Periodic Table of the Elements

IUPAC groups	1	2	3	4	5	6	7	8	9	10	11	12	13	14	15	16	17	18
Traditional groups (USA)	IA	IIA	IIIB	IVB	VB	VIB	VIIB		VIIIB		IB	IIB	IIIA	IVA	VA	VIA	VIIA	0

Period 1

| 1 H 1.008 | | | | | | | | | | | | | | | | | 2 He 4.003 |

Period 2

| 3 Li 6.941 | 4 Be 9.012 | | | | | | | | | | | 5 B 10.81 | 6 C 12.01 | 7 N 14.01 | 8 O 16.00 | 9 F 19.00 | 10 Ne 20.18 |

Period 3

| 11 Na 22.99 | 12 Mg 24.30 | | | | | | | | | | | 13 Al 26.98 | 14 Si 28.09 | 15 P 30.97 | 16 S 32.07 | 17 Cl 35.45 | 18 Ar 39.95 |

Period 4

| 19 K 39.10 | 20 Ca 40.08 | 21 Sc 44.96 | 22 Ti 47.88 | 23 V 50.94 | 24 Cr 52.00 | 25 Mn 54.94 | 26 Fe 55.85 | 27 Co 58.93 | 28 Ni 58.69 | 29 Cu 63.55 | 30 Zn 65.39 | 31 Ga 69.72 | 32 Ge 72.61 | 33 As 74.92 | 34 Se 78.96 | 35 Br 79.90 | 36 Kr 83.80 |

Period 5

| 37 Rb 85.47 | 38 Sr 87.62 | 39 Y 88.91 | 40 Zr 91.22 | 41 Nb 92.91 | 42 Mo 95.94 | 43 Tc [98.9] | 44 Ru 101.1 | 45 Rh 102.9 | 46 Pd 106.4 | 47 Ag 107.9 | 48 Cd 112.4 | 49 In 114.8 | 50 Sn 118.7 | 51 Sb 121.8 | 52 Te 127.6 | 53 I 126.9 | 54 Xe 131.3 |

Period 6

| 55 Cs 132.9 | 56 Ba 137.3 | 71 Lu 175.0 | 72 Hf 178.5 | 73 Ta 180.9 | 74 W 183.8 | 75 Re 186.2 | 76 Os 190.1 | 77 Ir 192.2 | 78 Pt 195.1 | 79 Au 197.0 | 80 Hg 200.6 | 81 Tl 204.4 | 82 Pb 207.7 | 83 Bi 209.0 | 84 Po [209.0] | 85 At [210.0] | 86 Rn [222.0] |

Period 7

| 87 Fr [223.0] | 88 Ra [226.0] | 103 Lr [260.1] | 104 Unq [261.1] | 105 Unp [262.1] | 106 Unh [263.1] | 107 Uns [262.1] | 108 Uno [265.1] | 109 Une [266.1] | | | | | | | | | |

Lanthanides

| 57 La 138.9 | 58 Ce 140.1 | 59 Pr 140.9 | 60 Nd 144.2 | 61 Pm [144.9] | 62 Sm 150.4 | 63 Eu 152.0 | 64 Gd 157.2 | 65 Tb 158.9 | 66 Dy 162.5 | 67 Ho 164.9 | 68 Er 167.3 | 69 Tm 168.9 | 70 Yb 173.0 |

Actinides

| 89 Ac [227.0] | 90 Th 232.0 | 91 Pa 231.0 | 92 U 238.0 | 93 Np [237.0] | 94 Pu [244.1] | 95 Am [243.1] | 96 Cm [247.1] | 97 Bk [247.1] | 98 Cf [251.1] | 99 Es [252.1] | 100 Fm [257.1] | 101 Md [258.1] | 102 No [259.1] |

Key

Atomic number	1
Symbol	H
Atomic Mass	1.008

Values in brackets are masses of most stable isotopes.

Physical Chemistry

Related Titles of Interest from the Benjamin/Cummings Series in the Life Sciences

W. M. Becker and D. W. Deamer
The World of the Cell, Second Edition (1991)

Benjamin/Maruzen
HGS Molecular Structure Models: General Chemistry (1969) *and Organic Chemistry Sets* (1988)

R. F. Boyer
Modern Experimental Biochemistry, Second Edition (1993)

I. S. Butler and J. F. Harrod
Inorganic Chemistry: Principles and Applications (1989)

N. A. Campbell
Biology, Third Edition (1993)

L. E. Hood, I. L. Weissman, W. B. Wood, and J. H. Wilson
Immunology, Second Edition (1984)

N. Kerner
Chemical Investigations (1986)

H. W. Knoche
Essentials of Organic Chemistry (1986)

G. M. Loudon
Organic Chemistry, Third Edition (1994)

B. Mahan and R. J. Meyers
University Chemistry, Fourth Edition (1987)

C. K. Mathews and K. E. van Holde
Biochemistry (1990)

G. Sackheim
Introduction to Chemistry for Biology Students, Fourth Edition (1991)

J. D. Watson, N. H. Hopkins, J. W. Roberts, J. A. Steitz, and A. M. Weiner
Molecular Biology of the Gene, Fourth Edition (1987)

W. B. Wood, J. H. Wilson, R. M. Benbow, and L. E. Hood
Biochemistry: A Problems Approach (1981)

C. Yoder and C. Schaeffer
Introduction to Multinuclear NMR (1987)

PHYSICAL CHEMISTRY

ROBERT G. MORTIMER

Rhodes College

The Benjamin/Cummings Publishing Company, Inc.

Redwood City, California · Menlo Park, California
Reading, Massachusetts · New York · Don Mills, Ontario
Wokingham, U.K. · Amsterdam · Bonn
Singapore · Tokyo · Madrid · San Juan

Sponsoring Editor: Anne Scanlan-Rohrer
Senior Production Editor: John Walker
Production Coordinator: Alyssa Wolf
Text Design: Mark Ong
Cover Design: John Martucci
Copyeditor: Mary Prescott
Proofreader: Anita Wagner
Illustrator: Ben Turner Graphics
Composition and Film: Polycomp Compositors Pte. Ltd.
Printing and Binding: R. R. Donnelly and Sons

Cover: The cloud-like formation in this image shows the quantum mechanical probability distribution for the electrons of a beryllium atom in an excited state with the $2px$, $2py$, and $2pz$ orbitals occupied. Data generated by Vital Images, Inc., Fairfield, Iowa, and reconstructed by the company's interactive volume rendering software system, Voxel View ®.

Library of Congress Cataloging-in-Publication Data

Mortimer, Robert G.
    Physical chemistry / Robert G. Mortimer.
        p.   cm.
    Includes bibliographical references and index.
    ISBN 0-8053-4560-4
    1. Chemistry, Physical and theoretical.   I. Title.
QD453.2.M67   1993
541.3—dc20   92-40711
                                                CIP

ISBN 0-8053-4560-4

2 3 4 5 6 7 8 9 10–DO–98 97 96 95 94 93

The Benjamin/Cummings Publishing Company, Inc.
390 Bridge Parkway
Redwood City, California 94065

To my teachers of physical chemistry

Norman Bauer
Norman Davidson
Henry Eyring
Joe Mayer
Robert Mazo
Lowell Tensmeyer

Preface

Physical chemistry is a fascinating field of study. It can reasonably be claimed that many parts of physics and all parts of chemistry are included within physical chemistry and its applications. Furthermore, it is the course in which most chemistry students first have the opportunity to synthesize what they have learned in mathematics, physics, and chemistry courses into a coherent pattern of knowledge.

This book is designed for a standard undergraduate two-semester course in physical chemistry. It is constructed about a central theme of systems, states, and processes. In any application of physical chemistry, one studies some physical object, such as a hydrogen atom or a mole of liquid water. This object is the system. Every system, whether a single atom or a macroscopic system of many molecules, has a set of states in which might be found, and a set of processes it might undergo to change its state. By calling attention to this fact and focussing attention on the commonalities between different branches of physical chemistry, the approach of this book attempts to assist students in structuring the variety of topics of physical chemistry into a coherent picture.

The topics of the traditional physical chemistry course can be grouped into several areas: (1) the study of the macroscopic properties of systems of many atoms or molecules; (2) the study of the processes which systems of many atoms or molecules can undergo; (3) the study of the properties of individual atoms and molecules; and (4) the study of the relationship between molecular and macroscopic properties.

The first eight chapters of the book present topics in thermodynamics, which is a comprehensive macroscopic theory of the behavior of material systems. These topics deal both with equilibrium macroscopic states and with a class of processes which begin and end with equilibrium states. Chapters 9 through 14 of the book present quantum mechanics and its application to the study of atomic and molecular structure. Chapter 15 presents a discussion of spectroscopy, which is the principal experimental tool for the study of atomic and molecular structure. Chapters 16 through 19 deal with nonequilibrium processes, discussed both from the macroscopic and molecular points of view. Chapter 20 presents statistical mechanics, which is the theory that connects molecular states and macroscopic states of systems of many molecules. The final two chapters present some applications of theories developed in earlier parts of the book to nonequilibrium processes and to the molecular structure of systems of many atoms or molecules.

Although the book is written from a unified point of view, the different portions of the book cover different parts of physical chemistry, as follows:

Chapter 1 Introduction to the description of systems

Chapters 2–8 Thermodynamics and its applications

Chapters 9–15 Quantum mechanics and its applications

Chapters 16–19 Nonequilibrium processes

Chapters 20–22 Statistical mechanics and its applications

The book is constructed so that several different sequences of these topic areas are possible with a minimum of adjustments. Four sequences which should be practical are:

1. As written
2. Chapters 1–8, Chapters 16–19, Chapters 9–15, Chapters 20–22
3. Chapter 1, Chapters 9–15, Chapters 2–8, Chapters 16–22
4. Chapter 1, Chapters 16–19, Chapters 2–8, Chapters 9–15, Chapters 20–22

Each chapter has a chapter preview, stating the principal facts and ideas that are presented in the chapter, as well as objectives for the student. There is a summary to assist in synthesizing the material of each chapter into a coherent whole and there are marginal notes throughout the chapters to assist the student in following the flow of topics in the chapter. Each chapter contains examples that illustrate various kinds of calculations, as well as exercises placed within the chapter. The exercises and problems at the end of the chapter are designed to provide practice in applying techniques and insights obtained through study of the chapter. Answers to all of the numerical exercises and to selected end-of-chapter problems are placed at the end of the book. A solutions manual, with complete solutions to all exercises and problems, is available from the publisher.

The book contains several appendixes, designed to improve the usefulness of the book. All of the tables of numerical data in the book are collected into Appendix A. Appendix C is a table of integrals, and Appendix G is a list of symbols used in the book. The other appendixes contain instructional material. Appendix B is a review of calculus needed for physical chemistry, Appendix D is a brief survey of classical mechanics, Appendix E is a proof of Euler's theorem, and Appendix F presents information about special mathematical functions encountered in quantum mechanics.

The author and the reviewers have made efforts to find all of the errors in the manuscript, but the author accepts full responsibility for them. The author welcomes feedback from students and instructors; please send your comments and suggestions to the author's attention.

Robert G. Mortimer
Department of Chemistry
Rhodes College
Memphis, TN 38112

Acknowledgments

The writing of this book began during a sabbatical leave from Rhodes College. Further support was received in the form of grants from the Faculty Development Committee of Rhodes College. It is a pleasure to acknowledge this support.

It has been my pleasure to have studied with dedicated and proficient teachers of physical chemistry to whom I have dedicated this book. I acknowledge their influence, example, and inspiration. I am also grateful to the students, whose efforts to understand the workings of the physical universe make teaching the most desirable of all professions.

While writing this book, I have benefitted from the expert advice of many reviewers:

Juana Acrivos	San Jose State University
James R. Barrante	Southern Connecticut State University
Robert K. Bohn	University of Connecticut
William R. Brennen	University of Pennsylvania
Dewey K. Carpenter	Louisiana State University, Baton Rouge
Jeff C. Davis, Jr.	University of South Florida
Norduff Debye	Towson State University
Jerald A. Devore	California State University, Long Beach
David L. Freeman	University of Rhode Island
Charles J. Fritchie, Jr.	Tulane University
L. Peter Gold	Pennsylvania State University
Elisheva Goldstein	California State Polytechnic University, Pomona
Charles Greenlief	Emporia State University
David K. Hoffman	Iowa State University
Donald J. Kouri	University of Houston

William R. Leenstra	University of Vermont
Lawrence L. Lohr	University of Michigan
C. Alden Mead	University of Minnesota
Clyde Metz	College of Charleston
Robert Nyland	Susquahana University
Lee G. Pederson	University of North Carolina
P. L. Polavarapu	Vanderbilt University
Carl F. Prenzlow	California State University, Fullerton
John L. Ragle	University of Massachusetts, Amherst
Marion B. Rhodes	University of Massachusetts, Amherst
Robert M. Rosenberg	Lawrence University
Robert E. Salomon	Temple University
Steven M. Schildcrout	Youngstown State University
R. P. Schmitt	Texas A & M University
Harris J. Silverstone	The Johns Hopkins University
Brian Swift	Marian College
John D. Vaughan	Colorado State University
Chia C. Yang	Arkansas Tech University

All these reviewers gave sound advice, and some of them went beyond the call of duty in searching out errors and unclarities and in suggesting remedies. The errors that remain are my responsibility, not theirs.

I wish to thank the editorial and production staff at The Benjamin/Cummings Publishing Company for their constant guidance and help during a rather long and complicated project. Anne Scanlan-Rohrer, Robin Heyden, Leslie With, John Walker, and Alyssa Wolf have devoted much of themselves to this book.

Finally, I must thank my wife, Dot, and my children, Julie, Paul, Jeannine, John and Dave, for enduring several years when the writing of this book occupied my time and my mind.

Contents

xii Contents

1 Systems, States and Processes: The Macroscopic Study of Gases and Liquids

OBJECTIVES

After studying this chapter, the student should:

1. understand the relationship between macroscopic behavior and its mathematical description,

2. be able to apply the methods of calculus to the macroscopic observables of a simple system,

3. be able to solve problems related to the macroscopic equilibrium properties of gases and liquids,

4. understand the critical point and be able to work problems related to the critical point,

5. understand the relationship between macroscopic and microscopic states.

PREVIEW

In this chapter we focus on systems, which are the objects that we study; on the states of systems, which are the possible circumstances in which the systems can be found; and on the processes that can change the states of systems. We discuss equilibrium states of gas and liquid systems. We introduce some mathematical tools and use them to describe the volumetric (pressure-volume-temperature) behavior of these systems, including the coexistence of phases (solid, liquid or gas).

PRINCIPAL FACTS AND IDEAS

1. The state of a system is specified by the values of a number of state variables.
2. In a simple one-phase system of one substance at equilibrium, if T, V, and n are chosen as independent variables, P is a dependent variable.
3. The language of mathematics is used to describe the macroscopic properties of systems.
4. Nonideal gases and liquids are described mathematically by various equations of state.
5. The coexistence of phases can be described mathematically.
6. Macroscopic states are determined by the nature of microscopic states.

1.1 Systems, States, and Processes

Physics has been defined as the study of the properties of matter that are shared by all substances, and chemistry has been defined as the study of the properties of individual substances. Physical chemistry comprises both of these studies. It is probably the most fundamental subdivision of chemistry. The theories and experiments that we discuss underlie much of organic chemistry, inorganic chemistry, biochemistry, and analytical chemistry.

Scientists use a **scientific method** to study the parts of the universe accessible to us, although most do not consciously apply such a routine method in their work. The first stage in the method is **description**, which means observing and reporting what a particular object does under certain conditions.

The second stage is **generalization**, or construction of empirical laws. This means concluding from experimental facts that all objects of a certain class exhibit some common behavior and finding a general statement that expresses that behavior. For example, the formula that expresses **Newton's second law** is

Newton's law is named for the British mathematician and physicist Isaac Newton, 1642–1727.

$$\mathbf{F} = m\mathbf{a} = m\frac{d\mathbf{v}}{dt} \tag{1.1-1}$$

where $\mathbf{F}$ is the force acting on an object having mass m, $\mathbf{v}$ is the object's velocity, and $\mathbf{a}$ is the object's acceleration. The force and the acceleration are both vectors, having direction as well as magnitude. Equation (1.1-1) is the most important relation in **Newtonian mechanics**, also known as **classical mechanics**.

The third stage in the scientific method is **explanation** or **theorizing**. This means contriving a set of assumptions, or hypotheses, about the nature of the physical universe or of a simpler "model system" in the hope of explaining its observed behavior.

The fourth stage is **testing the theory**. To do this, scientists deduce the consequences of the assumptions of the theory. This process is called deductive reasoning, or reasoning from the general case to the specific case. They then compare these predictions with experiment, usually carrying out new experiments to obtain relevant data. If the actual behavior and the predictions agree, the theory is likely to be accepted as an explanation of why systems behave as they do.

Systems

In physical chemistry we study a variety of material objects. The object being studied at a given moment is called the **system**. Our study is centered around three things: (1) the definition of the system and of its properties; (2) the values of variables that specify the **state of the system**, which means the condition of the system at a given time; and (3) the **processes** that can change the state of the system. In this entire book we focus on these three things, providing a common theme in all of the subject areas of physical chemistry.

The portion of the universe that is outside the system and that interacts with the system is called the **surroundings**. We must be specific about the boundary between system and surroundings. If we study a sample of liquid or gas confined in a container, we must decide whether the container is part of the system or part of the surroundings.

Figure 1.1 shows an example of a **macroscopic system** (a system of many molecules). A fluid (liquid or gas) is contained in a cylinder with a movable piston. The system consists of the fluid, and the cylinder and piston are part of the surroundings. There is a valve in the cylinder. When the valve is closed so that no matter can pass into or out of the system, we say that the system is a **closed system**. When the valve is open so that matter can pass into or out of the system, we say that the system is an **open system**. The cylinder-piston apparatus is immersed in a second part of the surroundings, a bath whose temperature can be controlled. If the system is completely insulated from its surroundings so that no heat can pass between system and surroundings, we call it an **adiabatic system** and say that it can undergo only adiabatic processes. If the system is completely separated from the rest of the universe so that no heat, work, or matter can be transferred, we call it an **isolated system**.

Sometimes, instead of discussing a real system, we will discuss a **model system**. A model system is designed to resemble some real system but is simpler than the real system, so it can be analyzed more easily. Model systems, which often exist only in our minds, are useful only if we can show that they at least approximately mimic the behavior of real systems.

Systems come in all sizes. The systems we discuss range from single atoms to mixtures of molecules of various substances. We will usually study either

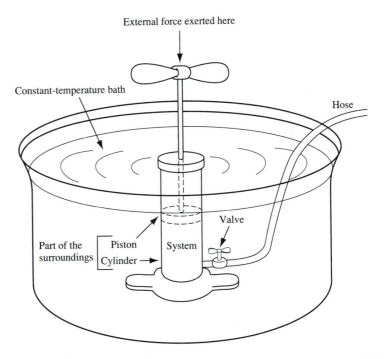

Figure 1.1. A Typical Fluid System Contained in a Cylinder with Variable Volume. Work can be done on the system by displacing the piston, heat can be transferred to the system from the bath in which it is immersed, and matter can be added to the system through the valve if it is open.

a macroscopic system containing many (perhaps 10^{23}) molecules or a **microscopic system** consisting of one atom or one molecule.

States and State Variables

The *state of a system* gives information about the circumstance in which the system is found. It is specified by giving values of a sufficient set of variables. The state of a system is a variable concept, depending on the kind of description we are using. There are two principal types of states: the *macroscopic state,* or *macrostate,* and the *microscopic state,* or *microstate.*

Macroscopic States

The macrostate is sometimes called the **thermodynamic state**. This state is specified by giving values of macroscopic variables, such as the pressure P, the temperature T, the volume V, the mass m, and the density ρ (equal to m/V). These variables are examples of **state variables** or **state functions**. A state variable depends only on the present state of a system, not on its history.

We do not need to know the value of every macroscopic variable to specify the macroscopic state of the system. It is an experimental fact that *when values of enough variables are specified, the values of the other variables are determined as dependent variables by the nature of the system.*

Consider a system consisting of a one-phase fluid sample of one substance at equilibrium. We specify the values of three variables: n, the amount of substance; T, the temperature; and P, the pressure. It is not necessary to specify the volume of the system. We know from experiments that the system will take on a volume determined by the values of T, P, and n and by the nature of the system. If these three variables are chosen as independent variables, the volume is a dependent variable whose value is given by a mathematical function determined by the nature of the system.

For example, if two ideal gases have the same values of T, P, and n, they must have the same value of V, given by V = nRT/P.

The statement that only three variables are required to specify a macrostate is restricted to equilibrium states. **Equilibrium** is a condition in which there is no tendency for any change to occur. There are systems in which no apparent change is observed, even over a long period of time, but which are not at equilibrium. For example, solid carbon exists in two allotropic forms, diamond and graphite. In order to be at chemical equilibrium at ordinary pressures, diamond must convert to graphite. However, the rate of conversion is very low, and diamond is said to be in a **metastable state**. A system in a metastable state is not at equilibrium and has a tendency to change toward equilibrium. However, a description that applies to equilibrium states can usually be applied to metastable states without significant error.

The advertisements that claim "A diamond is forever" are not strictly correct. Graphite is forever.

Processes

A **process** is an occurrence that changes the state of a system. A **reversible process** is a process whose direction can be reversed at any point by some infinitesimal change in the surroundings. Every process has a **driving force** that causes it to proceed. For example, a temperature difference is the driving force for a flow of heat. The driving force for a reversible process must be infinitesimal in magnitude. For example, if heat is flowing reversibly from a system to its surroundings, the temperatures of the system and the surroundings can differ only infinitesimally because an infinitesimal change in the temperature of the surroundings must be capable of reversing the direction of heat flow. The rate of a process is larger for larger magnitudes of its driving force, so a reversible process must occur infinitely slowly. The system has time to relax to equilibrium at each stage of a reversible process. *During a reversible process, the system passes through a sequence of equilibrium states.*

An **irreversible process** is a process that is not reversible. The system is not required to remain in equilibrium states during the process, and the process can occur in a finite time. When we discuss thermodynamics, we will discuss a class of irreversible processes in which the system is in an equilibrium state or metastable state when the process begins and is in an equilibrium state after the process ends, but passes through nonequilibrium states during the process.

1.2 The Behavior of Dilute Gases; The Ideal Gas

Robert Boyle was an English chemist who lived from 1627 to 1691.

A dilute gas is a gas at a fairly low pressure. It is found that all dilute gases have a common macroscopic behavior. The quantitative study of the behavior of the gases was begun by Boyle. The generalization of his results is expressed by **Boyle's law**, which holds for a fixed amount of any dilute gas at constant temperature:

$$PV = \text{constant} \qquad (1.2\text{-}1)$$

where P is the pressure and V is the volume.

Guillaume Amontons was a French chemist who lived from 1663 to 1705.

In 1701, Amontons studied the pressure of samples of air at fixed volumes and various temperatures. He found that the pressure of a gas was a linear function of the Celsius temperature such that the extrapolated pressure vanished at approximately $-273°C$ for all samples of all gases. He first defined the **absolute temperature scale**. The absolute temperature, T, is given by

$$T = t_C + 273.15 \, \text{K} \qquad (1.2\text{-}2)$$

where t_C is the Celsius (centigrade) temperature. The unit on this scale is the same size as the Celsius degree (°C) and is called the **kelvin** (abbreviated K). The results of Amontons are generalized to give the law of Amontons, which holds for a fixed amount of gas at constant volume:

$$\frac{P}{T} = \text{constant} \qquad (1.2\text{-}3)$$

The law of Gay-Lussac is named for the French chemist Joseph Gay-Lussac, 1778–1850.

Amontons died soon after proposing this law, and it did not become well known until it was republished by Gay-Lussac. It is commonly known as the law of Gay-Lussac.

Charles's law is named for the French physicist Jacques Alexandre Cesar Charles, 1746–1823.

The next empirical gas law was Charles's law, which holds for a fixed amount of gas at constant pressure:

$$\frac{V}{T} = \text{constant} \tag{1.2-4}$$

After this came Gay-Lussac's law of combining volumes, which was the first law to involve chemical change as well as **volumetric (P-V-T) behavior**. This law states that if the volumes of gases involved in chemical reactions are measured at the same temperature and pressure, the ratios of the volumes are equal to the ratios of small whole numbers.

John Dalton was an English chemist who lived from 1766 to 1844.

The **law of partial pressures** was published in 1803 by Dalton. The **partial pressure** of a gas is the pressure which that gas would exert at the same temperature if it were alone in the container. If P_1 is the partial pressure of gas number 1, P_2 the partial pressure of gas number 2, etc., and P is the total (observable) pressure, this law states that

$$P = P_1 + P_2 + \cdots$$

The hypothesis of Avogadro is named for the Italian chemist Amadeo Avogadro, 1776–1856.

The **hypothesis of Avogadro** is that N, the number of molecules in a sample of gas, is proportional to the volume at fixed temperature and pressure. This is not a purely empirical law because it employs the theoretical concept of molecules, introduced in the atomic theory of Dalton.

The Ideal Gas Law

The laws of Boyle and Charles and the hypothesis of Avogadro can be written as

$$V \propto \frac{1}{P} \quad \text{at constant } N \text{ and } T \tag{1.2-5}$$

$$V \propto T \quad \text{at constant } N \text{ and } P \tag{1.2-6}$$

$$V \propto N \quad \text{at constant } P \text{ and } T \tag{1.2-7}$$

These three proportionalities can be combined to give

$$V \propto \frac{NT}{P} \tag{1.2-8}$$

or

$$V = \frac{k_B NT}{P} \tag{1.2-9}$$

Boltzmann's constant is named for Ludwig Boltzmann, 1844–1906, an Austrian physicist who was one of the inventors of gas kinetic theory.

where k_B is a proportionality constant called Boltzmann's constant. Equation (1.2-9) is called the **ideal gas law**.

Units of Measurement; The SI Units

SI stands for Système International, the French name for the set of units.

In working with any equation like Equation (1.2-9), consistent units must be used. Such a set of units is the **international system of units**, or **SI units**. The SI units are mks units, in which lengths are measured in meters (m), masses in kilograms (kg), and time in seconds (s). Volume is measured in cubic meters (m³), temperature in kelvins (K), force in newtons (N), and

The joule is named for James Prescott Joule, 1818–1889, an English physicist who pioneered in the thermodynamic study of work, heat, and energy.

pressure in newtons per square meter (N m^{-2}), also known as pascals (Pa). From Newton's second law, Equation (1.1-1), the newton must have the units kg m s^{-2}, and 1 N is defined to equal 1 kg m s^{-2}. The SI unit of energy is the joule (J), defined so that $1\ J = 1\ N\ m = 1\ kg\ m^2\ s^{-2}$.

The value of Boltzmann's constant in SI units is

$$k_{\mathrm{B}} = 1.3807 \times 10^{-23}\ N\ m\ K^{-1} = 1.3807 \times 10^{-23}\ J\ K^{-1} \quad \textbf{(1.2-10)}$$

We will occasionally use other units, such as the calorie (cal). The calorie was originally defined as the amount of heat required to raise the temperature of 1 gram of water by 1°C. Its present definition is

$$\boxed{1\ cal = 4.184\ J \quad \text{(exactly, by definition)}} \quad \textbf{(1.2-11)}$$

Equations that will be used frequently are enclosed in boxes.

We will use several non-SI units of pressure. The **atmosphere** (abbreviated atm) is defined by

$$\boxed{1\ atm = 101325\ Pa \quad \text{(exactly, by definition)}} \quad \textbf{(1.2-12)}$$

The average barometric pressure near sea level is approximately equal to 1.00 atmosphere. The second non-SI unit of pressure is the **torr**, defined by

$$\boxed{760\ torr = 1\ atm \quad \text{(exactly, by definition)}} \quad \textbf{(1.2-13)}$$

A pressure equal to 1 torr will raise a column of mercury to a height of nearly 1 millimeter if the mercury is at a temperature of 273.15 K (0.00°C). Another unit of pressure is the **bar**, defined by

$$\boxed{1\ bar = 100000\ Pa \quad \text{(exactly, by definition)}} \quad \textbf{(1.2-14)}$$

The bar is approximately equal to 750.1 torr.

Chemists are accustomed to using a non-SI unit of volume, the **liter** (L), which is the same as a cubic decimeter:

$$\boxed{1\ L = 0.001\ m^3 = 1\ dm^3 \quad \text{(exactly, by definition)}} \quad \textbf{(1.2-15)}$$

The **formula unit** of a substance is the smallest particle (or set of particles) that retains the identity of the substance. It might be an atom, a molecule, or an electrically neutral combination of ions. The **mole** (mol) is the SI unit for the amount of a substance. A mole of any substance is a sample with the same number of formula units as the number of atoms contained in exactly 0.012 kg of the carbon-12 isotope. Therefore, the number of formula units, N, of a sample of any substance is proportional to the amount of the substance measured in moles, denoted by n:

$$N = N_A n \quad \textbf{(1.2-16)}$$

The proportionality constant N_A is called **Avogadro's constant** and is known from experiment to have the value

$$\boxed{N_A = 6.02214 \times 10^{23}\ mol^{-1}} \quad \textbf{(1.2-17)}$$

The **gas constant** R is defined by

$$R = N_A k = 8.3145 \text{ J K}^{-1} \text{ mol}^{-1}$$ **(1.2-18)**

The ideal gas law, Equation (1.2-9), can be written in terms of the amount of gas rather than the number of molecules:

$$PV = nN_A kT = nRT$$ **(1.2-19)**

▼ **EXAMPLE 1.1**

Find the pressure in Pa and in atm for an ideal gas if its temperature is 273.15 K (0°C), its volume is 22.400 L, and the amount of gas is 1.000 mol.

Solution

Although this example is elementary, it has similarities to more complicated problems. The steps used in this example are (1) finding the appropriate formula; (2) substituting the appropriate values into the formula, writing the units as well as the numerical values; (3) checking the units and inserting the appropriate labeled conversion factors to bring the units into consistency; and (4) carrying out the indicated arithmetic operations to obtain the final answer. An additional step, which is not explicitly shown, is looking at the final answer to see if it is reasonable.

$$P = \frac{(1.000 \text{ mol})(8.3145 \text{ J K}^{-1} \text{ mol}^{-1})(273.15 \text{ K})}{(22.400 \text{ L})(1 \text{ m}^3/1000 \text{ L})}$$

$$= 1.014 \times 10^5 \text{ J m}^{-3}$$

$$= 1.014 \times 10^5 \text{ N m}^2$$

$$= 1.014 \times 10^5 \text{ Pa}$$

$$= 101.4 \text{ kPa}$$

$$= (1.014 \times 10^5 \text{ Pa})\left(\frac{1 \text{ atm}}{101325 \text{ Pa}}\right)$$

$$= 1.001 \text{ atm}$$

▲

Exercise 1.1

Answers to numerical exercises appear in the back of the book.

Find the value of the gas constant in L atm K^{-1} mol^{-1}, in cal K^{-1} mol^{-1}, in cm^3 atm K^{-1} mol^{-1}, and in cm^3 bar K^{-1} mol^{-1}.

The ideal gas law is an illustration of the experimental fact stated earlier: for a system consisting of one phase and one substance, only three macroscopic state variables are independent. For a dilute gas, if the temperature, pressure, and amount of substance are specified, the volume is determined by Equation (1.2-19).

The ideal gas law is only approximately correct for real gases at nonzero pressures. Most gases deviate from the ideal gas law by 1% or less under ordinary conditions (temperatures near room temperature and pressures near 1 atmosphere). However, all gases come closer and closer to obeying the ideal gas law as the pressure is made smaller and smaller. A **dilute gas** is one whose pressure is small enough that it nearly obeys the ideal gas law.

We will represent all gases as ideal gases unless we have some specific need for a more accurate representation.

The Ideal Gas

An **ideal gas** is a gas that exactly obeys the ideal gas law for all values of the variables P, V, n, and T. An ideal gas is sometimes called a **perfect gas**.

However, defining an ideal gas does not cause such a system to exist in the real world—the ideal gas is an example of a model system.

Equation (1.2-19) is used to define the **ideal gas temperature scale**, with the condition that the pressure of a gas must be low enough so that the gas obeys this equation. We write

$$T = \lim_{P \to 0} \left(\frac{PV}{nR} \right) \tag{1.2-20}$$

We will regard a mathematical limit as an extrapolation.

The limit symbol in Equation (1.2-20) means that the quantity PV/nR is first evaluated at various fairly low pressures and its value is then extrapolated to zero pressure. The ideal gas temperature scale is the same as the absolute temperature scale of Amontons and Gay-Lussac.

1.3 Mathematics and the Macroscopic State of a Simple System

If we have a dilute gas at equilibrium and we choose values for n, T, and V for the system, then we have no choice about the value of P. The system "chooses" the value of P. The value of P can also be calculated from the ideal gas law

$$P = \frac{nRT}{V} \tag{1.3-1}$$

The important idea that system behavior can be described and predicted mathematically is expressed by saying that P is a **function** of T, V, and n, denoted by

$$P = f(T, V, n) \tag{1.3-2}$$

The pressure P is the **dependent variable**, and there are three **independent variables**: T, V, and n. The letter f stands for the functional relationship. It means that if you specify a value for each of the independent variables, the function will deliver a value for the dependent variable.

If a system is not a dilute gas, Equation (1.3-1) does not apply and Equation (1.3-2) would represent a different function. Whether a system is a gas, liquid, or solid, there is some function represented by Equation (1.3-2) that applies. Such a functional relation is called an **equation of state**. We can ordinarily obtain only approximate representations of equations of state for real systems.

A formula such as Equation (1.3-1) is not the only way to represent a function. Figure 1.2a shows a set of graphical curves that represent the dependence of P on V for an ideal gas at a fixed value of n (1.000 mol) and several fixed values of T. Figure 1.2b shows a perspective view of a graphical surface in three dimensions that represents the dependence of P on V and on T for a fixed value of n (1.000 mol). Just as the height of a curve in Figure 1.2a gives the value of P for a particular value of V, the height of the surface in Figure 1.2b gives the value of P for a particular value of T and a particular value of V.

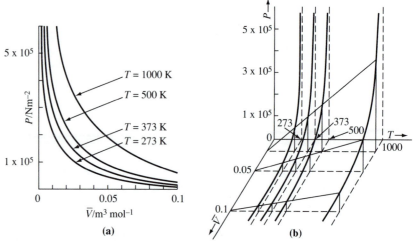

Figure 1.2. The Pressure of an Ideal Gas. (a) As a function of V at $n = 1$ and various constant values of T. The curves in this graph conform to the ideal gas equation of state, $PV = nRT$. **(b)** As a function of V and T at constant n. This three-dimensional graph gives the same information as the set of all curves in part (a).

Johannes Diderik van der Waals, 1837–1923, Dutch physicist who received the 1910 Nobel Prize in physics for his work on equations of state.

The **van der Waals equation of state** describes real gases more accurately than the ideal gas law:

$$\left(P + \frac{an^2}{V^2}\right)(V - nb) = nRT \tag{1.3-3}$$

To represent different gases we must choose different values of a and b for each gas.

The symbols a and b represent parameters that have different constant values for different substances. Table A1 gives values of the van der Waals parameters for several substances. We will see in Chapter 5 that the van der Waals equation can also be used as an approximate equation of state for a liquid.

▼   **EXAMPLE 1.2**

Use the van der Waals equation to predict the pressure of nitrogen gas at 273.15 K and a molar volume of 22.414 L. Compare with the pressure of an ideal gas at the same temperature and molar volume.

Solution

We solve the van der Waals equation for P as a function of T and the **molar volume**, denoted by $\bar{V}$ and defined to equal V/n. Some authors denote the molar volume by V_m and others denote it by $\tilde{V}$.

The solution of Example 1.2 follows the same four steps as that of Example 1.1.

$$P = \frac{RT}{\bar{V} - b} - \frac{a}{\bar{V}^2}$$

$$= \frac{(8.3145 \text{ J K}^{-1} \text{ mol}^{-1})(273.15 \text{ K})}{0.022414 \text{ m}^3 \text{ mol}^{-1} - 0.0000391 \text{ m}^3 \text{ mol}^{-1}}$$

$$- \frac{0.1408 \text{ Pa m}^6 \text{ mol}^{-2}}{(0.022414 \text{ m}^3 \text{ mol}^{-1})^2}$$

$$= 1.0122 \times 10^5 \text{ Pa}$$

$$= 0.9990 \text{ atm}$$

For the ideal gas,

$$P = \frac{RT}{\bar{V}} = \frac{(8.3145 \text{ J K}^{-1} \text{ mol}^{-1})(273.15 \text{ K})}{(0.022414 \text{ m}^3 \text{ mol}^{-1})}$$

$$= 1.0132 \times 10^5 \text{ Pa} = 1.0000 \text{ atm}$$

There is a difference of 0.1% between the two values. Some gases with larger molecules deviate more from ideal behavior than does nitrogen, but the behavior in this example is fairly typical of real gases under ordinary conditions.

▲

Exercise 1.2

The property of part a is essential, because all gases are found experimentally to approach ideal behavior for low enough pressures.

a. Show that in the limit that $\bar{V}$ becomes large, the van der Waals equation becomes identical to the ideal gas law.

b. Find the pressure of 1.000 mol of nitrogen at a volume of 0.500 L and a temperature of 298.15 K, using the van der Waals equation of state. Find the percent difference from the pressure of an ideal gas under the same conditions.

Instead of the notation of Equation (1.3-2), chemists usually write

$$P = P(T, V, n) \tag{1.3-4}$$

using the same letter for the function and for the dependent variable, even if a different set of independent variables is used. Furthermore, for a system other than an ideal gas, chemists use the same letter P to represent whatever combination of mathematical operations is needed to deliver a value of the pressure for that system. We follow these policies because we have many variables but only a limited supply of letters.

We can make another choice of independent variables. When we solve the ideal gas equation for V, we obtain

$$V = \frac{nRT}{P} \tag{1.3-5}$$

exchanging the roles of the dependent variable and one of the independent variables.

The fact that we can solve the ideal gas equation for different variables illustrates the fact that the independent variables that specify the state of the system can be chosen in different ways.

For a general one-phase, one-substance system we write

$$V = V(T, P, n) \tag{1.3-6}$$

We can also solve for T or n, choosing any one of our variables as the dependent variable.

Exercise 1.3

Assume that the volume of a liquid system is a linearly decreasing function of P, a linearly increasing function of T, and proportional to n. Write a formula expressing this functional relationship, using arbitrary symbols for constants and using the symbol V_0 for the volume at some reference temperature T_0, some reference pressure P_0, and some reference amount of substance n_0.

Macroscopic States and State Functions

We will be using a number of macroscopic state variables, including P, V, T, and n, as well as the mass m and the density $\rho = m/V$. We will define and discuss other macroscopic state variables in future chapters. One of the most important is the thermodynamic energy, or internal energy, U. We will use the internal energy now as an example of a macroscopic variable but will not specify its properties until Chapter 2.

There are two principal classes of macroscopic variables: **extensive variables**, which are proportional to the size of the system, and **intensive variables**, which are independent of the size of the system. For example, U, V, n, and m are extensive variables and P and T are intensive variables.

One way to determine whether a variable is extensive or intensive is to imagine combining a given system with a replica of that system, keeping P and T fixed. Any variable that has twice the value for the combined system as for the original system is extensive, and any variable that is unchanged is intensive.

The quotient of two extensive variables is an intensive variable. Examples are the density, $\rho = m/V$, the molar volume, $\bar{V} = V/n$, and the molar internal energy, $\bar{U} = U/n$. (We denote a molar quantity by a bar over the symbol for the extensive variable.)

We have already asserted as an experimental fact that for a one-phase, one-substance fluid system, the values of three variables from the set T, V, P, and n determine the value of the fourth variable. We now introduce an important assertion, which we regard as experimental fact: *All equilibrium macroscopic variables of a one-phase fluid system of one substance are functions of three independent variables, at least one of which must be extensive.* One independent variable must be extensive, because extensive variables cannot be functions only of intensive variables. This assertion follows from the fact that if all independent variables were intensive, the size of the system could be changed without changing the value of any independent variable. However, this change would change the values of extensive variables.

Specification of the values of three variables suffices to determine the values of all equilibrium macroscopic variables of a fluid one-phase, one-component system. Therefore, *the equilibrium macroscopic state of a one-component, one-phase fluid system is specified by giving the values of three macroscopic variables, of which at least one must be extensive.* If two one-component, one-phase fluid systems containing the same substance are at equilibrium and have the same values of three macroscopic variables (at least one of which is extensive), they must have the same values of all other macroscopic variables. They are in the same macroscopic state.

One-phase fluid systems that contain several substances have an additional independent variable for each substance beyond the first substance, so the total number of independent variables at equilibrium is $c + 2$, where c is the number of independent substances (called components). We discuss the definition of a component in Chapter 4, but for now consider it to be a substance whose amount can be varied independently of the other substances.

A one-phase system whose equilibrium state variables are functions of $c + 2$ variables is called a **simple system**. Fluid (liquid or gas) systems and strain-free solid systems are simple systems. A rubber band or a coil spring is not a simple system, because its length can be an independent variable in addition to T, P, and n. An electrochemical cell is another example of a nonsimple system. A system with more than one phase is technically not a simple system, since its energy depends on the areas of the interfaces between phases. However, as we will show in a later chapter, this surface energy is negligible in most cases and we generally consider such systems to be simple. We will provide another definition of a simple system in Chapter 2.

Different sets of independent variables can be chosen to specify the state of the system. For example, for the internal energy U of a simple one-substance, one-phase system we can write

$$U = U(T, V, n) \tag{1.3-7}$$

or we can write

$$U = U(T, P, n) \tag{1.3-8}$$

We do not always know the particular functions that are represented by these two equations, but we can sometimes get useful information without knowing them by using methods of calculus and measured values of derivatives.

Some facts about the calculus of functions of several variables are presented in Appendix B. A function of several independent variables has several derivatives. For example, the derivative of the internal energy with respect to the temperature, considering U to be a function of T, P, and n, is denoted by $(\partial U/\partial T)_{P,n}$. The subscripts indicate that P and n are held fixed (treated as constants) when the differentiation is performed. This kind of a derivative is called a **partial derivative**.

The derivative of the internal energy U with respect to the temperature, considering U to be a function of T, V, and n, is denoted by $(\partial U/\partial T)_{V,n}$. The subscripts now indicate that V and n are treated as constants. These two partial derivatives, $(\partial U/\partial T)_{P,n}$ and $(\partial U/\partial T)_{V,n}$, are not identical (see Appendix B).

A commonly measured quantity that is related to a partial derivative is the **isothermal compressibility** κ_T:

$$\boxed{\kappa_T = -\frac{1}{V}\left(\frac{\partial V}{\partial P}\right)_{T,n} \quad \text{(definition)}} \tag{1.3-9}$$

The factor $1/V$ is included to make the compressibility an intensive variable. T and n are fixed in the differentiation because measurements of the isothermal compressibility are taken on a closed system in a constant-temperature bath.

It is found experimentally that every system decreases in volume when the pressure on it is increased. The compressibility of any system is positive.

▼ **EXAMPLE 1.3**

The isothermal compressibility of liquid water at 298.15 K and 1.000 atm is equal to 4.524×10^{-5} bar$^{-1} = 4.524 \times 10^{-10}$ Pa^{-1}. Find the fractional change in the volume of a sample of water if its pressure is changed from 1.000 bar to 50.000 bar to 298.15 K.

Solution

The compressibility is so small in magnitude that we can write as an adequate approximation [see Equation (B-3)]:

In Example 1.3, the first step is to decide that an approximate formula can be used with adequate accuracy. The following steps are similar to those in the previous two examples.

$$\Delta V \approx \left(\frac{\partial V}{\partial P}\right)_{T,n} \Delta P = -V\kappa_T \, \Delta P$$

where we use the standard notation

$$\Delta V = V(\text{final}) - V(\text{initial})$$

with a similar equation for ΔP and any other variable. The fractional change is

$$\frac{\Delta V}{V} \approx -\kappa_T \, \Delta P = -(4.524 \times 10^{-5} \text{ bar}^{-1})(49.00 \text{ bar})$$

$$= -2.217 \times 10^{-3}$$

The **coefficient of thermal expansion** α is defined by

$$\boxed{\alpha = \frac{1}{V}\left(\frac{\partial V}{\partial T}\right)_{P,n} \quad \text{(definition)}}$$ (1.3-10)

Like the compressibility, the coefficient of thermal expansion is an intensive quantity. The coefficient of thermal expansion is usually, but not always, positive. Water has a negative coefficient of thermal expansion between 0°C and 3.98°C.

▼ **EXAMPLE 1.4**

The coefficient of thermal expansion of liquid water at 298.15 K and 1.000 atm is equal to 2.5705×10^{-4} K^{-1}. Find the fractional change in the volume of a sample of water at 1.000 atm if its temperature is changed from 298.15 K to 303.15 K.

Solution

In Example 1.4, as in Example 1.3, the first step is to decide that an approximate equation can be used.

Since the coefficient of thermal expansion is small,

$$\Delta V \approx \left(\frac{\partial V}{\partial T}\right)_{P,n} \Delta T = V\alpha \, \Delta T$$

The fractional change in volume is

$$\frac{\Delta V}{V} \approx \alpha \, \Delta T = (2.5705 \times 10^{-4} \text{ K}^{-1})(5.000 \text{ K})$$

$$= 1.29 \times 10^{-3}$$

Exercise 1.4

a. Find the fractional change in the volume of a sample of liquid water if its temperature is changed from 20.00°C to 30.00°C and its pressure is changed from 1.000 bar to 40.00 bar.

b. Find expressions for the isothermal compressibility and coefficient of thermal expansion for an ideal gas.

c. Find the value of the isothermal compressibility in atm^{-1}, in bar^{-1}, and in Pa^{-1} for an ideal gas at 298.15 K and 1.000 atm. Find the ratio of this value to that of liquid water at the same temperature and pressure, using the value from Table A3.

d. Find the value of the coefficient of thermal expansion of an ideal gas at 20°C and 1.000 atm. Find the ratio of this value to that of liquid water at the same temperature and pressure, using the value from Table A4.

The results of Examples 1.2, 1.3, and 1.4 are typical: Most gases are nearly ideal under ordinary conditions, and most liquids have a nearly constant volume under ordinary conditions. The volumes of most solids are even more nearly constant. We therefore recommend the following policy for making ordinary calculations: *Unless there is some reason to do otherwise, treat gases as though they are ideal and treat liquids and solids as though they have fixed volumes.*

1.4   Real Gases

Although most gases obey the ideal gas law to an accuracy of about 1% when near room temperature and 1 atmosphere, at pressures around 100 atmospheres one must seek a better description than the ideal gas law.

Several equations of state have been devised to describe the volumetric behavior (relation of pressure, temperature and volume) of real gases. We have already introduced the van der Waals equation in Equation (1.3-3). It can be written in the form

$$P = \frac{RT}{\bar{V} - b} - \frac{a}{\bar{V}^2} \tag{1.4-1}$$

where $\bar{V}$ is the molar volume, V/n.

Another equation of state is the **virial equation of state**:

$$\frac{P\bar{V}}{RT} = 1 + \frac{B_2}{\bar{V}} + \frac{B_3}{\bar{V}^2} + \frac{B_4}{\bar{V}^3} + \cdots \tag{1.4-2}$$

which is a power series in the independent variable $1/\bar{V}$. The B coefficients are called **virial coefficients**. The first virial coefficient, B_1, is equal to unity. The other virial coefficients must be taken as functions of temperature. Table A2 gives values of the second virial coefficient for several gases at several temperatures.

An equation of state that is a power series in P is also called a virial equation of state:

$$P\bar{V} = RT + A_2 P + A_3 P^2 + A_4 P^3 + \cdots \tag{1.4-3}$$

The coefficients A_2, A_3, etc. are called **pressure virial coefficients**. It can be shown that A_2 and B_2 are equal.

Exercise 1.5	Show that $A_2 = B_2$. Proceed by solving Equation (1.4-2) for P and substituting this expression for each P in Equation (1.4-3). Then use the fact that the coefficient of any power of $1/\bar{V}$ must be the same on both sides of the equation.

Table 1.1 displays several additional equations of state. Values of parameters for different gases may be found in Table A1. The values of the parameters must be determined from experimental data. Several of these equations of state are two-parameter equations of state and use the same letters, a and b, for their parameters. The parameters for a given gas do not necessarily have the same values in different equations.

Table 1.1. Some Equations of State

The Berthelot equation of state

$$\left(P + \frac{a}{T\bar{V}^2}\right)(\bar{V} - b) = RT$$

The Dieterici equation of state

$$Pe^{a/\bar{V}RT}(\bar{V} - b) = RT$$

The Redlich-Kwong equation of state

$$P = \frac{RT}{\bar{V} - b} - \frac{a}{T^{1/2}\bar{V}(\bar{V} + b)}$$

The Soave modification of the Redlich-Kwong equation of state

$$P = \frac{RT}{\bar{V} - b} - \frac{a\alpha(T)}{\bar{V}(\bar{V} + b)}$$

where $\alpha(T) = \{1 + m[1 - (T/T_c)^{1/2}]\}^2$, where m is a parameter and T_c is the critical temperature. See the article by Soave for values of the parameter m.

The Gibbons-Laughton modification of the Redlich-Kwong-Soave equation

The equation is the same as the Soave modification, but $\alpha(T)$ is

$$\alpha(T) = 1 + X[(T/T_c) - 1] + Y[(T/T_c)^{1/2} - 1]$$

where X and Y are parameters. See the article by Gibbons and Laughton for values of these parameters.

Other equations of state can be found in the book by Hirschfelder, Curtiss, and Bird, including the Beattie-Bridgeman equation, with five parameters, and the Benedict-Webb-Rubin equation, with eight parameters.
Table References:
O. Redlich and J. N. S. Kwong, *Chem. Rev.* **44**, 233 (1949).
G. Soave, *Chem. Eng. Sci.* **27**, 1197 (1972).
R. M. Gibbons and A. P. Laughton, *J. Chem. Soc. Faraday Trans. 2* **80**, 1019 (1984).
J. O. Hirschfelder, C. F. Curtiss, and R. B. Bird, *Molecular Theory of Gases and Liquids*, Wiley, New York, 1954, pp. 250ff.

The accuracy of the two-parameter equations of state has been evaluated.[1] The Redlich-Kwong equation seemed to perform better than the other two-parameter equations, and the van der Waals equation came in second best. The Gibbons-Laughton modification of the Redlich-Kwong equation (with four parameters) is more accurate than the two-parameter equations.

Equations of State for Mixtures

Discussions that are not essential to our major development are printed in small type. These parts of the book can be skipped without loss of continuity.

All equations of state can be used for gaseous mixtures of two or more substances if one uses parameters or coefficients that depend on composition. However, it is difficult to find experimental values of parameters for mixtures. In the absence of data, it is generally assumed that the a and b parameters in the van der Waals, Berthelot, and Dieterici equations of state for a two-component mixture obey the following "mixing rules" for a mixture of two gases:

$$a = a_1 x_1^2 + 2a_{12} x_1 x_2 + a_2 x_2^2 \qquad \textbf{(1.4-4)}$$

$$b = b_1 x_1^2 + 2b_{12} x_1 x_2 + b_2 x_2^2 \qquad \textbf{(1.4-5)}$$

where x_1 and x_2 are the mole fractions, defined for two components by

$$x_i = n_i/n = n_i/(n_1 + n_2) \qquad (i = 1, 2) \qquad \textbf{(1.4-6)}$$

Here a_1 and b_1 are the values of the parameters for substance 1 and a_2 and b_2 are the values for substance 2. The coefficients a_{12} and b_{12} might be evaluated from experimental data, but if no values can be found, one can use the empirical formulas:[2]

$$a_{12} = (a_1 a_2)^{1/2} \qquad \textbf{(1.4-7)}$$

$$b_{12} = [(b_1^{1/3} + b_2^{1/3})/2]^3 \qquad \textbf{(1.4-8)}$$

Graphical Presentation of Volumetric Data

Graphs of the **compression factor** Z are sometimes used to describe the behavior of gases:

$$\boxed{Z = \frac{P\bar{V}}{RT} \quad \text{(definition)}} \qquad \textbf{(1.4-9)}$$

Some older books use a different name, the "compressibility factor." We avoid this name because it might be confused with the compressibility, defined in Equation (1.3-9). Figure 1.3 shows the compression factor for nitrogen gas at several temperatures.

At fairly low temperatures, the compression factor decreases below unity for moderate pressures but rises above unity as the pressure is increased further. At higher temperatures, the compression factor is larger than unity for all pressures. The temperature at which the curve has zero slope at zero pressure is called the **Boyle temperature** and is the temperature at which the gas most nearly approaches ideality for moderate pressures.

---

[1] J. B. Ott, J. R. Goates, and H. T. Hall, Jr., *J. Chem. Educ.* **48**, 515 (1971); M. W. Kemp, R. E. Thompson, and D. J. Zigrang, *J. Chem. Educ.* **52**, 802 (1975).
[2] J. O. Hirschfelder, C. F. Curtiss, and R. B. Bird, *Molecular Theory of Gases and Liquids*, Wiley, New York, 1954, pp. 250ff.

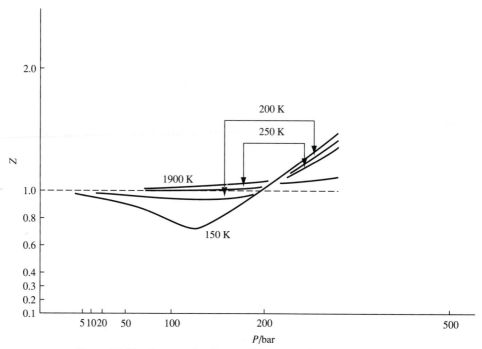

Figure 1.3. The Compression Factor of Nitrogen at Several Temperatures. The behavior of nitrogen is typical. At low temperatures, the value of Z is less than unity for moderate pressures but rises above unity for larger pressures. At higher temperatures, the value of Z is greater than unity for all pressures.

For a van der Waals gas, the compression factor is

$$Z = \frac{P\bar{V}}{RT} = \frac{\bar{V}}{\bar{V} - b} - \frac{a}{RT\bar{V}} = \frac{1}{1 - by} - \frac{ay}{RT} \qquad \textbf{(1.4-10)}$$

where we let $y = 1/\bar{V}$. The variable y is proportional to the pressure at constant temperature if the pressure is small enough.

Since a and b are both positive for all known gases, the first term on the right-hand side of Equation (1.4-10) gives a contribution to Z that is greater than unity, and the second term gives a negative contribution. The first term is constant if y is constant, and the second term is inversely proportional to the temperature if y is constant. For higher temperatures, the second term is relatively unimportant and the compression factor exceeds unity for all values of y. For temperatures below the Boyle temperature, the second term becomes relatively more important and a value of Z less than unity occurs if y is not too large. For large values of y, the denominator of the first term is small, making this term dominant even for lower temperatures. For any temperature below the Boyle temperature, there is thus a nonzero value of the pressure at which $Z = 1$.

▼ **EXAMPLE 1.5**

a. Find an expression for the Boyle temperature of a van der Waals gas.

b. Find the value of the Boyle temperature of nitrogen gas as predicted by the van der Waals equation.

Solution

a. Since y is proportional to P for small values of P, we seek the temperature at which

$$\left(\frac{\partial Z}{\partial y}\right)\Bigg|_{y=0} = 0 = \left(\frac{b}{(1-by)^2} - \frac{a}{RT}\right)\Bigg|_{y=0} = b - \frac{a}{RT}$$

so that

$$T_{\text{Boyle}} = a/Rb$$

b. For nitrogen,

$$T_{\text{Boyle}} = \frac{0.1408 \text{ Pa m}^6 \text{ mol}^{-2}}{(8.3145 \text{ J K}^{-1} \text{ mol}^{-1})(3.913 \times 10^{-5} \text{ m}^3 \text{ mol}^{-1})}$$

$$= 433 \text{ K}$$

The experimental value is approximately 327 K.

▲

The first step of part a is to apply the definition of the Boyle temperature. The second step is to carry out the differentiation, and the third step is to solve for T_{Boyle}. The plan of procedure is devised by determining exactly what the desired result is, analyzing the given information, and finding a method to go from the given information to the result.

Exercise 1.6

a. Find an expression for the Boyle temperature of a gas obeying the Dieterici equation of state.

b. Find the value of the Boyle temperature of nitrogen according to the Dieterici equation of state.

c. Find the expression for the molar volume at which $Z = 1$ for the van der Waals gas for a given temperature below the Boyle temperature. Hint: Find the nonzero value of y in Equation (1.4-10) that makes $Z = 1$.

d. Find the value of the molar volume and the pressure at which $Z = 1$ for nitrogen at 273.15 K, according to the van der Waals equation.

1.5 Real Liquids

For ordinary calculations, a sample of a liquid can be treated as if its volume is constant. If higher accuracy is needed, values of isothermal compressibilities and coefficients of thermal expansion can be used in approximate calculations such as those of Examples 1.3 and 1.4.

A few values of isothermal compressibilities for pure liquids at several temperatures and at two different pressures are given in Table A3. The values of the coefficient of thermal expansion for several substances are listed in Table A4. Each value applies only to a single temperature and a single pressure, but the dependence on temperature and pressure is not large and these values can be used over a fairly large range of temperature and pressure.

Exercise 1.7

a. Estimate the percent change in volume of a sample of carbon tetrachloride if it is heated from $-20.°\text{C}$ to $45°\text{C}$ at 1.000 atm.

b. Estimate the percent change in volume of a sample of carbon tetrachloride if it is pressurized at $55°\text{C}$ from 1.0 atm to 100.0 atm.

In order to indicate that trailing zeros are significant digits, we will place a decimal point after these digits. Zeros to the left of a decimal point are significant.

In addition to the coefficient of thermal expansion, there is a quantity called the **coefficient of linear thermal expansion** defined by

$$\alpha_L = \frac{1}{L}\left(\frac{\partial L}{\partial T}\right)_P \quad \text{(definition)} \tag{1.5-1}$$

where L is the length of the object. This coefficient is usually used for solids, whereas the coefficient of thermal expansion in Equation (1.3-10) is used more often for liquids. Unfortunately, the symbol α is usually used without the subscript L for the coefficient of linear thermal expansion, and the name "coefficient of thermal expansion" is also sometimes used for it. Since the units of both coefficients are the same (reciprocal temperature), there is an opportunity for confusion between them.

The linear coefficient is equal to one-third of the coefficient of thermal expansion, as can be seen from the following: Subject a cubic object of length L to an infinitesimal change in temperature dT. The new length of the object is

$$L(T + dT) = L(T) + \left(\frac{\partial L}{\partial T}\right)_P dT = L(T)(1 + \alpha_L\, dT)$$

The volume of the object is equal to L^3, so

$$\begin{aligned} V(T + dT) &= L(T)^3[1 + \alpha_L\, dT]^3 \\ &\doteq L(T)^3[1 + 3\alpha_L\, dT + 3(\alpha_L\, dT)^2 + (\alpha_L\, dT)^3] \end{aligned}$$

Since dT is infinitesimal, the last two terms are negligible:

$$V(T + dT) = L(T)^3(1 + 3\alpha_L\, dT) \tag{1.5-2}$$

The volume at temperature $T + dT$ is given by

$$\begin{aligned} V(T + dT) &= V(T) + (\partial V/\partial T)\, dT \\ &= V(T)[1 + \alpha\, dT] \end{aligned} \tag{1.5-3}$$

Comparison of Equations (1.5-3) and (1.5-2) shows that

$$\alpha = 3\alpha_L \tag{1.5-4}$$

The Tait Equation

There is an equation of state of liquids called the Tait equation.[3] The differential version is

$$\left(\frac{\partial V}{\partial P}\right)_T = \frac{-K}{P + L(T)} \tag{1.5-5}$$

where K is a constant parameter proportional to the size of the system and $L(T)$ is a function of T. These quantities are different for each substance. Equa-

---

[3] Hirschfelder, Curtiss, and Bird, *op. cit.*, p. 261.

tion (1.5-5) can be integrated at constant T to obtain

$$V - V_0 = -K \ln\left[\frac{P + L(T)}{P_0 + L(T)}\right] \tag{1.5-6}$$

where V_0 is the volume at the fixed pressure P_0 and ln denotes the natural logarithm (logarithm to the base $e = 2.72818\ldots$).

Exercise 1.8   Carry out the integration to obtain Equation (1.5-6).

Equation (1.5-6) is usually written in a different form:

$$k = \frac{V_0 - V}{V_0} = C \log_{10}\left[\frac{P + L(T)}{P_0 + L(T)}\right] \tag{1.5-7}$$

where k, the ratio of the decrease in volume to the initial volume, is called the **compression**, and $\log_{10}$ denotes the common logarithm (logarithm to the base 10). The constant C is

$$C = \frac{K}{V_0} \ln(10)$$

Gibson and Loeffler[4] found that for several derivatives of benzene C has the same value, 0.2159, and they determined values of L at different temperatures for these compounds. Table A5 gives Gibson and Loeffler's values.

An expression for the compressibility can be written from Equations (1.5-5) and (1.5-6):

$$\kappa_T = \frac{V_0 C}{V[P + L(T)] \ln(10)} \tag{1.5-8}$$

where V is to be calculated from Equation (1.5-6).

▼   **EXAMPLE 1.6**

a. Calculate the percent compression of chlorobenzene from a pressure of 1.00 bar to a pressure of 1000. bar at a temperature of 60°C.

b. Calculate the compressibility of chlorobenzene for a pressure of 1000. bar and a temperature of 65°C.

Solution

a. The compression is

$$k = 0.2159 \log_{10}\left(\frac{1000 + 961}{1 + 961}\right) = 0.0668 = 6.68\%$$

This corresponds to

$$\frac{V}{V_0} = 1 - 0.0668 = 0.9332$$

---

[4] R. E. Gibson and O. H. Loeffler, *J. Am. Chem. Soc.* **62**, 2515 (1939).

b. The compressibility is

$$\kappa_T = \frac{0.2159}{(0.9332)(2.303)(1000. + 961)} = 5.12 \times 10^{-5} \text{ bar}^{-1}$$

which nearly agrees with the value 5.10×10^{-5} bar^{-1} in Table A3 for a pressure of 1000. atm (1013.2 bar).

▲

Exercise 1.9

Calculate the compressibility of chlorobenzene at 1 atm and 65°C as predicted by the Tait equation.

1.6 The Coexistence of Phases and the Critical Point

Transitions from a gaseous state to a liquid state, from a liquid state to a solid state, etc. are called **phase transitions**, and the samples of matter in the different states are called phases. Such transitions take place abruptly, and different phases can coexist. For example, if a gas is at a temperature slightly above its condensation temperature at a certain fixed pressure, a small decrease in the temperature can produce coexisting liquid and gas phases, and a further small decrease in the temperature can produce a single liquid phase, greatly decreasing the volume of the system. Similarly, small increases in the pressure at constant temperature produce the same effects. This remarkable behavior seems to be an exception to the general rule that in nature small causes have small effects and large causes have large effects.

We will discuss the thermodynamics of phase equilibria in Chapter 6. It is found experimentally and understood thermodynamically that for any pure substance the pressure at which any two phases can coexist at equilibrium is a smooth mathematical function of the temperature (or equivalently, that the temperature is a smooth function of the pressure). Figure 1.4 shows schematic curves representing these functions for a typical substance. The curves are called **coexistence curves**, and the figure is called a **phase diagram**. The three curves shown are the solid-vapor (sublimation) curve at the bottom of the figure, the liquid-vapor (vaporization) curve at the upper right, and the solid-liquid (fusion or melting) curve at the upper left. The three curves meet at a point called the **triple point**. This point corresponds to the unique value of the pressure and unique value of the temperature at which all three phases can coexist.

The equilibrium temperature for coexistence of the liquid and solid at a pressure equal to 1 atmosphere is called the **normal melting temperature** or **normal freezing temperature**, and the equilibrium temperature for coexistence of the liquid and gas phases at a pressure equal to 1 atmosphere is called the **normal boiling temperature**. These temperatures are marked on Figure 1.4. If the triple point happens to lie higher in pressure than 1 atmosphere, the substance does not have a normal freezing temperature or a

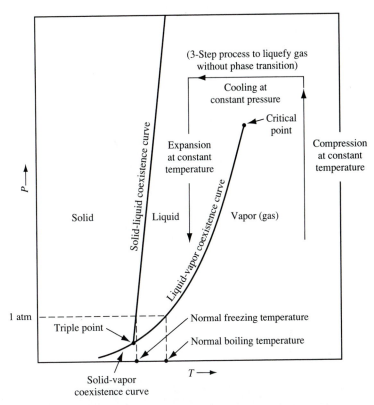

Figure 1.4. The Coexistence Curves for a Typical Pure Substance (schematic). This diagram is typical of many fluids with relatively simple molecules. The curves represent the coexistence of two phases, and the areas represent possible states for a single phase.

normal boiling temperature but instead has a **normal sublimation temperature**, at which the solid and gas coexist at a pressure equal to 1 atmosphere. Carbon dioxide is such a substance. Its triple point occurs at a pressure of 5.112 atm and a temperature of 216.55 K and its normal sublimation temperature is equal to 194.6 K ($-78.5°C$). Equilibrium liquid carbon dioxide can be observed only at pressures greater than 5.112 atm.

The Critical Point

A remarkable feature in Figure 1.4 is the termination of the liquid-vapor coexistence curve at a point called the **critical point**. The temperature, molar volume, and pressure at this point are called the **critical temperature**, denoted by T_c, the **critical molar volume**, denoted by $\bar{V}_c$, and the **critical pressure**, denoted by P_c. These three quantities are collectively called the **critical constants**, and Table A6 gives their values for several substances.

At temperatures higher than the critical temperature and pressures higher than the critical pressure, there is no transition between liquid and gas phases. It is possible to heat a gas to a temperature higher than the critical temperature, to compress it until its density is as large as that of a liquid,

Although solid iodine sublimes noticeably at 1 atmosphere pressure, it is possible to melt it at this pressure by raising the temperature to 113.5°C. Equilibrium liquid carbon dioxide does not exist at 1 atmosphere pressure at any temperature.

Fluids at supercritical temperatures are often referred to as gases, but such fluids can have liquid-like densities or gas-like densities with no phase transition between these densities. At high densities, multicomponent supercritical fluids can even exhibit phase separations like those of two immiscible liquids. Some industrial extractions, such as decaffeination of coffee, are carried out with supercritical fluids.

and then to cool it until it is a liquid without ever having passed through a phase transition. A path representing this kind of process is drawn in Figure 1.4.

There has been some speculation about whether the liquid-solid coexistence curve might also terminate at a critical point. Nobody has found such a critical point, and some people think that the presence of a lattice structure in the solid, which makes it qualitatively different from the liquid, makes the existence of such a point impossible. There is no such qualitative difference between liquid and gas, since both are disordered on the molecular level.

We now want to describe the pressure of a simple system as a function of T and V, the molar volume, over the entire three-phase region. Figure 1.5 shows schematically the pressure as a function of molar volume for several fixed temperatures in the fluid region, with one curve for each fixed temperature. These constant-temperature curves are called **isotherms**. No two isotherms can intersect each other.

For temperatures above the critical temperature, there is only one fluid phase and the isotherms are continuous smooth curves. For subcritical temperatures, the liquid-gas phase transition is represented by a horizontal line segment in the isotherm. This line segment, called a **tie line**, connects

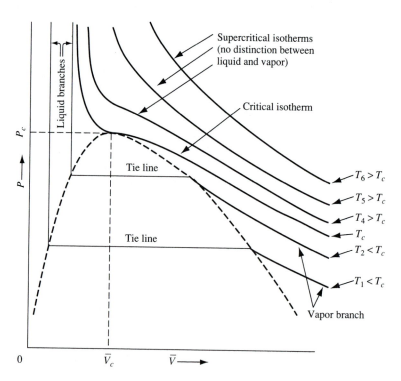

Figure 1.5. Isotherms for a Typical Pure Substance (schematic). The horizontal line segments (tie lines) for temperatures lower than the critical temperature represent the coexistence of the liquid and gas phases. Only the ends of the tie lines represent possible intensive states. The curves other than line segments represent possible intensive states of a single phase.

the two points representing the molar volumes of the coexisting liquid and gas phases. Tie lines are horizontal because the two phases must be at the same pressure to be at equilibrium.

Aside from the tie line, a subcritical isotherm consists of two smooth branches. The liquid branch is nearly vertical, because the liquid is nearly incompressible, and the gas branch of the curve is similar to the curve for an ideal gas.

As isotherms for subcritical temperatures closer and closer to the critical temperature are chosen, the tie lines become shorter and shorter until they shrink to zero length at the critical point. The isotherm that passes through the critical point must have a horizontal tangent line at the critical point, to avoid crossing any tie line. This point on the isotherm is an inflection point, with zero values of $(\partial P/\partial \bar{V})_T$ and $(\partial^2 P/\partial \bar{V}^2)_T$, corresponding to infinite compressibility.

At the critical point a fluid exhibits some unusual properties, such as strong scattering of light and infinite heat capacity, as well as infinite compressibility. Consider a sample of a pure fluid in a rigid closed container such that the average molar volume is equal to that of the critical state. If the temperature is raised through the critical value, the meniscus between the liquid and gas phases becomes diffuse and then disappears at the critical temperature. At this temperature, the liquid and gas phases become indistinguishable. Figure 1.6 illustrates this behavior in a system of carbon dioxide.[5]

The graphs in Figures 1.4 and 1.5 are projections of a single three-dimensional graph, with the pressure as the dependent variable and the temperature and molar volume as the independent variables, as shown in Figure 1.7. (The solid-liquid and solid-gas phase transitions are omitted from the diagram.)

As shown in Figure 1.7, only two independent variables are required to give the pressure as a dependent variable. This is true in general for intensive variables in a one-component fluid system. The **intensive state** of a system is the state as far as intensive variables are concerned. The size of the system is irrelevant. Specification of the intensive state requires one fewer independent variable than specification of the full macroscopic state. That is, for a one-phase simple system of c components, $c + 1$ intensive variables are independent.

Several isotherms (intersections of the surface with planes of constant T) are drawn on the surface in Figure 1.7. The liquid-gas equilibrium tie lines joining liquid and gas states are seen on the tongue-shaped region.

When the three-dimensional graph is viewed in a direction perpendicular to the T-P plane, each liquid-gas tie line is seen as a point, being parallel to the direction of viewing. The set of all of them makes up the gas-liquid coexistence curve seen in Figure 1.4. When the three-dimensional graph is viewed in a direction perpendicular to the $\bar{V}$-P plane, Figure 1.5 results.

----

[5] J. V. Sengers and A. L. Sengers, *Chem. Eng. News* **46**, 54 (June 10, 1968).

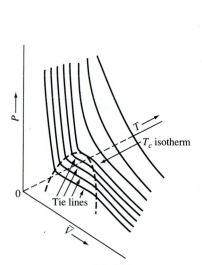

Figure 1.7. Surface Giving Pressure as a Function of Molar Volume and Temperature for a Typical Pure Substance in the Liquid-Vapor Region (schematic). The three-dimensional diagram gives the same information as the set of all isotherms, of which a few are shown in Figure 1.5.

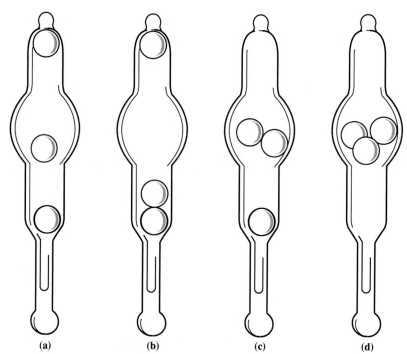

Figure 1.6. Liquid-Gas Equilibrium near the Critical Point. The three balls are slightly different in density, with densities close to the critical density of the fluid, carbon dioxide. **(a)** At a temperature slightly above the critical temperature. The density of the fluid depends slightly on height, due to gravity. **(b)** At the critical temperature, showing the scattering of light known as critical opalescence. **(c, d)** At subcritical temperatures, showing a definite meniscus. From J. V. Sengers and A. L. Sengers, *Chem. Eng. News*, June 10, 1968, p. 104. Used by permission of the copyright holder.

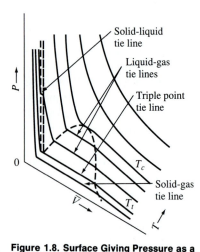

Figure 1.8. Surface Giving Pressure as a Function of Molar Volume and Temperature Showing All Three Phases (schematic). This is a more complete version of the surface of Figure 1.7, with the solid-liquid phase transition shown as well as the liquid-gas phase transition.

Figure 1.8 shows schematically a more complete view of the three-dimensional graph of Figure 1.7, including the solid-liquid and solid-gas phase transitions. There are three sets of tie lines, as labeled in the figure, corresponding to the three curves in Figure 1.4.

Because the entire fluid (liquid and gas) surface in Figure 1.7 is connected, it should be possible to obtain an equation of state that represents the entire surface accurately, making it unnecessary to use separate equations of state for the liquid and the gas. Except for the ideal gas equation, all of the gas-phase equations of state that we have discussed yield surfaces that resemble the true surface in the liquid region as well as in the gas region, although they do not represent the tie lines. The modified Redlich-Kwong-Soave equation of Gibbons and Laughton represents both the liquid and the gas fairly accurately, and the van der Waals equation is often used to give qualitative information. In Chapter 5 we will discuss the behavior of the equations of state in the two-phase region and the construction of the tie lines.

Location of the Critical Point for a van der Waals Gas

We seek the point at which

$$(\partial P/\partial \bar{V})_T = 0 \qquad \text{(1.6-1)}$$

$$(\partial^2 P/\partial \bar{V}^2)_T = 0 \qquad \text{(1.6-2)}$$

The first derivative of Equation (1.4-1) is

$$\left(\frac{\partial P}{\partial \bar{V}}\right)_T = -\frac{RT}{(\bar{V} - b)^2} + \frac{2a}{\bar{V}^3} \qquad \text{(1.6-3)}$$

and the second derivative is

$$\left(\frac{\partial^2 P}{\partial \bar{V}^2}\right)_T = \frac{2RT}{(\bar{V} - b)^3} - \frac{6a}{\bar{V}^4} \qquad \text{(1.6-4)}$$

Setting the right-hand side of each of these two equations equal to zero gives us two simultaneous algebraic equations, which are solved to give the values of the critical temperature T_c and the critical molar volume $\bar{V}_c$:

$$T_c = \frac{8a}{27Rb}, \qquad \bar{V}_c = 3b \qquad \text{(1.6-5)}$$

Exercise 1.10   Solve the simultaneous equations to verify Equation (1.6-5). One way to proceed is as follows: Obtain Equation (I) by setting the right-hand side of Equation (1.6-3) equal to zero and Equation (II) by setting the right-hand side of Equation (1.6-4) equal to zero. Solve equation (I) for T and substitute this expression into equation (II).

When the values of T_c and $\bar{V}_c$ are substituted into Equation (1.4-1), the value of the critical pressure for a van der Waals gas is obtained:

$$P_c = \frac{a}{27b^2} \qquad \text{(1.6-6)}$$

For a van der Waals gas, the compression factor at the critical point is

$$Z_c = \frac{P_c \bar{V}_c}{RT_c} = \frac{3}{8} = 0.375 \qquad \text{(1.6-7)}$$

Exercise 1.11   Verify Equations (1.6-6) and (1.6-7).

Equations (1.6-5) and (1.6-6) can be solved for a and b:

$$a = 3\bar{V}_c^2 P_c = \frac{9R\bar{V}_c T_c}{8} = \frac{27R^2 T_c^2}{64P_c} \qquad \text{(1.6-8)}$$

$$b = \frac{\bar{V}_c}{3} = \frac{RT_c}{8P_c} \qquad \text{(1.6-9)}$$

The van der Waals parameters a and b for a particular gas could also be obtained by fitting P-V-T data in any region. Most of the tabulated values seem to be from critical data.

There are two or three formulas for each parameter, because values for only two variables are needed to obtain values for a and b. Since no substance exactly fits the equation, different values can result from the different formulas. The most accurate fit is probably obtained using P_c and T_c as independent variables, since two-parameter equations of state do not usually give good values of the critical molar volume.

The values of the parameters for any two-parameter or three-parameter equation of state can be obtained from critical constants.

Exercise 1.12

a. Show that for the Dieterici equation of state

$$\bar{V}_c = 2b, \qquad T_c = \frac{a}{4bR}, \qquad P_c = \frac{a}{4b^2}\, e^{-2} \qquad \text{(1.6-10)}$$

b. Show that for the Dieterici equation of state, $Z_c = 2e^{-2} = 0.27067$.

c. Obtain the formulas giving the Dieterici parameters a and b as functions of P_c and T_c. Find the values of a and b for nitrogen.

The parameters a and b in the Redlich-Kwong equation of state can be obtained from the relations

$$a = \frac{R^2 T_c^{5/2}}{9(2^{1/3} - 1)P_c}, \qquad b = \frac{(2^{1/3} - 1)RT_c}{3P_c} \qquad \text{(1.6-11)}$$

Exercise 1.13

Find the values of a and b in the Redlich-Kwong equation of state for nitrogen.

The Law of Corresponding States

From Equation (1.6-7) we see that all substances that obey the van der Waals equation have the same value of the compression factor Z at the critical point, $Z_c = 0.375$. Any two-parameter equation of state gives a characteristic constant value of Z_c for all substances, although the different equations of state do not give the same value, as seen in Exercise 1.12. The Berthelot equation does give the same value as the van der Waals equation, 0.375. The experimental values for many different substances lie between 0.25 and 0.30.

There is even a greater degree of generality, which we now illustrate with the van der Waals equation of state. We define a new set of variables called reduced variables. The **reduced volume** is defined as the ratio of the molar volume to the critical molar volume:

$$V_r = \frac{V}{V_c} = \frac{\bar{V}}{\bar{V}_c} \qquad \text{(1.6-12)}$$

The **reduced pressure** is defined as the ratio of the pressure to the critical pressure:

$$P_r = \frac{P}{P_c} \qquad (1.6\text{-}13)$$

and the **reduced temperature** is defined as the ratio of the temperature to the critical temperature:

$$T_r = \frac{T}{T_c} \qquad (1.6\text{-}14)$$

Using the definitions in Equations (1.6-12), (1.6-13), and (1.6-14) and the relations in Equations (1.6-5) and (1.6-6), we obtain

$$P = \frac{aP_r}{27b^2}, \qquad \bar{V} = 3bV_r, \qquad T = \frac{8aT_r}{27Rb}$$

When these relations are substituted into Equation (1.4-1), the result is

$$\left(P_r + \frac{3}{V_r^2}\right)\left(V_r - \frac{1}{3}\right) = \frac{8T_r}{3} \qquad (1.6\text{-}15)$$

Exercise 1.14   Carry out the algebraic steps to obtain Equation (1.6-15).

In Equation (1.6-15), the parameters a and b have canceled out. The same equation of state, without adjustable parameters, applies to every substance that obeys the van der Waals equation of state, if the reduced variables are used instead of P, $\bar{V}$, and T. This is in agreement with an empirical law called the **law of corresponding states**:[6] *All substances obey the same equation of state in terms of the reduced variables.*

Figure 1.9 is a graph of the experimentally measured compression factor of a number of polar and nonpolar fluids as a function of reduced pressure at a number of reduced temperatures.[7] The agreement of the data for different substances with the law of corresponding states is better than the agreement of the data with any simple equation of state.

Exercise 1.15   All two-parameter equations of state conform to the law of corresponding states. Show this fact for the Dieterici equation of state by expressing it in terms of the reduced variables.

[6] Hirschfelder, Curtiss, and Bird, *op. cit.*, p. 235.
[7] G.-J. Su, *Ind. Eng. Chem.* **38**, 803 (1946).

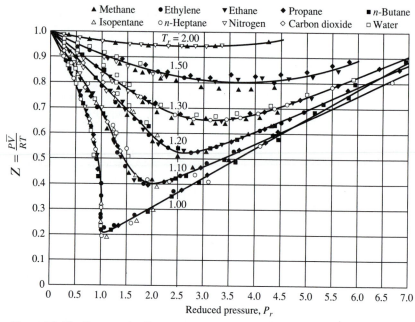

Figure 1.9. The Compression Factor as a Function of Reduced Pressure and Reduced Temperature for a Number of Gases. This figure shows the extent to which 10 substances obey the law of corresponding states. From G.-J. Su, *Ind. Eng. Chem.* **38**, 803 (1946). Used by permission of the copyright holder.

The Law of Rectilinear Diameters

Figure 1.10 illustrates an empirical law that is useful in locating the critical point from experimental data.[8] In this figure, the densities of the coexisting gas and liquid phases of hydrogen are plotted as a function of temperature. The arithmetic mean of these two quantities, which is called the **rectilinear diameter**, is also plotted. The **law of rectilinear diameters** states that the rectilinear diameter is a segment of a straight line. This empirical law is fairly well obeyed by nearly all fluids. Since the gas and liquid densities become equal at the critical point, the rectilinear diameter must pass through the critical point. Locating the rectilinear diameter aids in locating the critical point.

1.7   Microscopic States and Their Relationship to Macroscopic States

The second principal type of state is a **microscopic state** or **microstate**, which pertains to detailed mechanical properties, and which is sometimes called a **mechanical state**. If classical (Newtonian) mechanics can be used, the microstate is specified by giving the position and the velocity of every particle in

---

[8] Hirschfelder, Curtiss, and Bird, *op. cit.*, p. 362.

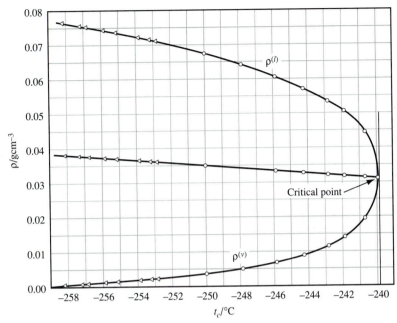

Figure 1.10. The Density of Hydrogen Liquid and Vapor as a Function of Temperature, with the Rectilinear Diameter. The rectilinear diameter closely approximates a straight line in this case. From Hirschfelder, Curtiss, and Bird, *op. cit.*, p. 362.

the system. Quantum mechanics is a more nearly accurate theory for systems containing particles of small mass than is classical mechanics, and we will discuss quantum mechanics beginning in Chapter 9. If quantum mechanics must be used, the microscopic state is specified in a different way, by specifying which member of a large list of possible states applies. This specification is usually done by giving the values of a set of quantum numbers. Roughly the same amount of information is needed to specify a quantum state as to specify the analogous classical state, and we illustrate the distinction between macrostates and microstates using classical mechanics.

The Relationship Between Macrostates and Microstates

If our system consists of a small number of atoms or molecules, the microstate is the only kind of state to be used. Macroscopic variables are not used because they apply only to aggregates of many molecules. For example, there is no such thing as the temperature of a single molecule. However, for a large system we can define both the macroscopic and microscopic state. For a system of many molecules, specification of the macrostate requires values of a few macroscopic variables, whereas specification of the microstate requires values of many mechanical variables.

Consider for example a sample of helium gas. The equilibrium macrostate can be specified by giving values for three variables—saying, for example, that the amount of helium in the system equals 1.000 mol, the pressure of the system of helium gas equals 1.000 atm, and the temperature of the system equals 298.15 K.

Even ignoring the motion of electrons inside the atoms and assuming that classical mechanics is an adequate description, specification of the microstate of our helium gas requires the values of three coordinates and three velocity components for each atom, a total of 3.6×10^{24} values. If these values were printed on ordinary computer printer paper with the six values for each atom on one line, the values for 66 atoms could be printed on one sheet. With 2600 sheets per box, 3.5×10^{18} boxes would be required for a system of 1.00 mol. At \$20 per box, this would cost \$70 billion billion for the paper, about 20 million times the 1992 federal debt of the United States. A computer printer that can print five sheets per minute would require 3.5×10^{15} years to print the values. Furthermore, the state represented by this huge mound of printed paper would apply only for an instant. As soon as any molecule moves or changes its velocity, a different state is occupied.

Just as with a macroscopic state, it is not necessary to specify the value of every mechanical variable to specify a microscopic state. In classical mechanics, if the values of the position and velocity of every particle are known, it is not necessary to specify the values of other variables, such as the angular momentum, the pressure, and the energy. They are state functions of the microscopic state.

We use the standard notation for a sum, a capital Greek sigma followed by an expression for the "general" term. The index for the first term is beneath the sigma, and the index for the last term is above the sigma.

The value of the **kinetic energy** is determined as a dependent variable by the velocities of the particles:

$$\mathscr{K} = \tfrac{1}{2}mv_1^2 + \tfrac{1}{2}mv_2^2 + \cdots + \tfrac{1}{2}mv_N^2 = \tfrac{1}{2}m \sum_{i=1}^{N} v_i^2$$
$$= \mathscr{K}(v_{x1}, v_{y1}, v_{z1}, v_{x2}, v_{y2}, v_{z2}, \ldots, v_{xN}, v_{yN}, v_{zN}) \qquad \textbf{(1.7-1)}$$

where v_{x1} is the x component of the velocity of particle number 1, etc. We have used the fact that all of the atoms of our system have the same mass.

The **potential energy** $\mathscr{V}$ is defined so that a velocity-independent force $\mathbf{F}_i$ on particle number i can be derived from it, as in Equation (D-5) of Appendix D:

$$F_{ix} = -\left(\frac{\partial \mathscr{V}}{\partial x_i}\right), \qquad F_{iy} = -\left(\frac{\partial \mathscr{V}}{\partial y_i}\right), \qquad F_{iz} = -\left(\frac{\partial \mathscr{V}}{\partial z_i}\right) \qquad \textbf{(1.7-2)}$$

where F_{ix} is the x component of the force $\mathbf{F}_i$, etc. The potential energy $\mathscr{V}$ depends on the positions of the particles of the system

$$\mathscr{V} = \mathscr{V}(x_1, y_1, z_1, x_2, y_2, z_2, \ldots, x_N, y_N, z_N) = \mathscr{V}(q) \qquad \textbf{(1.7-3)}$$

where q stands for the coordinates of all particles in the system.

An important property of the potential energy can be seen by inspection of Equation (1.7-2): If any constant is added to the value of $\mathscr{V}$, the forces are unchanged, because the derivative of a constant is equal to zero. Since the forces are the only physically meaningful effects of the potential energy, we have the following fact: *An arbitrary constant can be added to any potential energy without any physical effect.*

If we exclude the rest mass energy, the total energy of the system is the sum of the kinetic energy and the potential energy:

$$E = \mathcal{K} + \mathcal{V} \tag{1.7-4}$$

Since $\mathcal{K}$ is determined by the velocities of the molecules and $\mathcal{V}$ is determined by their positions, E is a function of positions and velocities of the particles and is therefore a state function of the microscopic state.

Now consider the macroscopic state of the same system. The position and velocity of the center of mass of a system are macroscopic variables and can be discussed macroscopically. We denote the kinetic energy of the macroscopic motion of the system by

$$\mathcal{K}_{cm} = \tfrac{1}{2} m_{sys} v_{cm}^2 \tag{1.7-5}$$

where m_{sys} is the mass of the system and v_{cm} is the speed of the center of mass of the system. We denote the potential energy of the entire system by $\mathcal{V}_g$. For example, near the surface of the earth, we can write

$$\mathcal{V}_g = m_{sys} g z_{cm} \tag{1.7-6}$$

where g is the acceleration due to gravity and z_{cm} is the vertical coordinate of the center of mass.

We define the internal kinetic and internal potential energy:

$$\mathcal{V}_{int} = \mathcal{V} - \mathcal{V}_g \tag{1.7-7a}$$

$$\mathcal{K}_{int} = \mathcal{K} - \mathcal{K}_{cm} \tag{1.7-7b}$$

The internal energy will be defined and discussed macroscopically in Chapter 2.

We define U, the **internal energy** or the **thermodynamic energy**:

$$U = \mathcal{K}_{int} + \mathcal{V}_{int} \tag{1.7-8}$$

Although $\mathcal{K}_{int}$ and $\mathcal{V}_{int}$ are obviously microscopic variables, we assert as an experimental fact that the internal energy is a macroscopic state function for a system at equilibrium. For a one-phase fluid system of one substance, this assertion means that the internal energy is a function of three independent variables:

$$U = U(T, V, n)$$

or

$$U = U(T, P, n)$$

etc.

The macroscopic and microscopic states of a given system are not independent. For example, we have seen that the value of the internal energy U, a macroscopic variable, is determined by the microscopic state of the system. We assert a more general relationship: *If the microscopic state is known, the macroscopic state is determined, but if the macroscopic state is known, we are almost completely ignorant of the microscopic state.* We argue now that a single macroscopic state corresponds to a large number of microscopic states: If a system is at equilibrium, its macrostate does not change over a period of time. However, since the molecules are moving rapidly (on the average at roughly the speed of sound), the microstate changes very rapidly and the system must pass through a very large number of microstates without changing the macroscopic state. The value of a macroscopic variable must represent an average of a corresponding microscopic variable over these many microstates.

Another argument for associating a macroscopic state with an average over microscopic states is the fact that macroscopic variables are measured with measuring instruments such as thermometers of macroscopic size and manometers of macroscopic size. Such large instruments always require a certain length of time to respond to a change in the value of the measured variable. During this *response time*, the system must pass through many microscopic states, and the measured value of the macroscopic variable must correspond to an average over all of the microscopic states the system occupied during the response time. We now briefly discuss the ways in which averages are calculated.

Averaging Procedures and Probability Distributions

If we have a set of numbers $w_1, w_2, w_3, w_4, \ldots, w_N$, the **mean** of this set is defined by

$$\langle w \rangle = \frac{1}{N} \sum_{i=1}^{N} w_i \quad \text{(definition)} \tag{1.7-9}$$

The mean is the most commonly used type of average.

Assume now that some of the w's are equal to each other. Let us arrange the members of our set so that all of the distinct values are at the first of the set, with $w_1, w_2, \ldots, w_M$ all different in value from each other. Every remaining member of the set is equal to one or another of the first M members. Let N_i be the total number of members of the set equal to w_i. The mean can now be written as a sum over only the distinct members of the set

$$\langle w \rangle = \frac{1}{N} \sum_{i=1}^{M} N_i w_i \tag{1.7-10}$$

There are fewer terms in this sum than in the sum of Equation (1.7-9) unless every N_i equals unity, but the mean value is unchanged. We must still divide by N, not by M, to get the correct mean value.

We define

$$p_i = \frac{N_i}{N} \tag{1.7-11}$$

The quantity p_i is the fraction of the entire set equal to w_i and is equal to the probability that a randomly chosen member of the set will be equal to w_i. The set of p_i values is called a **probability distribution**. We can now write

$$\langle w \rangle = \sum_{i=1}^{M} p_i w_i \tag{1.7-12}$$

From the definition of p_i in Equation (1.7-11), these probabilities are **normalized**, which means that they sum to unity.

$$\sum_{i=1}^{M} p_i = \frac{1}{N} \sum_{i=1}^{M} N_i = \frac{N}{N} = 1 \quad \text{(normalization)} \tag{1.7-13}$$

Exercise 1.16 A quiz was given to a class of 50 students. The scores were as follows:

Score	Number of Students
100	5
90	9
80	17
70	15
60	2
50	2

Find the mean score on the quiz without taking a sum of 50 terms.

We can also get the mean of a function of our values. If $h(w)$ is some function, its mean value is

$$\langle h \rangle = \sum_{i=1}^{M} p_i h(w_i) \tag{1.7-14}$$

Exercise 1.17 For the quiz scores in Exercise 1.16, find the mean of the squares of the scores

$$\langle w^2 \rangle = \sum_{i=1}^{M} p_i w_i^2$$

and the square root of this mean, called the **root-mean-square** score.

The Probability Distribution for Molecular States

Because we must average over very many molecular states, we require a probability distribution for microscopic states of a system. The microscopic states of the entire system can be specified by specifying the positions and velocities of the individual molecules. If the molecules do not exert forces on each other, the state of one molecule is independent of the states of the other molecules. This is nearly the case in a dilute gas, in which the molecules are relatively far apart.

For a system of independent molecules, we can average over the microstates of the system by averaging over the mechanical states of the individual molecules. However, the average over molecular states is not necessarily an average in which each molecular state occurs with equal probability. We now seek a probability distribution for these molecular states. We make two assumptions that will determine the mathematical form of the molecular probability distribution:

1. *The probability of a molecular state depends only on the energy of the molecular state.*
2. *The same probability distribution applies for all kinds of molecules.*

Let $p(\varepsilon)$ be the probability that a molecular state of energy ε will occur for a molecule randomly chosen from a large number of molecules. From the second assumption, there is only one function p for all types of molecules.

Consider a pair of molecules, which we call molecule 1 and molecule 2. Considering the pair of molecules as though it were a larger molecule, we assume that its probability density P depends on the sum of the energies of the two particles:

$$P = P(\varepsilon) = P(\varepsilon_1 + \varepsilon_2) \tag{1.7-15}$$

Since the molecules do not interact with each other, they are independent of each other. We apply a fact of probability theory: *The probability of the occurrence of two independent events is the product of the probabilities of the two events.* Therefore,

$$P(\varepsilon) = p(\varepsilon_1)p(\varepsilon_2) \tag{1.7-16}$$

The only function that satisfies Equation (1.7-16) is the exponential function, as we now show.

Differentiate Equation (1.7-16) with respect to ε_1.

$$\partial P / \partial \varepsilon_1 = (dp/d\varepsilon_1)p(\varepsilon_2) \tag{1.7-17}$$

By the chain rule, the left side of this equation is

$$\partial P / \partial \varepsilon_1 = (dP/d\varepsilon)(\partial \varepsilon/\partial \varepsilon_1) = dP/d\varepsilon \tag{1.7-18}$$

We substitute this expression into Equation (1.7-17) and divide by P.

$$\frac{1}{P(\varepsilon)}\frac{dP}{d\varepsilon} = \frac{1}{p(\varepsilon_1)}\frac{dp}{d\varepsilon_1} \tag{1.7-19}$$

An analogous equation can be written by differentiating with respect to ε_2 instead of ε_1. The left-hand sides of the two equations are identical, so

$$\frac{1}{p(\varepsilon_1)}\frac{dp}{d\varepsilon_1} = \frac{1}{p(\varepsilon_2)}\frac{dp}{d\varepsilon_2} \tag{1.7-20}$$

Here is our first use of separation of variables, a mathematical technique that we will use a number of times. The important step is one in which terms in an equation are set equal to constants.

In this equation, the variables ε_1 and ε_2 are separated. This means that ε_1 occurs only in one term and ε_2 occurs only in the other term. Since they are independent variables, we can keep ε_2 fixed while we allow ε_1 to vary. While ε_2 is held fixed, the right-hand side of the equation is fixed and the left-hand side must be a constant function of ε_1. A similar argument applies when ε_1 is held fixed. Each side of the equation must equal a constant, which we denote by c.

$$\frac{1}{p(\varepsilon_1)}\frac{dp}{d\varepsilon_1} = c \tag{1.7-21}$$

We multiply this equation by $d\varepsilon_1$ and carry out an indefinite integration:

$$\ln(p) = c\varepsilon_1 + A \tag{1.7-22}$$

where A is a constant of integration. This is the same as

$$p(\varepsilon_1) = e^A e^{c\varepsilon_1} \propto e^{c\varepsilon_1} \tag{1.7-23}$$

For now, we omit the evaluation of A and write the proportionality instead of the equality.

We must determine what the parameter c is. For now, we defer this analysis to Chapters 16 and 20 and assert without proof that

$$c = -\frac{1}{k_B T} \tag{1.7-24}$$

where k_B is Boltzmann's constant, introduced in Equation (1.2-10).

The probability distribution for molecular states in a dilute gas is

$$p(\varepsilon) \propto e^{-\varepsilon/k_B T} \tag{1.7-25}$$

where k_B is Boltzmann's constant and T is the absolute temperature. This probability distribution is called the **Boltzmann distribution**.

The Boltzmann probability distribution has a number of important properties:

1. States of higher energy are less probable than states of lower energy.
2. At higher temperatures, the difference in population between states of high energy and states of low energy decreases, until, as T approaches infinity, all states approach equal probability.
3. As T approaches zero on the Kelvin scale, only the states of lowest energy are populated.

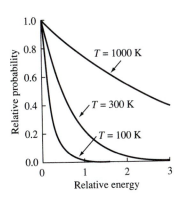

Figure 1.11. The Boltzmann Probability Distribution as a Function of Energy for Three Different Temperatures. This figure illustrates the properties of the Boltzmann distribution as described in the text, with higher probabilities for lower energies but with the difference becoming smaller at higher temperatures.

Figure 1.11 shows the Boltzmann probability distribution for a range of molecular energies and for three different temperatures, 100, 300, and 1000 K. The probability of a state at 0 K is arbitrarily set equal to 1 in this figure. We will find in Chapters 16 and 20 that the average kinetic energy of molecules of a monatomic gas is $3k_B T/2$ and is independent of the mass of the molecules. The value of this quantity at 300 K is 6.21×10^{-21} J, and this value corresponds to relative energy = 1. At 100 K, states of this energy are almost unpopulated, but at 1000 K they have a much higher probability.

EXAMPLE 1.7

Assume that the earth's atmosphere is at equilibrium at 298 K. Estimate the pressure at 8900 meters above sea level (roughly the altitude of Mount Everest). Assume that air is a single substance with a molar mass of 0.029 kg mol^{-1}.

Solution

We assume that the temperature is constant and that the kinetic energies of the molecules are the same on the average at all altitudes. The gravitational potential energy is $u = mgz$, where we take z to be the height above sea level.

$$u = \frac{(0.029 \text{ kg mol}^{-1})(9.80 \text{ m s}^{-2})(8900 \text{ m})}{6.022 \times 10^{23} \text{ mol}^{-1}}$$

$$= 4.2 \times 10^{-21} \text{ J}$$

If $P = 1.00$ atm at sea level and if we assume that air is an ideal gas:

$$P = (1.00 \text{ atm}) \exp\left[\frac{-4.20 \times 10^{-21} \text{ J}}{(1.3807 \times 10^{-23} \text{ J K}^{-1})(298 \text{ K})}\right]$$

$$= 0.36 \text{ atm}$$

Exercise 1.18   The altitude of Pike's Peak is 14110 ft. Estimate the barometric pressure on Pike's Peak on a winter day when the temperature is $-20°F$.

Summary

The state of a system is the numerical specification of the circumstance in which it is found. There are two important kinds of states of macroscopic systems: the macroscopic state, which pertains to variables of the system as a whole, and the microscopic state, which pertains to the mechanical variables of individual molecules.

Processes are the means by which the state of the system changes. The two principal kinds of processes are reversible processes and irreversible processes. The direction of a reversible process can be changed by an infinitesimal change in the surroundings. The system passes through equilibrium states during a reversible process.

The equilibrium macroscopic state of a one-phase simple system of one component is specified by the values of three independent state variables. All other macroscopic state variables are dependent variables, with values given by mathematical functions of the independent variables.

The volumetric (P-V-T) behavior of gases and liquids was described. A "calculation policy" was introduced. According to this policy, gases are treated as ideal unless there is some specific reason for a more accurate treatment. Solids and liquids are similarly treated as incompressible. For cases in which the ideal gas law is not an adequate description, several equations of state were presented.

Phase transitions were introduced. When two phases of a single substance are at equilibrium, the pressure is a function only of the temperature. A phase diagram for a pure substance contains three curves representing this dependence for the solid-liquid, solid-gas, and liquid-gas equilibria. These three curves meet at a point called the triple point. The liquid-vapor coexistence curve terminates at the critical point. Above the critical temperature, no gas-liquid phase transition occurs and there is only one fluid phase.

Microscopic states of macroscopic systems were introduced, and their relationship to macroscopic states was discussed. A macroscopic state of a system can be represented as an average over very many microscopic states of the system.

Additional Reading

C. Domb and M. S. Green, *Phase Transitions and Critical Phenomena*, Academic Press, New York

This is a series of volumes that present theories at a very high level for specialists in the field.

P. A. Egelstaff, *An Introduction to the Liquid State*, Academic Press, New York, 1967

A small book intended to introduce graduate students in physics to the theory and description of the liquid state.

J. O. Hirschfelder, C. F. Curtiss, and R. B. Bird, *Molecular Theory of Gases and Liquids*, Wiley, New York, 1954

This large and useful book contains a great deal of numerical data and carefully presents nearly all of the fluid theory known at the time of its publication. A second printing (1965) contains corrections and some added material.

R. C. Reid, J. M. Prausnitz, and Thomas K. Sherwood, *The Properties of Gases and Liquids*, 3rd ed., McGraw-Hill, New York, 1977

Intended for chemical engineers, this book contains quite a bit of data, as well as correlations that can be used to estimate numerical values when data are not available.

PROBLEMS

Problems for Section 1.2

1.19. a. A sample of oxygen gas is collected over water at $25°C$ at a total pressure of 748.5 torr, with a partial pressure of water vapor equal to 23.8 torr. If the volume of the collected gas is 454 mL, find the mass of the oxygen.

b. If the oxygen was produced by decomposition of $KClO_3$, find the mass of $KClO_3$.

*1.20. Find the volume of CO_2 gas produced from 100.0 g of $CaCO_3$ if the CO_2 is at a pressure of 746 torr and a temperature of 301.0 K.

Problems for Section 1.3

1.21. Show that the three partial derivatives obtained from $PV = nRT$ with n fixed conform to the cycle rule, Equation (B-15) of Appendix B.

1.22. For the van der Waals equation of state, obtain formulas for the partial derivatives $(\partial P/\partial T)_{V,n}$, $(\partial P/\partial V)_{T,n}$, and $(\partial P/\partial n)_{T,V}$.

1.23. Finish the equation

$$(\partial P/\partial V)_{T,n} = (\partial P/\partial V)_{U,n} + ?$$

Hint: See Equations (B-6) and (B-7) of Appendix B.

1.24. Take $z = ax \tan(y/b)$, where a and b are constants.

a. Find the partial derivatives $(\partial z/\partial x)_y$, $(\partial x/\partial y)_z$, and $(\partial y/\partial z)_x$.

b. Show that the derivatives of part a conform to the cycle rule, Equation (B-15) of Appendix B.

*1.25. a. Estimate the percent change in the volume of a sample of carbon tetrachloride if it is pressurized from 1.000 atm to 2.000 atm at $25°C$.

b. Estimate the percent change in the volume of a sample of carbon tetrachloride if its temperature is changed from $25°C$ to $35°C$.

1.26. Let $f(u) = \cos(au)$ and $u = x^2 + y^2$, where a is a constant. Using the chain rule, find $(\partial f/\partial x)_y$ and $(\partial f/\partial y)_x$.

1.27. For the virial equation of state:

a. Find the expressions for $(\partial P/\partial V)_{T,n}$ and $(\partial P/\partial T)_{V,n}$.

b. Show that $(\partial^2 P/\partial V \, \partial T)_n = (\partial^2 P/\partial T \, \partial V)_n$.

1.28. Derive an expression for the isothermal compressibility of a gas obeying the van der Waals equation of state. Hint: Use the reciprocal identity, Equation (B-8).

Problems for Section 1.4

1.29. Write expressions giving the compression factor Z as a function of temperature and molar volume for the van der Waals, Dieterici, and Redlich-Kwong equations of state.

1.30. a. For the van der Waals equation of state at temperatures below the Boyle temperature, find an expression for a value of the pressure other than $P = 0$ for which $P\bar{V} = RT$.

b. Find the value of this pressure for nitrogen gas at 273.15 K.

1.31. a. By differentiation, find an expression for the isothermal compressibility of a gas obeying the van der Waals equation of state.

*b. Find the value of the isothermal compressibility of nitrogen gas at 273.15 K.

Problems marked with an asterick have answers in the back of the book.

1.32. a. By differentiation, find an expression for the coefficient of thermal expansion of a gas obeying the van der Waals equation of state.

b. Find the value of the coefficient of thermal expansion of nitrogen gas at 273.15 K.

1.33. By differentiation, find an expression for the coefficient of thermal expansion of a gas obeying the Dieterici equation of state.

1.34. Manipulate the Dieterici equation of state into the virial form as in Equation (1.4-2). Use the identity

$$e^{-x} = 1 - x + \frac{x^2}{2!} - \frac{x^3}{3!} + \cdots + (-1)^n \frac{x^n}{n!} + \cdots$$

where $n! = n(n-1)(n-2)(n-3)\cdots(3)(2)(1)$. Write expressions for the second, third, and fourth virial coefficients.

1.35. Write an expression for the isothermal compressibility of a nonideal gas obeying the Redlich-Kwong equation of state.

1.36. Express A_3 in Equation (1.4-3) in terms of B_2 and B_3 in Equation (1.4-2).

*1.37. The experimental value of the compression factor $Z = P\bar{V}/RT$ for hydrogen gas at $T = 273.15$ K and $\bar{V} = 0.1497$ L mol^{-1} is 1.1336. Find the values of Z predicted by the van der Waals, Dieterici, and Redlich-Kwong equations of state for these conditions. Calculate the percent error for each.

1.38. a. Evaluate the parameters in the Dieterici equation of state for argon from critical point data.

b. Find the Boyle temperature of argon according to the Dieterici equation of state.

Problems for Section 1.5

1.39. The coefficient of linear expansion of borosilicate glass is equal to 3.2×10^{-6} K^{-1}.

a. Calculate the pressure of a sample of helium (assumed ideal) in a borosilicate glass vessel at 100°C if its pressure at 0°C is 1.000 atm. Compare with the value of the pressure calculated assuming that the volume of the vessel is constant.

b. Repeat the calculation of part a using the virial equation of state truncated at the B_2 term. The value of B_2 for helium is 11.8 cm^3 mol^{-1} at 0°C and 11.4 cm^3 mol^{-1} at 100°C.

*1.40. Assuming that the coefficient of thermal expansion of gasoline is roughly equal to that of benzene, estimate the fraction of your gasoline expense that could be saved by purchasing gasoline in the morning instead of the afternoon, assuming a temperature difference of 5°C.

1.41. The volume of a sample of a liquid is sometimes represented by

$$\bar{V}(t_C) = \bar{V}(0°C)(1 + \alpha' t_C + \beta' t_C^2 + \gamma' t_C^3)$$

where α', β', and γ' are constants and t_C is the Celsius temperature.

a. Find an expression for the coefficient of thermal expansion as a function of t_c.

b. Evaluate the coefficient of thermal expansion of benzene at 20.00°C, using $\alpha' = 1.17626 \times 10^{-3}$(°C)$^{-1}$, $\beta' = 1.27776 \times 10^{-6}$(°C)$^{-2}$, and $\gamma' = 0.80648 \times 10^{-8}$(°C)$^{-3}$. Compare your value with the value in Table A4.

1.42. The coefficient of thermal expansion of ethanol equals 1.12×10^{-3} K^{-1} at 20°C and 1.000 atm. Find the volume of 1.000 mol of ethanol at 10°C and at 30°C. The density at 20°C is equal to 0.7893 g cm^{-3}.

*1.43. Calculate the molar volume of liquid water at 100°C and 1.000 atm by the van der Waals equation of state. (A cubic equation must be solved. Get a numerical approximation to the solution by trial and error or other numerical method). Compare your answer with the correct value, 18.798 cm^3 mol^{-1}

Problems for Section 1.6

1.44. Show that the Redlich-Kwong equation of state conforms to the law of corresponding states.

1.45. a. Find formulas for the parameters a and b in the Soave and Gibbons-Laughton modifications of the Redlich-Kwong equation of state in terms of the critical constants. Show that information about the extra parameters is not needed.

b. Find the values of the parameters a and b for nitrogen.

1.46. Show that if the virial equation of state is truncated at the B_2 term, it cannot predict the existence of a critical point.

1.47. a. Show that if the virial equation of state is truncated at the third term,

$$\frac{P\bar{V}}{RT} = 1 + \frac{B_2}{\bar{V}} + \frac{B_3}{\bar{V}^2}$$

a critical point can occur.

b. Find the expressions for the critical pressure, the critical molar volume, and the critical temperature in terms of the virial coefficients.

c. Find the value of the compression factor $Z = P\bar{V}/RT$ at the critical point.

Problems for Section 1.7

*1.48. a. Find the probability of drawing the ace of spades from one deck of 52 cards and drawing the eight of diamonds from another deck of 52 cards.

b. Find the probability of drawing the ace of spades and the eight of diamonds (in that order) from a single deck of 52 cards.

c. Find the probability of drawing the ace of spades and the eight of diamonds (in either order) from a single deck of 52 cards.

1.49. Compute the odds for each of the possible values of the sum of the two numbers showing when two dice are thrown.

1.50. Assume that air is 80.% nitrogen and 20.% oxygen, by moles, at sea level. Calculate the percentages and the total pressure at an altitude of 20. km, assuming a temperature of $-20°C$ at all altitudes. Calculate the percent error in the total pressure introduced by assuming that air is a single substance with molar mass 0.029 kg mol^{-1}.

1.51. Calculate the difference in the density of air at the top and bottom of a vessel 1.00 m tall at 273.15 K at sea level.

*1.52. Estimate the difference in barometric pressure between the ground floor of a building and the forty-first floor, assumed to be 400. feet higher. State any assumptions.

General Problems

1.53. a. Manipulate the van der Waals equation of state into the virial form of Equation (1.4-2). Use the identity

$$\frac{1}{1-x} = 1 + x + x^2 + x^3 + \cdots$$

b. For each of the temperatures in Table A2 at which a value of the second virial coefficient is given for argon, calculate the value of the second virial coefficient from the values of the van der Waals parameters. Calculate the percent error for each value, assuming that the values in Table A2 are correct.

c. In terms of intermolecular forces, what does it mean when a second virial coefficient is positive? What does it mean when a second virial coefficient is negative? Draw a graph of the second virial coefficient of argon as a function of temperature, and comment on the temperature dependence.

d. Calculate the value of the third virial coefficient of argon at $0°C$ and at $50°C$, assuming the van der Waals equation is a correct description.

e. Calculate the value of the compression factor of argon at $0°C$ and a molar volume of 2.271 L mol^{-1}. Do it once using the ideal gas equation, once using the van der Waals equation, once using the virial equation of state truncated at the second virial coefficient and using the correct value of the second virial coefficient, once using the virial equation of state truncated at the second virial coefficient and using the value of the second virial coefficient from the van der Waals parameters, and once using the virial equation of state truncated at the third virial coefficient and using the values of the virial coefficients from the van der Waals parameters.

1.54. The volume of a sample of liquid water can be represented by the formula

$$\bar{V}(t_C) = \bar{V}(0°C)(1 + \alpha' t_C + \beta' t_C^2 + \gamma' t_C^3 + \delta' t_C^4)$$

where α', β', γ', and δ' are constants and t_C is the Celsius temperature.

a. Find an expression for the coefficient of thermal expansion as a function of t.

b. Two different sets of values are used. The first set is said to be valid from $0°C$ to $33°C$:

$$\alpha' = -6.4268 \times 10^{-5}(°C)^{-1},$$
$$\beta' = 8.505266 \times 10^{-6}(°C)^{-2},$$
$$\gamma' = -6.78977 \times 10^{-8}(°C)^{-3},$$
$$\delta' = 4.01209 \times 10^{-10}(°C)^{-4}$$

The second set is said to be valid from $0°C$ to $80°C$:

$$\alpha' = -5.3255 \times 10^{-5}(°C)^{-1},$$
$$\beta' = 7.615323 \times 10^{-6}(°C)^{-2},$$
$$\gamma' = -4.37217 \times 10^{-8}(°C)^{-3},$$
$$\delta' = 1.64322 \times 10^{-10}(°C)^{-4}$$

Calculate the volume of 1.000 g of liquid water at $25.000°C$ using the two sets of data. The density of liquid water at $0.000°C$ is 0.99987 g mL^{-1}. Compare your answers with the correct value, 1.00294 m Lg^{-1}.

c. Make a graph of the volume of 1.000 g of liquid from $0.000°C$ to $10.00°C$.

d. Find the temperature at which the density of liquid water is at a maximum (the temperature at which the volume is at a minimum) using each of the sets of data. The correct temperature of maximum density is $3.98°C$.

e. Derive a formula for the coefficient of thermal expansion of water. Calculate the value of this coefficient at $20°C$. Compare your value with the value in Table A4.

1.55. a. Calculate the values of the van der Waals parameters a and b for water, using the critical constants. Compare your values with those in Table A1.

b. Draw a graph of the isotherm (graph of P as a function of $\bar{V}$ at constant T) for water at the critical temperature, using the van der Waals equation of state.

c. Draw a graph of the vapor branch of the water isotherm for $350°C$ using the van der Waals equation of state. Use the fact that the vapor pressure of water at $350°C$ is equal to 163.16 atm to locate the point at which this branch ends.

d. Draw a graph of the water isotherm for $350°C$, using the van der Waals equation of state for the entire graph. Note that this equation of state gives a nonphysical "loop" instead of the tie line connecting the liquid and the vapor branches. This loop consists of a curve with a relative maximum and a relative minimum. The portion of the curve from the true end of the vapor branch to the maximum can represent metastable states (supercooled vapor). The portion of the curve from the end of the liquid branch to the minimum can also represent metastable states (superheated liquid). Find the location of the maximum and the minimum. What do you think about the portion of the curve between the minimum and the maximum?

e. For many temperatures, the minimum in the loop of the van der Waals isotherm is at negative values of the pressure. Such metastable negative pressures are said to be important in bringing water to the top of large trees, because a pressure of 1.000 atm can raise liquid water to a height of only 34 feet. What negative pressure would be required to bring water to the top of a giant sequoia tree of height 300 feet? Find the minimum negative pressure in the van der Waals isotherm for a temperature of 25°C.

f. Find the Boyle temperature of water vapor, using the van der Waals equation of state.

g. Draw a graph of the compression factor of water vapor as a function of pressure at the Boyle temperature, ranging from 0 bar to 500 bar, using the van der Waals equation of state. Instead of choosing equally spaced values of P to generate points for plotting, it is probably best to choose a set of values of $\bar{V}$ and then calculate both a value of P and a value of Z for each value of $\bar{V}$.

h. Draw an accurate graph of the compression factor of water at the critical temperature, ranging from 0 bar to 500 bar. Use the van der Waals equation of state. Tell how this graph is related to the graph of part b.

i. Calculate the density of liquid water at a temperature of 25°C and a pressure of 1000 bar, using the method of Example 1.3. The density of liquid water at this temperature and 1.000 bar is 0.997296 g mL^{-1}.

j. Calculate the density of liquid water at a temperature of 25°C and a pressure of 1000. bar, using the Tait equation.

1.56. Identify each statement as either true or false. If a statement is true only under special circumstances, label it as false.

a. All gases approach ideal behavior at sufficiently low pressures.

b. All gases obey the ideal gas equation of state within about 1% under all conditions.

c. Just as there is a liquid-vapor critical point, there must be a liquid-solid critical point.

d. For every macroscopic state of a system, there must correspond many microscopic states.

e. The state of a system is independent of the history of the system.

f. The macroscopic state of a simple one-phase system of one substance is specified by the value of three state variables.

g. Two gaseous systems with the same values of T, P, and n can have different volumes.

h. Negative pressures can occur in metastable systems.

i. Negative pressures can occur in equilibrium systems.

2 Work, Heat, and Energy: The First Law of Thermodynamics

PREVIEW

In this chapter, we begin our study of thermodynamics. We introduce the first law of thermodynamics and study some of its applications. This law, which we regard as experimental fact, is a version of the principle of conservation of energy and establishes work and heat as means of transferring energy.

OBJECTIVES

After studying this chapter, the student should:

1. understand the relationship of heat, work, and energy,
2. be able to calculate amounts of heat and work and energy changes for many nonchemical processes, using line integrals where appropriate,
3. be able to calculate enthalpy and energy changes for a class of chemical reactions.

PRINCIPAL FACTS AND IDEAS

1. Thermodynamics is a general macroscopic theory of the behavior of matter.
2. Thermodynamics is based on empirical laws.
3. The first law of thermodynamics is a version of the law of conservation of energy.
4. Work is one way of transferring energy.
5. Heat is another way of transferring energy.
6. The first law of thermodynamics defines U, the thermodynamic energy or internal energy.
7. The first law asserts that the internal energy is a state function.
8. The enthalpy is a variable whose change is equal to the amount of heat transferred in a constant-pressure process.

2.1 Work and the State of a System

Thermodynamics arose in the nineteenth century as a result of the strivings of scientists to understand the physical universe and the efforts of engineers to improve the efficiency of steam engines. This practical interest produced some important early advances in thermodynamics. In fact, someone has said, "Thermodynamics owes more to the steam engine than the steam engine owes to thermodynamics." However, thermodynamics quickly grew

43

to become the general theory of the macroscopic behavior of matter at equilibrium, and it is now a mature science.

Thermodynamics is an inherently macroscopic theory. It is possible to discuss thermodynamics without the concepts of atoms and molecules, and in fact its early development in the nineteenth century occurred before the atomic theory was universally accepted. Thermodynamics also differs from most of the theories of physics and chemistry in that it is based not on unproved assumptions but on empirical laws.

Thermodynamics is closely connected with the macroscopic state of a system, which we introduced in Chapter 1. In this chapter, we begin with easily defined and measured state variables such as n, T, V, and P. These state variables will be augmented by two experimental quantities, heat and work, which are not state variables.

The Definition of Mechanical Work

Nicolas Leonard Sadi Carnot, 1821–1894, a French engineer who was the first to consider quantitatively the interconversion of work and heat and who is credited with founding the science of thermodynamics.

The quantitative measurement of work was introduced by Carnot, who defined an amount of work as the height an object is lifted times the object's weight. This definition was extended by Coriolis, who provided the presently used definition of **work**: *The amount of work done on an object equals the force exerted on the object times the distance the object is moved in the direction of the force.* If the force on and the displacement of the object are both in the z direction, the work done in an infinitesimal displacement in the z direction is

Gaspard de Coriolis, 1792–1843, was a French physicist best known for discovering the Coriolis force.

$$dw = F_z \, dz \quad \text{(definition of work)} \tag{2.1-1}$$

Coriolis's definition of work becomes the same as that of Carnot if the force on the object is that due to gravity.

where dw is the quantity of work done on the object, F_z is the force exerted on the object in the z direction, and dz is an infinitesimal displacement of the object in the z direction.

The SI unit of work is the joule (abbreviated J), and the amount of work is expressed in joules if the force is expressed in newtons and the distance in meters:

$$1 \, \text{J} = 1 \, \text{N m} = 1 \, \text{kg m}^2 \, \text{s}^{-2} \tag{2.1-2}$$

If the force and the displacement are not in the same direction, dw can be written as the **scalar product** of two vectors:

$$dw = \mathbf{F} \cdot d\mathbf{r} = |\mathbf{F}||d\mathbf{r}| \cos(\alpha) \tag{2.1-3}$$

where $\mathbf{F}$ is the vector force and $d\mathbf{r}$ is the vector displacement. The angle between the vector $\mathbf{F}$ and the vector $d\mathbf{r}$ is denoted by α. The product $|d\mathbf{r}| \cos(\alpha)$ is the component of the displacement in the direction of the force, as shown in Figure 2.1. The scalar product of the vectors $\mathbf{F}$ and $d\mathbf{r}$, denoted by $\mathbf{F} \cdot d\mathbf{r}$, is defined as in Equation (B-28). No work is done if the object does not move or if the force and the displacement are perpendicular.

Equation (2.1-3) can be written in terms of cartesian components:[1]

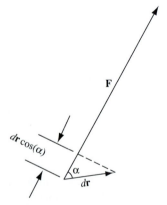

Figure 2.1. A Force and a Displacement. Only the component of the displacement in the direction of the force is effective in determining the amount of work.

---

[1] See Appendix B, Equation (B-27).

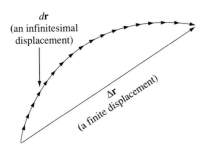

Figure 2.2. A Displacement Divided into Small Steps. If either the displacement or the force changes direction during the process, it must be divided into small steps such that the force and the displacement are constant during each step.

We will use quite a few line integrals in this and following chapters.

$$dw = F_x \, dx + F_y \, dy + F_z \, dz \qquad \text{(2.1-4)}$$

For a process that is not infinitesimal (a finite process), we can divide the displacement into small steps. This division is shown schematically in Figure 2.2. In the limit when the sizes of the steps approach zero, any force must be constant during a single step, and we can write a version of Equation (2.1-4) for the work done in each infinitesimal step. The work done on the object in the entire process is obtained by adding the work done in all of the steps making up the process. In the limit of infinitesimal steps, this sum becomes the integral

$$w = \int_c dw = \int_c \mathbf{F} \cdot d\mathbf{r} = \int_c (F_x \, dx + F_y \, dy + F_z \, dz) \qquad \text{(2.1-5)}$$

where we denote an amount of work by w.

A curve, or path, is defined by the motion of the object. We say that the integral is carried out along this curve, and the integral in Equation (2.1-5) is called a **line integral**. (See Appendix B). A line integral is sometimes denoted by adding the letter c below the integral sign, as in this equation. In a line integral, the integrand depends on more than one independent variable. In this case, it depends on x, y, and z. The curve specifies x as a function of y and z, y as a function of x and z, and z as a function of x and y. In the first term, if F_x depends on y and z, these relations are used to replace y and z in the expression for the integrand by functions of x, turning this term into an ordinary integral. Similarly, in the other terms the integrand is expressed as a function of only one variable by use of the relations given by the curve.

▼   **EXAMPLE 2.1**

An object is pushed in the z direction by a force $F_z = az + b$, where the constant a has the value 300.0 N m^{-1} and the constant b has the value 500.0 N. Find the work done in moving the object from $z = 0$ to $z = 10.0$ m.

Solution

$$w = \int_0^{10.0\,\text{m}} F_z \, dz = \int_0^{10.0\,\text{m}} (az + b)\, dz = \left(\frac{az^2}{2} + bz \right)\Bigg|_0^{10.0\,\text{m}}$$

$$= \frac{(300.0\ \text{N m}^{-1})(10.0\ \text{m})^2}{2} + (500.0\ \text{N})(10.0\ \text{m}) = 2.00 \times 10^4 \text{ J}$$

▲

The Work Done on a Closed Fluid System

Consider a closed fluid system confined in a vertical cylinder fitted with a piston, as in Figure 1.1. Let an external force F_{ext} be exerted downward in the z direction on the piston, including any force on the piston due to gravity or atmospheric pressure. Because F_{ext} is downward, we assign its value to be negative. Let the height of the piston be changed by the infinitesimal amount dz. If the piston moves downward, dz is negative. We assume that there is no friction, so all of the force is transmitted to the system. The amount of work done on the system is given by Equation (2.1-4):

$$dw = F_{ext}\, dz \tag{2.1-6}$$

We adopt the following convention: *A positive amount of work corresponds to work being done on the system by the surroundings. A negative amount of work corresponds to work being done on the surroundings by the system.* Equation (2.1-6) conforms to this convention. If dz is negative, then dw is positive, and work is done on the system.

Some older textbooks used the opposite convention for work; that is, positive value of w represented an amount of work done on the surroundings.

The amount of work done on the surroundings is the negative of the work done on the system:

$$dw_{surr} = -dw \tag{2.1-7}$$

In our notation, a quantity without a subscript applies to the system, and a quantity labeled with the subscript "surr" applies to the surroundings.

If the area of the piston is $\mathscr{A}$,

$$dw = \frac{F_{ext}}{\mathscr{A}}\, \mathscr{A}\, dz = -P_{ext} \mathscr{A}\, dz$$

$$\boxed{dw = -P_{ext}\, dV} \tag{2.1-8}$$

where P_{ext}, the **external pressure**, is defined as the magnitude of the external force on the piston divided by its area and $\mathscr{A}\, dz = dV$, the change in the volume of the system. The negative sign in Equation (2.1-8) comes from the fact that F_{ext} is negative, while P_{ext} is defined to be positive.

We have obtained Equation (2.1-8) only for a fluid system confined in a cylinder with a piston. However, we assert without proof that this equation holds for a fluid system of any shape. A system for which Equation (2.1-8) holds is called a **simple system**. Some systems are not simple. A solid system might also have work done on it by bending or stretching it, and this stress-strain work would have to be added to the expression in Equation (2.1-8). For a coil spring or a rubber band, the work done on the system is

$$dw = -P_{ext}\, dV + \tau\, dL \tag{2.1-9}$$

where τ is called the tension and dL denotes the change in the length of the system. Electrochemical cells, which we discuss in Chapter 8, are important nonsimple chemical systems.

Ordinarily, a fluid system or a strain-free solid system can be considered a simple system, but if surface tension is present the work done in creating new surface area must be included, so even a fluid system can be a nonsimple system. We discuss surface tension in a later chapter, and we will find that it is usually negligible for liquids and always negligible for gases. For the present, we neglect surface tension and consider all fluid systems to be simple systems.

Exercise 2.1

A sample of a gas is compressed from a volume of 10.00 L to a volume of 1.000 L at a constant external pressure of 1.000 atm (101325 N m^{-2}). Calculate the work done on the system.

In Chapter 1, we asserted as experimental fact that for a simple system which has one phase and contains a number c of independent substances (components), the equilibrium macroscopic state is specified by $c + 2$ variables (at least one of which is extensive). That is, $c + 2$ variables can be chosen as independent variables, and all other variables are dependent variables.

The three variables T, V, and n can be used to specify the equilibrium state of a simple one-phase, one-component system. Therefore, the location of a point in a three-dimensional space with T, V, and n axes specifies an equilibrium macroscopic state of the system, and conversely an equilibrium macroscopic state specifies the location of a point in this space. There is a **one-to-one correspondence** between an equilibrium state and a point in the space. The space is called a **state space** of the system, and the point corresponding to the state of the system is called the **state point**. Other choices of independent variables can also be made. For example, P, T, and n can be used, and a space with P, T, and n axes is then the state space.

Reversible Processes

In addition to the external force on the piston in contact with our sample system in Figure 1.1, the system exerts a force on the piston. If the system is at equilibrium and if there is no friction, the piston is stationary and the force on the piston due to the pressure of the system balances the external force:

$$F_{sys} = P\mathscr{A} = P_{ext}\mathscr{A} = -F_{ext} \quad \text{(equilibrium)} \qquad \textbf{(2.1-10a)}$$

or

$$\boxed{P = P_{ext} \quad \text{(equilibrium)}} \qquad \textbf{(2.1-10b)}$$

where P is the pressure of the system. During a reversible process, the system passes through equilibrium states, so for an infinitesimal reversible process

$$\boxed{dw_{rev} = -P\,dV \quad \text{(simple system)}} \qquad \textbf{(2.1-11)}$$

In a reversible process there can be no friction, and the process must take place infinitely slowly.

Reversible processes are also called **quasi-equilibrium processes** or **quasi-static processes**. Because the system passes through equilibrium states, the state point passes through a succession of points in the equilibrium state space, constituting a curve in the state space. Such a curve is represented schematically in Figure 2.3.

Reversible processes do not occur in the real world because they would take infinitely long to complete. However, some real processes differ from reversible processes only in ways that do not affect a particular calculation. For example, some real processes begin with a system in one equilibrium state and end with the system in another equilibrium state, although the system passes through nonequilibrium states during the process. This class of real processes can often be discussed by comparison with reversible processes having the same initial and final states. Thermodynamics has also

This is the only class of irreversible processes that we will discuss quantitatively until we reach Chapter 17.

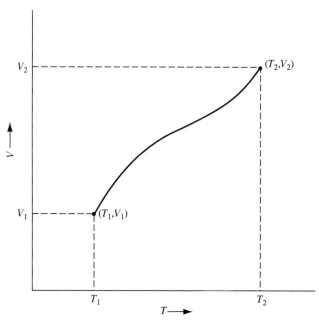

Figure 2.3. A Curve in State Space. During a reversible process the point representing the state of the system moves along a curve in the state space.

been extended to include irreversible processes by the use of additional assumptions,[2] but we do not discuss this branch of thermodynamics.

Work Done on an Ideal Gas

Let us carry out a finite reversible change in volume on a closed system consisting of n moles of an ideal gas. At any point in the process, the external pressure equals the pressure of the gas or differs from it only infinitesimally, so we can write the work done in the process as the integral of infinitesimal steps using Equation (2.1-11):

$$w_{\text{rev}} = \int_c dw_{\text{rev}} = -\int_c P \, dV = -nR \int_c \frac{T}{V} \, dV \quad \text{(ideal gas)} \quad \textbf{(2.1-12)}$$

This integral is a line integral, analogous to one term of the integral in Equation (2.1-5). Because the integrand depends on T as well as on V, the curve on which we integrate must specify a particular dependence of T on V. This dependence is used in the integrand function to replace T by the appropriate function of V.

---

[2] See, for example, S. R. deGroot and P. Mazur, *Nonequilibrium Thermodynamics*, North Holland, Amsterdam, 1962.

If the temperature is held fixed during the reversible volume change, the process is said to be **isothermal**. In this case, T is constant and can be factored out of the integral. If T were some function of V other than a constant function, it would have to be replaced by that function before the integration is carried out.

We now can write

$$w_{rev} = \int_c dw_{rev} = -\int_{V_1}^{V_2} \frac{nRT}{V} dV = -nRT \int_{V_1}^{V_2} \frac{1}{V} dV$$

$$w_{rev} = -nRT \ln\left(\frac{V_2}{V_1}\right) \quad \text{(ideal gas, isothermal process)} \qquad \textbf{(2.1-13)}$$

where V_1 is the initial value of the volume and V_2 is the final value.

EXAMPLE 2.2

a. Calculate the work done on a closed system consisting of 50.00 g of argon, assumed ideal, when it expands reversibly from a volume of 5.000 L to a volume of 10.00 L at a constant temperature of 298.15 K.

b. Calculate the integral of dP for the same process.

Solution

a.

$$w = -(50.00 \text{ g})\left(\frac{1 \text{ mol}}{39.948 \text{ g}}\right)(8.3145 \text{ J K}^{-1} \text{ mol}^{-1})(298.15 \text{ K}) \ln\left(\frac{10.00 \text{ L}}{5.000 \text{ L}}\right)$$

$$= -2151 \text{ J}$$

The negative sign indicates that work is done on the surroundings.

b. For a closed sample of ideal gas

$$dP = \left(\frac{\partial P}{\partial T}\right)_{V,n} dT + \left(\frac{\partial P}{\partial V}\right)_{T,n} dV = \frac{nR}{V} dT - \frac{nRT}{V^2} dV$$

At constant temperature, $dT = 0$ for each infinitesimal step of the process, so only the dV term contributes. In this term, we keep T constant and factor it out of the integral:

$$\Delta P = \int_{V_1}^{V_2} \left(\frac{\partial P}{\partial V}\right)_{T,n} dV = -nRT \int_{V_1}^{V_2} \frac{1}{V^2} dV = nRT\left(\frac{1}{V_2} - \frac{1}{V_1}\right)$$

$$= (50.00 \text{ g})\left(\frac{1 \text{ mol}}{39.948 \text{ g}}\right)(8.3145 \text{ J K}^{-1} \text{ mol}^{-1})(298.15 \text{ K})$$

$$\times \left(\frac{1}{0.01000 \text{ m}^3} - \frac{1}{0.005000 \text{ m}^3}\right)$$

$$= -3.104 \times 10^5 \text{ J m}^{-3} = -3.104 \times 10^5 \text{ Pa} = -3.063 \text{ atm}$$

Exact and Inexact Differentials

The differential of a function, such as dP, is called an **exact differential**. There is an important theorem of mathematics concerning the line integral of an exact differential (see Appendix B): *A line integral of a differential of a function (an exact differential) is equal to the function evaluated at the final*

This theorem implies that the path integral is path independent as long as the paths considered begin at the same point and end at the same point.

end of the integration curve minus the function evaluated at the initial end of the curve.

This theorem applies to the integral of dP in part b of Example 2.2. The integral is equal to the value of P at the end of the process minus the value of P at the beginning of the process:

$$\int_c dP = \Delta P = P_2 - P_1$$

Because a variety of different curves can have the same initial and final points, the line integral of an exact differential depends only on the starting point and the final point and is independent of the path between these points. It is said to be **path independent**. The converse of this theorem is also true. If the integral of a differential is path independent for all paths between the same end points, the differential must be an exact differential.

Work Is Not a State Function

This statement means that if the macroscopic equilibrium state of the system is specified, there is no such thing as a definite amount of work corresponding to the state of the system. Therefore, since w is not a function, dw is not the differential of a function. A differential such as dw, which is not the differential of a function, is called an **inexact differential**.

The line integral of an inexact differential such as dw depends on the path of integration as well as on the starting point and the final point. Two processes with the same initial and final states can correspond to different amounts of work done on the system. Let us illustrate this important fact by considering a reversible process with the same initial and final states as the process of Example 2.2, but with a different path.

▼ **EXAMPLE 2.3**

Calculate the work done on the ideal gas system of Example 2.2 if it is reversibly cooled at constant volume of 5.000 L from 298.15 K to 200.00 K, then reversibly expanded from 5.000 L to 10.00 L at a constant temperature of 200.0 K, and then reversibly warmed at a constant volume of 10.00 L from 200.0 K to 298.15 K.

Solution

In the cooling process at constant volume, $dV = 0$ for each infinitesimal step, so $w = 0$ for the cooling process. The same is true for the warming process, so the only nonzero contribution to w is from the isothermal expansion:

$$w = -(50.00 \text{ g}) \frac{1 \text{ mol}}{39.948 \text{ g}} (8.3145 \text{ J K}^{-1} \text{ mol}^{-1})(200 \text{ K}) \ln\left(\frac{10.00 \text{ L}}{5.000 \text{ L}}\right)$$

$$= -1443 \text{ J}$$

which is not equal to the amount of work in Example 2.2.

▲

A single case is enough to show a differential to be inexact, and this example shows that dw cannot be an exact differential.

Exercise 2.2
    a. Show that, because dw corresponds to Equation (B-19) of Appendix B with $N = 0$, dw does not satisfy the criteria of Equation (B-20) for an exact differential.
    b. Calculate the line integral of dP for the process of Example 2.3. Show that the integral is path independent for the two paths of Examples 2.2 and 2.3. The integral will have to be done in three sections.

If P_{ext} is a known function, w can be calculated directly for an irreversible process. For example, if the external pressure is constant,

$$w_{irrev} = -\int P_{ext}\, dV = -P_{ext}\int dV = -P_{ext}(V_2 - V_1)$$

$$\boxed{w_{irrev} = -P_{ext}\,\Delta V \quad \text{(constant } P_{ext})} \qquad \text{(2.1-14)}$$

Exercise 2.3
    a. Calculate the amount of work done on the surroundings if the isothermal expansion of Example 2.2 is carried out at a constant external pressure of 1.000 atm instead of reversibly, but with the same initial and final states as in Example 2.2. Why is less work done on the surroundings in the irreversible process than in the reversible process?
    b. What is the change in the pressure of the system for the irreversible process?

For any particular representation of a real gas, the expression for the work done in an isothermal reversible volume change can be obtained by integration.

▼   **EXAMPLE 2.4**

Obtain the formula for the work done per mole of gas during an isothermal reversible volume change of a real gas whose behavior is adequately represented by the truncated virial equation of state:

$$\frac{P\bar{V}}{RT} = 1 + \frac{B_2}{\bar{V}} \qquad \text{(2.1-15)}$$

where B_2, the second virial coefficient, depends only on the temperature, and where $\bar{V} = V/n$, the molar volume.

Solution

The amount of work per mole is

$$w_{rev} = -\int_c P\, d\bar{V} = -RT\left(\int_{\bar{V}_1}^{\bar{V}_2} \frac{1}{\bar{V}}\, d\bar{V} + \int_{\bar{V}_1}^{\bar{V}_2} \frac{B_2}{\bar{V}^2}\, d\bar{V}\right)$$

$$= -RT\ln\left(\frac{\bar{V}_2}{\bar{V}_1}\right) + RTB_2\left(\frac{1}{\bar{V}_2} - \frac{1}{\bar{V}_1}\right) \qquad \text{(2.1-16)}$$

▲

a. Obtain the equation analogous to Equation (2.1-16) for an arbitrary amount of gas, n moles.

b. Calculate the work done in the process of Example 2.2 if argon is the sample gas. The second virial coefficient of argon is equal to $-15.8 \, cm^3 \, mol^{-1}$ at 298.15 K.

c. Calculate the work done if the process is carried out irreversibly with $P_{ext} = $ constant = 1.000 atm.

2.2   Heat

Joseph Black, 1728–1799, was a Scottish chemist who discovered carbon dioxide ("fixed air") by heating calcium carbonate.

Black was the first to distinguish between the quantity of heat and the "intensity" of heat (temperature) and to recognize "latent heat" absorbed or given off in phase transitions. However, Black believed in the "caloric" theory of heat, which asserted that heat was an "imponderable" fluid called caloric. This fluid supposedly flowed spontaneously from a hotter object to a cooler object and could be extracted from an object by friction. This incorrect theory was not fully discredited until several decades after Black's death.

Heat Transferred During Temperature Changes

An amount of heat can be measured by determining the change in temperature that it produces in an object of known heat capacity. The **heat capacity**, C, of an object is defined such that

Equation (2.2-1) is considered by some not to be a logically satisfactory definition of heat, but it is the one used by Joule, Black, and other pioneers of thermodynamics. We will comment later on another definition.

$$dq = C \, dT \qquad (2.2\text{-}1)$$

where dq is an infinitesimal amount of heat transferred to the object and dT is a resulting infinitesimal change in temperature.

At first glance, Equation (2.2-1) might seem to indicate that C is a derivative of q with respect to T. However, this is not the case, because dq is an inexact differential, like dw. Just as there is no such thing as the work content of a system, there is no such thing as the heat content of a system. Therefore, the value of C depends on the way in which the system has its temperature changed. For example, the heat capacity at constant volume is generally different in value from the heat capacity at constant pressure.

The **specific heat** (better called the **specific heat capacity**), denoted by c, is defined as the heat capacity per unit mass, or C/m, where m is the mass of the object. The specific heat of a substance is an intensive quantity that is characteristic of the substance.

If an object is heated without any chemical reaction or phase change occurring, the quantity of heat transferred to the object is given by

$$q = \int_c dq = \int_{T_1}^{T_2} C \, dT \qquad (2.2\text{-}2)$$

where T_2 is the final temperature and T_1 is the initial temperature. If the heat capacity is constant, it can be factored out of the integral, giving

$$q = C(T_2 - T_1) = C \, \Delta T \qquad \text{(2.2-3)}$$

Just as with work, a positive value of q indicates heat transferred to the system and a negative value indicates heat transferred from the system to its surroundings.

Exercise 2.5

a. Find the amount of heat needed to heat 3.00 mol of liquid water from 20.00°C to 80.00°C. The specific heat of liquid water is nearly temperature independent and is nearly equal to $1.00 \text{ cal K}^{-1} \text{ g}^{-1} = 4.184 \text{ J K}^{-1} \text{ g}^{-1}$.

b. The specific heat of aluminum is $0.216 \text{ cal K}^{-1} \text{ g}^{-1}$. Find the final temperature if a piece of aluminum with mass 20.00 gram and at an initial temperature of 18.00°C is placed in 100.00 g of liquid water initially at 80.00°C. Assume that the water and aluminum are insulated from the rest of the universe.

▼ **EXAMPLE 2.5**

The molar heat capacity (heat capacity per mole) of water vapor at constant pressure of 1.000 atm is represented by

$$\bar{C}_P = 30.54 \text{ J K}^{-1} \text{ mol}^{-1} + (0.01029 \text{ J K}^{-2} \text{ mol}^{-1})T$$

Find the amount of heat required to raise the temperature of 2.000 mol of water vapor from 100.0°C to 500.0°C.

Solution

$$q = (2.000 \text{ mol}) \int_{373.15 \text{ K}}^{773.15 \text{ K}} [30.54 \text{ J K}^{-1} \text{ mol}^{-1} + (0.01029 \text{ J K}^{-2} \text{ mol}^{-1})T] \, dT$$

$$= (2.000 \text{ mol})\{(30.54 \text{ J K}^{-1} \text{ mol}^{-1})(400 \text{ K})$$

$$+ (0.01029 \text{ J K}^{-2} \text{ mol}^{-1})\tfrac{1}{2}[(773.15 \text{ K})^2 - (373.15 \text{ K})^2]\}$$

$$= (2.000 \text{ mol})(12216 \text{ J mol}^{-1} + 2359 \text{ J mol}^{-1}) = 2.915 \times 10^4 \text{ J}$$

▲

Heat Transferred During Phase Changes

This heat is sometimes called **latent heat** because it does not change the temperature of the system. The amount of latent heat per unit mass is characteristic of the substance and the phase transition. For example, the latent heat of fusion (melting) of water at its normal melting temperature, 0°C, and at a constant pressure of 1.000 atm equals 79.72 calories per gram, or 333.5 kilojoules per kilogram. The latent heat of vaporization (boiling) of water at 100°C and 1.000 atm equals 539.55 calories per gram, or 2257.5 kilojoules per kilogram.

Exercise 2.6

Find the maximum mass of liquid water that can be brought to 100.0°C from 20.00°C by contact with 100.0 g of steam at 100.0°C.

Internal Energy; The First Law

Although Lavoisier discredited the phlogiston theory of combustion, which held that combustion was the loss of an "imponderable fluid" called phlogiston, he was one of the principal promoters of the equally incorrect caloric theory of heat espoused by Black. The first experimental studies that discredited the caloric theory were done by Count Rumford. Rumford was at one time in charge of manufacturing cannons for the Elector of Bavaria, the ruler who made him a count. Rumford noticed that when a cannon was bored, a dull boring tool produced more heat than a sharp tool. He carried out a systematic set of experiments and showed that there was no apparent limit to the amount of heat that could be generated by friction. Rumford's results showed that "caloric" was not simply being extracted from the cannon. Work was being converted to heat.

From his experiments Rumford calculated a value for the "mechanical equivalent of heat," or the amount of heat to which a joule of work could be converted. His value was not very accurate. Better values were obtained by Mayer in 1842 and Joule in 1847. Mayer was apparently the first to espouse the law of conservation of energy, asserting that heat and work are just forms in which energy is transferred and that energy can neither be created nor destroyed.

Joule carried out experiments in which changes of state were produced either by doing work or by heating a system. His apparatus is depicted schematically in Figure 2.4. A falling mass turned a stirring paddle in a sample of water, raising the temperature by doing work on the liquid. The rise in temperature of the water was measured and the amount of work done by the falling mass was compared with the amount of heat required to produce the same change in temperature. Joule found that the ratio of the work required for a given change to the heat required for the same change was always the same, approximately 4.18 joules of work to 1 calorie of heat.

Exercise 2.7 Calculate the rise in temperature of 100.0 g of water if the falling weight of Figure 2.4 has a mass of 2.500 kg and drops by 0.600 m. Neglect friction in the pulleys, etc.

There was no detectable difference in the final state of the system if its temperature was raised by doing work on it, by heating it, or by some combination of work and heating. This indicates that heat and work are actually two different means of changing a single property of the system, the internal energy, U. It also indicates that U is a state variable. Based on the work of Rumford, Mayer, Joule, and many others since the time of Joule, we state the **first law of thermodynamics**: *For a closed system and any process that begins and ends with equilibrium states, ΔU is defined by*

$$\Delta U = q + w \qquad (2.3\text{-}1)$$

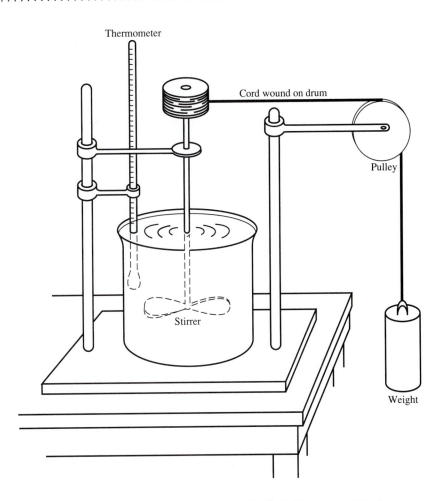

Figure 2.4. Joule's Apparatus for Determining the Mechanical Equivalent of Heat (schematic). The falling weight does work on the water in the container, changing its energy to an extent that can be determined from the temperature change.

where q is the amount of heat transferred to the system and w is the work done on the system. When so defined, ΔU is equal to the change in value of a state variable U, called the internal energy or the thermodynamic energy:

$$\Delta U = U_{\text{final}} - U_{\text{initial}} \tag{2.3-2}$$

In spite of the work of Mayer and Joule, initial credit for announcing the first law of thermodynamics went to Helmholtz.

A state variable is a mathematical function of the independent variables used to specify the state of the system. For a simple equilibrium one-phase system containing one component, we can write the internal energy as

$$U = U(T, V, n) \tag{2.3-3}$$

or

$$U = U(T, P, n) \tag{2.3-4}$$

Hermann Ludwig von Helmholtz, 1821–1894, was a German physicist and physiologist who studied the energy of muscle contraction. He was one of the first to propose that the energy for all processes on the earth ultimately came from solar radiation.

Because heat and work are both means of changing the value of the internal energy, they do not maintain separate identities after the transfer of energy is finished. The following analogy has been used:[3] Heat is analogous to rain falling on a pond, work is analogous to the influx of a stream into the pond, and energy is analogous to water in the pond. Evaporation (counted as negative rainfall) is analogous to heat flow to the surroundings, and efflux from the pond into a second stream is analogous to work done on the surroundings.

Once rain falls into the pond, it is no longer identifiable as rain, but only as water. Stream flow in the pond is also identifiable only as water, and not as stream flow. The amount of water in the pond is a well-defined quantity (a state variable), but one cannot measure separately how much rain and how much stream flow are in the pond.

Like rain, heat that has been transferred to a system is identifiable only as energy, and no longer as heat. Work that has been done on a system is no longer identifiable as work, but only as energy. There is no such thing as the heat content of a system in a given state and no such thing as the work content of a system in a given state. However, like the water in the pond, the energy of the system corresponding to a given equilibrium state is a state variable.

Conservation of Energy

The first law of thermodynamics is a **law of conservation of energy** for thermodynamic systems. The law of conservation of energy is a general law of physics to which there are no known exceptions. In fact, apparent violations of energy conservation led particle physicists to search for previously unknown particles that could be carrying energy away from a system. This search led to the discovery of the neutrino.[4] Even though no exceptions to the first law of thermodynamics have ever been verified, occasionally an unknown inventor announces a machine that will allegedly produce more energy than it takes in. Such machines are known as **perpetual motion machines of the first kind**.

It is the total energy of a system that is governed by the law of conservation of energy. The gravitational potential energy of the system, the kinetic energy of the center of mass of the system, and the rest-mass energy are usually excluded from the internal energy of a chemical system. If so, the total energy is

$$E_{\text{total}} = \tfrac{1}{2}mv_c^2 + \mathscr{V}_g + m_0c^2 + U \qquad (2.3\text{-}5)$$

where m is the mass of the system and m_0 is its rest mass, c is the speed of light, v_c is the speed of the system's center of mass, $\mathscr{V}_g$ is the gravitational potential energy of the system, and U is the internal energy of the system. Equation (2.3-1) applies only to a closed system whose center of mass is not

---

[3] Herbert B. Callen, *Thermodynamics*, Wiley, New York, 1960, p. 19.

[4] E. Fermi, *Z. Phys* **88**, 161 (1934).

accelerated and whose gravitational potential energy and rest mass do not change. These conditions apply to most laboratory systems. However, in the case of a rocket, in which energy of combustion is turned into kinetic energy of the whole system, we must consider the kinetic and potential energies of the entire system as well as the internal energy.

The Ideal Gas as an Example System

Thermodynamics applies to all systems, but it is convenient to have a system with simple properties to use for example derivations and calculations. The most commonly used example system is a dilute gas, approximately represented by the ideal gas. A mechanical model system that represents a monatomic ideal gas is a collection of point mass molecules, which have mass but zero size. Point mass molecules have no internal motions (no electronic motion, no vibration, and no rotation). We will analyze this model system mathematically in Chapters 16 and 20, once assuming that the molecules' motions are described by classical mechanics and once assuming that they are described by quantum mechanics. Because there are no intermolecular forces, the potential energy is constant and can be taken as equal to zero. The internal energy of this model system is equal to the kinetic energy of the molecules.

Although we often use an ideal gas as a simple example system, we must be careful not to apply ideal-gas equations to other kinds of systems.

For a model system of point mass molecules the internal energy is proportional to the amount of the gas and to the temperature:

$$U = \tfrac{3}{2}Nk_{\mathrm{B}}T = \tfrac{3}{2}nRT \quad \text{(gas of point-mass molecules)} \quad \textbf{(2.3-6)}$$

where N is the number of molecules and n is the amount of the gas in moles. If the factor $3/2$ is included, k_{B} is Boltzmann's constant and R is the gas constant (see Chapters 16 and 20). We state now as an assumption that for a dilute monatomic gas, if the electronic energy can be ignored, Equation (2.3-6) can be used. Experimental data for the inert gases (He, Ne, Ar, etc.) conform well to Equation (2.3-6) at ordinary temperatures.

For molecular gases, rotation and vibration of the molecules must be considered but the electronic motion can usually be ignored at room temperature. For many gases at ordinary temperatures, the energy of vibrational motion is nearly constant at its minimum value, which can be chosen to equal zero. We will find in Chapter 20 that, to a good approximation, the rotational energy of a diatomic gas near room temperature is equal to nRT. The rotational energy of a polyatomic gas is approximately equal to $3nRT/2$ if the molecules of the gas are nonlinear (the nuclei do not lie along a straight line) and approximately equal to nRT if the molecules are linear. Therefore, if the vibrational and electronic contributions can be ignored,

$$U \approx \tfrac{5}{2}nRT \quad \text{(dilute diatomic or linear polyatomic gas)} \quad \textbf{(2.3-7)}$$

$$U \approx 3nRT \quad \text{(dilute nonlinear polyatomic gas)} \quad \textbf{(2.3-8)}$$

Exercise 2.8

a. Find the value of the rest mass energy of 1.000 mol of argon gas, using Einstein's equation, $E = mc^2$.

b. Find the value of the internal energy of 1.000 mol of argon gas at 298.15 K.

Find the ratio of this energy to the rest mass energy of the system. Find the difference between the observed mass of the system at 298.15 K and at 0 K.

c. Explain why it would be difficult to use values of total energies for chemical purposes if the rest mass energy were included.

2.4   Calculation of Amounts of Heat and Energy Changes

Up to now, we have defined a change in energy as the sum of an amount of heat transferred plus an amount work done, assuming that both heat and work can be defined and measured satisfactorily. The definition of heat was through calorimetry, first defining the heat capacity. Some people think that this definition is logically inferior to the definition of work.

An Alternative Definition of Heat

The first law of thermodynamics provides a means of calculating quantities of heat without using the calorimetric definition of Equation (2.2-2). If separate means of calculating ΔU and w exist, the value of q can be calculated from

$$q = \Delta U - w \qquad (2.4\text{-}1)$$

If the first law is assumed as a postulate, Equation (2.4-1) can be used as the definition of the amount of heat transferred in a process.[5]

Although Equation (2.4-1) is considered by many to be the logically acceptable definition of heat, it requires that the first law of thermodynamics be accepted as a postulate.

An adiabatic process is one in which no heat is transferred, so $\Delta U_{\text{adiabatic}} = w_{\text{adiabatic}}$. In order to use Equation (2.4-1) to define q, we find an adiabatic process that has the same initial and final states as the process of interest. Because U is a state function,

$$\Delta U = w_{\text{adiabatic}} \qquad (2.4\text{-}2)$$
$$q = w_{\text{adiabatic}} - w \qquad (2.4\text{-}3)$$

Energy Changes in an Ideal Gas

The ideal gas gives simpler equations than almost any other system, so we frequently use it as an example system. However, equations that apply to an ideal gas do not usually apply to other systems. One must keep in mind which equations apply only to a particular system and which apply to all systems.

It is a property of ideal gases at equilibrium that the internal energy depends only on the amount of gas and the temperature and is independent of the volume or the pressure:

The ideal-gas property of Equation (2.4-4) is almost as important as PV = nRT.

$$\boxed{U = U(T, n) \quad \text{(ideal gas)}} \qquad (2.4\text{-}4)$$

---

[5] Callen, *op. cit.*, pp. 17ff.

We will prove this property in Chapter 4, assuming only that $PV = nRT$. For the present, we take it as an additional defined property of ideal gases. In an isothermal change of state of a closed ideal gas system, ΔU vanishes, so from Equation (2.3-1)

$$q = -w = nRT \ln\left(\frac{V_2}{V_1}\right) \quad \left(\begin{array}{l}\text{ideal gas; reversible} \\ \text{isothermal change}\end{array}\right) \qquad \textbf{(2.4-5)}$$

where V_2 is the final volume and V_1 is the initial volume.

▼ **EXAMPLE 2.6**

Find the amount of heat put into a system of 5.000 mol of argon (assumed ideal) in expanding reversibly and isothermally at 298.15 K from a volume of 50.00 L to 100.00 L.

Solution

$$w = -(5.000 \text{ mol})(8.3145 \text{ J K}^{-1} \text{ mol}^{-1})(298.15 \text{ K}) \ln\left(\frac{100.0 \text{ L}}{50.00 \text{ L}}\right) = -8591 \text{ J}$$

$$q = \Delta U - w = 0 - w = -w = 8591 \text{ J}$$

▲

Heat and work are not state functions, but U is a state function.

The amount of heat put into a system for a given change of state can depend on the path taken from the initial to the final state as well as on the initial and final states, just as was the case with an amount of work.

Exercise 2.9

Calculate the amount of heat that is put into the system of Example 2.6 if it expands irreversibly and isothermally at 298.15 K and at a constant external pressure of 1.000 atm (101325 Pa) from a volume of 50.00 L to a volume of 100.00 L. Hint: ΔU is the same as in Example 2.6.

Energy Changes in a General Closed Simple System

Energy changes can be calculated by carrying out a line integration of dU, either on the actual process or on another process with the same initial and final states. For a closed simple system,

$$\Delta U = \int_c \left(\frac{\partial U}{\partial T}\right)_{V,n} dT + \int_c \left(\frac{\partial U}{\partial V}\right)_{T,n} dV \quad \left(\begin{array}{l}\text{closed simple} \\ \text{system}\end{array}\right) \qquad \textbf{(2.4-6)}$$

where c indicates the curve in state space corresponding to the reversible process. This curve can be used to specify V as a function of T or to specify T as a function of V. These dependences are used in the integrands of the line integral of Equation (2.4-6).

The same value for ΔU will result if any curve having the same initial and final points is used for the line integration, so we can use any of these curves for the actual calculation. In addition, ΔU will have the same value for any irreversible process having the same equilibrium initial and final states. Keep in mind that q and w depend on the path of integration.

The Heat Capacity at Constant Volume

For an infinitesimal change,

$$dq = dU - dw = dU + P_{ext}\, dV$$

If V is constant, $dV = 0$ and $dw = 0$. Therefore,

$$dq = dU = \left(\frac{\partial U}{\partial T}\right)_{V,n} dT \quad (V \text{ constant, simple system}) \qquad \textbf{(2.4-7)}$$

Comparison of this equation with Equation (2.2-1) shows that

$$\boxed{C_V = \left(\frac{\partial U}{\partial T}\right)_{V,n}} \qquad \textbf{(2.4-8)}$$

where C_V is the **heat capacity at constant volume**. The constant-volume heat capacity is frequently measured for gases, but it is seldom measured for liquids and solids because it is difficult to maintain a liquid or solid system at constant volume when the temperature is varied.

The second partial derivative in Equation (2.4-6) gives the variation of the internal energy with volume at constant temperature. Again, the ideal gas is the simplest example system. Because U depends only on T and n for an ideal gas, as stated in Equation (2.4-4),

$$\boxed{\left(\frac{\partial U}{\partial V}\right)_{T,n} = 0 \quad (\text{ideal gas})} \qquad \textbf{(2.4-9)}$$

Therefore, for a closed ideal gas system

$$\boxed{dU = C_V\, dT \quad (\text{closed system, ideal gas})} \qquad \textbf{(2.4-10)}$$

For use in example calculations, we apply the properties of our model system of noninteracting molecules from Section 2.3. For a dilute monatomic gas in which electronic and vibrational energy can be neglected, Equations (2.3-6), (2.3-7), and (2.3-8) give, with Equation (2.4-8),

$$C_V \approx \tfrac{3}{2}nR \quad (\text{dilute monatomic gas}) \qquad \textbf{(2.4-11)}$$
$$C_V \approx \tfrac{5}{2}nR \quad (\text{dilute diatomic or linear polyatomic gas}) \quad \textbf{(2.4-12)}$$
$$C_V \approx 3nR \quad (\text{dilute nonlinear polyatomic gas}) \qquad \textbf{(2.4-13)}$$

These equations are satisfactory approximations for most substances but are not very accurate for some substances, such as Br_2 and CO_2.

Exercise 2.10　The Euler reciprocity relation, Equation (B-13), implies

$$\left(\frac{\partial^2 U}{\partial V\, \partial T}\right)_n = \left(\frac{\partial^2 U}{\partial T\, \partial V}\right)_n$$

Show that Equations (2.4-9) and (2.4-11) are consistent with this requirement.

The Joule Experiment

The first attempt to measure $(\partial U/\partial V)_{T,n}$ for real gases was made by Joule in 1853. His experiment, which became known as the **Joule experiment**, was carried out in an apparatus depicted schematically in Figure 2.5. A sample of a gas was placed in one side of the apparatus and the other side of the apparatus was evacuated. The entire apparatus was insulated from the surroundings so that the process would approximate an adiabatic process. The initial temperature, T_1, of the gas was measured. The stopcock was then opened to allow the gas to expand irreversibly into the vacuum, after which the final temperature, T_2, was measured.

To analyze this experiment, we define the contents of the apparatus to be the system. Because the surroundings are not affected, w is equal to zero, and because the process is adiabatic, q also vanishes. Therefore, from the first law, Equation (2.4-1), ΔU is equal to zero. If a change ΔT in temperature of the gas occurs, the derivative $(\partial U/\partial V)_{T,n}$ can be determined as follows: The Joule experiment is carried out several times with various volumes for the second chamber. The ratio $\Delta T/\Delta V$ is determined for each experiment and extrapolated to zero value of ΔV, where $\Delta V = V_2 - V_1$.

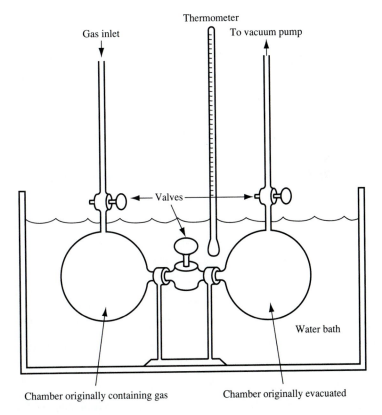

Figure 2.5. The Apparatus for the Joule Experiment (schematic). The gas is allowed to expand irreversibly into the evacuated side of the apparatus.

This extrapolation is equivalent to taking the mathematical limit

$$\mu_J = \lim_{\Delta V \to 0} \left(\frac{\Delta T}{\Delta V} \right) = \left(\frac{\partial T}{\partial V} \right)_{U,n} \tag{2.4-14}$$

The partial derivative $(\partial T/\partial V)_{U,n}$ is called the **Joule coefficient** and is denoted by μ_J. We have established that U is fixed in the Joule experiment, and n is fixed because the system is closed.

The Joule coefficient is related to $(\partial U/\partial V)_{T,n}$ by use of the cycle rule, Equation (B-15), and the reciprocal identity, Equation (B-8):

$$\left(\frac{\partial U}{\partial V} \right)_{T,n} = -\left(\frac{\partial T}{\partial V} \right)_{U,n} \left(\frac{\partial U}{\partial T} \right)_{V,n}$$

$$= -\mu_J C_V \tag{2.4-15}$$

The Joule coefficient of an ideal gas vanishes, as this equation shows.

Exercise 2.11　　Verify Equation (2.4-15).

The Joule coefficient for an ideal gas is equal to 0. Joule was unsuccessful in his attempt to measure nonzero values of $(\partial U/\partial V)_{T,n}$ for real gases because the changes in temperature that occurred were too small to be measured by the thermometers available at the time. Later versions of the experiment with better apparatus have given fairly good values. After the second law of thermodynamics has been introduced, we will present a better way to determine values of $(\partial U/\partial V)_{T,n}$.

Once values for C_V and for $(\partial U/\partial V)_{T,n}$ are obtained, ΔU can be calculated for any process that begins with one equilibrium state and ends with another equilibrium state by integrating Equation (2.4-6).

▼　　**EXAMPLE 2.7**　　For the truncated virial equation of state, Equation (1.4-2), it can be shown that

$$\left(\frac{\partial \bar{U}}{\partial \bar{V}} \right)_{T,n} = \frac{RT^2}{\bar{V}^2} \frac{dB_2}{dT} = \frac{RT^2 B_2'}{\bar{V}^2} \tag{2.4-16}$$

where R is the gas constant, $\bar{V}$ is the molar volume, and B_2' is introduced as an abbreviation for the derivative dB_2/dT. (See Exercise 4.13 for the derivation.) For argon gas at 298.15 K, B_2 is approximately equal to -15.8 cm^3 mol^{-1} and B_2' is approximately equal to 0.25 cm^3 mol^{-1} K^{-1}. The molar constant-volume heat capacity of argon gas is nearly constant and equal to $3R/2$.

a. Find ΔU, q, and w for a reversible isothermal expansion of 1.000 mol of argon at 298.15 K from a volume of 2.000 L to a volume of 20.00 L. Compare with values obtained assuming ideal gas behavior.

b. Find the value of the Joule coefficient for argon at 298.15 K and a molar volume of 20.000 L.

Solution

Our system consists of 1.000 mol, so we calculate the change in the molar energy, the work per mole, etc.

a.

$$\Delta \bar{U} = \int_c \left(\frac{\partial \bar{U}}{\partial \bar{V}} \right)_{T,n} d\bar{V} = \int_{\bar{V}_1}^{\bar{V}_2} \frac{RT^2 B_2'}{\bar{V}^2} d\bar{V}$$

$$= -RT^2 B_2' \left(\frac{1}{\bar{V}_2} - \frac{1}{\bar{V}_1} \right)$$

$$= -(8.3145 \text{ J K}^{-1} \text{ mol}^{-1})(298.15 \text{ K})^2 (0.25 \times 10^{-6} \text{ m}^3 \text{ mol}^{-1})$$

$$\times \left(\frac{1}{0.0200 \text{ m}^3 \text{ mol}^{-1}} - \frac{1}{0.00200 \text{ m}^3 \text{ mol}^{-1}} \right)$$

$$= 83 \text{ J mol}^{-1}$$

$$w = -\int_c P \, d\bar{V} = -RT \int_{\bar{V}_1}^{\bar{V}_2} \left(\frac{1}{\bar{V}} + \frac{B_2}{\bar{V}^2} \right) d\bar{V}$$

$$= -RT \left[\ln \left(\frac{\bar{V}_2}{\bar{V}_1} \right) - B_2 \left(\frac{1}{\bar{V}_2} - \frac{1}{\bar{V}_1} \right) \right]$$

$$= -(8.3145 \text{ J K}^{-1} \text{ mol}^{-1})(298.15 \text{ K}) \left[\ln \left(\frac{20.0 \text{ L}}{2.00 \text{ L}} \right) \right.$$

$$\left. -(-15.8 \text{ cm}^3 \text{ mol}^{-1}) \left(\frac{1}{20000 \text{ cm}^3 \text{ mol}^{-1}} - \frac{1}{2000 \text{ cm}^3 \text{ mol}^{-1}} \right) \right]$$

$$= -(2479 \text{ J mol}^{-1})[2.303 - 7.11 \times 10^{-3}] = -5690 \text{ J mol}^{-1}$$

$$q = \Delta U - w = 83 \text{ J mol}^{-1} - (-5690 \text{ J mol}^{-1}) = 5773 \text{ J mol}^{-1}$$

Compare these values with those obtained if ideal behavior is assumed: $\Delta U = 0$, $w = -5708 \text{ J mol}^{-1}$, and $q = -w = 5708 \text{ J mol}^{-1}$.

b.

$$\left(\frac{\partial U}{\partial V} \right)_{T,n} = (8.3145 \text{ J K}^{-1} \text{ mol}^{-1})(298.15 \text{ K})^2 \left(\frac{0.25 \times 10^{-6} \text{ m}^3 \text{ mol}^{-1} \text{ K}^{-1}}{(0.02000 \text{ m}^3 \text{ mol}^{-1})^2} \right)$$

$$= 462 \text{ J m}^{-3}$$

$$\mu_J = -\left(\frac{\partial U}{\partial V} \right)_{T,n} \frac{1}{C_V} = -\frac{462 \text{ J m}^{-3}}{12.17 \text{ J K}^{-1} \text{ mol}^{-1}} = -38 \text{ K (m}^3 \text{ mol}^{-1})^{-1}$$

Exercise 2.12

a. Find ΔU, q, and w for an irreversible isothermal expansion at 298.15 K of 1.000 mol of argon with the same initial and final molar volumes as in Example 2.7 but with a constant external pressure of 1.000 atm. Compare with the values obtained assuming ideal gas behavior and with values obtained in Example 2.7.

b. Find the change in temperature if 1.000 mol of argon initially at 298.15 K is expanded adiabatically into a vacuum so that its volume changes from 2.000 L to 20.00 L.

A change in internal energy for a nonisothermal process can be calculated by carrying out the line integral in Equation (2.4-6).

▼ **EXAMPLE 2.8**

Calculate ΔU for a process that takes 1.000 mol of argon from $T = 298.15$ K and $V = 2.000$ L to $T = 373.15$ K and $V = 20.000$ L.

Solution

We can choose any path with the proper end points. We integrate along the path shown in Example Figure 2.8. The path of the actual process must be used to calculate q and w, because they depend on the path.

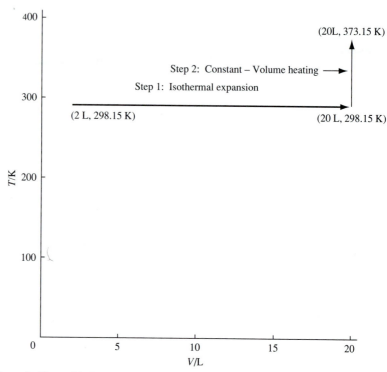

Example Figure 2.8. The Curve Representing the Path for the ΔU Line Integral. This is a reversible path consisting of an isothermal expansion and a constant-volume change in temperature.

The first line segment of the path is that of Example 2.7, so ΔU_1, the change in energy for that part, is equal to 83 J mol^{-1}. For the second line segment

$$\Delta U_2 = \int_{T_1}^{T_2} C_V \, dT = (1.000 \text{ mol}) \tfrac{3}{2} R (T_2 - T_1)$$

$$= (12.472 \text{ J K}^{-1})(75.00 \text{ K}) = 935.4 \text{ J}$$

Therefore,

$$\Delta U = \Delta U_1 + \Delta U_2 = 83 \text{ J} + 935 \text{ J} = 1018 \text{ J}$$

Exercise 2.13

a. Calculate q and w for the reversible process that follows the path used in Example 2.8.

b. Calculate q and w for the reversible process that takes the following path: the system is first heated from 298.15 K to 373.15 K at a constant volume of 2.000 L and is then expanded isothermally to a volume of 20.000 L.

c. Comment on the difference between the q and w values for parts a and b. What is the value of ΔU for the process of part b?

Adiabatic Processes

In an adiabatic process, q is equal to zero for the process, and dq is equal to zero for any infinitesimal step of the process. However, dw and w can be nonzero:

$$dU = dq + dw = dw \quad \text{(adiabatic process)} \qquad \text{(2.4-17)}$$

Because U depends only on n and T for an ideal gas, we can write as in Equation (2.4-10)

$$dU = C_V \, dT \quad \text{(ideal gas, reversible process)} \qquad \text{(2.4-18)}$$

$PV = nRT$ for an ideal gas, so

$$dw = -P \, dV = -\frac{nRT}{V} \, dV \quad \text{(ideal gas, reversible process)} \qquad \text{(2.4-19)}$$

Equating dU and dw (because $dq = 0$), we obtain

$$C_V \, dT = -\frac{nRT}{V} \, dV \quad \text{(ideal gas, reversible adiabatic process)} \qquad \text{(2.4-20)}$$

This is a differential equation that can be solved to obtain T as a function of V or V as a function of T if the dependence of C_V on T and V is known.

To a good approximation, C_V is a constant for many gases at ordinary temperatures. With this assumption, we can solve Equation (2.4-20) by separation of variables. We divide by T:

$$\frac{C_V}{T} \, dT = -\frac{nR}{V} \, dV \qquad \text{(2.4-21)}$$

We integrate Equation (2.4-21) from the initial state, denoted by V_1 and T_1, to the final state, denoted by V_2 and T_2:

$$C_V \ln\left(\frac{T_2}{T_1}\right) = -nR \ln\left(\frac{V_2}{V_1}\right)$$

We divide by C_V and take the exponential of both sides of this equation:

$$\boxed{\frac{T_2}{T_1} = \left(\frac{V_1}{V_2}\right)^{nR/C_V} = \left(\frac{V_1}{V_2}\right)^{R/\bar{C}_V} \quad \begin{pmatrix} \text{reversible adiabatic} \\ \text{process, ideal gas,} \\ C_V \text{ constant} \end{pmatrix}} \qquad \text{(2.4-22)}$$

If the initial values V_1 and T_1 are given, this equation gives T_2 as a function of V_2. It is an example of an important general fact: *For a reversible adiabatic process in a simple system, the final temperature is a single-valued function of the final volume for a given initial state.* That is, all of the possible final state points for reversible adiabatic processes lie on a single curve in the state space. This fact will be important in our discussion of the second law of thermodynamics in Chapter 3.

▼ **EXAMPLE 2.9**

A system consisting of 2.000 mol of neon, assumed ideal with C_V equal to $3nR/2$, expands adiabatically and reversibly from a volume of 5.000 L and a temperature of 373.15 K to a volume of 20.00 L. Find the final temperature.

Solution

$$T_2 = T_1\left(\frac{V_1}{V_2}\right)^{nR/C_V} = (373.15 \text{ K})\left(\frac{5.000 \text{ L}}{20.00 \text{ L}}\right)^{2/3} = 148.1 \text{ K}$$

Example Figure 2.9 represents the final temperature as a function of the final volume for this example:

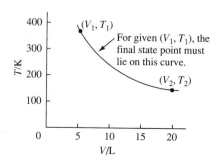

Example Figure 2.9. Final Temperature as a Function of Final Volume for the Adiabatic Expansion of an Ideal Gas. For each initial state, there is only one such curve.

▲

Exercise 2.14   **a.** Find the final temperature for the system of Example 2.9 if the final volume is 15.00 L.

b. Find the volume to which the system of Example 2.9 must be adiabatically and reversibly expanded in order to reach a temperature of 298.15 K.

An equation analogous to Equation (2.4-20) can be written for any representation of a real gas. For a gas obeying the van der Waals equation of state,

$$(P + a/\bar{V}^2)(\bar{V} - b) = RT \tag{2.4-23}$$

For this case, it can be shown (see Problem 4.33) that

$$\left(\frac{\partial \bar{U}}{\partial \bar{V}}\right)_{T,n} = \frac{a}{\bar{V}^2} \tag{2.4-24}$$

so that for a reversible process

$$d\bar{U} = C_V \, dT + \frac{a}{\bar{V}^2} \, d\bar{V} \tag{2.4-25}$$

$$dw = -P \, d\bar{V} = -\left(\frac{RT}{\bar{V} - b} - \frac{a}{\bar{V}^2}\right) d\bar{V} \tag{2.4-26}$$

Exercise 2.15   Show that Equation (2.4-26) is correct.

For an adiabatic process, $dU = dw$, so that when the right-hand sides of Equations (2.4-25) and (2.4-26) are equated and two identical terms canceled,

$$\bar{C}_V \, dT = -\frac{RT}{\bar{V} - b} \, d\bar{V} \qquad (2.4\text{-}27)$$

This equation is solved by separating the variables and integrating. If $\bar{C}_V$ is constant, the result is

$$\left(\frac{T_2}{T_1}\right) = \left(\frac{\bar{V}_1 - b}{\bar{V}_2 - b}\right)^{R/\bar{C}_V} \qquad \text{(van der Waals gas, } \bar{C}_V \text{ constant)} \qquad (2.4\text{-}28)$$

Exercise 2.16   Show that Equation (2.4-28) follows from Equation (2.4-27).

▼   **EXAMPLE 2.10**   Find the final temperature for the process of Example 2.9, using Equation (2.4-28) instead of Equation (2.4-22) but still assuming that $\bar{C}_V = 3R/2$.

Solution
From Table 1.1, $b = 3.219 \times 10^{-5} \, \text{m}^3 \, \text{mol}^{-1}$:

$$T_2 = (373.15 \text{ K}) \left(\frac{5.000 \times 10^{-3} \, \text{m}^3 \, \text{mol}^{-1} - 3.22 \times 10^{-5} \, \text{m}^3 \, \text{mol}^{-1}}{20.00 \times 10^{-3} \, \text{m}^3 \, \text{mol}^{-1} - 3.22 \times 10^{-5} \, \text{m}^3 \, \text{mol}^{-1}}\right)^{2/3}$$

$$= 147.6 \text{ K}$$

This value differs from ideal behavior by less than 0.5 K.   ▲

If the molar heat capacity of a van der Waals gas is not constant but can be represented by

$$\bar{C}_V = \alpha + \beta T \qquad (2.4\text{-}29)$$

the equation analogous to Equations (2.4-22) and (2.4-28) is

$$\frac{T_2}{T_1} \exp\left[\frac{\beta(T_2 - T_1)}{\alpha}\right] = \left(\frac{\bar{V}_1 - b}{\bar{V}_2 - b}\right)^{R/\alpha} \qquad (2.4\text{-}30)$$

Exercise 2.17   Show that Equation (2.4-30) is correct.

An equation analogous to Equation (2.4-22) exists for any simple system. For each such equation, there is a unique curve in the V-T plane containing all of the points that can be reached by adiabatic reversible processes from a given initial state.

Volume and the temperature are not the only choices for independent variables. For an ideal gas with constant heat capacity, we can substitute the ideal gas equation, $T = P\bar{V}/R$, into Equation (2.4-22) to obtain

$$\frac{P_2}{P_1} = \left(\frac{\bar{V}_1}{\bar{V}_2}\right)^{1 + R/\bar{C}_V} \qquad (2.4\text{-}31)$$

and

$$\frac{T_2}{T_1} = \left(\frac{P_2}{P_1}\right)^{R/(\bar{C}_V + R)} \qquad (2.4\text{-}32)$$

| Exercise 2.18 | The "Santa Ana" winds of the California coast are dry winds that begin in the mountains and drop to an altitude near sea level. Assume that the air is initially at a pressure of 0.81 atm (roughly the barometric pressure at 6000 feet above sea level) and a temperature of 25°C and that it is adiabatically and reversibly compressed to a pressure of 1.00 atm. Assume that air is an ideal gas with $\bar{C}_V = 5R/2$ = constant. Find the final temperature. This treatment ignores other factors that raise the temperature, such as frictional heating as the air passes along the ground. |

2.5   Enthalpy

However, the Guinness Book of World Records *lists the extreme barometric pressures as 1083.8 mbar (1.069 atm) at an altitude of 862 ft., and 870 mbar (0.859 atm) at sea level.*

Many chemical systems are contained in vessels that are open to the atmosphere and are thus maintained at a nearly constant pressure. For analysis of processes taking place under constant pressure conditions, we define a new variable, denoted by H and called the **enthalpy**:

$$H = U + PV \quad \text{(definition)} \tag{2.5-1}$$

The enthalpy is a state variable because U, P, and V are all state variables. It is one of a class of variables sometimes called **convenience variables**.

Consider a simple system whose pressure remains equal to a constant external pressure. From now on we will refer to these conditions simply as constant-pressure conditions, but we also mean that the pressure is equal to the external pressure. For any such process

$$dw = -P_{\text{ext}}\, dV = -P\, dV \quad \text{(constant pressure)} \tag{2.5-2}$$

This expression for dw is the same as that for reversible processes, Equation (2.1-11). We do not assert that all processes that occur at constant pressure are reversible, but only that the same expression for dw applies.

For a simple system at constant pressure,

$$w = \int_c dw = -\int_c P_{\text{ext}}\, dV = -\int_c P\, dV = -P(V_2 - V_1)$$
$$= -P\, \Delta V \quad \text{(simple system, constant pressure)} \tag{2.5-3}$$

where V_1 is the initial volume and V_2 is the final volume.

The heat transferred to the system is given by

$$dq = dU - dw = dU + P\, dV \quad \text{(constant pressure)} \tag{2.5-4}$$

From Equation (2.5-1)

$$dH = dU + P\, dV + V\, dP \tag{2.5-5}$$

At constant pressure, the $V\, dP$ term vanishes, so that

$$dq = dH \quad \text{(constant pressure)} \tag{2.5-6}$$

For a finite process

$$q = \Delta H \quad \text{(constant pressure)} \tag{2.5-7}$$

Although q is generally path dependent, it is path independent for constant-pressure processes. Since $dw = dU - dq$, w is also path independent for constant-pressure processes. Because enthalpy changes of constant-pressure processes are equal to amounts of heat, they are sometimes called "heats" of the processes.

The Heat Capacity at Constant Pressure

This heat capacity is given by

$$C_P = \lim_{\Delta T \to 0} \left(\frac{q}{\Delta T} \right)_{P \text{ const}} = \lim_{\Delta T \to 0} \left(\frac{\Delta H}{\Delta T} \right)_{P \text{ const}}$$

The limit in this equation is a partial derivative, so

Equation (2.5-8) can be regarded as the definition of C_P.

$$C_P = \left(\frac{\partial H}{\partial T} \right)_{P,n} \qquad \text{(2.5-8)}$$

The heat capacity at constant pressure is the most commonly measured heat capacity for solids and liquids.

We now obtain an expression for the difference between C_P and C_V. We first substitute Equation (2.5-1) into Equation (2.5-8):

$$C_P = (\partial H/\partial T)_{P,n} = (\partial U/\partial T)_{P,n} + P(\partial V/\partial T)_{P,n} \qquad \text{(2.5-9)}$$

There is no $V(\partial P/\partial T)$ term because P is held constant in the differentiation.

As an example of the variable-change identity, Equation (B-7) of Appendix B, we can write

$$(\partial U/\partial T)_{P,n} = (\partial U/\partial T)_{V,n} + (\partial U/\partial V)_{T,n}(\partial V/\partial T)_{P,n} \qquad \text{(2.5-10)}$$

We substitute this equation into Equation (2.5-9) and use the fact that $C_V = (\partial U/\partial T)_{V,n}$ to write

$$C_P = C_V + \left[\left(\frac{\partial U}{\partial V} \right)_{T,n} + P \right] \left(\frac{\partial V}{\partial T} \right)_{P,n} \qquad \text{(2.5-11)}$$

Equation (2.5-11) has a simple form for an ideal gas, because

$$(\partial U/\partial V)_{T,n} = 0 \quad \text{(ideal gas)}$$

and

$$(\partial V/\partial T)_{P,n} = nR/P \quad \text{(ideal gas)}$$

Remember not to use Equation (2.5-12) for liquids and solids.

so that

$$C_P = C_V + nR \quad \text{(ideal gas)} \qquad \text{(2.5-12)}$$

C_P is larger than C_V because in heating a system under constant-volume conditions, no work is done on the surroundings and all of the heat increases the energy of the system. Under constant-pressure conditions, some of the heat is turned into work against the external pressure as the system expands.

Equations (2.4-11)–(2.4-13) give, with Equation (2.5-12),

$$\bar{C}_P \approx 5R/2 \quad \text{(dilute monatomic gases)} \qquad \text{(2.5-13a)}$$

$$\bar{C}_P \approx 7R/2 \quad \text{(dilute diatomic or linear polyatomic gases} \atop \text{without electronic or vibrational excitation)} \quad \textbf{(2.5-13b)}$$

$$\bar{C}_P \approx 4R \quad \text{(dilute nonlinear polyatomic gases} \atop \text{without electronic or vibrational excitation)} \quad \textbf{(2.5-13c)}$$

The ratio of the constant-pressure heat capacity to the constant-volume heat capacity is denoted by γ:

$$\boxed{\gamma \approx \bar{C}_P/\bar{C}_V} \qquad \textbf{(2.5-14)}$$

The values in Equation (2.5-13) give

$$\gamma \approx 5/3 \quad \text{(dilute monatomic gas)} \qquad \textbf{(2.5-15a)}$$

$$\gamma \approx 7/5 \quad \text{(dilute diatomic or linear polyatomic gases} \atop \text{without electronic or vibrational excitation)} \quad \textbf{(2.5-15b)}$$

$$\gamma \approx 4/3 \quad \text{(dilute nonlinear polyatomic gases} \atop \text{without electronic or vibrational excitation)} \quad \textbf{(2.5-15c)}$$

Table A7 gives data on the molar constant-pressure heat capacity for several substances. For gases, the data are represented by the polynomial

$$\bar{C}_P = a + bT + cT^{-2} \qquad \textbf{(2.5-16)}$$

The temperature dependence in this formula is due to the contributions of vibrational and electronic motions. For liquids and solids near room temperature, heat capacities are nearly constant, so the values in the table can be used over a range of temperatures. Additional values are given in Table A9.

For many metallic solids and other solids in which one atom is the smallest formula unit, the molar heat capacity near room temperature is approximately equal to $3R = 24.94\ \text{J K}^{-1}\ \text{mol}^{-1}$. This is the **law of Dulong and Petit**. A more general property, shared by all substances, is that the heat capacity of any system approaches zero as the temperature approaches 0 K.

Pierre Louis Dulong, 1785–1838, was a French chemist originally trained as a physician. Alexis Therese Petit was a French physicist who lived from 1791–1820.

| Exercise 2.19 |

a. Evaluate $\bar{C}_P$ for oxygen gas at 298.15 K and at 500. K and find the percent differences between these values and $5R/2$.

b. Find the percent differences between the $\bar{C}_P$ values of copper and of iron and $3R$ at 298.15 K.

William Thomson, 1824–1907, later Lord Kelvin, was a Scottish mathematician and physicist who proposed the absolute temperature scale and made other important contributions. However, he stated in 1880 that all discoveries in physics had already been made and nothing remained to be done except to measure quantities to a few more significant digits.

The Joule-Thomson Experiment

This experiment was carried out by Joule and Thomson in a second attempt to demonstrate a difference in behavior between real gases and ideal gases. For an ideal gas, the enthalpy depends only on n and T, just as does U:

$$H = U + PV = U(T, n) + nRT = H(T, n)$$

$$\boxed{H = H(T, n) \quad \text{(ideal gas)}} \qquad \textbf{(2.5-17)}$$

Therefore,

$$
\left(\frac{\partial H}{\partial P}\right)_{T,n} = 0 \quad \text{(ideal gas)}
\tag{2.5-18}
$$

In order to determine the value of $(\partial H/\partial P)_{T,n}$ for real gases, Joule and Thomson used an apparatus equivalent to the one shown schematically in Figure 2.6. It consists of two cylinders that are fitted with pistons and separated by a porous plug. Each side has a manometer to measure the pressure and a thermometer to measure the temperature of the gas. The entire apparatus is adiabatically insulated from the surroundings. By pushing in on the left piston and pulling out on the right piston, a steady flow of gas can be maintained through the porous plug. If the pressure on each side is kept fixed, a time-independent nonequilibrium state will be attained with a different constant temperature on each side. A time-independent nonequilibrium state is called a **steady state**.

For each gas, several experiments were carried with different values of the pressure difference. The **Joule-Thomson coefficient** is defined as the extrapolated limit

$$
\mu_{\text{JT}} = \lim_{P_R \to P_L} \left(\frac{T_R - T_L}{P_R - P_L}\right) = \lim_{\Delta P \to 0} \left(\frac{\Delta T}{\Delta P}\right) \quad \text{(definition)}
\tag{2.5-19}
$$

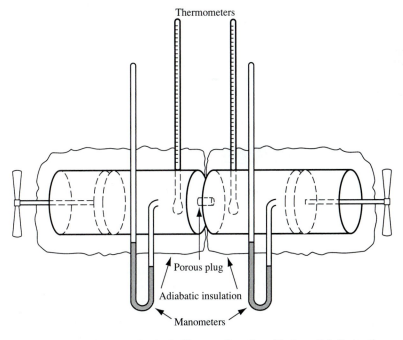

Figure 2.6. The Apparatus for the Joule-Thomson Experiment (schematic). During the experiment, one of the pistons is moved into its cylinder and the other piston is withdrawn in such a way that the pressure on each side remains constant as the gas flows irreversibly through the porous plug.

This set of assumptions allows us to use our technique of discussing non-equilibrium processes with equilibrium initial and final states.

where the subscripts R and L indicate right and left. The limit in Equation (2.5-19) is a partial derivative, but we must determine what variable is held fixed in the differentiation.

We choose as our system a sample of n moles of gas that flows through the porous plug after the steady state is established. We assume that all irreversible processes take place within the porous plug, so that the gas inside each chamber is at equilibrium and at a pressure equal to the external pressure on that side.

The work done on the gas on the left side is given by

$$w_L = -\int_{V_{L,1}}^{V_{L,2}} P_{L,\,ext}\, dV_L = -P_L(V_{L,2} - V_{L,1}) \qquad \textbf{(2.5-20)}$$

and the work done on the gas on the right side is given by

$$w_R = -\int_{V_{R,1}}^{V_{R,2}} P_{R,\,ext}\, dV_R = -P_R(V_{R,2} - V_{R,1}) \qquad \textbf{(2.5-21)}$$

where the subscript 1 denotes the initial value and the subscript 2 the final value of each quantity.

The process we consider is the complete transfer of the system from one side of the apparatus to the other. Before the transfer, the pressure P_1 of the system was equal to P_L, and the initial volume of the system must have been equal to the magnitude of the change in volume of the left side:

$$V_1 = V_{L,1} - V_{L,2} \qquad \textbf{(2.5-22)}$$

The final pressure P_2 must be equal to P_R, and the final volume V_2 must be equal to the change in volume of the right side:

$$V_2 = V_{R,2} - V_{R,1} \qquad \textbf{(2.5-23)}$$

From Equations (2.5-20)–(2.5-23), the total work done on the system is

$$w = w_L + w_R = P_1 V_1 - P_2 V_2 = -\Delta(PV) \qquad \textbf{(2.5-24)}$$

Note that $\Delta(PV)$ does not necessarily equal $P\,\Delta V$ or $V\,\Delta P$.

Exercise 2.20

a. Show that for any change in state
$$\Delta(PV) = P_1\,\Delta V + V_1\,\Delta P + (\Delta P)(\Delta V) \qquad \textbf{(2.5-25)}$$
b. When can $\Delta(PV)$ equal $P\,\Delta V$? When can it equal $V\,\Delta P$? When can it equal $P\,\Delta V + V\,\Delta P$?

Since the apparatus is adiabatically insulated from the laboratory, no heat is transferred to or from the laboratory. Also, no heat is transferred from the system to the apparatus after the steady state is established, because the chamber on the right is then at the same temperature as the gas that exits from the plug. Therefore,

$$q = 0 \qquad \textbf{(2.5-26)}$$
$$\Delta U = q + w = -\Delta(PV) \qquad \textbf{(2.5-27)}$$
$$\Delta H = \Delta U + \Delta(PV) = 0 \qquad \textbf{(2.5-28)}$$

The Joule-Thomson process therefore occurs at constant enthalpy. The limit that defines the Joule-Thomson coefficient is equal to a partial derivative at constant H and n:

$$\mu_{JT} = \left(\frac{\partial T}{\partial P}\right)_{H,n} \qquad \textbf{(2.5-29)}$$

We can use Equations (2.5-9) and (2.5-29) and the cycle rule, Equation (B-15) of Appendix B, to write an equation for the partial derivative in Equation (2.5-18):

$$\left(\frac{\partial H}{\partial P}\right)_{T,n} = -C_P\mu_{JT} \qquad \textbf{(2.5-30)}$$

Exercise 2.21   Show that Equation (2.5-30) is correct.

The Joule-Thomson coefficient of an ideal gas vanishes because $(\partial H/\partial P)_{T,n}$ vanishes for an ideal gas. Joule and Thomson found that the coefficient is measurably different from zero for ordinary gases at ordinary pressures. It depends on temperature, and it is positive at room temperature for most common gases except hydrogen and helium. Even for these gases, it is positive at temperatures below room temperature. This means that for some range of temperature any gas cools upon expansion through a porous plug. It also cools in an expansion through a nozzle or other aperture, which approximates the Joule-Thomson process. Expansion of a gas can be used to cool the gas enough to liquefy part of it. The final step in the production of liquid nitrogen and liquid helium is ordinarily carried out in this way.

▼   **EXAMPLE 2.11**   The Joule-Thomson coefficient of air at 300 K and 25 atm is equal to 0.173 K atm^{-1}. If a Joule-Thomson expansion is carried out from a pressure of 50.00 atm to a pressure of 1.00 atm, estimate the final temperature if the initial temperature is equal to 300 K.

Solution

$$\Delta T \approx \left(\frac{\partial T}{\partial P}\right)_{H,n} \Delta P = (0.173 \text{ K atm}^{-1})(49 \text{ atm}) = 8 \text{ K}$$
$$T_2 \approx 292 \text{ K}$$

▲

The molecular explanation for the fact that the Joule-Thomson coefficient is positive at sufficiently low temperature is that at low temperatures the attractive intermolecular forces are more important than the repulsive intermolecular forces. When the gas expands, work must be done to overcome the attractions. If no heat is added, the kinetic energy and the temperature must decrease. If the Joule-Thomson coefficient is negative, the repulsive intermolecular forces must be more important than the attractive forces.

2.6 Calculation of Enthalpy Changes for Nonchemical Processes

The simplest calculations of ΔH are for constant-pressure processes in closed systems. In this case, $\Delta H = q$. If no phase change or chemical reaction occurs,

$$
\begin{aligned}
\Delta H = q = \int_c dH &= \int_{T_1}^{T_2} \left(\frac{\partial H}{\partial T}\right)_{P,n} dT \\
&= \int_{T_1}^{T_2} C_P \, dT \qquad \begin{array}{l}\text{(constant } n \text{ and } P, \text{ no phase} \\ \text{transition, no chemical reaction)}\end{array}
\end{aligned}
\tag{2.6-1}
$$

▼   **EXAMPLE 2.12**

a. Find a formula for ΔH for the heating of a sample of a gas from temperature T_1 to temperature T_2 at constant pressure if $\bar{C}_P$ is represented by

$$\bar{C}_P = a + bT + cT^{-2}$$

b. Find ΔH and q for the heating of 2.000 mol of oxygen gas from 25.00°C to 100.00°C at 1.000 atm.

Solution

a.

$$
\begin{aligned}
\Delta H = q = n\int_{T_2}^{T_1} (a + bT + cT^{-2})\, dT \\
= n\left[a(T_2 - T_1) + \frac{b}{2}(T_2^2 - T_1^2) - c\left(\frac{1}{T_2} - \frac{1}{T_1}\right)\right]
\end{aligned}
$$

b. Using parameter values from Table A7,

$$
\begin{aligned}
\Delta H = q = (2.000 \text{ mol})&\left\{[30.0 \text{ J K}^{-1}\text{ mol}^{-1}][75.00 \text{ K}] \right. \\
&+ [4.18 \times 10^{-3} \text{ J K}^{-2}\text{ mol}^{-1}][(373.15 \text{ K})^2 - (298.15 \text{ K})^2] \\
&\left. - (1.7 \times 10^5 \text{ J K mol}^{-1})\left(\frac{1}{373.15 \text{ K}} - \frac{1}{298.15 \text{ K}}\right)\right\} \\
&= (2.000 \text{ mol})(2250 \text{ J mol}^{-1} + 210 \text{ J mol}^{-1} - 115 \text{ J mol}^{-1}) \\
&= 4690 \text{ J}
\end{aligned}
$$

▲

Enthalpy Changes for Reversible Phase Transitions

The values of enthalpy changes of constant-pressure phase transitions for many common substances can be found in published tables. Table A8 gives specific enthalpy changes (enthalpy changes per gram) for reversible fusion

It is best not to call enthalpy changes by the name "heat" because they are equal to amounts of heat only if the pressure is constant.

(melting) and vaporization (boiling) transitions for a number of pure substances at a constant pressure of 1.000 atm. The enthalpy changes for freezing or condensation processes are the negatives of these values. These enthalpy changes are sometimes called latent heats of fusion or of vaporization, or heats of fusion or vaporization.

▼ **EXAMPLE 2.13**

Find ΔH and q if 2.000 mol of liquid water at 0.00°C is reversibly frozen to ice at 0.00°C at a constant pressure of 1.000 atm.

Solution

$$q = \Delta H = (2.000 \text{ mol})(18.02 \text{ g mol}^{-1})(-333.5 \text{ J g}^{-1})$$
$$= -1.202 \times 10^4 \text{ J}$$

▲

Since H is a state variable, ΔH for any process is equal to ΔH for any other process having the same initial and final states.

▼ **EXAMPLE 2.14**

Calculate ΔH for the change of state of 1.000 mol of helium from a volume of 5.000 L and a temperature of 298.15 K to a volume of 10.000 L and a temperature of 373.15 K. Assume that $\bar{C}_P = 5R/2$ and that the gas is ideal.

Solution

For purposes of calculation, assume that the gas first expands isothermally to a pressure equal to the final pressure (step 1) and is then heated at constant pressure to its final temperature (step 2). Since the enthalpy depends on n and T only, ΔH for the first step vanishes. For the second step

$$\Delta H_2 = \int_{T_1}^{T_2} C_P \, dT = n\bar{C}_P \, \Delta T$$
$$= (1.000 \text{ mol})\tfrac{5}{2}(8.3145 \text{ J K}^{-1} \text{ mol}^{-1})(75.00 \text{ K})$$
$$= 1559 \text{ J}$$
$$\Delta H = \Delta H_1 + \Delta H_2 = 0 + 1559 \text{ J} = 1559 \text{ J}$$

▲

Although ΔH is the same for any process with the same initial and final states as the overall process in Example 2.14, q and w depend on the path of the particular process.

▼ **EXAMPLE 2.15**

Find q and w for the process used in the calculation of Example 2.14.

Solution

We first find V_2 and P_2, the volume and pressure at the end of step 1. From the ideal gas law,

$$V_2 = (10.00 \text{ L})\frac{298.15 \text{ K}}{373.15 \text{ K}} = 7.990 \text{ L}$$

$$P_2 = \frac{(1.000 \text{ mol})(8.3145 \text{ J K}^{-1} \text{ mol}^{-1})(298.15 \text{ K})}{0.007990 \text{ m}^3}$$

$$= 3.103 \times 10^5 \text{ Pa}$$

$$q_1 = (1.000 \text{ mol})(8.3145 \text{ J K}^{-1} \text{ mol}^{-1})(298.15 \text{ K}) \ln\left(\frac{7.990 \text{ L}}{5.000 \text{ L}}\right)$$

$$= 1162 \text{ J}$$

Because the pressure was constant during step 2,

$$q_2 = \Delta H_2 = 1559 \text{ J}$$

$$q = q_1 + q_2 = 1162 \text{ J} + 1559 \text{ J} = 2721 \text{ J}$$

Note that q is not equal to ΔH for the overall process, since the entire process is not a constant-pressure process. We calculate the work for the entire process. For the first step

$$w_1 = \Delta U_1 - q_1 = -q_1 = -1162 \text{ J}$$

For the second step, we let V_3 be the final volume

$$w_2 = -\int_{V_2}^{V_3} P \, dV = -P \int_{V_2}^{V_3} dV = -P \, \Delta V$$

$$= -(3.103 \times 10^5 \text{ Pa})(0.01000 \text{ m}^3 - 0.007990 \text{ m}^3)$$

$$= -623.6 \text{ J}$$

$$w = w_1 + w_2 = -1162 \text{ J} - 624 \text{ J} = -1786 \text{ J}$$

Exercise 2.22

a. Find ΔU for the process of Example 2.15.

b. Find ΔH, q, and w for the process in which the system of Examples 2.14 and 2.15 is first heated at constant volume from 298.15 K to 373.15 K and then expanded isothermally from a volume of 5.000 L to a volume of 10.000 L.

Enthalpy Changes for Irreversible Processes

We consider irreversible processes that begin with an equilibrium or metastable state and end with an equilibrium state. Since the enthalpy is a state variable, ΔH for any such process is equal to that for a reversible process with the same initial and final states.

EXAMPLE 2.16

Find ΔH and q if 2.000 mol of supercooled liquid water at $-15.00°C$ freezes irreversibly at a constant pressure of 1.000 atm to ice at $-15.00°C$. Assume the molar heat capacity of liquid water to be constant and equal to 76.1 J K^{-1} mol^{-1} and that of ice to be constant and equal to 37.15 J K^{-1} mol^{-1}.

Solution

Since H is a state function, ΔH is the same for every path having the same initial and final states. We calculate ΔH along the reversible path shown in Example Figure 2.16. Step 1 is the reversible heating of the system to 0.00°C, the equilibrium freezing temperature at 1.000 atm. Step 2 is the reversible freezing of the system at 0.00°C, and step 3 is the reversible cooling of the system to $-15.00°C$. Supercooled water is in a metastable state. However, the metastable state is sufficiently like an equilibrium state that no appreciable error is introduced by treating step 1 like a reversible process.

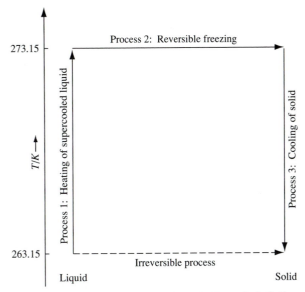

Example Figure 2.16. Irreversible and Reversible Paths for Example 2.16. The enthalpy change for the irreversible process is equal to the sum of the enthalpy changes for processes 1, 2, and 3.

$$\Delta H_1 = \int_{258.15\,K}^{273.15\,K} C_P(l)\,dT = C_P(l)\,\Delta T$$
$$= (2.000\ \text{mol})(76.1\ \text{J K}^{-1}\ \text{mol}^{-})(15.00\ \text{K}) = 2280\ \text{J}$$

The enthalpy change of step 2 is just like that of Example 2.13:

$$\Delta H_2 = -1.202 \times 10^4\ \text{J}$$

The enthalpy change of step 3 is similar to that of step 1.

$$\Delta H_3 = \int_{273.15\,K}^{258.15\,K} C_P(s)\,dT = C_P(s)\,\Delta T$$
$$= (2.000\ \text{mol})(37.15\ \text{J K}^{-1}\ \text{mol}^{-1})(-15.00\ \text{K}) = -1114\ \text{J}$$
$$\Delta H = \Delta H_1 + \Delta H_2 + \Delta H_3 = -1.085 \times 10^4\ \text{J}$$

Since the process is at constant pressure, $q = \Delta H$.

▲

2.7   Calculation of Enthalpy Changes for a Class of Chemical Reactions

This fact distinguishes chemical processes from physical processes. Thermodynamics applies equally to all processes, and any thermodynamic fact that applies to a physical process also applies to a chemical process and vice versa.

A chemical reaction involves the breaking of some chemical bonds and/or the formation of other chemical bonds. The breaking of bonds requires an input of energy. Therefore, nearly every constant-temperature chemical reaction is accompanied by energy and enthalpy changes. If the system gives off heat when the reaction takes place at constant temperature and pressure, the reaction is called **exothermic**, and if the system absorbs heat, the reaction is called **endothermic**.

In this chapter we will consider only chemical reactions of a restricted class, in which every reactant or product is either a gas, a pure liquid, or a pure solid. Three reactions in this class are:

$$2H_2(g) + O_2(g) \rightarrow 2H_2O(l) \tag{2.7-1}$$

$$CaCO_3(s) \rightarrow CaO(s) + CO_2(g) \tag{2.7-2}$$

$$N_2O_4(g) \rightarrow 2NO_2(g) \tag{2.7-3}$$

The label s refers to solid, l to liquid, and g to gas.

We now choose to write chemical reaction equations with the symbols for all substances on the right-hand side and with the $\rightarrow$ symbol replaced by an equals sign:

$$0 = 2H_2O(l) - 2H_2(g) - O_2(g) \tag{2.7-4}$$

$$0 = CaO(s) + CO_2(g) - CaCO_3(s) \tag{2.7-5}$$

$$0 = 2NO_2(g) - N_2O_4(g) \tag{2.7-6}$$

In order to have a way of writing a general chemical equation, we denote the stoichiometric coefficient of substance number i by v_i. If substance number i is a product, then $v_i > 0$, and if substance number j is a reactant, then $v_j < 0$. For example, if NO_2 is substance number 1 and N_2O_4 is substance number 2, then $v_1 = 2$ and $v_2 = -1$.

We number the substances from 1 to c, and any chemical reaction equation can now be written

$$0 = \sum_{i=1}^{c} v_i \mathscr{F}_i \tag{2.7-7}$$

where $\mathscr{F}_i$ is an abbreviation for the chemical formula of substance number i. The form of this equation will enable us to write a compact general formula for the enthalpy change of any chemical reaction.

We can express the enthalpy change for a reaction in our restricted class in terms of the enthalpies of pure substances. The enthalpy of a sample of any pure substance depends on the state of that sample. The **standard state** of a liquid or solid substance is specified to be the pure substance at a fixed pressure of exactly 1 bar (100000 Pa), which we designate by $P°$. The standard-state pressure was until recent years chosen as 1 atm (101325 Pa). The difference in numerical values is small, and the formulas involving $P°$ are the same with either choice. For highly accurate work, one must determine which standard pressure was used for a set of data.

The actual substance at pressure $P°$ is the standard state for solids and liquids, but the standard state for a gas is defined to be the *ideal gas* at pressure $P°$. This choice means that corrections must be made for the difference between the real gas and an ideal gas. These corrections are small, and we will learn how to calculate them in a later chapter.

If all substances are either pure condensed phases or ideal gases, if surface tension effects can be ignored, and if each substance is in an equilibrium or metastable state, the enthalpy of the system is a sum of contributions of the separate substances:

$$H = \sum_{i=1}^{c} n_i \bar{H}_i \tag{2.7-8}$$

where $\bar{H}_i$ is the molar enthalpy (enthalpy per mole) of substance number i and n_i is the amount (in moles) of that substance. If the substance is in its standard state, its molar enthalpy is denoted by $\bar{H}_i^\circ$.

Consider a chemical reaction that begins with all substances in equilibrium or metastable states at some particular temperature and pressure and ends with all substances in equilibrium states at the same temperature and pressure. The enthalpy change of the reaction is given by

$$\Delta H = H_{\text{final}} - H_{\text{initial}} = \sum_{i=1}^{c} \Delta n_i \bar{H}_i \qquad \textbf{(2.7-9)}$$

where Δn_i is the change in the amount of substance number i.

We say that *1 mole of reaction* occurs if

$$\Delta n_i = v_i$$

for each substance. That is, a number of moles of a product appears that is equal to its stoichiometric coefficient, and a number of moles of a reactant disappears that is equal to the magnitude of its stoichiometric coefficient. For 1 mole of reaction

$$\Delta H = \sum_{i=1}^{c} v_i \bar{H}_i \qquad \textbf{(2.7-10)}$$

This enthalpy change depends on the way in which the reaction equation is balanced. For example, if all stoichiometric coefficients are doubled, ΔH for the reaction doubles.

A **standard-state reaction** is one in which all substances are in their standard states both before and after the reaction. The enthalpy change for a standard-state reaction is denoted by ΔH°. If values for standard-state molar enthalpies were available, Equation (2.7-10) could be used to calculate ΔH° for a reaction. However, actual values for standard-state molar enthalpies are not found in tables. Instead, values of **standard-state enthalpy changes of formation** are tabulated. Some older tables use the name "heat of formation" or the name "enthalpy of formation."

The standard-state enthalpy change of formation of substance i, $\Delta \bar{H}_f^\circ(i)$, is defined to be the enthalpy change of the chemical reaction to form 1 mole of substance i in the specified phase from the appropriate elements in their most stable forms, with all substances in their standard states. The bar over the letter H denotes that the quantity is for the formation of 1 mole of substance i, and the superscript $^\circ$ denotes the standard state. Standard-state enthalpy changes of formation for a number of substances are listed in Table A9.

The enthalpy change for 1 mole of any standard-state reaction in our restricted class is given by

$$\boxed{\Delta H^\circ = \sum_{i=1}^{c} v_i \, \Delta \bar{H}_f^\circ(i)} \qquad \textbf{(2.7-11)}$$

We can show this equation to be correct as follows: Let process 1 convert the reactants into elements in their most stable form. The standard-state

When we give a value of ΔH, etc., for a reaction, it is always for 1 mole of the reaction as written.

Some authors use the superscript $\ominus$ to denote a standard state.

enthalpy change for process 1 is

$$\Delta H_1^\circ = H_{\text{elements}} - H_{\text{reactants}} = \sum_{i=1}^{c} v_i \, \Delta \bar{H}_f^\circ(i) \qquad \text{(2.7-12)}$$

(reactants only)

This process is equivalent to the reverse of all of the formation reactions multiplied by the magnitude of the stoichiometric coefficients. The sign in front of the sum in Equation (2.7-12) is positive because the stoichiometric coefficients are negative.

Let process 2 be the production of the products of the reaction of interest from the elements produced in process 1. The standard-state enthalpy change of process 2 is

$$\Delta H_2^\circ = \sum_{i=1}^{c} v_i \, \Delta \bar{H}_f^\circ(i) \qquad \text{(2.7-13)}$$

(products only)

Hess's law is named for Germain Henri Hess, 1802–1850, the Swiss-Russian chemist whose law first indicated that thermodynamics applies to chemistry.

We now invoke **Hess's law**, which states: *The enthalpy change of any process that is equivalent to the successive carrying out of two other processes is equal to the sum of the enthalpy changes of those two processes.* This law is a simple consequence of the fact that enthalpy is a state variable, so its change is path independent.

In our case

$$\begin{aligned} \Delta H &= H_{\text{products}} - H_{\text{reactants}} \\ &= H_{\text{elements}} - H_{\text{reactants}} + (H_{\text{products}} - H_{\text{elements}}) \\ &= \Delta H_1 + \Delta H_2 \end{aligned}$$

The products and the reactants of any reaction contain the same number of atoms of each element, because atoms are neither destroyed nor created in a chemical reaction. The two "element" terms cancel, and

$$\Delta H^\circ = \sum_{i=1}^{c} v_i \, \Delta \bar{H}_f^\circ(i) + \sum_{i=1}^{c} v_i \, \Delta \bar{H}_f^\circ(i) = \sum_{i=1}^{c} v_i \, \Delta \bar{H}_f^\circ(i) \quad \text{(2.7-14)}$$

(reactants only)   (products only)

where the final sum includes all substances involved in the reaction. This equation is the same as Equation (2.7-11).

▼   **EXAMPLE 2.17**

Find the standard-state enthalpy change of the reaction of Equation (2.7-6) at 298.15 K, using values of enthalpy changes of formation from Table A9.

Solution

$$\begin{aligned} \Delta H^\circ &= 2 \, \Delta \bar{H}_f^\circ(\text{NO}_2) + (-1) \, \Delta \bar{H}_f^\circ(\text{N}_2\text{O}_4) \\ &= (2)(33.18 \text{ kJ mol}^{-1}) + (-1)(9.16 \text{ kJ mol}^{-1}) \\ &= 57.20 \text{ kJ mol}^{-1} \end{aligned}$$

▲

We use dimensionless stoichiometric coefficients, so the enthalpy change for the reaction has the units of joules per mole (meaning per mole of the reaction *as written*). One can think of the units of v_i as moles of substance i per mole of reaction. If the way in which the reaction equation is balanced changes, the value of $\Delta H°$ changes. For example, if all stoichiometric co-efficients are doubled, the equation is still balanced but the value of ΔH doubles.

Exercise 2.23	Using values in Table A9, find the standard-state enthalpy change of the reaction of Equation (2.7-5) at 298.15 K.

If the formation reaction for a substance cannot actually be carried out, the enthalpy change of formation can be calculated from the enthalpy change of a combustion reaction or some other reaction that can be carried out.

▼   **EXAMPLE 2.18**

The standard-state enthalpy change of combustion of methane at 298.15 K equals -890.36 kJ mol^{-1}, with liquid water as one of the products. Find the enthalpy change of formation of methane at 298.15 K.

Solution

The balanced reaction equation is

$$0 = CO_2(g) + 2H_2O(l) - CH_4(g) - 2O_2(g)$$

so that

$$-890.36 \text{ kJ mol}^{-1} = \Delta\bar{H}_f°(CO_2) + 2\,\Delta\bar{H}_f°(H_2O)$$
$$+ (-1)\,\Delta\bar{H}_f°(CH_4) + (-2)\,\Delta\bar{H}_f°(O_2)$$

Since gaseous O_2 is the most stable form of oxygen at 298.15 K, $\Delta\bar{H}_f°(O_2) = 0$. Therefore, using values of the enthalpy changes of formation of the other substances from Table A9,

$$\Delta\bar{H}_f°(CH_4) = 890.36 \text{ kJ mol}^{-1} + (-393.509 \text{ kJ mol}^{-1})$$
$$+ 2(-285.830 \text{ kJ mol}^{-1}) + (-2)(0)$$
$$= -74.81 \text{ kJ mol}^{-1}$$

▲

Enthalpy Changes at Various Temperatures

An enthalpy change for a temperature other than the temperature found in a table can be calculated from heat capacity data. Assume that the enthalpy change for a given reaction can be calculated from enthalpy changes of formation at temperature T_1 and that the enthalpy change at temperature T_2 is desired.

Consider the processes shown in Figure 2.7. The reaction for which the reactants and products are at temperature T_2 is called process 2. This is the process whose enthalpy change we want to find. An alternative path-way for the same initial and final states consists of processes 3, 1, and 4. Process 3 is the change in temperature of the reactants from T_2 to T_1. Process 1 is the chemical reaction at temperature T_1. This is the process

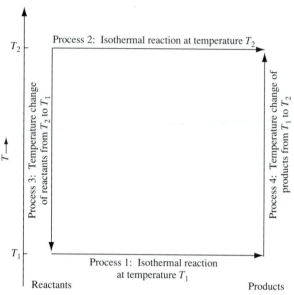

Figure. 2.7. The Process for Calculating $\Delta H(T_2)$ from $\Delta H(T_1)$. $\Delta H(T_2)$ is equal to the sum of the enthalpy changes of processes 3, 1, and 4.

whose enthalpy change is known. Process 4 is the change in temperature of the products from T_1 to T_2. Because enthalpy is a state variable,

$$\Delta H_2 = \Delta H_3 + \Delta H_1 + \Delta H_4 \tag{2.7-15}$$

For 1 mole of reaction, process 3 consists of changing the temperature of an amount of each reactant equal to the magnitude of its stoichiometric coefficient, so that

$$\Delta H_3 = -\int_{T_2}^{T_1} \sum_{i=1}^{c} v_i \bar{C}_P(i)\, dT = \int_{T_1}^{T_2} \sum_{i=1}^{c} v_i \bar{C}_P(i)\, dT \tag{2.7-16}$$
(reactants only in sum)

where $\bar{C}_P(i)$ is the molar heat capacity of substance i. The negative sign in front of the first integral comes from the fact that the stoichiometric coefficients of reactants are negative, and the second equality comes from interchanging the limits of integration, which changes the sign of a definite integral but not its magnitude.

Process 4 is the change in temperature of the products from T_1 to T_2, so that

$$\Delta H_4 = \int_{T_1}^{T_2} \sum_{i=1}^{c} v_i \bar{C}_P(i)\, dT \tag{2.7-17}$$
(products only in sum)

The sums in Equations (2.7-16) and (2.7-17) can be combined to give the expression

$$\Delta H(T_2) = \Delta H(T_1) + \int_{T_1}^{T_2} \Delta C_P\, dT \tag{2.7-18}$$

where the enthalpy change at each temperature is labeled with that temperature and

$$\Delta C_P = \sum_{i=1}^{c} v_i \bar{C}_P(i) \qquad \textbf{(2.7-19)}$$

All reactants and products are included in this sum. Since reactants have negative stoichiometric coefficients, ΔC_P is the heat capacity of the products minus the heat capacity of the reactants. If Equation (2.7-18) is applied to the standard-state reaction, then ΔC_P and each ΔH are replaced by the appropriate standard-state quantities.

▼   **EXAMPLE 2.19**

Using heat capacity data from Table A9 and assuming the heat capacities to be independent of temperature, find the standard-state enthalpy changes at 373.15 K for the reaction of Equation (2.7-6).

Solution

$$\Delta C_P^\circ = 2\bar{C}_P^\circ(NO_2) + (-1)\bar{C}_P^\circ(N_2O_4)$$
$$= 2(37.20 \text{ J K}^{-1} \text{ mol}^{-1}) + (-1)(77.28 \text{ J K}^{-1} \text{ mol}^{-1})$$
$$= -2.88 \text{ J K}^{-1} \text{ mol}^{-1} = -0.00288 \text{ kJ K}^{-1} \text{ mol}^{-1}$$

Using the value of $\Delta H^\circ(298.15 \text{ K})$ from Example 2.17,

$$\Delta H^\circ(298.15 \text{ K}) = 57.2 \text{ kJ mol}^{-1}$$
$$+ \int_{298.15 \text{ K}}^{373.15 \text{ K}} (-0.00288 \text{ kJ K}^{-1} \text{ mol}^{-1}) \, dT$$
$$= 57.2 \text{ kJ mol}^{-1} - 0.216 \text{ kJ mol}^{-1}$$
$$= 57.0 \text{ kJ mol}^{-1}$$

▲

Exercise 2.24   Find the value of the standard-state enthalpy change of the reaction of Equation (2.7-5) at 200°C. State any assumptions.

Reactions Other Than Standard-State Reactions

If the products and reactants are not at their standard states, the enthalpy change for a reaction can have a different value from that of the standard-state reaction. However, for the reactions in our present class, the effect is not large. The enthalpy of an ideal gas does not depend on the pressure, and real gases behave nearly like ideal gases for moderate pressures. The effect of moderate changes in pressures on the enthalpy of pure solids and liquids is also small. In a later chapter we will learn how to calculate these effects, but unless there is some need for really great accuracy, we will use the value of the standard-state enthalpy change for the enthalpy change of a reaction at ordinary pressures.

Adiabatic Reactions

Thus far, we have discussed reactions in which the final temperature is the same as the initial temperature. The enthalpy change will have a different value if the temperature of the system changes during the reaction. One case of interest is a chemical reaction that takes place adiabatically. If an adiabatic reaction takes place at constant pressure, the enthalpy change is zero.

In order to compute the final temperature of a system in which a chemical reaction takes place adiabatically, we consider the processes shown in Figure 2.8. Process 1 is the actual reaction, for which ΔH is equal to zero. Process 2 is the isothermal reaction. Process 3 is the change in temperature of the products plus any remaining reactants to the same final state as process 1. Since enthalpy is a state variable,

$$\Delta H_1 = \Delta H_2 + \Delta H_3 = 0 \qquad (2.7\text{-}20)$$

If a stoichiometric mixture is taken and the reaction proceeds to completion, there are no remaining reactants and we can write

$$\Delta H_3 = \int_{T_1}^{T_2} \sum_{i=1}^{c} v_i \bar{C}_P(i)\, dT = -\Delta H_2 \qquad (2.7\text{-}21)$$
$$\text{(products only in sum)}$$

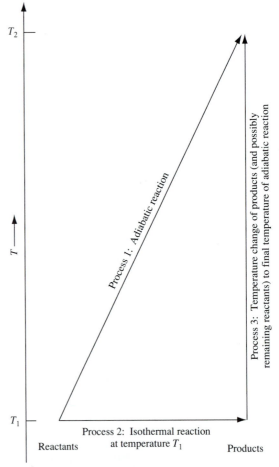

Figure 2.8. The Process for Calculating the Final Temperature of an Adiabatic Reaction.
The enthalpy change of process 1 is equal to the sum of the enthalpy changes of processes 2 and 3.

This equation can be solved for T_2. If there is not a stoichiometric mixture and the reaction proceeds to completion so that there are remaining amounts of some reactants, their heat capacities must be added to Equation (2.7-21). If the reaction does not proceed to completion, the equilibrium can shift as the temperature changes (see Chapter 7) and the calculation is more complicated. We will not discuss that case.

▼   **EXAMPLE 2.20**

Find the final temperature if the reaction of Equation (2.7-1) is carried out adiabatically at constant pressure, beginning at 298.15 K. Assume that a stoichiometric mixture is present before the reaction and that the reaction proceeds to completion. Assume that the heat capacity of water vapor is constant and equal to its value at 2000 K.

Solution

$$\Delta H°(298.15 \text{ K}) = 2 \, \Delta \bar{H}_f°(H_2O) - 0 - 0$$
$$= 2(-241.818 \text{ kJ mol}^{-1}) = -483.636 \text{ kJ mol}^{-1}$$
$$= -483636 \text{ J mol}^{-1}$$
$$C_P = (2)(51.18 \text{ J K}^{-1} \text{ mol}^{-1}) = 102.36 \text{ J K}^{-1} \text{ mol}^{-1}$$
$$\int_{298.15 \text{ K}}^{T_2} (102.36 \text{ J K}^{-1} \text{ mol}^{-1}) \, dT - 483636 \text{ J mol}^{-1} = 0$$
$$(102.36 \text{ J K}^{-1} \text{ mol}^{-1})(T_2 - 298.15 \text{ K}) = 483636 \text{ J mol}^{-1}$$
$$T_2 = 5023 \text{ K} \approx 5000 \text{ K}$$

▲

Exercise 2.25

Find the final temperature if a stoichiometric mixture of methane and oxygen is ignited at 298.15 K and allowed to react adiabatically at a constant pressure. Assume that the reaction proceeds to completion and that the heat capacities of the products are constant and equal to their values at 2000 K.

If heat capacities are represented by polynomials as in Table A7, a more nearly accurate final temperature can be calculated. However, this leads to a nonlinear equation. A quadratic equation can be solved by use of the quadratic formula, and a numerical approximation to the solution of a cubic equation, a quartic equation, etc. can be obtained by trial and error or by other techniques.[6]

Exercise 2.26

Using the parameters from Table A7, find the final temperature after adiabatic combustion of the stoichiometric mixture of hydrogen and oxygen in Example 2.20.

[6] See L. Gilman and R. H. McDowell, *Calculus*, 2nd ed., W. W. Norton, New York, 1978 pp. 465ff, or any other calculus text.

Energy Changes of Chemical Reactions

Energy changes of chemical reactions could be calculated from energy changes of formation in exactly the same way as enthalpy changes. However, tables of energy changes of formation are not commonly used because it is possible to calculate the energy change of a chemical reaction from the enthalpy change of the reaction. From the definition of the enthalpy, Equation (2.5-1),

$$\Delta U = \Delta H - \Delta(PV) \tag{2.8-1}$$

Ordinarily, $\Delta(PV)$ is much smaller than ΔH, so it can be calculated approximately while ΔH is calculated accurately. For example, if ΔH is 1000 times larger than $\Delta(PV)$ and if five significant digits are desired in ΔU, then five significant digits are required in ΔH but only two significant digits are needed in $\Delta(PV)$.

For the class of reactions that we considered in the previous section

$$PV = P[V(s) + V(l) + V(g)] \tag{2.8-2}$$

where $V(s)$ is the volume of all of the solid phases, $V(l)$ is the volume of all of the liquid phases, and $V(g)$ is the volume of the gas phase. Under ordinary conditions, the molar volume of a gas is several hundred to a thousand times larger than the molar volume of a solid or liquid. Therefore, if two or three significant digits suffice in the value of $\Delta(PV)$, we can ignore the volume of the solid and liquid phases and write

$$\Delta(PV) = \Delta[PV(g)] \tag{2.8-3}$$

If the products and reactants are at the same temperature and if we can use the ideal gas equation as an adequate approximation,

$$\Delta[PV(g)] = [\Delta n(g)]RT \tag{2.8-4}$$

where $\Delta n(g)$ is the change in the amount of gas, measured in moles. If 1 mole of reaction occurs, then

$$\Delta n(g) = \Delta v_g = \sum_{i=1}^{c} v_i \tag{2.8-5}$$
$$\text{(gases only)}$$

which defines the quantity Δv_g, equal to the number of moles of gas in the product side of the balanced chemical equation minus the numbers of moles of gas in the reactant side of the balanced equation.

▼   **EXAMPLE 2.21**   Find $\Delta(PV)$ and $\Delta U°$ for the reaction of Equation (2.7-1) at 298.15 K.

Solution

$$(\Delta v_g)RT = (-3)(8.3145 \text{ J K}^{-1} \text{ mol}^{-1})(298.15 \text{ K})$$
$$= -7437 \text{ J mol}^{-1} = -7.437 \text{ kJ mol}^{-1}$$
$$\Delta H° = 2\, \Delta \bar{H}_f°(H_2O) - 0 - 0 = -571.660 \text{ kJ mol}^{-1}$$
$$\Delta U° = -571.660 \text{ kJ mol}^{-1} - (-7.437 \text{ kJ mol}^{-1})$$
$$= -564.223 \text{ kJ mol}^{-1}$$

▲

Exercise 2.27   Using the fact that the molar volume of liquid water is 18 cm^3 mol^{-1} at 298.15 K, make a more accurate calculation of $\Delta(PV)$ for the reaction of Example 2.21.

Exercise 2.28   Find ΔU° for the reaction of Equation (2.7-2) at 298.15 K.

Calorimetry

The most common way to determine enthalpy changes of formation of combustible substances is to carry out the combustion reaction in a constant-volume calorimeter and calculate the enthalpy change of formation as in Example 2.18. Figure 2.9 depicts a **bomb calorimeter**. The reaction is carried out in a strong, rigid container, which is called the "bomb." A pellet of a solid reactant is placed in the bomb along with an excess of oxygen at a pressure of about 25 atm.

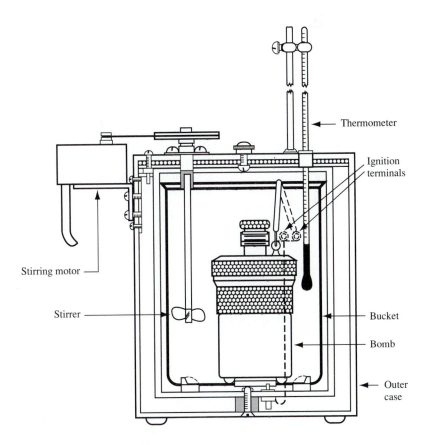

Figure 2.9. A Bomb Calorimeter. The bomb is immersed in a metal bucket containing a fixed amount of water. The system consists of the substances inside the bomb, and the bomb, bucket, and water are the surroundings. (Courtesy of Parr Instrument Co.)

The bomb is placed in a bucket filled with a measured amount of water, and the water is stirred until a nearly steady temperature is attained. The system, consisting of the solid reactant and the oxygen, is in a metastable state but is treated as though at equilibrium. The system is ignited electrically, and the bomb, water, and bucket are allowed to approach equilibrium at a new temperature several degrees higher than the initial temperature. The amount of heat transferred to the calorimeter (bomb, water, and bucket) is calculated from knowledge of the heat capacity of the calorimeter. If this heat capacity is independent of temperature

It is generally an excellent approximation to assume that the heat capacity of the calorimeter is constant.

$$q = -q_{cal} = -C_{cal}\,\Delta T \qquad \textbf{(2.8-6)}$$

Since the system is at constant volume, $w = 0$ and

$$q = \Delta U \qquad \textbf{(2.8-7)}$$

From the amount of solid reactant present, ΔU for 1 mole of reaction is calculated, allowing ΔH for 1 mole of reaction to be calculated from Equation (2.8-4). This value of ΔH is not quite equal to $\Delta H°$ because neither the final pressure nor the initial pressure is equal to the standard-state pressure and the gases present are not ideal. Also, the final temperature is not equal to the initial temperature. For ordinary work in an undergraduate physical chemistry laboratory, these differences are negligible compared with the experimental errors, but corrections can be made if needed for highly accurate work. Furthermore, in most calorimeters a wire fuse is used to ignite the sample, and the heat of combustion of the wire must be included in the calculation. We omit this contribution in the following example:

▼   **EXAMPLE 2.22**

A pellet of naphthalene of mass 1.234 g is burned in a bomb calorimeter with $C_{cal} = 14225$ J K^{-1} (assume constant). If the initial temperature is 298.150 K and the final temperature is 301.634 K, find ΔU and ΔH for 1 mole of reaction, with liquid water as one of the products.

Solution

The balanced reaction equation is

$$C_{10}H_8(s) + 12O_2(g) \rightarrow 10CO_2(g) + 4H_2O(l)$$
$$q = -q_{cal} = -(14225 \text{ J K}^{-1})(3.484 \text{ K}) = -4.956 \times 10^4 \text{ J}$$

For 1 mole of reaction

$$\Delta U = -\frac{4.956 \times 10^4 \text{ J}}{1.234 \text{ g}}\left(\frac{128.19 \text{ g}}{1 \text{ mol}}\right) = -5.148 \times 10^6 \text{ J mol}^{-1}$$

$$\Delta H = -5.148 \times 10^6 \text{ J mol}^{-1} + (-2)(8.3145 \text{ J K}^{-1}\text{ mol}^{-1})(298.15 \text{ K})$$
$$= -5.148 \times 10^6 \text{ J mol}^{-1} - 4958 \text{ J mol}^{-1}$$
$$= -5.143 \times 10^6 \text{ J mol}^{-1} = -5143 \text{ kJ mol}^{-1}$$

This value is in good agreement with the published value, $\Delta H° = -5156.8$ kJ mol^{-1}.

We can determine the error due to the inequality of the temperatures of the reactant and products. The initial amount of naphthalene was 0.00963 mol, giving 0.0963 mol of CO_2 and 0.0385 mol of H_2O. From the heat capacities in Table A7, the heat capacity of the products is 6.55 J K^{-1}. The heat required to bring the products back to 298.15 K is

$$q_2 = (6.55 \text{ J K}^{-1})(3.494 \text{ K}) = 22.9 \text{ J}$$

This means that $q_{\text{cal}} = 4.956 \times 10^4 \text{ J} + 22.9 \text{ J} = 4.958 \times 10^4 \text{ J}$ for the reaction with the products at the same temperature as the reactants, a difference of about 0.04%.

▲

Exercise 2.29 | What will be the rise in temperature of the calorimeter in Example 2.22 if a pellet of anthracene of mass 1.555 g is burned? The enthalpy change of combustion of anthracene is $-7114.5 \text{ kJ mol}^{-1}$.

The calorimetry of substances reacting in solutions is commonly carried out at constant pressure. In this case, q is equal to ΔH and no calculation of $\Delta(PV)$ is necessary to determine enthalpy changes of reactions.

Average Bond Energies

Chemical reactions involve the breaking and forming of chemical bonds. If it were possible to determine exactly the energy required to break every type of chemical bond, it would be possible to calculate standard-state energy changes of any gaseous chemical reaction. One problem is that breaking chemical bonds between the same pair of elements in different compounds requires different amounts of energy.

However, tables of average values have been constructed, from which estimates of energy changes can be made. Table A10 shows such values, given as the energy required to break 1 mole of the given bond. To estimate the energy change for a gas-phase reaction, one uses the relationship

$$\Delta U \approx \text{(sum of all bond energies in reactants)}$$
$$- \text{(sum of all bond energies in products)} \qquad \textbf{(2.8-8)}$$

The calculation can be simplified by omitting from both terms the bonds that occur in both reactants and products, since their contributions cancel.

In practice, this method is used only if data are not available for the enthalpy change of formation, because the results are only approximately correct. Average bond energies cannot be used for reactions involving solids or liquids, where intermolecular forces also make important contributions.

▼ **EXAMPLE 2.23**

Estimate the energy change of the reaction

$$C_2H_4(g) + H_2(g) \to C_2H_6(g)$$

Compare with the value obtained from enthalpy changes of formation.

Solution

Note that the 4 moles of C—H bonds that occur in both the products and the reactants are omitted from the calculation.

For 1 mole of reaction, 1 mole of C=C bonds and 1 mole of H—H bonds must be broken and 1 mole of C—C bonds and 2 moles of C—H bonds must be formed.

$$\Delta U = 613 \text{ kJ mol}^{-1} + 436 \text{ kJ mol}^{-1} - 348 \text{ J mol}^{-1} - 2(413 \text{ kJ mol}^{-1})$$
$$= -125 \text{ kJ mol}^{-1}$$

From enthalpy changes of formation at 298.15 K,

$$\Delta H_{298.15\,K} = \Delta \bar{H}_f^{\circ}(C_2H_6) - \Delta \bar{H}_f^{\circ}(C_2H_4) - \Delta \bar{H}_f^{\circ}(H_2)$$
$$= -84.68 \text{ kJ mol}^{-1} - (52.26 \text{ kJ mol}^{-1}) - 0$$
$$= -136.94 \text{ kJ mol}^{-1}$$
$$\Delta U_{298.15\,K} = -136.94 \text{ kJ mol}^{-1} - (-1)(8.3145 \text{ J K}^{-1} \text{ mol}^{-1})(298.15 \text{ K})$$
$$= -136.94 \text{ kJ mol}^{-1} + 2.48 \text{ kJ mol}^{-1}$$
$$= -134.46 \text{ kJ mol}^{-1}$$

The error of 9 kJ mol^{-1} comes from the fact that the C—H bonds in ethane are not exactly the same as the C—H bonds in ethylene.

The error of 9 kJ mol^{-1} in Example 2.23 is typical. It is larger than the difference between ΔH and ΔU, so the value of ΔU obtained from average bond energies is ordinarily used for ΔH without correction for the value of $\Delta(PV)$.

Exercise 2.30 Using average bond energy values, estimate ΔH for the reaction

$$CH_4(g) + 2O_2(g) \rightarrow CO_2(g) + 2H_2O(g)$$

Compare your value with the correct value of ΔH obtained from enthalpy changes of formation.

Summary

In this chapter, we have introduced thermodynamics and have discussed the first law of thermodynamics, which is a form of the law of conservation of energy. This law defines U, the internal energy of a system, through

$$\Delta U = q + w$$

where q is the heat added to the system and w is the work done on the system. The internal energy is a state function of the macroscopic state of the system.

The enthalpy, H, was defined through

$$H = U + PV$$

where P is the pressure and V is the volume of the system. At constant pressure, ΔH is equal to q for a simple system. Various calculations of ΔH values were carried out.

Additional Reading

H. B. Callen, *Thermodynamics*, Wiley, New York, 1960
   This carefully written text for physics students presents the basic principles of thermodynamics, treating the laws as postulates instead of experimental facts.

J. deHeer, *Phenomenological Thermodynamics*, Prentice-Hall, Englewood Cliffs, NJ, 1986

This is a modern text in chemical thermodynamics and a thorough theoretical survey.

J. G. Kirkwood and I. Oppenheim, *Chemical Thermodynamics*, McGraw-Hill, New York, 1961

A theoretically oriented text for graduate students in chemistry.

M. L. McGlashan, *Chemical Thermodynamics*, Academic Press, New York, 1979

This is a textbook for a chemical thermodynamics course at the graduate level.

K. S. Pitzer and L. Brewer, *Thermodynamics*, McGraw-Hill, New York, 1961

A revision of a classic graduate-level book of 1923 by G. N. Lewis and M. Randall, which brought chemical thermodynamics into graduate curricula.

F. T. Wall, *Chemical Thermodynamics*, 3rd ed., W. H. Freeman, San Francisco, 1974

This book is designed for a course at the senior or first-year graduate level and includes statistical mechanics.

PROBLEMS

Problems for Section 2.1

2.31. If dw_{rev} is written as

$$dw_{rev} = M\,dV + N\,dT$$

then for a closed fluid system $M = -P$ and $N = 0$. Show that the requirement for dw_{rev} to be an exact differential is not met.

2.32. The tension force for a spring that obeys Hooke's law is given by

$$\tau = \kappa(x - x_0)$$

where x is the length of the spring, x_0 is the equilibrium length, and κ is a constant called the spring constant. Obtain a formula for the work done on the spring if its length is changed reversibly from x_0 to x' at constant volume.

*2.33. a. Obtain a formula for the work done in isothermally compressing 1.000 mol of a van der Waals gas from a volume V_1 to a volume V_2.

b. Using the formula from part a, find the work done in compressing 1.000 mol of carbon dioxide from 20.00 L to 5.000 L at 298.15 K.

c. Calculate the work done on the surroundings if 1.000 mol of carbon dioxide expands isothermally but irreversibly from 5.000 L to 20.00 L at an external pressure of 1.000 atm.

Problems for Section 2.2

2.34. Calculate the amount of heat required to bring 1.000 mol of water from solid at 0.0°C to gas at 100.0°C.

2.35. Compute the rise in temperature of water that is brought to rest after falling over a waterfall 100. m high. Assume that no heat is transferred to the surroundings.

Problems for Section 2.3

2.36. Calculate q, w, and ΔU for melting 10.0 g of ice at 0.0°C and a constant pressure of 1.000 atm. The density of ice is 0.917 g/mL.

*2.37. Calculate q, w, and ΔU for vaporizing 5.000 mol of liquid water at 100.0°C to steam at 100.0°C at a constant pressure of 1.000 atm.

Problems for Section 2.4

2.38. Find the final pressure if 2.000 mol of nitrogen is expanded adiabatically and reversibly from a volume of 20.00 L to a volume of 50.00 L, beginning at a pressure of 2.500 atm. Assume nitrogen to be ideal with $\bar{C}_V = 5R/2$.

2.39. Find the final temperature and the final volume if 2.000 mol of nitrogen is expanded adiabatically and reversibly from standard temperature and pressure (STP) to a pressure of 0.500 atm. Assume nitrogen to be ideal with $\bar{C}_V = 5R/2$.

*2.40. 1.000 mol of carbon dioxide is expanded adiabatically and reversibly from 298.15 K and a molar volume of 5.000 L to a volume of 20.00 L.

a. Find the final temperature, assuming the gas to be ideal with $\bar{C}_V = 5R/2 = $ constant.

b. Find the final temperature, assuming the gas to be described by the van der Waals equation with $\bar{C}_V = 5R/2 = $ constant.

2.41. a. Find the final temperature, ΔU, q, and w for the reversible adiabatic expansion of O_2 gas from 373.15 K and a molar volume of 10.00 L to a molar volume of 30.00 L. Assume the gas to be ideal with $\bar{C}_V = 5R/2 = $ constant.

b. Repeat the calculation of part a for argon instead of oxygen. Assume that $\bar{C}_V = 3R/2 = $ constant.

c. Explain your answers for parts a and b in physical terms.

Problems for Section 2.5

2.42. Show that if $dU = dq + dw$, if dU is exact, and if dq is inexact, then dw must be inexact.

2.43. For constant-pressure processes, what is the function whose differential is equal to dw? Why is dw not equal to the differential of this function for processes in which pressure is not constant?

2.44. For a nonsimple system such as a spring or a rubber band governed by Equation (2.1-9), one must specify whether a heat capacity is measured at constant τ or at constant L, in addition to specifying constant P or constant V. Find a relation analogous to Equation (2.5-11) relating $C_{P,\tau}$ and $C_{P,L}$.

2.45. Express the exponents in Equations (2.4-25), (2.4-34), and (2.4-35) in terms of the heat capacity ratio γ, defined in Equation (2.5-14). Use Equation (2.5-12) to simplify your results.

*2.46. The Joule-Thomson coefficient of nitrogen gas at 50. atm and 0°C is equal to 0.044 K atm^{-1}.

a. Estimate the final temperature if nitrogen gas is expanded through a porous plug from a pressure of 60. atm to a pressure of 40. atm at 0°C.

b. Estimate the value of $(\partial \bar{H}/\partial P)_T$ for nitrogen gas at 50. atm and 0°C. State any assumptions.

Problems for Section 2.6

2.47. a. Calculate ΔH and ΔU for heating 1.00 mol of argon from 100. K to 300. K at a constant pressure of 1.00 atm. State any assumptions.

b. Calculate ΔH and ΔU for heating 1.00 mol of argon from 100 K to 300 K at a constant volume of 0.8206 L.

c. Explain the differences between the results of parts a and b.

2.48. Supercooled steam is condensed irreversibly but at a constant pressure of 1.000 atm and a constant temperature of 95°C. Find the molar enthalpy change.

Problems for Section 2.7

2.49. Calculate $\Delta H°$ and $\Delta U°$ for the following reactions at 298.15 K:

    a. $C_3H_8(g) + 5O_2(g) \rightarrow 3CO_2(g) + 4H_2O(l)$
*b. $2SO_2(g) + O_2(g) \rightarrow 2SO_3(g)$
    c. $SO_3(g) + H_2O(g) \rightarrow H_2SO_4(l)$
*d. $4CuO(s) \rightarrow 2Cu_2O(s) + O_2(g)$
    e. $2CO(g) + O_2(g) \rightarrow 2CO_2(g)$

2.50. Calculate $\Delta H°$ for the reactions of parts a, b, and e of the previous problem at 75°C.

2.51. Using the value of the enthalpy change of vaporization of water, find the enthalpy change of the reaction of part a of the previous problem at 298.15 K for the case that the water is vapor. Do not look up the enthalpy change of formation of water vapor.

*2.52. a. Find the values of the enthalpy changes of formation of methane, carbon dioxide, and liquid water at 373.15 K, using heat capacity values from Table A9.

b. Using the values from part a, find the standard-state enthalpy change for the following reaction at 373.15 K:

$$CH_4(g) + 2O_2(g) \rightarrow CO_2(g) + 2H_2O(l)$$

c. Find the standard-state enthalpy change of the reaction of part b, using Equation (2.7-18). Compare and comment on your answers for parts b and c.

Problems for Section 2.8

2.53. Calculate the value of $\Delta \bar{U}_f°$ for carbon dioxide at 298.15 K.

2.54. a. Calculate $\Delta H°$ for the following reaction at 298.15 K:

$$CH_4(g) + O_2(g) \rightarrow CO_2(g) + H_2O(l)$$

(Balance the reaction first.)

b. Calculate $\Delta U°$, neglecting the volume of liquid water.

c. Calculate $\Delta U°$ without neglecting the volume of liquid water, taking $V = 18.0$ cm^3 mol^{-1} for liquid water.

*2.55. a. Estimate the enthalpy change of the following reaction, using average bond energies:

$$2C_2H_6(g) + 7O_2(g) \rightarrow 4CO_2(g) + 6H_2O(g)$$

b. Compare the standard-state enthalpy change for this reaction at 298.15 K and at 373.15 K, using enthalpy changes of formation and heat capacity data.

General Problems

2.56. A sample of 1.000 mol of N_2 gas is expanded adiabatically from a volume of 10.00 L and a temperature of 400 K to a volume of 20.00 L. Assume that N_2 is ideal, with $\bar{C}_V = 5R/2$.

a. Find the final temperature if the expansion is carried out reversibly.

b. Find the final temperature if the expansion is carried out with a constant external pressure of 1.000 atm.

c. Find the final temperature if the gas expands into a vacuum.

d. Find ΔU and w for each of the processes of parts a, b, and c.

e. For the reversible expansion of part a, show that the value of w obtained from the integral

$$w_{\text{rev}} = -\int_c P \, dV$$

is the same as the value of w obtained from

$$w = \Delta U - q = \Delta U$$

2.57. It is shown in the theory of hydrodynamics[7] that the speed of sound in a fluid, v_s, is given by

$$v_s^2 = \frac{\bar{V} C_P}{M \kappa_T C_V} = \frac{\bar{V} \gamma}{M \kappa_T}$$

---

[7] H. Lamb, *Hydrodynamics*, 6th ed., Cambridge Univ. Press, New York, 1932.

where κ_T is the isothermal compressibility, $\bar{V}$ is the molar volume, and M is the molar mass.

a. Find the speed of sound in air at 298.15 K and 1.000 atm assuming a mean molar mass of 0.029 kg mol^{-1} and $\bar{C}_V = 5R/2$.

b. Find the speed of sound in helium at 298.15 K and 1.000 atm.

c. The speed of sound in ethane at 10.°C is equal to 308 m s^{-1}. Find the values of $\bar{C}_P$ and $\bar{C}_V$ for ethane at this temperature.

*2.58. a. The pressure on a sample of 1.000 mol of liquid water is increased from 1.00 bar to 100.00 bar at a constant temperature of 20.00°C. Find ΔH, ΔU, q, and w. State any assumptions.

b. The pressure on the same sample of water is increased adiabatically from 1.00 bar to 100.00 bar, beginning at 20.00°C. Find ΔH, ΔU, q, and w. State any assumptions.

2.59. a. Calculate the enthalpy change of combustion at 25°C of 1.000 mol of each of the following: (1) propane, $C_3H_8(g)$; (2) isooctane (2,2,4-trimethylpentane), $C_8H_{18}(l)$; (3) hexadecane (cetane), $C_{16}H_{34}(l)$. Assume that the combustion is complete, forming only CO_2 and H_2O, and that gaseous H_2O is formed. The enthalpy change of formation of cetane is -448 kJ mol^{-1}.

b. Liquefied propane is sold as a fuel for heating houses. Isooctane is a component of gasoline and is the substance used as the reference compound for "octane numbers." (It is "100 octane.") Cetane is a component of diesel fuel. In order to compare these fuels, compute the enthalpy change of combustion of 1.000 kg of each substance.

3 The Second and Third Laws of Thermodynamics: Entropy

PREVIEW

In this chapter we discuss the two additional laws that, together with the first law, form the entire basic theory of thermodynamics. These laws deal with the entropy, which is a subtler and more complicated state function than the internal energy.

OBJECTIVES

After studying this chapter, the student should:

1. understand the physical statements of the second law of thermodynamics as statements of what cannot happen,
2. understand the reason why a heat engine cannot have an efficiency as large as unity,
3. have a basic understanding of the physical meaning of the entropy of a system,
4. be able to calculate entropy changes for a variety of systems and processes,
5. have a basic understanding of the relationship between thermodynamic entropy and statistical entropy and be able to calculate simple statistical entropies.

PRINCIPAL FACTS AND IDEAS

1. The second law governs whether any macroscopic process can occur spontaneously.
2. There are two principal physical statements of the second law of thermodynamics: (1) It is impossible for a system in a cyclic process to turn heat completely into work. (2) Heat cannot flow spontaneously from a cooler to a hotter object if nothing else happens.
3. The second law provides a new state variable, the entropy.
4. A consequence of the second law is that the entropy of the universe cannot decrease.
5. The mathematical statement of the second law provides a means of calculating the entropy change of any process that begins and ends at equilibrium states.
6. Entropy is connected with lack of information through the definition of Boltzmann:

$$S_{st} = k_B \ln(\Omega)$$

where k_B is Boltzmann's constant and Ω is the number of mechanical states in which the system might exist, as far as we know.
7. The third law of thermodynamics allows the entropy of a pure crystalline substance to be set equal to zero at the absolute zero of temperature.
8. Zero temperature on the Kelvin scale is unattainable.

3.1 The Second Law of Thermodynamics and the Carnot Heat Engine

Like the first law of thermodynamics, the second law is based on experimental fact. It can be stated in a number of ways, both verbally and mathematically, and some of the statements do not seem closely related to other statements.

Physical Statements of the Second Law

There are two statements of the second law that we regard as generalizations of experimental fact. The **Kelvin statement of the second law of thermodynamics** involves **cyclic processes**, which are processes in which the final state of the system is the same as its initial state. This statement is: *It is impossible for a system to undergo a cyclic process whose sole effects are the flow of an amount of heat from the surroundings to the system and the performance of an equal amount of work on the surroundings.* In other words, *it is impossible for a system in a cyclic process to turn heat completely into work.*

Rudolf Julius Emmanuel Clausius, 1822–1888, was a German physicist and is generally considered to be the discoverer of the second law of thermodynamics.

The **Clausius statement of the second law of thermodynamics** is: *It is impossible for a process to occur that has the sole effect of removing a quantity of heat from an object at a lower temperature and transferring this quantity of heat to an object at a higher temperature.* In other words, *heat cannot flow spontaneously from a cooler to a hotter object if nothing else happens.*

The Kelvin and Clausius statements are called **physical statements of the second law**. No violation of either statement of the second law of thermodynamics has ever been observed in a properly done experiment. A machine that would violate the Kelvin statement of the second law and turn heat completely into work in a cyclic process is called a **perpetual motion machine of the second kind**.

The Clausius statement of the second law is closely related to ordinary experience.

However, heat can be completely turned into work if the process is not cyclic. For example, in an isothermal expansion of an ideal gas, since ΔU vanishes,

$$w_{surr} = -w = q \quad \text{(ideal gas, isothermal volume change)} \quad \text{(3.1-1)}$$

Since w_{surr} is the amount of work done on the surroundings by the system and q is the amount of heat put into the system, heat has been completely turned into work. However, the system has not been restored to its original state. The process is not cyclic, and there is no violation of the second law.

The Carnot Engine and the Carnot Cycle

The **Carnot heat engine** is an imaginary model machine that Carnot devised in 1824 to represent a steam engine. A simple steam engine is depicted schematically in Figure 3.1a. It has a cylinder with a piston connected to a crankshaft by a connecting rod. A boiler can inject high-pressure steam into the cylinder through an intake valve, and steam can be exhausted through an exhaust valve.

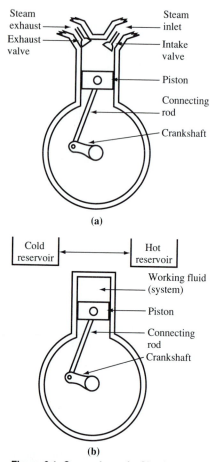

(a)

(b)

Figure 3.1. Comparison of a Simple Steam Engine and a Carnot Engine (schematic). (a) A simple steam engine. High-pressure steam enters through the intake valve and can push the piston, doing work on the surroundings. **(b)** A Carnot heat engine. The working fluid is a closed system that can accept heat from a hot reservoir, give up heat to a cool reservoir, and push on the piston.

Any cyclic process must begin and end at the same state. The state at which the isothermal compression ends must be chosen to accommodate this requirement.

This steam engine operates with a two-stroke cycle. The cycle begins with the piston at top dead center (the position of minimum volume in the cylinder) and with the intake valve open. High-pressure, high-temperature steam from the boiler enters the cylinder through the intake valve and pushes the piston, which turns the crankshaft. When the piston reaches bottom dead center (the position of maximum volume in the cylinder), the intake valve closes and the exhaust valve opens. The inertia of the crankshaft and its attached flywheel pushes the piston back toward top dead center, pushing the "spent" steam out through the exhaust valve. When top dead center is reached, the exhaust valve closes, the intake valve opens, and the engine is ready to repeat its cycle.

The Carnot engine is depicted in Figure 3.1b. This engine has no valves. The cylinder contains a "working fluid," which is our system. The cylinder, piston, reservoirs, etc. are the surroundings. The system is closed. Instead of passing steam into and out of the cylinder, the Carnot engine allows heat to flow from a "hot reservoir" into its working fluid and exhausts heat into a "cold reservoir" by conduction through the cylinder walls or cylinder head. The working fluid is presumably a gas but could be a liquid or a "gas" of photons. The Carnot engine is defined to operate reversibly and without friction, so there is no way to construct one in the real world.

The Carnot engine has a two-stroke cycle. The cycle begins at top dead center with the hot reservoir in contact with the cylinder. The expansion stroke is broken into two steps. The first step is an isothermal reversible expansion of the system at the temperature of the hot reservoir. We stop this expansion with the piston only partway toward bottom dead center and remove the hot reservoir from the cylinder. The second step is an adiabatic reversible expansion. We choose the beginning state of the second step so that this step ends with the system at the temperature of the cold reservoir and the piston at bottom dead center. The third step of the cyclic process is a reversible isothermal compression with the cylinder in contact with the cold reservoir. We end this step at such a volume that the fourth step, a reversible adiabatic compression, ends with the piston at top dead center and the system at the temperature of the hot reservoir. The engine is then ready to repeat the cycle.

Figure 3.2a shows the path followed by the state point of the system as the engine undergoes one cycle, using V and T as the independent variables. Figure 3.2b shows the same cycle using V and P as the independent variables. The state at the beginning of each step is labeled with the number of that step.

Because the second and fourth steps of the Carnot cycle are adiabatic,

$$q_2 = q_4 = 0 \qquad\qquad \textbf{(3.1-2)}$$

where the labels on the q's indicate the step. For the entire cycle,

$$q_{\text{cycle}} = q_1 + q_2 + q_3 + q_4 = q_1 + q_3 \qquad\qquad \textbf{(3.1-3)}$$

Because U is a state variable and the cycle begins and ends at the same state,

$$\Delta U_{\text{cycle}} = 0 \qquad\qquad \textbf{(3.1-4)}$$

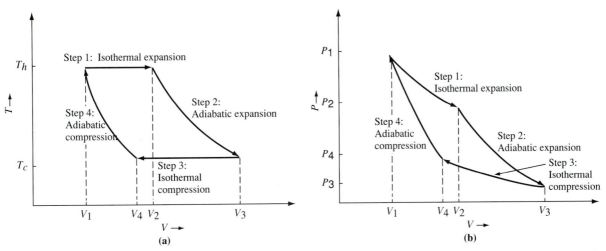

Figure 3.2. The Path of the State Point During a Carnot Cycle.
(a) In the V-T plane. A Carnot cycle must consist of an isothermal expansion, an adiabatic expansion, an isothermal compression, and an adiabatic compression that leads back to the initial state of the isothermal expansion. **(b)** In the V-P plane.

Therefore, since $\Delta U = q + w$,

$$w_{\text{cycle}} = -q_{\text{cycle}} = -q_1 - q_3 \qquad (3.1\text{-}5)$$

We define the efficiency, η_C, of the Carnot engine as the work done on the surroundings divided by the heat input at the hot reservoir. Since the heat exhausted at the cold reservoir is wasted, it is not reckoned as a negative part of the heat input.

$$\eta_C = \frac{w_{\text{surr}}}{q_1} = \frac{-w_{\text{cycle}}}{q_1} = \frac{q_{\text{cycle}}}{q_1}$$

$$= \frac{q_1 + q_3}{q_1} = 1 + \frac{q_3}{q_1} \qquad (3.1\text{-}6)$$

From the Kelvin statement of the second law, the efficiency must be less than unity. Therefore q_3 must be negative, since q_1 is positive. It is not possible to run a Carnot engine without exhausting some heat to a cool reservoir.

We now show that the ratio q_3/q_1 has the same value for all engines that produce work from heat, no matter what the working fluid is, as long as heat reservoirs at the same two temperatures are used. We show this assertion to be fact by considering a **Carnot heat pump**, which is our original Carnot heat engine run backward by another engine.

A heat pump is the reverse of a heat engine. Its input is work and its output is heat.

Figure 3.3 shows a reversible Carnot heat pump cycle, which is just the reverse of the cycle of Figure 3.2. We denote the amount of heat put into the hot reservoir by q'_4 and the amount of heat taken from the cold reservoir by q'_2. Then, since both cycles are reversible and we are considering the same Carnot engine run backward,

$$q'_4 = -q_1 \qquad (3.1\text{-}7)$$

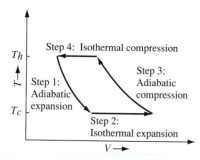

Figure 3.3. The Path of the State Point in the V-T Plane During a Carnot Heat Pump Cycle. This cycle is exactly the reverse of the Carnot engine cycle.

This is why it is less expensive to heat a house with a heat pump than with electric resistance heaters.

This technique is often used in the proof of mathematical theorems. However, we seek a contradiction with experimental fact, whereas a mathematician seeks a contradiction with an earlier theorem or with an axiom.

and

$$q_2' = -q_3 \tag{3.1-8}$$

Since the cycles are reversible, the amount of work done on the system in the reverse (heat pump) cycle, w', is equal to the amount of work done on the surroundings in the forward (engine) cycle if the same two reservoirs are used for both cycles:

$$w'_{\text{cycle}} = -w_{\text{cycle}} = w_{\text{surr}} \tag{3.1-9}$$

For a heat pump, the output is the heat delivered to the hot reservoir. The ratio of the output to the input is called the **coefficient of performance** and is analogous to an efficiency:

$$\eta_{\text{hp}} = \frac{|q_4'|}{w'_{\text{cycle}}} = -\frac{q_4'}{-q_2' - q_4'} = \frac{q_1}{q_1 + q_3}$$
$$= \frac{1}{1 + q_3/q_1} = \frac{1}{\eta_{\text{C}}} \tag{3.1-10}$$

This coefficient of performance equals the reciprocal of the Carnot efficiency because the input and output are reversed in their roles as well as their signs. The Carnot efficiency is always smaller than unity, so the Carnot heat pump coefficient of performance is always greater than unity. That is, the amount of heat delivered to the hot reservoir is always greater than the work put into the heat pump, because some heat has been transferred from the cold reservoir to the hot reservoir.

We now prove that any reversible heat engine has the same efficiency as the Carnot engine if it exchanges heat with the same two reservoirs. We proceed by assuming the opposite of what we want to prove and then showing that this assumption leads to a contradiction with fact and therefore must be incorrect.

Assume that a reversible heat engine does exist with a greater efficiency than a Carnot engine. This is the assumption that we want to show to be false. We call this engine a "superengine" and label its quantities with the letter *s*. We label the quantities for the original Carnot engine by the letter C. By our assumption, $\eta_s > \eta_{\text{C}}$, so

$$1 + \frac{q_3(s)}{q_1(s)} > 1 + \frac{q_3(\text{C})}{q_1(\text{C})} \tag{3.1-11}$$

Now use the superengine to drive the Carnot engine as a heat pump between the same two heat reservoirs as used by the superengine. If there is no loss to friction, all of the work done by the engine is transmitted to the heat pump:

$$w(s) = -w'(\text{C}) \tag{3.1-12}$$

From Equation (3.1-10), the amount of heat put into the hot reservoir by the Carnot heat pump is equal to

$$-q_4'(\text{C}) = \frac{w'(\text{C})}{1 + q_3(\text{C})/q_1(\text{C})}$$

The amount of heat removed from the hot reservoir by the superengine is

$$q_1(s) = -\frac{w(s)}{1 + q_3(s)/q_1(s)} < \frac{w'(C)}{1 + q_3(C)/q_1(C)}$$

Therefore

$$q_1(s) < -q_4'(C) \quad \text{(statement contrary to fact)}$$

and a larger amount of heat has been put into the hot reservoir by the heat pump than has been removed from it by the superengine. This contradicts the Clausius statement of the second law of thermodynamics, which means that our assumption must be false. *The efficiency of the second engine cannot be larger than that of the first Carnot engine.*

The second reversible heat engine also cannot have a smaller efficiency than the first Carnot engine. If it did, its coefficient of performance as a heat pump, which is the reciprocal of its efficiency as a heat engine, would be larger than that of a Carnot heat pump, and the second law could be violated by using the first engine to drive the second engine as a heat pump. We have shown that *the efficiency of a reversible heat engine operating with two heat reservoirs does not depend on the nature of the working fluid or on the details of its design.*

Exercise 3.1   Carry out the mathematics to show that a reversible engine cannot have a smaller efficiency than a Carnot engine if it uses the same heat reservoirs.

If a heat engine operates irreversibly, its coefficient of performance as a heat pump is not necessarily the reciprocal of its engine efficiency, since each step cannot necessarily be reversed. If the efficiency of the irreversible engine were greater than the Carnot efficiency, it could violate the second law if the engine were used to drive a Carnot heat pump. Therefore, an irreversible engine cannot have a higher efficiency than a Carnot engine. However, an irreversible engine can have a lower efficiency than a Carnot engine. This conclusion applies to real steam engines, internal combustion engines, etc. *Any real heat engine cannot be more efficient than a Carnot engine operating between the same two reservoirs.*

The Thermodynamic Temperature

The ratio q_3/q_1 has the same value for all heat engines operating between the same two reservoirs. It does not depend on the identities of the materials in the two reservoirs.

The **zeroth law of thermodynamics** states that if two objects, say A and B, are at thermal equilibrium with each other, and if B is at thermal equilibrium with a third object, say C, then A is also at thermal equilibrium with C. The temperature is the variable that has the same value in all objects at thermal equilibrium with each other, so we assert that any other reservoir at the same temperature could be substituted for one of our reservoirs without any change in our analysis. The value of the ratio q_3/q_1 therefore depends only on the temperatures of the reservoirs.

We define the **thermodynamic temperature**, θ, by the relation

$$\frac{\theta_c}{\theta_h} = \left|\frac{q_3}{q_1}\right| \quad \text{(definition)} \tag{3.1-13}$$

where θ_c is the thermodynamic temperature of the cold reservoir and θ_h is the thermodynamic temperature of the hot reservoir. We define the thermodynamic temperature to be positive, and the magnitude symbol is required because q_3 is negative. The Carnot efficiency is now given by

$$\eta_C = 1 - \frac{\theta_c}{\theta_h} \tag{3.1-14}$$

The thermodynamic temperature scale is not related to any particular kind of substance and is therefore more fundamental than the ideal gas temperature scale. Since only the ratio of the temperatures is defined, we can choose any convenient size for the unit of temperature. We now show that the thermodynamic temperature scale coincides with the ideal gas temperature scale if we choose the appropriate size for its unit.

Assume that the working fluid is an ideal gas with a constant heat capacity. If the two temperature scales coincide for this system, they must coincide for any system, since the ratio q_3/q_1 depends only on the temperatures of the reservoirs and not on the identity of the working fluid. For the first step of the Carnot cycle, from Equation (2.4-5)

Here is an example of a special case rigorously establishing a general principle. In some other discussions, we consider a special case and then assert without proof that the general principle is valid.

$$q_1 = nRT_h \ln(V_2/V_1) \tag{3.1-15}$$

Similarly, for the third step,

$$q_3 = nRT_c \ln(V_4/V_3) \tag{3.1-16}$$

We can now locate the states at which steps 1 and 3 terminate by using Equation (2.4-22):

$$\frac{T_c}{T_h} = \left(\frac{V_2}{V_3}\right)^{nR/C_V} \tag{3.1-17}$$

and

$$\frac{T_c}{T_h} = \left(\frac{V_1}{V_4}\right)^{nR/C_V} \tag{3.1-18}$$

These two equations imply that

$$\frac{V_1}{V_2} = \frac{V_4}{V_3} \tag{3.1-19}$$

When this relation is substituted into Equation (3.1-16), we have

$$q_3 = nRT_c \ln(V_1/V_2) = -nRT_c \ln(V_2/V_1) \tag{3.1-20}$$

This equation gives

$$w_{\text{cycle}} = -q_1 - q_3 = nR(-T_h + T_c) \ln(V_2/V_1) \tag{3.1-21}$$

and

$$\eta_C = \frac{w_{\text{surr}}}{q_1} = \frac{T_h - T_c}{T_h} = 1 - \frac{T_c}{T_h} \qquad (3.1\text{-}22)$$

Equation (3.1-14) holds for any working fluid, including an ideal gas. Therefore, by comparison of Equations (3.1-14) and (3.1-22),

$$\frac{\theta_c}{\theta_h} = \frac{T_c}{T_h} \qquad (3.1\text{-}23)$$

Since the two reservoirs can be at any temperatures so long as $T_h > T_c$, temperatures on the thermodynamic and ideal gas temperature scales are proportional to each other for any value of the temperature. We can choose the size of the temperature unit on a given scale, so we choose the unit on the thermodynamic temperature scale to be the kelvin, and the two scales then coincide. From now on, we will recognize the symbol T as the temperature on both the thermodynamic scale and the ideal gas scale.

| **Exercise 3.2** | Calculate the efficiency of a Carnot heat engine that represents a steam engine with its boiler at 500. K and its exhaust at 373.15 K. |

The thermodynamic temperature is a fundamental quantity, like mass, length, and time. We will later show that, as a consequence of the second and third laws of thermodynamics, the thermodynamic temperature must be positive and zero temperature is unattainable.

We can now express the Carnot heat pump coefficient of performance in terms of the temperature:

$$\eta_{\text{hp}} = \frac{1}{\eta_C} = \frac{1}{1 - T_c/T_h} \qquad (3.1\text{-}24)$$

If we are using a heat pump as a refrigerator (or air conditioner), the coefficient of performance is defined to be the heat removed from the cold reservoir divided by the work put into the refrigerator:

$$\eta_r = \frac{q_2'}{w_{\text{cycle}}} = \frac{q_2'}{-q_2' - q_4'} = -\frac{q_3}{q_1 + q_3}$$

$$\eta_r = \frac{1}{-q_1/q_3 - 1} = \frac{1}{T_h/T_c - 1} \qquad (3.1\text{-}25)$$

For Carnot heat pumps, the coefficient of performance is always greater than unity, and for Carnot refrigerators the coefficient of performance exceeds unity if $T_h/T_c < 2$.

| **Exercise 3.3** | **a.** Calculate the coefficient of performance of a reversible heat pump operating between a high temperature of 70.°F and a low temperature of 40.°F.
b. Calculate the coefficient of performance of a reversible refrigerator operating between an interior temperature of 4°C and an exterior temperature of 22°C. |

3.2 The Mathematical Statement of the Second Law; Entropy

The first law defines the internal energy as a state variable in a manner that is fairly easily visualized. On the other hand, the physical statement of the second law has no obvious connection with a state variable. Our task in this section is to establish such a connection, but even after we do this the connection will not be so easily visualized as the connection in the first law.

We will proceed by making a statement which defines a state function, but which at first seems to have little connection with the physical statements of the second law. Only after some rather subtle analysis will we be able to assert that it follows from the physical statements.

The mathematical statement of the second law of thermodynamics is as follows: *If the differential dS is defined as*

This equation defines the entropy, S, through its differential. Any constant can be added to the entropy without changing this definition.

$$dS = \frac{dq_{rev}}{T} \quad \text{(definition)}$$

(3.2-1)

where dq_{rev} is an infinitesimal amount of heat transferred in a reversible process and T is the temperature, then dS is an exact differential and S is a state variable, which is called the entropy.

For an irreversible process

$$dS > \frac{dq_{irrev}}{T_{surr}}$$

(3.2-2)

where T_{surr} is the temperature of the surroundings.

We now show that the mathematical statement of the second law of thermodynamics follows from the physical statements of the law. We first prove the assertion made in Section 2.4 that for any closed simple system there is only one curve in the system's state space that contains points accessible from a given state point by reversible adiabatic processes. We use the *V-T* plane as our state space.

Once again, we assume what we want to disprove. When we obtain a contradiction, we will have proved our assumption to be false.

We assume the opposite of what we want to prove, and then we show that this assumption leads to a contradiction with fact and so must be false. Curves in a state space representing adiabatic processes are called **adiabats**. We assume that there are two different reversible adiabats in the *V-T* plane of a closed simple system, such that the curves coincide at state number 1, as depicted in Figure 3.4. This is the assumption that we want to prove false. We have located two states of equal volume, labeled states number 2 and number 3, with state 3 at a higher

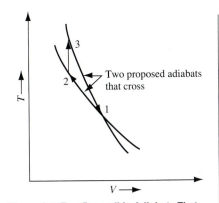

Figure 3.4. Two Reversible Adiabats That Cross (Assumption to Be Proved False). This diagram represents a situation that cannot happen without violating the second law of thermodynamics.

temperature than state 2. Now consider a reversible cyclic process $1 \rightarrow 2 \rightarrow 3 \rightarrow 1$, where step 2 from state 2 to state 3 is a constant-volume process.

Heat capacities are always positive. Therefore, $q_2 > 0$. Since steps 1 and 3 are adiabatic,

$$q_{\text{cycle}} = q_2 > 0 \tag{3.2-3}$$

Since $\Delta U = 0$ in any cyclic process,

$$w_{\text{surr}} = -w_{\text{cycle}} = -\Delta U_{\text{cycle}} + q_{\text{cycle}}$$
$$= q_{\text{cycle}} = q_2 \tag{3.2-4}$$

In our cyclic process, heat transferred to the system has been converted completely to work done on the surroundings, violating the second law of thermodynamics. The source of this incorrect result is the assumption that two reversible adiabats can cross. Therefore, only one reversible adiabat passes through state 1 or any other state.

The conclusion that two adiabats cannot cross shows that some heat must be exhausted to a cool reservoir in a Carnot cycle, as asserted in Section 3.1. There must be two isothermal or other nonadiabatic processes in the cycle to carry the state point of the system from one reversible adiabat to the other and back again.

The foregoing argument applies to a simple system, for which the volume is the important independent variable. A similar argument can be made for a nonsimple system. For example, in a magnetic system, the magnetization is the important independent variable. Adiabatic demagnetization is used to lower the temperature of magnetic systems, and the lowest temperatures have been achieved in this way. By an argument analogous to the foregoing one, it can be shown that irreversible demagnetization cannot lead to a lower temperature than reversible demagnetization.

The next part of our discussion is the proof that dq_{rev} possesses an integrating factor. That is, there exists a function y, the integrating factor, such that $y\, dq_{\text{rev}}$ is an exact differential (the differential of a function) even though dq_{rev} is an inexact differential. This theorem was proved by Carathéodory,[1] who showed that if there are points arbitrarily close to a given point in the V-T plane that could not be reached by processes for which $dq_{\text{rev}} = 0$, then dq_{rev} must possess an integrating factor. We have already shown that adiabatic reversible processes must lead to points on a single curve, so there are in fact points arbitrarily close to a given point that cannot be reached by reversible adiabatic processes. The hypotheses of Carathéodory's theorem are fulfilled.

We will give a nonrigorous outline of Carathéodory's proof.[2] The main idea of the proof is that if there is a single curve along which dq_{rev} vanishes, there is also a differential of a function, dS, that vanishes on the same curve. If dS and dq_{rev} vanish on the same curve, then in the vicinity of this curve

$$dS = y\, dq_{\text{rev}} \tag{3.2-5}$$

where y (the integrating factor) is a function that does not vanish in the vicinity of the curve.

Consider reversible adiabatic processes of a closed simple system starting from a particular initial state. Let the unique curve in the V-T state space that

[1] C. Carathéodory, *Math Ann.* **67**, 335 (1909).
[2] J. G. Kirkwood and I. Oppenheim, *Chemical Thermodynamics*, McGraw-Hill, New York, 1961, pp. 31ff; J. deHeer, *Phenomenological Thermodynamics*, Prentice-Hall, Englewood Cliffs, NJ, 1986, pp. 123ff.

represents these processes be represented mathematically by the function

$$T = f(V) \qquad\qquad (3.2-6)$$

Since n is fixed, it is not listed as an independent variable. Equation (2.4-22) is an example of such a function, holding for an ideal gas with constant heat capacity.

Equation (3.2-6) is the same as

$$0 = f(V) - T \quad \text{(valid only on the curve)} \qquad\qquad (3.2-7)$$

which applies only on the curve. Let S be defined by

$$S = S(T, V) = f(V) - T + C \qquad\qquad (3.2-8)$$

where C is a constant. Equation (3.2-8) applies for all values of T and V, not just values on the curve. Since f is a function of V, S is a function of T and V for our closed system and is therefore a state function. For reversible adiabatic processes, T is equal to $f(V)$ and S is equal to the constant C. Therefore, for adiabatic reversible processes

$$dS = 0 \quad \text{(reversible adiabatic processes)} \qquad\qquad (3.2-9)$$

Both dq_{rev} and dS vanish on the curve. Since reversible adiabatic processes cannot lead away from the curve, dq_{rev} vanishes only on the curve. Since $f(V)$ represents a unique curve, dS vanishes only on the curve, and we can write

$$dS = y \, dq_{rev} \qquad\qquad (3.2-10)$$

where y is a function that is nonzero in the vicinity of the curve. Since S is a function, y is an integrating factor, and we have proved the theorem of Carathéodory.

The line integral of the differential of a function (an exact differential) is equal to the value of the function at the final point minus the value of the function at the initial point. Such a line integral is therefore independent of the path between a given initial point and a given final point. The converse is also true. If a line integral of a differential between two given points is path independent, the differential is exact. If the initial and final points are the same point (a cyclic path), the line integral of an exact differential must vanish. Conversely, if a line integral vanishes for all cyclic paths (not just for a particular cyclic path), the differential must be exact.

This theorem is one that can be used in either direction.

We now show that $1/T$ is a possible choice for the integrating factor y by showing that

$$\oint \frac{dq_{rev}}{T} = 0 \qquad\qquad (3.2-11)$$

for all reversible cyclic processes. The symbol $\oint$ represents the line integral around a closed curve in the state space.

We begin with a Carnot cycle. From Equations (3.1-13) and (3.1-23)

$$\frac{q_1}{T_h} = -\frac{q_3}{T_c} \qquad\qquad (3.2-12)$$

Since T is constant on the isothermal segments and $dq_{rev} = 0$ on the adiabatic segments, the line integral for a Carnot cycle is

$$\oint \frac{dq_{rev}}{T} = \frac{q_1}{T_h} + \frac{q_3}{T_c} = 0 \qquad\qquad (3.2-13)$$

so Equation (3.2-11) is established for any Carnot cycle.

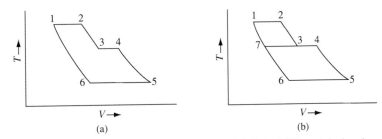

Figure 3.5. A Reversible Cycle of Isotherms and Adiabats. (a) The original cycle. **(b)** The cycle with an added process. The additional process from state 3 to state 7 allows us to show that this cycle is equivalent to two Carnot cycles.

Now consider the reversible cyclic process of Figure 3.5a. Steps 1, 3, and 5 are isothermal steps, and steps 2, 4, and 6 are adiabatic steps. We show that the line integral of Equation (3.2-11) vanishes for this cycle, as follows: Let point 7 lie on the curve from state 6 to state 1, at the same temperature as states 3 and 4, as shown in Figure 3.5b. We now carry out the reversible cyclic process $7 \rightarrow 1 \rightarrow 2 \rightarrow 3 \rightarrow 7$, which is a Carnot cycle (although we started in a different corner than previously). For this cycle, the line integral vanishes. We next carry out the cycle $7 \rightarrow 4 \rightarrow 5 \rightarrow 6 \rightarrow 7$. This is also a Carnot cycle, so the line integral around this cycle vanishes.

During the second cycle, on the way from state 7 to state 4, the path from state 7 to state 3 was traversed from left to right. During the first cycle, the path from state 3 to state 7 was traversed from right to left. When the two cyclic line integrals are added, these two paths exactly cancel each other, and if we leave them both out the sum of the two line integrals is unchanged. We now have a vanishing line integral for the cyclic process $7 \rightarrow 1 \rightarrow 2 \rightarrow 3 \rightarrow 4 \rightarrow 5 \rightarrow 6 \rightarrow 7$, which is the cycle of Figure 3.5a, except for starting in a different place. We have proved that Equation (3.2-11) holds for the cycle of Figure 3.5a.

We can similarly prove that Equation (3.2-11) holds for more complicated cyclic processes made up of isothermal and adiabatic reversible steps. For example, consider the process of Figure 3.6a, which can be divided into three Carnot cycles, just as that of Figure 3.5a was divided into two Carnot cycles.

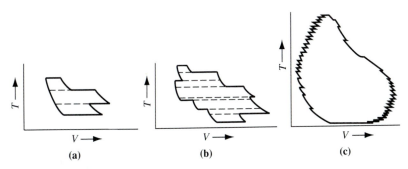

Figure 3.6. Reversible Cycles of Isotherms and Adiabats. (a) Cycle equivalent to three Carnot cycles. **(b)** Cycle equivalent to eight Carnot cycles. **(c)** Cycle equivalent to a large number of Carnot cycles.

Exercise 3.4 Show that the line integral of dq_{rev}/T vanishes around the cycle of Figure 3.6a.

Any cycle that is made up of reversible isothermal and adiabatic steps can be divided into Carnot cycles. If each Carnot cycle is traversed once, all of the paths in the interior of the original cycle are traversed twice, once in each direction, and therefore cancel out when all of the line integrals are added. The exterior curve is traversed once, and the integral of Equation (3.2-11) is shown to vanish around the cycle. For example, Figure 3.6b shows a more complicated cycle divided into Carnot cycles.

An arbitrary cycle can be divided into Carnot cycles.

In order to represent an arbitrary cycle, we construct isothermal and adiabatic steps that are smaller and smaller in size until the curve of the arbitrary cycle is more and more closely approximated, as crudely indicated in Figure 3.6c. In the limit that the sizes of the steps approach zero, any curve is exactly represented. The line integral of Equation (3.2-11) still vanishes, because even in the limit the cycle can be divided into Carnot cycles.

Since the line integral of Equation (3.2-11) vanishes for any reversible cycle, dq_{rev}/T must be an exact differential, and

$$dS = \frac{dq_{rev}}{T} \tag{3.2-14}$$

is the differential of a function S, which is called the entropy.

If S is a state function, it is a function of the independent variables that specify the state of the system. For example, for a simple system of one phase and one component, we can write

$$S = S(V, T, n) \tag{3.2-15}$$

Equation (3.2-14) provides a means of calculating the entropy change for any process that begins and ends at equilibrium states, even if the system passes through nonequilibrium states during the process. One can carry out the line integral of dS on any reversible path from the initial to the final state to obtain the entropy change. In the next section, we will carry out several such calculations.

Equation (3.2-1) defines only an infinitesimal change in the entropy. Any constant can be added to the value of the entropy without changing this equation or any of our discussion of the entropy. Only changes in the entropy are well defined. The third law of thermodynamics, which we will discuss later in this chapter, provides a conventional assignment of zero entropy to a particular state.

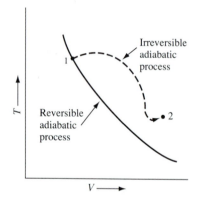

Figure 3.7. An Irreversible Adiabat. An irreversible adiabat does not proceed through a sequence of equilibrium states, even if the initial and final states are equilibrium states.

This assertion is the most important consequence of the second law and is often considered to be a statement of the second law.

We now want to show that if a system undergoes an irreversible adiabatic process, its entropy increases. We will then show that Equation (3.2-2) follows. Figure 3.7 shows schematically a change in state for an irreversible adiabatic process of a closed simple system in which the initial state (state 1) and the final state (state 2) are equilibrium states. During the process, the state of the system is not an equilibrium state and cannot be represented by a point in the V-T plane. The broken curve in the figure indicates that the state point leaves the V-T plane, going into some other mathematical space representing nonequilibrium states, and then returns to the V-T plane at the end of the process. In this kind of process, we must start with a system

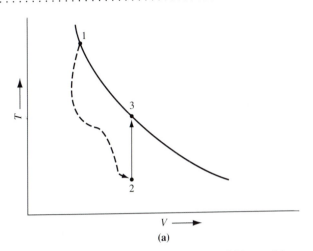

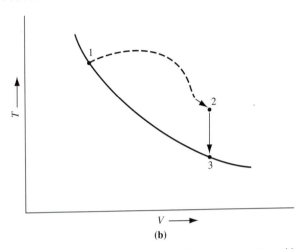

(a)   **(b)**

Figure 3.8. Reversible and Irreversible Adiabats. (a) Impossible case. An irreversible adiabatic process cannot lead to the low-temperature side of the reversible adiabat. **(b)** Possible case. The irreversible adiabatic process can lead to the high-temperature side of the reversible adiabat.

at equilibrium and must allow the system to come to an equilibrium state before we declare the process to be ended.

The figure also shows the unique curve corresponding to reversible adiabatic processes beginning at state 1. If the final state point of the irreversible process does not lie on the curve, it will be either above the curve or below it. The first case is shown in Figure 3.8a and the second case in Figure 3.8b.

Let us first assume the final state is below the curve, as in Figure 3.8a. State 3 is the state on the curve that has the same volume as state 2. After the irreversible process has occurred, we carry out a reversible constant-volume step from state 2 to state 3 (step 2). For a constant-volume process,

$$dq = C_V \, dT$$

The heat capacity of any system must be positive. Therefore, $q > 0$ for the step $2 \rightarrow 3$, because we hypothesized that the temperature of state 2 was lower than that of state 3.

After step 2, we carry out a reversible adiabatic step from state 3 to state 1 (step 3). Step 1 (the irreversible process) and step 3 are both adiabatic, so that

$$q_{\text{cycle}} = q_2 > 0 \tag{3.2-16}$$

Since U is a state variable

$$\Delta U_{\text{cycle}} = 0 \tag{3.2-17}$$

The work done on the surroundings in the cycle is

$$w_{\text{surr}} = -w_{\text{cycle}} = -\Delta U_{\text{cycle}} + q_2 = q_2 \tag{3.2-18}$$

Heat transferred to the system has been completely turned into work done on the surroundings in a cyclic process, which is a violation of the second law of thermodynamics. Therefore, the process is not possible: an irreversible adiabatic process cannot lead to a state below the reversible adiabatic curve.

Now consider the process of Figure 3.8b, in which state 2 is above the reversible adiabatic curve. Again, we carry out a constant-volume reversible step (step 2) from state 2 to state 3 and an adiabatic reversible step from state 3 to state 1. This time, because state 2 is at a higher temperature than state 3,

$$q_{\text{cycle}} = q_2 < 0 \tag{3.2-19}$$

so that

$$w_{\text{surr}} = -w_{\text{cycle}} = -\Delta U_{\text{cycle}} + q_2 < 0 \tag{3.2-20}$$

It is the system, not the surroundings, that is returned to its original state at the end of the process. The Kelvin statement of the second law applies only for a process in which the system is returned to its original state.

In this case, work done on the system has been turned completely into heat transferred to the surroundings. This does not violate the second law of thermodynamics because the surroundings do not undergo a cyclic process. The process of Figure 3.8b can occur, and an irreversible adiabatic process can lead to a state above the reversible adiabatic curve of V, the final temperature for an irreversible adiabatic process cannot be lower than for a reversible adiabatic process with the same final volume.

At this point, an argumentative reader might say, "Let me simply reverse the definition of system and surroundings. Whatever the author calls the surroundings, I will call the system, and whatever the author calls the system, I will call the surroundings. The second law then gives the opposite conclusion about which of the two processes is possible." The fallacy in this reasoning is that in the processes defined in Figure 3.8a and 3.8b, the surroundings undergo a noncyclic process. If the surroundings are redefined to be the system, heat has indeed been turned completely into work in the process of Figure 3.8b (the possible process), but this does not violate the second law because the surroundings (now called the system) have not undergone a cyclic process.

For a reversible adiabatic process, we integrate Equation (3.2-1) or Equation (3.2-14)

$$\Delta S_{\text{rev}} = \int_c \frac{dq_{\text{rev}}}{T} = 0 \quad \text{(reversible adiabatic process)} \tag{3.2-21}$$

This result is important. *A reversible adiabatic process does not change the entropy of the system.*

Now consider the irreversible adiabatic process from state 1 to state 2 in Figure 3.8b. Since S is a state variable,

$$\Delta S_{\text{cycle}} = \Delta S_1 + \Delta S_2 + \Delta S_3 = 0 \tag{3.2-22}$$

Since step 3 is reversible and adiabatic, $\Delta S_3 = 0$ and

$$\Delta S_1 = -\Delta S_2 \tag{3.2-23}$$

Since step 2 is reversible, we can integrate Equation (3.2-1) for this step:

$$\Delta S_2 = \int_{T_2}^{T_3} \frac{dq_{\text{rev}}}{T} = \int_{T_2}^{T_3} \frac{C_V}{T} \, dT < 0 \tag{3.2-24}$$

The inequality comes from the facts that the temperature and the heat capacity are both positive and the temperature of state 2 must be greater than that of state 3. We now have, from Equation (3.2-23), the conclusion that $\Delta S_1 > 0$. Therefore,

$$\Delta S_{irrev} > 0 \quad \text{(irreversible adiabatic processes)} \qquad \textbf{(3.2-25)}$$

Equation (3.2-26) is the mathematical statement of the most important consequence of the second law.

Combining Equations (3.2-21) and (3.2-25), we have

$$\Delta S \geq 0 \quad \text{(adiabatic processes)} \qquad \textbf{(3.2-26)}$$

where the equality holds for reversible processes. *For any adiabatic process, the entropy of the system cannot decrease.*

To study nonadiabatic processes in closed systems, we consider a closed system and its surroundings arranged as in Figure 3.9. The combination (system plus surroundings) is isolated from the rest of the universe, so the combination cannot exchange heat, work, or matter with the rest of the universe. However, the system can exchange heat and work with the surroundings. Assuming that surface tension effects between system and surroundings are negligible,

This design of the system plus surroundings allows us to use the properties of an adiabatically isolated system for the combination.

$$\Delta S_{combination} = \Delta S + \Delta S_{surr} \qquad \textbf{(3.2-27)}$$

Since the combination is isolated from the rest of the universe, it can undergo only adiabatic processes and Equation (3.2-26) applies to it. For an infinitesimal process

$$dS_{combination} = dS + dS_{surr} \geq 0 \qquad \textbf{(3.2-28)}$$

For reversible processes, the entropy change of the system and the entropy change of the surroundings cancel, and for irreversible processes the sum of dS and dS_{surr} must be positive. Since the combination is the only part of the universe involved in the process, the entropy of the universe cannot decrease. *In any reversible process, the entropy of the universe remains constant. In any irreversible process, the entropy of the universe increases.*

We rewrite Equation (3.2-28) in the form

$$dS \geq -dS_{surr} \qquad \textbf{(3.2-29)}$$

It is not necessary that dS be positive. However, if dS is negative, then dS_{surr} must be positive and large enough that the sum of the two entropy changes is not negative.

Since we are primarily interested in the processes of the system, the nature of the surroundings is not very important. We make the simplest possible assumption about the surroundings: we assume that the thermal conductivity of the surroundings is so large that the surroundings remain at equilibrium during any process and that the surroundings are so large that their temperature does not change. This means that we can apply Equation (3.2-1) or (3.2-14) to the surroundings:

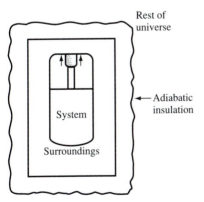

Figure 3.9. A System and Its Surroundings (a Combination) Adiabatically Insulated from the Rest of the Universe. This combination is used because all of the properties of an adiabatically isolated system can be ascribed to the combination.

$$dS_{surr} = \frac{dq_{surr}}{T_{surr}} = -\frac{dq}{T_{surr}} \qquad \textbf{(3.2-30)}$$

where the second equality comes from the fact that any heat transferred to the surroundings must come from the system. Equation (3.2-29) now gives

Equation (3.2-31) is the mathematical statement of the second law.

$$dS \geq \frac{dq}{T_{surr}}$$ (3.2-31)

The temperature of the surroundings, not the temperature of the system, occurs in Equation (3.2-31).

For a reversible process, Equation (3.2-31) is the same as Equation (3.2-1). For an irreversible process, it is the same as Equation (3.2-2). Equation (3.2-31) is the mathematical statement of the second law of thermodynamics, and the physical statements of Kelvin and Clausius can be considered to be alternative statements. In fact, if the mathematical statement is taken as a postulate, the physical statements can be derived from it. We do not carry out this analysis.[3]

Exercise 3.5   From the mathematical statement of the second law, show that heat cannot flow from a cooler object to a hotter object if nothing else happens. Hint: consider two objects that form a combination that is isolated from the rest of the universe.

The second law of thermodynamics is of far-reaching significance. Because no violations of it have ever been observed, we have no reason to doubt its universal applicability. However, if it is universally applicable, the ultimate fate of the universe will be to approach a state of equilibrium in which every object in the universe is at the same temperature. There will be no energy flow from stars to planets, and no life or any other macroscopic processes will be possible. This "heat death" of the universe will not occur for a very long time, but it is unavoidable if the second law is universally valid.

Some people have speculated that the second law is not universally valid but just a statement of what nearly always occurs. If this is the case, perhaps violations of the second law will be observed under some circumstances (e.g., if the universe some day begins to contract instead of expand). This idea is unsupported speculation, and we have every reason to apply the second law of thermodynamics to any process in any macroscopic system and no reason to assume that it might be violated.

3.3   **The Calculation of Entropy Changes**

Entropy changes can generally be calculated by carrying out the appropriate line integral of dS. For any process that begins at an equilibrium state (state 1) and ends at an equilibrium state (state 2), the entropy change

---

[3] Kirkwood and Oppenheim, *op. cit.*, p. 42.

is given by

$$\Delta S = S_{\text{final}} - S_{\text{initial}} = S_2 - S_1 = \int_c dS \qquad \text{(3.3-1)}$$

where S_2 is the entropy of state 2, the final state, and S_1 is the entropy of state 1, the initial state. The integral is a line integral and must be carried out along a curve in the equilibrium state space of the system beginning at the point representing state 1 and ending at the point representing state 2. The actual process does not have to be reversible as long as it has equilibrium initial and final states, but *the path on which the line integral is calculated must correspond to a reversible process.*

Equation (3.2-1) gives us an explicit expression for dS, enabling us to write for a closed system

$$\Delta S = \int_c \frac{dq_{\text{rev}}}{T} \quad \text{(closed system)} \qquad \text{(3.3-2)}$$

Equation (3.3-2) is a general working equation that we can use to calculate entropy changes for processes in closed systems. The rest of this section will be devoted to such calculations.

Isothermal Reversible Processes

The path of the actual process can be used for the path of integration. Since T is constant, we can factor $1/T$ out of the integral:

$$\Delta S = \int_c \frac{dq_{\text{rev}}}{T} = \frac{1}{T} \int_c dq_{\text{rev}} = \frac{q_{\text{rev}}}{T} \quad \text{(isothermal process)} \qquad \text{(3.3-3)}$$

If a process has a final temperature that is equal to its initial temperature, we can apply Equation (3.3-3) to it, even if the temperature of the system changes during the process and even if the process is not reversible. However, we must use q_{rev} for the isothermal reversible process in the formula, and not the value of q for the actual process if it differs from q_{rev}.

For isothermal reversible volume changes in a system consisting of an ideal gas, q_{rev} is given by Equation (2.4-5), so that

Equation (3.3-4) will be used frequently. However, be sure to apply it only to an ideal gas.

$$\Delta S = nR \ln\left(\frac{V_2}{V_1}\right) \quad \text{(ideal gas, isothermal process)} \qquad \text{(3.3-4)}$$

Since S is a state function, Equation (3.3-4) applies to any process of an ideal gas that begins and ends at equilibrium states at the same temperature, even if the temperature varies during the process and even if the process is irreversible.

▼   **EXAMPLE 3.1**

For the first part of this example, we find the appropriate formula and substitute values of variables directly into the formula. We write all units to make sure the units of our answer are correct. We write four significant digits because the smallest number of significant digits in the given information is four. In the second part of the example we use the fact that for a reversible process, $\Delta S + \Delta S_{surr} = \Delta S_{univ} = 0$.

Find ΔS and ΔS_{surr} for the reversible isothermal expansion of 3.000 mol of argon (assumed ideal) from a volume of 100.0 L to a volume of 500.0 L at 298.15 K.

Solution

$$\Delta S = (3.000 \text{ mol})(8.3145 \text{ J K}^{-1} \text{ mol}^{-1}) \ln\left(\frac{500.0 \text{ L}}{100.0 \text{ L}}\right)$$

$$= 40.14 \text{ J K}^{-1}$$

Since the process is reversible, $\Delta S + \Delta S_{surr} = 0$:

$$\Delta S_{surr} = -\Delta S = -40.14 \text{ J K}^{-1}$$

◥

Exercise 3.6   Find ΔS and ΔS_{surr} if 3.000 mol of argon is expanded reversibly and isothermally from a volume of 500.0 L to a volume of 2500. L at 298.15 K.

For a nonideal gas, the entropy change of a reversible isothermal volume change can be calculated from Equation (3.3-3) if an expression for q_{rev} is obtained.

▼   **EXAMPLE 3.2**

Find ΔS and ΔS_{surr} for the reversible expansion of 1.000 mol of argon from 2.000 L to 20.00 L at a constant temperature of 298.15 K (the process of Example 2.7a) if argon is represented by the truncated virial equation of state, as in that example.

Solution

Using the result of Example 2.7a,

Since we have already calculated q_{rev}, there is no need to calculate it again.

$$\Delta S = \frac{q_{rev}}{T} = \frac{5690 \text{ J}}{298.15 \text{ K}} = 19.08 \text{ J K}^{-1}$$

◥

Reversible Phase Changes

In Chapter 1, we asserted that two phases of a single substance can be at equilibrium with each other only at a particular temperature, which depends on the pressure. For example, liquid water and gaseous water can be at equilibrium with each other only at 100.00°C if the pressure is 1.000 atm (101325 Pa), and only at 25.00°C if the pressure is 23.756 torr (3167.2 Pa). If a phase change is carried out reversibly at constant pressure, the temperature is constant and Equation (3.3-3) applies. Since the pressure is constant, $q = \Delta H$, and

$$\Delta S = \frac{q_{rev}}{T} = \frac{\Delta H}{T} \quad \text{(reversible phase change)} \qquad (3.3\text{-}5)$$

▼ **EXAMPLE 3.3**

Find the entropy change of the system and of the surroundings if 3.000 mol of water freezes reversibly at 1.000 atm. The freezing temperature is 0.00°C at this pressure, and the specific enthalpy change of fusion is equal to 79.7 cal g^{-1} at this temperature.

Solution

$$\Delta S = \frac{(3.000 \text{ mol})(18.02 \text{ g mol}^{-1})(-79.7 \text{ cal g}^{-1})(4.184 \text{ J cal}^{-1})}{273.15 \text{ K}}$$

$$= -66.0 \text{ J K}^{-1}$$

Since the process is reversible,

$$\Delta S_{\text{surr}} = -\Delta S = 66.0 \text{ J K}^{-1}$$

▲

Exercise 3.7

Calculate ΔH, q, and ΔS for the reversible vaporization of 100.0 g of ethanol at 1.000 atm. The molar enthalpy change of vaporization is equal to 40.48 kJ mol^{-1} and the boiling temperature at 1.000 atm is 78.5°C.

Reversible Changes in Temperature

Another simple class of processes consists of temperature changes in closed systems without phase change or chemical reaction. If the pressure is constant, Equation (2.5-8) gives

$$dq = dH = C_P \, dT \quad \text{(closed system, constant pressure)} \quad \textbf{(3.3-6)}$$

so that Equation (3.3-2) becomes

$$\boxed{\Delta S = \int_{T_1}^{T_2} \frac{C_P}{T} \, dT \quad \text{(closed system, constant pressure)}} \quad \textbf{(3.3-7)}$$

where T_1 is the initial temperature and T_2 is the final temperature.

▼ **EXAMPLE 3.4**

Calculate ΔS and ΔS_{surr} for reversibly heating 2.000 mol of liquid water from 0.00°C to 100.00°C at a constant pressure of 1.00 atm.

Solution

The specific heat capacity of liquid water is nearly constant and equal to 1.000 cal K^{-1} g^{-1} = 4.184 J K^{-1} g^{-1}.

$$\Delta S = \int_{273.15 \text{ K}}^{373.15 \text{ K}} \frac{C_P}{T} \, dT = C_P \ln\left(\frac{373.15 \text{ K}}{273.15 \text{ K}}\right)$$

$$= (2.000 \text{ mol})(18.02 \text{ g mol}^{-1})(4.184 \text{ J K}^{-1} \text{ g}^{-1}) \ln\left(\frac{373.15 \text{ K}}{273.15 \text{ K}}\right)$$

$$= 47.03 \text{ J K}^{-1}$$

Once again, we use $\Delta S_{\text{univ}} = 0$.

$$\Delta S_{\text{surr}} = -\Delta S = -47.03 \text{ J K}^{-1}$$

▲

Exercise 3.8	For a gas whose molar heat capacity is represented by

$$\bar{C}_P = a + bT + cT^{-2}$$

derive a formula for ΔS if the temperature is changed from T_1 to T_2 at constant pressure.

For temperature changes at constant volume, Equation (2.4-7) gives

$$dq = dU = C_V\, dT \quad \text{(closed system, constant volume)} \tag{3.3-8}$$

so that Equation (3.3-2) becomes

$$\Delta S = \int_{T_1}^{T_2} \frac{C_V}{T}\, dT \quad \text{(closed system, constant volume)} \tag{3.3-9}$$

Exercise 3.9	Calculate the entropy change if 3.000 mol of argon gas is heated from 0.0°C to 250.0°C at a constant volume of 100.0 L. State any assumptions.

Irreversible Processes

Each irreversible process must be analyzed individually. Consider the system depicted in Figure 3.10a, which contains a large object at temperature T_1 and another large object at temperature T_2. These objects are insulated from the surroundings and from each other, except for a thin bar connecting the objects. If the objects are very large compared with the bar, the system will soon come to a **steady state** in which the properties of the system do not depend on time, although an irreversible process is taking place. Each large object will be essentially at equilibrium at a fixed temperature, and the bar will have a temperature that depends on position but not on time, even though heat is passing through the bar. Figure 3.10b shows the temperature of the bar as a function of position.

Consider a period of time Δt during which a quantity of heat q passes through the bar. Since the nonequilibrium state of the bar is time independent, the amount of heat entering one end of the bar is equal to the amount of heat leaving the other end. Also, since the nonequilibrium state of the bar is time independent, the entropy of the bar does not change and we do not have to study it.

Let us assume that $T_2 > T_1$. Since object 1 essentially remains at equilibrium, its entropy change is

$$\Delta S = \int \frac{dq_1}{T_1} = \frac{q_1}{T_1} \tag{3.3-10}$$

where q_1 is the amount of heat transferred in time Δt. Similarly, the entropy change of object 2 is

$$\Delta S_2 = \frac{q_2}{T_2} = -\frac{q_1}{T_2} \tag{3.3-11}$$

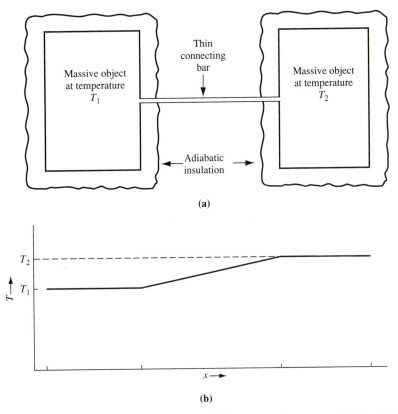

(a)

(b)

Figure 3.10. An Example Irreversible Process. (a) The system. It is assumed that the large objects at temperatures T_1 and T_2 remain at constant temperature, while the thin connecting bar, although not at equilibrium, remains in a steady state. **(b)** The temperature distribution in the system. It is assumed that this temperature distribution is time independent.

The entropy change of the system is

$$\Delta S = q_1 \left(\frac{1}{T_1} - \frac{1}{T_2} \right) = q_1 \frac{\Delta T}{T_1 T_2} \tag{3.3-12}$$

where $\Delta T = T_2 - T_1$. Because the system is adiabatically insulated from its surroundings, the entropy change of the surroundings vanishes and the entropy change of the universe is equal to the entropy change of the system.

Here we are using the fact that nothing happens to the surroundings, so S_{surr} is constant.

The time rate of change of the entropy of the universe is called the **entropy production**, because it is the rate at which new entropy is being generated. For our steady-state process, the entropy production is

$$\frac{dS_{\text{univ}}}{dt} = \left(\frac{dq}{dt} \right) \frac{\Delta T}{T_1 T_2} \tag{3.3-13}$$

where dq/dt is the rate at which heat is transferred.

Exercise 3.10   For the system of Figure 3.10, calculate the rate of entropy production if the first object is at 325 K, the second object is at 350. K, and 400. J of heat flows in 30. seconds.

Many of the irreversible processes that we discuss begin with the system in one equilibrium or metastable state and end with the system in another equilibrium state. *Because entropy is a state variable, we can calculate the entropy change of the system using a reversible process having the same initial and final states as the irreversible process.* If the entropy change of the surroundings is required, a separate calculation is necessary because the final state of the surroundings will not necessarily be the same in the reversible process as in the irreversible process.

We will repeatedly use a calculation process that differs from the actual process but has the same initial and final states. The change in any state function must be the same for both processes.

EXAMPLE 3.5

Calculate ΔS, ΔS_{surr}, and ΔS_{univ} if 2.000 mol of argon (assume ideal) expands isothermally at 298.15 K from a volume of 10.00 L to a volume of 40.00 L at a constant external pressure of 1.000 atm. Assume that the surroundings remain at thermal equilibrium at 298.15 K.

Solution

The initial and final states of the system are the same as for a reversible isothermal expansion, so

$$\Delta S = (2.000 \text{ mol})(8.3145 \text{ J K}^{-1} \text{ mol}^{-1}) \ln\left(\frac{40.00 \text{ L}}{10.00 \text{ L}}\right)$$

$$= 23.05 \text{ J K}^{-1}$$

To find ΔS_{surr}, we need q_{surr} for the actual process. Since $\Delta U = 0$ for an isothermal process in an ideal gas and P_{ext} is constant,

$$q_{surr} = -q = w = -P_{ext}\,\Delta V$$

$$= -(101325 \text{ N m}^{-2})(0.030 \text{ m}^3) = -3040 \text{ J}$$

$$\Delta S_{surr} = -3040 \text{ J}/(298.15 \text{ K}) = -10.20 \text{ J K}^{-1}$$

$$\Delta S_{univ} = 23.05 \text{ J K}^{-1} - 10.20 \text{ J K}^{-1} = 12.85 \text{ J K}^{-1}$$

Note that q for the actual process is not used in the ΔS calculations because $q_{rev} \neq q_{actual}$. However, because the surroundings remain at equilibrium, $q_{surr,\,rev} = q_{surr,\,actual}$.

Exercise 3.11

Find ΔS, ΔS_{surr}, and ΔS_{univ} if 2.000 mol of argon (assumed ideal) expands isothermally into a vacuum at 298.15 K, from a volume of 10.00 L to a volume of 40.00 L.

EXAMPLE 3.6

Calculate ΔS, ΔS_{surr}, and ΔS_{univ} if the process of Example 3.4 is carried out irreversibly at constant pressure with an arrangement similar to that of Figure 3.10, with the system and the surroundings connected only by a thin rod. Assume that the surroundings remain at equilibrium at 101.00°C as the system warms up.

We use the state function property in this example. We use an equation for the surroundings that applies only to isothermal processes because the surroundings remain at a fixed temperature as the temperature of the system changes.

Solution

Since entropy is a state variable, ΔS is the same as in Example 3.4:

$$\Delta S = 47.01 \text{ J K}^{-1}$$

$$\Delta S_{surr} = \int_c \frac{dq_{surr}}{T_{surr}} = \frac{-q}{T_{surr}}$$

$$= \frac{(2.000 \text{ mol})(18.01 \text{ g mol}^{-1})(4.184 \text{ J K}^{-1} \text{ g}^{-1})(100 \text{ K})}{374.15 \text{ K}}$$

$$\Delta S_{surr} = -40.28 \text{ J K}^{-1}$$
$$\Delta S_{univ} = \Delta S + \Delta S_{surr} = 6.73 \text{ J K}^{-1}$$

Metastable supercooled or superheated phases can undergo irreversible phase changes at constant pressure, and their entropy changes can be calculated by considering reversible processes with the same initial and final states, treating metastable states as though they were equilibrium states.

▼   **EXAMPLE 3.7**

Calculate the entropy change of the system, the surroundings, and the universe for the process of Example 2.16. Assume that the surroundings remain at equilibrium at $-15.00°C$.

Solution

We use the same reversible path as in Example 2.16:

$$\Delta S_1 = \int_{258.15 \text{ K}}^{273.15 \text{ K}} \frac{C_P(l)}{T} dT = C_P(l) \ln\left(\frac{273.15 \text{ K}}{258.15 \text{ K}}\right)$$
$$= (2.000 \text{ mol})(76.1 \text{ J K}^{-1} \text{ mol}^{-1}) \ln(1.0581) = 8.60 \text{ J K}^{-1}$$

$$\Delta S_2 = \frac{\Delta H_2}{T} = \frac{-1.202 \times 10^4 \text{ J}}{273.15 \text{ K}} = -44.01 \text{ J K}^{-1}$$

$$\Delta S_3 = \int_{273.15 \text{ K}}^{258.15 \text{ K}} \frac{C_P(s)}{T} dT = C_P(s) \ln\left(\frac{258.15 \text{ K}}{273.15 \text{ K}}\right)$$
$$= (2.000 \text{ mol})(37.15 \text{ J K}^{-1} \text{ mol}^{-1}) \ln(0.9451)$$
$$= -4.20 \text{ J K}^{-1}$$

$$\Delta S = \Delta S_1 + \Delta S_2 + \Delta S_3 = -39.61 \text{ J K}^{-1}$$

From Example 2.16, $\Delta H = q = -1.085 \times 10^4 \text{ J}$, so that $q_{surr} = 1.085 \times 10^4 \text{ J}$. Since the surroundings remain at equilibrium at constant temperature,

$$\Delta S_{surr} = \frac{q_{surr}}{T_{surr}} = \frac{1.085 \times 10^4 \text{ J}}{258.15 \text{ K}} = 42.03 \text{ J K}^{-1}$$
$$\Delta S_{univ} = \Delta S + \Delta S_{surr} = -39.61 \text{ J K}^{-1} + 42.03 \text{ J K}^{-1}$$
$$= 2.42 \text{ J K}^{-1}$$

The entropy change of mixing of a mixture of ideal gases can be calculated from the fact that each ideal gas acts as though it were alone in the volume containing the mixture (Dalton's law of partial pressure). Consider a mixture of several ideal gases, in which n_1 is the amount of substance 1, n_2 the amount of substance 2, etc.

To create the mixture, we take an initial state with each substance confined in a separate compartment of a container, as shown in Figure 3.11. We arrange the system so that each gas is at the same temperature T and the same pressure P by letting

$$V_i = \frac{n_i R T}{P} \qquad (i = 1, 2, 3, \ldots, s) \tag{3.3-14}$$

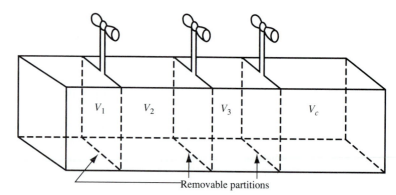

Figure 3.11. A System to Carry Out the Irreversible Mixing of Gases. The partitions can be removed so that the gases can mix by diffusion.

where V_i is the volume of compartment i and n_i is the amount of substance i in compartment i. The number of substances is denoted by c.

The gases are mixed by withdrawing the partitions between compartments, so that each gas expands irreversibly and isothermally into the entire volume. Since ideal gases act as though each were present by itself, the expansion is like an expansion into a vacuum, with zero external pressure and no work done. Since the initial and final states are equilibrium states, the entropy change of gas i is obtained from Equation (3.3-4):

$$\Delta S_i = n_i R \ln(V/V_i) \qquad (i = 1, 2, 3, \ldots, c) \tag{3.3-15}$$

where V is the total volume of the container

$$V = \sum_{i=1}^{c} V_i \tag{3.3-16}$$

The entropy change of the system is

$$\Delta S = \sum_{i=1}^{c} n_i R \ln(V/V_i) \tag{3.3-17}$$

The **mole fraction** of substance i is defined by

$$x_i = \frac{n_i}{n} \tag{3.3-18}$$

where n is the total amount of gas,

$$n = \sum_{j=1}^{c} n_j \tag{3.3-19}$$

From Equations (3.3-14), (3.3-16), (3.3-18), and (3.3-19) we have

$$x_i = \frac{V_i}{V} \tag{3.3-20}$$

so that the entropy change on mixing the ideal gases is

$$\Delta S = -R \sum_{i=1}^{c} n_i \ln(x_i) \tag{3.3-21}$$

All of the mole fractions must be less than or equal to unity, so that the entropy change of mixing is nonnegative.

▼ **EXAMPLE 3.8**

Find the entropy change of mixing of 1.000 mol of air. Assume that it is composed of 0.790 mol of nitrogen, 0.200 mol of oxygen, and 0.010 mol of argon.

Solution

$$\Delta S = -(8.3145 \text{ J K}^{-1} \text{ mol}^{-1})[(0.790 \text{ mol}) \ln(0.790)$$
$$+ (0.200 \text{ mol}) \ln(0.20) + (0.010 \text{ mol}) \ln(0.010)]$$
$$= 4.61 \text{ J K}^{-1}$$

▲

Equation (3.3-21) applies to the mixing of substances in other kinds of systems besides ideal gases, if the intermolecular interaction is unimportant. For example, it can be used to calculate the entropy change of mixing of isotopes of a single element.

Exercise 3.12

a. Find the entropy change of mixing for 1.000 mol of the normal mixture of Cl atoms, with 75.4% ^{35}Cl and 24.6% ^{37}Cl.
b. Find the entropy change of mixing in 0.500 mol of naturally occurring Cl_2. Note that there are three kinds of Cl_2 molecules if there are two isotopes.

3.4 ## Statistical Entropy

We have defined entropy as a macroscopic quantity, without molecular interpretation. However, we can define a statistical entropy that has essentially the same properties as the thermodynamic entropy. We will not discuss the statistical entropy in a fundamental way in this chapter, but will treat one model system in order to present some of its principal properties. We will return to the statistical entropy in Chapter 20.

The model system that we discuss is called the **lattice gas**. The name sounds like an oxymoron (a contradiction in terms), because a lattice is ordinarily a rigid solid structure. However, the lattice gas consists of particles that can move freely. It is the way we describe the mechanical states of the system that makes it a lattice gas.

Our system contains a fixed number N of noninteracting point mass molecules moving freely in a rectangular box of volume V. The molecules are assumed to obey classical mechanics, and the states of the molecules are specified by their locations and velocities.

We mentally divide the volume of the box into a number M of rectangular cells of equal size, as in Figure 3.12. Because the cell boundaries are imaginary, the molecules simply pass through them as they move. Instead of giving the values of three coordinates to specify the location of one particle, we specify which cell it occupies. This specification of location is called a **"coarse-grained" description**, and it gives less precise information about the location of the particles than would exact coordinate values. However, we can make the description more nearly exact by decreasing the size of the cells. We continue to specify the velocity of each particle in terms of the values of three velocity components.

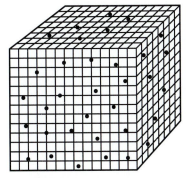

Figure 3.12. The Lattice Gas. This model system consists of a model gas in a volume divided into cells. The planes dividing the system into cells are imaginary and do not affect the motions of the molecules.

The **thermodynamic probability** Ω (capital Greek omega) is defined as the number of mechanical states (microstates) that a system might occupy, as far as we know. That is, the thermodynamic probability is a measure of our ignorance, or lack of information, about the mechanical state of the system, with larger values corresponding to less information.

For example, in the case of our lattice gas model system, we might know only that the system is at thermodynamic equilibrium with certain values of N, V, and T. This information specifies the equilibrium macroscopic state but gives very little information about the mechanical state of the system. Therefore, Ω has a very large value, since many mechanical states could correspond to the known macroscopic state (see Section 1.7).

We assume that the probability that a randomly chosen molecule has a particular velocity is independent of the probability that the same molecule has a particular position. Any velocity state of all the molecules can be combined with any coordinate state of the molecules, so for the set of molecules

The thermodynamic probability is also said to be a measure of randomness. Unlike Ω, "randomness" is not a well-defined quantity.

$$\Omega = \Omega_{\text{coord}}\Omega_{\text{vel}} \tag{3.4-1}$$

We now determine a formula for Ω_{coord} in our lattice gas model. Because the molecules are mass points, they do not interfere with each other and the presence of one molecule in a cell does not keep other molecules from occupying the same cell.

Equation (3.4-1) again uses probability theory: The probability that two independent events both occur is equal to the product of the probabilities of the individual events.

If we have only thermodynamic information, we know nothing about the positions of the molecules except that they are somewhere in the box. Since the cells are all of the same size, they are equally likely to contain a particular molecule. The number of possible coordinate states for a single molecule is equal to the number of cells, M. Any state of a second molecule can occur with any state of the first molecule, so the number of possible coordinate states for two molecules is M times M. Any state of a third molecule can occur with any state of the first pair of molecules, and so forth, so there is a factor of M for each molecule. For a system of N molecules,

$$\Omega_{\text{coord}} = M^N \tag{3.4-2}$$

If the molecules were not mass points, this equation would not be valid because a cell could fill up and not be able to accept any more molecules after a certain number.

▼   **EXAMPLE 3.9**

Calculate the value of Ω_{coord} for 1.000 mol of an ideal gas in a volume of 24.4 L if cells of 0.500 nm on a side are taken.

Solution

$$V_{\text{cell}} = (0.500 \times 10^{-9}\ \text{m})^3 = 1.25 \times 10^{-28}\ \text{m}^3$$

$$M = \frac{V}{V_{\text{cell}}} = \frac{0.0244\ \text{m}^3}{1.25 \times 10^{-28}\ \text{m}^3}$$

$$= 1.95 \times 10^{26}$$

$$\Omega_{\text{coord}} = M^N = (1.95 \times 10^{26})^{6.022 \times 10^{23}} = 10^{1.58 \times 10^{25}}$$

◢

Exercise 3.13	**a.** List the 36 possible states of two dice.

b. Determine how many possible states occur for four dice.

c. Determine how many possible states occur for two "indistinguishable" dice. (This means, for example, that there is no difference between a four on the first die with a five on the other and a five on the first die with a four on the other.) Explain why the correct answer is not equal to 18.

The **statistical entropy** was introduced by Boltzmann and is defined by

$$S_{st} = k_B \ln(\Omega) + S_0 \quad \text{(definition)} \qquad (3.4\text{-}3)$$

This definition is carved on Boltzmann's tombstone.

The choice $S_0 = 0$ is arbitrary. However, any other choice would be equally arbitrary, and our choice is the most convenient one.

where k_B is a constant to be evaluated later and S_0 is an arbitrary constant, which we take to equal zero. As with the definition of the thermodynamic probability, this equation applies to any kind of a system. Because the logarithm is a monotonic function of its argument, the statistical entropy is a measure of lack of information about the mechanical state of the system, just as is the thermodynamic probability.

Since Ω is defined as a number of possible mechanical states, it is equal to or greater than unity and the statistical entropy is nonnegative. If Ω is equal to unity, the statistical entropy is equal to zero. This circumstance corresponds to knowledge that the system definitely occupies a single mechanical state.

We now show that the statistical entropy and the thermodynamic entropy have the same behavior for isothermal volume changes in our lattice gas, which we assume to behave like an ideal gas. We will show in Chapter 16 that a model system of independent particles behaves like an ideal gas. Equation (3.3-4) gives the change in the thermodynamic entropy for an isothermal volume change in an ideal gas:

$$\Delta S = nR \ln(V_2/V_1) \qquad (3.4\text{-}4)$$

where V_2 is the final volume and V_1 the initial volume.

The change in the statistical entropy for any process is

$$\Delta S_{st} = k_B \ln(\Omega_2) - k_B \ln(\Omega_1) = k_B \ln(\Omega_2/\Omega_1)$$
$$= k_B \ln\left(\frac{\Omega_{coord(2)}\Omega_{vel(2)}}{\Omega_{coord(1)}\Omega_{vel(1)}}\right) \qquad (3.4\text{-}5)$$

The assertion that Ω_{vel} depends only on temperature is reasonable but not rigorously established. We will discuss the distribution of molecular velocities when we discuss the kinetic theory of gases.

The equilibrium velocity distribution in a gas of noninteracting molecules is independent of the coordinates of the particles. That is, any velocity can occur with any position. We assert that the velocity distribution depends only on the temperature and that the velocity factor in Ω depends only on the temperature. For an isothermal process,

$$\Omega_{vel(2)} = \Omega_{vel(1)} \qquad (3.4\text{-}6)$$

The velocity factors in Equation (3.4-5) cancel, and

$$\Delta S_{st} = k_B \ln\left(\frac{\Omega_{coord(2)}}{\Omega_{coord(1)}}\right) \qquad (3.4\text{-}7)$$

Using Equation (3.4-2) and the fact that N is fixed,

$$\Delta S_{st} = k_B \ln(M_2^N/M_1^N) = k_B \ln[(M_2/M_1)^N] = Nk_B \ln\left(\frac{M_2}{M_1}\right) \quad \text{(3.4-8)}$$

In order to maintain a given precision of position specification, the size of the cells must be kept constant, so the number of cells is proportional to the volume of the system:

$$M_2/M_1 = V_2/V_1$$

Therefore,

$$\Delta S_{st} = Nk_B \ln(V_2/V_1) \quad \text{(3.4-9)}$$

The change in the statistical entropy is identical to the change in thermodynamic entropy given in Equation (3.4-4) if we let

$$Nk_B = nR$$

which assigns a value to Boltzmann's constant:

$$\begin{aligned} k_B &= nR/N = R/N_A \\ &= \frac{8.3145 \text{ J K}^{-1}\text{ mol}^{-1}}{6.0221 \times 10^{23}\text{ mol}^{-1}} = 1.3807 \times 10^{-23}\text{ J K}^{-1} \end{aligned} \quad \text{(3.4-10)}$$

where N_A is Avogadro's constant.

We assert without further discussion that the equivalence between the statistical and thermodynamic entropies in a lattice gas and for isothermal expansions is typical of all systems and all processes. We write the general relation

We have established the relationship between S and S_{st} only for the lattice gas. To establish it for a general system is probably not possible.

$$\boxed{S = S_{st} + \text{constant}} \quad \text{(3.4-11)}$$

where the constant can be taken equal to zero.

▼   **EXAMPLE 3.10**   Find the value of the coordinate contribution to the statistical entropy for the value of Ω_{coord} in Example 3.9.

Solution

$$\begin{aligned} S_{st,\,coord} &= k_B \ln(10^{1.58 \times 10^{25}}) \\ &= (1.38 \times 10^{-23}\text{ J K}^{-1})(3.64 \times 10^{25}) \\ &= 502 \text{ J K}^{-1} \end{aligned}$$

▲

Exercise 3.14   Find the value of Ω for a system whose entropy is equal to 150 J K^{-1}.

Entropy and Randomness

We have established a connection between entropy and lack of information about the mechanical state of a system. It is commonly said that entropy is a measure of "randomness." This is an imprecise way of stating the connection between entropy and lack of information about the mechanical state, because disorder or randomness generally corresponds to lack of information about the mechanical state of the system.

3.5 The Third Law of Thermodynamics and Absolute Entropies

Walter Hermann Nernst, 1864–1941, was a German physical chemist who received the 1920 Nobel Prize in Chemistry for his work on the third law of thermodynamics, but who made numerous other contributions.

Max Karl Ernst Ludwig Planck, 1858–1947, was a German physicist who won the 1918 Nobel Prize in Physics for his pioneering work in quantum theory.

Gilbert Newton Lewis, 1875–1946, was an American chemist who first proposed that covalent chemical bonds arise from sharing of electrons according to the octet rule.

William Francis Giauque, 1895–1982, was an American chemist who discovered that ordinary oxygen consists of three isotopes. He received the 1949 Nobel Prize in Chemistry for proposing the process of adiabatic demagnetization to attain low temperatures.

Like the first and second laws of thermodynamics, the third law is a summary and generalization of experimental fact. The law was first stated by Nernst: *For certain isothermal chemical reactions of solids, the entropy change approaches zero as the thermodynamic temperature approaches zero.* Nernst based this statement on his analysis of experimental data of T. W. Richards. The statement of Nernst was sometimes called *Nernst's heat theorem.*

In 1911 Planck proposed extending Nernst's statement to assert that the entropies of individual substances actually approach zero as the temperature approaches zero. However, there is no experimental justification for this assertion.

In 1923 Lewis proposed the following acceptable statement of the third law: "*If the entropy of each element in some crystalline state be taken as zero at the absolute zero of temperature, every substance has a finite positive entropy—but at the absolute zero of temperature the entropy may become zero, and does so become in the case of perfect crystalline substances.*"[4]

The restriction to perfect crystals was made necessary by the discoveries of Giauque, a colleague of Lewis, who found that substances such as CO and NO fail to obey the third law in their ordinary crystalline forms. These substances easily form metastable crystals with some molecules in positions that are the reverse of the equilibrium positions, and ordinary crystals are in such metastable states. We discuss such systems later in this section.

Setting the entropies of the elements at zero temperature equal to zero is a convention, not necessarily a statement of a fundamental law as claimed by Planck.[5]

Exercise 3.15 Show that if the entropies of pure perfect crystalline elements are taken equal to nonzero constants at zero temperature, the entropy of a pure perfect crystalline compound at zero temperature is equal to the sum of the entropies of the appropriate numbers of moles of the elements at zero temperature.

---

[4] G. N. Lewis and M. Randall, *Thermodynamics and the Free Energy of Chemical Substances*, 1st ed., McGraw-Hill, New York, 1923, p. 448.
[5] M. L. McGlashan, *Chemical Thermodynamics*, Academic Press, New York, 1979, pp. 232ff.

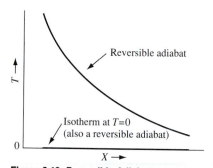

Figure 3.13. Reversible Adiabats Showing the Unattainability of Zero Temperature. The isotherm at $T = 0$ is a reversible adiabat as well as an isotherm. It therefore cannot intersect any reversible adiabat.

The Unattainability of Absolute Zero

We now show that it is impossible to bring a system to zero temperature in a finite number of operations. In Section 3.2, we showed that two reversible adiabats cannot cross. Since a reversible adiabat corresponds to constant entropy, the curve representing $T = 0$ is a reversible adiabat, as well as an isotherm (curve of constant temperature). This is depicted in Figure 3.13, in which the variable X represents an independent variable specifying the state of the system, such as the magnetization or the volume.

Just as a reversible adiabat gives the temperature of a simple system as a function of volume, in a general case the reversible adiabat gives the temperature as a function of X. Because two reversible adiabats cannot intersect, no other reversible adiabat can cross or meet the $T = 0$ isotherm. Therefore, no reversible adiabatic process can reduce the temperature of the system to zero. Furthermore, since irreversible adiabatic processes lead to the high-temperature side of the reversible adiabat (see Section 3.2), they cannot lead to lower temperatures than reversible adiabatic processes. Therefore, no adiabatic process can lead to zero temperature, except possibly by an infinite number of repetitions.

If no adiabatic process can lead to zero temperature, can some other kind of process lead to zero temperature? Unless a heat reservoir already exists at zero temperature, conduction of heat away from an object cannot do the job, because heat always flows from a hotter to a cooler object. We therefore conclude that no finite number of processes can cause a system to attain zero temperature, which is therefore called **absolute zero**. The unattainability of absolute zero is a consequence of both the second and third laws.

The lowest temperatures have been attained by **adiabatic demagnetization**. This process, invented by Giauque, consists of magnetizing an object isothermally. The magnetization process decreases the entropy because it aligns magnetic dipoles in the material, reducing the randomness of the system. Heat flows from the object to a heat reservoir during the magnetization. Once the object is magnetized, it is adiabatically insulated and then removed from the magnetic field. During the adiabatic demagnetization, which approximates a reversible process, the entropy remains nearly constant and the temperature drops. By carrying out this process repeatedly, temperatures of less than 0.000001 K (1 microkelvin) have been achieved in the nuclear spins of a magnetizable system, although an entire system has not been equilibrated at this temperature.

Absolute Entropies

We have agreed to set the entropy of pure crystalline substances equal to zero at zero temperature. The entropy change to bring a sample of a pure substance from zero temperature in a perfect crystalline form to some temperature of interest is called the **absolute entropy** of that system. We write for any system

$$S(T_1) = \int_0^{T_1} \frac{dq_{rev}}{T}$$

(3.5-1)

where $S(T_1)$ is the absolute entropy of the system at temperature T_1.

If there is no phase transition between $T = 0$ and $T = T_1$ (the system must be solid at temperature T_1) and if the final and initial states are at the same pressure,

$$S(T_1) = \int_0^{T_1} \frac{C_P}{T} dT \quad \text{(solid system)}$$

(3.5-2)

A similar equation can be written for a constant-volume integration.

Inspection of Equation (3.5-2) shows that the heat capacity must approach zero as the temperature approaches zero if the integral is not to diverge. Heat capacity data are difficult to obtain at very low temperatures, but all experimentally determined heat capacities tend toward zero as the temperature approaches zero.

An approximate theory of Debye[6] gives the result that for a crystal of a monatomic substance, the constant-volume heat capacity at low temperatures has a contribution from vibrational motions that is proportional to the cube of the temperature. An approximate theory for the motion of mobile electrons in metals gives a contribution proportional to the first power of the temperature. Thus,

The two terms in Equation (3.5-3) are the results of two different theories, both of which will be discussed in Chapter 22.

$$\bar{C}_V = aT^3 + bT \quad \text{(valid at low temperature)}$$

(3.5-3)

where a and b are parameters that can be determined from experimental data. For nonconductors, the parameter b is equal to zero. Equation (3.5-3) is quite reliable up to temperatures of about 15 K. Above this temperature, data are usually available. Since Equation (3.5-3) is only approximately valid and the difference between C_P and C_V is numerically small for solids, Equation (3.5-3) is usually used for C_P as well as C_V.

Exercise 3.16

a. Show that if Equation (3.5-3) is valid between zero temperature and some temperature T_1, the value of the molar entropy at T_1 is given by

$$\bar{S}(T_1) = \frac{aT_1^3}{3} = \frac{\bar{C}_V(T_1)}{3}$$

(3.5-4)

if b is equal to zero.

b. Find the expression for $\bar{S}(T_1)$ if b is not equal to zero.

---

[6] P. Debye, *Ann. Phys. (4)* **17**, 817 (1911); see Section 22.3.

If a phase transition occurs between zero temperature and the temperature of interest, Equation (3.5-2) must be modified to include the entropy change of the phase transition. If the substance is a liquid at temperature T_1, Equation (3.5-2) becomes

$$
S(T_1) = \int_0^{T_m} \frac{C_P(s)}{T} \, dT + \frac{\Delta H_{\text{fus}}}{T_m}
$$

$$
+ \int_{T_m}^{T_1} \frac{C_P(l)}{T} \, dT \quad \text{(liquid system)} \tag{3.5-5}
$$

where T_m is the reversible melting temperature, ΔH_{fus} is the enthalpy change of fusion (melting), $C_P(s)$ is the heat capacity of the solid, and $C_P(l)$ is the heat capacity of the liquid.

Exercise 3.17   Write the equation analogous to Equation (3.5-5) for a gaseous substance.

Calculation of Entropy Changes for Chemical Reactions

Absolute (third-law) entropies can be calculated from Equation (3.5-2), Equation (3.5-5), or the gas-phase analog, and tables of their values have been created. Some values for substances in their standard states are included in Table A9. These values can be used to calculate entropy changes for chemical reactions. For a reaction beginning with an equilibrium state or a metastable state and ending with an equilibrium state, the change in entropy is equal to the entropy of the products that appear minus the entropy of the reactants that disappear. For 1 mole of a reaction written as in Equation (2.7-7),

Compare Equation (3.5-6) with Equation (2.7-11), which contains enthalpy changes of formation. This equation contains absolute entropies, not entropy changes of formation.

$$
\Delta S = \sum_{i=1}^{c} v_i \bar{S}_i(T_1) \tag{3.5-6}
$$

if the temperature is equal to T_1 both before and after the reaction. The symbol $\bar{S}_i(T_1)$ stands for the absolute molar entropy of substance i at temperature T_1.

The **standard state for the entropy** is the same as the standard state for the enthalpy. For a solid or liquid, the standard state is the actual substance at pressure P°, equal to 1 bar (exactly). For a gas, the standard state is a "hypothetical ideal gas state" at pressure P°. That is, a correction must be made for the difference between the entropy of the real gas at pressure P° and the corresponding ideal gas at pressure P°. In Chapter 4, we will discuss how to make this correction. For most gases, the difference between the real gas and the ideal gas is small, and we can take the standard state as that of the real gas at pressure P° without serious error.

▼ **EXAMPLE 3.11**

Compute the standard-state entropy change for 1 mole of the reaction

$$0 = 2CO_2(g) - 2CO(g) - O_2(g)$$

if the product and reactants are at 298.15 K.

In this example, we substitute directly into Equation (3.5-6). Notice the units. The stoichiometric coefficients are taken as dimensionless, so the units of the entropy change of the reaction are J K^{-1} per mole of reaction.

Solution

Using values from Table A9,

$$\Delta S^\circ = (2)(213.64 \text{ J K}^{-1} \text{ mol}^{-1})$$
$$+ (-2)(197.564 \text{ J K}^{-1} \text{ mol}^{-1}) + (-1)(205.029 \text{ J K}^{-1} \text{ mol}^{-1})$$
$$= -172.88 \text{ J K}^{-1} \text{ mol}^{-1}$$

▲

Exercise 3.18

Assuming that the surroundings remain at equilibrium at 298.15 K, calculate the entropy change of the surroundings and of the universe for 1 mole of the reaction in Example 3.11.

In Chapter 2, we discussed the approximate calculation of energy changes of chemical reactions, using average bond energies. There is an analogous estimation scheme for the entropy changes of chemical reactions, in which contributions from bonds and contributions from groups of atoms are included.[7] We do not discuss this scheme, but the interested student is referred to the article by Benson and Buss.

Statistical Entropy and the Third Law of Thermodynamics

Some substances originally appeared not to obey the third law of thermodynamics, and these anomalies can be explained using statistical entropy. For example, consider a crystal of carbon monoxide. The absolute entropy of gaseous carbon monoxide determined by an integration such as in Equation (3.5-1) turned out to be too small to agree with values inferred from entropy changes of chemical reactions and absolute entropies of other substances.

Carbon monoxide molecules are nearly symmetrical in shape, with only a small dipole moment, so a carbon monoxide molecule fits into the crystal lattice almost as well with its ends reversed as in the equilibrium position. For this reason, metastable crystals can easily form with part of the molecules in the reversed position. (In fact, it is difficult to obtain a perfect crystal.) If we assume that the occurrence of reversed molecules is independent of the rest of the state of the crystal, we can write

This is why the statement of the third law is restricted to perfect crystals.

$$\Omega = \Omega_{\text{orient}}\Omega_{\text{rest}} \qquad (3.5-7)$$

---

[7] S. W. Benson and J. H. Buss, *J. Chem. Phys.* **29**, 546 (1958); S. W. Benson et al., *Chem. Rev.* **69**, 279 (1969).

where Ω_{orient} is the number of ways of orienting the molecules in ways compatible with our knowledge of the state of the system. The other factor, Ω_{rest}, is the number of possible states of the crystal if the orientation of the molecules is ignored.

At zero temperature, quantum statistical mechanics (which we will discuss in Chapter 20) predicts that the various vibrations of a crystal lattice all fall into a single lowest-energy state, as do the electronic motions. Therefore, if there is no entropy of isotopic mixing,

$$\lim_{T \to 0} \Omega_{rest} = 1 \tag{3.5-8}$$

If a metastable crystal exists in which, as far as we know, each molecule can occur with equal probability in either the equilibrium state or the reversed state, then

$$\Omega_{orient} = 2^N \tag{3.5-9}$$

where N is the number of molecules in the crystal. If Ω_{rest} is set equal to unity, the statistical entropy of the metastable crystal near zero temperature is

$$S_{st}(\text{metastable}) = k_B \ln(2^N) = N k_B \ln(2) \tag{3.5-10}$$

For 1 mole of carbon monoxide,

$$\bar{S}_{st}(\text{metastable}) = R \ln(2) = 5.76 \text{ J K}^{-1} \text{ mol}^{-1} \tag{3.5-11}$$

This value agrees with the amount by which carbon monoxide appears to deviate from the third law.

Exercise 3.19   Pretend that it somehow has been possible to synthesize 1.00 mol of $CaCO_3$ in which each carbonate ion has one ^{16}O atom, one ^{17}O atom, and one ^{18}O atom. Calculate the entropy of the metastable crystal near zero temperature, if nothing is known about the orientations of the carbonate ions.

Trouton's Rule

Trouton's rule is an empirical rule for entropy changes of vaporization. It states that for "normal" liquids, the molar entropy change of vaporization at the normal boiling temperature (at 1.000 atm) is approximately equal to $10.5R \approx 88 \text{ J K}^{-1} \text{ mol}^{-1}$.

In both ethanol and water, there is considerable molecular association through hydrogen bonding, and Trouton's rule underestimates the entropy change of vaporization by about 20 percent. Liquids without hydrogen bonding or other forms of molecular association are "normal" liquids as far as Trouton's rule is concerned. However, Trouton's rule also badly overestimates the entropy change of vaporization for hydrogen and helium.

Modifications of Trouton's rule have been proposed, including a version that uses entropy changes of vaporization to form gases with the same value of the molar volume instead of whatever volume corresponds to a pressure of 1 atm. The values for hydrogen and helium fall closer to those of other substances if this modified rule is used. There is also a method in which contributions for different groups of atoms in the molecule are considered.[8]

Summary

Kelvin's statement of the second law of thermodynamics is that heat put into a system that undergoes a cyclic process cannot be completely converted into work done on the surroundings. Clausius's statement of the same law is that heat cannot flow from a cooler to a hotter body if nothing else happens.

A mathematical statement of the second law is that S, the entropy, is a state function if we define

$$dS = \frac{dq_{rev}}{T}$$

It was shown that in any reversible process the entropy of the universe remains constant, whereas in any irreversible process the entropy of the universe must increase.

A general procedure for calculating an entropy change in a closed system is to carry out the line integral

$$\Delta S = \int_c \frac{dq_{rev}}{T}$$

over a curve in state space corresponding to a reversible path.

The statistical entropy is defined by

$$S_{st} = k_B \ln(\Omega) + S_0$$

where k_B is Boltzmann's constant and Ω is the thermodynamic probability, equal to the number of mechanical states that might be occupied by the system, as far as we know. The quantity S_0 is an arbitrary constant whose value we take to be zero.

The third law of thermodynamics states that if the entropies of all samples of pure perfect crystalline elements are arbitrarily taken as zero, then the entropies of all samples of pure perfect crystalline compounds can also be taken as zero. Entropies relative to the entropy at zero temperature are called absolute entropies.

A consequence of the second and third laws of thermodynamics is that no finite number of operations, either reversible or irreversible, can bring an object to zero temperature.

Additional Reading

J. D. Fast, *Entropy*, McGraw-Hill, New York, 1962
    A comprehensive discussion of thermodynamic and statistical entropy.

---

[8] D. Hoshino, K. Nagahama, and M. Hirata, *Ind. Eng. Chem. Fundam.* **22**, 430 (1983).

W. F. Magie, editor and translator, *The Second Law of Thermodynamics*, Harper and Brothers Publishers, New York, 1899
This little book consists of writings of Carnot, Clausius (in English translation), and Lord Kelvin. If you can find it, you can read the work of the pioneers.

D. R. Owen, *A First Course in the Mathematical Foundations of Thermodynamics*, Springer-Verlag, New York, 1984
This book discusses thermodynamics as a branch of mathematics.

See also the references listed in Chapter 2.

PROBLEMS

Problems for Section 3.1

ᶠ3.20. A Carnot engine contains as working fluid 0.100 mol of neon (assume ideal with $\bar{C}_V = 3R/2$). If $T_h = 500.$ K and $T_c = 373$ K, find the values of V_1, V_2, V_3, P_3, P_4, and V_4 if $P_1 = 20.00$ atm and $P_2 = 5.00$ atm. State any assumptions.

3.21. A Carnot engine contains as working fluid 0.100 mol of nitrogen (assume ideal with $\bar{C}_V = 5R/2$). If $T_h = 500.$ K and $T_c = 373$ K, find the values of V_1, V_2, V_3, P_3, P_4, and V_4 if $P_1 = 20.00$ atm and $P_2 = 50.00$ atm. Compare your results with those of Problem 3.20 and explain any differences.

3.22. A steam engine is 60% as efficient as a Carnot engine. If its boiler is at 200°C and its exhaust at 100°C, calculate the height to which it can lift a 1000-kg mass near the earth's surface if it burns 10.0 kg of coal. Pretend that the coal is pure graphite and that its enthalpy change of combustion is equal to that at 25°C.

3.23. Calculate the amount of electrical energy in kilowatt-hours necessary to freeze 100.0 kg of water in a reversible Carnot refrigerator with an interior temperature of $-10.$°C and an exterior temperature of 25°C. 1 watt = 1 J s^{-1}.

Problems for Section 3.2

*3.24. Calculate the entropy change for each of the four steps in the Carnot cycle of Problem 3.20, and show that the entropy change for the whole cycle vanishes.

3.25. A sample of 1.000 mol of helium gas (assume ideal with $\bar{C}_V = 3R/2$) expands adiabatically and irreversibly from a volume of 5.000 L and a temperature of 298.15 K to a volume of 10.00 L against an external pressure of 1.000 atm. Find the final temperature, ΔU, q, w, and ΔS for this process. Compare each quantity with the corresponding quantity for a reversible adiabatic expansion to the same final volume.

Problems for Section 3.3

3.26. a. Find the change in entropy for the vaporization of 2.000 mol of liquid water at 100°C and a constant pressure of 1.000 atm.

b. Find the entropy change for the heating of 2.000 mol of water vapor at a constant pressure of 1.000 atm from 100.°C to 200.°C. Use the polynomial representation in Table A7 for the heat cpacity of water vapor.

3.27. a. Calculate the entropy change for the isothermal expansion of 1.000 mol of helium gas (assume ideal) from a volume of 4.000 L to a volume of 8.000 L.

b. Calculate the entropy change for the isothermal expansion of 1.000 mol of helium gas (assume ideal) from a volume of 8.000 L to a volume of 12.000 L.

c. Explain in words why your answer in part b is not the same as that in part a, although the increase in volume is the same.

*3.28. a. Calculate the entropy change for the following reversible process: 1.000 mol of neon (assume ideal with $\bar{C}_V = 3R/2$) is expanded isothermally at 298.15 K from 2.000 atm pressure to 1.000 atm pressure and is then heated from 298.15 K to 398.15 K at a constant pressure of 1.000 atm. Integrate on the path representing the actual process.

b. Calculate the entropy change for the reversible process with the same initial and final states as in part a, but in which the gas is first heated at constant pressure and then expanded isothermally. Again, integrate on the path representing the actual process. Compare your result with that of part a.

c. Calculate the entropy change of the surroundings in each of parts a and b.

d. Calculate the entropy changes of the system and the surroundings if the initial and final states are the same as in parts a and b, but the gas is expanded irreversibly and isothermally against an external pressure of 1.000 atm and then heated irreversibly with the surroundings remaining essentially at equilibrium at 400. K.

Problems for Section 3.4

3.29. The calculation of the statistical entropy of a metastable disordered ice crystal is a famous problem that has not yet been solved exactly. Each water molecule has an oxygen atom with two hydrogens covalently bonded to it (at a smaller distance) and two hydrogens on other molecules hydrogen bonded to it (at a larger distance). A large number of coordinate states can be generated by moving the hydrogens around. For example, if one of the hydrogen-bonded hydrogens is brought closer to a given oxygen and covalently bonded to it, one of the covalently bonded hydrogens must move farther away and become hydrogen bonded. Make a crude first estimate of the statistical entropy of a disordered ice crystal containing 1.000 mol of water molecules by pretending that each oxygen can have its four hydrogens move independently of the others as long as two hydrogens are close to it and two are further from it.

3.30. Explain why the statistical entropy of the system changes in each of the following processes, and tell whether it increases or decreases.

    a. A sample of gas is heated at constant volume.
    b. A sample of gas is expanded at constant temperature.
    c. A sample of liquid water is heated.
    d. A sample of liquid water is frozen.
    e. A sample of liquid water is evaporated.

Problems for Section 3.5

3.31. Assign the following nonzero constant values for the standard-state molar entropies at 0 K:

    $C(s)$: $10.00 \text{ J K}^{-1} \text{ mol}^{-1}$
    $O_2(s)$: $20.00 \text{ J K}^{-1} \text{ mol}^{-1}$
    $H_2(s)$: $30.00 \text{ J K}^{-1} \text{ mol}^{-1}$

Accepting as experimental fact that the entropy changes of all reactions between pure solids at 0 K are equal to zero, assign values for the standard-state molar entropies of $CO_2(s)$ and $H_2O(s)$ at 0 K.

3.32. Calculate ΔS° at 298.15 K for each of the following reactions:

    *__a.__ $2H_2(g) + O_2(g) \rightarrow 2H_2O(l)$
    b. $2H_2(g) + O_2(g) \rightarrow 2H_2O(g)$
    *__c.__ $CaCO_3(s) \rightarrow CaO(s) + CO_2(g)$
    d. $CH_4(g) + 2O_2(g) \rightarrow CO_2(g) + 2H_2O(g)$

3.33. Using absolute entropy value from Table A9 and heat capacity values from Table A7 or Table A9, calculate $\bar{S}^\circ$ values for the substances at the indicated temperatures. If no polynomial representations are available, assume constant heat capacities.

    a. $H_2(g)$ at 200.0°C   **d.** $H_2O(g)$ at 100.0°C
    b. $O_2(g)$ at 200.0°C   **e.** $CH_4(g)$ at 200.0°C
    c. $H_2O(l)$ at 100.0°C   **f.** $H_2O(g)$ at 200.0°C

3.34. a. Using absolute entropy values from the previous problem, calculate $\Delta \bar{S}^\circ$ of vaporization for H_2O at 100.0°C.

    b. Calculate $\Delta \bar{S}^\circ$ of vaporization for H_2O at 25°C using Equation (3.3-5).

    c. Calculate $\Delta \bar{S}^\circ$ of vaporization for H_2O at 100.0°C using Equation (3.3-5).

    d. Comment on your values in light of Trouton's rule.

*__3.35.__ Using $\bar{S}^\circ$ values from Problem 3.33, calculate ΔS° at 200.°C for the reaction

$$0 = 2H_2O(g) - O_2(g) - 2H_2(g)$$

3.36. Tabulated entropy changes of formation could be used instead of absolute entropies to calculate entropy changes of chemical reactions.

    a. Using absolute entropies, calculate the standard-state entropy change of formation at 298.15 K for $CO(g)$, $O_2(g)$ and $CO_2(g)$.

    b. Calculate the standard-state entropy change at 298.15 K for the reaction

$$2CO(g) + O_2(g) \rightarrow 2CO_2(g)$$

using your values of entropy changes of formation from part a. Compare your result with that of Example 3.11.

*__3.37.__ Following are heat capacity data for pyridine.[9]

T (K)	$\bar{C}_P^\circ(s)/\text{J K}^{-1}\text{ mol}^{-1}$	T (K)	$\bar{C}_P^\circ(l)/\text{J K}^{-1}\text{ mol}^{-1}$
13.08	4.448	239.70	122.23
21.26	12.083	254.41	124.54
28.53	19.288	273.75	127.93
35.36	25.309	293.96	131.88
48.14	33.723	307.16	134.55
64.01	40.748		
82.91	46.413		
101.39	50.861		
132.32	58.325		
151.57	63.434		
167.60	68.053		
179.44	71.756		
193.02	76.467		
201.61	79.835		
212.16	84.446		
223.74	94.328		

A value of 8278.5 J mol^{-1} is reported for the enthalpy change of fusion at 231.49 K.

    a. Assuming that the Debye formula can be used between 0 K and 13.08 K, find the absolute entropy of solid pyridine at 231.49 K.

    b. Find the absolute entropy of liquid pyridine at 231.49 K and at 298.15 K.

    Use either a graphical or a numerical technique to approximate the integrals needed.[10]

---

[9] F. T. Gucker and R. L. Seifert, *Physical Chemistry*, W. W. Norton, New York, 1966, p. 445.
[10] See L. Gilman and R. H. McDowell, *Calculus*, 2nd ed., W. W. Norton, New York, 1978, pp. 479ff, or any other calculus text.

General Problems

3.38. Supercooled (metastable) water vapor commonly occurs in the atmosphere if dust particles are not present to begin condensation to the liquid. Sometimes small particles, such as tiny crystals of silver iodide, are released from airplanes in an attempt to begin condensation. This process is called cloud seeding. At a certain location, water vapor at 25°C has a metastable partial pressure of 30.0 torr. The equilibrium value at this temperature is 23.756 torr. Consider the air that is present to be the surroundings, and assume it remains at equilibrium at 25°C. A tiny particle is added to begin condensation. Calculate ΔS, ΔH, and ΔS_{surr} per mole of water that condenses. State any assumptions.

3.39. Without doing any detailed calculations, specify for each process whether each of the following quantities is positive, negative, or equal to zero: q, w, ΔU, ΔS, and ΔS_{surr}.

a. The system of Section 3.3, which consisted of two large objects and a small bar between them, is allowed to come to equilibrium from an initial state in which the two objects are at different temperatures. Assume that the objects have fixed volume.

b. A sample of water is boiled at 100.°C and 100. atm.

c. A sample of supercooled liquid water at $-10.$°C is allowed to equilibrate adiabatically at constant pressure after a tiny crystal of ice is dropped into it.

d. A sample of an ideal gas expands irreversibly and adiabatically into a vacuum, as in the Joule experiment.

e. A sample of gas is heated at constant volume.

f. A sample of gas is heated at constant pressure.

g. A sample of ideal gas expands reversibly at constant temperature.

3.40. Assume that an automobile engine burns 2,2,4-trimethylpentane (isooctane), forming only $CO_2(g)$ and $H_2O(g)$. The density of isooctane is 0.6909 g mL^{-1}, and 1.00 gallon is approximately equal to 3.76 L.

a. Find the amount of heat that can be obtained from combustion of 1.000 gallon of isooctane. Ignore the temperature dependence of ΔH of the combustion reaction.

b. If the combustion temperature is 2200.°C and the exhaust temperature is 800.°C, find the maximum height to which an automobile of 1000. kg can be lifted by combustion of 1.000 gallon of isooctane. Ignore all forms of friction. State any assumptions.

c. In some countries, antipollution laws require that the combustion temperature of automobile engines be lowered by exhaust gas recirculation, in an attempt to reduce the amount of nitrogen oxides produced. Repeat part b for a combustion temperature of 1800.°C.

*3.41. Which of the following statements are correct? If a statement is incorrect, give a counterexample.

a. The entropy of any system must increase when an irreversible process occurs.

b. The entropy of the surroundings must increase when an irreversible process occurs.

c. The entropy of the universe must increase when an irreversible process occurs.

d. The entropy of any system remains constant when a reversible process occurs.

e. The entropy of the surroundings remains constant when a reversible process occurs.

f. The entropy of the universe remains constant when a reversible process occurs.

g. The energy of an isolated system remains constant when a reversible process occurs in the system.

h. The energy of an isolated system decreases when a reversible process occurs in the system.

3.42. Make a graph of each of the following:

a. The molar entropy of a monatomic ideal gas with $\bar{C}_V = 3R/2 =$ constant as a function of temperature from 100. K to 300. K. Assume a constant pressure of 1.000 bar.

b. The molar entropy of an ideal gas as a function of molar volume from 1.000 L to 10.00 L at a constant temperature of 300. K.

c. The molar entropy of water from $-50.$°C to $+50.$°C at a constant pressure of 1.000 atm.

4 The Thermodynamics of Real Systems

PREVIEW

Chapters 2 and 3 contain all of the fundamental theory of thermodynamics. In this chapter we will take the abstract concepts of this theory and formulas and obtain concrete working formulas to apply to real systems and states.

OBJECTIVES

After studying this chapter, the student should:

1. understand the use of the second law of thermodynamics to determine whether or not a process is spontaneous,

2. be able to simplify thermodynamic formulas using appropriate tools,

3. be able to make various kinds of thermodynamic calculations,

4. be familiar with the way in which the Gibbs energy of a system depends on its state,

5. understand the meaning of partial molar quantities and be able to use them in calculations,

6. be able to use Euler's theorem and the Gibbs-Duhem relation in thermodynamic calculations.

PRINCIPAL FACTS AND IDEAS

1. Practical thermodynamics equations can be obtained from abstract principles.
2. The second law of thermodynamics provides the general criterion for spontaneous processes:

$$\Delta S_{\text{univ}} = \Delta S + \Delta S_{\text{surr}} \geq 0$$

3. The second law provides criteria for spontaneous processes in systems under specific circumstances.
4. The Gibbs and Helmholtz energies provide information about the maximum amount of work that can be done by a system in a given process.
5. The methods of calculus can provide useful thermodynamic relations, such as the Maxwell relations.
6. The description of open multicomponent systems introduces the chemical potential as an important variable.
7. Euler's theorem and the Gibbs-Duhem relation are useful formulas.

4.1 Criteria for Spontaneous Processes and for Equilibrium; The Gibbs and Helmholtz Energies

The second law of thermodynamics provides the fundamental criterion which must be obeyed for a process to take place in the real universe. This criterion is that the entropy of the universe cannot decrease in any process. We now investigate specific situations and express this fundamental criterion in different ways for different situations.

Criteria for Spontaneous Processes in Closed Systems

One way to avoid nonessential complications in our argument is to make a detailed analysis of the behavior of the surroundings unnecessary.

In this section, we express the criterion for possible processes in a closed system in terms of the state of a system in contact with surroundings, rather than in terms of the entropy of the universe. We assume that the surroundings are very large and have a very large thermal conductivity and a very large heat capacity, so that they remain at equilibrium at a fixed temperature.

Consider first a closed system that is not necessarily simple. The second law of thermodynamics implies that processes that are possible must obey the relation

$$dS \geq \frac{dq}{T_{surr}}$$ **(4.1-1)**

where T_{surr} is the temperature of the surroundings.

From the first law, $dq = dU - dw$, so that

$$dS \geq \frac{dU - dw}{T_{surr}}$$

which is the same as

$$\boxed{dU - dw - T_{surr}\, dS \leq 0}$$ **(4.1-2)**

An **isolated system** is a closed system that, in addition to being unable to exchange heat with the rest of the universe, cannot exchange work with it. That is, $dU = 0$, $dq = 0$, and $dw = 0$, so Equation (3.2-26) gives

$$dS \geq 0 \quad \text{(isolated system)}$$ **(4.1-3)**

Next, consider the special case that no work is done and that the entropy of the system is constant. A system known to be in a given mechanical state has constant statistical entropy and can fit this case. Since $dS = 0$ and $dw = 0$,

$$dU \leq 0 \quad (S \text{ constant}, dw = 0)$$ **(4.1-4)**

Next, consider the important case in which the temperature of the system is constant and equal to the temperature of the surroundings. We will refer to it simply as the case of constant temperature. Equation (4.1-2) becomes

$$dU - T\, dS - dw \leq 0 \quad (T \text{ constant})$$ **(4.1-5)**

If our system is simple, $dw = -P_{ext}\, dV$, and

$$dU - T\, dS + P_{ext}\, dV \leq 0 \quad (\text{simple system}, T \text{ constant})$$ **(4.1-6)**

There are two important cases of isothermal processes in closed simple systems. The first is the case of constant volume, for which

$$dU - T\, dS \leq 0 \quad (\text{simple system}, T \text{ and } V \text{ constant})$$ **(4.1-7)**

The **Helmholtz energy** is a convenience variable, named for Hermann Helmholtz and defined by

$$A = U - TS \quad \text{(definition)}$$ (4.1-8)

This quantity has been known to physicists as the *free energy* and as the *Helmholtz function* and has been known to chemists as the *work function* and as the *Helmholtz free energy*.

The differential of the Helmholtz energy is

$$dA = dU - T\,dS - S\,dT$$ (4.1-9)

so that if T is constant

$$dA = dU - T\,dS \quad \text{(constant } T\text{)}$$ (4.1-10)

Equation (4.1-7) is the same as

$$dA \leq 0 \quad \text{(simple system, } T \text{ and } V \text{ constant)}$$ (4.1-11)

Now consider the case in which the pressure of the system is constant and equal to the external pressure. In this case,

$$dU + P\,dV - T\,dS \leq 0 \quad \text{(simple system, } T \text{ and } P \text{ constant)}$$ (4.1-12)

When we specify constant pressure, we mean that the pressure is not only constant but also equal to the external pressure.

Another convenience variable is the **Gibbs energy**, named for the first American to gain a European reputation as an important theoretical scientist. It is defined by

$$G = U + PV - TS \quad \text{(definition)}$$ (4.1-13)

Josiah Willard Gibbs, 1839–1903, was an American physicist who made fundamental contributions to thermodynamics and statistical mechanics.

The Gibbs energy has also been called the *free energy*, the *Gibbs free energy*, the *Gibbs function*, and the *free enthalpy*.

The Gibbs energy is related to the enthalpy and the Helmholtz energy by the relations

$$G = H - TS = A + PV$$ (4.1-14)

From Equation (4.1-13), the differential of G is

$$dG = dU + P\,dV + V\,dP - T\,dS - S\,dT$$ (4.1-15)

If T and P are constant, this equation becomes

$$dG = dU + P\,dV - T\,dS \quad \text{(} T \text{ and } P \text{ constant)}$$ (4.1-16)

Equation (4.1-12) is the same as

$$dG \leq 0 \quad \text{(simple system, } T \text{ and } P \text{ constant)}$$ (4.1-17)

The criteria for finite processes are completely analogous to those for infinitesimal processes. For example, for a simple system at constant pressure and temperature, a spontaneous process must obey

$$\Delta G \leq 0 \quad \text{(simple system, } T \text{ and } P \text{ constant)} \qquad \textbf{(4.1-18)}$$

Equation (4.1-18) is the most useful criterion for the spontaneity of chemical reactions.

In the nineteenth century, Berthelot incorrectly maintained that all spontaneous reactions must be exothermic ($q < 0$). The incorrectness of Berthelot's conjecture was shown by Duhem, who established Equation (4.1-17).

From Equation (4.1-14), Equation (4.1-18) can be written

$$\Delta H - T\,\Delta S \leq 0 \quad \text{(simple system, } T \text{ and } P \text{ constant)} \qquad \textbf{(4.1-19)}$$

In many chemical reactions, the $T\,\Delta S$ term is numerically less important than the ΔH term and the incorrect criterion of Berthelot gives the correct prediction about the spontaneity of the reaction. However, in other cases the $T\,\Delta S$ term dominates and the criterion of Berthelot fails.

In general, the ΔH term dominates at low temperature (small values of T give small values of $T\,\Delta S$), and the $T\,\Delta S$ term becomes important and can dominate at sufficiently high temperature. The vaporization of a liquid is a simple example. Both ΔH and ΔS are positive. There is some temperature at which the vaporization is a reversible process and the two phases can coexist. At this temperature, $\Delta G = 0$. When T is smaller than this equilibrium temperature, the ΔH term dominates and $\Delta G > 0$ for the vaporization process. That is, the condensation is spontaneous and the equilibrium state is the liquid. At a higher temperature, the $T\,\Delta S$ term dominates and $\Delta G < 0$. The equilibrium state is then the gas.

We now have a set of criteria for possible processes in a closed simple system, given by Equations (4.1-3), (4.1-4), (4.1-11), and (4.1-17): *If the system is isolated, the entropy cannot decrease. If S is fixed and no work is done, U cannot increase. If T and V are fixed, A cannot increase. If T and P are fixed, G cannot increase.*

Criteria for a System to Be at Equilibrium

Let us now shift our discussion from the criteria for possible processes and seek the criteria for a system to be at macroscopic equilibrium. Consider first a closed simple system at constant P and T. All processes that do not violate the second law cause the Gibbs energy of the system to decrease. Therefore, when the system reaches equilibrium, the Gibbs energy must have reached the minimum value possible for that system at the given pressure and temperature.

Figure 4.1 represents the situation. The variable x schematically represents the extent to which a chemical reaction or some other process has occurred. The value of x at the minimum in the curve corresponds to the equilibrium state for the particular constant values of P and T, and other values of x must correspond to metastable or unstable states.

We nearly always assume that a state variable such as the Gibbs energy is a differentiable function of its independent variables, so that the depen-

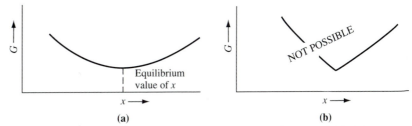

Figure 4.1. The Gibbs Energy as a Function of the Extent of a Process. (a) As it is assumed to be (schematic). We assume that the Gibbs energy is a differentiable function of the extent of the process. **(b)** As it assumed not to be (schematic). This function is not differentiable at its minimum.

dence is as shown in Figure 4.1a, not as shown in Figure 4.1b. There is a smooth minimum at which

$$\left(\frac{\partial G}{\partial x}\right)_{T,P} = 0 \quad \text{(at equilibrium, } P \text{ and } T \text{ constant)} \qquad \textbf{(4.1-20a)}$$

or

$$dG = 0 \quad \text{(at equilibrium, } P \text{ and } T \text{ constant)} \qquad \textbf{(4.1-20b)}$$

Equation (4.1-20) could correspond to either a maximum or a minimum. To ensure a minimum, we must also have

$$\left(\frac{\partial^2 G}{\partial x^2}\right)_{T,P} \geq 0 \quad \text{(at equilibrium)} \qquad \textbf{(4.1-21)}$$

Any spontaneous infinitesimal process in a system at equilibrium at constant pressure and temperature must conform to Equation (4.1-20).

For a closed simple system at constant temperature and volume, any spontaneous process must lower the value of the Helmholtz energy of the system. Therefore, at equilibrium the Helmholtz energy must be at a minimum, so that

$$\left(\frac{\partial A}{\partial x}\right)_{T,V} = 0 \quad \text{(at equilibrium, } T \text{ and } V \text{ constant)} \qquad \textbf{(4.1-22a)}$$

or

$$dA = 0 \quad \text{(at equilibrium, } T \text{ and } V \text{ constant)} \qquad \textbf{(4.1-22b)}$$

and to ensure a minimum, we must also have

$$\left(\frac{\partial^2 A}{\partial x^2}\right)_{T,V} \geq 0 \quad \text{(at equilibrium)} \qquad \textbf{(4.1-23)}$$

If a system is at equilibrium, both G and A are at minimum values at the same time, but with respect to different processes. If a system is at constant temperature and pressure, a spontaneous process could possibly increase A but must decrease G. If a system is at constant temperature and volume, a spontaneous process could possibly increase G but must decrease A.

Spontaneity and Equilibrium Criteria for Nonsimple Systems

If our system is not a simple system, we write

$$dw = -P_{ext}\,dV + dw_{net} \tag{4.1-24}$$

The term $-P_{ext}\,dV$ is called **compression work** or *P-V* **work**. The term dw_{net} is called the **net work**. It is any work that can be done in addition to compression work, such as electrical or stress-strain work.

Equation (4.1-5) for the case of constant temperature becomes

$$dU - T\,dS + P_{ext}\,dV - dw_{net} = dA - dw \le 0 \quad (T \text{ constant}) \tag{4.1-25}$$

which is the same as

$$dA \le dw \quad (T \text{ constant}) \tag{4.1-26}$$

In the case of constant pressure ($P = P_{ext}$) and constant temperature, Equation (4.1-25) is

$$dG - dw_{net} \le 0 \quad (T \text{ and } P \text{ constant}) \tag{4.1-27}$$

It is possible to increase the Gibbs energy of a nonsimple system at constant T and P by doing net work on the system. An example of this kind of process is electrolysis, in which a chemical reaction is caused to proceed in the nonspontaneous direction by forcing an electric current through an electrochemical cell.

The criteria for a nonsimple system to be at equilibrium are also different from those for a simple system: For a nonsimple system to be at equilibrium at constant T and V, the Helmholtz energy is not necessarily at a minimum value. If a system at equilibrium undergoes an infinitesimal change, instead of $dA = 0$ at constant T and V as in Equation (4.1-22), we must have

$$\boxed{dA = dw \quad (\text{at equilibrium, constant } T \text{ and } V)} \tag{4.1-28}$$

and instead of $dG = 0$ at constant T and P as in Equation (4.1-20),

$$\boxed{dG = dw_{net} \quad (\text{equilibrium, constant } T \text{ and } P)} \tag{4.1-29}$$

Work and the Helmholtz and Gibbs Energies

The Helmholtz energy and the Gibbs energy have important relationships to the work that can be done on the surroundings in a given process. Since $dw_{surr} = -dw$, Equation (4.1-25) is

$$\boxed{dw_{surr} \le -dA \quad (T \text{ constant})} \tag{4.1-30}$$

or for a finite process

$$w_{\text{surr}} \leq -\Delta A \quad (T \text{ constant}) \tag{4.1-31}$$

It is the work done on the surroundings that is limited by the second law of thermodynamics, not the work done on the system.

That is, the work that can be done on the surroundings in an isothermal process cannot exceed the negative of the Helmholtz energy change of the system, and only for a reversible process can it be equal to $-\Delta A$. Equation (4.1-31) holds for both a simple system and a nonsimple system. If work is to be done on the surroundings by any kind of a closed system at constant T, a process with negative ΔA must be found.

In the case that $P = P_{\text{ext}} = \text{constant}$ and $T = T_{\text{surr}} = \text{constant}$, Equation (4.1-27) is the same as

$$dw_{\text{surr(net)}} \leq -dG \quad (T, P \text{ constant}) \tag{4.1-32}$$

For a finite process

$$w_{\text{surr(net)}} \leq -\Delta G \quad (T, P \text{ constant}) \tag{4.1-33}$$

If the system is simple, Equation (4.1-33) becomes the same as Equation (4.1-18).

Just as the total work done on the surroundings at constant temperature is limited by $-\Delta A$, the net work done on the surroundings at constant temperature and pressure is limited by $-\Delta G$. That is, if T and P are constant, only a process corresponding to a negative change in the Gibbs energy can do net work on the surroundings, and the amount of net work done can be no greater than $-\Delta G$. If the process is reversible, the net work done on the surroundings is equal to $-\Delta G$.

Exercise 4.1 Write the relation governing the maximum amount of work done on the surroundings in a nonsimple system at constant T and V.

4.2 Fundamental Relations for Closed Simple Systems

As always, use equations only in situations in which they apply. These equations apply only for closed systems.

In this section, we obtain useful equations for the description of simple closed systems. For a closed simple system and for reversible processes, the first law is

$$dU = dq - P\,dV \tag{4.2-1}$$

and the second law is

$$dS = \frac{dq_{\text{rev}}}{T} \tag{4.2-2}$$

We combine these equations to obtain

$$dU = T\,dS - P\,dV \quad \left(\begin{array}{l}\text{simple closed system,} \\ \text{reversible processes}\end{array}\right) \qquad \text{(4.2-3)}$$

This equation holds for a closed simple system with any number of phases and any number of substances.

Our derivation of Equation (4.2-3) is restricted to reversible changes of state. However, in a branch of thermodynamics that concerns irreversible processes this equation is assumed to be valid, at least if the deviation from equilibrium is not large. This assumption is an additional hypothesis and does not follow from our analysis.

Let us first consider a system that contains only one phase, although it can contain any number of substances. Because the state of a simple one-phase closed system at equilibrium is specified by two independent variables (other than the amounts of the substances, which are fixed), we can consider U to be a function of S and V:

$$dU = \left(\frac{\partial U}{\partial S}\right)_{V,n} dS + \left(\frac{\partial U}{\partial V}\right)_{S,n} dV \quad \left(\begin{array}{l}\text{simple closed system,} \\ \text{reversible processes}\end{array}\right) \qquad \text{(4.2-4)}$$

where the single subscript n means that all substances present are at fixed amounts.

Comparison of Equations (4.2-3) and (4.2-4) gives us two important relations:

$$\left(\frac{\partial U}{\partial S}\right)_{V,n} = T \qquad \text{(4.2-5)}$$

$$\left(\frac{\partial U}{\partial V}\right)_{S,n} = -P \qquad \text{(4.2-6)}$$

The Maxwell Relations

From the Euler reciprocity relation, Equation (B-13) of Appendix B,

$$\left(\frac{\partial^2 U}{\partial S\,\partial V}\right)_n = \left(\frac{\partial^2 U}{\partial V\,\partial S}\right)_n \qquad \text{(4.2-7)}$$

Since the second derivative is the derivative of the first derivative,

$$\left(\frac{\partial^2 U}{\partial V\,\partial S}\right)_n = \left(\frac{\partial T}{\partial V}\right)_{S,n} \qquad \text{(4.2-8)}$$

$$\left(\frac{\partial^2 U}{\partial S\,\partial V}\right)_n = -\left(\frac{\partial P}{\partial S}\right)_{V,n} \qquad \text{(4.2-9)}$$

Therefore

$$\left(\frac{\partial T}{\partial V}\right)_{S,n} = -\left(\frac{\partial P}{\partial S}\right)_{V,n} \quad \text{(a Maxwell relation)} \quad (4.2\text{-}10)$$

James Clerk Maxwell, 1831–1879, great Scottish mathematician and physicist, made important contributions to many areas of physics. The Maxwell relations are remarkable. If it were not for their mathematical derivation, it would be hard to convince anyone that they are correct.

Equation (4.2-10) is one of a class of equations called **Maxwell relations**, after James Clerk Maxwell. Their principal utility is in replacing a hard-to-measure partial derivative with one that can be measured more easily. For example, it would be difficult to measure $(\partial P/\partial S)_{V,n}$ but much easier to measure $(\partial T/\partial V)_{S,n}$.

▼ **EXAMPLE 4.1**

From the relation in Equation (4.2-10), find an expression for $(\partial P/\partial S)_{V,n}$ for an ideal gas with constant heat capacity.

Solution

Equation (2.4-22) gives

$$T = T_1\left(\frac{V_1}{V}\right)^{nR/C_V}$$

where the subscripts are omitted on the final values of T and V.

$$\left(\frac{\partial P}{\partial S}\right)_{V,n} = -\left(\frac{\partial T}{\partial V}\right)_{S,n} = -T_1(V_1)^{nR/C_V}\left(\frac{nR}{C_V}\right)V^{-1-nR/C_V}$$

▲

Exercise 4.2

Find the value of $(\partial P/\partial S)_{V,n}$ for 1.000 mol of helium gas at 1.000 atm (101325 Pa) and 298.15 K.

To obtain the other three principal Maxwell relations, we write the differentials dH, dA, and dG:

$$dH = dU + P\,dV + V\,dP = T\,dS - P\,dV + P\,dV + V\,dP$$
$$= T\,dS + V\,dP \quad (4.2\text{-}11)$$

Therefore, as in Equations (4.2-5) and (4.2-6)

$$\left(\frac{\partial H}{\partial S}\right)_{P,n} = T \quad (4.2\text{-}12)$$

and

$$\left(\frac{\partial H}{\partial P}\right)_{S,n} = V \quad (4.2\text{-}13)$$

Note that T is now equal to two different partial derivatives, $(\partial U/\partial S)_{V,n}$ and $(\partial H/\partial S)_{P,n}$. Exploitation of such equalities is often useful.

By using the Euler reciprocity relation, we obtain a second Maxwell relation from Equations (4.2-12) and (4.2-13):

$$\left(\frac{\partial T}{\partial P}\right)_{S,n} = \left(\frac{\partial V}{\partial S}\right)_{P,n} \quad \text{(a Maxwell relation)} \qquad \textbf{(4.2-14)}$$

Exercise 4.3   Using Equation (4.2-14), find an expression for $(\partial V/\partial S)_{P,n}$ for an ideal gas with constant heat capacity. Evaluate the quantity for 1.000 mol of helium (assumed ideal) at 1.000 atm and 298.15 K. Take the molar heat capacity to equal $3R/2$.

The third Maxwell relation comes from the differential of the Helmholtz energy. Equations (4.1-8) and (4.2-3) give

$$dA = -S\,dT - P\,dV \quad \text{(closed system)} \qquad \textbf{(4.2-15)}$$

so that

$$\left(\frac{\partial A}{\partial T}\right)_{V,n} = -S \qquad \textbf{(4.2-16)}$$

$$\left(\frac{\partial A}{\partial V}\right)_{T,n} = -P \qquad \textbf{(4.2-17)}$$

and

$$\left(\frac{\partial S}{\partial V}\right)_{T,n} = \left(\frac{\partial P}{\partial T}\right)_{V,n} \quad \text{(a Maxwell relation)} \qquad \textbf{(4.2-18)}$$

Equations (4.1-13) and (4.2-3) give

$$dG = -S\,dT + V\,dP \qquad \textbf{(4.2-19)}$$

so that

$$\left(\frac{\partial G}{\partial T}\right)_{P,n} = -S \qquad \textbf{(4.2-20)}$$

$$\left(\frac{\partial G}{\partial P}\right)_{T,n} = V \qquad \textbf{(4.2-21)}$$

and

$$\left(\frac{\partial S}{\partial P}\right)_{T,n} = -\left(\frac{\partial V}{\partial T}\right)_{P,n} \quad \text{(a Maxwell relation)} \qquad \textbf{(4.2-22)}$$

A physical chemistry student should either memorize the Maxwell relations or be able to derive them quickly from the expressions for dU, dH, dA, and dG.

Equations (4.2-10), (4.2-14), (4.2-18), and (4.2-22) are the four principal Maxwell relations.

We can now derive Equation (3.3-4) in a different way. The entropy change for an isothermal volume change of an ideal gas is equal to

$$\Delta S = \int_{V_1}^{V_2} \left(\frac{\partial S}{\partial V}\right)_{T,n} dV = \int_{V_1}^{V_2} \left(\frac{\partial P}{\partial T}\right)_{V,n} dV$$

$$= \int_{V_1}^{V_2} \left(\frac{nR}{V}\right) dV = nR \ln\left(\frac{V_2}{V_1}\right) \tag{4.2-23}$$

▼   **EXAMPLE 4.2**

a. Find an expression for $(\partial S/\partial V)_{T,n}$ for a gas obeying the truncated virial equation of state

$$\frac{P\bar{V}}{RT} = 1 + \frac{B_2}{\bar{V}}$$

where B_2 is a function of T and $\bar{V}$ is the molar volume.

b. Evaluate the expression for 1.000 mol of argon in 25.00 L at 298.15 K. The values of B_2 and dB_2/dT may be found in Example 2.7.

c. Find an expression for ΔS for an isothermal volume change for n moles of a gas obeying the truncated virial equation of state in part a.

Solution

a.

$$\left(\frac{\partial S}{\partial V}\right)_{T,n} = \left(\frac{\partial P}{\partial T}\right)_{V,n} = \left[\frac{\partial(RT/\bar{V} + RTB_2/\bar{V}^2)}{\partial T}\right]_{V,n}$$

$$= \frac{R}{\bar{V}} + \frac{R}{\bar{V}^2}\left[B_2 + T\frac{dB_2}{dT}\right]$$

b.

$$\left(\frac{\partial S}{\partial V}\right)_{T,n} = \frac{8.3145 \text{ J K}^{-1} \text{ mol}^{-1}}{0.025 \text{ m}^3 \text{ mol}^{-1}} + \frac{8.3145 \text{ J}^{-1} \text{ mol}^{-1}}{(0.025 \text{ m}^3 \text{ mol}^{-1})^2}$$

$$\times [-15.8 \times 10^{-6} \text{ m}^3 \text{ mol}^{-1} + (298.15 \text{ K})(0.25 \times 10^{-6} \text{ m}^3 \text{ mol}^{-1} \text{ K}^{-1})$$

$$= 332.6 \text{ N m}^{-2} \text{ K}^{-1} + 0.78 \text{ N m}^{-2} \text{ K}^{-1}$$

$$= 333.4 \text{ N m}^{-2} \text{ K}^{-1} = 333.4 \text{ J K}^{-1} \text{ m}^{-3}$$

In Example 4.2 we see how a hard-to-measure quantity, $(\partial S/\partial V)_{T,n}$, is expressed in terms of more easily measured quantities. The solution of the example is straightforward, with substitution into the appropriate formula. Consistent units must be used.

The correction for gas nonideality, 0.78 J K^{-1} m^{-3}, is nearly negligible.

c. For an isothermal process in a closed system,

$$\Delta S = \int_c dS = \int_{V_1}^{V_2}\left(\frac{\partial S}{\partial V}\right)_{T,n} dV = \int_{V_1}^{V_2}\left[\frac{R}{\bar{V}} + \frac{R}{\bar{V}^2}\left(B_2 + T\frac{dB_2}{dT}\right)\right] dV$$

$$= n\int_{V_1}^{V_2}\frac{R}{\bar{V}} d\bar{V} + n\left(B_2 + T\frac{dB_2}{dT}\right)\int_{V_1}^{V_2}\frac{R}{\bar{V}^2} d\bar{V}$$

$$= nR\ln\left(\frac{\bar{V}_2}{\bar{V}_1}\right) - n\left(B_2 + T\frac{dB_2}{dT}\right)\left(\frac{1}{\bar{V}_2} - \frac{1}{\bar{V}_1}\right)$$

Exercise 4.4

a. Find an expression for $(\partial S/\partial P)_{T,n}$ for an ideal gas.

b. Evaluate the expression of part a for 2.000 mol of neon (assumed ideal) at 1.000 atm and 298.15 K.

c. Derive the expression for the entropy change for an isothermal pressure change of an ideal gas and calculate ΔS for the expansion of 2.000 mol of neon (assumed ideal) from 10.00 L to 40.00 L at 325 K.

4.3 Gibbs Energy Calculations

For a closed simple system at constant temperature, Equation (4.2-19) is

$$dG = V \, dP \quad \text{(simple system, } n \text{ and } T \text{ constant)} \tag{4.3-1}$$

where constant n means that the amounts of all substances present are constant.

Integration of this formula at constant n and T gives

$$G(T, P_2, n) - G(T, P_1, n) = \int_{P_1}^{P_2} V \, dP \tag{4.3-2}$$

The Gibbs Energy of an Ideal Gas

For an ideal gas, Equation (4.3-2) becomes

$$G(T, P_2, n) = G(T, P_1, n) + nRT \int_{P_1}^{P_2} \frac{1}{P} \, dP$$

$$= G(T, P_1, n) + nRT \ln\left(\frac{P_2}{P_1}\right) \quad \text{(ideal gas)} \tag{4.3-3}$$

In terms of the molar Gibbs energy, $\bar{G} = G/n$,

$$\boxed{\bar{G}(T, P_2) = \bar{G}(T, P_1) + RT \ln\left(\frac{P_2}{P_1}\right) \quad \text{(ideal gas)}} \tag{4.3-4}$$

The standard state for the Gibbs energy of an ideal gas is the same as for the entropy: a fixed pressure $P°$ (exactly 1 bar = 100000. Pa) and whatever temperature one is interested in. In the past, $P°$ was chosen as 1 atmosphere. This makes no difference to the formulas we write and makes only a small difference in numerical values. For highly accurate work, one must determine whether the 1 atm standard state or the 1 bar standard state has been used in a table of numerical values.

If state 1 is chosen to be the standard state and the subscript is dropped on P_2, Equation (4.3-4) is

$$\boxed{\bar{G}(T, P) = \bar{G}°(T) + RT \ln\left(\frac{P}{P°}\right) \quad \text{(ideal gas)}} \tag{4.3-5}$$

where $\bar{G}°(T)$ is the molar Gibbs energy of the gas in the standard state at temperature T.

▼ **EXAMPLE 4.3**

Obtain a formula to change from the 1-atm standard state to the 1-bar standard state.

Solution

Call the 1-atm standard-state pressure $P^{\circ\,\text{atm}}$, and use a similar symbol for the 1-atm standard-state molar Gibbs energies. We write

$$\bar{G}(T, P) = \bar{G}^{\circ}(T) + RT \ln(P/P^{\circ})$$
$$= \bar{G}^{\circ}(T) + RT \ln(P^{\circ\,\text{atm}}/P^{\circ}) + RT \ln(P/P^{\circ\,\text{atm}})$$

Example 4.3 is solved by manipulating the beginning equation into a form like the desired equation, followed by identifying the quantity $\bar{G}^{\circ\,\text{atm}}$ by comparison.

This equation has the correct form if

$$\bar{G}^{\circ\,\text{atm}} = \bar{G}^{\circ}(T) + RT \ln(P^{\circ\,\text{atm}}/P^{\circ}) = \bar{G}^{\circ}(T) + RT \ln(1.01325)$$
$$= \bar{G}^{\circ}(T) + (0.01316)RT$$

▲

Exercise 4.5

Find the difference between the two standard-state molar Gibbs energies of an ideal gas at 298.15 K.

The Gibbs Energy of a Real Gas; Fugacity

It is usually an adequate approximation to treat real gases as ideal gases. However, when corrections for nonideality are necessary, we write a new equation in the same form as Equation (4.3-5):

$$\boxed{\bar{G}(T, P) = \bar{G}^{\circ}(T) + RT \ln\left(\frac{f}{P^{\circ}}\right) \quad \text{(definition of } f)} \qquad \textbf{(4.3-6)}$$

The fugacity plays the same role for a nonideal gas as does the pressure for an ideal gas in determining its molar Gibbs energy.

The quantity f is called the **fugacity** of the gas and is defined by Equation (4.3-6). The fugacity has the dimensions of pressure, and the fugacity of an ideal gas is equal to its pressure.

The quantity $\bar{G}^{\circ}(T)$ is the molar Gibbs energy of the gas in its standard state, just as in Equation (4.3-5) for the ideal gas. However, the standard state of a real gas is not the real gas at pressure P°. It is a hypothetical state, defined to be the corresponding ideal gas at pressure P°. That is, the effects of gas nonideality are eliminated from the standard-state quantity.

The hypothetical ideal gas state is not the only possible choice for the standard state. The actual gas at pressure P° could have been chosen.

Since any gas approaches ideal gas behavior as its pressure approaches zero, we can obtain an expression for the difference between the molar Gibbs energy of the real gas at some given pressure P' and the standard-state (ideal gas) molar Gibbs energy:

$$\bar{G}_{\text{real}}(T, P') - \bar{G}^{\circ}(T)$$
$$= \bar{G}_{\text{real}}(T, P') - \lim_{P'' \to 0} [\bar{G}_{\text{real}}(T, P'') - \bar{G}_{\text{id}}(T, P'')] - \bar{G}^{\circ}(T) \qquad \textbf{(4.3-7)}$$

The real gas and the corresponding ideal gas are the same in the limit of low pressure. Therefore, the two terms inside the limit add to zero and can be included without changing the equation. From Equation (4.3-1), the first two terms on the right-hand side of Equation (4.3-7) can be combined:

$$(\text{first two terms}) = \int_0^{P'} \bar{V}_{\text{real}}\, dP$$

From Equation (4.3-3), the last two terms on the right-hand side of Equation (4.3-7) can be combined:

$$(\text{last two terms}) = \int_{P^\circ}^{0} \frac{RT}{P}\, dP$$

Equation (4.3-7) is therefore equivalent to integrating from the standard-state pressure P° down to zero pressure with the ideal gas and then integrating back up to pressure P', the pressure of interest, with the real gas. Because the real gas and the corresponding ideal gas are equivalent at zero pressure, this procedure is valid.

The integral to which the last two terms are equal can be broken into two parts as follows (we have also exchanged the limits and changed the sign):

$$(\text{last two terms}) = -\int_0^{P'} \frac{RT}{P}\, dP - \int_{P'}^{P^\circ} \frac{RT}{P}\, dP$$

$$= -\int_0^{P'} \frac{RT}{P}\, dP - RT \ln\left(\frac{P^\circ}{P'}\right)$$

The left-hand side of Equation (4.3-7) is equal to $RT \ln(f/P^\circ)$, so if f' denotes the fugacity at pressure P', we can combine the two integrals to write

$$RT \ln\left(\frac{f'}{P^\circ}\right) + RT \ln\left(\frac{P^\circ}{P'}\right) = \int_0^{P'} \left(\bar{V}_{\text{real}} - \frac{RT}{P}\right) dP \qquad \textbf{(4.3-8)}$$

which is the same as

$$\boxed{RT \ln\left(\frac{f'}{P'}\right) = \int_0^{P'} \left(\bar{V}_{\text{real}} - \frac{RT}{P}\right) dP} \qquad \textbf{(4.3-9)}$$

When the integral is written in this way, the integrand is a function that is generally small.

▼   **EXAMPLE 4.4**

Find an expression for the fugacity of a gas that obeys the equation of state

$$P\bar{V} = RT + A_2 P$$

where A_2 is a function of temperature. It was shown in Exercise 1.6 that A_2 is equal to B_2, the second virial coefficient.

Solution

$$RT \ln\left(\frac{f'}{P'}\right) = \int_0^{P'} \left(\frac{RT}{P} + A_2 - \frac{RT}{P}\right) dP = A_2 P'$$

or

$$f' = P' e^{A_2 P'/RT}$$

▲

| Exercise 4.6 | For argon at 273.15 K, $B_2 = -21.5$ cm^3 mol^{-1}. Find the value of the fugacity of argon gas at 5.000 atm and 273.15 K. |

The Gibbs Energy of Solids and Liquids

We now consider the pressure dependence of the Gibbs energy in a condensed phase (solid or liquid). As with the gas, we begin with Equation (4.3-1). The isothermal compressibility of a typical solid or liquid is near 10^{-9} Pa^{-1}, so a change in pressure of 10 atm (10^6 Pa) produces a change in volume of approximately a tenth of a percent. Thus, typical solids and liquids are nearly incompressible under ordinary pressure changes.

We assume the volume to be approximately constant in the integrand of Equation (4.3-2), giving

$$G(T, P_2, n) - G(T, P_1, n) = V(P_2 - P_1) \qquad (4.3\text{-}10)$$

The standard state of a substance in a condensed phase is chosen to be the actual pure substance at pressure P°. We can write for the molar Gibbs energy of a substance in a condensed phase at pressure P

$$\bar{G}(T, P) = \bar{G}^\circ(T) + \bar{V}(P - P^\circ) \qquad (4.3\text{-}11)$$

▼ **EXAMPLE 4.5**

Calculate $\bar{G} - \bar{G}^\circ$ for liquid water at 298.15 K and 10.00 bar.

Solution

$$\bar{G} - \bar{G}^\circ = (18.0 \times 10^{-6} \text{ m}^3 \text{ mol}^{-1})(9.00 \text{ bar})\left(\frac{100000 \text{ N m}^{-2}}{1 \text{ bar}}\right)$$

$$= 16.2 \text{ J mol}^{-1}$$

The difference can be neglected for most purposes.

▲

| Exercise 4.7 | Find $\bar{G} - \bar{G}^\circ$ for solid iron at 293.15 K and 1.100 bar. |

The Temperature Dependence of the Gibbs Energy

From Equation (4.2-19), if the pressure is constant and the system is closed,

$$dG = -S \, dT \quad (P \text{ and } n \text{ constant}) \qquad (4.3\text{-}12)$$

Integration of this equation at constant pressure gives

$$G(T_2, P, n) - G(T_1, P, n) = -\int_{T_1}^{T_2} S(T, P) \, dT \qquad (4.3\text{-}13)$$

The assignment of zero entropy at zero temperature is a convention, not a fundamental fact.

Equation (4.3-13) is not directly usable unless the actual value of the entropy is known. However, the calculation of "absolute" entropies is a convention, based on an assignment of zero entropy for elements at 0 K. A useful equation that is analogous to Equation (4.3-13) can be written for

ΔG of an isothermal process carried out once at temperature T_1 and once at temperature T_2.

Let $\Delta G(T)$ be the Gibbs energy change for an isothermal constant-pressure process at temperature T, and let $\Delta S(T)$ be the entropy change for the same process. We can write Equation (4.3-13) once for the initial state and once for the final state. The difference of these equations gives

$$\Delta G(T_2, P, n) - \Delta G(T_1, P, n) = -\int_{T_1}^{T_2} \Delta S(T, P)\, dT \qquad \textbf{(4.3-14)}$$

Although we are integrating the variable T, $\Delta S(T, P)$ and $\Delta G(T, P)$ are for isothermal processes.

We can integrate over values of T for isothermal processes taking place at temperature T. The integration corresponds to changing the temperature of the entire process during the integration process.

If ΔS is nearly independent of temperature between T_1 and T_2, Equation (4.3-14) becomes

$$\Delta G(T_2, P) - \Delta G(T_1, P) \approx -(T_2 - T_1)\,\Delta S \qquad \textbf{(4.3-15)}$$

This equation should be a usable approximation if the difference between T_2 and T_1 is not very large. An alternative equation is known as the Gibbs-Helmholtz equation:

$$\frac{\Delta G(T_2, P)}{T_2} - \frac{\Delta G(T_1, P)}{T_1} = -\int_{T_1}^{T_2} \frac{\Delta H(T, P)\, dT}{T^2} \qquad \textbf{(4.3-16)}$$

Exercise 4.8

a. Derive Equation (4.3-16).

b. At 373.15 K and 1.000 atm, the Gibbs energy change of vaporization of water is zero and the entropy change is 109 J K^{-1} mol^{-1}. Find the Gibbs energy change of vaporization of water at 383.15 K and 1.000 atm. What does the sign of your answer mean?

4.4   The Description of Multicomponent and Open Systems

The equilibrium macroscopic state of a simple closed system is specified by values of only two variables (in addition to the amounts of the substances present, which are fixed). At least one of these two variables must be an extensive variable. In Chapter 1, we asserted as an experimental fact that for a one-phase simple open system the number of variables required to specify the state is $c + 2$, where c stands for the number of independent substances, called **components**.

The number of components is not necessarily equal to the number of distinct chemical species present. In a one-phase system the number of components is equal to the number of substances whose amounts can be varied separately under the given conditions. It is also equal to the minimum number of substances from which the system can be prepared under the given conditions. If chemical reactions can come to equilibrium in the system, the substances present are not all independent because some can be produced from others by the chemical reactions. For example, if a gaseous

The conditions determine the number of components. For example, if a chemical reaction cannot equilibrate, inserting a catalyst that allows the chemical reaction to equilibrate changes the number of components.

system contains NO_2, it will also contain N_2O_4 at equilibrium, and at equilibrium the amount of N_2O_4 is not independent of the amount of NO_2. Generally speaking, each chemical reaction that can equilibrate reduces the number of components by one. We will return to the counting of components in a multiphase system in Chapter 6.

The Chemical Potential and Partial Molar Quantities

For a one-phase simple system containing c components, $c + 2$ independent variables are required to specify the state. We can make the choice

$$G = G(T, P, n_1, n_2, \ldots, n_c) \tag{4.4-1}$$

where n_i is the amount of substance i (measured in moles).

Equation (4.4-1) corresponds to the differential relation

$$dG = \left(\frac{\partial G}{\partial T}\right)_{P,n} dT + \left(\frac{\partial G}{\partial P}\right)_{T,n} dP + \sum_{i=1}^{c} \left(\frac{\partial G}{\partial n_i}\right)_{T,P,n'} dn_i \tag{4.4-2}$$

where the subscript n means that the amounts of all the components are fixed and the subscript n' means that the amounts of all components except component i are fixed.

The first two partial derivatives in Equation (4.4-2) are no different from the partial derivatives in Equations (4.2-20) and (4.2-21). In those equations, the amounts of all substances present were held fixed because the system was closed. In Equation (4.4-2), the amounts of all substances are held fixed because that is how partial derivatives are defined. Therefore, Equation (4.4-2) is the same as

$$dG = -S\, dT + V\, dP + \sum_{i=1}^{c} \mu_i\, dn_i \tag{4.4-3}$$

where μ_i is defined by

$$\mu_i = \left(\frac{\partial G}{\partial n_i}\right)_{T,P,n'} \tag{4.4-4}$$

The quantity μ_i is called the **chemical potential** of component i. Equation (4.4-3) is called the **Gibbs equation**, or sometimes the **fundamental relation of chemical thermodynamics**. It is the basis of the thermodynamic description of multicomponent systems.

Other independent variables can be used to specify the state of the system. For example,

$$dG = \left(\frac{\partial G}{\partial T}\right)_{V,n} dT + \left(\frac{\partial G}{\partial V}\right)_{T,n} dV + \sum_{i=1}^{c} \left(\frac{\partial G}{\partial n_i}\right)_{T,V,n'} dn_i \tag{4.4-5}$$

However, the partial derivatives $(\partial G/\partial T)_{V,n}$ and $(\partial G/\partial V)_{T,n}$ are not equal to any simple thermodynamic variables, as are the partial derivatives in Equation (4.4-2). We therefore say that the **natural independent variables** for the Gibbs energy are P, T, $n_1, n_2, \ldots, n_c$.

Exercise 4.9   Use an analogue of Equation B-7 of Appendix B to write a relation between $(\partial G/\partial n_i)_{T,V,n'}$ and μ_i.

The internal energy, the enthalpy, and the Helmholtz energy have their own sets of natural independent variables. For example, if we write

$$H = H(S, P, n_1, n_2, \ldots, n_c) \tag{4.4-6}$$

then

$$dH = \left(\frac{\partial H}{\partial S}\right)_{P,n} dS + \left(\frac{\partial H}{\partial P}\right)_{S,n} dP + \sum_{i=1}^{c} \left(\frac{\partial H}{\partial n_i}\right)_{S,P,n'} dn_i \tag{4.4-7}$$

Comparison with Equation (4.2-12) shows that

$$dH = T\,dS + V\,dP + \sum_{i=1}^{c} \left(\frac{\partial H}{\partial n_i}\right)_{S,P,n'} dn_i \tag{4.4-8}$$

The definitions of H, A, and G are still as they were, so that from Equation (4.4-3), Equation (4.4-8), and the relation $G = H - TS$,

$$dH = dG + T\,dS + S\,dT$$

$$\boxed{dH = T\,dS + V\,dP + \sum_{i=1}^{c} \mu_i\,dn_i} \tag{4.4-9}$$

Comparison of Equations (4.4-8) and (4.4-9) shows that

$$\mu_i = \left(\frac{\partial H}{\partial n_i}\right)_{S,P,n'} \tag{4.4-10}$$

Similarly,

$$\boxed{dU = T\,dS - P\,dV + \sum_{i=1}^{c} \mu_i\,dn_i} \tag{4.4-11}$$

and

$$\boxed{dA = -S\,dT - P\,dV + \sum_{i=1}^{c} \mu_i\,dn_i} \tag{4.4-12}$$

so that

$$\mu_i = \left(\frac{\partial U}{\partial n_i}\right)_{S,V,n'} \tag{4.4-13}$$

and

$$\mu_i = \left(\frac{\partial A}{\partial n_i}\right)_{T,V,n'} \tag{4.4-14}$$

Exercise 4.10   Carry out the mathematical steps to derive Equations (4.4-11) and (4.4-12).

It is not a problem that the chemical potential is equal to four different partial derivatives. Note that different variables are held fixed in each of the partial derivatives.

The chemical potential is thus equal to four different partial derivatives. Equation (4.4-4) is the most useful equality.

The chemical potential is an example of a **partial molar quantity** and is called the **partial molar Gibbs energy**. A partial molar quantity is a partial derivative of an extensive quantity with respect to the amount of one substance, keeping pressure, temperature, and the amounts of all other substances fixed. If the letter Y stands for any extensive quantity (U, H, A, G, S, V, etc.), the partial molar quantity for substance i is denoted by $\bar{Y}_i$ and defined by

$$\bar{Y}_i = \left(\frac{\partial Y}{\partial n_i}\right)_{T,P,n'} \quad \text{(definition)} \qquad \textbf{(4.4-15)}$$

All partial molar quantities are intensive quantities.

The partial derivatives in Equations (4.4-10), (4.4-13), and (4.4-14) are not partial molar quantities because P and T are not both held fixed in the differentiations.

▼   **EXAMPLE 4.6**

Find a relationship between the chemical potential and the partial molar enthalpy.

Solution

We begin with the relationship between G and H:

$$G = H - TS$$

Differentiation of both sides gives

$$\left(\frac{\partial G}{\partial n_i}\right)_{T,P,n'} = \left(\frac{\partial H}{\partial n_i}\right)_{T,P,n'} - T\left(\frac{\partial S}{\partial n_i}\right)_{T,P,n'}$$

or

$$\mu_i = \bar{H}_i - T\bar{S}_i$$

▲

The Partial Molar Quantities in a One-Component System

The equilibrium thermodynamic state of a simple one-component open system can be specified by T, P, and n, the amount of the single component. This gives the differential relation for a general extensive quantity Y:

$$dY = \left(\frac{\partial Y}{\partial T}\right)_{P,n} dT + \left(\frac{\partial Y}{\partial P}\right)_{T,n} dP + \left(\frac{\partial Y}{\partial n}\right)_{T,P} dn \qquad \textbf{(4.4-16)}$$

The derivative in the last term is the partial molar Y, since no other substances are present.

For a one-component system, the molar Y is defined by

$$\bar{Y} = \frac{Y}{n} \qquad \textbf{(4.4-17)}$$

The quantity $\bar{Y}$ is an intensive quantity. In a one-component system it depends on P and T but cannot depend on n because an intensive quantity cannot depend on an extensive quantity. Therefore,

$$\left(\frac{\partial Y}{\partial n}\right)_{T,P} = \left(\frac{\partial (n\bar{Y})}{\partial n}\right)_{T,P} = \bar{Y} \qquad \textbf{(4.4-18)}$$

Thus, in a one-component system, the molar quantity and the partial molar quantity are identical.

Partial Molar Quantities of an Ideal Gas

Just as with a pure substance, the partial molar volume of a one-component ideal gas is equal to the molar volume:

$$\bar{V} = \frac{V}{n} = \frac{RT}{P} \quad \text{(ideal gas)} \qquad \textbf{(4.4-19)}$$

The chemical potential is equal to the molar Gibbs energy only in a one-component system and in an ideal gas mixture.

The chemical potential is equal to the molar Gibbs energy. From Equation (4.3-5),

$$\mu = \bar{G} = \bar{G}^\circ(T) + RT \ln\left(\frac{P}{P^\circ}\right) \quad \text{(ideal gas)} \qquad \textbf{(4.4-20)}$$

Equation (4.4-20) is the same as

$$\mu = \mu^\circ + RT \ln\left(\frac{P}{P^\circ}\right) \quad \text{(ideal gas)} \qquad \textbf{(4.4-21)}$$

The standard state for the chemical potential of an ideal gas is the same as the standard state for the other thermodynamic functions: the ideal gas at pressure P° (1 bar).

The partial molar entropy is obtained by use of Equation (4.2-20):

$$\bar{S} = -(\partial \bar{G}/\partial T)_P = -(\partial \bar{G}^\circ/\partial T)_P + R \ln(P/P^\circ)$$

$$\bar{S} = \bar{S}^\circ - R \ln\left(\frac{P}{P^\circ}\right) \quad \text{(ideal gas)} \qquad \textbf{(4.4-22)}$$

The partial molar enthalpy of a one-component ideal gas is obtained from Equation (4.1-14):

$$\bar{H} = \bar{G} + T\bar{S}$$

$$= \bar{G}^\circ + RT \ln\left(\frac{P}{P^\circ}\right) + T\left[\bar{S}^\circ - R \ln\left(\frac{P}{P^\circ}\right)\right]$$

$$\bar{H} = \bar{G}^\circ + T\bar{S}^\circ = \bar{H}^\circ \quad \text{(ideal gas)} \qquad \textbf{(4.4-23)}$$

Remember that equations that are valid for an ideal gas do not necessarily apply to other systems.

The partial molar enthalpy of an ideal gas does not depend on pressure. We already knew this from Equation (2.5-18).

Exercise 4.11 Find the expression for the partial molar Helmholtz energy of a one-component ideal gas as a function of pressure.

According to Dalton's law of partial pressures, and also according to our ideal gas model of Chapter 16, each gas in a mixture of ideal gases behaves as if it were alone in the container. The chemical potential of a component of a mixture of ideal gases is independent of the presence of the other gases, so the chemical potential is equal to the molar Gibbs energy:

$$\mu_i = \mu_i^\circ + RT \ln\left(\frac{P_i}{P^\circ}\right) \quad \text{(ideal gas mixture)} \tag{4.4-24}$$

where μ_i° is the chemical potential of substance i in the standard state and P_i is the partial pressure of substance i.

Exercise 4.12 Calculate $\mu_i - \mu_i^\circ$ for argon gas in air at 298.15 K, assuming that the mole fraction of argon is 0.00934 and that the total pressure is 1.000 atm.

In a mixture of gases that cannot be assumed to be ideal, the situation is more complicated. We define f_i, the fugacity of component i, by the relation

$$\mu_i = \mu_i^\circ + RT \ln\left(\frac{f_i}{P^\circ}\right) \quad \text{(definition of } f_i) \tag{4.4-25}$$

where μ_i° is the same standard-state chemical potential as for the pure gas: the hypothetical ideal gas state at pressure P° and whatever temperature is desired. We will not discuss the evaluation of f_i in a mixture of nonideal gas because under most conditions gases can be considered to be ideal.

4.5 Additional Useful Thermodynamic Identities

We will generally obtain derivative equations from differential equations by pretending that we can divide by a differential of an independent variable, specifying what variables are held fixed. The process is mathematically indefensible but gives the correct derivative relation.

Our first identity of this section is an expression for $(\partial U / \partial V)_{T,n}$. We convert Equation (4.2-3) to a derivative equation by nonrigorously "dividing" by dV, specifying that T and n are held fixed:

$$\left(\frac{\partial U}{\partial V}\right)_{T,n} = T\left(\frac{\partial S}{\partial V}\right)_{T,n} - P\left(\frac{\partial V}{\partial V}\right)_{T,n}$$

The partial derivative of V with respect to V equals unity. We apply the Maxwell relation of Equation (4.2-18) to the first term to obtain

$$\left(\frac{\partial U}{\partial V}\right)_{T,n} = T\left(\frac{\partial P}{\partial T}\right)_{V,n} - P \qquad (4.5\text{-}1)$$

Equation (4.5-1) is called the **thermodynamic equation of state**. It can be used to show that $(\partial U/\partial V)_{T,n} = 0$ for an ideal gas, making it unnecessary to include this property as a separate part of the definition of an ideal gas. It is also possible to derive the ideal gas equation of state using Equation (4.5-1) and the assumption that $(\partial U/\partial V)_{T,n} = 0$.

Exercise 4.13

a. Show that $(\partial U/\partial V)_{T,n} = 0$ for an ideal gas, using only the equation of state, $PV = nRT$, and Equation (4.5-1).

b. For a gas obeying the truncated virial equation of state,

$$\frac{P\bar{V}}{RT} = 1 + \frac{B_2}{\bar{V}}$$

show that

$$\left(\frac{\partial U}{\partial V}\right)_{T,n} = RT^2 \frac{dB_2/dT}{\bar{V}^2}$$

Find the value of this derivative for 1.000 mol of argon at 1.000 atm and 298.15 K, using data in Example 2.7.

An equation for $(\partial H/\partial P)_{T,n}$ that is analogous to Equation (4.5-1) can be derived in a similar way. We convert Equation (4.2-11) to a derivative equation:

$$\left(\frac{\partial H}{\partial P}\right)_{T,n} = T\left(\frac{\partial S}{\partial P}\right)_{T,n} + V\left(\frac{\partial P}{\partial P}\right)_{T,n}$$

Equation (4.5-2) shows how a Maxwell relation can be used to replace a hard-to-measure quantity in a thermodynamic equation with a more easily measured quantity.

Using the Maxwell relation of Equation (4.2-22), we obtain

$$\left(\frac{\partial H}{\partial P}\right)_{T,n} = -T\left(\frac{\partial V}{\partial T}\right)_{P,n} + V \qquad (4.5\text{-}2)$$

Exercise 4.14

a. Show that for an ideal gas $(\partial H/\partial P)_{T,n} = 0$, using only the equation of state, $PV = nRT$, and Equation (4.5-2).

b. Find an expression for $(\partial H/\partial P)_{T,n}$ for a gas obeying the truncated virial equation of state:

$$P\bar{V} = RT + A_2 P$$

where it has been shown that $A_2 = B_2$.

Evaluate this derivative for 1.000 mol of argon at 1.000 atm and 298.15 K.

We can substitute Equation (4.5-1) into Equation (2.5-11) to obtain a relation between C_P and C_V.

$$C_P = C_V + \left[T \left(\frac{\partial P}{\partial T} \right)_{V,n} + P - P \right] \left(\frac{\partial V}{\partial T} \right)_{P,n} = C_V + T \left(\frac{\partial P}{\partial T} \right)_{V,n} \left(\frac{\partial V}{\partial T} \right)_{P,n}$$

We apply the cycle rule in the form

$$\left(\frac{\partial P}{\partial T} \right)_{V,n} \left(\frac{\partial T}{\partial V} \right)_{P,n} \left(\frac{\partial V}{\partial P} \right)_{T,n} = -1$$

to obtain

$$C_P = C_V - T \left(\frac{\partial P}{\partial V} \right)_{T,n} \left[\left(\frac{\partial V}{\partial T} \right)_{P,n} \right]^2 \tag{4.5-3}$$

Using the definition of the isothermal compressibility, Equation (1.3-9), and the coefficient of thermal expansion, Equation (1.3-10):

$$\boxed{C_P = C_V + \frac{TV\alpha^2}{\kappa_T}} \tag{4.5-4}$$

Since the coefficient of thermal expansion (which is occasionally negative) is squared and the compressibility is always positive, C_P is never smaller than C_V.

EXAMPLE 4.7

Example 4.7, Exercise 4.15, and Exercise 4.16 illustrate how small the difference $\bar{C}_P - \bar{C}_V$ is for condensed phases.

Calculate the constant-volume molar heat capacity of liquid water at 0.00°C and 1.000 atm. The constant-pressure heat capacity is equal to 75.983 J K^{-1} mol^{-1}. The coefficient of thermal expansion is equal to -68.14×10^{-6} K^{-1} (this is one of the few cases in which this quantity is negative). The compressibility is equal to 50.98×10^{-6} bar^{-1}. The molar volume is equal to 18.012 cm^3 mol^{-1}.

Solution

$$\bar{C}_P - \bar{C}_V = \frac{(273.15 \text{ K})(18.012 \times 10^{-6} \text{ m}^3)(-68.14 \times 10^{-6} \text{ K}^{-1})^2}{(50.98 \times 10^{-6} \text{ bar}^{-1})(1 \text{ bar}/100000 \text{ N m}^{-2})}$$

$$= 0.04481 \text{ J K}^{-1} \text{ mol}^{-1}$$

$$\bar{C}_V = 75.983 \text{ J K}^{-1} \text{ mol}^{-1} - 0.04481 \text{ J K}^{-1} \text{ mol}^{-1}$$

$$= 75.938 \text{ J K}^{-1} \text{ mol}^{-1}$$

Exercise 4.15

a. Find the value of $\bar{C}_V$ for liquid water at 25.00°C and 1.000 atm. The coefficient of thermal expansion is 2.5721×10^{-4} K^{-1}, the molar volume is 18.0687 cm^3 mol^{-1}, and the compressibility is 45.24×10^{-6} bar^{-1}. $\bar{C}_P$ is equal to 75.297 J K^{-1} mol^{-1}.

b. At 3.98°C, liquid water has a maximum density and the coefficient of thermal expansion vanishes. What is the difference between $\bar{C}_P$ and $\bar{C}_V$ at this temperature?

A number of organic liquids have fairly large coefficients of thermal expansion. In these cases, the difference between $\bar{C}_V$ and $\bar{C}_P$ can be fairly large (but still much smaller than with a gas). For almost all solids the difference between $\bar{C}_P$ and $\bar{C}_V$ is small.

Exercise 4.16 The constant-pressure specific heat capacity of metallic iron at 298.15 K and 1.000 atm is equal to 0.4498 J K^{-1} g^{-1}. The coefficient of thermal expansion is 3.55×10^{-5} K^{-1}, the density is 7.86 g cm^{-3}, and the isothermal compressibility is 6.06×10^{-7} atm^{-1}. Find the constant-volume specific heat capacity at 298.15 K.

Equations (2.5-8) and (2.4-8) give the relations

$$C_P = \left(\frac{\partial H}{\partial T}\right)_{P,n}, \qquad C_V = \left(\frac{\partial U}{\partial T}\right)_{V,n}$$

We can now obtain two additional relations from the expressions for dH and dU. We take Equation (4.2-11) for a closed system and convert it to a derivative relation, specifying that P is fixed:

$$\left(\frac{\partial H}{\partial T}\right)_{P,n} = T\left(\frac{\partial S}{\partial T}\right)_{P,n} + V\left(\frac{\partial P}{\partial T}\right)_{P,n}$$

The derivative of P with respect to anything at constant P is equal to zero, so that

$$C_P = \left(\frac{\partial H}{\partial T}\right)_{P,n} = T\left(\frac{\partial S}{\partial T}\right)_{P,n} \qquad (4.5\text{-}5)$$

Similarly,

$$C_V = \left(\frac{\partial U}{\partial T}\right)_{V,n} = T\left(\frac{\partial S}{\partial T}\right)_{V,n} \qquad (4.5\text{-}6)$$

Exercise 4.17 Use Equations (4.5-5) and (4.5-6) and the cycle rule to show that

$$\frac{C_P}{C_V} = \frac{\kappa_T}{\kappa_S} \qquad (4.5\text{-}7)$$

where κ_S is the adiabatic compressibility,

A reversible adiabatic process is a process at constant entropy.

$$\kappa_S = -\frac{1}{V}\left(\frac{\partial V}{\partial P}\right)_{S,n} \qquad (4.5\text{-}8)$$

We now obtain some equations similar to the Maxwell relations that can be used for multicomponent open systems. We begin with the Gibbs equation, Equation (4.4-3):

$$dG = -S\,dT + V\,dP + \sum_{i=1}^{c} \mu_i\,dn_i \qquad (4.5\text{-}9)$$

Using the Euler reciprocity relation, Equation (B-13) of Appendix B, we can write

$$-\left(\frac{\partial S}{\partial n_i}\right)_{T,P,n'} = \left(\frac{\partial \mu_i}{\partial T}\right)_{P,n}$$

which is the same as

$$-\bar{S}_i = \left(\frac{\partial \mu_i}{\partial T}\right)_{P,n} \qquad \textbf{(4.5-10)}$$

A second use of the Euler reciprocity relation gives

$$\left(\frac{\partial V}{\partial n_i}\right)_{T,P,n'} = \left(\frac{\partial \mu_i}{\partial P}\right)_{T,n}$$

which is the same as

$$\bar{V}_i = \left(\frac{\partial \mu_i}{\partial P}\right)_{T,n} \qquad \textbf{(4.5-11)}$$

4.6   Euler's Theorem and the Gibbs-Duhem Relation

Leonard Euler, 1707–1783, great Swiss mathematician, contributed to all branches of mathematics as well as to astronomy and physics.

Euler's theorem is a mathematical theorem that applies to homogeneous functions. A proof of this theorem may be found in Appendix E. A function of several independent variables, say $f(x_1, x_2, x_3, \ldots, x_c)$, is said to be **homogeneous of degree k** if

$$f(x_1, x_2, \ldots, x_c) = x_1^k g(x_2/x_1, x_3/x_1, \ldots, x_c/x_1) \qquad \textbf{(4.6-1)}$$

where g is some function of $c-1$ variables and k is an integer (it can equal zero).

Equation (4.6-1) is the same as

$$f(ax_1, ax_2, x_3, \ldots, ax_c) = a^k f(x_1, x_2, x_3, \ldots, x_c) \qquad \textbf{(4.6-2)}$$

where a is a positive constant. For example, if each independent variable is doubled, the new value of the function is equal to the old value times 2^k.

If T and P are held fixed, any extensive quantity is homogeneous of degree 1 in the amounts of the components, $n_1, n_2, \ldots, n_c$, and any intensive quantity is homogeneous of degree 0 in the amounts of the components. For example, if the amount of every component is doubled at constant T and P, the value of every extensive quantity doubles while the value of every intensive quantity remains unchanged.

If f is a homogeneous function of degree k, Euler's theorem states that

$$kf = \sum_{i=1}^{c} x_i \left(\frac{\partial f}{\partial x_i}\right)_{x'} \qquad \textbf{(4.6-3)}$$

where we use the subscript x' to stand for holding all of the x's constant except for x_i.

Let Y stand for any extensive quantity. Since Y is homogeneous of degree 1 in the n's if T and P are constant, Euler's theorem implies

$$Y = \sum_{i=1}^{c} n_i \bar{Y}_i \qquad (4.6\text{-}4)$$

where $\bar{Y}_i$ is the partial molar quantity for substance i. Of course, both sides of the equation must refer to the same state of the system.

Two examples of Equation (4.6-4) are

$$G = \sum_{i=1}^{c} n_i \mu_i \qquad (4.6\text{-}5)$$

and

$$V = \sum_{i=1}^{c} n_i \bar{V}_i \qquad (4.6\text{-}6)$$

Equation (4.6-4) is remarkable. It gives the value of the extensive quantity as a weighted sum of partial derivatives. An unbiased newcomer to thermodynamics would likely not believe this equation.

Euler's theorem can also be written in terms of the **mean molar quantity** $\bar{Y}$, defined by $\bar{Y} = Y/n$, where n is the total amount of all components.

$$\bar{Y} = \frac{1}{n} \sum_{i=1}^{c} n_i \bar{Y}_i = \sum_{i=1}^{c} x_i \bar{Y}_i \qquad (4.6\text{-}7)$$

where x_i is the mole fraction of component i.

A molar quantity, a partial molar quantity, and a mean molar quantity are all denoted by a capital letter with a bar over it. If it is not clear from the context, be sure to say which quantity you mean. Remember that the mean molar quantity never has a subscript because it does not pertain to a single component. The subscript is also commonly omitted on the molar quantity for a one-component system. The subscript should never be omitted from a partial molar quantity in a multicomponent system. In a one-component system, all three quantities are equal.

▼   **EXAMPLE 4.8**

In a solution of acetone (component 1) and chloroform (component 2) $x_1 = 0.531$ and $\bar{V}_1 = 74.2 \text{ cm}^3 \text{ mol}^{-1}$. If $\bar{V}$, the mean molar volume, is $77.0 \text{ cm}^3 \text{ mol}^{-1}$ at this composition, find $\bar{V}_2$ at this composition.

Solution

From Euler's theorem

$$\bar{V}_2 = \frac{\bar{V} - x_1 \bar{V}_1}{x_2}$$

$$= \frac{77.0 \text{ cm}^3 \text{ mol}^{-1} - (0.531)(74.2 \text{ cm}^3 \text{ mol}^{-1})}{0.469}$$

$$= 80.2 \text{ cm}^3 \text{ mol}^{-1}$$

▲

The Gibbs-Duhem Relation

From Euler's theorem, Equation (4.6-4), we write

$$dY = \sum_{i=1}^{c} n_i \, d\bar{Y}_i + \sum_{i=1}^{c} \bar{Y}_i \, dn_i \qquad \textbf{(4.6-8)}$$

and considering Y to be a function of T, P, and the n's, we write

$$dY = \left(\frac{\partial Y}{\partial T}\right)_{P,n} dT + \left(\frac{\partial Y}{\partial P}\right)_{T,n} dP + \sum_{i=1}^{c} \bar{Y}_i \, dn_i \qquad \textbf{(4.6-9)}$$

Equations (4.6-8) and (4.6-9) both represent an infinitesimal change in Y produced by an infinitesimal change in state, so we can equate the right-hand sides of the two equations:

This equality holds only if the same infinitesimal change is represented by Equations (4.6-8) and (4.6-9).

$$\boxed{\sum_{i=1}^{c} n_i \, d\bar{Y}_i = \left(\frac{\partial Y}{\partial T}\right)_{P,n} dT + \left(\frac{\partial Y}{\partial P}\right)_{T,n} dP} \qquad \textbf{(4.6-10)}$$

where we have canceled equal sums on the two sides of the equation.

Equation (4.6-10) is called the **generalized Gibbs-Duhem relation**. The original **Gibbs-Duhem relation** is a special case that applies to the Gibbs energy at constant T and P:

$$\boxed{\sum_{i=1}^{c} n_i \, d\mu_i = 0 \quad \text{(constant } T \text{ and } P\text{)}} \qquad \textbf{(4.6-11)}$$

This result is as remarkable as Euler's theorem. For example, in a two-component mixture, it specifies how much the chemical potential of one component must decrease if the chemical potential of the other component increases at constant temperature and pressure:

$$d\mu_1 = -\frac{x_2}{x_1} \, d\mu_2 \quad \text{(constant } T \text{ and } P\text{)} \qquad \textbf{(4.6-12)}$$

▼   **EXAMPLE 4.9**   A two-component ideal gas mixture at constant temperature and pressure has the partial pressure of gas number 1 changed by dP_1. Show that the expression for the chemical potential of a component of an ideal gas mixture, Equation (4.4-24), is compatible with Equation (4.6-12).

Solution

$$d\mu_1 = \left(\frac{\partial \mu_1}{\partial P_1}\right)_T dP_1 = \frac{RT}{P_1} \, dP_1$$

$$d\mu_2 = \left(\frac{\partial \mu_2}{\partial P_2}\right)_T dP_2 = \frac{RT}{P_2} \, dP_2$$

Since $P_1 + P_2 = P = \text{constant}$, $dP_2 = -dP_1$.

$$d\mu_2 = -\frac{RT}{P_2} \, dP_1$$

$$-\frac{x_2}{x_1} \, d\mu_2 = \frac{x_2 RT}{x_1 P_2} \, dP_1$$

In Example 4.9, a relatively simple method has been found to go from the expressions for $d\mu_1$ and $d\mu_2$ to the desired relation.

From Dalton's law of partial pressures, $x_2/P_2 = x_1/P_1$, so

$$-\frac{x_2}{x_1} d\mu_2 = \frac{RT}{P_1} dP_1 = d\mu_1 \quad \text{Q.E.D.}$$

The Gibbs-Duhem relation is often written as a derivative relation. It is necessary that the partial derivatives be taken with T and P constant. For a two-component system, we obtain the derivative relation by nonrigorously "dividing" by dx_1 and specifying that T and P are fixed:

$$x_1 \left(\frac{\partial \mu_1}{\partial x_1}\right)_{T,P} + x_2 \left(\frac{\partial \mu_2}{\partial x_1}\right)_{T,P} = 0 \tag{4.6-13}$$

Both derivatives must be with respect to the same mole fraction.

For a system with more than two components, the equation is

$$\sum_{i=1}^{c} x_i \left(\frac{\partial \mu_i}{\partial x_k}\right)_{T,P} = 0 \tag{4.6-14}$$

where the index k stands for any one of the components but must be the same in every term. Equation (4.6-14) is valid for any kind of changes in the mole fractions at constant T and P.

The Experimental Determination of Partial Molar Quantities

The partial molar volume is probably the most easily measured partial molar quantity, and we discuss it as an example. The most direct way to determine the partial molar volume is to measure the volume of the system as a function of the amount of the component of interest, keeping the pressure, temperature, and amounts of other substances fixed. If this volume can be represented by a polynomial or other functional form, the partial molar volume can be obtained by differentiation. If only the data points are available, the partial molar volume can be obtained by numerical means.[1]

▼   **EXAMPLE 4.10**

At constant temperature and pressure, the volume of a solution made from a fixed amount of solvent (component 1) and a variable amount of a solute (component 2) is represented by the polynomial

$$V = b_0 + b_1 n_2 + b_2 n_2^2 + b_3 n_2^3$$

where n_2 is the amount of component 2 in moles and the b's are constants at constant temperature and pressure. Find the partial molar volume of component 2 as a function of n_2.

Solution

$$\bar{V}_2 = b_1 + 2b_2 n_2 + 3b_3 n_2^2$$

---

[1] D. P. Shoemaker, C. W. Garland, and J. W. Nibler, *Experiments in Physical Chemistry*, 5th. ed., McGraw-Hill, New York, 1989, pp. 65ff.

The Method of Intercepts

This is a graphical method for determining partial molar quantities in a two-component system. In this method, the mean molar quantity is graphed as a function of one of the mole fractions for a fixed value of T and a fixed value of P. Figure 4.2 shows $\bar{V}$, the mean molar volume of a solution of ethanol (component 1) and water (component 2), as a function of x_1, the mole fraction of ethanol. From Euler's theorem, Equation (4.6-7),

$$\bar{Y} = x_1 \bar{Y}_1 + x_2 \bar{Y}_2 \qquad \textbf{(4.6-15)}$$

where Y stands for any extensive variable. Since $x_2 = 1 - x_1$ in a two-component system, this equation can be written

$$\bar{Y} = (\bar{Y}_1 - \bar{Y}_2)x_1 + \bar{Y}_2 \qquad \textbf{(4.6-16)}$$

Say that x_1' is a particular value of x_1 for which we desire the values of the partial molar quantities $\bar{Y}_1$ and $\bar{Y}_2$. At this value of x_1, we draw a tangent line to the curve, as shown in the figure. The intercepts of this line at the edges of the figure give the values of the two partial molar quantities for the composition $x_1 = x_1'$. The next part of this section contains a proof of the validity of this method, and may be skipped without loss of continuity.

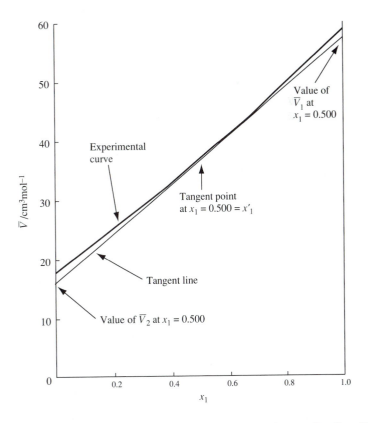

Figure 4.2. The Mean Molar Volume of an Ethanol-Water Solution as a Function of Mole Fraction of Ethanol. This diagram is used to carry out the method of intercepts graphically.

The value of the derivative $(\partial \bar{Y}/\partial x_1)_{T,P}$ gives the slope of the desired tangent line when evaluated at $x_1 = x_1'$. The derivative of Equation (4.6-15) with respect to x_1 is

$$(\partial \bar{Y}/\partial x_1)_{T,P} = \bar{Y}_1 + x_1(\partial \bar{Y}_1/\partial x_1)_{T,P} - \bar{Y}_2 + x_2(\partial \bar{Y}_2/\partial x_1)_{T,P} \quad \textbf{(4.6-17)}$$

where we have used the fact that $(\partial x_2/\partial x_1) = -1$. The second and fourth terms on the right-hand side of Equation (4.6-17) sum to zero by the analogue of Equation (4.6-13), giving

$$(\partial \bar{Y}/\partial x_1) = \bar{Y}_1 - \bar{Y}_2 \quad \textbf{(4.6-18)}$$

If this derivative is evaluated at $x_1 = x_1'$, it gives the slope of the tangent line at that point. If we let y stand for the ordinate of a point on the line, then

$$y = [\bar{Y}_1(x_1') - \bar{Y}_2(x_1')]x_1 + b \quad \textbf{(4.6-19)}$$

where b is the intercept of the tangent line at $x_1 = 0$ and we consider both of the partial molar quantities to be functions of x_1 and omit mention of the dependence on P and T.

The line and the curve must coincide at $x_1 = x_1'$, so from Equations (4.6-16) and (4.6-19)

$$\bar{Y}_2(x_1') + x_1'[\bar{Y}_1(x_1') - \bar{Y}_2(x_1')] = [\bar{Y}_1(x_1') - \bar{Y}_2(x_1')]x_1' + b$$

Canceling equal terms on both sides of the equation, we get

$$\bar{Y}_2(x_1') = b \quad \textbf{(4.6-20)}$$

One can repeat the entire argument with the roles of components 1 and 2 reversed to show that the intercept at the right side of the figure is equal to the value of $\bar{Y}_1$ at $x_1 = x_1'$. However, it can more easily be shown by evaluating the function represented by the line at $x_1 = 1$.

$$y(1) = [\bar{Y}_1(x_1') - \bar{Y}_2(x_1')] + \bar{Y}_2(x_1') = \bar{Y}_1(x_1') \quad \textbf{(4.6-21)}$$

Thus, the intercept at $x_1 = 1$ is equal to $\bar{Y}_1(x_1')$.

In order to gain better accuracy, it is better to graph a quantity of smaller magnitude than the mean molar quantity. Figure 4.3 shows a graph of the change in the mean molar quantity on mixing, defined as the quantity for 1 mol of solution minus the sum of the values of the quantity for the amounts of unmixed components in 1 mol of solution:

$$\boxed{\Delta \bar{Y}_{\text{mix}} = \bar{Y} - (x_1 \bar{Y}_1^* + x_2 \bar{Y}_2^*) \quad \text{(definition)}} \quad \textbf{(4.6-22)}$$

$$= \bar{Y} - [\bar{Y}_2^* + x_1(\bar{Y}_1^* - \bar{Y}_2^*)]$$

where $\bar{Y}_1^*$ is the molar quantity for pure component 1 and $\bar{Y}_2^*$ is the molar quantity for pure component 2. The intercepts of the line tangent to the curve at $x_1 = x_1'$ give the partial molar quantities evaluated at $x_1 = x_1'$, as follows:

$$\text{Right intercept} = \bar{Y}_1(x_1 = x_1') - \bar{Y}_1^* \quad \textbf{(4.6-23a)}$$

$$\text{Left intercept} = \bar{Y}_2(x_1 = x_1') - \bar{Y}_2^* \quad \textbf{(4.6-23b)}$$

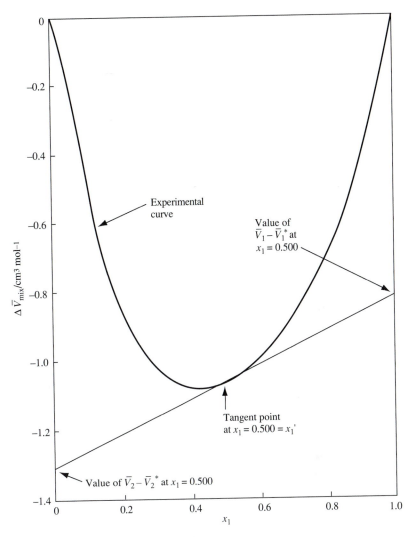

Figure 4.3. The Change in the Mean Molar Volume on Mixing for an Ethanol-Water Solution as a Function of Mole Fraction of Ethanol. This diagram is used to carry out a modified version of the method of intercepts, which gives greater accuracy.

Exercise 4.18   Show that Equations (4.6-23a) and (4.6-23b) are correct.

▼   **EXAMPLE 4.11**   From the intercepts in Figure 4.2, determine the partial molar volumes of ethanol (substance 1) and water (substance 2) if the mole fraction of ethanol is equal to 0.500. The molar volumes of the pure substances are $\bar{V}_1^* = 58.4 \text{ cm}^3 \text{ mol}^{-1}$, $\bar{V}_2^* = 18.02 \text{ cm}^3 \text{ mol}^{-1}$.

Solution

The tangent line is drawn in the figure.

$$\text{Right intercept} = \bar{V}_1 - \bar{V}_1^* = -0.83 \text{ cm}^3 \text{ mol}^{-1}$$
$$\text{Left intercept} = \bar{V}_2 - \bar{V}_2^* = -1.31 \text{ cm}^3 \text{ mol}^{-1}$$
$$\bar{V}_1 = 58.4 \text{ cm}^3 \text{ mol}^{-1} - 0.83 \text{ cm}^3 \text{ mol}^{-1} = 57.6 \text{ cm}^3 \text{ mol}^{-1}$$
$$\bar{V}_2 = 18.02 \text{ cm}^3 \text{ mol}^{-1} - 1.31 \text{ cm}^3 \text{ mol}^{-1} = 16.7 \text{ cm}^3 \text{ mol}^{-1}$$

Summary

In this chapter we have obtained two types of thermodynamic tools for the application of thermodynamics to real systems: criteria for spontaneous processes and useful formulas.

The second law of thermodynamics provides the general criterion for possible processes: no process can decrease the entropy of the universe. For a closed simple system at constant pressure and temperature the Gibbs energy G cannot increase, and for a closed simple system at constant temperature and volume the Helmholtz energy A cannot increase.

Several fundamental relations were obtained for simple closed systems. The first relations were for the differentials of the different energy-related state variables for closed simple systems. For example,

$$dG = -S\, dT - V\, dP$$

The Maxwell relations were derived from these differentials. For example,

$$\left(\frac{\partial S}{\partial P} \right)_{T,n} = -\left(\frac{\partial V}{\partial T} \right)_{P,n}$$

The Gibbs energy was expressed as the molar Gibbs energy of a pure substance in its standard state, $\bar{G}^\circ$, plus another term. For an ideal gas

$$\bar{G} = \bar{G}^\circ + RT \ln(P/P^\circ)$$

and for an incompressible solid or liquid

$$\bar{G} = \bar{G}^\circ + \bar{V}(P - P^\circ)$$

The description of open multicomponent systems is based on the Gibbs equation

$$dG = -S\, dT + V\, dP + \sum_{i=1}^{c} \mu_i\, dn_i$$

where μ_i is the chemical potential of component i.

A number of useful thermodynamic relations were obtained. The *thermodynamic equation of state* is

$$\left(\frac{\partial U}{\partial V} \right)_{T,n} = T\left(\frac{\partial P}{\partial T} \right)_{V,n} - P$$

For an extensive quantity, represented by Y, Euler's theorem is

$$Y = \sum_{i=1}^{c} n_i \bar{Y}_i$$

The generalized Gibbs-Duhem relation is

$$\sum_{i=1}^{c} n_i \, d\bar{Y}_i = \left(\frac{\partial Y}{\partial T}\right)_{P,n} dT + \left(\frac{\partial Y}{\partial P}\right)_{T,n} dP$$

PROBLEMS

Problem for Section 4.1

4.19. If a simple system is somehow maintained at constant S and P, show that its enthalpy cannot increase.

Problems for Section 4.2

4.20. The **fundamental equation** or **fundamental relation** of thermodynamics for a particular system is a formula giving

$$S = S(U, V, n) \quad \text{or} \quad U = U(S, V, n)$$

for that system. If this relation is known, all thermodynamic information about the system can be obtained from it. For an ideal monatomic gas with constant heat capacity,[2]

$$S = \frac{nS_0}{n_0} + nR \, \ln\left[\left(\frac{U}{U_0}\right)^{3/2}\left(\frac{V}{V_0}\right)\left(\frac{n}{n_0}\right)^{-5/2}\right]$$

where S_0, n_0, and V_0 are constants.

    a. Solve this equation for $U = U(S, V, n)$
    b. Use Equation (4.2-5) to obtain an expression for T.
    c. Use Equation (4.2-6) to obtain an expression for P.

4.21. Find an expression for $(\partial S/\partial P)_{T,n}$ for a gas obeying the truncated pressure virial equation of state

$$P\bar{V} = RT + A_2 P$$

4.22. If a gas is represented by the truncated virial equation of state

$$P\bar{V}/RT = 1 + B_2/\bar{V} + B_3/\bar{V}^2$$

where the virial coefficients depend on T, find an expression for the molar entropy change for an isothermal volume change of the gas.

4.23. a. Find an expression for $(\partial S/\partial V)_{T,n}$ for a gas obeying the van der Waals equation of state.

b. Find the value of ΔS for the isothermal expansion of 2.000 mol of argon from a volume of 20.00 L to a volume of 40.00 L at 298.15 K, assuming the van der Waals equation of state. Compare with the value for the same change in state assuming argon to be ideal.

4.24. a. Find an expression for $(\partial S/\partial V)_{T,n}$ for a gas obeying the Redlich-Kwong equation of state.

b. Find the value of ΔS for the isothermal expansion of 2.000 mol of argon from a volume of 20.00 L to a volume of 40.00 L at 298.15 K, assuming the Redlich-Kwong equation of state. Compare with the value for the same change in state assuming argon to be ideal.

Problems for Section 4.3

***4.25. a.** Find the value of ΔG for 1.000 mol of ideal gas if it is isothermally pressurized from 1.000 atm to 2.000 atm at 298.15 K.

b. Find the value of ΔG for 1.000 mol of ideal gas if it is isothermally pressurized from 2.000 atm to 3.000 atm at 298.15 K.

c. Explain in words why your answers for parts a and b are not equal although the changes in pressure are equal.

4.26. a. Find the value of ΔG if 1.000 mol of liquid water is pressurized at 0.00°C from 1.000 atm to 10.000 atm. State any assumptions.

b. Find the value of ΔG if 1.000 mol of solid water is pressurized at 0.00°C from 1.000 atm to 10.000 atm. State any assumptions.

c. Find the value of ΔG if 1.000 mol of solid water melts at 0.00°C and 1.000 atm.

d. Find the value of ΔG if 1.000 mol of solid water melts at 0.00°C and 10.000 atm.

4.27. a. Write an expression for ΔA for the isothermal expansion of an ideal gas from volume V_1 to volume V_2.

b. Find the value of ΔA for each of the processes in parts a and b of Problem 4.25. Explain the relationship of these values to the values of ΔG for the same processes.

***4.28.** Find the value of $\bar{G} - \bar{G}°$ for argon gas at 298.15 K and 1.000 atm, assuming the truncated pressure virial

---

    [2] H. B. Callen, *Thermodynamics*, Wiley, New York, 1960, pp. 26ff, 53ff.

equation of state

$$P\bar{V} = RT + A_2 P$$

with $A_2 = -15.8 \text{ cm}^3 \text{ mol}^{-1}$.

4.29. a. Write an expression for ΔA for the isothermal expansion from volume V_1 to volume V_2 for a gas obeying the truncated virial equation of state

$$P\bar{V}/RT = 1 + B_2/\bar{V}$$

b. Find the value of ΔA for the isothermal expansion of 1.000 mol of argon at 298.15 K from a volume of 10.000 L to a volume of 20.000 L. Compare with the result based on assuming argon to be an ideal gas.

4.30. a. Find an expression for ΔA for the isothermal expansion of a gas obeying the van der Waals equation of state from volume V_1 to volume V_2.

b. Find the value of ΔA for the isothermal expansion of 1.000 mol of argon at 298.15 K from a volume of 10.000 L to a volume of 20.000 L. Compare with your result for Problem 4.29.

Problems for Section 4.4

*4.31. Following are data on the density of ethanol-water solutions at 20.°C. Calculate the volume of a solution containing 0.700 mol of water and the appropriate amount of ethanol for each of the data points except for the 0% and 100% data points. Make a graph of this volume and determine the partial molar volume of ethanol for a solution with mole fraction of ethanol equal to 0.300.

Ethanol (% by mass)	Density (g mL^{-1})
0	0.99823
46.00	0.9224
48.00	0.9177
50.00	0.9131
52.00	0.9084
54.00	0.9039
56.00	0.8995
58.00	0.8956
100.00	0.7893

4.32. The partial specific volume of a system is defined as $(\partial V/\partial w_i)_{P,T,w'}$, where w_i is the mass of component i and the subscript w' stands for keeping the mass of every substance fixed except for substance i. All of the relations involving partial molar quantities can be converted to relations for partial specific quantities by consistently replacing n_i by w_i for every substance and by replacing x_i by the mass frac-

tion y_i:

$$y_i = w_i/w_{\text{total}}$$

for every substance. Using the data of Problem 4.31, find the partial specific volume of ethanol in the mixture with mass fraction 0.500.

Problems for Section 4.5

4.33. a. Find an expression for $(\partial U/\partial V)_{T,n}$ for a gas obeying the van der Waals equation of state.

b. Find the value of $(\partial U/\partial V)_{T,n}$ for argon at a molar volume of 0.244 m^3 mol^{-1}.

4.34. Show that

$$C_P = VT\alpha(\partial P/\partial T)_{S,n}$$

where α is the coefficient of thermal expansion.

*4.35. a. It is shown in the theory of hydrodynamics[3] that the speed of sound, v_s, is given by

$$v_s^2 = \frac{\bar{V}}{M\kappa_S}$$

where κ_S is the adiabatic compressibility defined in Equation (4.5-8), $\bar{V}$ is the molar volume, and M is the molar mass. Show that

$$v_s^2 = \frac{\bar{V}C_P}{M\kappa_T C_V}$$

where κ_T is the isothermal compressibility.

b. Find the speed of sound in air at 298.15 K and 1.000 atm, assuming a mean molar mass of 0.29 kg mol^{-1} and $\bar{C}_V = 5R/2$.

c. Find the speed of sound in helium at 298.15 K and 1.000 atm.

d. For parts a and b, find the ratio of the speed of sound to the mean speed of the gas molecules given by Equation (16.3-6).

4.36. Derive the following equation:

$$\left(\frac{\partial U}{\partial V}\right)_{T,n} = -P + \left[\left(\frac{\partial H}{\partial P}\right)_{T,n} - V\right]\left(\frac{\partial P}{\partial V}\right)_{T,n}$$

4.37. a. Find an expression for μ_{JT}, the Joule-Thomson coefficient, for a gas that obeys the truncated virial equation of state

$$P\bar{V} = RT + A_2 P + A_3 P^2$$

where the pressure virial coefficients depend on temperature.

b. Show that the Joule-Thomson coefficient in part a does not vanish in the limit of zero pressure, even though the Joule-Thomson coefficient of an ideal gas vanishes.

---

[3] H. Lamb, *Hydrodynamics*, 6th ed., Cambridge University Press, New York, 1932.

c. Evaluate the Joule-Thomson coefficient for argon at 298.15 K in the limit of zero pressure, assuming that $\bar{C}_P = 5R/2$. Use values in Example 2.7 and the fact that $A_2 = B_2$.

4.38. a. Find the change in enthalpy if 1.000 mol of liquid water is pressurized at 0.00°C from 1.000 atm to 10.000 atm. State any assumptions.

b. Find the change in enthalpy if 1.000 mol of solid water is pressurized at 0.00°C from 1.000 atm to 10.000 atm. State any assumptions.

c. From the results of parts a and b, find the value of the molar enthalpy change of fusion of water at 0.00°C and 10.00 atm.

Problems for Section 4.6

*4.39.** Determine which (if any) of the following functions are homogeneous with respect to all three independent variables x, y, and z. Find the degree of each homogeneous function. All letters except f, x, y, and z denote constants.

a. $f(x, y, z) = ax^2 + bxy + cy^2 + dyz + gxz + hz^2$
b. $f(x, y, z) = ax^2y^{-2} + b \ln(y/z) + c \tan(x^3y^{-3})$
c. $f(x, y, z) = ax^2 + b \cos(x^2y^{-2}) + cz^{-2}$
d. $f(x, y, z) = az^3 + bx^3 \cos(x^4y^{-4}) + c \exp(yz^{-1})$
e. $f(x, y, z) = ax^4 + bx^2yz \sin(x/y) + cy^3z \ln(z^2y^{-2})$

4.40. a. From the data of Problem 4.31, make a graph of the change in the mean molar volume on mixing as a function of ethanol mole fraction and determine the partial molar volume of each substance for the solution with ethanol mole fraction equal to 0.300, using the method of intercepts.

b. From the values of the partial molar volumes of water and ethanol in the solution of part a, find the volume of a solution containing 0.600 mol of ethanol and 1.400 mol of water. Compare this value with the value obtained by interpolating in the list of density values given in Problem 4.29.

General Problems

*4.41.** Which of the following statements are correct?

a. The minimum of the Gibbs energy of the system always corresponds to the maximum of the entropy of the system plus surroundings.

b. The minimum of the Helmholtz energy of a system always corresponds to the maximum of the entropy of the system plus surroundings.

c. The minimum of the Gibbs energy always corresponds to the minimum of the energy of the system plus surroundings.

d. If a system is at constant temperature and pressure, the minimum in the Gibbs energy of the system corresponds to the maximum of the entropy of the system plus surroundings.

e. If a system is at constant temperature and volume, the minimum in the Helmholtz energy corresponds to the maximum of the entropy of the system plus surroundings.

4.42. a. Find the value of the heat capacity at constant volume for 1.000 mol of liquid water at 20.00°C.

b. Find the value of the adiabatic compressibility of liquid water at 25.00°C.

c. Find the speed of sound in liquid water at 20.00°C, using the formula from Problem 4.35:

$$v_s^2 = \frac{\bar{V}C_P}{M\kappa_T C_V} = \frac{\bar{V}\gamma}{M\kappa_T}$$

d. Find the final volume of 1.000 mol of liquid water if it is compressed adiabatically from 1.000 bar and 25.00°C to a pressure of 100.00 bar. Assume that the adiabatic compressibility is constant.

4.43. For each of the following proposed processes, tell (1) whether the process is spontaneous, nonspontaneous, or reversible; (2) whether ΔG is positive, negative, or equal to zero; (3) whether ΔH is positive, negative, or equal to zero; (4) whether ΔS is positive, negative, or equal to zero; and (5) whether ΔH is smaller, equal to, or larger in magnitude than $T \Delta S$. Note that for spontaneous processes, the initial state must be metastable.

a. Liquid water is vaporized at 1.000 atm and 100°C.

b. Liquid water is vaporized at 1.000 atm and 105°C.

c. Liquid water is vaporized at 1.000 atm and 95°C.

d. Solid water melts at 1.000 atm and 5°C.

e. Solid water melts at 1.000 atm and 0°C.

f. Water vapor at 25°C and a partial pressure of 23.756 torr condenses to a liquid.

4.44. A nonideal gas is described equally well by two truncated virial equations of state:

$$\frac{P\bar{V}}{RT} = 1 + \frac{B_2}{RT} \quad \text{and} \quad P\bar{V} = RT + A_2P$$

where A_2 and B_2, the second virial coefficients, are functions of T and can be shown to equal each other (see Exercise 1.5).

a. Find expressions for the following:
    1. $\bar{G}(T, P') - \bar{G}^\circ(T)$
    2. $\bar{S}(T, P') - \bar{S}^\circ(T)$
    3. $\bar{H}(T, P') - \bar{H}^\circ(T)$
    4. $\bar{A}(T, P') - \bar{A}^\circ(T)$
    5. $\bar{U}(T, P') - \bar{U}^\circ(T)$
where P' is some pressure not necessarily equal to P°.

b. Evaluate each of the quantities in part a for carbon dioxide at 50°C, using data in Table A2.

4.45. Show that $(\partial T/\partial V)_{S,n} > 0$ unless $(\partial V/\partial T)_{P,n} < 0$. What is the sign of $(\partial T/\partial V)_{S,n}$ for water in the temperature range between 0.00°C and 3.98°C?

5 Phase Equilibrium

PREVIEW

In this chapter, we apply the principles of thermodynamic equilibrium to the equilibrium states of systems containing more than one phase.

PRINCIPAL FACTS AND IDEAS

1. The principles of thermodynamics determine equilibrium behavior in multiphase systems.
2. The fundamental fact of phase equilibrium is that at equilibrium the chemical potential of any substance must have the same value in all phases in which that substance appears.
3. The Gibbs phase rule gives the number of independent intensive variables in a multicomponent, multiphase system at equilibrium:

$$f = c - p + 2$$

where f is the number of independent intensive variables, c is the number of components, and p is the number of phases.
4. The Gibbs phase rule allows phase diagrams to be understood.
5. The Clausius and Clausius-Clapeyron equations govern the curves in phase diagrams.
6. Thermodynamics allows analysis of the stability of phases in systems.
7. Surface effects must be included in a complete thermodynamic treatment, but they are usually negligible.

OBJECTIVES

After studying this chapter, the student should:

1. understand the role of the chemical potential in phase equilibria and be able to solve problems involving phase equilibrium,
2. be able to apply the phase rule to the interpretation of phase equilibria,
3. be able to solve problems involving the Clapeyron and Clausius-Clapeyron equations,
4. understand the relationship of the Gibbs energy to phase equilibria,
5. understand the modifications to ordinary thermodynamics required by the inclusion of surface energy,
6. be able to solve problems involving surface tension.

5.1 The Fundamental Fact of Phase Equilibrium

A **phase** is a region of a system inside which intensive properties do not change abruptly as a function of position. The principal kinds of phases are solids, liquids, and gases, although plasmas (ionized gases), liquid crystals, and glasses are sometimes considered to be separate types of phases. Solid and liquid phases are called **condensed phases**, and a gas phase in contact with a condensed phase is often called a **vapor phase**.

Numerous elements, including carbon, exhibit **allotropy** in the solid phase. That is, there is more than one kind of solid phase of a single substance. Many compounds exhibit the same phenomenon, which is then called polymorphism. Most pure substances have only one liquid phase, but helium is a special case, exhibiting allotropy in liquid phases. Pure ^{4}He (the most abundant isotope) exists in two different liquid forms, and pure ^{3}He exists in three different liquid forms. A pure substance can have only one gas phase.

A mixture at equilibrium may contain several solid phases or several liquid phases. For example, by equilibrating mercury, a mineral oil, a methyl silicone oil, water, benzyl alcohol, and a perfluoro compound such as perfluoro (N-ethyl piperidine) at room temperature, one can obtain six co-existing liquid phases.[1] Each phase contains a large concentration of one substance and small concentrations of the other substances. Under ordinary conditions, only a single gas phase can exist in a single system. However, if certain gaseous mixtures are brought to supercritical temperatures and pressures, at which the distinction between gas and liquid disappears, two fluid phases can form without first making a gas-liquid phase transition.

Equilibrium Between Phases

Consider a two-phase simple closed system whose temperature and pressure can be controlled, as depicted in Figure 5.1. If the contribution of the surface area between the phases and between the system and its container is negligible, the Gibbs energy of the system is the sum of the Gibbs energies of the two phases:

$$G = G^{(I)} + G^{(II)} \tag{5.1-1}$$

where we denote the two phases by the superscripts (I) and (II).

We number the components from 1 to c. We assume that the system is closed, so that any substance moving out of one phase must move into the other phase:

$$dn_i^{(I)} = -dn_i^{(II)} \qquad (i = 1, 2, \ldots, c) \tag{5.1-2}$$

For an infinitesimal transfer of matter, the change in the Gibbs energy is given by

$$
\begin{aligned}
dG &= dG^{(I)} + dG^{(II)} \\
&= -S^{(I)}\,dT + V^{(I)}\,dP + \sum_{i=1}^{c} \mu_i^{(I)}\,dn_i^{(I)} \\
&\quad + \left(-S^{(II)}\,dT + V^{(II)}\,dP + \sum_{i=1}^{c} \mu_i^{(II)}\,dn_i^{(II)} \right)
\end{aligned}
\tag{5.1-3}
$$

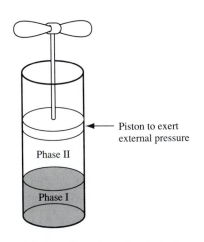

Figure 5.1. A Two-Phase Simple System. This system is closed to the surroundings, but each phase is open to the other.

Piston to exert external pressure

Phase II

Phase I

---

[1] J. Kochansky, *J. Chem. Educ.* **68**, 653 (1991).

We assume that both phases have the same values of T and P and thus the same values of dT and dP.

Let us maintain the system at equilibrium at constant T and P, so that dP and dT vanish, and carry out an infinitesimal transfer of matter between the phases. From our criteria for equilibrium in Chapter 4, dG must vanish for an equilibrium change at constant T and P. From Equations (5.1-2) and (5.1-3)

$$dG = \sum_{i=1}^{c} \mu_i^{(I)} \, dn_i^{(I)} + \sum_{i=1}^{c} \mu_i^{(II)} \, dn_i^{(II)}$$

$$= \sum_{i=1}^{c} (\mu_i^{(I)} - \mu_i^{(II)}) \, dn_i^{(I)} = 0 \qquad \textbf{(5.1-4)}$$

Our analysis does not depend on whether a suitable semipermeable membrane can exist in the real world. The important thing is that nothing in thermodynamics forbids the existence of such membranes. We can reach the same conclusion without membranes by arguing mathematically that each term must vanish separately because the dn's are linearly independent.

In principle it is possible, without violating the second law, to find a semipermeable membrane that will selectively allow one component to pass, but not the others. Therefore, each term of the sum in the right-hand side of Equation (5.1-4) must vanish individually, because all of the dn's but one, say dn_i, can be made to vanish. Because dn_i is not necessarily equal to zero, the other factor must vanish and

$$\boxed{\mu_i^{(I)} = \mu_i^{(II)} \quad \text{(system at equilibrium)}} \qquad \textbf{(5.1-5a)}$$

Equation (5.1-5a) is not restricted to a system with just two phases. If more than two phases are present at equilibrium, we can consider any pair of phases and conclude that the chemical potential of any substance has the same value in every phase into which it can pass. We write a second version of the equation:

$$\boxed{\mu_i^{(\alpha)} = \mu_i^{(\beta)} \quad \text{(system at equilibrium)}} \qquad \textbf{(5.1-5b)}$$

where the superscripts (α) and (β) designate any two phases of a multiphase system.

The properties of any equilibrium system are no different than they would be if the system arrived at its equilibrium state under some other conditions than the actual conditions. Therefore, Equation (5.1-5) is also valid for an open system and for a system which approached equilibrium without being at constant temperature and pressure.

Equation (5.1-5) is the *fundamental fact of phase equilibrium: At equilibrium, the chemical potential of any substance has the same value in all phases in which it occurs.*

Nonequilibrium Phases

Consider a two-phase simple system maintained at constant temperature and pressure. We assume that each phase is in a metastable state and is not yet at equilibrium with the other phase. The Gibbs energy of the system is still given by Equation (5.1-1).

The criterion for possible processes is now given by the inequality in Equation (4.1-17):

$$dG < 0 \qquad (T \text{ and } P \text{ constant}) \qquad \textbf{(5.1-6)}$$

Because dT and dP vanish and the system as a whole is closed, if a spontaneous transfer of matter occurs,

$$dG = \sum_{i=1}^{c} \mu_i^{(I)} \, dn_i^{(I)} + \sum_{i=1}^{c} \mu_i^{(II)} \, dn_i^{(II)}$$

$$= \sum_{i=1}^{c} (\mu_i^{(I)} - \mu_i^{(II)}) \, dn_i^{(I)} < 0 \qquad \textbf{(5.1-7)}$$

Once again, we argue that each term separately must be negative, since the introduction of semipermeable membranes would not violate the second law of thermodynamics. The two factors in each term of the sum in Equation (5.1-7) must be of opposite signs:

$$\mu_i^{(I)} > \mu_i^{(II)} \quad \text{implies that } dn_i^{(I)} < 0 \qquad \textbf{(5.1-8)}$$

$$\mu_i^{(I)} < \mu_i^{(II)} \quad \text{implies that } dn_i^{(I)} > 0 \qquad \textbf{(5.1-9)}$$

Therefore, *any substance moves spontaneously from a phase of higher value of its chemical potential to a phase of lower value of its chemical potential*. The term "chemical potential" was chosen by analogy with potential energy, because mechanical systems tend to move toward states of lower potential energy.

The spontaneous transfer of a substance from a phase of higher chemical potential to a phase of lower chemical potential is not restricted to systems at constant T and P.

Exercise 5.1 Argue that a substance will move spontaneously from any phase of higher value of its chemical potential to any other phase of lower value of its chemical potential in a system with each phase at constant T and V.

Transport of Matter in a Nonuniform Phase

Assume that a one-phase system is at the same temperature and pressure throughout but is of nonuniform composition. Imagine dividing the system into small regions or subsystems, each of which is small enough that the composition is almost uniform within it. We treat each subsystem in the same way as we treated one phase in obtaining Equations (5.1-8) and (5.1-9).

The analogue of the fundamental fact of phase equilibrium for nonuniform systems is obtained: *In a system with uniform temperature and pressure, any substance tends to move from a region of larger value of its chemical potential to a region of smaller value of its chemical potential.* Nonuniformity of the chemical potential is the driving force for diffusion.

5.2 The Gibbs Phase Rule

The equilibrium thermodynamic state of a one-phase simple system with c components is specified by the values of $c + 2$ thermodynamic variables, at least one of which must be an extensive variable. This statement means that $c + 2$ variables are independent variables, and all other equilibrium variables (both extensive and intensive) are independent variables. The **intensive equilibrium state** is the state of the system so far as only intensive variables are concerned. Only one extensive variable is required in addition to the intensive variables to give the full state of the system. Therefore, the number of variables required to specify the intensive state is smaller by one than the number required to specify the full state. The intensive state of a one-phase simple system is specified by the values of $c + 1$ variables, all of which are intensive. Thus $c + 1$ intensive variables are independent variables, and all other intensive variables are dependent variables. The size of a system can be varied without changing its intensive state.

One set of independent variables that can be used to specify the intensive state of a one-phase simple system consists of T, P, and $c - 1$ mole fractions. The mole fractions automatically obey the relation

$$\sum_{i=1}^{c} x_i = 1 \tag{5.2-1}$$

Therefore, no more than $c - 1$ of them can be used as independent variables.

Now consider a multiphase system consisting of p phases and c components. In counting phases, we count only regions that differ in their intensive properties from other regions. For example, crushed ice and liquid water are a two-phase system, just like a system of liquid water and a single ice cube.

The number of components in a system is equal to the number of substances present minus the number of relations that constrain the amounts of substances. There are three principal types of relations: those due to chemical equilibrium, those due to the requirement for electrical neutrality, and those due to the way the system was prepared (such as a specification that two substances are in their stoichiometric ratio). We will use the fact that for each chemical reaction that comes to equilibrium the number of components is reduced by one.

For example, a mixture of gaseous hydrogen, oxygen, and water vapor can remain unreacted for a very long time at room temperature because of the slowness of the reaction. We treat the metastable mixture as we would an equilibrium mixture in which no reaction is possible. There are three components. However, if a platinum catalyst is introduced into the system, chemical equilibrium is rapidly established, reducing the number of components to two (besides the catalyst). The amount of water vapor is determined by the amounts of hydrogen and oxygen and the nature of the chemical equilibrium. If the additional constraint is added that the hydrogen and oxygen are in the stoichiometric ratio of two moles to one, then the system has only one component. This would be the case if only water vapor were placed in the system and allowed to equilibrate in the presence of the catalyst. The fact that there is one component can be seen from the fact that the system can be produced from one substance.

▼ **EXAMPLE 5.1**   Determine the number of components in:
    a. An aqueous solution containing Na^+, Cl^-, and Br^-.
    b. An aqueous solution containing Na^+, K^+, Li^+, Cl^-, and Br^-.

c. A gaseous system containing NO_2 and N_2O_4 at chemical equilibrium with each other.

d. An aqueous solution containing Ca^{2+} ions and Cl^- ions.

Solution

a. There are three components, because the system can be produced from three pure substances: water, NaCl, and NaBr. The number can also be determined by counting up water and the three ions and subtracting unity for the condition of electrical neutrality.

b. There are five components (6 − 1). It would be possible to make the system with five substances: water, NaCl, KBr, KCl, and LiCl. No NaBr or LiBr is needed.

c. There is one component, because the amount of N_2O_4 is determined by the amount of NO_2 and the conditions of the equilibrium.

d. There are two components, because electrical neutrality imposes a relation on the amounts of Ca^{2+} and Cl^- ions.

▲

Exercise 5.2   Determine the number of components in:

a. An aqueous solution containing NaCl and KBr.

b. An aqueous solution containing Na^+, Cl^-, K^+, and Br^-.

c. A gaseous system containing PCl_5, PCl_3, and Cl_2 at chemical equilibrium with each other.

d. A solid mixture containing powdered graphite and powdered diamond without a catalyst or other means of converting one phase to the other.

e. A gaseous mixture containing carbon dioxide and water vapor.

f. A gaseous mixture containing carbon dioxide and water vapor, all of which was produced by combustion of a stoichiometric mixture of methane and oxygen (assume that the residual amounts of methane and oxygen are negligible).

If the p phases in a multiphase system of c components are separated from each other, there are $p(c + 1)$ independent intensive variables to specify the equilibrium intensive states of all phases ($c + 1$ variables for each phase). If the phases are placed in contact with each other, opened to each other, and allowed to equilibrate, thermal equilibrium means that all phases have the same temperature, mechanical equilibrium means that all phases have the same pressure, and phase equilibrium means that the chemical potential of every substance has the same value in every phase.

Each equality that did not exist when the phases were separated places a constraint on one variable, turning it into a dependent variable. Specifying that one variable has the same value in two phases means one equality, specifying that one variable has the same value in three phases means two equalities, etc., so that $p - 1$ equalities suffice for one variable and p phases. The number of variables that have equal values in all phases is $c + 2$ (P, T, and c chemical potentials), for a total of $(p - 1)(c + 2)$ constraints. This means that f, the **number of independent intensive variables** after equilibration of all phases, is equal to

$$f = p(c + 1) - (p - 1)(c + 2) = pc + p - pc + c - 2p + 2$$

$$\boxed{f = c - p + 2 \quad \text{(phase rule of Gibbs)}} \tag{5.2-2}$$

This equation is the **phase rule of Gibbs**. The number of independent intensive variables is also called the number of **degrees of freedom** or the **variance**, but these terms are also used for other quantities, so we avoid them.

For a one-component, one-phase system, the number of independent intensive variables at equilibrium is

$$f = 1 - 1 + 2 = 2 \quad \text{(one component, one phase)}$$

In a phase diagram such as that of Figure 1.4, an open area represents a single phase. Because there are two independent intensive variables, T and P can both be independent, and any point in the area can represent a possible intensive state of the system.

For one component and two phases, the number of independent intensive variables at equilibrium is

$$f = 1 - 2 + 2 = 1 \quad \text{(one component, two phases)}$$

If we fix the temperature, then every other variable is a dependent variable with a value determined by the nature of the system and that value of the temperature. Only one value of the pressure is possible, only one value for the density of each phase is possible, etc. For example,

$$P = P(T) \qquad (c = 1, p = 2) \tag{5.2-3}$$

Such a function is represented by a curve in a phase diagram such as that in Figure 1.4.

When a liquid and a vapor phase are equilibrated, the equilibrium pressure is called the **vapor pressure**. The vapor pressure of water at $100.00°C$ is 1.000 atm (760.00 torr) and at $25.00°C$ is 23.756 torr. Figure 5.2 shows the equilibrium vapor pressure of water as a function of temperature.

If one component and three phases are present, the number of independent intensive variables is zero. There is no choice of the temperature, pressure, density of each phase, or any other intensive variable, and the system is said to be at a **triple point**. The solid-liquid-vapor triple point of water occurs at a temperature of 273.16 K (this value defines the absolute temperature scale) and a pressure of 4.562 torr.

Figure 5.3 shows the phase diagram of water. The pressure scale is so compressed that the liquid-vapor curve of Figure 5.2 is too close to the horizontal axis to be visible. Nine different crystalline forms of water are shown, denoted by Roman numerals. There is no ice IV, because a metastable phase was mistaken for an equilibrium phase and given this number. When the error was discovered, the other forms were not renumbered.

In the novel *Cat's Cradle*,[2] a fictional form of ice that melts at $114°F$ is discovered. Because it is more stable than liquid water at room temperature,

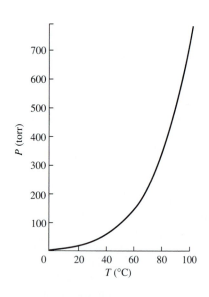

Figure 5.2. The Equilibrium Vapor Pressure of Water as a Function of Temperature. If both the liquid and the vapor phase are present, the equilibrium pressure is a function only of temperature. Data from R. C. Weast, *Handbook of Chemistry and Physics*, 69th ed., CRC Press, Boca Raton, FL, 1988–89, pp. D189–191.

---

[2] Kurt Vonnegut, *Cat's Cradle*, Delacorte Press, New York, 1963.

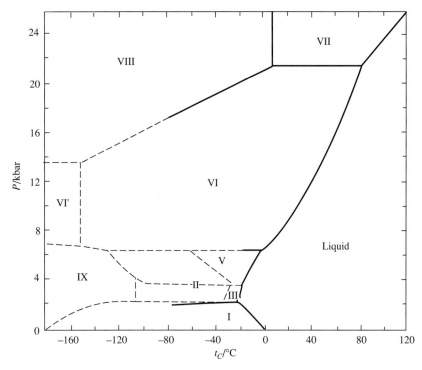

Figure 5.3. The Phase Diagram of Water. The liquid-vapor coexistence curve is too close to the horizontal axis to be seen. Of the solid phases that can equilibrate with the liquid, only ice I is less dense than the liquid. From B. Kamb, in E. Whalley, S. Jones, and L. Gold (eds.), *Physics and Chemistry of Ice*, University of Toronto Press, Toronto, 1973.

ultimately all of the water on the earth freezes to this form of ice, destroying life as we know it. In the late 1960s, it was thought for a time that there might be a second liquid phase of water, which was named "polywater" since it seemed to consist of polymers of water molecules.[3] This phase seemed to have a lower chemical potential than ordinary liquid water. Numerous experimental and theoretical studies of polywater were published before it was discovered that the small capillaries in which the polywater was supposedly prepared were leaching substances into the water, forming solutions. If it had been a real phase, polywater would have threatened life just as did the fictional form of ice in Vonnegut's novel.

Helium has different allotropic liquid phases and also has qualitatively different phase diagrams for different isotopes. Figure 5.4 shows the low-temperature phase diagrams of ^{4}He and ^{3}He. The diagram for ^{4}He has two triple points, one for the two liquid forms and vapor and one for the two liquid forms and the solid. The diagram for ^{3}He has three triple points.

---

[3] E. R. Lippincott et al., *Science* **164**, 1482 (1969); A. Cherkin, *Nature*, **224**, 1293 (1969).

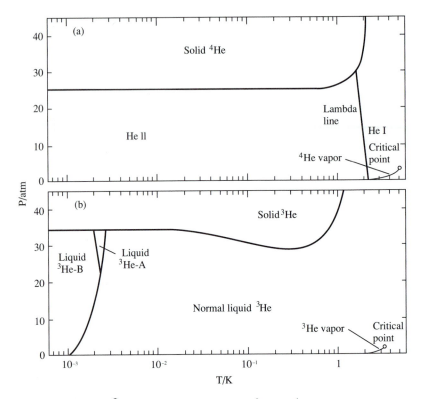

Figure 5.4. The Low-Temperature Phase Diagrams of ³He and ⁴He. From
P. V. E. McClintock, D. J. Meridith, and J. K. Wigmore, *Matter at Low Temperatures*, Wiley,
New York, 1984, p. 18.

Neither isotope exhibits coexistence between the solid and the vapor, and
the solid phases can exist only at pressures larger than 25 atm or 34 atm.

5.3 Phase Equilibria in a One-Component System

A one-component equilibrium system can consist of one phase, two phases,
or three phases. If more than one phase occurs, the equality of the chemical potential in the different phases imposes conditions, as specified by the
Gibbs phase rule. For example, when a one-component two-phase system
is at equilibrium, the pressure is a function of temperature, as in Equation (5.2-3):

$$P = P(T) \tag{5.3-1}$$

The two phases have equal values of the temperature, the pressure,
and the chemical potential of the single component. In a one-component

system, the chemical potential is equal to the molar Gibbs energy, so that

$$\bar{G}^{(I)} = \bar{G}^{(II)} \qquad (5.3\text{-}2)$$

Benoit-Pierre-Emile Clapeyron,
1799–1864, was a French engineer
who translated Carnot's cycle into the
language of calculus.

The Clapeyron Equation

This equation governs the dependence of P on T in a two-phase, one-component system. To derive the equation, we impose an infinitesimal change dT in the temperature of the system, maintaining equilibrium during the change. Since P is a function of T, the pressure will change by an amount dP that is determined by dT, and the molar Gibbs energies will undergo changes that are given by Equation (4.2-19):

$$d\bar{G}^{(I)} = -\bar{S}^{(I)}\, dT + \bar{V}^{(I)}\, dP \qquad (5.3\text{-}3\text{a})$$

$$d\bar{G}^{(II)} = -\bar{S}^{(II)}\, dT + \bar{V}^{(II)}\, dP \qquad (5.3\text{-}3\text{b})$$

The molar Gibbs energies remain equal to each other during the change, so $d\bar{G}^{(I)} = d\bar{G}^{(II)}$ and

$$-\bar{S}^{(I)}\, dT + \bar{V}^{(I)}\, dP = -\bar{S}^{(II)}\, dT + \bar{V}^{(II)}\, dP \qquad (5.3\text{-}4)$$

Nonrigorously "dividing" this equation by dT, we obtain the derivative equation

$$\boxed{\frac{dP}{dT} = \frac{\Delta\bar{S}}{\Delta\bar{V}} \quad \text{(Clapeyron equation)}} \qquad (5.3\text{-}5)$$

where $\Delta\bar{S} = \bar{S}^{(II)} - \bar{S}^{(I)}$ and $\Delta\bar{V} = \bar{V}^{(II)} - \bar{V}^{(I)}$. Equation (5.3-5) is the **Clapeyron equation**.

For a reversible phase change, Equation (3.3-5) gives

$$\Delta\bar{S} = \frac{\Delta\bar{H}}{T} \qquad (5.3\text{-}6)$$

so that the Clapeyron equation can be written

$$\frac{dP}{dT} = \frac{\Delta\bar{H}}{T\,\Delta\bar{V}} \qquad (5.3\text{-}7)$$

Using the Clapeyron equation, we can interpret the curves in a phase diagram.

▼   **EXAMPLE 5.2**

Interpret the vertical and horizontal line segments in the phase diagram of water, Figure 5.3.

Solution

A horizontal line corresponds to zero value of dP/dT, implying zero values of $\Delta\bar{H}$ and $\Delta\bar{S}$ for the phase transition. For example, between ice VI and ice VII there appears to be zero change in enthalpy and in entropy. The value of $\Delta\bar{V}$ is presumably nonzero. A vertical line corresponds to an undefined value for dP/dT, implying zero value for $\Delta\bar{V}$. For example, ice VII and ice VIII appear to have the same density.

▲

Exercise 5.3 For most substances, a solid-liquid coexistence curve has a positive slope. However, in the water phase diagram, the ice I–liquid curve has a negative slope. Explain this phenomenon. In the ^{3}He phase diagram, a horizontal region occurs in the solid–normal liquid curve and a region with a negative slope also occurs. Interpret these two phenomena (the explanation of the negative slope is not the same as that for water).

In order to have a representation for the function $P = P(T)$, we integrate the Clapeyron equation. We write Equation (5.3-5) in the form

$$dP = \frac{\Delta\bar{S}}{\Delta\bar{V}}\,dT \qquad (5.3\text{-}8)$$

Consider first a solid-liquid phase transition. Over a sufficiently small range of temperature, the quotient $\Delta\bar{S}/\Delta\bar{V}$ is nearly equal to a constant, and we can write for a sufficiently small value of $T_2 - T_1$,

$$P_2 - P_1 \approx \frac{\Delta\bar{S}}{\Delta\bar{V}}(T_2 - T_1) \qquad (T_2 - T_1 \text{ small}) \qquad (5.3\text{-}9\text{a})$$

$$\approx \frac{\Delta\bar{H}}{T\,\Delta\bar{V}}(T_2 - T_1) \qquad (T_2 - T_1 \text{ small}) \qquad (5.3\text{-}9\text{b})$$

In Equation (5.3-9b), T is a value of the temperature at which $\Delta\bar{H}$ and $\Delta\bar{V}$ were measured, such that $T_1 < T < T_2$.

▼ **EXAMPLE 5.3**

Estimate the pressure on a system of liquid and solid water if the equilibrium melting temperature is $-0.100°C$. The density of ice is 0.917 g cm^{-3}, the density of liquid water is 1.000 g cm^{-3}, and the molar enthalpy change of fusion is 6008 J mol^{-1}.

Solution

In Example 5.3 we again carry out an initial calculation of a quantity that we need to substitute into a formula. Note the use of the factor label method to express all quantities in consistent units.

$$\Delta\bar{V} = (18.015 \text{ g mol}^{-1})\left(\frac{1}{1.00 \text{ g cm}^{-3}} - \frac{1}{0.917 \text{ g cm}^{-3}}\right)\frac{1 \text{ m}^3}{10^6 \text{ cm}^3}$$

$$= -1.63 \times 10^{-6} \text{ m}^3 \text{ mol}^{-1}$$

$$P_2 - P_1 \approx \frac{6008 \text{ J mol}^{-1}}{(273.15 \text{ K})(-1.63 \times 10^{-6} \text{ m}^3 \text{ mol}^{-1})}(-0.100 \text{ K})$$

$$\approx 1.35 \times 10^6 \text{ J m}^{-3} = 1.35 \times 10^6 \text{ N m}^{-2} = 1.35 \times 10^6 \text{ Pa} = 13.3 \text{ atm}$$

$$P_2 \approx 14.3 \text{ atm}$$

If, instead of assuming that the quotient $\Delta\bar{S}/\Delta\bar{V}$ is approximately constant, one assumes that the quotient $\Delta\bar{H}/\Delta\bar{V}$ is approximately constant, one can integrate

$$dP = \frac{1}{\Delta\bar{V}}\frac{\Delta\bar{H}}{T}\,dT \qquad (5.3\text{-}10\text{a})$$

Equation (5.3-10b) is superior for large differences between T_2 and T_1 because T occurs explicitly in Equation (5.3-9b) and ΔH is likely to be more nearly independent of T than is ΔS.

to obtain

$$P_2 - P_1 \approx \frac{\Delta \bar{H}}{\Delta \bar{V}} \ln\left(\frac{T_2}{T_1}\right) \qquad (\Delta \bar{H} \text{ constant}) \qquad \textbf{(5.3-10b)}$$

Exercise 5.4

Compare the answer to Exercise 5.4 with the answer to Example 5.3 to see whether the assumption of constant ΔH gives different results from the assumption of constant ΔS.

Estimate the pressure of the system of Example 5.3, using Equation (5.3-10b) instead of Equation (5.3-9).

The Clausius-Clapeyron Equation

If one of the phases is a gas, we can write a useful modification of the Clapeyron equation. We assume (1) that the vapor phase is an ideal gas and (2) that the molar volume of the condensed phase (solid or liquid) is negligible compared to that of the vapor phase. These are both good approximations except in the vicinity of the critical point. Most gases obey the ideal gas equation to about 1% under ordinary conditions, and the molar volumes of vapor phases are typically several hundred times as large as the molar volumes of condensed phases.

We write for a vaporization (liquid-vapor transition):

$$\Delta \bar{V} = \bar{V}^{(g)} - \bar{V}^{(l)} \approx \bar{V}^{(g)} \approx \frac{RT}{P} \qquad \textbf{(5.3-11)}$$

Equation (5.3-7) now becomes the **derivative form** of the **Clausius-Clapeyron equation**:

$$\frac{dP}{dT} = \frac{P\,\Delta \bar{H}_{\text{vap}}}{RT^2} \qquad \textbf{(5.3-12)}$$

where $\Delta \bar{H}_{\text{vap}}$ is the molar enthalpy change of vaporization. For a sublimation (solid-vapor transition), Equation (5.3-12) applies except that $\Delta \bar{H}_{\text{vap}}$ is replaced by $\Delta \bar{H}_{\text{sub}}$, the molar enthalpy change of sublimation. We omit the subscript and apply the equation to either case.

We multiply Equation (5.3-12) by dT and divide by P:

$$\frac{1}{P}\frac{dP}{dT}\,dT = \frac{\Delta \bar{H}}{RT^2}\,dT \qquad \textbf{(5.3-13)}$$

Integrating Equation (5.3-13) with the assumption that $\Delta \bar{H}$ is constant gives the **integral form** of the **Clausius-Clapeyron equation**:

$$\ln\left(\frac{P_2}{P_1}\right) = -\frac{\Delta \bar{H}}{R}\left(\frac{1}{T_2} - \frac{1}{T_1}\right) \qquad \textbf{(5.3-14)}$$

If the enthalpy change of vaporization or sublimation depends on temperature, then $\Delta \bar{H}$ in this formula represents an average enthalpy change over the interval from T_1 to T_2.

| Exercise 5.5 | Carry out the steps to obtain Equation (5.3-14). |

▼ **EXAMPLE 5.4** Using the vapor pressure values for water given in Section 5.2, find the average enthalpy change of vaporization of water between 25.00°C and 100.00°C.

Solution

$$\Delta \bar{H}_{vap} = \frac{-RT_2T_1}{T_1 - T_2} \ln\left(\frac{P_2}{P_1}\right)$$

$$= \frac{(8.3145 \text{ J K}^{-1} \text{ mol}^{-1})(298.15 \text{ K})(373.15 \text{ K})}{75.00 \text{ K}} \ln\left(\frac{760.00 \text{ torr}}{23.756 \text{ torr}}\right)$$

$$= 4.274 \times 10^4 \text{ J mol}^{-1}$$

▲

| Exercise 5.6 | The molar enthalpy change of vaporization of benzene is 34.08 kJ mol^{-1}. From the fact that the normal boiling temperature is 80.1°C, find the vapor pressure of benzene at 50.0°C, assuming $\Delta \bar{H}_{vap}$ is constant. |

If the enthalpy change of a substance is not known and one wishes to estimate the vapor pressure of the liquid at one temperature from the known vapor pressure at another temperature, Trouton's rule (see Section 3.5) can be used as an approximation.

| Exercise 5.7 | The normal boiling temperature of chloroform is 61.7°C. Estimate the vapor pressure of chloroform at 50.0°C, using Trouton's rule. |

The vapor pressure we have discussed thus far is measured with no other substances present. We now investigate the effect of other substances. We assume that the amount of other substances that dissolves in the condensed phase is negligible, so the only effect of the other substances is to change the pressure on the condensed phase.

We denote the total pressure by P' and the vapor pressure by P. We apply the fundamental fact of phase equilibrium:

$$\bar{G}^{(l)} = \bar{G}^{(g)} \tag{5.3-15}$$

When the total pressure is changed from one value P'_1 to another value P'_2, the vapor pressure changes from one value P_1 to another value P_2. Since equilibrium is maintained, the change in the molar Gibbs energies of the two phases must be equal. Assuming the vapor phase to be ideal and using Equations (4.3-10) and (4.3-4),

$$\bar{V}^{(l)}(P'_2 - P'_1) = RT \ln(P_2/P_1) \tag{5.3-16}$$

Equation (5.3-16) can be solved for P_2:

$$P_2 = P_1 \exp\left[\frac{\bar{V}^{(l)}(P'_2 - P'_1)}{RT}\right] \tag{5.3-17}$$

▼ **EXAMPLE 5.5**

At 298.15 K the vapor pressure of water is 23.756 torr if the total pressure is equal to the vapor pressure. Calculate the vapor pressure of water if enough air is present in the vapor phase to give a total pressure of 1.000 atm.

Solution

The molar volume of water at this temperature is 18.05 cm^3 mol^{-1}.

$$\frac{\bar{V}^{(l)}(P'_2 - P'_1)}{RT} = \frac{(18.05 \times 10^{-6} \text{ m}^3 \text{ mol}^{-1})(760 \text{ torr} - 24 \text{ torr})}{(8.3145 \text{ J K}^{-1} \text{ mol}^{-1})(298.15 \text{ K})} \frac{101325 \text{ Pa}}{760 \text{ torr}}$$

$$= 7.14 \times 10^{-4}$$

$$P = (23.756 \text{ torr})e^{7.14 \times 10^{-4}} = 23.773 \text{ torr}$$

For most purposes, this change in vapor pressure is negligible.

▲

Exercise 5.8

Find the total pressure necessary to change the vapor pressure of water by 1.00% at 298.15 K.

5.4 The Gibbs Energy and Phase Transitions

Why is water a liquid at 1 atm and 373.14 K but a vapor at 1 atm and 373.16 K? Why should such a small change in temperature result in such a large change in structure? In this section, we investigate such questions.

The thermodynamic answer to this question involves the fact that at equilibrium at constant T and P, the Gibbs energy of the system is at a minimum. Figure 5.5 shows schematically the molar Gibbs energy (chemical potential) of liquid and gaseous water as a function of temperature at 1 atm pressure. If one phase has a more negative value of the molar Gibbs energy than the other phase, the system can lower its Gibbs energy by making the transition to the phase of lower molar Gibbs energy.

In order to construct Figure 5.5, we have written Equation (4.2-20) for molar quantities:

Since 1 mol is a fixed quantity, we can generally replace a partial derivative at fixed n by a derivative of a molar quantity without specification that n is fixed.

$$\left(\frac{\partial \bar{G}}{\partial T}\right)_P = -\bar{S} \tag{5.4-1}$$

The molar entropy of the water vapor is greater than the molar entropy of the liquid water, so the vapor curve in Figure 5.5 has a more negative slope than the liquid curve. The temperature at which the curves cross is the temperature of phase coexistence, because at this temperature the values

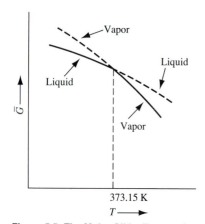

Figure 5.5. The Molar Gibbs Energy of Water as a Function of Temperature Near the Liquid-Vapor Phase Transition (schematic). This diagram, which applies at constant pressure, shows how minimization of the Gibbs energy determines the stable phase at any temperature.

of the chemical potential in the two phases are equal. Above this temperature, the vapor curve lies lower, so the vapor is the stable phase and the liquid must be metastable (indicated by a broken curve). Below the normal boiling temperature, the liquid is stable and the vapor is metastable.

We can reach the same conclusion in another way. The molar Gibbs energy is given by

$$\bar{G} = \bar{H} - T\bar{S} \tag{5.4-2}$$

At constant temperature, we can minimize $\bar{G}$ either by lowering $\bar{H}$ or by raising $\bar{S}$. The second term is more important at high temperature than at low temperature, because it is proportional to T. The phase of higher molar entropy is the more stable phase at high temperature, but the phase of lower molar enthalpy is the more stable phase at low temperature. The temperature of coexistence is the temperature at which these two tendencies balance each other.

Figure 5.6 shows the molar Gibbs energy of liquid and gaseous water as a function of pressure. The slope of the tangent to the curve is given by Equation (4.2-21):

$$\left(\frac{\partial \bar{G}}{\partial P}\right)_T = \bar{V} \tag{5.4-3}$$

The molar volume of the vapor is greater than that of the liquid phase, so the vapor curve has a more positive slope than the liquid curve. At a pressure greater than 1 atm at 373.15 K, the liquid is the stable phase, but at a pressure less than 1 atm, the vapor is the stable phase.

Exercise 5.9   Sketch rough graphs representing the molar Gibbs energy of water as a function of the temperature and as a function of the pressure in the vicinity of the solid-liquid phase transition. Liquid water has a smaller molar volume than solid water.

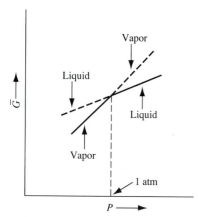

Figure 5.6. The Molar Gibbs Energy of Water as a Function of Pressure Near the Liquid-Vapor Phase Transition (schematic). This diagram, which applies at constant temperature, shows how minimization of the Gibbs energy determines the stable phase at any pressure.

For a fluid described by a particular equation of state, we can locate the pressure of liquid-vapor coexistence thermodynamically. Figure 5.7a shows the pressure as a function of molar volume at a fixed subcritical temperature as described by an equation of state such as the van der Waals equation. Instead of the tie line that actually describes the behavior of the fluid, there is an S-shaped curve (a "loop"). If we exchange the roles of the variables in this figure, we obtain Figure 5.7b. We want to locate two points, labeled a and e, on the curve in this figure, at which the chemical potentials in the two phases are equal, so that the phases can coexist at equilibrium. There are two regions, from a to b and from d to e, that can represent metastable phases. There is one region, from b to d, that corresponds to a negative compressibility. A real compressibility cannot be negative, so this portion of the curve cannot represent even a metastable system.

Since the curve corresponds to fixed temperature,

$$d\mu = d\bar{G} = \bar{V}\, dP \quad \text{(constant temperature)} \tag{5.4-4}$$

In order to have the molar Gibbs energy equal at points a and e, the integral of $d\bar{G}$ along the curve from point a to point e must vanish. This integral is written in

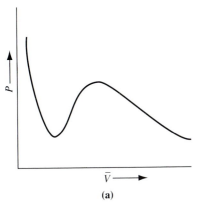

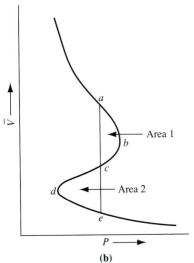

Figure 5.7. The Pressure and the Molar Volume for a Fluid Obeying an Equation of State Such as the van der Waals Equation (schematic). (a) The pressure as a function of molar volume. The region of positive slope cannot describe any real system, because it corresponds to negative compressibility. **(b)** The molar volume as a function of pressure. This diagram is used to determine the liquid state and the vapor state that have the same molar Gibbs energy, using the Maxwell construction.

the following way, because the curve does not represent a single-valued function:

$$\bar{G}(e) - \bar{G}(a) = \int_{P_a}^{P_b} \bar{V}\, dP + \int_{P_b}^{P_c} \bar{V}\, dP + \int_{P_c}^{P_d} \bar{V}\, dP + \int_{P_d}^{P_e} \bar{V}\, dP \quad \text{(5.4-5)}$$

The area to the right of the vertical line segment between points a and e is called area 1 and is equal to

$$\text{Area 1} = \int_{P_a}^{P_b} \bar{V}\, dP - \int_{P_c}^{P_b} \bar{V}\, dP = \int_{P_a}^{P_b} \bar{V}\, dP + \int_{P_b}^{P_c} \bar{V}\, dP \quad \text{(5.4-6)}$$

and the area to the left of this line segment, labeled area 2 in the figure, is given by

$$\text{Area 2} = \int_{P_d}^{P_c} \bar{V}\, dP - \int_{P_d}^{P_e} \bar{V}\, dP = -\int_{P_c}^{P_d} \bar{V}\, dP - \int_{P_d}^{P_e} \bar{V}\, dP \quad \text{(5.4-7)}$$

Comparison of Equation (5.4-5) with Equations (5.4-6) and (5.4-7) shows that when $\bar{G}(e) - \bar{G}(a) = 0$, area 1 and area 2 are equal. Adjustment of the locations of points a and e to make these areas equal is known as the **equal-area construction** and is due to Maxwell.

The van der Waals equation of state provides only a qualitatively correct description of the liquid-vapor transition when the equal-area construction is applied to it. The other common equations of state provide varying degrees of accuracy in describing the liquid-vapor transition when the equal-area construction is applied to them. Gibbons and Laughton obtained good agreement with experiment with their modification of the Redlich-Kwong equation of state (see Table 1.1).

Classification of Phase Transitions

A **first-order phase transition** is one in which at least one of the first derivatives of the molar Gibbs energy is discontinuous at the phase transition. That is, either the molar volume or the molar entropy is discontinuous. Usually both quantities are discontinuous, but in some first-order phase transitions, such as the transition between ice VI and ice VII, only one of these quantities will be discontinuous.

Ordinary phase transitions, such as vaporizations and freezings, are first-order transitions. In a first-order transition, the Gibbs energy has a cusp in at least one of the graphs like those of Figure 5.5 or 5.6. Figure 5.8 shows schematically the molar volume of gaseous and liquid water as a function of pressure, and Figure 5.9 shows the molar entropy as a function of temperature for a typical first-order transition.

Second derivatives of the Gibbs energy are also used in characterizing phase transitions. Equations (4.2-20), (4.2-21), (1.3-9), and (4.5-5) give

$$\left(\frac{\partial^2 \bar{G}}{\partial T^2} \right)_P = -\left(\frac{\partial \bar{S}}{\partial T} \right)_P = -\frac{\bar{C}_P}{T} \quad \text{(5.4-8)}$$

$$\left(\frac{\partial^2 \bar{G}}{\partial P^2} \right)_T = \left(\frac{\partial \bar{V}}{\partial P} \right)_T = -\bar{V}\kappa_T \quad \text{(5.4-9)}$$

where $\bar{C}_P$ is the molar heat capacity at constant pressure and κ_T is the isothermal compressibility.

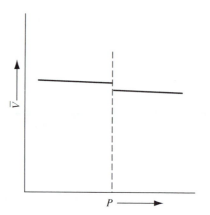

Figure 5.8. The Molar Volume as a Function of Pressure at a First-Order Phase Transition. There is a discontinuity in the molar volume at the phase transition.

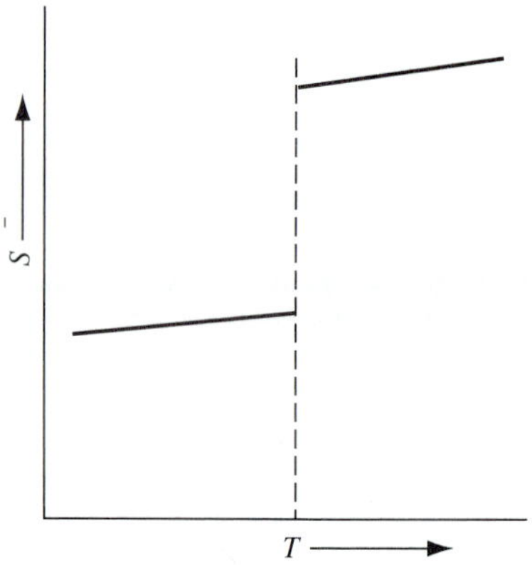

Figure 5.9. The Molar Entropy as a Function of Temperature at a First-Order Phase Transition. There is a discontinuity in the molar entropy at the phase transition.

If the molar volume or the molar entropy has a discontinuity, the heat capacity or the compressibility must have a singularity (a point at which it becomes infinite). Figure 5.10 shows the constant-pressure heat capacity as a function of temperature in the vicinity of a first-order phase transition, and Figure 5.11 shows the compressibility as a function of pressure in the vicinity of a first-order phase transition. The infinite value of the heat capacity at the first-order phase transition corresponds to the fact that a nonzero amount of heat produces no change in the temperature as one phase is converted to the other, and the infinite value of the compressibility corresponds to the fact that a finite volume change occurs as one phase is converted to the other with no change in the pressure.

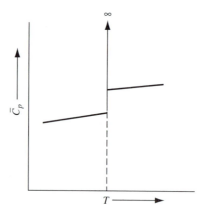

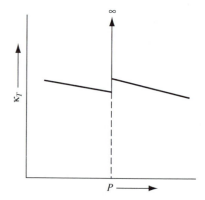

Figure 5.10. The Constant-Pressure Heat Capacity as a Function of Temperature at a First-Order Phase Transition. Because of the discontinuity in the entropy as a function of temperature, there is an infinite spike in the heat capacity at the phase transition.

Figure 5.11. The Isothermal Compressibility as a Function of Pressure at a First-Order Phase Transition (schematic). Because of the discontinuity in the molar volume as a function of pressure, there is an infinite spike in the compressibility at the phase transition.

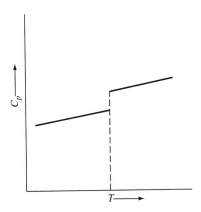

Figure 5.12. The Constant-Pressure Heat Capacity as a Function of Temperature at a Second-Order Phase Transition (schematic). In a second-order phase transition, there is no discontinuity in the molar entropy as a function of temperature, but there is a discontinuity in the heat capacity at the phase transition.

A **second-order phase transition** is one in which both of the first derivatives of the Gibbs energy are continuous but at least one of the second derivatives is discontinuous. Figure 5.12 shows the heat capacity in the vicinity of a second-order phase transition, and Figure 5.13 shows the compressibility in the vicinity of a second-order phase transition. In a second-order transition, neither the compressibility nor the heat capacity becomes infinite. Therefore, the molar entropies of the two phases must be equal and the molar volumes of the two phases must be equal.

Several other types of phase transitions occur in addition to those between solids and liquids, liquids and vapors, etc. These include normal-superconducting transitions in some metals and ceramics, paramagnetic-ferromagnetic transitions in some magnetic materials, and a type of transition that occurs in certain solid metal alloys, called an **order-disorder transition**. For example, beta brass, which is a nearly equimolar mixture of copper and zinc, has a low-temperature equilibrium state in which every copper atom in the crystal lattice is located at the center of a cubic unit cell, surrounded by eight zinc atoms at the corners of the cell. At 742 K, an order-disorder transition occurs from the ordered low-temperature state to a disordered state in which the atoms are randomly mixed in the same crystal lattice.

The order of a phase transition must be determined experimentally. To establish whether a phase transition is second order, the compressibility and the heat capacity must be measured carefully to determine whether they diverge at the phase transition. Second-order phase transitions are not common. The transition between normal and superconducting states is said to be the only well-established second-order transition. A phase transition that was once said to be second order is the transition between normal liquid helium and liquid helium II. (See Figure 5.4 for the phase diagram.) Later experiments indicated that the heat capacity of liquid helium appears to approach infinity at the transition, so the transition is not second order. However, the heat capacity rises smoothly toward infinity instead of rising abruptly as in a first-order transition. A plot of the heat capacity versus temperature has the general appearance of the Greek letter lambda, as shown in Figure 5.14, and the transition is called a **lambda transition**. The order-disorder transition in beta brass is also a lambda transition. A lambda transition is generally considered to be neither first order nor second order.

The Critical Point of a Liquid-Vapor Transition

The liquid-vapor critical point, which was introduced in Section 1.6, is the point beyond which the liquid-vapor phase transition does not occur. As can be seen in Figures 1.5 and 1.10, the molar volumes of the liquid and the vapor become more nearly equal as the temperature is increased toward the critical temperature. The discontinuity in Figure 5.8 gradually shrinks to zero, and two curves in Figure 5.6 approach each other more and more closely until there is only one curve at the critical point and above. Figure 5.15 shows schematically the molar volume as a function of pressure at the critical temperature.

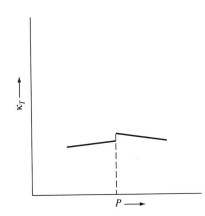

Figure 5.13. The Isothermal Compressibility as a Function of Pressure at a Second-Order Phase Transition. In a second-order phase transition, there is no discontinuity in the molar volume as a function of pressure, but there is a discontinuity in the compressibility at the phase transition.

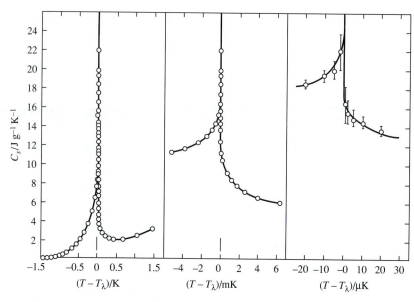

Figure 5.14. The Heat Capacity of Helium near the Lambda Transition. The heat capacity appears to become infinite, as in a first-order phase transition, but it rises smoothly instead of showing a spike at one point as does a first-order phase transition.

From H. E. Stanley, *Introduction to Phase Transitions and Critical Phenomena*, Oxford University Press, New York, 1971, p. 20.

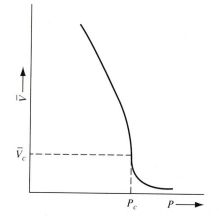

Figure 5.15. The Molar Volume near a Liquid-Vapor Critical Point. On the critical-temperature isotherm, the compressibility has an infinite value at the critical point, where the derivative $(\partial V/\partial P)_T$ is undefined.

The properties of the system in states near the critical point are abnormal. Quantities such as the compressibility and the heat capacity, which become infinite at the critical point, do not suddenly jump to infinite values as they do away from the critical point, but rise smoothly (and steeply) toward infinite values. In the vicinity of the critical point, the fluid scatters light strongly, a phenomenon called **critical opalescence**. Furthermore, as shown in Figure 1.6, if the system is near the critical state the meniscus between the liquid and vapor phases is diffuse, instead of appearing to be sharp like a mathematical plane.

Although it is not quantitatively accurate, the van der Waals equation of state predicts critical behavior that resembles that of a real fluid. For a van der Waals gas, Equation (1.6-3) gives the expression for $(\partial P/\partial \bar{V})_T$, and the reciprocal of this expression gives

$$\left(\frac{\partial \bar{V}}{\partial P}\right)_T = \frac{\bar{V}^3(\bar{V} - b)^2}{-RT\bar{V}^3 + 2a(\bar{V} - b)^2} \tag{5.4-10}$$

The compressibility is this derivative multiplied by $-1/\bar{V}$. We define the temporary variable $x = (\bar{V}/b) - 3$. By Equation (1.6-5), this variable vanishes at the critical point.

$$\left(\frac{\partial \bar{V}}{\partial P}\right)_T = \frac{b^3}{2a}\left[\frac{(3 + x)^3(2 + x)^2}{-(4/27)(3 + x)^3 + (2 + x)^2}\right] \tag{5.4-11}$$

where we have replaced T by its value at the critical point, given in Equation (1.6-5).

To find out what happens to the compressibility very close to the critical point, we investigate the behavior of $(\partial \bar{V}/\partial P)_T$ for small values of x. If x is small enough, x^2 is negligible compared with x, x^3 is negligible compared with x^2, etc. By multiplying out the factors in Equation (5.4-11) and keeping only the term of lowest power of x in each polynomial, we obtain a relation showing how $(\partial \bar{V}/\partial P)_T$ diverges at the critical point:

$$\left(\frac{\partial \bar{V}}{\partial P}\right)_T \approx -\frac{81b^3}{ax^2} \quad \text{(small } x \text{ only)} \tag{5.4-12a}$$

$$\kappa_T \approx \frac{27b^2}{ax^2} \tag{5.4-12b}$$

where we have used the fact that near the critical point $\bar{V} \approx 3b$. As x becomes small, the compressibility diverges as $1/x^2$.

Exercise 5.10   Verify Equations (5.4-11) and (5.4-12).

A number of properties either become infinite or vanish in characteristic ways as the critical point is approached. This behavior is described in terms of **critical exponents**, analogous to the exponent -2 in Equation (5.4-12). For example, the temperature dependence of the constant-volume heat capacity for fixed density equal to the critical density can be represented in the immediate vicinity of the critical point by the expression

$$C_V \propto (T - T_c)^{-\alpha} \quad \text{(for } T > T_c) \tag{5.4-13a}$$

$$C_V \propto (T_c - T)^{-\alpha'} \quad \text{(for } T < T_c) \tag{5.4-13b}$$

The exponents α and α' are the critical exponents for the heat capacity.

Similarly, the densities of the coexisting liquid and gas phases just below the critical point are represented by

$$\rho(l) - \rho(g) \propto (T_c - T)^{\beta} \tag{5.4-14}$$

The temperature dependence of the compressibility for fixed density equal to the critical density is represented by

$$1/\kappa_T \propto (T - T_c)^{-\gamma} \quad \text{(for } T > T_c) \tag{5.4-15a}$$

$$1/\kappa_T \propto (T_c - T)^{-\gamma'} \quad \text{(for } T < T_c) \tag{5.4-15b}$$

The dependence of the pressure on the density at constant temperature equal to the critical temperature is described by

$$P - P_c \propto |\rho - \rho_c|^{\delta - 1}(\rho - \rho_c) \tag{5.4-16}$$

Equation (5.4-16) has this form to ensure that the correct sign is produced for the left-hand side of the equation for any value of δ.

It appears that many fluids are well described by nearly equal values of the critical exponents, as would be expected from the law of corresponding states. Following are some experimental values, along with the values predicted by the van der Waals equation of state. We do not discuss the analysis that provides these predictions.[4]

	α'	β	γ'	δ	α	γ
CO_2	0.1	0.34	1.0	4.2	0.1	1.35
Xe	<0.2	0.35	1.2	4.4	—	1.3
van der Waals	0	0.5	1	3	0	1

5.5   Surface Structure and Thermodynamics

In our earlier discussions of thermodynamics, we stated that certain equations are valid only when surface tension can be neglected. For example, we assumed that the thermodynamic energy of a one-component fluid system depended on three variables, such as T, V, and n, but not on the surface area. The energy actually depends on the surface area, because molecules at the surface of the liquid have fewer nearest neighbors than molecules in the bulk of the liquid, although this contribution to the energy is generally small.

▼   **EXAMPLE 5.6**

For liquid carbon tetrachloride in contact with its vapor, estimate the surface energy per unit area, using the heat of vaporization to estimate the net attractive energy of the molecules.

Solution

Carbon tetrachloride molecules have a tetrahedral shape, which is nearly spherical. In a solid lattice, a spherical molecule can be surrounded by 12 molecules of the same size. Since the liquid is somewhat disordered and is less dense than the solid, we assume that each molecule in the interior of the liquid has about 10 nearest neighbors. Molecules at the surface of the liquid have no nearest neighbors on the vapor side, and we assume that a molecule at the surface has about seven nearest neighbors.

To estimate the surface energy, we first estimate the difference between the average attractive energy of a molecule and its neighbors in the bulk liquid and at the surface. The molar enthalpy change of vaporization of CCl_4 at 20°C is equal to 33.77 kJ mol^{-1}, so the molar energy change of vaporization is

$$\Delta \bar{U}_{vap} = \Delta \bar{H}_{vap} - \Delta(PV) \approx \Delta \bar{H}_{vap} - RT$$
$$= 33770 \text{ J mol}^{-1} - (8.3145 \text{ J K}^{-1} \text{ mol}^{-1})(293 \text{ K})$$
$$= 31330 \text{ J mol}^{-1}$$

---

[4] H. E. Stanley, *Introduction to Phase Transitions and Critical Phenomena*, Oxford University Press, New York, 1971, pp. 74ff.

The energy change of vaporization per molecule is

$$\frac{31330 \text{ J mol}^{-1}}{6.022 \times 10^{23} \text{ mol}^{-1}} = 5.203 \times 10^{-20} \text{ J}$$

When a molecule is brought to the surface, it gains energy equal to 30% of this value because it loses 3 of 10 nearest neighbors instead of losing all 10 as in vaporization. The surface energy per molecule is thus

$$\text{Surface energy per molecule} = (5.203 \times 10^{-20} \text{ J})(0.30)$$
$$= 1.6 \times 10^{-20} \text{ J}$$

We now estimate the surface area per molecule. The density of CCl_4 at 20°C is 1.594 g cm^{-3}, and its molar mass is 153.82 g mol^{-1}. This gives a molar volume of 96.5 cm^3 mol^{-1} and a volume per molecule of

$$\text{Volume per molecule} = \frac{96.5 \times 10^{-6} \text{ m}^3 \text{ mol}^{-1}}{6.022 \times 10^{23} \text{ mol}^{-1}} = 1.60 \times 10^{-28} \text{ m}^3$$

A sphere of this volume has a radius of 3.4×10^{-10} m, so we take this as the radius of the molecule.

$$\text{Area per molecule} = \pi r^2 = \pi (3.4 \times 10^{-10} \text{ m})^2$$
$$= 3.6 \times 10^{-19} \text{ m}^2$$

The surface energy per m^2 is

$$\gamma \approx \frac{1.6 \times 10^{-20} \text{ J}}{3.6 \times 10^{-19} \text{ m}^2} = 0.043 \text{ J m}^{-2}$$

This value agrees only roughly with the experimental value, 0.02695 J m^{-2} at 20°C.

The analysis in Example 5.6 is mostly a web of assumptions. It is important to be able to decide which assumptions are reasonable. Some people develop this skill by such exercises as estimating the number of piano tuners in New York City by knowing the total population and making reasonable assumptions.

▲

Exercise 5.11

a. Estimate the surface energy per square meter for liquid water, assuming that the principal intermolecular force is hydrogen bonding, with a bond energy of 20. kJ mol^{-1} for each hydrogen bond. Assume that a molecule in the interior of a sample of liquid water has four hydrogen-bonded nearest neighbors and a molecule in the surface has three. Compare your result with the experimental value at 25°C, 0.072 J m^{-2}.

b. For 1.00 mol of water contained in a beaker of diameter 5.00 cm, find the ratio of the surface energy of the upper surface to the energy required to vaporize 1.00 mol of water.

c. Explain why gases have negligible surface energy.

From the values of surface energies, it is apparent that the surface energy is insignificant unless a system has a very large surface area.

In some systems, there can be several different kinds of surfaces. For example, a system consisting of water and diethyl ether near room temperature may have a liquid phase that is mostly water, a liquid phase that is mostly diethyl ether, and a vapor phase. There is one surface between the two liquid phases and one surface between the upper liquid phase and the vapor phase, and each phase has a surface between itself and the container. The energy, Gibbs energy, and other energy-related functions depend separately on the area of each of these surfaces. We first consider a system with only one component and one kind of surface. The liquid-vapor surface is the easiest to visualize, as in Example 5.6.

We consider G to be a function of T, P, $\mathscr{A}$ (the surface area), and the amount of the substance, n:

$$dG = -S\,dT + V\,dP + \gamma\,d\mathscr{A} + \mu\,dn \tag{5.5-1}$$

where

$$\gamma = \left(\frac{\partial G}{\partial \mathscr{A}}\right)_{T,P,n} \tag{5.5-2}$$

The quantity γ is called the **surface tension** or **interfacial tension**. Since $U = G - PV + TS$,

$$dU = T\,dS - P\,dV + \gamma\,d\mathscr{A} + \mu\,dn \tag{5.5-3}$$

The surface tension can be interpreted as an internal energy per unit area or as a Gibbs energy per unit area. It is also equal to the reversible work per unit area required to produce new surface, as in Example 5.6. It can also be interpreted as a force per unit length. In SI units, γ can be expressed in either joules per square meter or newtons per meter. Consider a system such as that in Figure 5.16, which has a wire frame that protrudes from the surface of a sample of liquid. The film of liquid within the area of the frame has length equal to L. Let the frame be moved reversibly a distance dx, increasing the area. We assume that no substance is transferred to or from the vapor phase surrounding the system and that the volume of the system stays fixed. We have

$$dU = T\,dS + \gamma\,d\mathscr{A} \tag{5.5-4}$$

Since there are two sides to the liquid layer, the area increases by $2L\,dx$, and

$$dU = T\,dS + \gamma 2L\,dx \tag{5.5-5}$$

The arrangement shown in Figure 15.16 will not support a film of pure water, but works well with a detergent solution.

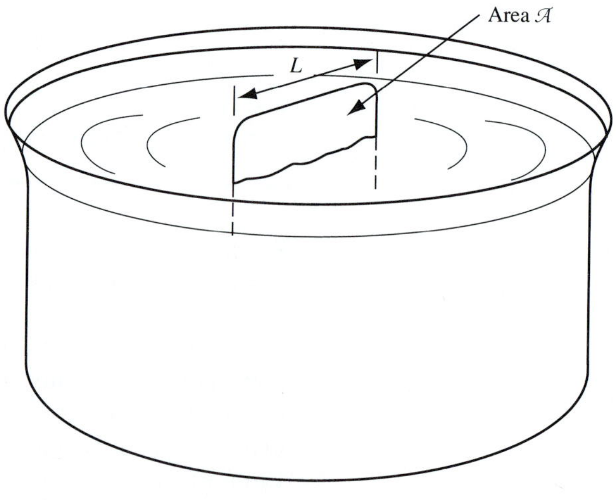

Figure 5.16. A Wire Frame of Adjustable Size to Illustrate Surface Tension. As the wire frame is raised, a thin film of liquid fills the area within the frame.

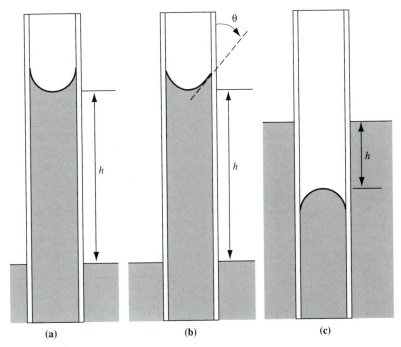

Figure 5.17. Capillary Rise or Depression of a Liquid in a Tube. (a) A liquid that wets the solid surface. The liquid surface (the meniscus) is tangent to the inner surface of the tube. This surface approximates a hemisphere. **(b)** A liquid that partially wets the solid surface. The contact angle, θ, is larger than 0, so the liquid surface is not tangent to the inner surface of the tube. The liquid surface approximates a portion of a hemisphere. **(c)** Capillary depression of mercury in a glass capillary. Because the liquid does not wet the glass surface, the contact angle is nearly equal to 180° and the meniscus is depressed below the surface of the bulk liquid.

By comparison with Equation (2.1-6),

$$dw_{\mathrm{rev}} = F_{\mathrm{rev}}\,dx = 2L\gamma\,dx = \gamma\,d\mathscr{A} \qquad \textbf{(5.5-6)}$$

and γ is recognized as the force exerted by the surface per unit length, justifying the name "surface tension." Table A11 gives values of the surface tension for several pure substances in contact with vapor of the same substance.

The surface of a liquid acts somewhat like a stretched elastic film, which has a tendency to reduce the area of the surface. Since the surface tension exerts a force on a liquid surface, a liquid can be attracted into or repelled from a tube of small diameter (or other small passages). For example, water wets a glass surface. The polar water molecules are attracted strongly enough to the polar and ionic groups on the glass surface that the surface is covered with a film of water, and water rises into a vertical glass tube. The liquid surface is tangent to the glass surface, as shown in Figure 5.17a. Figure 5.17b shows a case in which the liquid does not completely wet the surface of the solid. The liquid surface meets the solid surface at an angle, which is called the **contact angle**, denoted by θ. In the case of water on glass, the contact angle is nearly equal to zero.

The surface tension force is exerted tangent to the surface, so it is parallel to the tube if the contact angle is zero or 180°. At equilibrium the surface tension force balances the gravitational force on the liquid in the tube. If the radius of a vertical tube is r, the vertical surface tension force is equal to $2\pi r\gamma$. If the density of the liquid is ρ, the gravitational force on the column of liquid is $\pi r^2 h\rho g$, where g is the acceleration due to gravity and h is the height of the column. The height of the column is thus

$$h = \frac{2\gamma}{\rho g r} \tag{5.5-7}$$

Exercise 5.12

a. Find the height to which water at 20°C will rise in a capillary tube of diameter 0.35 mm.

b. Find the height to which the surface of water will rise in a tube with a diameter equal to 7.0 cm.

If a liquid does not completely wet the surface of a capillary tube, the surface tension force is not vertical but is exerted at an angle of θ from the vertical. Therefore, its upward component is $2\pi r\gamma \cos(\theta)$, and the height to which the liquid rises is

$$h = \frac{2\gamma \cos(\theta)}{\rho g r} \tag{5.5-8}$$

Mercury is attracted so weakly to glass that it forms a contact angle of nearly 180°. The surface tension force is downward, and a mercury meniscus is depressed in a glass capillary tube, as shown in Figure 5.17c.

Exercise 5.13

Assuming a contact angle of 180°, calculate the distance to which the mercury meniscus is depressed in a glass capillary tube of radius 1.00 mm.

Because the surface of a liquid acts like a stretched film, a sample of liquid adopts the shape of a sphere if the surface tension outweighs the effects of gravity. Because the surface of a liquid exerts a force on the liquid inside a small droplet, there is a difference between the pressures inside and outside the droplet. In order to obtain a formula for this pressure difference, we discuss the system depicted in Figure 5.18. The system is contained in a cylinder with a movable piston. There is a small droplet of liquid suspended in a vapor phase. If it is small enough, it can remain suspended in air for some time.

The piston is displaced reversibly so that the volume of the system changes. The work done is

$$dw_{\mathrm{rev}} = -P^{(g)} dV = -P^{(g)}(dV^{(g)} + dV^{(l)}) \tag{5.5-9}$$

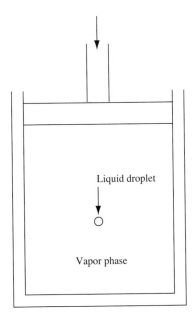

Figure 5.18. System to Illustrate the Pressure Difference in a Spherical Droplet. The droplet is assumed to be surrounded by the vapor phase and is assumed to be small enough that it does not immediately settle to the bottom of the vessel.

Laplace, Pierre Simon, Marquis de, 1749–1827, great French mathematician and astronomer, proposed that the solar system condensed from a rotating gas cloud.

where $P^{(g)}$ is the pressure of the gas phase, $P^{(l)}$ is the pressure of the liquid phase, $V^{(g)}$ is the volume of the gas phase, and $V^{(l)}$ is the volume of the liquid phase. Only the pressure of the gas phase enters, because only the gas phase is in contact with the piston.

We can write a different expression for the reversible work by adding the term of Equation (5.5-6) to the usual expression for dw_{rev} for the liquid phase. No such term is added for the vapor phase, which has negligible surface tension:

$$dw_{rev} = -P^{(g)} \, dV^{(g)} - P^{(l)} \, dV^{(l)} + \gamma \, d\mathscr{A} \qquad \textbf{(5.5-10)}$$

Equating the two expressions for dw_{rev} and canceling the term $P^{(g)} \, dV^{(g)}$ from both sides, we obtain

$$(P^{(l)} - P^{(g)}) \, dV^{(l)} = \gamma \, d\mathscr{A} \qquad \textbf{(5.5-11)}$$

Since the droplet is assumed to be spherical, the volume of the liquid and the area of the droplet are both functions of the radius of the droplet.

$$dV^{(l)} = d(\tfrac{4}{3}\pi r^3) = 4\pi r^2 \, dr \qquad \textbf{(5.5-12a)}$$

$$d\mathscr{A} = d(4\pi r^2) = 8\pi r \, dr \qquad \textbf{(5.5-12b)}$$

Substituting these relations into Equation (5.5-11) and canceling the common factor dr, we obtain

$$\boxed{P^{(l)} - P^{(g)} = \frac{2\gamma}{r}} \qquad \textbf{(5.5-13)}$$

Equation (5.5-13) is known as the **Laplace equation**. If the radius of the droplet becomes large, the difference in pressure between the inside and outside of the droplet approaches zero, as we would expect.

▼   **EXAMPLE 5.7**   Find the additional pressure inside a water droplet of radius 1.00 μm at 25°C.

Solution
At this temperature, the surface tension of water is 0.07197 N m^{-2}.

$$P^{(l)} - P^{(g)} = \frac{(2)(0.07197 \text{ N m}^{-1})}{1.00 \times 10^{-6} \text{ m}} = 1.44 \times 10^5 \text{ N m}^{-2}$$

$$= 1.42 \text{ atm}$$

▲

We have already learned that if the pressure on a sample of liquid is increased, the vapor pressure increases. We now derive a formula for the vapor pressure of a small droplet of a liquid and find that it increases as the size of the droplet decreases. We first rewrite Equation (5.5-1) for a planar surface:

$$dG = -S\,dT + V\,dP + \gamma\,d\mathscr{A} + \mu^{(p)}\,dn \qquad (5.5\text{-}14)$$

where the superscript (p) means that the value pertains to a planar surface. Let our system consist of a small spherical liquid droplet and the vapor at equilibrium with the droplet.

If the volume of the droplet is changed by an amount dV at constant T, constant P, and constant composition, then

$$dV = \bar{V}\,dn \qquad (5.5\text{-}15)$$

where $\bar{V}$ is the molar volume of the liquid. From Equation (5.5-12),

$$d\mathscr{A} = \frac{2}{r}\,dV = \frac{2\bar{V}}{r}\,dn \qquad (5.5\text{-}16)$$

Equation (5.5-14) can now be written

$$dG = -S\,dT + V\,dP + \left[\mu^{(p)} + \frac{2\bar{V}\gamma}{r}\right]dn \qquad (5.5\text{-}17)$$

The effective chemical potential is therefore

$$\mu = \mu^{(p)} + \frac{2\bar{V}\gamma}{r} \qquad (5.5\text{-}18)$$

If the vapor can be assumed ideal, the chemical potential and the partial vapor pressure are related by Equation (4.4-24):

$$\mu_i = \mu_i^\circ + RT\,\ln\left(\frac{P_i}{P^\circ}\right) \qquad (5.5\text{-}19)$$

When this equation is combined with Equation (5.5-18), we obtain

$$\boxed{\ln\left(\frac{P_i}{P_i^{(p)}}\right) = \frac{2\bar{V}\gamma}{rRT}} \qquad (5.5\text{-}20)$$

where $P_i^{(p)}$ is the partial vapor pressure for substance i at a planar surface.

▼ **EXAMPLE 5.8**

The vapor pressure of a planar surface of water at 298.15 K is 23.756 torr. Find the radius of a water droplet whose vapor pressure is 1.000 torr higher than this value at 298.15 K. At this temperature, the surface tension of water is 0.07197 N m^{-2}.

Solution

$$r = \frac{2(1.806 \times 10^{-5}\ \mathrm{m^3\ mol^{-1}})(0.07197\ \mathrm{J\ m^{-2}})}{(8.3145\ \mathrm{J\ K^{-1}\ mol^{-1}})(298.15\ \mathrm{K})\ \ln\!\left(\dfrac{24.756\ \mathrm{torr}}{23.756\ \mathrm{torr}}\right)}$$

$$= 2.543 \times 10^{-8}\ \mathrm{m} = 2.543 \times 10^{-5}\ \mathrm{mm}$$

▲

Because small droplets of a liquid have a larger vapor pressure than large droplets, under conditions of equilibrium with the vapor small droplets disappear while large droplets grow. This phenomenon is important in meteorology, because it is the mechanism by which raindrops grow large enough to fall.

The initial formation of a small droplet from the vapor is called **nucleation** and often requires a partial pressure of the liquid that is considerably larger than the equilibrium vapor pressure at a planar surface. This explains why clouds sometimes fail to form, even if water vapor is supersaturated. Most raindrops apparently nucleate on specks of solid material, and seeding of supersaturated air with small crystals of silver iodide is sometimes used in efforts to produce rain. **Homogeneous nucleation** (without a solid speck) requires spontaneous collection of a cluster of molecules, which is quite improbable unless the partial pressure of the substance greatly exceeds the equilibrium vapor pressure at a planar surface.

AgI is used for "cloud seeding" because it can be made into very small particles. It is possible to make 10^{15} particles of solid AgI from 1 gram of AgI by spraying a solution of AgI in acetone through a suitable nozzle, allowing the acetone to evaporate.

Just as the surface of a droplet produces a greater pressure inside the droplet than outside, the surface of a cavity inside a liquid produces a greater pressure inside the cavity (often carelessly called a "bubble"). The difference is that the vapor phase is outside a droplet and inside a cavity. The vapor pressure inside the cavity is therefore decreased below that of a planar surface.

Exercise 5.14   Show that the vapor pressure inside a cavity is given by

$$\ln\left(\frac{P_i}{P_i^{(p)}}\right) = -\frac{2\bar{V}\gamma}{rRT} \qquad \text{(5.5-21)}$$

Because the lessening of the vapor pressure inside a cavity is greater for a smaller cavity, the formation of a cavity requires a higher temperature than vaporization of the liquid from a planar surface. For this reason, a liquid can often be **superheated** well above its normal boiling temperature if its container is smooth and there are no dust particles in the liquid at which cavities can begin to form. In such a case, the liquid can suddenly boil ("bump") when the metastable superheated liquid finally begins to form cavities.

The surface energy of solids is generally even larger than that of liquids. However, since solids generally cannot flow and generally evaporate very slowly, these effects are not very noticeable.

5.6   Surfaces in Multicomponent Systems

In Example 5.6 we assumed that a surface had zero thickness: we calculated the energy as though one layer of molecules had the normal liquid on one side and the normal vapor on the other. A solid surface might be nearly like this crude model, but a liquid surface is more diffuse and it is perhaps appropriate to call it a surface region or even a surface phase. Figure 5.19a shows schematically an average density profile through a single-component liquid-vapor surface at equilibrium. The thickness of a liquid-vapor interfacial region is typically equal to several molecular diameters, perhaps nearly 1 nm.

Consider an arbitrary multicomponent two-phase system with either a liquid phase and a vapor phase or two liquid phases. Assuming that the boundary between the phases is planar and horizontal, we place three imaginary horizontal planes inside the interfacial region. The location of the center plane is denoted as z_0 in Figure 5.19a, and the other locations are called z_1 and z_2. We could speak of three phases, phase I, phase II, and a surface phase, extending from z_1 to z_2. However, this procedure has the two disadvantages that z_1 and z_2 are somewhat arbitrary and amounts of substances in the interfacial region are difficult to measure.

The standard procedure avoids these difficulties. The planes at z_1 and z_2 are abandoned, and a single dividing plane at z_0 is taken. The volume of each phase is assigned to be the volume that extends up to the dividing plane. The volume of the system is then exactly equal to the sum of the volumes of the two phases, and no volume is ascribed to the surface.

The concentration of each component in the homogeneous portion of each phase (the "bulk" portion) is extrapolated up to the surface plane, as shown in Figure 5.19a, which shows only one component. The amount of substance in each phase is assigned to be the amount that would occur if the concentration obeyed this extrapolation. We denote the amount of substance i thus assigned to phase I by $n_i^{(I)}$ and the amount assigned to phase II by $n_i^{(II)}$.

The shaded area to the right of the surface plane in Figure 5.19a represents the amount of substance that is present but not accounted for in

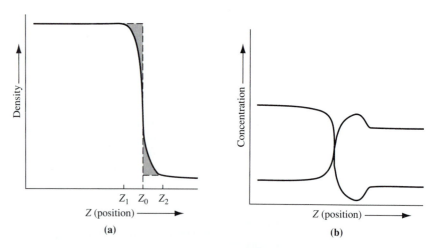

(a)   (b)

Figure 5.19. A Typical Density Profile Through a Surface Region (schematic). (a) A one-component system. The thickness of the surface region is several times a molecular diameter. The density profile shown in the diagram is an average profile. **(b)** A two-component system. The density profiles of the two substances are not required to have any simple relation to each other, because either or both of the substances can accumulate at the interface.

phase II by this convention. The shaded area to the left of the surface plane represents the amount of substance that is included in phase I by the convention but is actually not present.

For a one-component system, it is possible to place the plane in the interfacial region so that these two amounts of substance cancel, which would make the two shaded areas equal in size. In a multicomponent system, this cancellation can generally be achieved for only one substance. Figure 5.19b shows a schematic concentration profile for a two-component system, and it should be apparent that no placement of the plane will produce equal areas for both substances.

We now attribute an amount of each substance to the surface. We define for substance i

$$n_i^{(\sigma)} = n_i - n_i^{(\mathrm{I})} - n_i^{(\mathrm{II})} \quad \text{(definition)} \tag{5.6-1}$$

where n_i is the total amount of substance i and $n_i^{(\sigma)}$ is the amount of substance i attributed to the surface, called the **surface excess**. The surface excess of one substance (usually the solvent) can be made to vanish by appropriate placement of the plane, but the surface excess will then not generally vanish for the other substances. For a solute that accumulates at the surface, the surface excess is positive, and for a substance that avoids the surface, the surface excess is negative.

The amount of each substance attributed to each phase is the amount that would occur if each phase were a portion of a larger phase. Analogous values are assigned for all thermodynamic quantities. The values of the thermodynamic variables for each phase are equilibrium values for the volume assigned by the convention, as though this volume were an interior portion of a very large system (and thus with no surface contributions). They obey all of the equations of thermodynamics without surface contributions. For phase I,

$$dG^{(\mathrm{I})} = -S^{(\mathrm{I})}\, dT + V^{(\mathrm{I})}\, dP + \sum_{i=1}^{c} \mu_i\, dn_i \tag{5.6-2}$$

with a similar equation for phase II. The phases are at equilibrium, so T, P, and the μ's have the same values in all phases and require no superscripts.

Let us subtract Equation (5.6-2) and its analogue for phase II from the version of Equation (5.5-1) that applies to a multicomponent system:

$$d[G - G^{(\mathrm{I})} - G^{(\mathrm{II})}] = -(S - S^{(\mathrm{I})} - S^{(\mathrm{II})})\, dT + (V - V^{(\mathrm{I})} - V^{(\mathrm{II})})\, dP$$
$$+ \gamma\, d\mathscr{A} + \sum_{i=1}^{c} \mu_i\, d(n_i - n_i^{(\mathrm{I})} - n_i^{(\mathrm{II})}) \tag{5.6-3a}$$

which we rewrite as

$$dG^{(\sigma)} = -S^{(\sigma)}\, dT + \gamma\, d\mathscr{A} + \sum_{i=1}^{c} \mu_i\, dn_i^{(\sigma)} \tag{5.6-3b}$$

where $G^{(\sigma)}$ is called the **surface Gibbs energy**:

$$G^{(\sigma)} = G - G^{(\mathrm{I})} - G^{(\mathrm{II})} \tag{5.6-3c}$$

In Equation (5.6-3b), we have used the fact that $V^{(I)} + V^{(II)} = V$, so that the dP term vanishes. The surface Gibbs energy is the Gibbs energy that is assigned to the surface by our definition of the Gibbs energies of the phases. It depends on the location of the plane with which we divide the phases from each other. The definition of $S^{(\sigma)}$ is analogous to that of $G^{(\sigma)}$.

We assume that γ is an intensive variable, depending only on T, P, and the composition of the phases of the system. If so, when T and P are held fixed, G is a homogeneous function of $\mathscr{A}$ and the n's, so that Euler's theorem, instead of giving Equation (4.6-5), gives

$$G = \gamma\mathscr{A} + \sum_{i=1}^{c} \mu_i n_i \tag{5.6-4}$$

Exercise 5.15   Verify Equation (5.6-4).

Each phase obeys Euler's theorem without a surface term, so

$$G^{(I)} = \sum_{i=1}^{c} \mu_i n_i^{(I)} \tag{5.6-5}$$

with an analogous equation for phase II.

When Equation (5.6-5) and its analogue for phase II are subtracted from Equation (5.6-4), we obtain

$$G^{(\sigma)} = \gamma\mathscr{A} + \sum_{i=1}^{c} \mu_i n_i^{(\sigma)} \tag{5.6-6}$$

We can write an expression for $dG^{(\sigma)}$ from Equation (5.6-6):

$$dG^{(\sigma)} = \gamma \, d\mathscr{A} + \mathscr{A} \, d\gamma + \sum_{i=1}^{c} n_i^{(\sigma)} \, d\mu_i + \sum_{i=1}^{c} \mu_i \, dn_i^{(\sigma)} \tag{5.6-7}$$

Equating Equation (5.6-3b) and Equation (5.6-7) and dividing by $\mathscr{A}$, we obtain a surface version of the Gibbs-Duhem equation

$$0 = \frac{S^{(\sigma)}}{\mathscr{A}} \, dT + d\gamma + \sum_{i=1}^{c} \Gamma_i^{(\sigma)} \, d\mu_i \tag{5.6-8}$$

The surface excess per unit area, $\Gamma_i^{(\sigma)}$, is called the **surface concentration**:

$$\Gamma_i^{(\sigma)} = \frac{n_i^{(\sigma)}}{\mathscr{A}} \quad \text{(definition)} \tag{5.6-9}$$

If the temperature is constant,

$$d\gamma = -\sum_{i=1}^{c} \Gamma_i^{(\sigma)} \, d\mu_i \quad \text{(constant temperature)} \tag{5.6-10}$$

This equation can be interpreted as follows: if adding more of substance i (raising its chemical potential) decreases the surface tension, then $\Gamma_i^{(\sigma)}$ is positive and substance i accumulates at the interface. A substance that significantly lowers the surface tension and accumulates at the surface is called a **surfactant**. If raising the chemical potential of substance i increases the surface tension, then $\Gamma_i^{(\sigma)}$ is negative and substance i avoids the interface.

Many kinds of systems with interfaces are of intrinsic and practical interest. **Colloids** are suspensions of small solid particles in a liquid medium. **Aerosols** are suspensions of fine solid or liquid particles in a gas and are important in atmospheric chemistry and physics. Surface effects, including surface charges, dominate in determining the behavior of such systems, as well as some biological structures whose small size makes the surface important.

Summary

The fundamental fact of phase equilibrium is that

$$\mu_i^{(\alpha)} = \mu_i^{(\beta)}$$

where the subscript i denotes the substance and the superscript α or β denotes the phase.

The Gibbs phase rule is

$$f = c - p + 2$$

where f is the number of independent intensive variables, c is the number of components, and p is the number of phases.

The Clapeyron equation governs the curves in one-component pressure-temperature phase diagrams:

$$\frac{dP}{dT} = \frac{\Delta \bar{S}}{\Delta \bar{V}}$$

where P is the pressure at which two phases can coexist at equilibrium, $\Delta \bar{S}$ is the molar entropy change of the phase transition, and $\Delta \bar{V}$ is the molar volume change of the transition.

The integrated Clausius-Clapeyron equation is

$$\ln(P) = \frac{-\Delta \bar{H}}{RT} + \text{constant}$$

Inclusion of surface effects lead to the expression for dU:

$$dU = T\,dS - P\,dV + \gamma\,d\mathscr{A} + \mu\,dn$$

where γ is the surface tension and $\mathscr{A}$ is the interfacial surface area of the system. In most systems the effects of the surface energy are negligible.

Additional Reading

A. W. Adamson, *Physical Chemistry of Surfaces*, 2nd ed., Interscience Publishers, New York, 1967
This is a textbook for a graduate-level course in surface chemistry.

A. Findlay, *The Phase Rule and Its Applications*, 9th ed., edited by A. N. Campbell and N. O. Smith, Dover Publications, New York, 1951

This book is a standard source for information about phase equilibrium.

P. Gordon, *Principles of Phase Diagrams in Materials Systems*, McGraw-Hill, New York, 1968

This book is designed for materials scientists and presents a clear introduction to the thermodynamics of phase equilibrium. It contains a useful chapter on order-disorder transitions.

J. P. Hansen and I. R. McDonald, *Theory of Simple Liquids*, Academic Press, New York, 1976

Liquid theory is presented at an advanced level. Chapter 10 deals with phase transitions and critical phenomena.

C. Kittel and H. Kroemer, *Thermal Physics*, W. H. Freeman, San Francisco, 1980

A textbook for undergraduate students of physics. It contains a discussion of the theory of the van der Waals fluid, including critical phenomena, beginning on p. 288.

P. V. E. McClintock, D. J. Meredith, and J. K. Wigmore, *Matter at Low Temperatures*, John Wiley & Sons, New York, 1984

This is a readable and useful source for information about low-temperature properties, including phase diagrams of helium.

H. E. Stanley, *Introduction to Phase Transitions and Critical Phenomena*, Oxford University Press, New York, 1971

A general theoretical and experimental introduction to phase transitions, not only in fluid systems but also in magnetic systems.

PROBLEMS

Problems for Section 5.2

*5.16. For water at equilibrium at 1.000 atm and 273.15 K, find the values of $\bar{G}^{(l)} - \bar{G}^{(s)}$, $\bar{A}^{(l)} - \bar{A}^{(s)}$, $\bar{H}^{(l)} - \bar{H}^{(s)}$, $\bar{U}^{(l)} - \bar{U}^{(s)}$, and $\bar{S}^{(l)} - \bar{S}^{(s)}$.

5.17. For water at equilibrium at 1.000 atm and 373.15 K, find the values of $\bar{G}^{(g)} - \bar{G}^{(l)}$, $\bar{A}^{(g)} - \bar{A}^{(l)}$, $\bar{H}^{(g)} - \bar{H}^{(l)}$, $\bar{U}^{(g)} - \bar{U}^{(l)}$, and $\bar{S}^{(g)} - \bar{S}^{(l)}$. State any assumptions.

Problems for Section 5.3

5.18. Give the number of independent intensive variables for each of the following systems at equilibrium:

a. Ice and liquid water.

b. CO, O_2, and CO_2 in a single gas phase, with no catalyst present so that the chemical reaction cannot equilibrate, and with each substance added separately.

c. The system as in part b, but with a catalyst so that the chemical reaction can equilibrate.

d. The system as in part b, but with the system prepared by adding CO_2 only.

*5.19. Give the number of independent intensive variables for each of the following systems at equilibrium:

a. PCl_5, PCl_3, and Cl_2 in a one-phase gaseous system, with the chemical reaction among these substances at equilibrium and with each substance added separately.

b. The same substances as in part a, but with the system produced by placing only PCl_5 in the container.

c. Calcium carbonate heated so that the reaction equilibrates:

$$CaCO_3(s) \rightarrow CaO(s) + CO_2(g)$$

d. The same substances as in part c, except that calcium carbonate, calcium oxide, and carbon dioxide are added separately.

e. Ice VI, ice VII, and ice VIII.

5.20. A researcher exhibits a photo showing four phases, which he claims are ice I, ice II, liquid water, and water vapor. What is your comment?

Problems for Section 5.4

5.21. Find the pressure at which diamond and graphite co-exist at equilibrium at 298.15 K. The density of diamond is 3.52 g mL^{-1}, and that of graphite is 2.25 g mL^{-1}. The Gibbs energy change of formation of diamond is 2.90 kJ mol^{-1} at 298.15 K. State any assumptions.

5.22. The triple point of ammonia is at 196.2 K and 49.42 torr. The molar enthalpy change of vaporization at this temperature is 24.65 kJ mol^{-1}.

a. Find the normal boiling temperature of ammonia. State any assumptions.

b. The actual boiling temperature is $-33°C$. Find the average value of the molar enthalpy change of vaporization for the range between the triple point and the normal boiling temperature.

*5.23. a. The vapor pressure of ice is 1.950 torr at $-10.0°C$ and 4.579 torr at 0.0°C. Find the average enthalpy change of sublimation of ice for this range of temperature.

b. The vapor pressure of water is 23.756 torr at 25°C. Calculate the average enthalpy change of vaporization for the range of temperature from 0°C to 25°C. State any assumptions.

c. Find the enthalpy change of fusion of water. State any assumptions.

5.24. a. Find the pressure needed in an autoclave to attain a temperature of 125°C with liquid water and water vapor both present. Express the pressure in atmospheres and in psi (gauge), which means the pressure in excess of barometric pressure, measured in pounds per square inch. Assume that the barometric pressure is 1.00 atm, which is the same as 14.7 psi.

b. Calculate the freezing temperature of water at the pressure of part a.

5.25. At $-78.5°C$, the vapor pressure of solid carbon dioxide is 760. torr. The triple point is at 216.55 K and 5.112 atm. Find the average enthalpy change of sublimation.

5.26. The normal boiling point of oxygen is 90.18 K. The vapor pressure at 100.0 K is 2.509 atm. Find the enthalpy change of vaporization.

*5.27. The following data give the vapor pressure of liquid aluminum as a function of temperature:

T (K)	P (torr)	T (K)	P (torr)
1557	1	2022	100.
1760.	10.	2220.	400.
1908	40.	2329	760.

Using a graphical method or a linear least-squares procedure, find the enthalpy change of vaporization of aluminum.

5.28. Derive a modified version of the Clausius-Clapeyron equation using the relation

$$\Delta\bar{H}(T) = \Delta\bar{H}(T_1) + \Delta\bar{C}_P(T - T_1)$$

where $\Delta\bar{C}_P$ is assumed to be constant.

Problem for Section 5.5

5.29. a. Write a computer program to carry out the Maxwell equal-area construction, assuming the van der Waals equation of state. It is probably best to choose a trial value of the coexistence pressure and calculate the two areas, and then to carry out successive approximations until the areas are as nearly equal as you desire.

b. Using the equal-area construction, find the vapor pressure of water at 100.0°C according to the van der Waals equation of state.

c. Find the molar volumes of the coexisting liquid and vapor phases of water at 100.0°C according to the van der Waals equation of state and the equal-area construction.

d. Find the value of the compression factor, Z, for water vapor in coexistence with liquid water at 100.0°C.

Problems for Section 5.6

5.30. If the surface region is assumed to be two molecular diameters thick and the average density in the surface region is assumed to be half of that of the liquid, estimate the fraction of the molecules of a sample of water that are in the surface region if the water exists as (a) droplets of diameter 10.0 μm and (b) a "drop" containing 1.00 mol.

5.31. Explain why interfacial tensions between two liquid phases are generally smaller in magnitude than surface tensions between liquid and vapor phases.

*5.32. Estimate the surface tension of ethanol as was done for carbon tetrachloride in Example 5.6. The enthalpy change of vaporization is roughly 40 kJ mol^{-1}. Compare your answer to the correct value, and explain any discrepancy.

5.33. Calculate the capillary rise of pure water in a glass capillary of radius 0.200 mm at 298.15 K. Assume zero contact angle.

5.34. Give an alternative derivation of the capillary rise formula for zero contact angle beginning with the Laplace equation, Equation (5.5-13), and assuming that the meniscus is a hemisphere.

5.35. One method of measuring the surface tension of a liquid is to measure the force necessary to pull a fine wire ring out of the surface of the liquid. For water at 25°C, calculate the force for a ring 15 mm in diameter. Remember that there is a surface on both the inner and outer diameter of the ring.

*5.36. Calculate the vapor pressure of a droplet of ethanol with a radius of 0.0100 mm at 19°C. The vapor pressure at a planar surface is equal to 40.0 torr at this temperature.

5.37. Give an alternative derivation of the expression for the vapor pressure of a spherical droplet, Equation (5.5-20), using the Laplace equation, Equation (5.5-13), and the relation between pressure on the liquid phase and the vapor pressure, Equation (5.3-17).

General Problems

*5.38. **a.** The molar enthalpy change of vaporization of water is 44.01 kJ mol^{-1} at 298.15 K, and the vapor pressure of water at this temperature is 23.756 torr. Use the Clausius-Clapeyron equation to estimate the vapor pressure of water at 100°C. Compare your result with the actual value, 760.0 torr.

b. Use the modified Clausius-Clapeyron equation derived in Problem 5.28 to revise your estimate of the vapor pressure at 100°C, assuming that the heat capacities are constant. Comment on your result.

c. Assume that the heat capacities of the liquid and vapor phases are constant and equal to their values at 298.15 K. Find the value of $\Delta \bar{H}_{vap}$ at 100.0°C, and compare your value with the correct value, 40.66 kJ mol^{-1}.

d. Use the Clausius-Clapeyron equation to estimate the vapor pressure of water at the critical temperature, 647.4 K. The actual critical pressure is equal to 218.3 atm. Explain any discrepancy.

e. Use the modified Clausius-Clapeyron equation derived in Problem 5.28 to estimate the vapor pressure of water at 647.4 K. Comment on your result.

f. As the critical temperature is approached from below, all distinctions between the liquid phase and the vapor phase gradually disappear. Among other things, ΔH_{vap} must vanish at the critical point. Assume that the heat capacities of the liquid and vapor phases are constant and equal to their values at 298.15 K. Find the value of $\Delta \bar{H}_{vap}$ which this assumption predicts at 647.4 K. Find the temperature at which ΔH_{vap} vanishes according to this assumption, and compare your value with the actual critical temperature.

5.39. Identify the following statements as either true or false. If a statement requires some special condition to make it true, label it as false.

a. At equilibrium, a substance that occurs in two phases will have the same concentration in both phases.

b. The Clapeyron equation applies only to a phase transition involving a vapor phase.

c. The Clapeyron equation is an exact thermodynamic equation.

d. The Clausius-Clapeyron equation is an exact thermodynamic equation.

e. The Clausius-Clapeyron equation can be used for a solid-liquid phase transition.

f. The surface tension of liquid water is greater than that of liquid benzene because the hydrogen bonding between water molecules makes their intermolecular attraction larger than that of benzene.

g. It is impossible for four phases of a single substance to coexist at equilibrium.

h. It is impossible for four phases of a mixture of two substances to coexist at equilibrium.

6 Multicomponent Systems

PREVIEW

In this chapter, we study the equilibrium states of systems containing more than one component, applying the thermodynamic theory developed in Chapters 1–4 and the facts of phase equilibrium developed in Chapter 5.

OBJECTIVES

After studying this chapter, the student should:

1. be able to solve problems involving ideal and ideally dilute solutions,
2. be able to solve problems involving activities and activity coefficients in nonideal solutions,
3. understand the principles governing phase diagrams and be able to interpret phase diagrams for various kinds of systems,
4. be able to solve problems involving colligative properties.

PRINCIPAL FACTS AND IDEAS

1. All components of ideal solutions obey Raoult's law:
$$P_i = P_i^* x_i$$
where P_i^* is the vapor pressure of pure substance i and x_i is its mole fraction in the solution.
2. All solutes in ideally dilute solutions obey Henry's law:
$$P_i = k_i x_i$$
where k_i is the Henry's law constant and x_i is the mole fraction of solute. The solvent obeys Raoult's law.
3. Activities and activity coefficients describe deviations from ideal or ideally dilute behavior.
4. The activities of strong electrolyte solutes require special treatment.
5. Phase diagrams can be used to show the phase equilibria of multicomponent systems.
6. Colligative properties depend on concentrations of solutes but not on their identities.

6.1 Ideal Solutions

A liquid or solid phase containing two or more components is often called a solution. An **ideal solution** is defined as one in which the chemical potential of each component is given by the formula

$$\mu_i(T, P) = \mu_i^*(T, P) + RT \ln(x_i) \quad \text{(ideal solution)} \tag{6.1-1}$$

Much like an ideal gas, the ideal solution is a model system whose behavior is closely approximated by some real systems.

where $\mu_i^*(T, P)$ is the chemical potential of the pure substance i when it is at the temperature, T, and pressure, P, of the solution. The mole fraction of substance i in the solution is denoted by x_i.

The pure substance i at the pressure of the solution is taken as the **standard state** of substance i in the solution, and $\mu_i^*(T, P)$ is also denoted as μ_i°. We have already introduced a different kind of standard state for pure substance, which must be at pressure P° to be in their standard states, and we will introduce still more kinds of standard states. We will use the superscript $^\circ$ for any standard state. When necessary to avoid confusion, we use an additional label to specify which kind of standard state is meant; otherwise, the context determines which kind of standard state applies.

Raoult's law is named for Francois Marie Raoult, 1830–1901, a French chemist who was one of the founders of physical chemistry.

Ideal solutions exhibit two additional important properties, which follow from Equation (6.1-1).

*Property 1: The equilibrium partial vapor pressure of each component of an ideal solution very nearly obeys **Raoult's law**,*

$$P_i = P_i^* x_i \quad \text{(Raoult's law)} \tag{6.1-2}$$

Conformity to Raoult's law is sometimes used as the defining condition for an ideal solution. If so, Equation (6.1-1) can be shown as a consequence.

Here P_i is the **partial vapor pressure** of substance i, that is, the partial pressure of substance i in the vapor phase in equilibrium with the liquid or solid solution. P_i^* is the equilibrium vapor pressure of the pure substance i at the temperature and pressure of the solution, and x_i is the mole fraction of substance i in the solution (not in the vapor phase).

Property 2: The entropy change of mixing of an ideal solution is given by the same formula as in Equation (3.3-21) for a mixture of ideal gases.

Figure 6.1 illustrates Raoult's law in the case of toluene and benzene, which form a nearly ideal solution. It shows the partial vapor pressures of benzene and toluene and the total vapor pressure in a solution at 80°C, plotted as functions of the mole fraction of benzene.

To show that Raoult's law follows from Equation (6.1-1), we assume that the vapor phase in equilibrium with an ideal solution is an ideal gas mixture. The chemical potential of component i in the vapor phase is given by Equation (4.4-24),

$$\mu_i = \mu_i^{\circ(\text{gas})} + RT \ln\left(\frac{P_i}{P^\circ}\right) \tag{6.1-3}$$

where P° is the standard-state pressure (1 bar), $\mu_i^{\circ(\text{gas})}$ is the chemical potential of the gas in the standard state, and P_i is the partial pressure of component i in the vapor phase. From the fundamental fact of phase equilibrium, the chemical potential of component i has the same value in the solution and in the vapor, so

$$\mu_i^*(T, P) + RT \ln(x_i) = \mu_i^{\circ(\text{gas})} + RT \ln\left(\frac{P_i}{P^\circ}\right) \tag{6.1-4}$$

where P stands for the actual pressure exerted on the solution. Because P is not necessarily equal to P° or to the equilibrium vapor pressure of the pure substance, P_i^*, we next obtain an equation relating $\mu_i^*(T, P)$, the chemical potential of pure liquid component i at the actual pressure of

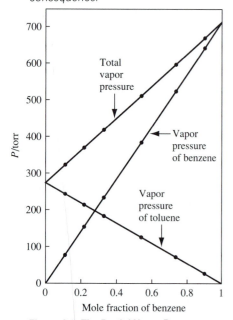

Figure 6.1. The Partial Vapor Pressures of Benzene and Toluene in a Solution at 80°C. These substances form a nearly ideal solution, as shown by conformity to Raoult's law. Drawn from data of M. A. Rosanoff, C. W. Bacon, and F. W. Schulze, *J. Am. Chem. Soc.* **36**, 1993 (1914).

the solution, to $\mu_i^\circ(T)$, the chemical potential of gaseous component i at pressure P°.

The chemical potential of pure liquid component i is equal to the chemical potential of gaseous component i at pressure P_i^*, by the fundamental fact of phase equilibrium, so

$$\mu_i(T, P_i^*) = \mu_i^{\circ(\text{gas})} + RT \ln\left(\frac{P_i^*}{P^\circ}\right) \tag{6.1-5a}$$

From Equation (5.4-4), if we assume that the liquid has fixed volume,

$$\mu_i^*(T, P_i^*) - \mu_i^*(T, P) = \bar{V}_{i(\text{liq})}^*(P_i^* - P) \tag{6.1-5b}$$

When Equations (6.1-5a) and (6.1-5b) are substituted into Equation (6.1-4) and the $\mu_i^{\circ(\text{gas})}$ terms are canceled, we obtain

$$\bar{V}_{i(\text{liq})}^*(P_i^* - P) + RT \ln(P_i^* x_i/P^\circ) = RT \ln(P_i/P^\circ) \tag{6.1-6}$$

Unless P and P_i^* are greatly different, the first term on the left-hand side of Equation (6.1-6) is much smaller than the other terms, like the correction in Example 5.4. If this term is neglected, we find that

$$P_i = P_i^* x_i \tag{6.1-7}$$

which is the same as Raoult's law, Equation (6.1-2). An ideal solution obeys Raoult's law except for a small pressure correction, which we neglect. There is an additional small correction for nonideality of the vapor, which we also neglect.

Exercise 6.1 The density of benzene is 0.87865 g cm^{-3} at $20.00°$C. Its equilibrium vapor pressure at this temperature is 57.8 torr. Find the value of the first term on the left-hand side of Equation (6.1-6) for benzene at $P = 1.000$ atm and compare it with the value of the second term, assuming a mole fraction of 0.500.

The Entropy of Mixing of an Ideal Solution

From Euler's theorem, Equation (4.6-4), and from Equation (6.1-1), the Gibbs energy of the solution is equal to

$$G_{(\text{soln})} = \sum_{i=1}^{c} n_i[\mu_i^* + RT \ln(x_i)] \tag{6.1-8}$$

The Gibbs energy of the unmixed components is a sum of contributions, one for each component. Since the molar Gibbs energy is the same as the chemical potential for a pure substance,

$$G_{(\text{unmixed})} = \sum_{i=1}^{c} n_i \mu_i^* \tag{6.1-9}$$

The **Gibbs energy change of mixing** is defined as the Gibbs energy change of forming the solution from the unmixed pure components at the same temperature and pressure:

$$\Delta G_{\text{mix}} = G_{(\text{soln})} - G_{(\text{unmixed})} = RT \sum_{i=1}^{c} n_i \ln(x_i) \qquad \textbf{(6.1-10)}$$

This is the same as the formula for the Gibbs energy change of mixing for a mixture of ideal gases.

Exercise 6.2   Write a formula for the Gibbs energy change of mixing for a mixture of ideal gases, and show that it is the same as Equation (6.1-10).

There are enough thermodynamic relations to give expressions for all thermodynamic variables of a system if one thermodynamic function is known as a function of its independent variables.

Once an expression for one thermodynamic quantity is obtained, we can obtain an expression for other thermodynamic quantities by using thermodynamic identities. For example, we can now obtain the entropy change of mixing for an ideal solution by using Equation (4.2-20).

$$-S_{(\text{soln})} = \left(\frac{\partial G_{(\text{soln})}}{\partial T} \right)_P \qquad \textbf{(6.1-11)}$$

Using Equation (6.1-8), this equation becomes

$$-S_{(\text{soln})} = \sum_{i=1}^{c} n_i [(\partial \mu_i^*/\partial T)_P + R \ln(x_i)] \qquad \textbf{(6.1-12)}$$

For the unmixed components,

$$-S_{(\text{unmixed})} = \sum_{i=1}^{c} n_i (\partial \mu_i^*/\partial T)_P \qquad \textbf{(6.1-13)}$$

so that

$$\Delta S_{\text{mix}} = -R \sum_{i=1}^{c} n_i \ln(x_i) \quad \text{(ideal solution)} \qquad \textbf{(6.1-14)}$$

which is the same as the formula for an ideal gas mixture, Equation (3.3-21).

The enthalpy change of mixing for a solution is defined as the change in enthalpy involved in forming the solution from the unmixed components at the same temperature and pressure. For an isothermal mixing process, from Equation (4.1-14)

$$\Delta H_{\text{mix}} = \Delta G_{\text{mix}} + T \, \Delta S_{\text{mix}} \qquad \textbf{(6.1-15)}$$

$$\Delta H_{\text{mix}} = RT \sum_{i=1}^{c} n_i [\ln(x_i) - \ln(x_i)] = 0 \quad \text{(ideal solution)} \qquad \textbf{(6.1-16)}$$

We can obtain a formula for the volume change of mixing from Equation (4.2-21), writing this equation for the solution and for the unmixed components. The result is

$$\Delta V_{\text{mix}} = 0 \quad \text{(ideal solution)} \qquad \textbf{(6.1-17)}$$

Exercise 6.3	Show that Equation (6.1-17) is correct.

▼ **EXAMPLE 6.1**

Calculate the Gibbs energy change of mixing, the entropy change of mixing, the enthalpy change of mixing, and the volume change of mixing for a solution of 1.200 mol of benzene and 1.300 mol of toluene at 20.00°C.

Solution

$$\Delta G_{\text{mix}} = RT[(1.200 \text{ mol}) \ln(0.4800) + (1.300 \text{ mol}) \ln(0.5200)]$$
$$= -4219 \text{ J}$$
$$\Delta S_{\text{mix}} = -R[(1.200 \text{ mol}) \ln(0.4800) + (1.300 \text{ mol}) \ln(0.5200)]$$
$$= 14.39 \text{ J K}^{-1}$$
$$\Delta H_{\text{mix}} = 0$$
$$\Delta V_{\text{mix}} = 0$$

▲

We can also obtain expressions for some partial molar quantities from the expression for the chemical potential. For example, to obtain the expression for the partial molar entropy, we use Equation (4.5-10):

$$\bar{S}_i = -(\partial \mu_i / \partial T)_{P,n} \qquad \textbf{(6.1-18)}$$

Applying this to Equation (6.1-1), we get

$$\bar{S}_i = -(\partial \mu_i^* / \partial T) - R \ln(x_i)$$

$$\boxed{\bar{S}_i = \bar{S}_i^* - R \ln(x_i) \quad \text{(ideal solution)}} \qquad \textbf{(6.1-19)}$$

Exercise 6.4	In the same way as Equation (6.1-19) was derived, obtain the relations for an ideal solution:

$$\boxed{\bar{V}_i = \bar{V}_i^* \quad \text{(ideal solution)}} \qquad \textbf{(6.1-20)}$$

$$\boxed{\bar{H}_i = \bar{H}_i^* \quad \text{(ideal solution)}} \qquad \textbf{(6.1-21)}$$

Although the entropy change of mixing, the Gibbs energy change of mixing, the enthalpy change of mixing, and the volume change of mixing for an ideal solution are given by the same formulas as the corresponding quantities for a mixture of ideal gases, an ideal solution does not resemble a mixture of ideal gases in its molecular structure. The ideal gas is a model system in which the molecules do not interact with each other. In a liquid or solid solution, a large intermolecular attraction holds the system together and a large intermolecular repulsion keeps the system from collapsing to a smaller volume.

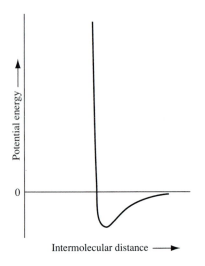

Figure 6.2. The Potential Energy of a Pair of Molecules (schematic). This potential energy leads to attractions at moderate intermolecular distances and repulsions at small intermolecular distances.

The intermolecular attraction and repulsion correspond to a potential energy that depends on intermolecular distance as shown in Figure 6.2, which depicts the potential energy of a pair of spherical molecules as a function of the distance between their centers. Since any potential energy $\mathscr{V}$ corresponds to a force pointing in the direction in which $\mathscr{V}$ decreases, there are an attraction at moderate distances and a repulsion at smaller distances. For nonspherical molecules, there is also a dependence of the potential energy on the relative orientations of the molecules, which we ignore for now.

It is found experimentally that Raoult's law applies best to mixtures of substances with molecules of similar size, shape, and polarity. For example, benzene and toluene form a nearly ideal solution because their molecules are similar and interact with each other in much the same way as with molecules of their own kind. For example, the curve analogous to that of Figure 6.2 that applies to one benzene molecule and one toluene molecule must be nearly the same as the curve that applies to two benzene molecules or to two toluene molecules. If a molecule of toluene is substituted for a molecule of benzene in the solution, the energy, enthalpy, and volume of the solution are almost unchanged. The similarity between the molecules also allows them to mix randomly in a solution, just as noninteracting molecules mix randomly in an ideal gas mixture, so the formulas for the entropy changes of mixing are identical.

Exercise 6.5   From the following list, pick pairs of substances that you think probably form nearly ideal solutions:

o-xylene	*m*-xylene
p-xylene	toluene
ethyl benzene	1-propanol
2-propanol	naphthalene
anthracene	phenanthrene
3-methylpentane	2-methylpentane

Two-Component Phase Diagrams of Ideal Solutions

In a phase diagram for one component, as in Figure 5.3, the location of a point in the diagram represents the values of T and P that specify the intensive state of the system. An equilibrium two-component system with one phase has three independent intensive variables. The full phase diagram therefore requires three dimensions. To make a two-dimensional phase diagram, we specify a fixed value for one variable.

There are two principal types of two-dimensional phase diagrams for a two-component solution: the pressure-composition and temperature-composition phase diagrams. In the **pressure-composition phase diagram**, the temperature is held fixed. The mole fraction (or mass fraction) of one component and the pressure are the two variables plotted. For a given solution, there is a different phase diagram of this type for each temperature.

For a solution obeying Raoult's law, the partial pressure of both components is given by Equation (6.1-2), so the total vapor pressure is

$$P_{tot} = P_1^* x_1 + P_2^* x_2 = P_2^* + (P_1^* - P_2^*) x_1 \qquad (6.1\text{-}22)$$

where we have used the relationship $x_2 = 1 - x_1$. This equation is represented by a straight line in the pressure-composition phase diagram.

Exercise 6.6	Show that the intercepts of the function in Equation (6.1-22) at $x_1 = 0$ and $x_1 = 1$ are equal to P_2^* and P_1^*.

The composition of the vapor phase at equilibrium with the liquid solution is not the same as the composition of the liquid solution. The mole fraction of component 1 in the vapor is given by Dalton's law of partial pressures:

$$x_{1(vap)} = \frac{P_1}{P_{tot}} = \frac{P_1^* x_1}{P_2^* + (P_1^* - P_2^*) x_1} \qquad (6.1\text{-}23)$$

▼   **EXAMPLE 6.2**

At 20.00°C the vapor pressure of pure benzene is 74.9 torr and that of pure toluene is 21.6 torr. Assuming ideality, find the partial vapor pressure of each component, the total vapor pressure, and the composition of the vapor at equilibrium with the solution of Example 6.1.

Solution

Call benzene component 1 and toluene component 2:

$$P_1 = (0.48)(74.9 \text{ torr}) = 36.0 \text{ torr}$$
$$P_2 = (0.52)(21.6 \text{ torr}) = 11.2 \text{ torr}$$
$$P_{tot} = P_1 + P_2 = 47.2 \text{ torr}$$
$$x_1 = \frac{P_1}{P_{tot}} = \frac{36.0 \text{ torr}}{47.2 \text{ torr}} = 0.763$$
$$x_2 = \frac{P_2}{P_{tot}} = \frac{11.2 \text{ torr}}{47.2 \text{ torr}} = 0.237$$

◢

The formula giving the total pressure as a function of $x_1^{(vap)}$ is

$$P_{tot} = \frac{P_1^* P_2^*}{P_1^* + x_1^{(vap)}(P_2^* - P_1^*)} \qquad (6.1\text{-}24)$$

The derivation of this formula is left as a problem (Problem 6.34).

Figure 6.3 shows the liquid-vapor pressure-composition phase diagram of benzene and toluene at 80°C. The line represents Equation (6.1-22), giving the pressure as a function of the benzene mole fraction in the liquid, and the other curve represents Equation (6.1-24), giving the pressure as a function of the mole fraction of benzene in the vapor.

The area below the curves represents possible equilibrium intensive states of the system when it is a one-phase vapor, and the area above the curves represents possible equilibrium states of the system when it is a one-phase

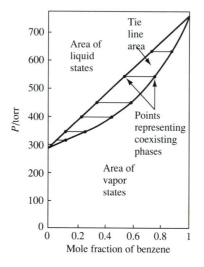

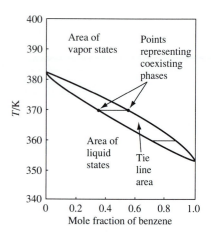

Figure 6.3. The Liquid-Vapor Pressure-Composition Phase Diagram of Benzene and Toluene at 80°C. The upper curve, which is the same as the upper curve in Figure 6.1, shows the pressure as a function of mole fraction in the liquid phase, and the lower curve shows the pressure as a function of mole fraction in the vapor phase. The two compositions that can coexist at a given pressure are given by the ends of the tie line at that pressure. Drawn from data of M. A. Rosanoff, C. W. Bacon, and F. W. Schulze, *J. Am. Chem. Soc.* **36**, 1993 (1914).

Figure 6.4. The Liquid-Vapor Temperature-Composition Phase Diagram of Benzene and Toluene at 1.000 bar. The lower curve shows the temperature as a function of the mole fraction in the liquid, and the upper curve shows the temperature as a function of the mole fraction in the vapor. Drawn from data of M. A. Rosanoff, C. W. Bacon, and F. W. Schulze, *J. Am. Chem. Soc.* **36**, 1993 (1914).

liquid. Points in the area between the curves do not represent possible intensive states of a single phase. This area represents two phases. Since the total pressure of the vapor and the pressure of the liquid solution at equilibrium with each other are the same, a horizontal line segment, or **tie line**, between the two curves connects the state points for the two phases at equilibrium with each other. Several tie lines are shown in the figure.

The second common type of two-dimensional phase diagram for a two-component system is the **temperature-composition diagram**, in which the variables are the mole fraction (or mass fraction) of one component and the temperature, with the pressure held fixed. Figure 6.4 shows the liquid-vapor temperature-composition diagram of benzene and toluene at 1.000 atm. The upper curve gives the boiling temperature at the given pressure as a function of the mole fraction of benzene in the vapor phase, and the lower curve gives the boiling temperature at the given pressure as a function of the mole fraction of benzene in the liquid phase.

Tie lines drawn between the two curves connect values of the mole fraction in the two phases at equilibrium with each other. Each tie line in this

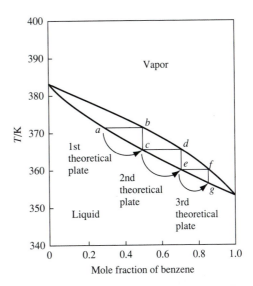

Figure 6.5. Diagram Representing a Constant-Pressure Distillation Process. Each theoretical plate corresponds to one step in the "staircase."

diagram must be the same as a tie line in one of the pressure-composition diagrams.

A constant-pressure distillation process can be described with the temperature-composition phase diagram, as depicted in Figure 6.5. Point a on the diagram in this figure represents the composition of a liquid solution that is being boiled in the still. Point b, at the other end of the tie line, represents the composition of the vapor at equilibrium with this liquid. This vapor condenses at the temperature represented by point c. A simple still in which this process can be carried out is said to have one **theoretical plate**.

A still can be made to produce a greater separation of the components by packing its column with glass beads or other objects. The liquid condenses on the glass beads partway up the column and then evaporates again, making only part of the column equivalent to one theoretical plate. For the process beginning at point a, a second evaporation at point c leads to a vapor with the composition at point d, and this vapor can condense still farther up the column, giving a liquid corresponding to point e. This process corresponds to two theoretical plates, with a temperature that changes as one moves up the column. Continuation of the process leads to a staircase pattern in the diagram, with one step for each theoretical plate. Three theoretical plates lead to a liquid with the composition at point g. A still with a large number of theoretical plates can produce a condensate that is almost entirely made up of the more volatile component. A "spinning-band" still has a rotating helical wire screen that wipes the walls of the column and can provide several hundred theoretical plates.

Exercise 6.7   From Figure 6.5, determine the composition and boiling temperature of the condensate produced from the liquid at point a by a still with three theoretical plates.

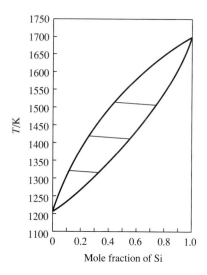

Figure 6.6. The Solid-Liquid Temperature-Composition Phase Diagram of Silicon and Germanium. Because both the solid and liquid phases are nearly ideal solutions, this diagram resembles the liquid-vapor phase diagram. From C. D. Thurmond, *J. Phys. Chem.* **57**, 827 (1953).

The solid-liquid phase diagrams for a two-component mixture are similar to liquid-vapor phase diagrams. Figure 6.6 shows the temperature-composition phase diagram of silicon and germanium, which form a nearly ideal solid solution. Note the similarity to Figure 6.4. Continuing this diagram to higher temperatures leads to the liquid-vapor transition region and a diagram with two areas of tie lines, as shown in Figure 6.7.

In order to represent the full equilibrium pressure-temperature-composition behavior of a two-component system, a three-dimensional graph is required. This is schematically represented in Figure 6.8, which represents the liquid-vapor equilibrium of a nearly ideal liquid solution. The compositions of the liquid and the vapor at equilibrium with each other are represented by two surfaces, and tie lines parallel to the composition axis connect coexisting states on these two surfaces. A pressure-composition diagram is created by passing a plane of constant temperature

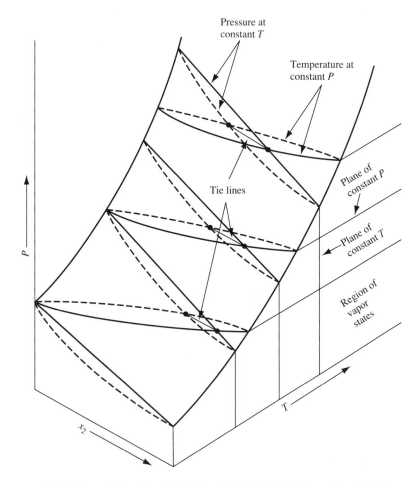

Figure 6.8. The Three-Dimensional Liquid-Vapor Phase Diagram for an Ideal Solution (schematic). This diagram shows how both the temperature-composition and the pressure-composition phase diagrams are obtained by passing planes through the same three-dimensional diagram.

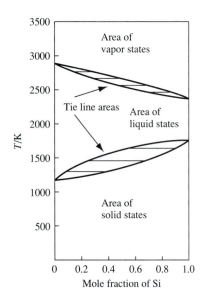

Figure 6.7. The Solid-Liquid and Liquid-Vapor Phase Diagram of Silicon and Germanium (schematic). This diagram shows both the solid-liquid and liquid-vapor phase transitions.

through the three-dimensional graph, and a temperature-composition diagram is created by passing a plane of constant pressure through the three-dimensional graph. The intersection of these two planes contains the common tie line of the two diagrams, as shown in the figure.

6.2   Henry's Law and Ideally Dilute Nonelectrolyte Solutions

Most liquid and solid solutions are not well described by Raoult's law. For example, Figure 6.9 shows the partial vapor pressures and total vapor pressure of a mixture of diethyl ether (component 1) and ethanol (component 2)

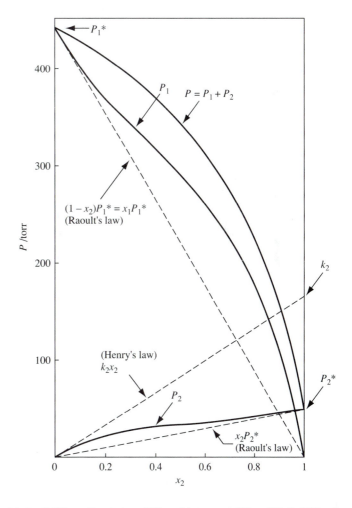

Figure 6.9. Partial Vapor Pressures of Ethanol (component 2) and Diethyl Ether in a Solution at 20°C, Showing Positive Deviation from Raoult's Law. In addition to the actual behavior, this diagram shows the Raoult's law behavior for both substances and the Henry's law behavior for ethanol. Drawn from data of J. Timmermans, *Physicochemical Constants of Binary Systems*, Vol. 2, Interscience, New York, 1959, p. 401.

at 20°C. In this system, the partial vapor pressures of both components are greater than predicted by Raoult's law. This behavior is called a **positive deviation** from Raoult's law and corresponds to greater repulsions between unlike molecules than between like molecules and/or to lesser attractions between unlike molecules than between like molecules. In the case of a **negative deviation**, the vapor pressure is smaller than predicted by Raoult's law. It is also possible (but less likely) for the deviation to be positive for one component and negative for another, as is the case with acetone and nitromethane at 318.5 K.

Exercise 6.8	At 318.15 K, acetone (component 1) has a negative deviation from Raoult's law in a solution with nitromethane (component 2), and nitromethane has a positive deviation from Raoult's law. What conclusions can you draw about 1-1, 1-2, and 2-2 molecular interactions?

Two features of Figure 6.9 are typical for nonionic substances. First, for small values of x_1 the curve representing P_1 is nearly straight, and for small values of x_2 the curve representing P_2 is nearly straight. Second, the curve representing P_1 nearly coincides with the line representing Raoult's law for values of x_1 near unity. The same is true of the curve representing P_2 for values of x_2 near unity.

These properties can be summarized: A nearly pure component approximately obeys Raoult's law, and a dilute component approximately obeys **Henry's law**, which is written

Henry's law is named for William Henry, 1774–1836, an English chemist who was a friend and colleague of John Dalton and who influenced Dalton's formulation of the atomic theory.

$$\boxed{P_i = k_i x_i \quad (x_i \ll 1) \quad \text{(Henry's law)}}$$

(6.2-1)

Here k_i is called the **Henry's law constant** for substance i. It is a constant for a given set of substances and a given temperature but depends on temperature and on the identity of the other substance or substances present. In this respect it differs from the Raoult's law constant, P_i^*, which does not depend on the identity of the other substance present.

If one component of a solution is present in a somewhat larger amount than the others, it is called the **solvent**. The other substances are called **solutes**. A small value of k_i for a given solvent corresponds to large solubility of substance i in that solvent.

A solution in which the solvent obeys Raoult's law and the solutes obey Henry's law is called an **ideally dilute solution**. Most nonelectrolyte solutions behave nearly like ideally dilute solutions at sufficiently low concentrations of the solutes.

The extrapolated line corresponding to Henry's law for ethanol in diethyl ether is shown in Figure 6.9. The intercept of this line with the edge of the graph is equal to the value of the Henry's law constant for ethanol, approximately 160 torr at the temperature of the figure. The Henry's law constant for ethanol in solution with some substance other than diethyl ether would have a different value.

▼   **EXAMPLE 6.3**

At 40°C, a solution of ethanol (component 2) in benzene (component 1) having a mole fraction of ethanol equal to 0.0200 has a partial vapor pressure of ethanol equal to 30.2 torr. Assuming that Henry's law holds at this composition, find the value of the Henry's law constant for ethanol in benzene.

Solution

$$k_2 = \frac{P_2}{x_2} = \frac{30.2 \text{ torr}}{0.0200} = 1.51 \times 10^3 \text{ torr}$$

▲

Exercise 6.9

Find the value of the Henry's law constant for benzene in ethanol at 40°C. The partial vapor pressure of benzene is 12.8 torr if the mole fraction of benzene is 0.0130. State any assumptions.

The Chemical Potential in an Ideally Dilute Solution

Consider an ideally dilute solution in which the partial vapor pressure of the solute is large enough to measure. We equilibrate it with a vapor phase, which we assume to be a mixture of ideal gases. Using Henry's law, Equation (6.2-1), for the partial vapor pressure of solute i, we have from the fundamental fact of phase equilibrium:

$$\mu_{i(\text{soln})} = \mu_{i(\text{vap})} = \mu_i^{\circ(\text{gas})} + RT \ln\left(\frac{k_i x_i}{P^\circ}\right)$$

$$\boxed{\mu_{i(\text{soln})} = \mu_i^{\circ(\text{H})} + RT \ln(x_i) \quad \text{(ideally dilute solute)}} \qquad \textbf{(6.2-2)}$$

where the standard-state chemical potential of substance i in the ideally dilute solution is given by

$$\mu_i^{\circ(\text{H})} = \mu_i^{\circ(\text{gas})} + RT \ln\left(\frac{k_i}{P^\circ}\right) \qquad \textbf{(6.2-3)}$$

Since $\mu_i^{\circ(\text{H})}$ is equal to the chemical potential of component i in the vapor phase when the partial pressure P_i is equal to k_i, it is equal to the chemical potential the pure liquid would have if it obeyed Henry's law for $x_i = 1$. We define this hypothetical pure liquid at the pressure of the solution to be the **standard state** for substance i in the ideally dilute solution. We refer to $\mu_i^{\circ(\text{H})}$ as the standard-state chemical potential of substance i for this particular standard state. This standard state depends on the identity of the solvent in our solution and on the pressure of the solution.

It may seem strange to have a standard state that corresponds to a hypothetical pure substance when we are discussing a dilute solute. The reason for this choice is that it allows us to write Equation (6.2-2) in the same form as Equation (6.1-1). We will write all equations for the chemical potential in ways that are similar to this equation, and will see the advantages of doing so.

An important case is the equilibration of two phases with each other, as in extraction processes. For example, if I_2 is dissolved in water, most of the I_2 can be extracted from the water by equilibrating this phase with carbon tetrachloride. For dilute solutions, the equilibrium mole fraction of I_2 in the water phase is proportional to the mole fraction of I_2 in the carbon tetrachloride phase. This is called **Nernst's distribution law**. The proportionality constant K_d is called the **distribution constant** and is given by

$$K_d = \frac{x_{i(\text{II})}^{(\text{eq})}}{x_{i(\text{I})}^{(\text{eq})}} \quad \text{(definition)} \tag{6.2-4}$$

where we use the subscript i for iodine and denote the carbon tetrachloride phase by II and the aqueous phase by I. The value of K_d depends on temperature and on the identities of the two solvents and ordinarily has a slight dependence on pressure.

We use the fundamental fact of phase equilibrium to derive Equation (6.2-4). The chemical potential of iodine in the two phases is given by

$$\mu_{i(\text{I})} = \mu_{i(\text{I})}^{\circ(\text{H})} + RT \ln(x_{i(\text{I})}) \tag{6.2-5a}$$

and

$$\mu_{i(\text{II})} = \mu_{i(\text{II})}^{\circ(\text{H})} + RT \ln(x_{i(\text{II})}) \tag{6.2-5b}$$

where we add a subscript to specify the phase because the standard-state chemical potentials in the two phases are not required to be equal. At equilibrium,

$$\mu_{i(\text{I})} = \mu_{i(\text{II})}$$

so that, after a few steps of algebra, we can write

$$K_d = \frac{x_{i(\text{II})}^{(\text{eq})}}{x_{i(\text{I})}^{(\text{eq})}} = \exp\left(\frac{\mu_{i(\text{I})}^{\circ(\text{H})} - \mu_{i(\text{II})}^{\circ(\text{H})}}{RT}\right)$$
$$= \frac{k_i^{(\text{I})}}{k_i^{(\text{II})}} \tag{6.2-6}$$

where $k_{i(\text{I})}$ and $k_{i(\text{II})}$ are the Henry's law constants for substance i in phases I and II, respectively. We see that K_d for a given solute depends only on temperature, pressure, and the identities of the two solvents.

Exercise 6.10

a. Carry out the steps to obtain Equation (6.2-6).

b. The distribution coefficient for iodine between water (phase I) and carbon tetrachloride (phase II) is approximately 0.0022 at 25.0°C. If a solution containing 0.0100 mol of iodine and 1.000 mol of water is equilibrated with 1.000 mol of carbon tetrachloride at this temperature, find the final mole fraction of iodine in each phase. Neglect any water that dissolves in the carbon tetrachloride phase and any carbon tetrachloride that dissolves in the aqueous phase.

The **molality** of component i in a solution is defined by

$$m_i = \frac{n_i}{w_1} \quad (i = 2, 3, \ldots, c) \quad \text{(definition)} \tag{6.2-7}$$

where n_i is the amount of component i in moles and w_1 is the mass of the solvent (component 1) in kilograms. The units of molality are mol kg^{-1}, also referred to as "molal."

For small concentrations, the molality is nearly proportional to the mole fraction. If M_1 is the molar mass of the solvent, then

$$w_1 = n_1 M_1 \tag{6.2-8}$$

so that the mole fraction of component i is given by

$$x_i = \frac{n_i}{n_1 + n_2 + \cdots + n_c}$$

$$= \frac{n_i}{(w_1/M_1) + n_2 + \cdots + n_c} \tag{6.2-9}$$

For a dilute solution

$$x_i \approx m_i M_1 \tag{6.2-10}$$

In this case, Henry's law becomes

$$P_i = k_i m_i M_1$$

$$= k_i^{(m)} m_i \tag{6.2-11}$$

where $k_i^{(m)} = k_i M_1$ is the **molality Henry's law constant** for substance i.

▼ **EXAMPLE 6.4**

a. From the value of k_2 for ethanol (substance 2) in Example 6.3, find the value of $k_2^{(m)}$.

b. Find the vapor pressure of a 0.0500 mol kg^{-1} solution of ethanol in benzene.

Solution

a.

$$k_2^{(m)} = (1.51 \times 10^3 \text{ torr})(0.07812 \text{ kg mol}^{-1})$$

$$= 118 \text{ torr (mol kg}^{-1})^{-1}$$

b.

$$P_2 = [118 \text{ torr (mol kg}^{-1})^{-1}](0.0500 \text{ mol kg}^{-1})$$

$$= 5.90 \text{ torr}$$

▲

For a dilute solution, the chemical potential can be expressed in terms of the molality in an equation similar to Equations (6.2-2) and (6.1-1). Using Equations (6.2-2) and (6.2-10),

$$\mu_i = \mu_i^{\circ(H)} + RT \ln(m_i M_1)$$

$$= \mu_i^{\circ(m)} + RT \ln(m_i/m^\circ) \quad \text{(dilute solution)} \tag{6.2-12}$$

where $\mu_i^{\circ(m)} = \mu_i^{\circ(H)} + RT \ln(M_1 m^\circ)$ and m° is defined to equal 1 mol kg^{-1} (exactly).

The quantity $\mu_i^{\circ(m)}$ is the chemical potential of substance i in its **molality standard state**—that is, in a hypothetical solution with m_i equal to 1 mol kg^{-1} and with Henry's law in the form of Equation (6.2-11) valid at this molality.

Exercise 6.11 Show that $\mu_i^{\circ(m)}$ is equal to the chemical potential of i in the vapor phase at equilibrium with a 1 mol kg^{-1} solution if Equation (6.2-11) is valid to this molality.

The **concentration** of component i is defined by

$$c_i = n_i/V \tag{6.2-13}$$

where V is the volume of the solution. If the volume is measured in liters instead of m^3, the concentration is called the **molarity**. Sometimes we use the symbol for the molarity:

$$c_i = [\mathscr{F}_i] \tag{6.2-14}$$

where $\mathscr{F}_i$ is an abbreviation for the formula of substance i.

Exercise 6.12 Assuming that the coefficient of thermal expansion of an aqueous solution of sucrose is the same as that of water, 2.0661×10^{-4} K^{-1}, determine the molarity at 25.0°C of a solution with a molarity of 0.1000 mol L^{-1} at 20.0°C.

Henry's law can also be expressed in terms of concentration. By Euler's theorem, Equation (4.6-4), the volume of a solution is

$$V = \sum_{i=1}^{c} n_i \bar{V}_i \approx n_1 \bar{V}_1 \tag{6.2-15}$$

where the approximate equality holds for a dilute solution. In this case

$$c_i \approx \frac{n_i}{n_1 \bar{V}_1} \approx \frac{x_i}{\bar{V}_1} \quad \text{(dilute solutions)} \tag{6.2-16}$$

Exercise 6.13 **a.** Show that for a dilute solution, Henry's law becomes
$$P_i = k_i \bar{V}_1 c_i = k_i^{(c)} c_i \tag{6.2-17}$$
b. Show that for a dilute solution, the chemical potential can be expressed in terms of the concentration:
$$\mu_i = \mu_i^{\circ(c)} + RT \ln(c_i/c^\circ) \tag{6.2-18}$$
where $\mu_i^{\circ(c)} = \mu_i^{\circ(H)} + RT \ln(c^\circ \bar{V}_1)$ and $c^\circ = 1$ mol per unit volume (exactly). Interpret the molarity standard state, at which the chemical potential is equal to $\mu_i^{\circ(c)}$.

In addition to the mole fraction, molality, and concentration, other measures are used to express the composition of solutions, including percentage by mass, percentage by volume, and parts per million by mass. For dilute solutions, Henry's law can be expressed in terms of any of these composition measures, since they are all proportional to the mole fraction for sufficiently dilute solutions.

Henry's law is also used to express the maximum solubility of gases in liquids. At equilibrium the amount of a gas dissolved in a liquid is proportional to the partial pressure of the gas. The only difference from the usual version of Henry's law is that instead of thinking of a solution producing a vapor phase at equilibrium with the solution, we think of a gas dissolving to produce a solution at equilibrium with the gas phase.

▼ **EXAMPLE 6.5**

At 25°C, water at equilibrium with air at 1.000 atm contains about 8.3 ppm of dissolved oxygen by mass. Compute the Henry's law constant. The mole fraction of oxygen in air is 0.203.

Solution

The mole fraction of oxygen in the water phase is

$$x_2 = \left(\frac{8.3\ \text{g}}{32.0\ \text{g mol}^{-1}}\right)\left(\frac{18.0\ \text{g mol}^{-1}}{1 \times 10^6\ \text{g}}\right) = 4.7 \times 10^{-6}$$

and the Henry's law constant is

$$k_2 = \frac{0.203\ \text{atm}}{4.7 \times 10^{-6}} = 4.3 \times 10^4\ \text{atm}$$

▲

If all solutes in a solution obey Henry's law over some range of composition near zero mole fractions of the solutes, then the solvent obeys Raoult's law over this range of composition. We show this for a two-component solution, leaving the proof for several components as a problem at the end of the chapter (Problem 6.43). We begin with the Gibbs-Duhem relation for constant pressure and temperature:

$$x_1(\partial\mu_1/\partial x_2)_{T,P} + x_2(\partial\mu_2/\partial x_2)_{T,P} = 0 \tag{6.2-19}$$

Note that P must be constant and, since the total vapor pressure depends on the composition of the solution, the vapor phase must have an additional gas or gases (such as air) present to keep the pressure constant. The small amount of air that dissolves in the solution can be ignored, and the small effect of the change in pressure on the chemical potentials in the solution is negligible (see Example 4.5).

Assume that component 2 obeys Henry's law over the range of composition from $x_2 = 0$ to $x_2 = x_2'$. In this range

$$(\partial\mu_2/\partial x_2)_{T,P} = RT[d\ \ln(x_2)/dx_2] = RT/x_2 \tag{6.2-20}$$

Since $x_1 + x_2 = 1$, $dx_2 = -dx_1$. Therefore, by Equation (6.2-19)

$$x_1(\partial\mu_1/\partial x_1)_{T,P} = x_2(\partial\mu_2/\partial x_2)_{T,P} = x_2 RT/x_2 = RT \tag{6.2-21}$$

Dividing this equation by x_1 and multiplying by dx_1, we obtain

$$(\partial\mu_1/\partial x_1)_{T,P}\ dx_1 = (RT/x_1)\ dx_1 \tag{6.2-22}$$

Integrating this equation from $x_1 = 1$ to $x_1 = x'_1 = 1 - x'_2$, we obtain

$$\mu_1(x'_1) - \mu_1(1) = RT[\ln(x'_1) - \ln(1)] = RT \ln(x'_1)$$

which is the same as

$$\mu_1(x_1) = \mu_1^* + RT \ln(x_1) \tag{6.2-23}$$

where we recognize $\mu_1(1)$ as μ_1^* and we have dropped the prime (') on the value of x_1. Equation (6.2-23) is the same as the ideal solution relation for component 1 and leads to Raoult's law for the solvent, as in Equation (6.1-7).

6.3	## The Activity and the Description of General Systems

Most solutions are neither ideal nor ideally dilute. To describe all possible cases, we define the **activity** a_i of substance i by

Equation (6.3-1) is to be used for all kinds of substances in all kinds of phases.

$$\boxed{\mu_i = \mu_i^\circ + RT \ln(a_i) \quad \text{(definition of } a_i)} \tag{6.3-1}$$

where μ_i° is the chemical potential of substance i in some standard state. Equation (6.3-1) is a general equation, to be used for pure substances and components of mixtures in solid, liquid, or gaseous phases. Notice the similarity of this equation to Equations (6.1-1), (6.2-2), (6.2-12), and (6.2-18).

The Activity of a Pure Solid or Liquid

The standard state of a pure solid or liquid is the substance at pressure P° (1 bar). This choice means that the activity of the pure substance is equal to unity if the pure substance is at pressure P°. If the substance is at a pressure other than P°, we assume the substance to be incompressible and write from Equation (4.3-11)

$$\mu_i(P') = \bar{G}_i(P') = \mu_i^\circ + \int_{P^\circ}^{P'} \bar{V}_i \, dP = \mu_i^\circ + \bar{V}_i^*(P' - P^\circ) \tag{6.3-2}$$

so that the activity at pressure P is given by

$$RT \ln(a_i) = \bar{V}_i^*(P - P^\circ) \tag{6.3-3a}$$

or

$$\boxed{a_i = \exp\left[\frac{\bar{V}_i(P - P^\circ)}{RT}\right] \quad \text{(pure solid or liquid)}} \tag{6.3-3b}$$

where we have dropped the prime (').

The exponent in Equation (6.3-3b) is generally quite small unless P differs greatly from P°, so the activity of a pure solid or liquid can be taken as nearly equal to unity under ordinary conditions.

▼   **EXAMPLE 6.6**

Find the activity of pure liquid water at a pressure of 2.000 atm and a temperature of 298.15 K.

Solution

The purpose of Example 6.6 is to show that the activity does not depend very much on the pressure for a condensed phase.

$$\ln(a) = \frac{(1.805 \times 10^{-5} \text{ m}^3 \text{ mol}^{-1})(1.00 \text{ atm})(101325 \text{ N m}^{-2} \text{ atm}^{-1})}{(8.3145 \text{ J K}^{-1} \text{ mol}^{-1})(298.15 \text{ K})}$$

$$= 0.0007378$$

$$a = e^{0.000736} = 1.000738$$

▲

Exercise 6.14

Find the pressure such that the activity of liquid water differs from unity by 1.000% at 298.15 K.

The Activity of an Ideal Gas

The standard state is the same as that used for the Gibbs energy of the gas: the pure ideal gas at whatever temperature we wish to discuss and at pressure P° (1 bar). Comparison of Equation (6.3-1) with Equation (4.4-24) shows that

$$\boxed{a_i = P_i/P^\circ \quad \text{(ideal gas)}} \qquad \text{(6.3-4)}$$

where P_i is the pressure (or partial pressure) of the gas. Since each gas in an ideal gas mixture behaves as though it were alone in the container, Equation (6.3-4) can also be used for components of ideal gas mixtures.

The Activity of a Nonideal Gas

The standard state is the hypothetical ideal gas at pressure P°, the same standard state as for the Gibbs energy of the gas. Comparison of Equation (6.3-1) with Equation (4.4-25) shows that the activity of a nonideal gas is

$$\boxed{a_i = f_i/P^\circ \quad \text{(any gas)}} \qquad \text{(6.3-5)}$$

The Activity Coefficient of a Nonideal Gas

The extent to which the activity of a nonideal gas deviates from that of an ideal gas is specified by the **activity coefficient**, γ, defined as the ratio

The activity coefficient for a nonideal gas is also known as the fugacity coefficient and is sometimes denoted by ϕ.

$$\boxed{\gamma_i = \frac{a_i(\text{real})}{a_i(\text{ideal})} = \frac{f_i}{P_i} \quad \text{(definition)}} \qquad \text{(6.3-6)}$$

The chemical potential of a nonideal gas can be written

$$\mu_i = \mu_i^\circ + RT \ln\left(\frac{\gamma_i P_i}{P^\circ}\right)$$ (6.3-7)

If the activity coefficient is greater than unity, the gas has a greater activity and a greater chemical potential than if it were ideal at the same temperature and pressure. If the activity coefficient is less than unity, the gas has a lower activity and a lower chemical potential than if it were ideal.

Activities in Ideal Solutions

For a component of an ideal liquid or solid solution, comparison of Equation (6.3-1) with Equation (6.1-1) shows that

$$\boxed{a_i = x_i \quad \text{(ideal solution)}}$$ (6.3-8a)

$$\boxed{\mu_i^\circ = \mu_i^* \quad \text{(ideal solution)}}$$ (6.3-8b)

The standard state of a substance in a solution must be in the same kind of phase as the solution. In this case, it must be a supercooled liquid.

An example of a nearly ideal system is a liquid mixture of benzene and naphthalene. The standard state of naphthalene for the liquid solution must be pure supercooled liquid naphthalene at the temperature and pressure of the solution. Since the chemical potential of the supercooled (metastable) liquid is higher than that of the solid at temperatures lower than the melting temperature, there is a range of mole fractions of naphthalene near unity at which the chemical potential of naphthalene in the solution would exceed that of solid naphthalene. Solutions in this range exceed the maximum solubility of naphthalene in benzene, and if they occur at all they are metastable and are called **supersaturated**. Figure 6.10 shows schematically the vapor pressure and chemical potential of naphthalene in a solution with benzene near room temperature.

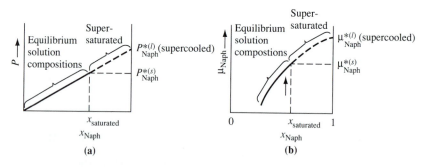

Figure 6.10. Naphthalene in an Ideal Solution (schematic). (a) The vapor pressure as a function of composition. The composition at which the vapor pressure is equal to that of solid naphthalene is the composition of maximum solubility (saturation). **(b)** The chemical potential as a function of composition. The composition at which the chemical potential is equal to that of solid naphthalene is the composition of maximum solubility.

The standard state of a component of an ideal solution is defined as the pure component at the pressure of the solution, rather than at some fixed pressure. However, as we have seen in Example 6.6, a moderate change in pressure does not change the chemical potential or the activity very much, so we will generally ignore the effect of changes in pressure on the standard state for condensed phases.

Activities and Activity Coefficients in Nonideal Solutions

The activities of the components of a nonideal solution are specified in different ways, depending on whether one of the components is designated as the solvent and depending on the variables used to specify the composition of the solution.

Convention I. All of the components are treated on an equal footing, with no substance designated as the solvent and with mole fractions used to specify the composition. The standard state for each component is the pure substance at the temperature and pressure of the solution, just as with an ideal solution.

For a solution of volatile substances at equilibrium, the chemical potential of substance i in the solution is equal to its chemical potential in the vapor phase. If we write the chemical potential in the solution in terms of the activity as in Equation (6.3-1) and assume the vapor to be an ideal gas,

$$\mu_i^{\circ(\mathrm{I})} + RT \ln[a_i^{(\mathrm{I})}] = \mu_i^{\circ(\mathrm{gas})} + RT \ln\left(\frac{P_i}{P^\circ}\right) \qquad \textbf{(6.3-9)}$$

where the superscript (I) specifies convention I. Since $\mu_i^{\circ(\mathrm{I})}$ is the chemical potential of the pure liquid, μ_i^*, it is equal to the chemical potential of the gaseous substance at a partial pressure equal to the equilibrium vapor pressure P_i^*. Equation (6.3-9) becomes

$$\mu_i^{\circ(\mathrm{gas})} + RT \ln\left(\frac{P_i^*}{P^\circ}\right) + RT \ln(a_i^{(\mathrm{I})}) = \mu_i^{\circ(\mathrm{gas})} + RT \ln\left(\frac{P_i}{P^\circ}\right) \qquad \textbf{(6.3-10)}$$

Canceling equal terms and taking antilogarithms in Equation (6.3-10) gives

$$\boxed{P_i = P_i^* a_i^{(\mathrm{I})}} \qquad \textbf{(6.3-11)}$$

which is just like Raoult's law, Equation (6.1-2), except that the activity $a_i^{(\mathrm{I})}$ occurs instead of the mole fraction x_i. The activity acts as an effective mole fraction in determining the partial vapor pressure of the substance.

In convention I, the activity is a kind of effective mole fraction to allow use of an equation similar to Raoult's law.

The **activity coefficient** $\gamma_i^{(\mathrm{I})}$ of substance i in a solution is defined as the ratio of the activity to the mole fraction. It is quite similar to the activity coefficient of a gas, defined in Equation (6.3-6). Both activity coefficients are equal to the ratio of the actual activity to the activity that would occur if the system were ideal.

For a volatile component, Equation (6.3-12) provides the easiest calculation of the activity coefficient: the actual vapor pressure divided by that predicted by Raoult's law.

From Equation (6.3-11), the activity coefficient is equal to the actual vapor pressure divided by the vapor pressure predicted by Raoult's law:

$$\gamma_i^{(I)} = \frac{a_i^{(I)}}{x_i} = \frac{P_i}{P_{i(\text{ideal})}} = \frac{P_i}{P_i^* x_i} \tag{6.3-12}$$

▼ **EXAMPLE 6.7**

Find the activity and the activity coefficient of 2,2,4-trimethylpentane (component 2) in ethanol at 25°C at a mole fraction of 0.2748, according to convention I. The partial vapor pressure is 48.31 torr and the vapor pressure of the pure liquid is 59.03 torr.

Solution

$$a_2^{(I)} = \frac{48.31 \text{ torr}}{59.03 \text{ torr}} = 0.8184$$

$$\gamma_2^{(I)} = 0.8184/0.2748 = 2.978$$

◢

Exercise 6.15

Find the activity and the activity coefficient of ethanol in the solution of Example 6.7, according to convention I. The partial vapor pressure is 46.91 torr and the vapor pressure of pure ethanol at 25°C is 49.31 torr.

Convention II. In this treatment one of the components is designated as the solvent, but mole fractions are used for all components. The solvent (which we always call component 1) is treated just as in convention I:

$$\gamma_1^{(II)} = \gamma_1^{(I)} = \frac{P_1}{P_i^* x_i} \quad (\text{solvent} = \text{component 1}) \tag{6.3-13}$$

Instead of specifying a solute's deviation from Raoult's law, we specify its deviation from Henry's law. This means that the standard state for the activity of a solute is a hypothetical pure liquid or solid with a vapor pressure equal to Henry's law constant, just as in the ideally dilute case.

From the fundamental fact of phase equilibrium, we write for a solute, substance i:

$$\mu_i^{\circ(II)} = \mu_i^{\circ(H)} = \mu_i^{\circ(\text{gas})} + RT \ln\left(\frac{k_i}{P^\circ}\right) \tag{6.3-14}$$

The activity takes on whatever value is needed to make Equation (6.3-1) valid:

$$\mu_i = \mu_i^{\circ(II)} + RT \ln(a_i^{(II)}) \tag{6.3-15}$$

Use of Equation (6.3-14) in Equation (6.3-15) gives

$$\mu_i^{\circ(\text{gas})} + RT \ln\left(\frac{k_i}{P^\circ}\right) + RT \ln(a_i^{(\text{II})}) = \mu_i^{\circ(\text{gas})} + RT \ln\left(\frac{P_i}{P^\circ}\right) \quad \textbf{(6.3-16)}$$

Canceling and taking antilogarithms lead to

$$P_i = k_i a_i^{(\text{II})} \quad (i = 2, 3, \ldots, c) \quad \textbf{(6.3-17)}$$

This equation is just like Henry's law, Equation (6.2-1), except for the occurrence of the activity instead of the mole fraction. If Henry's law is obeyed, the activity and the mole fraction are equal. If not, the activity is an "effective" mole fraction relating the chemical potential and the partial vapor pressure.

In convention II, the activity is a kind of effective mole fraction to allow use of an equation similar to Henry's law.

The activity coefficient is again defined as the ratio of the activity to the mole fraction. It is equal to the actual vapor pressure divided by the vapor pressure predicted by Henry's law:

$$\gamma_i^{(\text{II})} = \frac{a_i^{(\text{II})}}{x_i} = \frac{P_i}{k_i x_i} \quad \textbf{(6.3-18)}$$

Equation (6.3-15) can be written

$$\mu_i = \mu_i^{\circ(\text{II})} + RT \ln(\gamma_i^{(\text{II})} x_i) \quad \textbf{(6.3-19)}$$

Convention II is sometimes referred to as the application of the **solute standard state** to the solutes and of the **solvent standard state** to the solvent. Convention I is called the application of the **solvent standard state** to every component. There are two conventions because we would like to have activity coefficients nearly equal to unity as often as possible. If Henry's law is nearly obeyed by solutes, the use of convention II gives solute activity coefficients nearly equal to unity, whereas convention I could result in solute activity coefficients that deviate considerably from unity.

▼   **EXAMPLE 6.8**

The Henry's law constant for ethanol in diethyl ether at 20°C is 160. torr. Find the activity and activity coefficient of ethanol (component 2) in diethyl ether (component 1) at 20°C for a mole fraction of ethanol equal to 0.100, using both convention I and convention II. The partial pressure of ethanol at this composition and temperature is 12.45 torr, and the vapor pressure of pure ethanol at this temperature is 44.40 torr.

Solution

Convention I:

$$a_2^{(\text{I})} = \frac{P_2}{P_2^*} = \frac{12.45 \text{ torr}}{44.40 \text{ torr}} = 0.280$$

$$\gamma_2^{(\text{I})} = \frac{0.280}{0.100} = 2.80$$

Convention II:

$$a_2^{(\text{II})} = \frac{P_2}{k_2} = \frac{12.45 \text{ torr}}{160 \text{ torr}} = 0.0778$$

$$\gamma_2^{(\text{II})} = \frac{a_2}{x_2} = \frac{0.0778}{0.100} = 0.778$$

Example 6.8 shows that the activity coefficients in a fairly dilute solution are closer to unity when convention II is used than when convention I is used.

▲

| Exercise 6.16 | Using convention I, find the activity and activity coefficient for diethyl ether in the solution of Example 6.8. The partial vapor pressure of diethyl ether at this composition and pressure is 408.6 torr and the vapor pressure of pure diethyl ether at this temperature is 442.6 torr. |

Activities and activity coefficients of solutes are also defined to be used with molalities and concentrations instead of mole fractions. Once again, we require that Equation (6.3-1) be valid and choose standard states so that the activity is equal to the molality if Equation (6.2-10) is valid or to the concentration if Equation (6.2-15) is valid.

The Molality Description

From Equation (6.2-7), the molality of component i is

$$m_i = \frac{n_i}{w_1} = \frac{n_i}{n_1 M_1} = \frac{x_i}{x_1 M_1} \quad \text{(definition)} \tag{6.3-20}$$

where M_1 is the molar mass of the solvent and w_1 is the mass of the solvent. Using Equation (6.3-20) in Equation (6.3-19), we write

$$\mu_i = \mu_i^{\circ(\text{II})} + RT \ln(M_1 m^\circ) + RT \ln\left(\frac{\gamma_i^{(\text{II})} x_1 m_i}{m^\circ}\right) \tag{6.3-21}$$

where m° is defined to equal 1 mol kg^{-1} (exactly). This equation is in the form of Equation (6.3-1):

$$\mu_i = \mu_i^{\circ(m)} + RT \ln(a_i^{(m)}) \tag{6.3-22}$$

where

$$a_i^{(m)} = \frac{\gamma_i^{(\text{II})} x_1 m_i}{m^\circ} \tag{6.3-23}$$

and

$$\mu_i^{\circ(m)} = \mu_i^{\circ(\text{II})} + RT \ln(M_1 m^\circ) \tag{6.3-24}$$

As in the molality description of an ideally dilute solution, the standard state is the hypothetical solution having molality equal to 1 mol kg^{-1} and obeying the molality version of Henry's law, Equation (6.2-11).

We define the molality activity coefficient

$$\gamma_i^{(m)} = \gamma_i^{(\text{II})} x_1 \tag{6.3-25}$$

so that Equation (6.3-22) can be written

$$\mu_i = \mu_i^{\circ(m)} + RT \ln\left(\frac{\gamma_i^{(m)} m_i}{m^\circ}\right) \tag{6.3-26}$$

The standard state for the molality description of a nonideal solution is the same as that for the molality description of an ideally dilute solution.

Since x_1, the mole fraction of the solvent, is approximately equal to unity in dilute solutions, the molality activity coefficient and the mole fraction activity coefficient are nearly equal in dilute solutions.

Equation (6.3-26) is the same as Equation (6.2-12) except for the presence of the activity coefficient. All that is needed to convert an expression for an ideally dilute solution into one for an arbitrary solution is to insert the activity coefficient.

In all cases the effects of nonideality are included by inserting an activity coefficient into a formula.

The Concentration Description

The concentration is given by Equation (6.2-13):

$$c_i = \frac{n_i}{V} = \frac{x_i}{\bar{V}} \tag{6.3-27}$$

where $\bar{V}$ is the mean molar volume, V/n (n is the total amount of all substances). We want to write an equation of the form

$$\mu_i = \mu_i^{\circ(c)} + RT \ln\left(\frac{\gamma_i^{(c)} c_i}{c^\circ}\right) \tag{6.3-28}$$

so that the activity in the concentration description is

$$a_i^{(c)} = \frac{\gamma_i^{(c)} c_i}{c^\circ} \tag{6.3-29}$$

where c° is defined to equal 1 mol L^{-1} or 1 mol m^{-3}. Equation (6.3-28) is valid if

$$\mu_i^{\circ(c)} = \mu_i^{(II)} + RT \ln(\bar{V}_1^* c^\circ) \tag{6.3-30}$$

and

$$\gamma_i^{(c)} = \frac{\gamma_i^{(II)} \bar{V}}{\bar{V}_1^*} \tag{6.3-31}$$

The standard-state chemical potential is that of a hypothetical solution with a concentration of substance i equal to 1 mol L^{-1} (or 1 mol m^{-3}) and obeying Henry's law in the form of Equation (6.2-15). Correction is made for the fact that the mean molar volume is not equal to the molar volume of the solvent except in the limit of infinite dilution.

In all of our descriptions, the solvent is treated in the same way as in convention I. Its activity is always its mole fraction times its activity coefficient:

It is worth remembering that the solvent's activity and activity coefficient are the same in all descriptions.

$$a_1 = \gamma_1 x_1 \quad \text{(solvent, all descriptions)} \tag{6.3-32}$$

Since the activities and activity coefficients of the same solute in two different descriptions are not necessarily equal to each other, we have attached superscripts to specify which description is being used. We sometimes omit these superscripts, relying on the context to make clear which description is being used. Inspection of Equations (6.3-25) and (6.3-31) shows that all of the solute activity coefficients become equal to each other in the limit of infinite dilution.

The Gibbs-Duhem Integration

The activity of a nonvolatile solute cannot be determined by measuring its partial vapor pressure. However, all of our equations are still useful except for those containing the vapor pressure of the solute. For a two-component solution with a volatile solvent and a nonvolatile solute, the activity coefficient of the solute can be determined by measuring the vapor pressure of the solvent over a range of composition and then integrating the Gibbs-Duhem relation.

For constant pressure and temperature, the Gibbs-Duhem relation is given by Equation (4.6-11). When we substitute Equation (6.3-1) into this equation for the case of two components, we obtain

$$x_1 RT \, d[\ln(a_1)] + x_2 RT \, d[\ln(a_2)] = 0 \qquad (6.3\text{-}33)$$

Superscripts on activities and activity coefficients are generally omitted. One must tell from the context which description is being used.

We use convention II. For both components, $a_i = \gamma_i x_i$, where we omit superscripts. Using the fact that $x_i \, d[\ln(x_i)] = dx_i$, we obtain

$$x_1 RT \, d[\ln(\gamma_1)] + RT \, dx_1 + x_2 RT \, d[\ln(\gamma_2)] + RT \, dx_2 = 0$$

Since $x_1 + x_2 = 1$, $dx_1 + dx_2 = 0$, and two terms cancel. We divide by x_2 and obtain

$$d[\ln(\gamma_2)] = -\frac{x_1}{1 - x_1} \, d[\ln(\gamma_1)] \qquad (6.3\text{-}34)$$

The vapor pressure of the solvent (component 1) is measured over a range of compositions, beginning with pure solvent and extending to the composition at which we want the value of γ_2. From the vapor pressure, the activity coefficient of the solvent, γ_1, is calculated for a number of compositions in the range of interest. Equation (6.3-34) is then integrated from $x_1 = x_1''$ to $x_1 = x_1'$, a value of x within this range. x_2'' is taken nearly equal to 1, so that γ_2 is equal to unity at $x_1 = x_1''$

$$\ln[\gamma_2(x_1')] = \int_{x_1 = x_1''}^{x_i = x_i'} \frac{x_1}{1 - x_1} \, d[\ln(\gamma_1)] \qquad (6.3\text{-}35)$$

where we consider both γ_2 and γ_1 to be functions of x_1.

Unless the data are fit to some formula, this integral is approximated numerically. Before the advent of digital computers it was often done graphically, with the integrand $x_1/(1 - x_1)$ plotted on the vertical axis and $\ln(\gamma_1)$ plotted on the horizontal axis.

Thermodynamic Functions of Nonideal Solutions

From Euler's theorem and Equation (6.3-1), the Gibbs energy of a nonideal solution can be written in a way analogous to Equation (6.1-8) for an ideal solution:

$$G_{(\text{soln})} = \sum_{i=1}^{c} n_i [\mu_i^{\circ(l)} + RT \ln(a_i^{(l)})] \qquad (6.3\text{-}36)$$

Changes in any energy-related quantity are unambiguous, but the actual values of these quantities can always be modified by adding a constant to the potential energy.

We report only the difference between the variable in two specific states. For example, we report the **Gibbs energy change of mixing**, which is the value of the Gibbs energy change for the process that forms the solution from the pure components at the same temperature and pressure as the solution. Analogous mixing quantities are defined for other variables.

For convention I, the standard states are the pure components at the pressure of the solution, so Equation (6.3-1) gives the following expression for the Gibbs energy change of mixing:

$$\Delta G_{\text{mix}} = RT \sum_{i=1}^{c} n_i \ln(a_i^{(l)}) \qquad (6.3\text{-}37)$$

If we wish, we can rewrite this equation in terms of the activity coefficients:

$$\Delta G_{\text{mix}} = RT \sum_{i=1}^{c} n_i \ln(\gamma_i^{(\text{l})} x_i)$$

$$= RT \sum_{i=1}^{c} n_i \ln(x_i) + RT \sum_{i=1}^{c} n_i \ln(\gamma_i^{(\text{l})}) \qquad \text{(6.3-38)}$$

The first sum in the right-hand side of the final version of the equation is the same as for an ideal solution, so the second sum represents a correction for the nonideality of the solution. This correction is called the **excess Gibbs energy** and is denoted by G^E:

$$G^E = G(\text{actual}) - G(\text{ideal})$$

$$= RT \sum_{i=1}^{c} n_i \ln(\gamma_i^{(\text{l})}) \qquad \text{(6.3-39)}$$

Equation (6.3-38) can now be written

$$\boxed{\Delta G_{\text{mix}} = \Delta G_{\text{mix}}(\text{ideal}) + G^E} \qquad \text{(6.3-40)}$$

The **excess enthalpy, excess energy, excess entropy**, etc. can all be defined for a nonideal solution:

$$H^E = H(\text{actual}) - H(\text{ideal}) \qquad \text{(6.3-41a)}$$
$$U^E = U(\text{actual}) - U(\text{ideal}) \qquad \text{(6.3-41b)}$$
$$S^E = S(\text{actual}) - S(\text{ideal}) \qquad \text{(6.3-41c)}$$
$$V^E = V(\text{actual}) - V(\text{ideal}) \qquad \text{(6.3-41d)}$$

Exercise 6.17   Show that

$$S^E = -R \sum_{i=1}^{c} n_i \ln[\gamma_i^{(\text{l})}] - RT \sum_{i=1}^{c} n_i (\partial \ln(\gamma_i^{(\text{l})})/\partial T)_P \qquad \text{(6.3-42)}$$

The enthalpy change of mixing is often expressed in terms of the **heat of solution** or **enthalpy change of solution**. For a two-component solution, the **molar integral heat of solution** of component 1 in component 2 is defined by

$$\Delta \bar{H}_{\text{int, 1}} = \frac{\Delta H_{\text{mix}}}{n_1} \qquad \text{(6.3-43)}$$

and the molar integral heat of solution of component 2 is defined by

$$\Delta \bar{H}_{\text{int, 2}} = \frac{\Delta H_{\text{mix}}}{n_2} \qquad \text{(6.3-44)}$$

The same enthalpy change of mixing occurs in both equations, but it is divided by a different amount of substance in each case and assigned to a different component.

▼   **EXAMPLE 6.9**

If 2.000 mol of ethanol (substance 2) and 10.000 mol of water (substance 1) are mixed at a constant temperature of 298.15 K and a constant pressure of 1.000 atm, the enthalpy change is equal to -9.17 kJ. Find the molar integral heat of solution of ethanol in 5.000 mol of water and the molar integral heat solution of water in 0.200 mol of ethanol.

Solution

$$\Delta \bar{H}_{int, 2} = -\frac{9.17 \text{ kJ}}{2.00 \text{ mol}} = -4.58 \text{ kJ mol}^{-1}$$

$$\Delta \bar{H}_{int, 1} = -\frac{9.17 \text{ kJ}}{10.00 \text{ mol}} = -0.917 \text{ kJ mol}^{-1}$$

Example 6.9 shows how the same enthalpy change leads to different values of the integral enthalpy change of mixing for the two substances.

▲

Using the relation for a two-component solution

$$\Delta H_{mix} = n_1 \bar{H}_1 + n_2 \bar{H}_2 - (n_1 \bar{H}_1^* + n_2 \bar{H}_2^*) \qquad \text{(6.3-45)}$$

the integral heat of mixing of a component of a two-component solution can be written in terms of the partial molar enthalpies:

$$\Delta \bar{H}_{int, 2} = \frac{1}{n_2} [n_1 (\bar{H}_1 - \bar{H}_1^*) + n_2 (\bar{H}_2 - \bar{H}_2^*)]$$

$$= \frac{n_1}{n_2} (\bar{H}_1 - \bar{H}_1^*) + \bar{H}_2 - \bar{H}_2^* \qquad \text{(6.3-46)}$$

The integral heat of solution is the enthalpy change per mole of solute for the entire process of making the solution, starting with pure solvent and adding the pure solute to make the desired concentration. It is therefore a kind of average molar quantity for making a solution of the desired concentration. In the limit of zero concentration, it approaches a limit that depends only on the temperature, the pressure, and on the identity of the other substance.

The **differential heat of solution** is defined by

$$\Delta \bar{H}_{diff, 2} = \left(\frac{\partial \Delta H_{mix}}{\partial n_2} \right)_{T, P, n'} \qquad \text{(definition)} \qquad \text{(6.3-47)}$$

The differential heat of solution is the enthalpy change per mole of solute for adding an infinitesimal amount of solute to the solution (not changing its composition), or equivalently for adding 1 mole of solute to a very large amount of the solution (again not changing its composition). From Equations (6.3-45) and (6.3-47),

$$\Delta \bar{H}_{diff, i} = \bar{H}_i - \bar{H}_i^* \qquad \text{(6.3-48)}$$

Equation (6.3-48) is valid for any number of components, whereas Equation (6.3-46) is valid only for a two-component solution.

Exercise 6.18

Write the version of Equation (6.3-46) that applies to a solution of c components.

Tabulated Thermodynamic Properties for Solutions

In Chapters 2 and 4, we discussed enthalpy and Gibbs energy changes of formation for gases and pure substances. We now discuss enthalpy changes of formation and Gibbs energy changes of formation for components of solutions.

The **enthalpy change of formation** of substance i in a solution is defined as the enthalpy change to produce 1 mole of substance i from the necessary elements in their most stable forms and then to dissolve the 1 mole of substance i in a large amount of the solution of the specified composition. A large amount of solution is used so that the addition of solute does not change its composition. The enthalpy change of formation is therefore related to the differential heat of solution, not the integral heat of solution. The Gibbs energy change of formation is analogous; it is the change in Gibbs energy to produce 1 mole of the substance from the necessary elements in their most stable forms and then to dissolve it in a large amount of the solution of the specified composition. These quantities do *not* include the enthalpy change or the Gibbs energy change to produce the large amount of solution to which the substance is added.

The composition of the solution must be specified. The values of enthalpy changes and Gibbs energy changes of formation of solutes in Table A9 are for the hypothetical 1 mol kg^{-1} solution with activity coefficient set equal to unity, unless otherwise noted. The standard states for the chemical potential of solution components were defined to be at the pressure of the solution, whether or not that pressure was equal to $P°$. For the standard-state enthalpy change and Gibbs energy change of formation, the pressure is specified to equal $P°$. However, because enthalpies and Gibbs energies of liquids and solids do not depend very much on pressure, we will ignore the pressure dependence.

The standard-state enthalpy change of formation is defined by

$$\Delta \bar{H}_f^°(i, \text{soln}) = \bar{H}_i^° - H^°(\text{elements}) \quad \text{(definition)}$$

(6.3-49)

and the standard-state Gibbs energy of formation is given by

$$\Delta \bar{G}_f^°(i, \text{soln}) = \mu_i^° - G^°(\text{elements}) \quad \text{(definition)}$$

(6.3-50)

▼ **EXAMPLE 6.10**   The following are values for ethanol (component 2) at 298.15 K:

	$\Delta \bar{H}_f^°$ (kJ mol^{-1})	$\Delta \bar{G}_f^°$ (kJ mol^{-1})
Liquid	−277.69	−174.78
Aqueous	−288.3	−181.64

a. Find the differential heat of solution in the standard state.
b. Find the value of $\mu_2^{°(m)} - \bar{G}_2^*(\text{liq})$ for ethanol in H_2O.

Solution

a.

$$\Delta \bar{H}^{\circ}_{\text{diff}, 2} = \bar{H}^{\circ(m)}_2 - \bar{H}^{*}_{2(\text{liq})}$$
$$= \Delta \bar{H}^{\circ(m)}_f(2) - \Delta \bar{H}^{\circ}_f(2, \text{liq})$$
$$= -288.3 \text{ kJ mol}^{-1} - (-277.69 \text{ kJ mol}^{-1})$$
$$= -10.6 \text{ kJ mol}^{-1}$$

Note the difference in value between this quantity and the integral heat of solution in Example 6.9.

b.

$$\mu^{\circ(m)}_2 - \bar{G}^{*}_{2(\text{liq})} = \Delta \bar{G}^{\circ(m)}_f(2) - \Delta \bar{G}^{\circ}_f(2, \text{liq})$$
$$= -181.64 \text{ kJ mol}^{-1} - (-174.78 \text{ kJ mol}^{-1})$$
$$= -6.86 \text{ kJ mol}^{-1}$$

▲

6.4	

Activity Coefficients in Electrolyte Solutions

Some compounds ionize or dissociate to form electrically charged ions in a solution with water and some other solvents. The electrical forces between these ions in solution are strong and act over a distance that is large compared to molecular dimensions. The forces are said to be of *long range*.

The force between a macroscopic object of charge Q_1 and one of charge Q_2 is given by Coulomb's law

Coulomb's law is named for Charles Augustin de Coulomb, 1736–1806, the French physicist who discovered the law.

$$F = \frac{Q_1 Q_2}{4\pi\varepsilon r^2_{12}} \qquad (6.4\text{-}1)$$

where r_{12} is the distance between the centers of the objects and ε is a constant that depends on the substance (medium) between the charges and is called the **permittivity** of the medium. The permittivity of a vacuum is denoted by ε_0 and is equal to 8.85419×10^{-12} C^2 N^{-1} m^{-2}.

The ratio of the permittivity of a substance to that of a vacuum is called the **dielectric constant** and denoted by ε_{rel}. The dielectric constant of water at 25°C is equal to 78.54. Although Equation (6.4-1) was deduced for macroscopic charged objects, it is assumed to hold also for ions in a solution, at least for distances large enough that two ions have some water molecules between them.

Although the electrostatic forces between ions in water are weaker than those in a vacuum by a factor of about 80, these forces are of long range and have significant effects on the behavior of the solution. There are two principal equilibrium effects of the electrostatic forces: (1) the long-range electrostatic forces cause significant deviations from ideally dilute behavior even at low concentrations, and (2) the chemical equilibrium between the ions and the un-ionized or undissociated compound produces behavior qualitatively different from that of nonelectrolyte solutes. We will discuss the first effect in this section. The second effect will be discussed in Chapter 7 when we consider the thermodynamics of chemical equilibrium.

In an ideal solution, the effective molecular environment of the molecules of any substance must be composition independent, which can happen if the intermolecular forces between unlike molecules and like molecules are the same. In a sufficiently dilute solution, almost all solute molecules are surrounded only by molecules of the solvent. If the intermolecular forces are short-range forces, the molecular environment is independent of composition as long as the solution remains dilute, producing ideally dilute behavior. However, in an electrolyte solution the long-range electrostatic interaction between the ions makes fairly distant ions influence a given ion, so electrolyte solutions deviate from ideally dilute behavior except at extremely low concentrations.

The Debye-Hückel Theory

The Debye-Hückel theory is named for Peter J. W. Debye, 1884–1966, a Dutch-American physicist and chemist who received the Nobel Prize in chemistry in 1936 for his work on the dipole moments of molecules, and Erich Hückel, 1896–1980, a German chemist who is also known for an approximation scheme used in molecular quantum mechanics.

Debye and Hückel developed a molecular theory for the activity coefficients of an electrolyte solution. We will not discuss their theory except to mention its underlying ideas and its principal result.[1]

The Debye-Hückel theory begins with the assumption that there is a distance of closest approach a for a pair of ions. If r is the distance between the centers of two ions, it is assumed that $r < a$ cannot occur. Equation (6.4-1) is assumed to be valid for the potential energy of a pair of ions if $r > a$.

The main idea of the Debye-Hückel derivation was to imagine that the ions in a solution could somehow have their charges varied reversibly from zero to the actual values and to determine the consequences of this process from electrostatic theory and statistical mechanical theory. During the reversible charging process an "ion atmosphere" is created, consisting of an excess of ions of the opposite charge in the vicinity of any given ion, as these ions are attracted and ions of the same charge as the given ion are repelled.

An equation for the statistical average distribution of ions around the fixed ion was derived by combining the Poisson equation of electrostatics and the Boltzmann probability distribution of Equation (1.7-25) into an equation called the **Poisson-Boltzmann equation**.

Next, the reversible work of creating the ion atmosphere was calculated from electrostatic theory. According to Equation (4.1-29), the net work done on the system in a reversible process is equal to the change in the Gibbs energy. The reversible work of charging the ions gives the contribution to the Gibbs energy due to the electrostatic forces, which leads to equations for the chemical potential and the activity coefficient.

The first result of the Debye-Hückel theory is a formula for the statistical distribution of charge around a given ion in the solution:

$$n_j(r) = \frac{-z_j e^2 \exp[-\kappa(r-a)]}{4\pi\varepsilon r(1+\kappa a)} \qquad \textbf{(6.4-2)}$$

[1] An account of the development may be found in T. L. Hill, *Statistical Thermodynamics*, Addison-Wesley, Reading, MA, 1960, pp. 321ff. The original reference is P. Debye and E. Hückel, *Physik. Z.* **24**, 185 (1923).

where $n_j(r)$ represents the average net charge per unit volume at a distance r from a given ion of valence z_j and e is the charge on a proton, 1.6022×10^{-19} C. The quantity κ is

$$\kappa = e\left(\frac{2N_A\rho_1 I}{\varepsilon k_B T}\right)^{1/2} \tag{6.4-3}$$

where N_A is Avogadro's number, ρ_1 is the density of the solvent, k_B is Boltzmann's constant, and T is the temperature. The quantity I is the **ionic strength**, defined for a solution with s different charged species by

$$I = \frac{1}{2}\sum_{i=1}^{s} m_i z_i^2 \quad \text{(definition of ionic strength)} \tag{6.4-4}$$

where m_i is the molality of species i and z_i is its valence (number of proton charges per ion).

▼ **EXAMPLE 6.11**

Calculate the ionic strength of a solution that is 0.100 mol kg^{-1} in NaCl and 0.200 mol kg^{-1} in CaCl$_2$.

Solution

$$\begin{aligned}
I &= \tfrac{1}{2}[m(\text{Na}^+)(+1)^2 + m(\text{Ca}^{2+})(+2)^2 + m(\text{Cl}^-)(-1)^2] \\
&= \tfrac{1}{2}[(0.100 \text{ mol kg}^{-1}) + (0.200 \text{ mol kg}^{-1})(2^2) + (0.500 \text{ mol kg}^{-1})] \\
&= 0.700 \text{ mol kg}^{-1}
\end{aligned}$$

Notice the large contribution of a multiply charged species like Ca^{2+}.

▲

Exercise 6.19

Calculate the ionic strength of a solution that is 0.100 mol kg^{-1} in K$_2$SO$_4$ and 0.200 mol kg^{-1} in Na$_2$SO$_4$. Assume complete dissociation.

The expression for κ can be written

$$\kappa = \beta I^{1/2} \tag{6.4-5}$$

where β is a parameter that depends only on the temperature and the properties of the solvent:

$$\beta = e\left(\frac{2N_A\rho_1}{\varepsilon k_B T}\right)^{1/2} = e\left(\frac{2N_A\rho_1}{\varepsilon_{\text{rel}}\varepsilon_0 k_B T}\right)^{1/2} \tag{6.4-6}$$

The reciprocal of κ is called the **Debye length**. It is a measure of the effective range of net electrostatic interaction of the ions in the solution.

Exercise 6.20

Show that for water at 298.15 K

$$\beta = 3.282 \times 10^9 \text{ kg}^{1/2} \text{ mol}^{-1/2} \text{ m}^{-1} \quad \text{(water, 298.15 K)} \tag{6.4-7}$$

The density of water at 298.15 K is 997.14 kg m^{-3}.

The total net charge in a spherical shell of thickness dr and radius r centered on a given ion of valence z_j is given by

$$dQ_{\text{shell}} = 4\pi r^2 n_j(r)\, dr \qquad \text{(6.4-8)}$$

Figure 6.11 shows $|4\pi r^2 n_+(r)|$ around an anion of a 1-1 electrolyte at $0.0050\ \text{mol kg}^{-1}$. The maximum in the curve is at $r = 1/\kappa$, which for this case is approximately 43×10^{-10} m, illustrating the relatively long range of the net electrostatic forces.

Exercise 6.21   Show that the maximum in the curve of Figure 6.11 is at $r = 1/\kappa$ and verify the value of the distance to the maximum for a 1-1 electrolyte at $0.0050\ \text{mol kg}^{-1}$.

The final result for the activity coefficient of ions of valence z in the Debye-Hückel theory is

$$\ln(\gamma_i) = -\frac{z_i^2 \alpha I^{1/2}}{1 + \beta a I^{1/2}} \qquad \text{(6.4-9)}$$

The quantity α is a second function of temperature and of the properties of the solvent:

$$\alpha = (2\pi N_A \rho_1)^{1/2} \left(\frac{e^2}{4\pi \varepsilon k_B T} \right)^{3/2} \qquad \text{(6.4-10)}$$

Equation (6.4-9) is used for the activity coefficient in the mole fraction, molality, and molarity descriptions, since all three of these activity coefficients are nearly equal in a dilute solution.

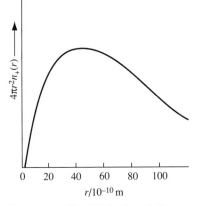

Figure 6.11. The Distribution of Charge Around a Central Ion According to the Debye-Hückel Theory. This diagram shows the charge density in spherical shells around a given ion.

Exercise 6.22   Show that for water at 298.15 K, the value of α is
$$\alpha = 1.171\ \text{kg}^{1/2}\ \text{mol}^{-1/2} \quad \text{(water, 298.15 K)} \qquad \text{(6.4-11)}$$

The chemical potentials and activity coefficients of individual ions cannot be measured, because the chemical potential is a partial derivative with respect to the amount of the given substance, keeping the amounts of other substances fixed. The amount of energy required to charge a system by adding ions of one charge without adding ions of the opposite charge is so large that it is not possible to add a significant amount of one kind of ion to a system without adding some ions of the other charge at the same time. For an electrolyte solute represented by the formula $M_{\nu_+} X_{\nu_-}$, where ν_+ and ν_- represent the numbers of cations and anions, respectively, in the formula of the compound, it is customary to define the **mean ionic activity coefficient**

$$\gamma_\pm = (\gamma_+^{\nu_+} \gamma_-^{\nu_-})^{1/\nu} \qquad \text{(6.4-12)}$$

where $\nu = \nu_+ + \nu_-$, the total number of ions in the formula.

From Equation (6.4-9) we can obtain

$$\ln(\gamma_\pm) = -z_+|z_-|\frac{\alpha I^{1/2}}{1 + \beta a I^{1/2}} \tag{6.4-13}$$

Exercise 6.23   Carry out the algebraic steps to obtain Equation (6.4-13).

▼   **EXAMPLE 6.12**   Calculate the value of $\gamma_\pm$ for a 0.0100 mol kg^{-1} solution of NaCl in water at 298.15 K. Assume that the distance of closest approach is 3.05×10^{-10} m. This distance gives a value of βa equal to 1.00 kg$^{1/2}$ mol$^{-1/2}$.

Solution

$$\ln(\gamma_\pm) = -\frac{(1.171 \text{ kg}^{1/2} \text{ mol}^{-1/2})(0.01 \text{ mol kg}^{-1})^{1/2}}{1 + (1.00 \text{ kg}^{1/2} \text{ mol}^{-1/2})(0.01 \text{ mol kg}^{-1})^{1/2}}$$

$$= -0.1065$$

$$\gamma_\pm = 0.899$$

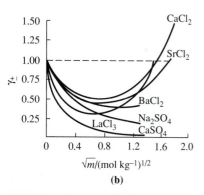

Figure 6.12. The Mean Ionic Activity Coefficients of Several Electrolyte Solutes as a Function of the Square Root of the Ionic Strength. According to the Debye-Hückel theory, the curves should be linear, as they are for a small region near zero ionic strength.

Equation (6.4-14) can be called an extended Debye-Hückel formula.

The Debye-Hückel theory has been shown experimentally to be an accurate **limiting law**. That is, it gives the correct behavior as the concentration is made very small. In practice, it is accurate enough for some purposes for ionic strengths up to about 0.01 mol kg^{-1} but fails badly for ordinary concentrations. Figure 6.12 shows experimental values of the mean ionic activity coefficient of several electrolyte solutes in water at 298.15 K as a function of $\sqrt{m}$.

Much work has been done to extend the Debye-Hückel theory, beginning in 1926 with a theory of Bjerrum[2] in which it was assumed that two ions of opposite charge closer to each other than a certain distance constituted an ion pair that could be treated as a single chemical species in chemical equilibrium with the dissociated ions. Later research is based largely on theoretical work of Mayer.[3] In his result, the Debye-Hückel result appears as the leading term of a series in powers and logarithms of I.

The work of Mayer gives some credibility to a semiempirical equation:

$$\ln(\gamma_\pm) = -z_+|z_-|\left(\frac{\alpha I^{1/2}}{1 + \beta a I^{1/2}} + bI\right) \tag{6.4-14}$$

where b is a parameter that is evaluated experimentally for each electrolyte solute. Use of an average value of 3.05×10^{-10} m for a and of an average value of b gives the **Davies equation**,[4] which for water as the solvent and

[2] N. Bjerrum, *Kgl. Danske Vidensk. Selskab.* **7**, no. 9 (1926).
[3] J. E. Mayer, *J. Chem. Phys.* **18**, 1426 (1950).
[4] C. W. Davies, *Ion Association*, Butterworth, London, 1962, pp. 35–52.

for a temperature of 298.15 K is

$$
\log_{10}(\gamma_{\pm}) = -0.510 z_{+} |z_{-}| \left[\frac{(I/m°)^{1/2}}{1 + (I/m°)^{1/2}} - 0.30 \frac{I}{m°} \right] \quad \textbf{(6.4-15)}
$$

where $m° = 1$ mol kg^{-1} (exactly). This equation has no adjustable parameters, and it is used when no experimental information is available for an ionic solute. In some cases it can give usable results for activity coefficients up to ionic strengths of 0.5 mol kg^{-1} or beyond, but it is ordinarily in error by several percent in this region. Table A12 gives experimental values of the mean ionic activity coefficients of several aqueous electrolytes at various concentrations. It also gives the predictions of the Debye-Hückel formula, Equation (6.4-13), with βa taken equal to unity, and of the Davies equation, Equation (6.4-15).

Exercise 6.24   Calculate the activity coefficient for the solution of Example 6.12 using the Davies equation. Find the percent difference between the result of the Davies equation and the Debye-Hückel limiting law.

6.5   Phase Diagrams for Nonideal Mixtures

In this section, we present and discuss some temperature-composition and pressure-composition phase diagrams for nonideal two-component and three-component solutions.

Liquid-Vapor Phase Diagrams of Nonideal Systems

Figure 6.13 shows a pressure-composition liquid-vapor phase diagram of ethanol and diethyl ether for a fixed temperature of 20°C. Figure 6.14 shows the temperature-composition phase diagram of ethanol and diethyl ether for a fixed pressure of 1.84 atm. The curve in Figure 6.13 representing total pressure as a function of liquid composition shows positive deviation from Raoult's law. The corresponding temperature curves in Figure 6.14 lie lower than the curves for an ideal solution, because if the vapor pressure is larger than that of an ideal solution, the solution will boil at a lower temperature than an ideal solution.

If the deviation from ideality is large enough, the curves in the phase diagram can exhibit a maximum or a minimum. Figure 6.15 shows the pressure-composition phase diagram of ethanol and benzene, in which there is a maximum in the vapor pressure curve. A strong negative deviation can give a minimum in the vapor pressure curve. Either a maximum or a minimum point in such a curve is called an **azeotrope**. The two curves representing liquid and vapor compositions are tangent at this point, so the two phases have the same composition at an azeotrope.

To show that the phases have the same composition and the same vapor pressure at the azeotrope, we write the Gibbs-Duhem relation,

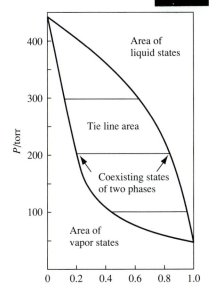

Figure 6.13. Pressure-Composition Phase Diagram for Diethyl Ether–Ethanol at 20°C. The lower curve shows the pressure as a function of mole fraction in the vapor, and the upper curve shows the pressure as a function of mole fraction in the liquid. Compare with Figure 6.3. Drawn from data in J. Timmermans, *Physicochemical Constants of Binary Systems*, Vol. 2, Interscience, New York, 1959, p. 401.

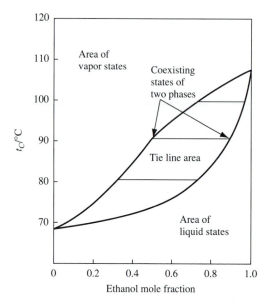

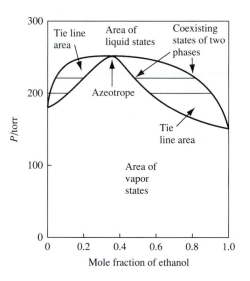

Figure 6.14. Temperature-Composition Phase Diagram for Diethyl Ether–Ethanol at 1.84 atm. The lower curve represents the temperature as a function of mole fraction in the liquid, and the upper curve represents the temperature as a function of mole fraction in the vapor. Compare with Figure 6.4 for a nearly ideal solution. Drawn from data in J. Timmermans, *Physicochemical Constants of Binary Systems*, Vol. 2, Interscience, New York, 1959, p. 401.

Figure 6.15. Liquid-Vapor Pressure-Composition Phase Diagram of Ethanol and Benzene, Showing an Azeotrope. In this diagram, a large positive deviation from ideality produces a maximum in the curves. Drawn from data in J. Timmermans, *Physicochemical Constants of Binary Systems*, Vol. 2, Interscience, New York, 1959, p. 61.

Equation (4.6-11), for the liquid phase at constant temperature and pressure. Using Equation (6.3-1) and dividing by RT, we obtain

$$x_1\, d[\ln(a_1)] + x_2\, d[\ln(a_2)] = 0 \qquad \textbf{(6.5-1)}$$

where x_1 and x_2 are the mole fractions in the liquid. In order to have constant pressure, an additional gas, such as air, must be present in the gas phase. The effect of this added gas on the total vapor pressure is small, as shown in Example 6.4.

If the gas phase (assumed ideal) is at equilibrium with the solution, the chemical potential of component 1 has the same value in each phase. Assuming convention I for the solution, we write

$$\mu_1^{\circ(\text{I})} + RT\,\ln(a_1) = \mu_1^{\circ(\text{gas})} + RT\,\ln(P_1/P^\circ)$$

For an equilibrium change in state at constant T and P

$$RT\, d[\ln(a_1)] = RT\, d[\ln(P_1/P^\circ)] \qquad \textbf{(6.5-2)}$$

When Equation (6.5-2) and the analogous equation for substance 2 are substituted in Equation (6.5-1), we obtain for the vapor at equilibrium with the solution

$$x_1 \, d\left[\ln\left(\frac{P_1}{P^\circ}\right)\right] + x \, d\left[\ln\left(\frac{P_2}{P^\circ}\right)\right] = 0$$

$$x_1 \frac{1}{P_1} \, dP_1 + x_2 \frac{1}{P_2} \, dP_2 = 0$$

We convert this equation into a derivative equation:

$$\frac{x_1}{P_1}\left(\frac{\partial P_1}{\partial x_1}\right)_{T,P} + \frac{x_2}{P_2}\left(\frac{\partial P_2}{\partial x_1}\right)_{T,P} = 0 \qquad \textbf{(6.5-3)}$$

The total vapor pressure is the sum of the partial vapor pressures.

$$P_{\text{vap}} = P_1 + P_2$$

At the azeotrope, the total vapor pressure is at a maximum or a minimum, so that

$$\left(\frac{\partial P_{\text{vap}}}{\partial x_1}\right)_{T,P} = \left(\frac{\partial P_1}{\partial x_1}\right)_{T,P} + \left(\frac{\partial P_2}{\partial x_1}\right)_{T,P} = 0 \qquad \textbf{(6.5-4)}$$

When this equation is substituted into Equation (6.5-3),

$$\frac{x_1}{P_1} = \frac{x_2}{P_2}$$

or

$$\frac{x_1}{x_2} = \frac{P_1}{P_2} = \frac{x_{1(\text{gas})}}{x_{2(\text{gas})}} \qquad \textbf{(6.5-5)}$$

where we have used the fact that in an ideal gas mixture, the mole fraction is proportional to the partial pressure (Dalton's law).

Deviations from ideality can be positive or negative. Figure 6.16 shows the temperature-composition phase diagram of acetone and chloroform, which have a negative deviation from ideality and which exhibit an azeotrope.

An azeotropic mixture is sometimes called a **constant-boiling mixture** because it distills without any change in composition. It is impossible to distill from one side of an azeotrope to the other. For example, ethanol and water have an azeotrope at 1.00 atmosphere pressure at an ethanol mole fraction equal to 0.90. Any mixture of ethanol and water can be distilled to this composition, but no further.

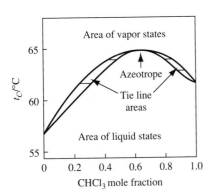

Figure 6.16. Liquid-Vapor Temperature-Composition Phase Diagram of Acetone and Chloroform. In this diagram a negative deviation from ideality produces a maximum in the curves. This maximum corresponds to a minimum in the pressure-composition phase diagram.

Exercise 6.25

The normal boiling temperature of water is 100°C, and that of ethanol is 78.3°C. At 1.000 atm, the azeotrope boils at 78.17°C.

a. Sketch the liquid-vapor temperature-composition phase diagram of ethanol and water.

b. By drawing a "staircase" of line segments representing distillation, as in Figure 6.5, show that a distillation process beginning with a mole fraction of ethanol less than 0.90 cannot give a distillate with an ethanol mole fraction greater than 0.90.

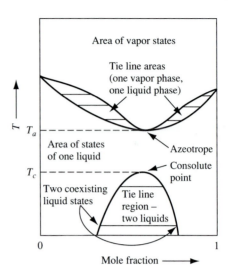

Figure 6.17. Temperature-Composition Phase Diagram of a System with Liquid-Liquid Phase Separation (schematic). A large positive deviation from ideality produces a phase separation into two liquid phases below a critical solution temperature T_c.

If a binary mixture has a sufficiently large positive deviation from ideality, there can be a separation into two liquid phases. Such phase separations are well known ("oil and water don't mix"). Figure 6.17 shows schematically the temperature-composition phase diagram of two hypothetical substances, A and B. Below the temperature labeled T_c is a region of tie lines in the center of the diagram. Only the compositions to the right or to the left of this region are possible equilibrium compositions of a single liquid phase at temperatures below T_c. Any horizontal tie line connects points representing the compositions of the two liquids that can be at equilibrium with each other at the temperature of the tie line.

The highest point in the tie-line region is called an **upper critical solution point**, or an **upper consolute point**. It has a number of properties similar to those of the gas-liquid critical point in Figure 1.5. For example, if a mixture has the same overall composition as that of the consolute point, it will be a two-phase system at a temperature below the consolute temperature. As its temperature is gradually raised, the meniscus between the phases becomes diffuse and disappears, just as the meniscus between the liquid and vapor phases disappears as the liquid-vapor critical point is approached, as shown in Figure 1.6.

There are mixtures, such as water and nicotine, that have both an upper and a lower consolute point, so that the boundary of the tie line region is a closed curve. For example, below the lower consolute point at 61°C water and nicotine mix in all proportions, and above the upper consolute temperature at 210°C they also mix in all proportions. Between these two temperatures, there is a tie line region in the diagram and the liquids are only partially miscible. The liquids do not mix at all compositions because the Gibbs energy of the system is lower when it separates into two phases than if it were in the (metastable) one-phase state.

If the positive deviation from ideality is even greater than that of Figure 6.17, the two-phase region can extend to the liquid-vapor region and produce a phase diagram like that of Figure 6.18, the temperature-composition phase diagram of furfural and water at 1.000 atm. The horizontal tie line at 97.90°C connects three points representing the compositions of two liquid phases and one gas phase that can coexist at equilibrium. For two components and three phases, only one intensive variable is independent. Since the pressure is fixed for this diagram, no other variable is independent and the compositions of all phases and the temperature are fixed for this state.

To purify furfural by the process of **steam distillation**, water is added to impure furfural and the two-phase mixture is boiled. If the impurities do not change the boiling temperature very much, the two-phase mixture boils

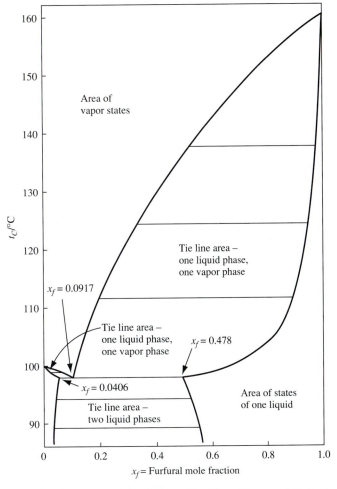

Figure 6.18. Liquid-Vapor Temperature-Composition Phase Diagram of Furfural and Water at 1.000 atm. A large positive deviation from ideality produces a phase separation into two liquid phases at all temperatures. After G. H. Mains, *Chem. Metall. Eng.* **26**, 779 (1922).

near 97.90°C at 1.000 atm. The vapor has a furfural mole fraction equal to 0.092, independent of the amounts of furfural and water, since all three phases have fixed compositions at fixed pressure. The vapor is condensed to a two-phase liquid, and the furfural layer is recovered and dried (at 20°C this layer has a furfural mole fraction of 0.78 before drying). The advantage of steam distillation is that it can be carried out at a lower temperature than an ordinary distillation.

Solid-Liquid Phase Diagrams

Solid-liquid phase diagrams are similar in appearance to liquid-vapor phase diagrams. Figure 6.19 shows the solid-liquid temperature-composition phase diagram of gold and copper.

Just as liquids are sometimes purified by distillation, solids are sometimes purified by **zone refining**, in which a rod-shaped piece of the solid is gradually passed through a ring-shaped furnace. A zone of the solid melts as it passes into the furnace and refreezes as it passes out of the furnace. This process is analogous to vaporization and recondensation of a liquid in distillation, except that the melting process gives a liquid of the same composition as the solid, making the initial process correspond to a vertical line segment in the diagram instead of a horizontal line segment. The liquid system is often not so easy to equilibrate as the vapor, but in many cases the solid that freezes out approximates the equilibrium composition at the other end of the tie line, being richer in the higher-melting component than the original solid. This is analogous to distillation in a still with one theoretical plate. A second pass through the furnace can lead to further purification.

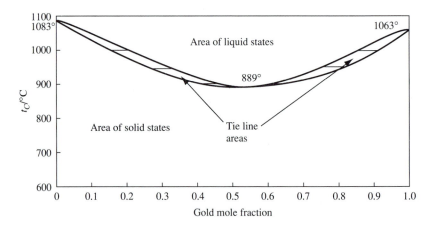

Figure 6.19. Solid-Liquid Temperature-Composition Phase Diagram of Gold and Copper.
Compare with Figure 6.16 for a liquid-vapor transition. From M. Hansen, *The Constitution of Binary Alloys*, McGraw-Hill, New York, 1958, p. 199.

Exercise 6.26 By drawing a "staircase" in Figure 6.19, determine what composition will result from three successive zone-refining passes starting with a gold-copper solid solution of gold mole fraction equal to 0.70. What would many successive zone refining passes lead to if the curves had a maximum instead of a minimum?

Figure 6.20 shows the solid-liquid temperature-composition phase diagram of silver and copper at 1.00 atm. The tie line at 779°C connects the state points representing two solid phases and one liquid phase that can be at equilibrium. The point representing the liquid phase at equilibrium with the two solid phases is called the **eutectic point**. If a liquid with the same composition as the eutectic is cooled, when it reaches the eutectic temperature two solid phases with the compositions represented by the ends of the tie line will freeze out. This two-phase solid mixture usually consists of very small grains of one phase imbedded in the other phase, so that it can look almost like a single phase unless examined through a microscope.

Exercise 6.27 For each one-phase region in the phase diagram of Figure 6.20, give the phase that can occur and give the number of independent intensive variables (excluding the pressure, which is fixed at 1 atm). For each two-phase (tie line) region give the phases that can be at equilibrium and give the number of independent intensive variables, excluding the pressure.

Solid-liquid phase diagrams like that of Figure 6.20 are constructed by analyzing experiments in which a mixture of known composition is heated

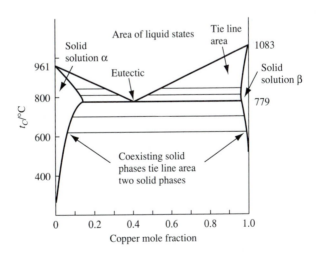

Figure 6.20. Solid-Liquid Temperature-Composition Phase Diagram of Silver and Copper. There is a sufficiently large deviation from ideality in the solid solutions that two different solid solution phases occur. From R. E. Dickerson, *Molecular Thermodynamics*, W. A. Benjamin, New York, 1969, p. 371.

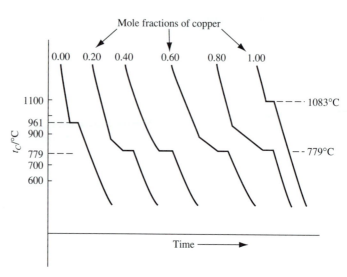

Figure 6.21. Cooling Curves for the Ag-Cu System. The compositions are those indicated in Figure 6.20. The first break in each curve represents the first appearance of a solid phase. The horizontal segments at 779°C represent the eutectic, when three phases occur. From R. E. Dickerson, *Molecular Thermodynamics*, W. A. Benjamin, New York, 1969, p. 371.

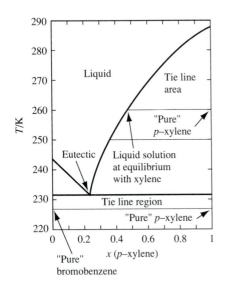

Figure 6.22. Solid-Liquid Temperature-Composition Phase Diagram of *p*-Xylene and Bromobenzene. There is a positive deviation from ideality that is sufficiently large to prevent any significant solid solubility. From M. L. McGlashan, *Chemical Thermodynamics*, Academic Press, New York, 1979, p. 268.

above its melting point and then allowed to cool slowly. Figure 6.21 shows **cooling curves** representing the temperature of such a mixture as a function of time for various mole fractions of copper. Consider the cooling curve for a mixture with copper of mole fraction 0.80. The temperature drops smoothly until a copper-rich solid solution begins to freeze out at about 950°C. At this point, the rate of cooling slows because the enthalpy of freezing is evolved as heat and the composition of the liquid changes as the solid is removed. When the eutectic temperature is reached at 779°C, a second solid solution, rich in silver, begins to freeze out. With three phases present, the temperature must remain constant. A horizontal portion of the cooling curve, called the "eutectic halt," results. Only after the system is entirely frozen can the temperature drop further.

Figure 6.22 shows the solid-liquid temperature-composition phase diagram of *para*-xylene and bromobenzene at 1 atm. The solids are almost completely insoluble in each other, so the regions of solid solubility are too small to show in the figure and the solids that freeze out are essentially pure substances.

Solid-Liquid Phase Diagrams with Compounds

Sometimes two substances form solid-state compounds. Figure 6.23 shows the solid-liquid temperature-composition phase diagram of aniline (A) and phenol (P), which form a one-to-one compound in the solid state and are completely miscible in the liquid phase.

The phase diagram resembles two phase diagrams set side by side, and that is essentially what it is. The vertical line at mole fraction 0.5 repre-

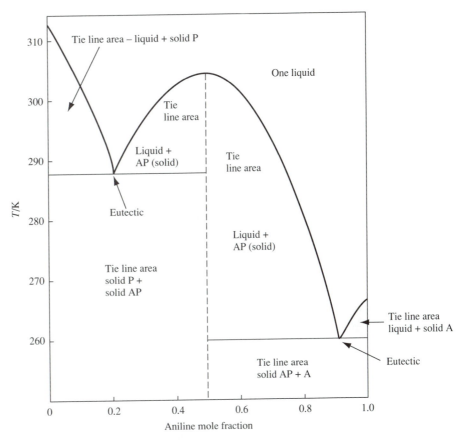

Figure 6.23. Solid-Liquid Temperature-Composition Phase Diagram of Phenol and Aniline. In this diagram we see the occurrence of a stoichiometric compound of the two substances, containing the same number of moles of each substances.

sents the compound $C_6H_5OH \cdot C_6H_5NH_2$, which we abbreviate AP. The right half of the diagram is the phase diagram for the two substances AP and A. The left half is the phase diagram for the two substances P and AP. Each diagram contains a eutectic point, below which the two nearly pure solid substances can coexist. Either A or P can coexist with the compound AP, but they cannot coexist with each other at equilibrium. The compound AP exists only in the solid state. When it melts, an equimolar liquid solution of aniline and phenol results.

   If more than one solid compound occurs, a solid-liquid phase diagram can be quite complicated. Figure 6.24 shows the temperature-composition phase diagram of copper and lanthanum, which have four different compounds. There are only two maxima in the diagram. The two compounds $LaCu_4$ and $LaCu$ do not melt in the same way as does the compound of aniline and phenol and the compounds $LaCu_2$ and $LaCu_6$. For example, at 551°C LaCu melts to form a liquid phase with a lanthanum mole fraction of 0.57 plus solid $LaCu_2$, as indicated by the tie line. This phenomenon is called **incongruent melting** because the liquid phase does not have the same

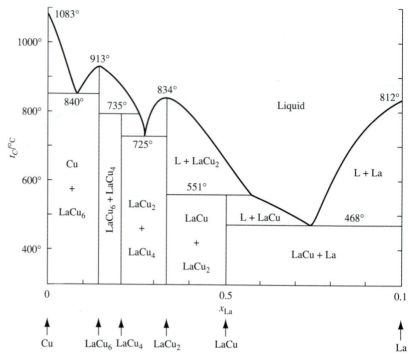

Figure 6.24. Solid-Liquid Temperature-Composition Phase Diagram of Copper and Lanthanum. There are four compounds. There is no eutectic between LaCu and $LaCu_2$. The compound LaCu melts incongruently to give solid $LaCu_2$ and a liquid solution. From R. E. Dickerson, *Molecular Thermodynamics*, W. A. Benjamin, New York, 1969, p. 379.

composition as the solid phase from which it arose. The point representing the composition of the liquid at 551°C is called a **peritectic point**.

Exercise 6.28   For each area in Figure 6.24, tell what phase or phases occur and what the number of independent intensive variables is.

Three-Component Phase Diagrams

For three components and one phase, $f = 3 - 1 + 2 = 4$. We have already had to keep one variable fixed in order to draw a two-dimensional phase diagram for two components. For three components, we must keep two variables fixed to have a two-dimensional phase diagram.

For a composition-composition phase diagram at a fixed temperature and a fixed pressure, a diagram in an equilateral triangle is customarily used, as depicted in Figure 6.25. A theorem of plane geometry asserts that the sum of the three perpendicular distances to the sides is the same for any point inside an equilateral triangle. Therefore, these three perpendicular distances can be used to represent the three mole fractions, which must add to unity. Each vertex of the triangle represents a different pure component, and the perpendicular distance from the opposite side to any point inside the triangle equals the mole fraction of that component.

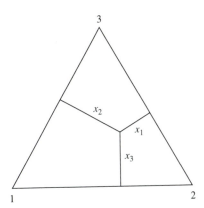

Figure 6.25. Mole Fractions of a Three-Component System Plotted in an Equilateral Triangle. The three perpendicular distances from any interior point to the three sides of an equilateral triangle have a constant sum.

Figure 6.26 shows the liquid-liquid composition-composition phase diagram of water, acetone, and ethyl acetate at 30°C and 1.00 atm. The tie lines in the two-phase region connect points representing the compositions of two coexisting liquid phases. Since all points in the diagram correspond to the same temperature and pressure, the tie lines must remain in the plane of the diagram, but they are not required to be parallel to any side of the triangle or to each other. Their directions must be determined experimentally.

In order to display a temperature-composition phase diagram at constant pressure, a three-dimensional space must be used. Figure 6.27 shows a perspective view of a partial solid-liquid phase diagram for bismuth, tin, and lead. These substances actually dissolve appreciably in each other in the solid phases, but the solid solution regions are artificially shrunk to zero to simplify the diagram.

Each face of the triangular prism is a two-component temperature-composition diagram with a single eutectic. The bismuth-lead diagram also has a peritectic point, but we will ignore this in describing the phases that can occur. The interior of the prism represents compositions in which all three components are present. There is a **triple eutectic** at 96°C, a lower temperature than any of the three two-component eutectics. The surface shown in the diagram is the lower boundary of the three-dimensional one-phase liquid region. This surface has three grooves that lead down to the triple eutectic from the three two-component eutectics.

Below the surface lie regions corresponding to two, three, or four phases. Figure 6.28 shows schematically a composition-composition phase diagram obtained by passing a plane through the prism slightly above the triple eutectic. The central roughly triangular region is a one-phase liquid region inside which all points represent possible compositions of a single liquid phase. Each tie line from an edge of this region connects to a corner of the diagram. This means that only one solid phase freezes out from a composition corresponding to a point on the edge of the region. At a composition corresponding to a corner of the region, two components freeze out, as in

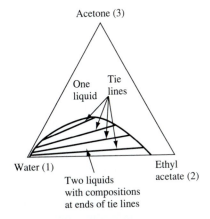

Figure 6.26. Liquid Composition-Composition Phase Diagram of Water, Acetone, and Ethyl Acetate at 1 atm and 30°C. The region at the bottom of the diagram corresponds to separation of two liquid phases, as shown by the ends of the tie lines.

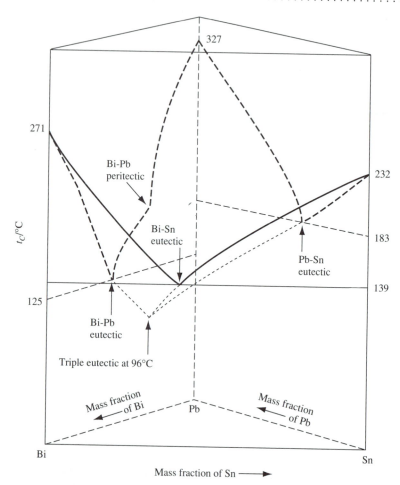

Figure 6.27. Perspective View of Three-Dimensional Temperature-Composition Phase Diagram of Bismuth, Tin, and Lead at 1 atm. The diagram has been simplified by omission of the regions of solid solubility. Each face of the triangular prism is a two-component temperature-composition phase diagram with a eutectic. There is also a peritectic point in the Bi-Pb phase diagram. From F. T. Gucker and R. L. Seifert, *Physical Chemistry*, W. W. Norton, New York, 1966, p. 773.

a two-component eutectic. The triangular regions along the sides of the triangle are three-phase regions. Just as the two ends of a tie line give the possible compositions of two phases, the corners of these regions give the compositions of three coexisting phases (the liquid with a composition corresponding to the corner of the liquid region and the two solids). There are no tie lines in these regions. Each region is a "tie triangle."

At the triple eutectic temperature, the liquid region has shrunk to a point from which three tie lines extend, one to each corner, indicating that at the triple eutectic three solid phases freeze out, each consisting of a single component. Four phases are at equilibrium, and if the pressure is fixed the temperature and the compositions of the four phases are fixed, since for three components and four phases there is only one independent intensive variable.

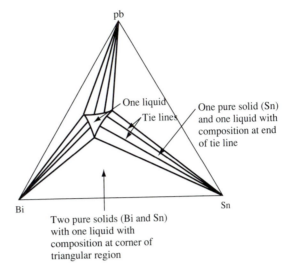

Figure 6.28. Schematic Solid-Liquid Composition Phase Diagram of Bismuth, Tin, and Lead at 100°C and 1 atm. This temperature is slightly above the triple eutectic temperature. The diagram is produced by passing a horizontal plane through the three-dimensional diagram shown in Figure 6.27. The small, roughly triangular liquid area is the only one-phase area in the diagram.

6.6 Colligative Properties

Colligative properties are properties that depend on the concentration of a solute but not on its identity. The name comes from a Latin word meaning "tied together" and is used because of the common dependence of the properties on solute concentration. The four principal colligative properties are freezing point depression, boiling point elevation, vapor pressure lowering, and osmotic pressure. Discussing these four phenomena is like performing variations on a theme. The theme on which we will see four variations is the fundamental fact of phase equilibrium.

We will use the fundamental fact of phase equilibrium in discussing each of the colligative properties.

Freezing Point Depression

Consider a system consisting of two phases at equilibrium with each other: a liquid solution with a single solute and the pure solid solvent. We assume that the solute is completely insoluble in the frozen solvent.

From the fundamental fact of phase equilibrium,

$$\mu_{1(\text{liq})} = \mu^*_{1(\text{solid})} \tag{6.6-1}$$

where the solvent is called component 1. We assume that the solution is ideally dilute, so that

$$\mu^*_{1(\text{liq})} + RT \ln(x_1) = \mu^*_{1(\text{solid})} \tag{6.6-2}$$

There are two phases and two components, so by the Gibbs phase rule there are two independent intensive variables. If we regard pressure and temperature as independent, the mole fraction of the solvent is a dependent variable.

We divide Equation (6.6-2) by T and then differentiate it with respect to temperature, keeping the pressure fixed:

$$\left(\frac{\partial \mu^*_{1(\text{liq})}/T}{\partial T}\right)_P + R\left(\frac{\partial \ln(x_1)}{\partial T}\right)_P = \left(\frac{\partial \mu^*_{1(\text{solid})}/T}{\partial T}\right)_P \tag{6.6-3}$$

We obtain an identity by a general thermodynamic derivation:

$$\left(\frac{\partial(\mu^*_i/T)}{\partial T}\right)_P = \frac{-T\bar{S}^*_i - \mu^*_i}{T^2} = -\frac{\bar{H}^*_i}{T^2} \tag{6.6-4}$$

Use of this identity in Equation (6.6-3) gives

$$R\left(\frac{\partial \ln(x_1)}{\partial T}\right)_P = \frac{\bar{H}^*_{1(\text{liq})} - \bar{H}^*_{1(\text{solid})}}{T^2} = \frac{\Delta\bar{H}^*_{1(\text{fus})}}{T^2} \tag{6.6-5}$$

where $\Delta\bar{H}^*_{1(\text{fus})}$ is the molar enthalpy change of fusion (melting) of the pure solvent.

In Equation (6.6-5) we see that x_1 is a dependent variable in the present case.

We multiply Equation (6.6-5) by dT and integrate both sides of the equation from the normal melting temperature of the pure solvent T_{m1} to some lower temperature T'.

$$R\int_{T_{m1}}^{T'}\left(\frac{\partial \ln(x_{1(\text{soln})})}{\partial T}\right)_P dT = \int_{T_{m1}}^{T'}\frac{\Delta\bar{H}_{1(\text{fus})}}{T^2} dT \tag{6.6-6}$$

To a good approximation, the enthalpy change of fusion is constant over a small range of temperature. The result of the integration is therefore

$$R \ln[x_{1(\text{soln})}(T')] = -\Delta\bar{H}_{1(\text{fus})}\left(\frac{1}{T'} - \frac{1}{T_{m1}}\right) \tag{6.6-7}$$

where $x_{1(\text{soln})}(T')$ is the mole fraction of the solvent in the solution that is at equilibrium with the pure solvent at temperature T', and we have used the fact that the mole fraction of component 1 at equilibrium at temperature T_{m1} is equal to unity.

Our present system (solid pure component 1 plus liquid solution) is the same as the system that corresponds to one of the liquid composition curves in a solid-liquid phase diagram such as Figure 6.22, so Equation (6.6-7) is the equation for this curve if Raoult's law holds. In this case, the curve is independent of the identity of the second component (except that it stops when it meets the other curve at the eutectic point).

For dilute solutions, Equation (6.6-7) is simplified by using the approximation

$$\ln(x_1) = \ln(1 - x_2) \approx -x_2 \tag{6.6-8}$$

Exercise 6.29 Find the percentage error for the approximation of Equation (6.6-8) for $x_2 = 0.100$, 0.0100, 0.00100, and 0.000100.

Use of the approximation of Equation (6.6-8) gives

$$x_2 \approx \Delta \bar{H}_{1(\text{fus})} \frac{T_{m1} - T}{RT_{m1}T} \qquad \textbf{(6.6-9a)}$$

where we drop the prime symbol on the equilibrium temperature. Equation (6.6-9a) is accurate only for dilute solutions, in which case T is approximately equal to T_{m1}. We write as a further approximation

$$x_2 \approx \left(\frac{\Delta \bar{H}_{1(\text{fus})}}{RT_{m1}^2} \right) \Delta T_f \qquad \textbf{(6.6-9b)}$$

where we use the symbol ΔT_f for $T_{m1} - T$, the freezing point depression (a positive quantity).

Equation (6.6-9b) is often rewritten in terms of the molality, using Equation (6.3-20) to relate the molality and the mole fraction for a dilute solution The result is

$$\Delta T_f = K_{f1} m_2 \qquad \textbf{(6.6-10)}$$

where m_2 is the molality of the solute (component 2) and M_1 is the molar mass of the solvent (measured in kilograms). The quantity K_{f1} is called the **freezing point depression constant**.

Remember that several approximations were made to obtain Equations (6.6-9) and (6.6-10).

$$K_{f1} = \frac{M_1 R T_{m1}^2}{\Delta \bar{H}_{1(\text{fus})}} \qquad \textbf{(6.6-11)}$$

It has a different value for each solvent but is independent of the identity of the solute. Equation (6.6-11) is still valid if there are several solutes, in which case m_2 is replaced by the sum of the molalities of all solutes. If a solute dissociates or ionizes, the total molality of all species must be used.

Exercise 6.30 The molar enthalpy change of fusion of water is 6.01 kJ/mol. Show that the value of the freezing point depression constant for water is 1.86 K kg mol^{-1}.

EXAMPLE 6.13 Find the freezing point depression of a solution of 10.00 g of sucrose in 1.000 kg of water.

Solution

$$m_2 = \frac{10.00 \text{ g sucrose}(1 \text{ mol sucrose}/342.30 \text{ g sucrose})}{1.000 \text{ kg}}$$

$$= 0.02921 \text{ mol kg}^{-1}$$

$$\Delta T = (1.86 \text{ K kg mol}^{-1})(0.02921 \text{ mol kg}^{-1}) = 0.0543 \text{ K}$$

Boiling Point Elevation

Consider a volatile solvent (component 1) and a nonvolatile solute (component 2). This solution is at equilibrium with a vapor phase, which is pure component 1. We assume that our solution is ideally dilute, that the gas phase is ideal, and that our system is at a fixed pressure P.

The fundamental fact of phase equilibrium gives

$$\frac{\mu^*_{1(\text{liq})}}{T} + R\ln(x_{1(\text{liq})}) = \frac{\mu_{1(\text{vap})}}{T} \tag{6.6-12}$$

where we have divided the equation by T. Differentiating this equation with respect to T at constant P and using the identity of Equation (6.6-4), we get

$$R\left(\frac{\partial \ln(x_{1(\text{liq})})}{\partial T}\right)_P = -\frac{\bar{H}_{1(\text{vap})}}{T^2} + \frac{\bar{H}_{1(\text{liq})}}{T^2} = -\frac{\Delta\bar{H}_{1(\text{vap})}}{T^2} \tag{6.6-13}$$

where $\Delta\bar{H}_{1(\text{vap})}$ is the molar enthalpy change of vaporization of the pure liquid component 1. This equation is similar to Equation (6.6-5) except that here the right-hand side is a negative quantity, whereas in Equation (6.6-5) it is a positive quantity. The reason is that going from the liquid phase to the vapor phase is an endothermic process, whereas going from the liquid phase to the solid phase is an exothermic process.

We now multiply Equation (6.6-13) by dT and integrate from T_{b1}, the normal boiling temperature of component 1, to a higher temperature, T'. Over a small interval of temperature, the enthalpy change of vaporization is nearly constant, giving an equation analogous to Equation (6.6-7):

$$R\ln(x_{1(\text{liq})}) = \Delta\bar{H}_{1(\text{vap})}\left(\frac{1}{T'} - \frac{1}{T_{b1}}\right) \tag{6.6-14}$$

Equation (6.6-14) can be simplified in the case of small boiling point elevations by using the same approximations as were used in Equation (6.6-9b), with the result

$$x_2 = \frac{\Delta\bar{H}_{1(\text{vap})}}{RT^2_{b1}}\Delta T_b \tag{6.6-15}$$

where $\Delta T_b = T - T_{b1}$ is the boiling point elevation (a positive quantity).

When Equation (6.6-15) is solved for the boiling point elevation and written in terms of the molality, the result is analogous to Equation (6.6-10):

$$\Delta T_b = K_{b1}m_2 \tag{6.6-16}$$

where the **boiling point elevation constant** for component 1 is given by

Notice how similar the boiling point elevation is to the freezing point depression.

$$K_{b1} = \frac{M_1 RT^2_{b1}}{\Delta\bar{H}_{1(\text{vap})}} \tag{6.6-17}$$

Again, this quantity is different for each solvent but does not depend on the identity of the solute or solutes. If more than one solute is present, the molality m_2 is replaced by the sum of the molalities of all solutes.

Exercise 6.31

a. Find the boiling point elevation constant for water. The molar enthalpy change of vaporization is 40.67 kJ mol^{-1}.

b. Find the boiling temperature at 1.000 atm of a solution of sucrose with 10.00 g of sucrose in 1.000 kg of water.

Vapor Pressure Lowering

For a nonvolatile solute and a volatile solvent that obeys Raoult's law, the total vapor pressure is equal to the partial vapor pressure of the solvent, given by

$$P_{\text{vap(total)}} = x_1 P_1^* \qquad \text{(6.6-18)}$$

where P_1^* is the vapor pressure of the pure solvent (component 1). The lowering of the vapor pressure is given by

$$\Delta P_{\text{vap}} = P_1^* - x_1 P_1^* = P_1^*(1 - x_1) = P_1^* x_2 \qquad \text{(6.6-19)}$$

Exercise 6.32	**a.** Calculate the vapor pressure at 100.0°C of the solution in Exercise 6.31b. **b.** From Equation (6.6-19), obtain an expression for the vapor pressure lowering of a dilute solution in terms of the molality.

Osmotic Pressure

This colligative property involves the equilibrium of a liquid solution and its pure liquid solvent on opposite sides of a semipermeable membrane that allows only the solvent to pass. The chemical potential of the solvent is made to have equal values in the two phases by having the two phases at different pressures. A simple apparatus in which this equilibrium can be accomplished is shown in Figure 6.29. The pressure of the solution is increased above that of the pure solvent by the gravitational force on the solution in the left column. Let the pressure on the pure solvent on one side of the semipermeable membrane be called P and the pressure on the solution of the other side be called $P + \Pi$. The difference Π is the **osmotic pressure**.

At equilibrium, we must have

$$\mu_1^*(T, P) = \mu_1(T, P + \Pi) = \mu_1^*(T, P + \Pi) + RT \ln(x_1) \qquad \text{(6.6-20)}$$

From Equation (4.3-10), if the molar volume of the pure liquid is nearly independent of pressure

$$\mu_1^*(T, P + \Pi) - \mu_1^*(T, P) = \int_P^{P+\Pi} \bar{V}_1^* \, dP = \Pi \bar{V}_1^* \qquad \text{(6.6-21)}$$

which gives

$$\Pi \bar{V}_1^* = -RT \ln(x_1) = -RT \ln(1 - x_2)$$
$$\approx RT x_2 \qquad \text{(6.6-22)}$$

where the approximation of Equation (6.6-8) for dilute solutions has been applied. It is customary to rewrite Equation (6.6-22) in another form. For a dilute solution of two components

$$x_2 = \frac{n_2}{n_1 + n_2} \approx \frac{n_2}{n_1}$$

and

$$V = n_1 \bar{V}_1 + n_2 \bar{V}_2 \approx n_1 \bar{V}_1$$

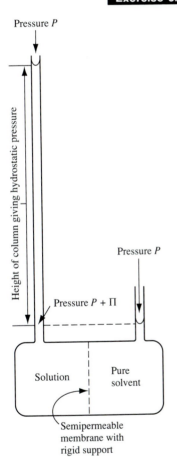

Pressure P

Height of column giving hydrostatic pressure

Pressure P

Pressure $P + \Pi$

Solution

Pure solvent

Semipermeable membrane with rigid support

Figure 6.29. An Osmometer (schematic). The system is equilibrated by the hydrostatic pressure of the column of solution in the left tube.

The van't Hoff equation is named for Jacobus Henricus van't Hoff, 1852–1911, a Dutch physical chemist who won the 1901 Nobel Prize in chemistry for his work on osmotic pressure. He was also the first person to propose the tetrahedral carbon atom.

Use of these two equations gives

$$\Pi = n_2 RT/V = c_2 RT \qquad \text{(6.6-23)}$$

where c_2 is the concentration of the solute. Equation (6.6-23) is remarkably similar to the ideal gas equation of state. It is known as the **van't Hoff equation**.

Exercise 6.33

For solutions of moderate concentration, the osmotic pressure is quite large. According to the van't Hoff equation, a 1.00 mol L^{-1} solution has an osmotic pressure of 24.4 atm at 25°C, corresponding to the hydrostatic pressure of a column of water over 800 feet high.

a. Verify Equation (6.6-23).
b. Find the osmotic pressure at 298.15 K of a solution of 10.00 g of sucrose in enough water to make 1.000 liter of solution.

Summary

For each component of an ideal solution,

$$\mu_i(T, P) = \mu_i^*(T, P) + RT \ln(x_i)$$

Each component of an ideal solution obeys Raoult's law

$$P_i = P_i^* x_i$$

where x_i is the mole fraction of component i in the solution.

The partial vapor pressure of a sufficiently dilute solute is governed by Henry's law:

$$P_i = k_i x_i$$

A solution in which Henry's law is obeyed by the solutes is called an ideally dilute solution.

The activity a_i is defined by the general relation

$$\mu_i = \mu_i^\circ + RT \ln(a_i)$$

where μ_i° is the chemical potential in some standard state.

According to two different mole fraction descriptions, called convention I and convention II, the activity is given by

$$a_i = \gamma_i x_i$$

The activity of a solute in the molality description is given by

$$a_i^{(m)} = \frac{\gamma_i^{(m)} m_i}{m^\circ}$$

The Debye-Hückel theory provides an accurate limiting law for the activity coefficients of electrolyte solutes. A semiempirical equation, the Davies equation, can provide usable estimates of electrolyte activity coefficients at larger concentrations.

Two-component pressure-composition and temperature-composition phase diagrams give information about phases present at equilibrium. For a three-component system, a composition-composition diagram at constant temperature and pressure is plotted in an equilateral triangle and a temperature-composition diagram is plotted in a (three-dimensional) triangular prism.

The four principal colligative properties are freezing point depression, boiling point elevation, vapor pressure lowering, and osmotic pressure. In each case, the magnitude of the effect in an ideally dilute solution is determined by the concentration of the solute but is independent of its identity.

Additional Reading

Allen M. Alper, ed., *Refractory Materials*. Vol. 6-1, *Phase Diagrams: Materials Science and Technology*, Academic Press, New York, 1970
This useful book contains a general introduction to the thermodynamics of phase diagrams and their interpretation, as well as some examples of phase diagrams. Other volumes in the same series also contain information about phase diagrams.

A. Findlay, *The Phase Rule and Its Applications*, 9th ed. edited by A. N. Campbell and N. O. Smith, Dover, New York, 1951
This is a standard source of information about phase equilibrium.

P. Gordon, *Principles of Phase Diagrams in Materials Systems*, McGraw-Hill, New York, 1968
Designed for materials scientists, this book presents a clear introduction to the thermodynamics of phase equilibrium and contains a useful chapter on order-disorder transitions.

J. P. Hansen and I. R. McDonald, *Theory of Simple Liquids*, Academic Press, New York, 1976
A general presentation of liquid theory at an advanced level. Chapter 10 deals with phase transitions and critical phenomena.

M. Hansen, *The Constitution of Binary Alloys*, McGraw-Hill, New York, 1958
A number of phase diagrams and other information about two-component systems of metals are presented.

C. Kittel and H. Kroemer, *Thermal Physics*, W. H. Freeman, San Francisco, 1980
A textbook for undergraduate students of physics at about the level of physical chemistry. It contains a discussion of the theory of the van der Waals fluid, including critical phenomena, beginning on p. 288.

P. V. E. McClintock, D. J. Meredith, and J. K. Wigmore, *Matter at Low Temperatures*, Wiley, New York, 1984

A readable and useful source for information about low-temperature properties, including phase diagrams of helium.

J. Nyvlt, *Solid-Liquid Phase Equilibria*, Elsevier Scientific Publishing, Amsterdam, 1977
This book has a practical orientation toward the construction of phase diagrams. It contains a number of useful tables of solubilities.

H. E. Stanley, *Introduction to Phase Transitions and Critical Phenomena*, Oxford University Press, New York, 1971
A general theoretical and experimental introduction to phase transitions, not only in fluid systems but also in magnetic systems.

J. Timmermans, *Physicochemical Constants of Binary Systems in Concentrated Solutions*, 4 vols., Interscience, New York, 1959–1960
This work is a compendium of data on two-component mixtures, including partial vapor pressures, boiling temperatures, freezing temperatures, refractive indices, densities, etc. Volumes I and II contain information on pairs of organic compounds, Volume III on pairs of metallic substances and of one metallic substance with water or an organic compound, and Volume IV on pairs of inorganic substances and on inorganic substances with organic substances.

D. D. Wagman, W. H. Evans, V. B. Parker, R. H. Schumm, I. Halow, S. M. Bailery, K. L. Churney, and R. H. Schumm, The NBS table of chemical thermodynamic properties. Selected values for inorganic and C_1 and C_2 organic substances in SI units, *J. Phys. Chem. Ref. Data* **11** (Suppl. 2) (1982)

M. W. Chase, Jr., C. A. Davies, J. R. Downey, Jr., D. J. Frurip, R. A. McDonald, and A. N. Syverud, JANAF thermochemical tables, third edition, parts I and II, *J. Phys. Chem. Ref. Data* **14** (Suppl. 1) (1985).
These volumes are a standard source of thermochemical data. They include data on solutes in aqueous and other solutions, among other data, and are available as bound volumes.

J. Wisniak, *Phase Diagrams* (Physical Sciences Data, 10), Elsevier Science Publishers, Amsterdam, 1981
A two-volume set containing phase diagrams.

J. Wisniak and A. Tamir, *Mixing and Excess Thermodynamic Properties. A Literature Source Book* (Physical Sciences Data, 1), Elsevier Science Publishers, Amsterdam, 1978
This book and two supplements, which appeared in 1982 and 1986, present thermodynamic data on mixtures.

PROBLEMS

Problems for Section 6.1

6.34. Derive Equation (6.1-24).

6.35. Find ΔS_{mix}, ΔG_{mix}, ΔH_{mix}, and ΔV_{mix} if 100.0 g of benzene and 50.0 g of naphthalene are mixed at 50.0°C. State any assumptions.

*6.36. Assume that a 1.000 molal solution of naphthalene in benzene is ideal. Calculate the value of the activity coefficient in the molality description and in the concentration description.

6.37. Assume that carbon tetrachloride and 1,1,1-trichloroethane (methyl chloroform) form an ideal solution. Look up the vapor pressures of the pure compounds at 25°C and plot a pressure-composition phase diagram for this temperature (four points besides the end points should give an adequate plot).

6.38. Assume that carbon tetrachloride and 1,1,1-trichloroethane form an ideal solution. Look up the normal boiling temperatures and the enthalpy changes of vaporization of the pure substances and plot an accurate temperature-composition phase diagram for 1.000 atm (four points besides the end points should give an adequate plot).

Problems for Section 6.2

6.39. Deep-sea divers can suffer a condition known as the "bends" when they breathe ordinary air, because nitrogen gas dissolves in blood at high pressure and is released as bubbles in the bloodstream when the diver decompresses.

   a. Calculate the amount of nitrogen dissolved in 5.000 L of blood (roughly the volume in an adult human) at equilibrium with air (78 mol % nitrogen) at a depth of 200 m, assuming that Henry's law constant for nitrogen in blood is 7.56×10^4 atm, the value for nitrogen in water at 20.0°C.

   b. Calculate the volume of this amount of nitrogen as a gas at 1.000 atm and 20.0°C.

*6.40. a. From the value in Exercise 6.10, find the distribution coefficient for iodine between water and carbon tetrachloride at 25°C, using the concentration description. The density of carbon tetrachloride is 1.59 g cm^{-3}.

   b. Iodine is equilibrated between water and carbon tetrachloride at 25°C. The final concentration of iodine in the carbon tetrachloride phase is 0.0765 mol L^{-1}. Find the volume of a sodium thiosulfate solution with 0.0100 mol L^{-1} required to titrate 50.00 mL of the aqueous phase (2 moles of thiosulfate are required to react with 1 mole of I_2).

6.41. From the value in Exercise 6.10, find the distribution coefficient for iodine between water and carbon tetrachloride at 25°C, using the molality description.

6.42. From the Henry's law constant for ethanol in benzene at 40°C, calculate the Henry's law constant for ethanol in benzene at this temperature if mass fractions are used instead of mole fractions to describe the system.

6.43. Show that the solvent in a solution of several solutes obeys Raoult's law at a certain composition if all solutes obey Henry's law for all compositions between this composition and the pure solvent. Hint: carry out the integration for fixed proportions of solutes, so that the mole fractions of the solutes remain proportional to each other.

Problems for Section 6.3

6.44. Find the activity of pure liquid water at 1000. bar and 25°C.

*6.45. Find the activity of graphite and that of diamond at the coexistence pressure at 298.15 K.

6.46. Find the activity coefficient of gaseous nitrogen at 298.15 K and 10.00 atm, using whatever information you need from earlier chapters of this book.

6.47. Naphthalene is a solid near room temperature, but it forms a nearly ideal liquid solution with benzene or with toluene. Show that the mole fraction of naphthalene in any solution that is equilibrated with solid naphthalene (a saturated solution) has the same mole fraction of naphthalene, independent of the solvent, if the solution is ideal.

6.48. At 35.2°C, the vapor pressure of pure acetone is 344.5 torr and that of pure chloroform is 293 torr. At this temperature, a solution of 0.7090 mol of acetone and 0.2910 mol of chloroform has a total vapor pressure of 286 torr and a mole fraction of acetone in the vapor of 0.8062.

   a. Using convention I, find the activity and activity coefficient of each component.

   b. Find the Gibbs energy change of mixing and the excess Gibbs energy for the solution.

   c. Henry's law constant for chloroform in acetone at this temperature is equal to 145 torr. Considering acetone to be the solvent, find the activity and activity coefficient for each component according to convention II.

*6.49. a. From data in Problem 6.48, find the activity coefficient of chloroform in acetone for the solution of that problem, using the molality description and regarding acetone as the solvent.

   b. Find the activity coefficient of acetone in the solution of Problem 6.48, using the molality description and regarding acetone as the solvent.

6.50. a. From data on enthalpy changes of formation in Table A9, find the standard-state differential heat of solution of KOH at 298.15 K. Note that tabulated values for strong electrolytes in solution are listed for the separate ions.

   b. From data on Gibbs energy changes of formation in Table A9, find the value of $\mu^{\circ(m)} - \bar{G}^*(\text{solid})$ for KOH at 298.15 K.

6.51. The maximum solubility of iodine in water at 1.000 atm and 298.15 K is 1.42×10^{-3} mol kg^{-1}. Find the value of $\mu^{\circ(m)} - \mu^{*(\text{solid})}$ for I_2. State any assumptions.

Problems for Section 6.4

6.52. a. Look up the dielectric constant and density of methanol. Calculate the values of the Debye-Hückel parameters α and β for methanol at 298.15 K.

   b. Calculate $\gamma_{\pm}$ for a 0.0100 mol kg^{-1} NaCl solution in methanol at 298.15 K. Calculate the percent difference between this value and that in water at the same temperature and molality.

*6.53. a. Calculate $\gamma_\pm$ for a 0.0050 mol kg^{-1} KCl solution at 298.15 K using the Debye-Hückel formula, Equation (6.4-13), and using $\beta a = 1.00$ kg$^{1/2}$ mol$^{-1/2}$. Repeat the calculation for a 0.0050 molal FeSO$_4$ solution at the same temperature, using the same value of βa.

b. Repeat the calculations of part a using the Davies equation. Calculate the percent difference between each value in part a and the corresponding value in part b.

6.54. a. Make a plot of $\gamma_\pm$ as a function of the molality for a 1-1 electrolyte at 298.15 K, using the Davies equation. Compare your graph with that of Figure 6.12 and comment on any differences.

b. Make a plot of $\gamma_\pm$ as a function of the molality for a 1-2 electrolyte at 298.15 K, using the Davies equation. Compare your graph with that of Figure 6.12 and comment on any differences.

c. Make a plot of $\gamma_\pm$ as a function of the molality for a 2-2 electrolyte at 298.15 K, using the Davies equation. Compare your graph with that of Figure 6.12 and comment on any differences.

Problems for Section 6.5

6.55. Describe what happens if one begins with a fairly small amount of water at 99.0°C and gradually adds furfural until one has a mixture with a mole fraction of furfural equal to 0.90. The temperature is maintained at 99.0°C. Give the approximate mole fractions at which phase transitions occur.

6.56. Describe what happens if one begins with a vapor phase containing water and furfural with a water mole fraction equal to 0.500 and gradually cools this mixture from 150°C to 85°C. Give the approximate temperatures at which phase transitions occur.

6.57. Sketch cooling curves for mixtures of phenol (P) and aniline (A) with aniline mole fractions of 0.26 (at the eutectic between A and AP), 0.40, 0.50, and 0.95. Describe what happens at each break in each curve.

6.58. Sketch the solid-liquid temperature-composition phase diagram of Na and K. There is a single compound, Na$_2$K, which melts incongruently at 6.6°C to give essentially pure Na and a solution with a sodium mole fraction of 0.42. The melting temperature of Na is 97.5°C, and that of K is 63°C. There is a eutectic at -12.5°C and a sodium mole fraction of 0.15. There is no appreciable solid solubility. Label each area in the diagram with the number of independent intensive variables corresponding to that phase.

6.59. Sketch the solid-liquid and liquid-vapor temperature-composition phase diagram of titanium and uranium. The two substances form a nearly ideal liquid solution with a uranium boiling temperature of 1133°C and a titanium boiling temperature of 1660°C. The melting temperature of uranium is 770°C, and that of titanium is 882°C. There is a compound, TiU$_2$, which melts at 890°C. The eutectic between the com-

pound and uranium is at uranium mole fraction 0.95 and 720°C, and the eutectic between titanium and the compound is at uranium mole fraction 0.28 and 655°C. Label each area with the number of independent intensive variables.[5]

6.60. a. Sketch a cooling curve for a mixture of lanthanum and copper with a mole fraction of lanthanum equal to 0.30. Label each break in the curve and tell what phase or phases are freezing out for each portion of the curve.

b. Repeat part a for a lanthanum mole fraction of 0.53.

6.61. Sketch the titanium-nickel temperature-composition phase diagram. The melting temperature of nickel is 1453°C and that of titanium is 1675°C. There are three compounds: TiNi$_3$, melting near 1370°C; TiNi, melting near 1300°C; and Ti$_2$Ni, which melts incongruently near 990°C. There are fairly sizable regions of solid solubility except that nothing dissolves in TiNi$_3$. The three eutectics, in order of increasing titanium mole fraction, are near 1300°C, 1110°C, and 940°C. Incongruent melting of Ti$_2$Ni gives a solid solution that is mostly TiNi and a solution with a titanium mole fraction near 0.70. Label each area with the phase or phases present and the number of independent intensive variables.

6.62. Problem Figure 6.62 shows a phase diagram of lithium sulfate, ammonium sulfate, and water. For each area in the diagram, tell what phase or phases are present and the number of independent intensive variables.

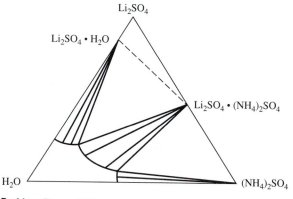

Problem Figure 6.62

Problems for Section 6.6

6.63. Find the freezing point depression of a solution of 1.000 g of KCl in 1.000 kg of water. Assume that the KCl

---

[5] The actual diagram might be more complicated. See M. Hansen, *The Constitution of Binary Alloys*, McGraw-Hill, New York, 1958, pp. 1238ff.

dissociates completely, so that 2 moles of dissolved ions result from 1 mole of KCl.

6.64. a. Calculate the freezing temperature at 1.000 atm of a solution of 5.00 g of sucrose in 95.00 g of water, using Equation (6.6-9b).

b. Calculate the freezing temperature of the solution of part a using Equation (6.6-7).

c. Modify Equation (6.6-7) by discontinuing the assumption that the solution is ideally dilute, including an activity coefficient for the water. From the actual freezing point depression of 2.047°C, find the activity coefficient of water in the solution of part a.

*6.65. a. Calculate the boiling temperature of the solution of Problem 6.63, using Equation (6.6-14).

b. Repeat the calculation using Equation (6.6-15) and calculate the percent difference between the boiling point elevation and that of part a.

c. Repeat the calculation using Equation (6.6-16) and calculate the difference between the boiling point elevation and that of part a.

6.66. The density of ethylene glycol is 1.1088 g cm^{-3} and its melting temperature is $-11.5°C$. Calculate the freezing temperature of a mixture of 1.00 liter of ethylene glycol and 10.00 liter of water. State any assumptions.

6.67. a. The vapor pressure of pure water at 25°C is 23.756 torr if the total pressure on the liquid is just that due to the vapor. Calculate the vapor pressure of water at 25°C if enough oxygen gas is added to the vapor to give a total pressure of 1.000 atm. Neglect any oxygen dissolved in the water.

b. Calculate the mole fraction of dissolved oxygen in the water under the conditions of part a. The Henry's law constant is in Example 6.5. Calculate the effect of this dissolved oxygen on the vapor pressure.

c. Repeat parts a and b with 100.00 atm total pressure instead of 1.000 atm.

6.68. a. Calculate the osmotic pressure at 20.00°C of a solution of 5.000 g of sucrose in 95.00 g of water, using Equation (6.6-23). The density of the solution is 1.0194 g cm^{-3}. Calculate the height of a column of the solution sufficient to equilibrate an osmometer like that of Figure 6.29.

b. Repeat the calculation using the version of Equation (6.6-22) containing the natural logarithm.

c. Repeat the calculation, assuming the same value for the activity coefficient of water as found in part c of Problem 6.64.

*6.69. Assuming complete dissociation and assuming that the water activity coefficient equals unity, calculate the osmotic pressure at 25.00°C of a solution of 2.000 g of KCl in 1.000 kg of water. The density of the solution is 1.002 g cm^{-3}.

General Problems

*6.70. Identify each statement as true or false (if a statement is true only under some special circumstance, count it as false):

a. For a small enough concentration of a solute, its activity according to convention I and its activity according to convention II are nearly equal.

b. For a small enough concentration of a solute, its activity coefficient according to convention I and its activity coefficient according to convention II are nearly equal.

c. For a small enough concentration of a solute, its activity coefficient according to convention I and its molality activity coefficient are nearly equal.

d. For a small enough concentration of a solute, its activity coefficient according to convention II and its molality activity coefficient are nearly equal.

e. In an ideally dilute solution, the mole fraction and the activity of a solute are equal.

f. In an ideally dilute solution, the mole fraction and the activity of the solvent are equal.

g. In an ideal solution, the mole fraction of every component is equal to its activity.

h. In an ideal solution the mole fraction and the activity coefficient of every component are equal.

i. If two substances form an ideal solution, they are miscible in all proportions.

j. The mole fraction of naphthalene in a saturated solution is the same in all solvents with which naphthalene forms ideal solutions.

6.71. The **lever rule** can be used to determine the relative amounts of material in the two coexisting phases represented by the ends of a tie line in a two-component phase diagram, as in Problem Figure 6.71. The rule is

$$L^{(\alpha)} n^{(\alpha)} = L^{(\beta)} n^{(\beta)}$$

where $n^{(\alpha)}$ is the total amount of both substances in phase α and $n^{(\beta)}$ is the total amount of both substances in phase β. The distances $L_1^{(\beta)}$ and $L_1^{(\alpha)}$ are labeled on the diagram. The mole fraction x_1 is the overall mole fraction

$$x_1 = \frac{n_1^{(\alpha)} + n_1^{(\beta)}}{n_1^{(\alpha)} + n_1^{(\beta)} + n_2^{(\alpha)} + n_2^{(\beta)}}$$

Derive the lever rule.

6.72. The freezing temperature of CH_2OHCH_2OH, ethylene glycol (the main ingredient of automobile antifreeze), is $-11.5°C$. Its density is 1.1088 g cm^{-3}, and its enthalpy change of fusion is 11.23 kJ mol^{-1}.

a. Find the freezing point depression constant for solutions with ethylene glycol as the solvent.

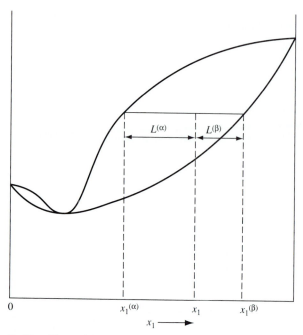

Problem Figure 6.71

b. Find the freezing temperature of a solution of 10.00 g of water in 1.000 kg of ethylene glycol.

c. Find the freezing temperature of a solution of 10.00 g of ethylene glycol in 1.000 kg of water.

d. Assuming an ideal liquid solution, draw the water–ethylene glycol temperature-composition phase diagram, using Equation (6.6-7) to calculate the curves. Assume that the enthalpy changes of fusion are constant. Calculate at least three points on each curve leading to the eutectic. What is the temperature of the eutectic point?

e. A common practice is to mix equal parts of water and antifreeze by volume. Find the freezing temperature of this mixture, using your graph from part d. What proportions by volume would give the lowest freezing temperature?

7 Chemical Equilibrium

PREVIEW

Through the work of Duhem, Gibbs, and others it was established in the nineteenth century that application of the principles of thermodynamics leads to the macroscopic understanding of chemical equilibrium. This chapter is devoted to the analysis that gives this understanding.

OBJECTIVES

After studying this chapter, a student should:

1. understand the thermodynamic origin of equilibrium constant expressions,

2. be able to carry out equilibrium calculations on various kinds of chemical systems,

3. be able to apply the principle of Le Châtelier,

4. understand some models for the coupling of chemical reactions in biological systems.

PRINCIPAL FACTS AND IDEAS

1. The principles of thermodynamics determine the state of chemical equilibrium for any system.
2. The familiar equilibrium constant expression of elementary chemistry is equal to a constant at constant temperature and pressure, when modified to include activity coefficients.
3. The principle of Le Châtelier can predict how an equilibrium chemical system responds to changes in temperature or composition.
4. The coupling of biochemical reactions can be understood through thermodynamics and the use of postulated mechanisms.

7.1 Gibbs Energy Changes and Equilibria of Chemical Reactions; The Equilibrium Constant

Equation (2.7-7) is a "generic" chemical equation that can stand for any chemical reaction:

$$0 = \sum_{i=1}^{c} v_i \mathscr{F}_i \qquad (7.1\text{-}1)$$

where the formula of substance i is denoted by $\mathscr{F}_i$ and its stoichiometric coefficient is denoted by v_i. The stoichiometric coefficients of products are positive, and those of reactants are negative. For example, the equation

261

for the combustion of propane is written

$$0 = 3CO_2(g) + 4H_2O(l) - C_3H_8(g) - 5O_2(g) \qquad \textbf{(7.1-2)}$$

The inconvenience of writing chemical equations in this unfamiliar way is more than outweighed by the resulting ability to write thermodynamic equations in compact forms that apply to any chemical reaction.

We consider reactions in closed systems at constant pressure and temperature and assume that the system can be considered to be in a metastable state during the reaction, so that the Gibbs energy and other thermodynamic variables of the system have well-defined values as equilibrium is approached. The Gibbs energy of the system must decrease until it reaches a minimum value at equilibrium.

This criterion for equilibrium at constant temperature and pressure was derived in Chapter 4.

If an infinitesimal amount of reaction takes place at constant T and P, the change in the Gibbs energy is given by Equation (4.4-3):

$$dG = \sum_{i=1}^{c} \mu_i \, dn_i \quad \text{(constant } T \text{ and } P) \qquad \textbf{(7.1-3)}$$

Use of any substance in Equation (7.1-4) gives the same value of ξ if only one chemical reaction can occur.

To express dG in terms of the amount of reaction that takes place, we define the **progress variable** ξ by

$$n_i = n_i(\text{initial}) + v_i \xi \qquad \textbf{(7.1-4)}$$

where n_i is the amount of substance i.

The progress variable has the dimensions of moles. We say that if ξ increases by one mole, one mole of reaction has occurred: v_i moles of i have appeared if i is a product, and $|v_i|$ moles of i have disappeared if i is a reactant. The meaning of one mole of reaction changes if the balancing of the reaction equation is changed. If all stoichiometric coefficients are doubled, the equation is still balanced but one mole of reaction now corresponds to twice the amount of each substance.

For an infinitesimal amount of reaction $d\xi$, we can write a relation that is valid for every substance involved in the reaction:

$$dn_i = v_i \, d\xi \qquad \textbf{(7.1-5)}$$

In Equation (7.1-6) we have used the fact that a common factor in every term can be factored out of a sum.

Equation (7.1-3) now becomes

$$dG = \sum_{i=1}^{c} \mu_i v_i \, d\xi = \left(\sum_{i=1}^{c} v_i \mu_i \right) d\xi \quad (T \text{ and } P \text{ constant}) \qquad \textbf{(7.1-6)}$$

and we can identify

$$\boxed{\left(\frac{\partial G}{\partial \xi} \right)_{T,P} = \sum_{i=1}^{c} v_i \mu_i} \qquad \textbf{(7.1-7)}$$

The quantity $(\partial G/\partial \xi)_{T,P}$ is the **rate of change of Gibbs energy per mole of reaction**. Since it is a derivative of an extensive quantity with respect to an extensive quantity, it is an intensive quantity. The forward reaction corresponds to an increase in the value of ξ $(d\xi > 0)$.

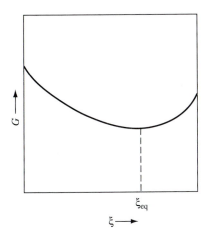

Figure 7.1. The Gibbs Energy of a Reacting System as a Function of the Progress Variable. The minimum in the curve corresponds to the equilibrium state, and other portions of the curve correspond to metastable states.

Equation (6.3-1) is general. It applies to any description of a solution or a pure substance.

Since $dG = (\partial G/\partial \xi)_{T,P}\, d\xi$, the condition that $dG < 0$ implies that

$$\left(\frac{\partial G}{\partial \xi}\right)_{T,P} < 0 \quad \text{(forward reaction spontaneous)} \qquad \textbf{(7.1-8)}$$

if the forward reaction is spontaneous. If the backward reaction is spontaneous,

$$\left(\frac{\partial G}{\partial \xi}\right)_{T,P} > 0 \quad \text{(backward reaction spontaneous)} \qquad \textbf{(7.1-9)}$$

If the equilibrium state has been attained, there is no tendency for reaction to occur and

$$\left(\frac{\partial G}{\partial \xi}\right)_{T,P} = \sum_{i=1}^{c} v_i \mu_i = 0 \quad \text{(equilibrium)} \qquad \textbf{(7.1-10)}$$

The situation is as represented in Figure 7.1. A system in any nonequilibrium state will spontaneously react to approach the equilibrium state at the minimum, beginning from either side of the minimum.

We now express the chemical potential in terms of the activity, using Equation (6.3-1):

$$\mu_i = \mu_i^\circ + RT \ln(a_i) \qquad \textbf{(7.1-11)}$$

where μ_i° is the chemical potential of substance i in whatever standard state applies.

The rate of change of the Gibbs energy per mole of reaction is

$$\left(\frac{\partial G}{\partial \xi}\right)_{T,P} = \sum_{i=1}^{c} v_i \mu_i^\circ + RT \sum_{i=1}^{c} v_i \ln(a_i) \qquad \textbf{(7.1-12)}$$

If the reaction could be carried out with all of the reactants and products maintained in their standard states, the activities would equal unity, the second sum in Equation (7.1-12) would vanish, and the change in G for one mole of reaction would be equal to

$$\Delta G^\circ = \int_0^1 \left[\left(\frac{\partial G}{\partial \xi}\right)_{T,P\,\big|\text{standard state}}\right] d\xi = \int_0^1 \left(\sum_{i=1}^{c} v_i \mu_i^\circ\right) d\xi = \left(\sum_{i=1}^{c} v_i \mu_i^\circ\right) \int_0^1 d\xi$$

$$\Delta G^\circ = \sum_{i=1}^{c} v_i \mu_i^\circ \quad \text{(standard-state reaction)} \qquad \textbf{(7.1-13)}$$

The quantity ΔG° is called the **standard-state Gibbs energy change** for one mole of reaction.

If the standard-state Gibbs energy change is negative, the forward reaction will be spontaneous under standard conditions, and if it is positive, the backward reaction will be spontaneous under standard conditions. Although it is not generally possible to carry out the standard-state reaction, the spontaneity of the standard-state reaction gives an indication of the spontaneity under real conditions.

The Evaluation of the Standard-State Gibbs Energy Change

The Gibbs energy change of formation of a substance is the Gibbs energy change to produce one mole of the substance from the appropriate elements in their most stable forms. The Gibbs energy change for the standard-state reaction can be calculated from standard-state Gibbs energy changes of formation:

The derivation of Equation (7.1-14) is exactly like that of Equation (2.7-11). This derivation can be reviewed if necessary.

$$\Delta G^\circ = \sum_{i=1}^{c} v_i \, \Delta \bar{G}_f^\circ(i) \qquad \text{(7.1-14)}$$

This equation is an exact analogue of Equation (2.7-11) for the enthalpy change of the reaction, and it follows from the fact that the Gibbs energy is a state function, as is the enthalpy. Values of standard-state Gibbs energy changes of formation for a number of substances are included in Table A9. The table includes values of the quantity $-(\bar{G}^\circ - \bar{H}_{298}^\circ)/T$, which can also be used to calculate ΔG° for a reaction. This function is included in the table because it is found experimentally to be a more slowly varying function with temperature than is $\Delta \bar{G}_f^\circ$. Interpolation of its values gives greater accuracy than does interpolation of $\Delta \bar{G}_f^\circ$ values.

Exercise 7.1

a. Using Gibbs energy changes of formation from Table A9, calculate ΔG° at 298.15 K for the reaction

$$PCl_5(g) \rightarrow PCl_3(g) + Cl_2(g)$$

b. Calculate ΔG° for the same reaction using values of $-(\bar{G}^\circ - \bar{H}_{298}^\circ)/T$ and values of $\Delta \bar{H}_f^\circ$. Comment on the comparison of your answer with that of part a.

When the same quantity can be calculated in different ways, it can provide a test of the consistency of the data used.

The Gibbs energy change of a constant-temperature reaction can also be calculated from

$$\Delta G = \Delta H - T \, \Delta S \qquad \text{(7.1-15)}$$

where the enthalpy change is calculated from enthalpy changes of formation by Equation (2.7-11) and the entropy change is calculated from third-law ("absolute") entropies, using Equation (3.5-6).

Exercise 7.2

a. Calculate the standard-state enthalpy change and entropy change at 298.15 K for the reaction of Exercise 7.1.

b. Calculate the standard-state Gibbs energy change of the reaction of Exercise 7.1 using Equation (7.1-15). Compare with the answer of Exercise 7.1.

The Gibbs Energy Change at Fixed Composition

Using the identity that a sum of logarithms is equal to the logarithm of a product, we write Equation (7.1-12) in the form

$$\left(\frac{\partial G}{\partial \xi}\right)_{T,P} = \Delta G^\circ + RT \ln(Q) \qquad \textbf{(7.1-16)}$$

where

The notation Π denotes a product of factors, just as Σ denotes a sum of terms.

$$Q = a_1^{v_1} a_2^{v_2} \cdots a_c^{v_c} = \prod_{i=1}^{c} a_i^{v_i} \qquad \textbf{(7.1-17)}$$

The quantity Q is called the **activity quotient**. It is called a quotient because some of its factors have negative exponents.

Exercise 7.3   Carry out the steps to obtain Equations (7.1-16) and (7.1-17).

If it is possible to keep the activity of every substance fixed during a chemical reaction, by having either a very large system or a small amount of reaction or by adding reactants and removing products, the change in Gibbs energy for one mole of reaction can be written

$$\Delta G\big|_{\text{fixed comp}} = \int_0^1 [\Delta G^\circ + RT \ln(Q)]\, d\xi = [\Delta G^\circ + RT \ln(Q)] \int_0^1 d\xi$$

In deriving Equation (7.1-18), we have factored a constant factor out of the integral.

$$\Delta G\big|_{\text{fixed comp}} = \Delta G^\circ + RT \ln(Q) \qquad \textbf{(7.1-18)}$$

Exercise 7.4   Calculate the value of $(\partial G/\partial \xi)_{T,P}$ for the reaction of Exercise 7.1 if the partial pressure of PCl_5 is 0.100 atm, the partial pressure of PCl_3 is 0.900 atm, and the partial pressure of Cl_2 is 0.500 atm.

The Thermodynamics of Chemical Equilibrium

At equilibrium, Equation (7.1-16) becomes

$$0 = \Delta G^\circ + RT \ln(Q_{\text{eq}}) \qquad \textbf{(7.1-19)}$$

For reactions involving pure substances and gases, ΔG° depends only on temperature. For reactions in solutions, it depends on temperature and pressure. The value of Q_{eq} is therefore equal to a constant at constant temperature (or constant temperature and pressure in the case of a solution

The equilibrium constant is not a true constant, but depends on temperature and sometimes on pressure. It does not depend on composition.

reaction). It is called the **equilibrium constant**, K:

$$K = Q_{eq} = e^{-\Delta G^\circ / RT} = \prod_{i=1}^{c} a_{i(eq)}^{v_i}$$

(7.1-20)

where $a_{i(eq)}$ denotes the equilibrium value of a_i. Equation (7.1-20) is valid for any kind of reacting system.

Like the standard-state Gibbs energy change, the equilibrium constant is a quick indication of the spontaneity of a reaction. If the equilibrium constant is greater than unity, the equilibrium activities of the products will be greater than those of the reactants; if the equilibrium constant is smaller than unity, the equilibrium activities of the reactants will be larger.

7.2   Reactions Involving Gases and Pure Substances

For our present discussion of chemical equilibrium, we will assume that all gases can be assumed to be ideal and that all liquids and solids can be assumed to have constant volumes. The expressions for the activities of these substances are relatively simple. Reactions involving such substances are therefore fairly easy to discuss.

Gaseous Reactions

For reactions in which all substances are ideal gases the activities are given by

$$a_i = P_i/P^\circ \quad \text{(ideal gas)}$$

(7.2-1)

Equation (7.1-20) for the equilibrium constant is

$$K_P = \prod_{i=1}^{c} \left(\frac{P_{i(eq)}}{P^\circ} \right)^{v_i} \quad \text{(ideal gas reaction)}$$

(7.2-2)

It is advantageous to use dimensionless equilibrium constants so that equations like Equation (7.1-20) can be written with consistent units. The arguments of logarithms and exponentials should not have units.

where the subscript P indicates that all activities are expressed in terms of partial pressures. Some chemistry texts omit the P° divisors, giving K_P the dimensions of pressure raised to the appropriate power, and others define two different equilibrium constants, one with the divisors and the other without them.

▼   **EXAMPLE 7.1**

Consider the reaction

$$0 = 2NO_2(g) - N_2O_4(g)$$

a. Calculate the value of ΔG° at 298.15 K.

b. Calculate the value of K_P at 298.15 K.

c. Calculate the equilibrium pressure of a system that initially consists of 1.000 mol of dinitrogen tetroxide and is confined in a fixed volume of 24.46 L at 298.15 K.

Solution

a. From the Gibbs energy changes of formation

$$\Delta G^\circ = 2\,\Delta\bar{G}_f^\circ(NO_2) + (-1)\,\Delta\bar{G}_f^\circ(N_2O_4)$$
$$= (2)(51.258 \text{ kJ mol}^{-1}) + (-1)(97.787 \text{ kJ mol}^{-1})$$
$$= 4.729 \text{ kJ mol}^{-1}$$

b.
$$K_P = \exp\left[\frac{-4729 \text{ J mol}^{-1}}{(8.3145 \text{ J K}^{-1}\text{ mol}^{-1})(298.15 \text{ K})}\right] = 0.1484$$

We sometimes label partial pressures, amounts, and activities with the formula for the substance inside parentheses rather than with a subscript. There is no difference between P_i and $P(i)$ or between n_i and $n(i)$.

c. Let α be the degree of dissociation (the fraction of the initial N_2O_4 that dissociates). We assume ideal gas behavior:

$$P(NO_2) = \frac{n(NO_2)RT}{V} = \frac{(1.000 \text{ mol})(2\alpha)RT}{V}$$

$$P(N_2O_4) = \frac{n(N_2O_4)RT}{V} = \frac{(1.000 \text{ mol})(1-\alpha)RT}{V}$$

We can now write

$$K_P = 0.1484 = \frac{(2\alpha)^2}{1-\alpha}\frac{(1.000 \text{ mol})RT}{P^\circ V}$$
$$= \frac{4\alpha^2(1.013)}{1-\alpha}$$

which is the same as

$$4.054\alpha^2 + 0.1484\alpha - 0.1484 = 0$$

Use of the quadratic formula gives

$$\alpha = \frac{-0.1484 \pm \sqrt{(0.1484)^2 + (4)(4.054)(0.1484)}}{2(4.054)}$$
$$= 0.174$$

Whenever an algebraic equation with more than one root is encountered in chemistry, physical reasoning is used to decide which root is the appropriate one.

We must discard the negative root of the quadratic equation because a negative value of α is not physically possible if no nitrogen dioxide is initially present. The total amount of gas is $(1.000 \text{ mol})(1 - \alpha + 2\alpha) = 1.174$ mol, so the total pressure is

$$P = \frac{(1.174 \text{ mol})(0.08206 \text{ L atm K}^{-1}\text{ mol}^{-1})(298.15 \text{ K})}{24.46 \text{ L}}$$
$$= 1.174 \text{ atm}$$
$$= 1.190 \times 10^5 \text{ Pa}$$

◢

Reactions Involving Pure Condensed Phases and Gases

If the pressure of the system does not differ very much from P°, Equation (6.3-3b) gives for the activity of a pure liquid or solid

$$a_i \approx 1 \quad \text{(pure liquid or solid near } P^\circ\text{)} \qquad \textbf{(7.2-3)}$$

Substances that are pure liquids or solids contribute a factor nearly equal to unity to the equilibrium constant expression. To a good approximation, they can be omitted from it.

▼   **EXAMPLE 7.2**

a. Write the equilibrium constant quotient for the reaction

$$CaCO_3(s) \rightarrow CaO(s) + CO_2(g)$$

b. Find the equilibrium constant at 298.15 K and the pressure of CO_2 at equilibrium with $CaO(s)$ and $CaCO_3(s)$ at 298.15 K.

Solution

a. Since the activities of the solids are nearly equal to unity,

$$K = \frac{a_{eq}(CaO)a_{eq}(CO_2)}{a_{eq}(CaCO_3)} \approx a_{eq}(CO_2) \approx \frac{P_{eq}(CO_2)}{P°}$$

b. From the Gibbs energies of formation

$$\Delta G° = (1)(-603.501 \text{ kJ mol}^{-1}) + (1)(-394.389 \text{ kJ mol}^{-1}$$
$$+ (-1)(-1128.79 \text{ kJ mol}^{-1})$$
$$= 130.90 \text{ kJ mol}^{-1}$$

$$K_P = \exp\left[-\frac{130900 \text{ J mol}^{-1}}{(8.3145 \text{ J K}^{-1} \text{ mol}^{-1})(298.15 \text{ K})}\right] = 1.17 \times 10^{-23}$$

This equilibrium constant is strongly temperature dependent and becomes much larger at higher temperatures.

$$P_{eq}(CO_2) = (P°)(1.17 \times 10^{-23}) = 1.17 \times 10^{-23} \text{ bar} \approx 1.15 \times 10^{-23} \text{ atm}$$

▲

Reactions that appear to proceed to completion are not fundamentally different from other reactions. They just have very large equilibrium constants.

▼   **EXAMPLE 7.3**

a. Write the activity quotient for the combustion of propane, Equation (7.1-2), at 298.15 K.

b. Find the equilibrium constant for this reaction.

Solution

a.
$$Q = \frac{[P(CO_2)/P°]^3(1)^4}{[P(C_3H_8)/P°](P(O_2)/P°)^5}$$

b. $\Delta G° = 3\Delta\bar{G}_f°(CO_2) + 4\Delta\bar{G}_f°(H_2O) + (-1)\Delta\bar{G}_f°(C_3H_8) + (-5)\Delta\bar{G}_f°(O_2)$
$$= (3)(-394.389 \text{ kJ mol}^{-1}) + (4)(-237.141 \text{ kJ mol}^{-1})$$
$$+ (-1)(-23.27 \text{ kJ mol}^{-1}) + (-5)(0)$$
$$= -2108.46 \text{ kJ mol}^{-1}$$

$$K = \exp\left[\frac{2108460 \text{ J mol}^{-1}}{(8.3145 \text{ J K}^{-1} \text{ mol}^{-1})(298.15 \text{ K})}\right] = e^{850.5} = 2 \times 10^{369}$$

▲

Exercise 7.5

a. Find the final partial pressure of propane and oxygen if a stoichiometric mixture of propane and oxygen comes to equilibrium at 298.15 K and a total pressure of 760. torr.

b. Find the volume containing one molecule of propane at the equilibrium of part a. Comment on the magnitude of your result.

Chemical Equilibrium in Solution

For a component of a liquid or solid solution, the activity is given in one of several different ways. In discussing chemical equilibrium, one must specify which description is being used because activities and standard states are different for different descriptions. The molality description is most commonly used for aqueous solutions, but convention II is also used for non-electrolyte solutes in nonaqueous solutions.

In the molality description, the activity of a solute is given by Equation (6.3-23):

$$a_i = \gamma_i m_i / m^\circ \qquad (7.3\text{-}1)$$

where m_i is the molality of substance i, m° is equal to 1 mol kg^{-1} by definition, and γ_i is the activity coefficient of substance i. We omit superscripts on the activity coefficients.

The activity of the solvent is given by Equation (6.3-32):

$$a_1 = \gamma_1 x_1 \qquad (7.3\text{-}2a)$$

where x_1 is the mole fraction of the solvent and γ_1 is its activity coefficient. In a dilute solution, both the activity coefficient and the mole fraction of the solvent are nearly equal to unity, so

$$a_1 \approx 1 \qquad (7.3\text{-}2b)$$

It is necessary to determine from the context which description is being used. For dilute aqueous solutions, the molality and molarity (concentration in mol L^{-1}) descriptions are nearly equivalent.

Gibbs Energy Changes and Equilibrium Constants

The Gibbs energy change for a standard-state solution reaction can be calculated from tables of the standard-state Gibbs energy changes of formation. Table A9 gives values for some solutes in the molality standard state.

▼ **EXAMPLE 7.4**

Using tabulated Gibbs energy changes of formation, find the standard-state Gibbs energy change at 298.15 K for the reaction

$$2CO(aq) + O_2(aq) \rightleftharpoons 2CO_2(aq)$$

Solution

From values in Table A9,

$$\Delta G^\circ = 2\,\Delta \bar{G}_f^\circ(CO_2) + (-2)\,\Delta \bar{G}_f^\circ(CO) + (-1)\,\Delta \bar{G}_f^\circ(O_2)$$
$$= 2(-385.98 \text{ kJ mol}^{-1}) + (-2)(-119.90 \text{ kJ mol}^{-1}) + (-1)(16.4 \text{ kJ mol}^{-1})$$
$$= -548.56 \text{ kJ mol}^{-1}$$

This value compares with -512.682 kJ mol^{-1} for the gas-phase reaction at the same temperature.

The difference between the value of ΔG° in Example 7.4 and that for the gas-phase reaction is due primarily to the effects of solvation and the difference in the standard states.

▲

Exercise 7.6

Find the value of ΔG° for the reaction of Example 7.4 using mole fractions according to convention II instead of the molality description. State any assumptions.

The equilibrium constant for a reaction involving only solutes is given in the molality description by

$$K = Q_{eq} = \prod_{i=2}^{c} \left(\frac{\gamma_i m_{i(eq)}}{m^\circ} \right)^{\nu_i} \tag{7.3-3}$$

where γ_i is the activity coefficient and $m^\circ = 1$ mol kg^{-1}. Component 1 (the solvent) is omitted from the product because it is not involved in the reaction. The equilibrium constant for a reaction involving the solvent is given by

$$K = Q_{eq} = (\gamma_1 x_1)^{\nu_1} \prod_{i=2}^{c} \left(\frac{\gamma_i m_{i(eq)}}{m^\circ} \right)^{\nu_i} \tag{7.3-4}$$

For a dilute solution, x_1 and γ_1 are both nearly unity, so the activity factor for the solvent can be omitted to a good approximation.

For a dilute solution in which the activity coefficients are nearly equal to unity, we often write an approximate equilibrium constant in which we omit all activity coefficients as well as the mole fraction of the solvent:

$$K_m = \prod_{i=2}^{c} \left(\frac{m_{i(eq)}}{m^\circ} \right)^{\nu_i} \tag{7.3-5}$$

The lower limit $i = 2$ means that component 1 (the solvent) is excluded, whether or not it is involved in the reaction. Some authors also omit the m° factors, giving an equilibrium constant that has the dimensions of molality raised to some power. Of course, when the activity coefficients differ considerably from unity, the use of K_m can be a poor approximation.

The value of the equilibrium constant for a solution reaction can be calculated from the Gibbs energy change of the standard-state reaction, using Equation (7.1-20), which applies to any kind of reaction. Once the equilibrium constant is evaluated, the equilibrium composition can be calculated for any particular case, if information about activity coefficients is available, either from experimental data or from theoretical estimates.

▼   **EXAMPLE 7.5**

a. Find the equilibrium constant for the reaction of Example 7.4 at 298.15 K.

b. Find $(\partial G/\partial \xi)_{T,P}$ for the case in which the molality of carbon monoxide is 0.00010 mol kg^{-1}, that of carbon dioxide is 0.00015 mol kg^{-1}, and that of oxygen is 0.00020 mol kg^{-1}. Assume activity coefficients equal to unity.

c. Find the equilibrium composition for initial molalities of 0.00015 mol kg^{-1} for carbon dioxide, 0.00020 mol kg^{-1} for oxygen, and zero for carbon monoxide. Assume activity coefficients equal to unity.

Solution

The reaction is

$$2CO(aq) + O_2(aq) \rightleftharpoons 2CO_2(aq)$$

From Example 7.4, $\Delta G^\circ = -548.56$ kJ mol^{-1} at 298.15 K.

a.

$$K = e^{-\Delta G^\circ/RT} = \exp\left[\frac{548560 \text{ J mol}^{-1}}{(8.3145 \text{ J K}^{-1} \text{ mol}^{-1})(298.15 \text{ K})} \right]$$

$$= 1.27 \times 10^{96}$$

b.

$$(\partial G/\partial \xi)_{T,P} = \Delta G^\circ + RT \ln(Q)$$
$$= -548560 \text{ J mol}^{-1}$$
$$+ (8.3145 \text{ J K}^{-1} \text{ mol}^{-1})(298.15 \text{ K}) \ln\left[\frac{(0.00015)^2}{(0.00010)^2(0.00020)}\right]$$
$$= -548560 \text{ J mol}^{-1} + 23000 \text{ J mol}^{-1} = -525000 \text{ J mol}^{-1}$$
$$= -525 \text{ kJ mol}^{-1}$$

c. Let $m(CO)/m^\circ = 2x$, so that $m(O_2)/m^\circ = 0.00020 + x$ and $m(CO_2)/m^\circ = 0.00015 - 2x$:

$$K = 1.27 \times 10^{96} = \frac{[m(CO_2)/m^\circ]^2}{[m(CO)/m^\circ]^2 m(O_2)/m^\circ} = \frac{(0.00015 - 2x)^2}{(2x)^2(0.00020 + x)}$$

Since x will be very small,

$$x^2 \approx \frac{(0.00015)^2}{4(1.27 \times 10^{96})(0.00020)} = 2.21 \times 10^{-101}$$
$$x \approx 4.7 \times 10^{-51}, \qquad m(CO) \approx 9.4 \times 10^{-51} \text{ mol kg}^{-1}$$

7.4 Equilibria in Solutions of Strong Electrolytes

The discussion in Section 7.3 applies to solutions containing electrolyte solutes as well as nonelectrolyte solutes, and it provides us with the means of discussing some unique properties of the activities of strong electrolytes.

We first consider a volatile strong electrolyte such as hydrogen chloride at equilibrium with a vapor phase. From the fundamental fact of phase equilibrium

$$\mu(HCl, aq) = \mu(HCl, g) \tag{7.4-1}$$

The HCl ionizes in solution

$$HCl(aq) \rightleftharpoons H^+(aq) + Cl^-(aq) \tag{7.4-2}$$

Writing $H^+(aq)$ does not express any particular assumption about hydration of the hydrogen ions. Most aqueous hydrogen ions are apparently bonded to a water molecule, forming H_3O^+, the hydronium ion, but others are apparently attached to water dimers, producing $H_5O_2^+$, etc. The symbol $H^+(aq)$ stands for all of these species taken together.

The condition for equilibrium of the reaction of Equation (7.4-2) is, from Equation (7.1-10),

$$\mu(HCl, aq) = \mu(H^+, aq) + \mu(Cl^-, aq) \tag{7.4-3}$$

We use the ordinary molality description for the H^+ and Cl^- ions. For the aqueous HCl, we use a new molality description in which m° (1 mol kg^{-1}) is replaced by m', a constant molality that is unspecified. This policy is necessary because $m(HCl)$, the equilibrium molality of unionized HCl, is too small to be measured.

We write the chemical potentials in terms of activity coefficients and molalities:

$$\mu^\circ(\text{HCl}, aq) + RT \ln[a(\text{HCl})]$$
$$= \mu^\circ(\text{HCl}, aq) + RT \ln[\gamma(\text{HCl})m(\text{HCl})/m']$$
$$= \mu^\circ(\text{H}^+) + RT \ln[\gamma(\text{H}^+)m(\text{H}^+)/m^\circ]$$
$$+ \mu^\circ(\text{Cl}^-) + RT \ln[\gamma(\text{Cl}^-)m(\text{Cl}^-)/m^\circ] \qquad \textbf{(7.4-4)}$$

where we omit the label "aq" on the ion quantities. In this equation, the standard-state chemical potentials of H^+ and Cl^- refer to the hypothetical solution with molality m° (1 mol kg^{-1}) and activity coefficient equal to unity, but that of HCl refers to the hypothetical solution with molality m' and unit activity coefficient.

The value of $m(\text{HCl})$ is too small to measure, and the value of m' will be of similar magnitude. The conventional procedure is to avoid use of the value of m' and to specify that

$$\mu^\circ(\text{HCl}, aq) = \mu^\circ(\text{H}^+) + \mu^\circ(\text{Cl}^-) \quad \text{(convention)} \qquad \textbf{(7.4-5)}$$

which uniquely determines m', even though we still do not know its value. Canceling the terms occurring in Equation (7.4-5) and taking antilogarithms, we obtain from Equation (7.4-4)

$$a(\text{HCl}, aq) = \gamma(\text{HCl})m(\text{HCl})/m'$$
$$= (\gamma(\text{H}^+)m(\text{H}^+)/m^\circ)(\gamma(\text{Cl}^-)m(\text{Cl}^-)/m^\circ)$$
$$= \gamma_+\gamma_- m_+ m_-/m^{\circ 2} \qquad \textbf{(7.4-6)}$$

where we have simplified the notation, replacing $m(\text{H}^+)$ by m_+ and $m(\text{Cl}^-)$ by m_-, with similar notation for the γ's.

Exercise 7.7   Show that Equation (7.4-6) is equivalent to Equation (7.4-4).

Equation (7.4-5) is the same as requiring that

$$\Delta G^\circ = 0 \qquad \textbf{(7.4-7)}$$

for the ionization reaction of Equation (7.4-2), which means that the equilibrium constant for the reaction is given by

The fact that the value of m' is unknown keeps us from making equilibrium calculations with Equation (7.4-8).

$$K = \frac{(\gamma_+ m_+/m^\circ)(\gamma_- m_-/m^\circ)}{\gamma(\text{HCl})m(\text{HCl})/m'} = 1 \qquad \textbf{(7.4-8)}$$

Equation (7.4-8) does *not* mean that the molality of un-ionized HCl is roughly equal to the product of the molalities of the ions, because m' is not equal to m°.

We assume that the vapor is an ideal gas and write

$$\mu(\text{HCl}, g) = \mu^\circ(\text{HCl}, g) + RT \ln\left[\frac{P(\text{HCl})}{P^\circ}\right] \qquad \textbf{(7.4-9)}$$

From Equations (7.4-1) and (7.4-4)

$$\mu°(HCl, g) + RT \ln(P(HCl)/P°)$$
$$= \mu_+° + \mu_-° + RT \ln(\gamma_+\gamma_- m_+ m_-/m°^2) \qquad \textbf{(7.4-10)}$$

The **gross molality** m is defined as the total amount of HCl per kilogram of solvent, including both ionized and unionized forms. If no H^+ ions and Cl^- ions are added to the solution from other solutes and if the amount of H^+ ions from water can be neglected, m_+ and m_- are both equal to m if ionization is complete. Equation (7.4-10) is equivalent to

$$P(HCl) = k^{(m)}\gamma_+\gamma_- m^2 = k_\pm^{(m)} m^2 \gamma_\pm^2 \qquad \textbf{(7.4-11)}$$

where

$$k_\pm^{(m)} = \left(\frac{P°}{m°^2}\right)\exp\left[\frac{\mu°(H^+) + \mu°(Cl^-) - \mu°(HCl, g)}{RT}\right] \qquad \textbf{(7.4-12)}$$

and $\gamma_\pm$ is the mean ionic activity coefficient of Equation (6.4-12):

$$\gamma_\pm = (\gamma_+\gamma_-)^{1/2} \qquad \textbf{(7.4-13)}$$

For concentrations of HCl sufficiently small that the activity coefficients nearly equal unity, the partial vapor pressure of HCl is nearly proportional to the square of the molality, not proportional to the molality as in Henry's law. Figure 7.2 shows the partial vapor pressures of HCl as a function of molality in an aqueous solution at 298.15 K as a function of the molality. Table A13 gives the values of the partial pressure for larger molalities for the same temperature. The table represents experimental data, and the graph represents values calculated from activity coefficient values determined by other techniques.

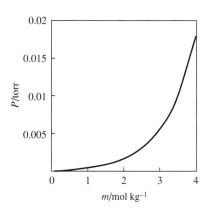

Figure 7.2. The Partial Vapor Pressure of HCl as a Function of Molality. In contrast to Henry's law, the curve in this diagram is quadratic instead of linear for small molality.

▼   **EXAMPLE 7.6**

The partial vapor pressure of a 5.00 mol kg^{-1} aqueous solution of HCl at 298.15 K is equal to 6.97×10^{-5} atm.[1] The mean ionic activity coefficient is 2.38. Find the value of $k_\pm^{(m)}$ and of $\mu°(H^+) + \mu°(Cl^-) - \mu°(HCl, g)$.

Solution

$$k_\pm^{(m)} = \frac{P(HCl)}{\gamma_\pm^2 m^2} = \frac{6.97 \times 10^{-5} \text{ atm}}{(2.38)^2(5.00 \text{ mol kg}^{-1})^2}$$
$$= 4.92 \times 10^{-7} \text{ atm kg}^2 \text{ mol}^{-2}$$

$$\mu°(H^+) + \mu°(Cl^-) - \mu°(HCl, g)$$
$$= RT \ln\left(\frac{k_\pm^{(m)} m°^2}{P°}\right)$$
$$= (8.3145 \text{ J K}^{-1} \text{ mol}^{-1})(298.15 \text{ K})$$
$$\times \ln\left[\frac{(4.92 \times 10^{-7} \text{ atm})(1 \text{ mol kg}^{-1})^2}{1 \text{ bar}}\left(\frac{1.01325 \text{ bar}}{1 \text{ atm}}\right)\right]$$
$$= -3.60 \times 10^4 \text{ J mol}^{-1} = -36.0 \text{ kJ mol}^{-1}$$

▲

[1] S. J. Bates and H. D. Kirschman, *J. Am. Chem. Soc.* **41**, 1991 (1919).

Equations (7.4-4) through (7.4-7) are valid for nonvolatile electrolytes such as NaCl and AgCl as well as volatile electrolytes such as HCl or HNO_3. Since the partial vapor pressure of NaCl or AgCl is too small to measure, we cannot apply Equations (7.4-8) through (7.4-12) to these substances. We must find a different way to evaluate their activity coefficients than the use of the vapor pressure. We discuss electrochemical methods in Chapter 8.

For a general univalent-univalent electrolyte represented by the formula MX, we write the same equation as Equation (7.4-6):

$$a(\text{MX}, aq) = \gamma(\text{M}^+)m(\text{X}^-)\gamma(\text{X}^-)m(\text{X}^-)/m^{\circ 2} = \gamma_\pm^2 m_+ m_-/m^{\circ 2} \quad \textbf{(7.4-14)}$$

For an electrolyte that is not univalent-univalent, the expression is different. For $CaCl_2$, we have

$$\text{CaCl}_2 \rightarrow \text{Ca}^{2+} + 2\text{Cl}^- \quad \textbf{(7.4-15)}$$

Setting ΔG° for this reaction equal to zero as in Equation (7.4-5) or (7.4-7) leads to

$$a(\text{CaCl}_2) = (\gamma_+ m_+/m^\circ)(\gamma_- m_-/m^\circ)^2 \quad \textbf{(7.4-16)}$$

Exercise 7.8   Show that Equation (7.4-16) is correct.

For $CaCl_2$, the mean ionic activity coefficient is

$$\gamma_\pm = (\gamma_+ \gamma_-^2)^{1/3} \quad \textbf{(7.4-17)}$$

and the mean ionic molality is defined to be

$$m_\pm = (m_+ m_-^2)^{1/3} \quad \textbf{(7.4-18)}$$

so that

$$a(\text{CaCl}_2) = (\gamma_\pm m_\pm)^3 \quad \textbf{(7.4-19)}$$

If there is no other source of Ca^{2+} or Cl^- than the $CaCl_2$,

$$m_+ = m, \qquad m_- = 2m \quad \textbf{(7.4-20)}$$

where m is the gross molality (total amount of $CaCl_2$ per kilogram of solvent), so that for $CaCl_2$:

$$m_\pm = m(2^2 1^1)^{1/3} = m v_\pm \quad \textbf{(7.4-21)}$$

where the second equality defines the quantity $v_\pm$ for $CaCl_2$ or a similar 1-2 electrolyte. We now write

$$a(\text{CaCl}_2) = (\gamma_\pm m_\pm/m^\circ)^3 = (\gamma_\pm v_\pm m/m^\circ)^3 \quad \textbf{(7.4-22)}$$

The analogous quantities are defined for any electrolyte. If the formula of the electrolyte is represented by $M_{v_+}X_{v_-}$ and if there is no other source of either ion,

$$a(M_{v_+}X_{v_-}) = (\gamma_\pm m_\pm/m^\circ)^{(v_+ + v_-)} \quad \textbf{(7.4-23)}$$

where, as in Equation (6.4-12),

$$\gamma_\pm = (\gamma_+^{v_+} \gamma_-^{v_-})^{1/(v_+ + v_-)} \quad \textbf{(7.4-24)}$$

and

$$m_\pm = (m_+^{\nu_+} m_-^{\nu_-})^{1/(\nu_+ + \nu_-)} = m\nu_\pm = m(\nu_+^{\nu_+} \nu_-^{\nu_-})^{1/(\nu_+ + \nu_-)} \qquad \textbf{(7.4-25)}$$

For example, the mean ionic molality of a 3-1 electrolyte such as $CrCl_3$ is $27^{1/4}m$, and that of a 2-2 electrolyte such as $ZnSO_4$ is $2m$.

Exercise 7.9

a. Verify Equations (7.4-23) and (7.4-24).

b. Write expressions for $\gamma_\pm$ and $m_\pm$ for $Mg_3(PO_4)_2$ in terms of m, the gross molality, and the activity coefficients of the ions.

Tabulated Thermodynamic Values for Electrolytes

Gibbs energy changes of formation, activities, activity coefficients, etc. for separate ions cannot be measured because of the near impossibility of adding ions of one charge to a system without adding ions of the opposite charge. In spite of this, it is desirable to tabulate values of enthalpy changes of formation, Gibbs energies of formation, and absolute entropies for ions instead of the neutral substances in order to have a shorter table.

This arbitrary convention does not change the value of ΔG, ΔS, etc. for a chemical reaction. The purpose of the convention is to make a shorter table of values.

It is customary to make an arbitrary division of the Gibbs energy change of formation, enthalpy change of formation, entropy, etc. between the ions of one electrolyte and then to make all other values consistent with this division. The arbitrary choice that is made is to assign zero value to the Gibbs energy change of formation, enthalpy change of formation, entropy, etc. of the hydrogen ion in its standard state in aqueous solution. This convention is equivalent to assigning the entire Gibbs energy change of formation of aqueous HCl to the chloride ion, etc.

▼   **EXAMPLE 7.7**

Using tabulated values in Table A9, calculate $\Delta H°$, $\Delta G°$, and $\Delta S°$ for the reaction

$$0 = H_2O + NaCl - NaOH(aq) - HCl(aq)$$

Solution

We write the net ionic equation

$$0 = H_2O - OH^-(aq) - H^+(aq)$$

$$\Delta H° = \Delta\bar{H}_f°(H_2O) + (-1)\,\Delta\bar{H}_f°(OH^-) + (-1)\,\Delta\bar{H}_f°(H^+)$$
$$= (-285.830\text{ kJ mol}^{-1}) + (-1)(-229.994\text{ kJ mol}^{-1}) + (-1)0$$
$$= -55.836\text{ kJ mol}^{-1}$$

$$\Delta G° = \Delta\bar{G}_f°(H_2O) + (-1)\,\Delta\bar{G}_f°(OH^-) + (-1)\,\Delta\bar{G}_f°(H^+)$$
$$= (-237.141\text{ kJ mol}^{-1}) + (-1)(-157.244\text{ kJ mol}^{-1}) + (-1)0$$
$$= -79.897\text{ kJ mol}^{-1}$$

In Example 7.7, omission of the "spectator ions" makes no difference because the standard state of each ion is independent of the other ions present.

$$\Delta S° = \bar{S}°(H_2O) + (-1)\bar{S}°(OH^-) + (-1)\bar{S}°(H^+)$$
$$= (69.95\text{ J K}^{-1}\text{ mol}^{-1}) + (-1)(-10.75\text{ J K}^{-1}\text{ mol}^{-1}) + (-1)0$$
$$= 80.70\text{ J K}^{-1}\text{ mol}^{-1}$$

▲

<table>
<tr><td>

</td></tr>
</table>

| 7.5 | **Acid-Base Equilibrium Calculations** |

It is customary to specify the acidity of a solution in terms of the pH, which has been variously defined. Chemists would like to be able to use the definition

An operational definition of pH will be introduced in Chapter 8.

$$\text{pH} = -\log_{10}[a(\text{H}^+)] = -\log_{10}\left[\frac{\gamma(\text{H}^+)m(\text{H}^+)}{m^\circ}\right] \quad \text{(7.5-1)}$$

where $\log_{10}$ denotes the logarithm to the base 10 (the common logarithm). A version of the definition of Equation (7.5-1) or of the analogous concentration definition, which omits the activity coefficient, is commonly found in elementary chemistry textbooks.

The pH as defined in Equation (7.5-1) cannot be measured exactly because of the impossibility of exactly measuring $\gamma(\text{H}^+)$. However, we will assume that $\gamma(\text{H}^+)$ can be obtained to an adequate approximation and will use Equation (7.5-1).

▼   **EXAMPLE 7.8**

a. Find the equilibrium constant at 298.15 K for the reaction

$$\text{HA} \rightleftharpoons \text{H}^+ + \text{A}^-$$

where HA stands for acetic acid, CH_3COOH.

b. Find the molality of hydrogen ions and the pH in a solution prepared from 0.100 mol of acetic acid and 1.000 kg of water and maintained at 298.15 K. Use the Davies equation to estimate activity coefficients.

Solution

a. From Gibbs energy changes of formation in Table A9,

$$\Delta G^\circ = \Delta \bar{G}_f^\circ(\text{A}^-) + \Delta \bar{G}_f^\circ(\text{H}^+) - \Delta \bar{G}_f^\circ(\text{HA})$$
$$= -369.31 \text{ kJ mol}^{-1} + 0 - (-396.46 \text{ kJ mol}^{-1}) = 27.15 \text{ kJ mol}^{-1}$$

$$K = \exp\left[\frac{-27150 \text{ J mol}^{-1}}{(8.3145 \text{ J K}^{-1}\text{ mol}^{-1})(298.15 \text{ K})}\right] = 1.75 \times 10^{-5}$$

b. We assume that hydrogen ions from water can be neglected and that the activity coefficient of unionized acetic acid is equal to unity.

We use a method of successive approximations. For our first approximation, we assume that $\gamma_\pm$ is equal to unity:

$$K = 1.75 \times 10^{-5} = \frac{[m(\text{H}^+)/m^\circ][m(\text{A}^-)/m^\circ]}{m(\text{HA})/m^\circ} = \frac{x^2}{0.100 - x}$$

where we let $x = m(\text{H}^+)/m^\circ = m(\text{A}^-)/m^\circ$. This equation is the same as

$$x^2 + 1.75 \times 10^{-5}x - 1.75 \times 10^{-6} = 0$$

Use of the quadratic formula gives

$$x = 1.31 \times 10^{-3}, \qquad m(\text{H}^+) = 1.31 \times 10^{-3} \text{ mol kg}^{-1}$$

where a negative root has been discarded.

We now use the Davies equation to obtain a value for the activity coefficients of the hydrogen ions and acetate ions at the molality we have found. From Equation (6.4-15) with I equal to 1.31×10^{-3} mol kg^{-1}, we find that $\gamma_\pm$ is equal to 0.959.

We now obtain a second approximation by substituting this value of $\gamma_{\pm}$ into the equilibrium expression:

$$K = \frac{(0.959)^2 x^2}{0.100 - x}$$

This equation is solved, with the result that

$$x = 1.37 \times 10^{-3}, \qquad m(H^+) = 1.37 \times 10^{-3} \text{ mol kg}^{-1}$$

Using the value of γ from the Davies equation,

$$pH = 2.88$$

Since further iterations would not change the activity coefficient appreciably, we stop with this second approximation.

In Example 7.8 we use a procedure of successive approximations. This is necessary because we have no way to determine the activity coefficients until the ionic strength is known.

Exercise 7.10

a. Find the standard-state Gibbs energy change and the equilibrium constant at 298.15 K for the ionization of water, which is the reverse reaction from that of Example 7.7.

b. Calculate the molalities of hydrogen and hydroxide ions in pure water at 298.15 K. Use the Davies equation to estimate activity coefficients.

The calculation of pH in aqueous solutions is not always as simple as it was in Example 7.8. If a weak acid is quite dilute, the hydrogen ions that come from water ionization cannot be neglected, and this complicates the calculation.

▼   **EXAMPLE 7.9**

Find the pH of a solution made from 1.00×10^{-7} mol of acetic acid and 1.000 kg of water at 298.15 K.

Solution

We must solve two simultaneous equations:

$$K = 1.75 \times 10^{-5} = \frac{[\gamma(H^+)m(H^+)/m^\circ][\gamma(A^-)m(A^-)/m^\circ]}{\gamma(HA)m(HA)/m^\circ}$$

The assumption that the activity coefficient of HA is equal to unity is acceptable because the solution is sufficiently dilute that only charged substances behave nonideally.

and

$$K_w = 1.0 \times 10^{-14} = [\gamma(H^+)m(H^+)/m^\circ][\gamma(OH^-)m(OH^-)/m^\circ] \quad \textbf{(7.5-2)}$$

We assume that γ_{HA} is equal to unity. Since all of the ions are univalent, we assume that all of their activity coefficients are equal, as would be predicted by the Davies equation. We represent them by γ.

We choose the temporary symbols

$$x = \frac{m(H^+)}{m^\circ}, \qquad y = \frac{m(OH^-)}{m^\circ}$$

Since there are only two equations, we must be sure that there are only two independent unknown quantities to be determined. Otherwise, no solution would be possible.

We have two equations, and although we have four molalities only two are independent. We express the other variables in terms of x and y. For each OH^- ion, an H^+ ion is released from water, and for each A^- ion an H^+ ion is released from the acid, so that

The two simultaneous equations are solved by solving one of them for y and substituting this solution into the other equation.

$$\frac{m(A^-)}{m^\circ} = x - y, \qquad \frac{m(HA)}{m^\circ} = \frac{m}{m^\circ} - (x - y)$$

where m is the gross molality.

From the water ionization equilibrium expression,

$$y = \frac{K_w}{\gamma^2 x}$$

After $m(A^-)$ and $m(HA)$ are eliminated and this expression is substituted into the acid equilibrium expression, it becomes

$$K = \frac{\gamma^2 x[x - K_w/(\gamma^2 x)]}{m/m^\circ - x + K_w/(\gamma^2 x)}$$

When this equation is multiplied out, we obtain a cubic equation:

$$\gamma^2 x^3 + Kx^2 - [(m/m^\circ)K + K_w]x - KK_w/\gamma^2 = 0$$

For our first approximation, we assume that the activity coefficient is equal to unity and solve for x. Use of a computer program to obtain a numerical approximation to the root gives $m(H^+) = 1.61 \times 10^{-7}$ mol kg^{-1}, only 61% higher than in pure water. Use of the Davies equation gives a value of 0.999999 for γ, so no repetition of the calculation is necessary. The value of the pH is 6.79.

If the hydrogen ions from water are ignored, the method of Example 7.8 gives $m(H^+) = 9.9 \times 10^{-8}$ mol kg^{-1}, which is in error by 38% and is smaller than the value of $m(H^+)$ ions in pure water.

▲

Example 7.9 is more complicated than equilibrium calculations we have done up to now, because we must include the hydrogen ions from water as well as from the acid.

Another possible complication is that some acids are polyprotic, which means that one molecule of the acid ionizes successively to give more than one hydrogen ion.

Exercise 7.11   Find the pH of a solution made from 1.000 kg of water and 0.100 mol of phosphoric acid, H_3PO_4, for which the three acid dissociation constants are

$$K_1 = 7.52 \times 10^{-3}, \qquad K_2 = 6.23 \times 10^{-8}, \qquad K_3 = 2.2 \times 10^{-13}$$

Three simultaneous equations must be solved. It is best to seek simplifying approximations, such as neglecting the H^+ ions from the third ionization in discussing the first two ionizations. The validity of such approximations should be checked at the end of the calculation.

Buffer Solutions

In titrating a weak acid with a strong base, it is found that the pH changes only slowly when about half of the acid has been neutralized. Figure 7.3a shows the titration curve of 0.100 molal acetic acid with 0.100 molal sodium hydroxide, and Figure 7.3b shows the titration curve of 0.100 molal hydrochloric acid with 0.100 molal sodium hydroxide. Near the neutral pH value of 7, the curve has a large positive slope in both diagrams, but near a pH value of 5 the acetic acid curve has a much smaller slope than the hydrochloric acid curve.

A solution that resists changes in pH is called a **buffer solution**. A buffer solution can be produced by adding both the weak acid and its salt or by partially titrating the acid. Consider a solution with n_a mol of a weak acid and n_s mol of a salt of this acid added to 1.000 kg of water. The acid will ionize slightly and the anion will hydrolyze slightly. These effects will nearly

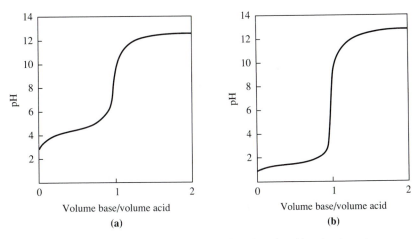

Figure 7.3. Titration Curves. (a) Curve for a solution of acetic acid and a strong base. **(b)** Curve for a solution of hydrochloric acid. For the acetic acid solution, there is a buffering region around pH 5.

cancel, and we can write to a good approximation

$$m(\text{HA}) = \frac{n_a}{1.000 \text{ kg}}, \qquad m(\text{A}^-) = \frac{n_s}{1.000 \text{ kg}}$$

Assuming γ_{HA} to equal unity (a good approximation), the equilibrium expression is

$$K = \frac{[\gamma(\text{H}^+)m(\text{H}^+)/m^\circ][\gamma(\text{A}^-)m(\text{A}^-)/m^\circ]}{m(\text{HA})/m^\circ} = \frac{a(\text{H}^+)\gamma(\text{A}^-)n_s}{n_a} \quad \textbf{(7.5-3)}$$

If $\gamma(\text{A}^-)$ is also assumed to equal unity, this equation can be written

$$\boxed{\text{pH} = \text{p}K_a + \log_{10}\left(\frac{n_s}{n_a}\right) \quad \begin{pmatrix}\text{Henderson-Hasselbalch}\\ \text{equation}\end{pmatrix}} \quad \textbf{(7.5-4)}$$

where

$$\boxed{\text{p}K_a = -\log_{10}(K) \quad \text{(definition)}} \qquad \textbf{(7.5-5)}$$

and where $\log_{10}$ denotes the common logarithm (logarithm to the base 10). Equation (7.5-4) is known as the **Henderson-Hasselbalch equation**. This equation is used frequently in making buffer solutions.

For greater accuracy than the Henderson-Hasselbalch equation affords, one must use Equation (7.5-3) with experimental or theoretical values of the activity coefficients.

▼   **EXAMPLE 7.10**   **a.** Using the Henderson-Hasselbalch equation, calculate the pH of a solution made from 0.400 mol of acetic acid and 0.600 mol of sodium acetate in 1.000 kg of water.

b. Repeat the calculation of part a using Equation (7.5-3) and the Davies equation.

Solution

a. Using the value $K_a = 1.75 \times 10^{-5}$,

$$pH = 4.757 + \log_{10}\left(\frac{0.600 \text{ mol}}{0.400 \text{ mol}}\right) = 4.933$$

b. Let m_a be the molality of acid and m_s the molality of the anion that would occur if the acid did not ionize and the salt did not hydrolyze. We assume that the salt dissociates completely, that $\gamma(H^+)$, $\gamma(A^-)$, and $\gamma(OH^-)$ are equal and can be approximated by the Davies equation, and that $\gamma(HA)$ is equal to unity.

The hydrolysis of the salt is represented by

$$A^- + H_2O \rightleftharpoons HA + OH^-$$

We let $m(OH^-)/m^\circ = y$ and $m(H^+)/m^\circ = x$.

$$K = \frac{\gamma^2 x(m_s/m^\circ + x - y)}{m_a/m^\circ - x + y}$$

When y is replaced by $K_w/\gamma^2 x$, we obtain

$$\gamma^2 x^3 + (\gamma^2 m_s/m^\circ + K)x^2 - (K_w + Km_a/m^\circ)x - KK_w/\gamma^2 = 0 \qquad \textbf{(7.5-6)}$$

The ionic strength of the solution is equal to $m_s + x$, so the activity coefficient γ can be approximated by setting $I = m_s$ in the Davies equation for the initial calculation and can be approximated more nearly exactly by successive approximation.

Equation (7.5-6) is solved by successive approximation. The result is that $x = 2.13 \times 10^{-5}$, $\gamma = 0.740$, $a(H^+) = 1.576 \times 10^{-5}$, and $pH = 4.802$. The pH value in part a is in error by 0.131.

Example 7.10 is designed to show the degree of accuracy of the Henderson-Hasselbalch equation. In part b several assumptions are necessary, and deciding what assumptions are reasonable is an important part of the solution.

▲

Exercise 7.12

a. Verify Equation (7.5-6).

b. Substitute the numerical result of part b of Example 7.10 into the equilibrium expression, Equation (7.5-3), to verify its correctness.

Biological Buffering

The principal buffering system in mammalian blood is that of carbonic acid, which can equilibrate with its two anions and with gaseous and aqueous carbon dioxide through the reactions:

$$CO_2(g) \rightleftharpoons CO_2(aq) \qquad \textbf{(7.5-7a)}$$
$$CO_2(aq) + H_2O \rightleftharpoons H_2CO_3(aq) \qquad \textbf{(7.5-7b)}$$
$$H_2CO_3(aq) \rightleftharpoons H^+ + HCO_3^- \qquad \textbf{(7.5-7c)}$$
$$HCO_3^- \rightleftharpoons H^+ + CO_3^{2-} \qquad \textbf{(7.5-7d)}$$

Exercise 7.13

The Henry's law constant for CO_2 in water at 25°C is 1.24×10^6 torr. The equilibrium constant for reaction (7.5-7b) is 2.58×10^{-3}, that for reaction (7.5-7c) is

1.70×10^{-4}, and that for reaction (7.5-7d) is 4.69×10^{-11}. It is not posssible by chemical analysis to distinguish $CO_2(aq)$ from H_2CO_3. If the combined concentration of $CO_2(aq)$ and H_2CO_3 is used in the equilibrium expression, a different value for the first acid ionization is found. [The molality of $CO_2(aq)$ is larger than that of H_2CO_3.]

a. Find the first ionization constant of carbonic acid, using the combined molalities of $CO_2(aq)$ and H_2CO_3 in the equilibrium expression instead of the molality of H_2CO_3. Explain why reaction (7.5-7c) can act as an effective buffering reaction near the normal pH of blood, around pH 7.4.

b. Find the pH of a solution produced by equilibrating water with carbon dioxide gas at 760 torr at 25°C. Assume all activity coefficients are equal to unity.

c. Repeat the calculation of part a, using the Davies equation to estimate activity coefficients.

The pH of human blood can be raised by removing dissolved carbon dioxide from the blood through hyperventilation (rapid breathing). The reaction of Equation (7.5-7b) is catalyzed by an enzyme, carbonic anhydrase, so that equilibrium is established rapidly. If the molality of dissolved carbon dioxide is lowered below its normal value, the reaction of Equations (7.5-7b) and (7.5-7c) shifts to the left, raising the pH of the blood and producing a feeling of "lightheadedness." A person who has hyperventilated is often told to breathe in and out of a paper bag. Since exhaled air is about 4% carbon dioxide, this practice increases the level of carbon dioxide in the blood and alleviates the symptoms. The body also attempts to correct the situation by increasing the excretion of bicarbonate ion in urine.

7.6   Temperature and Pressure Dependence of Equilibrium Constants. The Principle of Le Châtelier

The temperature derivative of $\Delta G°$ at constant pressure is given by differentiation of Equation (7.1-13):

$$\left(\frac{\partial \Delta G°}{\partial T}\right)_P = \sum_{i=1}^{c} v_i \left(\frac{\partial \mu_i°}{\partial T}\right)_{P,n} \tag{7.6-1}$$

The derivatives on the right-hand side of this equation are at fixed n, since each standard state corresponds to a specified fixed composition. If the reaction involves only pure substances and gases, $\Delta G°$ does not depend on pressure, and the left-hand side of Equation (7.6-1) can be replaced by an ordinary derivative. For a solution reaction, the standard states are at the pressure of the solution, so $\Delta G°$ does depend on pressure, although this dependence is usually small.

From Equation (4.5-10)

$$(\partial \mu_i / \partial T)_{P,n} = -\bar{S}_i \tag{7.6-2}$$

Substituting the standard-state version of Equation (7.6-2) into Equation (7.6-1) gives

$$\left(\frac{\partial \, \Delta G^\circ}{\partial T}\right)_P = -\sum_{i=1}^{c} v_i \bar{S}_i^\circ = -\Delta S^\circ \tag{7.6-3}$$

where ΔS° is the entropy change for one mole of the standard-state reaction. We now use Equation (7.6-3) to write

$$\left[\frac{\partial (\Delta G^\circ / T)}{\partial T}\right]_P = \frac{1}{T}\left(\frac{\partial \, \Delta G^\circ}{\partial T}\right)_P - \frac{1}{T^2}\Delta G^\circ$$

$$\boxed{\left[\frac{\partial (\Delta G^\circ / T)}{\partial T}\right]_P = -\frac{\Delta S^\circ}{T} - \frac{\Delta G^\circ}{T^2} = -\frac{\Delta H^\circ}{T^2}} \tag{7.6-4}$$

where we have used Equation (7.1-15). Equation (7.6-4) is the differential version of the **Gibbs-Helmholtz equation**. This equation holds only for a reaction that begins and ends at the same temperature, because $\Delta(TS) = T\,\Delta S$ only if T has the same initial and final values. The value of ΔG can be calculated only for a process that begins and ends at the same temperature, since otherwise

$$\Delta G = \Delta H - T_1\,\Delta S - S_1\,\Delta T - \Delta S\,\Delta T \tag{7.6-5}$$

where $\Delta G = G_2 - G_1$, $\Delta H = H_2 - H_1$, etc. This expression cannot be evaluated because the actual value of the entropy of the system is not known. The third law implies only that the entropy can consistently be chosen to equal zero at zero temperature.

Exercise 7.14

a. Show that Equation (7.6-5) is correct.

b. Show that

$$\left[\frac{\partial (\Delta G^\circ / T)}{\partial (1/T)}\right]_P = \Delta H^\circ \tag{7.6-6}$$

Equation (7.1-20) can be written

$$\ln(K) = -\Delta G^\circ / RT \tag{7.6-7}$$

From this equation and the Gibbs-Helmholtz equation, we can write

$$\left[\frac{\partial \ln(K)}{\partial T}\right]_P = \frac{\Delta H^\circ}{RT^2} \tag{7.6-8}$$

or

$$\left[\frac{\partial \ln(K)}{\partial (1/T)}\right]_P = -\frac{\Delta H^\circ}{R} \tag{7.6-9}$$

This equation is called the **van't Hoff equation**.

In the case of a reaction involving only pure substances and gases, all of the partial derivatives of ΔG° and $\ln(K)$ in the above equations can be replaced by ordinary derivatives.

Exercise 7.15   Verify Equations (7.6-8) and (7.6-9).

If the value of ΔH° is known as a function of temperature, Equation (7.6-8) can be integrated to obtain the value of K at one temperature from the value at another temperature:

$$\ln\left[\frac{K(T_2)}{K(T_1)}\right] = \int_{T_1}^{T_2} \frac{\Delta H^\circ}{RT^2} dT \qquad (7.6\text{-}10)$$

which is equivalent to

$$\frac{\Delta G^\circ(T_2)}{T_2} - \frac{\Delta G^\circ(T_1)}{T_1} = -\int_{T_1}^{T_2} \frac{\Delta H^\circ}{T^2} dT \qquad (7.6\text{-}11)$$

where $\Delta G^\circ(T_2)$ is the value of ΔG° at temperature T_2 and $\Delta G^\circ(T_1)$ is the value of ΔG° at temperature T_1.

If ΔH° is temperature independent, Equation (7.6-10) becomes

$$\boxed{\ln\left[\frac{K(T_2)}{K(T_1)}\right] = -\frac{\Delta H^\circ}{R}\left(\frac{1}{T_2} - \frac{1}{T_1}\right)} \qquad (7.6\text{-}12)$$

which is the integrated van't Hoff equation. Equation (7.6-11) becomes

$$\boxed{\frac{\Delta G^\circ(T_2)}{T_2} - \frac{\Delta G^\circ(T_1)}{T_1} = \Delta H^\circ\left(\frac{1}{T_2} - \frac{1}{T_1}\right)} \qquad (7.6\text{-}13)$$

Equations (7.6-12) and (7.6-13) depend on the assumption that ΔH° is temperature independent.

which is the integral version of the Gibbs-Helmholtz equation.

Exercise 7.16   Carry out the integrations to obtain Equations (7.6-12) and (7.6-13).

▼   **EXAMPLE 7.11**   Assuming tht ΔH° is temperature independent, calculate the value of the equilibrium constant and of ΔG° at 100.°C for the reaction

$$0 = 2NO_2(g) - N_2O_4(g)$$

Solution

$$\Delta H^\circ = 2\,\Delta\bar{H}_f^\circ(NO_2) - \Delta\bar{H}_f^\circ(N_2O_4)$$
$$= 2(33.095 \text{ kJ mol}^{-1}) + (-1)(9.179 \text{ kJ mol}^{-1}) = 57.011 \text{ kJ mol}^{-1}$$
$$\ln\left[\frac{K(373.15)}{K(298.15)}\right] = -\frac{57011 \text{ J mol}^{-1}}{8.3145 \text{ J K}^{-1}\text{ mol}^{-1}}\left(\frac{1}{373.15 \text{ K}} - \frac{1}{298.15 \text{ K}}\right)$$
$$= 4.622$$

Using the value of $K(298.15)$ from Example 7.1,

$$K(373.15) = K(298.15)e^{4.622} = (0.148)(101.7) = 15.10$$

$$\Delta \bar{G}^\circ = -RT \ln(K)$$

$$= -(8.3145 \text{ J K}^{-1} \text{ mol}^{-1})(373.15 \text{ K}) \ln(15.10)$$

$$= -8420 \text{ J mol}^{-1} = -8.42 \text{ kJ mol}^{-1}$$

If the assumption of constant ΔH° is not sufficiently accurate, the next simplest assumption is that the heat capacities are constant, so that

$$\Delta H^\circ(T) = \Delta H^\circ(T_1) + \Delta C_P(T - T_1) \tag{7.6-14}$$

When Equation (7.6-14) is substituted into Equation (7.6-8) and an integration is carried out from T_1 to T_2, the result is

$$\ln\left[\frac{K(T_2)}{K(T_1)}\right] = -\frac{\Delta H^\circ(T_1)}{R}\left(\frac{1}{T_2} - \frac{1}{T_1}\right) + \frac{\Delta C_P^\circ}{R}\left[\ln\left(\frac{T_2}{T_1}\right) + \frac{T_1}{T_2} - 1\right] \tag{7.6-15}$$

Exercise 7.17

a. Verify Equation (7.6-15).

b. Using heat capacity data from Table A9 and assuming the heat capacities to be temperature independent, evaluate K and ΔG° for the reaction of Example 7.11 at 100°C. Calculate the percent difference between your value for K and that in Example 7.11.

▲

The Principle of Le Châtelier

This principle is named for French chemist Henri Louis Le Châtelier, 1850–1936.

The behavior of a system at chemical equilibrium when subjected to a change in temperature illustrates the **principle of Le Châtelier**, a general qualitative principle. This principle states that, if possible, a system will respond to a "stress" placed upon it by reacting in the direction that minimizes the effect of that stress on intensive properties of the system.[2]

The transfer of heat to the system is the applied stress (a stress is generally an extensive quantity). From Equation (7.6-8), we see that the equilibrium constant for an endothermic reaction ($\Delta H > 0$) has a positive temperature derivative. The stress causes the reaction to shift toward the right (producing more products), thus absorbing part of the heat put into the system. The temperature of the system rises by a smaller amount than if the reaction were somehow "frozen" and could not shift its reaction equilibrium. This moderation of the temperature rise is the lessening of the effect of the stress referred to in the statement of the principle.

Similarly, for an exothermic reaction ($\Delta H < 0$), transfer of heat to the system will cause the equilibrium to shift toward the left, again causing the

---

[2] See J. A. Campbell, *J. Chem. Educ.* **62**, 231 (1985) for an interesting rule for predicting the direction of the shift in a reaction equilibrium produced by a change in temperature if the sign of ΔH° is not known and also for references to articles discussing the correct statement of the principle of Le Châtelier.

temperature of the system to rise by a smaller amount than if the reaction were frozen.

The principle of Le Châtelier can also be applied to a shift in equilibrium produced by changing the volume of a system at constant temperature. Let us write an equilibrium expression for a gaseous reaction in terms of the mole fractions, given by Dalton's law as

$$x_i = P_i/P_{tot} \tag{7.6-16}$$

where P_{tot} is the total pressure. Equation (7.2-2) becomes

$$K_P = \prod_{i=1}^{c} \left(\frac{x_i P_{tot}}{P^\circ} \right)^{v_i} = \left(\frac{P_{tot}}{P^\circ} \right)^{\Delta v} \prod_{i=1}^{c} (x_i)^{v_i}$$

$$= \left(\frac{P_{tot}}{P^\circ} \right)^{\Delta v} K_x \tag{7.6-17}$$

Since the v's are negative for reactants, this sum is the difference between the number of moles of products and the number of moles of reactants.

where Δv is the sum of the stoichiometric coefficients:

$$\Delta v = \sum_{i=1}^{c} v_i \tag{7.6-18}$$

equal to the net change in the number of moles of gas in the system if 1 mol of reaction occurs.

The equilibrium constant K_P for a gaseous reaction cannot depend on pressure.

The quantity K_x is not a true equilibrium constant, since it depends on pressure for a gaseous reaction.

$$K_x = (P_{tot}/P^\circ)^{-\Delta v} K_P \tag{7.6-19}$$

The conclusions drawn here follow from the fact that K_P is constant at constant temperature. The pressure variation of K_x is therefore the same as the variation of the first factor in the right-hand side of Equation (7.6-19).

If the pressure on the system is increased, K_x will increase and the mole fractions of the products will increase if the products consist of fewer moles of gas than the reactants. Similarly, they will decrease if the products consist of more moles of gas than the reactants. In either case, a reduction of volume (the stress) will increase the pressure by a smaller amount than if the reaction were "frozen," in agreement with the principle of Le Châtelier.

Exercise 7.18   For the reaction of Example 7.1, calculate the degree of dissociation if the volume is reduced to 12.23 L at 298.15 K. Interpret the results in terms of the principle of Le Châtelier.

The principle of Le Châtelier can also be applied to adding an additional amount of a reactant or product to an equilibrium reaction mixture.

▼   **EXAMPLE 7.12**   For the system of Example 7.1, find the effect of adding an additional 0.500 mol of NO_2.

Solution

Let

$$n(N_2O_4) = (1.000 \text{ mol})(1.000 - \alpha)$$

$$n(NO_2) = (1.000 \text{ mol})(2.000\alpha + 0.500)$$

$$0.148 = \frac{(2.000\alpha + 0.500)^2}{1.000 - \alpha} (1.000 \text{ mol}) \frac{RT}{P^\circ V}$$

$$\frac{RT}{P^\circ V} = \frac{(8.3145 \text{ J K}^{-1} \text{ mol}^{-1})(298.15 \text{ K})}{(100000 \text{ Pa})(0.02446 \text{ m}^3)} = 1.0135 \text{ mol}^{-1}$$

$$0.148 = \frac{(2.027\alpha + 0.5067)^2}{1.000 - \alpha}$$

In using the quadratic formula, we disregarded a root that gives a physically impossible value of α.

which gives

$$4.0540\alpha^2 + 2.1750\alpha + 0.1054 = 0$$

The quadratic formula gives

$$\alpha = -0.0539$$

In Example 7.12 the original definition of α is maintained. Because of the addition of NO₂, there is nothing wrong with a negative value of α.

The final amount of N_2O_4 is 1.0539 mol. The mole fraction of NO_2 is 0.271, instead of 0.506, the value it would have if no shift occurred, in agreement with the principle of Le Châtelier.

▲

Exercise 7.19   Verify the result of Example 7.12 by substituting the value of α into the equilibrium constant expression.

7.7 Chemical Reactions and Biological Systems

An important feature of biochemical reactions of metabolism and respiration is the **coupling** of pairs of reactions, or the driving of a nonspontaneous reaction by the progress of a spontaneous reaction. This coupling is used both in driving useful reactions and in regenerating reactants for the spontaneous reactions. For example, the hydrolysis of adenosine triphosphate (ATP) to form adenosine diphosphate (ADP) and phosphoric acid (P) is shown in Figure 7.4. Since ATP, ADP, and phosphoric acid are all weak polyprotic acids, they exist as various anions in aqueous solution, as well as in the forms shown in Figure 7.4. This reaction equation is abbreviated

$$ATP + H_2O \rightleftharpoons ADP + P \qquad (7.7\text{-}1)$$

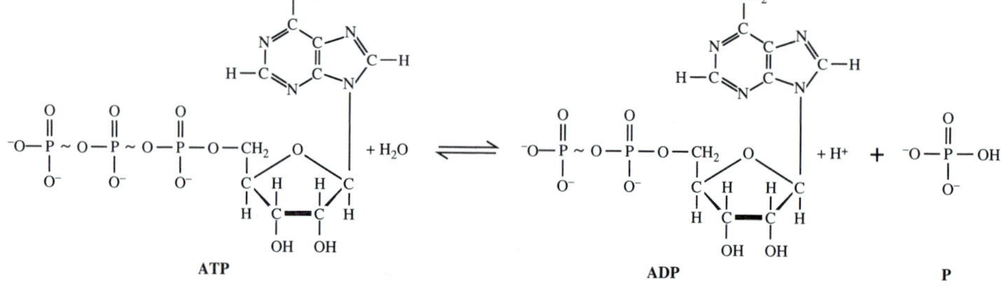

Figure 7.4. The Hydrolysis of Adenosine Triphosphate. This reaction is one of the principal energy delivery systems in biological organisms. The common abbreviation for each species is shown below its formula.

where the symbols in Equation (7.7-1) stand for whatever ionized and un-ionized forms of ATP, ADP, and phosphoric acid occur. At neutral pH, the most abundant form of ATP is a triply negative anion with a single proton on the acid groups.

The anions of ATP and ADP have a strong tendency to form complexes with positive ions, especially cations with multiple charges, such as Mg^{2+} or Ca^{2+}. It is customary to define a modified standard-state reaction in which the substances in the reaction equation are at unit activities but the hydrogen ions and any complexing cations are at specified activities not necessarily equal to unity. The symbol $\Delta G^{\circ\prime}$ is used for the Gibbs energy change of such a modified standard-state reaction.

For the reaction of Equation (7.7-1), using the concentration description, $\Delta G^{\circ\prime}$ is equal to -29.3 kJ mol^{-1} at 298.15 K with pH equal to 7.00 and pMg equal to 4.00. The pH is defined on the concentration description and the pMg is defined by analogy with pH:

$$pMg = -\log_{10}[a(Mg^{2+})] = -\log_{10}[\gamma(Mg^{2+})c(Mg^{2+})/c^{\circ}] \quad \textbf{(7.7-2)}$$

This value of $\Delta G^{\circ\prime}$ is for the combined reactions of whatever unionized, anionic, and complexed forms occur under the specified conditions. For example, the standard state for ATP is the state with the sum of the concentrations of all forms of ATP equal to 1 mol L^{-1} and with all activity coefficients equal to unity.

▼ **EXAMPLE 7.13**

a. Find the equilibrium constant for the reaction of Equation (7.7-1).
b. Find the equilibrium concentrations of ADP and ATP at pH 7.00 and pMg 4.00 if all of the phosphoric acid present comes from the hydrolysis of ATP and if the initial concentration of ATP is 0.0100 mol L^{-1}. Approximate all activity coefficients by unity.

Solution

a.

$$K = \exp\left[\frac{2.93 \times 10^4 \text{ J mol}^{-1}}{(8.3145 \text{ J K}^{-1} \text{ mol}^{-1})(298.15 \text{ K})}\right] = 1.36 \times 10^5$$

b. Let $x = c_{eq}(ATP)/c^{\circ}$

$$1.36 \times 10^5 = \frac{(0.0100 - x)^2}{x}$$

$$x = \frac{(0.0100 - x)^2}{1.36 \times 10^5} \approx \frac{(0.0100)^2}{1.36 \times 10^5} \approx 7.35 \times 10^{-10}$$

$$c_{eq}(ATP) \approx 7.35 \times 10^{-10} \text{ mol L}^{-1}$$

In the solution to Example 7.13, the value of $\Delta G^{\circ\prime}$ must be expressed in consistent units, which means joules instead of kilojoules.

▲

Exercise 7.20

Find the value of $(\partial G/\partial \xi)_{T,P}$ for the case that $c(ATP) = 0.0200$ mol L^{-1} and $c(ADP) = c(P) = 0.0100$ mol L^{-1} at pH 7.00 and pMg = 4.00. Approximate all activity coefficients by unity.

The hydrolysis of ATP is a spontaneous reaction under the conditions occurring in biological systems as well as under the modified standard-state conditions. This reaction is coupled to various other reactions that would otherwise not be spontaneous. That is, the spontaneous hydrolysis of ATP drives the nonspontaneous reactions, causing them to proceed. For example, the reaction

$$P + \text{glucose} \rightarrow \text{glucose-6-phosphate} + H_2O \qquad (7.7\text{-}3)$$

is driven by the reaction of Equation (7.7-1).

The ADP formed in the reaction of Equation (7.7-1) is "recycled." That is, other substances undergo spontaneous reactions that are coupled to the reaction of Equation (7.7-1), driving it from right to left. A reaction that drives the regeneration of ATP is the hydrolysis of phosphoenolpyruvic acid

$$
\begin{array}{c}
O \\
| \\
HO-P-OH \\
| \\
O \quad\; O \\
| \quad\; \| \\
CH_2{=}C{-}C{-}OH
\end{array}
$$

(abbreviated PEP).

The hydrolysis of PEP is sufficiently spontaneous to produce ATP from ADP. The sum of the two reactions is equivalent to a spontaneous reaction:

(A) $PEP + H_2O \rightleftharpoons Py + P$ $\Delta G^{\circ\prime} = -53.6 \text{ kJ mol}^{-1}$

(B) $ADP + P \rightleftharpoons ATP + H_2O$ $\Delta G^{\circ\prime} = +29.3 \text{ kJ mol}^{-1}$

(C) $ADP + PEP \rightleftharpoons ATP + Py$ $\Delta G^{\circ\prime} = -24.3 \text{ kJ mol}^{-1}$

where Py stands for pyruvic acid

$$
\begin{array}{c}
O \quad\; O \\
\| \quad\; \| \\
CH_3{-}C{-}C{-}OH
\end{array}
$$

and/or pyruvate ion.

We must ask how the two reactions can be combined as a single reaction. The hydrolysis of PEP produces phosphoric acid, which is a reactant in the regeneration of ATP from ADP. According to the principle of Le Châtelier, the phosphoric acid would shift the equilibrium of the regeneration reaction, producing more ATP. In the next example, we show that the effect of this shift is much too small to regenerate a significant amount of ATP.

▼ **EXAMPLE 7.14**

Calculate the concentration of ATP produced at 298.15 K by the equilibrium shift due to the principle of Le Châtelier if the initial concentrations of PEP and ADP are 0.0100 mol L^{-1}.

Solution

The equilibrium constant for reaction A is

$$K_A = \exp\left[\frac{5.36 \times 10^4 \text{ J mol}^{-1}}{(8.3145 \text{ J K}^{-1} \text{ mol}^{-1})(298.15 \text{ K})}\right] = 2.46 \times 10^9$$

Example 7.14 shows that the principle of Le Châtelier is insufficient to explain the regeneration of ATP by the hydrolysis of PEP.

This reaction proceeds essentially to completion, giving an equilibrium concentration of phosphate that is nearly equal to 0.0100 mol L^{-1}.

The equilibrium constant for reaction B is the reciprocal of that for the reaction of Equation (7.7-1), or 7.35×10^{-6}. We let $x = c_{eq}(ATP)/c^\circ$. The equilibrium expression gives

$$x = (7.35 \times 10^{-6})(0.0100 - x)^2 \approx (7.35 \times 10^{-6})(0.0100)^2 = 7.35 \times 10^{-10}$$

$$c_{eq}(ATP) \approx 7.35 \times 10^{-10} \text{ mol L}^{-1}$$

▲

To explain the coupling of reactions, one must have a mechanism that in some molecular sense makes one reaction out of two reactions. A proposed mechanism for the coupling of these two reactions involves two steps and an enzyme:[3]

$$(1) \qquad \text{E} + \text{PEP} \rightarrow \text{EP} + \text{Py}$$

$$(2) \qquad \frac{\text{EP} + \text{ADP} \rightarrow \text{E} + \text{ATP}}{\text{ADP} + \text{PEP} \rightarrow \text{ATP} + \text{Py}}$$

where E represents the enzyme pyruvate kinase. An enzyme generally has an **active site**, a cavity into which a reactant molecule can fit. Once in the active site, the reactant molecule is rendered more reactive. The important aspect of the proposed mechanism is that the phosphate is not simply released into the solution: it is held in the active site of the enzyme until it reacts with an ADP molecule. Since the first step is not repeated until the second step occurs, the hydrolysis of PEP does not occur without regeneration of ATP, and the two reactions are combined into a single reaction.

Exercise 7.21

a. Find the equilibrium constant at 298.15 K for the combined reaction

$$\text{ADP} + \text{PEP} \rightleftharpoons \text{ATP} + \text{P}$$

b. Find the equilibrium ATP concentration for the initial concentrations of Example 7.14, treating the combined reaction as a single reaction.

The coupling of the spontaneous hydrolysis of ATP to drive other reactions is similar to the coupling that regenerates ATP.

Exercise 7.22

a. Write a possible mechanism for the coupling of the spontaneous hydrolysis of ATP to drive the phosphorylation of glucose, Equation (7.7-3).

b. For the combined reaction, $\Delta G^{\circ\prime} = -9.6$ kJ mol^{-1}. Find $\Delta G^{\circ\prime}$ for the phosphorylation of glucose and find the equilibrium constant for the combined reaction.

---

[3] K. J. Laidler, *Physical Chemistry with Biological Applications*, Benjamin/Cummings, Menlo Park, CA, 1978, pp. 246ff.

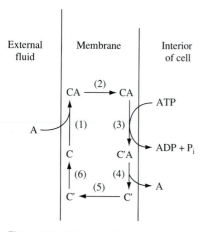

Figure 7.5. A Proposed Mechanism for Active Transport Through a Membrane. Every step of this mechanism is spontaneous, but the mechanism is able to transport substance A from a region of lower chemical potential to a region of higher chemical potential. The ATP must be provided from outside the mechanism to make it operate.

Processes other than nonspontaneous chemical reactions are also coupled to the hydrolysis of ATP. Figure 7.5 depicts a proposed mechanism for the **active transport** of a hypothetical substance A through a biological membrane from a solution of low concentration and low chemical potential of A (on the left in the figure) to a region of high concentration and high chemical potential of A (on the right).[4] This transport is opposite in direction to the spontaneous transport of A, which is from a higher to a lower value of the chemical potential (see Chapter 6).

The mechanism assumes that inside the membrane there is a carrier substance which has two forms and that substance A is able to pass through the surfaces of the membrane. The first form of the carrier, denoted by C, has a tendency to form a complex with substance A, while the second form, denoted by C', has no such tendency. The conversion of C to C' is nonspontaneous and is coupled to the hydrolysis of ATP.

Step (1) of the mechanism is the combination of C with A at the left surface of the membrane. This is followed by step (2), spontaneous transport of the complex CA through the membrane from left to right, to a region where the concentration and chemical potential of CA are small. Step (3) is the conversion of C to C', which is coupled to the hydrolysis of ATP. The transported molecule A is still attached to C when C is converted to C', but A is released in step (4) as soon as the conversion to C' is complete. Because of the dissociation of C'A, the concentration of CA is kept small on the right side of the membrane, which makes step (2) spontaneous.

After molecule A is released from C', the C' molecules move spontaneously from right to left in step (5), because they are converted in step (6) back to form C at the left side of the membrane by an enzyme located there, keeping the concentration of C' small at the left side of the membrane. The C molecules at the left side of the membrane are now available to complex again with A molecules, and the process can be repeated.

The overall process transports A molecules from a lower to a higher chemical potential because of step (3), which consumes ATP. Although the chemical potential of A increases, the Gibbs energy of the entire system decreases due to the negative Gibbs energy change of hydrolysis of ATP.

Just as the active transport of a substance through a membrane can be driven by hydrolysis of ATP, the regeneration of ATP can also be driven by spontaneous transport of hydrogen ions through a membrane.[5]

Summary

For a reaction at equilibrium at constant temperature and pressure,

$$0 = \left(\frac{\partial G}{\partial \xi}\right)_{T,P} = \sum_{i=1}^{c} \nu_i \mu_i$$

---

[4] *ibid.*, pp. 487ff.

[5] E. D. P. DeRobertis and E. M. F. DeRobertis, Jr., *Cell and Molecular Biology*, Saunders College Publishing Co., Philadelphia, 1980, pp. 267ff.

which leads to the constancy of K, the equilibrium constant:

$$K = \prod_{i=1}^{c} a_{eq}(i)^{\nu_i}$$

where $a_{eq}(i)$ is the equilibrium value of the activity of substance i.

The equilibrium constant is related to the Gibbs energy change of the standard-state reaction:

$$K = e^{-\Delta G°/RT}$$

The equilibrium constant for a reaction involving ideal gases is

$$K_P = \prod_{i=1}^{c} (P_{eq}(i)/P°)^{\nu_i} \quad \text{(gaseous reaction only)}$$

The equilibrium constant for a reaction in solution

$$K = (\gamma_1 x_1)^{\nu_1} \prod_{i=1}^{c} \left[\frac{\gamma_i m_{eq}(i)}{m°} \right]^{\nu_i}$$

where the solvent is designated as substance 1. For dilute solution, the solvent factor can be omitted from the equilibrium expression.

The Gibbs-Helmholtz equation for the temperature dependence of an equilibrium constant is

$$\left(\frac{\partial \ln(K)}{\partial T} \right)_P = \frac{\Delta H°}{RT^2}$$

The principle of Le Châtelier asserts that in general a system will react to lessen the effect of a stress on an intensive variable, if it can do so. This effect was illustrated by considering the shift in equilibrium caused by changing the temperature or the pressure on a system and by adding a reactant or product to the system.

PROBLEMS

Problem for Section 7.1

7.23. A hypothetical gaseous isomerization reaction

$$A \rightleftharpoons B$$

has $\Delta G° = -6.000$ kJ mol^{-1} at 298.15 K. Draw a graph of G as a function of the progress variable ξ for the case that the initial amount of A is 1.000 mol and that of B is 0. The total pressure is kept fixed at 1.000 bar and the temperature is kept fixed at 298.15 K. Locate the minimum in G on the graph and calculate the equilibrium constant from the value of ξ at the minimum.

Problems for Section 7.2

*__7.24.__ Find the equilibrium constant for the reaction of Problem 7.23 and compare it with the value found graphically in Problem 7.23.

7.25. Find the standard-state Gibbs energy change at 298.15 K for each of the reactions:

   a. $2SO_2(g) + O_2(g) \rightarrow 2SO_3(g)$
   b. $2NO(g) + O_2(g) \rightarrow N_2O_4(g)$
   c. $S(\text{cr, rhombic}) + O_2(g) \rightarrow SO_2(g)$

7.26. Using formation values and third-law entropies, find $\Delta H°$, $\Delta G°$, and $\Delta S°$ for each of the following reactions at 298.15 K. Calculate $\Delta H° - T \Delta S°$ and compare it with $\Delta G°$ to check the consistency of the data.

   *__a.__ $2HgO(s) \rightarrow 2Hg(l) + O_2(g)$
   b. $CaCO_3(s) + 2HCl(g) \rightarrow CaCl_2(s) + H_2O(l) + CO_2(g)$
   c. $2Mg(s) + O_2(g) \rightarrow 2MgO(s)$

7.27. Find the equilibrium constant at 298.15 K for each of the reactions of the previous problem.

7.28. a. Find the equilibrium constant at 298.15 K for the reaction

$$PCl_5(g) \rightleftharpoons PCl_3(g) + Cl_2(g)$$

b. Find the total pressure if 0.100 mol of PCl_5 is placed in a vessel with a volume of 2.000 L at 298.15 K and allowed to equilibrate. Assume ideal gas behavior.

*7.29. a. Find the equilibrium constant at 298.15 K for the reaction

$$N_2(g) + 3H_2(g) \rightleftharpoons 2NH_3(g)$$

b. Find the equilibrium composition of a system originally consisting of 1.000 mol of N_2 and 3.000 mol of H_2, maintained at 298.15 K and 1.000 atm. Neglect gas non-ideality.

7.30. a. Find $\Delta G°$ at 298.15 K for the gas-phase reaction

$$2SO_2(g) + O_2(g) \rightleftharpoons 2SO_3(g)$$

b. Find K_P for the reaction at 298.15 K.

c. If a stoichiometric mixture of SO_2 and O_2 is allowed to come to equilibrium at 298.15 K and 1.000 atm, find the partial pressure of SO_2.

7.31. Find $\Delta G°$ and K_P at 298.15 K for each of the gas-phase reactions:

a. $N_2O + O \rightleftharpoons 2NO$
b. $H_2O + \frac{7}{2}H_2S + 3NO_2 \rightleftharpoons 3NH_3 + \frac{7}{2}SO_2$
c. $H_2 + CO_2 \rightleftharpoons H_2O + CO$

7.32. If a reacting gaseous system is held at constant T and P instead of constant T and V, under some circumstances addition of one of the reactants or products can shift the equilibrium to produce more of the added substance.

a. Derive an expression for the derivative $(\partial \ln(K_x)/\partial n_i)_{T,P}$, where n_i is the amount of substance i.

b. Under what circumstances could addition of N_2 to an equilibrium system containing gaseous N_2, H_2, and NH_3 shift the equilibrium to produce more N_2?

Problems for Section 7.4

7.33. Find the partial vapor pressure of HCl over a 0.85 molal aqueous HCl solution at 298.15 K. State any assumptions.

7.34. Find $\Delta H°$, $\Delta S°$, and $\Delta G°$ at 298.15 K for each of the following reactions, using formation data. Compare the value of $\Delta H° - T \Delta S°$ with that of $\Delta G°$ to test the consistency of the data.

*a. $Cl_2(g) + 2I^-(aq) \rightarrow 2Cl^-(aq) + I_2(s)$
b. $2Ag^+(aq) + Ca(s) \rightarrow 2Ag(s) + Ca^{2+}(aq)$
c. $2Fe^{2+}(aq) + Sn^{4+} \rightarrow 2Fe^{3+}(aq) + Sn^{2+}(aq)$

7.35. a. Find the equilibrium constant (solubility product constant) at 298.15 K for the reaction

$$AgCl(s) \rightleftharpoons Ag^+ + Cl^-$$

b. Find the molality of Ag^+ produced by equilibrating solid AgCl with pure water at 298.15 K. Use the Davies equation to approximate activity coefficients.

7.36. The solubility product constant of lithium carbonate is 1.7×10^{-3} at 25°C.

a. Find the solubility in mol kg^{-1} of lithium carbonate in water at this temperature. Use the Davies equation to estimate activity coefficients.

b. Find the solubility in mol kg^{-1} of lithium carbonate in a solution of 0.200 mol of sodium carbonate in 1.000 kg of water at 25°C. Use the Davies equation.

c. Find the solubility in mol kg^{-1} of lithium carbonate in a solution of 0.200 mol of sodium sulfate in 1.000 kg of water at 25°C. Use the Davies equation.

*7.37. The solubility of Ag_2SO_4 in pure water at 25°C is 0.0222 mol kg^{-1}. Find the value of the solubility product constant. Use the Davies equation to estimate activity coefficients.

Problems for Section 7.5

7.38. For each of the following weak acids, find the pH of a solution made from 0.100 mol of the acid and 1.000 kg of water at 298.15 K. Do each calculation twice: once assuming that $\gamma_{\pm}$ equals unity and once using the Davies equation to estimate $\gamma_{\pm}$. In each case, decide whether hydrogen ions from water must be included:

	Acid	K_a
a.	Naphthalene sulfonic	2.7×10^{-1}
b.	Periodic	2.3×10^{-2}
c.	Chloroacetic	1.40×10^{-3}
d.	o-Phenylbenzoic	3.47×10^{-4}
e.	Benzoic	6.46×10^{-5}

7.39. Find the pH at 298.15 K of a solution made from 0.0100 mol of aspartic acid and 1.000 kg of water. Do the calculation twice, once assuming that $\gamma_{\pm}$ equals unity and once using the Davies equation to estimate $\gamma_{\pm}$. For this acid, $K_1 = 1.38 \times 10^{-4}$ and $K_2 = 1.51 \times 10^{-10}$. State any assumptions.

7.40. Calculate the pH of a solution of 0.0100 mol of cacodylic acid in 1.000 kg of water at 298.15 K. The acid dissociation constant is equal to 6.4×10^{-7}. Include the hydrogen ions from water and use the Davies equation to estimate activity coefficient.

7.41. Find the molalities of H_2CO_3 and HCO_3^- in an aqueous solution at 298.15 K that has been equilibrated with gaseous CO_2 at 0.0400 atm and in which the pH is equal to 7.40.

*7.42. How much solid NaOH must be added to 0.100 mol of cacodylic acid in 1.000 kg of water at 298.15 K to make a buffer solution with pH equal to 7.00? See Problem 7.40 for K.

   a. Use the Henderson-Hasselbalch equation.

   b. Use Equation (7.5-3) and the Davies equation.

7.43. Consider an acetic acid–acetate buffer solution made from 0.060 mol of acetic acid and 0.040 mol of sodium acetate in 1.000 kg of water and maintained at 298.15 K.

   a. Find the pH of the buffer, using the Henderson-Hasselbalch equation.

   b. Find the pH of the buffer using the Davies equation to estimate activity coefficients.

   c. Find the change in pH if 0.030 mol of sodium hydroxide is added.

   d. Find the change in pH if 0.030 mol of sodium hydroxide is added to a solution of hydrochloric acid and sodium chloride in 1.000 kg of water if the solution has the same pH and ionic strength as the buffer solution in part a.

7.44. A solution of ammonium benzoate is made from 0.0100 mol of ammonium benzoate and 1.000 kg of water and is maintained at 25°C. The acid ionization constant K_a for benzoic acid is 6.46×10^{-5} at 25°C, and the base dissociation constant K_b for ammonia is 1.774×10^{-5} at 25°C. Aguirre-Ode[6] gives an approximate formula (assuming that all activity coefficients equal unity):

$$\frac{m(H^+)}{m^\circ} = \left(\frac{K_a K_w}{K_b}\right)^{1/2} \left(\frac{K_b + m/m^\circ}{K_a + m/m^\circ}\right)^{1/2}$$

where m is the gross molality. Find the pH using this formula.

Problems for Section 7.6

7.45. Multiply Equation (7.6-9) by $d(1/T)$ and carry out an integration to obtain a formula analogous to Equation (7.6-12). Comment on your result.

*7.46. a. Find the equilibrium constant for the reaction

$$N_2 + 3H_2 \rightleftharpoons 2NH_3$$

at 400 K. State any assumptions.

   b. Find the equilibrium composition of a system originally consisting of 1.000 mol of N_2 and 3.000 mol of H_2, maintained at 400 K and 1.000 atm. Neglect gas nonideality. Compare your answer with that of Problem 7.29.

7.47. Find the total pressure so that at equilibrium at 298.15 K the partial pressure of NO_2 is twice as large as the partial pressure of N_2O_4.

7.48. a. Find $\Delta H°$, $\Delta G°$, and K_P at 298.15 K for the reaction

$$I_2(g) \rightleftharpoons 2I(g)$$

   b. Assuming that $\Delta H° =$ constant, find K at 1000. K from the values at 298.15 K.

   c. Assuming that $\Delta H° =$ constant, find the temperature at which $K = 1.000$.

   d. Assuming that $\Delta C_P° =$ constant, find K at 1000. K.

7.49. The solubility product constant for MgF_2 is 7.1×10^{-9} at 18°C and 6.4×10^{-9} at 27°C. Find the values of $\Delta G°$, $\Delta H°$, and $\Delta S°$ at 25°C.

7.50. Find the value of K_P at 400 K for the reaction

$$2SO_2(g) + O_2(g) \rightleftharpoons 2SO_3(g)$$

State any assumptions.

7.51. a. For a general reaction in solution, derive an equation for $[\partial \ln(K)/\partial P]_T$. Explain the relation of your equation to the principle of Le Châtelier.

   b. Assuming that partial molar volumes are pressure independent, derive an equation for $\ln[K(T, P_2)/K(T, P_1)]$, where P_1 and P_2 are two different pressures.

   c. Typical values of the compressibilities of liquids are near 10^{-9} Pa^{-1}. Estimate the inaccuracy of the result of part b due to the neglect of compressibilities.

   d. For a hypothetical reaction with $\Delta V° = 50.$ ml mol^{-1} and $K = 0.15$ at 298.15 K and 1.000 atm, find the value of the equilibrium constant at 10.00 atm.

Problems for Section 7.7

*7.52. Creatine phosphate is another substance besides phosphoenolpyruvate that can regenerate ATP. If the temperature is 25°C, the pH is 7.00, and the pMg is 4.00, the value of $\Delta G°'$ for its hydrolysis is -43.1 kJ mol^{-1}. Assuming that a mechanism exists to couple the reactions, find the equilibrium constant for the combined reaction to regenerate ATP.

7.53. It is proposed that spontaneous transport of hydrogen ions through a membrane can drive the regeneration of ATP.[7] Assuming the existence of a suitable mechanism, calculate the minimum difference in pH on the two sides of the membrane that would be required to drive this regeneration at 298.15 K.

General Problems

7.54. Identify the following statements as true or false. If a statement requires some special circumstance to make it true, label it as false.

---

[6] F. Aguirre-Ode, *J. Chem. Educ.* **64**, 957 (1987).

[7] E. D. P. DeRobertis and E. M. F. DeRobertis, *op. cit.*

a. Equilibrium constants are true constants.

b. Equilibrium constants depend on temperature but do not depend on pressure.

c. Equilibrium constants for reactions involving only gases depend on temperature but do not depend on pressure.

d. Dilution of a solution of a weak acid at constant temperature increases the degree of ionization of the acid.

e. Dilution of a solution of a weak acid can lower the pH of the solution.

f. According to the principle of Le Châtelier, changing the pressure on the system always causes a gas-phase reaction to shift its equilibrium composition.

g. According to the principle of Le Châtelier, changing the temperature always causes the equilibrium composition of a system to shift.

7.55. The Haber process[8] produces ammonia directly from hydrogen gas and nitrogen gas. Since the reaction proceeds very slowly, a catalyst is used. The catalyst used in industrial manufacture is a mixture of iron oxide and potassium aluminate. The reaction is

$$N_2(g) + 3H_2(g) \rightleftharpoons 2NH_3(g)$$

a. Find $\Delta G°$ for this reaction at 298.15 K.

b. Find K_P for this reaction at 298.15 K.

c. Find K_x for this reaction at 298.15 K and 1.000 bar.

d. Find $\Delta H°$ for this reaction at 298.15 K.

e. Using the principle of Le Châtelier, specify the conditions of temperature and pressure (high or low temperature, high or low pressure) that would increase the yield of ammonia from the process.

f. The process is actually carried out at high pressure (around 500 bar) and fairly high temperature (around 500°C). Comment on this practice. Why might this temperature be used instead of room temperature?

g. Calculate K_P at 500.°C, using values of $\Delta G°$ and $\Delta H°$ at 298.15 K and assuming that $\Delta H°$ is constant.

h. Calculate K_P at 500.°C, using values of

$$-(\bar{G}° - \bar{H}°_{298})/T$$

from Table A9 and interpolating. Compare with your result from part g.

i. If equilibrium is attained at a total pressure of 1.000 bar at 298.15 K, calculate the partial pressure of each substance if a stoichiometric mixture of hydrogen and nitrogen is introduced into the system at the start of the reaction.

j. Find the value of K_x at 298.15 K and 500. bar.

k. Find the value of K_x at 500.°C and 1.000 bar. Find the value of K_x at 500.°C and 500. bar.

l. If equilibrium is attained at a total pressure of 500. bar at 298.15 K, calculate the partial pressure of each substance if a stoichiometric mixture of hydrogen and nitrogen is introduced into the system at the start of the reaction.

m. If equilibrium is attained at a total pressure of 500. bar at 500.°C, calculate the partial pressure of each substance if a stoichiometric mixture of hydrogen and nitrogen is introduced into the system at the start of the reaction.

7.56. The following reaction is known as the "water-gas reaction."

$$H_2(g) + CO_2(g) \rightleftharpoons H_2O(g) + CO(g)$$

The reaction has been used industrially as a source of carbon monoxide, which is used as a reducing agent in obtaining metallic iron from its ore. The system is allowed to approach equilibrium, using an iron catalyst.

a. Find the value of $\Delta G°$ and of K_P for this reaction at 298.15 K and 1000. K.

b. Find $\Delta H°$ for this reaction at 298.15 K and 1000. K.

c. What experimental conditions will favor the maximum yield of carbon monoxide?

*d. Find $\Delta G°$ and K_P at 1259 K, interpolating between values of $-(\bar{G}° - \bar{H}°_{298})/T$ in Table A9. Compare your value of K_P with the accepted value of 1.60.

*e. Find $\Delta G°$ and K_P for this reaction at 1259 K, using the values at 1000. K and assuming that $\Delta H°$ is constant between 1000. K and 1259 K. Compare your result with the result of part d.

7.57. For a sparingly soluble 1-1 electrolyte, the molality at saturation is given by

$$m_{sat}/m° = \sqrt{K_{sp}}$$

where K_{sp} is the solubility product constant, if it can be assumed that activity coefficients are equal to unity. Derive a modified version of this equation, assuming the Debye-Hückel expression for activity coefficients.

8 The Thermodynamics of Electrochemical Systems

PREVIEW

In this chapter, we apply the principles and tools of thermodynamics to electrochemical systems. Electrochemistry is a broad and rather complicated field, but equilibrium electrochemistry can be understood through thermodynamics. Our discussion in this chapter will be limited to the equilibrium thermodynamics of electrochemical cells.

PRINCIPAL FACTS AND IDEAS

1. Thermodynamic relations can be specialized to give useful information about electrochemical systems.
2. The effects of the electric potential must be included in the chemical potentials of substances with charged particles.
3. In an electrochemical cell, a flow of current is accompanied by the progress of a chemical reaction.
4. In an electrochemical cell, the chemical reaction that occurs is physically divided into two half-reactions.
5. Electrochemical experiments can be used to obtain thermodynamic information about chemical reactions.

OBJECTIVES

After studying this chapter, the student should:

1. understand how the thermodynamics of a nonsimple system is applied to electrochemical cells,

2. be able to calculate cell voltages for standard conditions and other conditions using standard reduction potentials and the Nernst equation,

3. be able to solve problems relating equilibrium constants and Gibbs energy changes to electrochemically measured quantities.

8.1 The Chemical Potential and the Electric Potential

Electrical forces on charged objects are described in terms of the **electric field E**, which is a vector quantity defined by the relation

$$\mathbf{E} = \frac{\mathbf{F}}{Q_t} \tag{8.1-1}$$

If the presence of the test charge disturbs the other charges that produce the field, the limit of Equation (8.1-1) is taken as $Q_t \to 0$.

where $\mathbf{F}$ is the electrostatic force on a "test charge" Q_t. If the force is measured in newtons and the charge is measured in coulombs, the units of $\mathbf{E}$ are volts m^{-1}.

Exercise 8.1   Using the definition 1 volt = 1 joule per coulomb, show that the SI units of E are volt m^{-1}.

By inspection of Equation (6.4-1), the electric field around an isolated charge Q_1 in a medium of permittivity ε is

$$\mathbf{E} = \frac{\mathbf{r}}{|\mathbf{r}|} \frac{Q_1}{4\pi\varepsilon|\mathbf{r}|} = \frac{\mathbf{r}}{r} \frac{Q_1}{4\pi\varepsilon r} \tag{8.1-2}$$

where $\mathbf{r}$ is the vector from the charge to the location of interest and either $|\mathbf{r}|$ or r denotes the magnitude of the vector $\mathbf{r}$. The factor $\mathbf{r}/r$, which is also denoted by $\mathbf{e}_r$, is a vector of magnitude unity that points directly away from the charge Q_1. For a collection of charges, the electric field is a vector sum of terms like that of Equation (8.1-2).

The **electric potential** φ is defined by specifying that the electric field is equal to the negative gradient of the electric potential

The gradient of a scalar function is defined in Equation (B-37) of Appendix B.

$$\mathbf{E} = -\nabla\varphi \tag{8.1-3a}$$

For example, the x component of $\mathbf{E}$ is

$$E_x = -\frac{\partial\varphi}{\partial x} \tag{8.1-3b}$$

The electric potential is a potential energy per unit charge, just as the electric field is a force per unit charge.

The difference in electric potential between point 2 and point 1 is equal to the reversible work per coulomb required to move a test charge Q_t from point 1 to point 2, assuming that gravitational and other forces are absent:

$$\varphi(\mathbf{r}_2) - \varphi(\mathbf{r}_1) = \frac{w_{rev(1\to2)}}{Q_t} \tag{8.1-4}$$

The electric potential has the units of volts, which is the same as joules per coulomb. An electric potential difference is called a voltage or an **electromotive force**, abbreviated **e.m.f.** As with any potential energy, Equation (8.1-4) gives only a difference, not an actual value. An arbitrary constant can be added to the electric potential without any physical effect, as with any potential energy. It is customary to choose the value of the electric potential to be zero at a location that is infinitely distant from all charges.

Electric potentials relative to this zero are sometimes called "absolute" potentials.

The electrostatic contribution to the potential energy of a particle of charge Q is

$$\mathscr{V}_{\text{electrostatic}} = Q\varphi \tag{8.1-5}$$

If we include this potential energy in the thermodynamic energy U, it is also included in the Gibbs energy. Therefore, the chemical potential of an ionic species i is given by

$$\mu_i = \mu_{i(\text{chem})} + N_A e z_i \varphi \qquad \text{(8.1-6)}$$

where e is the proton charge, N_A is Avogadro's number, and z_i is the valence of the ion (the number of proton charges on the ion: positive for a cation and negative for an anion). The quantity $\mu_{i(\text{chem})}$, the chemical part of the chemical potential, is assumed to be independent of the electric potential, so that it depends only on temperature, pressure, and composition. If the substance is uncharged or if the electric potential has a zero value, $\mu_{i(\text{chem})}$ is equal to the entire chemical potential.

Unfortunately, $\mu_{i(\text{chem})}$ has sometimes been called the "chemical potential." The chemical potential including the electric potential has then been called the "electrochemical potential." The chemical potential including the electric potential term is the true chemical potential. It is this quantity that obeys the Gibbs-Duhem relation and the fundamental fact of phase equilibrium. We will refer to it as the chemical potential and will call $\mu_{i(\text{chem})}$ the **chemical part of the chemical potential**.

The charge on a mole of protons is denoted by F and called **Faraday's constant**.

Faraday's constant is named for Michael Faraday, 1791–1867, the great English physicist and chemist who discovered the laws of electrolysis and invented the first electric generator.

$$F = N_A e = 96485 \text{ C mol}^{-1} \qquad \text{(8.1-7)}$$

Equation (8.1-6) can be written

$$\mu_i = \mu_{i(\text{chem})} + z_i F \varphi \qquad \text{(8.1-8)}$$

Exercise 8.2   Verify Equation (8.1-7).

Equation (8.1-8) applies to electrons, just as it applies to any other type of charged particle. For an electron, Equation (8.1-8) becomes

$$\mu_e = \mu_{e(\text{chem})} - F \varphi \qquad \text{(8.1-9)}$$

The chemical potential of a single charged species cannot be measured, because charged particles cannot be added without adding counterions at the same time. We use the analogue of Equation (7.4-3) to write the chemical potential of a neutral electrolyte solute:

$$\begin{aligned} \mu_i &= v_+ \mu_+ + v_- \mu_- \\ &= v_+ \mu_{+(\text{chem})} + v_- \mu_{-(\text{chem})} + (v_+ z_+ + v_- z_-) F \varphi \end{aligned} \qquad \text{(8-1-10)}$$

Electrical neutrality implies that $v_+ z_+ + v_- z_- = 0$. The v's are all positive, but the z's are negative for anions.

From the fact that the compound is electrically neutral, the last term in Equation (8.1-10) vanishes:

$$\mu_i = v_+ \mu_{+(\text{chem})} + v_- \mu_{-(\text{chem})} = \mu_{i(\text{chem})} \qquad \text{(8.1-11)}$$

Exercise 8.3   Show that the chemical potential of aqueous $CaCl_2$ has no dependence on the electric potential.

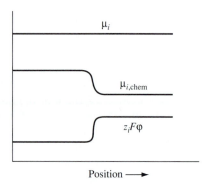

Figure 8.1. The Equilibrium Chemical Potential, the Chemical Part of the Chemical Potential, and the Electric Potential at the Interface Between Two Metallic Phases (schematic). Since the chemical potential must be uniform, if the chemical part of the chemical potential is not uniform, the electric potential is not uniform.

If two different homogeneous conducting phases are placed in close contact and allowed to come to equilibrium, the chemical potential of electrons must have the same value in both phases. However, because the phases are of different materials, the chemical part of the electron's chemical potential can have different values in the two phases. Therefore, the electric potential can have different values in the two phases at equilibrium, as depicted in Figure 8.1.

Within a phase of uniform composition at constant temperature and pressure, the chemical part of the chemical potential must be uniform. In this case, a gradient in the electric potential drives an electron current, and we can measure an electric potential difference by the tendency of electrons to flow. To make a current of electrons flow between two phases of different compositions, there must exist a gradient in the chemical potential of electrons.

8.2 Electrochemical Cells at Equilibrium

An electrochemical cell is a device in which the passage of an electric current is accompanied by the progress of a chemical reaction. There are three principal types of electrochemical cells. In an **electrolytic cell**, a current is passed by an external driving force, causing an otherwise nonspontaneous chemical reaction to proceed. In a **galvanic cell**, the progress of a spontaneous chemical reaction causes the electric current to flow, doing work on the surroundings.

Galvanic cells are named after Luigi Galvani, 1737–1798, the Italian anatomist who showed that electricity caused frog muscles to contract and that dissimilar metals in contact with the muscle tissue could produce an electric current.

An **equilibrium electrochemical cell** is at the state between an electrolytic cell and a galvanic cell. The tendency of a spontaneous reaction to push a current through the eternal circuit must be balanced by an external source of e.m.f. that exactly cancels this tendency. If this **counter e.m.f.** is increased beyond the equilibrium value the cell becomes an electrolytic cell, and if it is decreased below the equilibrium value the cell becomes a galvanic cell.

Electrochemical cells always contain several phases. There are two or more phases such as metals or graphite that can conduct a current of electrons. These conducting phases are called **electrodes** and are placed in contact with an electrolyte solution.

Under certain conditions, solvated electrons can occur at low concentrations in solutions,[1] but we will consider uncombined electrons to be insoluble in electrolyte solutions. The zero-valent metal of a typical electrode is also insoluble in liquid solutions, and the components of a liquid solution are insoluble in the metal of the electrode. We cannot use the fundamental fact of phase equilibrium, because no substance occurs in both phases. However, if the material of the electrode can oxidize to form an ion that occurs in the electrolyte solution or can accept electrons from a chemical reaction in the solution, a chemical reaction can come to equilibrium at the phase boundary, and this fact will provide an equilibrium condition on the chemical potentials.

The fundamental fact of phase equilibrium can be applied only to substances occurring in both phases being considered.

Figure 8.2 schematically depicts a particular electrochemical cell. A figure such as Figure 8.2 is called a **cell diagram**, showing the phases that occur and how they are connected. This cell is one of a class called **cells without liquid junction**, which means that it contains a single solution, so there is no junction between two different liquid solutions.

The electrode at the left in Figure 8.2 is called a **hydrogen electrode**. It is a piece of platinum that has been "platinized" (plated with porous platinum) to increase its surface area. Hydrogen gas at a specified pressure is bubbled through the solution around the electrode and is adsorbed on the platinum, where it can undergo the oxidation process

$$H_2(g) \rightarrow 2H^+ + 2e^- \qquad \textbf{(8.2-1)}$$

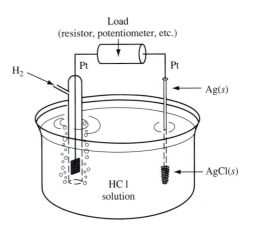

Figure 8.2. An Electrochemical Cell. An electrochemical cell always consists of several phases, and at least one phase is an ionic solution. This cell also has a gas phase, and the adsorbed gas reacts at the left electrode surface.

---

[1] L. Kevan and B. Webster (eds.), *Electron-Solvent and Anion-Solvent Interactions*, Elsevier, New York, 1976.

The solution in the cell is an aqueous solution of hydrochloric acid. As in this equation, we will label substances in gaseous phases by (g) and those in solid phases by (s), but will not usually label substances in aqueous solution. The process of Equation (8.2-1) is called a **half-reaction**, since it cannot take place without another process to accept the electrons produced. This half-reaction is an oxidation half-reaction. The electrode for which an oxidation half-reaction is written is called the **anode**.

The electrode at the right in the cell of Figure 8.2 is called the **silver–silver chloride electrode**. It is a piece of silver that is coated with solid silver chloride. The reduction half-reaction

$$AgCl(s) + e^- \rightarrow Ag(s) + Cl^- \qquad \textbf{(8.2-2)}$$

You can remember the names "anode" and "cathode" by noting that "anode" and "oxidation" both begin with vowels and that "cathode" and "reduction" both begin with consonants.

can occur at this electrode. A piece of platinum (the **terminal**) is attached to the silver of the right electrode so that an electric potential difference can be measured between two terminals of the same material. The electrode for which a reduction half-reaction is written is called the **cathode**.

We multiply the reduction half-reaction equation by 2 and add the two half-reaction equations to obtain the **cell reaction equation**:

$$H_2(g) + 2AgCl(s) + 2e^-(R) \rightarrow 2H^+ + 2Ag(s) + 2Cl^- + 2e^-(L) \quad \textbf{(8.2-3)}$$

A cell reaction is generally a chemical reaction that could take place outside the cell as well as in the cell.

We have labeled the electrons at the right (R) and left (L) electrodes. In an electrochemical cell the half-reactions take place in different locations, so the electrons must be transported from one electrode to the other through an external circuit if the reaction occurs.

We can rewrite the cell reaction equation using only electrons and uncharged substances:

$$H_2(g) + 2AgCl(s) + 2e^-(R) \rightarrow 2HCl + 2Ag(s) + 2e^-(L) \quad \textbf{(8.2-4)}$$

Equations (8.2-3) and (8.2-4) are equivalent, but the second equation makes it easier to write the chemical potentials because the electric potential terms of neutral electrolytes will cancel, as in Equation (8.1-11).

Several conventions have been adopted to make the description of electrochemical cells systematic. The first convention is: *The left electrode in a cell diagram is assigned to be the anode, and the right electrode is assigned*

One way to remember this convention is to note that "reduction" and "right" both begin with the letter "r."

to be the cathode. This choice for the right and left electrodes dictates the direction in which we write the cell reaction equation. It can happen that the spontaneous direction of the reaction is opposite to the way we write it.

A **cell symbol** communicates the same information as the cell diagram. In this symbol the phases of the cell are listed, beginning with the terminal of the left electrode and proceeding through the cell to the terminal of the right electrode. The symbol for each phase is separated from the next by a vertical line.

The cell symbol for our present cell is

$$Pt|H_2(g)|HCl(aq)|AgCl(s)|Ag|Pt$$

Since the platinum wire at the right terminal is present only to provide a phase at which to measure the potential, it is sometimes omitted from the cell symbol.

Sometimes the concentration or the activity of the electrolyte solute involved in the reaction is included in the symbol. For example, if the molality of the HCl in solution is 0.500 mol kg^{-1} and the pressure of the hydrogen gas is 0.990 atm, the cell symbol can be written

$$Pt|H_2(0.990 \text{ atm})|HCl(m = 0.500 \text{ mol kg}^{-1})|AgCl(s)|Ag|Pt$$

A similar specification of composition is used for the electrode that is a solid solution or an amalgam (solution in liquid mercury).

Let us leave our galvanic cell on "open circuit" (with the two terminals not connected to a circuit). We allow the cell to stabilize at constant temperature and pressure, with a fixed partial pressure of hydrogen at the anode. The state that is reached is metastable, because bringing the reactants to the same location or short-circuiting the terminals would permit a reaction to occur. Some authors call such a state a "partial equilibrium." We treat this metastable state of our system as though it were an equilibrium state.

We now connect the cell to an external circuit in order to measure the electric potential difference between its terminals. We want to make this measurement without disturbing the equilibrium state of the cell. One way to do this is with a **potentiometer**, a device in which an adjustable counter e.m.f. opposes the electric potential difference of the cell. The counter e.m.f. is adjusted until it is just sufficient to stop the flow of electrons in the external circuit, as indicated by a galvanometer. The state of the cell is now the same as if the cell were an open circuit, and the counter e.m.f. is equal in magnitude to the potential difference between the electrodes. The value of this potential difference is called the cell's **reversible potential difference** or its **reversible voltage**.

To an excellent approximation, the cell reaction is now thermodynamically reversible. If the counter e.m.f. is made slightly smaller than its equilibrium value, the cell functions as a galvanic cell and a current flows while the reaction proceeds in the spontaneous direction. If the counter e.m.f. is made slightly larger, the cell functions as an electrolytic cell and a current flows in the opposite direction.

From Equation (8.1-10), an infinitesimal amount of reaction $d\xi$ in our system at equilibrium corresponds to

$$0 = \left(\frac{\partial G}{\partial \xi}\right)_{T,P} d\xi = [2\mu(\text{HCl}) + 2\mu(\text{Ag}) - \mu(\text{H}_2) - 2\mu(\text{AgCl})$$
$$+ 2\mu(\text{e}^-(\text{L})) - 2\mu(\text{e}^-(\text{R}))] \, d\xi \qquad \textbf{(8.2-5)}$$

which can be written

$$0 = \left(\frac{\partial G_{\text{chem}}}{\partial \xi}\right)_{T,P} d\xi + \left(\frac{\partial G_{\text{electron}}}{\partial \xi}\right)_{T,P} d\xi \qquad \textbf{(8.2-6)}$$

where G_{chem} includes the chemical potentials of substances other than electrons. All of these substances are neutral, so only the chemical parts of these chemical potentials are included. If the reaction took place outside an electrochemical cell, $(\partial G_{\text{chem}}/\partial \xi)_{T,P}$ would be the rate of change of Gibbs energy for the reaction.

Since all other substances were written as uncharged species, the electric potential occurs only in the chemical potential expression of the electrons. Both terminals are made of platinum and are at the same temperature and pressure, so the chemical parts of the chemical potential of the electrons cancel, and we can write from Equation (8.1-6):

$$0 = \left(\frac{\partial G_{\text{chem}}}{\partial \xi}\right)_{T,P} + 2FE \qquad (8.2\text{-}7)$$

where E is the difference in electric potential between the right and left terminals and is called the potential difference of the cell. We introduce our second convention: *The potential difference of a cell is defined as the electric potential of the terminal of the right electrode minus that of the left electrode*:

$$\boxed{E = \Delta\varphi = \varphi(\text{R}) - \varphi(\text{L})} \qquad (8.2\text{-}8)$$

Both terminals must be made of the same material.

Since electrons are negatively charged, a positive value of E means that the chemical potential of the electron is larger in the left terminal, and electrons move spontaneously from the left terminal to the right terminal if a wire is connected between the terminals. In this case, oxidation occurs at the left electrode and the cell reaction proceeds spontaneously in the direction in which we wrote it. If the cell potential difference is negative, the reverse of the cell reaction proceeds spontaneously.

We now use Equation (6.3-1) to write the chemical potential of each substance other than electrons in the form:

$$\mu_i = \mu_i^\circ + RT \ln(a_i) \qquad (8.2\text{-}9)$$

We can now write

$$0 = 2FE = -\Delta G^\circ - RT \ln(Q) \qquad (8.2\text{-}10)$$

where

$$\Delta G^\circ = 2\mu^\circ(\text{HCl}) + 2\mu^\circ(\text{Ag}) + \mu^\circ(\text{H}_2) + 2\mu^\circ(\text{AgCl}) \qquad (8.2\text{-}11)$$

and Q is the activity quotient:

$$Q = [a(\text{HCl})]^2 [a(\text{Ag})]^2 [a(\text{H}_2)]^{-1} [a(\text{AgCl})]^{-2}$$
$$= \frac{[a(\text{HCl})]^2 [a(\text{Ag})]^2}{[a(\text{H}_2)][a(\text{AgCl})]^2} \qquad (8.2\text{-}12)$$

This quantity is the same as the activity quotient of Chapter 7. The only difference is that in Chapter 7 we applied it to the case of equilibrium outside a cell, and Q_{eq} was equal to the equilibrium constant. Now it can take on other values.

The **standard-state potential difference** E° for this cell is defined:

$$\boxed{E^\circ = -\frac{\Delta G^\circ}{2F}} \qquad (8.2\text{-}13)$$

It is the potential difference (voltage) that would be measured if all substances were in their standard states (with unit activities). Equation (8.2-10) is now

$$E = E° - \frac{RT}{2F} \ln(Q) \tag{8.2-14}$$

The Nernst equation is named for Hermann Walther Nernst, the German physical chemist who was mentioned in Chapter 3 for his work on the third law of thermodynamics.

Equation (8.2-14) is the **Nernst equation** for this cell.

Since the Ag and AgCl are both pure solids, their activities are equal to or nearly equal to unity and can be omitted from the product Q. Treating hydrogen as an ideal gas and using Equation (7.4-6) for the activity of HCl in terms of molalities,

$$E = E° - \frac{RT}{2F} \ln\left[\left(\frac{\gamma(H^+)^2 m(H^+)^2 \gamma(Cl^-)^2 m(Cl^-)^2}{m°^4}\right)\left(\frac{P(H_2)}{P°}\right)^{-1}\right] \tag{8.2-15}$$

$$= E° - \frac{RT}{2F} \ln\left[\frac{\gamma_\pm^4 m^4 / m°^4}{P(H_2)/P°}\right] \tag{8.2-16}$$

where we assume that $m(H^+)$ and $m(Cl^-)$ are equal and denote them by m.

▼ **EXAMPLE 8.1**

Find the potential difference of our cell at 298.15 K if the hydrogen pressure is 744 torr and the HCl is at 0.500 mol kg^{-1} with mean ionic activity coefficient equal to 0.757. The standard-state potential difference is 0.2223 V.

Solution

$$E = 0.2223 \text{ V}$$
$$- \left[\frac{(8.3145 \text{ J K}^{-1} \text{ mol}^{-1})(298.15 \text{ K})}{(2)(96485 \text{ C mol}^{-1})}\right] \ln\left[\frac{(0.757)^4(0.500)^4}{(744 \text{ torr})/(750 \text{ torr})}\right]$$
$$= 0.2223 \text{ V} - (-0.0498 \text{ V}) = 0.2721 \text{ V}$$

▲

So far, we have discussed a particular cell. Now consider a general cell without a liquid junction and with a cell reaction that can be written with neutral substances and electrons:

$$0 = \sum_{i=1}^{c} v_i \mathscr{F}_i + ne^-(R) - ne^-(L) \tag{8.2-17}$$

where n is the number of electrons in the reaction equation.

The Nernst equation is

$$E = E° - \frac{RT}{nF} \ln(Q) \tag{8.2-18}$$

where

The activity quotient is exactly the same as if the reaction took place outside a cell.

$$Q = \prod_{i=1}^{c} a_i^{v_i} \tag{8.2-19}$$

The activity quotient Q does not include the activity of the electron.

Exercise 8.4

a. Write the cell symbol, the cell reaction equation, and the Nernst equation for the cell with the half-reactions

$$2Hg(l) + 2Cl^- \rightarrow Hg_2Cl_2(s) + 2e^-$$

$$Cl_2(g) + 2e^- \rightarrow 2Cl^-$$

b. Find the potential difference of the cell at 298.15 K if $P(Cl_2) = 0.950$ atm and $a(Cl^-) = 0.500$. $E° = 1.091$ V.

If the cell reaction equation is modified by multiplying all of the stoichiometric coefficients by the same constant, say C, the Nernst equation is unchanged, because Q will be raised to the power C while the n factor in the denominator in front of $\ln(Q)$ will be increased by the same factor C, canceling the effect of the exponent C.

Exercise 8.5

Multiply the cell reaction equation in Equation (8.2-4) by 1/2. Write the Nernst equation for the new reaction equation and show that it is the same as Equation (8.2-16).

Because the standard states of solutes are hypothetical states, the standard-state potential difference of any cell cannot be measured directly. If the activity coefficients are known, Q can be calculated, and $E°$ can be calculated using Equation (8.2-18). If the activity coefficients are not known, an extrapolation to zero concentration can be used, because activity coefficients approach unity in this limit.

For example, consider the cell of Figure 8.2. We keep the pressure of the hydrogen equal to $P°$ and measure E at various molalities. For small enough values of m, $\gamma_\pm$ will be given by the Debye-Hückel formula. Let us use the augmented version in Equation (6.4-14). For HCl, the ionic strength is equal to the molality m if no other electrolytes are present. If Equation (6.4-14) is substituted into Equation (8.2-16) and if we assume that $\beta a \approx 1$,

$$E = E° + \frac{2RT}{F}\left[\frac{\alpha m^{1/2}}{1 + (m/m°)^{1/2}} + bm + \ln\left(\frac{m}{m°}\right) \right]$$

which can be rewritten

$$E + \frac{2RT}{F}\left[\frac{\alpha m^{1/2}}{1 + (m/m°)^{1/2}} + \ln\left(\frac{m}{m°}\right) \right] = E° + \frac{2RT}{F}bm \quad \textbf{(8.2-20)}$$

The left-hand side of Equation (8.2-20) contains only measurable quantities. If this function is plotted as a function of m, the plot should be linear in the region in which Equation (8.2-20) is valid (the region near $m = 0$), so extrapolation to the $m = 0$ axis is straightforward. The intercept is equal to $E°$, and the slope of the tangent line near the axis is equal to $2RTb/F$.

An alternative method can also be used. For small values of m, not only is the $m/m°^{1/2}$ term negligible compared to unity but also the bm term is negligible compared to the $\alpha m^{1/2}$ term. If these terms are omitted,

Equation (8.2-20) becomes

$$E + \frac{2RT}{F} \ln\left(\frac{m}{m^\circ}\right) = E^\circ - \frac{2RT}{F} \alpha m^{1/2} \qquad \textbf{(8.2-21)}$$

If the left-hand side of this equation is plotted as a function of $m^{1/2}$, the plot should be nearly linear in the region near $m = 0$ for small values of $m^{1/2}$, allowing accurate extrapolation to obtain the value of E° from the intercept.

8.3 Half-Cell Potentials and Cell Potentials

Consider the cell of Figure 8.3. This cell has a hydrogen electrode on the left side and a calomel electrode on the right side. The calomel electrode contains liquid mercury with solid calomel [mercury(I) chloride, Hg_2Cl_2] in contact with it. A platinum wire extends from the pool of mercury and acts as a terminal. The solution in this cell is an aqueous solution of hydrochloric acid, just as in the cell of Figure 8.2.

The reduction half-reaction of the calomel electrode is

$$Hg_2Cl_2(s) + 2e^- \rightarrow 2Hg(l) + 2Cl^- \qquad \textbf{(8.3-1)}$$

and the oxidation half-reaction of the hydrogen electrode is the same as in Equation (8.2-1). It is found that $E^\circ = 0.268$ V for this cell.

Exercise 8.6

  a. Write the cell reaction equation for the cell of Figure 8.3.
  b. Write the Nernst equation for the cell of Figure 8.3.
  c. Write an equation analogous to Equation (8.2-20) that could be used to determine the value of E° for the cell of Figure 8.3.

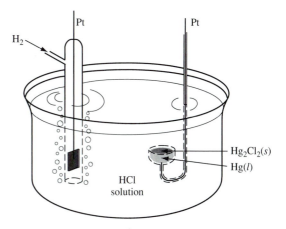

Figure 8.3. An Electrochemical Cell with a Hydrogen Electrode and a Calomel Electrode.
In this cell, the half-reaction at the calomel electrode involves substances in three phases: the liquid Hg phase, the solid Hg_2Cl_2 phase, and the solution.

Consider the cell that is obtained by interchanging the right and left half-cells of Figure 8.2. The cell reaction equation is reversed:

$$2HCl + 2Ag(s) + 2e^-(R) \rightarrow H_2(g) + 2AgCl(s) + 2e^-(L) \quad \textbf{(8.3-2)}$$

The standard-state potential difference of this cell is the negative of that of the cell of Figure 8.2, or -0.2223 volt.

Figure 8.4 shows a cell containing the silver–silver chloride electrode on the left and the calomel electrode on the right. It is possible to calculate the value of $E°$ for this cell without making a measurement, as follows:

We construct a double cell, as depicted in Figure 8.5. It consists of two complete cells, one of which is the "reversed" version of the cell in Figure 8.2

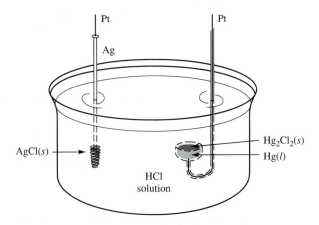

Figure 8.4. An Electrochemical Cell with a Silver–Silver Chloride Anode and a Calomel Cathode. In this cell, both half-reactions involve three phases.

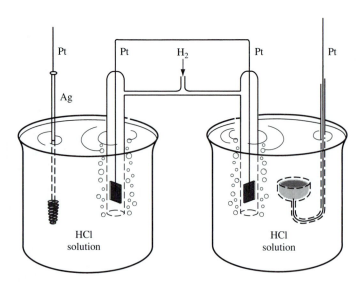

Figure 8.5. A Double Cell. This cell is used to illustrate the fact that potentials of cells with one hydrogen electrode can be used as half-cell potentials.

and one of which is the cell of Figure 8.3. Hydrogen gas at the same pressure is fed into both hydrogen electrodes and the two HCl solutions are at the same concentration. A wire is connected between the two hydrogen electrodes and maintains them at the same electric potential. We can write

$$E(\text{double cell}) = \varphi_R(\text{right cell}) - \varphi_L(\text{left cell})$$
$$= \varphi_R(\text{right cell}) - \varphi_L(\text{right cell}) + \varphi_R(\text{left cell}) - \varphi_L(\text{left cell})$$

The second equality follows from the fact that the two hydrogen electrodes are in the same state. The second equality is the same as

$$E(\text{double cell}) = E(\text{right cell}) + E(\text{left cell})$$

Because this relation is true for an arbitrary state, it is true for the standard state:

$$E^\circ(\text{double cell}) = E^\circ(\text{right cell}) + E^\circ(\text{left cell})$$
$$= 0.268 \text{ V} + (-0.2223 \text{ V}) = 0.046 \text{ V} \qquad \textbf{(8.3-3)}$$

The voltage of the cell in Figure 8.4 is also 0.046 V, because the state of the left electrode of the left cell in Figure 8.5 is not different from the state of the left electrode of the cell in Figure 8.2 and the state of the right electrode of the right cell in Figure 8.5 is not different from the state of the right electrode of the cell in Figure 8.4.

We adopt the following convention: *The standard-state potential difference of a cell consisting of a hydrogen electrode on the left and any other electrode on the right is called the* **standard reduction potential** *of the right electrode, or of the right half-cell. It is also sometimes called the* **standard half-cell potential** *or* **electrode potential**.

The standard-state potential difference of the cell of Figure 8.4 can now be written

$$E^\circ = E^\circ(\text{right half-cell}) - E^\circ(\text{left half-cell}) \qquad \textbf{(8.3-4)}$$
$$= 0.268 \text{ V} - 0.222 \text{ V} = 0.046 \text{ V} \qquad \textbf{(8.3-5)}$$

The negative sign in Equation (8.3-4) comes from the fact that the left cell in Figure 8.5 has its hydrogen electrode on the right, so its standard-state potential difference is the negative of that conventionally assigned to its left electrode.

The procedure that led to Equation (8.3-4) for the cell of Figure 8.4 can be applied to any pair of electrodes that can be combined to make a galvanic cell: *To obtain the standard-state potential difference of any cell, subtract the standard reduction potential of the left half-cell from the standard reduction potential of the right half-cell.* Using this procedure we can make a fairly short table of standard reduction potentials and use it to calculate potential differences for a large number of cells.

| Exercise 8.7 | Show that from a table of N half-cell potentials, the potential differences for $N(N-1)/2$ cells can be calculated if each half-cell can be combined with every other half-cell to make a cell. |

The convention that assigns standard reduction potentials relative to the standard hydrogen electrode is arbitrary. Efforts have been made to determine theoretically the "absolute" potential of electrodes (relative to the potential at a point infinitely distant from all charges). All of these approaches require assumptions and the use of nonthermodynamic theories. One work cites a value of -4.43 V (absolute) for the standard hydrogen electrode.[2] Others have come up with values ranging from this value to -4.73 V. We will use only half-cell potentials relative to the standard hydrogen electrode.

If you are not certain whether an old table gives reduction potentials or oxidation potentials, look for an active metal like sodium or potassium. If the table gives reduction potentials, the half-cell potential of such a metal will be negative.

Table A14 gives values for standard reduction potentials (in the molality description) for a number of half-cells. Longer versions of such tables are available in handbooks. Unfortunately, some older works use the convention opposite to the presently accepted one and give values that are the negative of the reduction potentials (these are called oxidation potentials).

▼   **EXAMPLE 8.2**

Write the cell reaction equation and find the standard-state potential difference of the cell

$$Pt|Cl_2|FeCl_2(aq)|Fe|Pt$$

Solution

The cell reaction equation is (canceling the electrons)

$$2Cl^- + Fe^{2+} \rightarrow Cl_2(g) + Fe(s)$$

The standard-state potential difference of the cell is

$$E° = -0.409 \text{ V} - (+1.3583 \text{ V}) = -1.767 \text{ V}$$

A positive value of E corresponds to a reaction that is spontaneous as written.

The negative sign means that the standard-state cell reaction would proceed spontaneously in the reverse direction.

▲

Cells with Liquid Junctions

Some pairs of half-cells cannot be combined into a cell with a single liquid solution. That is, a single solution in contact with both electrodes cannot be made to contain all of the ionic species. Either a precipitate would form or one of the dissolved species would react directly with an electrode or another dissolved species.

For example, consider the cell with the half-reactions:

$$Zn(s) \rightarrow Zn^{2+} + 2e^-$$
$$Cu^{2+} + 2e^- \rightarrow Cu(s)$$

If the Cu^{2+} ion were contained in the solution in contact with the zinc electrode, the oxidation and reduction half-reactions would take place at the interface between the zinc electrode and the solution without transferring electrons through an external circuit.

---

[2] H. Reiss and A. Heller, *J. Phys. Chem.* **89**, 4207 (1985).

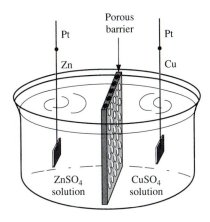

Figure 8.6. The Daniell Cell. This is not a reversible cell because it contains a liquid junction between two different electrolyte solutions.

However, a galvanic cell can be constructed to use these half-reactions, as shown in Figure 8.6. Cells such as this are known as **Daniell cells** and were once used to power telephones and railroad signals. The cell container has two compartments. The left compartment contains a zinc electrode and a solution of zinc sulfate, and the right compartment contains a copper electrode and a solution of copper(II) sulfate. The barrier between the compartments is porous, so that ions can diffuse through the liquid filling its pores but the two solutions cannot mix by flowing together.

The cell has also been constructed without a porous barrier. Concentrated $CuSO_4$ is placed in the bottom of the cell container and a less dense dilute $ZnSO_4$ solution is layered above it. The cell is not allowed to stand on open circuit. A small current is always allowed to flow, so that Cu^{2+} ions are moving downward toward the copper electrode in the bottom of the container. This motion overcomes the tendency of the Cu^{2+} ions to diffuse upward, and the solutions do not mix appreciably.

The liquid interface between the two solutions is called a **liquid junction**. A cell with a liquid junction is also called a **cell with transference**. In a cell symbol, a liquid junction is sometimes represented by a vertical broken line, but it can be represented by an unbroken vertical line like that of any other phase boundary.

The presence of the liquid junction makes it impossible to have a persistent metastable state of the cell, because ions can diffuse through the liquid junction even if the cell is on open circuit. A cell with a liquid junction is not reversible because reversing the current does not reverse these diffusion processes. The irreversible processes make a contribution to the potential difference of the cell, called the **liquid junction potential**.

▼   **EXAMPLE 8.3**

a. Write the cell reaction equation and the cell symbol for the Daniell cell.

b. Write the Nernst equation for the Daniell cell.

c. Find the standard-state potential difference of the Daniell cell, neglecting the liquid junction potential. In which direction will the cell reaction proceed spontaneously under standard conditions?

Solution

a. Canceling the electrons,

$$Zn(s) + Cu^{2+} \rightarrow Zn^{2+} + Cu(s)$$
$$Zn|Zn^{2+}|Cu^{2+}|Cu$$

(We have omitted the platinum terminal wire symbols.)

b.

$$E = E° + E_{LJ} - \frac{RT}{2F} \ln\left[\frac{a(Zn^{2+})}{a(Cu^{2+})}\right]$$

where E_{LJ} represents the liquid junction potential.

c. Neglecting E_{LJ},

$$E° = 0.3402 \text{ V} - (-0.7628 \text{ V}) = 1.1030 \text{ V}$$

The reaction would proceed spontaneously as written if the terminals were connected by a short circuit.

▲

Exercise 8.8 Find the potential difference of a Daniell cell if the activity of zinc ions is 0.250 and the activity of copper ions is 0.550 (both on the molality scale). Neglect the liquid junction potential.

Because the liquid junction is not an equilibrium or metastable system, it cannot be studied thermodynamically. However, nonthermodynamic analyses have been carried out.[3] The resulting equations contain activity coefficients and transference numbers, whose values are usually not accurately known. The **transference number** of an ion is the fraction of the current that is carried by that type of ion. Ions with the largest concentrations generally have the largest transference numbers and can dominate in determining the liquid junction potential.

A common procedure for minimizing the liquid junction potential is to use a **salt bridge**, as shown schematically in Figure 8.7. The salt bridge contains a third electrolyte that forms a liquid junction with each cell solution. This electrolyte solution is usually very concentrated, and dominates the liquid junction potential. The electrolyte most commonly used is saturated KCl, chosen because the potassium ions and chloride ions have nearly equal transference numbers at equal concentrations. The KCl is often suspended in an agar gel for ease in handling. For solutions that are incompatible with chloride ions, such as those containing silver or lead ions, ammonium nitrate can be used.

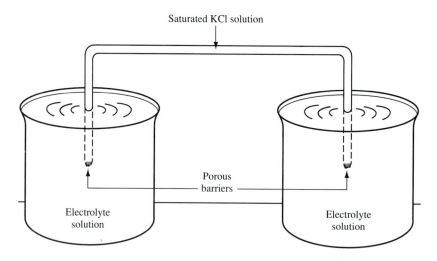

Figure 8.7. A Salt Bridge. This device is designed to minimize the liquid junction potential between two electrolyte solutions by creating two liquid junctions whose liquid junction potentials nearly cancel.

[3] J. G. Kirkwood and I. Oppenheim, *Chemical Thermodynamics*, McGraw-Hill, New York, 1961, pp. 211ff.

In a cell symbol, a salt bridge is represented by a double vertical line. The cell symbol for the Daniell cell with a salt bridge would be

$$Zn|Zn^{2+}||Cu^{2+}|Cu$$

For a saturated KCl salt bridge, the transference numbers of the potassium and chloride ions are far larger than those of the ions in the cell solutions, because of the large concentrations of the potassium and chloride ions. Two ions of opposite charge make opposing contributions to the liquid junction potential when diffusing in the same direction, and if their transference numbers are equal their contributions cancel. The liquid junction potentials at the two ends of the salt bridge are therefore fairly small. They are opposite in sign and nearly cancel. Calculations indicate that the net liquid junction potential with a salt bridge is generally no larger in magnitude than a few millivolts and can be negligible in some cells.

Another approach to handling the liquid junction potential is to design cells that avoid it. For example, Figure 8.8 schematically depicts a **concentration cell**. In a concentration cell, there are two solutions of the same electrolyte, but with different concentrations.

The cell symbol of the cell of Figure 8.8 can be written

$$Pt|H_2(P_1)|HCl(m_1)||HCl(m_2)|H_2(P_2)|Pt$$

where P_1 and P_2 represent two pressures of the hydrogen, and m_1 and m_2 represent two molalities of the HCl solutions.

If the hydrogen is at the same pressure in both sides of the cell, the reaction for this concentration cell is

$$HCl(m_2) \rightarrow HCl(m_1) \tag{8.3-6}$$

The standard states for the half-cells are the same, so $E°$ vanishes for any concentration cell. The Nernst equation for our concentration cell is,

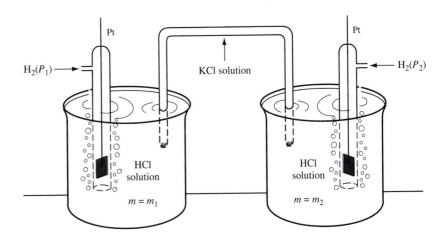

Figure 8.8. A Concentration Cell. The two parts of this cell are identical except for the concentrations of the electrolyte solutions.

assuming zero liquid junction potential and complete ionization,

$$E = -\frac{RT}{F} \ln\left[\frac{a_1(\text{HCl})}{a_2(\text{HCl})}\right]$$

$$= -\frac{RT}{F} \ln\left[\frac{(\gamma_{\pm 1} m_1/m^\circ)^2}{(\gamma_{\pm 2} m_2/m^\circ)^2}\right] = -\frac{2RT}{F} \ln\left[\frac{(\gamma_{\pm 1} m_1/m^\circ)}{(\gamma_{\pm 2} m_2/m^\circ)}\right] \quad \textbf{(8.3-7)}$$

where the extra subscript on $\gamma_\pm$ indicates the cell in which that value applies. If the activity coefficient of HCl at one molality is known, this equation can be used to determine the activity coefficient at the other molality if the liquid junction potential can be evaluated.

In order to eliminate the liquid junction potential, the cell of Figure 8.8 can be replaced by the double cell of Figure 8.9. This cell is reversible. Assuming that the pressure of hydrogen is equal to P° in both cells, the Nernst equation for the left-hand cell is

$$E(\text{left cell}) = 0.2223 \text{ V} - \frac{2RT}{F} \ln(\gamma_{\pm 1} m_1/m^\circ)$$

and that of the right-hand cell is

$$E(\text{right cell}) = -0.2223 \text{ V} + \frac{2RT}{F} \ln(\gamma_{\pm 2} m_2/m^\circ)$$

so that the potential difference of the double cell is equal to that of Equation (8.3-7) without the liquid junction potential.

Another possibility is to build just one cell, like half of the double cell, and then to measure its potential difference once when filled with a solution of a given molality and once when filled with a solution of another molality.

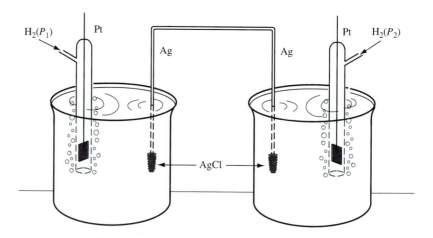

Figure 8.9. A Double Cell Equivalent to a Concentration Cell. This is used to replace the concentration cell with a cell having no liquid junctions.

Exercise 8.9	Find the potential difference of the concentration cell of Figure 8.8 if in solution 1 the molality is 0.500 mol kg^{-1} with mean ionic activity coefficient equal to 0.757, and in solution 2 the molality is 0.200 mol kg^{-1} with mean ionic activity coefficient equal to 0.767.

8.4 The Determination of Activity Coefficients of Electrolytes

In the case of a volatile electrolyte such as HCl, the mean ionic activity coefficient can be determined for large concentrations from the partial vapor pressure, using Equation (7.4-11).

EXAMPLE 8.4

Using information in Example 7.6, find the value of $\gamma_{\pm}$ for HCl at 10.00 mol kg^{-1} and 298.15 K, in which state the partial vapor pressure is equal to 4.20 torr.[4]

Solution

$$\gamma_{\pm}^2 = \frac{P_{HCl}}{k_{\pm}^{(m)} m^2}$$

$$= \frac{(4.20 \text{ torr})\left(\dfrac{1 \text{ atm}}{760 \text{ torr}}\right)}{(4.92 \times 10^{-7} \text{ atm kg}^2 \text{ mol}^{-2})(10.00 \text{ mol kg}^{-1})^2} = 112$$

$$\gamma_{\pm} = 10.6$$

For a two-component solution with a volatile solvent, values of the vapor pressure of the solvent can be used to obtain values of the activity coefficient of the solvent. If this is done for several values of the solvent mole fraction between unity and the composition of interest, integration of the Gibbs-Duhem relation can given the value of the activity coefficient of the solute.

The vapor pressure of the solvent is not usually measured directly but is determined indirectly using the **isopiestic method**. In this procedure, the solution of interest and a solution of a well-studied reference solute are placed in the same closed container at a fixed temperature, as schematically shown in Figure 8.10. For aqueous solutions, KCl is usually used as the reference solute because accurate water activity coefficient data are available for KCl solutions. The solutions are left undisturbed at constant temperature until enough solvent has evaporated from one solution and condensed in the other solution to equilibrate the solvent in the two solutions. The

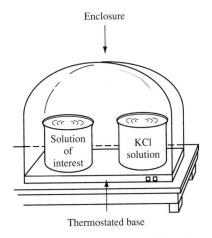

Figure 8.10. Apparatus for the Isopiestic Method of Determining Activity Coefficients. This system is allowed to equilibrate so that the chemical potential of the volatile solvent has the same value in both solutions. The activity coefficient of the solute is determined by a Gibbs-Duhem integration.

[4] S. J. Bates and H. D. Kirschman, *op. cit.*

solutions are then analyzed to determine the mole fractions of the solvent in both solutions.

At equilibrium, the activity of solvent (substance 1) in the solution of interest (phase II) is equal to the activity of solvent in the reference solution (phase I). Therefore

Because the chemical potential of water has the same value in both phases, the activity has the same value.

$$\gamma_{1(II)} = \frac{a_{1(I)}}{x_{1(II)}} = \frac{\gamma_{1(I)} x_{1(I)}}{x_{1(II)}} \tag{8.4-1}$$

The activity of the solvent is often expressed in terms of the **osmotic coefficient** ϕ, defined by

$$\phi = -\frac{\ln(a_1)}{M_1 v m_2} = \frac{\mu_1^\circ - \mu_1}{RTM_1 v m_2} \quad \text{(definition)} \tag{8.4-2}$$

where a_1 is the activity of the solvent, M_1 is the molar mass of the solvent, v is the sum $v_+ + v_-$ for the solute, and m_2 is the gross molality of the solute. If the solute dissociates completely, $v m_2$ is equal to the sum of the molalities of the ions.

From Equation (8.4-2)

$$\mu_1 = \mu_1^\circ - RTM_1 v m_2 \phi \tag{8.4-3}$$

The chemical potential of the solute can be written

$$\mu_2 = \mu_2^\circ + vRT \ln(v_\pm \gamma_2 m_2 / m^\circ) \tag{8.4-4}$$

For constant pressure and temperature, the Gibbs-Duhem relation for a two-component system is given by Equation (4.6-11)

$$n_1 \, d\mu_1 + n_2 \, d\mu_2 = 0 \tag{8.4-5}$$

Since the molality m_2 is equal to the amount of substance 2 divided by the mass of substance 1 measured in kilograms

$$n_2 = m_2 n_1 M_1 \tag{8.4-6}$$

Use of Equations (8.4-3), (8.4-4) and (8.4-6) in Equation (8.4-5) gives

$$-n_1 vRTM_1[m_2 \, d\phi + \phi \, dm_2] + m_2 n_1 M_1 vRT[d \ln(\gamma_2) + d \ln(m_2)] = 0$$

Cancellation of the common factor and use of the identity

$$d \ln(m) = (1/m) \, dm$$

gives

$$-m_2 \, d\phi - \phi \, dm_2 + m_2 \, d \ln(\gamma_2) + dm_2 = 0 \tag{8.4-7}$$

which is the same as

$$d \ln(\gamma_2) = d\phi + \frac{\phi - 1}{m_2} dm_2 \tag{8.4-8}$$

Equation (8.4-8) can be integrated from $m_2 = 0$ to $m_2 = m_2'$, a particular value of m_2:

$$\int_{m_2=0}^{m_2=m_2'} d \ln(\gamma_2) = \int_{m_2=0}^{m_2=m_2'} d\phi + \int_0^{m_2'} \frac{\phi - 1}{m_2} dm_2 \tag{8.4-9}$$

The integral on the left-hand side yields zero at its lower limit, since the activity coefficient approaches unity as m_2 approaches 0. It can be shown that ϕ approaches unity as m_2 approaches zero, so that

$$\ln[\gamma(m_2')] = \phi(m_2') - 1 + \int_0^{m_2'} \frac{\phi - 1}{m_2}\, dm_2 \qquad \text{(8.4-10)}$$

If values of ϕ are measured over the range of molalities between 0 and m_2', numerical integration of this equation gives the value of γ_2 at m_2'.

Exercise 8.10	Using the relation $x_1 = 1 - x_2 = 1 - n_2/n_1$ at high dilution, show that ϕ approaches 1 as m_2 approaches 0.

Activity coefficients of electrolyte solutes can be determined from electrochemical cell measurements. For example, consider the cell of Figure 8.2, for which the Nernst equation is

$$E = E^\circ - \frac{RT}{2F} \ln\left[\frac{a(HCl)^2}{P(H_2)/P^\circ}\right] \qquad \text{(8.4-11)}$$

where $E^\circ = 0.2223$ V. Let us maintain the pressure of hydrogen gas at P° and measure the voltage at a molality of HCl equal to m:

$$\ln[a(HCl)] = 2\ln(\gamma_\pm m/m^\circ) = \frac{F(E^\circ - E)}{RT} \qquad \text{(8.4-12)}$$

EXAMPLE 8.5

In a cell such as that of Figure 8.2, a voltage of 0.3524 V was measured with a solution having a molality of HCl equal to 0.1000 mol kg^{-1} and a hydrogen pressure equal to 1.000 bar. Find the value of the activity and of the mean ionic activity coefficient of HCl, assuming hydrogen is an ideal gas.

Solution

From Equation (8.4-12)

$$\ln(\gamma_\pm m/m^\circ) = \frac{(96485\ \text{C mol}^{-1})(0.2223\ \text{V} - 0.3524\ \text{V})}{2(8.3145\ \text{J K}^{-1}\ \text{mol}^{-1})(298.15\ \text{K})}$$

$$= -2.532$$

$$\gamma_\pm = \frac{e^{-2.532}}{0.1000} = 0.795$$

For comparison, the Davies equation gives $\gamma_\pm = 0.781$.

Determination of pH

Because the activity of a single ion cannot be measured, we cannot correctly use the definition of the pH given in Equation (8.4-1). Instead we define the pH by

$$pH = -\log_{10}[a'(H^+)] \qquad \text{(8.4-13)}$$

Just writing this definition does not give us a definition of a' or a way to measure it.

where $a'(H^+)$ is the closest approximation to the activity of hydrogen ions that can be obtained.

In Example 8.5, Equation (8.4-12) was used to calculate the activity of HCl. We can write

$$a(H^+)a(Cl^-) = a(HCl) \tag{8.4-14}$$

If we can assume that $a(H^+)$ is approximately equal to $a(Cl^-)$, then we have an approximate value of $a(H^+)$, and thus of the pH.

Exercise 8.11

Find the pH of the solution in the cell of Example 8.5, using the assumption stated above.

In order to measure the pH conveniently, it is customary to modify the cell of Figure 8.2 as shown in Figure 8.11. The calomel electrode is fitted with a porous plug in which a liquid junction is formed. The solution in the container of the calomel electrode is a KCl solution saturated with calomel. If the concentration of KCl is 1.0000 mol L^{-1}, the electrode is called the **normal calomel electrode**. If the KCl solution is saturated, the electrode is called the **saturated calomel electrode**. The half-cell potential of the normal calomel electrode is 0.2802 V, and that of the saturated calomel electrode is 0.2415 V.

We place the calomel electrode and the hydrogen electrode in the solution whose pH we wish to measure, as shown in the figure. The solution does not have to be HCl, since the chloride ions needed to react at the calomel electrode are in the solution of the calomel electrode.

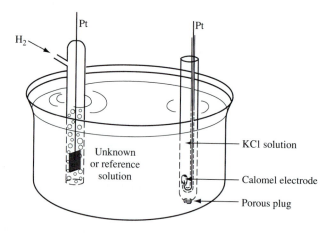

Figure 8.11. A Cell for the Measurement of pH. The potential of this cell is a well-defined function of the pH of the solution except for the liquid junction potential.

The Nernst equation for the cell of Figure 8.11 is

$$E = 0.2415 \text{ V} + E_{LJ} - \frac{RT}{F} \ln \left[\frac{a(H^+)a(Cl^-)}{(P(H_2)/P^\circ)^{1/2}} \right] \qquad \textbf{(8.4-15)}$$

where $a(H^+)$ is measured in one solution and $a(Cl^-)$ is measured in the other. Since the KCl solution on the calomel electrode side of the liquid junction is presumably much more concentrated than the solution on the other side, the magnitude of the liquid junction potential should be fairly small and should not change much when a different solution is placed in the cell.

The activity of the chloride ion is not known accurately, since it is a single-ion activity, so we do not try to use Equation (8.4-15) directly. Instead, we first put into the cell a reference solution (solution I) with a pH value that we regard as reliable and measure the voltage of the cell at a known hydrogen pressure. Then we remove this solution and put a solution whose pH we wish to determine (solution II) into the cell, and measure the voltage with the same hydrogen pressure as before. We assume that the liquid junction potential and the activity of the chloride ion are the same in both cases and write

$$E^{(II)} - E^{(I)} = -\frac{RT}{F} \ln[a(H^+, II)] + \frac{RT}{F} \ln[a(H^+, I)] \qquad \textbf{(8.4-16)}$$

where $E^{(I)}$ and $E^{(II)}$ are the two voltages measured.

Equation (8.4-16) is the same as

$$\boxed{pH^{(II)} - pH^{(I)} = \frac{F}{RT \ln(10)} (E^{(II)} - E^{(I)})} \qquad \textbf{(8.4-17)}$$

If a reference solution of known pH is not available, we cannot use this procedure.

We can regard Equation (8.4-17) as an operational definition of pH, equivalent to Equation (8.4-13). The pH of solution I must be known in advance.

Exercise 8.12   Show that Equation (8.4-17) is correct.

The **glass electrode** has come into common use to replace the hydrogen electrode, which is somewhat dangerous. This electrode is schematically depicted in Figure 8.12. It consists of a silver–silver chloride electrode inside a thin-walled glass bulb filled with a buffered solution of nearly constant pH. To measure the pH of an unknown solution, a glass electrode and a calomel electrode are immersed in the solution. The special glass of which the bulb is made allows hydrogen ions to establish an electrical potential difference across the glass in order to equilibrate the hydrogen ions inside the membrane with those outside the membrane.

Although the voltage of the cell depends on the nature of the glass membrane and on the hydrogen ion activity inside the bulb, the dependence of the voltage on the pH of the unknown solution is the same as in the

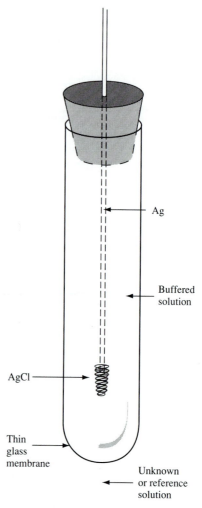

Figure 8.12. A Glass Electrode (schematic). This electrode is used to replace a hydrogen electrode in a pH measuring device.

The first commonly available pH meter was made possible in the 1930s when Arnold Beckman, a chemistry professor at the California Institute of Technology, invented an amplifier that allowed the cell to operate the voltage-measuring device and display. Professor Beckman left Caltech and founded a company that sold the pH meters and the famous Beckman DU spectrophotometer, which used the same amplifier.

cell of Figure 8.11, so Equation (8.4-17) can be used. pH meters, which are found in almost every chemistry laboratory, consist of the pair of electrodes described above, a voltage-measuring device, an analog or digital display, and a circuit that gives the pH directly without requiring the operator to substitute numbers into Equation (8.4-17). Since the temperature occurs in Equation (8.4-17), most pH meters have a control knob with which one can set the temperature.

Exercise 8.13   Calculate the difference between the cell voltages that occur for pH readings of 7.00 and 13.50 at 298.15 K.

Thermodynamic Information from Electrochemistry

Thermodynamic information about many chemical reactions that occur outside electrochemical cells can be obtained from electrochemical measurements. For a general reaction written in the form of Equation (8.2-17), the analogue of Equation (8.2-7) is

$$-nFE = \left(\frac{\partial G_{\text{chem}}}{\partial \xi}\right)_{T,P} \tag{8.5-1}$$

where n is the number of electrons in the cell reaction equation.

For the standard-state reaction,

$$-nFE^\circ = \Delta G^\circ_{\text{chem}} \tag{8.5-2}$$

The Gibbs energy change in our electrochemical equations is for the reaction outside the cell, and this value is what we need.

The equilibrium constant for the reaction can be calculated from Equation (7.1-20):

$$K = Q_{\text{eq}} = e^{-\Delta G^\circ_{\text{chem}}/RT} = e^{nFE^\circ/RT} \tag{8.5-3}$$

▼   **EXAMPLE 8.6**

a. Find the value of K for the reaction of Equation (8.2-4) at 298.15 K.

b. Find the equilibrium value of the hydrogen pressure when the molality of HCl is 0.500 mol kg^{-1} and its mean ionic activity coefficient is 0.757.

Solution

a.

$$K = \exp\left[\frac{2(96485 \text{ C mol}^{-1})(0.2223 \text{ V})}{(8.3145 \text{ J K}^{-1} \text{ mol}^{-1})(298.15 \text{ K})}\right]$$

$$= e^{17.30} = 3.28 \times 10^7$$

b.

$$3.28 \times 10^7 = \frac{\gamma_{\pm}^4 m_{\text{eq}}^4/m^{\circ 4}}{P_{\text{eq}}(\text{H}_2)/P^\circ}$$

$$P(\text{H}_2)_{\text{eq}} = P^\circ\left[\frac{(0.757)^4(0.500)^4}{3.28 \times 10^7}\right] = (6.26 \times 10^{-10})P^\circ$$

$$= 6.26 \times 10^{-10} \text{ bar} = 4.69 \times 10^{-7} \text{ torr}$$

▲

Exercise 8.14   Find the equilibrium constant for the reaction of Exercise 8.4 at 298.15 K.

From Equation (6.6-3), we can write an expression for the entropy change of a reaction outside an electrochemical cell (we now omit the subscript "chem"):

$$\left(\frac{\partial S}{\partial \xi}\right)_{T,P} = -\left[\frac{\partial}{\partial T}\left(\frac{\partial G}{\partial \xi}\right)_{T,P}\right]_P = nF\left(\frac{\partial E}{\partial T}\right)_P \tag{8.5-4}$$

For the standard-state reaction, this equation becomes

$$\Delta S^\circ = -\left(\frac{\partial \Delta G^\circ}{\partial T}\right)_P = nF\left(\frac{\partial E^\circ}{\partial T}\right)_P \tag{8.5-5}$$

The enthalpy change of a reaction is given by

$$\left(\frac{\partial H}{\partial \xi}\right)_{T,P} = \left(\frac{\partial G}{\partial \xi}\right)_{T,P} + T\left(\frac{\partial S}{\partial \xi}\right)_{T,P} = -nFE + nFT\left(\frac{\partial E}{\partial T}\right)_P \tag{8.5-6}$$

For the standard-state reaction,

$$\Delta H^\circ = -nFE^\circ + nFT\left(\frac{\partial E^\circ}{\partial T}\right)_P \tag{8.5-7}$$

From Equation (7.6-13), in the case that ΔH° can be assumed temperature independent, we can also write

$$\frac{E^\circ(T_2)}{T_2} - \frac{E^\circ(T_1)}{T_1} = -\frac{\Delta H^\circ}{nF}\left(\frac{1}{T_2} - \frac{1}{T_1}\right) \tag{8.5-8}$$

▼

EXAMPLE 8.7

Assuming that ΔH° is constant, find E° at 323.15 K for the Daniell cell. Neglect the liquid junction potential.

Solution

We have already found that $E^\circ = 1.103$ V at 298.15 K. We find from enthalpy changes of formation in Table A9 that $\Delta H^\circ = -218.66$ kJ mol^{-1}. From Equation (8.5-8)

$$\frac{E^\circ(323.15 \text{ K})}{323.15 \text{ K}} = \frac{1.103 \text{ V}}{298.15 \text{ K}} - \frac{-218660 \text{ J mol}^{-1}}{(2)(96485 \text{ C mol}^{-1})}\left(\frac{1}{323.15 \text{ K}} - \frac{1}{298.15 \text{ K}}\right)$$

$$= 3.699 \times 10^{-3} \text{ V K}^{-1} - 2.94 \times 10^{-4} \text{ V K}^{-1}$$

$$= 3.405 \times 10^{-3} \text{ V K}^{-1}$$

$$E(323.15 \text{ K}) = (3.405 \times 10^{-3} \text{ V K}^{-1})(323.15 \text{ K}) = 1.100 \text{ V}$$

▲

A reaction that is not an oxidation-reduction reaction can often be written as a sum of an oxidation half-reaction and a reduction half-reaction.

A chemical reaction does not have to be an oxidation-reduction reaction for us to apply Equations (8.5-2), (8.5-3), (8.5-5), and (8.5-8) to the reaction. It is necessary only to be able to write the reaction as the sum of two half-cell reactions.

▼

EXAMPLE 8.8

Find the solubility product constant of AgI at 298.15 K from electrochemical data.

Solution

The reaction equation is

$$AgI(s) \rightarrow Ag^+ + I^-$$

This reaction equation can be written as the sum of the half-reaction equations:

$$AgI(s) + e^- \rightarrow Ag(s) + I^- \qquad E^\circ = -0.1519 \text{ V}$$
$$Ag(s) \rightarrow Ag^+ + e^- \qquad E^\circ = -0.7986 \text{ V}$$

The standard-state voltage of the cell with the given cell reaction is -0.9505 V, and the solubility product constant is, from Equation (8.5-3),

$$K_{sp} = \exp\left[\frac{(1)(96485 \text{ C mol}^{-1})(-0.9505 \text{ V})}{(8.3145 \text{ J K}^{-1} \text{ mol}^{-1})(298.15 \text{ K})}\right]$$
$$= e^{-36.99} = 8.6 \times 10^{-17}$$

Summary

In this chapter, we have discussed the thermodynamics of electrochemical cells. An electrochemical cell can function as an electrolytic cell, in which an externally imposed voltage produces a chemical reaction, or as a galvanic cell, in which a spontaneous chemical reaction produces a current in an external circuit. An equilibrium electrochemical cell is at the state between these two conditions.

The chemical potential of a charged species was separated into two contributions:

$$\mu_i = \mu_{i(\text{chem})} + z_i F \varphi$$

where z_i is the valence of the charged species i, F is Faraday's constant, and φ is the electric potential.

The Nernst equation is

$$E = E^\circ - \frac{RT}{nF} \ln(Q)$$

where Q is the activity quotient for the reaction and E° is the reversible standard-state cell voltage.

Standard-state half-cell reduction potentials can be used to obtain the standard-state voltage for any cell that can be made from two half-cells in the table, using the relation

$$E^\circ = E^\circ(\text{right}) - E^\circ(\text{left})$$

where $E^\circ(\text{right})$ and $E^\circ(\text{left})$ are the half-cell potentials for the two electrodes in their standard states.

Thermodynamic functions can be determined electrochemically, using the relations

$$\Delta G^\circ_{\text{chem}} = -nFE^\circ$$
$$K = e^{-\Delta G^\circ_{\text{chem}}/RT} = e^{nFE^\circ/RT}$$

where $\Delta G^\circ_{\text{chem}}$ refers to the reaction outside the cell.

Additional Reading

A. W. Adamson, *Physical Chemistry of Surfaces*, 4th ed., Wiley, New York, 1982

This is one of the relatively few books devoted entirely to surface chemistry. Several chapters apply to the surfaces of electrodes.

A. J. Bard and L. R. Faulkner, *Electrochemical Methods—Fundamentals and Applications*, Wiley, New York, 1980

This book is designed for a course in analytical chemistry at the senior or first-year graduate level. Electrochemistry is practiced primarily by analytical chemists, and one must look in textbooks of analytical chemistry like this for many of the fundamentals of electrochemistry.

J. O'M. Bockris and A. K. Reddy, *Modern Electrochemistry*, Plenum, New York, 1970

The senior author of this book is one of the most prominent practitioners of electrochemistry.

W. C. Gardiner, Jr., *Rates and Mechanisms of Chemical Reactions*, W. A. Benjamin, New York, 1969

This is intended for an undergraduate course in chemical kinetics at the level of a general physical chemistry course. It includes a discussion of the rates of electrode processes.

J. Goodisman, *Electrochemistry: Theoretical Foundations*, Wiley, New York, 1987

A modern presentation of equilibrium and nonequilibrium electrochemistry from a physical chemical point of view and a very useful source.

H. A. Laitinen and W. E. Harris, *Chemical Analysis—An Advanced Text and Reference*, 2nd ed., McGraw-Hill, New York, 1975

This book is designed for a course in analytical chemistry at the senior or first-year graduate level and includes a discussion of electrochemistry and electrochemical analysis.

PROBLEMS

Problem for Section 8.1

*8.15. Assume that the electric field in a region near a phase boundary in an electrochemical cell is equal to

$$1.0 \times 10^7 \text{ V m}^{-1}$$

Find the magnitude of the concentration gradient of a univalent ion that is necessary to make the total chemical potential uniform if the concentration of the ion is 0.100 mol L^{-1} and the temperature is 298.15 K.

Problems for Section 8.2

8.16. a. Find the reversible cell voltage for the cell of Figure 8.3 at 298.15 K if the partial pressure of hydrogen is 712 torr and the molality of the HCl solution is 0.100 mol kg^{-1}. Assume that the hydrogen gas is ideal and use the Davies equation to estimate the activity coefficient of the HCl.

b. Repeat the calculation of part a using the value of the activity coefficient from Table A12.

8.17. a. Using the extrapolation of Equation (8.2-20), find $E°$ for the cell of Figure 8.2. The following are (contrived) data for the cell voltage at 298.15 K with $P(H_2) = P°$:

m (mol kg^{-1})	E (volt)	m (mol kg^{-1})	E (volt)
0.100	0.3523	0.600	0.2624
0.200	0.3186	0.700	0.2539
0.300	0.2985	0.800	0.2456
0.400	0.2838	0.900	0.2395
0.500	0.2722	1.000	0.2332

b. Using the extrapolation of Equation (8.2-21), find $E°$ for the cell of Figure 8.3.

Problems for Section 8.3

*8.18. a. Write the cell symbol for the cell with the half-reactions

$$Pb^{2+} + 2e^- \rightarrow Pb(s)$$
$$Ag(s) + Cl^- \rightarrow AgCl(s) + e^-$$

Why is a salt bridge needed? What substance would you not use in the salt bridge?

b. Draw a sketch of the cell of part a.

c. Find the value of E for the cell if the activity of Pb^{2+} is 0.100 and that of Cl^- is 0.200 on the molality scale. State any assumptions.

8.19. Instead of the oxidation half-reaction used in Example 8.1, use the reverse of the half-reaction

$$2H_2O + 2e^- \rightarrow H_2 + 2OH^- \qquad E° = -0.8277 \text{ V}$$

to calculate the reversible cell potential. Since the OH^- ion is not in its standard state, the Nernst equation must be applied.

8.20. The lead storage battery used in automobiles has an anode made of lead, a cathode containing lead oxide, PbO_2, and an electrolyte solution of sulfuric acid. Solid lead sulfate, $PbSO_4$, is formed.

a. Write the cell reaction of the lead storage battery.

b. Find $E°$ for a six-cell lead storage battery at 25°C.

c. Find E for a six-cell lead storage battery at 25°C if each cell contains aqueous sulfuric acid with 24.4% sulfuric acid by weight and density 1.2023 g cm^{-3}. The mean ionic activity coefficient of sulfuric acid under these conditions is 0.150.

d. The voltage regulator in most automobiles is set to charge the battery at about 13.5 V. Calculate the activity of sulfuric acid necessary to make the reversible voltage equal to 13.5 V.

8.21. Find the reversible cell voltage at 298.15 K of a concentration cell in which both half-cells have calomel electrodes and HCl solutions. In one solution the molality is 2.50 mol kg^{-1} and in the other solution the molality is 0.100 mol kg^{-1}. Carry out the calculation once assuming that activity coefficients are equal to unity and once using activity coefficient values from the graph of Figure 6.12.

Problems for Section 8.4

8.22. Following are data on the activity of water in calcium chloride solution at 25°C.[5]

$m(CaCl_2)(\text{mol kg}^{-1})$	$a(H_2O)$
0.1	0.99540
0.2	0.99073
0.3	0.98590
0.4	0.98086
0.5	0.97552
0.6	0.96998
0.7	0.96423
0.8	0.95818
0.9	0.95174
1.0	0.94504

Find the activity coefficient of calcium chloride at 1.000 mol kg^{-1}, using a Gibbs-Duhem integration.

*8.23. If a solution of NaOH with a molality of 0.100 mol kg^{-1} is placed in a cell with a standard hydrogen electrode and a normal calomel electrode at 298.15 K, find the cell voltage and the pH of the solution. Neglect the liquid junction potential. State any other assumptions.

Problems for Section 8.5

8.24. a. Using half-cell potentials, find the equilibrium constant for the reaction at 298.15 K.

$$I_2(aq) + I^- \rightleftharpoons I_3^-$$

b. Find the final concentrations if 100. mL of a 0.00100 mol kg^{-1} solution of I_2 and 100. mL of a 0.00100 molal solution of KI are mixed and allowed to equilibrate. Use the Davies equation to estimate activity coefficients.

8.25. Find the solubility product constants for

*a. Hg_2Cl_2 ***c.** $PbSO_4$
b. Ag_2S **d.** $AgBr$

8.26. a. Find the equilibrium constant at 298.15 K for the reaction

$$2Ce^{4+} + 2Cl^- \rightleftharpoons 2Ce^{3+} + Cl_2(g)$$

b. Find the equilibrium composition if a 0.0500 mol kg^{-1} solution of CeF_4 is added to an equal volume of a 0.0500 mol kg^{-1} solution of NaCl at 298.15 K. Assume that the chlorine gas which is formed is maintained at pressure $P°$ and use the Davies equation to estimate activity coefficients.

8.27. Find the equilibrium constant at 298.15 K for the reaction

$$I_2(s) + 2Cl^- \rightleftharpoons 2I^- + Cl_2(g)$$

8.28. Fuel cells have been used in space vehicles to provide electrical energy. A fuel cell is a galvanic cell into which reactants are brought continuously, so that no recharging is necessary. A hydrogen-oxygen fuel cell can have either an acidic or a basic electrolyte solution.

---

[5] Robinson and Stokes, *op. cit.*, p. 478.

a. Write the half-reaction equations and the cell equation for a hydrogen-oxygen fuel cell with a KOH electrolyte.

b. Calculate the value of $E°$ for the fuel cell of part a and write its Nernst equation. To what circumstance does the standard state correspond?

c. Write the half-cell reaction equations and the cell reaction for a hydrogen fuel cell with an acidic electrolyte. Find its $E°$.

d. Find the reversible cell voltage of a hydrogen-oxygen fuel cell at 298.15 K if the hydrogen and oxygen are both at 150. torr (roughly the total pressure maintained in a manned space vehicle with a pure oxygen atmosphere) and if the electrolyte solution is an HCl solution at 2.50 mol kg^{-1}.

General Problems

*8.29. **a.** A hydrogen-oxygen fuel cell is operating reversibly at constant $P = P°$ and with unit activity of all substances. Using data from Table A9, calculate the maximum amount of work that could be done if 2.000 mol of hydrogen gas and 1.000 mol of oxygen gas are consumed at 298.15 K to form liquid water.

b. Using the Carnot efficiency formula, Equation (3.1-22), find the maximum work that can be done with a heat engine operating reversibly between 1000.°C and 100.°C if 2.000 mol of hydrogen gas and 1.000 mol of oxygen gas are reacted. Assume that the water is produced as vapor, that the enthalpy change is approximately equal to that at 298.15 K,

and that all of the enthalpy change is transferred to the engine. Compare your answer with the result of part a.

c. Calculate the minimum amounts of wasted energy in the form of heat for parts a and b of this problem.

8.30. A fuel cell with an acidic electrolyte using methane as a fuel has the anode half-reaction equation:

$$CH_4(g) + 2H_2O(l) \rightarrow CO_2(g) + 8H^+ + 8e^-$$

a. Write the cathode half-reaction equation and the cell reaction equation for the methane fuel cell. Calculate $\Delta G°$ and $E°$ for the cell reaction.

b. Fuel cells have a thermodynamic advantage over energy delivery systems in which fuel is burned in an internal combustion engine or a turbine to drive an electric generator. Calculate the maximum amount of work that can be obtained from the combustion of 100.0 kg of methane in a turbine or other heat-driven engine with an upper temperature of 2000. K and an exhaust temperature of 1200. K.

c. Calculate the maximum energy in kilowatt-hours that can be obtained from the work in part b, assuming one has an electric generator that is 100% efficient.

d. Calculate the maximum energy in kilowatt-hours that can be obtained from a fuel cell that consumes 100.0 kg of methane. Assume that all products and reactants are in their standard states. Comment on the advisability of using fuel cells for municipal electric power generation. What disadvantages might such use have?

9 The Principles of Quantum Mechanics: I. The Schrödinger Equation

PREVIEW

In this chapter, we introduce quantum mechanics, which is a general theory for the mechanical states of any system. Quantum mechanics has superseded classical (Newtonian) mechanics as the accepted description of these states.

PRINCIPAL FACTS AND IDEAS

1. Classical mechanics ascribes exact positions and trajectories of particles.
2. The old quantum theory contained quantization in hypotheses.
3. The "matter waves" of de Broglie led to quantum mechanics.
4. The Schrödinger equation is a wave equation that governs the states of systems according to quantum mechanics.
5. Quantum mechanics contains the concept of wave-particle duality: objects can exhibit wave-like properties as well as particle-like properties.
6. The time-independent Schrödinger equation can be solved for some example systems and produces quantization as a natural part of the solution.

OBJECTIVES

After studying this chapter, the student should:

1. understand the solution of classical equations of motion,
2. understand the classical wave equation and its solutions and be able to solve problems involving classical wave phenomena,
3. understand the way in which quantization is introduced in the old quantum theory and be able to solve problems related to this theory,
4. understand the relation of the Schrödinger equation to the classical wave equation and understand the boundary conditions imposed on its solutions,
5. understand and be able to use the method of separation of variables to solve a class of differential equations,
6. be able to solve problems related to the Schrödinger equation.

9.1 Classical Mechanics

Classical physics consists of theories that existed prior to 1900. It includes thermodynamics, classical mechanics, and the electrodynamics of Maxwell. Classical mechanics is based on the laws of motion of Newton and is also called **Newtonian mechanics**. Appendix D presents a survey of classical mechanics.

The Classical Mechanics of the Harmonic Oscillator

The harmonic oscillator is a model system that represents a mass suspended by a spring from a stationary object, as shown in Figure 9.1. Let the coordinate z be the distance from the mass's equilibrium position to its actual position, taken as positive if the mass is above its equilibrium position and negative if it is below its equilibrium position. We assume that the mass moves only in the z direction, so the coordinates x and y do not enter into the description of the motion of the mass.

The force on the mass suspended by a spring is well described for fairly small values of z by **Hooke's law**.

$$F_z = -kz \qquad (9.1\text{-}1)$$

Named for Robert Hooke, 1635–1703, one of Newton's contemporaries and rivals.

where k is a positive constant called the **force constant**. The negative sign in Equation (9.1-1) indicates that the force is always toward the equilibrium position.

The harmonic oscillator is a model system that is defined to obey Hooke's law exactly for all values of the coordinate. We will assert that a real mass on a spring behaves very much like a harmonic oscillator as long as the magnitude of z is not too large. We will also use this model to represent molecular vibrations and vibrations in a crystal.

Once again, we have a model system that resembles a real system but is simpler so that mathematical analysis is possible.

An **equation of motion** is a differential equation that can be solved to give z as a function of t. From Newton's second law, Equation (D-1), the equation of motion of the harmonic oscillator is

$$-kz = m\frac{d^2z}{dt^2} \qquad (9.1\text{-}2)$$

The **general solution** of such an equation is a family of functions that includes nearly every solution of the equation. A general solution of Equa-

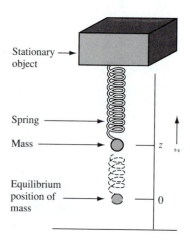

Figure 9.1. A System Represented by a Harmonic Oscillator. This system is a mass on a spring.

tion (9.1-2) can be written

$$z(t) = A \sin\left(\sqrt{\frac{k}{m}}\, t\right) + B \cos\left(\sqrt{\frac{k}{m}}\, t\right) \qquad (9.1\text{-}3)$$

where A and B are arbitrary constants.

Exercise 9.1	Verify by substitution that the function of Equation (9.1-3) satisfies Equation (9.1-2).

A velocity component is the time derivative of the corresponding co-ordinate, so Equation (9.1-3) also delivers the velocity, which has only a z component:

$$v_z(t) = \frac{dz}{dt} = \sqrt{\frac{k}{m}}\left[A \cos\left(\sqrt{\frac{k}{m}}\, t\right) - B \sin\left(\sqrt{\frac{k}{m}}\, t\right) \right] \qquad (9.1\text{-}4)$$

To make the general solution apply to a specific case, we apply **initial conditions**, which are the values of the position and velocity at an initial time for the case of interest. Consider the case that at time $t = 0$

$$z(0) = C, \qquad v_z(0) = 0 \qquad (9.1\text{-}5)$$

where C is a known constant.

Since $\sin(0)$ equals zero and $\cos(0)$ equals unity, our velocity expression can vanish for $t = 0$ only if $A = 0$ and $B = C$. Our solution now contains no arbitrary constants:

$$z(t) = C \cos\left(\sqrt{\frac{k}{m}}\, t\right) \qquad (9.1\text{-}6)$$

$$v_z(t) = -\sqrt{\frac{k}{m}}\, C \sin\left(\sqrt{\frac{k}{m}}\, t\right) \qquad (9.1\text{-}7)$$

Initial conditions are different from boundary conditions in that they apply only at one time.

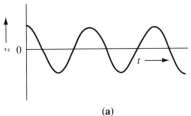

(a)

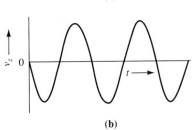

(b)

Figure 9.2. The Behavior of a Harmonic Oscillator. (a) The position as a function of time according to classical mechanics. **(b)** The velocity as a function of time according to classical mechanics. This motion is called uniform harmonic motion. The position and velocity both vary sinusoidally.

Figure 9.2a shows the position as a function of time, and Figure 9.2b shows the velocity as a function of time. The motion the classical harmonic oscillator undergoes is called **simple harmonic motion**. It is a **periodic motion**, repeating the same pattern over and over. The oscillator is moving at the greatest speed when it is near the origin and with zero speed when it is farthest from the origin, when $|z| = C$. The constant C is called the **amplitude of the oscillation**.

The **probability density** for finding the particle is defined so that if z' is a particular value of z

$$\begin{array}{l} \text{Probability of finding the particle} \\ \text{between } z' \text{ and } z' + dz \text{ at time } t \end{array} = f(z', t)\, dz \qquad (9.1\text{-}8)$$

The probability density is a probability per unit length on the z axis. It depends on time because the mass is moving. Since our solution gives a definite value of the position as a function of time, the probability density f at a given time is a sharply spiked function, as shown in Figure 9.3a. The spike moves as governed by Equation (9.1-6).

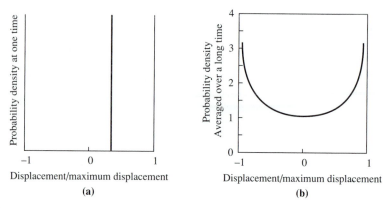

Figure 9.3. The Probability Density for the Position of a Harmonic Oscillator According to Classical Mechanics. (a) Instantaneous probability density. At a given instant, the particle can be located exactly. **(b)** Probability density averaged over a long time. The probability density rises at the ends of the classically allowed region, due to the slowness of the particle's motion near the ends of the region.

The probability density averaged over a long time can be obtained from the solution and is shown in Figure 9.3b. The probability density is greatest in the regions where the mass is moving most slowly and smallest in the regions where the mass is moving most rapidly, as expected.

The length of time required for the oscillator to go from a certain state to the next repetition of that state is called the **period** of the oscillator and denoted by τ. It is the length of time required for the argument of the sine function in Equation (9.1-6) or the cosine function in Equation (9.1-7) to change by 2π:

$$\sqrt{\frac{k}{m}}\,\tau = 2\pi$$

or

$$\tau = 2\pi\sqrt{\frac{m}{k}} \tag{9.1-9}$$

The **frequency** v of the oscillator is the reciprocal of the period, or the number of oscillations per second:

$$v = \frac{1}{2\pi}\sqrt{\frac{k}{m}} \tag{9.1-10}$$

The frequency is larger if the force constant is larger (the spring is stiffer) and smaller if the mass is larger.

EXAMPLE 9.1

An object of mass 0.250 kg is suspended from a spring with $k = 5.55$ N m^{-1}. Find the period and the frequency.

Solution

$$\tau = (2\pi)\sqrt{\frac{0.250 \text{ kg}}{5.55 \text{ N m}^{-1}}} = 1.33 \text{ s}$$

$$v = \frac{1}{1.33 \text{ s}} = 0.750 \text{ s}^{-1} = 0.750 \text{ hertz}$$

▼ **EXAMPLE 9.2**

A typical chemical bond is similar to a spring with a force constant near 500. N m^{-1}. Estimate the frequency of oscillation of a hydrogen atom at one end of such a spring with the other end held fixed.

Solution

This frequency is typical of vibrational frequencies of molecules.

$$v = \frac{1}{2\pi}\sqrt{\frac{500. \text{ N m}^{-1}}{1.674 \times 10^{-27} \text{ kg}}} = 8.70 \times 10^{13} \text{ s}^{-1}$$

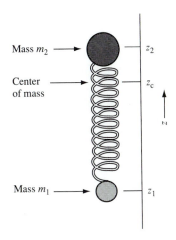

Mass m_2 —→ z_2

Center of mass —→ z_c

Mass m_1 —→ z_1

Figure 9.4. A Second System Represented by a Harmonic Oscillator. This system consists of two masses connected by a spring.

A second version of the harmonic oscillator represents a system of two masses connected by a spring, as shown in Figure 9.4. We assume that the two masses do not rotate about each other, and we extend our z coordinate through the two masses. The state of the system can be represented by the z coordinate of each mass and the velocity of each mass. Instead of the coordinates of the two masses, we can alternatively use the z coordinate of the center of mass and the relative z coordinate (see Appendix D). The relative coordinate is

$$z_{rel} = z_2 - z_1 \qquad (9.1\text{-}11)$$

where z_1 and z_2 are the z coordinates of mass 1 and mass 2.

In this version of the harmonic oscillator, the force on particle 2 is given by

$$F_2 = -k(z_{rel} - z_{rel,0}) \qquad (9.1\text{-}12)$$

where $z_{rel,0}$ is the equilibrium length of the spring. As shown in Appendix D, z_{rel} moves exactly as would the coordinate of a fictitious particle under the influence of the same force as in Equation (9.1-12). The fictitious particle has a mass equal to μ, the **reduced mass** of the two particles, defined by

$$\mu = \frac{m_1 m_2}{m_1 + m_2} \qquad (9.1\text{-}13)$$

where m_1 and m_2 are the masses of the two objects.

The consequence of the similarity between the two kinds of harmonic oscillators is that we can use the formulas for the first kind of oscillator for the second type by replacing the mass m in the formulas by μ and replacing z by z_{rel}.

Exercise 9.2   The frequency of vibration of a diatomic hydrogen molecule with two ^{1}H atoms is 1.32×10^{14} s^{-1}. Find the value of the force constant. What would the frequency of vibration be if one hydrogen nucleus was replaced by a deuterium nucleus?

Since the mechanical state of the system is specified in classical mechanics by values of the coordinates and velocity components, all other mechanical quantities are state functions of these variables. The kinetic energy of our harmonic oscillator is determined by the velocity:

$$\mathscr{K} = \frac{1}{2} m v^2 = \frac{k}{2} C^2 \left[\sin\left(\sqrt{\frac{k}{m}} t \right) \right]^2 = \frac{k}{2} C^2 \sin^2\left(\sqrt{\frac{k}{m}} t \right) \quad (9.1\text{-}14)$$

The potential energy and the corresponding force are determined by the coordinate and obey Equation (D-5) of Appendix D:

$$-d\mathscr{V}/dz = F_z$$

The potential energy of the harmonic oscillator is

$$\mathscr{V}(z) = \frac{1}{2} k z^2 \quad (9.1\text{-}15)$$

with the choice $\mathscr{V}(0) = 0$. Figure 9.5a shows the potential energy of the harmonic oscillator as a function of z, and Figure 9.5b shows the force due to this potential energy.

For our particular initial conditions, the potential energy is given as a function of time by

$$\mathscr{V} = \frac{k}{2} C^2 \cos^2\left(\sqrt{\frac{k}{m}} t \right) \quad (9.1\text{-}16)$$

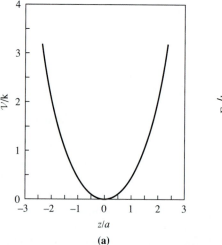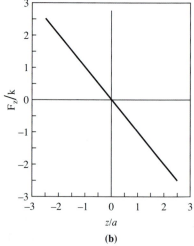

Figure 9.5. Mechanical Variables of a Harmonic Oscillator. (a) The potential energy. **(b)** The force on the oscillator. The force is given as the negative derivative of the potential energy.

The total energy, E, is given by

$$E = \mathscr{K} + \mathscr{V} = \frac{1}{2} kC^2 \left[\sin^2\left(\sqrt{\frac{k}{m}}\, t\right) + \cos^2\left(\sqrt{\frac{k}{m}}\, t\right) \right] \quad \textbf{(9.1-17)}$$

According to a trigonometric identity, the factor inside the square brackets equals unity for all values of t:

$$E = \frac{1}{2} kC^2 = \frac{1}{2} kz_t^2 \quad \textbf{(9.1-18)}$$

Conservation of energy not only is an important physical principle but also can be useful in solving problems.

The total energy is constant, an example of **conservation of energy**.

The constant z_t is the magnitude of z at the **turning point**, the greatest magnitude z can take on for a given energy. There are a negative turning point and a positive turning point, at which the potential energy has its maximum value, the kinetic energy vanishes, and the oscillator must begin its return toward $z = 0$.

9.2   Properties of Waves in Classical Mechanics

There are various wave phenomena in classical physics, including sound, light, waves on the surface of bodies of water, and vibrations of the strings in musical instruments. In any wave, there is some oscillating displacement. For example, in a water wave the displacement is the distance to a point on the surface from the equilibrium position of this part of the surface. A region of positive displacement is called a **crest**, and a region of negative displacement is called a **trough**. A location where the displacement of a wave equals zero is called a **node**. Most waves are periodic, with a number of crests and troughs having the same shape. The distance from one crest to the next is called the **wavelength**. The **period** is the time for one oscillation at a fixed point. That is, it is the time for the oscillating object to return to a given state. The **frequency** is the reciprocal of the period, or the number of oscillations per unit time. A wave is generally much longer than one wavelength, but in any event a wave is inherently **delocalized**, or not capable of existing at a single point in space.

There are two principal simple types of waves. A **traveling wave** propagates, or moves along, like the waves on the surface of a large body of water. A **standing wave** does not move along, but stays in the same location with stationary nodes. An example of a standing wave is the fundamental vibration of a string in a musical instrument, in which the length of the string is equal to one-half wavelength and there is a node at each end of the string. The displacement of such a wave is the instantaneous distance from a point in the oscillating string to its equilibrium position.

Figure 9.6 represents some features of traveling and standing waves. It shows how the traveling wave in Figure 9.6a moves to the right without changing shape, whereas the standing wave in Figure 9.6b oscillates up and down between stationary nodes while the displacement at each node remains zero.

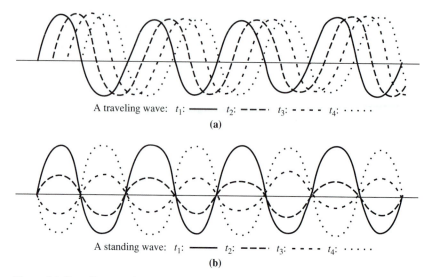

A traveling wave: t_1: —— t_2: – – –· t_3: – – – – t_4: · · · · ·

(a)

A standing wave: t_1: —— t_2: – – –· t_3: – – – – t_4: · · · · ·

(b)

Figure 9.6. Traveling and Standing Waves. (a) A traveling wave at times $t_1 < t_2 < t_3 < t_4$. The nodes of the traveling wave move along (from left to right in this diagram). **(b)** A standing wave at times $t_1 < t_2 < t_3 < t_4$. The nodes of the standing wave remain at fixed positions.

Water waves and string vibrations are **transverse waves**: the displacement is perpendicular (transverse) to the direction of propagation of the traveling wave or to the length of the string. In **longitudinal waves**, the displacement is parallel to the direction of propagation. An example of a longitudinal wave is a type of sound wave in solids in which the particles vibrate in the direction of propagation.

One important property of waves is **interference**. When two waves come to the same location, their displacements add. When two crests or two troughs coincide, a displacement of large magnitude results. This addition is called **constructive interference**. When a crest of one wave and a trough of another wave coincide, they partially or completely cancel each other. This cancellation is called **destructive interference**. Constructive and destructive interferences are depicted qualitatively in Figure 9.7a, which shows the sum of two waves of slightly different wavelengths. At some points the resulting wave has a displacement equal to the sum of the amplitudes, and at other places destructive interferences produce a sum of zero value.

A property that arises from interference is **diffraction**. For example, if a water wave encounters a post, a wave will be reflected away from the post which has crests that are arcs of circles. The reflected waves from a row of equally spaced posts can interfere to produce a diffracted wave with straight crests, which travels in a direction different from that of the incident wave. Figure 9.7b illustrates diffraction by a set of equally spaced posts or other scattering centers. The straight lines represent the crests of a plane wave moving from left to right. The arcs represent the crests of diffracted waves moving outward from the scattering centers. At a considerable distance from the scattering centers, these crests combine to produce a diffracted plane wave. The wave nature of light was established experimentally when interference and diffraction of light were observed.

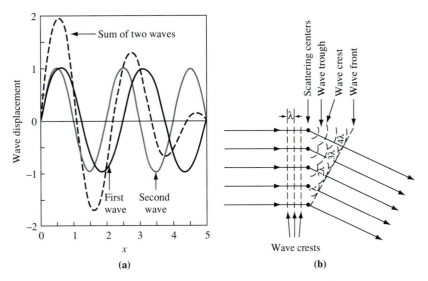

Figure 9.7. The Superposition of Two Waves of Different Wavelengths. (a) Constructive and destructive interference at one time. **(b)** The diffraction of waves by a row of scattering centers. The direction of the diffracted wave depends on the wavelength.

Waves in a Flexible String

The flexible string is a model system designed to represent a real vibrating string like one in a musical instrument. It has the properties that (1) it is uniform, with a mass per unit length equal to ρ; (2) there is a tension force of magnitude T pulling at each end of the string; (3) the equilibrium position of the string is a segment of a straight line; and (4) the string is perfectly flexible—that is, no force can be put on any part of the string by bending it. The force exerted on one portion of the string by an adjacent portion is tangent to the string at the point dividing the portions and has magnitude equal to T.

We consider a string of length L, with its ends fixed on the x axis of a cartesian coordinate system and the left end of the string at the origin. There is no displacement from the equilibrium position at $x = 0$ and $x = L$. At some initial time, we displace the string into some position in the x-z plane other than its equilibrium position and release it to vibrate freely. It will remain in the x-z plane as it vibrates.

The state of the string is specified by giving the displacement and velocity at each point of the string as a function of t. Functions of x and t are thus required:

$$z = z(x, t) \tag{9.2-1}$$

$$v_z = v_z(x, t) = \frac{\partial z}{\partial t} \tag{9.2-2}$$

We assume that our string undergoes only small displacements, so that the total length of the string remains nearly constant and the tension force T is nearly constant.

Derivation of the Wave Equation for a Flexible String

We consider a small portion of the string lying between x and $x + \Delta x$, as shown in Figure 9.8. The force on the left end of the string segment is called $\mathbf{F}_1$ and the force on the right end is called $\mathbf{F}_2$. Both forces have magnitude T and are tangent to the curve representing the position of the string. If the string is curved, the forces will not cancel.

We denote the angles between the x axis and the two tangent lines by α_1 and α_2. For small displacements, the net force on the string segment will lie in the z direction:

$$F_z = F_{2z} + F_{1z} = T \sin(\alpha_2) - T \sin(\alpha_1) \approx T[\tan(\alpha_2) - \tan(\alpha_1)]$$

$$\approx T\left[\left(\frac{\partial z}{\partial x}\right)\bigg|_{x+\Delta x} - \left(\frac{\partial z}{\partial x}\right)\bigg|_{x}\right] \tag{9.2-3}$$

where the subscripts on the derivatives denote the positions at which they are evaluated. We have used the fact that the sine and the tangent are nearly equal for small angles, which corresponds to our case of small displacements. The measure of the angle in radians is also nearly equal to the sine and the tangent for this small angle.

This equality of sine, tangent, and measure of the angle in radians for small angles is sometimes useful.

Exercise 9.3   Check the validity of the above assertion by evaluating the sine and the tangent of $1.0000°$ (0.017453 radian).

We have also used the fact that the first derivative is equal to the tangent of the angle between the horizontal and the tangent line.

The mass of the string segment is $\rho \Delta x$ and its acceleration is $\partial^2 z / \partial t^2$, so from Newton's second law

$$T\left[\left(\frac{\partial z}{\partial x}\right)\bigg|_{x+\Delta x} - \left(\frac{\partial z}{\partial x}\right)\bigg|_{x}\right] = \rho \Delta x \frac{\partial^2 z}{\partial t^2} \tag{9.2-4}$$

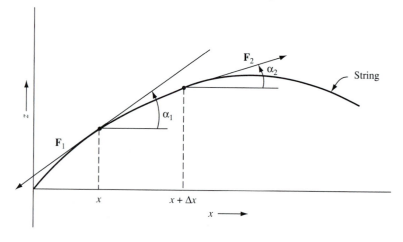

Figure 9.8. The Position of a Flexible String and the Forces on a Segment of the String.
This arbitrary conformation of the string as a function of length along the string is used to describe the forces on a small segment of the string.

We divide both sides of this equation by $\rho\,\Delta x$ and then take the limit as Δx approaches zero. The quotient of differences in Equation (9.2-4) becomes a second derivative:

$$\lim_{\Delta x \to 0}\left[\frac{(\partial z/\partial x)|_{x+\Delta x} - (\partial z/\partial x)|_{x}}{\Delta x}\right] = \frac{\partial^2 z}{\partial x^2}$$

This equation is substituted into Equation (9.2-4) to obtain the classical wave equation.

Solution of the Wave Equation

The classical wave equation of the string is

$$c^2 \frac{\partial^2 z}{\partial x^2} = \frac{\partial^2 z}{\partial t^2} \tag{9.2-5}$$

where we let $T/\rho = c^2$. We will show later that c is equal to the speed of propagation of the wave in the string.

Equation (9.2-5) is a partial differential equation, and its solution is a function of x and t, describing the displacement of the string at each point along the string for each value of the time. We solve this equation by **separation of variables**, which is a step-by-step procedure that we will use several times to solve one type of partial differential equation.

1. The first step is to assume a trial solution that is a product of factors, each depending on one of the independent variables:

   $$z(x, t) = \psi(x)\zeta(t) \tag{9.2-6}$$

   The function ψ does not depend on t, and the function ζ does not depend on x.

2. The second step is to substitute the trial solution into the differential equation and carry out the algebra needed to obtain terms that depend separately on the two independent variables. We substitute Equation (9.2-6) into Equation (9.2-5):

   $$c^2\zeta \frac{d^2\psi}{dx^2} = \psi \frac{d^2\zeta}{dt^2}$$

   We have used the fact that a constant can be factored out of a derivative, and we have written ordinary derivatives because ζ and ψ depend on single variables. We divide by $\psi\zeta c^2$:

   $$\frac{1}{\psi(x)}\frac{d^2\psi}{dx^2} = \frac{1}{c^2\zeta(t)}\frac{d^2\zeta}{dt^2} \tag{9.2-7}$$

   The variables are separated in this equation. That is, each term depends on only one independent variable.

3. The third step is to set each term equal to a constant. This is because x and t are independent variables. For example, t can be held fixed while x is allowed to vary. The left-hand side of Equation (9.2-7) is a function of x, but it must be equal to the right-hand side of the

Setting each term equal to a constant transforms one partial differential equation into two ordinary differential equations.

equation, which is equal to a constant if we keep t fixed. The left-hand side is therefore a constant function of x. Similarly, the right-hand side is a constant function of t.

4. The fourth step is to obtain and solve an ordinary differential equation for each factor of the trial solution. Equating each side of Equation (9.2-7) to a constant gives the two equations:

$$\frac{1}{\psi(x)} \frac{d^2\psi}{dx^2} = \text{constant} = -\kappa^2 \qquad \textbf{(9.2-8)}$$

$$\frac{1}{c^2\zeta(t)} \frac{d^2\zeta}{dt^2} = -\kappa^2 \qquad \textbf{(9.2-9)}$$

We assign the symbol $-\kappa^2$ to the constant because κ will turn out to be a real quantity with this assignment.

Multiplying Equation (9.2-8) by ψ and Equation (9.2-9) by $c^2\zeta$, we now have the ordinary differential equations

$$\frac{d^2\psi}{dx^2} + \kappa^2\psi(x) = 0 \qquad \textbf{(9.2-10)}$$

$$\frac{d^2\zeta}{dt^2} + \kappa^2 c^2\zeta(t) = 0 \qquad \textbf{(9.2-11)}$$

These are the same as Equation (9.1-2) except for the symbols used. We have the general solutions

$$\psi(x) = B \cos(\kappa x) + D \sin(\kappa x) \qquad \textbf{(9.2-12)}$$
$$\zeta(t) = F \cos(\kappa ct) + G \sin(\kappa ct) \qquad \textbf{(9.2-13)}$$

where B, D, F, and G are arbitrary constants. The product of these two functions is a wave function that satisfies the wave equation, Equation (9.2-5).

When we apply the boundary conditions, we are choosing specific members of the family of functions represented by the general solution.

We must now conform our solution to the **boundary condition** that z vanishes at $x = 0$ and $x = L$. The factor ψ must vanish at these points, since it contains all of the x dependence of z. The condition that ψ vanishes at $x = 0$ requires that $B = 0$, since $\sin(0) = 0$ and $\cos(0) = 1$. The condition that ψ vanishes at $x = L$ places restrictions of the value of κ. The sine function vanishes if its argument is an integral multiple of π, so that

$$\kappa L = n\pi \qquad \textbf{(9.2-14)}$$

where n is some integer. Thus,

$$\psi(x) = D \sin(n\pi x/L) \qquad \textbf{(9.2-15)}$$

Now that we have satisfied the boundary conditions, we still have some constants whose values we do not know. The constants F and G are chosen to match whatever initial conditions we have in a particular case. Let us take the case (1) that the string is passing through its equilibrium position ($z = 0$ for all x) at the time $t = 0$ and (2) that the maximum displacement is equal to A. Condition (1) requires that $F = 0$, since $\sin(0) = 0$ and $\cos(0) = 1$. If the maximum displacement is equal to A,

$$z(x, t) = \psi(x)\zeta(t) = A \sin\left(\frac{n\pi x}{L}\right) \sin\left(\frac{n\pi ct}{L}\right) \qquad \textbf{(9.2-16)}$$

The solution function $z(x, t)$ is called a **wave function**. We call A the **amplitude** of the wave.

The z component of the velocity of any point of the string is given by the function

$$v_z = \frac{\partial z}{\partial t} = \left(\frac{n\pi c}{L}\right) A \sin\left(\frac{n\pi x}{L}\right) \cos\left(\frac{n\pi c t}{L}\right) \qquad \textbf{(9.2-17)}$$

Each point of the string that is not at a node undergoes uniform harmonic motion.

The wave function in Equation (9.2-16) represents a different standing wave for each value of n. Figure 9.9a shows the wave function for $n = 1$, Figure 9.9b shows the wave function for $n = 2$, etc. In each graph, the solid curve represents the position of the string when $\sin(n\pi c t/L) = 1$, and the broken curve represents the position when $\sin(n\pi c t/L) = -1$. The string oscillates between these two positions.

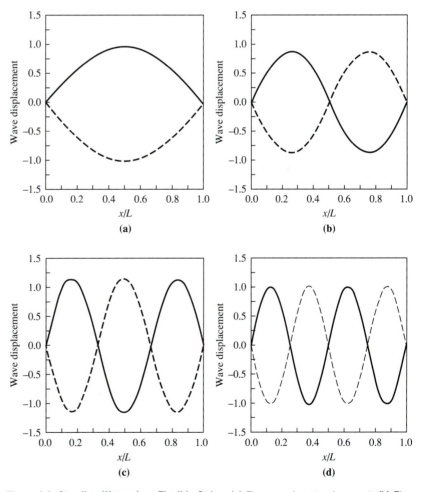

Figure 9.9. Standing Waves in a Flexible String. (a) The wave function for $n = 1$. **(b)** The wave function for $n = 2$. **(c)** The wave function for $n = 3$. **(d)** The wave function for $n = 4$. These are the first few standing waves that satisfy the condition that the ends of the string are fixed. The nodes are fixed. Between the nodes the string oscillates.

Exercise 9.4

a. Show by substitution that Equation (9.2-16) satisfies Equation (9.2-5).
b. What is the effect on the wave function of replacing n by its negative?
c. What is the relationship between the value of n and the number of nodes?

The wavelength is the length of one period of the sine function, which is the distance for the argument of the sine function to change by 2π. For a standing wave in a string of length L, the wavelength λ is given by

$$n\lambda = 2L \quad \text{or} \quad \lambda = \frac{2L}{n} = \frac{2\pi}{\kappa} \tag{9.2-18}$$

The wavelength takes on only values from a discrete set. We say that it is quantized.

Because the ends of the string are fixed, the wavelength can take on only values from a discrete set, instead of being allowed to take on any value.

The **period** of the motion is the time for the string to return to a given initial state. It is denoted by τ. It is the time for the argument of $\sin(n\pi ct/L)$ to change by 2π, so that

$$2\pi = \frac{n\pi c\tau}{L} \quad \text{or} \quad \tau = \frac{2L}{nc} \tag{9.2-19}$$

The **frequency** v is the number of oscillations per unit time, or the reciprocal of the period:

$$v = \frac{nc}{2L} = \frac{n}{2L}\sqrt{\frac{T}{\rho}} \tag{9.2-20}$$

We have a different frequency for each value of n. For a fixed value of n, the frequency can be increased by increasing the tension force, by decreasing the length of the string, or by decreasing the mass per unit length of the string.

We have obtained a set of solutions, one for each positive integer value of n. Not all frequencies and wavelengths are possible. From Equations (9.2-18) and (9.2-20), the frequencies and wavelengths are **quantized**. That is, the possible frequencies and wavelengths form discrete sets of values with gaps between them. The standing wave with $n = 1$ is called the **fundamental** or **first harmonic**, the standing wave with $n = 2$ is called the **first overtone** or **second harmonic**, etc.

A string does not usually move as described by a single harmonic. For example, in a musical instrument the frequencies of a number of harmonics can be heard simultaneously. To represent this behavior mathematically, we write

The Fourier series is named for Jean Baptiste Joseph Fourier, 1768–1830, famous French mathematician and physicist. This Fourier sine series is a series of coordinate sine functions with time-dependent coefficients. Fourier cosine series are also defined, and a general series is a sum of a sine series and a cosine series.

$$z(x, t) = \sum_{n=1}^{\infty} a_n \sin\left(\frac{n\pi x}{L}\right) \tag{9.2-21}$$

The sum in Equation (9.2-21) is a **Fourier sine series**. It is a **linear combination** (sum of functions multiplied by coefficients) of all of the members of the set of solutions to our partial differential equation. Fourier cosine series are linear combinations of cosine functions, and a more general Fourier series contains both sine and cosine terms. Fourier was able to show that any periodic function obeying a few conditions can be represented by a Fourier series (an exact representation might require all of the infinitely many terms to be nonzero).

The **Fourier coefficients** $a_1, a_2, \ldots$ can be constants if a function only of x is represented. We must let these coefficients depend on t to satisfy the wave equation. With our initial condition that the string was passing through its equilibrium position at $t = 0$, the following sum is a solution:

$$z(x, t) = \sum_{n=1}^{\infty} A_n \sin\left(\frac{n\pi x}{L}\right) \sin\left(\frac{n\pi ct}{L}\right) \qquad \textbf{(9.2-22)}$$

Each term represents a harmonic, and each harmonic has its own frequency. The constants $A_1, A_2, \ldots$ can have any values. Some of the coefficients can vanish, and if only one coefficient is nonzero we recover our single-frequency solution of Equation (9.2-16). Some or all of the harmonics can occur at once, with constructive and destructive interference that continually changes because the different harmonics have different frequencies. Figure 9.10 shows the summation of three harmonics. Figure 9.10a shows the $n = 1$, $n = 2$, and $n = 3$ terms in Equation (9.2-22) at time $t = L/(4c)$, and Figure 9.10b shows the same three terms at $t = 3L/(4c)$. For the figure we have chosen $A_1 = 1$, $A_2 = 0.2$, and $A_3 = 0.1$.

Our solution in Equation (9.2-22) is only one possibility. All of our harmonics were "in phase" at $t = 0$. That is, they all made zero contribution to the position of the string. The harmonics do not remain in phase, since they have different frequencies, and other solutions could have different phases for the harmonics at $t = 0$.

Exercise 9.5   Show by substitution that the series in Equation (9.2-22) satisfies Equation (9.2-5).

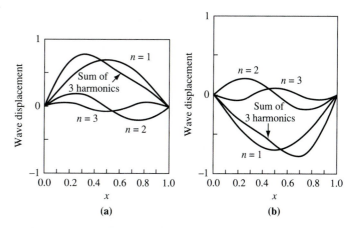

Figure 9.10. The Superposition of Three Harmonics of a Flexible String. (a) At time $t = L/(4c)$. **(b)** At time $t = 3L/(4c)$. By the principle of superposition, a sum of the three harmonics satisfies the wave equation. Constructive and destructive interference produces waves that are not sinusoidal and do not maintain a fixed shape.

The fact that a linear combination of solutions can be a solution to the wave equation is called the **principle of superposition**.

In a string of finite length, stationary nodes are required at the ends of the string, which prevents the occurrence of ordinary traveling waves, in which the nodes move. However, if the length of a string is infinite, we can have traveling waves. A traveling wave that satisfies Equation (9.2-5) is described by the wave function

$$z(x, t) = A \sin(\kappa x - \kappa c t) \qquad \textbf{(9.2-23)}$$

Exercise 9.6   Show by substitution that the function in Equation (9.2-23) satisfies Equation (9.2-5).

We can find the speed of our traveling wave by following the motion of one of the nodes. At time $t = 0$, there is a node at $x = 0$. As time passes, this node will be located at the point where $x - ct = 0$. Thus

$$x(\text{node}) = ct \qquad \textbf{(9.2-24)}$$

The node is moving toward the positive end of the x axis with a speed equal to c, as stated earlier. Since $c = \sqrt{T/\rho}$, increasing the mass per unit length decreases the speed, and increasing the tension force increases the speed.

Exercise 9.7   What change would have to be made in the mass per unit length to quadruple the speed of propagation? What change would have to be made in the tension force to double the speed of propagation?

If the function were

$$z(x, t) = A \sin(\kappa x + \kappa c t) \qquad \textbf{(9.2-25)}$$

the wave would be traveling toward the negative end of the x axis with the same speed c.

In one wavelength, the argument of the sine function changes by 2π for fixed t, so the same relationship occurs as in Equation (9.2-18) for the standing wave:

$$\kappa = \frac{2\pi}{\lambda} \qquad \textbf{(9.2-26)}$$

The relationship between the frequency and the wavelength can be obtained by observing that in time t, the length of the wave "train" that passes a fixed point is

$$\text{Length} = ct$$

The number of wavelengths in this wave train is equal to

$$\text{Number} = \frac{ct}{\lambda}$$

In time t, the number of oscillations is equal to

$$\text{Number} = vt$$

so that $vt = ct/\lambda$, or

$$v = \frac{c}{\lambda}$$

(9.2-27)

Equation (9.2-27) is a very useful equation that we will apply repeatedly to light waves.

Equation (9.2-27) is the general relation between wavelength and frequency.

Although we have derived our wave formulas for waves in a flexible string, the same formulas hold for other waves.

Exercise 9.8

The speed of sound in air at sea level and room temperature is approximately equal to 338 m s^{-1}. Find the wavelength of a sound wave with a frequency of 440 s^{-1}. (This frequency is the standard frequency of "A" above middle "C" in a musical instrument.)

Standing waves can be created by interference between traveling waves moving in opposite directions. Consider two traveling waves of equal amplitude, moving in opposite directions:

$$z_R(x, t) = A \sin(\kappa x + \kappa c t)$$ (9.2-28a)
$$z_L(x, t) = A \sin(\kappa x - \kappa c t)$$ (9.2-28b)

If the two waves interfere, the resulting displacement is

$$z_{tot} = A[\sin(\kappa x - \kappa c t) + \sin(\kappa x + \kappa c t)]$$ (9.2-29)

Trigonometric identities give

$$\sin(\kappa x + \kappa c t) = \sin(\kappa x) \cos(\kappa c t) + \cos(\kappa x) \sin(\kappa c t)$$
$$\sin(\kappa x + \kappa c t) = \sin(\kappa x) \cos(-\kappa c t) + \cos(\kappa x) \sin(-\kappa c t)$$
$$= \sin(\kappa x) \cos(\kappa c t) - \cos(\kappa x) \sin(\kappa c t)$$

where additional identities

$$\sin(x) = -\sin(-x)$$

and

$$\cos(x) = \cos(-x)$$

have been used. Use of these equations gives

$$z_{tot} = 2A \sin(\kappa x) \cos(\kappa c t)$$ (9.2-30)

which is a standing wave similar to the standing wave in Equation (9.2-16). The fact that we now have a cosine instead of a sine for the time dependence corresponds to starting our clock at a different time.

The stationary nodes are produced by destructive interference of the two traveling waves, and the oscillations between the nodes are produced by interference that is alternately constructive and destructive. The relationship between standing and traveling waves allows us to use the concept of the speed of propagation for a standing wave and also permits us to use Equation (9.2-27) for a standing wave.

| **Exercise 9.9** | Sketch a graph of the two traveling waves and their sum at a moment when their nodes coincide and also at a slightly later time, showing that the node in their sum stays at the same place. |

| **Exercise 9.10** | Carry out the mathematical steps to obtain Equation (9.2-30). |

The Classical Wave Theory of Light

In 1865 Maxwell developed a mathematical theory of electromagnetism. In this theory, there are four important vector quantities: the electric field $\mathbf{E}$, the electric displacement $\mathbf{D}$, the magnetic field strength $\mathbf{H}$, and the magnetic induction $\mathbf{B}$. The dependence of these quantities on time and position is described by **Maxwell's equations**, which Maxwell deduced from empirical laws. He found that the electric and magnetic fields can oscillate like waves. These oscillations constitute electromagnetic radiation. Visible light is an example of such radiation, as are ultraviolet radiation, X-rays, radio waves, microwaves, etc., which differ from visible light only in having different wavelengths.

At first it was thought that light consisted of oscillations in a medium called "the luminiferous ether" which presumably filled all of space. The assumption that such a medium exists was abandoned after Michelson and Morley demonstrated that the speed of light has the same value for observers moving with different velocities.

Albert A. Michelson, 1852–1931, an American physicist, was the first American to win a Nobel Prize in science (in 1907). Edward W. Morley, 1838–1923, was an American chemist.

In a medium with zero electrical conductivity (a perfect insulator or a vacuum), the following equations follow from Maxwell's equations.[1]

$$\nabla^2\mathbf{E} - \varepsilon\mu(\partial^2\mathbf{E}/\partial t^2) = 0 \qquad \textbf{(9.2-31)}$$

$$\nabla^2\mathbf{H} - \varepsilon\mu(\partial^2\mathbf{H}/\partial t^2) = 0 \qquad \textbf{(9.2-32)}$$

where ε is called the **permittivity** of the medium and μ is called the **permeability** of the medium. The permittivity was already introduced in Chapters 6 and 8 in connection with the forces between charged particles. The values of these quantities for a vacuum are denoted by ε_0 and μ_0. These values are given inside the back cover. The operator ∇^2 is called the **Laplacian**. In cartesian coordinates,

The Laplacian is named for Pierre-Simon Laplace, 1749–1827, a famous French mathematician who pioneered the use of noncartesian coordinates.

$$\nabla^2 f = (\partial^2 f/\partial x^2) + (\partial^2 f/\partial y^2) + (\partial^2 f/\partial z^2) \qquad \textbf{(9.2-33)}$$

where f is an arbitrary function.

Equations (9.2-31) and (9.2-32) are wave equations similar to Equation (9.2-5). Their solutions represent electromagnetic radiation. The simplest electromag-

---

[1] J. C. Slater and N. H. Frank, *Electromagnetism*, McGraw-Hill, New York, 1947, pp. 90ff.

netic wave is a **plane wave**, which is a wave propagating in a fixed direction with each crest or trough located in a moving plane perpendicular to this direction. Let us take the direction of propagation as the positive y axis. If the electric field **E** oscillates only in the z direction, we have the case of **plane polarization**. One can show from Maxwell's equations[2] that in this case the magnetic field oscillates only in the x direction. There is another independent plane-polarized wave propagating in the y direction, in which the electric field oscillates in the x direction and the magnetic field oscillates in the z direction.

In our case, Equations (9.2-31) and (9.2-32) are

$$\frac{\partial^2 E_z}{\partial y^2} - \frac{1}{c^2}\frac{\partial^2 E_z}{\partial t^2} = 0 \qquad (9.2\text{-}34)$$

$$\frac{\partial^2 H_x}{\partial y^2} - \frac{1}{c^2}\frac{\partial^2 H_x}{\partial t^2} = 0 \qquad (9.2\text{-}35)$$

An additional condition from Maxwell's equations makes these equations interdependent:

$$\frac{E_z}{H_x} = \pm\sqrt{\frac{\mu}{\varepsilon}} \qquad (9.2\text{-}36)$$

The electric field cannot oscillate without oscillation of the magnetic field, and vice versa. If we solve Equation (9.2-34), we can use Equation (9.2-36) to obtain the solution of Equation (9.2-35).

In Equations (9.2-34) and (9.2-35), we have let

$$c = \frac{1}{\sqrt{\varepsilon\mu}} \qquad (9.2\text{-}37)$$

Equations (9.2-31) and (9.2-32) have the same form as Equation (9.2-5), in which c is the speed of propagation of the wave. The theory of Maxwell correctly predicts the value of the speed of light.

▼ **EXAMPLE 9.3**

Calculate the speed of light in a vacuum.

Solution

From the table of Fundamental Constants (inside back cover),

$$c_{\text{vacuum}} = \frac{1}{\sqrt{(8.8542 \times 10^{-12}\ \text{C}^2\ \text{N}^{-1}\ \text{m}^{-2})(4\pi \times 10^{-7}\ \text{J s}^2\ \text{C}^{-2}\ \text{m}^{-1})}}$$

$$= 2.9979 \times 10^8\ \text{m s}^{-1}$$

▲

Exercise 9.11 Show that the units in Equation (9.2-37) are the same on both sides.

_____

[2] Slater and Frank, *op. cit.*

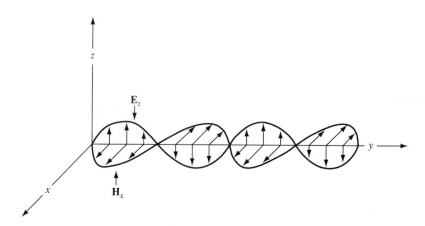

Figure 9.11. The Electric and Magnetic Fields in an Electromagnetic Wave. The wave is propagating to the right in the y direction. The electric field is oscillating in the z direction, and the magnetic field is oscillating in the x direction. The nodes of the electric field and those of the magnetic field coincide.

This interaction of oscillating charges and electromagnetic radiation is the basis of spectroscopy, in which absorbed or emitted radiation is used to determine molecular energy levels.

A traveling-wave solution to Equations (9.2-34) and (9.2-35) is

$$E_z(y, t) = E_0 \sin(\kappa y - \kappa ct) \qquad \textbf{(9.2-38)}$$

$$H_x(y, t) = H_0 \sin(\kappa y - \kappa ct) \qquad \textbf{(9.2-39)}$$

where E_0 and H_0 are constants that obey Equation (9.2-36). Only one of these constants, say E_0, is needed to represent the amplitude of the electromagnetic wave. The constant κ can take on any real value.

Figure 9.11 shows $\mathbf{E}$ and $\mathbf{H}$ as functions of y at time $t = 0$ with $\mathbf{E}$ and $\mathbf{H}$ plotted in the directions in which they point. As time passes, the traveling wave moves to the right without changing its shape or wavelength.

By Equation (9.2-18), we can also write the traveling wave as

$$E_z(y, t) = E_0 \sin[2\pi(y - ct)/\lambda] \qquad \textbf{(9.2-40)}$$

$$H_x(y, t) = H_0 \sin[2\pi(y - ct)/\lambda] \qquad \textbf{(9.2-41)}$$

As a traveling electromagnetic wave passes, a fixed observer sees oscillating electric and magnetic fields. Since oscillating electric and magnetic fields put oscillating forces on charged particles such as the electrons and nuclei in molecules, molecules can absorb electromagnetic radiation. The converse is also true. Oscillating electric charges emit electromagnetic radiation.

Electromagnetic waves contain energy and transport it from one place to another. The radiant energy density (radiant energy per unit volume) is given by[3]

$$U = \frac{\varepsilon E^2 + \mu H^2}{2} \qquad \textbf{(9.2-42)}$$

and the energy flux (flow of energy per unit area per unit time) is given by **Poynting's vector**

$$\mathbf{S} = \mathbf{E} \times \mathbf{H} \qquad \textbf{(9.2-43)}$$

The cross product of two vectors is discussed in Appendix B.

where $\times$ represents the vector cross product.

---

[3] Slater and Frank, *op. cit.*, pp. 99ff.

For our plane polarized plane wave, Poynting's vector points in the y direction, with

$$S_y = E_z H_x \qquad (9.2\text{-}44)$$

For our plane wave, there is a relationship between the energy density and the energy flux. Using Equation (9.2-36), we can rewrite Equation (9.2-42) as

$$U = \frac{\mu H_x^2 + \mu H_x^2}{2} = \mu H_x^2 \qquad (9.2\text{-}45)$$

Again using Equation (9.2-36), we can write Equation (9.2-44) as

$$S_y = \sqrt{\mu/\varepsilon}\, H_x^2 \qquad (9.2\text{-}46)$$

From Equation (9.2-37), we see that $c\sqrt{\varepsilon\mu}$ equals unity, so that

$$S_y = \mu H_x^2 c = Uc \qquad (9.2\text{-}47)$$

Thus the flow of energy per unit area per second equals the amount of energy contained in a volume with unit cross-sectional area and a length equal to the speed of light times 1 second.

An electromagnetic wave cannot penetrate a perfect conductor. A finite non-zero electric field would produce an infinite current inside the object. Because the wave cannot simply disappear, it is completely reflected, producing a node at the surface of the conductor. Real conductors, which have finite conductivities, reflect only part (often 95–98%) of incident electromagnetic waves.

Since electromagnetic waves must have nodes at perfectly conducting walls and nearly vanish at a real conducting wall, reflection between walls in a cavity can produce standing electromagnetic waves, similar to the standing wave in Equation (9.2-30).

9.3 The Old Quantum Theory

Near the end of the nineteenth century, classical physics was generally accepted as an accurate description of the real physical universe. However, several important phenomena were discovered that could not be explained by classical physics. Three of these were explained early in the twentieth century by new theories: Planck's theory of blackbody radiation, Einstein's theory of the photoelectric effect, and Bohr's theory of the hydrogen atom. These theories are the major parts of what is called the "old quantum theory." They were based on assumptions of **quantization**—the idea that the value of a physical quantity can equal one of a discrete set of values, but not any of the values between those in the discrete set.

The concept of quantization was introduced into these three theories as an assumption, whereas it occurs naturally in the quantum theory of Schrödinger.

Planck's Theory of Blackbody Radiation

If an object has a temperature somewhat above 500°C, it glows visibly with a red color. At higher temperatures, it glows orange, yellow, white, or even blue if the temperature is high enough. At a given temperature, an object with a lower reflectivity glows more intensely at every wavelength, so that a **blackbody**, a model system that reflects no radiation at any wavelength, has the maximum possible emissivity at every wavelength.

The best laboratory approximation to a blackbody is not an object, but a small hole in a hollow box. If the inside of the box (the "cavity") is made fairly nonreflective, any light falling on the hole from outside will be absorbed as it is reflected around in the box. Measurements on the light emitted through the hole when such a box was heated showed that the amount of light emitted and its spectral distribution depended only on the temperature of the walls of the box. Figure 9.12 shows the **spectral radiant emittance** η as a function of wavelength for several temperatures. The spectral radiant emittance is defined such that $\eta(\lambda)\,d\lambda$ is the energy per unit time per unit area emitted in the wavelengths between λ and $\lambda + d\lambda$. The maximum spectral radiant emittance is at shorter wavelengths for higher temperatures. The visible part of the electromagnetic spectrum, which ranges from about 400 nm to 750 nm, is labeled in the figure. For temperatures near room temperature, nearly all of the radiation is in the infrared region. This is the radiation that is involved in the greenhouse effect in the earth's atmosphere. At 2000 K, only the red part of the visible spectrum

The greenhouse effect is the absorption in the upper atmosphere of infrared radiation emitted by the earth. It is principally due to CO_2, H_2O, CH_4, and various chlorofluorocarbons.

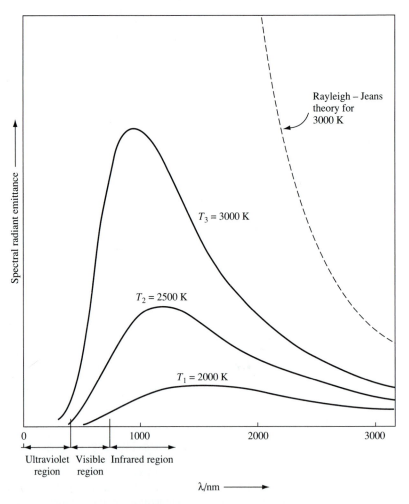

Figure 9.12. The Spectral Radiant Emittance of a Blackbody for $T_1 = 2000$ K, $T_2 = 2500$ K, and $T_3 = 3000$ K. The maximum in the curve shifts to shorter wavelengths as the temperature is raised. The Rayleigh-Jeans curve and the experimental curve coincide for sufficiently long wavelengths.

(around 650 to 750 nm) is represented, but at higher temperatures the other wavelengths are also represented. At around 6000 K the maximum in the curve is in the middle (green) portion of the visible region.

The **total radiant emittance** (emission per unit area per unit time, summed over all wavelengths) is equal to the area under the curve. The **Stefan-Boltzmann law** relates the total radiant emittance to the absolute temperature of the blackbody:

$$\text{Total energy flux} = \sigma T^4 \qquad \textbf{(9.3-1)}$$

The **Stefan-Boltzmann constant** σ has the value

$$\sigma = 5.67051 \times 10^{-8} \text{ J m}^{-2} \text{ s}^{-1} \text{ K}^{-4} = 5.67051 \times 10^{-8} \text{ W m}^{-2} \text{ K}^{-4}$$

This theory was developed by John William Strutt, third Baron Rayleigh, 1842–1919, winner of the 1904 Nobel Prize in physics, and Sir James Jeans, 1877–1946, British astronomer and physicist.

Rayleigh and Jeans constructed a classical theory of blackbody radiation. They defined as their system the set of standing electromagnetic waves that could exist inside a cavity. For a rectangular cavity, they counted up the possible standing waves of various wavelengths that could exist in the cavity with nodes at the walls and then computed the average energy of each standing wave.

Assuming nodes at the walls, the number of possible standing waves with wavelength in the region between λ and $\lambda + d\lambda$ is given by[4]

$$\text{Number} = g(\lambda)\,d\lambda = \frac{8\pi}{\lambda^4}\,V\,d\lambda \qquad \textbf{(9.3-2)}$$

where V is the volume of the system. A factor of 2 has been included to account for the fact that there are two possible perpendicular directions of plane polarization for a given direction of propagation.

Rayleigh and Jeans obtained an average energy of a single standing wave, using the Boltzmann probability distribution of Equation (1.7-25). According to the Boltzmann probability distribution, the probability of a state with energy per unit volume equal to U is

$$p(U) \propto e^{-U/k_B T} = \exp\left(\frac{\varepsilon_0 E^2 + \mu_0 H^2}{k_B T}\right)$$

where we have used Equation (9.2-42). Averaging over values of E and H, Rayleigh and Jeans obtained a result for $\langle U \rangle$, the mean energy of a single standing wave:

$$\langle U \rangle = k_B T \qquad \textbf{(9.3-3)}$$

where k_B is Boltzmann's constant and T is the absolute temperature.

Exercise 9.12

Derive Equation (9.3-3). The average is computed from the formula

$$\langle U \rangle = \frac{\displaystyle\int_{-\infty}^{\infty} \int_{-\infty}^{\infty} U e^{-U/k_B T}\, dE\, dH}{\displaystyle\int_{-\infty}^{\infty} \int_{-\infty}^{\infty} e^{-U/k_B T}\, dE\, dH}$$

where U is given by Equation (9.2-42). The integration is analogous to the summation in Equation (1.7-10) or Equation (1.7-12). The denominator in this formula is said to provide "normalization."

---

[4] N. Davidson, *Statistical Mechanics*, McGraw-Hill, New York, 1962, p. 214. See also Section 22.4.

The result of Equation (9.3-3) is independent of wavelength, so the energy per unit volume of the standing waves with wavelengths in the range $d\lambda$ is

$$w(\lambda)\,d\lambda = \frac{8\pi k_B T}{\lambda^4}\,d\lambda \tag{9.3-4}$$

The spectral radiant emittance is analogous to the result expressed in Equation (9.2-47). We omit the derivation. The result is

$$\eta(\lambda)\,d\lambda = \frac{c}{4}\,w(\lambda)\,d\lambda = \frac{2\pi c k_B T}{\lambda^4}\,d\lambda \tag{9.3-5}$$

where c is the speed of light.

Equation (9.3-5) agrees well with experiment for large values of the wavelength (much larger than visible wavelengths) but predicts that the spectral radiant emittance becomes large without bound in the limit of short wavelength. This prediction disagrees with the experimental curves in Figure 9.12, and this failure of the Rayleigh-Jeans theory was called the "ultraviolet catastrophe."

In 1900, Planck devised a new theory of blackbody radiation. After attempting to explain blackbody radiation within classical physics, he came up with a theory based on assumptions that at the outset had no direct evidence to support them. Planck hypothesized *quantization*, which defines a new kind of model system for cavity radiation. The following statements are a simplified version of his assumptions:[5]

1. The walls of the cavity contain oscillating electric charges. Each such oscillator has a characteristic fixed frequency of oscillation, but many oscillators are present and every frequency is represented.
2. The standing waves in the cavity are equilibrated with the oscillators in such a way that the average energy of standing waves of a given frequency equals the average energy of the oscillators of that frequency.
3. The energy of a wall oscillator is *quantized*. That is, it is capable of assuming only one or another of the values

$$E = 0,\, h\nu,\, 2h\nu,\, 3h\nu,\, 4h\nu, \ldots,\, nh\nu, \ldots \tag{9.3-6}$$

where ν is the frequency of the oscillator and h is a new constant, now known as **Planck's constant**. The quantity n, which can take on any nonnegative integral value, is our first example of a **quantum number**. Figure 9.13 shows this energy quantization. The figure resembles a ladder, and the quantization has been compared to a ladder. A person can stand on any rung but nowhere between the rungs. The energy can take on any of the values in Equation (9.3-6) but no value between these values.
4. The probability of any energy is given by the Boltzmann probability distribution, Equation (1.7-25).

From assumption 4, the probability of a state with energy $nh\nu$ is

$$p(nh\nu) \propto \exp(-nh\nu/k_B T) \tag{9.3-7}$$

Calculation of the mean energy of an oscillator using this probability gives (see Problem 9.46):

$$\langle E \rangle = \frac{h\nu}{\exp(h\nu/k_B T) - 1} \tag{9.3-8}$$

Max Planck, 1858–1947, a German physicist, received the Nobel Prize in physics in 1918 for this work.

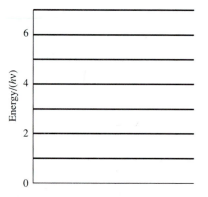

Figure 9.13. The Quantized Energies of an Oscillator as Postulated by Planck. The horizontal line segments are plotted at the heights of the assumed energy values, 0, $h\nu$, $3h\nu$, $4h\nu$, $5h\nu$, $6h\nu$, $7h\nu$, etc.

The introduction of quantization as an assumption was the crucial part of Planck's theory.

---

[5] M. Jammer, *The Conceptual Development of Quantum Mechanics*, McGraw-Hill, New York, 1966, pp. 10ff.

The mean energy of the standing waves with wavelength in the interval $d\lambda$ is assumed to equal this mean oscillator energy, so the energy in this frequency interval is this mean energy times the number of such standing waves, given by Equation (9.3-2):

$$w(\lambda)\, d\lambda = \frac{8\pi h\nu}{\lambda^4(e^{h\nu/k_B T} - 1)}\, d\lambda = \frac{8\pi hc}{\lambda^5(e^{hc/\lambda k_B T} - 1)}\, d\lambda \qquad \textbf{(9.3-9)}$$

where Equation (9.2-27) has been used. The spectral radiant emittance is given by a derivation similar to that leading to Equation (9.3-5):

$$\eta(\lambda)\, d\lambda = \frac{2\pi hc^2}{\lambda^5(e^{hc/\lambda k_B T} - 1)}\, d\lambda \qquad \textbf{(9.3-10)}$$

This formula agrees accurately with experimental measurements of blackbody radiation at all temperatures and wavelengths if an optimum value of the constant h is taken. By fitting data available at the time, Planck was able to get a value of h nearly equal to the presently accepted value, 6.6261×10^{-34} J s.

▼ **EXAMPLE 9.4**

Find the relation of the wavelength of maximum spectral radiant emittance to the temperature.

Solution

One method of numerical solution is the method of bisection, a methodical trial-and-error procedure in which two values are first found such that a root lies between them. The center of the interval between these values is the next trial value. Next one tries the middle of that half which is found to contain the root, etc. Another method of solving such an equation is the Newton-Raphson method, which is discussed in many elementary calculus courses.

We set the derivative of the function of Equation (9.3-10) equal to zero:

$$\frac{d\eta}{d\lambda} = 2\pi hc^2 \frac{(hc/\lambda k_B T)e^{hc/\lambda k_B T} - 5(e^{hc/\lambda k_B T} - 1)}{\lambda^6(e^{hc/\lambda k_B T} - 1)} = 0$$

This expression can vanish only if the numerator vanishes:

$$\frac{hc}{\lambda_{max}k_B T} = 5(1 - e^{-hc/\lambda_{max}k_B T})$$

This equation must be solved by numerical approximation. The result is the **Wien displacement law**,

$$\lambda_{max} = (0.2014)\frac{hc}{k_B T} = \frac{2.898 \times 10^{-3}\ \text{m K}}{T}$$

The wavelength of maximum spectral radiant emittance is inversely proportional to the temperature.

▲

Exercise 9.13

Find the temperature that corresponds to a wavelength of maximum spectral emittance in the red part of the visible spectrum at 700. nm.

Planck's formula agrees with the result of Rayleigh and Jeans for large wavelengths.

Exercise 9.14

Show that in the limit as $\lambda \to \infty$, Equation (9.3-10) agrees with Equation (9.3-5).

The Planck formula also agrees with the Stefan-Boltzmann law.

Exercise 9.15   Use the definite integral

$$\int_0^\infty \frac{x^3}{e^x - 1}\, dx = \frac{\pi^4}{15}$$

to derive the Stefan-Boltzmann law, Equation (9.3-1). Calculate the theoretical value of the Stefan-Boltzmann constant.

Einstein's Theory of the Photoelectric Effect

When a piece of metal inside an evacuated glass tube is illuminated with light of sufficiently short wavelength, electrons are emitted. According to classical physics, one would expect a more intense beam of light to eject electrons of greater energy. However, it was found that electrons were not ejected at all unless the wavelength of the light was at least as small as a threshold wavelength. Also, the maximum energy of the ejected electrons depended not on the intensity of the light but only on its wavelength, with shorter wavelengths producing electrons of larger energy. There was no explanation for this behavior in classical physics.

In 1905, Einstein published a theory for the photoelectric effect, based on the hypothesis that the energy in a beam of light consists of discrete "quanta" and that each quantum has an energy determined by its frequency or wavelength:

$$E = h\nu = \frac{hc}{\lambda} \qquad \textbf{(9.3-11)}$$

Albert Einstein, 1879–1955, a German-Swiss-American physicist, received the 1921 Nobel Prize in physics for this work, but made fundamental contributions in almost every area of theoretical physics.

where h is Planck's constant and ν is the frequency of the light. Equation (9.3-11) is known as the **Planck-Einstein relation**. The quanta of light are called **photons**.

Einstein's hypothesis that light of a given frequency has energy that comes in indivisible quanta of a fixed size is similar to Newton's hypothesis that light consists of particles, which he called "corpuscles." However, Einstein's photons are not simply like little bullets, as Newton pictured his corpuscles. They are much more complicated. Maxwell pictured light as a wave whose amplitude can take on any value and whose energy is therefore not quantized in any way. Maxwell's theory is needed to explain the well-known diffraction and interference phenomena exhibited by light, but it was unable to explain the photoelectric effect.

We obtain the quantitative explanation for the photoelectric effect from Equation (9.3-11). The energy of an electron ejected from the metal is equal to the energy of the photon minus the energy required to detach the electron from the metal. The **work function**, W, is the minimum energy to detach an electron from a given substance. The maximum electron energy is

$$E_{\max}(\text{electron}) = h\nu - W = \frac{hc}{\lambda} - W \qquad \textbf{(9.3-12)}$$

Robert A. Millikan, 1868–1953, American physicist who received the Nobel Prize in physics in 1923 for his measurement of the charge on the electron.

In 1916 Millikan made accurate measurements of the photoelectric effect that agreed well with Equation (9.3-12).

Since light exhibits a particle-like nature in some experiments and wave-like properties in other experiments, we speak of a **wave-particle duality**. This terminology means that light sometimes acts like a wave and sometimes acts like a particle. We cannot adequately answer the question "What is light *really* like?" We must content ourselves with the wave description when it explains the observations of a particular experiment and with the particle description when it explains the observations of another experiment.

Exercise 9.16 The work function of nickel equals 5.0 eV. Find (a) the threshold wavelength for nickel and (b) the maximum electron speed for a wavelength of 200. nm.

Bohr's Theory of the Hydrogen Atom

Johannes Robert Rydberg, 1854–1919, Swedish physicist.

If moist hydrogen gas at low pressure is placed in a tube with a metal electrode at each end and an alternating electric voltage is applied to the electrodes, a reddish-violet glow (a **discharge**) is produced. The major part of this discharge comes from hydrogen atoms formed by dissociation of hydrogen molecules. When light is viewed in a spectroscope, the different wavelengths are **dispersed**, or sent in different directions. Only a few wavelengths arise from hydrogen atoms. Four wavelengths are present in the visible light and other lines occur in the ultraviolet and in the infrared. When only a few wavelengths are present, each wavelength produces an image of the slit of the spectroscope, which resembles a line segment. Such a set of separated lines is called a **line spectrum** and the slit images are called **spectral lines**.

Rydberg was able to represent the wavelengths of all of the spectral lines of hydrogen atoms with a single empirical formula:

$$\frac{1}{\lambda} = \mathscr{R}_{\mathrm{H}}\left(\frac{1}{n_1^2} - \frac{1}{n_2^2}\right) \tag{9.3-13}$$

where n_1 and n_2 are two positive integers and $\mathscr{R}_{\mathrm{H}}$ is **Rydberg's constant** for the hydrogen atom, equal to 1.09677581×10^7 m^{-1}. Using classical physics, no explanation for this relationship could be found.

Ernest Rutherford, first Baron Rutherford of Nelson, 1871–1937, a British physicist originally from New Zealand, won the 1908 Nobel Prize in chemistry and coined the terms alpha, beta, and gamma radiation.

In 1911, Rutherford scattered alpha particles from a thin piece of gold foil. From the way in which the alpha particles were scattered, he concluded that atoms consisted mostly of nearly empty space but with a very small, dense, positive nucleus. If all of the positive charge was concentrated in the nucleus, the electrons in the atom had to be orbiting around the nucleus. However, according to the electrodynamics of Maxwell, an orbiting electron would emit electromagnetic radiation, losing energy and falling onto the nucleus. Classical physics was unable to explain either the line spectrum of the hydrogen atom or its continuing existence.

Niels Henrik David Bohr, 1885–1962, a Danish physicist, received the Nobel Prize in physics in 1922 for this work and was responsible for much of the accepted physical interpretation of quantum mechanics.

In 1913, Bohr published a theory of the hydrogen atom, based on hypotheses that had no prior justification. A simplified version of Bohr's hypotheses are:

1. The hydrogen atom consists of a positive nucleus of charge e and an electron of charge −e moving around it in a circular orbit. The charge e had been measured by Millikan and is now known to have the value 1.6022×10^{-19} C.
2. The angular momentum (see Appendix D) of the electron is quantized: Its magnitude can take on one of the values $h/2\pi$, $2h/2\pi$, $3h/2\pi$, $4h/2\pi$, etc., where h is Planck's constant. Figure 9.14 schematically shows the quantization of the angular momentum.

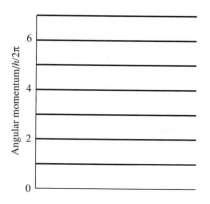

Figure 9.14. The Quantized Angular Momentum Values of Electronic Motion in a Hydrogen Atom as Postulated by Bohr. The horizontal line segments are plotted at the heights of the assumed angular momentum values, $h/2\pi$, $2h/2\pi$, $3h/2\pi$, $4h/2\pi$, $5h/2\pi$, $6h/2\pi$, $7h/2\pi$, etc.

3. Maxwell's equations do not apply. Radiation is emitted or absorbed only when a sudden transition is made from one quantized value of the angular momentum to another.

4. The wavelength of emitted or absorbed light is given by the Planck-Einstein relation, Equation (9.3-11), with the energy of the photon equal to the difference in energy of the initial and final states of the atom.

5. In all other regards, classical mechanics is valid.

We now derive the consequences of Bohr's hypotheses. For simplicity, we pretend that the nucleus is stationary. This assumption can be removed if desired (see Appendix D) by replacing the mass of the orbiting electron by the reduced mass of the electron and the nucleus, which differs only slightly from the electron's mass.

To maintain a circular orbit, there must be a centripetal force on the electron:

$$F_r = -\frac{mv^2}{r} \tag{9.3-14}$$

where v is the speed of the electron, m is its mass, and r is its distance from the nucleus. [See Equation (D-13) of Appendix D.]

Exercise 9.17   Find the centripetal force on an object of mass 1.00 kg if you swing it on a rope so that the radius of the orbit is 3.00 m and the time required for one orbit is 1.00 second (a speed of 9.43 m s^{-1}).

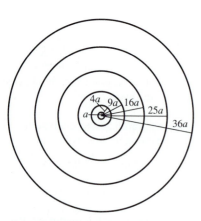

Figure 9.15. The Quantized Bohr Orbits. The radius of an electron orbit can take on only the values a, $4a$, ..., n^2a, ... where n is a positive integer.

The centripetal force is provided by the electrostatic attraction of the positive nucleus for the negative electron:

$$\frac{mv^2}{r} = \frac{e^2}{4\pi\varepsilon_0 r^2} \tag{9.3-15}$$

where ε_0 is the permittivity of the vacuum.

The angular momentum of the electron in a circular orbit is given by Equation (D-15) of Appendix D. It is quantized, according to assumption 2:

$$L = mrv = \frac{nh}{2\pi} \qquad (n = 1, 2, 3, \ldots) \tag{9.3-16}$$

where the quantum number n is a positive integer.

Equation (9.3-16) is solved for the speed v and the result is substituted into Equation (9.3-15). The resulting equation is solved for r to give

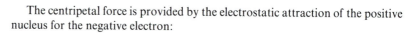

$$r = \frac{h^2 4\pi\varepsilon_0}{4\pi^2 m e^2} n^2 = a_0 n^2 \tag{9.3-17}$$

where a_0 is called the **Bohr radius** for an infinitely massive nucleus and is equal to 5.29×10^{-11} m. The radius of the electron's orbit is quantized: the distance of the electron from the nucleus can be one of the values given by Equation (9.3-17), but not any value between these. Figure 9.15 depicts the first few Bohr orbits with the quantized orbit radii given by Equation (9.3-17).

Exercise 9.18   **a.** Obtain Equation (9.3-17) from Equations (9.3-15) and (9.3-16).
b. Using the accepted values of the physical constants, verify the value of the Bohr radius.

The energy of the electron is also quantized. The potential energy for an electron in an orbit of radius r is

$$\mathscr{V} = -\frac{e^2}{4\pi\varepsilon_0 r} \tag{9.3-18}$$

where we choose a value of zero for the potential energy at infinite separation of the electron and the nucleus.

Exercise 9.19   Using the analogue of Equation (D-6) of Appendix D for the r direction, show that Equation (9.3-18) leads to the force expression on the right-hand side of Equation (9.3-15).

The kinetic energy is given by

$$\mathscr{K} = \frac{1}{2}mv^2 = \frac{1}{2}\frac{e^2}{4\pi\varepsilon_0 r} \tag{9.3-19}$$

where Equation (9.3-15) has been used to replace v^2. Comparison of Equations (9.3-18) and (9.3-19) shows that the kinetic energy is equal to half the magnitude of the potential energy. This is one of the consequences of the **virial theorem** of mechanics,[6] and it holds for any system acted upon only by electrostatic forces.

The total energy of the hydrogen atom is

$$E = E_n = \mathscr{K} + \mathscr{V} = -\frac{2\pi^2 m e^4}{(4\pi\varepsilon_0)^2 h^2 n^2} \tag{9.3-20}$$

where we have used Equation (9.3-17) for the value of r. We label each possible value of the energy with a value of n, the quantum number. Figure 9.16 depicts the first few energy levels. Each horizontal line segment is placed at a height proportional to the energy value.

The energy of an emitted or absorbed photon is equal to the difference between two quantized energies of the atom:

$$E(\text{photon}) = E_{n_2} - E_{n_1} = \frac{2\pi^2 m e^4}{(4\pi\varepsilon_0)^2 h^2}\left(\frac{1}{n_1^2} - \frac{1}{n_2^2}\right) \tag{9.3-21}$$

Figure 9.17 depicts the first few transitions corresponding to emission of photons.
Using the Planck-Einstein relation for the energy of the photon,

$$\frac{1}{\lambda} = \frac{E_{n_2} - E_{n_1}}{hc} = \frac{2\pi^2 m e^4}{(4\pi\varepsilon_0)^2 h^3 c}\left(\frac{1}{n_1^2} - \frac{1}{n_2^2}\right) \tag{9.3-22}$$

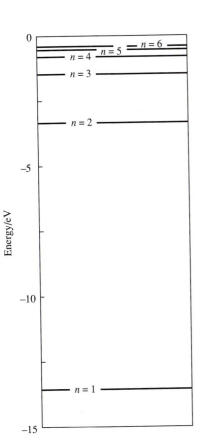

Figure 9.16. The Quantized Electron Energies by the Bohr Theory. The energy values are all negative, since an energy value of zero corresponds to enough energy to remove the electron from the atom.

[6] Ira N. Levine, *Quantum Chemistry*, 3rd ed., Allyn & Bacon, Boston, 1983, pp. 393ff.

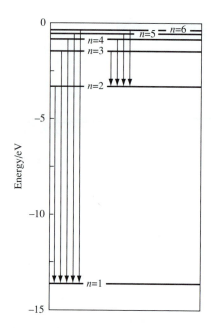

Figure 9.17. The Transitions Between Bohr Theory Energies for the Hydrogen Atom. These are some of the transitions that lead to the wavelengths given by the Rydberg formula.

This is exactly like the formula of Rydberg, Equation (9.3-13), with the constant $\mathscr{R}$ given by the expression in front of the square bracket. The first set of transitions shown in Figure 9.17, in which the lower-energy state (n_1 state) is the $n = 1$ state, corresponds to the series of wavelengths known as the Lyman series. The second set of transitions, in which $n_1 = 2$, is the Balmer series. The next series, which is not shown, is the Paschen series.

When the values of the physical constant are substituted into the expression of Equation (9.3-22), we get

$$\mathscr{R}_\infty = 1.097373 \times 10^7 \text{ m}^{-1} \qquad \textbf{(9.3-23)}$$

This value is labeled with the subscript ∞, corresponding to the assumption that the nucleus is stationary, as it would be if infinitely massive. If we replace the mass of the electron by the reduced mass of the electron and proton, Equation (D-27) of Appendix D, in order to correct for the actual motion of the nucleus, we get

$$\mathscr{R}_\text{H} = 1.09678 \times 10^7 \text{ m}^{-1} \qquad \textbf{(9.3-24)}$$

which is in good agreement with the experimental value.

The value of Rydberg's constant is for wavelengths measured in a vacuum. Wavelengths measured in air are slightly shorter than vacuum wavelengths, so the value of $\mathscr{R}$ in air is larger by a factor of the refractive index of air, equal to 1.00027 for visible wavelengths.

Exercise 9.20   Calculate the wavelength and frequency of the light emitted when n changes from 3 to 2. What color does this correspond to?

De Broglie Waves and the Schrödinger Equation

The theories of Planck, Einstein, and Bohr, known as the old quantum theory, were soon supplanted by a more successful quantum theory. This theory, sometimes known as "wave mechanics," was developed in the 1920s by de Broglie, Schrödinger, and others.

De Broglie Waves

Prince Louis Victor de Broglie, 1892–1977, who was a graduate student in 1923, won the Nobel Prize in physics in 1929 for this work.

In 1923, de Broglie was trying to find a physical justification for Bohr's hypothesis of quantization of angular momentum. In classical physics, one thing that is quantized is the wavelength of standing waves, given for example by Equation (9.2-18). De Broglie sought a way to relate this to Bohr's theory of the hydrogen atom, and came up with the idea that a moving particle such as an electron might somehow by accompanied by a "fictitious wave."[7] At the time, de Broglie thought of the particle as moving classically, but somehow being constrained in its motion so that it moved together with its accompanying wave.

We have already seen in Section 9.3 that light exhibits a wave-particle duality. According to Einstein's theory of relativity, a photon of energy E has a mass m such that

$$E = mc^2 \tag{9.4-1}$$

where c is the speed of light. This photon mass is not a rest mass such as a particle exhibits when it is stationary with respect to the laboratory in which its mass is determined.

If the Planck-Einstein relation, Equation (9.3-11), is used for the energy and mc is replaced by the momentum p, Equation (9.4-1) becomes

$$\frac{hc}{\lambda} = pc \quad \text{or} \quad p = \frac{h}{\lambda} \tag{9.4-2}$$

where λ is the wavelength.

De Broglie deduced that the velocity of the wave accompanying a particle is the same as the velocity of the particle if the wavelength is given by the same formula as in Equation (9.4-2):

$$\lambda = \frac{h}{p} = \frac{h}{mv} \tag{9.4-3}$$

We omit de Broglie's argument, which is more complicated than simply saying that Equation (9.4-3) is analogous to Equation (9.4-2).

The quantization assumption of Bohr's theory arises naturally from Equation (9.4-3) if one assumes that the circumference of a circular electron

---

[7] Jammer, *op. cit.*, pp. 243ff.

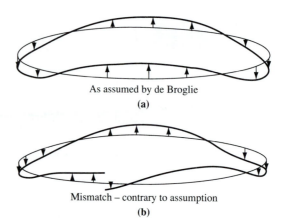

As assumed by de Broglie

(a)

Mismatch – contrary to assumption

(b)

Figure 9.18. De Broglie Waves Around a Closed Orbit. (a) An integral number of wavelengths on the circumference. Only if the circumference is an integral multiple of the wavelength. **(b)** Not an integral number of wavelengths on the circumference. In this case, a unique point occurs at which there is a discontinuity in the wave.

orbit in a hydrogen atom is equal to an integral number of wavelengths. This assumption means that the wave repeats itself with the same phase (with crests in the same positions) on each trip around the orbit, as depicted in Figure 9.18a. The situation in Figure 9.18b, in which a mismatch occurs, is assumed to be implausible, since a second circuit around the orbit would be out of phase with the first circuit.

For a circular orbit

$$2\pi r = n\lambda = \frac{nh}{mv} \tag{9.4-4}$$

This equation is the same as

$$mvr = \frac{nh}{2\pi} \tag{9.4-5}$$

which is Equation (9.3-16), the hypothesis of Bohr.

Although he had established his wave-particle relation only for the motion of electrons in the hydrogen atom, de Broglie hypothesized that this relation holds for all motions of all particles. This proposal of **matter waves** was revolutionary. In classical physics, a particle was supposed to be a localized massive object with its center of mass at a single point, and a wave was supposed to be delocalized type of motion. The idea that a particle is tied to a wave seemed implausible, and when de Broglie presented his doctoral thesis containing this proposal, the examining committee refused to believe that it might correspond to physical reality.

Only for objects of very small mass, such as electrons, can matter waves actually be observed.

▼ **EXAMPLE 9.5**   Calculate the de Broglie wavelength of a baseball of mass 5.1 ounces thrown at 95 miles per hour.

Solution

$$\lambda = \frac{6.6261 \times 10^{-34}\,\text{J s}}{(5.1\,\text{oz})(95\,\text{mi/h}^{-1})} \left(\frac{16\,\text{oz}}{1\,\text{lb}}\right)\left(\frac{1\,\text{lb}}{0.4536\,\text{kg}}\right)\left(\frac{3600\,\text{s}}{1\,\text{h}}\right)\left(\frac{1\,\text{mi}}{1609\,\text{m}}\right)$$

$$= 1.1 \times 10^{-34}\,\text{m}$$

De Broglie suggested at his final oral examination that electron diffraction by crystals could verify his theory. In 1927, after the development of the Schrödinger wave equation for matter waves, Davisson and Germer[8] accidentally grew a single crystal while heating a piece of nickel. When they irradiated this piece of nickel with a beam of electrons they observed diffraction effects, verifying the existence of de Broglie's matter waves.

Exercise 9.21   Find the speed of electrons with a de Broglie wavelength equal to 2.15×10^{-10} m, the lattice spacing in a nickel crystal.

The notion of a wave moving along with a particle as it traces out a classical trajectory has been abandoned, and we now speak of a wave-particle duality for electrons and other particles, with the wave-like properties inherently belonging to the object and not to an accompanying wave.

This wave-particle duality is illustrated by a hypothetical experiment in an evacuated system.[9] A beam of electrons is allowed to stream from a heated tungsten wire toward a partition with two slits in it, as depicted in Figure 9.19a. We assume that some means is used to give all of the electrons the same speed, and we assume that both slits are the same distance from the source. At some distance from the other side of the partition is a screen coated with a material such as zinc sulfide, which glows when an electron strikes it.

When an electron passes through a slit, its direction can be changed by interaction with the edges of the slit, and not all electrons reach the screen at the points lying on straight lines from the source through the slits. A distributed glowing pattern of bands is observed on the screen when an intense beam of electrons passes through the slits. This pattern is schematically depicted in Figure 9.19b, where the intensity of the glow is plotted as a function of position on the screen.

The pattern is explained by the constructive and destructive interference of waves appearing to pass through the two slits. If the difference in the path lengths from the two slits to a given point on the screen equals an integral number of wavelengths, there is constructive interference and a glowing band. Between the bands, there is destructive interference and little

---

[8] C. J. Davisson and L. H. Germer, *Phys. Rev.* **30**, 705 (1927).
[9] R. P. Feynman, R. B. Leighton, and M. Sands, *The Feynman Lectures on Physics*, Vol. 3, Addison-Wesley, Reading, MA, 1965, Ch. 1.

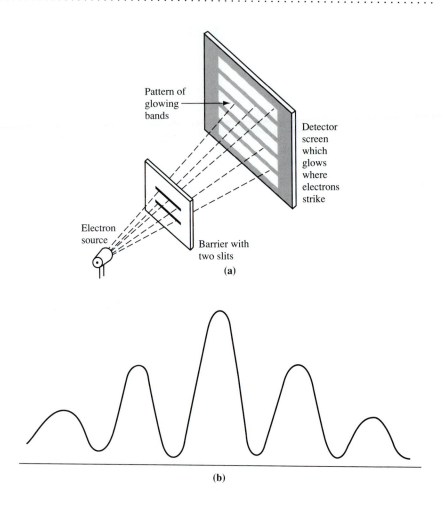

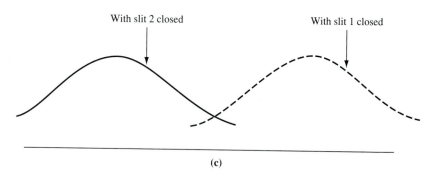

Figure 9.19. A Hypothetical Experiment with Electrons Passing Through Two Slits.
(a) The apparatus. **(b)** The intensity of the glow due to electrons arriving at the screen
in part (a) with both slits open. This diagram shows the diffraction pattern which proves
that the particles have a wave-like character. **(c)** The intensity of the glow due to electrons
arriving at the screen in part (a) with one slit open at a time. If only one slit is open at
a time, no diffraction pattern occurs.

or no glow. Exactly the same behavior is observed with light of a single wavelength instead of electrons.

If the intensity of the source is decreased to the point that electrons pass through the slits one at a time, it can be observed that each electron lands at a single point on the screen. There is a tiny localized flash when each electron arrives. Furthermore, if the screen is moved right up to the slits, each electron is found to pass through a single slit. Still, if the flashes are recorded and summed up, exactly the same pattern of diffraction bands appears as with an intense beam of electrons.

The experiment is now modified. The first slit is covered while the electrons continue to pass through the second slit. The diffraction pattern is no longer observed. There is a single band distributed about the point on the screen in a straight line through the open slit. Now the first slit is uncovered and the second slit is covered. Another single band is observed. The sum of these two single bands shows no interference effect, as shown schematically in Figure 9.19c.

Our observations are summarized and interpreted as follows: The path of any electron from the source to the screen cannot be specified when no attempt is made to detect its location along the path. Only when the screen is placed at the slits is it possible to say which slit the electron passes through. When the screen is some distance from the slits, there is no way to say whether the electron went through slit 1 or slit 2, and wave-like interference properties are observed as though the electron passed through both slits in a delocalized fashion.

A confused observer might plan to follow the path of an electron by shining a beam of light through the apparatus and detecting light scattered by the moving electrons. This procedure should show which slit each electron passes through. However, the resolution of an optical apparatus is no smaller than the wavelength of the light used. Accurate determination of the path would require light of a wavelength smaller than the distance between the slits. The energy of photons is inversely proportional to their wavelength, so if the slits are close together, the photons have high energy. If the energy of one photon is transferred to the electron when the light is reflected by the electron, the velocity of the electron will be greatly changed and it will no longer contribute to the same diffraction pattern as before.

The position of the electron can be determined only by doing something to it, like stopping it with a screen or allowing a photon to be scattered by it. If it strikes the screen, we can say where it hit the screen but cannot say exactly how it got there. Since a wave-like diffraction pattern is created by many electrons striking the screen, we must accept the fact that objects like electrons exhibit wave-like properties as well as particle-like properties.

Erwin Schrödinger, Austrian physicist who shared the 1933 Nobel Prize in physics with P. A. M. Dirac. Dirac pioneered the development of relativistic quantum mechanics, much as Schrödinger pioneered the nonrelativistic version which we discuss.

Peter Debye, 1884–1966, Dutch-American physicist and chemist who won the 1936 Nobel Prize in chemistry for his work on dipole moments (and the Debye of the Debye-Hückel theory).

The Schrödinger Equation

In 1925 Schrödinger was studying wave equations for classical waves. At this time, Debye reportedly invited Schrödinger to give a seminar on de Broglie's dissertation. With the proper mathematical tools at hand, Schrödinger managed to guess the correct wave equation for matter waves.

In 1926, he published a series of four papers.[10] The first three papers presented the time-independent version of the Schrödinger equation and applied it to the hydrogen atom, rotation and vibration of diatomic molecules, and the effect of an external electric field on energy levels. The time-dependent version of the equation was more difficult to obtain and was reported in the fourth paper at the end of 1926.

In the formal theory of quantum mechanics, the Schrödinger wave equation is taken as a postulate. It cannot be derived rigorously. In order to demonstrate a relationship with the classical wave equation, we obtain this equation nonrigorously for the case of a particle that moves parallel to the x axis. The reader who wishes to accept the Schrödinger equation as a postulate can skip to Equation (9.4-9).

Nonrigorous Derivation of the Schrödinger Equation

For a wave moving parallel to the x axis, the classical coordinate wave equation of (9.2-10) can be written

$$\frac{d^2\psi}{dx^2} + \frac{4\pi^2}{\lambda^2}\psi = 0 \qquad \textbf{(9.4-6)}$$

where we have used Equation (9.2-18) to replace the wave constant κ in terms of the wavelength λ. Use of the de Broglie relation, Equation (9.4-3), to replace λ gives

$$\frac{d^2\psi}{dx^2} + \frac{4\pi^2}{h^2}m^2v^2\psi = 0 \qquad \textbf{(9.4-7)}$$

This equation now represents a matter wave moving along the x axis.

We eliminate the speed v from our equation, using the relation

$$E = \mathscr{K} + \mathscr{V} = \frac{1}{2}mv^2 + \mathscr{V}(x) \qquad \textbf{(9.4-8)}$$

where $\mathscr{K}$ is the kinetic energy, $\mathscr{V}(x)$ is the potential energy, and E is the total energy.

The result of the preceding analysis is the **time-independent Schrödinger equation** for one-dimensional motion:

$$-\frac{\hbar^2}{2m}\frac{d^2\psi}{dx^2} + \mathscr{V}(x)\psi = E\psi \qquad \textbf{(9.4-9)}$$

where ψ is the coordinate wave function, or time-independent wave function. For an isolated particle, E is a constant. We introduce the symbol $\hbar$ ("h-bar"):

$$\hbar = \frac{h}{2\pi} \qquad \textbf{(9.4-10)}$$

The use of this quantity simplifies the form of many equations.

---

[10] The time-independent equations are presented in *Ann. Physik* **79**, 361 (1926); **79**, 489 (1926); and **80**, 437 (1926). The time-dependent equation is presented in *Ann. Physik* **81**, 109 (1926).

We abbreviate the left-hand side of Equation (9.4-9):

$$\hat{H} = -\frac{\hbar^2}{2m}\frac{d^2}{dx^2} + \mathcal{V}(x) \tag{9.4-11}$$

The time-independent Schrödinger equation can be written

$$\boxed{\hat{H}\psi = E\psi} \tag{9.4-12}$$

The quantity $\hat{H}$ is a **mathematical operator**, since it stands for carrying out mathematical operations. It is called the **Hamiltonian operator**. We discuss mathematical operators in Chapter 10.

The Time-Dependent Schrödinger Equation

For motion in the x direction, the time-dependent Schrödinger equation is

$$\boxed{\hat{H}\Psi = i\hbar\frac{\partial\Psi}{\partial t}} \tag{9.4-13}$$

where i is the imaginary unit

$$i = \sqrt{-1} \tag{9.4-14}$$

The function Ψ is the **time-dependent wave function**, or the displacement of the matter wave as a function of position and time. In this chapter and the next, we will use a capital psi (Ψ) for a time-dependent wave function and a lowercase psi (ψ) for a coordinate wave function.

The time-independent Schrödinger equation, Equation (9.4-12), can be obtained from the time-dependent equation by separation of variables. We assume a trial solution of the same type as with the classical wave function:

$$\Psi(x, t) = \psi(x)\zeta(t) \tag{9.4-15}$$

Once again, we use the method of separation of variables. Compare our present use of it with our use in Section 9.2.

where $\psi(x)$ will be found to be the same function as the coordinate wave function and ζ is a function of t.

We substitute (9.4-15) into Equation (9.4-13) and divide by $\psi(x)\zeta(t)$, obtaining

$$\frac{1}{\psi}\hat{H}\psi = \frac{i\hbar}{\zeta}\frac{d\zeta}{dt} \tag{9.4-16}$$

The variables x and t are separated in this equation. Each side is equal to a constant, which we denote by E:

$$\frac{1}{\psi}\hat{H}\psi = E$$

and

$$\frac{i\hbar}{\zeta}\frac{d\zeta}{dt} = E$$

Multiplication of the first equation by ψ and of the second equation by ζ gives

$$\hat{H}\psi = E\psi \tag{9.4-17}$$

and

$$\frac{d\zeta}{dt} = \frac{E}{i\hbar} \zeta \qquad (9.4\text{-}18)$$

Equation (9.4-17) is the same as the time-independent Schrödinger equation, Equation (9.4-12), so ψ is the same coordinate wave function as in that equation.

Equation (9.4-18) has the solution

$$\zeta(t) = Ce^{Et/i\hbar} = Ce^{-iEt/\hbar} \qquad (9.4\text{-}19)$$

where C is a constant. If we take $C = 1$, the full wave function is

$$\Psi(x, t) = \psi(x)e^{-iEt/\hbar} \qquad (9.4\text{-}20)$$

If we have a solution to the time-independent Schrödinger equation, including knowledge of the value of the energy E, we can immediately write a solution to the time-dependent equation by multiplying the coordinate wave function by the function ζ. As we will see, solution of the time-independent Schrödinger equation will deliver not only the coordinate wave function ψ but also the value of the energy. For this reason, it is often necessary to solve only the time-independent equation.

The time-dependent Schrödinger equation also has solutions that do not consist of a coordinate function multiplied by a time function, including some that represent traveling waves, and the time-independent Schrödinger equation does not necessarily apply to such solutions. The time-dependent equation applies to all cases.

The coordinate wave function can in many cases be chosen to be a real function. However, even in this case the full wave function is a complex function. The function ζ can be written

$$e^{-iEt/\hbar} = \cos(-Et/\hbar) + i \sin(-Et/\hbar)$$
$$= \cos(Et/\hbar) - i \sin(Et/\hbar) \qquad (9.4\text{-}21)$$

By an even function, we mean that $f(-x) = f(x)$. *By an odd function, we mean that* $(-x) = -f(x)$.

A standing wave corresponds to a stationary state.

where we have used the fact that the cosine is an even function and the sine is an odd function. The real part and the imaginary part oscillate with the same frequency, but out of phase. If the coordinate wave function is real, the real and imaginary parts of the full wave function have stationary nodes, representing a standing wave.

The Schrödinger Equation in Three Dimensions

For a single particle moving in three dimensions, the Hamiltonian operator is

$$\hat{H} = -\frac{\hbar^2}{2m}\left(\frac{\partial^2}{\partial x^2} + \frac{\partial^2}{\partial x^2} + \frac{\partial^2}{\partial x^2}\right) + \mathscr{V}(x, y, z)$$
$$= -\frac{\hbar^2}{2m}\nabla^2 + \mathscr{V}(x, y, z) \qquad (9.4\text{-}22)$$

The potential energy $\mathscr{V}$ can now depend on x, y, and z. The operator ∇^2 is the Laplacian operator defined in Equation (9.2-33).

The three-dimensional Schrödinger equations are still of the same form as Equations (9.4-12) and (9.4-13). However, since the Hamiltonian operator depends on all three coordinates, the coordinate wave function can depend on x, y, and z.

The Schrödinger Equation for a Multiparticle System

If the system consists of a number, n, of point-mass particles moving in three dimensions, the potential energy can depend on $3n$ coordinates. The Hamiltonian operator for such a system of n particles is

$$\hat{H} = -\sum_{j=1}^{n} \frac{\hbar^2}{2m_j} \nabla_j^2 + \mathcal{V}(q) \qquad \text{(9.4-23)}$$

The only differences between the Schrödinger equations for one system and another are the number of coordinates involved and the form of the potential energy function.

where ∇_j^2 is the Laplacian operator for the coordinates of particle j and q stands for the coordinates of all n particles. The Schrödinger equations are still written in the same form as Equations (9.4-12) and (9.4-13). The coordinate wave function is now a function of $3n$ coordinates.

Here is the general recipe for constructing a Schrödinger equation.

To treat a particular system, we find the potential energy function that applies to that system and write the Hamiltonian operator with that potential energy function and with a Laplacian term for every particle. We then write the Schrödinger equation with that Hamiltonian operator and a wave function depending on the coordinates of all particles.

Just as with the Schrödinger equation of a single particle, a solution to the Schrödinger equation for a system of many particles gives a solution to the time-dependent Schrödinger equation when multiplied by the time-dependent function of Equation (9.4-19).

Exercise 9.22   Carry out the steps to show that equations analogous to Equations (9.4-17) and (9.4-18) hold for a system of n particles.

Eigenvalue Equations

The time-independent Schrödinger equation is one of a class of equations called **eigenvalue equations**. The word eigenvalue is a "semitranslation" of the German word *Eigenwert*; a full translation is "characteristic value." An eigenvalue equation has on one side an operator operating on a function and on the other side a constant (the **eigenvalue**) multiplying the same function, which is called the **eigenfunction**. In the time-independent Schrödinger equation, the eigenvalue is the value of the energy, E, and is called the **energy eigenvalue**. The coordinate wave function is called the **energy eigenfunction**.

Eigenvalue equations such as the Schrödinger equation generally have a set of many solutions and not just a single solution.

A given eigenvalue equation generally has a set of many different solutions, each corresponding to a specific eigenvalue. A single eigenvalue can correspond to several eigenfunctions, but a single eigenfunction can correspond to only one eigenvalue. Two common cases occur: (1) the eigenvalue can take on any value within some range of values (a **continuous spectrum** of eigenvalues); (2) the eigenvalue can take on values only from a discrete set, with the values between the allowed values not permittted (a **discrete spectrum** of eigenvalues). The occurrence of a discrete spectrum of eigenvalues corresponds to quantization.

In addition to satisfying the Schrödinger equation, the wave function must satisfy other conditions. Since it represents a wave, we assume it has the properties that are shared by all waves; that is, the wave function is (1) single-valued, (2) continuous, and (3) finite.

9.5   The Particle in a Box; The Free Particle

In this section we solve the time-independent Schrödinger equation for the two simplest cases. This analysis will show how the form of the wave function is determined and how the energy quantization enters.

The Particle in a One-Dimensional Box

The particle in a one-dimensional box is a model system with the following properties:

1. The system consists of a single point-mass particle that can move parallel to the x axis.
2. The particle moves without friction but is confined to a finite segment of the x axis, from $x = 0$ to $x = a$. Inside this interval (the box) there is no force on the particle. Since the particle is absolutely confined in the box, we call it a "hard" box.

The macroscopic system most nearly represented by our model system would be a particle sliding in a tight-fitting (but frictionless) tube with closed ends, or a bead sliding on a frictionless wire between barriers. The principal chemical system represented by this model is an electron moving in a conjugated system of single and double bonds. The model only crudely represents this system, since the electron actually interacts with the other electrons and the nuclei in the molecule, but we will discuss its use in Chapter 13.

We will require a potential energy function for our Schrödinger equation. Since the particle experiences no force inside the box, its potential energy is constant there and we choose the value zero for this constant. If the potential energy function has a positive constant value for all positions outside the box, the particle must have a total energy at least as large as this value in order to escape from the box. To represent absolute confinement within the box, we say that this positive constant value is made to approach infinity.

Figure 9.20a shows the position of the particle as a function of time according to classical mechanics, and Figure 9.20b shows the velocity of the particle as a function of time, with the same time scale in both graphs. This behavior is derived by obtaining an equation of motion from Newton's

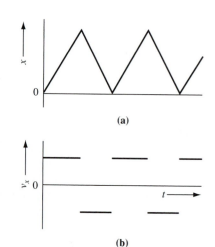

Figure 9.20. Mechanical Variables of a Particle in a Hard One-Dimensional Box. **(a)** The position according to classical mechanics. **(b)** The velocity of a particle according to classical mechanics. This diagram shows that the particle in a box moves back and forth at constant speed.

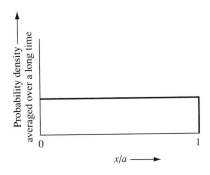

Figure 9.21. The Probability Density of the Position of a Particle in a Box as a Function of x According to Classical Mechanics, Averaged over a Long Time. The classical prediction is that the particle is equally likely to be found in any part of the box.

second law of motion and solving it, much as we solved it for the harmonic oscillator.

Figure 9.21 shows the probability density for finding the particle as a function of position in the box, averaged over a long time. This figure is analogous to Figure 9.3b for the harmonic oscillator. The average probability density is uniform, indicating that on the average the probability of finding the particle at one position in the box is equal to the probability of finding the particle at any other position in the box.

The time-independent Schrödinger equation for our system is

$$-\frac{\hbar^2}{2m}\frac{d^2\psi}{dx^2} + \mathcal{V}(x)\psi(x) = E\psi(x) \qquad \textbf{(9.5-1)}$$

where $\psi(x)$ is the coordinate wave function (energy eigenfunction) and E is the energy eigenvalue. We divide the x axis into three regions and solve separately in each region:

Region I: $x < 0$

Region II: $0 \leq x \leq a$

Region III: $a < x$

We have already stated the assumption that the full wave function is continuous. The coordinate wave function must also be continuous. After obtaining the solutions for the three regions, we will match them up at the boundaries between the regions, requiring ψ to be continuous at these boundaries.

In regions I and III, Equation (9.5-1) becomes

$$\frac{d\psi^2}{dx^2} - \lim_{\mathcal{V}\to\infty} \frac{2m\mathcal{V}}{\hbar^2}\psi = -\frac{2mE}{\hbar^2}\psi \qquad \textbf{(9.5-2)}$$

The left-hand side of Equation (9.5-2) would be infinite unless ψ vanishes, so we write

$$\psi^{(\text{I})}(x) = \psi^{(\text{III})}(x) = 0 \qquad \textbf{(9.5-3)}$$

This result is reasonable because the particle is confined to region II. We will always assert that a wave function vanishes in a region where there is no particle and vice versa.

For region II we have

$$\frac{d^2\psi^{(\text{II})}}{dx^2} = -\frac{2mE}{\hbar^2}\psi^{(\text{II})} = -\kappa^2\psi^{(\text{II})} \qquad \textbf{(9.5-4)}$$

where κ is given by

$$\kappa^2 = \frac{2mE}{\hbar^2} \qquad \textbf{(9.5-5)}$$

Equation (9.5-4) is the same as Equation (9.2-10) except for the symbols used. Its solution is

$$\psi^{(\text{II})}(x) = B\cos(\kappa x) + C\sin(\kappa x) \qquad \textbf{(9.5-6)}$$

For ψ to be continuous at $x = 0$ and $x = a$, we must have the boundary conditions

$$\psi^{(II)}(0) = \psi^{(I)}(0) = 0, \qquad \psi^{(II)}(a) = \psi^{(III)}(a) = 0 \qquad \text{(9.5-7)}$$

Notice the similarity to the boundary conditions for the vibrations of a string of finite length in Section 9.2.

For $\psi^{(II)}(0)$ to vanish, the constant B must vanish, because cos(0) equals unity and sin(0) equals zero. Thus

$$\psi^{(II)}(x) = C \sin(\kappa x) \qquad \text{(9.5-8)}$$

The condition that $\psi^{(II)}(a)$ vanishes imposes a condition on κ, as in Equation (9.2-15). The sine function vanishes only when its argument is an integral multiple of π, so that

$$n\pi = \kappa a \quad \text{or} \quad \kappa = \frac{n\pi}{a} \qquad \text{(9.5-9)}$$

where n is an integer. Our solution for region II is

$$\psi_n(x) = C \sin(n\pi x/a) \qquad \text{(9.5-10)}$$

where we now omit the superscript (II). There is a different energy eigenfunction for each integral value of n, the quantum number.

The energy eigenvalues are quantized. From Equation (9.5-5),

$$E = E_n = \frac{\hbar^2 \kappa^2}{2m} = \frac{\hbar^2 n^2 \pi^2}{2ma^2} = \frac{h^2 n^2}{8ma^2} \qquad \text{(9.5-11)}$$

There is a different energy eigenvalue for each energy eigenfunction. This case is called the **nondegenerate** case. In the **degenerate case**, more than one energy eigenfunction corresponds to a given value of E.

Figure 9.22a shows the energy eigenvalues, represented by horizontal line segments at heights proportional to their energy values. Figure 9.22b shows the wave functions (energy eigenfunctions). Each wave function is plotted on a separate axis, placed at a height in the diagram corresponding to its energy eigenvalue. Equation (9.5-10) resembles Equation (9.2-15), and each wave function in Figure 9.22b resembles one of the standing waves in Figure 9.9. As in that case, we eliminate negative values of n, because replacing a value of n by its negative does not change the energy eigenvalue and is equivalent to simply changing the sign of C. We also eliminate $n = 0$, since $n = 0$ implies $\psi = 0$.

The quantization of the energy eigenvalues is a consequence of solving the Schrödinger equation and requiring the solution to conform to the boundary condition that it vanish at the boundaries of the box. Unlike the quantization hypotheses of the old quantum theory, quantization has arisen in a natural way from the mathematical analysis of the eigenvalue equation. Quantization of energy occurs in the same way in other quantum mechanical systems, and always occurs when a particle is confined to a finite region.

The energy of a classical wave is determined by the amplitude of the wave (it is proportional to the square of the amplitude) and is independent

A quantum number is an integer or some other value that can be used to specify a specific wave function from a set of wave functions. This is our third example of a quantum number, and we will see numerous examples in future chapters.

The similarity between the classical wave in a string and the quantum mechanical wave function is striking. Notice the similarity between the effect of the boundary conditions in this case and in the case of a classical wave in a string of finite length, discussed in Section 9.2.

$\psi = 0$ means that no particle is present.

Confinement of a particle in a finite region always leads to quantization of its energy and to the existence of a zero-point energy.

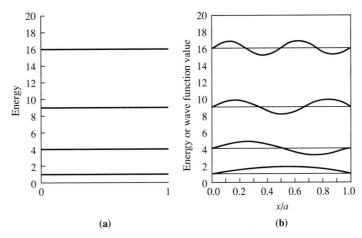

Figure 9.22. The Solutions to the Schrödinger Equation for a Particle in a One-Dimensional Box. (a) The energy eigenvalues. **(b)** The energy eigenfunctions. Compare the information about the motion of the particle in this figure with that in Figure 9.20.

of the wavelength. The amplitude of the wave described by the Schrödinger equation is equal to the constant C in Equation (9.5-10). The value of C does not affect the energy, since it is not determined by the Schrödinger equation. Later, we will introduce a normalization convention that assigns convenient values to such constants.

EXAMPLE 9.6

Find the energy of an electron in a box of length 1.000 nm (10.00 Å or 1000 pm) for $n = 1$.

Solution

$$E = \frac{(6.6261 \times 10^{-34} \text{ J s})^2 (1)^2}{(8)(9.109 \times 10^{-31} \text{ kg})(1.000 \times 10^{-9} \text{ m})^2} = 6.025 \times 10^{-20} \text{ J}$$

Exercise 9.23

How does the energy for a given value of n change if the length of the box is doubled? How does it change if the mass of the particle is doubled?

There is an important relationship between the energy of a matter wave and its wavelength. In the case of zero potential energy, all of the energy is kinetic energy and we can write

$$E = \frac{1}{2} mv^2 = \frac{p^2}{2m} \tag{9.5-12}$$

where we use the definition of the momentum, which is the mass times the velocity. From Equation (9.4-3) and Equation (9.5-12),

$$\lambda = \frac{h}{p} = \frac{h}{\sqrt{2mE}} \tag{9.5-13}$$

which is the same as

The energy of a de Broglie wave is inversely proportional to the square of its wavelength.

$$E = \frac{h^2}{2m\lambda^2}$$ (9.5-14)

The energy increases rapidly as the wavelength decreases.

Exercise 9.24

a. Argue that for a given value of n, the value of the wavelength for a particle in a box of length a is $2a/n$.

b. Show that the same formula for the energy as in Equation (9.5-11) is obtained by substituting the result of part a into Equation (9.5-14).

A larger number of nodes corresponds to a shorter de Broglie wavelength, to a larger speed, and to a larger value of the energy.

As the value of n increases, the energy increases, the wavelength decreases and the number of nodes increases. Furthermore, the more nodes, the higher the energy.

The energy in Equation (9.5-11) is kinetic energy, since we set the potential energy inside the box equal to zero. Because we do not allow $n = 0$, the minimum possible kinetic energy is positive and is called the **zero-point energy**. It is not possible for the particle in a box to have zero kinetic energy. This remarkable result is very different from classical mechanics, which always allows a particle to be at rest. All systems in which particles are confined in finite regions exhibit zero-point energies. In some systems, in addition to zero-point kinetic energy, there is also a minimum potential energy larger than the classically allowed minimum.

Exercise 9.24 illustrates the fact that we have mentioned repeatedly: Adding a constant to the potential energy has no physical effect.

If the potential energy inside the box is assigned a nonzero constant value $\mathscr{V}_0$ instead of zero, the energy eigenfunction is unchanged and the energy eigenvalue is increased by the value of $\mathscr{V}_0$, as shown in the next exercise.

Exercise 9.25

a. Solve the time-independent equation for the particle in a one-dimensional box with constant potential $\mathscr{V}_0$ in the box. Show that the energy eigenvalue is

$$E_n = \mathscr{V}_0 + \frac{h^2 n^2}{8ma^2}$$

but that the wave function is unchanged.

b. The result of part a is generally true. That is, adding a constant to the potential energy adds the same constant to the energy eigenvalues. Write the time-independent Schrödinger equation for a general system of n particles from Equation (9.4-23), and show that this statement is correct.

Just as in the Bohr theory of the hydrogen atom, energy can be gained or lost by a particle in a box only in amounts (**quanta**) of the appropriate sizes. If the particle is charged, this energy can be absorbed or emitted as radiant energy. The energy of a photon that is emitted or absorbed by a charged particle in a box will be equal to the difference in energy of the initial and final states of the particle.

Exercise 9.26	Calculate the wavelength and frequency of the photon emitted if an electron in a one-dimensional box of length 10.0 Å (1.000×10^{-9} m) makes a transition from $n = 2$ to $n = 1$ and the energy difference is entirely converted into the energy of the photon.

Equations (9.4-20) and (9.4-21) can be used to obtain the full wave function for a particle in a one-dimensional hard box.

$$\Psi_n(x, t) = C \sin\left(\frac{n\pi x}{a}\right) e^{-iE_n t/\hbar} \qquad (9.5\text{-}15)$$

It is generally possible to choose a real energy eigenfunction for a particle confined in a finite region, but the time-dependent wave function is always complex.

Exercise 9.27	Calculate the frequency of the matter wave for the $n = 1$ and $n = 2$ states of an electron in a box of length 1.000 nm. Compare these frequencies with the photon frequency in Exercise 9.26. Do you think there is any simple relationship between these frequencies?

Consider the difference between the solution of a classical equation of motion and our solution of the time-independent Schrödinger equation. In our classical solution, the particle is located at a point at a given time, and this point moves as shown in Figure 9.20. Any value of the energy is possible. In the solution of the time-independent Schrödinger equation, we obtain a set of coordinate wave functions (energy eigenfunctions) and a set of quantized energy eigenvalues. We can write the corresponding set of solutions to the time-dependent equation from Equation (9.5-15).

The state of the particle can be specified by specifying which wave function and energy eigenvalue correspond to the state of the particle. We recognize two cases:

1. The wave function of the system is known to be one or another of the energy eigenfunctions times the appropriate time-dependent factor, as in Equation (9.5-15). Chemists are usually interested in this case. For example, when a photon is absorbed or emitted by a molecule, we can identify the initial and final molecule states as states that correspond to energy eigenfunctions.
2. The wave function is some other function. Such a function must obey the time-dependent Schrödinger equation and the same boundary conditions as the energy eigenfunctions. It can be represented by a linear combination analogous to that of Equation (9.2-23):

$$\Psi(x, t) = \sum_{n=1}^{\infty} A_n \psi_n(x) e^{-iE_n t/\hbar} \qquad (9.5\text{-}16)$$

where the constants A_1, A_2,... are a set of time-independent constants. The first case is included in this case if one of the A's equals unity and the others all vanish. As in the classical case, this equation expresses the principle of superposition.

| **Exercise 9.28** | Show that the function of Equation (9.5-16) satisfies the time-dependent Schrödinger equation for the particle in a one-dimensional hard box. |

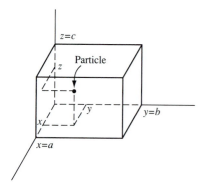

Figure 9.23. A Particle in a Three-Dimensional Box. This system contains a single particle that moves freely as long as it remains in the box.

The Particle in a Three-Dimensional Box

We now solve the Schrödinger equation for a model system consisting of a single point mass particle confined in a hard three-dimensional rectangular box. This system will be used later in the book to represent the translational motion of a gas molecule in a container and also the motion of an electron confined in such a region.

We place a cartesian coordinate system with its origin at the lower left rear corner of the box and with the coordinate axes perpendicular to the box walls. Denote the length of the box in the x direction by a, the length in the y direction by b, and the length in the z direction by c. To represent a hard box, we choose the potential energy function

$$\mathscr{V} = \begin{cases} 0 & \text{if } 0 < x < a \text{ and } 0 < y < b \text{ and } 0 < z < c \\ \mathscr{V}_0 & \text{otherwise (outside the box)} \end{cases} \quad \textbf{(9.5-17)}$$

and then take the limit that $\mathscr{V}_0$ approaches $+\infty$. Figure 9.23 depicts the system and shows the three cartesian coordinates specifying the position of the particle.

The Hamiltonian operator is given by Equation (9.4-22). Let us divide our space into two regions: region I, outside the box, and region II, inside the box. For reasons exactly the same as in the one-dimensional case, the coordinate wave function vanishes in region I. In region II, the time-independent Schrödinger equation is

$$\frac{\partial^2 \psi}{\partial x^2} + \frac{\partial^2 \psi}{\partial y^2} + \frac{\partial^2 \psi}{\partial z^2} = -\frac{2mE}{\hbar^2}\psi \quad \textbf{(9.5-18)}$$

Once again, the method of separation of variables. This time we have three coordinates and assume a trial function with three factors.

We solve Equation (9.5-18) by the method of separation of variables, as in Section 9.2. We write the trial function

$$\psi(x, y, z) = X(x)Y(y)Z(z) \quad \textbf{(9.5-19)}$$

The operator on the left-hand side of the Equation (9.5-18) has terms containing only one variable each, which implies that this trial function will separate the variables, as we will see. Substitution of the trial function into the Schrödinger equation (9.5-18) and division by XYZ gives

$$\frac{1}{X}\frac{d^2X}{dx^2} + \frac{1}{Y}\frac{d^2Y}{dy^2} + \frac{1}{Z}\frac{d^2Z}{dz^2} = -\frac{2mE}{\hbar^2} \quad \textbf{(9.5-20)}$$

after cancellation of factors. Since x, y, and z are independent variables, we can, if we wish, keep y and z fixed while allowing x to vary. Every term except the first then equals a constant, so the first term must be a constant function of x. We write

$$\frac{1}{X}\frac{d^2X}{dx^2} = K_x = -\frac{2mE_x}{\hbar^2} \tag{9.5-21}$$

where K_x is a constant and this equation defines another constant, E_x. The second and third terms of Equation (9.5-20) must also be constant functions:

$$\frac{1}{Y}\frac{d^2Y}{dy^2} = K_y = -\frac{2mE_y}{\hbar^2} \tag{9.5-22}$$

$$\frac{1}{Z}\frac{d^2Z}{dz^2} = K_z = -\frac{2mE_z}{\hbar^2} \tag{9.5-23}$$

where E_y and E_z are newly defined constants. To satisfy Equation (9.5-18),

$$E = E_x + E_y + E_z \tag{9.5-24}$$

We multiply Equation (9.5-21) by the function X:

$$\frac{d^2X}{dx^2} = -\frac{2mE_x}{\hbar^2}X \tag{9.5-25}$$

When we transcribe a solution to a mathematical equation, we use the fact that the mathematics of an equation is independent of the symbols used for the variables in the equation.

This equation is identical to Equation (9.5-4) except for the symbols used, and it has the same boundary conditions, so we can transcribe the solution of the one-dimensional problem:

$$X_{n_x}(x) = C_x \sin(n_x \pi x/a) \tag{9.5-26}$$

$$E_x = \frac{h^2}{8ma^2} n^2 \tag{9.5-27}$$

where we use the symbol n_x for the quantum number and C_x is a constant.

The Y and Z equations are identical except for the symbols used, so we can write their solutions:

$$Y_{n_y}(y) = C_y \sin(n_y \pi y/b) \tag{9.5-28}$$

$$Z_{n_z}(z) = C_z \sin(n_z \pi z/c) \tag{9.5-29}$$

$$E_y = \frac{h^2}{8mb^2} n_y^2 \tag{9.5-30}$$

$$E_z = \frac{h^2}{8mc^2} n_z^2 \tag{9.5-31}$$

Here n_y and n_z are positive integers that are not necessarily equal to n_x.

The energy eigenfunction is

$$\psi_{n_x n_y n_z}(x, y, z) = C \sin(n_x \pi x/a) \sin(n_y \pi y/b) \sin(n_z \pi z/c) \tag{9.5-32}$$

where we let $C = C_x C_y C_z$. The energy eigenvalue is

$$E_{n_x n_y n_z} = \frac{h^2}{8m}\left(\frac{n_x^2}{a^2} + \frac{n_y^2}{b^2} + \frac{n_z^2}{c^2}\right) \tag{9.5-33}$$

We attach the three quantum numbers n_x, n_y, and n_z to the symbols ψ and E. If the state of the system corresponds to an energy eigenfunction, the state is specified by giving the values of the three quantum numbers.

Consider a cubical box, which has $a = b = c$. This case leads to energy values that are degenerate. The energy eigenvalue for a particle in a hard cubical box is

$$E_{n_x n_y n_z} = \frac{h^2}{8ma^2}(n_x^2 + n_y^2 + n_z^2) \tag{9.5-34}$$

A set of states with equal energies is called an **energy level**, and the number of states making up the energy level is called the **degeneracy** of the energy level. For example, the two sets of quantum numbers (1, 2, 3) and (3, 2, 1) both correspond to the same energy.

▼   **EXAMPLE 9.7**

For an electron in a cubical box of side 1.00×10^{-9} m, find the energy and the degeneracy of the level in which the state corresponding to (1, 2, 3) occurs.

Solution

The energy eigenvalue is

$$E_{123} = \frac{14h^2}{8ma^2} = \frac{(14)(6.6261 \times 10^{-34} \text{ J s})^2}{(8)(9.109 \times 10^{-31} \text{ kg})(1.00 \times 10^{-9} \text{ m})^2}$$
$$= 8.43 \times 10^{-19} \text{ J}$$

There are six permutations of the three distinct numbers: (1, 2, 3), (2, 3, 1), (3, 1, 2), (3, 2, 1), (1, 3, 2) and (2, 1, 3). There are no other sets of three integers whose squares add up to 14, so the degeneracy is 6.

▲

Exercise 9.29

For an electron of Example 9.7, find the energies and degeneracies of all energy levels of lower energy than that in Example 9.7.

The Free Particle in One Dimension

The free particle is a system consisting of a mass point on which no forces act. The potential energy of the particle is a constant, which we set equal to zero. If the particle can move only parallel to the x axis, the time-independent Schrödinger equation is

$$-\frac{\hbar^2}{2m}\frac{d^2\psi}{dx^2} = E\psi \tag{9.5-35}$$

Equation (9.5-35) is the same as Equation (9.5-4) for the motion of a particle in a hard box. The general solution to Equation (9.5-35) is the same as the general solution to Equation (9.5-4), given by Equation (9.5-6). We write this solution (the energy eigenfunction) in a different way:

$$\psi(x) = De^{i\kappa x} + Fe^{-i\kappa x} \tag{9.5-36}$$

where the constant κ is given by Equation (9.5-5).

| **Exercise 9.30** | Use the identity |

$$e^{ix} = \cos(x) + i \sin(x) \tag{9.5-37}$$

to find the relations between the constants B, C, D, and F that cause Equation (9.5-6) and Equation (9.5-36) to represent the same function.

The boundary conditions are different from our previous case but still determine the constants in the wave function.

The boundary conditions are different from the case of the particle in a box, since there are now no walls at which the wave function must vanish. We must still conform our solution to the assumptions that the wave function be continuous and finite. The finiteness condition requires that κ be real. To show this fact, let

$$\kappa = a + ib$$

where a and b are real. The solution is now

$$\psi(x) = D e^{iax} e^{-bx} + F e^{-iax} e^{bx} \tag{9.5-38}$$

If b is positive, the second term grows without bound for large positive values of x. If b is negative, the first term grows without bound if x becomes large and positive. In order to keep the wave function finite, b must vanish, so that κ is real. There is no restriction on the values of D and F except that they must be finite.

The energy eigenvalues are given by Equation (9.5-5):

$$E = \frac{\hbar^2 \kappa^2}{2m} \tag{9.5-39}$$

There is no restriction on the values of the parameter κ except that it must be real, so E can take on any real nonnegative value. The energy is not quantized and there is no zero point energy.

Our wave functions can represent traveling waves. If F vanishes, the full wave function is

$$\Psi(x, t) = D e^{i\kappa x - iEt/\hbar} = D e^{i(\kappa x - Et/\hbar)} \tag{9.5-40}$$

where E is given by Equation (9.5-39). Separating the real and imaginary parts, we obtain

$$\Psi(x, t) = D[\cos(\kappa x - Et/\hbar) + i \sin(\kappa x - Et/\hbar)] \tag{9.5-41}$$

Comparing this with Equation (9.2-23) shows that both the real and imaginary parts are traveling waves propagating to the right with a speed given by

$$c = \frac{\hbar \kappa}{2m} \tag{9.5-42}$$

| **Exercise 9.31** | Show that Equation (9.5-42) is correct. |

Exercise 9.32   Show that the function

$$\Psi(x, t) = Fe^{-i\kappa x - iEt/\hbar} \tag{9.5-43}$$

represents a traveling wave moving to the left, and find its speed.

If D and F happen to be equal, we obtain a standing wave.

$$\psi(x) = D(e^{i\kappa x} + e^{-i\kappa x}) = 2D\cos(\kappa x) \tag{9.5-44}$$

Exercise 9.33   Use Equation (9.5-37) to verify Equation (9.5-44).

The full wave function corresponding to Equation (9.5-44) is

$$\Psi(x, t) = 2D\cos(\kappa x)e^{-iEt/\hbar} \tag{9.5-45}$$

Exercise 9.34   Show that if $D = -F$, a standing wave results. How does it compare with that of Equation (9.5-45)?

If the constants D and F are both nonzero but have unequal magnitudes, the full wave function becomes

$$\Psi(x, t) = De^{i(\kappa x - Et/\hbar)} + Fe^{-i(\kappa x + Et/\hbar)} \tag{9.5-46}$$

which represents a combination of traveling waves, one moving to the right and one moving to the left. This behavior is rather different from that found in classical mechanics, in which one state always corresponds to only one kind of behavior.

The Free Particle in Three Dimensions

The time-independent Schrödinger equation for a free particle moving in three dimensions is

$$\frac{\partial^2\psi}{\partial x^2} + \frac{\partial^2\psi}{\partial y^2} + \frac{\partial^2\psi}{\partial z^2} = -\frac{2mE}{\hbar^2}\psi \tag{9.5-47}$$

This is the same as Equation (9.5-18) for a particle inside a three-dimensional box. As with the particle in a box, we can solve the equation by separation of variables. We do not give the general solution. For the special case of a traveling wave moving in a definite direction, the energy eigenfunction is

$$\psi(x, y, z) = De^{i\kappa_x x}e^{i\kappa_y y}e^{i\kappa_z z} \tag{9.5-48}$$

where

$$\kappa_x^2 = \frac{2mE_x}{\hbar^2}, \qquad \kappa_y^2 = \frac{2mE_y}{\hbar^2}, \qquad \kappa_z^2 = \frac{2mE_z}{\hbar^2} \tag{9.5-49}$$

The vector $\boldsymbol{\kappa}$ with components κ_x, κ_y, and κ_z is called the **wave vector**. The direction in which this vector points is the direction in which the traveling wave moves.

The energy eigenvalue is given by

$$E = E_x + E_y + E_z = \frac{\hbar^2}{2m}(\kappa_x^2 + \kappa_y^2 + \kappa_z^2) \tag{9.5-50}$$

The three components of the wave vector can take on any real values. The energy is not quantized.

9.6

The Harmonic Oscillator

In order to make the one-dimensional Hamiltonian operator in Equation (9.4-9) apply to the harmonic oscillator, we write in the potential energy function of Equation (9.1-15). The Schrödinger equation is

$$\hat{H}\psi = -\frac{\hbar^2}{2m}\frac{d^2\psi}{dz^2} + \frac{1}{2}kz^2\psi = E\psi \tag{9.6-1}$$

The letter used for any variable is unimportant.

where our coordinate is now called z instead of x.

Let us define the constants

$$b = \frac{2mE}{\hbar^2}, \qquad a = \frac{\sqrt{km}}{\hbar} \tag{9.6-2}$$

so that the Schrödinger equation is now

$$\frac{d^2\psi}{dz^2} + (b - a^2z^2)\psi = 0 \tag{9.6-3}$$

Charles Hermite, 1822–1901, a great French mathematician who made many contributions to mathematics, including the proof that e (2.71828...) cannot solve any algebraic equation.

This differential equation is the same as one that was solved by Hermite. We present a brief outline of his solution.

The Solution of the Hermite Equation

The first step is to find an **asymptotic solution**, which is a solution that applies for very large magnitudes of z. If z has a very large magnitude, b will be negligible compared with a^2z^2, so that

$$\frac{d^2\psi}{dz^2} - a^2z^2\psi \approx 0 \quad \text{(for large magnitudes of } z) \tag{9.6-4}$$

The solution to this equation (the asymptotic solution) is

$$\psi_\infty \approx e^{\pm az^2/2} \tag{9.6-5}$$

Exercise 9.35 Show by taking the second derivative of the function in Equation (9.6-5) that it satisfies Equation (9.6-4) for large values of z. Hint: Neglect a term that is small compared to another term for large values of z.

There are two possible signs in the exponent in Equation (9.6-5). If the positive sign is taken, the function will grow without bound as $|z|$ becomes large, violating our boundary condition that the wave function is finite. We therefore reject the positive sign.

Our asymptotic solution satisfies Equation (9.6-4), but also satisfies Equation (9.6-3) for one particular value of b.

Exercise 9.36 Find the value of b and the corresponding value of E that make the function of Equation (9.6-5) satisfy Equation (9.6-3).

In order to find a general solution to Equation (9.6-3), we choose a trial solution of the form

$$\psi(z) = \psi_\infty(z)S(z) = e^{-az^2/2}S(z) \tag{9.6-6}$$

where $S(z)$ is a power series

$$S(z) = c_0 + c_1 z + c_2 z^2 + c_3 z^3 + \cdots = \sum_{n=1}^{\infty} c_n z^n \tag{9.6-7}$$

with constant coefficients $c_1, c_2, c_3, \ldots$. We might have tried to represent the solution by a power series instead of by a power series multiplied by the asymptotic solution, but this turns out to be intractable.[11] The second derivative of our trial solution is

$$\frac{d^2\psi}{dz^2} = e^{-az^2/2}\left[\frac{d^2S}{dz^2} - 2az\frac{dS}{dz} + (a^2z^2 - a)S\right]$$

Substitution of this expression into Equation (9.6-3) gives (after cancellation of two terms)

$$e^{-az^2/2}\left[\frac{d^2S}{dz^2} - 2az\frac{dS}{dz} + (b-a)S\right] = 0 \tag{9.6-8}$$

The exponential factor does not vanish for any finite real value of z, so the quantity in square brackets must vanish. From Equation (9.6-7), the first two derivatives of S are

$$\frac{dS}{dz} = \sum_{n=1}^{\infty} nc_n z^{n-1}$$

$$\frac{d^2S}{dz^2} = \sum_{n=2}^{\infty} n(n-1)c_n z^{n-2} = \sum_{j=0}^{\infty} (j+2)(j+1)c_{j+2} z^j$$

where we let $j = n - 2$. The index n or the index j can be called a "dummy index." The symbol used for it is unimportant; it just stands for successive integral values. We can therefore replace j by n without changing the sum, even though n now has a different meaning than in the original sum. Also, we can add an $n = 0$ term to the expression for dS/dz without any change, since the $n = 0$ term has a factor of zero. The result is

$$\sum_{n=0}^{\infty} [(n+2)(n+1)c_{n-2} - 2anc_n + (b-a)c_n]z^n = 0 \tag{9.6-9}$$

The quantity in the square brackets must vanish for each value of n, since every power of z on the right-hand side of the equation has a zero coefficient, and

---

[11] Ira N. Levine, *op. cit.*, pp. 58ff.

If two power series are equal to each other for all values of the independent variable, the two coefficients of a given power in the two series must be equal.

every power of z must have the same coefficient on both sides of the equation. Therefore,

$$c_{n+2} = \frac{2an + a - b}{(n+2)(n+1)} c_n \qquad (n = 0, 1, 2, \ldots) \qquad \textbf{(9.6-10)}$$

Equation (9.6-10) is called a **recursion relation**. Given a value of c_n, it provides a value for c_{n+2}. For example, if c_1 is given, then c_3 is determined, and thus c_5, etc. If c_0 is known, then c_2 is determined, and thus c_4, etc. If we pick any value for c_0 and any other value for c_1 and let the recursion relation pick the other values, then Equation (9.6-6) gives a solution to the Schrödinger equation.

Our solution must be finite for all values of z, including the limit as $|z|$ approaches infinity. However, if the series is permitted to have infinitely many terms, it is found that the series becomes large very rapidly for large values of $|z|$. It overcomes the rapidly decreasing gaussian factor, and the wave function fails to remain finite for large values of $|z|$.[12] Therefore, the series cannot have infinitely many terms. It obeys the Schrödinger equation if it does, but it does not obey the boundary condition that the wave function remain finite for all values of z.

If the series does not have an infinite number of terms, then it might have one term, it might have two terms, it might have three terms, etc. Each of these cases can occur. However, we cannot simply require all coefficients past a certain point in the series to vanish if this violates the recursion relation. The function would then fail to satisfy the Schrödinger equation. We must have a termination of the series that satisfies the recursion relation.

We cannot arbitrarily terminate the series. We must have a termination that satisfies the differential equation, which means that it must satisfy the recursion relation.

Let us say that c_{v+2} is a vanishing coefficient, where c_v does not vanish. The numerator in the right-hand side of Equation (9.6-10) must vanish for $n = v$:

$$2av + a - b = 2av + a - \frac{2mE}{\hbar^2} = 0 \qquad \textbf{(9.6-11)}$$

If v is an even integer, all of the odd-numbered coefficients must vanish, because there is no second recursion relation to terminate the part of the series containing odd-numbered coefficients. Therefore, a single solution will contain even powers of z or odd powers of z but not both.

Equation (9.6-11) dictates the value of the energy eigenvalue for the harmonic oscillator:

$$E = \left(\frac{\hbar^2}{2m}\right)(2av + a) = \frac{\hbar^2 a}{m}\left(v + \frac{1}{2}\right)$$

$$= \frac{h}{2\pi}\sqrt{\frac{k}{m}}\left(v + \frac{1}{2}\right) = hv\left(v + \frac{1}{2}\right) \qquad \textbf{(9.6-12)}$$

where v is the frequency of the oscillator predicted by classical mechanics, given by Equation (9.1-10), and where the quantum number v can take on any nonnegative integral value. The energy is quantized, with a discrete spectrum of equally spaced values.

---

[12] This assertion is not obvious. See Levine, *ibid.*, p. 61.

Since no lower energy is possible, E_0 is the zero point energy.

$$E_0 = \frac{1}{2}hv \quad \text{(zero point energy)} \qquad \textbf{(9.6-13)}$$

Figure 9.24 shows the energy levels superimposed on a graph of the potential energy function. The horizontal axis represents the coordinate z, but the quantized energy levels do not depend on z. They are constants.

Once again, solution of the Schrödinger equation and application of the boundary conditions imposed on the wave function have resulted in a quantization of energy. The wave functions decrease rapidly as $|z|$ becomes large. The matter wave is effectively confined to a region fairly close to $z = 0$, although the wave function does not completely vanish until $|z| \to \infty$.

Exercise 9.37 Find the frequency of a photon with energy equal to the difference in energy between the $v = 0$ state and the $v = 1$ state. How does this frequency compare with the classical frequency of the oscillator? How do you interpret this comparison?

The polynomial in z in the energy eigenfunction is called a **Hermite polynomial**. These polynomials are discussed in Appendix F. For $v = 0$,

$$\psi_0 = S_0 e^{-az^2/2} = c_0 e^{-az^2/2} \qquad \textbf{(9.6-14)}$$

A conventional choice for the value of c_0 gives

$$\psi_0 = \left(\frac{a}{\pi}\right)^{1/4} e^{-az^2/2} \qquad \textbf{(9.6-15)}$$

For $v = 1$, the energy eigenfunction is

$$\psi_1 = \left(\frac{4a^3}{\pi}\right)^{1/4} z e^{-az^2/2} \qquad \textbf{(9.6-16)}$$

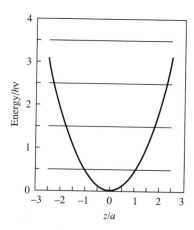

Figure 9.24. Harmonic Oscillator Energy Eigenvalues. This is really two diagrams: one to show how the potential energy depends on position, and one to show the quantized energy levels of the system as given by the Schrödinger equation.

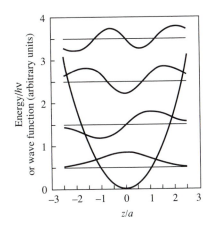

Figure 9.25. Harmonic Oscillator Wave Functions. This diagram shows the first four energy eigenfunctions for the harmonic oscillator, superimposed on a diagram of the energy eigenvalues.

For $v = 2$, the wave function is

$$\psi_2 = \left(\frac{a}{4\pi}\right)^{1/4}(2az^2 - 1)e^{-az^2/2} \qquad (9.6\text{-}17)$$

Other energy eigenfunctions can be generated from formulas for the Hermite polynomials in Appendix F.

Figure 9.25 shows the energy eigenfunctions for $v = 0$, $v = 1$, $v = 2$, and $v = 3$. Each wave function is plotted on a separate axis, which is placed at a height representing the energy eigenvalue. Compare these graphs with those for the particle in a one-dimensional hard box in Figure 9.22b. The general pattern of the nodes is the same. In addition to the nodes at infinite $|z|$, the lowest-energy wave function has no nodes, the next-lowest-energy wave function has one node, etc. The quantum number v is equal to the number of nodes (not counting the nodes at infinite $|z|$). This pattern occurs for any kind of system: the more nodes, the higher the energy.

Contrast the classical and quantum mechanical solutions for the harmonic oscillator. The classical solution gives the position and velocity of the oscillator as a function of time, as shown in Figure 9.2. The quantum mechanical solution gives a set of coordinate wave functions (energy eigenfunctions), as shown in Figure 9.25. When these are multiplied by the appropriate time factors from Equation (9.4-19), we have the displacement of a matter wave as a function of position and time. The classical state is described by specifying the position and velocity of the particle, but the quantum state is described by specifying which wave function applies.

EXAMPLE 9.8   Calculate the frequency of oscillation of the matter wave corresponding to the $v = 0$ state of the oscillating hydrogen atom of Example 9.2.

Solution

From Equation (9.4-19) the time-dependent factor of the wave function is

$$\zeta(t) = e^{-iE_0 t/\hbar} = e^{-ih\nu t/2\hbar} = e^{-\pi i\nu t}$$

Inspection of this equation shows that the frequency is $\nu/2$, where ν is the classical frequency of the oscillator, or 4.35×10^{13} s^{-1}. It is interesting that this frequency is not the same as that of the classical oscillator.

Exercise 9.38   Find the frequency of oscillation of the matter wave for the $v = 1$ state. Compare this frequency and the frequency from Example 9.8 with the frequency of the photon in Exercise 9.37.

There is an interesting difference between the wave functions of the harmonic oscillator and those of the particle in a hard box. The wave function for the particle in a box vanishes outside the classically permitted region (the box). That is, the matter wave extends only over the region in which the particle can move according to classical mechanics. Although the classical analysis says that the oscillating mass of a harmonic oscillator with a given energy cannot go past a turning point, the matter wave extends to infinity, although with a rapidly decreasing displacement for large values of z.

▼   **EXAMPLE 9.9**   Find the classical amplitude of oscillation for a hydrogen atom attached to a chemical bond as in Example 9.2. Take the energy as equal to that of the $v = 0$ quantum state.

Solution

For $v = 0$, the turning point is given by

$$z_t^2 = \frac{hv}{k} = \frac{h}{2\pi} \sqrt{\frac{1}{km}}$$

$$= \frac{6.6261 \times 10^{-34} \text{ J s}}{2\pi\sqrt{(500. \text{ N m}^{-1})(1.674 \times 10^{-27} \text{ kg})}}$$

$$= 1.15 \times 10^{-22} \text{ m}^2$$

$$z_t = 1.07 \times 10^{-11} \text{ m} = 0.107 \text{ Å}$$

▲

Exercise 9.39   Find the classical amplitude of oscillation of a hydrogen molecule with an energy equal to that of the $v = 0$ quantum state. Express it as a percent of the bond length, 0.74×10^{-10} m.

Summary

In this chapter we have introduced quantum mechanics, the theory on which the modern understanding of chemistry is based. We began with a solution of a classical equation of motion for the harmonic oscillator, obtaining formulas for the position and velocity of the mass as functions of time. The classical wave equation for a flexible string was also discussed. The position and velocity of each point of the string were prescribed for all times by the solution of the equation of motion.

The "old quantum theory" consists of theories with arbitrary assumptions of quantization, devised to explain phenomena that classical physics could not explain. Three parts of this theory were discussed: the blackbody radiation theory of Planck, the photoelectric effect theory of Einstein, and the hydrogen atom theory of Bohr.

De Broglie sought a physical justification for Bohr's assumption of quantization and hypothesized that all particles move with a wave-like character. According to this wave-particle duality, electrons and other objects have some of the properties of classical waves and some of the properties of classical particles. Schrödinger developed a wave equation for these matter waves. The time-independent equation is an eigenvalue equation given by

$$\hat{H}\psi = E\psi$$

where E is the energy of the system, ψ is a wave function, and $\hat{H}$ is the Hamiltonian operator. The time-dependent Schrödinger equation is

$$\hat{H}\Psi = ih\frac{\partial \Psi}{\partial t}$$

By assuming that the wave function Ψ is a product of a coordinate factor ψ and a time factor ζ, the coordinate factor is found to obey the time-independent Schrödinger equation.

The time-independent Schrödinger equation was written and solved for three example systems: the particle in a hard box, the free particle, and the harmonic oscillator. Sets of energy eigenfunctions and energy eigenvalues were obtained, and in the cases of the particle in a box and the harmonic oscillator, a discrete spectrum of energies was found, corresponding to energy quantization.

Two new phenomena were found to occur. First, the particle in a box and harmonic oscillator were both found to exhibit a zero point energy. Second, the harmonic oscillator was found to have a nonzero wave function in regions which classical mechanics predicts that the particle cannot enter.

Additional Reading

J. C. Davis, Jr., *Advanced Physical Chemistry*, Ronald Press, New York, 1965
  One of very few existing textbooks for a general physical chemistry course following the standard one-year undergraduate course. It contains a clear and fairly detailed discussion of nonrelativistic quantum mechanics.

M. W. Hanna, *Quantum Mechanics in Chemistry*, 3rd ed., Benjamin/Cummings, Menlo Park, 1981
  This fairly small book, available in paperback, is designed as a supplementary text for an undergraduate course in physical chemistry or as a text for part of a course. It is quite clear and gives somewhat more detail than most physical chemistry textbooks.

I. N. Levine, *Quantum Chemistry*, 3rd ed., Allyn & Bacon, Newton, MA, 1983
  This is a widely used text for a one-semester quantum chemistry course at the senior or first-year graduate level. It is clear and authoritative.

J. P. Lowe, *Quantum Chemistry*, Academic Press, New York, 1978
  A book for a course in quantum chemistry at the senior or first-year graduate level. It presents basic concepts, although it is oriented mostly toward molecular orbital theory.

F. Mandl, *Quantum Mechanics*, Butterworths Scientific Publications, London, 1975
  This theoretically inclined book for physics students at the graduate level presents the basic concepts in a careful and rather complete way. Many concepts are unfortunately buried in a welter of detail.

D. A. McQuarrie, *Quantum Chemistry*, University Science Books, Mill Valley, CA, 1983
A text for a standard course in quantum mechanics at the senior or first-year graduate level.

F. L. Pilar, *Elementary Quantum Chemistry*, McGraw-Hill, New York, 1968
This is a book for a one-year course in quantum chemistry at the first-year graduate level. It contains some things not readily found elsewhere in books written for chemists. Most undergraduates will feel that the word "elementary" does not belong in the title.

PROBLEMS

Problems for Section 9.2

*9.40. If a violin string is tuned so that its fundamental frequency is 440. hertz, find the frequency of each of the first three overtones.

9.41. Assume that the fundamental and the first overtone are simultaneously excited in a flexible string such that

$$z(y, t) = A_1 \sin(\pi y/L) \sin(\pi ct/L) + A_2 \sin(2\pi y/L) \sin(2\pi ct/L)$$

where

$$A_1 = 2A_2$$

Draw a rough graph representing the shape of the string at time $t = L/(4c)$ and at time $t = 3L/(4c)$. Comment on the differences in the two shapes.

9.42. Show that the relationship $v = c/\lambda$ for a traveling wave also holds for the standing wave in Equation (9.2-30).

*9.43. For liquid water at 25°C, the permittivity depends on frequency and for fairly low frequency is equal to 6.954×10^{-10} C^2 N^{-1} m^{-2}. The refractive index is defined as the ratio

$$\text{Refractive index} = n = c_{\text{vacuum}}/c_{\text{medium}}$$

The refractive index of water is 1.33 for visible light. Find the speed of light in water and the permittivity for frequencies corresponding to visible light. To four significant digits, the permeability is the same as that of a vacuum for these frequencies.

Problems for Section 9.3

9.44. a. Find the temperature of a blackbody with a maximum in its spectral radiant emittance curve at a wavelength of 430. nm.

b. Assume that the surface temperature of the sun is 5800. K and that it radiates like a blackbody. Find the wavelength of maximum spectral radiant emittance.

*9.45. Interstellar space is filled with isotropic radiation that corresponds to blackbody radiation with a temperature of 2.736 K. This fact substantiates the "big bang" theory of cosmology. Find the wavelength of maximum spectral radiant emittance of blackbody radiation at this temperature

and draw a graph of the spectral radiant emittance curve for this temperature.

9.46. Show that the normalized probability of a state of a Planck oscillator with energy $E = nh\nu$ is equal to

$$p(nh\nu) = (1 - e^{-h\nu/k_B T})e^{-nh\nu/k_B T}$$

and that the mean energy of an oscillator in Planck's theory is

$$\langle E \rangle = \frac{h\nu}{e^{h\nu/k_B T} - 1}$$

9.47. a. Derive an expression for the period of the electronic motion in the Bohr theory (the time required for an electron to make one circuit around a Bohr orbit).

b. Find the period and the frequency (the reciprocal of the period) for $n = 1$ and for $n = 1,000,000$.

*9.48. The Balmer series of hydrogen atom spectral lines corresponds to transitions from higher values of n to $n = 2$ in the Bohr energy expression. Find the wavelengths of all lines in the Balmer series that lie in the visible region.

9.49. Find the wavelengths of the first six lines in the hydrogen atom spectrum corresponding to transitions to $n = 1$. In what region of the electromagnetic spectrum do these lines lie?

9.50. A positronium atom is a hydrogen-like atom with a nucleus consisting of a positron (an antiparticle with charge e and mass equal to that of the electron).

a. Find the value of the Bohr radius for positronium.

b. Find the energy of the $n = 1$ state of positronium, and find the ratio of this energy to that of a hydrogen atom.

c. Find the radius of the circle in which each particle moves around the center of mass.

Problems for Section 9.4

*9.51. In Chapter 16, we will find that the root-mean-square speed of molecules of mass m in a gas is given by $v_{\text{rms}} = \sqrt{3k_B T/m}$. Calculate the de Broglie wavelength of an argon atom moving with a speed equal to the root-mean-square speed of argon atoms at 300. K.

9.52. Thermal neutrons are neutrons with a distribution of speeds nearly like the equilibrium distribution for gas molecules. In Chapter 16 we will find that the most probable speed of gas molecules of mass m is $\sqrt{2k_BT/m}$. Find the de Broglie wavelength of a neutron moving at the most probable speed for 300. K. Would thermal neutrons be useful for diffraction experiments to determine crystal lattice spacings?

9.53. Find the de Broglie wavelength of a 1500.-kg automobile moving at 65 miles per hour.

Problems for Section 9.5

9.54. Derive a formula for the kinetic energy of a particle with de Broglie wavelength equal to $2a/n$ and show that this is the same as the energy of a particle in a hard one-dimensional box of length a with quantum number n.

*9.55. The particle in a one-dimensional box is sometimes used as a model for pi electrons in a conjugated bond system (alternating double and single bonds).

a. Find the first three energy levels for a pi electron in 1,3-butadiene. Assume a carbon-carbon bond length of 1.39×10^{-10} m and assume that the box is hard and consists of the three carbon-carbon bonds plus an additional length of 1.39×10^{-10} m at each end.

b. The molecule has four pi electrons. Assume that two are in the state corresponding to $n = 1$ and that two are in the state corresponding to $n = 2$. Find the frequency and wavelength of the light absorbed if an electron makes a transition from $n = 2$ to $n = 3$.

9.56. **a.** Sketch a graph of the product of ψ_1 and ψ_2, the first two energy eigenfunctions of a particle in a one-dimensional box, and argue from the graph that the two functions are orthogonal, which means that $\int_{-\infty}^{\infty} \psi_1(x)^* \psi_2(x)\, dx = 0$.

b. Work out the integral and show that ψ_1 and ψ_2 are orthogonal.

9.57. Think of a baseball on its way from the pitcher's mound to home plate as being a particle in a box of length 60 feet. Assume that the baseball has a mass of 5.1 ounces. If the baseball has a speed of 95 miles per hour, find its kinetic energy and the value of the quantum number n corresponding to this value of E. Find the number of nodes in the wave function and find the wavelength corresponding to this many nodes in a length of 60 feet. Compare this wavelength with the de Broglie wavelength in Example 9.5.

Problems for Section 9.6

*9.58. Find the classical turning point for a harmonic oscillator that has the same energy as the $v = 1$ quantum mechanical energy. Calculate the probability that a quantum mechanical oscillator is farther from its equilibrium position than this value.

9.59. Using the recursion relation, Equation (9.6-10), obtain the energy eigenfunctions ψ_3, ψ_4, and ψ_5 for the harmonic oscillator.

9.60. A two-dimensional harmonic oscillator has the potential energy function

$$\mathscr{V} = \mathscr{V}(x, y) = \frac{k}{2}(x^2 + y^2)$$

a. Write the time-independent Schrödinger equation and find its solutions, using the one-dimensional harmonic oscillator solutions.

b. Find the energy eigenvalues and degeneracies for the first 10 energy levels.

*9.61. The harmonic oscillator is used as a model for molecular vibrations, considering the nuclei to be masses connected by spring-like chemical bonds. The molecule vibrates like a harmonic oscillator with mass equal to the reduced mass of the nuclei of the molecule.

a. Calculate the reduced mass of the nuclei of an HBr molecule.

b. The vibrational frequency of the HBr molecule is $v = 7.944 \times 10^{13}$ s^{-1}. Find the force constant k.

9.62. A harmonic oscillator potential energy function is modified so that

$$\mathscr{V} = \begin{cases} kz^2/2 & \text{if } |z| < z' \\ \infty & \text{if } |z| > z' \end{cases}$$

where z' is some positive constant that is greater than the classical turning point for the energies we will consider.

a. Tell qualitatively how this will affect the classical solution.

b. Tell qualitatively how this will affect the quantum mechanical solution.

c. Will tunneling occur? Draw a rough sketch of the first two wave functions.

General Problems

9.63. Consider an automobile with a coil spring at each wheel. If a mass of 50. kg is suspended from one such spring, the spring lengthens by 0.025 m. The "unsprung weight" (the effective mass of the wheel and suspension components) of one wheel is 40. kg. The mass of the part of the automobile supported by the springs is 1400. kg.

a. Find the force constant for each spring.

b. Assuming that all four springs are identical and that one-fourth of the supported mass is supported at each wheel, find the distance that each spring is compressed from its equilibrium length when the automobile is resting on its wheels.

c. Find the potential energy of each spring when the automobile is resting on its wheels.

d. Find the period and the frequency of oscillation of a wheel when it is hanging freely.

e. If the automobile is suddenly lifted off its wheels, find the speed of the wheel when the spring passes through its equilibrium length if no shock absorber is present to slow it down.

f. Find the energy of a quantum of energy of an oscillating wheel according to quantum mechanics.

g. Find the value of the quantum number when the energy of the oscillating wheel is equal to the energy of part c.

h. Find the wavelength of the electromagnetic radiation whose photons have energy equal to $h\nu$, where ν is the frequency of oscillation of part d.

*9.64. Calculate the de Broglie wavelength of an electron moving with the kinetic energy corresponding to the $n = 4$ state of a hydrogen atom according to the Bohr theory. Show that this wavelength is equal to one-fourth of the circumference of the fourth Bohr orbit.

9.65. Assume that the motion of the earth around the sun is described by the Bohr hydrogen atom theory. The electrostatic attraction is replaced by the gravitational attraction, given by the formula

$$F = \frac{Gm_1 m_2}{r^2}$$

where G is the gravitational constant, equal to $6.67 \times 10^{-11} \ \mathrm{m^3 \ s^{-2} \ kg^{-1}}$ and m_1 and m_2 are the masses of the two objects. The mass of the earth is 5.983×10^{24} kg, and the mass of the sun is larger by a factor of 332958. The earth's orbit is slightly elliptical, but pretend that it is circular with a radius of 1.4967×10^{11} m. Assume that the sum is stationary (as it would be if it were infinitely massive).

a. Find the value of the Bohr radius.

b. Find the value of the quantum number corresponding to the size of the earth's actual orbit.

c. Find the kinetic energy, the potential energy, and the total energy of the earth's orbital motion.

d. Find the ratio of the reduced mass of the earth-sun system to the mass of the earth.

*9.66. Identify each statement as either true or false. If a statement is true only under special circumstances, label it as false.

a. A de Broglie wave can be identified as a transverse wave.

b. A de Broglie wave can be identified as a longitudinal wave.

c. The oscillating quantity in a de Broglie wave cannot be physically identified.

d. The Bohr theory of the hydrogen atom is a hybrid theory, maintaining elements of classical mechanics along with quantization.

e. Several different eigenvalues can correspond to the same eigenfunction.

f. Several different energy eigenfunctions can correspond to the same energy eigenvalue.

g. Light can be identified as a wave in a pervasive medium.

h. Light exhibits both wave-like and particle-like properties.

i. If the length of its box is made to approach infinity, a particle in a box behaves like a free particle.

j. A free particle cannot be described by a standing-wave type of wave function, but must be described by a traveling-wave type of wave function.

10 The Principles of Quantum Mechanics: II. The Postulates of Quantum Mechanics

OBJECTIVES

After studying this chapter, the student should:

1. understand the idea of founding quantum mechanical theory on postulates,

2. be able to solve a variety of problems involving mathematical operators,

3. be able to construct a quantum mechanical operator from the classical expression for the corresponding operator,

4. be able to calculate expectation values for mechanical variables and determine whether the variable has a unique value for the state in question,

5. understand the determination of the state of a system by a set of measurements.

PREVIEW

In this chapter, we present and discuss the theoretical foundation of quantum mechanics.

PRINCIPAL FACTS AND IDEAS

1. The formulation of quantum mechanics is based on a set of postulates.
2. The first two postulates establish the role of the wave function in quantum mechanics.
3. The third postulate of quantum mechanics establishes a connection between each mechanical variable and a mathematical operator.
4. The fourth postulate provides the means of obtaining information about the values of mechanical variables.
5. The fifth postulate concerns the determination of the state of a system by experimental measurements.

10.1 The First Two Postulates of Quantum Mechanics

The Schrödinger equation is central to quantum mechanics in much the same way that Newton's second law of motion is central to classical mechanics. However, Schrödinger did not originally derive his equation from other principles, and it is not based on experimental fact. We therefore take the time-dependent Schrödinger equation as a **postulate**. A postulate is a fundamental assumption on which a theory is based, and the consequences of the postulates must be compared with experiment to validate the theory.

385

Werner Karl Heisenberg, 1901–1976, was a German physicist who invented matrix mechanics, a form of quantum mechanics equivalent to the Schrödinger formulation, and who discovered the uncertainty principle, for which he received the 1932 Nobel Prize in physics.

The way in which the state is specified is crucial to quantum mechanics.

Just as Newton's second law is augmented by two other laws, the Schrödinger equation is augmented by additional postulates. Schrödinger, Heisenberg, and others devised several postulates that form a consistent logical foundation for quantum mechanics. We will state five postulates in a form similar to that of Mandl[1] and Levine.[2]

The first two postulates have already been discussed and used in Chapter 9, but without calling them postulates. We restate them as follows:

Postulate I. *All information that can be obtained about the state of a mechanical system is contained in a wave function Ψ, which is a continuous, finite, and single-valued function of time and of the coordinates of the particles of the system.*

Postulate II. *The wave function Ψ obeys the time-dependent Schrödinger equation*

$$\hat{H}\Psi = i\hbar \frac{\partial \Psi}{\partial t} \tag{10.1-1}$$

where $\hat{H}$ is the Hamiltonian operator, $\hbar = h/2\pi$, and h is Planck's constant.

In quantum mechanics, since the available information about the state is contained in a wave function, we can specify the state by specifying which wave function applies to the system at a given instant. Information about values of energy, momentum, etc. must be obtained from this wave function rather than from values of coordinates and velocities as in classical mechanics.

There is a **one-to-one relationship** between the state of the system and its wave function. That is, to each state there corresponds one wave function, and to each wave function there corresponds one state. The terms "state" and "wave function" are often used interchangeably, and the wave function is sometimes referred to as the **state function**.

The idea that the values of some mechanical variables can be uncertain even if the state is completely known is a strange idea for a newcomer to quantum mechanics.

Specification of the state in quantum mechanics usually gives less information about the mechanical variables of the system than in classical mechanics, where specification of the state allows precise calculation of the values of all mechanical variables. In some cases we will be able to predict with certainty from a known wave function what result an error-free measurement of a mechanical variable will give. In other cases we will find that only statistical predictions can be made, even in the absence of experimental error.

There are solutions to the time-dependent Schrödinger equation that are not of this form, but we will not consider them.

It is not necessary to establish the time-independent Schrödinger equation by a separate postulate. It can be derived from the time-dependent equation, as shown in Chapter 9. If

$$\Psi = \psi(q)\zeta(t) \tag{10.1-2}$$

---

[1] F. Mandl, *Quantum Mechanics*, Butterworths Scientific Publications, London, 1957, pp. 60ff.

[2] I. N. Levine, *Quantum Chemistry*, 3rd ed., Allyn & Bacon, Newton, MA, 1983, pp. 136ff.

where q stands for all the coordinates of the particles in the system, then the coordinate wave function ψ satisfies the time-independent Schrödinger equation (is an energy eigenfunction).

10.2 Mathematical Operators

A **mathematical operator** is a symbol standing for a mathematical operation. When the symbol for an operator is written to the left of the symbol for a function, the operation is to be applied to that function. For example, d/dx is a derivative operator; z is a multiplication operator, standing for multiplication by the variable z; and c is also a multiplication operator, standing for multiplication by the constant c. The result of operating on a function with an operator is another function. If $f(x)$ is the function on which we operate and $g(x)$ is the resulting function, in most cases $g(x)$ is a different function from $f(x)$. Figure 10.1 shows an example of a function, $f(x) = \ln(x)$, and $g(x) = 1/x$, the result of operating with the derivative operator, d/dx.

The only function that is equal to its own derivative is e^x.

Linear Operators

An operator $\hat{A}$ is linear if

$$\hat{A}[f(q) + g(q)] = \hat{A}f(q) + \hat{A}g(q)$$

(10.2-1a)

and if

$$\hat{A}[cf(q)] = c\hat{A}f(q)$$

(10.2-1b)

where c is a constant and f and g are arbitrary functions of the variables on which $\hat{A}$ acts. The independent variable or variables are represented by the single symbol q.

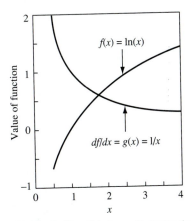

Figure 10.1. A Function and Its Derivative. This figure illustrates the fact that operating on a function generally yields a different function.

Hermitian Operators

An operator $\hat{A}$ is **hermitian** if

$$\int f^*\hat{A}g \, dq = \int (\hat{A}f)^*g \, dq = \int (\hat{A}^*f^*)g \, dq \qquad (10.2\text{-}2)$$

The letter q stands for all coordinates. For example, if q represents the coordinates of two particles that can move in three dimensions, the integral is a sixfold integral, and dq stands for $dx_1 dy_1 dz_1 dx_2 dy_2 dz_2$. The integral in this equation is a definite integral, extending over all values of the coordinates. If the particles can move in all of space, the integration limits are $-\infty$ to ∞ for each coordinate. We will often omit the limits of integration, but all of our integrals in quantum mechanics are definite integrals. The functions f and g must obey boundary conditions which ensure that the integral converges.

In Equation (10.2-2), f^* denotes the complex conjugate of the function f and $\hat{A}^*$ denotes the complex conjugate of the operator $\hat{A}$. Any complex quantity z can be written in the form

$$z = x + iy \qquad (10.2\text{-}3)$$

Definition of real part and imaginary part. Note that y is a real number, although it is called the imaginary part of z.

where x and y are real quantities and $i = \sqrt{-1}$. The quantity x is called the **real part** of z and y is called the **imaginary part** of z. The **complex conjugate** of z is defined as

$$z^* = x - iy \quad \text{(definition of } z^*) \qquad (10.2\text{-}4)$$

The fact that a complex conjugate can be taken by changing the sign in front of every i often saves us the trouble of separating the imaginary part from the real part of a complex expression.

That is, the complex conjugate of z has the same real part as z and an imaginary part that is the negative of that of z. A real quantity is equal to its complex conjugate.

It is a fact that any complex expression can be turned into its complex conjugate by changing the sign in front of every i in the expression. The following exercise is one example.

Exercise 10.1

Using the identity

$$e^{ix} = \cos(x) + i\sin(x)$$

show that

$$(e^{ix})^* = e^{-ix} \qquad (10.2\text{-}5)$$

An operator can be separated into a real part and an imaginary part, and its complex conjugate is obtained in the usual way.

The quantum mechanical operators that we will use are hermitian operators. It can be shown that hermitian operators are linear, but a linear operator is not necessarily hermitian.

▼ **EXAMPLE 10.1**

Show (a) that the operator d/dx is linear and (b) that it is not hermitian.

Solution

a. Linearity:

$$\frac{d}{dx}(f+g) = \frac{df}{dx} + \frac{dg}{dx}$$

$$\frac{d(cf)}{dx} = c\frac{df}{dx} \qquad \text{Q.E.D.}$$

b.

$$\int_{-\infty}^{\infty} f(x)^* \frac{dg}{dx}\, dx = f(x)^* g(x)\Big|_{-\infty}^{\infty} - \int_{-\infty}^{\infty} g(x)\frac{df^*}{dx}\, dx$$

If f and g obey the same boundary conditions as wave functions, they will vanish at the limits of integration, and we have

$$\int_{-\infty}^{\infty} f(x)^* \frac{dg}{dx}\, dx = -\int_{-\infty}^{\infty} \frac{df^*}{dx} g(x)\, dx$$

which is the negative of what we would require for a hermitian operator.

▲

Exercise 10.2

a. Show that the multiplication operator x is linear and hermitian.
b. Show that the operator $i(d/dx)$ is linear and hermitian.
c. Show that any hermitian operator is linear.

Operator Algebra

There is a kind of algebra in which we manipulate the operators themselves without specifying the functions on which they might operate. We write equations containing the operators, but not containing the symbols for the functions that are operated on.

For example, the operator that always produces the same function as the one on which it operates is called the **identity operator** and is denoted by $\hat{E}$:

$$\hat{E}f(q) = f(q) \qquad (10.2\text{-}6)$$

Equation (10.2-6) can be written as an operator equation:

$$\hat{E} = 1 \qquad (10.2\text{-}7)$$

That is, the identity operator is equivalent to the multiplication operator 1 (multiplication by unity). An **operator equation** means that the operators on the two sides of the equation always produce the same result when applied to any well-behaved function.

The **product of two operators** is defined as successive application of the operators and is denoted by writing the two operator symbols adjacent to each other:

$$\hat{C}f(q) = \hat{A}\hat{B}f(q) = \hat{A}(\hat{B}f(q)) = \hat{A}g(q) \qquad (10.2\text{-}8a)$$

It is important that the rightmost operator operates first. Since we are accustomed to reading text from left to right, there is opportunity for confusion here.

where $g(q)$ is the function produced when $\hat{B}$ operates on $f(q)$. The operator written on the right operates first. The first equality in Equation (10.2-8a) can be written in operator form:

$$\hat{C} = \hat{A}\hat{B} \tag{10.2-8b}$$

Operator multiplication is **associative**, so that

$$\hat{A}\hat{B}\hat{C} = (\hat{A}\hat{B})\hat{C} = \hat{A}(\hat{B}\hat{C}) \tag{10.2-9}$$

Operator multiplication is not necessarily **commutative**. It can happen that

$$\hat{A}\hat{B} \neq \hat{B}\hat{A} \quad \text{(in some cases)} \tag{10.2-10}$$

If $\hat{A}\hat{B} = \hat{B}\hat{A}$, the operators $\hat{A}$ and $\hat{B}$ are said to **commute**.

We define the **commutator** of two operators $\hat{A}$ and $\hat{B}$, which is denoted by $[\hat{A}, \hat{B}]$ and is equal to the two-term operator

$$\boxed{[\hat{A}, \hat{B}] = \hat{A}\hat{B} - \hat{B}\hat{A} \quad \text{(definition of commutator)}} \tag{10.2-11}$$

If two operators commute, their commutator vanishes.

▼ **EXAMPLE 10.2**

Find the commutator $\left[x, \dfrac{d}{dx} \right]$.

Solution

In Example 10.1 it is necessary to allow the commutator to operate on an arbitrary function in order to proceed with the solution.

We let the commutator act on an arbitrary function, $f(x)$:

$$\left[x, \frac{d}{dx} \right] f(x) = x\frac{df}{dx} - \frac{d(xf)}{dx} = x\frac{df}{dx} - x\frac{df}{dx} - f(x) = -f(x)$$

As an operator equation,

$$\left[x, \frac{d}{dx} \right] = -\hat{E} = -1$$

▲

Exercise 10.3

Find the commutator $\left[x^2, \dfrac{d}{dx} \right]$.

The following facts are useful: (1) Every operator commutes with itself. (2) Multiplication operators commute with each other. (3) A constant multiplication operator commutes with all other hermitian operators. (4) A derivative operator almost never commutes with a multiplication operator containing the same independent variable. (5) Operators that act on different variables commute with each other.

▼ **EXAMPLE 10.3**

Find the operator $\hat{C} = (\hat{\mathcal{H}} + \hat{\mathcal{V}})^3$ if $\hat{\mathcal{H}}$ and $\hat{\mathcal{V}}$ are two operators that do not commute.

Solution

$$C = (\hat{\mathscr{H}} + \hat{\mathscr{V}})^3 = (\hat{\mathscr{H}} + \hat{\mathscr{V}})(\hat{\mathscr{H}} + \hat{\mathscr{V}})(\hat{\mathscr{H}} + \hat{\mathscr{V}})$$
$$= (\hat{\mathscr{H}} + \hat{\mathscr{V}})(\hat{\mathscr{H}}^2 + \hat{\mathscr{H}}\hat{\mathscr{V}} + \hat{\mathscr{V}}\hat{\mathscr{H}} + \hat{\mathscr{V}}^2)$$
$$= \hat{\mathscr{H}}^3 + \hat{\mathscr{H}}^2\hat{\mathscr{V}} + \hat{\mathscr{H}}\hat{\mathscr{V}}\hat{\mathscr{H}} + \hat{\mathscr{H}}\hat{\mathscr{V}}^2 + \hat{\mathscr{V}}\hat{\mathscr{H}}^2 + \hat{\mathscr{V}}\hat{\mathscr{H}}\hat{\mathscr{V}} + \hat{\mathscr{V}}^2\hat{\mathscr{H}} + \hat{\mathscr{V}}^3$$

The order of the operators in every term must be maintained, since the operators $\hat{\mathscr{H}}$ and $\hat{\mathscr{V}}$ do not commute.

Terms like $\hat{\mathscr{H}}\hat{\mathscr{V}}\hat{\mathscr{H}}$ and $\hat{\mathscr{V}}\hat{\mathscr{H}}^2$ are different from each other if the two operators do not commute.

▲

Exercise 10.4

a. Find the operator $(\hat{\mathscr{H}} + \hat{\mathscr{V}})^4$ if $\hat{\mathscr{H}}$ and $\hat{\mathscr{V}}$ do not commute.
b. Find the operator $(\hat{A} + \hat{B})^4$ if $\hat{A}$ and $\hat{B}$ do commute.

Properties of Hermitian Operators

Hermitian operators have some properties that are useful in quantum mechanics:

Property 1: Two hermitian operators are not required to commute with each other.

Exercise 10.5

Show that the two hermitian operators x and $i(d/dx)$ do not commute and find their commutator.

Property 2: A hermitian operator has a set of eigenfunctions.

We have already seen the Hamiltonian operators for the particle in a box and for the harmonic oscillator have their own sets of eigenfunctions. Other hermitian operators also have their own sets of eigenfunctions. For an operator $\hat{A}$ we write

$$\hat{A}f_j(q) = a_j f_j(q) \tag{10.2-12}$$

The eigenfunction f_j must be a function of the same independent variables as the variables on which the operator $\hat{A}$ operates.

where a_j is the eigenvalue corresponding to the eigenfunction $f_j(q)$. There will generally be a set of infinitely many eigenfunctions: $f_1(q), f_2(q), f_3(q), \ldots$.

One and only one eigenvalue corresponds to a given eigenfunction, but more than one eigenfunction can correspond to a given eigenvalue. The number of eigenfunctions that have eigenvalues equal to a certain value is called the **degeneracy** of that eigenvalue.

The eigenvalues of an operator can be discrete (separated from each other) or continuous (take on any real value in some range). The discrete eigenvalues are quantized and are said to form a **discrete spectrum**. The continuous eigenvalues are said to form a **continuous spectrum**.

▼   **EXAMPLE 10.4**

Find the eigenfunctions and eigenvalues of the operator d^2/dx^2.

Solution

$$d^2f/dx^2 = af(x)$$

Example 10.4 is solved by a standard technique for solving a linear differential equation with constant coefficients: a trial solution of the form e^{bx} is assumed and substituted into the differential equation.

This equation is solved by assuming the trial solution

$$f(x) = e^{bx}$$

where b is a constant. Substitution of this function into the original equation and division by e^{bx} gives the **characteristic equation**:

$$b^2 = a_j$$

The eigenvalue equals b^2. If we apply the same boundary conditions that wave functions obey, the eigenfunction must remain finite. If x can take on any real value, the finiteness condition requires that b must be imaginary, without a real part. If we let $b = ik$, then

$$f = Ce^{ikx}$$

where C is a constant and k is a real constant. The eigenvalue equals $-k^2$. The spectrum of eigenvalues is continuous, with k able to take on any real value.

The solution of a linear differential equation can always be multiplied by a constant that can take on any value and still be a solution.

▲

Exercise 10.6

a. Argue that b in Example 10.4 cannot have a real part if the above boundary conditions are applied.

b. Find the eigenvalues and eigenfunctions of the operator d^4/dx^4.

Property 3: The eigenvalues of a hermitian operator are real.

The eigenvalues of certain hermitian operators will be equal to measured values of mechanical variables, which must be real.

To establish property 3, we take Equation (10.2-12), multiply both sides by the complex conjugate of the eigenfunction, and integrate over all values of the coordinates:

$$\int f_j^* \hat{A} f_j^* \, dq = a_j \int f_j^* f_j \, dq \tag{10.2-13}$$

We now apply the definition of a hermitian operator, Equation (10.2-2), to the left-hand side of this equation:

$$\int f_j^* \hat{A} f_j \, dq = \int (\hat{A}^* f_j^*) f_j \, dq$$

From the complex conjugate of Equation (10.2-12), we can replace $\hat{A}^* f_j^*$ by $a_j^* f_j^*$:

$$\int f_j^* \hat{A} f_j \, dq = \int (\hat{A}^* f_j^*) f_j \, dq = a_j^* \int f_j^* f_j \, dq \tag{10.2-14}$$

The right-hand side of Equation (10.2-14) must equal the right-hand side of Equation (10.2-13), so

$$a_j^* = a_j \tag{10.2-15}$$

A quantity equals its complex conjugate if and only if it is real.

In this derivation we have found the necessary definitions and properties and combined them into a fairly simple and straightforward proof. In this kind of proof, it is necessary to keep in mind where we start from and what we need to show and to find a path from the starting point to the finish.

Property 4: Two eigenfunctions of a hermitian operator with different eigenvalues are orthogonal to each other.

Two functions f and g are **orthogonal** to each other if

$$\int f^*g \, dq = \int g^*f \, dq = 0 \quad \left(\begin{array}{c}\text{definition of}\\ \text{orthogonality}\end{array}\right) \qquad \textbf{(10.2-16)}$$

where the integrations extend over all values of the coordinates. The two integrals in Equation (10.2-16) are the complex conjugates of each other, so that if one vanishes, so does the other.

We prove property 4 as follows: Multiply the eigenvalue equation, Equation (10.2-12), by f_k^*, the complex conjugate of a different eigenfunction, and integrate:

$$\int f_k^* \hat{A} f_j \, dq = a_j \int f_k^* f_j \, dq \qquad \textbf{(10.2-17)}$$

Now apply the hermitian property to the left-hand side of this equation:

$$\int f_k^* \hat{A} f_j \, dq = \int (\hat{A}^* f_k^*) f_j \, dq = a_k^* \int f_k^* f_j \, dq$$

$$= a_k \int f_k^* f_j \, dq \qquad \textbf{(10.2-18)}$$

where we have replaced a_k^* by a_k because we know a_k to be real. The left-hand sides of Equations (10.2-17) and (10.2-18) are equal, so the difference of the right-hand sides vanishes:

$$(a_j - a_k) \int f_k^* f_j \, dq = 0$$

If the two eigenvalues are not equal to each other, the integral must vanish, and we have proved the orthogonality of f_k and f_j:

$$\int f_k^* f_j \, dq = 0$$

If two eigenfunctions have equal eigenvalues, they are not necessarily orthogonal to each other.

▼ **EXAMPLE 10.5**

Show that the first two eigenfunctions of the Hamiltonian operator for the particle in a hard one-dimensional box are orthogonal to each other.

Solution

Since the wave function vanishes outside the region $0 \le x \le a$, we can change the limits of the integral to 0 and a:

$$\int_{-\infty}^{\infty} \psi_1(x)\psi_2(x) \, dx = \frac{2}{a} \int_0^a \sin\left(\frac{\pi x}{a}\right) \sin\left(\frac{2\pi x}{a}\right) dx$$

$$= \frac{2}{\pi} \int_0^{\pi} \sin(y) \sin(2y) \, dy = 0$$

If you can argue without calculation that an integral vanishes, you can sometimes save some time. The principal such argument is that the integrand function is an odd function about the center of the interval of integration.

where we have looked the integral up in Appendix C. One can also make a graph of the integrand and argue that the positive and negative contributions to the integral cancel each other.

▲

Exercise 10.7 Show that the first two energy eigenfunctions of the harmonic oscillator are orthogonal to each other.

Property 5: Two commuting hermitian operators can have a set of common eigenfunctions.

This statement means that a set of functions $f_{jk}(q)$ can be found such that

$$\hat{A}f_{jk}(q) = a_j f_{jk}(q)$$
$$\hat{B}f_{jk}(q) = b_k f_{jk}(q)$$

where $\hat{A}$ and $\hat{B}$ are two operators that commute and a_j and b_k are eigenvalues. Two indices are needed to enumerate all of the functions in the set, because several functions can have the same eigenvalue for $\hat{A}$ but different eigenvalues for $\hat{B}$. We omit the proof of this property.

10.3   Postulate III. Mathematical Operators in Quantum Mechanics

In the next two sections, we establish by postulate how hermitian operators are used in quantum mechanics to obtain information about the values of mechanical variables.

The third postulate is:

Postulate III. *To every mechanical variable, there is a hermitian mathematical operator in one-to-one correspondence.*

This postulate is suggested by the pattern seen in the time-independent Schrödinger equation:

$$\hat{H}\psi = E\psi$$

The operator $\hat{H}$ is the only operator associated with the variable E in such an equation, and E is the only variable associated with $\hat{H}$. That is, there is a **one-to-one correspondence** between $\hat{H}$ and E. We postulate that any other mechanical variable, such as momentum, angular momentum, or position, has its own operator in one-to-one correspondence with it.

Determination of the Operator Corresponding to a Particular Variable

We use the correspondence between the Hamiltonian operator and the energy as a pattern for establishing operators for other mechanical variables.

We write the classical expression for the energy (the "classical Hamiltonian") and associate it with the Hamiltonian operator. The classical

Hamiltonian must be written as a function of momentum, not velocity (see Appendix D). In cartesian coordinates, the momentum is the mass times the velocity. For one particle moving parallel to the x axis,

$$\frac{p_x^2}{2m} + \mathscr{V}(x) \leftrightarrow -\frac{\hbar^2}{2m}\frac{d^2}{dx^2} + \mathscr{V}(x) \tag{10.3-1}$$

where the symbol $\leftrightarrow$ means "is in one-to-one correspondence with."

The potential energy function $\mathscr{V}(x)$ occurs on both sides in the same way, so we postulate that the operator for the potential energy is the operator for multiplication by the potential energy function.

$$\hat{\mathscr{V}} \leftrightarrow \mathscr{V}(x) \tag{10.3-2}$$

We extend this assumption and postulate that: *Any function of coordinates corresponds to the operator for multiplication by that function.*

If the potential energy is canceled from the two sides of Equation (10.3-1), the remaining terms indicate that the operator for the kinetic energy $\mathscr{K}$ is

$$\hat{\mathscr{K}} = \frac{1}{2m}\hat{p}_x^2 = -\frac{\hbar^2}{2m}\frac{d^2}{dx^2} \tag{10.3-3}$$

The operator for the square of the x component of the momentum is therefore

$$\hat{p}_x^2 = -\hbar^2\frac{d^2}{dx^2} \tag{10.3-4}$$

The operator for the square of a momentum component must be the square of the operator for that momentum component. The square of an operator means operating twice with the operator. Therefore,

Equation (10.3-5) is the fundamental relation for the construction of operators that correspond to mechanical variables.

$$\hat{p}_x = -i\hbar\frac{d}{dx} = \frac{\hbar}{i}\frac{d}{dx} \tag{10.3-5}$$

Since any quantity has two square roots, the opposite sign could also have been taken. The sign in Equation (10.3-5) gives the momentum the correct sign when a particle is moving in a known direction (see Problem 10.32).

We complete the third postulate by the additional assumption that the pattern of Equation (10.3-5) holds for all momentum components and all functions of momentum components: *The quantum mechanical operator for any mechanical variable is obtained by (1) expressing the quantity classically in terms of cartesian coordinates and cartesian momentum components and (2) replacing the momentum components by $\hbar/i$ times the derivative with respect to the appropriate coordinate.*

This argument is not the only way to establish the proper relationship between a mechanical variable and its operator. Another way is to require the commutation relation of Equation (10.3-11), which leads to Equation (10.3-5).

If another coordinate system is required in a particular problem, the expression for every operator is constructed in cartesian coordinates and then transformed to the other coordinate system. Even then, for a complicated variable, the above recipe must be augmented by the requirement that the operator be hermitian, and the operator must be verified by comparing its action with experimental fact.

We can now justify the form of the Hamiltonian operator for motion in three dimensions, Equation (9.4-22). In three dimensions

$$\mathscr{H} = \frac{p_x^2 + p_y^2 + p_z^2}{2m} \tag{10.3-6}$$

and Equation (9.4-22) follows.

▼ **EXAMPLE 10.6**

Construct the operator for the z component of the angular momentum of one particle about the origin of a cartesian coordinate system.

Solution

The angular momentum is defined in Appendix D as the vector product

$$\mathbf{L} = \mathbf{r} \times \mathbf{p}$$

The expression for the z component is

$$L_z = xp_y - yp_x \tag{10.3-7}$$

The operator is

$$\hat{L}_z = \frac{\hbar}{i}\left[x\frac{\partial}{\partial y} - y\frac{\partial}{\partial x}\right] \tag{10.3-8}$$

▲

Exercise 10.8

Construct the operator for L_z^2.

▼ **EXAMPLE 10.7**

For motion in the x-y plane, transform the expression for $\hat{L}_z$ to plane polar coordinates.

Solution

The necessary relations are

$$\phi = \arctan(y/x), \qquad \rho^2 = (x^2 + y^2)$$

If f is an arbitrary function of x and y and is also expressible as a function of ρ and ϕ, then

$$x\frac{\partial f}{\partial y} = x\frac{\partial f}{\partial \phi}\frac{\partial \phi}{\partial y} + x\frac{\partial f}{\partial \rho}\frac{\partial \rho}{\partial y} = x\frac{\partial f}{\partial \phi}\frac{x}{x^2 + y^2} + x\frac{\partial f}{\partial \rho}\frac{y}{(x^2 + y^2)^{1/2}}$$

$$y\frac{\partial f}{\partial x} = y\frac{\partial f}{\partial \phi}\frac{\partial \phi}{\partial x} + y\frac{\partial f}{\partial \rho}\frac{\partial \rho}{\partial x} = y\frac{\partial f}{\partial \phi}\frac{-y}{x^2 + y^2} - y\frac{\partial f}{\partial \rho}\frac{x}{(x^2 + y^2)^{1/2}}$$

Since the second terms cancel,

$$\hat{L}_z f = \frac{\hbar}{i}\left[\frac{\partial f}{\partial \phi}\frac{x^2}{x^2 + y^2} - \frac{\partial f}{\partial \phi}\frac{-y^2}{x^2 + y^2}\right] = \frac{\hbar}{i}\frac{\partial f}{\partial \phi}$$

so that

$$\hat{L}_z = \frac{\hbar}{i}\frac{\partial}{\partial \phi} \tag{10.3-9}$$

Example 10.7 illustrates the procedure that must be used if an operator is needed in other than cartesian coordinates. The operator is first written in cartesian coordinates and then transformed to another coordinate system.

This expression also holds for spherical polar coordinates and for cylindrical polar coordinates in three dimensions, although we do not prove this fact.

The expression for the operator $\hat{L}_z$ in Equation (10.3-9) will be important in our later discussions of atomic and molecular wave functions. We will also use the expression for $\hat{L}^2$ in spherical polar coordinates, which we present without proof. This operator is closely related to the Laplacian operator expressed in spherical polar coordinates, which we present in the next chapter.

$$\hat{L}^2 = -\hbar^2 \left[\frac{1}{\sin(\theta)} \frac{\partial}{\partial \theta} \sin(\theta) \frac{\partial}{\partial \theta} + \frac{1}{\sin^2(\theta)} \frac{\partial^2}{\partial \phi^2} \right] \qquad \textbf{(10.3-10a)}$$

This equation can also be written

$$\hat{L}^2 = -\hbar^2 \left[\frac{\partial^2}{\partial^2 \theta} + \cot(\theta) \frac{\partial}{\partial \theta} + \frac{1}{\sin^2(\theta)} \frac{\partial^2}{\partial \phi^2} \right] \qquad \textbf{(10.3-10b)}$$

The commutation relations between operators are useful.

▼   **EXAMPLE 10.8**   Find the commutator $[\hat{x}, \hat{p}_x]$.

Solution

Operate on an arbitrary function $f(x)$:

$$[\hat{x}, \hat{p}_x]f = \frac{\hbar}{i} \left[x \frac{\partial f}{\partial x} - \frac{\partial (xf)}{\partial x} \right] = -\frac{\hbar}{i} f$$

so that

Example 10.8 is solved by operating on an arbitrary function. The result does not depend on what this function is.

$$[\hat{x}, \hat{p}_x] = -\frac{\hbar}{i} = i\hbar \qquad \textbf{(10.3-11)}$$

Equation (10.3-11) is an important result. Some authors find the form of $\hat{p}_x$ by postulating that this commutation relation must hold, instead of deducing the form of $\hat{p}$ by inspection of the Hamiltonian operator.

▲

Exercise 10.9   Show that $[\hat{L}_x, \hat{p}_y] = i\hbar\hat{p}_z$.

10.4   Postulate IV. Expectation Values

The first postulate of quantum mechanics asserts that if we know what the wave function of a system is, we know its state. Any information about the values of mechanical variables must be obtained from the wave function.

In classical mechanics, knowledge of the state allows simple calculation of the value of any mechanical variable. In quantum mechanics, the calculation is less direct and can lead to predictions with uncertainties.

The fourth postulate provides the methods for obtaining this information:

Postulate IV. (a) *If a mechanical variable A is measured without experimental error, the only possible measured values of a variable A are eigenvalues of the operator $\hat{A}$ that corresponds to A.*

(b) *The expectation value for the error-free measurement of a mechanical variable A can be calculated from the formula*

$$\langle A \rangle = \frac{\int \Psi^* \hat{A} \Psi \, dq}{\int \Psi^* \Psi \, dq} \tag{10.4-1}$$

where $\hat{A}$ is the operator corresponding to the variable A and $\Psi = \Psi(q, t)$ is the wave function corresponding to the state of the system immediately prior to the measurement.

The definite integrals in Equation (10.4-1) extend over all values of the coordinates.

Prediction of the Outcomes of Measurements

The first part of the fourth postulate is independent of the state of our system. It allows us to determine the list of possible values for any variable by solving the eigenvalue equation for that variable. If there is a discrete spectrum of eigenvalues, the variable is quantized (can take on values from a discrete list).

The **expectation value** is a prediction of the mean value of a set of many values selected randomly from a population. The second part of the postulate hypothesizes that if a set of many error-free measurements of the variable A is undertaken, the mean of the set will be equal to the result obtained from Equation (10.4-1).

This part of the fourth postulate gives us a way to obtain as much information as we can obtain about the values of all mechanical variables for a given state of a system. For certain systems, certain states, and certain variables, it is possible to make a precise prediction of the outcome of a measurement from knowledge of the wave function. This case will be called case I.

For some systems, some states, and some variables, even though the state (and thus the wave function) of the system are known, it is impossible to make a certain prediction of the outcome of a single error-free measurement. The outcomes of individual measurements will be distributed over various values (all of which must be eigenvalues of the operator). Only statistical predictions can be made. This case will be called case II.

Normalization

The denominator in Equation (10.4-1) is the same whether we are calculating the expectation value of the angular momentum, the energy, or any other variable. There is a way to eliminate this denominator: Every wave function that satisfies the Schrödinger equation has a constant multiplying it, such as the constant C in Equation (9.5-10). Any value of this constant can be chosen.

Exercise 10.10

a. Show that if a wave function Ψ satisfies the time-dependent Schrödinger equation

$$\hat{H}\Psi = i\hbar \frac{\partial \Psi}{\partial t}$$

then the function $C\Psi$ also satisfies it, where C is any constant.
b. Show that if a wave function ψ satisfies the time-independent Schrödinger equation

$$\hat{H}\psi = E\psi$$

then the function $C\psi$ also satisfies it, where C is any constant.

We can choose a value of a constant multiplying a wave function such that

$$\int \Psi^*\Psi \, dq = 1 \quad \text{(definition of normalization)} \qquad \textbf{(10.4-2)}$$

The wave function Ψ is then said to be **normalized**. The advantage of using normalized wave functions is that the denominator in equations like (10.4-1) can be omitted.

Exercise 10.11

Carry out the integration to show that the harmonic oscillator wave function in Equation (9.6-15) is normalized.

Position Measurements

For almost any state, position measurements belong to case II. Consider the position of a particle that moves parallel to the x axis. If the wave function of the system just before each one of a set of position measurements is $\Psi(x, t)$, the expectation value of x is

$$\langle x \rangle = \int \Psi(x, t)^* x \Psi(x, t) \, dx \qquad \textbf{(10.4-3)}$$

if the wave function Ψ is normalized. We will not discuss the eigenfunctions of the position operator in detail, but all values of x can be eigenvalues and are thus possible outcomes of the position measurement. See Problem 10.28 for some information about these eigenfunctions and eigenvalues.
Since the multiplication operator x commutes with multiplication by Ψ^*, we can write

$$\langle x \rangle = \int x\Psi(x, t)^*\Psi(x, t) \, dx = \int x|\Psi(x, t)|^2 \, dx \qquad \textbf{(10.4-4)}$$

where we use the fact that any quantity times its complex conjugate is equal to the square of the absolute magnitude of the quantity.

Exercise 10.12
For a complex quantity $z = x + iy$, show that
$$|z|^2 = z^*z = x^2 + y^2$$
The real nonnegative quantity z^*z is the square of the magnitude of z. The **magnitude** or **absolute value** of a complex quantity z is

$$\boxed{|z| = \sqrt{z^*z}}$$ (10.4-5)

EXAMPLE 10.9

Find the expectation value for the position of a particle in a one-dimensional hard box of length a if the coordinate wave function is known to be the energy eigenfunction with $n = 1$.

Solution

The normalized particle-in-a-box wave function is

$$\psi_n(x) = \sqrt{\frac{2}{a}} \sin\left(\frac{n\pi x}{a}\right)$$ (10.4-6)

The expectation value is

$$\langle x \rangle = \frac{2}{a} \int_0^a \sin\left(\frac{\pi x}{a}\right) e^{iE_1 t/\hbar} x \sin\left(\frac{\pi x}{a}\right) e^{-iE_1 t/\hbar} \, dx$$

Example 10.9 is solved by substituting the expression for the wave function into the formula for the expectation value and carrying out the integral, using the method of substitution to obtain an integral that was looked up in a table (see Appendix C). Note how the fact that the wave function vanishes outside the box was used to eliminate the rest of the x axis from the range of integration.

We have used the fact that the integrand vanishes except in the region between $x = 0$ and $x = a$. We used a normalized wave function and used Equation (10.2-5) to write the complex conjugate of $\exp(-iEt/\hbar)$. The two complex exponentials cancel each other. We have

$$\langle x \rangle = \frac{2}{a} \int_0^a \sin\left(\frac{\pi x}{a}\right) x \sin\left(\frac{\pi x}{a}\right) \, dx = \frac{2}{a} \int_0^a x \sin^2\left(\frac{\pi x}{a}\right) \, dx$$
$$= \frac{2}{a}\left(\frac{a}{\pi}\right)^2 \int_0^\pi y \sin^2(y) \, dy = \frac{a}{2}$$

The predicted mean position of the particle is the middle of the box. This is a reasonable result, since the box is symmetrical about its center. That is, the two halves of the box are just like mirror images of each other.

Exercise 10.13
Show that the particle-in-a-box energy eigenfunction given in Equation (10.4-6) is normalized.

Since the time-dependent factors canceled in Example 10.9, the coordinate wave function (energy eigenfunction) could have been used in the original integral instead of the full wave function. This behavior occurs with any expectation value if the operator is independent of the time and if the wave function is the product of an energy eigenfunction and a time factor. A state corresponding to such a wave function is called a **stationary state**. For stationary states, all expectation values of variables with time-independent operators are time independent, and coordinate wave functions can be used to calculate expectation values.

Exercise 10.14 Find $\langle x \rangle$ for a particle in a hard one-dimensional box of length a for the $n = 2$ state.

Probability Densities

We have asserted that position measurements belong to case II for almost any kind of wave function. That is, we can predict the mean of a set of many repeated measurements of the position of a particle, but individual members of the set can have different values. We now want to study the probabilities of different outcomes of the position measurement.

In Chapter 1 we discussed the computation of a mean value, $\langle w \rangle$, from a set of values $w_1, w_2, w_3, \ldots$. Equation (1.7-12) is

$$\langle w \rangle = \sum_{i=1}^{M} p_i w_i \qquad \textbf{(10.4-7)}$$

where p_i is the probability that a randomly chosen member of the set is equal to w_i.

In the case of a position measurement, the outcome of a measurement can take on any value in some range. The position operator has a continuous spectrum of eigenvalues. Instead of the probability that a measurement has a given value, we must define the probability that a measurement lies in a given range, as in Equation (9.1-8):

$$\text{Probability that } u \text{ lies} \atop \text{between } u' \text{ and } u' + du = f(u')\, du \qquad \textbf{(10.4-8)}$$

where du is a small (infinitesimal) range of values of u and u' is a value of u in this range. The function $f(u)$ is a **probability density**, or a probability per unit length on the u axis. Figure 10.2 schematically depicts a probability density. The probability of a small interval of width du is given by the product $f(u')\, du$, which is equal to the area that is shaded in the figure.

If u_1 is the smallest value of u and u_2 is the largest value, the total probability is equal to the integral

$$\text{Total probability} = \int_{u_1}^{u_2} f(u)\, du = 1 \qquad \textbf{(10.4-9)}$$

which is equal to the area under the curve representing the probability density. To normalize the probability density, we scale it so that the total probability equals unity, as in the second equality of Equation (10.4-9).

The mean value of u is given by an equation analogous to Equation (10.4-7):

$$\langle u \rangle = \int_{u_1}^{u_2} u f(u)\, du \qquad \textbf{(10.4-10)}$$

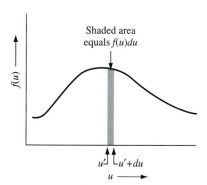

Shaded area equals $f(u)du$

Figure 10.2. An Example Probability Density (Probability Distribution). The function $f(u)$ is defined such that the probability that the variable u lies in a range du is equal to $f(u)\, du$.

Just as in Equation (1.7-14), we can take the mean of a function of the independent variable u:

$$\langle h(u) \rangle = \int_{u_1}^{u_2} h(u) f(u) \, du \tag{10.4-11}$$

If we compare Equation (10.4-4) with Equation (10.4-10), we see that the probability of finding the particle between x and $x + dx$ is equal to

$$\text{Probability} = |\Psi(x, t)|^2 \, dx \tag{10.4-12}$$

The square of the magnitude of the wave function is here identified as the probability density for finding the particle or particles of the system.

The square of the magnitude of the wave function, $|\Psi(x, t)|^2$, is a probability density. In this case, it is a probability per unit length on the x axis. At any location where the square of the wave function is nonzero, there is some probability of finding the particle. Case II obviously applies. We now have a connection between the wave function (the displacement of a matter wave) and the probability of finding the particle. This connection corresponds to our earlier assertion that a wave function equal to zero corresponds to the absence of a particle.

For the motion of one particle in three dimensions, the probability that the particle lies between x and $x + dx$ in the x direction, between y and $y + dy$ in the y direction, and between z and $z + dz$ in the z direction is analogous to that in Equation (10.4-12):

$$\text{Probability} = |\Psi(x, y, z, t)|^2 \, dx \, dy \, dz \tag{10.4-13}$$

The probability density in this case is a probability per unit volume in three dimensions.

Exercise 10.15   For a particle in a three-dimensional hard box, the eigenfunction of the Hamiltonian operator is given by Equation (9.5-32). For the $n_x = 1$, $n_y = 1$, $n_z = 1$ state, find the probability that the particle is in a small rectangular region in the center of the box such that the length of the region in each direction is equal to 1.000% of the length of the box in that direction. Proceed as though the region were infinitesimal.

For the motion of two particles in three dimensions, the wave function depends on $x_1, y_1, z_1, x_2, y_2,$ and z_2. The probability that the first particle is between x_1 and $x_1 + dx_1$, between y_1 and $y_1 + dy_1$, and between z_1 and $z_1 + dz_1$ and that simultaneously the second particle is between x_2 and $x_2 + dx_2$, between y_2 and $y_2 + dy_2$, and between z_2 and $z_2 + dz_2$ is

$$\text{Probability} = |\Psi(x_1, y_1, z_1, x_2, y_2, z_2, t)|^2 \, dx_1 \, dy_1 \, dz_1 \, dx_2 \, dy_2 \, dz_2$$
$$= |\Psi(\mathbf{r}_1, \mathbf{r}_2, t)|^2 \, d^3\mathbf{r}_1 \, d^3\mathbf{r}_2 \tag{10.4-14}$$

That is, $|\Psi|^2$ is a probability per unit six-dimensional volume. Similarly, for a system with n particles, the square of the magnitude of the wave function is a probability density in a space with $3n$ dimensions.

If a wave function is normalized, its probability density is also normalized. The total probability of all positions is equal to the integral of the square of the magnitude of the wave function over all values of the coordinates, which equals unity for a normalized wave function.

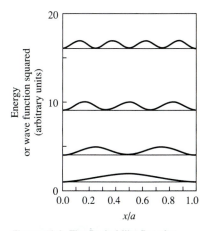

Figure 10.3. The Probability Density for Positions of a Particle in a One-Dimensional Hard Box. This diagram shows the squares of the energy eigenfunctions (probability densities) for the first four states of a particle in a one-dimensional hard box.

In quantum mechanics, there are points of zero probability density for finding a particle. This behavior does not occur in the same way in classical mechanics.

For a stationary state, the probability density is time independent:

$$\psi(x)^* e^{iEt/\hbar} \psi(x) e^{-iEt/\hbar} = \psi^*(x)\psi(x) = |\psi(x)|^2 \qquad \textbf{(10.4-15)}$$

The analogue of Equation (10.4-15) can be written for a wave function that depends on more than one coordinate.

Figure 10.3 shows the probability density (square of the magnitude of the wave function) for four energy eigenfunctions of a particle in the box. These four graphs are placed at heights proportional to the energy eigenvalue corresponding to each wave function. The lowest curve of the figure shows the probability density for $n = 1$, the next curve shows the probability density for $n = 2$, the third curve from the bottom shows the probability density for $n = 3$, and the upper curve shows the probability density for $n = 4$. Note how different the probability densities are from one described by classical mechanics. If the state of a classical particle in a box is known, its position is known, and the probability at a given time would be nonzero at only one point, as in Figure 10.4a. The classical probability density averaged over a long time would be uniform, with all parts of the box being equally probable, as shown in Figure 10.4b, which is the same as Figure 9.21.

The quantum mechanical probability density for a particle in a box is not uniform, and there are points at which the probability density vanishes. However, if a very large value of n is taken, these points become closer and closer together as schematically shown in Figure 10.5, which is drawn for $n = 10$. A measurement of x has some experimental uncertainty, which corresponds to measuring a local average probability density over a small range equal to the experimental uncertainty. For very large values of n, the local average probability density is effectively uniform, as in the classical case, since the width of the oscillations in the curve becomes smaller than the experimental uncertainty. This behavior conforms to the **correspondence principle**, which states that for sufficiently large energies and masses, the behavior predicted by quantum mechanics approaches the behavior predicted by classical mechanics.

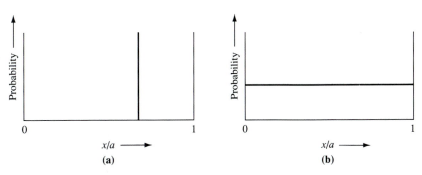

Figure 10.4. The Probability Density for Positions of a Particle in a One-Dimensional Hard Box According to Classical Mechanics. (a) The instantaneous probability. At a given time, there is no uncertainty about the position of the particle. **(b)** The probability averaged over a long time. The average probability density is uniform, with all parts of the box equally probable.

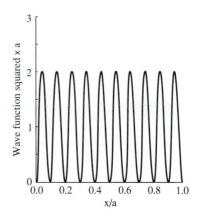

Figure 10.5. The Probability Density for Positions of a Particle in a One-Dimensional Hard Box for $n = 10$. This probability density resembles the classical probability density more closely than does that of the lower-energy states.

▼ **EXAMPLE 10.10**

For a particle in a one-dimensional box in the state corresponding to $n = 2$, find the probability that the particle will be found in each of the regions making up thirds of the box.

Example 10.10 is solved by integrating the probability density (the square of the wave function) over the range of interest.

Solution

For $0 < x < a/3$,

$$\text{Probability} = \int_0^{a/3} \psi(x)^2 \, dx = \frac{2}{a} \int_0^{a/3} \sin^2\left(\frac{2\pi x}{a}\right) dx$$

$$= \frac{2}{a}\frac{a}{2\pi} \int_0^{2\pi/3} \sin^2(y) \, dy = \frac{1}{\pi}\left[\frac{y}{2} - \frac{1}{4}\sin(2y)\right]\Bigg|_0^{2\pi/3}$$

$$= \frac{1}{\pi}\left[\frac{\pi}{3} - \frac{1}{4}\sin\left(\frac{4\pi}{3}\right)\right] = 0.402249$$

The right one-third of the box will have the same probability as the left one-third. The probability of finding the particle in the center region will be

$$\text{Probability} = 1 - 2(0.402249) = 0.195501$$

The node in the wave function at $x = a/2$ results in a smaller probability for the middle third of the box than for the left or right third of the box.

▲

Exercise 10.16

From inspection of Figure 10.3, estimate the probability of finding the particle in the left one-third of the box for the $n = 1$ state. After making this estimate, make an accurate calculation of the probability.

Figure 10.6 shows the probability density for the first few energy eigen-functions of the harmonic oscillator. Each graph is placed at a height in the figure proportional to its energy eigenvalue, and the potential energy is also plotted. The vertical axis is used for two different variables, as in

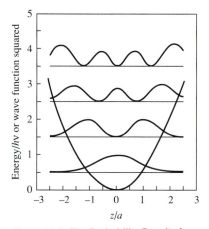

Figure 10.6. The Probability Density for the First Few Energy Eigenstates of the Harmonic Oscillator. Each wave function graph is placed at the height corresponding to its energy eigenvalue, and the potential energy is also plotted. The point where the potential energy curve crosses the energy level line is the classical turning point.

Figure 10.3. The behavior of the harmonic oscillator probability density is qualitatively like that of the particle in a box, and the number of nodes follows the same pattern. However, for the harmonic oscillator, the probability density does not vanish outside a finite region but extends to infinite values of the coordinate. However, it approaches zero so rapidly that large magnitudes of the coordinate make only a small contribution to the total probability.

Each of the probability densities in Figure 10.6 can be compared with the classical probability density in Figure 9.3b. The probability density for $v = 3$ looks a little bit like the classical probability density, with the greatest probability at the ends of the classically allowed region. The resemblance will be greater for larger values of v, similar to the behavior in Figure 10.5 for the particle in a box.

Distinguishing Case I from Case II

A common measure of the "spread" of a probability distribution is the **standard deviation**. The standard deviation for the measurement of an arbitrary variable A is denoted by σ_A and defined by

$$\sigma_A = (\langle A^2 \rangle - \langle A \rangle^2)^{1/2} \quad \text{(definition)} \qquad \textbf{(10.4-16)}$$

The square of the standard deviation is called the **variance**.

In case I, all outcomes for repeated measurements will be equal and the standard deviation will be equal to zero. In case II, the outcomes will be distributed and the standard deviation will be nonzero.

▼   **EXAMPLE 10.11**

Find the standard deviation for the position of a particle in a hard one-dimensional box of length a for the $n = 1$ state.

Solution

The time-dependent factors of the wave function will cancel, so we work from the beginning with the coordinate wave function:

$$\langle x^2 \rangle = \frac{2}{a} \int_0^a \sin\left(\frac{\pi x}{a}\right) x^2 \sin\left(\frac{\pi x}{a}\right) dx$$

$$= \frac{2}{a}\left(\frac{a}{\pi}\right)^3 \int_0^\pi y^2 \sin(y)\, dy = a^2\left(\frac{1}{3} - \frac{1}{2\pi^2}\right) = 0.282673 a^2$$

From Example 10.9 we have $\langle x \rangle = a/2$, so that

$$\sigma_x = [0.282673 a^2 - (a/2)^2]^{1/2} = 0.180757 a$$

The fact that the standard deviation is nonzero shows that for this system and this wave function, case II applies to the position of the particle.

Carl Freidrich Gauss, 1777–1855, great German mathematician and physicist.

The Normal (Gaussian) Probability Distribution

A probability distribution that occurs in a number of applications in the physical and social sciences is the **gaussian distribution**, also called the

normal distribution:

$$f(u) = \frac{1}{\sqrt{2\pi}\sigma} e^{-(u-\mu)^2/2\sigma^2}$$

(10.4-17)

where σ is the standard deviation and μ is the mean value of u. The independent variable (statisticians call it the **random variable**) u is assumed to range over all real values from $-\infty$ to $+\infty$. A graph of this probability distribution is shown in Figure 10.7a. This graph is often called a bell-shaped curve. The probability density for the $v = 0$ state of a harmonic oscillator is a gaussian probability density.

In a gaussian probability distribution, 68.3% of the members of the statistical population lie within one standard deviation of the mean. This probability is represented in Figure 10.7b as a shaded area. The standard deviation is a convenient measure of the width of a probability distribution. Most ordinary distributions have roughly the same value (approximately 2/3) for the probability that the random variable lies within one standard deviation of the mean.

It is a common occurrence for the probability that a value lies within one standard deviation from the mean to be approximately 2/3.

▼   **EXAMPLE 10.12**

Calculate the probability that a particle in a one-dimensional hard box of length a will be found within one standard deviation of its mean position if the wave function is the $n = 1$ energy eigenfunction.

Solution

$$\text{Probability} = \frac{2}{a} \int_{0.319243a}^{0.6807566a} \sin^2\left(\frac{\pi x}{a}\right) dx = \frac{2}{\pi} \int_{1.00293}^{2.13866} \sin^2(y)\, dy$$

$$= \frac{2}{\pi} \left[\frac{y}{2} - \frac{\sin(2y)}{4} \right]\Big|_{1.00293}^{2.13866} = 0.65017$$

This value is reasonably close to 2/3 and to 0.683.

▲

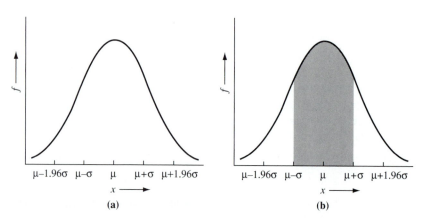

Figure 10.7. The Gaussian (Normal) Probability Distribution. (a) The graph showing the famous bell-shaped curve. **(b)** The probability that the variable deviates no more than one standard deviation from its mean. The shaded area represents this probability, given that the total area under the curve equals unity (normalization).

Exercise 10.17 Calculate the probability that a particle in a one-dimensional hard box will be found within one standard deviation of its mean position for the $n = 2$ state.

Uncertainty in the Measurement of a Variable

If case II applies, we will use the standard deviation, which we can evaluate from Equation (10.4-16), as a measure of the width of the probability distribution, or of the **uncertainty** of the variable. We can predict that a single measurement will have roughly a two-thirds probability of lying within one standard deviation of the expectation value.

Consider now the probability density for the harmonic oscillator. In Section 9.6, we commented on the fact that the wave function of the harmonic oscillator is nonzero for positions past the classical turning points.

▼   **EXAMPLE 10.13**

Calculate the probability that the harmonic oscillator will be found in the classically forbidden region for the $v = 0$ state.

Solution

For a classical energy equal to the quantum mechanical energy for $v = 0$, the turning point is given by

$$z_t^2 = \frac{hv}{k} = \frac{h}{k}\frac{1}{2\pi}\sqrt{\frac{k}{m}} = \frac{\hbar}{\sqrt{km}} = \frac{1}{a}$$

$$|z_t| = \sqrt{\frac{1}{a}}$$

The probability that the harmonic oscillator is in the classically permitted region is

$$\text{Probability} = \left(\frac{a}{\pi}\right)^{1/2}\int_{-\sqrt{1/a}}^{\sqrt{1/a}} e^{-az^2}\,dz = 2\left(\frac{a}{\pi}\right)^{1/2}\int_{0}^{\sqrt{1/a}} e^{-az^2}\,dz$$

where we have used the fact that, since the integrand is an even function, the integral over half the interval is equal to half of the integral over the entire interval. This integral cannot be worked out in closed form. However, it is related to the error function, for which tables of values are available:[3]

$$\text{erf}(x) = \frac{2}{\sqrt{\pi}}\int_{0}^{x} e^{-y^2}\,dy \quad \text{(definition)}$$

If we make the substitution $y = \sqrt{a}\,z$, we have

$$\text{Probability} = \frac{2}{\sqrt{\pi}}\int_{0}^{1} e^{-y^2}\,dy = \text{erf}(1) = 0.8427$$

where the value of the error function was obtained from the table in Appendix C. The probability that the oscillating particle is farther away from its equilibrium position than the classical turning point is thus $1.0000 - 0.8427 = 0.1573$, or 15.73% (7.86% past each end of the classically permitted region). This probability is represented by the two shaded areas in Figure 10.8.

Example 10.13 is solved as were several examples by integrating the probability density over the range of interest. In this case, a tabulated value of the error function must be used to evaluate the integral. The result illustrates tunneling.

---

[3] M. Abramowitz and I. A. Stegun (eds.), *Handbook of Mathematical Functions with Formulas, Graphs, and Mathematical Tables*, U.S. Government Printing Office, Washington, DC, 1964. See Appendix C for a table of values.

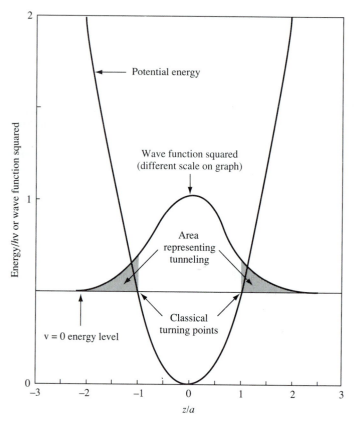

Figure 10.8. The Probability Density of a Harmonic Oscillator in Its Ground State, Showing Tunneling. The shaded areas represent the probability that the oscillator will be in classically forbidden regions.

Exercise 10.18   Calculate the value of the ratio

$$\text{Ratio} = \frac{|\psi_0(z_t)|^2}{|\psi_0(0)|^2}$$

Explain in words what this ratio represents.

Penetration into a classically forbidden region is called *tunneling*. The name was chosen because a tunnel into a hillside allows access to a location under an inaccessible location of high gravitational potential energy.

We have used the position of a particle as an example of a mechanical variable. We must also consider the momentum, the energy, and the angular momentum.

▼   **EXAMPLE 10.14**   For a particle in a hard one-dimensional box, find $\langle p_x \rangle$, $\langle p_x^2 \rangle$ and σ_{p_x} for the $n = 1$ state.

Solution

$$\langle p_x \rangle = \frac{2}{a} \int_0^a \sin\left(\frac{\pi x}{a}\right) \frac{\hbar}{i} \frac{d}{dx} \sin\left(\frac{\pi x}{a}\right) dx$$

$$= \frac{2\hbar}{\pi i} \int_0^a \sin\left(\frac{\pi x}{a}\right) \cos\left(\frac{\pi x}{a}\right) dx = 0$$

$$\langle p_x^2 \rangle = \frac{2}{a} \int_0^a \sin\left(\frac{\pi x}{a}\right) (-\hbar^2) \frac{d^2}{dx^2} \sin\left(\frac{\pi x}{a}\right) dx$$

$$= \frac{2}{a} \hbar^2 \left(\frac{\pi}{a}\right)^2 \int_0^a \sin\left(\frac{\pi x}{a}\right) \sin\left(\frac{\pi x}{a}\right) dx = \frac{\hbar^2 \pi^2}{a^2}$$

$$\sigma_{p_x} = [\langle p_x^2 \rangle - \langle p_x \rangle^2]^{1/2} = \langle p_x^2 \rangle^{1/2} = \frac{\hbar \pi}{a}$$

Example 10.14 is solved by substituting into the expectation value formula for p_x and p_x^2. The nonzero value of the standard deviation shows that for this system and this state, case II applies.

Heisenberg's Uncertainty Principle

We use σ_x and σ_{p_x} as measures of the uncertainty in predictions of position and momentum. Their product is a measure of the combined uncertainty of the two variables and is called an **uncertainty product**. From Examples 10.11 and 10.14, the value of this product for the $n = 1$ state of the particle in a one-dimensional hard box is

$$\sigma_x \sigma_{p_x} = (0.180756a) \frac{\hbar \pi}{a} = 0.56786\hbar = 0.09038h$$

The coordinate x and the momentum p_x are said to be "conjugate variables," in the sense of Equation (D-17) of Appendix D. Any coordinate has a "conjugate momentum," and the conjugate momentum for any cartesian coordinate is the corresponding component of the ordinary momentum.

The **Heisenberg uncertainty principle** is a general statement of the combined uncertainties of two conjugate variables: *The product of the uncertainties of two conjugate variables is equal to or larger than $h/4\pi$, where h is Planck's constant.*

If we use the symbols Δx and Δp_x for the uncertainties of a coordinate and its conjugate momentum, then

$$\Delta x \, \Delta p_x \geq \frac{h}{4\pi} \qquad \textbf{(10.4-18)}$$

Equation (10.4-18) corresponds to the use of the standard deviation as the measure of uncertainty. There are other measures of the uncertainty of a statistical prediction besides the standard deviation. For example, for a gaussian probability distribution, a prediction at the 95% probability level is made by using 1.96 times the standard deviation. At this level of probability, the right-hand side of Equation (10.4-18) would be replaced by a larger value. The actual value of the uncertainty product depends on the nature of the system and on the state considered. The uncertainty product for the $n = 1$ state of the particle in a box, $0.09038h$, is slightly larger than $h/(4\pi)$. The uncertainty product for the $v = 0$ state of the harmonic oscillator is equal to $h/(4\pi)$.

Coordinates and momenta are not the only variables that have nonzero uncertainty products. The commutator of the operators of two conjugate variables is nonzero, as we have already seen for the commutator $[x, p_x]$. Any two variables whose operators do not commute must have a nonzero uncertainty product. There is a general relation

$$\Delta A \; \Delta B \geq \left| \frac{1}{2} \int \psi^* [\hat{A}, \hat{B}] \psi \; dq \right| \tag{10.4-19}$$

where $[\hat{A}, \hat{B}]$ is the commutator of $\hat{A}$ and $\hat{B}$. We omit the proof of Equation (10.4-19).[4] From the result of Problem 10.35, one can see that two components of the angular momentum obey a type of uncertainty relation, at least for states for which the eigenvalue of the third component is not equal to zero.

Exercise 10.19 Use Equations (10.4-19) and (10.3-11) to obtain the uncertainty relation for x and p_x.

The uncertainty principle is a rather subtle concept and deserves more discussion than we give it in this book. However, the main idea is that it requires that case II applies to conjugate pairs of variables. There is no such thing as a state for which both of the variables of a conjugate pair of variables belong to case I.

▼ **EXAMPLE 10.15**   Find $\langle p_x \rangle$ and σ_{p_x} for a free particle in a state corresponding to the wave function of Equation (9.5-40).

Solution

We cancel the time-dependent factors to obtain

$$\langle p_x \rangle = \frac{\dfrac{\hbar}{i} D^* D \displaystyle\int e^{-i\kappa x} \dfrac{d}{dx} e^{i\kappa x} \, dx}{D^* D \displaystyle\int e^{-i\kappa x} e^{i\kappa x} \, dx}$$

$$= \hbar\kappa \, \frac{D^* D \displaystyle\int e^{-i\kappa x} e^{i\kappa x} \, dx}{D^* D \displaystyle\int e^{-i\kappa x} e^{i\kappa x} \, dx} = \hbar\kappa \, \frac{D^* D \displaystyle\int dx}{D^* D \displaystyle\int dx}$$

We have not stated the limits on the integrals, which must cover the entire range of possible coordinate values. In this case the limits are $-\infty$ and ∞. We specify that the limits are $-L$ and L, with the intention of taking the limit that $L \to \infty$. We cancel the integrals in the last quotient of integrals before taking the limit and obtain

$$\langle p_x \rangle = \hbar\kappa$$

---

[4] Levine (*Quantum Chemistry*, 4th ed., Allyn & Bacon, 1983, p. 188) assigns the proof as a homework problem. A lot of hints are included, but it is a fairly long proof.

To calculate σ_{p_x}, we require $\langle p_x^2 \rangle$, which is given by a similar quotient of integrals except that the second derivative is taken:

$$\langle p_x^2 \rangle = \hbar^2 \kappa^2$$

The square of the standard deviation is

$$\langle \sigma_{p_x} \rangle^2 = \langle p_x^2 \rangle - \langle p_x \rangle^2 = \hbar^2 \kappa^2 - \hbar^2 \kappa^2 = 0$$

The uncertainty in one of a pair of conjugate variables can vanish if the uncertainty in the other variable is infinite.

The zero value for the standard deviation of the momentum in Example 10.15 does not violate the uncertainty principle. Even though the uncertainty in the momentum vanishes, the wave function is nonzero over an infinite range of values of x and the uncertainty in x is infinite.

Exercise 10.20   Write the integral to calculate σ_x for the free particle of Example 10.15 and argue that its value is infinite.

From the expectation value of the momentum of a free particle, we can now justify the apparently arbitrary choice of sign made in Equation (10.3-5). It appeared at that time that either $i\hbar \, d/dx$ or $-i\hbar \, d/dx$ could have been chosen as the operator for p_x. The free-particle wave function in Equation (9.5-40) represents a traveling wave moving toward the positive end of the x axis. This wave function contains a time factor, $\exp(-iEt/\hbar)$. The negative sign in this factor results from having a positive sign in the right-hand side of the time-dependent Schrödinger equation, Equation (9.4-13). If $i\hbar \, d/dx$ had been chosen for the $\hat{p}_x$ operator, a negative value for $\langle p_x \rangle$ would have resulted, indicating motion in the wrong direction.

Exercise 10.21   Show that taking the opposite sign for the momentum operator leads to a wave moving in the wrong direction.

The Time-Energy Uncertainty Relation

Like position and momentum, energy and time also obey an uncertainty relation:

$$\Delta E \, \Delta t \geq \frac{h}{4\pi} \qquad\qquad \textbf{(10.4-20)}$$

The time-energy uncertainty relation is different from that of position and momentum, since time is not a mechanical variable that can be expressed in terms of the state of the system. Time does not correspond to a quantum mechanical operator. Although the time-dependent Schrödinger equation has the Hamiltonian operator on one side and the time derivative operator on the other, this does not imply that $i\hbar \, \partial/\partial t$ can be used as an operator for the energy, although such an operator relationship would lead to a commutator that would establish Equation (10.4-20).[5]

The standard interpretation of the time-energy uncertainty relation is that if Δt is the time during which the system is known to be in a given state (the "lifetime" of the state), then there is a minimum uncertainty ΔE in the energy of the state as given by Equation (10.4-20).

Exercise 10.22   If the energy of a system is to be measured to an uncertainty of 1.0×10^{-21} J, find the minimum time during which the system must be in the state at the measured energy.

We have stated that case I applies when the state of the system just prior to a measurement corresponds to an eigenfunction of the operator for the variable. This fact must be related to the time-energy uncertainty principle.

EXAMPLE 10.16

For a particle in a one-dimensional hard box, find $\langle E \rangle$ and σ_E for the state corresponding to the $n = 1$ energy eigenfunction.

Solution

$$\sigma_E = (\langle E^2 \rangle - \langle E \rangle^2)^{1/2} \tag{10.4-21}$$

Using the normalized wave function

$$\langle E \rangle = \int \psi_i^* \hat{H} \psi \, dx = \int \psi_1^* E_1 \psi_1 \, dx = E_1 \int \psi_1^* \psi_1 \, dx$$

$$= E_1 = \frac{h^2}{8ma^2}$$

since the wave function is an eigenfunction of $\hat{H}$. Also

$$\langle E^2 \rangle = \int \psi_1^* \hat{H} \psi_1 \, dx = \int \psi_1^* E_1^2 \psi_1 \, dx$$

$$= E_1^2 \int \psi_1^* x \psi_1 \, dx = E_1^2 \tag{10.4-22}$$

where we have used the fact that $\hat{H}^2$ means operation twice with $\hat{H}$. The standard deviation vanishes:

$$\sigma_E = (E_1^2 - E_1^2)^{1/2} = 0 \tag{10.4-23}$$

The solution to Example 10.16 shows that for a state corresponding to an energy eigenfunction, case I applies to the energy.

The result of Example 10.16 illustrates an important general fact: *If the wave function is an eigenfunction of the operator corresponding to the variable being measured, the outcome of an error-free measurement is completely predictable.* The only value that will occur is the eigenvalue corresponding to the given eigenfunction.

---

[5] Y. Aharanov and D. Bohm, *Phys. Rev.* **122**, 1649 (1961).

Exercise 10.23	**a.** For a general system whose wave function ψ_j is an eigenfunction of the operator $\hat{A}$ with eigenvalue a_j, show that $\langle A \rangle = a_j$ and that the standard deviation, σ_A, vanishes.
	b. For a one-dimensional harmonic oscillator, find $\langle E \rangle$ and σ_E for the state corresponding to the $v = 1$ energy eigenfunction.

There is zero uncertainty in the energy if the system is in a state corresponding to an energy eigenfunction. However, the time-energy uncertainty principle must be obeyed. Therefore, a system must be observed for an infinite length of time to be completely certain it is in such a state.

Although we have discussed energy eigenfunctions to the exclusion of other kinds of wave functions, there is no requirement that the wave function actually corresponding to the state of a system be an energy eigenfunction.

▼   **EXAMPLE 10.17**   For a one-dimensional harmonic oscillator, find $\langle E \rangle$ and σ_E if the state just prior to the measurements is the state corresponding to the normalized wave function

$$\psi = \sqrt{\frac{1}{2}} (\psi_0 + \psi_1)$$

where ψ_0 and ψ_1 are the first two energy eigenfunctions, given in Equations (9.6-15) and (9.6-16).

Solution

Since ψ is normalized

$$
\begin{aligned}
\langle E \rangle &= \frac{1}{2} \int_{-\infty}^{\infty} (\psi_0^* + \psi_1^*) \hat{H} (\psi_0 + \psi_1) \, dx \\
&= \frac{1}{2} \int_{-\infty}^{\infty} (\psi_0^* + \psi_1^*)(E_0 \psi_0 + E_1 \psi_1) \, dx \\
&= \frac{1}{2} \left[E_0 \int_{-\infty}^{\infty} \psi_0^* \psi_0 \, dx + E_1 \int_{-\infty}^{\infty} \psi_0^* \psi_1 \, dx \right. \\
&\quad \left. + E_0 \int_{-\infty}^{\infty} \psi_1^* \psi_0 \, dx + E_1 \int_{-\infty}^{\infty} \psi_1^* \psi_1 \, dx \right] \\
&= \frac{1}{2} (E_0 + 0 + 0 + E_1) = \frac{1}{2}(E_0 + E_1) = h\nu
\end{aligned}
$$

We have used the normalization of the energy eigenfunctions, the fact that the two energy eigenfunctions are orthogonal to each other, and the expression for the energy eigenvalues.

$$
\begin{aligned}
\langle E^2 \rangle &= \frac{1}{2} \int_{-\infty}^{\infty} (\psi_0^* + \psi_1^*) \hat{H}^2 (\psi_0 + \psi_1) \, dx \\
&= \frac{1}{2} \int_{-\infty}^{\infty} (\psi_0^* + \psi_1^*)(E_0^2 \psi_0 + E_1^2 \psi_1) \, dx \\
&= \frac{1}{2} (E_0^2 + E_1^2) = \frac{5}{4}(h\nu)^2
\end{aligned}
$$

We have omitted some steps. We now have

$$\sigma_E = (\langle E^2 \rangle - \langle E \rangle^2)^{1/2} = \left[\frac{5}{4} (h\nu)^2 - (h\nu)^2 \right]^{1/2} = \frac{h\nu}{2}$$

For the state corresponding to this wave function, case II applies to the energy.

▲

Exercise 10.24

For a particle in a one-dimensional hard box, find $\langle E \rangle$ and σ_E for the coordinate wave function

$$\psi = \sqrt{\frac{1}{3}}\, \psi_1 + \sqrt{\frac{2}{3}}\, \psi_2$$

where ψ_1 and ψ_2 are the first two energy eigenfunctions.

Example 10.17 and Exercise 10.24 are examples of a general case. We assume that we can write any wave function as a **linear combination** (sum of functions multiplied by constant coefficients) of the eigenfunctions of a hermitian operator if the eigenfunctions obey the same boundary conditions:

$$\psi = \sum_{j=1}^{\infty} c_j \varphi_j \qquad\qquad \textbf{(10.4-24)}$$

where φ_1, φ_2, φ_3,... are the set of eigenfunctions of $\hat{A}$, having eigenvalues a_1, a_2, a_3,.... The wave function ψ is said to be **expanded** in terms of the set of functions φ_1, φ_2, φ_3,.... This set of functions is called the **basis set**. The coefficients c_1, c_2, c_3,... are called the **expansion coefficients** and must be chosen to have different values to represent different wave functions. If a set of functions allows a sum such as the sum in Equation (10.4-24) to represent an arbitrary function obeying the same boundary conditions, it is called a **complete set**. The sine and cosine functions in a Fourier series are an example of a complete set of functions for representing periodic functions. If the functions in a complete set are normalized and orthogonal to each other we call them a **complete orthonormal set**. Although a general proof is lacking, it is assumed that the set of all eigenfunctions of any hermitian operator forms a complete set for representing functions obeying the same boundary conditions.

We now obtain a formula for the expectation value of $\hat{A}$, assuming also that ψ is normalized:

$$\langle A \rangle = \int \psi^* \hat{A} \psi \, dq$$

We do not indicate the limits of the integration, which must be over all possible values of the coordinates that are abbreviated by q. We substitute the expansion of Equation (10.4-24) into the expression for $\langle A \rangle$.

$$\langle A \rangle = \int \sum_{j=1}^{\infty} c_j^* \varphi_j^* \hat{A} \sum_{k=1}^{\infty} c_k \varphi_k \, dq = \sum_{j=1}^{\infty} \sum_{k=1}^{\infty} c_j^* c_k a_k \int \varphi_j^* \varphi_k \, dq$$

We have used the eigenfunction property, factored the constants out of the integrals, and exchanged the order of integrating and summing.

Since the functions φ_1, φ_2,... are an orthonormal set, only terms in which $j = k$ will be nonzero, and the integrals in these terms will equal unity. We write

$$\int \varphi_j^* \varphi_k \, dq = \delta_{jk} = \begin{cases} 1 & \text{if } j = k \\ 0 & \text{if } j \neq k \end{cases} \qquad \text{(10.4-25)}$$

This equation defines the quantity δ_{jk}, which equals unity when its two indices are equal and equals zero otherwise. It is called the **Kronecker delta**.

When the sum over k is performed, only the $j = k$ term will be nonzero, and we have

$$\langle A \rangle = \sum_{j=1}^{\infty} \sum_{k=1}^{\infty} c_j^* c_k a_k \, \delta_{jk} = \sum_{j=1}^{\infty} c_j^* c_j a_j = \sum_{j=1}^{\infty} |c_j|^2 a_j \qquad \text{(10.4-26)}$$

Comparison of Equation (10.4-26) with Equation (1.7-12) shows that $\langle A \rangle$ is given in the same way as a mean value is given from individual values and their probabilities. We therefore assert that individual measurements of A can give as a result only one or another of the eigenvalues of the operator $\hat{A}$ and that the probability that the eigenvalue a_j will occur is

Equation (10.4-27) is sometimes useful because it allows calculation of the probabilities of all of the eigenvalues.

$$p_j = |c_j|^2 \qquad \text{(10.4-27)}$$

10.5 Postulate V. Determining the State of a System

The fifth and final postulate gives the rule for determining the mechanical state of a quantum mechanical system:

> **Postulate V.** *Immediately after a measurement of the mechanical variable A in which the outcome was the eigenvalue a_j, the state of the system corresponds to a wave function that is an eigenfunction of $\hat{A}$ with eigenvalue equal to a_j.*

Postulate V says nothing about the state of the system prior to the measurement, because the act of measurement can change the state of the system. To illustrate this change of state, we consider determining the position of a particle by allowing it to scatter electromagnetic radiation. This procedure would be similar to determining the position of an airplane with a radar apparatus.

The act of measurement determines the state of the system immediately after the measurement. The concept that the measurement process can change the state of the system does not occur in classical mechanics.

However, when a macroscopic object such as an airplane reflects an electromagnetic wave, the effect on the object is negligible. When an object of small mass such as an electron scatters light, the situation is very different. If the position of an electron is to be determined to an accuracy of 0.1 nm, radiation with a wavelength of no more than 0.1 nm is needed.

▼ **EXAMPLE 10.18**

Compare the energy of a photon of wavelength 0.10 nm with the lowest kinetic energy of an electron in a one-dimensional hard box of length 1.0 nm.

Solution

$$E(\text{photon}) = h\nu = \frac{hc}{\lambda} = \frac{(6.6261 \times 10^{-34} \text{ J s})(3.00 \times 10^8 \text{ m s}^{-1})}{1.00 \times 10^{-10} \text{ m}}$$

$$= 1.99 \times 10^{-15} \text{ J}$$

The solution to Example 10.18 shows numerically why the act of measurement can change the state of an electron.

From the result of Example 9.6

$$E(\text{electron}) = 6.0 \times 10^{-20} \text{ J}$$

▲

In Example 10.18, the photon energy is about 30000 times as large as the kinetic energy of the electron, so that the energy transferred to the electron in a measurement can be larger than the original kinetic energy of the electron.

Another argument implies that the measurement process can change the state of the system. Assume that a particle in a box of length a is in a state corresponding to one of the energy eigenfunctions. The square of the wave function is the probability density for finding the particle, and this quantity is nonzero nearly everywhere in the box. However, a single position measurement will give a single well-defined outcome, such as the location of a flash of light at a screen. An immediate repetition of the measurement would have to give a position very near the first position, since there would be no time for the particle to move appreciably. The wave function immediately after the first measurement must be a function that is nonzero only in the immediate vicinity of the measured position. The act of measurement has changed the wave function. Figure 10.9a shows the wave function just before the measurement of position. Figure 10.9b shows the wave function immediately after the position measurement. Figure 10.9c shows the wave function after a fairly short time has elapsed. The wave function has begun to evolve back into a delocalized wave function. We could follow this evolution by solving the time-dependent Schrödinger equation.

This analysis is based on physical reasoning rather than mathematical reasoning, but it has valid mathematical consequences.

Information About the State Prior to a Measurement

A single measurement gives us information about the state after the measurement. Some information about the original wave function of a quantum mechanical

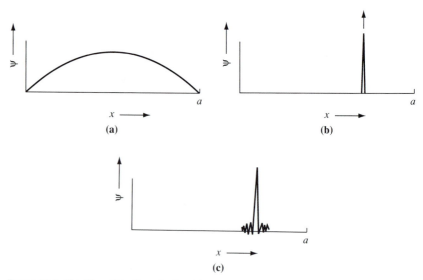

Figure 10.9. The Wave Function of a Particle in a Box. (a) Before a position measurement (schematic). The probability density is nonzero over the entire box (except for the end points). **(b)** Immediately after the position measurement (schematic). In a very short time, the particle cannot have moved far from the position given by the measurement, and the probability density must be a sharply peaked function. **(c)** Shortly after a position measurement (schematic). After a short time, the probability density can be nonzero over a larger region.

system can be obtained by repeated measurements, assuming we have a way to put the system back into the original state before each measurement.

Consider first the nondegenerate case, such that each eigenfunction of the operator $\hat{A}$ corresponding to a variable A has a distinct eigenvalue. Since the set of eigenfunctions is assumed to be a complete set, we can write the wave function prior to the measurement as a linear combination:

$$\psi(\text{prior}) = \sum_{j=1}^{\infty} c_j \varphi_j \tag{10.5-1}$$

where $\varphi_1, \varphi_2, \ldots$ is the set of eigenfunctions of $\hat{A}$ such that

$$\hat{A}\varphi_j = a_j \varphi_j \tag{10.5-2}$$

To obtain information about the coefficients $c_1, c_2, c_3, \ldots$, we carry out many measurements of A, ensuring that the system is in the same state prior to each measurement. By Equation (10.4-22), the magnitude of the c coefficients is determined by the probabilities of the different eigenvalues in the set of measurements:

$$|c_j| = \sqrt{p_j} \tag{10.5-3}$$

where p_i is the fraction of the measurements that give results equal to the eigenvalue A_i. This equation is as far as we can go with measurements of A. We would have to know the real and imaginary parts of each expansion coefficient to know the wave function exactly, and measurements of mechanical variables cannot give us this information.

The case that a single measurement can determine what the state of the system was prior to the measurement is exceptional.

If only one of the p's happens to be nonzero in the set of repeated measurements, we can say what the state was prior to the measurement. If p_i equals unity, we can assert that $|c_i|$ equals unity and that all the other c's vanish. The state prior to the measurement must have been the state corresponding to φ_i.

Measurement of a Complete Set of Variables Can Put the System into a Known State

If several eigenfunctions of $\hat{A}$ have the same eigenvalue, measurements of A give even less information about the prior state. Let a_i be an eigenvalue, and let g_i be the number of different eigenfunctions whose eigenvalues equal this value. If these are numbered from 1 to g_i, the probability that the value a_i will occur in a measurement of A is

$$p_i = \sum_{k=1}^{g_i} |c_k|^2 \tag{10.5-4}$$

The measurement has put the system into a state corresponding to a linear combination of all eigenfunctions with eigenvalues equal to the observed values.

The fifth postulate says only that the wave function after the measurement is an eigenfunction with eigenvalue a_i. Such an eigenfunction can be

$$\psi(\text{after}) = \sum_{j=1}^{g_i} c_j(\text{after})\varphi_j \tag{10.5-5}$$

so, even if every p_i is determined by a set of many measurements of A, not even the magnitudes of the individual c's can be determined.

Exercise 10.25   Argue that the linear combination in Equation (10.5-5) is an eigenfunction of $\hat{A}$ with eigenvalue a_i.

Only the functions with the same value for the eigenvalue are included in the sum, but we do not know what the coefficients $c_1, c_2, c_3, \ldots, c_g$ are.

Assume that there is another mechanical variable B whose operator $\hat{B}$ commutes with $\hat{A}$. There can be a set of common eigenfunctions, φ_{jk}:

$$\hat{A}\varphi_{jk} = a_j\varphi_{jk} \quad \text{and} \quad \hat{B}\varphi_{jk} = b_k\varphi_{jk} \qquad \textbf{(10.5-6)}$$

where we now use two indices to specify a given eigenfunction. Assume that all the eigenfunctions that have the same eigenvalue a_j have different eigenvalues b_k. If a measurement of A gives the result $a_{j'}$ where j' is a particular value of j, and a measurement of B gives the result b_k, then

$$\psi(\text{after}) = \varphi_{j'k'} \qquad \textbf{(10.5-7)}$$

We say that in this case the variables A and B form a **complete set of commuting observables**. Successive measurement of all of a complete set of commuting variables puts the system into a state corresponding to a known wave function that is an eigenfunction of all the operators corresponding to the set of variables. For most systems, two variables are not enough to form a complete set of commuting observables. Enough variables must be included so that the operator for the last variable measured has only nondegenerate eigenvalues among the possible states.

The energy eigenfunctions are so important that we usually refer to them as "wave functions."

Because of the great importance of the energy, the eigenfunctions of the Hamiltonian operator (the energy eigenfunctions) are the most commonly studied wave functions. Nearly every useful complete set of commuting observables includes the energy, but other sets could also be chosen. If a system is put into a state corresponding to an energy eigenfunction, its time-dependent wave function is a product of the energy eigenfunction and a time-dependent factor, and the coordinate factor will not change after the measurement because the wave function is that of a stationary state.

Summary

In this chapter we have presented postulates that are the theoretical basis of quantum mechanics. The first two postulates establish a one-to-one correspondence between the mechanical state of a system and a wave function and establish the Schrödinger equation, which governs the wave functions.

The third postulate is that there is a hermitian mathematical operator in one-to-one correspondence to each mechanical variable for a given system. The recipe for writing the operator for a given variable is: (1) write the classical expression for the variable in terms of cartesian coordinates and momentum components and (2) replace each momentum component by the relation

$$p_{x_j} \leftrightarrow \frac{\hbar}{i}\frac{\partial}{\partial x}$$

and its analogues.

The fourth postulate provides the means for predicting values of mechanical variables from operators and the wave function of the system. The first part of the postulate is that the only possible outcomes of a measurement of a variable are the eigenvalues of the operator corresponding to that variable. The second part is that the expectation value of the variable A is given by

$$\langle A \rangle = \frac{\int \Psi^* \hat{A} \Psi \, dq}{\int \Psi^* \Psi \, dq}$$

By studying the standard deviation of A, given by

$$\sigma_A = [\langle A^2 \rangle - \langle A \rangle^2]^{1/2}$$

it was established that if the state just before a measurement of A corresponds to an eigenfunction of $\hat{A}$, the only possible outcome of the measurement is the eigenvalue corresponding to that eigenfunction. In this case, a measurement is completely predictable, similar to the case in classical mechanics. If the wave function is not an eigenfunction of $\hat{A}$, the standard deviation gives a measure of the spread of the distribution of results.

The fifth postulate states that in a measurement of A, if the result is a_j, one of the eigenvalues of $\hat{A}$, then the state of the system immediately after the measurement corresponds to a wave function that is a linear combination of only the eigenfunctions whose eigenvalues equal a_j.

Measuring a complete set of commuting observables on the same system suffices to put the system into a state that is completely known, even though only partial information is available about the state of the system prior to the measurements.

Additional Reading

M. Jammer, *The Conceptual Development of Quantum Mechanics*, McGraw-Hill, New York, 1966
This is a book for physicists, and it presents the foundations of quantum mechanics in historical context.

M. Jammer, *The Philosophy of Quantum Mechanics*, Wiley, New York, 1974
This book is what its title suggests. It includes discussions of some controversial aspects, including paradoxes that arise in certain interpretations of the postulates.

See also the references at the end of Chapter 9.

PROBLEMS

Problems for Section 10.3

10.26. Find an expression for the commutator $[x, d^2/dx^2]$.

10.27. Find a simplified expression for the operator

$$\left[\frac{1}{x} + \frac{d}{dx} \right]^2$$

10.28. The eigenfunctions of coordinate operators are **Dirac delta functions**, defined such that $\delta(x - a) \to \infty$ if $x = a$ and

$\delta(x - a) = 0$ if $x \neq a$, and such that

$$\int_b^c \delta(x - a) \, dx = 1$$

if $b < a < c$.

a. Show that $\delta(x - a)$ is an eigenfunction of x.
b. What is the eigenvalue of $\delta(x - a)$?

10.29. a. Show that the operator x is linear and hermitian.

b. Show that the operator for multiplication by a function, $h(x)$, is linear and hermitian.

*10.30. a. Find the eigenfunctions and eigenvalues of $\hat{p}_x = -i\hbar(\partial/\partial x)$.

b. Are the energy eigenfunctions for a particle in a hard one-dimensional box eigenfunctions of this operator? If so, find the eigenvalues.

c. Are the energy eigenfunctions for a free particle eigenfunctions of this operator? If so, find the eigenvalues.

10.31. A function of an operator is defined through the Taylor series representing the function. For example, the exponential of a Hamiltonian operator is defined as the series

$$e^{\hat{H}} = 1 + \hat{H} + \frac{1}{2!}\hat{H}^2 + \frac{1}{3!}\hat{H}^3 + \cdots$$

where the operator powers are defined in the usual way, as successive operations of the operator.

a. Write the formula for the result when $e^{\hat{H}}$ operates on an eigenfunction of $\hat{H}$.

b. Write the expression for the first three terms of $e^{(\hat{A}+\hat{B})}$, where $\hat{A}$ and $\hat{B}$ are two operators that do not necessarily commute.

c. Find the expression for $\sin(\hat{H})$.

Problems for Section 10.4

10.32. Show that the momentum operator $(\hbar/i)\,\partial/\partial x$ gives the correct sign for $\langle p_x \rangle$ for a traveling wave given by

$$\Psi = e^{i\kappa x}e^{-iEt/\hbar}$$

*10.33. a. Find the eigenfunctions $\Phi(\phi)$ of the operator for the z component of the angular momentum, $\hat{L}_z = -i\hbar(\partial/\partial\phi)$.

b. Since $\phi = 0$ and $\phi = 2\pi$ refer to the same location, the boundary condition

$$\Phi(0) = \Phi(2\pi)$$

is imposed. Find the eigenvalues of $\hat{L}_z$.

10.34. a. Draw sketches of the first two energy eigenfunctions of a particle in a one-dimensional hard box of length a. Without doing the integral explicitly, argue from the graphs that the two functions are orthogonal.

b. Draw sketches of the first two energy eigenfunctions of a harmonic oscillator. Without doing the integral explicitly, argue from the graphs that the two functions are orthogonal.

10.35. Show that the two operators $\hat{L}_z$ and $\hat{L}_y$ do not commute, and find their commutator.

Problems for Section 10.5

*10.36. a. Calculate $\langle p_x^2 \rangle$ for each of the first three energy eigenfunctions for the particle in a one-dimensional hard box. Hint: Use the energy eigenvalues to avoid detailed calculations.

b. Obtain a formula (a function of n) for $\langle p_x^2 \rangle$ for a general energy eigenfunction for a particle in a one-dimensional hard box.

10.37. a. Calculate the uncertainty product $\sigma_x\sigma_{p_x}$ for a harmonic oscillator for the $v = 0$ state.

b. Without doing any calculations, state whether the uncertainty product for the $v = 1$ state would be smaller than, equal to, or larger than the product for the $v = 0$ state.

c. Calculate the uncertainty product $\sigma_x\sigma_{p_x}$ for a harmonic oscillator for the $v = 1$ state.

10.38. Using Equation (10.4-19), find an uncertainty relation for $\hat{L}_z$ and $\hat{L}_y$. The necessary commutator is to be found in Problem 10.35. Use a wave function that is a product of three harmonic oscillator wave functions with $v = 0$ for each:

$$\psi = \psi_0(x)\psi_0(y)\psi_0(z)$$

Problems for Section 10.6

*10.39. a. Find $\langle E \rangle$ for the coordinate wave function

$$\psi(x) = \sqrt{\frac{1}{3}}\,\psi_0 + \sqrt{\frac{1}{3}}\,\psi_1 + \sqrt{\frac{1}{3}}\,\psi_2$$

where ψ_0, ψ_1, and ψ_2 are the three lowest-energy harmonic oscillator energy eigenfunctions.

b. Find σ_E for the wave function in part (a).

c. Tell what values would occur in a set of many measurements of E, given that the system is in the state corresponding to the wave function of part a immediately before each measurement. Give the probability of each value.

10.40. a. A measurement of the energy of a particle in a three-dimensional cubic hard box gives a value $14h^2/(8ma^2)$. Tell what eigenfunctions are included in the linear combination representation of the wave function after the measurement.

b. How could the particle be put into a known state?

10.41. The energy of a particle in a one-dimensional hard box of length a is measured repeatedly with the particle restored to a specific but unknown state before each measurement. The results are summarized as follows:

Value	Probability
$h^2/(8ma^2)$	0.25
$4h^2/(8ma^2)$	0.375
$9h^2/(8ma^2)$	0.125
$16h^2/(8ma^2)$	0.25

What can you say about the state prior to the measurement?

General Problems

*10.42. a. Obtain a formula for the uncertainty product $\sigma_x \sigma_{p_x}$ for each of the first three energy eigenfunctions of a particle in a one-dimensional box. Comment on any trend you see in these values.

b. Evaluate the uncertainty product for each of the first three energy eigenfunctions of an electron in a box of length 10.0 Å (1.00×10^{-9} m).

10.43. a. Obtain a formula for the uncertainty σ_x for a particle in a one-dimensional box of length a for a state corresponding to a general energy eigenfunction ψ_n.

b. Find the limit of the formula of part a as $n \to \infty$.

c. Obtain a formula for the uncertainty σ_{p_x} for a particle in a one-dimensional box of length a for a state corresponding to a general energy eigenfunction ψ_n.

d. Obtain a formula for the uncertainty product $\sigma_x \sigma_{p_x}$ for a particle in a one-dimensional box of length a corresponding to a general energy eigenfunction ψ_n.

e. For the baseball in Problem 9.57, find the value of the uncertainty σ_x.

f. For the baseball in Problem 9.57, find the value of the uncertainty σ_{p_x}.

g. For the baseball in Problem 9.57, find the value of the uncertainty product $\sigma_x \sigma_{p_x}$.

*10.44. Label each statement as either true or false. If a statement is true only under certain circumstances, label it as false.

a. Every wave function satisfies the time-independent Schrödinger equation.

b. Every wave function satisfies the time-dependent Schrödinger equation.

c. Knowledge of the time-independent wave function provides all available information about mechanical variables of a system.

d. Knowledge of the time-dependent wave function provides all available information about mechanical variables of a system.

e. Measurement of the energy of a particle in a one-dimensional box determines the state of the system.

f. Measurement of the energy of a particle in a three-dimensional box determines the state of the system.

g. The uncertainty in the position of a particle can vanish only if nothing is known about its momentum.

h. The uncertainty in the momentum of a particle can vanish only if nothing is known about its position.

i. If a system is known to be in a state corresponding to an energy eigenfunction, the time during which the system is known to be in that state is infinite.

j. If a free particle moving in one dimension is known to be in a state corresponding to an eigenfunction of the momentum operator, nothing can be said about the position of the particle.

11 The Electronic States of Atoms: I. The Hydrogen Atom and the Simple Orbital Approximation for Multielectron Atoms

PREVIEW

In this chapter, we begin our application of quantum mechanics to atoms. We begin with the simplest atom, the hydrogen atom, which has a single electron. We discuss other atoms in the crudest approximation, the simple orbital approximation or zero-order orbital approximation. In this approximation, the electrons in multielectron atoms move independently in the same way that the electron of a hydrogen atom moves.

PRINCIPAL FACTS AND CONCEPTS

1. The Schrödinger equation for the hydrogen atom is an example of the "central-force problem."
2. In the central-force problem, the angular momentum of the system can have definite values when the system is in a state corresponding to an energy eigenfunction.
3. The Schrödinger equation for the hydrogen atom can be solved exactly, giving electronic wave functions called orbitals.
4. Electrons have intrinsic (spin) angular momentum in addition to the angular momentum of orbital motion. Spin orbitals describe both space and spin behavior.
5. In the simple orbital approximation, each electron in a multielectron atom occupies a hydrogen-like spin orbital.

6. The wave function for a multielectron atom is assumed to be anti-symmetric. That is, the wave function changes sign if the coordinates of two electrons are exchanged.

7. In an orbital wave function, every electron must occupy a different spin orbital (the Pauli exclusion principle).

11.1 The Central-Force Problem. Angular Momentum

We begin with a discussion of systems of two particles in which the potential energy of the system depends only on the distance between the particles. Such systems are called central-force systems, and the hydrogen atom is a member of this class of systems.

The Hydrogen Atom System

A hydrogen atom consists of a single electron and a nucleus containing a single proton. A **hydrogen-like atom** is like a hydrogen atom except for having a nucleus with a number of protons equal to Z. Any result for the hydrogen-like atom can be applied to a hydrogen atom by setting Z equal to 1, to an He^+ ion by setting Z equal to 2, etc.

Figure 11.1 depicts the hydrogen-like atom system. We let the cartesian coordinates of the nucleus be denoted by x_p, y_p, and z_p and let the cartesian coordinates of the electron be denoted by x_e, y_e, and z_e. The potential energy of the hydrogen-like atom depends only on the distance between the particles. It is given by Coulomb's law, which we introduced in Chapter 8:

$$\mathscr{V}(r) = -\frac{Ze^2}{4\pi\varepsilon_0 r} \tag{11.1-1}$$

where ε_0 is the permittivity of the vacuum and r is the distance between the particles:

$$r = (x^2 + y^2 + z^2)^{1/2} \tag{11.1-2}$$

where x, y, and z are called the relative coordinates:

$$x = x_e - x_p \tag{11.1-3a}$$

$$y = y_e - y_p \tag{11.1-3b}$$

$$z = z_e - z_p \tag{11.1-3c}$$

The coordinates of the center of mass are

$$X = \frac{m_e x_e + m_p x_p}{M} \tag{11.1-4a}$$

$$Y = \frac{m_e y_e + m_p y_p}{M} \tag{11.1-4b}$$

$$Z = \frac{m_e z_e + m_p z_p}{M} \tag{11.1-4c}$$

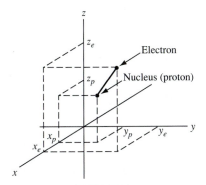

Figure 11.1. The System Consisting of a Nucleus and an Electron. This figure shows the cartesian coordinates of both particles, with a line segment drawn between the particles.

where the sum of the masses is denoted by M:

$$M = m_p + m_e \qquad \textbf{(11.1-5)}$$

The Central-Force System

Instead of beginning with the hydrogen atom, we discuss first a two-particle system for which $\mathscr{V}$ is an unspecified function of the relative coordinates. In this case the motion of the two-particle system can be separated into two simpler problems. We will further assume that $\mathscr{V}$ depends only on r, the distance between the two particles. Such a potential energy function corresponds to a **central force**, which points in the direction from one particle to the other. Our results will apply to a hydrogen atom as well as any other system in which $\mathscr{V}$ depends only on r. A second example of a central-force system is a model system that represents a rotating diatomic molecule.

The Central-Force Hamiltonian

We construct the Hamiltonian operator for our system according to the rules of Section 10.3. For a two-particle system, the kinetic energy can be expressed in terms of the velocity of each particle, or alternatively in terms of the velocity of the center of mass and the relative velocity, as in Equation (D-26) of Appendix D:

We here express the kinetic energy in terms of the center-of-mass velocity and the relative velocity. We are not required to make this change, but we could not solve the problem if we did not do it.

$$\mathscr{K} = \frac{m_p}{2}\left(v_{px}^2 + v_{py}^2 + v_{pz}^2\right) + \frac{m_e}{2}\left(v_{ez}^2 + v_{ey}^2 + v_{ez}^2\right)$$

$$= \frac{M}{2}\left(V_x^2 + V_y^2 + V_z^2\right) + \frac{\mu}{2}\left(v_x^2 + v_y^2 + v_z^2\right) \qquad \textbf{(11.1-6)}$$

where v_p and v_e are the velocities of the proton and electron, respectively, and V and v are the velocity of the center of mass and the relative velocity, respectively. The x component of each of these vectors is given by

$$v_{px} = dx_p/dt \qquad \textbf{(11.1-7a)}$$

$$v_{ex} = dx_e/dt \qquad \textbf{(11.1-7b)}$$

$$V_x = \frac{dX}{dt} \qquad \textbf{(11.1-7c)}$$

$$v_x = \frac{dx}{dt} \qquad \textbf{(11.1-7d)}$$

with similar equations for the y and z components. The **reduced mass** is denoted by μ:

$$\mu = \frac{m_p m_e}{m_p + m_e}$$

To write the Hamiltonian operator, we must express the kinetic energy in terms of momenta. The momenta conjugate to the center-of-mass coordinates X, Y, and Z are

Conjugate momenta are discussed in Appendix D.

$$P_x = MV_x, \qquad P_y = MV_y, \qquad P_z = MV_z \qquad \textbf{(11.1-8a)}$$

The momenta conjugate to the relative coordinates x, y, and z are

$$p_x = \mu v_x, \qquad p_y = \mu y_y, \qquad p_z = \mu v_z \qquad \textbf{(11.1-8b)}$$

The classical Hamiltonian function is

$$H_{cl} = \frac{1}{zM}(P_x^2 + P_y^2 + P_z^2) + \frac{1}{z\mu}(p_x^2 + p_y^2 + p_z^2) + \mathscr{V}(r) \quad \textbf{(11.1-9)}$$

| **Exercise 11.1** | Show that Equation (11.1-9) is correct. |

By Postulate III, the quantum mechanical Hamiltonian operator is obtained from Equation (11.1-9) by Equation (10.3-5) and similar equations. The result is

To write the Hamiltonian operator, we make the replacements analogous to Equation (10.3-5).

$$\hat{H} = -\frac{\hbar}{2M}\left(\frac{\partial^2}{\partial X^2} + \frac{\partial^2}{\partial Y^2} + \frac{\partial^2}{\partial Z^2}\right) - \frac{\hbar^2}{2\mu}\left(\frac{\partial^2}{\partial x^2} + \frac{\partial^2}{\partial y^2} + \frac{\partial^2}{\partial z^2}\right) + \mathscr{V}(r)$$

$$= -\frac{\hbar^2}{2M}\nabla_c^2 - \frac{\hbar^2}{2\mu}\nabla_r^2 + \mathscr{V}(r) \quad \textbf{(11.1-10)}$$

The symbol ∇_c^2 stands for the sum of the three second-derivative operators for the center-of-mass coordinates in the first line of Equation (11.1-10) and is called the **center-of-mass Laplacian**. The symbol ∇_r^2 stands for the **relative Laplacian**.

The first term in the Hamiltonian operator is the **center-of-mass Hamiltonian**:

$$\hat{H}_c = -\frac{\hbar^2}{2M}\nabla_c^2 \quad \textbf{(11.1-11)}$$

and the other two terms constitute the **relative Hamiltonian**:

$$\hat{H}_r = -\frac{\hbar^2}{2\mu}\nabla_r^2 + \mathscr{V}(r) \quad \textbf{(11.1-12)}$$

When a differential equation contains two terms such that one set of independent variables occurs in one term but not in the other, the equation can be solved by separation of variables.

The center-of-mass coordinates do not occur in the relative Hamiltonian, and the relative coordinates do not occur in the center-of-mass Hamiltonian, because the potential energy $\mathscr{V}$ depends only on r. This fact will enable us to separate the variables in the Schrödinger equation.

We now write the time-independent Schrödinger equation

$$(\hat{H}_c + \hat{H}_r)\Psi = E\Psi \quad \textbf{(11.1-13)}$$

In order to separate the variables, we assume the trial function

Previously we used Ψ for time-dependent wave functions. There are not enough letters in the Greek and Latin alphabets to use each symbol for only one thing.

$$\Psi = \chi(X, Y, Z)\psi(x, y, z) \quad \textbf{(11.1-14)}$$

We will use both Ψ and ψ for coordinate wave functions in the next several chapters. Generally, Ψ will be used for wave functions of entire atoms or molecules and ψ will be used for simpler wave functions.

By the method of separation of variables used in Sections 9.1 and 9.4, Equation (11.1-14) leads to the two equations

$$\hat{H}_c\chi = E_c\chi \quad \textbf{(11.1-15)}$$

$$\hat{H}\psi = E_r\psi \quad \textbf{(11.1-16)}$$

where the energy eigenvalue E is given by

$$E = E_c + E_r \quad \textbf{(11.1-17)}$$

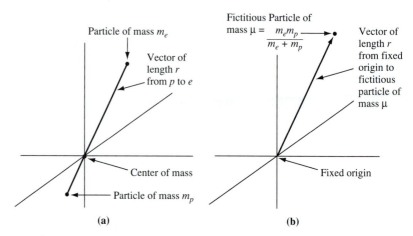

Figure 11.2 Figure Illustrating the Equivalence Between the Motion of a Particle of Mass μ Around a Fixed Center and the Relative Motion of Two Particles. (a) The actual two-particle system. **(b)** The fictitious particle of mass μ. The two vectors move in exactly the same way.

Exercise 11.2   Carry out the steps to obtain Equations (11.1-15)–(11.1-17).

Equation (11.1-15) for the center of mass is the same as the Schrödinger equation for a free particle of mass M, Equation (9.5-47). The energy eigenfunctions and energy eigenvalues are given by Equations (9.5-48) and (9.5-49), with replacement of the symbol for the mass by M. The center-of-mass motion is usually called **translation** of the atom. We will now consider the relative motion. The translational motion will be considered in Chapter 14.

Equation (11.1-16) for the relative motion is mathematically equivalent to the problem of a particle of mass μ moving at distance r from a fixed origin under the effect of the potential energy $\mathcal{V}(r)$. Figure 11.2 depicts this equivalence. The vector from one particle (labeled p) to the other (labeled e) in Figure 11.2a is equal to the vector from a fixed origin to the fictitious particle of mass μ in Figure 11.2b. If one of the particles is much heavier than the other, as is the case in any hydrogen-like atom, the reduced mass is nearly equal to the mass of the lighter object. The center of mass is much closer to the heavier particle than to the other, and the motion is nearly the same as though the heavier particle were stationary with the lighter particle moving around it. The relative motion in a hydrogen-like atom is very nearly the same as electronic motion around a stationary nucleus.

One particle moves relative to the other in the same way as a fictitious particle of mass μ moves relative to a fixed origin.

Exercise 11.3   The mass of the electron is 9.1094×10^{-31} kg and the mass of the proton is 1.673×10^{-27} kg.

a. Calculate the ratio of the reduced mass of the hydrogen atom to the mass of the electron.

b. For a hydrogen atom with the electron at a distance 1.000×10^{-10} m from the nucleus, find the distance from the center of mass to the nucleus and to the electron. Hint: This calculation is easier if the particles are temporarily assumed to be on the x axis.

Solution of the Relative Schrödinger Equation

Since the potential energy function depends on r, which is a fairly complicated function of x, y, and z, it simplifies the problem to transform to spherical polar coordinates, in which r is one of the coordinates. These coordinates are shown in Figure 11.3.

Using the expression for the Laplacian operator in these coordinates,[1] the relative Schrödinger equation is

$$
\hat{H}_r\psi = -\frac{\hbar^2}{2\mu r^2}\left[\frac{d}{dr}\left(r^2\frac{d\psi}{dr}\right) + \frac{1}{\sin(\theta)}\frac{\partial}{\partial\theta}\left(\sin(\theta)\frac{\partial\psi}{\partial\theta}\right) + \frac{1}{\sin^2(\theta)}\frac{\partial^2\psi}{\partial\phi^2}\right]
$$
$$
+ \mathscr{V}(r)\psi
$$
$$
= E_r\psi \tag{11.1-18}
$$

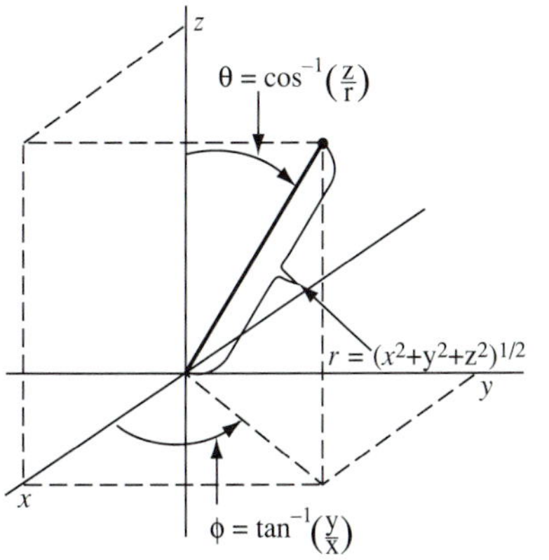

Figure 11.3. Spherical Polar Coordinates. These coordinates are used to simplify the solution of the Schrödinger equation.

Comparison of this equation with Equation (10.3-10) shows that the operator for the square of the angular momentum is contained in the Hamiltonian operator:

$$
-\frac{\hbar^2}{2\mu r^2}\frac{d}{dr}\left(r^2\frac{d\psi}{dr}\right) + \frac{1}{2\mu r^2}\hat{L}^2\psi + \mathscr{V}(r)\psi = E_r\psi \tag{11.1-19}
$$

This equation can be solved by separation of variables. We assume the trial solution

$$
\psi(r, \theta, \phi) = R(r)Y(\theta, \phi) \tag{11.1-20}
$$

The operator in Equation (11.1-18) does not contain three terms with only one coordinate in each term, so it is not obvious that separation of variables will work until we try it.

Since the operator $\hat{L}^2$ does not contain r, substitution of the trial solution into Equation (11.1-19) gives

$$
-\frac{\hbar^2}{2\mu}\left[Y\frac{1}{r^2}\frac{d}{dr}\left(r^2\frac{dR}{dr}\right) + R\hat{L}^2Y\right] + (\mathscr{V} - E_r)RY = 0 \tag{11.1-21}
$$

We multiply this equation by $2\mu r^2/\hbar^2$ and divide by RY. This separates r from the other variables, giving

$$
-\frac{1}{R}\frac{d}{dr}\left(r^2\frac{dR}{dr}\right) + \frac{2\mu r^2}{\hbar^2}(\mathscr{V} - E_r) + \frac{1}{\hbar^2}\frac{1}{Y}\hat{L}^2Y = 0 \tag{11.1-22}
$$

The Angular Factors in the Wave Function

The final term on the left-hand side of Equation (11.1-22) contains no r and the other terms contain no θ or ϕ. The last term must be a constant function of θ and ϕ, which we set equal to the constant K. Multiplication

---

[1] See Equation (B-42) of Appendix B.

by $\hbar^2 Y$ gives the equation

$$\hat{L}^2 Y = \hbar^2 K Y \qquad (11.1\text{-}23)$$

which is also the eigenvalue equation for the square of the angular momentum. The function Y is the eigenfunction for the $\hat{L}^2$ operator, as well as a factor in the eigenfunction of the Hamiltonian operator.

Equation (11.1-23) can be written

$$-\hbar^2 \left[\frac{1}{\sin(\theta)} \frac{\partial}{\partial \theta} \left[\sin(\theta) \frac{\partial Y}{\partial \theta} \right] + \frac{1}{\sin^2(\theta)} \frac{\partial^2 Y}{\partial \phi^2} \right] = \hbar^2 K Y \quad (11.1\text{-}24)$$

To separate the variables θ and ϕ, we assume the trial solution

$$Y = \Theta(\theta) \Phi(\phi) \qquad (11.1\text{-}25)$$

Substitution into Equation (11.1-24) followed by division by $\Theta(\theta)\Phi(\phi)$ and multiplication by $\sin^2(\theta)$ gives

$$\frac{\sin(\theta)}{\Theta} \frac{d}{d\theta} \left[\sin(\theta) \frac{d\Theta}{d\theta} \right] + \frac{1}{\Phi} \frac{d^2\Phi}{d\phi^2} = -K \sin^2(\theta) \qquad (11.1\text{-}26)$$

The last term on the left-hand side of this equation depends only on ϕ, so it must be a constant function of ϕ, which we call $-m^2$. If this choice for the constant is made, m will turn out to be a real integer. Multiplication by Φ gives the equation

$$\frac{d^2\Phi}{d\phi^2} = -m^2\Phi \qquad (11.1\text{-}27)$$

Except for the symbols used, Equation (11.1-27) is exactly the same as several equations already encountered, and its general solution can be written as in Equation (9.5-36)

$$\Phi = Ae^{im\phi} + Be^{-im\phi} \qquad (11.1\text{-}28)$$

where A and B are constants.

The variable ϕ ranges from 0 to 2π radians. Since $\phi = 0$ and $\phi = 2\pi$ refer to the same location for fixed values of r and θ, we must impose the condition

$$\Phi(0) = \Phi(2\pi) \qquad (11.1\text{-}29)$$

so that the wave function will be continuous at this value of ϕ. This condition is satisfied only if m is real and equal to an integer.

Exercise 11.4   Use the identity

$$e^{ix} = \cos(x) + i\sin(x)$$

to show that m is real and equal to an integer.

There are no further conditions that restrict possible values of the constants A and B in Equation (11.1-28). However, two standard forms of the function are customarily used. For the first form, we choose the values so that Φ is an eigenfunction of $\hat{L}_z$, given by Equation (10.3-9):

$$\hat{L}_z = \frac{\hbar}{i}\frac{\partial}{\partial \phi} \qquad (11.1\text{-}30)$$

We operate on Φ with $\hat{L}_z$:

$$\hat{L}_z\Phi = \frac{\hbar}{i}(imAe^{im\phi} - imBe^{-im\phi}) \qquad (11.1\text{-}31)$$

We have an eigenfunction of $\hat{L}_z$ with eigenvalue $\hbar m$ if B is chosen to equal zero. It can be shown that $\hat{L}^2$ and $\hat{L}_z$ commute and that both commute with $\hat{H}_r$, so these three operators are all possible members of a complete set of commuting observables for a central-force problem.

Exercise 11.5   Show that $\hat{H}_r$, $\hat{L}^2$, and $\hat{L}_z$ all commute.

The complex Φ functions are eigenfunctions of the $\hat{L}_z$ operator. Not all of the real functions are eigenfunctions.

With $B = 0$, the normalized Φ function is

$$\Phi = \Phi_m = \frac{1}{\sqrt{2\pi}}e^{im\phi} \qquad (11.1\text{-}32)$$

where we label the members of the set of functions with the quantum number m.

For the second standard form, we choose the constants A and B so that Φ is a real function. If A and B are equal,

$$\Phi = \Phi_{mx} = A(e^{im\phi} + e^{-im\phi}) = 2A\cos(m\phi)$$

$$\Phi_{mx} = \frac{1}{\sqrt{\pi}}\cos(m\phi) \qquad (11.1\text{-}33)$$

If $B = -A$, then

$$\Phi = \Phi_{my} = A(e^{im\phi} - e^{-im\phi}) = 2iA\sin(m\phi)$$

$$\Phi_{my} = \frac{1}{\sqrt{\pi}}\sin(m\phi) \qquad (11.1\text{-}34)$$

where the values of A and B are chosen for normalization.

Exercise 11.6   Show that Φ_{mx} and Φ_{my} are not eigenfunctions of $\hat{L}_z$ except for $m = 0$.

We now seek the Θ factor. After replacement of the constant term by $-m^2$ and multiplication by Θ, Equation (11.1-26) becomes

$$\sin(\theta)\frac{d}{d\theta}\left[\sin(\theta)\frac{d\Theta}{d\theta}\right] - m^2\Theta + K\sin^2(\theta)\Theta = 0 \qquad \textbf{(11.1-35)}$$

The equation is named for Adrien-Marie Legendre, 1752–1833, a famous French mathematician.

This equation can be transformed into the **associated Legendre equation** by the change of variables:

$$y = \cos(\theta), \qquad P(y) = \Theta(\theta) \qquad \textbf{(11.1-36)}$$

The associated Legendre equation and its solutions are given in Appendix F. The solutions are called **associated Legendre functions** and are derivatives of polynomials known as **Legendre polynomials**.

For a solution to exist which obeys the relevant boundary conditions, the constant K must be equal to $\ell(\ell+1)$ where ℓ is an integer at least as large as $|m|$. (We do not prove this fact.) There is one solution for each set of values of the two quantum numbers ℓ and m:

$$\Theta(\theta) = \Theta_{\ell m}(\theta) \qquad \textbf{(11.1-37)}$$

The solutions are the same for a given value of m and its negative:

$$\Theta_{\ell m}(\theta) = \Theta_{\ell, -m}(\theta) \qquad \textbf{(11.1-38)}$$

where we insert a comma to avoid confusing two subscripts having values ℓ and $-m$ with a single subscript having a value $\ell - m$.

The Y functions in Equation (11.1-25) are called **spherical harmonic functions**. Each one is a product of a Θ function and a Φ function having the same value of m as the Θ function. Table 11.1 gives the normalized spherical harmonic functions for $\ell = 0$, $\ell = 1$, and $\ell = 2$. Additional functions can be derived from formulas in Appendix F.

Angular Momentum Values

The classical angular momentum **L** of a particle of mass m about a fixed center is defined in Equation (D-14) of Appendix D:

$$\mathbf{L} = m\mathbf{r} \times \mathbf{v} \qquad \textbf{(11.1-39)}$$

The vector product of two vectors is defined in Equation (D-33) of Appendix D.

where **r** is the position vector with the fixed center taken as the origin, **v** is the velocity of the particle, and × represents the vector product of the two vectors. Figure 11.4 shows the vectors **r**, **v**, and **L** for a particle orbiting in a plane. The angular momentum is perpendicular to the plane containing **r** and **v**. In a central-force problem, the orbit of the moving mass remains in this plane, and the angular momentum remains in a fixed direction.

In the central-force problem, the relative coordinate moves like a single particle of mass μ moving about a fixed center, so the classical angular momentum of relative motion is

$$\mathbf{L} = \mu\mathbf{r} \times \mathbf{v} \qquad \textbf{(11.1-40)}$$

The total angular momentum of a set of particles is the vector sum of the individual angular momenta. If the set of particles constitutes a rotating rigid symmetrical body such as a gyroscope spinning on its axis, the

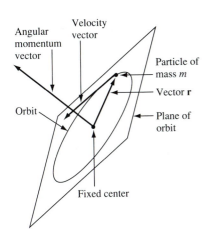

Angular momentum vector

Velocity vector

Orbit

Particle of mass m

Vector $\mathbf{r}$

Plane of orbit

Fixed center

Figure 11.4. Diagram Illustrating the Definition of the Angular Momentum Vector. The angular momentum vector, **L**, is perpendicular to the vector **r** and the vector **v** and has magnitude equal to $m|\mathbf{r}||\mathbf{v}|$.

Table 11.1. Spherical Harmonic Functions $\psi_{\ell m}(\theta, \phi) = \Theta_{\ell m}(\theta)\Phi_m(\phi)$

Complex Φ functions, eigenfunctions of $\hat{L}_z$.

$$\Phi_m(\phi) = \frac{1}{\sqrt{2\pi}} e^{im\phi}$$

Real Φ functions, not necessarily eigenfunctions of $\hat{L}_z$.

$$\Phi_{mx}(\phi) = \frac{1}{\sqrt{\pi}} \cos(m\phi)$$

$$\Phi_{my}(\phi) = \frac{1}{\sqrt{\pi}} \sin(m\phi)$$

Θ functions:

$$\Theta_{00}(\theta) = \frac{\sqrt{2}}{2}$$

$$\Theta_{10}(\theta) = \frac{\sqrt{6}}{2} \cos(\theta), \qquad \Theta_{11}(\theta) = \Theta_{1,-1}(\theta) = \frac{\sqrt{3}}{2} \sin(\theta)$$

$$\Theta_{20}(\theta) = \frac{\sqrt{10}}{4} (3 \cos^2(\theta) - 1)$$

$$\Theta_{21}(\theta) = \Theta_{2,-1}(\theta) = \frac{\sqrt{15}}{2} \sin(\theta) \cos(\theta)$$

$$\Theta_{22}(\theta) = \Theta_{2,-2}(\theta) = \frac{\sqrt{15}}{4} \sin^2(\theta)$$

Additional Θ functions can be obtained from Appendix F.

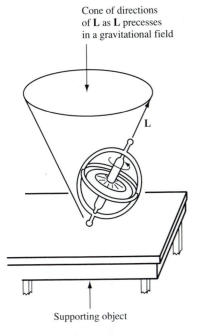

Cone of directions of **L** as **L** precesses in a gravitational field

L

Supporting object

Figure 11.5. A Simple Gyroscope. If a torque is place on a gyroscope, as by supporting one end of it in a gravitational field, the gyroscope will precess; that is, its axis of rotation will move around a cone with a vertical axis.

angular momentum vector of every particle points along that axis and has magnitude proportional to its rate of rotation. Figure 11.5 shows a simple gyroscope. If a gyroscope stands on one end of its axis at an angle in a gravitational field, the gravitational torque, instead of making the gyroscope fall on its side, makes the axis move (precess) around a vertical cone, as shown in the figure.

Equation (11.1-23) gives us the eigenvalues of the operator for the square of the angular momentum:

$$\hat{L}^2 Y_{\ell m} = \hat{L}^2 \Theta_{\ell m} \Phi_m = \hbar^2 \ell(\ell + 1)\Theta_{\ell m}\Phi_m \qquad (\ell = 0, 1, 2, \ldots) \quad \textbf{(11.1-41)}$$

Since the eigenvalues of an operator are the only possible outcomes of measurements of the corresponding mechanical quantity, the square of the angular momentum is quantized, taking on values

$$\boxed{L^2 = 0, 2\hbar^2, 6\hbar^2, 12\hbar^2, 20\hbar^2, \ldots} \quad \textbf{(11.1-42)}$$

The magnitude of the angular momentum vector can taken on the quantized values

$$\boxed{L = |\mathbf{L}| = 0, \sqrt{2}\,\hbar, \sqrt{6}\,\hbar, \sqrt{12}\,\hbar, \sqrt{20}\,\hbar, \ldots} \quad \textbf{(11.1-43)}$$

The quantization of angular momentum arises from the fact that the constant K had to equal $\ell(\ell + 1)$ for Equation (11.1-35) to be the same as the associated Legendre equation and have known solutions. Compare this mathematically generated quantization with the quantization of the magnitude of angular momentum in the Bohr theory of the hydrogen atom, where it was assumed as a hypothesis that the angular momentum could take on values $\hbar$, $2\hbar$, $3\hbar$, ... Not only is the origin of the quantization different, but the values are different from those of the Bohr theory.

The magnitudes of the quantized values for L_z are the same as the assumed values that Bohr took for the magnitude of **L**.

The function Φ_m in Equation (11.1-31) is an eigenfunction of $\hat{L}_z$ with eigenvalue $\hbar m$, so the possible values of L_z are

$$L_z = m\hbar = 0, \pm\hbar, \pm 2\hbar, \pm 3\hbar, \ldots \pm \ell\hbar \qquad \textbf{(11.1-44)}$$

The list is restricted by the condition that $|m|$ cannot be any larger than the integer ℓ.

In order to specify completely the direction of the angular momentum vector, L_x and L_y are needed as well as L_z. However, $\hat{L}_x$, $\hat{L}_y$, and $\hat{L}_z$ do not commute with each other, so they cannot have common eigenfunctions. No more than one component of the angular momentum can have a predictable value for a given state, and the exact direction of the angular momentum vector cannot be determined. This is a consequence of the uncertainty principle, since both position and vector momentum in the same direction would have to be specified with arbitrary accuracy in order to specify exact values of two or three different components, as seen in the expression for the $\hat{L}_z$ operator in Equation (10.3-8).

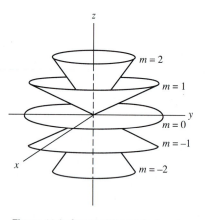

Figure 11.6. Cones of Possible Angular Momentum Directions for $\ell = 2$. These cones are similar to the cones of precession of a gyroscope and represent possible directions for the angular momentum vector. The z component is arbitrarily chosen as the one component that can have a definite value.

The situation is depicted in Figure 11.6, which shows the case that $\ell = 2$, for which m can take on the values 2, 1, 0, -1, and -2. The magnitude of L^2 is $6\hbar^2$, and the possible values of L_z are $2\hbar$, $\hbar$, 0, $-\hbar$, and $-2\hbar$. The maximum value of L_z is always smaller than the magnitude of L unless **L** is equal to zero, so that **L** cannot point along the z axis. (In the case of $\ell = 2$, $2\hbar$ is smaller than $\sqrt{6}\,\hbar$.) The angular momentum vector can point anywhere on the five cones drawn in the figure. If the wave function is known to correspond to a particular value of m, then it is known which cone applies, but the direction on that cone is not known. For any value of ℓ, there are $2\ell + 1$ cones, one for each possible value of m. Notice the similarity of each cone in Figure 11.6 to the cone of directions around which the gyroscope axis precesses in Figure 11.5.

Since only one component of the angular momentum can be a member of a complete set of commuting observables, only one component can have predictable values. We choose L_z for this component because it is more convenient in spherical polar coordinates.

There is nothing unique about the z direction. If we wished, we could choose $\hat{L}_x$ or $\hat{L}_y$ as a member of a set of commuting observables instead of $\hat{L}_z$. In that event, the Φ functions would be different and would correspond to cones in Figure 11.6 oriented around either the x axis or the y axis. In fact, the Φ_{mx} function in Equation (11.1-33) is an eigenfunction of $\hat{L}_x$ and the Φ_{my} function in Equation (11.1-34) is an eigenfunction of $\hat{L}_y$. We use $\hat{L}_z$ because its expression in spherical polar coordinates is simple and because the spherical harmonic functions are eigenfunctions of $\hat{L}_z$.

Exercise 11.7   Transform the expression for Φ_{1x} to cartesian coordinates. Show that this function is an eigenfunction of the operator $\hat{L}_x$ and find its eigenvalue.

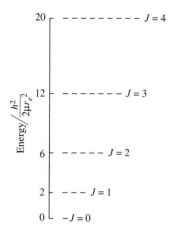

Figure 11.7. Energy Levels of the Rigid Rotor. The energy levels are quantized, as shown. Each level is at a height proportional to its energy. The degeneracies are shown by placing a line segment for each state of the level at the same height in the diagram.

Using the letters J and M instead of ℓ and m makes no difference to our results.

The degeneracy of the energy level is obtained by counting up the number of values M can take on: a number of positive values equal to J, a number of negative values equal to J, plus the zero value, for a total of 2J + 1 values.

The Rigid Rotor

The rigid rotor is a model system that approximately represents the rotation of a diatomic molecule. It is a central-force system consisting of two masses with fixed interparticle distance. Since r is fixed, the potential energy has a constant value if the system is a central-force system. We choose this value to equal zero:

$$r = r_e \text{ (fixed)}, \qquad \mathscr{V} = 0 \text{ (fixed)} \qquad \textbf{(11.1-45)}$$

For this case, Equation (11.1-19) for the radial factor R is simplified. Since r is constant in this model, the derivative with respect to r is omitted and the Schrödinger equation becomes

$$R \frac{1}{2\mu r^2} \hat{L}^2 Y = E_r R Y \qquad \textbf{(11.1-46)}$$

Since both sides of the equation are multiplied by the radial factor R, we assume that this factor does not vanish and cancel it. The equation is now the same as the angular momentum eigenvalue equation, so the energy eigenfunction is

$$\psi_{\text{rot}} = Y_{JM}(\theta, \phi) = \Theta_{JM}(\theta)\Phi_M(\phi) \qquad \textbf{(11.1-47)}$$

where we call the eigenfunction ψ_{rot}. We use different letters, J and M, for the quantum numbers, but they are the same as ℓ and m and obey the same relations as ℓ and m.

Using Equation (11.1-41), the energy eigenvalue is

$$E = E_J = \frac{\hbar^2}{2\mu r_e^2} J(J + 1) \quad \text{(rigid rotor)} \qquad \textbf{(11.1-48)}$$

There is one energy eigenfunction for each set of values of J and M. The number of values M can assume for a given value of J is $2J + 1$, so the degeneracy (the number of states in the energy level) is

$$g_J = 2J + 1 \qquad \textbf{(11.1-49)}$$

Figure 11.7 shows the first few energy levels with their degeneracies. If the rigid rotor has ends that have electrical charges (if it has a dipole moment), it can emit or absorb electromagnetic radiation.

Exercise 11.8

a. The internuclear distance of the carbon monoxide molecule is 1.128×10^{-10} m. Assume that it rotates like a rigid rotor and find the value of the rotational energy in the $J = 0$ and $J = 1$ energy levels.

b. Find the frequency and wavelength of the radiation absorbed if a carbon monoxide molecule makes a transition from the $J = 0$ state to one of the $J = 1$ states.

The Hydrogen-like Atom

In Section 11.1 we wrote the energy eigenfunction for any central-force problem as

$$\psi(r, \theta, \phi) = R(r)Y_{\ell m}(\theta, \phi) = R(r)\Theta_{\ell m}(\theta)\Phi_m(\phi) \qquad (11.2\text{-}1)$$

The spherical harmonic functions $Y_{\ell m}(\theta, \phi) = \Theta_{\ell m}(\theta)\Phi_m(\phi)$ are the same functions for any central-force problem.

To complete the solution of the time-independent Schrödinger equation for a particular central-force problem, we must write and solve the equation for R, the **radial factor**. This equation is obtained from Equation (11.1-22) by replacing $\hat{L}^2 Y$ by $\hbar^2 \ell(\ell + 1)Y$ according to Equation (11.1-41). Multiplication of the resulting equation by R gives

$$-\frac{d}{dr}\left(r^2 \frac{dR}{dr}\right) + \frac{2\mu r^2}{\hbar^2}(\mathcal{V} - E)R + \ell(\ell + 1)R = 0 \qquad (11.2\text{-}2)$$

where we now drop the subscript r from the symbol for the relative energy. Since the nucleus is much more massive than the electron, the relative energy is essentially the energy of the electron's motion around the nucleus.

Equation (11.2-2), which applies to any central-force problem, is now made to apply to a hydrogen-like atom by replacing $\mathcal{V}(r)$ by the expression in Equation (11.1-1). If this is done and if the derivative term is expanded into two terms, we have

$$-r^2 \frac{d^2R}{dr^2} - 2r\frac{dR}{dr} - \frac{2\mu r^2}{\hbar^2}\left(E - \frac{Ze^2}{4\pi\varepsilon_0 r}\right)R + \ell(\ell + 1)R = 0 \quad (11.2\text{-}3)$$

The choice of the zero of potential energy is arbitrary. Our present choice is the most convenient one, although it makes the total energy negative for the bound states.

The potential energy function is chosen to equal zero when the electron and nucleus are infinitely far apart. We consider only negative values of the energy eigenvalue E, corresponding to **bound states**, in which the system does not have sufficient relative energy for the electron to escape from the nucleus. There are also nonbound states, called **scattering states**, in which the relative energy is positive and the electron can escape from the nucleus. It is found that the energy for these nonbound states is not quantized but can take on any positive value. We do not discuss these states.[2]

We now make the following substitutions:

$$\alpha^2 = -\frac{2\mu E}{\hbar^2}, \qquad \beta = \frac{\mu Ze^2}{4\pi\varepsilon_0 \alpha \hbar^2}, \qquad \rho = 2\alpha r \qquad (11.2\text{-}4)$$

This equation is named for Edmund Laguerre, 1834–1866, a famous French mathematician.

The resulting equation is divided by ρ^2, giving an equation that is known as the **associated Laguerre equation**:

$$\frac{d^2R}{d\rho^2} + \frac{2}{\rho}\frac{dR}{d\rho} - \frac{R}{4} + \frac{\beta R}{\rho} + \ell(\ell + 1)\frac{R}{\rho^2} = 0 \qquad (11.2\text{-}5)$$

where we use the letter R for the function of ρ that is equal to $R(r)$.

---

[2] H. A. Bethe and E. E. Salpeter, *Quantum Mechanics of One- and Two-Electron Systems*, Plenum, New York, 1977, pp. 21ff, pp. 32ff.

The solution is written as

$$R(\rho) = G(\rho)e^{-\rho/2} \tag{11.2-6}$$

where $G(\rho)$ is a power series

$$G(\rho) = \sum_{j=0}^{\infty} a_j \rho^j \tag{11.2-7}$$

with constant coefficients $a_1, a_2, a_3, \ldots$.

This series must terminate after a finite number of terms in order to keep the wave function from becoming infinite for large values of ρ, violating our boundary conditions. We omit the details of the analysis.[3] This termination requires that the parameter β in Equation (11.2-4) be equal to an integer n, which must be at least as large as $\ell + 1$. The minimum value of n is unity, and this value occurs only for $\ell = 0$. Solving the second equality in Equation (11.2-4) for α, we obtain

$$\alpha = \frac{\mu Z e^2}{4\pi\varepsilon_0 \hbar^2 n} \tag{11.2-8}$$

From the first relation in Equation (11.2-4), the energy is quantized:

$$E = E_n = -\frac{\hbar^2 \alpha^2}{2\mu} = -\frac{\mu Z^2 e^4}{2(4\pi\varepsilon_0 \hbar n)^2} \qquad (n = 1, 2, 3, \ldots) \tag{11.2-9}$$

The value of the energy is determined by the value of the quantum number n. Once again, as in the case of the particle in a box and the harmonic oscillator, we have an energy that is quantized by the nature of the Schrödinger equation and its boundary conditions, and not by arbitrary assumption as in the Bohr theory. However, the energy expression in Equation (11.2-9) is identical to that of the Bohr theory.

Exercise 11.9   Substitute the values of the constants into Equation (11.2-9) to show that the energy of relative motion of a hydrogen atom can take on the values

$$E = E_n = -\frac{(2.1787 \times 10^{-18} \text{ J})Z^2}{n^2} = -\frac{(13.60 \text{ eV})Z^2}{n^2} \tag{11.2-10}$$

where 1 eV is the energy required to move one electron through an electric potential difference of 1 volt and is equal to 1.602×10^{-19} J.

The parameter a, called the **Bohr radius** of the hydrogen atom, is defined by

$$a = \frac{\hbar^2 4\pi\varepsilon_0}{\mu e^2} = 5.2947 \times 10^{-11} \text{ m} = 0.52947 \text{ Å} \tag{11.2-11}$$

---

[3] Frank L. Pilar, *Elementary Quantum Chemistry*, McGraw-Hill, New York, 1968, pp. 151ff.

We use the symbol a_0 for the Bohr radius if the nucleus is considered stationary so that the mass of the electron appears in the formula instead of μ. a_0 is equal to unit of length called the bohr.

where Å represents the angstrom unit, 10^{-10} m. The value of a is the same as the radius of the smallest orbit in the Bohr theory of the hydrogen atom. When we express the energy in terms of this parameter, we get

$$E = E_n = -\frac{\hbar^2 \alpha^2}{2\mu} = -\frac{Z^2 e^2}{2(4\pi\varepsilon_0)an^2} \qquad (11.2\text{-}12)$$

Exercise 11.10   Verify Equations (11.2-11) and (11.2-12).

In the (fictitious) limit that the nucleus is infinitely heavy compared to the electron, the electron moves about the stationary nucleus, and the reduced mass becomes

$$\lim_{m_p \to \infty} \mu = \lim_{m_p \to \infty} \left(\frac{m_e m_p}{m_e + m_p} \right) = m_e \qquad (11.2\text{-}13)$$

Equation (11.2-11) becomes

$$\lim_{m_p \to \infty} a = a_0 = \frac{\hbar^2 4\pi\varepsilon_0}{m_e e^2} = 5.29177 \times 10^{-11} \text{ m} \qquad (11.2\text{-}14)$$

The relative motion of the electron and the nucleus is generally called electronic motion because the nucleus moves very little compared with the electron.

For ordinary purposes, the distinction between relative motion of the nucleus and electron about their center of mass and electronic motion about a stationary nucleus is numerically unimportant, because the nucleus is so much more massive than the electron. We usually refer to the relative motion as electronic motion.

Exercise 11.11   Calculate the percentage error in the hydrogen atom Bohr radius and the hydrogen atom energy introduced by replacing the reduced mass by the mass of the electron.

The Radial Factor of the Hydrogen Atom Wave Functions

The polynomial G in Equation (11.2-6) is expressed as a function of ρ, where from Equations (11.2-4) and (11.2-8)

$$\rho = 2\alpha r = \frac{2Zr}{na} \qquad (11.2\text{-}15)$$

This polynomial is related to the **associated Laguerre functions**. Appendix F describes these functions and the **Laguerre polynomials** of which they are derivatives.

There is a different R factor for each set of values of the quantum numbers ℓ and n. Table 11.2 gives the R functions for $\ell = 1$, 2, and 3, and others can be written from the formulas for associated Laguerre functions given in Appendix F. There is a different spherical harmonic function Y for each set of values of ℓ and m, as given in Table 11.1, so a particular energy eigenfunction is obtained by multiplying an R factor by a spherical harmonic

Table 11.2. Radial Factors for Hydrogen-like Energy Eigenfunctions

$$R_{10}(r) = R_{1s}(r) = \left(\frac{Z}{a}\right)^{3/2} 2e^{-Zr/a}$$

$$R_{20}(r) = R_{2s}(r) = \frac{1}{2\sqrt{2}}\left(\frac{Z}{a}\right)^{3/2}\left(2 - \frac{Zr}{a}\right)e^{-Zr/2a}$$

$$R_{21}(r) = R_{2p}(r) = \frac{1}{2\sqrt{6}}\left(\frac{Z}{a}\right)^{3/2}\left(\frac{Zr}{a}\right)e^{-Zr/2a}$$

$$R_{30}(r) = R_{3s}(r) = \frac{1}{9\sqrt{3}}\left(\frac{Z}{a}\right)^{3/2}\left[6 - \frac{4Zr}{a} + \left(\frac{2Zr}{3a}\right)^2\right]e^{-Zr/3a}$$

$$R_{31}(r) = R_{3p}(r) = \frac{2}{27\sqrt{6}}\left(\frac{Z}{a}\right)^{3/2}\left(\frac{4Zr}{a} - \frac{2Z^2r^2}{3a^2}\right)e^{-Zr/3a}$$

$$R_{32}(r) = R_{3d}(r) = \frac{1}{9\sqrt{30}}\left(\frac{Z}{a}\right)^{3/2}\left(\frac{2Zr}{3a}\right)^2 e^{-Zr/3a}$$

Additional functions can be obtained from Appendix F.

function having the same value of ℓ. If the complex Φ functions are used, an energy eigenfunction is specified by giving the values of the quantum numbers n, ℓ, and m. To each of these eigenfunctions there corresponds a stationary state of the orbital electronic motion of the atom, with predictable values of the energy, the magnitude of the angular momentum, and the z component of the angular momentum.

The rules the quantum numbers obey can be restated:

$$n = 1, 2, 3,\ldots \tag{11.2-16a}$$
$$\ell = 0, 1, 2,\ldots, n - 1 \tag{11.2-16b}$$
$$m = 0, \pm 1, \pm 2,\ldots, \pm 1 \tag{11.2-16c}$$

The quantum number n is called the **principal quantum number**. The quantum number ℓ has been called the **azimuthal quantum number** but might better be called the **angular momentum quantum number**. The quantum number m has been called the **magnetic quantum number** but might better be called the **angular momentum projection quantum number**.

Since the energy depends only on the value of the principal quantum number, the energy levels are degenerate except for the $n = 1$ level.

▼   **EXAMPLE 11.1**

Find an expression for the degeneracy of the hydrogen atom energy levels.

Solution

For a given value of n, the possible values of ℓ range from 0 to $n - 1$. For a given value of ℓ, the values of m range from $-\ell$ to ℓ. The number of possible values of m is $2\ell + 1$, since m can have any of ℓ positive values, any of ℓ negative values, or be equal to zero. The degeneracy is

$$g_n = \sum_{\ell=0}^{n-1}(2\ell + 1) = 2n\frac{0 + n - 1}{2} + n = n^2$$

The degeneracy for different values of ℓ is called an "accidental degeneracy" because it occurs only for the Coulomb potential, not for other central-force potentials.

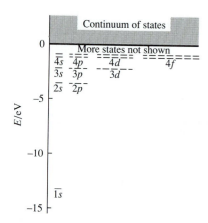

Figure 11.8. Energy Levels of the Hydrogen Atom. The bound-state energy levels are quantized.

We will not use the older K, L, M notation.

The letters s, p, d, and f came from the spectroscopic terms "sharp," "principal," "diffuse," and "fundamental," but these names have no connection with the present usage.

where we have used the fact that the sum of a set of successive integers is the mean of the first and the last times the number of members of the set.

Figure 11.8 shows the diagram of the electronic energy levels of a hydrogen-like atom. Each state is represented by a horizontal line segment at the appropriate height for its energy. There is also a continuous spectrum of unbound states of positive energy. The characteristic pattern for the numbers of bound states is that (1) increasing the value of n by unity makes one more value of ℓ available and (2) increasing the value of ℓ by unity makes two more values of m available.

The electronic energy levels of the hydrogen-like atom are called **shells** because the expectation value of the distance of the electron from the nucleus is approximately the same for all states with the same value of n and is larger for larger values of n. The $n = 1$ level consists of a single state, with $\ell = 0$ and $m = 0$, and is called the first shell. The energy level with $n = 2$ is called the second shell, etc. The shells are also labeled K, L, M, etc. instead of 1, 2, 3, etc.

Within a given shell, the states with a given value of ℓ constitute a **sub-shell**. For example, the second shell has two different values of ℓ: zero and unity. The $\ell = 0$ state of any shell is called the **s subshell** of that shell, and the three states in a shell with $\ell = 1$ are collectively called the **p subshell** of that shell. The third shell also has a **d subshell** consisting of the five $\ell = 2$ states. The fourth shell also has an **f subshell** consisting of the seven $\ell = 3$ states. As further subshells appear, they are given the letters g, h, i, etc. (alphabetical after f). There are n subshells in the nth shell. For example, the seventh shell has seven subshells: the $7s, 7p, 7d, 7f, 7g, 7h$, and $7i$ subshells.

Exercise 11.12   Give the value of each of the three quantum numbers for each state of the fourth shell.

Formulas from Tables 11.1 and 11.2 can be used to construct expressions for the first few bound-state electronic energy eigenfunctions of the hydrogen atom. Others can be constructed from formulas in Appendix F.

EXAMPLE 11.2

Write the formula for ψ_{211}.

Solution

From Equation (11.1-32)

$$\Phi_1 = \frac{1}{\sqrt{2\pi}} e^{i\phi}$$

From Table 11.1

$$\Theta_{11} = \sqrt{\frac{3}{4}} \sin(\theta)$$

From Table 11.2

$$R_{21}(\rho) = \left(\frac{Z}{a}\right)^{3/2} \frac{1}{2\sqrt{6}} \rho e^{-\rho/2}$$

Our energy eigenfunction is

Note that ρ has a different meaning in each shell.

$$\psi_{211} = \left(\frac{Z}{a}\right)^{3/2} \frac{1}{8\sqrt{\pi}} \rho e^{-\rho/2} \sin(\theta) e^{i\phi} = \left(\frac{Z}{a}\right)^{3/2} \frac{Zr/a}{8\sqrt{\pi}} e^{-Zr/2a} \sin(\theta) e^{i\phi}$$

▲

The real Φ functions are often more useful in describing chemical bonding and the complex Φ functions are more useful in discussing angular momentum eigenvalues.

The real Φ functions can also be chosen, giving real ψ functions. For a given function $\Theta_{\ell m}(\theta)$, either Φ_{mx} or Φ_{my} can be chosen instead of choosing between Φ_m and Φ_{-m}. In either case, there are two functions, so the number of energy eigenfunctions is the same. However, except for $m = 0$, the real energy eigenfunctions are not eigenfunctions of $\hat{L}_z$ and do not correspond to predictable values of L_z.

For $\ell = 1$, an energy eigenfunction with a Φ_{1x} function is called a *px* function and an energy eigenfunction with a Φ_{1y} function a *py* function. The $l = 1$, $m = 0$ function is called the *pz* function. The nomenclature for $\ell = 2$ is more complicated.

Instead of giving the value of the subscript ℓ, we can give the letter corresponding to the value of ℓ. The 210 function can be called the *2p0* function, etc. Table 11.3 contains formulas representing hydrogen-like energy eigenfunctions for the first three shells. Both the real and complex *2p* functions are included, but only the real *3p* and *3d* functions are given. The *2p0* function is the same as the *2pz* function, and the *2p* functions containing Φ_{1x} and Φ_{1y} are called the *2px* and *2py* functions, with similar notation for the *3p* functions, etc. The notation for the real *d* functions is more complicated, as can be seen from the table. The first few complex functions can be written from Tables 11.1 and 11.2, and additional functions can be written from formulas in Appendix F.

Single-electron wave functions are called **orbitals**. This name is derived from the word "orbit," but orbitals do not represent classical motion along a precise trajectory, as the word orbit often implies. It is important to have a grasp of the qualitative properties of the hydrogen-like orbitals. Figure 11.9 shows graphs of the R functions for the first three shells. The number of nodes in the R function increases by unity if n is increased by unity for fixed ℓ and decreases by unity if ℓ is increased by unity for fixed n.

The Θ_{00} function and the Φ_0 function for the s subshells are equal to constants. Therefore, the orbitals in the s subshells depend only on r. We say that they are **spherically symmetric** functions. The other Θ and Φ functions are more complicated, especially in the case of the complex Φ functions, for which we must plot a real and an imaginary part unless we choose to plot only the magnitude of the orbital. Figure 11.10 shows graphs of several of these functions. The three spherical harmonic functions that occur in the *2p* subshell are exactly the same as the three spherical harmonic functions that occur in the *3p* subshell or any other *p* subshell, and those of the *3d* subshell are the same as those of any other *d* subshell, etc.

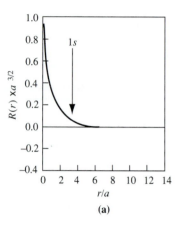

(a)

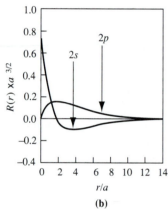

(b)

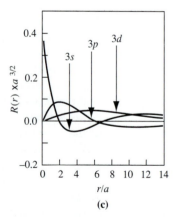

(c)

Figure 11.9. Radial Factors for Hydrogen-like Energy Eigenfunctions. (a) $n = 1$. **(b)** $n = 2$. **(c)** $n = 3$. The pattern of numbers of nodal surfaces is important. As n is increased with ℓ fixed, the number of spherical nodal surfaces increases. As ℓ is increased with n fixed, the number of spherical nodal surfaces decreases.

Table 11.3. Real Hydrogen-like Energy Eigenfunctions

$$\psi_{10} = \psi_{1s} = \frac{1}{\sqrt{\pi}}\left(\frac{Z}{a}\right)^{3/2}e^{-Zr/a}$$

$$\psi_{20} = \psi_{2s} = \frac{1}{4\sqrt{2\pi}}\left(\frac{Z}{a}\right)^{3/2}\left(2 - \frac{Zr}{a}\right)e^{-Zr/2a}$$

$$\psi_{21x} = \psi_{2px} = \frac{1}{4\sqrt{2\pi}}\left(\frac{Z}{a}\right)^{3/2}\left(\frac{Zr}{a}\right)e^{-Zr/2a}\sin(\theta)\cos(\phi)$$

$$\psi_{21y} = \psi_{2py} = \frac{1}{4\sqrt{2\pi}}\left(\frac{Z}{a}\right)^{3/2}\left(\frac{Zr}{a}\right)e^{-Zr/2a}\sin(\theta)\sin(\phi)$$

$$\psi_{300} = \psi_{3s} = \frac{1}{18\sqrt{3\pi}}\left(\frac{Z}{a}\right)^{3/2}\left[6 - \frac{4Zr}{a} + \left(\frac{2Zr}{3a}\right)^2\right]e^{-Zr/3a}$$

$$\psi_{310} = \psi_{3pz} = \frac{\sqrt{2}}{81\sqrt{\pi}}\left(\frac{Z}{a}\right)^{3/2}\left(\frac{6Zr}{a} - \frac{Z^2r^2}{a^2}\right)e^{-Zr/3a}\cos(\theta)$$

$$\psi_{31x} = \psi_{3px} = \frac{\sqrt{2}}{81\sqrt{\pi}}\left(\frac{Z}{a}\right)^{3/2}\left(\frac{6Zr}{a} - \frac{Z^2r^2}{a^2}\right)e^{-Zr/3a}\sin(\theta)\cos(\phi)$$

$$\psi_{31y} = \psi_{3py} = \frac{\sqrt{2}}{81\sqrt{\pi}}\left(\frac{Z}{a}\right)^{3/2}\left(\frac{6Zr}{a} - \frac{Z^2r^2}{a^2}\right)e^{-Zr/3a}\sin(\theta)\sin(\phi)$$

$$\psi_{320} = \psi_{3dz^2} = \frac{1}{81\sqrt{6\pi}}\left(\frac{Z}{a}\right)^{3/2}\left(\frac{Zr}{a}\right)^2e^{-Zr/3a}[3\cos^2(\theta) - 1]$$

$$\psi_{3dxz} = \frac{\sqrt{2}}{81\sqrt{\pi}}\left(\frac{Z}{a}\right)^{3/2}\left(\frac{Zr}{a}\right)^2e^{-Zr/3a}\sin(\theta)\cos(\theta)\cos(\phi)$$

$$\psi_{3dyz} = \frac{\sqrt{2}}{81\sqrt{\pi}}\left(\frac{Z}{a}\right)^{3/2}\left(\frac{Zr}{a}\right)^2e^{-Zr/3a}\sin(\theta)\cos(\theta)\sin(\phi)$$

$$\psi_{dx^2-y^2} = \frac{1}{81\sqrt{2\pi}}\left(\frac{Z}{a}\right)^{3/2}\left(\frac{Zr}{a}\right)^2e^{-Zr/3a}\sin^2(\theta)\cos(2\phi)$$

$$\psi_{3dxy} = \frac{1}{81\sqrt{2\pi}}\left(\frac{Z}{a}\right)^{3/2}\left(\frac{Zr}{a}\right)^2e^{-Zr/3a}\sin^2(\theta)\sin(2\phi)$$

It is difficult to draw a graph representing a function of three independent variables, and it is also difficult to visualize the qualitative properties of the orbital by looking at three separate graphs for the R, Θ, and Φ functions. Therefore, we introduce the **orbital region**, which we define as the region in space where the magnitude of the orbital function is larger than some specified small value.

Since the square of the orbital function is the probability density, the orbital region is also the region which there is a fairly large probability of finding the electron. A common policy is to choose a constant magnitude of the orbital at the boundary of the orbital regions such that 90% of the total probability of finding the electron is inside the orbital region. Alternatively, one could include in the orbital region all points at which the

The notation Re(Φ_1) stands for the real part of Φ_1 and the notation Im(Φ_1) stands for the imaginary part of Φ_1.

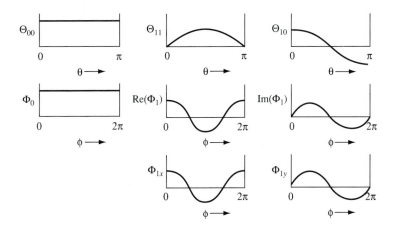

Figure 11.10. Some Factors of Spherical Harmonic Functions. These are the factors by which the radial functions must be multiplied to give the energy eigenfunctions.

magnitude of the orbital function is at least as large as some fraction, say 5%, of its maximum value. For qualitative discussions, either policy is suitable.

Pictures of orbital regions are seen in almost all elementary chemistry and organic chemistry textbooks. The orbital and the orbital region are two different things. The orbital is a one-electron wave function, while the orbital region is a three-dimensional region in space where the orbital is larger in magnitude than some small value.

This distinction between an orbital and its orbital region is often not emphasized in elementary chemistry courses but is important.

Figure 11.11 shows several orbital regions. The sign of the orbital function is indicated for the real orbitals. Notice the differences between the orbital regions for the complex $2p$ and the real $2p$ orbitals. For the complex orbitals, we must take the magnitude of the complex exponential $e^{im\phi}$ or $e^{-im\phi}$, which is a constant, whereas for the real orbitals we have either $\sin(m\phi)$ or $\cos(m\phi)$. The compactness of the orbital regions of the real p functions often makes them more useful than the complex p orbitals in discussing chemical bonding.

The orbital regions can be approximately constructed from the pattern of the nodes in the R, Θ and Φ functions. Since a node in any of the three factors leads to a vanishing value of the product, a node in any factor leads to a nodal surface on which its independent variable is constant. If there is a node in the R factor, the nodal surface is a sphere. If there is a node in the Θ factor, the nodal surface is a cone, or a plane if the node occurs at $\theta = \pi/2$ (90°). If there is a node in a real Φ factor, the nodal surface is a half-plane with edge at the z axis, which is always paired with another half-plane to make a nodal plane containing the z axis. The nodes in the real and imaginary parts of a complex Φ function are not in the same places, but each is just like that of a real Φ function.

A node in R gives a spherical nodal surface, a node in Θ gives a conical nodal surface, and a node in a real Φ function gives a planar nodal surface.

The number of nodal surfaces is the same for all orbitals of a shell and is equal to $n - 1$. The first shell has no nodes in its single orbital (the $1s$ orbital). The second shell has one node in each of its wave functions.

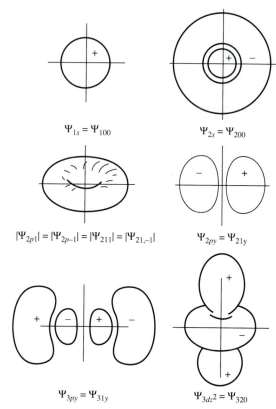

$$\Psi_{1s} = \Psi_{100} \qquad \Psi_{2s} = \Psi_{200}$$

$$|\Psi_{2p1}| = |\Psi_{2p-1}| = |\Psi_{211}| = |\Psi_{21,-1}| \qquad \Psi_{2py} = \Psi_{21y}$$

$$\Psi_{3py} = \Psi_{31y} \qquad \Psi_{3d_z^2} = \Psi_{320}$$

Figure 11.11. Cross Sections of Some Orbital Regions for Hydrogen-like Orbitals. The orbital region is the region in space inside which the orbital function differs significantly from zero.

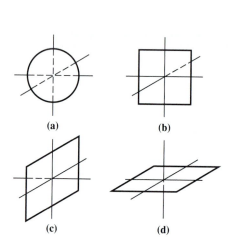

(a)   **(b)**

(c)   **(d)**

Figure 11.12. The Nodal Surfaces of the Real Energy Eigenfunctions of the Second Shell. (a) The nodal sphere of the 2s wave function. **(b)** The nodal plane of the 2px wave function. **(c)** The nodal plane of the 2py wave function. **(d)** The nodal plane of the 2pz wave function. Each of these surfaces represents the points in space where the wave function vanishes.

▼   **EXAMPLE 11.3**

Describe the nodal surfaces for the real orbitals of the second ($n = 2$) shell.

Solution

In each of the orbitals there is one node. For the 2s function, there is a node in the R function, producing a single spherical nodal surface. The 2pz function has a node in the Θ function at $\theta = \pi/2$, producing a nodal plane in the x-y plane. The 2px function has nodes in the Φ function at $\phi = \pi/2$ and at $3\pi/2$, producing a nodal plane in the y-z plane. Similarly, the 2py orbital has a nodal plane in the x-z plane. Figure 11.12 depicts the nodal surfaces in the real orbitals of the second shell. ▲

Exercise 11.13

Describe the nodal surfaces for the real orbitals of the 3d subshell.

The Connection Between the Number of Nodes and the Energy Eigenvalue

There is a general qualitative connection between the number of nodes an energy eigenfunction possesses and its energy eigenvalue. With a particle in a one-dimensional box, the number of nodes was $n - 1$, where n was the single quantum number. The energy was proportional to the square of n. Thus, the more nodes, the higher the energy. In the real hydrogen-like orbitals, the number of nodal surfaces is $n - 1$, where n is the principal quantum number, and the energy is also higher for larger values of n. For almost any bound system, we have the same relation: the more nodes, the higher the energy. This correlates with the fact that the de Broglie wavelength has a smaller value when there are more nodes. By Equation (9.4-3) the de Broglie wavelength is inversely proportional to the speed and thus has a smaller value when the kinetic energy is larger.

Once again, we see the connection: the more nodes, the higher the energy.

Normalization of the Hydrogen-like Orbitals

For motion of one particle in three dimensions, normalization using cartesian coordinates means

$$\int_{-\infty}^{\infty} \int_{-\infty}^{\infty} \int_{-\infty}^{\infty} \psi(x, y, z)^* \psi(x, y, z)\, dx\, dy\, dz = 1 \qquad \textbf{(11.2-17)}$$

The **element of volume** $dx\, dy\, dz$ in cartesian coordinates has been abbreviated by $d^3\mathbf{r}$ or by dq, and we will use the same abbreviations for other coordinate systems.

In spherical polar coordinates, normalization means

$$\int_0^{\infty} \int_0^{\pi} \int_0^{2\pi} \psi(r, \theta, \phi)^* \psi(r, \theta, \phi) r^2 \sin(\theta)\, d\phi\, d\theta\, dr = 1 \qquad \textbf{(11.2-18)}$$

The factor $r^2 \sin(\theta)$, which is called a **Jacobian**, is required to complete the element of volume:

$$d^3\mathbf{r} = r^2 \sin(\theta)\, d\phi\, d\theta\, dr \qquad \textbf{(11.2-19)}$$

This equation follows from the fact that an infinitesimal length in the r direction is dr, an infinitesimal arc length in the θ direction is $r\, d\theta$, and an infinitesimal arc length in the ϕ direction is $r \sin(\theta)\, d\phi$. Since the lengths are infinitesimal, there is no distinction between arc lengths and linear lengths. The element of volume is the product of these mutually perpendicular infinitesimal lengths, giving Equation (11.2-19).

Since our energy eigenfunctions for electronic motion are written as a product of three factors, the normalization integral can be factored:

This factoring of the normalization integral is done for our convenience.

$$\int_0^{\infty} R^* R r^2\, dr \int_0^{\pi} \Theta^* \Theta \sin(\theta)\, d\theta \int_0^{2\pi} \Phi^* \Phi\, d\phi = 1 \qquad \textbf{(11.2-20)}$$

Expectation Values for the Hydrogen Atom

The normalizations we have introduced make each factor of the integral in Equation (11.2-20) equal unity. These separate normalizations simplify the calculation of most expectation values.

▼   **EXAMPLE 11.4**

Calculate $\langle 1/r \rangle$ and $\langle \mathcal{V} \rangle$, where $\mathcal{V}$ is the potential energy, for a hydrogen-like atom in the 1s state.

Solution

In Example 11.4, the separate normalizations of the three factors in the wave function make the calculation easier. If the quantity whose expectation value we desire depends on only one variable, two of the three factors can be omitted in the calculation.

$$\left\langle \frac{1}{r} \right\rangle = \int_0^\infty R^* \frac{1}{r} R r^2 \, dr \int_0^\pi \Theta^* \Theta \, \sin(\theta) \, d\theta \int_0^{2\pi} \Phi^* \Phi \, d\phi \qquad \text{(11.2-21)}$$

By our separate normalizations, the second and third integrals both equal unity, so

$$\left\langle \frac{1}{r} \right\rangle = \int_0^\infty R_{10}^* \frac{1}{r} R_{10} r^2 \, dr = 4\left(\frac{Z}{a}\right)^3 \int_0^\infty e^{-2Zr/a} r \, dr$$

$$= 4(Z/a)^3 (a/2Z)^2 (1) = Z/a \qquad \text{(11.2-22)}$$

$$\langle \mathcal{V} \rangle = -\frac{Ze^2}{4\pi\varepsilon_0} \left\langle \frac{1}{r} \right\rangle = -\frac{Z^2 e^2}{4\pi\varepsilon_0 a} \qquad \text{(11.2-23)}$$

▲

Since the θ and the ϕ integrals in Equation (11.2-21) both equal unity, they can be omitted from the outset and only the function R used to make the calculation. The same is true in calculating the expectation value of any function of r.

As seen in Equation (11.2-23), the expectation value of the potential energy is exactly twice the total energy of Equation (11.2-12). Therefore, the kinetic energy is half as large as the magnitude of the potential energy and is equal in magnitude to the total energy (the kinetic energy must be positive while the total energy and the potential energy are negative). This behavior occurs in all systems of particles interacting only with the Coulomb potential energy and is a consequence of the **virial theorem** of mechanics.[4]

The Radial Distribution Function

Unfortunately, a different quantity in fluid theory is also called the radial distribution function. See Chapter 22.

The **radial distribution function**, f_r, is defined as the probability per unit value of r of finding the electron at a distance r from the nucleus. That is,

$$f_r \, dr = \begin{array}{l} \text{Probability that the particle} \\ \text{lies at a distance from the} \\ \text{nucleus between } r \text{ and } r + dr \end{array} \qquad \text{(11.2-24)}$$

The locations that lie at distances from the nucleus between r and $r + dr$ constitute a spherical shell of radius r and thickness dr, as shown in Figure 11.13a. The total probability of finding the electron in this shell is obtained by integrating over θ and ϕ:

$$f_r \, dr = \left(\int_0^\pi \int_0^{2\pi} \psi(r, \theta, \phi)^* \psi(r, \theta, \phi) r^2 \, \sin(\theta) \, d\phi \, d\theta \right) dr \quad \text{(11.2-25)}$$

---

[4] Ira N. Levine, *Quantum Chemistry*, Allyn & Bacon, Boston, 1983, pp. 393ff.

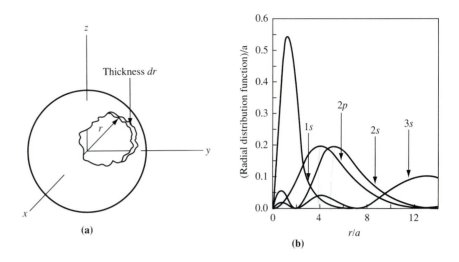

Figure 11.13. The Probability Distribution for Electron-Nucleus Distances. (a) A spherical shell of radius r and thickness dr. This shell is centered on the origin and contains all of the points that are at distances from the origin between r and $r + dr$. **(b)** Radial distribution functions for hydrogen-like orbitals. The radial distribution function is the probability density for finding the particle at a certain distance from the origin, irrespective of direction.

The result of Equation (11.2-26) arises from the separate normalizations of the three factors.

where r is not integrated. The integral can be factored

$$f_r \, dr = R^*R\left(\int_0^\pi \Theta^*\Theta \, \sin(\theta) \, d\theta \int_0^{2\pi} \Phi^*\Phi \, d\phi\right) r^2 \, dr$$
$$= R^*R r^2 \, dr \qquad \qquad \textbf{(11.2-26)}$$

The expectation value of a quantity depending only on r can be computed using the radial distribution function. For example,

$$\left\langle \frac{1}{r} \right\rangle = \int_0^\infty \frac{1}{r} R^*R r^2 \, dr = \int_0^\infty \frac{1}{r} f_r \, dr \qquad \textbf{(11.2-27)}$$

Figure 11.13b shows graphs of the radial distribution function for several energy eigenfunctions. All the states of a given subshell have the same radial distribution function because they have the same radial factor in their wave functions.

Since the radial distribution function is proportional to r^2, it vanishes at the nucleus and goes through one or more relative maxima. The location of these relative maxima can be determined by the methods of calculus, setting the first derivative of f_r equal to zero. The maximum value corresponds to the most probable value of the electron-nucleus distance.

Exercise 11.14

a. Calculate the expectation value $\langle r \rangle$ for a hydrogen-like atom in the 1s state. Why is this not equal to $\langle 1/r \rangle^{-1}$?

b. Calculate $\langle r^2 \rangle$ for a hydrogen-like atom in the 1s state. Why is this not equal to $\langle r \rangle^2$?

c. Calculate the most probable value of r for a hydrogen-like atom in the $1s$ state. Why is this not equal to $\langle r \rangle$?

The Time-Dependent Wave Function of the Hydrogen Atom

Having found the energy eigenvalues and energy eigenfunctions, we can now write the time-dependent wave function, using the analogue of Equation (9.4-20):

$$\Psi_{n\ell m}(r, \theta, \phi, t) = \psi_{n\ell m}(r, \theta, \phi)e^{-iE_n t/\hbar} \qquad \textbf{(11.2-28)}$$

This represents a stationary state. The probability density for finding the electron is time independent and the expectation value of any time-independent variable is time independent.

Exercise 11.15 Show that the expectation value $\langle r \rangle$ is exactly the same as in Example 11.4 when the time-dependent wave function Ψ_{100} is used instead of the time-independent wave function ψ_{100}.

Intrinsic Angular Momentum of the Electron

It is found experimentally that in addition to the angular momentum that we have discussed, electrons have another angular momentum that is not included in the Schrödinger description. To obtain adequate agreement with experiment, we must add this feature to our theory. We refer to the angular momentum included in the Schrödinger description as the **orbital angular momentum** and the additional angular momentum as the **intrinsic angular momentum** or the **spin angular momentum**.

The z component of the intrinsic angular momentum takes on one of only two possible values, $\hbar/2$ and $-\hbar/2$. We denote the intrinsic angular momentum by **S** and write

$$\boxed{S_z = \pm\frac{\hbar}{2}} \qquad \textbf{(11.2-29a)}$$

We assign a new quantum number, m_s, for the z component of the intrinsic angular momentum, with the values

$$\boxed{m_s = \pm\frac{1}{2}} \qquad \textbf{(11.2-29b)}$$

Angular momenta are vectors, and the total angular momentum of an electron is the vector sum of the orbital and intrinsic angular momenta. If we denote the total angular momentum by **J**, then J_z has values given by

$$J_z = m\hbar + m_s\hbar \qquad (11.2\text{-}30)$$

where m is the same quantum number as in our original treatment.

The pattern of values of m_s is analogous to that of Equation (11.1-44) if we assign a quantum number s for the square of the intrinsic angular momentum with a fixed value of $1/2$. The square of the intrinsic angular momentum has the fixed value

$$s^2 = \hbar^2 \frac{1}{2}\left(\frac{1}{2} + 1\right) \qquad (11.2\text{-}31)$$

following the same pattern as Equation (11.1-42).

The quantum mechanics of Schrödinger is nonrelativistic. That is, it cannot be expected to be correct when particles have speeds near the speed of light. There is a relativistic version of quantum mechanics, based on the Dirac equation rather than the Schrödinger equation. The intrinsic angular momentum occurs naturally in relativistic quantum mechanics, which we do not discuss.

All angular momenta have possible values and quantum numbers that follow the same pattern. The principal difference between the intrinsic angular momentum and the orbital angular momentum is that the intrinsic angular momentum can have half-integral values for quantum numbers.

▼ **EXAMPLE 11.5**

Calculate the expectation value of the square of the speed of the electron in a hydrogen atom in the 1s state, and from this calculate the root-mean-square speed. Compare this to the speed of light.

Solution

We can obtain this quantity from the expectation value of the kinetic energy:

$$\langle \mathcal{K} \rangle = E_1 - \langle \mathcal{V} \rangle = -2.18 \times 10^{-18}\,\text{J} + \frac{e^2}{4\pi\varepsilon_0}\left\langle \frac{1}{r} \right\rangle$$

$$= -2.18 \times 10^{-18}\,\text{J} + 2(2.18 \times 10^{-18}\,\text{J}) = 2.18 \times 10^{-18}\,\text{J}$$

$$\langle v^2 \rangle = \frac{2\langle \mathcal{K} \rangle}{m} = \frac{2(2.18 \times 10^{-18}\,\text{kg m}^2\,\text{s}^{-2})}{9.11 \times 10^{-31}\,\text{kg}} = 4.79 \times 10^{12}\,\text{m}^2\,\text{s}^{-2}$$

$$v_{\text{rms}} = \langle v^2 \rangle^{1/2} = 2.19 \times 10^6\,\text{m s}^{-1}$$

which is smaller than the speed of light by a factor of 100. ▲

Example 11.5 is designed to show that relativistic corrections are not important for the hydrogen atom.

There are three principal differences between the orbital angular momentum and the intrinsic angular momentum. First, the orbital angular momentum occurred naturally in the nonrelativistic Schrödinger theory, while the intrinsic angular momentum is arbitrarily added to the theory to make it agree with experiment. Second, the intrinsic angular momentum has only one possible magnitude, while the orbital angular momentum has variable (but quantized) magnitude. Third, this single magnitude corresponds to a quantum number that is a half-integer instead of an integer. Figure 11.14 shows the two cones of possible directions of the intrinsic angular momentum.

Here are some important differences between orbital and intrinsic angular momenta.

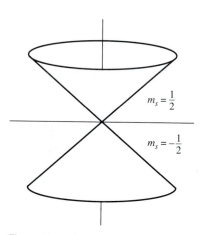

Figure 11.14. Cones of Spin Angular Momentum Directions for One Electron. Compare this diagram with that of Figure 11.5. The z component of an orbital angular momentum can take on various values, depending on the value of ℓ. The z component of the spin angular momentum can take on only one of two possible values.

Find the angle between the z axis and the intrinsic angular momentum for $m_s = +1/2$.

Solution

The angle θ in Figure 11.14 is

$$\theta = \arccos\left[\frac{\hbar/2}{\hbar\sqrt{(1/2)(3/2)}}\right] = \arccos\left(\frac{1/2}{\sqrt{3/4}}\right)$$

$$= \arccos(0.57735) = 54.7356\ldots \text{ degrees} = 0.9553166\ldots \text{ radians}$$

It is natural to seek a classical interpretation for the intrinsic angular momentum. Although we have previously treated the electron as a mass point and it is not clear what its internal structure is, it is customary to ascribe the intrinsic angular momentum to rotation of the electron about its own axis, calling it **spin angular momentum**. The assumed motion is analogous to the rotation of the earth on its axis as it revolves about the sun. This spin interpretation is very commonly used, but we could proceed without any mental picture of spinning motion. However, there is value in having a mental picture of what is going on, even if it is not completely justified. We will use the spin interpretation of intrinsic angular momentum, visualizing the direction of the intrinsic angular momentum as the direction of the axis of rotation of the electron.

We now have twice as many possible states of electronic motion in a hydrogen-like atom as we did before, because for every set of values of the quantum numbers n, ℓ, and m there are two possible values of m_s. We will call the state for $m_s = +1/2$ the "spin-up" state and the state for $m_s = -1/2$ the "spin-down" state, corresponding to the direction of the intrinsic angular momentum vector.

There are two different ways to include spin in our notation. The first is to attach another subscript for the new quantum number, replacing $n\ell m$ by $n\ell mm_s$. There is no need to include the value of s, since it is fixed. The new orbital is now called a **spin orbital**. The second way is to multiply the old orbital by a **spin function**, which is called α for $m_s = +1/2$ and β for $m_s = -1/2$. The original orbital is now called a **space orbital** and the product is called a spin orbital. The spin function is thought of as a function of some unspecified spin coordinates, which we do not explicitly represent. The two ways of writing a spin orbital are equivalent:

$$\psi_{n\ell m,\, 1/2} = \psi_{n\ell m}\alpha, \qquad \psi_{n\ell m,\, -1/2} = \psi_{n\ell m}\beta \qquad \textbf{(11.2-32)}$$

We define operators for the spin angular momentum, analogous to the orbital angular momentum operators. We do not write any explicit mathematical forms for them, but assign their properties by definition. The spin functions α and β are defined as eigenfunctions of $\hat{S}^2$, the operator for the square of the spin angular momentum:

$$\hat{S}^2\alpha = \hbar^2(1/2)(3/2)\alpha \qquad \textbf{(11.2-33)}$$

$$\hat{S}^2\beta = \hbar^2(1/2)(3/2)\beta \qquad \textbf{(11.2-34)}$$

The eigenvalue properties in Equations (11.2-33)–(11.2-36) are assigned by definition to give the proper pattern of eigenvalues.

They are also defined as eigenfunctions of $\hat{S}_z$, the operator for the z component of the spin angular momentum:

$$\hat{S}_z \alpha = +\frac{\hbar}{2}\alpha \qquad (11.2\text{-}35)$$

$$\hat{S}_z \beta = -\frac{\hbar}{2}\beta \qquad (11.2\text{-}36)$$

Use of this formalism modifies the Schrödinger theory of the electron so that it agrees adequately with experiment for many purposes. Further modifications can be made to include additional aspects of relativistic quantum mechanics. These modifications result in small differences between the energies of spin-up and spin-down states for states of nonzero orbital angular momentum. We will not discuss the **spin-orbit coupling** that produces this effect, although it is numerically important in heavy atoms.[5]

11.3 The Helium Atom in the "Zero-Order" Orbital Approximation

From now on, we will be forced to discuss approximations instead of exact solutions to Schrödinger equations.

The hydrogen-like atom is the only atom for which the Schrödinger equation can be solved exactly. This does not invalidate the Schrödinger quantum theory for other atoms, since approximate treatments of other atoms give accurate agreement with experimental energy values. It does mean that the only way to proceed is with approximations.

In treating the hydrogen-like atom, it was possible to separate the center-of-mass motion and relative motion exactly. However, no serious numerical error was introduced into the treatment of the relative motion by assuming that the nucleus was stationary. With the helium atom, which has two electrons and a nucleus, although the center of mass still moves like a free particle, the same separation cannot be performed exactly for both electrons.

We assume that the nucleus is stationary when we study the electronic motion. This is a good approximation because the helium nucleus is even more massive than the hydrogen nucleus. The translation of the center of mass of the atom is still a separate problem and is exactly represented as the motion of a free particle with mass equal to the sum of the masses of the nucleus and all electrons. We defer discussion of this translational motion to Chapter 14 and proceed to the problem of electronic motion.

The Hamiltonian of a Helium-like Atom

We define a "helium-like" atom with a nucleus containing Z protons and with two electrons, so that $Z = 2$ represents the He atom, $Z = 3$ represents the Li^+ ion, etc. The system is shown in Figure 11.15. Assuming a stationary

---

[5] Pilar, *op. cit.*, pp. 301ff; K. Balasubramanian, *J. Phys. Chem.* **93**, 6585 (1989).

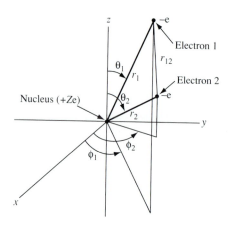

Figure 11.15. The Helium Atom System. The occurrence of two electrons makes it impossible to obtain an exact solution to the Schrödinger equation for this system.

nucleus, the expression for the classical energy is

$$H_{cl} = \frac{1}{2m} p_1^2 + \frac{1}{2m} p_2^2 + \frac{1}{4\pi\varepsilon_0}\left(-\frac{Ze^2}{r_2} - \frac{Ze^2}{r_1} + \frac{e^2}{r_{12}}\right) \quad \text{(11.3-1)}$$

where $\mathbf{p}_1$ is the vector momentum of electron 1, $\mathbf{p}_2$ is the vector momentum of electron 2, m is the electron mass, and the distances are as labeled in Figure 11.15.

We create the Hamiltonian operator for the helium-like atom by making the replacements of Equation (10.3-5) and its analogues:

$$\hat{H} = -\frac{\hbar^2}{2m}(\nabla_1^2 + \nabla_2^2) + \frac{1}{4\pi\varepsilon_0}\left(-\frac{Ze^2}{r_2} - \frac{Ze^2}{r_1} + \frac{e^2}{r_{12}}\right) \quad \text{(11.3-2)}$$

where ∇_1^2 and ∇_2^2 are the Laplacian operators for electrons 1 and 2.

The "Zero-Order" Orbital Approximation

The Hamiltonian operator of Equation (11.3-2) gives a time-independent Schrödinger equation that has not been solved exactly. The **zero-order approximation** is obtained by neglecting the last term in this operator. It assumes that the electrons' repulsion for each other is negligible compared with their attraction for the nucleus. It is not a good approximation, but it makes a good starting point for better approximations.

The approximate Hamiltonian operator is now

$$\hat{H}^{(0)} = -\frac{\hbar^2}{2m}\nabla_1^2 - \frac{Ze^2}{4\pi\varepsilon_0 r_1} - \frac{\hbar^2}{2m}\nabla_2^2 - \frac{Ze^2}{4\pi\varepsilon_0 r_2} \quad \text{(11.3-3)}$$

where we add a superscript (0) to distinguish the approximate "zero-order" Hamiltonian from the correct Hamiltonian. It is a hydrogen-like Hamiltonian operator for electron 1 plus a hydrogen-like Hamiltonian operator for electron 2:

$$\hat{H}^{(0)} = \hat{H}_{HL}(1) + \hat{H}_{HL}(2) \quad \text{(11.3-4)}$$

where the subscript HL stands for hydrogen-like and we abbreviate the coordinates of a particle by writing only the particle index.

The approximate time-independent Schrödinger equation is

$$\hat{H}^{(0)}\Psi^{(0)}(1,2) = [\hat{H}_{HL}(1) + \hat{H}_{HL}(2)]\Psi^{(0)}(1,2) = E^{(0)}\Psi^{(0)}(1,2) \quad \text{(11.3-5)}$$

where we attach a superscript (0) to the zero-order wave function and eigenvalue. This wave function is a function of the coordinates of both particles:

$$\Psi^{(0)} = \Psi^{(0)}(r_1, \theta_1, \phi_1, r_2, \theta_2, \phi_2) = \Psi^{(0)}(1,2) \quad \text{(11.3-6)}$$

where we use spherical polar coordinates and again abbreviate the coordinates by giving only the subscripts.

The Hamiltonian operator $\hat{H}^{(0)}$ has two terms, each of which contains one set of variables that do not occur in the other term. Because of this, Equation (11.3-5) can be solved by separation of variables, using the trial solution

The wave function must depend on all of the coordinates that occur in the Hamiltonian operator.

$$\Psi^{(0)}(1,2) = \psi_1(r_1, \theta_1, \phi_1)\psi_2(r_2, \theta_2, \phi_2) = \psi_1(1)\psi_2(2) \quad \text{(11.3-7)}$$

where ψ_1 and ψ_2 are two orbitals (functions of the coordinates of one electron). The subscript tells which orbital is meant, and the number inside parentheses tells which particle's coordinates are the independent variables for the orbital. The approximate two-electron wave function is a product of two orbitals and is called an **orbital wave function**.

We substitute the trial solution into Equation (11.3-5) and use the fact that $\psi_1(1)$ is treated as a constant when $\hat{H}_{HL}(2)$ operates and $\psi_2(2)$ is treated as a constant when $\hat{H}_{HL}(1)$ operates. The result is

$$\psi_2(2)\hat{H}_{HL}(1)\psi_1(1) + \psi_1(1)\hat{H}_{HL}(2)\psi_2(2) = E^{(0)}\psi_1(1)\psi_2(2)$$

Division of this equation by $\psi_1(1)\psi_2(2)$ gives

$$\frac{1}{\psi_1(1)}\hat{H}_{HL}(1)\psi_1(1) + \frac{1}{\psi_2(2)}\hat{H}_{HL}(2)\psi_2(2) = E^{(0)} \qquad \textbf{(11.3-8)}$$

We have separated the variables. Each term on the left-hand side of the equation contains only a set of variables not occurring in the other term, and the right-hand side is a constant. Therefore, the first must be equal to a constant, E_1, and the second must be equal to a constant, E_2. We now have

$$\hat{H}_{HL}(1)\psi_1(1) = E_1\psi_1(1) \qquad \textbf{(11.3-9)}$$
$$\hat{H}_{HL}(2)\psi_2(2) = E_2\psi_2(2) \qquad \textbf{(11.3-10)}$$

where

$$E_1 + E_2 = E^{(0)} \qquad \textbf{(11.3-11)}$$

One motivation for discussing the hydrogen-like atom instead of simply discussing the hydrogen atom was to have orbitals that we can use in equations like Equation (11.3-9) and (11.3-10).

Equations (11.3-9) and (11.3-10) are exactly the same as two hydrogen-like Schrödinger equations. Therefore E_1 and E_2 are hydrogen-like energies, called **orbital energies**. The total electronic energy is

$$E^{(0)}_{n_1 n_2} = E_{n_1}(HL) + E_{n_2}(HL) = -(13.6 \text{ eV})(Z^2)\left(\frac{1}{n_1^2} + \frac{1}{n_2^2}\right) \qquad \textbf{(11.3-12)}$$

where n_1 and n_2 are two values of the principal quantum number for a hydrogen-like atom.

The orbitals $\psi_1(1)$ and $\psi_2(2)$ are hydrogen-like orbitals:

$$\Psi^{(0)}(1, 2) = \psi_1(1)\psi_2(2) = \psi_{n_1 \ell_1 m_1 m_{s1}}(1)\psi_{n_2 \ell_2 m_2 m_{s2}}(2) \qquad \textbf{(11.3-13)}$$

The subscripts indicate the quantum numbers for the two orbitals. The values of a given quantum number for the two orbitals are not necessarily equal, so we add a subscript on each subscript to distinguish them from each other. The notation with separate spin functions can also be used.

Probability Densities for Two Particles

Here is a two-particle probability density, which is a function in a six-dimensional space.

For a system of two particles whose wave function is $\Psi(1, 2)$, the probability of finding particle 1 in the volume element $d^3\mathbf{r}_1$ and finding particle 2 in the volume element $d^3\mathbf{r}_2$ is given by

$$\text{Probability} = \Psi^*(1, 2)\Psi(1, 2)\, d^3\mathbf{r}_1\, d^3\mathbf{r}_2 = |\Psi(1, 2)|^2\, d^3\mathbf{r}_1\, d^3\mathbf{r}_2 \qquad \textbf{(11.3-14)}$$

where we temporarily ignore the spins of the particles. The square of the magnitude of the wave function is a probability density in a six-dimensional space.

In the simple orbital wave function, the probability densities of the two particles are independent of each other.

For the orbital wave function of Equation (11.3-13), the probability density for two particles is the product of two one-particle probability densities:

$$|\Psi(1, 2)|^2 = |\psi_1(1)|^2 |\psi_2(2)|^2 \qquad \textbf{(11.3-15)}$$

If this probability density is normalized,

$$\int |\Psi(1, 2)|^2 \, d^3\mathbf{r}_1 \, d^3\mathbf{r}_2 = 1 \qquad \textbf{(11.3-16)}$$

We consider the inclusion of spin functions later.

Indistinguishability of Identical Particles

Although we have obtained a function that satisfies our approximate Schrödinger equation and the appropriate boundary conditions, it must be modified to conform to a condition that is required to obtain agreement with experiment: *Identical particles are inherently indistinguishable from each other.*

The inherent indistinguishability of identical particles is a feature of quantum mechanics that is different from classical mechanics.

This condition is an additional hypothesis that must be tested by comparing its consequences with experimental fact. However, it is sometimes argued that the uncertainty principle makes exact trajectories impossible to specify, so if two identical particles approach each other closely, it might not be possible to tell which is which after the encounter. Figure 11.16 shows two encounters that might be distinguished from each other if classical mechanics is valid but might not be distinguished assuming that quantum mechanics is valid.

Since identical particles are indistinguishable, we must not build anything into our theory that would allow us to distinguish one particle from another of the same kind. For example, in our helium atom wave function, the probability of finding electron 1 at location 1 and electron 2 at location 2 must equal the probability of finding electron 1 at location 2 and electron 2 at location 1. Any difference in these two probabilities would give an illusory means of distinguishing the particles.

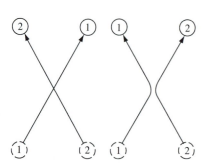

Figure 11.16. Two Encounters of Classical Particles. Since classical mechanics includes exact trajectories, we can "track" each particle exactly.

The probability density in Equation (11.3-14) must remain unchanged if the locations of the two particles are interchanged:

$$\Psi(1, 2)^*\Psi(1, 2) = \Psi(2, 1)^*\Psi(2, 1) \qquad \textbf{(11.3-17)}$$

We say that the probability density $\Psi^*\Psi$ must be **symmetric** with respect to interchange of the two particles' locations.

If the probability density were not symmetric, it would offer an (incorrect) possibility of distinguishing two identical particles. The probability density of two particles does not have to be symmetric if the particles are not identical.

With real functions, there are only two ways to satisfy Equation (11.3-17). First, the wave function might be symmetric with respect to interchange of the particles:

$$\Psi(1, 2) = \Psi(2, 1) \quad \text{(one possibility)} \qquad \textbf{(11.3-18)}$$

or second, the wave function might be **antisymmetric** with respect to interchange of the particles (change sign if the locations of the particles are switched):

$$\Psi(1, 2) = -\Psi(2, 1) \quad \text{(another possibility)} \qquad \textbf{(11.3-19)}$$

Although our wave functions are not required to be real, we consider only these two possibilities. Furthermore, if we have wave functions for more than two particles, we will consider only these possibilities if the locations of any two particles are interchanged.

Particles that obey Equation (11.3-18) are called **bosons**, and particles that obey Equation (11.3-19) are called **fermions**. Electrons are found experimentally to be fermions, so our approximate two-electron wave function must be modified to obey Equation (11.3-19). Protons and neutrons are also fermions. Atomic nuclei containing an even number of nucleons (protons and neutrons) are bosons, and so are photons.

The simplest way to modify the wave function is to add a second term that is the negative of the first term with the orbital labels interchanged, giving for an orbital wave function

$$\Psi(1, 2) = C[\psi_1(1)\psi_2(2) - \psi_2(1)\psi_1(2)] \tag{11.3-20}$$

where C is a constant whose value can be chosen to normalize the function and where the orbital symbols represent spin orbitals. We say that the wave function has been **antisymmetrized**. With this wave function it is not possible to say which electron occupies which spin orbital, because the labels are in one order in the first term of the antisymmetric wave function and in the other order in the second term.

Exercise 11.16 By explicit manipulation, show that the function of Equation (11.3-20) obeys Equation (11.3-19).

The Pauli Exclusion Principle

An important fact about fermions can be seen in Equation (11.3-20). If the orbitals ψ_1 and ψ_2 are the same function, the two-particle wave function is the difference of two identical terms and vanishes. A vanishing wave function cannot represent any state of the system. Any physically meaningful two-electron orbital wave function cannot contain the same spin orbital more than once in any term.

We will later construct orbital wave functions for more than two electrons. When antisymmetrized, these will consist of a sum of terms with different signs. Each term will be a product of spin orbitals, one for each electron. The **Pauli exclusion principle** is a generalization of our observations for two electrons and can be stated: *In an orbital wave function, the same spin orbital cannot occur more than once in each term.*

A spin orbital that occurs in an orbital wave function is said to be "occupied" by an electron. Another statement of the Pauli exclusion principle is: *In an orbital wave function, no two electrons can occupy the same spin orbital.*

The probability density for the antisymmetrized wave function of Equation (11.3-20) is

$$\Psi(1, 2)^*\Psi(1, 2) = |C|^2[|\psi_1(1)|^2|\psi_2(2)|^2 + |\psi_2(1)|^2|\psi_1(2)|^2$$
$$- \psi_1(1)^*\psi_2(1)\psi_2(2)^*\psi_1(2) - \psi_2(1)^*\psi_1(1)\psi_1(2)^*\psi_2(2)] \tag{11.3-21}$$

The value of the normalizing constant C is obtained by integrating the expression of Equation (11.3-21) over all values of the coordinates and

The origin of the Pauli exclusion principle is the inherent indistinguishability of identical particles. It was discussed by Wolfgang Pauli, 1900–1958, who received the 1945 Nobel Prize in physics.

setting the result equal to unity. Each term in Equation (11.3-21) gives an integral that factors into a product of two one-particle integrals. Each of the first two terms gives unity if the orbitals are normalized. Each of the last two terms gives zero if the orbitals are orthogonal to each other. In this case, we have

$$1 = |C|^2[1 + 1] = 2|C|^2$$

or if C is taken to be real and positive,

$$C = \sqrt{\frac{1}{2}} \qquad \text{(11.3-22)}$$

The probability of finding particle 1 in the volume element $d^3\mathbf{r}_1$ irrespective of the location of particle 2 is given by integrating the probability density in Equation (11.3-14) or Equation (11.3-21) over all positions of particle 2:

$$\begin{array}{c}\text{Probability of finding} \\ \text{particle 1 in } d^3\mathbf{r}_1\end{array} = \left[\int \Psi(1, 2)^*\Psi(1, 2)\, d^3\mathbf{r}_2\right] d^3\mathbf{r}_1 \quad \text{(11.3-23)}$$

If the two-electron wave function is a one-term orbital wave function such as that of Equations (11.3-7) and (11.3-14), the orbital function for electron 1 factors out of the integral:

$$\begin{array}{c}\text{Probability of finding} \\ \text{particle 1 in } d^3\mathbf{r}_1\end{array} = \psi_1(1)^*\psi_1(1)\, d^3\mathbf{r}_1 \int \psi_2(2)^*\psi_2(2)\, d^3\mathbf{r}_2$$

If the orbital function ψ_2 is normalized, the integral equals unity and

Here we have again the fact that if the wave function is not antisymmetrized, each electron's probability density is just that of the occupied orbital.

$$\begin{array}{c}\text{Probability of finding} \\ \text{particle 1 in } d^3\mathbf{r}_1\end{array} = \psi_1(1)^*\psi_1(1)\, d^3\mathbf{r}_1 = |\psi_1(1)|^2\, d^3\mathbf{r}_1 \quad \text{(11.3-24)}$$

The probability density for electron 1 is just that of its own orbital, independent of electron 2. An analogous equation can be written for electron 2.

If the antisymmetrized wave function is used, the expression in Equation (11.3-21) must be integrated over the coordinates of particle 2 to obtain the probability density for particle 1. Only the first two terms survive, due to the orthogonality of the orbitals, and the result is

Since we cannot specify which orbital is occupied by the electron, the probability density is the average of that of the occupied orbitals.

$$\begin{array}{c}\text{Probability of finding} \\ \text{particle 1 in } d^3\mathbf{r}_1\end{array} = |C|^2[|\psi_1(1)|^2 + |\psi_2(1)|^2]\, d^3\mathbf{r}_1 \quad \text{(11.3-25)}$$

This probability is the average of what would occur if electron 1 occupied orbital 1 and what would occur if it occupied orbital 2.

An exactly analogous expression can be written for electron 2. The total probability of finding some electron in a volume $d^3\mathbf{r}$ is the sum of the probabilities for the two electrons:

Because the electrons cannot be distinguished, the probability density for the second electron is exactly equal to that of the first electron.

Probability of finding an electron in $d^3\mathbf{r}$

$$= 2|C|^2[|\psi_1(\mathbf{r})|^2 + |\psi_2(\mathbf{r})|^2]\, d^3\mathbf{r} \quad \text{(11.3-26)}$$

where we have labeled the coordinate without specifying a particular electron. When this probability density is multiplied by $-e$, the charge density (charge per unit volume) due to the electrons.

The Ground State of the Helium Atom

The lowest-energy state of a system is called the **ground state**. Since the subshell of lowest orbital energy, the $1s$ subshell, contains two spin orbitals, our approximate ground state wave function is

$$\Psi_{1s1s}^{(0)}(1, 2) = \Psi_{1s, 1/2; 1s, -1/2}(1, 2)$$
$$= C[\psi_{100, 1/2}(1)\psi_{100, -1/2}(2) - \psi_{100, -1/2}(1)\psi_{100, 1/2}(2)] \quad \textbf{(11.3-27)}$$

where two sets of orbitals subscripts are necessary on the symbol for the orbital wave function because it contains two spin orbitals. This wave function satisfies the Pauli exclusion principle, since there are two different spin orbitals.

Let us write our antisymmetrized orbital wave function in the other notation, expressing the spin orbitals as products of space orbitals and spin functions:

$$\Psi^{(0)}(1, 2) = C[\psi_{100}(1)\alpha(1)\psi_{100}(2)\beta(2) - \psi_{100}(1)\beta(1)\psi_{100}(2)\alpha(2)]$$
$$= C\psi_{100}(1)\psi_{100}(2)[\alpha(1)\beta(2) - \beta(1)\alpha(2)] \quad \textbf{(11.3-28)}$$

Equations (11.3-27) and (11.3-28) are the same equation written in two different notations.

where we have factored the space orbitals out because they are the same in both terms. The antisymmetric wave function is now a product of a symmetric **space factor** and an antisymmetric **spin factor**.

The value of the normalization constant C must be such that

$$1 = C^*C \int \Psi^{(0)}(1, 2)^*\Psi^{(0)}(1, 2) \, dq_1' \, dq_2' \quad \textbf{(11.3-29)}$$

where the coordinates of both particles are integrated. Since we have introduced spin functions, an integration over the independent variables of the spin functions as well as over the space coordinates is indicated. We regard dq_1' and dq_2' as representing both space and spin coordinates:

$$dq_1' = d^3\mathbf{r}_1 \, d^s(1) \quad \textbf{(11.3-30)}$$

where $d^3\mathbf{r}_1$ is the volume element in ordinary space and $d^s(1)$ is the "volume element" of the unspecified spin coordinates.

We do not explicitly integrate over the unspecified spin coordinates, but define the spin functions α and β to be normalized and orthogonal to each other:

Since the spin "coordinates" are unspecified, the integrations in Equations (11.3-31) and (11.3-32) are just devices to express the normalization and orthogonality of the spin functions. There are other devices to express the same thing.

$$\int \alpha(1)^*\alpha(1) \, d^s(1) = \int \beta(1)^*\beta(1) \, d^s(1) = 1 \quad \text{(by definition)} \quad \textbf{(11.3-31)}$$

and

$$\int \beta(1)^*\alpha(1) \, d^s(1) = \int \alpha(1)^*\beta(1) \, d^s(1) = 0 \quad \text{(by definition)} \quad \textbf{(11.3-32)}$$

We use these definitions when an integration over spin coordinates is indicated.

The two-electron wave function in Equation (11.3-28) is substituted into the normalization integral of Equation (11.3-29). The integral can be factored, since the space and spin coordinates of each particle occur in separate factors:

$$1 = C^*C \int \psi_{100}(1)^* \psi_{100}(1)\, d^3\mathbf{r}_1 \int \psi_{100}(2)^* \psi_{100}(2)\, d^3\mathbf{r}_2$$

$$\times \int [\alpha(1)\beta(2) - \beta(1)\alpha(2)]^* [\alpha(1)\beta(2) - \beta(1)\alpha(2)]\, d^s(1)\, d^s(2) \quad \textbf{(11.3-33)}$$

Since the hydrogen-like orbitals are normalized, the integrals over the space coordinates equal unity, and we have, after multiplying out the terms and factoring the spin integrals:

$$1 = C^*C \left\{ \left[\int \alpha(1)^*\alpha(1)\, d^s(1) \int \beta(2)^*\beta(2)\, d^s(2) \right. \right.$$

$$+ \int \beta(1)\beta(1)\, d^s(1) \int \alpha(2)^*\alpha(2)\, d^s(2)$$

$$- \int \alpha(1)^*\beta(1)\, d^s(1) \int \beta(2)^*\alpha(2)\, d^s(2)$$

$$\left. \left. - \int \beta(1)^*\alpha(1)\, d^s(1) \int \alpha(2)^*\beta(2)\, d^s(2) \right] \right\} \quad \textbf{(11.3-34)}$$

where we have factored the double integrals. Each of the first two terms in the final equation above gives unity because of the defined normalization of the spin functions. The last two terms give zero because of the defined orthogonality of the spin functions, so if we choose C to be real and positive,

$$C = \sqrt{\frac{1}{2}} \quad \textbf{(11.3-35)}$$

The energy eigenvalue for our zero-order ground-state wave function is

$$E_{1s1s}^{(0)} = E_1(\text{HL}) + E_1(\text{HL}) = 2(-13.60 \text{ eV})Z^2 \quad \textbf{(11.3-36)}$$

For helium $Z = 2$, so that

$$E_{1s1s}^{(0)} = -108.8 \text{ eV}$$

This approximate energy eigenvalue is seriously in error—the experimental value is -79.0 eV. Since 1 eV is equivalent to 96.5 kJ mol^{-1}, an error of 30 eV is very large, larger than chemical bond energies.

Excited States of the Helium Atom

States of higher energy than the ground state of a system are called **excited states**. In our zero-order orbital approximation for the helium atom, the total electronic energy eigenvalue for any state is equal to the sum of the orbital energy eigenvalues, as in Equation (11.3-12). The electronic wave function contains two spin orbitals, which must be different functions. For excited states, there are two cases: (1) both electrons occupy the same space orbital with different spin functions, and (2) the two electrons occupy different space orbitals, with either the same or different spin functions.

For orbital wave functions, a statement of which orbitals are occupied is called the **electron configuration**. The **detailed configuration** is specified by writing the designation of each occupied space orbital with a right superscript giving the number of electrons occupying that space orbital. This superscript can equal either 1 or 2. The **subshell configuration** is specified by writing the designation of each subshell with a right superscript giving the number of electrons occupying orbitals of that subshell. The configuration of the ground state of helium is $(1s)^2$ (subshell and detailed configurations are the same with s subshells). Two different excited configurations are $(1s)^1(2s)^1$ and $(1s)^1(2p0)^1$ (detailed) or $(1s)^1(2p)^1$ (subshell). A superscript equal to unity is often omitted, so $(1s)(2s)$ has the same meaning as $(1s)^1(2s)^1$.

If both electrons occupy the same space orbital, a wave function for an excited state is similar to that of the ground state. For configuration $(2s)^2$, there is only one state:

$$\Psi_{2s2s} = \frac{1}{\sqrt{2}} \, \psi_{2s}(1)\psi_{2s}(2)[\alpha(1)\beta(2) - \beta(1)\alpha(2)] \qquad \textbf{(11.3-37)}$$

For configuration $(1s)^1(2s)^1$, there are four states. We can count them up simply by saying that each electron has two choices, spin up and spin down. However, we must write wave functions that are antisymmetric. Four possible wave functions are

$$\Psi_1 = \frac{1}{\sqrt{2}} \, [\psi_{1s}(1)\psi_{2s}(2) - \psi_{2s}(1)\psi_{1s}(2)]\alpha(1)\alpha(2) \qquad \textbf{(11.3-38a)}$$

$$\Psi_2 = \frac{1}{\sqrt{2}} \, [\psi_{1s}(1)\psi_{2s}(2) - \psi_{2s}(1)\psi_{1s}(2)]\beta(1)\beta(2) \qquad \textbf{(11.3-38b)}$$

$$\Psi_3 = \frac{1}{2} \, [\psi_{1s}(1)\psi_{2s}(2) - \psi_{2s}(1)\psi_{1s}(2)][\alpha(1)\beta(2) + \beta(1)\alpha(2)] \qquad \textbf{(11.3-38c)}$$

$$\Psi_4 = \frac{1}{2} \, [\psi_{1s}(1)\psi_{2s}(2) + \psi_{2s}(1)\psi_{1s}(2)][\alpha(1)\beta(2) - \beta(1)\alpha(2)] \qquad \textbf{(11.3-38d)}$$

The product of an antisymmetric factor and a symmetric factor is an antisymmetric function; the product of two antisymmetric factors or of two symmetric factors is a symmetric function.

Each of these wave functions is a product of a space function and a spin function. Other kinds of functions can be written, but this type of wave function is most easily constructed and analyzed. A symmetric space part is combined with an antisymmetric spin part, or vice versa, in each function. All of these functions are eigenfunctions of the $\hat{L}^2$ and $\hat{S}^2$ operators, although we do not prove that fact.

Exercise 11.17 Show that Ψ_3 and Ψ_4 satisfy the zero-order Schrödinger equation and find the energy eigenvalue. Show that these functions are normalized if the orbitals are normalized.

Angular Momentum in the Helium Atom

States of atoms are often characterized by their values of orbital and spin angular momenta. Angular momenta are vectors, and the sum of two angular momenta is a vector sum. A helium atom is not a rigid body, so the orbital angular momenta and spin angular momenta of the two electrons are not necessarily parallel.

It is a theorem of both classical and quantum mechanics that the total angular momentum of an isolated system is **conserved**. That is, if no external forces act on the system, its total angular momentum does not change in time. A conserved quantity is called a **constant of the motion**. The quantum mechanical operator of a constant of the motion always commutes with the Hamiltonian operator. A quantum number determining the value of a conserved quantity is called a **good quantum number**, and the quantity itself is sometimes referred to by the same name.

Let ℓ_1 and s_1 be the orbital and spin angular momenta of electron 1 and ℓ_2 and s_2 be the orbital and spin angular momenta of electron 2. We will now use lowercase letters for angular momenta of single electrons and capital letters for angular momenta of atoms. The total orbital and spin angular momenta of the helium atom are vector sums:

$$\mathbf{L} = \ell_1 + \ell_2 \tag{11.3-39}$$

$$\mathbf{S} = \mathbf{s}_1 + \mathbf{s}_2 \tag{11.3-40}$$

and the total angular momentum of the atom is

$$\mathbf{J} = \mathbf{L} + \mathbf{S} \tag{11.3-41}$$

If relativistic effects are included, only J^2 and J_z are good quantum numbers. However, $\hat{L}^2$, $\hat{L}_z$, $\hat{S}^2$, and $\hat{S}_z$ commute with our nonrelativistic Hamiltonian, and we can choose energy eigenfunctions corresponding to definite values of L^2, L_z, S^2, and S_z. The angular momenta of the individual electrons do not commute with the zero-order Hamiltonian and do not have definite values.[6]

The eigenfunctions and eigenvalues of the $\hat{L}^2$, $\hat{L}_z$, $\hat{S}^2$, and $\hat{S}_z$ operators follow the same pattern as other angular momenta:

$$\hat{L}^2\Psi = \hbar^2 L(L+1)\Psi \tag{11.3-42}$$

$$\hat{L}_z\Psi = \hbar M_L\Psi \tag{11.3-43}$$

$$\hat{S}^2\Psi = \hbar^2 S(S+1)\Psi \tag{11.3-44}$$

$$\hat{S}_z\Psi = \hbar M_S\Psi \tag{11.3-45}$$

All angular momenta follow the same pattern of eigenvalues for the square and one component.

The orbital angular momentum quantum number L is a nonnegative integer. The orbital angular momentum projection quantum number M_L ranges from $+L$ to $-L$, just as m ranges from $+\ell$ to $-\ell$ in Equation (11.1-44). The spin angular momentum quantum number S is a nonnegative integer or half-integer and the spin angular momentum projection quantum number M_S ranges from $+S$ to $-S$.

Figure 11.17 illustrates how angular momentum vectors can add vectorially to produce some particular values of L, M_L, S, and M_S. In each diagram, the tail of the second vector is placed at the head of the first vector, as in the geometric representation of vector addition.

Russell-Saunders Coupling

The assumption that L and S are both good quantum numbers is called **Russell-Saunders coupling**. Each set of states corresponding to a particular value of L and a particular value of S is called a **term**. A **Russell-Saunders term symbol** is assigned to each term. The principal part of the symbol is

---

[6] *ibid.*, pp. 294ff.

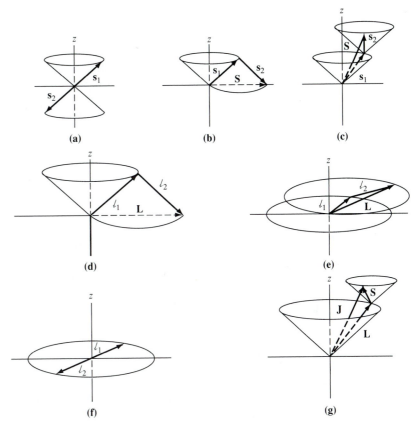

Figure 11.17. Examples of Vector Addition of Angular Momenta. These examples show how angular momenta add vectorially to give sums which follow the standard pattern for an angular momentum. **(a)** Two electron spins: $m_{s1} = 1/2$, $m_{s2} = -1/2$, $M_S = 0$, $S = 0$. **(b)** Two electron spins: $m_{s1} = 1/2$, $m_{s2} = -1/2$, $M_S = 0$, $S = 1$. **(c)** Two electron spins: $m_{s1} = 1/2$, $m_{s2} = 1/2$, $M_S = 1$, $S = 1$. **(d)** Two p electrons' orbital angular momenta: $\ell_1 = 1$, $m_\ell = 1$, $\ell_2 = 1$, $m_2 = 1$, $M_L = 0$, $L = 1$. **(e)** Two p electrons' orbital angular momenta: $\ell_1 = 1$, $m_\ell = 0$, $\ell_2 = 1$, $m_2 = 0$, $M_L = 0$, $L = 2$. **(f)** Two p electrons' orbital angular momenta: $m_1 = 0$, $\ell_1 = 1$, $m_2 = 0$, $\ell_2 = 1$, $M_L = 0$, $L = 0$. **(g)** Total orbital and spin angular momenta: $L = 1$, $M_L = 1$, $S = 1/2$, $M_S = 1/2$, $M_J = 0$, $J = 3/2$.

a letter giving the value of L, as follows:

Value of L	Symbol
0	S
1	P
2	D
3	F
4	G
etc.	

From this point on the symbols are in alphabetic order.

A left superscript with the value $2S + 1$ is attached. For our two sets of states we will show that we have ^3S and ^1S ("triplet S" and "singlet S"). The value of the left superscript is called the **multiplicity** of the term and is equal to the number of values of M_S that occur. A right subscript giving the value of J is also sometimes used.

We now investigate the values of the quantum numbers for particular configurations. Since the angular momenta are vectors, their operators are expressed as vector sums:

$$\hat{L}^2 = (\hat{\ell}_{x1} + \hat{\ell}_{x2})^2 + (\hat{\ell}_{y1} + \hat{\ell}_{y2})^2 + (\hat{\ell}_{z1} + \hat{\ell}_{z2})^2 \quad \text{(11.3-46)}$$

$$\hat{L}_z = \hat{\ell}_{z1} + \hat{\ell}_{z2} \quad \text{(11.3-47)}$$

$$\hat{S}^2 = (\hat{s}_{x1} + \hat{s}_{x2})^2 + (\hat{s}_{y1} + \hat{s}_{y2})^2 + (\hat{s}_{z1} + \hat{s}_{z2})^2 \quad \text{(11.3-48)}$$

$$\hat{S}_z = \hat{s}_{z1} + \hat{s}_{z2} \quad \text{(11.3-49)}$$

The $\hat{L}^2$ and $\hat{S}^2$ expressions are not easily usable because they contain terms that do not commute with each other, and we will not obtain explicit expressions for them.[7] However, we can operate on specific products of orbitals with $\hat{L}_z$ and $\hat{S}_z$ to find the values of M_L and M_S and from these infer the values of L and S:

$$\hat{L}_z \psi_{n_1 \ell_1 m_1 m_{s1}}(1)\psi_{n_2 \ell_2 m_2 m_{s2}}(2)$$
$$= (\hat{L}_{z1} + \hat{L}_{z2})\psi_{n_1 \ell_1 m_1 m_{s1}}(1)\psi_{n_2 \ell_2 m_2 m_{s2}}(2)$$
$$= \hbar(m_1 + m_2)\psi_{n_1 \ell_1 m_1 m_{s1}}(1)\psi_{n_2 \ell_2 m_2 m_{s2}}(2) \quad \text{(11.3-50)}$$

so that

$$\boxed{M_L = m_1 + m_2} \quad \text{(11.3-51)}$$

Equations (11.3-51) and (11.3-52) allow us to determine the values of M_L and M_S.

A similar equation can be written for $\hat{S}_z$, giving

$$\boxed{M_S = m_{s1} + m_{s2}} \quad \text{(11.3-52)}$$

▼   **EXAMPLE 11.7**   Find the values of the quantum numbers M_L and M_S for each of the wave functions in Equation (11.3-38a)–(11.3-38d).

Solution

$$\hat{L}_z \Psi_1 = (\hat{\ell}_{z1} + \hat{\ell}_{z2})\Psi_1$$
$$= \frac{1}{\sqrt{2}}[(\hat{\ell}_{z1}\psi_{1s}(1)\psi_{2s}(2) - \psi_{2s}(1)(\hat{\ell}_{z2}\psi_{1s}(2)]\alpha(1)\alpha(2)$$
$$= 0 + 0 = 0$$

---

[7] Levine, *op. cit.*, pp. 268ff.

so that $M_L = 0$. All the other wave functions also contain only s orbitals, so $M_L = 0$ for all of them.

$$\hat{S}_z \Psi_1 = (\hat{s}_{z1} + \hat{s}_{z2})\Psi_1$$

$$= \sqrt{\frac{1}{2}} [\psi_{1s}(1)\psi_{1s}(2) - \psi_{1s}(1)\psi_{1s}(2)][\hat{s}_{z1}\alpha(1)\alpha(2) + \alpha(1)\hat{s}_{z2}\alpha(2)]$$

$$= \sqrt{\frac{1}{2}} [\psi_{1s}(1)\psi_{1s}(2) - \psi_{1s}(1)\psi_{1s}(2)]\left[\frac{\hbar}{2}\alpha(1)\alpha(2) + \alpha(1)\frac{\hbar}{2}\alpha(2)\right]$$

$$= \left(\frac{\hbar}{2} + \frac{\hbar}{2}\right)\Psi_1 = \hbar\Psi_1$$

so that Ψ_1 is an eigenfunction of $\hat{S}_z$ with eigenvalue $\hbar$, corresponding to $M_S = 1$.

By a similar calculation,

$$\hat{S}_z \Psi_2 = \left(-\frac{\hbar}{2} - \frac{\hbar}{2}\right)\Psi_2 = -\hbar\Psi_2$$

so that $M_S = -1$ for Ψ_2.

$$\hat{S}_z \Psi_3 = \frac{1}{2}[\psi_{1s}(1)\psi_{2s}(2) - \psi_{2s}(1)\psi_{1s}(2)]$$

$$\times [\hat{s}_{z1}\alpha(1)\beta(2) + \hat{s}_{z1}\beta(1)\alpha(2) + \alpha(1)\hat{s}_{z2}\beta(2) + \beta(1)\hat{s}_{z2}\alpha(2)]$$

$$= \frac{1}{2}[\psi_{1s}(1)\psi_{2s}(2) - \psi_{2s}(1)\psi_{1s}(2)]$$

$$\times \left[\left(\frac{\hbar}{2}\right)\alpha(1)\beta(2) + \left(\frac{-\hbar}{2}\right)\beta(1)\alpha(2) + \alpha(1)\left(\frac{-\hbar}{2}\right)\beta(2) + \beta(1)\left(\frac{\hbar}{2}\right)\alpha(2)\right]$$

$$= 0$$

so that $M_S = 0$ for Ψ_3. A similar calculation leads to $M_S = 0$ for Ψ_4.

◢

Since the spin operators do not act on the space factors, the space factors could be omitted from the calculations of the eigenvalues of $\hat{S}_z$ in Example 11.7. Similarly, the spin factors could be omitted from the calculations of the eigenvalues of $\hat{L}_z$. The values of M_L and M_S that occur can be counted up without writing the wave functions.

The wave functions in Equation (11.3-38) are eigenfunctions of the $\hat{S}^2$ operator. We state the eigenvalues without proof:

$$\hat{S}^2\alpha(1)\alpha(2) = 2\hbar^2\alpha(1)\alpha(2) \qquad \textbf{(11.3-53a)}$$

$$\hat{S}^2\beta(1)\beta(2) = 2\hbar^2\beta(1)\beta(2) \qquad \textbf{(11.3-53b)}$$

$$\hat{S}[\alpha(1)\beta(2) + \beta(1)\alpha(2)] = 2\hbar^2[\alpha(1)\beta(2) + \beta(1)\alpha(2)] \qquad \textbf{(11.3-53c)}$$

$$\hat{S}^2[\alpha(1)\beta(2) - \beta(1)\alpha(2)] = 0 \qquad \textbf{(11.3-53d)}$$

The first three functions correspond to $S = 1$ and the fourth corresponds to $S = 0$. States with $S = 1$ are called **triplet states**. States with $S = 0$ are called **singlet states**.

We can infer the values of L and S from the values of M_L and M_S and the pattern of quantum numbers, which is that M_L ranges from L to $-L$ and M_S from S to $-S$. The only value of M_L that occurs is zero, so the only possible value of L is zero. There are two states with $M_S = 0$, one with

The symmetric spin factor in Equation (11.3-53c) belongs to the triplet, and the antisymmetric spin factor in Equation (11.3-53d) is the singlet.

$M_S = 1$, and one with $M_S = -1$. The maximum value of S is 1. Three of the wave functions, with $M_S = 1$, $M_S = 0$, and $M_S = -1$ must correspond to $S = 1$. There is one additional state, with $M_S = 0$. This can correspond only to $S = 0$. These wave functions are identified in Equation (11.3-53), but we have enumerated them without using these expressions.

Example 11.7 was a simple case, with zero orbital angular momentum. Other cases are more complicated. For example, the $(1s)(2p)$ subshell configuration consists of three detailed configurations: $(1s)(2p1)$, $(1s)(2p0)$, and $(1s)(2p, -1)$.

▼   **EXAMPLE 11.8**

Enumerate the states in the $(1s)(2p)$ configuration.

Solution

With the $1s$ orbital and one of the p orbitals, we can construct either a symmetric space function or an antisymmetric space function. For example, using the $2p1$ orbital we have

$$\Psi_s = \frac{1}{\sqrt{2}} \left[\psi_{1s}(1)\psi_{2p1}(2) + \psi_{2p1}(1)\psi_{1s}(2) \right] \tag{11.3-54}$$

and

$$\Psi_a = \frac{1}{\sqrt{2}} \left[\psi_{1s}(1)\psi_{2p1}(2) - \psi_{2p1}(1)\psi_{1s}(2) \right] \tag{11.3-55}$$

An antisymmetric spin factor must be combined with the symmetric space factor so that the wave function will be antisymmetric. The singlet spin factor is the only antisymmetric spin factor, so the symmetric space factor leads to only one state, with $S = 0$. The three triplet spin factors are all symmetric, so the antisymmetric space factor can combine with any of these, leading to three states, with M_S equal to 1, 0 and -1 but with $S = 1$ for all three.

The $2p0$ orbital and the $2p, -1$ orbital each combine with the $1s$ orbital to make two space factors, one symmetric and one antisymmetric. Each gives four states, as above, giving an additional eight states, for a total of 12 states.

The space factors of Equations (11.3-54) and (11.3-55) both correspond to $M_L = 1$. The analogous factors containing the $2p0$ orbital correspond to $M_L = 0$, and those containing the $2p, -1$ orbital correspond to $M_L = -1$. These three values of M_L corresponds to $L = 1$ with no states left over, so only P terms occur.

Each triplet spin factor combines with each of the three antisymmetric space factors to give the nine states of the ^3P term, and the singlet spin factor combines with each of the symmetric space factors to give the three states of the ^1P term.

▲

Example 11.8 is solved by straight-forward application of the pattern that angular momentum magnitudes and projections always follow.

Exercise 11.18   By explicit operation with $\hat{L}_z = \hat{L}_{z1} + \hat{L}_{z2}$, show that the eigenvalues of the two space factors in Equation (11.3-54) and Equation (11.3-55) both equal $\hbar$, corresponding to $M_L = 1$.

In order to finish characterizing our electronic states, we give the values of J, the quantum number for the total angular momentum, and M_J, the quantum number for its z component. Since $\mathbf{J}$ is the sum of $\mathbf{L}$ and $\mathbf{S}$,

$$\hat{J}_z = \hat{L}_z + \hat{S}_z \qquad (11.3\text{-}56)$$

Therefore,

$$\boxed{M_J = M_L + M_S} \qquad (11.3\text{-}57)$$

The possible values of J can be deduced by using the rule that for each value of J, the values of M_J range from $+J$ to $-J$.

Since the largest value of M_J equals the largest value of M_L plus the largest value of M_S, the largest value of J is

Equations (11.3-58) and (11.3-59) are simple results of the fact that $\mathbf{L}$ and $\mathbf{S}$ add as vectors. The magnitude of $\mathbf{J}$ and the quantum number J are both positive.

$$J_{\max} = L + S \qquad (11.3\text{-}58)$$

The smallest value of J is

$$J_{\min} = |L - S| \qquad (11.3\text{-}59)$$

J must be nonnegative.

Exercise 11.19

Hint: First make a list of all sets of quantum numbers that can occur. In counting up the terms that occur, find the states with the largest values of the quantum numbers first. Delete enough states from the list to account for all the values of the projection quantum number, and then examine the states that are left.

Tabulate the M_L and M_S values of the 12 states of Example 11.8. Show that the following terms occur:

$$^1P_1, \quad ^3P_1, \quad ^3P_3, \quad ^3P_0$$

11.4   Atoms with More Than Two Electrons

Beginning with the helium atom, we must take account of electron-electron repulsion. Our discussion of larger atoms will be similar to that of the helium atom, beginning with the zero-order treatment, which neglects this repulsion completely. In Chapter 12 we will describe without any detail the approximate inclusion of this repulsion.

The Lithium Atom in Zero Order

The Hamiltonian operator for a lithium-like atom with stationary nucleus is

$$\hat{H} = -\frac{\hbar^2}{2m}(\nabla_1^2 + \nabla_2^2 + \nabla_3^2)$$

$$+ \frac{1}{4\pi\varepsilon_0}\left(-\frac{Ze^2}{r_1} - \frac{Ze^2}{r_2} - \frac{Ze^2}{r_3} + \frac{e^2}{r_{12}} + \frac{e^2}{r_{13}} + \frac{e^2}{r_{23}}\right) \qquad (11.4\text{-}1)$$

where $Z = 3$ represents Li, $Z = 4$ represents Be$^+$, etc.

In any atom, the zero-order problem always corresponds to neglect of the electron-electron repulsions.

Just as in the helium atom treatment, we obtain the zero-order Hamiltonian by discarding the electron-electron repulsion terms, giving

$$\hat{H}^{(0)} = \hat{H}_{HL}(1) + \hat{H}_{HL}(2) + \hat{H}_{HL}(3) \qquad (11.4\text{-}2)$$

The zero-order Schrödinger equation is solved by separation of variables, assuming a trial function which is a product of three orbitals:

$$\Psi^{(0)} = \psi_1(1)\psi_2(2)\psi_3(3) \qquad (11.4\text{-}3)$$

The zero-order problem always leads to the simple orbital wave function (a one-term product of orbitals) and to an energy that is a sum of orbital energies.

The three orbitals obey hydrogen-like Schrödinger equations and are thus hydrogen-like orbitals, so the zero-order wave function without antisymmetrization but with inclusion of spin is

$$\Psi^{(0)} = \psi_{n_1\ell_1 m_1 m_{s1}}(1)\psi_{n_2\ell_2 m_2 m_{s2}}(2)\psi_{n_3\ell_3 m_3 m_{s3}}(3) \qquad (11.4\text{-}4)$$

The subscripts on the subscripts indicate that the quantum numbers do not necessarily have the same value for each orbital.

The electronic energy of the atom is the sum of three hydrogen-like energy eigenvalues:

$$E^{(0)} = E^{(0)}_{n_1 n_2 n_3} = E_{n_1}(HL) + E_{n_2}(HL) + E_{n_3}(HL)$$

$$= -Z^2(13.60 \text{ eV})\left(\frac{1}{n_1^2} + \frac{1}{n_2^2} + \frac{1}{n_3^2}\right) \qquad (11.4\text{-}5)$$

Exercise 11.20   Carry out the steps to obtain Equations (11.4-4) and (11.4-5).

Antisymmetrization

The orbital wave function of Equation (11.4-4) can be antisymmetrized by adding one term with each permutation of the orbital labels for a fixed order of particle labels, or equivalently by adding one term with each permutation of particle labels for a fixed order of orbital labels. Each term that is generated from the first term by one permutation of a pair of indexes has a negative sign, and each term generated by two permutations of pairs of indexes has a positive sign. The antisymmetrized function is

There are six terms in this antisymmetrized wave function because there are 3! = 6 ways to permute three objects. The second, third, and fourth terms are obtained from the first term by a single binary permutation (interchange of two things), and the last two terms are obtained by two such permutations.

$$\Psi = \frac{1}{\sqrt{6}} [\psi_1(1)\psi_2(2)\psi_3(3) - \psi_2(1)\psi_1(2)\psi_3(3) - \psi_1(1)\psi_3(2)\psi_2(3)$$

$$- \psi_3(1)\psi_2(2)\psi_1(3) + \psi_3(1)\psi_1(2)\psi_2(3) + \psi_2(1)\psi_3(2)\psi_1(3)] \qquad (11.4\text{-}6)$$

where we abbreviate the quantum numbers by including only the subscripts on them.

Exercise 11.21   Show that the function produced by exchanging particle labels 1 and 3 in Equation (11.4-6) is the negative of the original function. Choose another permutation and show the same thing.

A considerable difference between the helium and lithium ground states is that there is no way to construct an antisymmetric three-electron wave function using only 1s orbitals, since there are only two 1s spin orbitals.

▼ **EXAMPLE 11.9**

If orbitals ψ_1 and ψ_3 are the same function, show that the wave function of Equation (11.4-6) vanishes.

Solution

$$
\begin{aligned}
\Psi = \frac{1}{\sqrt{6}} [&\psi_1(1)\psi_2(2)\psi_1(3) - \psi_2(1)\psi_1(2)\psi_1(3) - \psi_1(1)\psi_1(2)\psi_2(3) \\
&- \psi_1(1)\psi_2(2)\psi_1(3) + \psi_1(1)\psi_1(2)\psi_2(3) + \psi_2(1)\psi_1(2)\psi_1(3)] \\
= \; &0
\end{aligned}
$$

▲

Exercise 11.22

Show that the wave function of Equation (11.4-6) is normalized if the orbitals are normalized and orthogonal to each other. The normalization integral is an integral over the coordinates of all three electrons. Each term will factor, but there will be 36 terms. Look for a way to write the result of integrating each term without having to write all the integrands, using the orthogonality and normalization of the orbitals.

Slater Determinants

The Slater determinant is named after John C. Slater, 1900–1976, a prominent American physicist who made various contributions to atomic and molecular quantum theory.

Another notation can be used to write the wave function of Equation (11.4-6). A **determinant** is a quantity derived from a square matrix by a certain set of multiplications, additions, and subtractions. If orbitals are used for the elements of a square matrix, the determinant of that matrix is a function of the coordinates on which the orbitals depend. The wave function of Equation (11.4-6) is equal to the determinant

$$
\Psi = \frac{1}{\sqrt{6}} \begin{vmatrix} \psi_1(1) & \psi_1(2) & \psi_1(3) \\ \psi_2(1) & \psi_2(2) & \psi_2(3) \\ \psi_3(1) & \psi_3(2) & \psi_3(3) \end{vmatrix} \qquad \textbf{(11.4-7)}
$$

which is called a **Slater determinant**.

Exercise 11.23

Use the following rules for expanding a three-by-three determinant to show that the function of Equation (11.4-7) is the same as that of Equation (11.4-6):

$$
\begin{vmatrix} a_{11} & a_{12} & a_{13} \\ a_{21} & a_{22} & a_{23} \\ a_{31} & a_{32} & a_{33} \end{vmatrix} = a_{11} \begin{vmatrix} a_{22} & a_{23} \\ a_{32} & a_{33} \end{vmatrix} - a_{12} \begin{vmatrix} a_{21} & a_{23} \\ a_{31} & a_{33} \end{vmatrix} + a_{13} \begin{vmatrix} a_{21} & a_{22} \\ a_{31} & a_{32} \end{vmatrix}
$$

$$
\begin{aligned}
= \; &a_{11}(a_{22}a_{33} - a_{23}a_{32}) - a_{12}(a_{21}a_{33} - a_{23}a_{31}) \\
&+ a_{13}(a_{21}a_{32} - a_{22}a_{31})
\end{aligned}
$$

Two properties of determinants are related directly to the properties of antisymmetrized orbital wave functions:

1. If one exchanges two columns or two rows of a determinant, the resulting determinant is the negative of the original determinant.

Exchanging the locations of two particles is equivalent to exchanging two columns, so the Slater determinant exhibits the necessary antisymmetry.

2. If two rows or two columns of a determinant are identical, the determinant vanishes. If two electrons occupy identical spin orbitals, two rows of the determinant in Equation (11.4-7) are identical and the determinant vanishes, in agreement with the Pauli exclusion principle.

For the ground-state configuration in zero order, we must choose three spin orbitals with the minimum possible sum of orbital energies, since the zero-order energy is equal to the sum of the orbital energies, from Equation (11.4-5). However, the spin orbitals must all be different so that the wave function can be antisymmetrized without vanishing. This policy for choosing the ground-state configuration is called the **Aufbau principle**, from the German word for "building up." For the lithium atom, we choose the two $1s$ spin orbitals and one spin orbital from the second shell. In zero order, all of the $2s$ and $2p$ orbitals have the same energy, but we anticipate that higher-order calculations will give a lower energy for the $2s$ subshell than for the $2p$ subshell and choose one of the $2s$ spin orbitals. The zero-order energy of the ground state is

$$E_{gs}^{(0)} = E_{1s1s2s}^{(0)} = 2E_1(\text{HL}) + E_2(\text{HL})$$

$$= (-13.60 \text{ eV})\left(2Z^2 + \frac{Z^2}{4}\right) = -275.4 \text{ eV} \qquad \textbf{(11.4-8)}$$

where we have used the lithium value $Z = 3$ in computing the numerical value. This value is seriously in error, as was the zero-order value for helium. It differs from the experimental value of -203.5 eV by 35%.

The antisymmetrized zero-order wave function is

$$\Psi^{(0)} = \frac{1}{\sqrt{6}} \begin{vmatrix} \psi_{1s}(1)\alpha(1) & \psi_{1s}(2)\alpha(2) & \psi_{1s}(3)\alpha(3) \\ \psi_{1s}(1)\beta(1) & \psi_{1s}(2)\beta(2) & \psi_{1s}(3)\beta(3) \\ \psi_{2s}(1)\alpha(1) & \psi_{2s}(2)\alpha(1) & \psi_{2s}(3)\alpha(3) \end{vmatrix} \qquad \textbf{(11.4-9)}$$

The $2s$ spin-down orbital could have been chosen instead of the $2s$ spin-up orbital. We therefore have two states of equal energy instead of a single ground state. This **doubly degenerate** ground level corresponds to a doublet term with $S = 1/2$, since the possible values of M_S are $+1/2$ and $-1/2$. Since $M_L = 0$, the value of L is 0, the only value of J is $1/2$, and the term symbol is $^2S_{1/2}$.

Excited states of the lithium atom can correspond to various choices of orbitals. The values of M_L and M_S for these excited states can be calculated by algebraic addition. Using the rules that M_L ranges from $+L$ to $-L$ and M_S ranges from $+S$ to $-S$, one can deduce the values of S and L that occur and can assign term symbols. Higher-order calculations must be used to determine the order of the energies of the excited states.

Exercise 11.24   Consider the excited-state configuration $(1s)(2s)(3s)$ for a lithium atom.
     a. Show that quartet states with $S = 3/2$ can occur.
     b. Write the term symbols for all terms that occur.
     c. Find the zero-order energy eigenvalue for this configuration.

Atoms with More Than Three Electrons

The treatment of the other atoms is similar to the helium and lithium treatments. For an atom with atomic number Z (Z protons in the nucleus and Z electrons), the stationary-nucleus Hamiltonian operator is

$$\hat{H} = -\frac{\hbar^2}{2m}\sum_{i=1}^{Z}\nabla_i^2 + \frac{e^2}{4\pi\varepsilon_0}\left(-\sum_{i=1}^{Z}\frac{Z}{r_i} + \sum_{i=2}^{Z}\sum_{j=1}^{i-1}\frac{1}{r_{ij}}\right) \quad \textbf{(11.4-10)}$$

where r_i is the distance from the nucleus to the position of the ith electron and r_{ij} is the distance from the ith electron to the jth electron.

The first two sums in Equation (11.4-10) are sums of hydrogen-like one-particle Hamiltonian operators, and the double sum is a sum of terms like those we have been unable to handle exactly with helium and lithium. Just as before, we neglect these terms in carrying out the zero-order solution. The zero-order Hamiltonian operator is

The zero-order solution still corresponds to neglect of the electron-electron repulsions and still leads to a wave function that is a product of orbitals and an energy that is a sum of orbital energies.

$$\hat{H}^{(0)} = \sum_{i=1}^{Z}\hat{H}_{HL}(i) \quad \textbf{(11.4-11)}$$

The corresponding time-independent Schrödinger equation can be solved by separation of variables, using the trial function

$$\Psi^{(0)} = \psi_1(1)\psi_2(2)\psi_3(3)\psi_4(4)\cdots\psi_Z(Z) = \prod_{i=1}^{Z}\psi_i(i) \quad \textbf{(11.4-12)}$$

where the symbol Π stands for a product of factors, just as Σ stands for a sum of terms. Since the terms in the Hamiltonian are hydrogen-like Hamiltonians, the factors $\psi_1(1)$, $\psi_2(2)$, $\psi_3(3)$, etc. are all hydrogen-like orbitals and the energy eigenvalue is a sum of hydrogen-like orbital energies:

$$\psi_i(i) = \psi_{n_i\ell_i m_i m_{s1}}(i) \quad \textbf{(11.4-13)}$$

$$E^{(0)} = E_{n_1}(HL) + E_{n_2}(HL) + \cdots = \sum_{i=1}^{Z}E_{n_i}(HL) \quad \textbf{(11.4-14)}$$

where n_i, ℓ_i, etc. are values of the quantum numbers for hydrogen-like orbitals. Just as with the helium and lithium atoms, the zero-order wave functions and energies of Equations (11.4-13) and (11.4-14) are very poor approximations.

Just as we antisymmetrized two-electron and three-electron wave functions to satisfy the indistinguishability requirements, we must antisymmetrize the orbital wave function of Equation (11.4-12). This is done by writing a Slater determinant with one row for each spin orbital and one column for each electron:

$$\Psi = \frac{1}{\sqrt{Z!}}\begin{vmatrix} \psi_1(1) & \psi_1(2) & \psi_1(3) & \psi_1(4) & \cdots & \psi_1(Z) \\ \psi_2(1) & \psi_2(2) & \psi_2(3) & \psi_2(4) & \cdots & \psi_2(Z) \\ \psi_3(1) & \psi_3(2) & \psi_3(3) & \psi_3(4) & \cdots & \psi_3(Z) \\ \psi_4(1) & \psi_4(2) & \psi_4(3) & \psi_4(4) & \cdots & \psi_4(Z) \\ \vdots & \vdots & \vdots & \vdots & \vdots & \vdots \\ \psi_Z(1) & \psi_Z(2) & \psi_Z(3) & \psi_Z(4) & \cdots & \psi_Z(Z) \end{vmatrix} \quad \textbf{(11.4-15)}$$

where the $1/\sqrt{Z!}$ factor normalizes the wave function, assuming that all orbitals are normalized and orthogonal to each other, and where we have abbreviated the quantum numbers. The Pauli exclusion principle must be followed. No two spin orbitals can be the same, or two rows of the determinant would be identical, causing the wave function to vanish.

The values of M_L, M_S, L, and S can be computed in the same way as with the helium and lithium atoms. Because the hydrogen-like orbitals in the same shell all have the same energy, many of the terms are degenerate in zero order but will have different energies when better approximations are used.

In the next chapter we will discuss further approximations beyond the simple orbital approximation, which will give better approximations to atomic energies. We will primarily remain with the orbital approximation but will go beyond the zero-order results. We will find that the orbitals in different subshells in the same shell do not correspond to the same energy and will use the facts about the orbital energies to understand the periodic chart of the elements.

Summary

The time-independent Schrödinger equation for a general central-force system was separated into a one-particle Schrödinger equation for the motion of the center of mass of the two particles and a one-particle Schrödinger equation for the motion of one particle relative to the other.

The Schrödinger equation for the relative motion was solved by separation of variables in spherical polar coordinates, assuming the trial function

$$\psi(r, \theta, \phi) = R(r)Y(\theta, \phi) = R(r)\Theta(\theta)\Phi(\phi)$$

The angular functions $Y_{\ell m}(\theta, \phi)$ are a set of functions called spherical harmonic functions. These functions are also eigenfunctions of the operator for the square of the orbital angular momentum and its z component, with eigenvalues given by the quantum numbers ℓ and m.

The equation for the radial factor $R(r)$ in the relative wave function was solved exactly for the hydrogen atom, giving a set of wave functions with two quantum numbers: n, the principal quantum number, and ℓ, the angular momentum quantum number from the spherical harmonic functions.

The energy eigenvalues were found to depend only on the principal quantum number:

$$E = E_n = -\frac{(13.60 \text{ eV})Z^2}{n^2}$$

where Z is the number of protons in the nucleus.

An intrinsic electronic angular momentum of the electron was introduced. This angular momentum was ascribed to a spinning motion of the electron in addition to its orbital motion. It corresponds to fixed magnitude and two possible z projections, $\hbar/2$ and $-\hbar/2$.

In the zero-order approximation, the energy eigenfunctions of the helium atom were products of hydrogen-like orbitals, one for each electron.

Antisymmetrizing these orbital wave functions to conform to the physical indistinguishability of the electrons produced the Pauli exclusion principle, which states that no two electrons can occupy the same orbital in any orbital wave function.

In zero order, the wave functions for multielectron atoms are products of hydrogen-like orbitals. By utilizing the Pauli exclusion principle, possible electron configurations and term symbols can be computed.

Additional Reading

Hans A. Bethe and Edwin E. Salpeter, *Quantum Mechanics of One- and Two-Electron Atoms*, Plenum Publishing, New York, 1977
This book by two well-respected physicists gives a complete and careful treatment.

Michael A. Morrison, Thomas F. Estle, and Neal F. Lane, *Quantum States of Atoms, Molecules, and Solids*, Prentice-Hall, Englewood Cliffs, NJ, 1976
This is a comprehensive account of the applications of quantum mechanics to these systems.

PROBLEMS

Problems for Section 11.1

11.25. Using formulas in Appendix F, write the formulas for the spherical harmonic functions Y_{43} and Y_{42}. Omit the normalizing factors.

11.26. Sketch graphs of the functions and of their squares: (a) $\Theta_{10}(\theta)$, (b) $\Theta_{11}(\theta)$, (c) $\Theta_{20}(\theta)$, (d) $\Theta_{21}(\theta)$, (e) $\Theta_{22}(\theta)$.

11.27. Sketch the nodal surfaces for the spherical harmonic functions: (a) Y_{10}, (b) Y_{22}, (c) Y_{31}, (d) Y_{21}.

Problems for Section 11.2

11.28. a. Sketch the nodal surfaces for the first five energy eigenfunctions for a particle in a three-dimensional spherical hard box. Hint: Use the analogy with the hydrogen atom functions.

b. Sketch the orbital regions for the first few energy eigenfunctions for a particle in a three-dimensional spherical hard box.

*11.29. Calculate the percent difference in energy between an ordinary hydrogen atom and a deuterium atom in the ground state.

11.30. A positronium atom is a hydrogen-like atom consisting of an electron and a positron (an antielectron with charge $+e$ and mass equal to the electron mass). Find the energy of a positronium atom in the 1s state. Describe the

classical motion of the two particles about the center of mass. Find the value of the Bohr radius for positronium.

*11.31. Calculate the angle between the z axis and each of the cones of possible directions of the orbital angular momentum for $\ell = 2$.

11.32. Find the ratio of the magnitude of the orbital angular momentum to the maximum value of its z component for each of the cases $\ell = 1, 2, 3$, and 4.

*11.33. Calculate $\langle r \rangle$ for the 2s and 2p states of a hydrogen-like atom. Comment on your answer.

11.34. a. Find the value of the distance b such that there is a 95% chance that an electron in a hydrogen atom in the 1s state is no farther from the nucleus than distance b.

b. Find the ratio of the 1s wave function at $r = b$ to the same function at $r = 0$.

c. Repeat parts a and b for 90% probability instead of 95%.

11.35. For a hydrogen atom in a 1s state, find the probability that the electron is no farther from the nucleus than (a) a, (b) $2a$, (c) $3a$ where a is the Bohr radius.

*11.36. Calculate $\langle z \rangle$ and σ_z for the electron in a hydrogen atom in the 1s state. Explain the meaning of the values.

11.37. Calculate the expectation values of p_x and of p_x^2 for the electron in a hydrogen atom in the 1s state. Why does $\langle p_x^2 \rangle$ not equal $\langle p_x \rangle^2$?

11.38. Calculate the expectation value and the standard deviation of L_z for a hydrogen atom in the $2px$ state. Explain what the values mean.

*11.39. Find the most probable value of the electron's distance from the nucleus for a hydrogen atom in the $1s$, the $2s$, and the $2p$ states.

11.40. a. Draw a rough picture of the nodal surfaces of each of the real $3d$ orbitals. From these, draw rough pictures of the orbital regions.

b. Do the same for the complex $3d$ orbitals (eigenfunctions of $\hat{L}_z$).

Problems for Section 11.3

11.41. Draw a graph of a probability density for finding any electron at a distance r from the nucleus of a helium atom in the $(1s)(2s)$ configuration, using the zero-order wave function of Equation (11.3-38a).

*11.42. If the spin angular momentum vector for one electron with spin up lies in the y-z plane and the spin angular momentum vector for another electron with spin down also lies in the y-z plane, find the magnitude and direction of their vector sum. To what values of S and M_S (if any) does this vector sum correspond?

11.43. Find the possible term symbols for the subshell configuration $(2p)(3d)$ for the helium atom. Which will probably have the lowest energy?

Problems for Section 11.4

11.44. Find the possible term symbols for the ground state of the chlorine atom.

*11.45. Find the ground-state electron configuration and the possible term symbols for the following atoms:

   a. C   **b.** Se   **c.** Ar   **d.** Mg

11.46. Find the possible term symbols for the following configurations of the Be atom:

a. $(1s)^2(2s)^2$ (ground state)
b. $(1s)^2(2s)(3s)$ (an excited state)
c. $(1s)(2s)(3s)(4s)$ (an excited state)

General Problems

11.47. From the pattern of nodal surfaces observed in the subshells we have discussed, predict the following:

a. The number of nodal spheres in the $6s$ wave function.

b. The number of nodal spheres in a $6p$ wave function.

c. The number of nodal planes containing the z axis in the real part of the $6d0$ (620) wave function.

d. The number of nodal cones in the real part of the $6p1$ (611) wave function.

11.48. Consider the beryllium atom, Be.

a. Write the Hamiltonian operator, assuming a stationary nucleus.

b. Write the zero-order Hamiltonian operator (excluding the electron-electron repulsion terms).

c. Write the ground-state wave function in the simple orbital approximation, without antisymmetrization.

d. Write the antisymmetrized ground-state wave function as a Slater determinant.

e. Consider the ground-state configuration $(1s)^2(2s)^2$. Determine the values of S, L, M_L, and M_S that can occur. Write the Russell-Saunders term symbols for all terms that can occur.

f. Consider the subshell configuration $(1s)^2(2s)(2p)$. Determine the values of S, L, M_L, and M_S that can occur. Write the Russell-Saunders term symbols for all terms that can occur.

g. Consider the subshell configuration $(1s)^2(2p)(3p)$. Determine the values of S, L, M_L, and M_S that can occur. Write the Russell-Saunders term symbols for all terms that can occur.

*11.49. Identify the following statements as either true or false. If a statement is true only under special circumstances, label it as false.

a. The angular factors Θ and Φ are the same functions for the hydrogen atom wave functions and those of any other central-force problem.

b. In a central-force problem, the motion of the center of mass and the relative motion can be treated separately only to a good approximation.

c. Every atom is spherical in shape.

d. The x or y axis could be chosen as the unique direction for angular momentum components instead of the z axis.

e. The energy eigenvalues for the H atom in quantum mechanics are identical to those in the Bohr theory.

f. The angular momentum eigenvalues for the H atom in quantum mechanics are identical to those in the Bohr theory.

g. There is a one-to-one correspondence between the states of the H atom in the Bohr theory and the states of the H atom in quantum mechanics.

h. Electrons in a multielectron atom move exactly like electrons in a hydrogen-like atom with the appropriate nuclear charge.

12 The Electronic States of Atoms: II. Higher-Order Approximations for Multielectron Atoms

PREVIEW

In this chapter, we discuss the three principal approximation methods by which we are able to go beyond the zero-order orbital approximation. Using these approximation schemes, we can explain many of the properties of atoms of different elements and can understand the periodic table of the elements.

OBJECTIVES

After studying this chapter, the student should:

1. understand the basic ideas behind the variation method, the perturbation method, and the self-consistent field method,
2. be able to solve simple problems using the variation and perturbation methods,
3. understand the Aufbau principle and its relationship to the periodic table of the elements,
4. be able to solve problems related to electron configurations and term symbols of multielectron atoms.

PRINCIPAL FACTS AND IDEAS

1. The interelectron repulsions must be included in approximation methods beyond the zero-order orbital approximation.
2. The variation theorem allows calculation of upper bounds to ground-state energies.
3. The perturbation method allows approximate calculations of interelectron repulsion energies.
4. The self-consistent field method allows calculations of average interelectron repulsion energies.
5. The electronic structure of multielectron atoms can be described in terms of the approximation schemes.
6. The structure of the periodic table of the elements can be understood in terms of higher-order orbital approximations.

The Variation Method and Its Application to the Helium Atom

In Chapter 11, we presented the zero-order orbital theory of the electronic structure of atoms. The interelectron repulsions were neglected, with the result that the electrons were independent of each other and occupied states corresponding to hydrogen-like orbitals. There are three principal approximation schemes by which we are able to go beyond the zero-order orbital approximation. The first scheme is the *variation method*, which is based on the *variation theorem*:

The Variation Theorem

The expectation value of the energy for a state corresponding to a wave function ψ is given by

$$\langle E \rangle = \frac{\int \psi^* \hat{H} \psi \, dq}{\int \psi^* \psi \, dq} \tag{12.1-1}$$

where $\hat{H}$ is the correct Hamiltonian operator for a system and the coordinates of the particles of the system are abbreviated by q.

The **variation theorem** states that for any conceivable wave function ψ obeying the same boundary conditions as the correct ground-state energy eigenfunction

$$\boxed{\langle E \rangle \geq E_{gs} \quad \text{(variation theorem)}} \tag{12.1-2}$$

where E_{gs} is the correct ground-state energy eigenvalue of the system. $\langle E \rangle$ is equal to E_{gs} if and only if the function ψ is the same function as the correct ground-state energy eigenfunction; otherwise $\langle E \rangle$ is higher than E_{gs}.

The variation theorem can also be stated as follows: *If we use any wave function other than the correct ground-state energy eigenfunction to calculate the expectation value of the energy, the result must be higher than the correct ground-state energy.* We omit the proof of the theorem, which is assigned in Problem 12.10.

The variation theorem suggests the **variation method** for finding an approximate ground-state energy. When this method is applied, the expectation value $\langle E \rangle$ is called the **variational energy** and is denoted by W. First, we choose a family of possible approximate wave functions. Second, we find the member of the family that gives a lower (more negative) value of W than any other member of the family. Since W can never be too negative, the function that gives the minimum energy gives a better approximation to the ground-state energy than any other member of the family of functions. The theorem does not guarantee that this function is a better approximation to the correct wave function than any other member of the family for any purpose other than the calculation of the energy.

It is reasonable to hope that if the variational energy is close to the correct energy, the wave function is close to the correct wave function.

A typical application of the variation method uses a family of functions that can be represented by a single formula containing one or more variable parameters. Such a family of functions is called a **variation function** or a

variation trial function. The variational energy W is calculated as a function of the parameters, and the minimum value of W is found.

Application of the Variation Method to the Helium Atom[1]

Let us first use the zero-order orbital wave function of Equation (11.3-13) as a variation trial function. This is a single function, so no minimization can be done.

The wave function is normalized so that the denominator in Equation (12.1-1) equals unity, and the variational energy is

The entire Hamiltonian, not a zero-order approximate Hamiltonian, must be used in calculating the variational energy.

$$W = \frac{1}{2} \int \psi_{100}(1)^* \psi_{100}(2)^* [\alpha(1)\beta(2) - \alpha(1)\beta(2)]^* \hat{H} \psi_{100}(1)\psi_{100}(2)$$
$$\times [\alpha(1)\beta(2) - \alpha(1)\beta(2)] \, dq_1' \, dq_2' \tag{12.1-3}$$

where $\hat{H}$ is the correct Hamiltonian operator and dq_1' and dq_2' indicate integration over space and spin coordinates.

Because the Hamiltonian operator is independent of the spin coordinates, the integration over the spin coordinates gives a factor of 2, which cancels the factor $1/2$ that came from the normalizing constant. We could have omitted the spin factor and the spin integration from the beginning. We obtain

$$W = \int \psi_{100}(1)^* \psi_{100}(2)^* \left[\hat{H}_{HL}(1) + \hat{H}_{HL}(2) \right.$$
$$\left. + \frac{e^2}{4\pi\varepsilon_0 r_{12}} \right] \psi_{100}(1)\psi_{100}(2) \, dq \tag{12.1-4}$$

where the symbol dq stands for $d^3\mathbf{r}_1 \, d^3\mathbf{r}_2$.

The $\hat{H}_{HL}(1)$ and $\hat{H}_{HL}(2)$ terms in the Hamiltonian operator give energy eigenvalues for a hydrogen-like atom.

Exercise 12.1   Show that the $H_{HL}(1)$ term in Equation (12.1-4) yields a contribution to W equal to $E_1(HL)$ and that the $\hat{H}_{HL}(2)$ term yields an equal contribution.

We now have

$$W = 2E_1(HL) + \int \psi_{100}(1)^* \psi_{100}(2)^* \left(\frac{e^2}{4\pi\varepsilon_0 r_{12}} \right) \psi_{100}(1)\psi_{100}(2) \, dq \tag{12.1-5}$$

Evaluation of the integral in this equation is difficult, and we do not carry it out.[2] The result is that the integral equals

$$\frac{5Ze^2}{8(4\pi\varepsilon_0 a)} = -\frac{5}{8} \langle \mathscr{V} \rangle_{HL(1s)} = -\frac{5}{4} E_1(HL) \tag{12.1-6}$$

[1] J. C. Davis, Jr., *Advanced Physical Chemistry*, Ronald Press, New York, 1965, pp. 221ff.
[2] I. N. Levine, *Quantum Chemistry*, 3rd ed., Allyn & Bacon, Boston, 1983, pp. 202ff.

where $\langle \mathcal{V} \rangle_{HL(1s)}$ is the expectation value of the potential energy for the hydrogen-like atom in its ground state. The variational energy is, using Equation (11.2-12),

$$W = -2\frac{Z^2 e^2}{2(4\pi\varepsilon_0 a)} + \frac{5Ze^2}{8(4\pi\varepsilon_0 a)}$$

$$= -108.8 \text{ eV} + 34.0 \text{ eV} = -74.8 \text{ eV} \qquad \textbf{(12.1-7)}$$

where we have used the value $Z = 2$ in the calculation of the numerical value. This result is more positive than the correct value of -79.0 eV, as the variation theorem guaranteed. The error is approximately 4 eV. This result is much better than the value of -108.8 eV obtained with the zero-order approximation.

This improvement in the energy value was not obtained by changing the wave function. Our wave function is still the zero-order wave function obtained by complete neglect of the interelectron repulsion. The improvement came from using the complete Hamiltonian operator, including the interelectron repulsion energy, in calculating the variational energy.

Let us now use a variational trial function that represents a family of functions. We replace the nuclear charge Z in the hydrogen-like 1s orbitals by a variable parameter, Z'. The modified space orbital is

$$\psi'_{100} = \psi'_{100}(Z') = \frac{1}{\sqrt{\pi}}\left(\frac{Z'}{a}\right)^{3/2} e^{-Z'r/a} \qquad \textbf{(12.1-8)}$$

where a is the Bohr radius. We indicate that the orbital depends on the value of Z', and label the variation trial orbital with a prime (').

The variation trial function is

$$\varphi = \varphi(Z') = \psi'(1)\psi'(2)\frac{1}{\sqrt{2}}\left[\alpha(1)\beta(2) - \beta(1)\alpha(2)\right] \qquad \textbf{(12.1-9)}$$

where we omit the subscripts on the orbital symbols.

The concept of shielding is used to give a qualitative explanation of the things that we are now discussing.

There is a physical motivation for choosing this variation function. As an electron (the "first" electron) moves about the nucleus, there is some probability that the second electron will be somewhere between the first electron and the nucleus, "shielding" the first electron somewhat from the full nuclear charge and causing it to move as though the nucleus had a smaller charge. Therefore, a value of Z' smaller than 2 should produce a better approximation than $Z' = 2$.

The wave function of Equation (12.1-9) is substituted into Equation (12.1-1) to calculate the variational energy. The correct number of protons, $Z = 2$, must be used in the Hamiltonian operator. The variational energy is

$$W = \int \psi'(1)^* \psi'(2)^* \left[\hat{\mathscr{K}}(1) - \frac{Ze^2}{4\pi\varepsilon_0 r_1} + \hat{\mathscr{K}}(2) \right.$$

$$\left. - \frac{Ze^2}{4\pi\varepsilon_0 r_2} + \frac{e^2}{4\pi\varepsilon_0 r_{12}} \right] \psi'(1)\psi'(2)\, dq \qquad \textbf{(12.1-10)}$$

where $\hat{\mathscr{K}}$ is the kinetic energy operator for one electron.

The first term in the Hamiltonian operator operates only on the coordinates of electron 1, so that

$$\iint \psi(1)^*\psi(2)^*\hat{\mathscr{K}}(1)\psi'(1)\psi'(2)\, dq$$
$$= \int \psi(1)^*\hat{\mathscr{K}}(1)\psi'(1)\, d^3\mathbf{r}_1 = Z'^2\langle\mathscr{K}_{\mathrm{H}}\rangle_{1s} = -\frac{Z'^2}{2}E_1(\mathrm{H}) \quad \textbf{(12.1-11)}$$

where $\langle\mathscr{K}_{\mathrm{H}}\rangle_{1s}$ is the expectation value of the kinetic energy of the hydrogen (not hydrogen-like) atom in the $1s$ state. We have used the fact that the integral over the coordinates of particle 2 can be factored out and the assumption that the orbital $\psi'(2)$ is normalized so that this integral equals unity. The factor Z'^2 comes from the fact that orbital $\psi'(1)$ is the $1s$ orbital for an effective nuclear charge equal to $Z'\mathrm{e}$. The final equality comes from Equation (11.2-23) and the discussion following Example 11.4.

The second term in the Hamiltonian operator in Equation (12.1-10) gives

$$-\int \psi'(1)^*\psi'(2)^* \frac{Z\mathrm{e}^2}{4\pi\varepsilon_0 r_1} \psi'(1)\psi'(2)\, dq$$
$$= ZZ'\langle\mathscr{V}\rangle_{\mathrm{H}(1s)} = 2ZZ'E_1(\mathrm{H}) \quad \textbf{(12.1-12)}$$

where $\langle\mathscr{V}\rangle_{\mathrm{H}(1s)}$ is the expectation value of the potential energy of a hydrogen (not hydrogen-like) atom in the $1s$ state. We have a factor of Z from the original factor Z in the Hamiltonian and a factor of Z' from the $1s$ orbital with a nuclear charge of $Z'\mathrm{e}$. The final equality comes from Equation (11.2-23).

Exercise 12.2　Show that Equation (12.1-12) is correct.

The next two terms in the Hamiltonian operator in Equation (12.1-10) are just like the first two, except that the roles of particles 1 and 2 are interchanged. After the integrations are done, this interchange makes no difference and these two terms give contributions equal to those of the first two terms.

The final term is the same as in Equation (12.1-6) except that the orbitals are for the nuclear charge of $Z'\mathrm{e}$ instead of $Z\mathrm{e}$, so that its contribution is

$$\int \psi'(1)^*\psi'(2)^*\left(\frac{\mathrm{e}^2}{4\pi\varepsilon_0 r_{12}}\right)\psi'(1)\psi'(2)\, dq = -\frac{5}{4}Z'E_1(\mathrm{H}) \quad \textbf{(12.1-13)}$$

The variational energy is a function of a parameter because our variation function was a family of functions expressed by a formula with a parameter.

The final result is

$$W = E_1(\mathrm{H})\left(-2Z'^2 + 4ZZ' - \frac{5}{4}Z'\right) \quad \textbf{(12.1-14)}$$

Exercise 12.3　Verify Equation (12.1-14).

We find the minimum value of W by differentiating with respect to the variable parameter Z' and setting this derivative equal to zero:

$$0 = E_1(\text{H})\left(-4Z' + 4Z - \frac{5}{4}\right)$$

This equation is satisfied by

$$Z' = Z - \frac{5}{16} \qquad \textbf{(12.1-15)}$$

A theorem of electrostatics asserts that a spherically symmetric distribution of charge produces an electric field just like that of a point charge as long as the field is measured outside the charge distribution.

For $Z = 2$, $Z' = 27/16 = 1.6875$. Our optimized helium atom wave function corresponds to a shielding of the nucleus so that an electron sees an effective nuclear charge of 1.6875 instead of 2 protons. This effective charge is equivalent to saying that the second electron has a 31.25% probability of being between the nucleus and the first electron. Figure 12.1 shows the zero-order $1s$ orbital (with $Z = 2$) and the variational orbital we have just obtained, with $Z' = 1.6875$. The variable on the horizontal axis is the distance from the nucleus divided by the Bohr radius a.

The minimum value of W is

$$W = (-13.60 \text{ eV})\left[-2(1.6875)^2 + 4(2)(1.6875) - \frac{5}{4}(1.6875)\right]$$

$$= -77.5 \text{ eV} \qquad \textbf{(12.1-16)}$$

This value differs from the experimental value of -79.0 eV by 1.5 eV, an error of 2%. This error corresponds to 145 kJ mol^{-1}. Our result is still not accurate enough for quantitative chemical purposes.

More nearly accurate values can be obtained by choosing more complicated variation functions. For example, Hylleraas used the variation function[3]

$$\varphi = Ce^{-Z''r_1/a}e^{-Z''r_2/a}(1 + br_{12}) \qquad \textbf{(12.1-17)}$$

which is not an orbital wave function, because of the dependence of the final factor on r_{12}, the distance between the electrons. This function gave a variational energy equal to -78.7 eV with a value of Z'' equal to 1.849 and a value of b equal to 0.364. The energy is in error by 0.3 eV, or about 0.4%. More elaborate variational functions have been used and have given excellent agreement with experiment.[4]

The presence of the factor $(1 + br_{12})$ introduces **electron correlation** into the wave function. That is, it introduces a dependence on the interelectron distance. In a one-term orbital wave function, the probability density of each electron is independent of the position of any other electrons, as in Equation (11.3-24), and there is no electron correlation. Electrons repel each other, and the correct wave function must depend on interelectron distances.

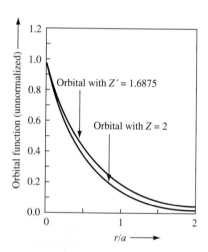

Figure 12.1. Zero-Order and Variationally Obtained Orbitals for the Ground State of the Helium Atom. This diagram shows how the two orbitals compare, with the effect of shielding making the variationally obtained orbital extend farther from the nucleus.

[3] E. A. Hylleraas, *Z. Physik* **65**, 209 (1930).
[4] T. Koga, *J. Chem. Phys.* **94**, 5530 (1991).

The wave function of Equation (12.1-17) represents electron correlation by giving a larger probability density for larger separations of the electrons. This inclusion of explicit dependence on interelectron distance is called **dynamical correlation**. An antisymmetrized orbital wave function also exhibits correlation because it vanishes if two electrons are at the same location and has a small magnitude if they are near each other. This effect is called **statistical correlation**.

Exercise 12.4	Consider the antisymmetrized orbital wave function

$$\Psi = C[\psi_1(1)\psi_2(2) - \psi_2(1)\psi_1(2)]$$

where ψ_1 and ψ_2 are any two distinct space orbitals. Show that the wave function vanishes if both electrons are at the same location.

12.2   The Perturbation Method and Its Application to the Helium Atom

This method is the second of the three most commonly used approximation schemes. It is based on the use of power series. If a function $f(x)$ obeys certain conditions that ensure convergence, it can be represented by the power series

$$f(x) = a_0 + a_1 x + a_2 x^2 + a_3 x^3 + \cdots \qquad \text{(12.2-1)}$$

where the coefficients $a_0, a_1, a_2, \ldots$ are independent of x. For a completely accurate representation, infinitely many terms must generally be included, but for fairly small values of x an adequate approximation can often be obtained with a partial sum of only a few terms.

The perturbation method is applied to a problem in which the Hamiltonian operator can be separated into two terms

$$\hat{H} = \hat{H}^{(0)} + \hat{H}' \qquad \text{(12.2-2)}$$

such that $\hat{H}^{(0)}$ gives a Schrödinger equation that can be solved:

$$\hat{H}^{(0)}\Psi^{(0)} = E^{(0)}\Psi^{(0)} \qquad \text{(12.2-3)}$$

This equation is called the **unperturbed equation** or **zero-order equation**. The wave function $\Psi^{(0)}$ and the energy eigenvalue $E^{(0)}$ are called the **zero-order wave function** and the **zero-order energy eigenvalue**. The zero-order orbital approximation that we used in Chapter 11 is such a problem.

The term $\hat{H}'$ in the Hamiltonian operator is called the **perturbation**. The best results are obtained if the perturbation term is small compared to other terms in the Hamiltonian operator. We will treat the interelectron repulsion term in the helium atom Hamiltonian as a perturbation. The condition of smallness is not met in the helium atom, for which the interelectron repulsion is nearly as large as the attraction of an electron for the nucleus, so our results will not be very accurate.

The first step in the perturbation method is to construct a new Hamiltonian operator in which the perturbation term is multiplied by a fictitious parameter, λ:

$$\hat{H} = \hat{H}^{(0)} + \lambda \hat{H}' \qquad \text{(12.2-4)}$$

The new Schrödinger equation is

$$\hat{H}(\lambda)\Psi(\lambda) = E(\lambda)\Psi(\lambda) \qquad \text{(12.2-5)}$$

where the energy eigenvalue and the energy eigenfunction now depend on λ.

When we let λ equal unity, this new problem becomes the same as the original problem. If we could solve the new problem, its solution would become that of the original problem by letting $\lambda = 1$. It seems at first that we are further complicating an already intractable problem by introducing a new independent variable. If we cannot solve the problem for λ equal to unity, the problem for variable λ would seem even less likely to be solved. However, we do not always have to obtain a complete solution as a function of λ in order to obtain useful information.

Consider a particular energy eigenfunction, say Ψ_n, and its energy eigenvalue, E_n. We will now assume that the state corresponding to this eigenfunction is nondegenerate. That is, there is only one eigenfunction whose eigenvalue is equal to E_n. This assumption is true for the ground state of the helium atom but not for the higher-energy states, which require a more complicated discussion.

We assume that the energy eigenvalues and energy eigenfunctions can be represented by a power series in λ:

We use superscripts on the coefficients instead of subscripts because we already have subscripts on our eigenvalues and eigenfunctions.

$$E_n = E_n^{(0)} + E_n^{(1)}\lambda + E_n^{(2)}\lambda^2 + \cdots \qquad \text{(12.2-6)}$$
$$\Psi_n = \Psi_n^{(0)} + \Psi_n^{(1)}\lambda + \Psi_n^{(2)}\lambda^2 + \cdots \qquad \text{(12.2-7)}$$

The idea of the perturbation method is to obtain only a few coefficients (often just two) in Equations (12.2-6) and (12.2-7) and to hope that a partial sum containing these terms gives an adequate approximation to the entire series when we let $\lambda = 1$. Figure 12.2 shows schematically a typical energy eigenvalue as represented by the first two partial sums of the series for values of λ between zero and unity.

We must assume that as λ varies between zero and unity, the states change gradually, with no new states appearing or old states disappearing, so that there is a one-to-one correspondence between the correct states and the approximate zero-order states. We must also assume that the power series converges. It is not ordinarily possible to prove that either of these conditions is met, but comparison with experiment indicates that there is no reason to avoid using the method.

We would like to base everything we do on things that have been established with certainty, but this goal is almost never achieved.

When the power series of Equations (12.2-6) and (12.2-7) are substituted into the time-independent Schrödinger equation we obtain

$$(\hat{H}^{(0)} + \lambda \hat{H}')(\Psi_n^{(0)} + \Psi_n^{(1)}\lambda + \Psi_n^{(2)}\lambda^2 + \cdots)$$
$$= (E_n^{(0)} + E_n^{(1)}\lambda + E_n^{(2)}\lambda^2 + \cdots)(\Psi_n^{(0)} + \Psi_n^{(1)}\lambda + \Psi_n^{(2)}\lambda^2 + \cdots) \qquad \text{(12.2-8)}$$

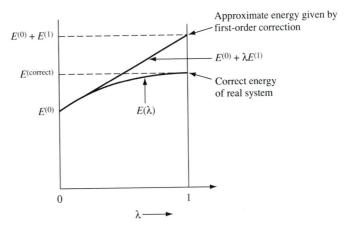

Figure 12.2. An Energy Eigenvalue as a Function of λ for a Hypothetical System. This diagram shows how the fictitious parameter λ is used in the perturbation method. Only the value $\lambda = 1$ has physical meaning, corresponding to the actual system. $\lambda = 0$ corresponds to a soluble problem with part of the Hamiltonian operator omitted (the zero-order problem).

If two power series are equal, one can differentiate both series n times and then set the independent variable equal to zero. The coefficient of the nth-degree term is all that survives on each side.

When two power series in the same independent variable are equal to each other for all values of the independent variable, the coefficients of the same power in the two series are equal. We multiply out the products in Equation (12.2-8) and then equate the constant terms on the two sides of the equation, after which we equate the coefficients of the linear terms, etc. The constant (zero-order) terms obey the relation

$$\hat{H}^{(0)}\Psi_n^{(0)} = E_n^{(0)}\Psi_n^{(0)} \tag{12.2-9}$$

which is the same as Equation (12.2-3) except that we have now specified a particular energy eigenfunction and eigenvalue.

The coefficients of the linear (first-order) terms obey the relation

$$H'\Psi_n^{(0)} + H^{(0)}\Psi_n^{(1)} = E_n^{(1)}\Psi_n^{(0)} + E_n^{(0)}\Psi_n^{(1)} \tag{12.2-10}$$

Exercise 12.5   Verify Equations (12.2-9) and (12.2-10).

Any function obeying the same boundary conditions as a complete set of functions can be represented by a linear combination such as Equation (12.2-11). It is believed that the eigenfunctions of any hermitian operator form a complete set.

To obtain an equation for $E_n^{(1)}$, the first-order correction to the energy, we express the first-order correction to the wave function as a linear combination of the unperturbed (zero-order) wave functions:

$$\Psi_n^{(1)} = \sum_{j=1}^{\infty} a_{nj}\Psi_j^{(0)} \tag{12.2-11}$$

where we assume that there are infinitely many zero-order wave functions. Since $\hat{H}^{(0)}$ is a hermitian operator, its eigenfunctions presumably form a complete set, allowing this to be an exact representation if all of the infinitely many eigenfunctions are included. This representation is substituted into Equation (12.2-10), giving

$$\hat{H}'\Psi_n^{(0)} + \hat{H}^{(0)} \sum_{j=1}^{\infty} a_{nj}\Psi_j^{(0)} = E_n^{(1)}\Psi_n^{(0)} + E_n^{(0)} \sum_{j=1}^{\infty} a_{nj}\Psi_j^{(0)} \tag{12.2-12}$$

Since $\Psi_j^{(0)}$ is an eigenfunction of $\hat{H}^{(0)}$ with eigenvalue $E_j^{(0)}$,

$$\hat{H}'\Psi_n^{(0)} + \sum_{j=1}^{\infty} a_{nj}E_j^{(0)}\Psi_j^{(0)} = E_n^{(1)}\Psi_n^{(0)} + E_n^{(0)}\sum_{j=1}^{\infty} a_{nj}\Psi_j^{(0)} \quad \textbf{(12.2-13)}$$

We now multiply each term of Equation (12.2-13) by $\Psi_n^{(0)*}$ and integrate over all coordinates on which the wave function depends. Since the zero-order energy eigenfunctions are orthogonal to each other if they belong to different eigenvalues, every integral in the sums vanishes except the one for which $n = j$. This integral equals unity if the zero-order wave functions are normalized. We now have only one term surviving in each sum:

The technique of multiplying by a member of an orthonormal set of functions and integrating is often very useful in obtaining an equation for coefficients in a linear combination of functions.

$$\int \Psi_n^{(0)*}\hat{H}'\Psi_n^{(0)}\,dq + a_{nn}E_n^{(0)} = E_n^{(1)} + a_{nn}E_n^{(0)}$$

where we abbreviate the coordinates of the system by q. The second term on each side cancels.

The result for the first-order correction to the energy eigenvalue is

$$\boxed{E_n^{(1)} = \int \Psi_n^{(0)*}\hat{H}'\Psi_n^{(0)}\,dq} \quad \textbf{(12.2-14)}$$

For the ground state of the helium atom,

$$\hat{H}^{(0)} = \hat{H}_{\mathrm{HL}}(1) + \hat{H}_{\mathrm{HL}}(2) \quad \textbf{(12.2-15)}$$

$$\hat{H}' = \frac{e^2}{4\pi\varepsilon_0 r_{12}} \quad \textbf{(12.2-16)}$$

The zero-order Hamiltonian operator is just like that of Equation (11.2-2), so $E_{1s1s}^{(0)}$ is given by Equation (11.2-12) and $\Psi_{1s1s}^{(0)}$ is given by Equation (11.3-13). Integration over the spin coordinates in Equation (12.2-14) yields

$$E_{1s1s}^{(1)} = \int \psi_{100}(1)^*\psi_{100}(2)^*\left(\frac{e^2}{4\pi\varepsilon_0 r_{12}}\right)\psi_{100}(1)\psi_{100}(2)\,d^3\mathbf{r}_1\,d^3\mathbf{r}_2 \quad \textbf{(12.2-17)}$$

The first-order correction to the energy depends only on the zero-order wave function, so it is not surprising that the result to first order is the same as the variation result using the zero-order wave function.

This result is the same as the integral in Equation (12.1-5), so our perturbation method result to first order is the same as the result we obtained with the variation method using the unmodified zero-order wave function as our variation function:

$$E_{1s1s}^{(0)} + E_{1s1s}^{(1)} = -108.8 \text{ eV} + 34.0 \text{ eV} = -74.8 \text{ eV} \quad \textbf{(12.2-18)}$$

The first-order correction to the wave function and the second-order correction to the energy eigenvalue are more complicated than the first-order correction to the energy eigenvalue, and we do not discuss them. No exact calculation of the second-order correction to the energy of the helium atom has been made, but a calculation made by a combination of the perturbation and variation methods gives an accurate upper bound:[5]

$$E_{1s1s}^{(2)} = -4.3 \text{ eV} \quad \textbf{(12.2-19)}$$

---

[5] C. W. Scherr and R. E. Knight, *Rev. Mod. Phys.* **35**, 436 (1963).

Thus the second-order value of the energy is -79.1 eV, within 0.1 eV of the experimental value, -79.0 eV. Approximate calculations through thirteenth order have been made and have given values that agree with experiment nearly as well as the best results of the variation method.[6]

12.3   The Self-Consistent Field Method

The third general approximation method is the **self-consistent field** (SCF) method introduced in 1928 by Hartree.[7] The goal of the self-consistent field method is similar to that of the variation method. It is to find the best possible orbitals. However, the search is not restricted to any particular family of functions, but allows the form of the orbital functions to be varied. It is therefore capable of finding the best possible orbital approximation. The SCF method is extensively used in modern quantum chemistry.

We illustrate this method by applying it to the ground state of the helium atom. Just as in the variation method and the perturbation method, we begin with the zero-order solution of Equation (11.3-27). The ground state of helium is a singlet state, and the antisymmetrization is in the spin factor of the wave function. We can proceed with the space factor of the wave function, omitting the spin factor, since the Hamiltonian contains no spin dependence.

The zero-order orbitals satisfied Equations (11.3-9) and (11.3-10), which omit the potential energy of electron-electron repulsion. We add a correction term to Equation (11.3-9) to represent this potential energy. If electron 2 were fixed at location $\mathbf{r}_2$,

$$-\frac{\hbar^2}{2m}\nabla_1^2\psi(1) - \frac{Ze^2}{4\pi\varepsilon_0 r_1}\psi_1(1) + \frac{e^2}{4\pi\varepsilon_0 r_{12}}\psi_1(1) = E_1\psi_1(1) \qquad \textbf{(12.3-1)}$$

where r_{12} is the distance between the fixed position of electron 2 and the variable position of electron 1 and where E_1 is a new orbital energy.

If electron 2 is not at a fixed position but occupies the normalized orbital $\psi_2(2)$, then its probability of being found in the volume element $d^3\mathbf{r}_2$ is

$$\text{Probability} = \psi_2(2)^*\psi_2(2)\,d^3\mathbf{r}_2 = |\psi_2(2)|^2\,d^3\mathbf{r}_2 \qquad \textbf{(12.3-2)}$$

We now replace the electron-electron term in the Hamiltonian of Equation (12.3-1) by a weighted average over all positions of electron 2, obtaining

$$-\frac{\hbar^2}{2m}\nabla_1^2\psi(1) - \frac{Ze^2}{4\pi\varepsilon_0 r_1}\psi_1(1) + \left[\int\frac{e^2}{4\pi\varepsilon_0 r_{12}}|\psi_2(2)|^2\,d^3\mathbf{r}_2\right]\psi_1(1)$$
$$= E_1\psi_1(1) \qquad \textbf{(12.3-3)}$$

After the integration, the integral term depends only on the coordinates of electron 1, so this equation can be solved if the orbital for electron 2 is a known function. For the ground state of the helium atom, both $\psi_1(1)$ and $\psi_2(2)$ are $1s$ orbitals, so both orbitals in Equation (12.3-3) are the same unknown function ψ_{1s}. This type of equation is called an **integrodifferential equation**.

In the self-consistent field method, the integrodifferential equation is solved by successive approximations, a method also called **iteration**. The first step is to

[6] *ibid.*
[7] D. R. Hartree, *Proc. Cambridge Philos. Soc.* **24**, 89, 111, 426 (1928).

If the second electron is assumed to be governed by the zero-order solution, solving the wave equation for the first electron should give a better result than neglecting the repulsion of the second electron entirely. Repetition of the method with successive approximations converges to the best possible orbital approximation.

replace the orbital under the integral by the zero-order function we obtained earlier. The $\psi_{1s}(1)$ orbital that results from solving this equation is called the first-order solution $\psi_{1s}^{(1)}(1)$. The equation it obeys is

$$-\frac{\hbar^2}{2m}\nabla_1^2\psi_{1s}^{(1)}(1) - \frac{Ze^2}{4\pi\varepsilon_0 r_1}\psi_{1s}^{(1)}(1) + \left[\int\frac{e^2}{4\pi\varepsilon_0 r_{12}}|\psi_{1s}^{(0)}(2)|^2\,d^3\mathbf{r}_2\right]\psi_{1s}^{(1)}$$

$$= E_{1s}^{(1)}\psi_{1s}^{(1)}(1) \tag{12.3-4}$$

where $E_{1s}^{(1)}$ is a new approximation to the orbital energy.

It is found that the integral term depends only on r_1, not on θ_1 and ϕ_1, so only a one-dimensional problem needs to be solved. It is ordinarily not possible to solve Equation (12.3-4) analytically, but an accurate numerical representation of $\psi_{1s}^{(1)}(1)$ can be obtained.

The next iteration (repetition) in the method of successive approximations is carried out by replacing $\psi_{1s}^{(0)}(2)$ under the integral sign by $\psi_{1s}^{(1)}(2)$ and denoting the new unknown function by $\psi_{1s}^{(2)}(1)$. This equation is solved, and the resulting solution is used under the integral for the next iteration, and so forth. The equation for the jth iteration is

$$-\frac{\hbar^2}{2m}\nabla_1^2\psi_{1s}^{(j)}(1) - \frac{Ze^2}{4\pi\varepsilon_0 r_1}\psi_{1s}^{(j)}(1) + \left[\int\frac{e^2}{4\pi\varepsilon_0 r_{12}}|\psi_{1s}^{(j-1)}(2)|^2\,d^3\mathbf{r}_2\right]\psi_{1s}^{(j)}(1)$$

$$= E_{1s}^{(j)}\psi_{1s}^{(j)}(1) \tag{12.3-5}$$

When additional iterations produce only negligible changes in the orbital function and the energy, we say that the integral term provides a self-consistent contribution to the force on electron 1, or a **self-consistent field**. At this point, the iteration is stopped.

In the SCF method, the expectation value of the energy is not the sum of the orbital energies, because the entire potential energy of electron-electron repulsion has been included in Equation (12.3-3) for each orbital. Since both orbitals are obtained from this equation, the sum of the two orbital energies includes the interelectron repulsion energy twice. We correct for this double inclusion by subtracting the expectation value of the interelectron repulsion energy from the sum of the orbital energies. If n iterations have been carried out, the expectation value of the energy is

$$E(\text{atom}) = 2E_{1s}^{(n)} - \int\frac{e^2}{4\pi\varepsilon_0 r_{12}}|\psi_{1s}^{(n)}(1)|^2|\psi_{1s}^{(n)}(2)|^2\,d^3\mathbf{r}_1\,d^3\mathbf{r}_2$$

$$= 2E_{1s}^{(n)} - J_{1s1s} \tag{12.3-6}$$

The integral J_{1s1s} is called a **Coulomb integral** because it represents an approximate expectation value of a Coulomb (electrostatic) repulsion energy.

Clementi and Roetti expressed the unknown orbitals as a linear combination of **Slater-type orbitals** (STOs), which are exponential radial functions similar to $1s$ orbitals with variable effective nuclear charges multiplied by powers of r, combined with the correct spherical harmonic angular functions. Using this expression instead of a numerical representation to evaluate the integrals in the self-consistent-field method, they obtained an energy for the ground state of the helium atom equal to -77.9 eV.[8]

---

[8] E. Clementi and C. Roetti, *At. Data. Nucl. Data Tables* **14**, 177 (1974).

The self-consistent field method can produce the best possible orbital wave function. The difference between the best energy calculated with an orbital wave function and the correct nonrelativistic energy is called the **correlation energy**. In the next chapter we will describe the configuration interaction method for improving on orbital wave functions.

12.4

Excited States of the Helium Atom

The variation method in its original version applies only to the ground state. However, an **extended variation theorem** states that the calculated variational energy will be no lower than the correct energy of the first excited state if the variation trial function is orthogonal to the correct ground-state energy eigenfunction. It will be no lower than the energy of the second excited state if the variation trial function is orthogonal to both the ground state and the first excited state, etc.[9] Unfortunately, the correct ground-state energy eigenfunction is not generally known, so a family of functions exactly orthogonal to it cannot be chosen. In some calculations a family of functions has been chosen that is orthogonal to an approximate ground-state variation function. With luck, this family of functions is nearly orthogonal to the correct ground-state function, and the minimum variational energy from this family might be a good approximation to the energy of the first excited state.

The perturbation method as described earlier in this chapter must be modified to apply to excited states of the helium atom. In the form we outlined, it does not apply to a zero-order state that has the same energy as other zero-order states of the system (the degenerate case). The excited energy levels of the helium atom are degenerate in zero order. For example, the zero-order orbital energies of the $2s$ and $2p$ hydrogen-like orbitals are all equal, so all of the zero-order states of the $(1s)(2s)$ and $(1s)(2p)$ helium configurations have the same energy. A version of the perturbation method has been developed to handle the degenerate case. We will describe this method only briefly and present some results for the first excited states of the helium atom.[10]

The first task of the degenerate perturbation method is to find the **correct zero-order wave functions**, the ones that are in one-to-one correspondence with the exact wave functions. "Correct zero-order function" means that as λ is increased from zero to unity, each correct zero-order function smoothly turns into one of the exact functions without getting mixed up with other functions. We express the correct zero-order wave functions as linear combinations of the degenerate "initial" zero-order wave functions:

$$\Psi_{n(\text{new})}^{(0)} = \sum_{j=1}^{g} c_{nj}\Psi_{j}^{(0)} \qquad (12.4\text{-}1)$$

[9] Levine, *op. cit.*, pp. 176ff.
[10] *ibid.*, pp. 209ff.

In order to find the c coefficients that define the correct zero-order functions, one must solve a set of homogeneous linear simultaneous equations. An equation that must be satisfied for a nontrivial solution to exist is called a **secular equation**.[11] We do not discuss secular equations. However, solution of the secular equation gives the first-order corrections to the energies as well as allowing solution of the equations for the c coefficients for each correct zero-order function. It is found that the wave functions of Equation (11.3-38) are the correct zero-order functions for the $(1s)(2s)$ configuration and that three sets of similar functions are the correct zero-order functions for the $(1s)(2p)$ configuration.

Figure 12.3 shows the results of calculations to first order and to third order for the energies of the four levels that result.[12] We observe the follow-

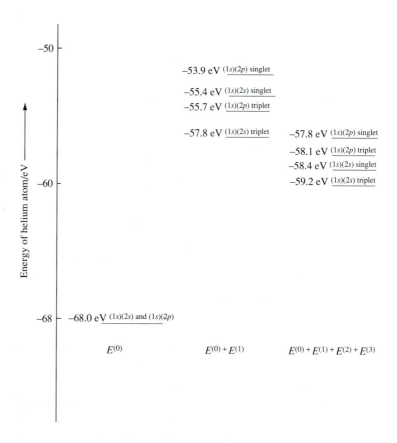

Figure 12.3. Approximate Energies of Helium Excited States. These excited states are the same as those found in perturbation theory. It is assumed that perturbation theory is capable of finding all of the actual states, even if the correct energies cannot be calculated exactly.

[11] *ibid.*, p. 189.
[12] *ibid.*, pp. 223–224.

Here are some experimental facts that we can use in understanding the periodic table.

ing facts: (1) Each triplet state has a lower energy than the corresponding singlet state. (2) The $(1s)(2s)$ configuration gives states of lower energy than the $(1s)(2p)$ configuration. That is, the orbital energies of the $2p$ subshell are higher than the orbital energies of the $2s$ subshell. The same behavior is found by experiment to be generally true for various atoms, and it is also found in higher shells that the orbital energies of a d subshell lie higher than the orbital energies of the p subshell in the same shell, etc.

It is possible to explain the difference in the subshell energies on the basis of shielding. According to an optimized variational wave function, an electron in a ground-state helium atom moves as though the nuclear charge were reduced, due to the shielding of the positive nuclear charge by the negative charge of the other electron. An electron in a $2s$ orbital spends more time close to the nucleus than one in a $2p$ orbital, as shown in the radial distribution functions of Figure 11.13b. An electron in a $2s$ orbital will experience less shielding and its energy will be lower than one in a $2p$ orbital, in agreement with the results shown in Figure 12.3.

It is possible to explain the difference in the energies of the singlet and triplet states on the basis of statistical correlation. The singlet state wave functions all have symmetric space factors because the spin factors are antisymmetric. Statistical correlation is not found in symmetric space factors. In the triplet state wave functions, the spin factor is symmetric, so the space factor is antisymmetric, giving statistical correlation. The electrons have lower probability of being found close together than of being far apart when the system is in a triplet state. Since close proximity of two electrons corresponds to higher potential energy, a triplet state has lower energy than a singlet state with the same space orbitals. We will use these explanations involving shielding and statistical correlation again in discussing multielectron atoms.

> Our analysis is oversimplified. It also is found that the antisymmetric space factor corresponds to a lower probability that the electrons will be far apart, as well as to a lower probability that they will be close together. There is also a higher probability that the electrons will be found close to the nucleus, and this fact may be a controlling factor.[13]

Exercise 12.6   Sketch a qualitative energy level diagram for the excited states of the $(1s)(3s)$, $(1s)(3p)$, and $(1s)(3d)$ configurations of the helium atom.

The self-consistent field method must also be modified in order to treat excited states of the helium atom, because two different orbital functions are involved. Two simultaneous integrodifferential equations must be solved by iteration. Furthermore, an antisymmetrized wave function requires two terms in the space factor of the wave function. The original self-consistent

---

[13] *ibid.*, p. 280.

field method of Hartree did not provide for antisymmetrization. The method was modified by Fock[14] to include antisymmetrization.

12.5   Atoms with More Than Two Electrons

Higher-Order Approximations for the Lithium Atom

In Chapter 11, we wrote the zero-order wave function for the ground state of the lithium atom. An application of the variation method to the lithium atom ground state uses an orbital wave function containing hydrogen-like orbitals with variable effective nuclear charges. The minimum in the variational energy, -201.2 eV, is found to occur with effective nuclear charges of 2.686 protons for the $1s$ orbitals and 1.776 protons for the $2s$ orbital.[15] This variational energy differs from the correct value, -203.5 eV, by 1%.

The difference in the two effective nuclear charges corresponds to the fact that an electron occupying a $2s$ orbital is on the average farther from the nucleus than an electron occupying a $1s$ orbital, so there is a larger probability that other electrons are found between it and the nucleus than is the case with a $1s$ electron. The effective charge for the $2s$ orbital corresponds to 1.224 electrons being found between the nucleus and the $2s$ electron, while a $1s$ electron appears to have 0.314 electron between itself and the nucleus. The $2s$ electron is more effectively shielded from the nuclear charge by the other electrons than are the $1s$ electrons.

The effective nuclear charge seen by the $1s$ electrons is nearly the same as seen by the $1s$ electrons in a helium-like atom with three protons in the nucleus, since the minimum in the variational energy of Equation (12.1-16) occurs at $Z' = 2.6875$ if $Z = 3$. A $1s$ electron is shielded primarily by the other $1s$ electron and sees almost no shielding due to the $2s$ electron. Since the $2s$ electron on the average is found farther away from the nucleus than the $1s$ electron, this result is reasonable.

In further variational calculations the $2p$ orbital is found to be higher in energy than the $2s$ orbitals, so the ground configuration is $(1s)^2(2s)$, not $(1s)^2(2p)$. The $2p$ electron is more effectively screened from the nuclear charge than is a $2s$ electron. However, an electron in a $2p$ orbital is not on the average farther from the nucleus than one in a $2s$ orbital for the same nuclear charge. (See Problem 11.33.) It is not just the average distance from the nucleus but the entire radial probability distribution that determines the effectiveness of the shielding. An electron in a $2s$ orbital has a considerable probability of being found close to the nucleus, where the shielding is least effective. Figure 11.9 shows that the $2s$ orbital is nonzero for $r = 0$, whereas the $2p$ orbitals vanish for $r = 0$. Figure 11.13b shows that the radial probability distribution for the $2s$ orbital has a "hump" close to the nucleus,

---

[14] V. Fock, *Z. Phys.* **61**, 126 (1930).
[15] Levine, *op. cit.*, pp. 252ff.

which the $2p$ orbital does not have. It is said that the $2s$ orbital is more "penetrating" toward the nucleus than are the $2p$ orbitals.

The **ionization potential** can be used to obtain an estimate of the effective nuclear charge for the outermost electron. The first ionization potential is defined as the energy required to remove one electron from an isolated neutral atom. If the orbitals for the other electrons are not changed much by the removal of one electron, the ionization potential is nearly equal to the energy of the orbital occupied by the outermost electron. In the case of lithium, the effective charge seen by the $1s$ electrons is nearly unaffected by the presence of the $2s$ electron, so this condition is fairly well met.

▼   **EXAMPLE 12.1**

Find the effective nuclear charge seen by the $2s$ electron in lithium from the ionization potential, which is 124 kcal mol^{-1}.

Solution

The ionization potential in electron volts is

$$IP = \frac{(124000 \text{ cal mol}^{-1})(4.184 \text{ J cal}^{-1})}{96485 \text{ J mol}^{-1} \text{ eV}^{-1}} = 5.38 \text{ eV}$$

The energy of the $2s$ orbital is given by Equation (11.2-10b) as

$$E_2 = -\frac{(13.6 \text{ eV})Z'^2}{4}$$

where Z' is the effective nuclear charge. Setting this energy equal to 5.38 eV gives $Z' = 1.26$, which is only roughly in agreement with the value of 1.776 obtained by the variational calculation.

▲

The perturbation method can also be applied to the lithium atom. The first-order correction to the ground-state energy is 83.5 eV, resulting in an energy through first order equal to -192.0 eV. This value is considerably less accurate than the value obtained by the simple variational calculation.[16]

The Hartree-Fock method is the most successful of the three common approximation methods. It leads to a ground-state energy of -202.3 eV, differing from the correct value by only 0.6%.[17] This error is presumably the correlation energy.

One way to include dynamical electron correlation in an orbital wave function is to construct a wave function that is a linear combination of several Slater determinants corresponding to different configurations. For example, for the ground state of the lithium atom, one could use

$$\Psi = c_1\Psi_{1s1s2s} + c_2\Psi_{1s2s2s} + c_3\Psi_{1s1s3s} + \cdots \tag{12.5-1}$$

where c_1, c_2, c_3, etc are variable parameters and the Ψ's represent Slater determinant wave functions with the given configurations. The variational

[16] *ibid.*
[17] F. L. Pilar, *Elementary Quantum Chemistry*, McGraw-Hill, New York, 1968, p. 336.

energy is minimized with respect to these parameters. Although it is not obvious from inspection of Equation (12.5-1) that Ψ includes dynamical correlation, it does in fact depend on interelectron distances.

The procedure of adding orbital wave functions corresponding to excited configurations is called **configuration interaction**, abbreviated CI. Unfortunately, the process converges slowly, so many configurations must be used to get good accuracy. Using large computers, atomic and molecular calculations have been made with as many as a million configurations.

Atoms with More Than Three Electrons

The higher-order approximate treatment of the other atoms is similar to the helium and lithium treatments. All three approximation schemes can be applied. The most accurate work has been done with the Hartree-Fock method. The optimum orbitals appear to be in one-to-one correspondence to the hydrogen-like orbitals. As with helium, the optimum energy is expressed as a sum of orbital energies with suitable Coulomb integral corrections.

Figure 12.4 shows approximate orbital energies in neutral atoms, obtained by an approximation scheme called the Thomas-Fermi method. This method gives orbital energies that generally agree with those from the Hartree-Fock method. Notice that logarithmic scales are used in the figure. Several things are apparent: First, the orbitals in the same shell but in different subshells have different energies, with higher values of ℓ corresponding to higher energies; second, all of the orbitals in a given subshell have the same energy; third, the energies depend strongly on the nuclear charge, with some pairs of curves crossing and recrossing as a function of the nuclear charge.

The energy differences between subshells in the same shell can be ascribed to differences in shielding. An electron in an s orbital spends more time close to the nucleus than an electron in a p orbital and is less effectively shielded from the nucleus by other electrons, giving it a lower orbital energy. Similarly, an electron in a p orbital is less effectively shielded than an electron in a d orbital, etc. All of the orbitals in a subshell have the same orbital energy because they all contain the same radial factor.

Using Figure 12.4 or some equivalent source of orbital energies, it is now possible to determine the ground-level configuration for any neutral atom, using the Aufbau principle. For the first 18 elements, the subshell energies lie in the increasing order $1s$, $2s$, $2p$, $3s$, $3p$. For example, the configuration of the ground state of argon is $(1s)^2(2s)^2(2p)^6(3s)^2(3p)^6$. From Figure 12.4 we see that beyond atomic number 15 the $3d$ orbital energy is higher than that of $4s$. Therefore, elements 19, potassium, and 20, calcium, in their ground states have the $4s$ orbitals occupied in preference to the $3d$ orbitals. Beyond atomic number 23, the figure shows the $4s$ energy above the $3d$ energy. However, it is found experimentally that most of the transition elements from scandium, element 23, through zinc, element 30, have two electrons occupying the $4s$ spin orbitals in their ground levels, although chromium, element 24, and copper, element 29, have only one $4s$ electron.

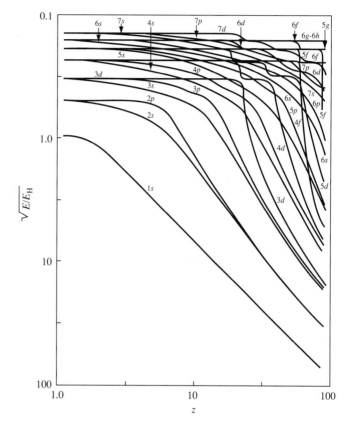

Figure 12.4. Approximate Orbital Energies in Neutral Atoms. The axes in this diagram are logarithmic. The order of occupation of subshells can only approximately be determined from this diagram. From R. Latter, *Phys. Rev.* **99**, 510 (1955).

Assuming the energies in the figure to be essentially correct, we must conclude that other factors besides orbital energy, principally the repulsions between electrons, are important in determining the ground-level configuration.

The correct ground-level configuration for most elements can be obtained from the scheme of Figure 12.5, which shows the **diagonal mnemonic device** or the **diagonal rule**. To determine the order of orbitals for the Aufbau principle, one draws diagonals from upper right to lower left through the subshell symbols in the array, as shown. One then traverses these diagonals, moving top to bottom from one diagonal to the next but traversing each diagonal from upper right to lower left, as shown by the path in the figure. The number of spin orbitals in each subshell is listed at the top of the figure, so one can tell when enough subshells have been chosen to be occupied by the electrons of a given atom.

The diagonal mnemonic device is equivalent to the "$n + \ell$ rule," which states that subshells of a given value of $n + \ell$ are occupied before those of the next higher value of $n + \ell$ and that within a given value of $n + \ell$, the subshells are occupied in order of increasing n.

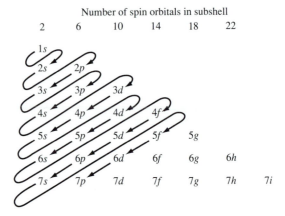

Figure 12.5. The Diagonal Mnemonic Device for Applying the Aufbau Principle to Neutral Atoms. Following the arrows in the diagram gives the order of occupation of subshells in neutral atoms. Exceptions are given in Table 12.1.

EXAMPLE 12.2

Give the ground-level configuration of (a) Al and (b) Mn.

Solution

    a. Al (13 electrons): $(1s)^2(2s)^2(2p)^6(3s)^2(3p)$

    b. Mn (25 electrons): $(1s)^2(2s)^2(2p)^6(3s)^2(3p)^6(3d)^5(4s)^2$

Configurations are often abbreviated by giving an inert gas configuration as part of them. The Mn configuration is sometimes abbreviated $[Ar](3d)^5(4s)^2$, where $[Ar]$ stands for the ground-level configuration of argon.

Exercise 12.7

Give the ground-level configuration of the elements: (a) Se (b) Nb (c) Pb

Table 12.1 lists the known exceptions to the diagonal rule through element 103. There has been some discussion about whether lanthanum and actinium are also exceptions to the rule. Lanthanum has been assigned a $5d$ electron in the ground level, and actinium has been assigned a $6d$ electron. However, from an analysis of spectroscopic observations, it has been concluded that lanthanum has a $4f$ electron in its ground state and that actinium has a $5f$ electron, as predicted by the diagonal rule.[18]

The diagonal mnemonic rule does not necessarily apply to ions, since the shielding is different for ions than for neutral atoms. For example, the iron atom has six $3d$ electrons and two $4s$ electrons, in conformity with the diagonal rule. The Ni^{2+} ion, with the same number of electrons, has eight $3d$ electrons and no $4s$ electrons. The correct electron configuration for pos-

---

[18] W. B. Jensen, *J. Chem. Educ.* **59**, 635 (1982).

Table 12.1. Exceptions to the Diagonal Mnemonic Rule

Atomic No.	Symbol	Ground Configuration
24	Cr	$[\text{Ar}](3d)^5(4s)^1$
29	Cu	$[\text{Ar}](3d)^{10}(4s)^1$
41	Nb	$[\text{Kr}](4d)^4(5s)^1$
42	Mo	$[\text{Kr}](4d)^5(5s)^1$
44	Ru	$[\text{Kr}](4d)^7(5s)^1$
45	Rh	$[\text{Kr}](4d)^8(5s)^1$
46	Pd	$[\text{Kr}](4d)^9(5s)^1$
47	Ag	$[\text{Kr}](4d)^{10}(5s)^1$
64	Gd	$[\text{Xe}](4f)^7(5d)^1(6s)^2$
65	Tb	$[\text{Xe}](4f)^8(5d)^1(6s)^2$
66	Dy	$[\text{Xe}](4f)^9(5d)^1(6s)^2$
67	Ho	$[\text{Xe}](4f)^{10}(5d)^1(6s)^2$
68	Er	$[\text{Xe}](4f)^{11}(5d)^1(6s)^2$
78	Pt	$[\text{Xe}](4f)^{14}(5d)^9(6s)^1$
79	Au	$[\text{Xe}](4f)^{14}(5d)^{10}(6s)^1$
90	Th	$[\text{Rn}](6d)^2(7s)^1$
91	Pa	$[\text{Rn}](5f)^2(6d)^1(7s)^2$
92	U	$[\text{Rn}](5f)^3(6d)^1(7s)^2$
93	Np	$[\text{Rn}](5f)^4(6d)^1(7s)^2$
94	Cm	$[\text{Rn}](5f)^7(6d)^1(7s)^2$
103	Lw	$[\text{Rn}](5f)^{14}(6d)^1(7s)^2$

itive ions can usually be obtained by finding the configuration of the neutral atom and then first removing electrons from the outer shell, instead of removing the last electrons added according to the diagonal rule.

For elements with partially filled subshells, the detailed configuration and the values of the quantum numbers L and S of the ground level can be predicted, using rules due to Hund. **Hund's first rule** is: *For the same value of L, the level with the largest value of S has the lowest energy*. **Hund's second rule** is: *For a given value of S, the level with the largest value of L has the lowest energy*.

Hund's first rule has precedence over his second rule.

Hund's second rule is applied only after the first rule has been applied. These rules are quite reliable for ground levels, but less reliable for other levels.[19] There is also a third rule, which states that for subshells that are more than half filled, higher values of J correspond to lower energies, and for subshells that are less than half filled, lower values of J correspond to lower energies.

---

[19] Levine, *op. cit.*, p. 280.

With several electrons, the operators for the squares of the total orbital and spin angular momentum are more complicated than with two electrons, and we will not discuss them. However, the operators for the z components are sums of the one-electron operators:

$$\hat{L}_z = \sum_{i=1}^{Z} \hat{L}_{iz}, \qquad \hat{S}_z = \sum_{i=1}^{Z} \hat{S}_{iz} \qquad \qquad \text{(12.5-2)}$$

The consequence is

$$M_L = \sum_{i=1}^{Z} m_i \qquad \qquad \text{(12.5-3)}$$

$$M_S = \sum_{i=1}^{Z} m_{si} \qquad \qquad \text{(12.5-4)}$$

For any given detailed configuration, the possible values of M_L and M_S can be determined by algebraic addition. The addition is simplified by the fact that contributions to both M_L and M_S from filled subshells vanish. The possible values of L and S and the Russell-Saunders term symbols can be found from the fact that M_L ranges from $-L$ to $+L$ and M_S ranges from $-S$ to $+S$. The ground-level term can then be determined from Hund's rules.

▼   **EXAMPLE 12.3**

Using Hund's first and second rules, find the ground-level term symbol for the nitrogen atom.

Solution

The ground-level configuration is $(1s)^2(2s)^2(2p)^3$. Applying Hund's first rule, we seek the largest value S can have. The filled $1s$ and $2s$ subshells make no net contribution to L or S because the subshells are filled and the electrons are paired. The electrons in the $2p$ subshell can have their spins parallel if they occupy different space orbitals, so the largest value of M_S is $+3/2$ and the smallest is $-3/2$. Therefore, the largest value of S is $3/2$, and this will be the ground-level value.

Since we are looking for values of M_L, we use the space orbitals that are eigenfunctions of the $\hat{L}_z$ operators, the ψ_{2p1}, ψ_{2p0}, and $\psi_{2p,-1}$ orbitals. Each of these is occupied by one electron, so $M_L = 1 + 0 - 1 = 0$, and the only value of L is zero. The term symbol is ^4S (quartet S). There is no need to apply Hund's second rule, since only one value of L occurs.

In the solution to Example 12.3, the filled subshells can be omitted. To find the lowest-energy term, we need to find the largest value of M_S, since this value is equal to the largest value of S.

▲

Exercise 12.8   Find the ground-level term symbols for (a) B, (b) C, and (c) O.

The explanation of Hund's first rule is the same as the explanation for the fact that the triplet levels were lower in energy than the singlet levels in helium, discussed in Section 12.4. The higher values of S correspond to more electrons occupying states of parallel spins, which means that they occupy a larger number of space orbitals. This occupation lowers the probability that the electrons will be found close together, thus lowering the potential energy.

The Periodic Table of the Elements

Dmitri Mendeleev, 1834–1907, Russian chemist who primarily correlated valence and atomic mass, and Julius Lothar Meyer, 1830–1895, German chemist who primarily correlated atomic volume and atomic mass.

The periodic table was invented independently by Mendeleev and Meyer. Both noticed that if the elements were listed in increasing order of atomic mass, there was a repetition, or periodicity, of chemical and physical properties. For example, lithium, sodium, potassium, rubidium, and cesium all form oxides with the formula M_2O and chlorides with the formula MCl, whereas beryllium, magnesium, calcium, strontium, and barium all form oxides with the formula MO and chlorides with the formula MCl_2.

Inside the front cover of this book is a modern periodic table. The elements are listed in order of atomic number, instead of atomic mass, except that some elements are listed separately at the bottom of the table. There are several ways of numbering the columns, and the two most common ways are shown. One scheme, which is supposed to become the standard scheme, is to number the 18 columns from 1 to 18. The other is to number the columns 1A through 8A and 1B through 8B, as indicated. Three columns are grouped together as column 8B. This numbering corresponds closely to the numbering scheme used by Mendeleev, although the A and B columns were not distinguished in his table, which had only eight columns.

The elements in the columns labeled A are called **representative elements**, and those in the columns labeled B are called **transition elements**. The two sets of 14 elements at the bottom of the chart are called **inner transition elements**. Figure 12.6 shows a periodic table in which all elements are listed in order of increasing atomic number.

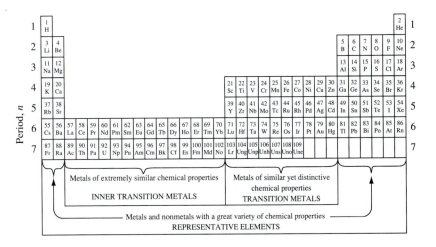

Figure 12.6. A Periodic Table of the Elements. This table is different from the periodic table inside the front cover in that all elements are listed in order of atomic number. The inner transition elements are given their own columns in the body of the table, instead of being placed under the body of the table.

The elements following uranium (U, element 92) in the table do not occur in the earth's crust and have been synthesized in nuclear reactors. The elements after lawrencium, element 103, have been given temporary names: element 104 is unnilquadrium, element 105 is unnilpentium, etc., based on the Latin version of the atomic numbers. These names will be replaced when the discoverers of the elements are officially recognized and choose permanent names.

Many elements were unknown at the time of Mendeleev. These included the inert gases, most of the inner transition elements, and others scattered about the table, such as scandium, gallium, and germanium. However, Mendeleev had sufficient confidence in the periodicity principle that he left blank spaces in the table for undiscovered elements.

Mendeleev listed the elements in order of atomic mass because the concept of atomic number was unknown. There are cases in which a larger atomic mass occurs before a smaller one (Ar and K, Co and Ni, Te and I). However, Mendeleev had an incorrect value for the atomic mass of tellurium; he listed Fe, Co, and Ni together in his column 8; and argon had not been discovered. He might have been unaware of these reversals of order.

The form of the periodic table was first explained by Niels Bohr, who also introduced the modern ("long") form of the chart with 18 columns. The similarity of chemical properties of the elements in a given column is due to the similarity of their electron configurations in the outermost shell (the **valence shell**). For example, sodium and potassium both easily lose one electron because sodium has only one electron in its valence shell (the third shell) and potassium has only one electron in its valence shell (the fourth shell).

The eight columns of representative elements occur as two columns on the left and six columns on the right, corresponding to the two spin orbitals of an s subshell and the six spin orbitals of a p subshell. The transition elements occur in 10 columns, corresponding to the 10 spin orbitals of a d subshell, and the inner transition elements occur in 14 columns, corresponding to the 14 spin orbitals of an f subshell.

The general chemical behavior of an element can be predicted from its first ionization potential and its electron affinity. The **electron affinity** is the energy required to remove the extra electron from a singly charged negative ion of the element. It is therefore equal to the amount of energy given off in forming a negative ion and is positive if a gaseous atom spontaneously attracts an electron. Elements with relatively high ionization potential also have relatively high electron affinities (except for the inert gases). Those with relatively small values of the ionization potential tend to lose electrons when combining chemically. Elements with high electron affinities tend to gain electrons when combining chemically.

Figure 12.7 shows the first ionization potential of the elements as a function of atomic number. The elements with the highest ionization potentials are the inert gases, which have eight electrons in the valence shell (except for helium). A similar graph of the electron affinity would show that the elements of column 7A, the halogens, have the greatest electron affinity. In other words, if the halogen achieves the same configuration as an inert gas by gaining an electron, it becomes relatively stable.

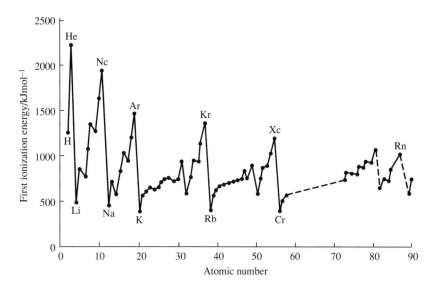

Figure 12.7. First Ionization Potentials of the Elements. The trends in the table and their relation to position in the periodic chart are understood in terms of electron configurations. Moving from left to right in one row of the periodic chart corresponds to increasing the number of protons in the nucleus, holding the electrons more tightly, but without increasing the number of occupied shells. Moving from top to bottom in the periodic chart corresponds to moving to shells farther from the nucleus, lessening the attraction of the electrons to the nucleus, which is shielded by the electrons in the inner shells.

The elements with the lowest ionization potentials are the elements in column 1A, the alkali metals, which have a single electron in the s subshell of the valence shell. It is relatively easy to remove an electron from an atom of an alkali metal, giving the inert gas configuration in the shell just below the valence subshell.

Figure 12.7 shows several additional elements, such as beryllium, nitrogen, magnesium, phosphorus, zinc, and mercury, which have higher ionization potentials than their immediate neighbors. All of these elements have ground-level configurations with all subshells completely filled (beryllium, magnesium, zinc, and mercury), or with all subshells filled except for a half-filled valence subshell (nitrogen and phosphorus). We conclude that not only is a filled subshell relatively stable, but also a half-filled subshell is relatively stable. This behavior can also be seen in some of the exceptions to the diagonal rule, such as chromium and copper. The stabilization of having the d subshell half full in chromium and full in copper is apparently great enough to produce a ground-state configuration with only one electron in the $4s$ subshell and either five or ten in the $3d$ subshell.

This stability of filled subshell and half-filled subshell also determines the ground-state configurations of Mo, Ag, and a number of other elements in the same columns.

By Hund's first rule, a subshell that is half full or less than half full in the ground level will have each electron occupying a different space orbital, in order to have parallel spins, resulting in the state of maximum M_S. A half-filled subshell therefore has one electron occupying each space orbital and has the same electron charge distribution as a full subshell (except for having only half as much total charge). **Unsöld's theorem** asserts that the charge distribution in a filled hydrogen-like subshell is spherically symmetric (independent of θ and ϕ). This theorem also must hold for a half-filled subshell.

Exercise 12.9   For hydrogen-like orbitals, show that

$$|\psi_{2p1}|^2 + |\psi_{2p0}|^2 + |\psi_{2p-1}|^2$$

is independent of θ and ϕ, as asserted by Unsöld's theorem.

A number of additional chemical and physical properties, including atomic size, melting temperature, and electronegativity, can be correlated with electron configuration and thus with position in the periodic table.

Summary

In this chapter, we introduced three approximation schemes. The first was the variation method, in which a variation trial function is chosen to minimize the approximate ground-state energy. A simple orbital variation trial function was found to correspond to a reduced, or shielded, nuclear charge in the helium atom. This result was interpreted to mean that each electron in a helium atom shields the other electron from the full charge of the nucleus. A better approximation corresponds to introduction of electron correlation, a dependence of the wave function on the interelectronic distance.

The next approximation method discussed was the perturbation method. This method is based on representations of the energy eigenvalues and energy eigenfunctions as power series in λ, a fictitious parameter that has the value unity in the actual case. A term in the correct Hamiltonian that prevents an exact solution (the perturbation) is arbitrarily multiplied by this parameter. In the helium atom treatment, the interelectronic repulsive potential energy was treated as the perturbation term in the Hamiltonian operator. The method gave useful results for excited states.

The third approximation scheme was the self-consistent field method of Hartree and Fock. In this method, an optimum orbital wave function is sought without restricting the search to a single family of functions. In the helium atom, the interelectronic repulsive energy is represented by assuming the probability density for the second electron to be given by an earlier approximate orbital and solving the resulting integrodifferential equation by iteration.

In higher-order orbital approximations, the energies of the orbitals in multielectron atoms depend on the angular momentum quantum number as well as on the principal quantum number, increasing as ℓ increases. The ground state is identified by the Aufbau principle, choosing orbitals that give the lowest sum of the orbital energies consistent with the Pauli exclusion principle.

Hund's first rule is that the largest value of S corresponds to the lowest energy in a configuration. The second rule is that for fixed value of S, the largest value of L, the quantum number for the total orbital angular momentum, corresponds to the lowest energy. The first rule correlates with the fact that the larger values of S correspond to lower probability for small interelectron distances, lowering the potential energy.

The form of the periodic table is determined by electron configurations. Elements with the same number of electrons in the outer (valence) shell have similar chemical properties. For example, all of the inert gases have eight electrons in the outer shell, corresponding to the stable configuration with fully occupied s and p subshells.

Additional Reading

Michael A. Morrison, Thomas F. Estle, and Neal F. Lane, *Quantum States of Atoms, Molecules, and Solids*, Prentice-Hall, Englewood Cliffs, NJ, 1976

See also the references at the end of Chapters 9–11.

PROBLEMS

Problems for Section 12.1

12.10. Prove the variational theorem. Hint: Assume that all of the energy eigenfunctions and energy eigenvalues are known, and write the variation function as a linear combination of the energy eigenfunctions:

$$\varphi = \sum_{i=0}^{\infty} c_i \psi_i$$

Substitute this expression into the formula for the variational energy and use eigenfunction and orthogonality properties.

*12.11. Calculate the variational energy of a particle in a hard one-dimensional box of length a, with:

  a. the trial function $\varphi(x) = Ax(a - x)$,

  b. the trial function $\varphi(x) = Ax^2(a - x)^2$,

  c. the trial function $\varphi(x) = \begin{cases} Ax & \text{if } 0 < x < a/2 \\ A(a - x) & \text{if } a/2 < x < a. \end{cases}$

For trial function c, determine if the trial function is an acceptable choice. Calculate the percent error for each trial function.

12.12. Calculate the variational energy of a harmonic oscillator using the trial function $\psi(x) = A/(b^2 + x^2)$, where b is a variable parameter. Minimize the energy and find the percent error from the correct ground-state energy.

Problems for Section 12.2

12.13. Using first-order perturbation, calculate the energy of the ground state of an anharmonic oscillator with potential energy $\mathscr{V} = kz^2/2 + bz$, where k and b are constants.

*12.14. Using first-order perturbation, calculate the energy of the ground state of an anharmonic oscillator with potential energy function $\mathscr{V} = kz^2/2 + cz^4$, where k and c are constants. $s_0 = (\frac{\hbar}{\cdots})^{1/4}$??

12.15. Using first-order perturbation, calculate the energy of the ground state of a system with the potential energy

$$\mathscr{V}(z) = \begin{cases} \infty & \text{if } z < -a \text{ or } a < z \\ kz^2/2 & \text{if } -a < z < a \end{cases}$$

Problems for Section 12.4

12.16. Using the values for the orbital exponents (effective number of protons in the nucleus) in Section 12.4, draw graphs of the radial distribution function for each orbital in the ground state of the Li atom. Draw a graph of the total radial distribution function.

*12.17. Write the ground-state electron configurations by subshells for the following elements: (a) Fe, (b) Rn, (c) Tc, (d) Rb.

12.18. Using Hund's rules, write the ground-term symbol for each of the elements in Problem 12.17.

12.19. Write all of the term symbols for the ground configurations of the following elements: (a) P, (b) Ca, (c) Cu, (d) Cl.

12.20. Explain why each of the following elements has a ground configuration different from that predicted by the diagonal mnemonic device: (a) Mo, (b) Ag, (c) Pd.

*12.21. The ionization potential (energy to remove one electron) of a sodium atom in its ground state is 5.1 eV. Use this to calculate a value for the effective nuclear charge felt by the $3s$ electron in a sodium atom in its ground state. State any assumptions.

12.22. Which of the elements in the first two rows of the periodic table have electronic charge distributions that are spherically symmetric?

*12.23. For each of the first 18 elements of the periodic table, give the number of unpaired electrons in the ground state.

12.24. Use the expression for the time-dependent wave function to show that the real hydrogen-like energy eigenfunctions correspond to standing waves, whereas the complex hydrogen-like energy eigenfunctions correspond to traveling waves. Tell how the traveling waves move. Show that both types of energy eigenfunctions correspond to stationary states.

General Problems

12.25. Prove an extended variational theorem: If the trial function φ is orthogonal to ψ_0, the correct ground-state wave function, the variational energy cannot be lower than the correct energy of the first excited state, E_1. See the hint in Problem 12.10.

12.26. Each element in the second row of the periodic table has a higher ionization potential than the element to its left, except for nitrogen, which has a higher ionization potential than either carbon or oxygen, and beryllium, which has a higher ionization potential than either boron or lithium. Explain both the general trend and these exceptions to the general trend.

12.27. The ionization potential generally decreases toward the bottom of a column of the periodic table. Explain this fact in terms of effective nuclear charge and electron shielding.

12.28. a. Using first-order perturbation, obtain a formula for the ground-state energy of a particle in a one-dimensional hard box with an additional potential energy $\hat{H}' = bx$, where b is a negative constant.

b. Obtain a formula for the ground-state energy of the particle of part a, using the variation method and the zero-order wave function as a trial function. Compare your answer with that of part a.

c. Obtain a formula for the ground-state energy of the particle of part a, using the variation method and the trial function $\varphi = Ax^2(a - x)$. Compare your answer with that of part a.

d. Obtain a formula for the ground-state energy of the particle of part a, using the variation method and the trial function $\varphi = Ax^n(a - x)$.

*e. Assume that the particle is an electron in a one-dimensional box of length 1.00 nm and that the ends of the box are charged so that $b = -1.60 \times 10^{-11}$ J m^{-1}. Evaluate the energy according to parts a and c.

*f. Evaluate the energy according to part d and find the optimum value of n.

*12.29. Identify each statement as either true or false. If a statement is true only under special circumstances, label it as false.

a. The orbital energy of a $4s$ subshell is always lower than that of a $3d$ subshell in the same atom.

b. The ground state of every inert gas corresponds to a filled valence shell.

c. The inert gases are the only elements with spherically symmetric electron charge distributions.

d. An electron configuration that contains only filled subshells can correspond to only one term symbol.

e. Orbital occupations that do not correspond to Hund's first rule cannot occur.

f. The self-consistent field theory can deliver the best possible orbital wave function for a multielectron atom.

g. A second-order perturbation result is always more nearly correct than a first-order result.

h. An electron configuration with two unpaired electrons cannot correspond to a doublet term symbol.

i. An antisymmetrized orbital wave function contains no electron correlation.

j. An antisymmetrized orbital wave function contains no dynamical correlation.

k. If a variational energy equals the correct ground-state energy, the variational trial function must be equal to the correct ground-state wave function.

13 The Electronic States of Molecules

PREVIEW

In this chapter, we study the electronic states of molecules, based on the Born-Oppenheimer approximation. In this approximation, the nuclei are assumed to remain stationary as the electrons move. We will primarily use orbital wave functions, in which the states of one electron are independent of the states of the other electrons and correspond to molecular orbitals.

PRINCIPAL FACTS AND IDEAS

1. In the Born-Oppenheimer approximation, the nuclei are assumed to be stationary when the electronic states are studied.
2. The Schrödinger equation for the hydrogen molecule ion, H_2^+, can be solved in the Born-Oppenheimer approximation without further approximations.
3. Molecular orbitals can be represented approximately as linear combinations of atomic orbitals (LCAO-MOs).
4. The electronic states of homonuclear diatomic molecules can be described with a common set of LCAO-MOs.
5. The valence bond method is an alternative to the molecular orbital method.
6. Heteronuclear diatomic molecules are described with molecular orbitals that differ from those of homonuclear diatomic molecules.
7. Qualitative descriptions of the electronic states of molecules can be obtained by using general criteria for forming good bonding LCAO molecular orbitals.
8. The electronic structure of polyatomic molecules can be described with LCAO molecular orbitals.
9. Various advanced techniques exist for molecular orbital calculations.

OBJECTIVES

After studying this chapter, the student should:

1. be familiar with and understand the properties of the electronic states of the hydrogen molecule ion,
2. be able to construct approximate wave functions and electron configurations for homonuclear diatomic molecules,
3. be able to use general properties of molecular orbitals, including criteria for formation of good bonding orbitals, to predict the qualitative properties of electronic states of heteronuclear diatomic molecules,
4. be able to describe qualitatively the bonding in a fairly small polyatomic molecule, including bond angles, bond polarities, and the dipole moment of the molecule, using the criteria for formation of good bonding orbitals,
5. be able to describe qualitatively the bonding in fairly small polyatomic molecules, using the valence bond method.

13.1

The Born-Oppenheimer Approximation; The Hydrogen Molecule Ion

The Schrödinger equation cannot be solved exactly for a system of more than two particles. For this reason, we had to resort to approximation schemes for all atoms except hydrogen. One of the approximations we used in studying the electronic motion was that the nucleus was stationary. The center of mass actually moves like a particle with mass equal to the total atomic mass, but the electrons move much more rapidly and adjust to each new position of the nucleus.

Just as in atoms, the center of mass of a molecule moves independently of the relative motion, and we will discuss this translational motion in Chapter 14. Our study of the relative motion in molecules is based on the **Born-Oppenheimer approximation.**[1] This approximation is the use of a Schrödinger equation for the electrons, with the nuclei assumed to be stationary. Fixed bond distances and bond angles are assumed, although we can change the bond distances or angles and repeat the solution for a new set of nuclear positions.

The nuclei of a molecule do rotate and vibrate. However, the electrons in a typical molecule move around the molecule several hundred times or a thousand times while the nuclei vibrate once and even more times while the nuclei rotate once. Each small shift in the position of the nuclei occurs slowly compared to electronic motion and is immediately followed by rapid equilibration of the electrons to a new electronic wave function and energy eigenvalue, which are nearly what they would have been if the nuclei had always been at their new location. The Born-Oppenheimer approximation is therefore a good approximation.

The Born-Oppenheimer approximation enables us to solve the Schrödinger equation repeatedly for different positions of the nuclei. Having done so, we can construct graphs of the energy eigenvalues as a function of nuclear positions. For example, Figure 13.1 shows schematically the ground-state electronic energy of a diatomic molecule as a function of internuclear distance, R.

Use of the Born-Oppenheimer approximation does not keep us from studying vibrations and rotations. The electronic energy acts as a potential energy for nuclear motion, since it depends on the nuclear positions and not on their velocities. We will discuss translation, rotation, and vibration of molecules in Chapter 14.

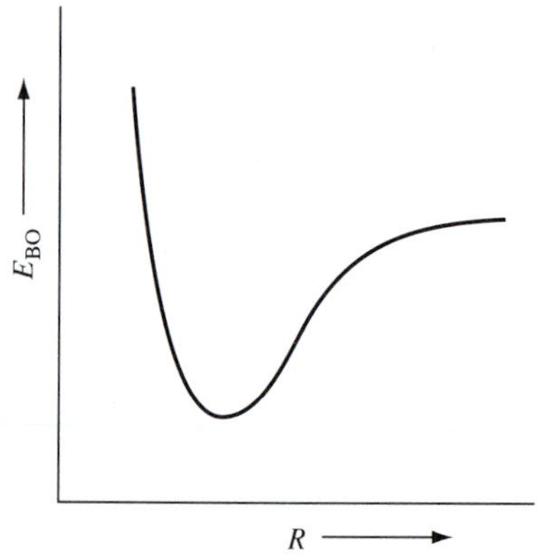

Figure 13.1. Born-Oppenheimer Energy as a Function of Internuclear Distance for a Diatomic Molecule (schematic). This energy is the total energy of the molecule in the Born-Oppenheimer approximation. It consists of the electronic energy (kinetic plus potential), plus the energy of repulsion of the nuclei.

The Hydrogen Molecule Ion

The simplest molecular system is the hydrogen molecule ion, H_2^+. The system consists of two nuclei and a single electron, as shown in Figure 13.2. The fixed location of one nucleus is denoted by A and the fixed location

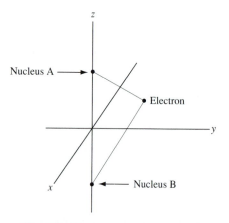

Figure 13.2. Hydrogen Molecule Ion (H_2^+) System. This system is the simplest molecule, having only one electron. It is the only molecule for which the Schrödinger equation can be solved in the Born-Oppenheimer approximation without any additional approximations.

[1] Max Born and J. Robert Oppenheimer, *Ann. Phys.* **84**, 457 (1927).

of the other by B. We choose our coordinate system so that the z axis passes through the two nuclei with the origin of coordinates midway between the nuclei.

In the Born-Oppenheimer approximation, the Hamiltonian operator for the hydrogen molecule ion is

$$\hat{H} = -\frac{\hbar^2}{2m}\nabla^2 + \frac{e^2}{4\pi\varepsilon_0}\left(\frac{1}{R} - \frac{1}{r_A} - \frac{1}{r_B}\right) \qquad \textbf{(13.1-1)}$$

where ∇^2 is the Laplacian operator for the electron's coordinates, m is the electron mass, R is the internuclear distance, r_A is the distance from the electron to the nucleus at A, and r_B is the distance from the electron to the nucleus at B. There are no kinetic energy terms for the nuclei, which are considered to be fixed.

Since the internuclear distance R is constant in the Born-Oppenheimer approximation,

$$\frac{e^2}{4\pi\varepsilon_0 R} = \mathscr{V}_{nn} = \text{constant} \qquad \textbf{(13.1-2)}$$

where $\mathscr{V}_{nn}$ is the internuclear potential energy. We write

$$\hat{H} = \hat{H}_{el} + \mathscr{V}_{nn} \qquad \textbf{(13.1-3)}$$

$$\hat{H}_{el} = -\frac{\hbar^2}{2m}\nabla^2 + \frac{e^2}{4\pi\varepsilon_0}\left(-\frac{1}{r_A} - \frac{1}{r_B}\right) \qquad \textbf{(13.1-4)}$$

The electronic Schrödinger equation is

$$\hat{H}_{el}\psi_{el} = E_{el}\psi_{el} \qquad \textbf{(13.1-5)}$$

A constant added to a Hamiltonian operator does not change the energy eigenfunctions and results in adding that constant to the energy eigenvalues (see Exercise 9.25). We can write

$$E_{BO} = E_{el} + \mathscr{V}_{nn} \qquad \textbf{(13.1-6)}$$

where E_{BO} is the total energy eigenvalue with stationary nuclei, called the Born-Oppenheimer energy. This energy eigenvalue is a constant for constant R, but if we solve repeatedly for different values of R we can obtain its value as a function of R.

The variables can be separated in Equation (13.1-5) by transforming to coordinates called confocal polar elliptical coordinates. We will not discuss the solution,[2] but will present some facts about the ground state and first excited state. We will refer to the energies and orbitals of these states as the "exact Born-Oppenheimer" energies and orbitals. They contain no approximations other than the Born-Oppenheimer approximation.

Figure 13.3 shows the Born-Oppenheimer energy as a function of the internuclear distance for these two states. The lower curve has a minimum at $R = 1.06 \times 10^{-10}$ m. This value of R is denoted by R_e and is called the

In the Born-Oppenheimer approximation, the Schrödinger equation for the H_2^+ molecule ion is a one-body problem. The only Schrödinger equations that have been solved analytically are one-body problems and problems that can be separated into one-body problems.

---

[2] D. R. Bates, K. Ledsham and A. L. Stewart, *Philos. Trans. Roy. Soc. A* **246**, 215 (1963).

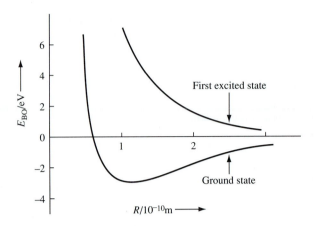

Figure 13.3. The Born-Oppenheimer Energy of the Ground State and First Excited State of the Hydrogen Molecule Ion as a Function of Internuclear Distance. This figure shows the results of solving the Schrödinger equation repeatedly for different internuclear distances in the Born-Oppenheimer approximation and then drawing a smooth curve through the energy eigenvalues as a function of internuclear distance. Both the ground state and the first excited state are shown.

equilibrium internuclear distance. For large values of R, the energy approaches a constant value. The difference between the value of the energy at $R = R_e$ and this constant value is denoted by D_e and is called the **dissociation energy**. It is equal to 2.8 eV. It is the energy to break the bond, starting from the minimum in the Born-Oppenheimer energy.

The first excited state has an energy that decreases monotonically as R increases. If the molecule is in the ground electronic state it is stable, with a chemical bond. If it is in the first excited state, it is not bonded and will dissociate, forming a hydrogen atom and a positive hydrogen ion.

The eigenfunctions of the Hamiltonian in Equation (13.1-5) are one-electron wave functions that correspond to electronic motion around both nuclei. They are called **molecular orbitals**. Figure 13.4 shows qualitatively the orbital regions for the ground state and first excited state. There is some similarity between the ground-state orbital region and the ground-state orbital region for the hydrogen atom, and between the orbital region of the first excited state and that of the $2pz$ orbital of the hydrogen atom.

If the mathematical limit is taken as R approaches zero, a hypothetical single atom is obtained, called the **united atom**. The united atom for H_2^+ is the He^+ ion, with a single electron and $Z = 2$. In this limit the ground-state molecular orbital turns smoothly into the united-atom $1s$ orbital of He^+ and the first excited-state molecular orbital turns smoothly into the united-atom $2pz$ orbital of He^+.

The ground-state orbital has no nodes, while the first excited state orbital has one nodal surface between the nuclei. In general, a molecular orbital without a nodal surface between the nuclei corresponds to an electronic energy with a minimum value. Such an orbital is called a **bonding molecular orbital**. An orbital with a nodal surface between the nuclei generally corresponds to an electronic energy that decreases monotonically as R increases. Such an orbital is called an **antibonding orbital**.

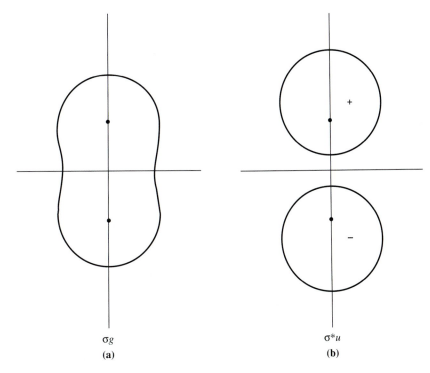

σ*g
(a)

σ*u
(b)

Figure 13.4. The Orbital Regions for the Hydrogen Molecule Ion (schematic). (a) The ground state. **(b)** The first excited state. These sketches show the important qualitative properties of these two orbitals. The ground state corresponds to a standing wave with no nodes, and the first excited state corresponds to a standing wave with a nodal plane between the nuclei.

With the hydrogen atom, we found that the orbital angular momentum operators $\hat{L}^2$ and $\hat{L}_z$ commute with the electronic Hamiltonian. Therefore, energy eigenfunctions could be chosen which were eigenfunctions of these two operators. In the hydrogen molecule ion, $\hat{L}^2$ does not commute with the electronic Hamiltonian. This fact is not obvious; it has to do with the fact that not all directions are equivalent for the electron because there are two fixed nuclei. However, $\hat{L}_z$ does commute with the electronic Hamiltonian operator if the nuclei are placed on the z axis. The energy eigenfunctions can be eigenfunctions of $\hat{L}_z$, although not necessarily of $\hat{L}^2$.

All angular momentum magnitudes and projections on one axis follow the same pattern.

The eigenvalues of $\hat{L}_z$ follow the same pattern as in the atomic case:

$$\hat{L}_z\psi = \hbar m\psi \qquad \textbf{(13.1-7)}$$

where the quantum number m is an integer and ψ represents an energy eigenfunction. For molecular orbitals, we define a nonnegative quantum number λ:

$$\lambda = |m| \qquad \textbf{(13.1-8)}$$

A nonzero value of λ corresponds to two states, since m can be either positive or negative. That is, each level for $\lambda \neq 0$ has a degeneracy equal to 2 (is "doubly degenerate").

Specifying values of λ with σ, π, δ, etc. is analogous to the s, p, d, f, ... notation for atomic orbitals.

Atomic orbitals corresponding to $\ell = 0$ were called s orbitals, orbitals with $\ell = 1$ were called p orbitals, etc. For molecular orbitals, we use the designation σ for orbitals corresponding to $\lambda = 0$, π for $\lambda = 1$, δ for $\lambda = 2$, etc. Both the ground-state orbital and the first excited-state orbital of the hydrogen molecule ion are σ orbitals.

Symmetry Properties of the Orbitals and Symmetry Operators

An important class of operators can commute with the Born-Oppenheimer electronic Hamiltonian operator for a molecule and can be used to characterize molecular orbitals. These operators are **symmetry operators**, which move points from one location to another in three-dimensional space. Each symmetry operator is classified and named by the way it moves a point.

The **inversion operator**, $\hat{\imath}$, is defined to move a point on a line through the origin of coordinates to a location at the same distance from the origin as the original location. If the cartesian coordinates of the original location are (x, y, z), we denote the operation of $\hat{\imath}$ on this point by writing

$$\hat{\imath}(x, y, z) = (-x, -y, -z) \qquad (13.1\text{-}9)$$

A **symmetry element** is a point, line, or plane with respect to which a symmetry operation is performed. The symmetry element for the inversion operator is the origin. Since there is only one origin, there is only one inversion operator.

Point symmetry operators are symmetry operators that leave a point at its original location if that location is at the origin. The inversion operator is an example of a point symmetry operator. The symmetry elements of point symmetry operators always include the origin. We consider only point symmetry operators.

A **reflection operator** is defined to move a point along a line perpendicular to a specified plane to a location at the same distance from the plane as the original location. It is said to "reflect" the point through the plane, which is the symmetry element. The reflection operator $\hat{\sigma}_h$ reflects through a horizontal plane:

$$\hat{\sigma}_h(x, y, z) = (x, y, -z) \qquad (13.1\text{-}10)$$

There is only one horizontal plane through the origin, so there is only one $\hat{\sigma}_h$ operator among the point symmetry operators.

A symmetry operator that reflects through a vertical plane is denoted by $\hat{\sigma}_v$. Since there are infinitely many vertical planes containing the origin, there are infinitely many $\hat{\sigma}_v$ operators. It is convenient to attach subscripts or other labels to distinguish them from each other.

Exercise 13.1

Find the coordinates of the points resulting from the operations:
a. $\hat{\imath}(1, 2, 3)$
b. $\hat{\sigma}_h(4, -2, -2)$
c. $\hat{\sigma}_{vyz}(7, -6, 3)$, where $\hat{\sigma}_{vyz}$ is the reflection operator that reflects through the yz plane.

Rotation operators cause a point to move as it would if it were part of a rigid body rotating about a specified line (an axis), which is the symmetry element. That is, the point moves around a circle centered on the axis of rotation and perpendicular to it. Infinitely many lines pass through the origin, and for each rotation axis there can be rotations by different angles. There are infinitely many different rotation operators. We consider only rotation operators that produce a full rotation (360°) when applied an integral number of times. By convention, all rotations are counterclockwise when viewed from the positive end of the rotation axis.

A rotation operator that produces one full rotation when applied n times is denoted by $\hat{C}_n$. It is sometimes convenient to add subscripts to denote the axis. For example, the $\hat{C}_{4z}$ operator rotates by 90° about the z axis, and its effect on a point at (x, y, z) is

$$\hat{C}_{4z}(x, y, z) = (-y, x, z) \tag{13.1-11}$$

Figure 13.5 shows the effect of the operators $\hat{\imath}$, $\hat{\sigma}_h$, and $\hat{C}_{4z}$ on a point in the first octant.

Exercise 13.2

Find the following locations:
  a. $\hat{C}_{2x}(1, 2, 3)$ (the axis of rotation is the x axis)
  b. $\hat{C}_{3y}(1, 1, 1)$ (the axis of rotation is the y axis)

Ordinary mathematical operators operate on functions, not on isolated points. We define a mode of operation so that symmetry operators also operate on functions such as orbital functions. Let $f(x, y, z)$ be some function of the coordinates x, y, and z, and let $\hat{O}$ be some symmetry operator that carries a point at (x, y, z) to a location (x', y', z'). The function $\hat{O}f$ is defined as the function that has the same value at the location (x', y', z') as the function f has at the location (x, y, z). We denote this new function by

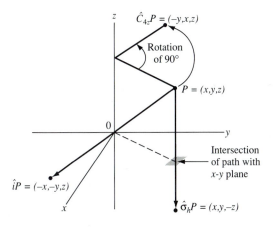

Figure 13.5. The Effect of the Symmetry Operators $\hat{\imath}$, $\hat{\sigma}_h$, and $\hat{C}_{4z}$. These mathematical operators move a point in three-dimensional space in ways defined in the text.

g, so that

$$g(x, y, z) = \hat{O}f(x, y, z) \qquad \text{(13.1-12a)}$$

$$g(x', y', z') = f(x, y, z) \qquad \text{(13.1-12b)}$$

A function can be an eigenfunction of a symmetry operator. The only eigenvalues that occur are $+1$ and -1.

▼   **EXAMPLE 13.1**   Show that the hydrogen-like $1s$ orbital is an eigenfunction of the inversion operator $\hat{i}$. Find the eigenvalue.

Solution

$$\psi_{1s} = \sqrt{\frac{1}{\pi}} \left(\frac{Z}{a}\right)^{3/2} e^{-Zr/a}$$

where

$$r = (x^2 + y^2 + z^2)^{1/2}$$

When x is replaced by $-x$, y is replaced by $-y$, and z is replaced by $-z$, the value of r is unchanged, so

$$\hat{i}\psi_{1s} = \psi_{1s}$$

The $1s$ function is an eigenfunction of the inversion operator with eigenvalue 1.

▲

Exercise 13.3   **a.** Determine the spherical polar coordinates of $\hat{i}P$ and $\hat{\sigma}_h P$ if P represents a point whose location is (r, θ, ϕ).
b. Show that the ψ_{2pz} hydrogen-like orbital is an eigenfunction of the $\hat{\sigma}_h$ operator with eigenvalue -1.

Symmetry operators will be useful for studying the electronic states not only of diatomic molecules but also of polyatomic molecules. The symmetry operators that are important for a given molecule are determined by the positions of the nuclei. When the electronic Schrödinger equation of a diatomic molecule is solved repeatedly in the Born-Oppenheimer approximation, a bond length of minimum Born-Oppenheimer energy can be found. Similarly, when the electronic Schrödinger equation of a polyatomic molecule is solved in the Born-Oppenheimer approximation, a conformation of minimum energy can be found, with fixed bond lengths and angles. This is the equilibrium conformation, and its symmetry properties are the ones of greatest importance.

We use symmetry operators in two ways: First, we apply them to the nuclei in their equilibrium positions. If a symmetry operator moves every nucleus to a location previously occupied by a nucleus of the same kind, we say that the symmetry operator "belongs" to the molecule. Second, we apply them to the electrons, leaving the nuclei fixed in their equilibrium positions. A symmetry operator that belongs to the molecule will not change the value of the potential energy when it is applied to the electrons with the nuclei fixed. It will bring every electron to a point where it is the same distance from each nucleus as it was in its original position, or it is the same distance from a different nucleus of the same kind. The operation of the inversion operator on the electron of an H_2^+ molecule ion is illustrated in Figure 13.6. A symmetry operator that belongs to the molecule will commute with the electronic Hamiltonian operator, and the electronic wave function can be an eigenfunction of this symmetry operator.

When we are dealing with a molecule, we must specify how the molecule is oriented. The standard practice is to place the highest-order symmetry axis of the molecule on the vertical axis (z axis). For example, the H_2^+ molecule is placed with the nuclei on the z axis. The horizontal plane is perpendicular to the bond axis, and all vertical planes contain the bond axis.

Exercise 13.4   Show that the symmetry operators $\hat{\imath}$, $\hat{\sigma}_h$, $\hat{C}_{nz}$, and $\hat{C}_{2a}$ belong to the H_2^+ molecule, where n is any positive integer and a stands for any axis in the x-y plane. Show also that if these operators are applied to the electron position with fixed nuclei, the potential energy is unchanged.

The ground electronic orbital of H_2^+ is an eigenfunction of the symmetry operators in Exercise 13.4 with all eigenvalues equal to unity. The first excited state is also an eigenfunction of these operators, but the eigenvalues of $\hat{\imath}$, $\hat{\sigma}_h$, and $\hat{C}_{2a}$ are equal to -1.

An eigenfunction having an eigenvalue of $\hat{\imath}$ equal to $+1$ is denoted by a subscript g (from the German *gerade*, meaning "even") and an eigenvalue of $\hat{\imath}$ equal to -1 is denoted by a subscript u (from the German *ungerade*, meaning "odd"). An eigenvalue of $\hat{\sigma}_h$ equal to -1 is denoted by an asterisk (*). Orbitals with asterisks are antibonding, since they have a nodal plane through the origin perpendicular to the bond axis. No superscript or subscript is used to denote an eigenvalue of $\hat{\sigma}_h$ equal to $+1$, corresponding to a bonding orbital.

The ground-state orbital of the hydrogen molecule ion is denoted by $\psi_{1\sigma g}$, and the first excited state can be denoted by $\psi_{2\sigma^*u}$. The subscripts 1 and 2 are chosen because orbitals are generally numbered from the lowest to the highest orbital energy.

13.2   LCAO-MOs—Molecular Orbitals That Are Linear Combinations of Atomic Orbitals

The exact Born-Oppenheimer solution to the Schrödinger equation for the hydrogen molecule ion gives complicated orbitals expressed in an unfamiliar coordinate system, which we did not explicitly display. It will be convenient to have some easily expressed approximate molecular orbitals. We define approximate molecular orbitals that are **linear combinations of atomic orbitals**, abbreviated LCAO-MO.

If $f_1, f_2, f_3, \ldots$ are a set of functions, then g is a linear combination of these functions if

$$g = c_1 f_1 + c_2 f_2 + c_3 f_3 + \cdots \qquad \textbf{(13.2-1)}$$

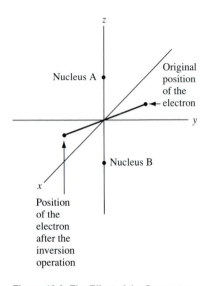

z

Original
position
of the
electron

Nucleus A

y

Nucleus B

x

Position
of the
electron
after the
inversion
operation

Figure 13.6. The Effect of the Symmetry Operator $\hat{\imath}$ on the Electron of the H_2^+ Ion. This symmetry operator moves the electron so that its distance from nucleus B is the same as its original distance from nucleus A.

where the c's are a set of constant coefficients. The set of f functions is called the **basis set**. If the linear combination can be an exact representation of an arbitrary function obeying the same boundary conditions as the basis set, the basis set is said to be a **complete set**. We will not attempt to use a complete set of functions, but will begin with a linear combination of two atomic orbitals centered on two different nuclei.

As in Figure 13.2, let r_A be the distance from the nucleus at location A to the electron, and let r_B be the distance from the nucleus at location B to the electron. Take two hydrogen-like orbitals, one with r_A as its independent variable and one with r_B as its independent variable. We use the abbreviations

$$\psi_{1sA} = \psi_{1s}(r_A) \tag{13.2-2a}$$

$$\psi_{1sB} = \psi_{1s}(r_B) \tag{13.2-2b}$$

The orbital ψ_{1sA} has its orbital region centered at location A and the orbital ψ_{1sB} has its orbital region centered at location B. We now form a molecular orbital that is a linear combination of atomic orbitals (LCAO-MO):

$$\psi_{MO} = c_A \psi_{1sA} + c_B \psi_{1sB} \tag{13.2-3}$$

There are now two ways to proceed. The first procedure is to regard ψ_{MO} as a variational trial function and minimize the variational energy as a function of c_A and c_B to find the approximate ground state. The result is that the variational energy is minimized when $c_A = c_B$. An approximation to the first excited state is obtained when $c_A = -c_B$.[3]

The other procedure is to choose values of c_A and c_B so that the approximate orbital is an eigenfunction of the same symmetry operators as the exact orbitals. The ground-state exact Born-Oppenheimer orbital is an eigenfunction of the inversion operator with eigenvalue $+1$. In order to obtain an LCAO-MO with this eigenvalue, we must choose

$$c_A = c_B \tag{13.2-4}$$

Since the origin is midway between the two nuclei, inversion from any point leads to a point that is the same distance from nucleus B as the original point was from nucleus A and vice versa. The first term in the linear combination becomes equal to the original value of the second term and vice versa, if the two c's are equal.

In order to obtain a molecular orbital with the same symmetry properties as the exact Born-Oppenheimer orbital of the first excited state, we must choose

$$c_A = -c_B \tag{13.2-5}$$

Equations (13.2-4) and (13.2-5) give the desired symmetry properties because of the similarity of the two atomic orbitals.

The symmetry properties are sufficiently fundamental that choosing the molecular orbitals to be eigenfunctions of symmetry operators leads to the same LCAO-MOs as the variation procedure.

---

[3] J. C. Davis, Jr., *Advanced Physical Chemistry*, Ronald Press, New York, 1965, p. 404.

Exercise 13.5
a. Argue that $c_A = c_B$ leads to an eigenvalue of $+1$ for the $\hat{\sigma}_h$ operator and for the $\hat{C}_{2a}$ operator.
b. Argue that $c_A = -c_B$ leads to an eigenvalue of -1 for the $\hat{\sigma}_h$ operator and for the $\hat{C}_{2a}$ operator.

We introduce the symbols for our two LCAO-MOs:

$$\psi_{\sigma g 1s} = C_g(\psi_{1sA} + \psi_{1sB}) \tag{13.2-6}$$

$$\psi_{\sigma^* u 1s} = C_u(\psi_{1sA} - \psi_{1sB}) \tag{13.2-7}$$

where the $1s$ subscripts indicate the atomic orbitals from which the LCAO-MOs were constructed. The values of the constants C_g and C_u can be chosen to normalize the molecular orbitals.

Figure 13.7 schematically shows the orbital regions for the $\sigma g 1s$ LCAO-MO and the $\sigma^* u 1s$ LCAO-MO, as well as the orbital regions for the $1s$ atomic orbitals. The intersection of the two atomic orbital regions is called the **overlap region**. This is the region where both atomic orbitals differ significantly from zero. For the $\sigma g 1s$ orbital, the two atomic orbitals combine with the same sign in the overlap region, producing an orbital region characteristic of a bonding orbital, with no nodal surfaces. For the $\sigma^* u 1s$ orbital, the atomic orbitals combine with opposite signs in the overlap region, canceling to produce a nodal surface between the nuclei, characteristic of an antibonding orbital. This addition and cancellation are very similar to constructive and destructive interference of waves.

The analogy between addition of atomic orbitals and interference of waves is interesting, but one should not make too much of it.

Figure 13.8 shows the electronic energy for each of these LCAO molecular orbitals along with the exact Born-Oppenheimer energies. The value of D_e for the $\sigma g 1s$ orbital is equal to 1.76 eV, with R_e equal to 1.32×10^{-10} m. As we expect from the variation theorem, the approximate

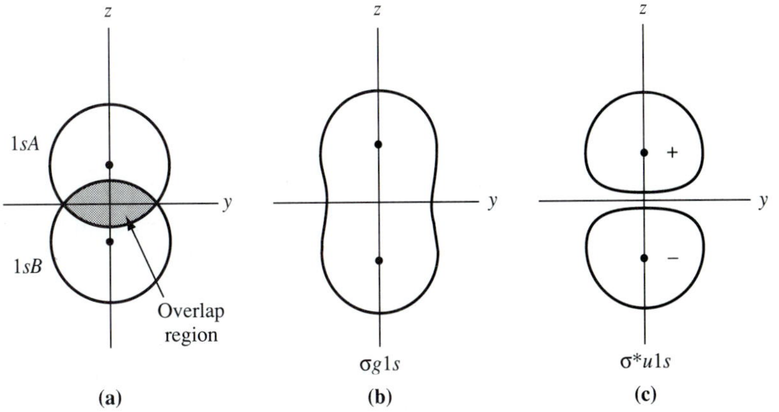

Figure 13.7. The Orbital Region for the $\sigma g 1s$ and $\sigma^* u 1s$ LCAO Molecular Orbitals.
(a) The overlapping orbital regions of the $1sA$ and $1sB$ atomic orbitals. **(b)** The orbital region of the $\sigma g 1s$ LCAO-MO. **(c)** The orbital region of the $\sigma^* u 1s$ LCAO-MO. The orbital regions of the LCAO molecular orbitals have the same general features as the "exact" Born-Oppenheimer orbitals whose orbital regions were depicted in Figure 13.4.

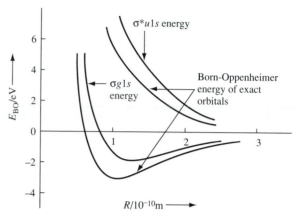

E_{BO}/eV

$\sigma^* u 1s$ energy

$\sigma g 1s$ energy

Born-Oppenheimer energy of exact orbitals

$R/10^{-10}m$

Figure 13.8. The Orbital Energies for the $\sigma g 1s$ and $\sigma^* u 1s$ LCAO Molecular Orbitals. This diagram shows qualitatively how the Born-Oppenheimer energies of the LCAO molecular orbitals compare with the Born-Oppenheimer energies of the "exact" orbitals. The approximate orbital energies must lie above the corresponding exact energies for all internuclear distances.

The variation theorem implies that our approximate energy cannot lie below the correct energy for any value of the internuclear distance, so the dissociation energy must be smaller for the approximate wave function than for the correct wave function.

energies lie above the exact energies for all values of R. The energies can be improved greatly by "scaling" the atomic orbitals, that is, by replacing the atomic number Z in the orbital exponent by a variable parameter, as we did with the helium atom in Chapter 12.

LCAO-MOs can be constructed that are linear combinations of more than two atomic orbitals. For the ground state of the hydrogen molecule ion we can write

$$\psi_{MO} = c_{1sA}\psi_{1sA} + c_{1sB}\psi_{1sB} + c_{2sA}\psi_{2sA}$$
$$+ c_{2sB}\psi_{2sB} + c_{2pzA}\psi_{2pzA} + c_{2pzB}\psi_{2pzB} \qquad (13.2\text{-}8)$$

When the variational energy is minimized with respect to the c coefficients, a better (lower) value is obtained than with the $\sigma g 1s$ orbital. However, we will use linear combinations of only two atomic orbitals as much as possible, since we will content ourselves with qualitative description rather than quantitative calculation.

Putting orbitals with different symmetry properties in the same linear combination yields a function that has no simple symmetry properties and is not useful in describing any molecule.

The $2px$ and $2py$ atomic orbitals are not included in Equation (13.2-8) because they have different symmetry about the bond axis than does the exact ground-state orbital. If they were included with nonzero coefficients, the LCAO-MO would not be an eigenfunction of the same symmetry operators as the exact orbitals.

Exercise 13.6 Argue that the $2px$ and $2py$ atomic orbitals are eigenfunctions of the $\hat{C}_{2z}$ operator with eigenvalue -1, while the $2pz$ orbital is an eigenfunction with eigenvalue $+1$. Argue that a linear combination of all three of these orbitals is not an eigenfunction of the $\hat{C}_{2z}$ operator.

Further excited states of the hydrogen molecule ion are approximated by LCAO-MOs using excited-state hydrogen-like orbitals. For example, two linear combinations of $2s$ orbitals that are eigenfunctions of the proper symmetry operators are

$$\psi_{\sigma g 2s} = C_g[\psi_{2s}(r_A) + \psi_{2s}(r_B)] = C_g(\psi_{2sA} + \psi_{2sB}) \qquad \textbf{(13.2-9)}$$

$$\psi_{\sigma^* u 2s} = C_u[\psi_{2s}(r_A) - \psi_{2s}(r_B)] = C_u(\psi_{2sA} - \psi_{2sB}) \qquad \textbf{(13.2-10)}$$

As seen in Figure 13.8, the fact that the $\sigma g 2s$ orbital is a bonding orbital does not mean that its energy is lower than that of the $\sigma^* u 1s$ antibonding orbital, because they dissociate to different atomic states.

The $\sigma g 2s$ orbital is a bonding orbital, and the $\sigma^* u 2s$ orbital is an antibonding orbital. However, the $\sigma g 2s$ orbital energy is higher than that of $\sigma^* u 1s$, because the molecule dissociates from the $\sigma g 2s$ state to a hydrogen nucleus and a hydrogen atom in the $2s$ state, as shown schematically in Figure 13.9.

Exercise 13.7	Draw sketches of the orbital regions for the functions in Equations (13.2-9) and (13.2-10). Argue that the designations σg and $\sigma^* u$ are correct.

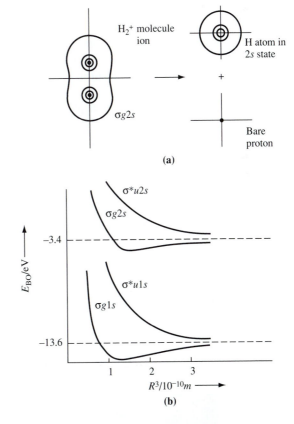

(a)

(b)

Figure 13.9. Dissociation of a Hydrogen Molecule Ion in the $\sigma g 1s$ and $\sigma g 2s$ States. **(a)** The orbital regions before and after dissociation. **(b)** The energy levels of the LCAO molecular orbitals as a function of internuclear distance. If a molecule in the $\sigma g 2s$ state dissociates, an atom in a $2s$ state and a bare nucleus result. A molecule in the $\sigma g 1s$ state dissociates to form an atom in the $1s$ state and a bare nucleus.

To normalize the $\sigma g1s$ orbital, we write, abbreviating the $1sA$ and $1sB$ subscripts by A and B:

$$1 = |C_g|^2 \int (\psi_A + \psi_B)^*(\psi_A + \psi_B) \, d^3\mathbf{r} \qquad \text{(13.2-11)}$$

Since the $1s$ atomic orbitals are real functions, the complex conjugate symbol can be omitted. We will choose the normalization constant C_g to be real, so that

$$1 = C_g^2 \int (\psi_A^2 + 2\psi_A\psi_B + \psi_B^2) \, d^3\mathbf{r} \qquad \text{(13.2-12)}$$

The atomic orbitals ψ_A and ψ_B are normalized, so that the first and last terms in the integral will each yield unity when the integration is done. The second term gives an integral denoted by S:

$$\int \psi_A^* \psi_B \, d^3\mathbf{r} = S \qquad \text{(13.2-13)}$$

The integral S is called the **overlap integral** because the major contribution to its integrand is from the overlap region. In other regions, at least one of the factors in the integrand is small. Since the $1s$ orbitals are positive everywhere, the overlap integral for two $1s$ orbitals is positive. Its value depends on R, approaching zero if the two nuclei are very far apart and approaching unity when R approaches zero. Similar overlap integrals can be defined for other pairs of atomic orbitals, and it is convenient to attach two subscripts to the symbol S to indicate which two orbitals are involved. The overlap integral in Equation (13.2-13) would be denoted as $S_{1s,1s}$. For normalized atomic orbitals, the values of overlap integrals range from -1 to $+1$.

The property of approaching unity if the nuclei approach each other applies only to overlap integrals for two orbitals of the same type. An overlap integral between a 1s and a 2pz orbital on different nuclei would approach zero as the nuclei approached each other, because it approaches an orthogonality integral.

We now have

$$1 = C_g^2(1 + 2S + 1) \qquad \text{(13.2-14)}$$

so the normalized LCAO-MO is

$$\psi_{\sigma g1s} = \frac{1}{\sqrt{2 + 2S}}(\psi_A + \psi_B) \qquad \text{(13.2-15)}$$

Exercise 13.8 | Show that the normalization constant for the $\sigma^* u1s$ LCAO-MO is

$$C_u = \frac{1}{\sqrt{2 - 2S}} \qquad \text{(13.2-16)}$$

13.3   Homonuclear Diatomic Molecules

Homonuclear diatomic molecules, in which the two nuclei are of the same kind, are easier to discuss than heteronuclear diatomic molecules. We will base our discussion on the H_2^+ molecular orbitals in much the same way as we based our discussion of atoms in Chapters 11 and 12 on the hydrogen atom atomic orbitals.

The Hydrogen Molecule

Figure 13.10 shows the hydrogen molecule system, consisting of two nuclei at locations A and B and two electrons at locations 1 and 2. The hydrogen molecule, with its two electrons, bears the same relationship to the hydrogen molecule ion as the helium atom does to the hydrogen atom, and our treatment of it resembles that of the helium atom.

The H_2 molecule is strongly analogous to the He atom. Compare the discussion of the H_2 molecule with that of the He atom in Chapters 11 and 12.

We apply the Born-Oppenheimer approximation, assuming the nuclei to be fixed. We pass the z axis of our coordinate system through the nuclei, with the origin at their center of mass. The distances between the particles are labeled as shown in the figure. The Born-Oppenheimer Hamiltonian operator is

$$\hat{H} = -\frac{\hbar^2}{2m}(\nabla_1^2 + \nabla_2^2) + \frac{e^2}{4\pi\varepsilon_0}\left(\frac{1}{R} - \frac{1}{r_{1A}} - \frac{1}{r_{1B}} - \frac{1}{r_{2A}} - \frac{1}{r_{2B}} + \frac{1}{r_{12}}\right)$$

$$= \hat{H}_{HMI}(1) + \hat{H}_{HMI}(2) + \frac{e^2}{4\pi\varepsilon_0}\left(\frac{1}{R} + \frac{1}{r_{12}}\right) \tag{13.3-1}$$

where $\hat{H}_{HMI}(1)$ and $\hat{H}_{HMI}(2)$ are hydrogen-molecule-ion electronic Hamiltonian operators as in Equation (13.1-1). In the second version of this equation, the nuclear repulsion term (proportional to $1/R$) must be included because the HMI Hamiltonian operators do not contain this term. This term is a constant in the Born-Oppenheimer approximation. It can be omitted during the solution of the electronic Schrödinger equation and then added to the resulting electronic energy eigenvalue to obtain the total Born-Oppenheimer energy, as in Equation (13.1-6).

The final term in the Hamiltonian operator, representing interelectron repulsion, prevents separation of the equation into two one-electron equations. We neglect this term, just as we did in the helium atom treatment, obtaining the zero-order electronic Hamiltonian operator:

$$\hat{H}_{el}^{(0)} = \hat{H}_{HMI}(1) + \hat{H}_{HMI}(2) \tag{13.3-2}$$

This approximate Hamiltonian leads to a separation of variables, with a trial function that is a product of two orbitals

$$\Psi^{(0)} = \psi_1(1)\psi_2(2) \tag{13.3-3}$$

The orbitals $\psi_1(1)$ and $\psi_2(2)$ are hydrogen-molecule-ion orbitals, and the zero-order Born-Oppenheimer energy is the sum of two hydrogen-molecule-ion electronic energies plus the $1/R$ term:

$$E^{(0)} = E_{HMI}(1) + E_{HMI}(2) + \frac{e^2}{4\pi\varepsilon_0 R} \tag{13.3-4}$$

If we wanted, we could use the exact Born-Oppenheimer orbitals in Equation (13.3-3). However, the zero-order orbital approximation is crude enough that no appreciable further damage is done by using the LCAO-MO orbitals instead of the exact Born-Oppenheimer orbitals.

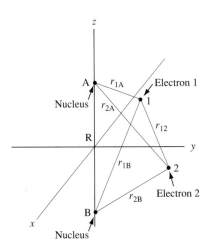

Figure 13.10. The Hydrogen Molecule System. This system is analogous to the helium system in having two electrons. Our treatment of it is similar to the treatment of the helium atom, except for using molecular orbitals instead of atomic orbitals.

The Aufbau principle is the same for molecules and atoms.

To find the ground-state wave function, we apply the Aufbau principle, choosing the lowest-energy LCAO-MOs compatible with the Pauli exclusion principle. Including the spin factor, the antisymmetrized and normalized zero-order ground state wave function is

$$\Psi^{(0)} = \frac{1}{\sqrt{2}}\,\psi_{\sigma g 1s}(1)\psi_{\sigma g 1s}(2)[\alpha(1)\beta(2) - \beta(1)\alpha(2)] \qquad \text{(13.3-5)}$$

Electron configurations are assigned just as with atoms. This function corresponds to the configuration $(\sigma g 1s)^2$.

The zero-order wave function corresponds to dissociation energy $D_e = 2.65$ eV and equilibrium internuclear distance $R_e = 0.84 \times 10^{-10}$ m, compared with the experimental values of 4.75 eV and 0.741×10^{-10} m. More accurate values can be obtained with configuration interaction or by using the perturbation method, variation method, or self-consistent field method. For example, introduction of variable orbital exponents into the atomic orbitals in the LCAO-MO wave function gives $D_e = 3.49$ eV and $R_e = 0.732 \times 10^{-10}$ m with an orbital exponent corresponding to an apparent nuclear charge of 1.197 protons. A Hartree-Fock calculation gives $D_e = 3.64$ eV and $R_e = 0.74 \times 10^{-10}$ m.[4] Since the Hartree-Fock orbitals are the best possible orbitals (no error except for the correlation error), the correlation error is equal to 1.11 eV.

The wave function of Equation (13.3-5) corresponds to two electrons moving in one bonding molecular space orbital whose orbital region encompasses both nuclei. This situation is called a **single covalent bond**, with one pair of shared electrons.

A configuration interaction wave function using the configurations $(\sigma g 1s)^2$ and $(\sigma^* u 1s)^2$ is

$$\Psi_{CI} = C_{CI}[\psi_{\sigma g 1s}(1)\psi_{\sigma g 1s}(2) + c_u \psi_{\sigma^* u 1s}(1)\psi_{\sigma^* u 1s}(2)]$$
$$\times [\alpha(1)\beta(2) - \beta(1)\alpha(2)] \qquad \text{(13.3-6)}$$

where the value of c_u is chosen to minimize the variational energy and the value of C_{CI} is chosen to normalize the function. With the optimum value of the parameter c_u, this function gives $D_e = 4.02$ eV and $R_e = 0.75 \times 10^{-10}$ m. Including a single additional configuration has thus removed about a third of the correlation error.[5]

Excited states of the hydrogen molecule correspond to configurations other than $(\sigma g 1s)^2$. Some of these states are not bound states. For example, for the configuration $(\sigma g 1s)(\sigma^* u 1s)$, one electron is in a bonding orbital and one in an antibonding orbital. The antibonding effect of one electron approximately cancels the bonding effect of the other, and the molecule will dissociate into two hydrogen atoms if placed in such a state.

Although there is no net repulsion between the nuclei, a molecule with as many bonding orbitals as antibonding orbitals will dissociate, since the dissociation will decrease the Gibbs energy of the system by increasing the entropy.

---

[4] Ira N. Levine, *Quantum Chemistry*, 3rd ed., Allyn & Bacon, Boston, 1983, pp. 355ff.
[5] *ibid.*

Molecular Term Symbols

Much as with atoms, term symbols are used to designate electronic states of homonuclear diatomic molecules. In the atomic case, the main part of the term symbol was determined by the orbital angular momentum: the letter S stood for $L = 0$, the letter P for $L = 1$, D for $L = 2$, etc. The energy eigenfunctions of molecules are not necessarily eigenfunctions of $\hat{L}^2$, since this operator does not commute with the electronic Hamiltonian operator. However, $\hat{L}_z$ does commute with the electronic Hamiltonian operator, and an energy eigenfunction can be an eigenfunction of $\hat{L}_z$:

$$\hat{L}_z \Psi = \hbar M_L \Psi \tag{13.3-7}$$

where M_L is the same quantum number as in the atomic case and equals an integer. In the orbital approximation, M_L is equal to the algebraic sum of the values of m for each orbital, just as with atoms. For example, in the ground state of diatomic hydrogen, both of the electrons occupy sigma orbitals, so the value of M_L is zero.

For molecules, we define a non-negative quantum number Λ, equal to $|M_L|$. The main part of the molecular term symbol is a capital Greek letter, assigned as follows:

Value of Λ:	0	1	2	3...
Symbol:	Σ	Π	Δ	Φ...

Since our nonrelativistic electronic Hamiltonian operator contains no spin, all of the spin angular momentum operators commute with the electronic Hamiltonian operator, and the energy eigenfunctions can be eigenfunctions of the same spin operators as the energy eigenfunctions of atoms. The quantum number for the square of the total spin angular momentum, S, is equal to a nonnegative integer or half-integer, and a left superscript equal to $2S + 1$ is used, just as in the atomic case.

There will often be several degenerate states with the same values of Λ and S. These states make up a term, and the same term symbol applies to all of them. The ground state of the hydrogen molecule is nondegenerate, and constitutes a $^1\Sigma$ (singlet sigma) term.

The energy eigenfunctions of homonuclear diatomic molecules can be chosen to be eigenfunctions of the symmetry operators belonging to the molecule, just as can individual orbitals. If the wave function is an eigenfunction of the inversion operator with eigenvalue unity, a right subscript g is attached to the term symbol. If it is an eigenfunction of the inversion operator with eigenvalue -1, a right subscript u is attached. With sigma terms, if the wave function is an eigenfunction of a $\hat{\sigma}_v$ operator with eigenvalue $+1$, a right superscript $+$ is added, and if it is an eigenfunction of this operator with eigenvalue -1, a right superscript $-$ is added.

The excited configuration $(\sigma g1s)(\sigma^*u1s)$ can correspond to two different terms. The eigenvalue of the inversion operator is -1 for both terms, since one orbital is g and the other is u. Because the spins do not have to be paired, one term is a triplet and the other is a singlet. The term symbols for the configuration are $^1\Sigma_u^+$ and $^3\Sigma_u^+$.

The Valence Bond Method

Orbital wave functions are not the only type of approximate wave functions that have been used for the hydrogen molecule. In 1927, Heitler and London[6] introduced a type of approximate wave function for the ground state of the hydrogen molecule called the **valence bond function**:

$$\Psi_{VB} = C[\psi_{1sA}(1)\psi_{1sB}(2) + \psi_{1sB}(1)\psi_{1sA}(2)][\alpha(1)\beta(2) - \beta(1)\alpha(2)] \quad \textbf{(13.3-8)}$$

where ψ_{1sA} and ψ_{1sB} are the $1s$ atomic orbitals centered on the nuclei at locations A and B, respectively, and C is a normalizing factor. This wave function expresses the sharing of electrons in a different way from an orbital wave function. It contains one term in which electron 1 occupies an atomic orbital centered on nucleus A while electron 2 occupies an atomic orbital centered on nucleus B, and another term in which the locations are switched. The two terms make a symmetric space factor, which must be multiplied by an antisymmetric spin factor.

When the valence bond function of Equation (13.3-8) is used to calculate the variational energy without using variable orbital exponents or other variable parameters, the values $D_e = 3.20$ eV and $R_e = 0.80 \times 10^{-10}$ m are obtained. These are in better agreement with experiment than the values obtained from the simple LCAO-MO wave function of Equation (13.3-5).

If we express the molecular orbitals in the wave function of Equation (13.3-5) in terms of atomic orbitals, the space factor is

$$C_g^2[\psi_{1sA}(1) + \psi_{1sB}(1)][\psi_{1sA}(2) + \psi_{1sB}(2)]$$
$$= C_g^2[\psi_{1sA}(1)\psi_{1sA}(2) + \psi_{1sB}(1)\psi_{1sB}(2)$$
$$+ \psi_{1sA}(1)\psi_{1sB}(2) + \psi_{1sB}(2)\psi_{1sA}(2)] \quad \textbf{(13.3-9)}$$

where the normalizing constant is called C_g.

The last two terms on the right-hand side of Equation (13.3-9) are the same as the space factor of the valence bond wave function. These terms correspond to sharing of electrons and are called **covalent terms**. The other two terms are called **ionic terms**, since one term has both electrons on nucleus A and the other has them on nucleus B. The simple LCAO-MO wave function gives the ionic terms equal weight with the covalent terms, while the simple valence bond function omits them completely. Because electrons repel each other, the ionic terms should be less prominent in an optimum wave function than the covalent terms. From the numerical results, it is apparent that it is better to omit them completely than to include them with the same weight as the covalent terms. In fact, this omission corresponds to taking approximate account of electron correlation.

[6] W. Heitler and F. London, *Z. Phys.* **44**, 455 (1927).

It is even better to include the ionic terms with reduced weight. A modified valence bond wave function is

$$\Psi_{MVB} = c_{VB}\Psi_{VB} + c_I\Psi_I \qquad (13.3\text{-}10)$$

where c_{VB} and c_I are variable parameters and Ψ_I contains only the ionic terms:

It is nearly always true that a more flexible variation function can give a better (lower) energy than a less flexible function.

$$\Psi_I = [\psi_{1sA}(1)\psi_{1sA}(2) + \psi_{1sB}(1)\psi_{1sB}(2)][\alpha(1)\beta(2) - \beta(1)\alpha(2)] \quad (13.3\text{-}11)$$

This wave function gives better results than either the simple LCAO-MO function or the simple valence bond function. In fact, it can be made identical to the optimized configuration interaction function of Equation (13.3-6).

Exercise 13.9   By expressing the function of Equation (13.3-6) in terms of atomic orbitals, show that it can be made the same as the function of Equation (13.3-10). Express the parameters c_{VB} and c_I in terms of C_{CI} and c_u.

Because an improved valence bond wave function can be the same as a configuration interaction LCAO-MO wave function, the distinction between the valence bond method and the LCAO-MO method at least partially disappears when improvements are made in the simple functions. However, the valence bond method has become less popular than the molecular orbital method. One reason is that it is possible to make further improvements in the molecular orbital method by adding more configurations, but improvements beyond addition of ionic terms to the simple valence bond functions are more difficult.

Diatomic Helium

The simple LCAO-MO method explains why He_2 does not exist in its ground state. The zero-order Born-Oppenheimer Hamiltonian operator for diatomic helium consists of four hydrogen molecule-like Hamiltonian operators (with interelectron repulsions and constant internuclear repulsion omitted):

If the nuclear repulsion terms are omitted from each HMIL Hamiltonian, then the nuclear repulsion term must be added to the energy eigenvalue at the end of the calculation.

$$\hat{H}^{(0)} = \hat{H}_{HMIL}(1) + \hat{H}_{HMIL}(2) + \hat{H}_{HMIL}(3) + \hat{H}_{HMIL}(4) \quad (13.3\text{-}12)$$

The one-electron Hamiltonians on the right-hand side of this equation are called "hydrogen molecule ion–like" (HMIL) because they include two protons in each nucleus instead of one. This molecular Hamiltonian leads to separation of variables with a trial function that is a product of four hydrogen molecule ion–like orbitals. We apply the Aufbau principle to find the ground state. Including spin functions, the nonantisymmetrized zero-order ground-state wave function is

Even when we do not antisymmetrize a function, we must write it so that it could be antisymmetrized without vanishing. That is, we must obey the Pauli exclusion principle.

$$\Psi^{(0)} = \psi_{\sigma g1s}(1)\psi_{\sigma g1s}(2)\psi_{\sigma^*u1s}(3)\psi_{\sigma^*u1s}(4)\alpha(1)\beta(2)\alpha(3)\beta(4) \quad (13.3\text{-}13)$$

This wave function can be antisymmetrized without vanishing.

Exercise 13.10　Antisymmetrize the function of Equation (13.3-13) by writing it as a 4-by-4 Slater determinant.

The wave function of Equation (13.3-13) corresponds to the configuration $(\sigma g 1s)^2(\sigma^* u 1s)^2$. Two electrons occupy bonding orbitals and two occupy antibonding orbitals. The repulsive effect of the antibonding orbitals roughly cancels the attractive effect of the bonding orbitals. The molecule, if ever formed in this electronic state, would dissociate into two helium atoms. However, the molecule has been observed in excited states.

We define a **bond order**:

$$\text{BO} = \text{bond order} = \frac{1}{2}(n_{\text{bonding}} - n_{\text{antibonding}}) \qquad \textbf{(13.3-14)}$$

where n_{bonding} is the number of electrons occupying bonding orbitals and $n_{\text{antibonding}}$ is the number of electrons occupying antibonding orbitals. The division by 2 is included to make our bond order conform to the traditional definition of a single bond as having a pair of shared electrons, a double bond as having two pairs, etc. The bond order of the diatomic helium molecule in its ground state is zero, and that of the diatomic hydrogen molecule in its ground state is unity.

Homonuclear Diatomic Molecules with More Than Four Electrons

We adopt these policies in order to get the simplest possible wave functions that include realistic bonding.

For homonuclear diatomic molecules of elements with more than four electrons, we will require additional molecular orbitals beyond the two space orbitals we have already used. We now construct a set of simple LCAO-MOs, following the policies:

1. Each LCAO-MO is a combination of two atomic orbitals of the same type with one centered on each of two nuclei.
2. Each LCAO-MO is an eigenfunction of the symmetry operators belonging to the molecule.

For a given element, we will use the atomic orbitals for that element, not those for hydrogen, because these orbitals correspond to the appropriate number of protons in each nucleus.

Two independent linear combinations can be made from two independent basis functions, etc. The number of independent functions remains constant as linear combinations are formed.

Two independent LCAO molecular orbitals can be constructed from two atomic orbitals. One of them will be a bonding orbital, for which the two atomic orbitals will add in the overlap region, and the other will be an antibonding orbital, for which the two atomic orbitals will cancel in the center of the overlap region. A bonding LCAO-MO and an antibonding LCAO-MO are made from the $2s$ atomic orbitals in exactly the same way as the $\sigma g 1s$ and the $\sigma^* u 1s$ orbitals. These are called the $\sigma g 2s$ and the $\sigma^* u 2s$ orbitals.

The configuration of diatomic lithium in the ground state is $(\sigma g 1s)^2(\sigma^* u 1s)^2(\sigma g 2s)^2$, and the configuration of diatomic beryllium is $(\sigma g 1s)^2(\sigma^* u 1s)^2(\sigma g 2s)^2(\sigma^* u 2s)^2$. The bond order for Li_2 is unity and the bond order for Be_2 is zero, explaining why Be_2 does not exist in its ground state.

There are six $2p$ atomic space orbitals on two nuclei, and six LCAO molecular space orbitals can be constructed from them. We use the real atomic orbitals, ψ_{2px}, ψ_{2py}, and ψ_{2pz}, in order to make the orbital regions more compact, but the complex atomic orbitals that are eigenfunctions of the $\hat{L}_z$ operator could also be used.

In order to obtain eigenfunctions of the symmetry operators, the two $2pz$ orbitals can be combined, but not a $2pz$ with a $2px$, etc. We obtain the LCAO molecular orbitals:

$$\psi_{\sigma g2pz} = C(\psi_{2pzA} - \psi_{2pzB}) \tag{13.3-15a}$$

$$\psi_{\sigma^*u2pz} = C(\psi_{2pzA} + \psi_{2pzB}) \tag{13.3-15b}$$

$$\psi_{\pi u2px} = C(\psi_{2pxA} + \psi_{2pxB}) \tag{13.3-15c}$$

$$\psi_{\pi^*g2px} = C(\psi_{2pxA} - \psi_{2pxB}) \tag{13.3-15d}$$

$$\psi_{\pi u2py} = C(\psi_{2pyA} + \psi_{2pyB}) \tag{13.3-15e}$$

$$\psi_{\pi^*g2py} = C(\psi_{2pyA} - \psi_{2pyB}) \tag{13.3-15f}$$

where the normalizing constant C can have a different value in each case.

Figure 13.11 shows the orbital regions of the atomic orbitals and the LCAO molecular orbitals. The $2pz$ atomic orbitals produce a sigma molecular orbital, since they correspond to $m = 0$ (they are the same as

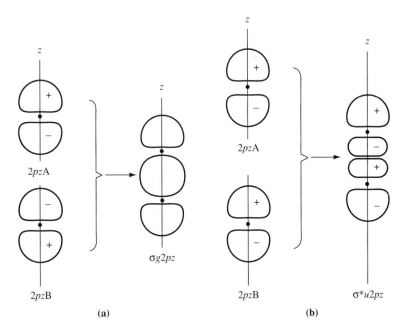

Figure 13.11. Orbital Regions for 2p Atomic Orbitals and LCAO Molecular Orbitals Made from Them. (a) The $\sigma g2pz$ LCAO-MO. **(b)** The σ^*u2pz LCAO-MO. **(c)** The $\pi u2py$ LCAO-MO. **(d)** The π^*g2py LCAO-MO. This diagram shows how the orbital regions for bonding LCAO-MOs arise from addition of atomic orbitals in the overlap regions and how the orbital regions for antibonding LCAO-MOs arise from cancellation of atomic orbitals in the overlap regions.

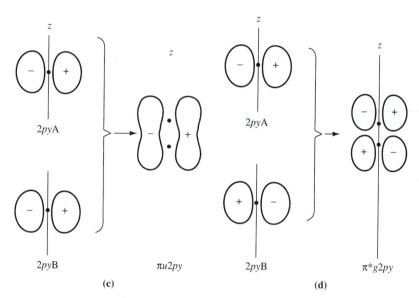

Figure 13.11. (*Continued*)

210 orbitals). A difference of the $2pz$ atomic orbitals produces the bonding orbital, while a sum produces an antibonding orbital. This behavior comes from the fact that the orbital function has different signs in the two lobes of the atomic orbital region. Some authors reverse the sign of one orbital or reverse the direction of the z axis for one orbital in order to write a positive sign in the bonding orbital. The important thing is that the atomic orbitals must add and not cancel in the overlap region to make a bonding orbital.

The LCAO-MOs made from the $2px$ and $2py$ atomic orbitals are called π orbitals, because the $2px$ and $2py$ atomic orbitals are linear combinations of the 211 and 21, -1 orbitals. The bonding pi orbitals are "u" instead of "g" because the two lobes of the $2p$ atomic orbitals have opposite signs.

Figure 13.12 shows a **correlation diagram**, in which the energies of the atomic orbitals and the LCAO molecular orbitals are shown schematically, with line segments connecting the LCAO-MOs and the atomic orbitals from which they were constructed. From Figure 12.4 it is apparent that the atomic orbital energies depend on nuclear charges. The LCAO-MO energies also depend on nuclear charge, as well as on internuclear distance. The energy scale in the figure is only qualitative and is shifted up or down for different elements.

Although the $2px$, $2py$, and $2pz$ atomic orbitals are at the same energy, the overlap for the $\sigma 2p$ LCAO-MOs is different from that of the $\pi 2p$ LCAO-MOs. The $\sigma 2p$ energy is not equal to the $\pi u2px$ and $\pi u2px$ energies, although the $\pi u2px$ and $\pi u2py$ energies are equal to each other. The $2p1$ and $2p, -1$ orbitals also lead to degenerate π orbitals, which we call the $\pi u2p1$ and $\pi u2p, -1$ LCAO-MOs.

The order of the LCAO-MO energies in the figure is correct for elements up to nitrogen for distances near the equilibrium internuclear distance.

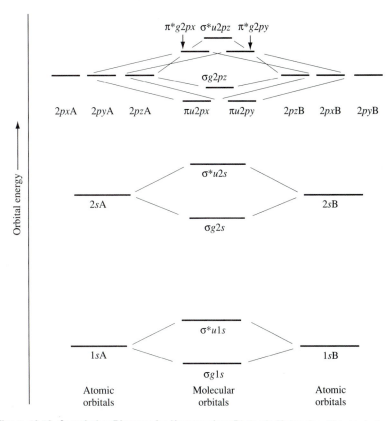

Figure 13.12. Correlation Diagram for Homonuclear Diatomic Molecules. This kind of diagram is an orbital energy level diagram in which the connections (correlations) between an LCAO-MO and the atomic orbitals making up the linear combination are displayed.

However, for oxygen and fluorine the $\sigma g2pz$ bonding orbital is lower in energy than the $\pi u2px$ and $\pi u2py$ bonding orbitals.

We can now write the ground-state configurations in Table 13.1 for all of the homonuclear diatomic molecules of elements of the first and second rows of the periodic chart, using the LCAO molecular orbitals in Figure 13.12 and the Aufbau principle. We could have used the $\pi u2p1$ and $\pi u2p, -1$ instead of the $\pi u2px$ and $\pi u2py$ orbitals, and the value of Λ could more easily have been determined if we had done so.

In boron and oxygen, the final two electrons were assigned to different space orbitals with parallel spins, in accordance with Hund's first rule, which is the same for molecules as for atoms. This agrees with the experimental fact that both diatomic boron and diatomic oxygen have triplet ground levels (two unpaired electrons).

Since each space orbital is listed separately, these configurations are detailed configurations. The detailed configurations convey all of the information conveyed by a nonantisymmetrized wave function. Since the $\pi u2px$ and $\pi u2py$ orbitals are degenerate, they are analogous to a subshell and could be lumped together, as could the π^*g2px and π^*g2py orbitals. For

Hund's first rule is the same for molecules as for atoms. Hund's second and third rules do not apply, since L and J are not good quantum numbers for molecules.

Table 13.1. Ground-State Electron Configurations of Diatomic Molecules

Molecule	Configuration	Bond Order
H_2	$(\sigma g1s)^2$	1
He_2	$(\sigma g1s)^2(\sigma *u1s)^2$	0
Li_2	$(\sigma g1s)^2(\sigma *u1s)^2(\sigma g2s)^2$	1
Be_2	$(\sigma g1s)^2(\sigma *u1s)^2(\sigma g2s)^2(\sigma *u2s)^2$	0
B_2	$(\sigma g1s)^2(\sigma *u1s)^2(\sigma g2s)^2(\sigma *u2s)^2(\pi u2px)(\pi u2py)$	1
C_2	$(\sigma g1s)^2(\sigma *u1s)^2(\sigma g2s)^2(\sigma *u2s)^2(\pi u2px)^2(\pi u2py)^2$	2
N_2	$(\sigma g1s)^2(\sigma *u1s)^2(\sigma g2s)^2(\sigma *u2s)^2(\pi u2px)^2(\pi u2py)^2(\sigma g2pz)^2$	3
O_2	$(\sigma g1s)^2(\sigma *u1s)^2(\sigma g2s)^2(\sigma *u2s)^2(\sigma g2pz)^2(\pi u2px)^2(\pi u2py)^2(\pi *g2px)(\pi *g2py)$	2
F_2	$(\sigma g1s)^2(\sigma *u1s)^2(\sigma g2s)^2(\sigma *u2s)^2(\sigma g2pz)^2(\pi u2px)^2(\pi u2py)^2(\pi *g2px)^2(\pi *g2py)^2$	1
Ne_2	$(\sigma g1s)^2(\sigma *u1s)^2(\sigma g2s)^2(\sigma *u2s)^2(\sigma g2pz)^2(\pi u2px)^2(\pi u2py)^2(\pi *g2px)^2(\pi *g2py)^2(\sigma *u2pz)^2$	0

example, the ground-state configuration of diatomic oxygen would be written

$$(\sigma g1s)^2(\sigma *u1s)^2(\sigma g2s)^2(\sigma *u2s)^2(\sigma g2p)^2(\pi u2p)^4(\pi *g2p)^2$$

This is analogous to the subshell configurations in Chapter 12 and does not explicitly show that the final two antibonding electrons occupy different orbitals.

▼ **EXAMPLE 13.2**

Write an orbital wave function without antisymmetrization for the diatomic boron molecule in its ground level.

Solution

$$\Psi = \psi_{\sigma g1s}(1)\alpha(1)\psi_{\sigma g1s}(2)\beta(2)\psi_{\sigma *u1s}(3)\alpha(3)\psi_{\sigma *u1s}(4)\beta(4)$$
$$\times \ \psi_{\sigma g2s}(5)\alpha(5)\psi_{\sigma g2s}(6)\beta(6)\psi_{\sigma *u2s}(7)\alpha(7)\psi_{\sigma *u2s}(8)\beta(8)$$
$$\times \ \psi_{\pi u2px}(9)\alpha(9)\psi_{\pi u2py}(10)\alpha(10)$$

The wave function shown in the example is one of the triplet wave functions making up the ground level. Another has spin-down spin functions instead of spin up for the final two factors, and a third would have paired spins but total spin magnitude equal to $\sqrt{2}\hbar$, as in Figure 11.17b.

Exercise 13.11

Write an orbital wave function without antisymmetrization for
   a. Diatomic oxygen in its ground level.
   b. Diatomic helium in the excited-state configuration $(\sigma g1s)^2(\sigma *u1s)(\sigma g2s)$. What is the bond order for this molecule? Do you think it could exist?

Term Symbols for Homonuclear Diatomic Molecules

The symmetry operators must be applied to all factors. Two u orbitals make a g wave function, etc. The same vertical reflection plane must be used for all orbitals.

Term symbols can be written for states of homonuclear diatomic molecules from inspection of their configurations. For example, in the ground level of diatomic boron, the electron spins and orbital angular momentum projections occur in canceling pairs in the filled subshells. Only the electrons in pi orbitals make a contribution. By Hund's first rule, these electrons occupy different space orbitals with unpaired spins. The orbital angular momentum projections cancel, making a triplet sigma term. Both orbitals have eigenvalue -1 for the inversion operator, so their product has eigenvalue $+1$. A vertical mirror plane in the x-z plane will given eigenvalue $+1$ for the $\pi u2px$ orbital and eigenvalue -1 for the $\pi u2py$ orbital, so the term symbol is $^3\Sigma_g^-$.

An Alternative Set of Wave Functions

We are never able to write a unique "correct" wave function as we did with the hydrogen atom. We are just trying to write a wave function that resembles the unobtainable correct wave function closely enough for the purposes at hand. Since the attractive effect of an electron in a bonding orbital and the repulsive effect of an electron in an antibonding orbital nearly cancel, the variational energy will be almost unchanged if we replace pairs of bonding and antibonding orbitals by the atomic orbitals from which the LCAO-MOs were constructed. Electrons occupying atomic orbitals are counted as nonbonding and are omitted from bond order calculations.

▼   **EXAMPLE 13.3**

Give the configuration for the ground state of diatomic carbon using nonbonding orbitals as much as possible.

Solution

The configuration is $(1sA)^2(1sB)^2(2sA)^2(2sB)^2(\pi u2px)^2(\pi u2py)^2$. The bond order is still equal to 2, with four bonding electrons, no antibonding electrons, and eight nonbonding electrons.

▲

Exercise 13.12

Using as many nonbonding orbitals as possible, give the ground-state configurations of diatomic boron and diatomic fluorine. Write the corresponding orbital wave function without antisymmetrization for each of these molecules.

Homonuclear Diatomic Molecules in the Valence-Bond Approximation

For some homonuclear diatomic molecules, satisfactory simple valence bond wave functions can be constructed. In the simple valence bond approximation, two bonding electrons occupy a bonding factor constructed from two atomic orbitals on different nuclei, as in Equation (13.3-8). We make maximum use of nonbonding orbitals, since our simple valence bond theory has nothing analogous to antibonding orbitals. We make no attempt to describe the ground states of diatomic boron and diatomic oxygen because our simple bonding factors do not accommodate unpaired electrons.

▼   **EXAMPLE 13.4**   Write a simple valence bond wave function for diatomic carbon in its ground state.

Solution

$$\Psi = \psi_{1sA}(1)\alpha(1)\psi_{1sA}(2)\beta(2)\psi_{1sB}(3)\alpha(3)\psi_{1sB}(4)\beta(4)$$
$$\times \ \psi_{2sA}(5)\alpha(5)\psi_{2sA}(6)\beta(6)\psi_{2sB}(7)\alpha(7)\psi_{2sB}(8)\beta(8)$$
$$\times \ \sqrt{\frac{1}{2}}\ [\psi_{2pxA}(9)\psi_{2pxB}(10) + \psi_{2pxB}(9)\psi_{2pxA}(10)]$$
$$\times \ [\alpha(9)\beta(10) - \beta(9)\alpha(10)]$$
$$\times \ \sqrt{\frac{1}{2}}\ [\psi_{2pyA}(11)\psi_{2pyB}(12) + \psi_{2pyB}(11)\psi_{2pyA}(12)]$$
$$\times \ [\alpha(11)\beta(12) - \beta(11)\alpha(12)] \qquad \textbf{(13.3-16)}$$

▲

Since a bonding factor corresponds to a single covalent bond, the bond order for diatomic carbon in the valence bond approximation is equal to 2, just as in the LCAO-MO method. We have used bonding factors made with p orbitals to represent the double bond. These bonding factors represent two pi bonds. In the next section we will discuss hybrid atomic orbitals, with which we can represent a double bond as a sigma bond and a pi bond.

The wave function in Example 13.4 is partially antisymmetrized, since the bonding factors and associated spin factors provide antisymmetrization between electrons 9 and 10 and also between 11 and 12. If an improved valence bond wave function is desired, ionic terms can be added to the covalent terms in Equation (13.3-16), just as in Equation (13.3-10) for the hydrogen molecule.

Exercise 13.13   **a.** Write a simple valence bond wave function for the ground state of diatomic fluorine.

b. Write a modified valence bond wave function for the ground state of diatomic fluorine, including ionic terms.

Excited States of Homonuclear Diatomic Molecules

Excited states are represented in the LCAO-MO approximation by configurations other than the one arrived at using the Aufbau principle. Term symbols can be written for an excited configuration.

Write the term symbols that can occur for the diatomic beryllium configuration $(\sigma g 1s)^2(\sigma^*u1s)^2(\sigma g2s)^2(\sigma^*u2s)(\sigma u2p)$. Which term will have the lowest energy? What is the bond order for this configuration? Do you think the molecule could exist in this configuration? Can you write a valence bond wave function equivalent to this configuration?

If a molecule absorbs energy so that it makes a transition from its ground state to a state corresponding to a wave function obtained by deleting one nonbonding orbital and adding one antibonding pi orbital, the transition is called an $n \rightarrow \pi^*$ (*n* to pi-star) transition, and we say that a nonbonding electron has been promoted to the π^* orbital. Similarly, if an electron is promoted from a pi bonding orbital to a pi antibonding orbital, the transition is called a $\pi \rightarrow \pi^*$ transition.

13.4 Heteronuclear Diatomic Molecules

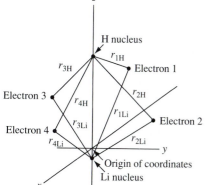

Figure 13.13. The LiH System. This system has two nuclei of different elements and four electrons. The origin of coordinates is placed at the center of mass. The system is less symmetrical than a homonuclear diatomic molecule.

The simplest commonly encountered heteronuclear diatomic molecule is lithium hydride, LiH. This molecule is a system of two nuclei and four electrons, like diatomic helium. We say that lithium hydride and diatomic helium are **isoelectronic** (have the same number of electrons). Figure 13.13 shows the LiH system.

The zero-order Born-Oppenheimer Hamiltonian operator for the LiH molecule in zero order (with the internuclear repulsion term and the interelectron repulsion terms omitted) is

$$\hat{H}^{(0)} = \hat{H}_1(1) + \hat{H}_1(2) + \hat{H}_1(3) + \hat{H}_1(4) \qquad \textbf{(13.4-1)}$$

where $H_1(i)$ is the one-electron Hamiltonian operator for electron i:

$$\hat{H}_1(i) = -\frac{\hbar^2}{2m}\nabla_i^2 + \frac{e^2}{4\pi\varepsilon_0}\left(\frac{1}{r_{iH}} + \frac{3}{r_{iLi}}\right) \qquad \textbf{(13.4-2)}$$

Here r_{iH} is the distance from the hydrogen nucleus to electron i and r_{iLi} is the distance from the lithium nucleus to electron i, as labeled in Figure 13.13. H_1 is not a hydrogen molecule ion–like operator because the two nuclear charges are different.

The zero-order Hamiltonian operator of Equation (13.4-1) gives a Schrödinger equation in which the variables can be separated by assuming the wave function:

$$\Psi = \psi_1(1)\psi_2(2)\psi_3(3)\psi_4(4) \qquad \textbf{(13.4-3)}$$

where each factor is a molecular orbital.

We seek LCAO approximations to the orbitals in Equation (13.4-3). We cannot obtain the values of the coefficients from symmetry arguments, because the only symmetry operators that commute with the Hamiltonian operator are rotations about the z axis and reflections through vertical planes. However, optimum values of the coefficients in the LCAO molecular orbitals can be obtained by the variation method or by the Roothaan modification of the Hartree-Fock method, in which LCAO molecular orbitals are used instead of numerical representations of the molecular orbitals.[7]

This is a minimal basis set, containing as few atomic orbitals as possible.

We take a basis set consisting of four space orbitals: the lithium $1s$, lithium $2s$, lithium $2pz$, and hydrogen $1s$ atomic orbitals. These orbitals all have cylindrical symmetry about the bond axis (the z axis). Any linear combination of them will be an eigenfunction of the symmetry operators that commute with the Hamiltonian operator. They also correspond to $m = 0$, so any linear combination of these atomic orbitals is a sigma orbital.

The $2px$ and $2py$ lithium orbitals have different symmetry about the bond axis and have $m = \pm 1$, so that the LCAO-MOs would not be eigenfunctions of the rotation operators about the bond axis and of the $\hat{L}_z$ operator if these atomic orbitals were included.

Four independent LCAO-MOs can be made from our basis set:

$$\psi_{i\sigma} = c^{(i)}_{1s\text{Li}}\psi_{1s\text{Li}} + c^{(i)}_{2s\text{Li}}\psi_{2s\text{Li}} + c^{(i)}_{2pz\text{Li}}\psi_{2pz\text{Li}} + c^{(i)}_{1s\text{H}}\psi_{1s\text{H}} \quad \textbf{(13.4-4)}$$

where i is an index ranging from 1 to 4, used to specify which one of the LCAO-MOs is meant. We have added a σ subscript to indicate that all of these orbitals are sigma orbitals. An additional subscript is added to the atomic orbitals to show which atom the atomic orbital is taken from.

We do not discuss the calculations that lead to optimum values of the coefficients, but Table 13.2 shows the results of the Hartree-Fock-Roothaan procedure for the four LCAO-MOs. These orbitals are numbered 1σ, 2σ, etc., with the lowest index corresponding to the lowest energy, etc. Figure 13.14 is a correlation diagram showing schematically the atomic and molecular orbital energies.

By the Aufbau principle, the ground-state wave function requires only two of the four space orbitals:

$$\Psi_{gs} = \psi_{1\sigma}(1)\alpha(1)\psi_{1\sigma}(2)\beta(2)\psi_{2\sigma}(3)\alpha(3)\psi_{2\sigma}(4)\beta(4) \quad \textbf{(13.4-5)}$$

Table 13.2. Results of Hartree-Fock-Roothaan Calculation for the LiH Ground State at an Internuclear Distance Equal to 159 pm

MO	$c_{1s\text{H}}$	$c_{2s\text{Li}}$	$c_{2pz\text{Li}}$	$c_{1s\text{H}}$
1	0.9996	−0.0000	−0.0027	0.0035
2	0.0751	0.3288	−0.2048	−0.7022
3	−0.0115	0.7432	0.6601	0.1236
4	−0.1256	0.8769	−1.0107	1.2005

From A. M. Karo, *J. Chem. Phys.* **30**, 1241 (1959).

[7] C. C. J. Roothaan, *Rev. Mod. Phys.* **23**, 69 (1951).

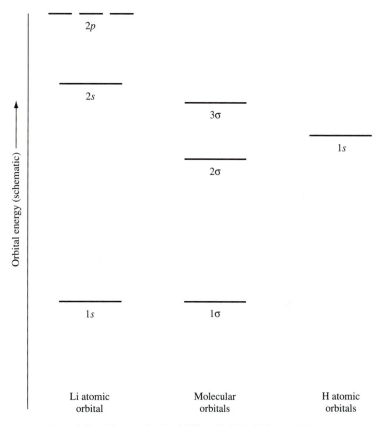

Figure 13.14. Correlation Diagram for the Lithium Hydride Molecule. Three molecular orbitals are shown. Only two of them are occupied in the ground state. The 1σ molecular orbital is essentially nonbonding, and the 2σ molecular orbital is essentially bonding. The molecule has a single bond, in contrast to the He_2 molecule, which has the same number of electrons but has bond order zero.

Let us now construct an approximate wave function that is qualitatively the same as that of Equation (13.4-5) but can be constructed without calculation. The 1σ LCAO-MO is almost exactly the same as the $1s$ lithium orbital, so we replace this LCAO-MO by the $1s$ lithium orbital, which is a nonbonding orbital. This will make little difference to the calculated molecular energy.

The coefficient of the $1s$ lithium orbital in the 2σ LCAO-MO is smaller than the others, and the coefficients of the $2s$ and $2pz$ lithium orbitals are roughly the same size and of opposite signs. This suggests a way to go back to our previous policy of including only two atomic orbitals in our LCAO-MOs, with only minimal damage to the wave function: We omit the $1s$ lithium orbital and replace the $2s$ and $2pz$ orbitals by a new kind of atomic orbital:

$$\psi_{2sp1} = N_1(-\psi_{2s} + \psi_{2pz}) \tag{13.4-6}$$

The designation $2sp$ is used because the new atomic orbital is a linear combination of the $2s$ and a $2p$ orbital, and the subscript 1 indicates that this is the first orbital of this type. This orbital is one of a class of orbitals called **hybrid orbitals**, which are linear combinations of atomic orbitals centered on the same nucleus.

There is another independent $2sp$ hybrid orbital:

$$\psi_{2sp2} = N_2(-\psi_{2s} - \psi_{2pz}) \tag{13.4-7}$$

Exercise 13.15

a. Using the fact that the $2s$ and $2pz$ orbitals are normalized and orthogonal to each other, show that N_1 and N_2 both equal $\sqrt{1/2}$.

b. Show that the $2sp1$ and $2sp2$ orbitals are orthogonal to each other.

Figure 13.15 shows cross sections of the orbital regions of the $2s$ and $2pz$ orbitals and of the $2sp1$ and $2sp2$ hybrid orbitals, assuming hydrogen-like atomic orbitals. Consider the $2sp1$ hybrid orbital. Since the $2s$ orbital is negative in the outer part of its orbital region, the $2s$ and the $2pz$ orbitals

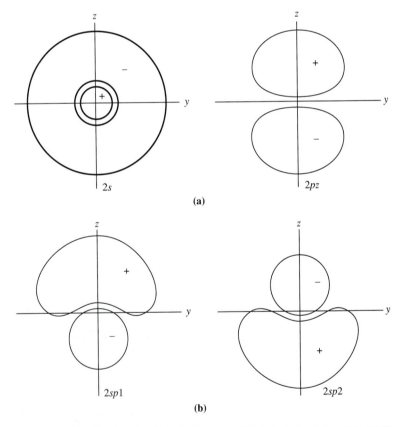

Figure 13.15. Orbital Regions for Nonhybridized and Hybridized Atomic Orbitals. (a) The nonhybridized orbital regions. **(b)** The hybridized orbital regions. This figure shows that the hybrid orbitals which are formed from one $2s$ atomic orbital and one $2p$ atomic orbital are directional. That is, each hybrid orbital has a lobe that extends farther from the nucleus and can give a larger overlap region. These two orbitals have directions 180° from each other.

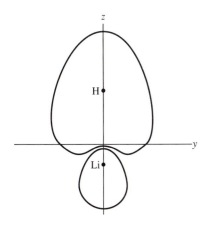

Figure 13.16. Cross Section of the Orbital Region of the Bonding LiH Orbital Made with a Hybrid Lithium Orbital (schematic). This diagram shows that the LCAO-MO has the typical bonding-orbital properties: it has no node between the nuclei, and it can be seen by inspection that it is made up from atomic orbitals that have the same symmetry about the bond axis.

add in the direction of the positive z axis and partially cancel in the direction of the negative z axis. The orbital regions of the hybrid orbitals are "directional." That is, each orbital region is localized, extending farther in one direction than in other directions. The orbital regions of the hybrid orbitals also extend slightly farther from the nucleus than do the orbital regions of the $2s$ and $2p$ orbitals. The orbital region of the $2sp2$ hybrid orbital extends in the opposite direction from that of the $2sp1$ orbital.

Using hybrid orbitals, we can to a fairly good approximation write two occupied LCAO-MOs as linear combinations of no more than two atomic orbitals:

$$\psi_{1\sigma} = \psi_{1s\text{Li}} \tag{13.4-8a}$$

$$\psi_{2\sigma} = c_{2sp1\text{Li}}\psi_{2sp1\text{Li}} + c_{1s\text{H}}\psi_{1s\text{H}}$$

$$\approx -0.47\psi_{2sp1\text{Li}} - 0.88\psi_{1s\text{H}} \tag{13.4-8b}$$

In the second line of Equation (13.4-8b), the values of the coefficients were chosen to maintain approximately the same relative weights of the atomic orbitals as in the Hartree-Fock orbital and to normalize the LCAO-MO. Figure 13.16 shows schematically the orbital region of this LCAO-MO. The 3σ molecular orbital has roughly equal coefficients for the $2s$ and $2p$ atomic orbitals, so it can be approximately represented as a linear combination of the $2sp2$ hybrid orbital and the hydrogen $1s$ orbital.

Exercise 13.16

a. Estimate the coefficients to represent the 3σ molecular orbital as a linear combination of the $2sp2$Li hybrid orbital and the $1s$H orbital. Is this molecular orbital a bonding, an antibonding, or a nonbonding orbital?

b. Estimate the coefficients to represent the 4σ molecular orbital as a linear combination of the $2sp1$Li hybrid orbital and the $1s$H orbital. Is this molecular orbital a bonding, an antibonding, or a nonbonding orbital?

The approximate orbitals of Equation (13.4-8) are not as good as the Hartree-Fock orbitals, but we can use them for a qualitative description of the bonding of the LiH molecule in its ground state. Lithium hydride exists because the two lowest-energy lithium hydride space orbitals are a nonbonding orbital and a bonding orbital, whereas in helium the two lowest-energy space orbitals are a bonding orbital and an antibonding orbital. Occupying the two lowest-energy space orbitals with four electrons gives a bond order of 1 for lithium hydride, whereas it would give a bond order of 0 for diatomic helium.

In the 2σ LCAO-MO, the coefficient of the $1s$H orbital is roughly twice as large as the coefficient of the $2sp1$Li orbital. There is thus a larger probability of finding an electron in the vicinity of the hydrogen nucleus than in the vicinity of the lithium nucleus if it occupies this space orbital. We say that the shared electrons are not equally shared, and we call this kind of chemical bond a **polar covalent bond**. It is intermediate between a covalent bond and an ionic bond. The ends of the molecule possess net charges, giving the molecule an **electric dipole moment**.

An arbitrary collection of charges can be represented as the sum of a monopole, a dipole, a quadrupole, an octopole, a hexadecapole, etc. The **monopole** is the algebraic sum of all of the charges and vanishes for a neutral molecule. The dipole moment of a neutral molecule is defined by

$$\boldsymbol{\mu} = \sum_{i=1}^{n} Q_i \mathbf{r}_i \qquad \text{(13.4-9)}$$

where Q_i is the ith charge and $\mathbf{r}_i$ is its position vector. The dipole moment is a measure of the charge separation in a molecule. A molecule with a positive end and a negative end has a nonzero dipole moment. The dipole moment points from the negative to the positive end of the dipole.

If a molecule contains only two electric charges $+Q$ and $-Q$ separated by a distance r, the magnitude of the dipole moment is given by

$$|\boldsymbol{\mu}| = \mu = Qr \qquad \text{(13.4-10)}$$

and the dipole moment points from the negative charge to the positive charge. A larger dipole moment corresponds either to a larger distance between the charges or to larger charges.

Exercise 13.17   For two charges of equal magnitude and opposite signs at arbitrary locations, show that the magnitude of the vector in Equation (13.4-9) is the same as that given by Equation (13.4-10).

A simple **quadrupole** consists of four charges of equal magnitude at the corners of a square such that the charges are alternately positive and negative around the square. Most molecules have nonzero quadrupole moments. However, we will not discuss quadrupoles and will say nothing about octopoles, hexadecapoles, etc.

Exercise 13.18   A simple quadrupole consists of a charge $+Q$ at the origin, a charge $-Q$ at the point $(1, 0)$, a charge $+Q$ at $(1, 1)$, and a charge $-Q$ at $(0, 1)$. Show that the dipole moment vanishes.

The classical expression in Equation (13.4-9) contains no momenta, so it is equal to its quantum mechanical operator:

$$\hat{\boldsymbol{\mu}} = \boldsymbol{\mu} \qquad \text{(13.4-11)}$$

The expectation value of the electric dipole of a molecule in a state Ψ is

$$\langle \boldsymbol{\mu} \rangle = \int \Psi^* \hat{\boldsymbol{\mu}} \Psi \, dq \qquad \text{(13.4-12)}$$

where the integration over q stands for integration over all of the coordinates of all particles in the molecule. In the Born-Oppenheimer approximation, the nuclei are fixed and we do not integrate over their coordinates. Since the operator contains no spin dependence, spin functions and spin integrations can be omitted.

With a one-term orbital wave function in which the orbitals are orthogonal to each other, each electron makes its contribution to the dipole independently [see Equation (11.3-24), which also holds for molecular orbitals]. In an antisymmetrized orbital wave function, the total electron probability density is the same as in the one-term function, as shown in Equation (11.3-26). The integral in Equation (13.4-12) is equal to a sum of one-electron integrals:

$$\hat{\boldsymbol{\mu}} = \sum_{i=1}^{n} Q_i \mathbf{r}_i$$

$$\langle \boldsymbol{\mu} \rangle = \boldsymbol{\mu}_{\text{nuc}} - e \sum_{i=1}^{n_e} \int \psi_i(i)^* \mathbf{r}_i \psi_i(i) \, d^3\mathbf{r}_i \tag{13.4-13}$$

where ψ_i is the ith occupied space orbital and n_e is the number of electrons. For most molecules, each space orbital is occupied by two electrons and occurs twice in this equation.

The contribution of the nuclei is computed with the nuclei at fixed positions:

$$\boldsymbol{\mu}_{\text{nuc}} = \sum_{A=1}^{n_n} e Z_A \, \mathbf{r}_A$$

where n_n is the number of nuclei and $\mathbf{r}_A$ is the position vector of the Ath nucleus, which contains Z_A protons.

The formula in Equation (13.4-13) is the same as if the ith electron were a classical "smeared-out" charge with a density distribution equal to $\psi_i^* \psi_i$. It is a theorem of electrostatics that a spherically symmetrical distribution of charge has an effect outside the charge distribution as though the charge were concentrated at the center of symmetry. Therefore, an electron moving in an undistorted atomic orbital of an s subshell contributes just as though it were at the nucleus. Although it does not follow from the theorem, electrons moving in other undistorted, nonhybridized atomic orbitals also contribute to the dipole as though they were at the nucleus. Electrons in the symmetrical LCAO-MOs constructed for the homonuclear diatomic molecules make an equal contribution at each end of the molecule. Electrons moving in LCAO molecular orbitals that have unequal coefficients make a larger contribution to the negative charge at the end of the molecule with the coefficient of larger magnitude, so the LiH molecule has a sizable dipole moment, with the hydrogen end negative.

The SI unit in which dipole moments are measured is the coulomb-meter. The bond length in LiH is experimentally measured to be 1.596×10^{-10} m, so that if the bond were purely ionic with an undistorted Li^+ ion and an undistorted H^- ion,

$$\mu_{\text{ionic}} = (1.6022 \times 10^{-19} \text{ C})(1.596 \times 10^{-10} \text{ m})$$
$$= 2.56 \times 10^{-29} \text{ C m}$$

There is a common unit named the **debye**, which is defined by

$$1 \text{ debye} = 3.335641 \times 10^{-30} \text{ C m} \tag{13.4-14}$$

The debye unit is named for Peter Debye, of the Debye-Hückel theory, whose 1936 Nobel Prize was for his work on dipole moments. The debye unit was defined in terms of the cgs unit of charge, the esu, such that $1 \text{ debye} = 10^{-10}$ esu Å. The charge on the proton is 4.80×10^{-10} esu, so molecular dipole moments are near 1–5 debye.

The dipole moment of the LiH molecule would equal 7.666 debye if the bond were completely ionic.

▼   **EXAMPLE 13.5**

Estimate the dipole moment of the LiH molecule from the orbitals of Equation (13.4-8.)

Solution

The lithium nucleus has charge 3e, and there are two nonbonding electrons that contribute as though they were at the lithium nucleus. The probability density of an electron in the 2σ orbital is

$$|\psi_{2\sigma}|^2 = (0.47)^2|\psi_{2sp1Li}|^2 + (0.47)(0.88)\psi^*_{2sp1Li}\psi_{1sH} + (0.88)^2|\psi_{1sH}|^2$$

We make some reasonable approximations: Each atomic orbital is assumed to contribute at the nucleus of the atom, and an overlap term is neglected. The contributions at each nucleus are added up and multiplied by the bond length to obtain the dipole moment.

We neglect the second term in this expression because it is appreciably nonzero only in the overlap region. The third term will make its contribution to the integral in Equation (13.4-13) as though the electron were centered at the hydrogen nucleus. The $2sp$ hybrid orbital does not have its center of charge exactly at the lithium nucleus, but we approximate the contribution of the first term as though it did. The net charge at the lithium nucleus is

$$Q_{Li} = 3e - 2e - 2(0.47)^2e = 0.56\ e = 8.9 \times 10^{-20}\ C$$

The net charge at the hydrogen nucleus is

$$Q_H = e - 2(0.88)^2e = -0.56\ e = -8.9 \times 10^{-20}\ C$$

with a bond length of 1.595×10^{-10} m,

$$\mu = (8.9 \times 10^{-20}\ C)(1.595 \times 10^{-10}\ m) = 1.42 \times 10^{-29}\ C\ m$$

$$= 4.3\ \text{debye}$$

This dipole moment is about 60% as large as the value for a purely ionic bond, indicating a bond that is roughly 60% ionic in character. The experimental value of the LiH dipole moment is 5.88 debye.

▲

The properties of the LCAO-MOs from the Hartree-Fock treatment of lithium hydride conform to a general pattern of results that can be expressed in the following rules:

This pattern expresses what generally is found in the results of Hartree-Fock calculations. We will use it to guide our qualitative discussions.

1. *Two atomic orbitals on different nuclei must have roughly equal orbital energies to form a good bonding LCAO-MO.* If the energies are different, the lower-energy (bonding) LCAO-MO constructed from them will have a coefficient of larger magnitude for the atomic orbital of lower energy, and the higher-energy (antibonding) LCAO-MO will have a coefficient of larger magnitude for the higher-energy atomic orbital. For greatly different energies, the lower-energy LCAO-MO will be almost the same as the lower-energy atomic orbital, making it nearly a nonbonding orbital.
2. *Two atomic orbitals on different nuclei must have a fairly large overlap region to form a good bonding LCAO-MO.*
3. *If two atomic orbitals on different nuclei do not have the same symmetry about the bond axis, they will not form a good bonding LCAO-MO.* In fact, they cannot form an eigenfunction of the proper symmetry operators for a diatomic molecule.

These three rules are generalizations that are valid for most molecules. We will use them as a means of predicting what would probably happen if a calculation were carried out, enabling us to compose a fairly reliable qualitative description of bonding in a given molecule without performing calculations.

The 1s lithium orbital and the 1s hydrogen orbital do not form a good bonding LCAO-MO for LiH because of the large difference in energies and because the overlap region of these two orbitals is small (the lithium 1s orbital is not in the valence shell). The 2sp hybrid orbital has better overlap with the 1s hydrogen orbital than either the 2s or 2pz orbital, since its orbital region is directional. The 2px and 2py orbitals do not have the same symmetry about the bond axis as the other orbitals, so they are not used to make LCAO-MOs. Furthermore, these orbitals have an overlap region with a 1s hydrogen orbital that is composed of two parts, with the product of the two orbital functions having a positive sign in one part and a negative sign in the other part. This produces a small (or zero) value of the overlap integral and counts as poor overlap, even though the overlap region is not small.

▼ **EXAMPLE 13.6**

Example 13.6 is the kind of discussion that our crude level of analysis permits. To begin such a discussion, you need a rough energy level diagram for all the atomic orbitals. The relative energies of atomic orbitals on different atoms can be estimated from a knowledge of electronegativities. For example, we know that the fluorine 2sp hybrids lie lower than the hydrogen 1s orbitals because fluorine is more electronegative than hydrogen.

Describe the bonding in hydrogen fluoride.

Solution

We use the fluorine 1s as a nonbonding orbital. The fluorine 2px and 2py atomic orbitals do not have the same symmetry around the z axis (the bond axis) as the hydrogen 1s orbital, so they are also used as nonbonding orbitals. We make two hybrid orbitals with the fluorine 2s and spz orbitals, calling them 2sp1 and 2sp2. We make a bonding orbital with the 2sp1 hybrid orbital and the hydrogen 1s orbital. We denote this LCAO-MO by 1σ. We use the 2sp2 orbital as a nonbonding orbital.

Since the energy of the fluorine 2sp1 hybrid is lower than that of the hydrogen 1s orbital, this bonding orbital will have a coefficient of larger magnitude for the fluorine atomic orbital than for the hydrogen orbital. We say that the bonding orbital is "heavy" on the fluorine end. The accompanying antibonding orbital, denoted by $2\sigma^*$, remains vacant in the ground state. The antibonding orbital is heavy on the hydrogen end.

The energies of the LCAO-MOs and hybrid orbitals are shown schematically in the correlation diagram of Figure 13.17. We have placed the line segment for the bonding orbital closer to the fluorine side to indicate that this orbital is heavy on the fluorine end.

From the Aufbau principle, the ground-state electron configuration is

$$(1s\text{F})^2(2sp2\text{F})^2(2px\text{F})^2(2py\text{F})^2(1\sigma)^2$$

Since the bonding orbital is heavy on the fluorine end, the bond is polar with the fluorine end negative.

▲

Exercise 13.19 Describe the bonding in LiF. How much ionic character do you think there will be in the bond?

Carbon monoxide is an interesting case. A good qualitative description of the chemical bonding is obtained by constructing LCAO-MOs similar

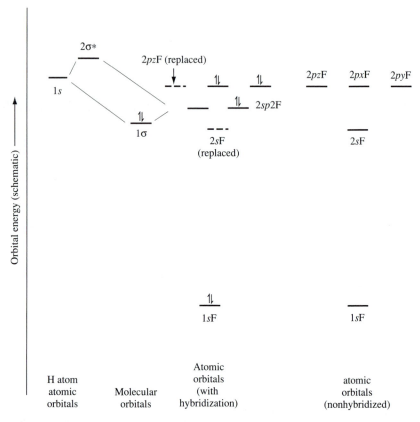

Figure 13.17. Correlation Diagram for Hydrogen Fluoride. This correlation diagram shows that in the simple approximation used, the bonding orbital is made as a linear combination of the 1s hydrogen orbital and one of the 2sp hybrid fluorine orbitals. The antibonding orbital is made from the same two atomic orbitals. The other orbitals used in the description are nonbonding fluorine orbitals.

to those of N_2, which is isoelectronic with CO. Since oxygen's effective nuclear charge is greater than that of carbon, all the bonding LCAO-MOs will have somewhat larger coefficients for the oxygen orbitals than for the carbon orbitals, and the antibonding LCAO-MOs will have somewhat larger coefficients for carbon than for oxygen.

The electron configuration of the molecule is

$$(\sigma 1s)^2(\sigma^*1s)^2(\sigma 2s)^2(\sigma^*2s)^2(\pi 2px)^2(\pi 2py)^2(\sigma 2pz)^2$$

We have dropped the designations g and u, since the orbitals are not eigenfunctions of the inversion operator, but keep the * to designate an antibonding orbital. The bond order is 3, the same as nitrogen.

The bonding and antibonding effects in the lowest-energy four orbitals approximately cancel, as in the homonuclear diatomics, so we can almost equally well use the alternative configuration

$$(1sC)^2(1sO)^2(2sC)^2(2sO)^2(\pi 2px)^2(\pi 2py)^2(\sigma 2pz)^2$$

The carbon end of CO is negative because the oxygen atom has more protons than the carbon atom. Two of the shared electrons can be considered to come from the oxygen atom, constituting a "coordinate covalent bond."

The dipole moment of the CO molecule is rather small, about 0.1 debye. The fact that the coefficients of the oxygen orbitals are greater in the bonding orbitals causes the bonding electrons to contribute a net negative charge at the oxygen end of the molecule. However, the oxygen nucleus is more positive than the carbon nucleus by two protons, and this contributes a positive charge at the oxygen end that slightly more than cancels the charge due to the electrons. The carbon end is negative.

Exercise 13.20 Assuming the alternative electron configuration with eight electrons occupying nonbonding orbitals, and assuming that the coefficients of the atomic orbitals in all three of the bonding molecular orbitals are the same, estimate the values of the coefficients corresponding to the experimental value of the dipole moment of the CO molecule.

Exercise 13.21 Give a qualitative description of the bonding of the BN molecule. Compare it with diatomic carbon.

The Valence Bond Method for Heteronuclear Diatomic Molecules

A valence bond wave function such as that of Equation (13.3-8) or Equation (13.3-16) corresponds to equally shared electrons in completely covalent bonds. Addition of the ionic terms as in Equation (13.3-11) does not correspond to ionic bonding, since one ionic term corresponds to bond polarity in one direction and the other to bond polarity in the opposite direction.

In the lithium hydride molecule, the hydrogen end of the molecule is negative, and a good approximation of the bonding can be obtained in the valence bond method by including only one ionic term, the ionic term corresponding to this polarity. We write the modified valence bond function for LiH:

$$\Psi_{MVB} = c_{VB}\Psi_{VB} + c_I\Psi_I \tag{13.4-15}$$

where

$$\Psi_{VB} = \psi_{1sLi}(1)\alpha(1)\psi_{1sLi}(2)\beta(2)[\psi_{2sp1Li}(3)\psi_{1sH}(4) + \psi_{1sH}(3)\psi_{2sp1Li}(4)]$$
$$\times [\alpha(3)\beta(4) - \beta(3)\alpha(4)] \tag{13.4-16}$$

and

$$\Psi_I = \psi_{1sLi}(1)\alpha(1)\psi_{1sLi}(2)\beta(2)\psi_{1sH}(3)\psi_{1sH}(4)[\alpha(3)\beta(4) - \beta(3)\alpha(4)] \tag{13.4-17}$$

Only one ionic term is added, since only one end of the molecule needs to be assigned a negative charge.

The wave function Ψ_I corresponds to purely ionic bonding, describing a positive lithium ion and a negative hydride ion, and the wave function Ψ_{VB} corresponds to purely covalent bonding. The values of the coefficients c_{VB} and c_I could be determined by the variation method and would describe the optimum mixture of covalent and ionic bonding. If $|c_I| > |c_{VB}|$, we say that the bond is primarily ionic, and if $c_{VB} = 0$ we have a purely ionic bond.

Exercise 13.22   Calculate the values of c_{VB} and c_I that make the wave function Ψ_{MVB} nearly equivalent to the LCAO-MO wave function in Equation (13.4-5) if the orbitals of Equation (13.4-8) are used. Find the percent ionic character, defined as

$$[c_I^2/(c_{VB}^2 + c_I^2)] \times 100\%$$

Electronegativity

An empirical parameter called the electronegativity can be used to estimate the degree of inequality of electron sharing in a bond between atoms of two elements. This quantity was introduced by Pauling. Pauling observed that polar covalent bonds are generally stronger (have larger bond dissociation energies) than purely covalent bonds and used this to define the electronegativity.

Linus Pauling, 1901–   , American chemist who won the 1954 Nobel Prize in chemistry for his work on molecular structure.

If the electronegativity of element A is denoted by X_A and that of element B is denoted by X_B, Pauling defined

$$|X_A - X_B| = (0.102 \text{ mol}^{1/2} \text{ kJ}^{-1/2})(\Delta E_{AB})^{1/2} \qquad \textbf{(13.4-18)}$$

where ΔE_{AB} is the difference between the average bond energy of an A-B bond and the mean of the average bond energies of A-A and B-B bonds:

$$\Delta E_{AB} = E_{AB} - \frac{1}{2}(E_{AA} + E_{BB}) \qquad \textbf{(13.4-19)}$$

This definition gives values for the electronegativity that are nearly self-consistent. Since only the difference in electronegativity is defined, the value for one element is chosen arbitrarily and the other values are relative to it. Choosing a value of 2.1 for hydrogen makes all electronegativities positive, ranging from 0.7 to 4.0.

There are other definitions that are used to obtain values of the electronegativity. Table A15 gives the electronegativity values for several elements. Fluorine is the most electronegative element, followed by oxygen and chlorine. The alkali metals are the least electronegative. In any row of the periodic chart, the electronegativity increases from left to right, and in any column it decreases from top to bottom. Electronegativities of the inert gases are not defined.

The trends in electronegativity can be understood on the basis of the observation that when a bonding molecular orbital is constructed from atomic orbitals of different energies, optimizing the values of the coefficients gives a coefficient of larger magnitude for the lower-energy atomic orbital than for the other atomic orbital. For example, with atomic orbitals in the same shell, a larger effective nuclear charge must correspond to a larger electronegativity, since a larger effective nuclear charge corresponds to a lower atomic orbital energy, as indicated in Equation (11.2-12). As one moves from left to right across a row of the periodic chart, the nuclear charge and the effective nuclear charge increase, corresponding to the observed increase in the electronegativity.

| Exercise 13.23 | Using average bond energies from Table A10, calculate the differences in electronegativity between (a) H and F, (b) C and O, and (c) C and Cl. Compare with the values in Table A15. |

A rule of thumb is that if the difference between the electronegativities of two elements is greater than 1.7, a bond between those elements will be primarily ionic. A difference of less than 1.7 corresponds to a polar covalent bond, and a pure covalent bond requires a difference of zero.

| Exercise 13.24 | Classify the bonds between the following pairs of elements as purely covalent, polar covalent, and primarily ionic: (a) Li and H, (b) C and O, (c) N and Cl, and (d) H and F, (e) Li and F, and (f) F and F. |

13.5   Polyatomic Molecules

In this section we will treat chemical bonding in polyatomic molecules, describing qualitative features of approximate molecular orbital electronic wave functions but not reporting on any calculations.

In zero order, interelectron repulsion terms are neglected and the Born-Oppenheimer Hamiltonian operator for any molecule consists of one-electron operators plus internuclear repulsion terms, which are treated as constants. In this approximation and in higher-order orbital approximations such as that of the Hartree-Fock method, the solution to the Schrödinger equation is a product of spin orbitals, one for each electron. For the wave function to survive antisymmetrization, the Pauli exclusion principle must be applied, with no more than one electron occupying each spin orbital (no more than two electrons occupying each space orbital). For the ground state, the wave function is constructed according to the Aufbau principle, choosing the lowest-energy set of orbitals consistent with the Pauli exclusion principle.

The Aufbau principle is the same for molecules as for atoms.

As much as possible, we will use simple LCAO molecular orbitals constructed according to the policies applied to diatomic molecules in Section 13.4: each bonding LCAO-MO will be constructed of two atomic orbitals centered on different atoms such that the orbitals (1) have orbital energies that are fairly close in value, (2) overlap significantly, and (3) have the same symmetry about the bond axis.

Let us begin with the H_2O molecule. This molecule is known from experiment to have a bond angle of $104.5°$ and a bond length of 0.958×10^{-10} m. The molecule has a dipole moment of magnitude 1.85 debye bisecting the bond angle, with the oxygen end negative.

Since the molecule is a 10-electron system, at least five space orbitals are required for an orbital wave function. Figure 13.18 is a correlation diagram showing schematically the energies of the seven lowest-energy atomic space

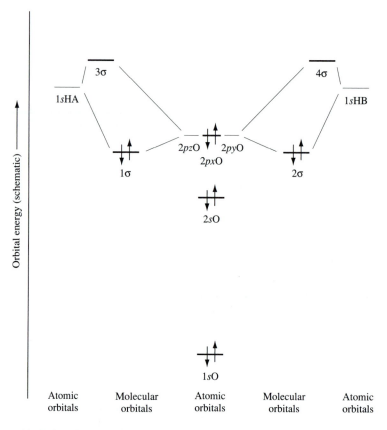

Figure 13.18. Correlation Diagram for the Water Molecule. Here we see two bonding LCAO-MOs, each made from a hydrogen 1s orbital and one of the 2p oxygen orbitals. This description is unsatisfactory, because the bond angle is predicted to be 90° in order to satisfy the criteria for bonding LCAO-MOs. The actual bond angle is around 105°.

orbitals of one oxygen atom and two hydrogen atoms. For our first attempted description, we use these unmodified atomic orbitals as our basis set. The 1s oxygen orbital is far lower in energy than the other orbitals, so we use it as a nonbonding orbital. All of the oxygen orbitals from the second shell have appropriate energies to be used in bonding LCAO-MOs with the hydrogen 1s orbitals. We first try to use the 2p orbitals in constructing bonding LCAO-MOs.

The atomic orbitals in an LCAO-MO must have the same symmetry about the bond axis. Since the $2pz$ orbital is cylindrically symmetric about the z axis, we place a hydrogen nucleus denoted by A on the z axis and construct the bonding molecular orbital

$$\psi_{1\sigma} = c_O \psi_{2pzO} + c_H \psi_{1sA} \tag{13.5-1}$$

where ψ_{1sA} is the 1s hydrogen orbital centered on hydrogen nucleus A and the coefficients c_O and c_H are constants. We place a hydrogen atom denoted

In this discussion, we are using the facts about LCAO molecular orbitals that were presented in the LiH discussion in the previous section.

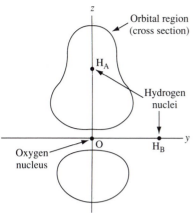

Figure 13.19. The Nuclear Framework of the Water Molecule Using the Molecular Orbitals of Equations (13.5-1) and (13.5-2). This figure shows the positions of the nuclei, as well as the cross section of the orbital region of the 1σ orbital. We see the unsatisfactory description with a bond angle of 90°.

by B on the y axis and construct the LCAO-MO:

$$\psi_{2\sigma} = c_O\psi_{2py O} + c_H\psi_{1sB} \qquad (13.5\text{-}2)$$

Because of the similarity of the $2py$ and $2pz$ orbitals, the coefficients in Equation (13.5-2) will be equal to those in Equation (13.5-1). Figure 13.19 shows the nuclear framework in the required position, as well as the orbital region of the 1σ orbital.

The 1σ orbital is a sigma orbital because it is a linear combination of two atomic orbitals corresponding to $m = 0$. The 2σ orbital bears the same relationship to the y axis as the 1σ orbital does to the z axis, so by analogy the projection of the orbital angular moment on the y axis for this orbital also vanishes, and it is also a sigma orbital. Whenever an orbital is cylindrically symmetrical about its bond axis, it is a sigma orbital or can be a term in a sigma orbital that is a linear combination.

Since the energy of the oxygen $2p$ orbitals is lower than that of the hydrogen $1s$ orbitals, optimization of the coefficients will give for the bonding orbital

$$|c_O| > |c_H| \qquad (13.5\text{-}3)$$

An electron occupying either the 1σ or 2σ orbital will more likely be found near the oxygen nucleus than near the hydrogen nucleus, and the bonds are polar covalent bonds.

Of the 10 electrons, two can occupy the oxygen $1s$ space orbital, four can occupy the $2s$ and $2px$ oxygen orbitals, and four can occupy the bonding LCAO-MOs. The ground-state configuration is thus

$$(1sO)^2(2sO)^2(2pxO)^2(1\sigma)^2(2\sigma)^2$$

There are two single bonds, with a bond angle of 90°. The molecule is polar with the oxygen end negative.

The approximate wave function corresponding to this electronic configuration needs to be improved, even with our limited goal of qualitative description. First, the bond angle of 90° is in poor agreement with the experimental value of 105°. Second, a water molecule tends to form two equivalent hydrogen bonds between its oxygen atom and hydrogen atoms on other water molecules, indicating that the two pairs of nonbonding electrons in the valence shell should occupy equivalent orbitals, instead of occupying the oxygen $2s$ and $2px$ orbitals.

Furthermore, this result disagrees with the **valence shell electron pair repulsion** (VSEPR) theory. According to this elementary theory, the electron pairs in the valence shell of an atom should arrange themselves so that they are as far from each other as possible. The two bonding electron pairs and the two nonbonding electron pairs should arrange themselves in a tetrahedral shape, with a bond angle near 109°.

A better simple wave function for the water molecule is obtained by creating a set of hybrid orbitals that are linear combinations of the $2s$ space orbital and all three of the $2p$ space orbitals. These orbitals are called the

$2sp^3$ hybrid orbitals:

$$\psi_{2sp^31} = \frac{1}{2}(-\psi_{2s} + \psi_{2px} + \psi_{2py} + \psi_{2pz}) \qquad \textbf{(13.5-4a)}$$

$$\psi_{2sp^32} = \frac{1}{2}(-\psi_{2s} + \psi_{2px} - \psi_{2py} - \psi_{2pz}) \qquad \textbf{(13.5-4b)}$$

$$\psi_{2sp^33} = \frac{1}{2}(-\psi_{2s} - \psi_{2px} + \psi_{2py} - \psi_{2pz}) \qquad \textbf{(13.5-4c)}$$

$$\psi_{2sp^34} = \frac{1}{2}(-\psi_{2s} - \psi_{2px} - \psi_{2py} + \psi_{2pz}) \qquad \textbf{(13.5-4d)}$$

Exercise 13.25   Using the fact that the $2s$, $2px$, $2py$, and $2pz$ atomic orbitals are all normalized and all orthogonal to each other, choose one of the hybrid orbitals in Equation (13.5-4) and show that it is normalized. Choose a pair of orbitals in Equation (13.5-4) and show that they are orthogonal to each other.

The localization, or directional character, of the sp^3 hybrid orbital regions is deduced here. This character is important in determining bond directions.

The orbital regions of the $2sp^3$ hybrid orbitals are directional. Figure 13.20a shows schematically how the three $2p$ orbitals combine in the $2sp^31$ hybrid orbital. Since the $2s$ orbital is negative in the outer part of its orbital region, it is entered in the linear combinations with a negative coefficient so that it will make a positive contribution in this region. In the figure, vectors are drawn along the positive coordinate axes, in the directions of the largest positive values of the $2p$ orbitals. The direction of the largest positive value of the hybrid orbital lies between these directions, in this case in the direction toward the upper right front corner of a cube centered at the origin and parallel to the coordinate planes. Although we do not prove this fact, this vector is an axis of cylindrical symmetry of the hybrid orbital. The symmetry axes of the other three hybrid orbitals are shown in the figure. Each region is cylindrically symmetrical about an axis pointing toward one of four alternate corners of the cube.

Exercise 13.26   Pick one of the $2sp^3$ hybrid orbitals other than $2sp^31$ and argue that its orbital region is directed as shown in Figure 13.20a.

Since connecting the four alternate corners of a cube with line segments constructs a regular tetrahedron, the angle between any two of the axes shown in Figure 13.20a is called the **tetrahedral angle** and equals 109 degrees, 28 minutes, 16.39... seconds.

Exercise 13.27   Using the theorem of Pythagoras and values of trigonometric functions, show that the angle between alternate diagonals of a cube is 109 degrees, 28 minutes, 16.39... seconds.

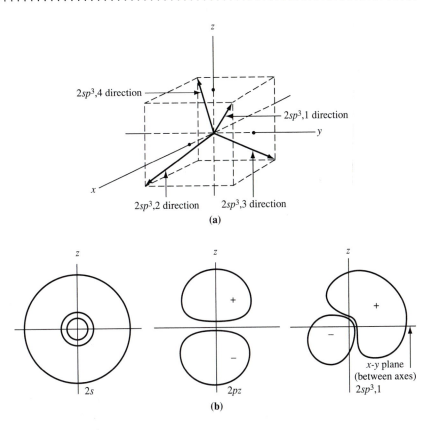

Figure 13.20. The sp^3 Hybrid Orbitals. (a) The addition of s and p atomic orbitals to form sp^3 hybrid orbitals (represented by vectors in the directions of positive contributions). This figure shows the directions of maximum extension of the four sp^3 hybrid orbitals, which point to alternate corners of a cube centered at the origin. **(b)** The orbital regions of $2s$, $2p$, and $2sp^3$ orbitals. The orbital region of the sp^3 hybrid orbital resembles that of the sp hybrids, except for the direction of maximum extension. All three of the $2p$ orbitals are included in each linear combination, so the directions lie as shown in part (a).

If we start with a certain number of independent functions, we can make the same number of independent linear combinations.

We place the hydrogen atoms on the symmetry axes of the hybrid orbitals in order to satisfy our criteria for making good LCAO bonding molecular orbitals.

A cross section of the orbital region of a $2sp^3$ orbital in a plane containing its axis of symmetry is shown in Figure 13.20b. For comparison, cross sections of the orbital regions for the $2s$ and $2pz$ nonhybridized orbitals are also shown. The orbital region of the hybrid orbital extends farther in the direction of its symmetry axis than that of either the $2s$ or $2p$ orbital, making it possible to form a more strongly bonding LCAO-MO using a $2sp$ orbital than using a $2p$ orbital, because of the additional overlap.

We can now construct an approximate wave function for the water molecule, using the $2sp^3$ hybrid orbitals, as shown in the correlation diagram of Figure 13.21. In this diagram, the nonhybridized atomic orbitals are shown with broken lines. We place the two hydrogens on the symmetry axes for two $2sp^3$ hybrid orbitals: hydrogen A on the axis of the $2sp^3 2$ orbital and hydrogen B on the axis of the $2sp^3 3$ hybrid orbital. We form two

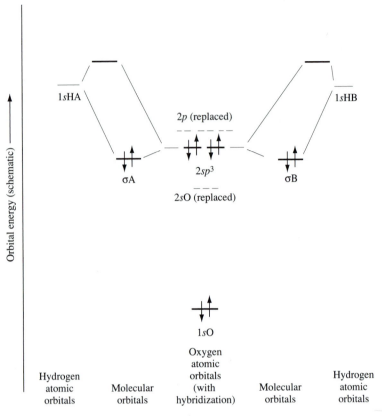

Figure 13.21. Correlation Diagram for the Water Molecule Using 2sp³ Hybrid Orbitals.
This description is the most nearly satisfactory description that can be obtained with the
hybrid orbitals which we have defined and with the policy that no more than two atomic
orbitals be used in one LCAO-MO. The bond angle is 109.5°, the angle between two of
the directions shown in Figure 13.20a.

bonding LCAO molecular orbitals:

$$\psi_{\sigma A} = c_O \psi_{2sp^3 2} + c_H \psi_{1sA} \tag{13.5-5a}$$

$$\psi_{\sigma B} = c_O \psi_{2sp^3 3} + c_H \psi_{1sB} \tag{13.5-5b}$$

Since the hybrid oxygen orbitals are somewhat lower in energy than the
hydrogen $1s$ orbitals, we expect that

*We know that the oxygen orbitals are
lower in energy than the hydrogen
orbitals because of the experimental
fact that oxygen is more electronegative than hydrogen.*

$$|c_O| > |c_H| \tag{13.5-6}$$

That is, the bonding orbitals will be heavy at the oxygen end. Along with
the bonding orbitals σA and σB, there are antibonding orbitals σ^*A and
σ^*B, which are heavy on the hydrogen end. These antibonding orbitals
remain vacant in the ground state.

By the Aufbau principle, the electron configuration of the ground state
is $(1sO)^2(2sp^3 1)^2(2sp^3 4)^2(\sigma A)^2(\sigma B)^2$. The configuration is also denoted in
the energy level diagram of Figure 13.21 by arrows pointing up and down
to represent electrons occupying a space orbital with spin up and spin down.

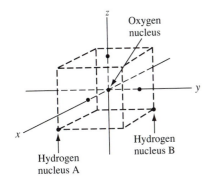

Figure 13.22. The Positions of the Nuclei in the Description of the Water Molecule Using sp^3 Hybrid Orbitals. This figure shows the placement of the nuclei to obtain proper bonds using only two atomic orbitals in the LCAO molecular orbitals.

As with the earlier approximate wave function, there are two single bonds and two pairs of valence shell nonbonding electrons, which occupy nonbonding sp^3 hybrid orbitals. The bond angle is equal to the tetrahedral angle, 109°, and is in reasonably good agreement with the experimental bond angle of 105°. The molecule is polar with the oxygen end negative, in agreement with experiment. Figure 13.22 shows the water molecule structure according to our description.

Even better descriptions of the water molecule can be constructed by using LCAO-MOs that are linear combinations of all atomic orbitals in our basis set and by using larger basis sets. Hybrid orbitals are then not needed. Their use would only restrict the flexibility needed in finding the optimum orbitals.[8]

We can now understand the **Lewis octet rule.** According to this rule, atoms tend to have eight electrons in their valence shells, counting both bonding and nonbonding electrons. For the water molecule and other molecules made from elements of the second row of the periodic chart, the valence shell is the second shell, in which eight spin orbitals (four space orbitals) occur. When linear combinations of functions are made, the number of independent linear combinations is the same as the number of basis functions used. Two linear combinations can be made from two atomic orbitals. From two atomic orbitals, one bonding LCAO-MO and one antibonding LCAO-MO will result. If the antibonding orbitals remain vacant, the total number of occupied spin orbitals around a given nucleus of a second-row element equals eight, including nonbonding atomic orbitals and bonding molecular orbitals. This fact corresponds to the octet rule of Lewis.

▼   **EXAMPLE 13.7**

Give a qualitative description of the bonding in methane, CH_4. Does the molecule have a dipole moment?

Solution

Place each hydrogen nucleus on a symmetry axis of one of the $2sp^3$ hybrid orbitals. Form four bonding LCAO molecular orbitals, one from each $2sp^3$ hybrid and the corresponding hydrogen $1s$ orbital. Occupy the $1s$ carbon space orbital with two nonbonding electrons and occupy each bonding molecular space orbital with two electrons, making four single bonds, each at the tetrahedral angle with the other three bonds. Since the carbon $2sp^3$ hybrid orbitals are slightly lower in energy than the hydrogen $1s$ orbitals, the bonds are slightly polar. However, the dipole contributions from the four bonds cancel each other because of the symmetry of the molecule, and the molecule has zero dipole moment.

In the solution of Example 13.7, we have deduced the relative energies of the orbitals from the fact that carbon is slightly more electronegative than hydrogen.

▲

Exercise 13.28

Give a qualitative description of the bonding in ammonia, NH_3, including a statement about the dipole moment of the molecule.

---

[8] Levine, *op. cit.*, p. 424ff.

Molecules with Double and Triple Bonds

In Section 13.3 we discussed several diatomic molecules with multiple bonds. For example, C_2 had a double bond and N_2 had a triple bond. In our approximate wave functions for these molecules, some of the shared electrons occupied pi LCAO-MOs that were constructed from $2p$ atomic orbitals. We can construct similar pi LCAO-MOs for polyatomic molecules.

Let us begin with the ethyne (acetylene) molecules, C_2H_2. The Lewis electron dot formula for this molecule is

$$H\!:\!C\!:\!:\!:\!C\!:\!H$$

and the molecule is known to be linear. The acetylene molecule is isoelectronic with N_2, which also has a triple bond. In N_2, two of the bonding electrons occupy the $\sigma g2pz$ orbital, two occupy the $\pi u2px$ orbital, and two occupy the $\pi u2py$ orbital. The other eight electrons occupied $\sigma g1s$, σ^*u1s, $\sigma g2s$, and σ^*u2s orbitals in one LCAO-MO description. In the description with nonbonding atomic orbitals, the other eight electrons occupy the two $1s$ and two $2s$ orbitals.

Let us make a wave function for acetylene similar to the second description. Instead of the $2s$ carbon orbitals, we make a bonding LCAO-MO from each $2s$ carbon orbital and the $1s$ hydrogen orbital on the adjacent hydrogen. Call one of these σA and the other one σB. Assuming the C—C bond axis to be the z axis, the electron configuration of the molecule is

$$(1sA)^2(1sB)^2(\sigma A)^2(\sigma B)^2(\sigma g2pz)^2(\pi u2px)^2(\pi u2py)^2$$

There are two C—H single bonds and a C—C triple bond, consisting of a sigma bond and two pi bonds, just as in the diatomic nitrogen molecule configuration of Section 13.3. The carbon-carbon sigma bonding orbital is constructed from $2pz$ orbitals, while the two carbon-hydrogen sigma bonding orbitals are constructed from carbon $2s$ orbitals and hydrogen $1s$ orbitals. The C—H bonds can point in any direction, since the $2s$ orbitals are not directional.

A better description is obtained by constructing two $2sp$ hybrids on each carbon, using the $2s$ and $2pz$ orbitals. The orbital region of a $2sp$ hybrid is directional and extends farther from the nucleus than the $2p$ or $2s$ non-hybridized orbitals, giving greater overlap and better bonding LCAO-MOs than the sigma LCAO-MOs in our first description.

We construct a carbon-carbon sigma bonding orbital from the two $2sp$ hybrid orbitals that overlap between the carbons. Call this the 1σ bonding orbital. On each carbon, this leaves a $2sp$ hybrid orbital whose orbital region is directed away from the C—C bond but along the same axis. With each of these and a hydrogen $1s$ orbital we form a carbon-hydrogen sigma bonding orbital, placing the hydrogens on the same axis as the C—C bond (the z axis). Call these two bonding orbitals $\sigma A'$ and $\sigma B'$. The configuration is now

$$(1sA)^2(1sB)^2(\sigma A')^2(\sigma B')^2(1\sigma)^2(\pi u2px)^2(\pi u2py)^2$$

This is similar to the earlier configuration, but the wave function corresponding to it provides lower energy (greater bonding energy) because of

The $\pi u2p1$ and $\pi u2p-1$ molecular orbitals could also have been used instead of the real orbitals.

Here we are using the fact that more extensive overlap generally leads to a stronger bond. This is why the hybrid orbitals are useful in our crude description.

the greater overlap of the $2sp$ hybrid orbitals, and it correctly predicts that the molecule is linear. The triple bond still consists of a sigma bond and two pi bonds.

Exercise 13.29 Describe the bonding in diatomic nitrogen using orbitals similar to those in the second description of acetylene.

Better wave functions can be constructed by making linear combinations of all atomic orbitals in our basis set and finding the coefficients by the Hartree-Fock method. Hybrid orbitals are not needed in the basis set, and the LCAO-MOs are linear combinations of all atomic orbitals in the basis set and have orbital regions that extend over the entire molecule.

After the LCAO-MOs are optimized, we can take certain linear combinations of them, called "energy-localized" orbitals, that have orbital regions concentrated between pairs of atoms. It is found that the three energy-localized bonding molecular orbitals of the triple bond are equivalent. That is, they all have the same energy and the same shape, with the orbital regions lying 120° from each other around the bond axis. These orbitals are sometimes called "banana orbitals" because of their shape.[9]

In addition to the sp and sp^3 hybrid orbitals, there are hybrid orbitals constructed from an s orbital and two p orbitals. The $2s$, $2px$, and $2py$ orbitals can be used to construct three independent hybrid orbitals that are called the $2sp^2$ orbitals:

$$\psi_{2sp^21} = -\sqrt{\frac{1}{3}}\psi_{2s} + \sqrt{\frac{2}{3}}\psi_{2px} \qquad \textbf{(13.5-7a)}$$

$$\psi_{2sp^22} = -\sqrt{\frac{1}{3}}\psi_{2s} - \sqrt{\frac{1}{6}}\psi_{2px} + \sqrt{\frac{1}{2}}\psi_{2py} \qquad \textbf{(13.5-7b)}$$

$$\psi_{2sp^23} = -\sqrt{\frac{1}{3}}\psi_{2s} - \sqrt{\frac{1}{6}}\psi_{2px} - \sqrt{\frac{1}{2}}\psi_{2py} \qquad \textbf{(13.5-7c)}$$

The orbital regions of these $2sp^2$ hybrids are directional and lie 120° apart from each other in the x-y plane. Figure 13.23a shows the direction of the symmetry axes for the three $2sp^2$ orbitals, and Figure 13.23b shows a cross section of the orbital region for one of them.

The $2sp^2$ hybrid orbitals can be used to construct approximate wave functions for molecules containing double bonds. For example, ethene (ethylene) has the structural formula

$$\begin{array}{ccc} H & & H \\ \diagdown & & \diagup \\ & C{=}C & \\ \diagup & & \diagdown \\ H & & H \end{array}$$

The molecule is planar with a C—C—H bond angle of 122°.

---

[9] Levine, *op. cit.*, p. 453.

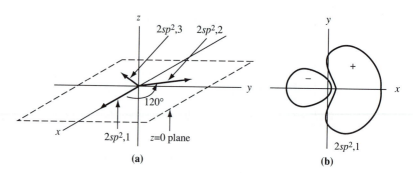

Figure 13.23. The sp^2 Hybrid Orbitals. (a) Directions of the symmetry axes of the orbital regions of the $2sp^2$ hybrid orbitals. **(b)** A cross section of an orbital region. Since only the $2px$ and $2py$ atomic orbitals are used in these hybrid orbitals, the directions of maximum extension lie in the x-y plane, 120° apart. Again, the orbital region resembles those of the other hybrid orbitals except for direction.

We can construct three sigma bonding orbitals around each carbon from the $2sp^2$ hybrids, and we can construct a pi bonding orbital from non-hybridized $2p$ orbitals. Let the plane of the molecule be the x-y plane and orient the x axis through the double bond. The hybrids in Equation (13.5-7) are appropriate for both carbons if we reverse the direction of the x axis for one carbon to make the larger lobes of two $2sp^2$ orbital regions on different carbons point toward each other.

Denote the C—C sigma bonding orbital by 1σ. The C—C pi bonding orbital is a sum of the two $2pz$ orbitals. Denote this orbital by 1π. Denote the C—H bonding orbitals by 2σ, 3σ, 4σ, and 5σ. The electronic configuration of the molecule in the ground state is

$$(1sA)^2(1sB)^2(1\pi)^2(1\pi)^2(2\sigma)^2(3\sigma)^2(4\sigma)^2(5\sigma)^2$$

Since the nuclei must be placed on the symmetry axes of the hybrid orbitals to have good sigma bonds, our C—C—H bond angles equal 120°, in fairly good agreement with the experimental value of 122°.

If LCAO-MOs are made that are linear combinations of all the atomic orbitals considered, the Hartree-Fock self-consistent field method gives the correct bond angle of 122°, and formation of energy-localized orbitals represents the double bond as two equivalent banana bonds.[10]

Exercise 13.30 Describe the bonding in diatomic carbon using orbitals such as those used in ethylene. Do you think this description would be superior to that in Section 13.3?

The Valence Bond Description of Polyatomic Molecules

In the examples we have considered so far, no antibonding orbitals were needed. In such cases, we can obtain a simple valence bond wave function

[10] Levine, *op. cit.*, p. 450ff.

Here is a way to construct a valence bond description from the LCAO-MO description for any molecule without unpaired electrons. We use the same criteria for good bonding as in the LCAO-MO treatment.

by replacing each pair of bonding molecular spin orbitals with a bond factor such as that of Equation (13.3-8). The same types of hybrid orbitals can be used. The criteria for forming a good valence bond bonding factor are the same as those for forming a good bonding molecular orbital: The two atomic orbitals should have the same symmetry around the bond axis, they should have roughly equal energies, and they should have considerable overlap. The nuclei are placed as in the molecular orbital treatment.

For the water molecule, a partially antisymmetrized (but unnormalized) valence bond wave function is

$$
\begin{aligned}
\Psi_{VB} = {}& \psi_{1sO}(1)\alpha(1)\psi_{1sO}(2)\beta(2)\psi_{2sp^31}(3)\alpha(3)\psi_{2sp^31}(4)\beta(4) \\
& \times \psi_{2sp^34}(5)\alpha(5)\psi_{2sp^34}(6)\beta(6) \\
& \times [\psi_{2sp^32}(7)\psi_{1sA}(8) + \psi_{1sA}(7)\psi_{2sp^32}(8)] \\
& \times [\alpha(7)\beta(8) - \beta(7)\alpha(8)] \\
& \times [\psi_{2sp^33}(9)\psi_{1sB}(10) + \psi_{1sB}(9)\psi_{2sp^33}(10)] \\
& \times [\alpha(9)\beta(10) - \beta(9)\alpha(10)] \qquad\qquad \textbf{(13.5-8)}
\end{aligned}
$$

where the subscript $1sA$ stands for the $1s$ orbital on one hydrogen and $1sB$ stands for the $1s$ orbital on the other hydrogen. The hydrogens are placed on the symmetry axes of the oxygen hybrid orbitals.

This wave function corresponds to nonpolar covalent bonds. Ionic terms could be added as in Equation (13.4-15), writing

In Equation (13.5-9) we include only one ionic term, since oxygen is more electronegative than hydrogen.

$$
\Psi_{MVB} = c_{VB}\Psi_{VB} + c_I\Psi_I \qquad\qquad \textbf{(13.5-9)}
$$

where Ψ_I is the completely ionic wave function and the value of the coefficients c_{VB} and c_I would be determined by minimizing the variational energy.

Exercise 13.31   Write the expression for Ψ_I in Equation (13.5-9).

The ionic character of the bond can also be represented by placing an ionic term in each bonding factor. The first bonding factor would become

$$
\psi_{2sp^32}(7)_{1sA}(8) + \psi_{1sA}(7)\psi_{2sp^32}(8) + c\psi_{2sp^32}(7)\psi_{2sp^32}(8) \quad \textbf{(13.5-10)}
$$

where we omit the spin factor. Only one ionic term is included, with both electrons on the oxygen atom, which is the more electronegative atom. The value of the coefficient c would be optimized by minimizing the variational energy of the molecule. The space factor in Equation (13.5-9) is symmetric, so an antisymmetric spin factor as in Equation (13.5-8) would be used.

To give a quick description of the bonding in the water molecule using the valence bond method, it is necessary only to specify that two nonbonding electrons occupy the oxygen $1s$ space orbital, four nonbonding electrons occupy two oxygen $2sp^3$ hybrid space orbitals, and four electrons occupy two bonding factors, each constructed from an oxygen $2sp^3$ hybrid and a hydrogen $1s$ orbital, so that the bond angle equals $109°$.

Exercise 13.32   Using the valence bond method, give a qualitative description of the bonding in (a) ammonia and (b) methane.

Valence bond descriptions of multiple bonds are also similar to the LCAO-MO description. Each bonding factor replaces a pair of LCAO spin orbitals, but nonbonding orbitals are the same in both methods. There are no analogues to antibonding orbitals in the simple valence bond method, so molecules that require antibonding orbitals cannot be well described in this method.

▼   **EXAMPLE 13.8**

In the solution to Example 13.8, our valence bond description is similar to the LCAO-MO description except for the use of bonding factors instead of pairs of bonding orbitals. Hybrid orbitals were used in valence bond wave functions before being used in LCAO molecular orbitals.

Describe the bonding in the ethyne (acetylene) molecule, using the valence bond method.

Solution

Four electrons occupy the two carbon $1s$ space orbitals. Two electrons occupy a carbon-carbon sigma bonding factor made from a $2sp$ hybrid on each carbon, and four electrons occupy two carbon-hydrogen sigma bonding factors made from a $2sp$ hybrid on a carbon and a $1s$ on a hydrogen. The final four electrons occupy two carbon-carbon pi bonding factors, one made from the $2px$ on each carbon and one made from the $2py$ orbital on each carbon, making a triple carbon-carbon bond. The molecule is linear because of the $180°$ angle between the symmetry axes of the $2sp$ hybrids on each carbon.

▲

Exercise 13.33   Using the valence bond method, describe the bonding in the propene (propylene) molecule. Give the bond angles around each carbon.

Other Types of Hybrid Orbitals

In addition to the sp, sp^2, and sp^3 hybrid orbitals, there are three common types that include d orbitals, either from the same shell as the s and p orbitals or from the next lower shell.

If one d space orbital is included in addition to the four space orbitals of the s and p subshells, five hybrid space orbitals can be constructed, which are called dsp^3 hybrids if the d orbital is from the next lower shell or sp^3d hybrids if the d orbital is from the same shell.[11] These orbitals have directional orbital regions that point toward the apices of a trigonal bipyramid. The symmetry axis of one orbital points along the positive z axis, that of another points along the negative z axis, and three point in the x-y plane in directions $120°$ from each other. These directions are the same as the trigonal bipyramidal directions assigned to five electron pairs in the VSEPR

---

[11] Keith F. Purcell and John C. Kotz, *Inorganic Chemistry*, W. B. Saunders, Philadelphia, 1977, p. 104.

theory. The sp^3d orbitals can be used to construct sigma LCAO molecular orbitals or valence bonding factors for molecules, such as iodine trifluoride, that have five pairs of electrons in the valence shell of a central atom.

Exercise 13.34   Write the LCAO-MO electron configuration of the iodine trifluoride molecule. Choose arbitrary subscripts for the five sp^3d hybrids and the sigma bonding orbitals, but specify the direction of the orbital region of each.

Hybrid orbitals can also be formed from two d orbitals, one s orbital, and three p orbitals.[12] These orbitals are called d^2sp^3 if the d subshell is from the shell below that of the s and p orbitals and sp^3d^2 if the d subshell is from the same shell. These six orbitals have symmetry axes pointing along the positive and negative coordinate axes. These directions point toward the apices of a regular octahedron, just as do the directions chosen for six pairs of electrons in the VSEPR theory. When these hybrids are used in a molecule wave function, all of the bond angles are equal to $90°$.

The five types of hybrid orbitals we have defined suffice to give all the electronic geometries predicted by the VSEPR theory for up to six pairs of valence shell electrons. In addition, the dsp^2 (or sp^2d) hybrids made from one d orbital, one s orbital, and two p orbitals give a square planar electronic geometry.

Exercise 13.35   Using either the LCAO-MO method or the valence bond method, describe the bonding in the following molecules, including a specification of the shape of the molecule.
  a. Sulfur hexafluoride, SF_6.
  b. Xenon tetrafluoride, XeF_4.
  c. Phosphorus trichloride, PCl_3.

Delocalized Bonding

In all of the cases considered thus far, we have used LCAO molecular orbitals constructed from no more than two atomic orbitals, or have used a one-term valence bond wave function. In some molecules, this kind of description is unsatisfactory. The deficiency is usually that the approximate wave function has different kinds of bonds in two locations where the actual molecule has two equivalent bonds. In the molecular orbital method, this deficiency can be remedied by using delocalized LCAO-MOs, which are linear combinations of atomic orbitals centered on three or more nuclei. In the valence bond method, it can be remedied by the use of a technique called **resonance**.

---

[12] *ibid.*, p. 105.

We illustrate both of these techniques by their application to the benzene molecule, C_6H_6, which is known from experiment to be hexagonal in shape, with six equivalent carbon-carbon bonds. Consider first the valence bond approach.

Two structural formulas corresponding to different valence bond wave functions can be written:

These structures are called **resonance structures**. It is customary to write a double-headed arrow between resonance structures, as shown. Let the valence bond wave functions corresponding to the two structures be called Ψ_I and Ψ_{II}. Each one of these wave functions consists of orbital factors for nonbonding electrons and sigma and pi two-electron bonding factors for bonding electrons. If the plane of the molecule is the x-y plane, $2sp^2$ hybrids made from $2px$ and $2py$ orbitals are used to form sigma bonding factors. Rotation of the x and y axes of each carbon is necessary to make the symmetry axes point along sides of the hexagon. The nonhybridized $2pz$ orbitals are used for three pi bonding factors. The only difference between the two structures is the location of the three pi bonding factors.

We write the wave function as a linear combination:

$$\Psi = c_I\Psi_I + c_{II}\Psi_{II} \tag{13.5-11}$$

The linear combination of Equation (13.5-11) expresses the concept of resonance. The wave function is a combination of the wave functions for the different resonance structures. More than two structures can be included, and coefficients of different magnitudes can be used to vary the contribution of the different resonance structures.

where c_I and c_{II} are coefficients whose values can be found by minimizing the variational energy or by other arguments. This is the mathematical expression of resonance. Neither formula alone represents the structure of the molecule, but since the wave function is a linear combination of both wave functions, we say that we have a "blending" of the two structural formulas.

Since our two resonance structures differ only by the double bond locations, the two coefficients in this case will be equal to each other, giving equal weight to each resonance structure. Various other resonance structures have been constructed for benzene, including some with "long bonds" across the ring. Terms corresponding to such resonance structures enter with small coefficients, and we omit them in our crude treatment.

The difference between the variational energy calculated with a wave function including resonance and that calculated with a single resonance structure is called the **resonance energy**, but the same term is sometimes applied to the difference between the correct ground-state energy and that calculated with a single resonance structure. An experimental estimate of this resonance energy for benzene is obtained from the difference between the enthalpy change of hydrogenation of benzene and three times the enthalpy change of hydrogenation of ethylene. The value of this estimate is 150 kJ mol^{-1}.

In the LCAO-MO method, we do not need the concept of resonance. However, we must abandon our simple policy of making LCAO-MOs from only two atomic orbitals. We will make LCAO-MOs as linear combinations of atomic orbitals centered on more than two nuclei. Such molecular orbitals are called **delocalized orbitals**. In the case of benzene, we proceed as follows: We first construct ordinary (localized) sigma bonding orbitals for carbon-carbon bonds and carbon-hydrogen bonds. Since the molecule is hexagonal, all bond angles are equal to 120°. We choose the x-y plane for the plane of the molecule, so that $2sp^2$ carbon hybrid orbitals are the appropriate atomic orbitals for the sigma LCAO-MOs, turning the x and y axes for each carbon as necessary. The $2pz$ orbitals remain nonhybridized.

The molecule has 42 electrons. Twelve of these will occupy the six nonbonding carbon $1s$ space orbitals. Twelve will occupy the six carbon-carbon sigma bonding space orbitals, and twelve will occupy the six carbon-hydrogen sigma bonding space orbitals. This leaves six electrons and six nonhybridized $2pz$ carbon space orbitals. Since six independent linear combinations can be made from six independent functions, we can construct six delocalized LCAO-MOs from these six $2pz$ orbitals:

The orbital of Equation (13.5-12) is delocalized, which simply means that it is a linear combination of atomic orbitals on more than two nuclei. An electron occupying such an orbital moves in an orbital region encompassing several atoms.

$$\varphi_i = c_1^{(i)}\psi_1 + c_2^{(i)}\psi_2 + c_3^{(i)}\psi_3 + c_4^{(i)}\psi_4 + c_5^{(i)}\psi_5 + c_6^{(i)}\psi_6 \quad \textbf{(13.5-12)}$$

where i is an index used to specify which of the delocalized LCAO-MOs is meant and the six nonhybridized $2pz$ orbitals are abbreviated $\psi_1, \psi_2, \ldots, \psi_6$. Since all six atomic orbitals are included in each LCAO-MO, an electron occupying such an orbital moves around the entire ring of carbon atoms. These LCAO-MOs are called pi orbitals, although we cannot talk of an angular momentum projection on a single bond axis.

The coefficients in Equation (13.5-12) are determined by applying the variational method, minimizing the orbital energy. However, the usual approach does not begin with the zero-order Born-Oppenheimer Hamiltonian operator, which is a sum of known one-electron Hamiltonian operators. Instead, it begins by assuming a sum of "effective" one-electron Hamiltonian operators that apply only to the electrons that will occupy the delocalized pi orbitals and include electron-electron and electron-nucleus interactions in some kind of unspecified average way. No specific expression for these one-electron operators is given, but a **semiempirical approach** is taken, using experimental data to give approximate values of certain integrals.

In our approximation, the energy of the molecule is the sum of orbital energies. If we separately minimize the orbital energies, we can minimize the energy of the molecule.

The orbital variation energy for the ith delocalized orbital is

$$W_i = \frac{\int \varphi_i^* \hat{H}_1^{\text{eff}} \varphi_i \, d^3\mathbf{r}}{\int \varphi_i^* \varphi_i \, d^3\mathbf{r}} \quad \textbf{(13.5-13)}$$

where $\hat{H}_1^{\text{eff}}$ is the effective one-electron Hamiltonian operator for one electron and $d^3\mathbf{r}$ stands for the volume element of this electron.

When the expression of Equation (13.5-12) is substituted into Equation (13.5-13), W is given as a function of the coefficients c_1, c_2, etc. and the minimum in W is found by differentiating W with respect to each of the c coefficients and setting these derivatives equal to zero. This procedure gives a set of simultaneous equations that can in principle be solved for the coefficients.

This secular equation is different from that of degenerate perturbation theory. We do not discuss either one.

The simultaneous equations are a set of linear homogeneous equations that must obey a certain condition in order to have a nontrivial solution.[13] This condition is in the form of an equation, called a **secular equation**, in which a certain determinant is set equal to zero. It can be solved to find the permissible values of W, the orbital energy. The number of values of the orbital energy W is equal to the number of orbitals in the linear combinations, although some of the values may equal each other. One set of c coefficients is obtained for each value of W.

The Hückel method is named for Ewald Hückel, the coinventor of the Debye-Hückel theory of electrolyte solutions.

The simultaneous equations are not usually solved exactly. Various approximation schemes are used. The simplest set of approximations constitutes the **Hückel molecular orbital method**. The expression for W contains two types of integrals:

$$H_{ab} = \int \psi_a^* \hat{H}_1^{\text{eff}} \psi_b \, d^3\mathbf{r} \qquad \textbf{(13.5-14)}$$

$$S_{ab} = \int \psi_a^* \psi_b \, d^3\mathbf{r} \qquad \textbf{(13.5-15)}$$

where ψ_a and ψ_b are two of the $2pz$ atomic orbitals. The integral S_{ab} is an **overlap integral**, and the integral H_{ab} is called a **matrix element** of the Hamiltonian. The simple Hückel theory is defined by assuming that all of the overlap integrals vanish for $a \neq b$; that H_{ab} has one value, α, when $a = b$ and another value, β, when a and b represent orbitals on atoms bonded to each other; and that H_{ab} vanishes when atoms a and b are not bonded to each other.

These assumptions allow solution of the secular equation and the simultaneous equations for the coefficients. We do not discuss the solution.[14] For benzene, the lowest orbital energy is (note that β is negative)

$$W = \alpha + 2\beta \qquad \textbf{(13.5-16)}$$

corresponding to equal values for all six c's:

$$\varphi_1 = \sqrt{\frac{1}{6}} (\psi_1 + \psi_2 + \psi_3 + \psi_4 + \psi_5 + \psi_6) \qquad \textbf{(13.5-17)}$$

Figure 13.24a shows the orbital energies of this space orbital and the other five pi LCAO-MOs, obtained by solution of the secular equation and the simultaneous equations for the c's in the Hückel approximation.[15]

---

[13] Levine, *op. cit.*, pp. 188ff.

[14] Frank L. Pilar, *Elementary Quantum Chemistry*, McGraw-Hill, New York, 1968, pp. 584ff.

[15] Levine, *op. cit.*, pp. 474–475.

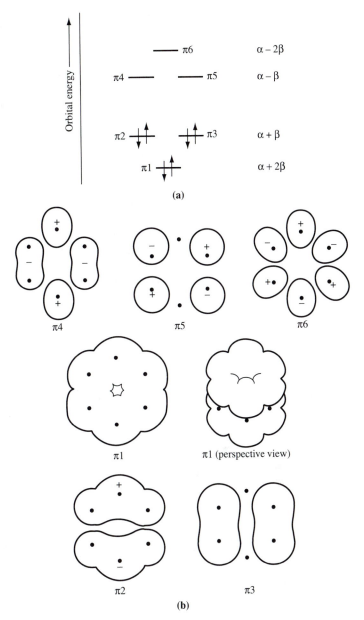

Figure 13.24. Results of the Hückel Molecular Orbital Calculation for Benzene. (a) Orbital energies. It is interesting that if a diagram of the hexagonal shape of the six carbons is placed over the orbital energy diagram, each corner corresponds to one of the orbital energies. **(b)** The orbital regions. Each orbital region is shown from above (from the direction of the positive end of the z axis), except that the orbital region for the $\pi 1$ orbital is shown twice, once from above and once from an angle. Orbital regions exist below the plane of the nuclei and above the plane of the nuclei. The sign shown applies to the region above the plane. The region directly below has the opposite sign, because there is a nodal plane in the plane of the nuclei. Note that the energies correlate with the numbers of nodes in the same way as with other systems: the more nodes, the higher the energy.

Three of the energies are obtained as relative minima in the energy, and three are obtained as relative maxima. Notice the interesting fact that the pattern of the energy levels has the same shape as the molecule. This correspondence occurs in the Hückel solution for all single-ring aromatic molecules. Figure 13.24b shows a view of the orbital regions of the six LCAO-MOs, looking perpendicular to the plane of the molecule. Slightly different orbital regions occur if one takes a linear combination of $\psi_{\pi 2}$ and $\psi_{\pi 3}$ or of $\psi_{\pi 4}$ and $\psi_{\pi 5}$.

Again the general relation: the more nodes, the higher the energy.

The general relation between energy and number of nodes is followed, with no nodes in the lowest-energy orbital, one nodal plane in the next two orbitals (which are degenerate), two nodal planes in the next, and three nodal planes in the highest-energy orbital. Without doing any calculations, we might have guessed the number of energy levels and the number of states in each from the following facts: (1) no more than three nodal planes can be drawn between the atoms in the six-membered ring, (2) there is only one way to have no nodes, (3) there are two perpendicular directions in which a single nodal plane can be drawn, (4) there are two simple ways to draw two nodal planes between the atoms, and (5) there is only one way to draw three nodal planes between the atoms. However, there is no way to determine the spacing between the energy levels without solving the secular equation.

Six electrons occupy the pi orbitals, so in the ground state each of the lowest three space orbitals is occupied by two electrons, as shown by arrows in Figure 13.24a. In the first excited state, an electron in one of the highest occupied orbitals is promoted to one of the lowest unoccupied orbitals, increasing the energy of the molecule by $2|\beta|$. This transition can be observed spectroscopically.

Since we have no explicit expression for the effective one-electron Hamiltonian operator, we have no way to calculate the α and β integrals. Their values are ordinarily deduced from experiment, and the Hückel method is therefore called a semiempirical method.

Exercise 13.36   From the fact that benzene absorbs strongly at wavelengths near 180 nm, estimate the value of β.

▼   **EXAMPLE 13.9**   Describe the delocalized pi molecular orbitals for *trans*-1,3-butadiene.

Solution

The molecule is planar, with bond angles near 120°. We choose the x-y plane for the plane of the molecule and construct $2sp^2$ hybrids for the sigma bonds. After the nonbonding electrons and sigma bonding electrons are taken care of, there are four electrons and four nonhybridized $2pz$ carbon orbitals left. The delocalized orbitals are linear combinations of all four of these orbitals.

Without doing any calculations, we can see that there are four possible numbers of nodes between the atoms: no nodes, one node, two nodes, and three nodes. There is only one symmetrical way to have each number of nodes. Therefore, the orbital

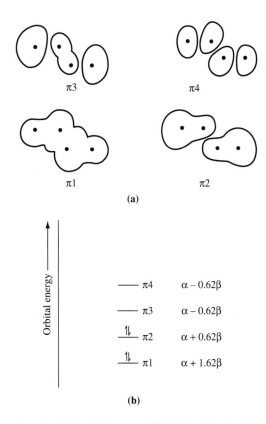

Figure 13.25. Results of the Hückel Molecular Orbital Calculation for 1,3-Butadiene.
(a) The orbital regions for the 1,3-butadiene Hückel molecular orbitals. These orbital regions are shown in much the same way as those for the benzene molecule shown in Figure 13.24.
(b) The orbital energies. Once again, we see the correlation between number of nodes and energy: the more nodes, the higher the energy.

regions of the four delocalized orbitals must be as in Figure 13.25a, looking perpendicular to the plane of the molecule. Since each orbital has a different number of nodes from the others, they are nondegenerate. From the general relation between numbers of nodes and orbital energies, the orbital energies must be as in Figure 13.25b, which gives the energies from the Hückel calculation. The two lowest orbitals are occupied in the ground state and there are no unpaired electrons. This configuration is indicated by arrows in the figure.

Exercise 13.37   **a.** Describe the pi LCAO-MOs in the cyclobutadiene molecule, assuming a square structure. Use $2sp^2$ hybrid orbitals for the sigma bonds, although they do not quite fit. There are two ways to make a single node in an LCAO-MO (either horizontal or vertical in these dot formulas). Give the electron configuration of the ground state of the molecule.

b. Describe the bonding using the valence bond method, using the resonance structures[16]

Free-Electron Molecular Orbitals

The free-electron molecular orbital (FEMO) method is a simple way of representing delocalized molecular orbitals. The electrons that move over several bond lengths are assumed to move in a one-dimensional box with an appropriate length. Repulsions between delocalized electrons are neglected, so the factor of the wave function for the delocalized electrons is a product of single-electron functions that can be antisymmetrized.

Let us discuss 1,3-butadiene, which has already been treated in Example 13.9. We assume the sigma bond framework to be as before. The carbon-carbon bond lengths are 146 pm for the center bond and 134 pm for the others.[17] From Figure 13.25a, we see that the orbital regions extend beyond the end carbon nuclei, so we assign the length of the box in which the pi electrons move to be the sum of the three bond lengths plus one additional bond length at each end. The additional bond length at each end will be taken as the average of the two bond lengths, giving a box length of 694 pm.

The energy levels and energy eigenfunctions of a particle in a box are given by Equations (9.5-10) and (9.5-11):

$$E = E_n = \frac{h^2 n^2}{8ma^2}, \qquad \psi = \psi_n = \sqrt{\frac{2}{a}} \sin\left(\frac{n\pi x}{a}\right) \qquad \text{(13.5-18)}$$

where a is the length of the box and n is a quantum number equal to a positive integer. Since we have four electrons, the Aufbau principle gives the ground-state wave function (including only the pi electrons) as

$$\Psi_{\text{gs}} = \psi_1(1)\alpha(1)\psi_1(2)\beta(2)\psi_2(3)\alpha(3)\psi_2(4)\beta(4) \qquad \text{(13.5-19)}$$

The ground-state pi electron energy is

$$E\pi = \frac{h^2}{8ma^2}(1 + 1 + 4 + 4) = \frac{10h^2}{8ma^2} \qquad \text{(13.5-20)}$$

EXAMPLE 13.10

Calculate the wavelength of the light absorbed when 1,3-butadiene makes the transition from the ground state to the first excited state.

---

[16] This structure is apparently not a correct representation of the actual shape of the molecule, which is rectangular but not square. See Levine, *op. cit.*, p. 476.
[17] K. Kuchitsu, F. Tsutomu, and Y. Morino, *J. Mol. Struct.* **37**, 2074 (1962).

Solution

This transition is the promotion of one electron from $n = 2$ to $n = 3$, so

$$\Delta E = \frac{h^2}{8ma^2}(9 - 4) = \frac{(5)(6.6261 \times 10^{-34}\ \text{J s})^2}{(8)(9.109 \times 10^{-31}\ \text{kg})(6.94 \times 10^{-10}\ \text{m})^2}$$

$$= 6.25 \times 10^{-19}\ \text{J}$$

$$\lambda = \frac{hc}{\Delta E} = \frac{(6.6261 \times 10^{-34}\ \text{J s})(2.9979 \times 10^8\ \text{m s}^{-1})}{6.25 \times 10^{-19}\ \text{J}}$$

$$= 3.18 \times 10^{-7}\ \text{m} = 318\ \text{nm}$$

In view of the crudity of the free-electron model, this degree of agreement is much better than we should expect.

This is in fair agreement with the experimental value of 217 nm. Better agreement could be attained by adding less than a full bond length at each end of the carbon-carbon chain.

▲

Exercise 13.38   Using the same bond lengths as with 1,3-butadiene, find the reciprocal wavelength of the longest-wavelength electronic transition of 1,3,5-hexatriene. Compare with the experimental value, 24000 cm^{-1}.

13.6   More Advanced Treatments of Molecular Electronic Structure

Our discussion of molecular electronic structure has been extremely crude compared with current quantum chemistry research. In the past several decades, modern digital computers have made possible calculations that previously could only be dreamed of, and a great deal of research effort has been expended in quantum chemical calculations with the goal of calculating wave functions, molecular geometries, and molecular energies accurately enough for chemical purposes.

It is difficult to characterize briefly the research being done in such an active field as quantum chemistry and even more difficult to predict what will be done in the future. However, chemists should try to be aware of the general trends in the field and of the degree of success currently achieved.

Much of the current research in quantum chemistry is based on the LCAO-MO method. Two approaches are commonly used:

1. The **semiempirical approach** begins with an assumed Hamiltonian operator that is a sum of effective one-electron Hamiltonian operators, much as in the Hückel method (interelectron repulsion terms are included in some versions of this approach). Since the one-electron operators are generally not explicitly expressed, empirical data must be used to assign values to matrix elements and other integrals.

2. The **ab initio** ("from the beginning") approach begins with the correct nonrelativistic Hamiltonian and requires no inputs of empirical information, although nuclei are often placed according to experimental data on bond lengths and angles.

Dewar and Storch have written a review article comparing the results of different semiempirical and ab initio methods in calculating enthalpy changes of reactions.[18] At the time of this article, no method had given chemically adequate accuracy for anything but the smallest molecules.

There are many versions of the semiempirical LCAO-MO approach, each with its own set of approximations. The typical calculation is the minimization of orbital energies. Since linear combinations of all of the basis functions are used, this kind of calculation involves the computation of many integrals, some of which are analogous to the α and β integrals of the Hückel method. Most of the approximations of the different methods are assumptions that certain integrals vanish. Some methods are named according to the types of integrals that are assumed to vanish, such as the CNDO method (complete neglect of differential overlap), the INDO method (intermediate neglect of differential overlap), the NDDO method (neglect of diatomic differential overlap), and the MINDO method (modified intermediate neglect of differential overlap). Other methods are named for approximations of different sorts, such as the PCILO method (perturbative configuration interaction using localized orbitals) and the DIM method (diatomic in molecules). There are enough other methods that the situation is reminiscent of acronyms of government agencies or of alphabet soup.

The principal ab initio methods are based on the Hartree-Fock-Roothaan self-consistent field procedure, which uses LCAO-MOs constructed from some basis set:

$$\psi = \sum_{i=1}^{n} c_i f_i \qquad \text{(13.6-1)}$$

where n is the number of functions in the basis set $f_1, f_2, \ldots$.

A "minimal" basis set contains one function for every inner-shell and valence shell orbital of each atom. A minimal basis set is the smallest basis set that can be used. Much larger basis sets have been used. When the LCAO-MOs are substituted into the Hartree-Fock equations, a set of equations called the Roothaan equations is obtained.[19] The Roothaan equations are solved by iteration, and the results are the orbital energies and the values of the c coefficients for each of the LCAO-MOs.

Various kinds of basis functions are in common use. An important criterion is the speed with which computers can evaluate the integrals occurring in the calculation. It is found that **Slater-type orbitals** (STOs) require less computer time than hydrogen-like orbitals. These orbitals contain the same spherical harmonic functions as the hydrogen-like orbitals, but their radial factors are exponential functions multiplied by powers of r instead of by polynomials. There are rules for guessing appropriate values for the exponents.[20]

[18] M. J. S. Dewar and D. M. Storch, *J. Am. Chem. Soc.* **107**, 3898 (1985).
[19] Levine, *op. cit.*, pp. 376ff.
[20] *ibid.*, pp. 542–543.

In addition to Slater-type orbitals, **gaussian orbitals** have been widely used. In these functions, the radial factor is

$$R(r) = e^{-br^2} \tag{13.6-2}$$

where b is a parameter to be evaluated in some way. The correct spherical harmonic functions are used for the angular factors. Such gaussian functions are not very good representations of radial factors, but they allow even more rapid computer evaluation of integrals than Slater-type orbitals, because the product of two gaussian functions centered on different nuclei is another gaussian function centered somewhere else. Some basis sets contain multiple gaussian functions chosen to simulate Slater-type orbitals. For example, in the STO-3G basis set, each Slater-type orbital is represented approximately by a linear combination of three gaussian functions.

Since the best single-configuration orbital wave function still contains the correlation error, configuration interaction is used to improve the energies. This means that the wave function, instead of being a single Slater determinant corresponding to a particular electronic configuration, is a linear combination of Slater determinants, each corresponding to a different configuration. Up to a million configurations have been used, but present-day computers still require a very long time to do such a calculation.

Because the energy change in a chemical reaction is a fairly small fraction of the total energy of the molecules, the total energies must be calculated to very high accuracy in order to approximate the energy change of a reaction. For example, if the energy change in a reaction is approximately 500 kJ mol^{-1} (about 5 eV molecule^{-1}), the total energies of the molecules might be several thousand eV, or several hundred thousand kJ mol^{-1}. Achieving an error of 10% in the energy change of reaction requires the maximum error in the molecular energies to be smaller than a tenth of a percent, unless some errors fortuitously cancel.

However, some things are inherently very difficult to observe experimentally, and any quantum chemical calculation giving information about such a process gives useful information. For example, the elementary steps in a chemical reaction might take place in 10^{-13} second, making experimental study nearly impossible with present techniques. Quantum chemical calculations giving the energies and geometries (bond angles and lengths) of reactive intermediates might be a source of such information.

Molecular Mechanics

A branch of calculational chemistry has achieved some degree of success in predicting geometries of molecules and intermediates without direct study of the electronic wave functions of a molecule. In this method, called **molecular mechanics**, approximations to Born-Oppenheimer energies as a function of bond lengths and bond angles are calculated, using various expressions for the interaction of different atoms and groups of atoms in molecules. Such potential energy functions are built into a computer program, which then carries out the process of finding the conformations of minimum potential energy. With many atoms in a molecule, this minimization is a difficult problem, and various techniques exist for its solution.[21]

Summary

In this chapter we have discussed the quantum theory of molecules, using the Born-Oppenheimer approximation, which is the assumption that the nuclei are stationary as the electrons move. In this approximation, the time-independent Schrödinger equation for the hydrogen molecule ion, H_2^+, can be solved exactly to give energy eigenvalues and orbitals dependent on the internuclear distance R.

Linear combination of atomic orbitals, called LCAO-MOs, provide an approximate representation of molecular orbitals for H_2^+. The ground-state LCAO-MO, called the $\sigma g1s$ function, is a sum of the $1s$ atomic orbitals for each nucleus and is called a bonding orbital. The first excited-state LCAO-MO, called the σ^*u1s function, is an antibonding orbital.

An approximate wave function for a diatomic molecule is a product of two LCAO molecular orbitals similar to those of H_2^+. The ground state of the H_2 molecule corresponds to the configuration $(\sigma g1s)^2$. A wave function with a single configuration can be improved on by adding terms corresponding to different configurations.

Most information about an LCAO-MO wave function is contained in the specification of the electron configuration, which is constructed for the ground state by the Aufbau principle much as with atoms. Molecular term symbols can be assigned much as with atoms.

In the valence bond method, a bonding factor is included in the wave function to represent electron sharing between nuclei by containing two "covalent" terms, with each electron occupying an orbital on one nucleus in one term and on the other nucleus in the other term. Ionic terms, with both electrons on the same nucleus, can also be included.

Heteronuclear diatomic molecules require unsymmetrical molecular orbitals. An approximate wave function for the LiH molecule was constructed using hybrid orbitals, which are linear combinations of atomic orbitals on the same nucleus. A bonding molecular orbital made from a $2sp$ hybrid on the Li nucleus and a $1s$ orbital on the H nucleus provides an adequate description of the bonding in LiH.

The criteria for a good bonding LCAO-MO were presented: A good bonding LCAO-MO is formed from a pair of atomic orbitals with nearly equal energies, with considerable overlap, and with the same symmetry about the bond axis.

LCAO molecular orbitals for polyatomic molecules were constructed as linear combinations of only two atomic orbitals, conforming to the three

---

[21] U. Burkert and N. L. Allinger, *Molecular Mechanics*, ACS Monograph 177, American Chemical Society, Washington, DC, 1982.

general criteria that predict good bonding molecular orbitals. In the case of the water molecule, sp^3 hybrid atomic orbitals could produce a satisfactory wave function with a bond angle of $109°$.

The sp^2 hybrid orbitals were useful in constructing LCAO molecular orbitals for molecules with double bonds, such as ethene, The sp hybrid orbitals were used in molecules with triple bonds, such as ethyne.

Delocalized bonding was described in the valence bond method by use of the concept of resonance. In the molecular orbital description, delocalized LCAO-MOs are used. For example, in the benzene molecule, six of the electrons occupy delocalized orbitals.

Additional Reading

John A. Pople and David L. Beveridge, *Approximate Molecular Orbital Theory*, McGraw-Hill, New York, 1970
This small book is designed to aid senior and first-year graduate students in learning Hartree-Fock semiempirical molecular orbital methods.

Donald J. Royer, *Bonding Theory*, McGraw-Hill, New York, 1968
This small book is designed for advanced undergraduates. It has a clear discussion of Hückel molecular orbitals.

See also the references at the end of Chapters 9–12.

PROBLEMS

Problems for Section 13.1

*13.39. What is the symmetry operator that is equivalent to the operator product $\hat{i}\hat{\sigma}_h$? Is this the same as the product $\hat{\sigma}_h\hat{i}$?

13.40. Write the function that is equal to $\hat{C}_{4z}\psi_{2px}$.

Problems for Section 13.2

13.41. Predict what will be formed if a hydrogen molecule ion in the state corresponding to the σ^*g2px LCAO-MO dissociates.

13.42. Sketch the orbital region of the $\sigma g3s$ LCAO-MO for a homonuclear diatomic molecule.

13.43. For a homonuclear diatomic molecule, sketch the orbital regions for the six LCAO molecular orbitals that can be formed from the $3p$ atomic orbitals.

13.44. For a heteronuclear diatomic molecule, sketch the orbital region for the LCAO-MO

$$\psi = c_1\psi_{2sA} + c_2\psi_{2pzA} + c_1\psi_{2sB} - c_2\psi_{2pzB}$$

where c_1 and c_2 are roughly but not exactly the same size and are both positive. Take the z axis as the bond axis. Determine whether this orbital will be an eigenfunction of $\hat{i}$ and of $\hat{\sigma}_h$. Give the eigenvalues if it is an eigenfunction.

13.45. Sketch the orbital region of the (unusable) LCAO-MO for a diatomic molecule

$$\psi = c(\psi_{1sA} + \psi_{2pyB})$$

The z axis is the bond axis.

*13.46. By inspection of the orbital regions, predict which united-atom orbital will result from each of the following LCAO-MOs of the hydrogen molecule ion: (a) $\pi u2px$ (b) $\pi u2py$ (c) $\sigma g2pz$

Problems for Section 13.3

13.47. Give the term symbol or symbols for each of the excited configurations of the hydrogen molecule: (a) $(\sigma g1s)$ $(\pi u2p1)$, (b) $(\pi u2p1)(\sigma^*g2p, -1)$, (c) $(\sigma g2p0)^2$

13.48. Give the term symbol for the ground-state configuration of each of the second-row homonuclear diatomic molecules (lithium through fluorine).

*13.49. Excited states of diatomic neon exist, although the ground state has bond order zero. Give the configuration and term symbol for an excited state that might exist.

Problems for Section 13.4

13.50. Describe the bonding in the possible molecule LiB. Do you think the molecule can exist?

13.51. Using the modified valence bond method, describe the bonding of the NaCl molecule. Predict whether the coefficient of the covalent term or the ionic term will be larger. Look up the electronegativity of sodium if necessary.

15.52. By analogy with the $2sp$ hybrid orbitals, sketch the orbital region of the two $3sp$ hybrid orbitals.

13.53. Using the molecular orbital method, describe the bonding of the NaCl molecule. Predict what the bonding molecular orbitals will look like if optimized.

13.54. Describe the bonding in the carbon monoxide molecule using the valence bond method. Include ionic terms.

*13.55. The dipole moment of the HCl molecule in its ground state equals 1.1085 debye and the internuclear distance equals 127.455 pm. Estimate the percent ionic character and the values of the coefficients of the covalent and ionic terms in the modified valence bond wave function.

13.56. Using average bond energies from Table A10, calculate the electronegativity differences for H and C, H and N, and C and N. Compare with the values in Table A15.

Problems for Section 13.7

13.57. Show that the $2sp^2$ hybrid orbitals of Equation (13.5-7) are normalized and orthogonal to each other.

13.58. Using the valence bond method with resonance, describe the bonding in the carbonate ion, CO_3^{2-}.

*13.59. Tell which of the following molecules and ions will have nonzero dipole moments: (a) SO_4^{2-}, (b) BF_3, (c) $HClO_4$, (d) CO_2, (e) $H_2C{=}C{=}CH_2$, (f) NO_2.

13.60. Show that the dipole moment of a molecule such as carbon tetrachloride vanishes. Hint: There are four polar bonds of equal magnitude directed along the four tetrahedral directions. Place the bonds on alternate diagonals of a cube and use cartesian components.

13.61. Using hybrid orbitals and LCAO-MOs, describe the bonding and molecular shape of each of the molecules or ions in Problem 13.59.

13.62. Using the valence bond method, describe the bonding and molecular shape of (a) 1,3,5-hexatriene, (b) NO_2^-, (c) CH_3 (methyl radical).

13.63. Using the valence bond method, with resonance where appropriate, describe the bonding in the molecules (a) H_2CO, (b) HNO_3, (c) SO_2.

*13.64. The methylene radical, CH_2, is found experimentally to be linear, and CCl_2 is found to be bent, with a bond angle near 120°. Describe the bonding in both molecules and explain the difference in shape. Which molecule will have unpaired electrons?

13.65. Describe the bonding of the carbonate ion using LCAO molecular orbitals. Place the nuclei in the x-y coordinate plane and make delocalized orbitals with the nonhybridized pz orbitals on all four atoms, trying to guess where the nodes are in the lowest-energy delocalized orbitals.

13.66. Crystal field theory is an approximate theory for complex ions with a transition metal atom in the center and several atoms or groups (ligands) bonded around it. The ligands are approximately represented as point charges.

   a. If six negative charges are octahedrally arranged about an iron(II) ion, tell which of the real $3d$ orbitals will have their energies raised by a greater amount and which will have their energies raised by a lesser amount.

   b. If the energy difference in part a is small, the $3d$ orbitals will be occupied as though they were at the same energy, and if the energy difference is large, the lower-energy $3d$ orbitals will be preferentially occupied in the ground state. In each case, use Hund's first rule to determine the number of unpaired $3d$ electrons in the iron.

13.67. Use the energy level expressions for benzene and for 1,3-butadiene to obtain two different values for the parameter β, using the fact that the longest-wavelength ultraviolet absorption is at wavelength 180 nm in benzene and at 217 nm in 1,3-butadiene. Compare these values with an accepted value of -2.71 eV[22] and explain why the values do not agree.

*13.68. The motion of the pi electrons around the benzene molecule is sometimes represented as de Broglie waves moving around a circular ring. Take the carbon-carbon distance in the ring as 139 pm, and take the lowest-energy electron state to have a de Broglie wavelength equal to the circumference of the ring, the next to have a de Broglie wavelength equal to half the circumference, etc. Find the energy and wavelength of the photons absorbed in the longest-wavelength ultraviolet absorption and compare with the value given in Problem 13.67.

13.69. In the free-electron molecular orbital model, the electrons actually move in three dimensions. For 1,3-butadiene represent the electrons as particles in a three-dimensional box with a length in the x direction equal to 694 pm, width in the y direction equal to 268 pm, and height in the z direction equal to the width. Find the wavelength of the photons absorbed in the longest-wavelength absorption due to changes in the quantum numbers n_y and n_z. Explain why the representation as a one-dimensional box can be used successfully to understand the near-ultraviolet spectrum.

13.70. Find the correct fraction of a bond length to add to each end of the carbon-carbon chain in 1,3-butadiene to give agreement with the observed wavelength of light absorbed in the longest-wavelength absorption.

[22] Donald J. Royer, *Bonding Theory*, McGraw-Hill, New York, 1968, p. 162.

13.71. Write an approximate LCAO-MO wave function without antisymmetrization for the formaldehyde molecule, using no more than two atomic orbitals in an LCAO-MO. Place the C—O bond on the z axis and the H atoms in the y-z plane. Use hybrid orbitals where appropriate. Specify which atomic orbitals make up each LCAO-MO and identify each with an index and the designation sigma or pi. Give the shape of the molecule.

13.72. Calculations[23] indicate that the C—H bond in methane has the opposite polarity from that predicted by the electronegativity difference, with each bond having a dipole moment possibly as large as 1.67 debye with carbon positive. If each sigma C—H bond is represented by

$$C(\psi_{2sp^3C} + c_H\psi_{1sH})$$

estimate the value of the coefficient c_H. Neglect the overlap integral in the calculation.

*13.73. The dipole moment of chloromethane is 1.87 debye. Assume that the C—H bonds has length 111 pm and the C—Cl bond has length 178 pm. Estimate the net charge on each atom. State any assumptions.

13.74. Obtain access to any of the common computer programs that will solve the Hückel molecular orbital problem for various molecules. You will have to find out how the necessary information is put into the computer.

   a. Run the program for benzene and for 1,3,5-hexatriene. Compare the results and explain the differences.

   b. Run the program for cyclobutadiene and for 1,3-butadiene. Compare the results and explain the differences.

[23] A. E. Reed and F. Weinhold, *J. Chem. Phys.* **84**, 2428 (1986).

General Problems

13.75. Describe the bonding in the HF molecule using sp^3 hybrids for the fluorine instead of the sp hybrids used in Example 13.6. Which description fits better with the VSEPR theory?

13.76. Consider the two molecules BH_3 and NH_3.

   a. Describe the bonding in each, using the simple LCAO-MO approach of using two atomic orbitals in each LCAO-MO. Use the appropriate hybrid atomic orbitals in the LCAO molecular orbitals.

   b. Explain why the molecules have different shapes.

   c. In the gas phase, a mixture of these two substances undergoes a Lewis acid-base reaction to form an adduct with a coordinate covalent bond. Describe the bonding in this molecule, using the simple LCAO-MO approach with the appropriate hybrid atomic orbitals.

   d. Compare the adduct of part c to ethane, C_2H_6.

   e. Would the adduct of part c be polar or nonpolar? Justify your answer in terms of electronegativity and in terms of orbital energy levels.

*13.77. Identify each statement as either true or false. If a statement is true only under special circumstances, label it as false.

   a. The total electronic angular momentum of a diatomic molecule is a good quantum number.

   b. The orbital angular momentum of the electrons of a diatomic molecule is a good quantum number.

   c. The component on the bond axis of the orbital angular momentum of an electron in a diatomic molecule is a good quantum number.

   d. Hybrid orbitals are used to allow reasonable approximate molecular orbitals to be written as linear combinations of only two atomic orbitals.

14 Translational, Rotational, and Vibrational States of Atoms and Molecules

OBJECTIVES

After studying this chapter, the student should:

1. understand the separation of the translational and electronic energy of atomic substances and be able to solve problems related to the translational energy,

2. understand the translational, rotational, and vibrational energy levels of diatomic molecules and be able to solve problems related to these energy levels,

3. understand the equilibrium populations of diatomic energy levels and be able to solve problems related to these populations,

4. understand the rotational and vibrational energy levels of polyatomic molecules and be able to solve problems related to these energy levels.

PREVIEW

In the previous three chapters, we have focused on the electronic motion in atoms and molecules. We now study the motions of the nuclei, which according to the Born-Oppenheimer approximation carry the electrons along with them. The only nuclear motion in atoms is translation, but in molecules we also have rotation and vibration.

PRINCIPAL FACTS AND IDEAS

1. Atoms have only translational and electronic energy.
2. The translational energy of an atom is like that of a structureless particle.
3. Molecules have rotational and vibrational energy in addition to translational and electronic energy.
4. The Born-Oppenheimer approximation is used to separate the electronic motion from the rotational and vibrational motion.
5. Translation, rotation, and vibration of diatomic molecules can be discussed separately from the electronic structure.
6. The equilibrium populations of the energy levels of molecules are governed by the Boltzmann probability distribution.
7. The rotational and vibrational energies of polyatomic molecules can be discussed in the Born-Oppenheimer approximation.

14.1 **Translational Motions of Atoms**

In Chapter 11, we began our study of the states of atoms. The wave function of a hydrogen atom is given by Equation (11.1-14)

$$\Psi = \chi(X, Y, Z)\psi_r(x, y, z) = \psi_{\text{tr}}(X, Y, Z)\psi_{\text{el}}(x, y, z) \qquad \textbf{(14.1-1)}$$

where χ is the center-of-mass factor in the total wave function of the atom. We now denote this factor by ψ_{tr} and call it the **translational factor**. This factor represents the motion of a particle with mass equal to the total mass of the atom. The factor ψ_r is the **relative factor**, which corresponds to motion of the electron and the nucleus around the center of mass. Because the nucleus is much more massive than the electron, it corresponds nearly to electronic motion about the nucleus. We now denote this factor by ψ_{el} and call it the **electronic factor**.

The separation of Equation (14.1-1) can also be used for other atoms as a good approximation. The translational factor is still the same as that of a particle with the same mass as the total mass of the atom. The relative factor is obtained with the assumption that the nucleus is stationary; it is the electronic wave function that we studied in Chapters 11 and 12.

The total energy of a hydrogen atom is given by Equation (11.1-17)

$$E_{\text{total}} = E_c + E_r = E_{\text{tr}} + E_{\text{el}} \qquad \textbf{(14.1-2)}$$

where E_{tr} is the translational energy (previously called E_c, the center-of-mass energy), and E_r is the relative energy, which we now call the electronic energy E_{el}. This is the atomic electronic energy that we studied in Chapters 11 and 12. We also apply Equation (14.1-2) to atoms other than hydrogen. It is exactly correct for the hydrogen atom and a good approximation for other atoms.

We now study the translational energy, which is the same for any atom. If the atom is not confined in a container or subject to any other external force, its center of mass obeys the time-independent Schrödinger equation of a free particle, Equation (9.5-47). The unnormalized solution of this Schrödinger equation for the center-of-mass motion in a particular direction is given by Equation (9.5-48), and the translational energy is given by Equation (9.5-50) and is not quantized.

If the atom is confined in a container, we assume that the container is a rectangular box oriented parallel to the cartesian coordinate axes with one corner at the origin. Let its length in the x direction be a, its length in the y direction be b, and its length in the z direction be c. If the atom were a point mass confined in this box, the potential energy would be represented by Equation (9.5-17):

$$\mathscr{V}_{\text{ext}} = \mathscr{V}_{\text{ext}}(x, y, z) = \begin{cases} 0 & \text{if } 0 < x < a, 0 < y < b, \text{ and } 0 < z < c \\ \infty & \text{otherwise} \end{cases} \qquad \textbf{(14.1-3)}$$

This potential function does not exactly apply to an atom. The center of the atom cannot move completely up to the wall because of the presence of the electrons in the atom. However, we will consider only a box much larger than the size of an atom, and Equation (14.1-3) will be a very good approximation.

The solution of the time-independent Schrödinger equation for the translation of the center of mass of an atom in a box is the same as that of a point mass in a box, Equation (9.5-32):

$$\psi_{\text{tr}} = \sqrt{\frac{8}{abc}} \sin\left(\frac{n_x \pi x}{a}\right) \sin\left(\frac{n_y \pi y}{b}\right) \sin\left(\frac{n_z \pi z}{c}\right) \tag{14.1-4}$$

where the quantum numbers n_x, n_y, and n_z are three nonnegative integers which we now call the **translational quantum numbers**. The translational energy is given by Equation (9.5-33):

$$E_{\text{tr}} = \frac{h^2}{8M}\left(\frac{n_x^2}{a^2} + \frac{n_y^2}{b^2} + \frac{n_z^2}{c^2}\right) \tag{14.1-5}$$

where M is the total mass of the atom. The translational energy levels, although quantized, lie very close together for a box of macroscopic size, and in the limit that the box becomes infinitely large, the energy levels approach each other like those of a free particle.

▼   **EXAMPLE 14.1**

The energy difference between the ground state and first excited electronic level of a hydrogen atom is 10.2 eV. Compare this energy difference with the spacing between the ground state and first excited translational level of a hydrogen atom in a cubical box 0.100 m on a side.

Solution

The ground translational state corresponds to $n_x = n_y = n_z = 1$, which we denote by (111). The first excited level consists of (112), (121), and (211), so the energy difference is

$$\Delta E = E_{112} - E_{111} = \frac{h^2}{8ma^2}(6 - 3)$$

$$= \frac{(6.6261 \times 10^{-34}\text{ J s})^2}{8(1.674 \times 10^{-27}\text{ kg})(0.100\text{ m})^2}(3)\frac{1\text{ eV}}{1.6022 \times 10^{-19}\text{ J}}$$

$$= 6.14 \times 10^{-20}\text{ eV}$$

which is smaller than the electronic excitation energy by a factor of 10^{-20}.

Example 14.1 is solved by straightforward substitution of values into the formula for the energy levels. Its purpose is to show the large difference between the energy level spacings.

▲

Exercise 14.1   Calculate the difference in energy between the ground state and the first excited translational level of a xenon atom in a box 0.100 m on each side. Express it in joules and in electron volts. Compare it with the excitation energy to the first excited electronic level of the xenon atom, 8.315 eV.

Even though translational energy levels are very close together, an atom can possess considerable translational energy. In Chapter 16, we will find that in a system of many point mass molecules the average translational energy is equal to

$$\langle E_{tr} \rangle = \frac{3k_B T}{2} \tag{14.1-6}$$

where k_B is Boltzmann's constant, introduced in Chapter 1 and equal to 1.3807×10^{-23} J K^{-1}, and T is the absolute temperature.

▼   **EXAMPLE 14.2**

For a hydrogen atom in the box of Example 14.1 with translational energy equal to $3k_B T/2$ at 300 K, find the value of the translational quantum numbers, assuming them to be equal to each other.

Solution

$$\frac{3k_B T}{2} = \frac{3}{2}(1.3807 \times 10^{-23} \text{ J K}^{-1})(300 \text{ K}) = 6.21 \times 10^{-21} \text{ J}$$

$$= \frac{h^2}{8Ma^2}(3n_x^2)$$

Example 14.2 is solved by a straight-forward substitution. Its purpose is to show how large typical values of the translational quantum numbers are.

$$n_x = \left[\frac{8(1.674 \times 10^{-27} \text{ kg})(0.100 \text{ m})^2(6.21 \times 10^{-21} \text{ J})}{3(6.6261 \times 10^{-34} \text{ J s})^2} \right]^{1/2}$$

$$= (6.314 \times 10^{17})^{1/2} = 7.95 \times 10^{8}$$

▲

Exercise 14.2

a. For a xenon atom in a cubical box with side 0.100 m, find the values of the translational quantum numbers (assumed equal) if the energy is equal to $3k_B T/2$ at 300 K.
b. Find the change in energy if one of the translational quantum numbers is increased by unity from its value in part a.

14.2   The Nonelectronic States of Molecules

Diatomic molecules are simpler to discuss than polyatomic molecules, so we discuss them first. However, we start our discussion with some analysis that applies to all molecules.

Equations for All Molecules

In order to study the motions of the electrons in molecules, we introduced the Born-Oppenheimer approximation in Chapter 13. In this approximation, the Schrödinger equation for the electronic motion is constructed with the nuclei assumed to be stationary.

We assume that the Schrödinger equation for the electrons has been solved in the Born-Oppenheimer approximation. If the internuclear repulsions are omitted from the electronic Hamiltonian, the Born-Oppenheimer energy for a given electronic state is the electronic energy eigenvalue plus the energy of internuclear repulsions, as in Equation (13.1-6):

$$E_{BO} = \mathcal{V}_{nn} + E_{el} \tag{14.2-1}$$

where E_{el} is the electronic energy eigenvalue and $\mathscr{V}_{nn}$ is the internuclear potential energy:

$$\mathscr{V}_{nn} = \frac{e^2}{4\pi\varepsilon_0} \sum_{A=2}^{n} \sum_{B=1}^{A-1} \frac{Z_A Z_B}{r_{AB}} \tag{14.2-2}$$

where we use the indexes A and B to specify two nuclei. Z_A is the number of protons in nucleus A, Z_B is the number of protons in nucleus B, r_{AB} is the distance between these two nuclei, and n is the total number of nuclei. The limits of the sums are chosen so that the indexes A and B are never equal to each other and so that each pair of nuclei is included only once.

Any energy that depends only on positions is a potential energy.

Since the Born-Oppenheimer energy depends on nuclear positions but not on their velocities, it acts like a potential energy for nuclear motion. We denote it now by $\mathscr{V}$:

$$\mathscr{V}(\mathbf{r}_1, \mathbf{r}_2, \ldots, \mathbf{r}_n) = E_{BO} \tag{14.2-3}$$

where $\mathbf{r}_1, \mathbf{r}_2, \ldots, \mathbf{r}_n$ are the nuclear position vectors.

The Hamiltonian operator for nuclear motion is

$$\hat{H}_{nuc} = -\hbar^2 \sum_{A=1}^{n} \frac{1}{2m_A} \nabla_A^2 + \mathscr{V}(\mathbf{r}_1, \mathbf{r}_2, \ldots, \mathbf{r}_n) + \mathscr{V}_{ext} \tag{14.2-4}$$

where $\nabla_1^2, \nabla_2^2, \ldots, \nabla_n^2$ are the Laplacian operators for the nuclear positions. The term $\mathscr{V}_{ext}$ represents the potential energy confining the molecule to a container. For a molecule that can move anywhere in space, this term can be set equal to zero. If the molecule is confined in a container, we use the expression of Equation (14.1-3) for $\mathscr{V}_{ext}$. As with atoms, this use is an approximation, since the closeness of approach of the center of mass of a molecule to the wall of a box will depend on the rotational and vibrational state of the molecule. For a box of macroscopic size, Equation (14.1-3) will be a very good approximation.

Diatomic Molecules

We now use lowercase r instead of R for the internuclear distance, because we will use R for the radial factor of the wave function.

The Born-Oppenheimer energy of a diatomic molecule depends only on the internuclear distance, which we now denote by r. The nuclear Hamiltonian operator for a diatomic molecule is

$$\hat{H}_{nuc} = -\hbar^2 \left(\frac{1}{2m_1} \nabla_1^2 + \frac{1}{2m_2} \nabla_2^2 \right) + \mathscr{V}(r) + \mathscr{V}_{ext} \tag{14.2-5}$$

where m_1 and m_2 are the nuclear masses. Since $\mathscr{V}_{ext}$ does not depend on the relative coordinates, $\hat{H}_{nuc}$ is the Hamiltonian operator for a central-force system and can be transformed into the same form as Equation (11.1-10):

$$\hat{H}_{nuc} = -\frac{\hbar^2}{2M} \nabla_c^2 + \mathscr{V}_{ext} - \frac{\hbar^2}{2\mu} \nabla_r^2 + \mathscr{V}(r) = \hat{H}_c + \hat{H}_r \tag{14.2-6}$$

The center-of-mass (translational) Hamiltonian is

$$\hat{H}_c = \hat{H}_{tr} = -\frac{\hbar^2}{2M} \nabla_c^2 + \mathscr{V}_{ext} \tag{14.2-7}$$

The relative Hamiltonian is

$$\hat{H}_r = -\frac{\hbar^2}{2\mu}\nabla_r^2 + \mathscr{V}(r) \qquad \textbf{(14.2-8)}$$

Here ∇_c^2 is the Laplacian operator for the center of mass of the two nuclei and ∇_r^2 is the Laplacian operator for their relative coordinates. M is the sum of the masses of the two nuclei, and μ is their reduced mass:

$$M = m_1 + m_2, \qquad \mu = \frac{m_1 m_2}{m_1 + m_2} \qquad \textbf{(14.2-9)}$$

All of the analysis of the central-force problem in Section 11.1 applies to the Hamiltonian operator of Equation (14.2-6). The variables separate to give two separate Schrödinger equations

$$\hat{H}_{\text{tr}}\psi_{\text{tr}} = E_{\text{tr}}\psi_{\text{tr}} \qquad \textbf{(14.2-10)}$$

and

$$\hat{H}_r\psi_r = E_r\psi_r \qquad \textbf{(14.2-11)}$$

The translational Schrödinger equation, Equation (14.2-10), is the same as for atoms. The translational wave function for a diatomic molecule in a rectangular box will be given by Equation (14.1-4) and the translational energy levels by Equation (14.1-5).

The relative Schrödinger equation for a diatomic molecule is given in spherical polar coordinates by Equation (11.1-18). This equation is solved by the trial function of Equation (11.1-20):

$$\psi_r(r, \theta, \phi) = R(r)\Theta(\theta)\Phi(\phi) = R(r)Y(\theta, \phi) \qquad \textbf{(14.2-12)}$$

The angular factors are the same spherical harmonic functions as for the rigid rotor and the hydrogen atom, with the same quantization of the angular momentum. The eigenvalues for the square of the angular momentum and its z projection are given by Equations (11.1-42) and (11.1-44):

$$\hat{L}^2 Y_{JM} = \hbar^2 J(J + 1)Y_{JM} \qquad \textbf{(14.2-13)}$$
$$\hat{L}_z Y_{JM} = \hbar M Y_{JM} \qquad \textbf{(14.2-14)}$$

where we now call the quantum numbers J and M instead of ℓ and m, to follow customary usage. The quantum number J in this usage is *not* the same as the quantum number J for the total electronic angular momentum in Chapters 11 and 12.

In an exact treatment, the orbital and spin angular momenta of the electrons would combine with the rotational angular momentum and the spin angular momentum of the nuclei to make up the total angular momentum, which would be the only exactly conserved quantity. However, in the Born-Oppenheimer approximation, the states with definite values of J and M are good approximations to actual states.

The radial factor in the wave function of Equation (14.2-12) is different from that of the hydrogen atom. In our present notation, Equation (11.2-2) for the radial factor R is

$$-\frac{d}{dr}r^2\frac{dR}{dr} + \frac{2\mu r^2}{\hbar^2}\hbar(\mathscr{V} - E)R + J(J + 1)R = 0 \qquad \textbf{(14.2-15)}$$

where we drop the subscript on the relative energy E_r. We need an explicit expression for $\mathscr{V}(r)$ in order to proceed. The function $\mathscr{V}(r)$ is a function such as the one depicted in Figure 13.1. As a first approximation, we express $\mathscr{V}(r)$ as a Taylor series in the variable $x = r - r_e$, where r_e is the value of r at the minimum in $\mathscr{V}$:

$$\mathscr{V}(r) = \mathscr{V}_e + \left(\frac{d\mathscr{V}}{dr}\right)_e x + \frac{1}{2!}\left(\frac{d^2\mathscr{V}}{dr^2}\right)_e x^2 + \cdots \quad \textbf{(14.2-16)}$$

where the subscript e means that the quantity is evaluated at $r = r_e$.

The function $\mathscr{V}$ is at a minimum at $r = r_e$, so the first derivative vanishes. To a fairly good approximation, we truncate the series at the quadratic term and write

$$\mathscr{V}(r) = \mathscr{V}_e + \frac{1}{2!}\left(\frac{d^2\mathscr{V}}{dr^2}\right)_e x^2 = \mathscr{V}_e + \frac{1}{2}kx^2 \quad \textbf{(14.2-17)}$$

where $\mathscr{V}_e$ is the value of $\mathscr{V}$ at the minimum and where we denote the second derivative evaluated at $r = r_e$ by k, called the **force constant**. It is the same kind of force constant as we introduced for the harmonic oscillator in Chapter 9. This approximation will be more nearly correct for small magnitudes of x (values of r that do not differ much from r_e).

A multiple chemical bond (a double or triple bond) generally has a greater dissociation energy than a single bond and thus a deeper minimum in $\mathscr{V}$. It is also likely to have a correspondingly larger value of k. Also, we have a different value of k, of $\mathscr{V}_e$, and of r_e for each bound electronic state. Figure 14.1 shows schematically the function $\mathscr{V}$ and the truncated power series representations for two electronic states of a diatomic molecule. The potential function of Equation (14.2-17) is called a **harmonic potential** because it leads to vibrational energy levels identical to those of a harmonic oscillator.

The radial Schrödinger equation is now

$$\frac{d}{dr}r^2\frac{dR}{dr} - J(J+1)R + \frac{2\mu r^2}{\hbar^2}\left(E - \mathscr{V}_e - \frac{kx^2}{2}\right)R = 0 \quad \textbf{(14.2-18)}$$

We define a new dependent variable

$$S(r) = rR(r) \quad \textbf{(14.2-19)}$$

When this variable is substituted into Equation (14.2-18), the result is

$$\frac{\hbar^2}{2\mu}\left[\frac{d^2S}{dr^2} - \frac{J(J+1)S}{r^2}\right] + \left(E - \mathscr{V}_e - \frac{kx^2}{2}\right)S = 0 \quad \textbf{(14.2-20)}$$

To express this equation in terms of x, we write

$$\frac{1}{r^2} = \frac{1}{(r_e + x)^2} = \frac{1}{r_e^2}\left(1 - \frac{2x}{r_e} + \frac{3x^2}{r_e^2} + \cdots\right) \quad \textbf{(14.2-21)}$$

If x is quite small, it is a fairly good approximation to keep only the first term of this series:

$$\frac{1}{r^2} \approx \frac{1}{r_e^2}$$

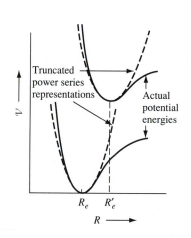

Figure 14.1. Vibrational Potential Energy for a Typical Diatomic Molecule in Two Electronic States. This figure shows not only the realistic curve representing the Born-Oppenheimer energy (which acts as the vibrational potential energy) but also the parabolic curve representing the quadratic truncated power series representation. The equilibrium internuclear distances for the two electronic states are not equal to each other; they show the general pattern that the excited state has the longer equilibrium internuclear distance.

Truncated power series representations

Actual potential energies

R_e R_e'

$R \longrightarrow$

Equation (14.2-20) becomes

$$-\frac{\hbar^2}{2\mu}\frac{d^2S}{dx^2} + \frac{kx^2}{2}S = \left[E - \mathscr{V}_e - \frac{\hbar^2}{2\mu r_e^2}J(J+1)\right]S \quad \textbf{(14.2-22)}$$

Exercise 14.3   Carry out the mathematical steps to obtain Equation (14.2-22) from Equation (14.2-18).

Equation (14.2-22) is the same as the harmonic oscillator Schrödinger equation of Chapter 9, Equation (9.6-1), except for the presence of a constant term subtracted from the energy eigenvalue. The function S is the same as the harmonic oscillator energy eigenfunction, given by Equations (9.6-15) etc., and the energy eigenvalue E is the harmonic oscillator energy eigenvalue of Equation (9.6-12) plus the constant term (see Exercise 9.25):

$$E = E_{Jv} = h v_e\left(v + \frac{1}{2}\right) + \frac{\hbar^2}{2I_e}J(J+1) + \mathscr{V}_e \quad \textbf{(14.2-23)}$$

where v_e is the oscillator frequency predicted by classical mechanics,

$$v_e = \frac{1}{2\pi}\sqrt{\frac{k}{\mu}} \quad \textbf{(14.2-24)}$$

The quantum number v is equal to $0, 1, 2, \ldots$; it is the same as the harmonic oscillator quantum number of Chapter 9. We use the symbol I_e for the equilibrium **moment of inertia** of the diatomic molecule:

$$I_e = \mu r_e^2 \quad \textbf{(14.2-25)}$$

In all of these variables, the subscript e is used to recognize the fact that we have replaced r by r_e in Equation (14.2-22).

We have now completed our solution of the Schrödinger equation for the motion of a diatomic molecule. The wave function for the relative motion of the nuclei is now

$$\psi_r = \Theta_{JM}(\theta)\Phi_M(\phi)\frac{S_v(r - r_e)}{r} = \psi_{\text{rot},JM}\psi_{\text{vib},v} \quad \textbf{(14.2-26)}$$

where we replace the radial factor R by S/r. The factor ψ_{rot} is called the rotational wave function:

$$\psi_{\text{rot},JM} = \Theta_{JM}(\theta)\Phi_M(\phi) = Y_{JM}$$

It is the same as the spherical harmonic function of Section 11.1, and has the same two quantum numbers, which we now call J and M instead of ℓ and m.

The radial factor R is the vibrational wave function. The vibrational wave function is a harmonic oscillator wave function divided by r, the internuclear distance.

We will often sketch vibrational wave functions as though they were harmonic oscillator functions. This practice is acceptable if r remains nearly equal to r_e.

$$\psi_{\text{vib},v} = R = \frac{S_v}{r} = \frac{\psi_{\text{HO}}}{r}$$

In the case of atoms, the wave function was a product of a translational factor and an electronic factor. The total energy was the translational energy plus the electronic energy. For a diatomic molecule, the total wave function is given by

$$\psi_{\text{tot}} = \psi_{\text{tr}}\psi_r = \psi_{\text{tr}}\psi_{\text{rot}}\psi_{\text{vib}}\psi_{\text{el}} \tag{14.2-27}$$

For a diatomic molecule, the total energy is the translational energy plus the relative energy in Equation (14.2-23). The second term in the right-hand side of Equation (14.2-23) is the same as the energy of a rigid rotor, Equation (11.1-48), and we call it the rotational energy:

$$E_{\text{rot}} = \frac{\hbar^2}{2I_e} J(J + 1) \qquad (J = 0, 1, 2, \ldots) \tag{14.2-28a}$$

The rotational energy does not depend on the value of the quantum number M. For a given value of J, the values of M range from J to $-J$ in unit steps, so there are $2J + 1$ states with the energy given in Equation (14.2-28a). If we denote this degeneracy by g_J, we have

$$g_J = 2J + 1 \tag{14.2-28b}$$

The $J = 0$ energy level consists of one state, the $J = 1$ level consists of three states, etc.

The first term in the right-hand side of Equation (14.2-23) is the energy of a harmonic oscillator, and we call it the vibrational energy:

$$E_{\text{vib}} = h\nu_e\left(v + \frac{1}{2}\right) \qquad (v = 0, 1, 2, 3, \ldots) \tag{14.2-29}$$

The energy levels of a harmonic oscillator are nondegenerate, so there is only one vibrational state for each value of v.

The final term is the Born-Oppenheimer energy at the minimum, and we call it the electronic energy.

$$E_{\text{el}} = \mathcal{V}_e = \mathcal{V}(r_e) \tag{14.2-30}$$

This electronic energy is a different constant for each electronic state. It is equal to the Born-Oppenheimer energy at $r = r_e$. The rest of the Born-Oppenheimer energy was taken as the potential energy of vibration.

The center-of-mass (translational) energy is given by Equation (14.1-5):

$$E_{\text{tr}} = E_{n_x n_y n_z} = \frac{h^2}{8M}\left(\frac{n_x^2}{a^2} + \frac{n_y^2}{b^2} + \frac{n_z^2}{c^2}\right) \tag{14.2-31}$$

This separation of energy is an approximation. Among other things that are omitted is the fact that rotation and vibration interact with each other.

We can write the total energy as a sum of translational, vibrational, rotational, and electronic contributions:

$$E_{\text{tot}} = E_{\text{tr}} + E_{\text{vib}} + E_{\text{rot}} + E_{\text{el}} \tag{14.2-32}$$

The translational energy is independent of the electronic state, but the quantities ν_e, I_e, and E_{el} are different for each electronic state. The electronic energy E_{el} is a different constant for each electronic state. It is often chosen to equal zero for the electronic ground state, which makes it equal to a different positive constant for each excited level.

The spacings between the different quantized values of the energy contributions are widely different.

▼   **EXAMPLE 14.3**

These energy level spacings are typical. Remember the relative sizes of these spacings.

For a carbon monoxide molecule in a cubic box 0.100 m on a side, compare the spacing between the ground state and the first excited level for translation, rotation, vibration, and electronic motion. The equilibrium bond length is 1.128×10^{-10} m and the vibrational frequency is 6.505×10^{13} s^{-1} for the ground electronic state. The excitation energy to the minimum in the first excited electronic state is 6.036 eV.

Solution

We assume the most common isotopes, ^{12}C and ^{16}O. For translation,

$$E_{211} - E_{111} = \frac{h^2}{8Ma^2}(6-3) = \frac{(6.6261 \times 10^{-34} \text{ J s})^2(3)}{8(4.469 \times 10^{-26} \text{ kg})(0.100 \text{ m})^2}$$

$$= 3.68 \times 10^{-40} \text{ J} = 2.30 \times 10^{-21} \text{ eV}$$

For rotation,

$$E_1 - E_0 = \frac{\hbar^2}{2I_e}(2-0) = \frac{h^2}{8\pi^2\mu r_e^2}(2-0)$$

$$= \frac{(6.6261 \times 10^{-34} \text{ J s})^2(2)}{8\pi^2(1.138 \times 10^{-26} \text{ kg})(1.128 \times 10^{-10} \text{ m})^2}$$

$$= 7.68 \times 10^{-23} \text{ J} = 4.79 \times 10^{-4} \text{ eV}$$

For vibration,

$$E_1 - E_0 = h\nu_e\left(\frac{3}{2} - \frac{1}{2}\right) = h\nu_e$$

$$= (6.6261 \times 10^{-34} \text{ J s})(6.505 \times 10^{13} \text{ s}^{-1})$$

$$= 4.310 \times 10^{-20} \text{ J} = 0.269 \text{ eV}$$

The vibrational energy level spacing is smaller than the electronic level spacing by a factor of about $\frac{1}{20}$, the rotational level spacing is smaller by a factor of $\frac{1}{600}$ than the vibrational level spacing, and the translational level spacing is smaller than the rotational level spacing by a factor of 10^{-17}.

▲

The results of Example 14.3 are quite typical. For most molecules, the four kinds of energy levels have spacings that are very different from each other, with the translational levels much closer together than any of the others. Figure 14.2 shows schematically (not to scale) the spacings of the energy levels for a typical diatomic molecule. The vibrational levels are superimposed on the graph of the potential function $\mathscr{V}$, and the spacings between the rotational levels have been enlarged so as to be seen. The translational levels are too closely spaced to be shown in this diagram.

Exercise 14.4

Compare the energy level spacings for a Cl_2 molecule in a cubic box 0.100 m on a side. Its vibrational frequency is 1.694×10^{13} s^{-1} and its equilibrium bond length for the ground electronic state is 1.988×10^{-10} m. The energy of the first excited electronic state observed in the gas phase is 2.208 eV above that of the ground electronic state.

The energy level expression of Equation (14.2-23) is only a first approximation. The power series expression for the vibrational potential function

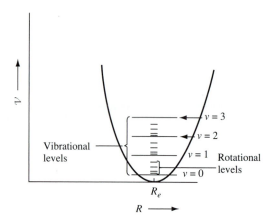

Figure 14.2. Energy Levels for a Typical Diatomic Molecule (schematic). The vibrational energy levels are superimposed on the vibrational potential energy with the same zero of energy, so that the zero-point energy can be identified as the height from the minimum in the curve to the $v = 0$ line segment. The rotational energy levels are superimposed for each vibrational state. These energy levels are drawn farther apart from each other than is the case for a typical molecule. Their actual spacing is very much smaller than the vibrational energy level spacing.

$\mathscr{V}$ was truncated at the quadratic term. If one additional term of this expansion is kept, the equation can be transformed into the harmonic oscillator equation, and the energy expression is[1]

$$E = E_{vJ} = h v_e \left(v + \frac{1}{2} \right) + \frac{\hbar^2}{2 I_e} J(J + 1)$$
$$- \frac{\hbar^4}{8 \pi^2 v_e^2 I_e^3} J^2 (J + 1)^2 + E_{\mathrm{el}} \qquad \textbf{(14.2-33)}$$

If a diatomic molecule has a dipole moment (that is, if the ends of the molecule are charged), the molecule can absorb or emit electromagnetic radiation. The difference in energy between two states divided by Planck's constant is equal to the frequency of the photon absorbed or emitted when a transition occurs between two states. For this reason, energies divided by Planck's constant are sometimes used. Equation (14.2-33) becomes

$$\frac{E_{vJ}}{h} = v_e \left(v + \frac{1}{2} \right) + B_e J(J + 1) - \mathscr{D} J^2 (J + 1)^2 \qquad \textbf{(14.2-34)}$$

where we define the parameters

$$B_e = \frac{\hbar}{4 \pi I_e} = \frac{h}{8 \pi^2 I_e}, \qquad \mathscr{D} = \frac{4 B_e^3}{v_e^2} \qquad \textbf{(14.2-35)}$$

The term containing the parameter $\mathscr{D}$ represents the nonconstancy of the internuclear distance. It is independent of the vibrational quantum num-

---

[1] Jeff C. Davis, Jr., *Advanced Physical Chemistry*, Ronald Press, New York, 1965, p. 285.

ber v and becomes more important for larger values of the rotational quantum number J. Figure 14.3 shows the rotational energy levels of the CO molecule with and without this correction term. The corrected levels are more closely spaced for larger values of J than the uncorrected levels. Inspection of Equations (14.2-34) and (14.2-35) shows that this shift would correspond to decreased values of B_e and increased values of I_e for larger values of J. Since I_e is proportional to r_e^2, this shift corresponds to larger values of r_e^2 due to centrifugal stretching.

A second approximation that went into Equation (14.2-23) was the assumption of the harmonic potential function of Equation (14.2-17). Several alternative formulas are used as approximate representations of the vibrational potential energy. The most commonly used is the **Morse function**

$$\mathscr{V}(r) - \mathscr{V}(r_e) = D_e[1 - e^{-a(r-r_e)}]^2 \qquad \textbf{(14.2-36)}$$

where D_e is the **dissociation energy**, or the energy required to dissociate the molecule from the state of minimum $\mathscr{V}$. The parameter a determines the curvature of the function and is equal to $k/(2D_e)$. The values of these parameters must be determined for each molecule. Figure 14.4a depicts the Morse

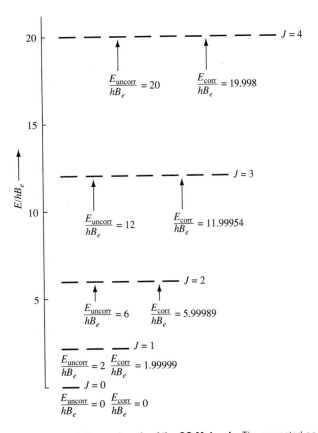

Figure 14.3. The Rotational Energy Levels of the CO Molecule. The corrected and uncorrected energy levels are so nearly equal that this figure represents both sets of levels. The values of the energy eigenvalues divided by hB_e are given in the figure.

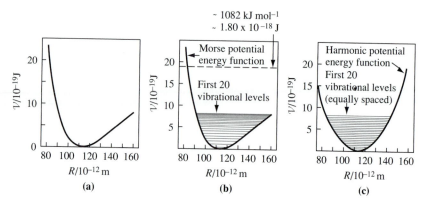

Figure 14.4. The Vibrational Potential Energy and Energy Levels of the CO Molecule.
(a) The vibrational potential energy represented by the Morse function. **(b)** The first
20 vibrational energy levels for the Morse potential. **(c)** The first 20 vibrational energy
levels for the harmonic potential. The Morse potential parameters for CO are
$D_e = 11.2$ eV $= 1.80 \times 10^{-18}$ J and $a = 2.2994 \times 10^{10}$ m^{-1}. The force constant for the
harmonic potential is $k = 1900$ N m^{-1}.

potential function for the CO molecule. The Schrödinger equation for the
Morse potential has been solved.[2]

Exercise 14.5

Using the general relation between potential energy and force, Equation (D-6) of
Appendix D, obtain a formula for the force on a nucleus in a diatomic molecule
described by the Morse potential function. Show that D_e is equal to the difference
in potential energy between the minimum and the value for large r. Show that r_e is
the value of r at the minimum and show that there is no force if $r = r_e$.

Corrections for anharmonicity and for interaction between vibration and
rotation can be added to the energy level expression, giving instead of
Equation (14.2-34)

$$E_{vJ} = hv_e\left(v + \frac{1}{2}\right) - hv_e x_e\left(v + \frac{1}{2}\right)^2 + hB_e J(J + 1)$$

$$- h\mathcal{D}J^2(J + 1)^2 - h\alpha\left(v + \frac{1}{2}\right)J(J + 1) \qquad \textbf{(14.2-37)}$$

The parameters α and x_e are given for the Morse potential by[3]

$$\alpha = \frac{3h^2 v_e}{16\pi^2 \mu r_e^2 D_e}\left(\frac{1}{ar_e} - \frac{1}{a^2 r_e^2}\right) \qquad \textbf{(14.2-38)}$$

$$x_e = \frac{hv_e}{4D_e} \qquad \textbf{(14.2-39)}$$

---

[2] P. M. Morse, *Phys. Rev.* **34**, 57 (1929).
[3] Davis, *op. cit.*, p. 351.

where D_e is the dissociation energy of the molecule, not the parameter $\mathscr{D}$. The parameters α and x_e have different values for different diatomic molecules, as does the parameter $\mathscr{D}$.

The term containing the parameter x_e expresses the effect of the anharmonicity of the vibrational potential energy and is independent of the rotational quantum number. Figure 14.4b shows the vibrational energy levels of the CO molecule, including the anharmonicity correction. Figure 14.4c shows the uncorrected energy levels. The uncorrected energy levels are equally spaced, like those in Figure 9.24. The corrected energy levels are more closely spaced for larger values of the vibrational quantum number. The dissociation energy D_e is sometimes related to v_e and x_e by determining the energy at which the energy difference between two successive levels shrinks to zero (see Problem 14.21).

The term containing the parameter α contains both the vibrational and rotational quantum numbers and expresses the interaction of vibration and rotation. The physical origin of this interaction can be seen in Figure 14.4b, which shows that for larger values of v the classically allowed region of the vibrational coordinate moves to the right in the figure, so that the moment of inertia is larger for larger values of v.

All three of the correction terms cause the energy levels to be lower for large values of the quantum numbers than the uncorrected energy levels. For some purposes, the terms in α, x_e, and $\mathscr{D}$ can be omitted and the expression of Equation (14.2-23) can be used.

EXAMPLE 14.4

Calculate the energy of the $v = 2$, $J = 2$ level of the CO molecule using the values of the parameters: $v_e = 6.50488 \times 10^{13}$ s^{-1}, $x_e = 6.124 \times 10^{-3}$, $B_e = 5.7898 \times 10^{10}$ s^{-1}, $\mathscr{D} = 1.83516 \times 10^5$ s^{-1}, $\alpha = 5.24765 \times 10^8$ s^{-1}.

Solution

$$E/h = v_e(5/2) + v_e x_e(5/2)^2 + B_e(2)(3) - \mathscr{D}(36) - \alpha(5/2)(6)$$
$$= (6.50488 \times 10^{13} \text{ s}^{-1})(5/2) + (3.9836 \times 10^{11} \text{ s}^{-1})(5/2)^2$$
$$+ (5.7898 \times 10^{10} \text{ s}^{-1})(6) - (1.83516 \times 10^5 \text{ s}^{-1})(36)$$
$$- (5.24765 \times 10^8 \text{ s}^{-1})(5/2)(6)$$
$$= 1.6262 \times 10^{14} \text{ s}^{-1} + 2.4897 \times 10^{12} \text{ s}^{-1} + 3.4739 \times 10^{11} \text{ s}^{-1}$$
$$- 6.6066 \times 10^6 \text{ s}^{-1} - 7.8715 \times 10^9 \text{ s}^{-1}$$
$$= 1.6545 \times 10^{14} \text{ s}^{-1}$$
$$E = (6.6261 \times 10^{-34} \text{ J s})(1.6545 \times 10^{14} \text{ s}^{-1})$$
$$= 1.0963 \times 10^{-19} \text{ J}$$

To five significant digits, the $\mathscr{D}$ term is insignificant and the α term is fairly small. The $v_e x_e$ term is large compared with the main rotational term (the J term) but is fairly small compared with the main vibrational term (the v_e term).

The energy levels are also sometimes given in terms of energies divided by hc, where c is the speed of light. This quantity has the dimensions of reciprocal wavelength, and its difference for two levels is equal to the reciprocal of the wavelength of the photon emitted or absorbed in the transition between these levels. The energies are commonly given in reciprocal centimeters, sometimes called **wave numbers**.

$$\frac{E_{vJ}}{hc} = \tilde{v}_e\left(v + \frac{1}{2}\right) + \tilde{v}_e x_e\left(v + \frac{1}{2}\right)^2 + \tilde{B}_e J(J + 1)$$

$$- \tilde{\mathcal{D}}J^2(J + 1)^2 - \tilde{\alpha}\left(v + \frac{1}{2}\right)J(J + 1) \tag{14.2-40}$$

Unfortunately, many spectroscopic references use the same letters without a tilde to represent the parameters in Equation (14.2-40). The parameters in this equation have the units of reciprocal length (usually cm^{-1}). Those in Equation (14.2-37) have the units of s^{-1}.

The parameters in this equation are marked with a tilde ($\tilde{}$) to distinguish them from the parameters in Equation (14.2-36). They are equal to the other parameters divided by c, the speed of light. Table A16 gives the values of parameters for several diatomic molecules.

Exercise 14.6

a. Using the expression of Equation (14.2-40), find the wavelength and frequency of the light absorbed when carbon monoxide molecules make the transition from the $v = 0$, $J = 0$ state to the $v = 1$, $J = 1$ state.

b. Find the wavelength for the same transition, neglecting the terms in $\tilde{\alpha}$ and $\tilde{\mathcal{D}}$.

c. Find the wavelength for the same transition, neglecting the terms in x_e, $\tilde{\alpha}$, and $\tilde{\mathcal{D}}$.

Nuclear Spins and Wave Function Symmetry

For the special case of homonuclear diatomic molecules, we return to the wave function of Equation (14.2-27). We determine the consequence of complying with the requirement that the wave function not pretend to distinguish between two identical nuclei, which are indistinguishable, just as two electrons are indistinguishable from each other. Although electrons are fermions, some nuclei are fermions and others are bosons. We assume that the interchange of two fermion nuclei must change the sign of the molecular wave function and that the interchange of two boson nuclei must leave the wave function unchanged. The consequence is that for some homonuclear diatomic molecules, only rotational wave functions with odd values of J can occur. For the other diatomic molecules, only even values of J can occur. For heteronuclear diatomic molecules, all values of J can occur, since the nuclei are distinguishable.

For our purposes, a nucleus can be considered to be made up of protons and neutrons, collectively called **nucleons**. Protons and neutrons have a spin quantum number of $1/2$, as do electrons, so they are fermions. If a nucleus contains an odd number of nucleons, the nucleus is a fermion, since exchanging two such nuclei changes the sign of the wave function once for each nucleon (an odd number of times). If a nucleus contains an even number of nucleons, it is a boson. Exchanging two such nuclei changes the sign of the wave function an even number of times, restoring the original sign.

The wave function for nuclear motion in Equation (14.2-26) must be multiplied by a nuclear spin wave function to be a complete wave function. The nuclear spin angular momentum has the same general properties as any angular momentum. It takes on the values

$$|\mathbf{I}|^2 = \hbar^2 I(I + 1) \tag{14.2-41}$$

Once again, we see that all angular momenta follow the same pattern of eigenvalues.

where the vector $\mathbf{I}$ represents the spin angular momentum, which is the vector sum of the spin angular momenta of all the nucleons in the nucleus. The nuclear spin quantum number I is an integer if the nucleus is a boson and a half-integer if the nucleus is a fermion. Its value depends on how the nucleon angular momenta add vectorially in the particular nuclear state. A given nucleus can have different spin states, just as an atom can have different electronic spin states. However, very large energies are required to raise nuclei to excited states, and chemists ordinarily encounter nuclei only in their ground states. Table A17 lists the spin quantum number for the nuclear ground state of several common nuclei.

The projection of the spin angular momentum on the z axis takes on the values

$$I_z = \hbar M_I \tag{14.2-42}$$

where M_I is a quantum number ranging from $+I$ to $-I$ in integral steps. If I is an integer, so is M_I, and if I is a half-integer, so is M_I.

If $I = 0$, as in the case of ^{16}O, the nuclei are bosons, containing an even number of nucleons. There is a single spin function, and the total wave function must be symmetric with respect to exchange of the nuclei, since they are bosons. If $I = 1/2$, as in ^1H, the nuclear spin wave functions are like those for electrons, with α representing spin up and β representing spin down. The triplet nuclear spin functions for such a diatomic molecule are symmetric:

$$\alpha(1)\alpha(2), \qquad \beta(1)\beta(2), \qquad \sqrt{\frac{1}{2}}\left[\alpha(1)\beta(2) + \beta(1)\alpha(2)\right]$$

and the singlet spin function is antisymmetric:

$$\sqrt{\frac{1}{2}}\left[\alpha(1)\beta(2) - \beta(1)\alpha(2)\right]$$

The space part of the total wave function must be antisymmetric if it is combined with a symmetric nuclear spin function and must be symmetric if it is combined with the antisymmetric spin function, since the nuclei are fermions.

For values of I greater than 1/2, the spin functions are more complicated than those of two hydrogen nuclei, and we do not discuss them. The total wave function must still be symmetric if I is an integer or antisymmetric if I is a half-integer.

The space factor of the total wave function of a homonuclear diatomic molecule could be symmetrized or antisymmetrized by making a two-term wave function. However, our one-term function is generally an eigenfunction of spatial symmetry operators, and this makes it either symmetric or antisymmetric with respect to exchange of particles. For example, if the entire wave function is operated on by the inversion operator and then the electronic factor is operated again by the inversion operator, the effect is to exchange the nuclei but not the electrons. Similarly, if the entire molecule rotates by 180° around an axis perpendicular to the internuclear axis and the electrons are then reflected through a plane perpendicular to the axis of rotation and finally inverted, the nuclei are exchanged and the electrons are back where they started. We can determine the effect of such operations on each factor of the total wave function.

Electronic wave functions are denoted by g if they are eigenfunctions of the inversion operator with eigenvalue $+1$ and by u if the eigenvalue is -1. Functions with eigenvalue $+1$ are said to have **even parity**, and those with eigenvalue -1 are said to have **odd parity**. A superscript $+$ is used to denote eigenfunctions of $\hat{\sigma}_v$ with eigenvalue $+1$, and a superscript $-$ is used to denote eigenfunctions of $\hat{\sigma}_v$ with eigenvalue -1.

The rotational factor of the wave function of a diatomic molecule is a spherical harmonic function. The sperical harmonic functions for even values of J are eigenfunctions of the inversion operator with eigenvalue unity (are "gerade") and for odd values of J are eigenfunctions with eigenvalue negative unity (are "ungerade").

Exercise 14.7 Show that the spherical harmonic function Y_{00} is an eigenfunction of the inversion operator with eigenvalue 1, while the spherical harmonic function Y_{11} is an eigenfunction with eigenvalue -1. In spherical polar coordinates the inversion operator replaces θ by $\pi - \theta$ and replaces ϕ by $\pi + \phi$.

The vibrational factor is a harmonic oscillator wave function divided by r and depends only on r, which is a scalar quantity that remains unchanged under inversion. The vibrational factor is an eigenfunction of the inversion operator with eigenvalue unity for all values of the quantum number v.

For any diatomic molecule, we must choose wave functions such that the outcome of a set of symmetry operations that exchanges the nuclei changes the sign of the wave function for fermion nuclei and does not change the sign of the wave function for boson nuclei.

▼ **EXAMPLE 14.5** Find the permitted wave functions for H_2 in its electronic ground state.

Solution

The ground electron configuration is $(\sigma g1s)^2$. The electronic wave function is a product of two gerade space orbitals and is thus itself gerade. The nuclear wave function must be antisymmetric. If the nuclear spin function is the antisymmetric singlet function, the rotational wave function must be gerade, which means that J must be an even integer. If the nuclear spin function is one of the triplet functions, which are symmetric, the rotational wave function must be ungerade, which means that J must be an odd integer.

The form of hydrogen with triplet nuclear spin states and odd values of J is called **ortho hydrogen**, and the singlet, even-J form is called **para hydrogen**. Since the interconversion between the singlet and triplet states is extremely slow in the absence of a magnetic field or a catalyst, these two forms of hydrogen can be separated. ▲

In the para form the nuclear spins point in opposite directions, just as do two groups para to each other on a benzene ring.

In Example 14.5 we have seen that for each of the forms of hydrogen, only half of the rotational wave functions occur. This is a general occurrence for homonuclear diatomic molecules, since a given electronic wave function will always demand either a symmetric or antisymmetric nuclear function. Therefore, for a given electronic state and a given nuclear spin state in a homonuclear diatomic molecule, either odd values of J will occur or even values of J will occur, but not both.

For any given homonuclear molecule, either even J occurs or odd J occurs, but not both. This restriction does not apply to a heteronuclear molecule.

Exercise 14.8 For ^{200}Hg, $I = 0$ and the electronic ground state is g and $+$. What values of J can occur with $^{200}Hg_2$?

14.3   **Rotation and Vibration in Polyatomic Molecules**

Rotation and vibration are more complicated in polyatomic molecules than in diatomic molecules, and we consider only a first approximation. To study the rotation of a polyatomic molecule, we pretend that the molecule is somehow prevented from vibrating. To study the vibration of a polyatomic molecule we pretend that the molecule is prevented from rotating. We could have adopted this policy to obtain Equation (14.2-23) for the energy of a diatomic molecule, where we found the energy was that of a nonrotating harmonic oscillator plus that of a rigid (nonvibrating) rotor. Our result for the polyatomic molecules will be at the same level of approximation as Equation (14.2-23).

Rotation of Polyatomic Molecules

We assume that all bond lengths and bond angles of a polyatomic molecule are locked at their equilibrium values, so that the molecule rotates as a rigid body. For example, we assume that a methane molecule rotates with all bond angles held at the tetrahedral angle of 109° and all bond lengths held fixed at the equilibrium length of 1.11×10^{-10} m. This is a fairly good approximation for small values of the rotational and vibrational energies.

The rotation of a rigid body is described classically in terms of **moments of inertia** taken relative to three mutually perpendicular axes. For an object consisting of n point masses, the moment of inertia about an axis is defined to be

$$I_{\text{axis}} = \sum_{i=1}^{n} m_i r_{i(\text{axis})}^2 \tag{14.3-1}$$

where m_i is the mass of the ith point mass and $r_{i(\text{axis})}$ is the perpendicular distance from this point mass to the axis. For example, the moment of inertia about the z axis is

$$I_z = \sum_{i=1}^{n} m_i(x_i^2 + y_i^2) \tag{14.3-2}$$

The other two moments of inertia are defined similarly.

There are three additional quantities called **products of inertia**. For example,

$$I_{xy} = I_{yx} = \sum_{i=1}^{n} m_i x_i y_i \tag{14.3-3}$$

The other two products of inertia, I_{yz} and I_{xz}, are defined analogously. For both the moments of inertia and the products of inertia, we neglect the masses of the electrons and include only the nuclei in the sums.

There is a theorem that for any rigid object, it is possible to choose a set of perpendicular axes with the origin at the center of mass of the object such that all products of inertia vanish. Such axes are called **principal axes**, and the moments of inertia relative to them are called **principal moments of inertia**.

For a symmetrical molecule, it is usually possible to assign a set of principal axes by inspection. One first decides on the symmetry operators that belong to the molecule. Principal axes are then obtained by placing the axes along symmetry elements as much as possible. If there is an axis of symmetry that is at least a threefold rotation axis, a set of principal axes is obtained simply by choosing this rotation axis as one of the axes. If there is a twofold rotation axis, a set of principal axes is obtained by choosing this axis as one of the principal axes and placing the other two axes in planes of symmetry.

Since the principal axes are defined relative to the molecule and rotate with it, it is customary to call the axes by the letters A, B, and C instead of x, y, and z. By convention, the axes are ordered so that

$$I_A \leq I_B \leq I_C \tag{14.3-4}$$

If all three of its principal moments are equal, an object is called a **spherical top**. The name "top" is apparently chosen because of the rotating toys by that name. Any kind of smooth spherical ball such as a billiard cue ball is a spherical top, and a tetrahedral molecule such as methane or an octahedral molecule such as sulfur hexafluoride is also a spherical top. Any mutually perpendicular axes passing through the center of mass are principal axes.

If two of the principal moments are equal, the object is called a **symmetric top**. A **prolate symmetric top** has a unique moment of inertia that is smaller than the other two. An American football and a rugby ball are prolate symmetric tops if the lacing is ignored. An **oblate symmetric top** has a unique moment that is larger than the other two. A discus is an oblate symmetric top. Any molecule with at least a threefold rotation axis is a symmetric top (or possibly a spherical top).

If all three principal moments of inertia are unequal, the object is called an **asymmetric top**. A boomerang is an asymmetric top, as is a bent triatomic molecule such as sulfur dioxide or water.

▼   **EXAMPLE 14.6**

Example Figure 14.6. The Boron Trifluoride Molecule and Its Principal Axes. These principal axes are identified by inspection. All must pass through the center of mass, and one must coincide with the threefold rotational axis. The A and B axes could be placed anywhere in the horizontal plane with no change in the moments of inertia.

Show that BF_3, which is a trigonal planar molecule, is an oblate symmetric top.

Solution

Orient the molecule as in Example Figure 14.6, with one BF bond on the A axis. Let the bond length be called a.

$$I_A = 2m_F[a\sin(120°)]^2 = 2m_F a^2 \left[\frac{1}{2}\sqrt{3} \right]^2 = \frac{3}{2}m_F a^2$$

$$I_B = m_F a^2 + 2m_F[a\cos(120°)]^2 = \frac{3}{2}m_F a^2 = I_A$$

$$I_C = 3m_F a^2 > I_A$$

▲

| **Exercise 14.9** | Show that methane is a spherical top. |

The total angular momentum **L** of an object is the vector sum of the angular momenta of all the particles making up the object. The classical rotational energy of a rigid object is given in terms of **L** and its principal moments of inertia by

$$E_{\text{classical}} = \frac{L_A^2}{2I_A} + \frac{L_B^2}{2I_B} + \frac{L_C^2}{2I_C} \tag{14.3-5}$$

where L_A, L_B, and L_C are the instantaneous components of the vector **L** on the A, B, and C axes.

We can write the quantum mechanical rotational energy from Equation (14.3-5) by using the general properties of angular momentum. Consider first the case of a diatomic molecule or linear polyatomic molecule, for which I_A vanishes and the other two moments are equal. There can be no component of angular momentum on the A axis because there are no nuclei that are not on this axis (we are excluding the electrons from our moments of inertia). Equation (14.3-5) becomes

$$E_{\text{classical}} = \frac{1}{2I_B}(L_B^2 + L_C^2) = \frac{L^2}{2I_B} \tag{14.3-6}$$

From Equation (11.1-42) we have the values that a general angular momentum can assume in quantum mechanics. Substitution of this formula into Equation (14.3-6) gives

$$E_{\text{qm}} = E_J = \frac{\hbar^2}{2I_B} J(J+1) \tag{14.3-7}$$

which agrees with Equation (14.2-28a). The possible values of L_z are given by Equation (11.1-44).

$$L_z = \hbar M \qquad (M = 0, \pm 1, \pm 2, \ldots, \pm J) \tag{14.3-8}$$

The energy does not depend on the quantum number M, so the energy level for a particular value of J has degeneracy $2J + 1$ (one state for each value of M).

Consider next a spherical top, for which $I_A = I_B = I_C$. For this case

$$E_{\text{classical}} = \frac{1}{2I_A}(L_A^2 + L_B^2 + L_C^2) = \frac{1}{2I_A} L^2 \tag{14.3-9}$$

The quantum mechanical energy is

$$E_{\text{qm}} = E_J = \frac{\hbar^2}{2I_A} J(J+1) \tag{14.3-10}$$

This formula is the same as that for the energy of the linear molecule, with the energy depending only on the quantum number J. The degeneracy is not the same. In both cases, there is a quantum number M that specifies the projection of the angular momentum on one of the coordinate axes, which

The number of quantum numbers is always equal to the number of classical coordinates.

we choose as the z axis. When there are three independent variables in the expression for the classical kinetic energy, such as the three components of the angular momentum, there are three quantum numbers. In this case, the third quantum number is for the projection of the angular momentum on one of the principal axes, say the A axis. This projection can take on the values

$$L_A = \hbar K \qquad (K = 0, \pm 1, \pm 2, \ldots, \pm J) \qquad \textbf{(14.3-11)}$$

The quantum number K has the same range of values as M. For a given value of J, there is one state for each value of M and for each value of K, so the degeneracy is

$$g_J = (2J + 1)^2 \qquad \text{(spherical top)} \qquad \textbf{(14.3-12)}$$

The energy levels of symmetric tops and asymmetric tops are quite complicated, and we do not discuss them.[4] In all cases, the three quantum numbers J, M, and K occur. For an asymmetric top, the energy levels can depend on the values of all three quantum numbers.

In Section 14.2, we found that only half of the nonnegative integral values of the rotational quantum number J occurred for a homonuclear diatomic molecule. In the case of polyatomic molecules, the situation is more complicated. We do not discuss the problem but assert that the fraction of the states that can occur is $1/\sigma$, where σ is called the **symmetry number** of the molecule. It is defined as the number of equivalent orientations of the molecule. That is, it is the number of orientations in which a model of the nonvibrating molecule can be placed and have each nuclear location occupied by a nucleus of the same kind as in the first orientation.

For example, the symmetry number of a boron trifluoride molecule is 6 (three positions with one side of a model upward and three more positions with the other side upward). For this molecule, only one-sixth of the conceivable sets of values of J, K, and M can occur. The symmetry number of methane is 12 (three positions with each of the four hydrogens up). Only one-twelfth of the conceivable sets of values of J, M, and K can occur for methane. The symmetry number of any homonuclear diatomic molecule is 2, corresponding to the result that only half of the conceivable values of J can occur. The symmetry number of a heteronuclear diatomic molecule equals unity, as does that of many asymmetric tops, so that all values of the rotational quantum numbers can occur in these cases.

Exercise 14.10	Find the symmetry numbers of the molecules:

    a. chloroform, $CHCl_3$
    b. water, H_2O
    c. benzene, C_6H_6
    d. dichloromethane, CH_2Cl_2

---

[4] G. Herzberg, *Infrared and Raman Spectra*, Van Nostrand Reinhold, New York, 1945, pp. 42ff.

Vibrations of Polyatomic Molecules

In a polyatomic molecule there are several bond lengths and bond angles that can oscillate about their equilibrium values. However, each bond length or bond angle does not oscillate independently of the others. An observed vibration generally consists of concerted motions of all or most of the nuclei.

The first problem in analyzing the vibrations is to determine the vibrations that the molecule would undergo if governed by classical mechanics. This is a complicated process, and we will give only a cursory summary.[5] We begin with each nucleus in its equilibrium position and assume that the molecule is not rotating. For the first nucleus, let q_1 be its displacement from its equilibrium position in the x direction, q_2 be its displacement in the y direction, and q_3 be its displacement in the z direction. For the second nucleus, let q_4, q_5, and q_6 be similar displacements, and so on. There are $3n$ such variables if there are n nuclei.

We assume that the Born-Oppenheimer electronic energy (the vibrational potential energy) depends quadratically on these displacements:

$$\mathscr{V} = \mathscr{V}_e + \sum_{i=1}^{3n} \sum_{j=1}^{i} b_{ij} q_i q_j \qquad \textbf{(14.3-13)}$$

where the b's are constants. Equation (14.3-13) will be a good approximation for small values of the q's. For a diatomic molecule, it is equivalent to the harmonic potential, and we refer to it as a harmonic potential energy function. Cross-terms do occur in which i and j are not equal.

We now find a transformation to a new set of coordinates such that each new coordinate can oscillate independently without transferring energy to make the other coordinates oscillate. In order to do this, we must find new coordinates such that cross-terms do not occur in the potential energy function. These new coordinates are linear combinations of the q's:

$$w_i = \sum_{i=1}^{3n} c_{ij} q_j \qquad \textbf{(14.3-14)}$$

where the c's are constants.

A total of $3n$ coordinates is required to locate all of the n nuclei in three-dimensional space. Three coordinates can be used to specify the location of the center of mass of the molecule. For a linear (or diatomic) molecule, two angular coordinates specify the orientation of the molecule. For a nonlinear molecule, three angular coordinates are required to specify the orientation, as shown in Figure 14.5. The electronic energy is independent of these 5 (or 6) coordinates. For a linear molecule, we need to include only $3n - 5$ of the new coordinates in Equation (14.3-14). For a nonlinear molecule, $3n - 6$ of them need to be included.

We do not discuss the procedure, but it is found that $3n - 5$ or $3n - 6$ new coordinates can be specified such that the potential energy can be

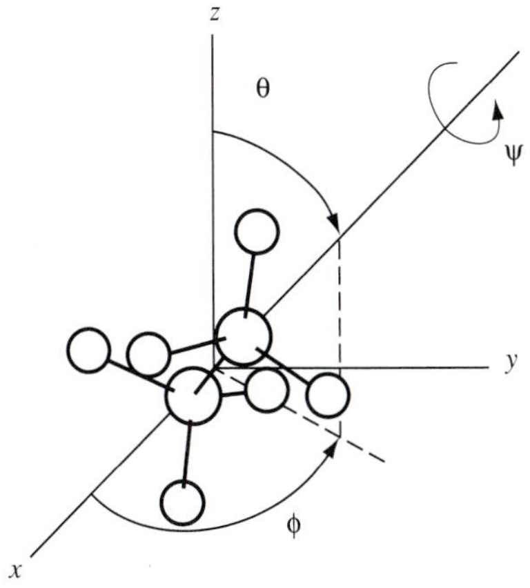

Figure 14.5. Three Angles to Specify the Orientation of a Nonlinear Polyatomic Molecule. Two angles are required to specify the orientation of one molecular axis. The angles θ and ϕ are used for these two angles. The angle ψ is used to specify the angle of rotation about this axis.

---

[5] E. B. Wilson, Jr., J. C. Decius, and P. C. Cross, *Molecular Vibrations*, McGraw-Hill, New York, 1955.

written

$$\mathcal{V} = \mathcal{V}_e + \frac{1}{2} \sum_{i=1}^{3n-5(6)} \kappa_i w_i^2 \tag{14.3-15}$$

The κ's are constants (effective force constants for the new coordinates). The upper limit of the sum indicates that there are $3n - 5$ terms for a linear molecule and $3n - 6$ terms for a nonlinear molecule. These new coordinates are called **normal coordinates**.

The vibrational energy can be written

$$E_{\text{vib}} = \mathcal{K}_{\text{vib}} + \mathcal{V}_{\text{vib}} = \frac{1}{2} \sum_{i=1}^{3n-5(6)} \left[M_i \left(\frac{dw_i}{dt} \right)^2 + \kappa_i w_i^2 \right] \tag{14.3-16}$$

where the M's are constants (effective masses for the new coordinates). The motions of the normal coordinates are called **normal modes** of motion. Equation (14.3-16) is a sum of harmonic oscillator energy expressions without cross-terms. Since each normal coordinate is a linear combination of cartesian coordinates of the nuclei, the motion of a normal coordinate corresponds to a concerted motion of some or all of the nuclei.

Normal modes oscillate independently in both classical mechanics and quantum mechanics.

According to classical mechanics, each normal coordinate oscillates independently with a characteristic classical frequency given by

$$v_i = \frac{1}{2\pi} \sqrt{\frac{\kappa_i}{M_i}} \tag{14.3-17}$$

When the quantum mechanical Hamiltonian operator is written, there are $3n - 5$ or $3n - 6$ terms, each one a harmonic oscillator Hamiltonian operator. The variables can be separated, and the vibrational Schrödinger equation is solved by a vibrational wave function that is a product of $3n - 5$ or $3n - 6$ factors:

$$\psi_{\text{vib}} = \psi_1(w_1)\psi_2(w_2)\cdots = \prod_{i=1}^{3n-5(6)} \psi_i(w_i) \tag{14.3-18}$$

where each factor is a harmonic oscillator wave function. The energy is a sum of harmonic oscillator energy eigenvalues:

$$E_{\text{vib}} = \sum_{i=1}^{3n-5(6)} hv_i \left(v_i + \frac{1}{2} \right) \tag{14.3-19}$$

where $v_1, v_2, \ldots$ are vibrational quantum numbers, one for each normal mode, and $v_1, v_2, \ldots$ are the classical frequencies. The quantum numbers are equal to nonnegative integers and are independent of each other.

Figure 14.6 shows schematically the motion for the four normal modes of carbon dioxide (linear) and the three normal modes of sulfur dioxide (nonlinear), with the frequencies divided by the speed of light (generally called "frequencies"). The arrows in the diagrams show the direction of motion of each nucleus during one half of the cycle. Each nucleus moves in the opposite direction during the other half of the cycle.

Vibrational frequencies are usually given as the frequency divided by the speed of light. The common unit is cm^{-1}.

In each of the triatomic molecules, there is a normal mode in which both bonds shorten and lengthen together. Such a mode is called a **symmetric stretch**. There is also a mode in which one bond lengthens while the other

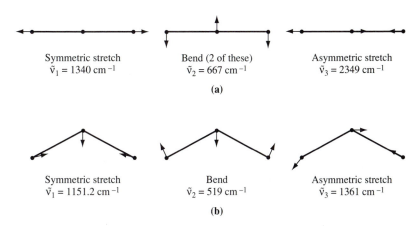

Figure 14.6. Vibrational Normal Modes. (a) Carbon dioxide. The linear CO_2 molecule has four normal modes. These two bends have the same frequency and are shown together. **(b)** Sulfur dioxide. The nonlinear SO_2 molecule has three normal modes, as shown. The general pattern is followed by both molecules: the bend has the lowest frequency, the symmetric stretch has an intermediate frequency, and the asymmetric stretch has the highest frequency.

shortens, called an **asymmetric stretch**. There is also a mode in which the bond lengths do not change much but the bond angle oscillates, called a **bend**. A linear triatomic molecule such as carbon dioxide can bend in two perpendicular directions, so there are two bending modes with the same frequency. However, a nonlinear triatomic molecule can bend only in the plane of the molecule. A "bend" perpendicular to the plane of the molecule has no restoring force to make it oscillate and therefore corresponds to a rotation.

There is a common pattern of frequencies. Asymmetric stretches usually have the highest frequency, symmetric stretches are a little lower in frequency, and bends are quite a bit lower in frequency.

This pattern of frequencies occurs for most triatomic molecules.

Even a moderate-size molecule has numerous normal modes, and we do not attempt to describe all of them for other molecules. For example, benzene, with 12 nuclei, has 30 normal modes, including a "breathing mode" in which the ring alternately contracts and swells and a "pseudorotation" in which a kind of puckered wave moves around the ring. Advanced techniques, including group theory, are used in studying the normal modes of polyatomic molecules.[6]

In some large molecules, a few of the normal modes correspond to fairly large oscillations of one bond length or bond angle while other bond lengths and bond angles remain nearly constant. Often, the frequency of such a normal mode is nearly the same for the same pair of elements in different compounds. For example, most hydrocarbons exhibit a C—H stretching frequency in the range 2850–3000 cm^{-1}, and compounds with an O—H

---

[6] Ira N. Levine, *Molecular Spectroscopy*, Wiley, New York, 1975, pp. 427ff.

bond usually exhibit an O—H stretching frequency in the range 3600–3700 cm^{-1}. Table A18 lists a few such characteristic frequencies, and organic chemistry textbooks give longer lists.

14.4   The Equilibrium Populations of Molecular States

We have derived formulas for the energy eigenvalues for vibrational, rotational, and translational energy for diatomic and polyatomic molecules. These eigenvalues correspond to the states that are available to a single isolated diatomic molecule.

In dilute gas of many molecules, each of these molecular states will be occupied by some number of molecules. At thermal equilibrium, the states are occupied by numbers of molecules proportional to the Boltzmann factor of Equation (1.7-25):

$$\text{Population of state of energy } E \propto e^{-E/k_{\mathrm{B}}T} \qquad \textbf{(14.4-1a)}$$

where k_{B} is Boltzmann's constant, equal to 1.3807×10^{-23} J K^{-1}, and T is the absolute temperature. We now use the letter E for the energy of the molecular state instead of the letter ε used in Chapter 1.

Here we use E for a molecular energy, instead of ε as in Chapter 1. In Chapter 20 we will switch back.

If an energy level has degeneracy g, the states in the level have equal populations, so each level has a population proportional to the degeneracy times the Boltzmann factor:

$$\text{Population of energy level} \propto g e^{-E/k_{\mathrm{B}}T} \qquad \textbf{(14.4-1b)}$$

Rotational Energy Levels of Diatomic Molecules

To the approximation of Equation (14.2-28), a diatomic molecule has a rotational energy eigenvalue that is independent of the vibrational energy. We can therefore consider the rotational energy levels separately. A rotational level has an energy eigenvalue

$$E_J = hB_e J(J + 1) = hc\tilde{B}_e J(J + 1) \qquad \textbf{(14.4-2)}$$

and a degeneracy

$$g_J = 2J + 1 \qquad \textbf{(14.4-3)}$$

Therefore,

$$\begin{array}{l}\text{Population of rotational} \\ \text{energy level } J\end{array} \propto (2J + 1)e^{-E_J/k_{\mathrm{B}}T} \qquad \textbf{(14.4-4)}$$

▼ **EXAMPLE 14.7**

a. Using the value of B_e for CO in Example 14.4, find the ratio of the population of one of the $J = 2$ states to that of the $J = 0$ state at 298 K.

b. Find the ratio of the population of the $J = 2$ level to that of the $J = 0$ state at this temperature.

c. Find the level with the largest population at this temperature.

Solution

a. Let $N(J, M)$ be the number of molecules with quantum numbers equal to J and M. The population of all of the $J = 2$ states is the same. For the $J = 2$, $M = 0$ state:

$$\frac{N(2, 0)}{N(0, 0)} = e^{-(E_2 - E_0)/k_B T} = e^{-6hB_e/k_B T}$$

$$= \exp\left[\frac{-2.302 \times 10^{-22} \text{ J}}{(1.3807 \times 10^{-23} \text{ J K}^{-1})(298 \text{ K})}\right] = e^{-0.05595}$$

$$= 0.9456$$

b. Let $N(J)$ be the population of level J:

$$N(J) = (2J + 1)N(J, M)$$

$$\frac{N(2)}{N(0)} = \frac{5N(2, 0)}{N(0, 0)} = 5(0.9456) = 4.728$$

c. The level of maximum population is the level with the maximum value of $(2J + 1)\exp(-E_J/k_B T)$. The quickest way to find this is to treat J as a continuously variable quantity, differentiating the function to be maximized with respect to J, setting this derivative equal to zero, and solving for J. A nonintegral value of J will result, but rounding to the nearest integer will give the desired value.

Here is a mathematical trick. We find a value as though the quantum number J could take on any real value and then round off to an integer.

$$\frac{d}{dJ}\left[(2J + 1)\exp\left[\frac{-hc\tilde{B}_e J(J + 1)}{k_B T}\right]\right] = \exp\left[\frac{-hc\tilde{B}_e J(J + 1)}{k_B T}\right]\left[2 - \frac{hc\tilde{B}_e}{k_B T}(2J + 1)^2\right]$$

This expression vanishes when J has the value J_{mp}, where

$$(2J_{mp} + 1)^2 = \frac{2k_B T}{hc\tilde{B}_e} = 214.5$$

$$J_{mp} = 6.82 \approx 7$$

From this example, it is apparent that for a typical molecule at room temperature, several of the rotational energy levels are significantly populated. Molecules with a smaller reduced mass have more widely spaced rotational energy levels, and have smaller populations for rotational excited levels than molecules with larger reduced masses.

Exercise 14.11   Find the rotational level with the largest population for HF molecules at 298 K. The internuclear distance is 0.9168×10^{-10} m.

Since the vibrational levels of a diatomic molecule are nondegenerate, we have for the population of a vibrational energy level with quantum number v

$$\text{Population of vibrational level} \propto e^{-E_v/k_B T} \qquad \text{(14.4-5)}$$

▼   **EXAMPLE 14.8**   Find the ratio of the population of the $v = 1$, $J = 0$ state to the $v = 0$, $J = 0$ state for the CO molecule at 298 K. Neglect the anharmonicity correction $v_e x_e$.

Solution

From the values in Example 14.4,

$$\frac{E_{10} - E_{00}}{k_{\rm B}T} = \frac{(6.6261 \times 10^{-34} \text{ J s})(2.9979 \times 10^{10} \text{ cm s}^{-1})(2129.8 \text{ cm}^{-1})}{(1.3807 \times 10^{-23} \text{ J K}^{-1})(298.15 \text{ K})} = 10.28$$

$$\frac{N(1, 0)}{N(0, 0)} = e^{-10.28} = 3.4 \times 10^{-5}$$

▲

Since the vibrational energy levels are much more widely spaced than the rotational energy levels, the population of the excited vibrational states is very small at room temperature for typical diatomic molecules.

Exercise 14.12   Find the ratio of the population of the $v = 1$ vibrational level to that of the $v = 0$ vibrational level for the Br_2 molecule at 298.15 K. The required information is given in Table A16.

It is a fact of probability theory that the probability of the occurrence of two independent events is the product of the probabilities of the two events. If we denote the probability of a vibrational level v by $p_{\rm vib}(v)$ and the probability of a rotational level by $p_{\rm rot}(J)$, the probability that these two levels are occupied by the same molecule is

This fact from probability theory is very useful.

$$p_{\rm vib, rot}(v, J) = p_{\rm vib}(v)p_{\rm rot}(J) \tag{14.4-6}$$

▼   **EXAMPLE 14.9**   For CO at 298.15 K, find the ratio of the population of the $v = 1$, $J = 2$ level to the $v = 0$, $J = 0$ level.

Solution

From the previous two examples,

$$\text{Probability} = (4.728)(3.4 \times 10^{-5}) = 1.6 \times 10^{-4}$$

▲

Exercise 14.13   Use the energy of the level with $v = 1$, $J = 2$ directly in the Boltzmann formula of Equation (14.4-1b) to obtain the result of Example 14.9 by a different calculation.

The probability of an electronic energy level is also governed by the Boltzmann probability distribution. Since most electronic states are even more widely spaced than vibrational states, excited electronic states are almost unpopulated at room temperature.

Exercise 14.14   Calculate the ratio of one of the states of the first excited electronic level of the Cl_2 molecule to the ground state at 298 K. The energy of the first excited level is 2.208 eV above the ground state.

Summary

The motion of the nuclei of molecules consists of translational, rotational, and vibrational motions. To a good approximation, these three types of motion make separate contributions to the molecular energy of a diatomic molecule. The translational energy for a molecule confined in a box is the same as that of a point mass particle in the same box.

To a first approximation, the vibrational energy of a diatomic molecule is that of a harmonic oscillator, the rotation is that of a rigid rotor, and correction terms can be included if necessary. The rotational energy of a polyatomic molecule is taken to be that of a rigid rotating body. The vibrational energy is taken to be that of normal modes, each of which oscillates like a harmonic oscillator.

The populations of rotational, vibrational, and electronic states of molecules are described by the Boltzmann probability distribution.

PROBLEMS

Problems for Section 14.1

*14.15. Repeat the calculation of Example 14.1 for a box 1.00 m on a side. Compare your answer with that of Example 14.1 and comment on any qualitative difference.

14.16. Find the values of the three translational quantum numbers (assume equal) of a xenon atom in a cubical box 0.100 m on a side if the translational energy is 8.315 eV, the excitation energy to the first excited electronic level.

Problems for Section 14.2

*14.17. Using information on the normal H_2 molecule, predict the reciprocal wavelengths of the light absorbed by HD in making the transition from $J = 2$ to $J = 3$. The isotope D is deuterium, ^2H.

14.18. Find a formula for the rotational frequency (number of revolutions per second) of a rigid diatomic molecule assuming that classical mechanics holds but that the angular momentum has the magnitude $\hbar\sqrt{J(J+1)}$. Compare this with the frequency of a photon absorbed when a quantum mechanical molecule makes a transition from J to $J + 1$. Show that the two frequencies are nearly equal for large values of J.

14.19. Using information on the normal H_2 molecule, find the frequencies of vibration of the HD and D_2 molecules, where D is deuterium, ^2H. Compare these with the vibrational frequency of normal H_2.

*14.20. Calculate the percent change in the rotational and vibrational energies of an HCl molecule if (a) ^{35}Cl is replaced by ^{37}Cl and (b) ^1H is replaced by ^2H (assume $H^{35}Cl$).

14.21. The dissociation energy is sometimes approximated by determining the point at which two successive vibrational energy levels have the same energy when the x_e correction is included. Estimate the value of D_e for the CO molecule using this approach. Compare your value with that in Table A16.

*14.22. a. From the vibrational frequency in Table A16, find the value of the force constant for the HF molecule.

b. From the vibrational frequency in Table A16, find the value of the force constant for the H_2 molecule.

14.23. We will find in Chapter 15 that when electromagnetic radiation is emitted or absorbed by a transition between values of the rotational quantum number J, the value of J changes by ± 1.

a. Find the frequency and wavelength of radiation absorbed in the $J = 0$ to $J = 1$ transition for CO.

b. Find the frequency and wavelength of radiation absorbed in the $J = 1$ to $J = 2$ transition for CO.

Problems for Section 14.3

14.24. Classify each of the following molecules as linear, spherical top, prolate symmetric top, oblate symmetric top, or asymmetric top:

a. H_2O	**g.** C_6H_6
b. CO_2	**h.** C_2H_6
c. CH_3Cl	**i.** C_2H_5Cl
d. $CHCl_3$	**j.** C_2H_4
e. CH_2Cl_2	**k.** C_2H_2
f. CCl_4	**l.** trans-$C_2H_2F_2$

*14.25. Determine the number of vibrational normal modes for each of the molecules in Problem 14.24.

14.26. Without doing any calculations, assign principal axes for the molecules in parts a–h, j, and k of Problem 14.24.

*14.27. Calculate the three principal moments of inertia for the water molecule, assuming a bond length of 96 pm and a bond angle of 105°. You must first find the location of the center of mass in the molecule. Pick a product of inertia and show that it vanishes.

14.28. Calculate the principal moments of inertia for the chloroform molecule, assuming tetrahedral bond angles, a C—H bond length of 111 pm, and a C—Cl bond length of 178 pm. Classify the molecule as oblate symmetric top, prolate symmetric top, spherical top, or asymmetric top.

14.29. At high temperatures, the two methyl groups making up the ethane molecule rotate nearly freely with respect to each other about the C—C bond instead of undergoing a torsional vibration relative to each other. Under these conditions, how many vibrational normal modes does ethane have?

14.30. Give the number of vibrational normal modes for each molecule:

    a. C_6H_6 **f.** C_2H_2
    b. C_2N_2 **g.** SF_6
    c. C_2H_4 **h.** BH_3
    d. C_2H_6 **i.** NH_3
    e. C_8H_{18}

Problems for Section 14.4

14.31. Find the ratio of the populations of the $v = 1, J = 1$ state and the $v = 0, J = 0$ state at 298 K for (a) normal H_2, (b) HD, and (c) D_2, where D is deuterium, ^2H.

*14.32. For a temperature of 298 K, find the ratio of the population of the $v = 1$ vibrational state to the population of the $v = 0$ vibrational state of (a) H_2 and (b) I_2. (c) Explain physically why the values in parts a and b are so different.

14.33. a. Find the rotational level of maximum population for H_2 at 298 K.

    b. Find the rotational level of maximum population for N_2 at 298 K.

General Problems

*14.34. If a molecule is confined in a very small box, its energy levels can be spaced widely enough that the spacing can be observed with light absorbed or emitted by transitions between translational levels. Assume that a CO molecule is confined in a matrix of solid argon at 75 K and that the center of the CO molecule can move in a cubical region that is 4.0×10^{-10} m on a side.

    a. Find the energies and degeneracies of the first three translational energy levels.

    b. Find the wavelength and frequency of the light absorbed if a molecule makes a transition from the lowest energy level to the next energy level.

    c. Find the ratio of the populations of the first two energy levels.

14.35. a. From information in Table A16, find the value of the force constant for each of the molecules: N_2, O_2, and F_2. From the LCAO-MO treatment in Chapter 13, find the bond order for each molecule. Comment on the relative sizes of these distances.

    b. From information in Table A16, find the value of the internuclear distance in each of the molecules: N_2, O_2, and F_2. Comment on the relative sizes of these internuclear distances.

    c. From information in Table A16, find the value of the force constant for HF and HI. Comment on the relative sizes of these force constants.

    d. From information in Table A16, find the value of the internuclear distance for HF and HI. Comment on the relative sizes of these distances.

*14.36. Identify each of the following statements as either true or false. If a statement is true only under specific circumstances, label it as false.

    a. The behavior of a molecule confined in a container will be noticeably different from the behavior of a free molecule.

    b. Although part of the electronic energy in the Born-Oppenheimer approximation is kinetic energy, this energy acts as a potential energy for nuclear motion.

    c. Principal axes can be chosen for any object.

    d. A linear triatomic molecule has more distinct frequencies of vibration than a bent triatomic molecule.

    e. A methane molecule has nine vibrational normal modes.

    f. An SeO_2 molecule has four vibrational normal modes.

    g. A normal oxygen molecule, $^{16}O_2$, has rotational levels qualitatively different from those of the isotopically substituted oxygen molecule $^{16}O^{18}O$.

    h. The corrections for anharmonicity in the vibrational energy levels of a diatomic molecule cause the energy levels to be farther apart than the uncorrected energy levels.

    i. The corrections for centrifugal stretching in the rotational energy levels of a diatomic molecule cause the energy levels to be closer together than the uncorrected energy levels.

    j. The corrections for the interaction of rotation and vibration cause the energy levels to be closer together than the uncorrected energy levels.

15 Spectroscopy and Photochemistry

PREVIEW

Spectroscopy is the most useful technique for obtaining information about the spacings of energy levels, and thus about molecular structure.

PRINCIPAL FACTS AND IDEAS

1. Spectroscopy is the study of the interactions of matter and electromagnetic radiation.
2. Emission/absorption spectroscopy is based on the Bohr frequency rule:

$$E_{\text{photon}} = h\nu = \frac{hc}{\lambda} = E_{\text{upper}} - E_{\text{lower}}$$

3. Selection rules predict which transitions between pairs of levels will occur with absorption or emission of radiation.
4. Concentrations can be determined spectroscopically using the Beer-Lambert law.
5. The spectroscopy of atoms involves electronic energy levels.
6. Transitions between rotational states of molecules produce spectra in the microwave region.
7. Transitions between vibrational states of molecules produce spectra in the infrared region.
8. Transitions between electronic states of molecules produce spectra in the visible and ultraviolet regions.
9. Photochemistry is closely related to spectroscopy.
10. Other types of optical spectroscopy, such as Raman spectroscopy, can supplement emission/absorption spectroscopy.
11. Magnetic resonance spectroscopy involves transitions between states that have different energies in a magnetic field.
12. Some modern spectrometers use Fourier transform techniques.

OBJECTIVES

After studying this chapter, the student should:

1. understand the Bohr frequency rule and be able to solve problems related to it,
2. understand the origin of selection rules and be able to use to interpret spectra.
3. be able to solve problems using the Beer-Lambert law,
4. understand the principles of photochemistry,
5. understand the principles of Raman spectroscopy and be able to predict which normal modes of a polyatomic molecule will be Raman active,
6. be able to deduce molecular structural information from infrared, microwave, and Raman spectral features,
7. understand the principles of electron spin resonance and nuclear magnetic resonance spectroscopy,
8. be able to deduce molecular structural information from electron spin resonance and nuclear magnetic resonance spectra,
9. understand the principles used in Fourier transform spectroscopy.

593

| 15.1 | **Spectroscopic Study of Energy Levels** |

In Chapters 11 through 14, we studied the energies and wave functions of atoms and molecules. We found that to a good approximation, the energy could be separated into separate contributions: For atoms, these contributions are electron and translational energies. For molecules, they are electronic, vibrational, rotational, and translational energies. This separation allows these different energy levels to be studied separately by studying the wavelengths of photons emitted or absorbed in transitions between energy levels.

According to the Planck-Einstein relation of Equation (9.3-11), the energy of a photon is

$$E_{photon} = h\nu = \frac{hc}{\lambda} \tag{15.1-1}$$

where h is Planck's constant, ν is the frequency of the radiation, c is the speed of propagation of electromagnetic radiation, and λ is the wavelength. Since wavelengths of radiation can be measured very accurately, photon energies can be determined with great accuracy.

The fundamental fact of spectroscopy is that if a photon is emitted or absorbed by an atom or molecule, the atom or molecule makes a transition between levels whose difference in energy is equal to the energy of the photon. This is expressed by the **Bohr frequency rule**,

The Bohr frequency rule. expressed in Equation (15.1-2), makes spectroscopy the most useful tool for the experimental study of energy levels.

$$E_{photon} = h\nu = \frac{hc}{\lambda} = E_{upper} - E_{lower} \tag{15.1-2}$$

where E_{upper} and E_{lower} are the energy values for the upper and lower levels of the atom or molecule. This statement includes the assumption that only one photon is absorbed or emitted at a time. Multiphoton transitions are increasingly being studied, but we will not discuss these processes.[1]

The spectrum of electromagnetic radiation is divided into several regions:

Name of Region	Wavelengths
Gamma radiation	< 10 pm
X-radiation	10 pm – 10 nm
Ultraviolet radiation	10 nm – 400 nm
Visible radiation	400 nm – 750 nm
Infrared radiation	750 nm – 1 mm
Microwave radiation (including radar)	1 mm – 10 cm
Radio-frequency radiation (including AM, FM, TV)	10 cm – 10 km

---

[1] See for example C. K. Rhodes, *Science* **229**, 1345 (1985).

Typical spacings between electronic energy levels are generally in the range 2–10 eV, corresponding to photon energies in the visible and ultraviolet region. Spacings between vibrational energy levels correspond to photon energies in the infrared region, and spacings between rotational energy levels correspond to photon energies in the microwave region. Spacings between translational energy levels of molecules in macroscopic containers are too small to observe spectroscopically.

Absorption of photons can also break chemical bonds or cause transitions to reactive excited states, which is the basis of photochemistry. Typical chemical bond energies are near 4 to 5 eV (400 to 500 kJ mol^{-1}), and photons with this energy lie in the ultraviolet region. We will discuss photochemistry in Section 15.6.

▼ **EXAMPLE 15.1**

Find the frequency and wavelength of a photon with enough energy to break a chemical bond whose bond energy is 4.0 eV.

Solution

$$ v = \frac{E}{h} = \frac{(4.0 \text{ eV})[1.602 \times 10^{-19} \text{ J(eV)}^{-1}]}{6.6261 \times 10^{-34} \text{ J s}} = 9.7 \times 10^{14} \text{ s}^{-1} $$

$$ \lambda = \frac{3.0 \times 10^{8} \text{ m s}^{-1}}{9.7 \times 10^{14} \text{ s}^{-1}} = 3.1 \times 10^{-7} \text{ m} = 310 \text{ nm} $$

As shown in Example 15.1, ultraviolet radiation has enough energy per photon to break chemical bonds.

▲

Chemical bond energies are usually expressed in kilojoules per mole. A mole of photons is called an **einstein**.

Exercise 15.1

Find the energy per photon and per einstein for
a. microwave radiation with $\lambda = 1.0$ cm.
b. X-radiation with $\lambda = 1.0 \times 10^{-10}$ m.

Until now, our discussion of quantum mechanics has included only the study of stationary states of atomic and molecular systems. We must now study a time-dependent process, the evolution of the state of a system containing matter plus radiation. The standard approximate treatment uses a time-dependent version of perturbation theory, treating the radiation as a classical wave.[2] We give only a brief summary.

In the classical description, electromagnetic radiation consists of an oscillating electric field and an oscillating magnetic field (see Section 9.2). If the radiation is plane polarized, the electric field oscillates in a plane containing the direction of propagation and the magnetic field oscillates in the plane perpendicular to this plane and also containing the direction of propagation, as in Figure 9.6. Because an electric field puts a force on any charged particle and a magnetic field puts a force on a moving charged

---

[2] Jeff C. Davis, *Advanced Physical Chemistry*, Ronald Press, New York, 1965, pp. 243ff.

particle, both of these fields interact with the nuclei and electrons of an atom or molecule, and both can cause absorption or emission of energy.

The principal type of transition produced by the electric field is called an **electric dipole transition**, and a similar transition due to the magnetic field is called a **magnetic dipole transition**. The electric field is many times more effective in causing atomic and molecular transitions than the magnetic field, and we discuss only the electric dipole transitions.

In the time-dependent perturbation method, the Hamiltonian operator of the system (atom or molecule) is written in a way similar to Equation (12.2-4):

$$\hat{H} = \hat{H}^{(0)} + \hat{H}' \tag{15.1-3}$$

where $\hat{H}^{(0)}$ is the complete time-independent Hamiltonian operator of the atom or molecule in the absence of radiation, and $\hat{H}'$ is the time-dependent perturbation. In our case, this term describes the interaction between the atom or molecule and the electric field of the radiation.

The time-dependent wave function of the atom or molecule is written as a linear combination as in Equation (12.2-11) except that the coefficients $a_1, a_2, \ldots$ are now time dependent:

$$\Psi = \sum_j a_j(t)\psi_j^{(0)} \tag{15.1-4}$$

Here $\psi_j^{(0)}$ is an eigenfunction of $\hat{H}^{(0)}$ and is thus time independent. The summation extends over all of the eigenfunctions, but we do not explicitly write limits for the summation. In order to observe transitions, the initial condition is chosen:

$$a_j(0) = \delta_{jn} = \begin{cases} 1 & \text{if } j = n \\ 0 & \text{if } j \neq n \end{cases} \tag{15.1-5a}$$

The symbol δ_{jn} is the **Kronecker delta**, which was defined in Chapter 13.

At time zero the wave function is

$$\Psi(0) = \Psi^{(0)} = \psi_n^{(0)} \tag{15.1-5b}$$

By starting the system off in a state that is a single energy eigenfunction, we make transitions easily identifiable.

Only state n is occupied at time zero. If at later times another coefficient, a_j, becomes nonzero, there is a nonzero probability of a transition from state n to the state corresponding to $\psi_j^{(0)}$.

A formula that gives the coefficients as a function of time is derived. We do not carry out the analysis, but cite its principal result. If the radiation is polarized in the z direction, $a_j(t)^*a_j(t)$, which gives the probability of a transition to state j, is proportional to the intensity of the radiation of the proper wavelength and to the square of the integral

$$(\mu_z)_{jn} = \int \psi_j^{(0)*}\hat{\mu}_z\psi_n^{(0)}\,dq \tag{15.1-6}$$

The square of the transition dipole moment must be positive, and it is this quantity to which the transition probability is proportional.

where $\hat{\mu}_z$ is the z component of the operator for the electric dipole of the atom or molecule, as in Equation (13.4-9). The integration is over all the coordinates of the system, abbreviated by the symbol q. The quantity $\boldsymbol{\mu}_{jn}$ is called the **matrix element of the dipole moment** or the **transition dipole moment**, and the integral in Equation (15.1-6) is its z component.

For two states that have a nonzero transition dipole moment, the theory predicts that a transition between them can occur with the absorption or emission of a photon. Such a transition is called an **allowed transition**. A transition between two states that have a zero transition dipole moment is not predicted to occur and is called a **forbidden transition**. A rule that tells which transitions are allowed is called a **selection rule**. The selection rules we derive are generally obtained with approximate wave functions and with nonrelativistic Hamiltonian operators. Furthermore, only the electric dipole transitions are considered. Most of our selection rules are not exactly obeyed. Some forbidden transitions do occur, but nearly always with lower probabilities than allowed transitions.

It is not surprising that the dipole moment operator is involved, since an electric field can induce an electric dipole in a system of charged particles, much as the electric field in electromagnetic radiation induces an oscillating current in a radio or television receiving antenna. In such an antenna, the frequency of the radiation must be the same as the resonant frequency of the circuit containing the antenna for absorption to occur. In spectroscopy, the frequency of the radiation must be such that the photon energy is equal to the difference in energy between two levels of the atom or molecule.

Inspection of Equation (15.1-6) shows that the value of the transition dipole moment is unchanged if the initial and final states' wave functions are switched. This means that incident radiation will induce transitions from a lower-energy state to a higher-energy state with the same probability as it will induce transitions in the other direction. A transition that lowers the energy of the atom or molecule corresponds to emission of a photon. This emission is called **stimulated emission**, since it is stimulated, or caused to occur, by the radiation field. Radiation emitted in stimulated emission has the same wavelength as the incident photons, moves in the same direction, and is in phase with the incident radiation. That is, the crests and troughs of its waves are at the same places as those of the waves of the incident radiation. With many atoms or molecules emitting radiation, we can get a strong beam of unidirectional radiation with all waves in phase. We say that such radiation is **coherent**. Such radiation is emitted by lasers, which amplify electromagnetic radiation by adding photons to an incident beam by stimulated emission.

"Laser" is an acronym for light amplification by stimulated emission of radiation.

Transitions resulting in emission of photons can also occur in the absence of stimulating radiation. Such emission is called **spontaneous emission**. The probability of such transitions is also proportional to the square of the transition dipole moment, but is independent of the intensity of any radiation. Because there is no inducing radiation to specify a direction and phase, spontaneously emitted radiation is usually emitted in all directions and is not coherent.

Optical spectroscopy is of two principal types. In **absorption spectroscopy**, the attenuation of an incident beam due to absorption is observed. In **emission spectroscopy**, spontaneously emitted radiation from excited atoms or molecules is observed. In both absorption and emission spectroscopy, one must be able to measure the wavelength of the radiation and to observe the intensity of the radiation as a function of wavelength.

Two principal factors determine the intensity of observed absorption or emission due to a given transition. The first is the inherent probability that the transition will take place, which is proportional to the square of the transition dipole moment. The second is the number of atoms or molecules that occupy the initial state. If a system of many atoms or molecules is at thermal equilibrium, the number of atoms or molecules occupying a given state of energy E is proportional to the Boltzmann factor of Equation (1.7-25):

$$\text{Number} \propto e^{-E/k_B T} \tag{15.1-7}$$

where k_B is Boltzmann's constant, equal to 1.3807×10^{-23} J K^{-1}, and T is the absolute temperature.

The number of atoms or molecules occupying excited states drops rapidly with increasing energy, so states of energy values much higher than $k_B T$ will be occupied by very few atoms or molecules at equilibrium. Absorptive or emissive transitions from such states cannot be observed unless the states are significantly populated. For emission spectroscopy, this is commonly done with an electric arc, spark, or other discharge. A flame can also be used.

A nonequilibrium population distribution in an arc or flame can approximate an equilibrium distribution for a high temperature.

Exercise 15.2   For hydrogen atoms at thermal equilibrium at 298 K, find the ratio of the number of hydrogen atoms in one of the $n = 2$ states to the number in one of the $n = 1$ states. Take $E = 0$ for the $n = 1$ level (the ground level), so that $E = 10.2$ eV for the $n = 2$ state.

The simplest apparatus with which to observe emission spectra is a spectroscope. The emitted light from a flame or electric discharge tube is passed through a narrow slit to make a localized beam. It is then **dispersed**, which means that the different wavelengths present are spread out. A triangular prism of transparent material was first used for this purpose, since light is slowed to a characteristic speed when passing through a transparent material. Light that strikes a boundary between two materials is refracted, or has its path bent, if it strikes the boundary at an angle other than a right angle. Different wavelengths are refracted by different amounts, so the radiation is dispersed when it passes through a prism whose sides are not parallel.

Diffraction gratings are also used to disperse the radiation. Figure 15.1a shows schematically how radiation is dispersed by a prism and Figure 15.1b shows how it is dispersed by a transmission grating, which can be considered to pass radiation only through a set of parallel, equally spaced slots. Figure 15.1b is very similar to Figure 9.7b. In the radiation that passes through the grating, crests that are in constructive interference produce a beam in a direction determined by the wavelength of the radiation. Different wavelengths come out at different angles. A reflection grating functions in a similar way.

In a simple spectroscope, the wavelengths at which emission takes place are observed by viewing the locations of bright images of the slit. If only

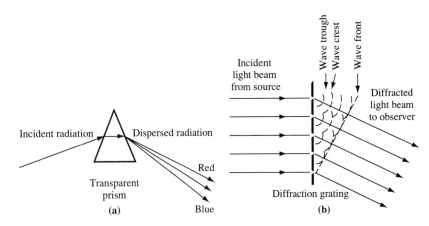

Figure 15.1. The Dispersion of Electromagnetic Radiation. (a) Prism. Because the refractive index depends on wavelength, different wavelengths are refracted through different angles. **(b)** Transmission grating. Because constructive interference is necessary to give the diffracted beam, different wavelengths are diffracted through different angles. The diagram shows first order, in which there is a difference of one wavelength in the paths from adjacent scattering centers. Second order corresponds to two wavelengths difference, etc.

narrow bands of wavelengths are emitted, the slit images look like line segments, and these features are called **spectral lines**.

In a **spectrograph**, the screen on which the spectral lines are viewed is replaced by a photographic film or plate, and a permanent record of the spectrum is obtained, allowing accurate measurement of the line positions. Figure 15.2 shows a simulation of the visible portion of the emission spectrum of atomic hydrogen at low pressure.

Absorption spectroscopy is usually carried out in a **spectrophotometer** such as the one shown schematically in Figure 15.3. The light is dispersed by a prism or grating, collimated into a beam of nearly parallel rays, and then passed through a cell containing the sample. Only a narrow band of wavelengths passes at one time. The wavelength can be chosen by turning the prism or grating. A photocell or other detector determines the intensity

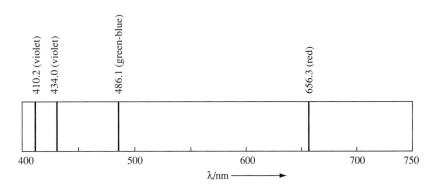

Figure 15.2. The Visible Portion of the Hydrogen Atom Emission Spectrum (simulated). Each wavelength represented produces an image of the slit of the spectrograph. If only discrete wavelengths are present, as in this case, the spectrum is called a line spectrum.

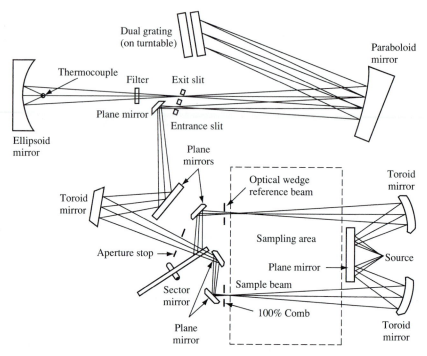

Figure 15.3. Schematic Diagram of a Filter-Grating, Double-Beam Infrared Spectro-photometer. This diagram shows a scanning instrument. The wavelength passed is determined by the angular position of the grating, which is mounted on a turntable and is generally rotated automatically by a motor. Courtesy of Perkin-Elmer Corporation.

of transmitted radiation. In a single-beam instrument, the cell containing the sample substance and a "blank" cell not containing the sample are placed alternately in the beam. In a double-beam instrument, the beam is divided and passed simultaneously through the sample cell and the blank cell.

The intensity of a collimated beam of radiation is defined as the energy passing a unit area per unit time. We define a total intensity, I_{tot}, which is the energy of all wavelengths per unit area per unit time, and an intensity per unit wavelength interval, $I(\lambda)$, such that the energy carried by radiation in the infinitesimal wavelength range between λ and $\lambda + d\lambda$ is

$$\frac{\text{Energy in range } d\lambda}{\text{per unit time per unit area}} = I(\lambda)\, d\lambda \qquad \textbf{(15.1-8)}$$

A variable that is commonly plotted to represent an absorption spectrum is the transmittance, T, usually expressed in percent:

$$T(\lambda) = \frac{I(\lambda)_{\text{out}}}{I(\lambda)_{\text{in}}} \times 100\% \quad \text{(definition)} \qquad \textbf{(15.1-9)}$$

where $I(\lambda)_{\text{out}}$ is the intensity after the light beam passes through the cell and $I(\lambda)_{\text{in}}$ is the incident intensity.

Figure 15.4 shows an absorption spectrum for a sample in liquid solution, plotted as a function of $1/\lambda$, measured in cm^{-1}. This spectrum shows

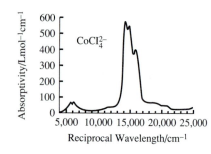

Figure 15.4. The Absorption Spectrum of 0.001 mol L^{-1} CoCl$_4^{2-}$ in 10 mol L^{-1} HCl Solution. This continuous spectrum arises because of the broadening of the spectral lines due to solvent interaction. From Russel S. Drago, *Physical Methods in Chemistry*, W. B. Saunders, Philadelphia, 1977, p. 392.

absorption over broad angles of reciprocal wavelength, because the sample substance is dissolved in a liquid solvent. The constant interactions with solvent molecules shift the energy levels of the sample molecules, making some of them absorb at different wavelengths from others. If the absorption spectrum of the same substance were taken in the gas phase instead of in a solvent, the regions of absorption would be much narrower. However, every spectral line has an inherent nonzero width. In addition to solvent interactions, there are other origins of the spread, or finite width, of a spectral line, including Doppler shifts from the translation of gaseous molecules. We will not discuss the theory of line widths.[3]

No spectrograph or spectrophotometer can disperse radiation so completely that a single wavelength is separately detected. Therefore, a spectral line with a very narrow inherent width appears to have a larger width due to the limitations of the instrument. The difference in wavelength of the most closely spaced narrow spectral lines that an instrument can resolve is called the **resolution** of the instrument.

Absorption spectroscopy can be used to measure the concentration of a substance that absorbs radiation. Figure 15.5 shows a cell containing an absorbing substance with concentration c. Consider a beam of light in the x direction, and consider a thin slab of unit area within the cell, lying between x and $x + dx$. The volume of the slab is dx times unit area, so the amount of absorbing substance in this portion of the slab equals $c\,dx$ times unit area. Let the intensity of light in the small range of wavelengths $d\lambda$ be denoted by $I(\lambda, x)$. This intensity depends on x because the light becomes less intense the farther it travels into the cell, due to absorption. The amount of light absorbed in the slab per unit time is proportional to the intensity

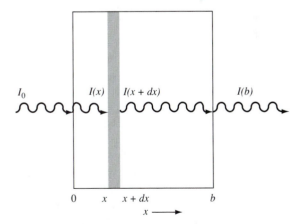

Figure 15.5. Absorption of Light in a Thin Slab. The attenuation of the light due to the absorbing substance in the small slab is assumed proportional to the concentration of the substance and to the thickness of the slab.

[3] Ira N. Levine, *Quantum Chemistry*, Allyn & Bacon, Boston, 1983, pp. 133ff.

of light and to the amount of absorbing substance, so

$$-dI = k(\lambda)Ic\,dx \tag{15.1-10}$$

The proportionality factor k is a function of wavelength but not of concentration. It is strongly different from zero only for wavelengths such that the photon energy equals the energy difference between a significantly occupied molecular energy level and a higher level.

Equation (15.1-10) is a differential equation that can be solved for the intensity of light as a function of position. Division by I separates the variables, giving

$$-\frac{dI}{I} = k(\lambda)c\,dx \tag{15.1-11}$$

This equation is solved by a definite integration, letting $x = 0$ be the front of the cell and $x = b$ the back of the cell. Assuming that the concentration is uniform,

$$-\ln\left[\frac{I(\lambda, b)}{I(\lambda, 0)}\right] = \ln\left[\frac{I(\lambda, 0)}{I(\lambda, b)}\right] = k(\lambda)cb \tag{15.1-12}$$

The **absorbance** $A(\lambda)$ (formerly called the optical density) is defined as the common logarithm of the same ratio as in Equation (15.1-12). It is governed by the **Beer-Lambert Law**:

$$\boxed{A(\lambda) = \log_{10}\left[\frac{I(\lambda, 0)}{I(\lambda, b)}\right] = a(\lambda)bc \quad \text{(Beer-Lambert law)}} \tag{15.1-13}$$

where $a(\lambda)$ is the **absorptivity** (formerly called the extinction coefficient):

$$a(\lambda) = \frac{k(\lambda)}{\ln(10)} = \frac{k(\lambda)}{2.302585\ldots} \tag{15.1-14}$$

If the concentration is measured in mol L^{-1}, the absorptivity is called the **molar absorptivity**. The Beer-Lambert law is well obeyed by many substances at small concentrations. At higher concentrations, deviations occur. For careful determinations of concentrations, a nonlinear calibration curve is used, corresponding to an absorptivity that depends on concentration.

▼   **EXAMPLE 15.2**

A solution of a certain dye has a molar absorptivity of 1.8×10^5 L mol^{-1} cm^{-1} at a wavelength of 606 nm. Find the concentration of a solution of this dye that has an absorbance at this wavelength equal to 1.65 in a cell 1.000 cm in length.

Solution

$$c = \frac{A}{ab} = \frac{1.65}{(1.8 \times 10^5 \text{ L mol}^{-1} \text{ cm}^{-1}(1.000 \text{ cm})} = 9.2 \times 10^{-6} \text{ mol L}^{-1}$$

▲

Exercise 15.3 A solution of a certain dye has a concentration of 0.000100 mol/L and gives an absorbance of 1.234 in a cell of length 1.000 cm at a wavelength of 587 nm. Find the molar absorptivity at this wavelength.

15.2 Spectra of Atoms

Translational levels of atoms are too closely spaced to observed spectroscopically, unless the atoms are confined in very small volumes. Therefore, the only atomic energy levels that can ordinarily be observed are electronic levels. For a hydrogen atom, when the orbitals are substituted in the integral of Equation (15.1-6) the following selection rules are found:[4]

Hydrogen Atom Selection Rules

$$\Delta m = 0, \pm 1 \qquad\qquad \textbf{(15.2-1a)}$$

$$\Delta \ell = \pm 1 \qquad\qquad \textbf{(15.2-1b)}$$

There is no restriction on the values Δn can take on. Since $\Delta \ell = \pm 1$, a hydrogen atom in a state of an s subshell can make a transition only to a state in a p subshell, an atom in a p subshell state can make a transition only to an s subshell state or a d subshell state, etc. However, except for small relativistic corrections, all states in the same shell have the same energy in hydrogen-like atoms, and a simple spectrum is obtained, as was shown in Figure 15.1. Figure 15.6 shows the transitions that take place. The line segments representing each pair of states between which transitions can occur are connected with other line segments.

Exercise 15.4 From the expression for the energy of a hydrogen atom in Equation (12.2-19), find the wavelength and frequency of the photons emitted by a hydrogen atom undergoing the $n = 2 \rightarrow n = 1$ transition, the $n = 3 \rightarrow n = 2$ transition, and the $n = 4 \rightarrow n = 3$ transition. In what spectral range (visible, ultraviolet, or infrared) does each lie?

For multielectron atoms, most spectra can be understood in terms of orbital wave functions. If such wave functions are used to calculate transition dipole moments, the resulting selection rules are[5]

Selection Rules for Multielectron Atoms

$$\Delta L = \pm 1 \qquad\qquad \textbf{(15.2-2a)}$$

$$\Delta S = 0 \qquad\qquad \textbf{(15.2-2b)}$$

$$\Delta J = 0, \pm 1 \qquad (0 \rightarrow 0 \text{ not allowed}) \qquad \textbf{(15.2-2c)}$$

$$\Delta M_J = 0, \pm 1 \qquad (0 \rightarrow 0 \text{ not allowed for } \Delta J = 0) \qquad \textbf{(15.2-2d)}$$

---

[4] Davis, *op. cit.*, pp. 256–257.
[5] *ibid.*

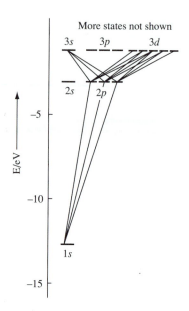

Figure 15.6. The Energy Levels of the Hydrogen Atom and the Allowed Transitions Between Them. This type of diagram is known as a Grotrian diagram. It is constructed by interpreting the spectrum of the substance.

where J is the quantum number for the total electronic angular momentum and M_J is the quantum number for its z component. The most important of these selection rules is the one that forbids transitions between singlet and triplet states, etc. The selection rules of Equation (15.2-2) are only approximately correct. "Forbidden" transitions do occur, but their probabilities are small.

15.3 Rotational and Vibrational Spectra of Diatomic Molecules

Transitions between rotational, vibrational, and electronic states of diatomic molecules can be observed. All three kinds of transition can occur simultaneously, but the selection rules also allow transitions in which only the rotational and vibrational states change, as well as transitions in which only the rotational states change.

Rotational Spectra of Diatomic Molecules

If only the rotational state changes, the transition dipole moment integral is

$$(\boldsymbol{\mu})_{J''M'',J'M'} = \int Y^*_{J''M''}\boldsymbol{\mu}Y_{J'M'}\sin(\theta)\,d\theta\,d\phi \qquad \textbf{(15.3-1)}$$

The Y functions are the rotational wave functions (spherical harmonic functions) and $\boldsymbol{\mu} = \boldsymbol{\mu}(r, \theta, \phi)$ is the dipole moment operator of the molecule in

The rule of Equation (15.3-2b) implies that homonuclear diatomic molecules have no rotational spectrum, since for them $\mu = 0$. If $\mu = 0$, the transition dipole moment vanishes and the transition is forbidden.

the Born-Oppenheimer approximation. We do not present the analysis of this integral.[6] The selection rules that result are

$$\Delta J = \pm 1 \text{ for nonzero permanent dipole moment} \qquad \textbf{(15.3-2a)}$$

all ΔJ's forbidden for zero permanent dipole moment   **(15.3-2b)**

The selection rule of Equation (15.3-2b) can be understood qualitatively by considering what would happen if the molecules and the radiation obeyed classical mechanics: no rotational force (torque) could be put on an uncharged molecule by an oscillating electric field unless the molecule possessed opposite charges at different locations (a dipole moment). A way to determine whether an interaction occurs is to see whether the motion presents to the radiation a periodically varying dipole. If a diatomic molecule has a permanent dipole moment, rotation without vibration does present a periodically varying dipole to the radiation, since the dipole's direction is changing, even though its magnitude is not.

Rotational transitions give photon wavelengths in the microwave region. The radiation sources in microwave spectrometers are klystron tubes, which were originally designed for radar apparatuses. Hollow metal waveguides are used to carry the radiation to the sample cell, which is a hollow metal cavity. The resonant radiation in the cavity is sampled to detect absorption.

From the selection rule, the photon energy for an allowed transition is

$$E_{\text{photon}} = h\nu = \frac{hc}{\lambda} = E_{v,\,J+1} - E_{vJ} \qquad \textbf{(15.3-3)}$$

where J is the rotational quantum number for the lower-energy state and the vibrational quantum number v has the same value for both states. The reciprocal wavelength is

$$\tilde{\nu} = \frac{1}{\lambda} = \frac{1}{hc}(E_{v,\,J+1} - E_{vJ}) \qquad \textbf{(15.3-4)}$$

The use of cm^{-1} is so thoroughly established that the SI unit, m^{-1}, is very seldom used.

where we introduce the symbol $\tilde{\nu}$ for the reciprocal wavelength. This quantity is usually expressed in cm^{-1} and is sometimes called a "frequency," because it is proportional to the frequency of the light through the relation $c/\lambda = \nu$. The unit cm^{-1} is also referred to as **wavenumber**, and the term "wavenumber" is sometimes applied to the reciprocal wavelength itself.

Equation (14.2-40) gives the required energy levels. For a first approximation, we neglect the terms in $\tilde{\alpha}$ and $\tilde{D}$, obtaining

$$\tilde{\nu} = \frac{1}{\lambda} = \tilde{B}_e[(J+1)(J+2) - J(J+1)$$

$$= \tilde{B}_e(J^2 + 3J + 2 - J^2 - J) = 2\tilde{B}_e(J+1) \qquad \textbf{(15.3-5)}$$

Since J can take on values $0, 1, 2, \ldots$, this corresponds to a set of equally spaced spectral lines with reciprocal wavelengths equal to $2\tilde{B}_e$, $4\tilde{B}_e$, ... Figure 15.7a shows the energy levels and transitions, and Figure 15.7b shows a simulated spectrum.

---

[6] Ira N. Levine, *Molecular Spectroscopy*, Wiley, New York, 1975, pp. 162ff.

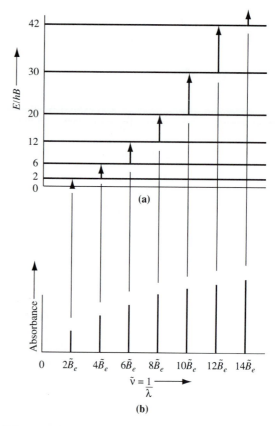

Figure 15.7. A Microwave Spectrum. (a) The allowed transitions. **(b)** The simulated spectrum. The positions of the spectral lines are correlated with the transitions that produce the lines. The intensities are related to the populations of the rotational levels.

Exercise 15.5   Find an expression for the reciprocal wavelengths of the microwave spectrum of a diatomic molecule including all the terms in Equation (14.2-40).

▼ **EXAMPLE 15.3**   The splitting between the spectral lines in the CO spectrum is 3.8626 cm^{-1}. Find the value of r_e, the equilibrium internuclear distance.

Solution

The splitting is equal to $2\tilde{B}_e$, so $\tilde{B}_e = 1.9313$ cm^{-1}. We assume that the lines are due to the principal isotopes, ^{12}C and ^{16}O, with masses 12.0000 amu and 15.994915 amu, so that

$$\mu = \left[\frac{(12.0000 \text{ amu})(15.994915 \text{ amu})}{12.0000 \text{ amu} + 15.9949154 \text{ amu}}\right]\left(\frac{0.001 \text{ kg}}{6.022 \times 10^{23} \text{ amu}}\right)$$

$$= 1.1385 \times 10^{-26} \text{ kg}$$

Example 15.3 is solved by straight-forward substitution into a formula obtained by solving another formula for r_e. The first part is the calculation of the reduced mass. We use the atomic masses of the most abundant isotopes, since the spectrum is produced by individual molecules. If an element consists of a mixture of isotopes, the lines for the different isotopes can sometimes be resolved from each other.

$$r_e^2 = \frac{h}{8\pi^2 \mu \tilde{B}_e c}$$

$$= \frac{6.6261 \times 10^{-34} \text{ J s}}{8\pi^2 (1.1385 \times 10^{-26} \text{ kg})(1.9313 \text{ cm}^{-1})(2.9979 \times 10^{10} \text{ cm s}^{-1})}$$

$$= 1.2731 \times 10^{-20} \text{ m}^2$$

$$r_e = 1.1283 \times 10^{-10} \text{ m} = 112.83 \text{ pm} = 1.1283 \text{ Å}$$

Exercise 15.6   The equilibrium internuclear distance of HCl is 1.275×10^{-10} m. Find the spacing between the lines in the microwave spectrum for both $H^{35}Cl$ and $H^{37}Cl$. The chlorine atomic masses are 34.96885 amu and 36.96590 amu.

The intensity of a given line in a spectrum is determined by the magnitude of the transition dipole moment for the transition or transitions producing the line and by the number of molecules occupying the initial state or states for the spectral line.

Since M can equal $J, J - 1, J - 2, \ldots, -J + 1, -J$, the rotational levels have a degeneracy of $2J + 1$. If all of the states were equally occupied and the transition dipole moments were equal, the spectral lines would have intensities proportional to $2J + 1$.

However, the states are not equally occupied. At equilibrium, the rotational states have populations proportional to the Boltzmann factor of Equation (1.7-25). Therefore,

The population of a rotational level is a product of the degeneracy of the level and the Boltzmann factor for a single state of the level. The degeneracy increases as J increases, and the Boltzmann factor decreases as J increases. The population rises to a maximum and then decreases as J increases.

$$\text{Population of energy level } J \propto (2J + 1)e^{-E_J/k_B T} \qquad \textbf{(15.3-6)}$$

In Example 14.7 of the previous chapter, we found for example that the most populated rotational level for the CO molecule at room temperature is the $J = 7$ level.

If the transition dipole moments for different rotational transitions in the same molecule are roughly equal, the level with the largest population is the one with the largest absorption intensity. The spacings between rotational energy levels for typical molecules are such that roughly 5 to 50 of the levels are significantly populated near room temperature.

Exercise 15.7   Find the rotational level with the largest population for HCl molecules at 298 K. The internuclear distance equals 1.275×10^{-10} m.

Vibration-Rotation Spectra of Diatomic Molecules

When transitions are observed between vibrational levels, the emitted or absorbed radiation is in the infrared region of the spectrum. However, vibrational transitions do not occur without rotational transitions, since $\Delta J = 0$ is forbidden. The infrared spectrum is a vibration-rotation spectrum.

The vibrational selection rules are derived in the Born-Oppenheimer approximation by evaluating the transition dipole moment integral

$$(\mu)_{v'v''} = \int \psi_{v'}^* \hat{\mu}(x) \psi_{v''} \, dx \tag{15.3-7}$$

where $\psi_{v'}$ and $\psi_{v''}$ are two vibrational wave functions, expressed in terms of $x = r - r_e$, and $\hat{\mu}(x)$ is the operator for the molecular dipole moment. Since the vibrational wave functions approach zero rapidly for large magnitudes of x, taking the limits of the integral as infinite produces no significant numerical error but contributes to the fact that our vibrational selection rules are only approximately correct.

We represent the classical expression for the dipole moment by the truncated Taylor series

$$\mu(x) = \mu(0) + (d\mu/dx)_0 x + \cdots \tag{15.3-8}$$

where $\mu(0)$ is the dipole moment at $x = 0$ and the subscript 0 on the derivative means that it is evaluated at $x = 0$. Using harmonic oscillator wave functions, the selection rule is

$$\Delta v = 0, \; \pm 1 \text{ for nonzero dipole moment} \tag{15.3-9a}$$

$$\text{All } \Delta v \text{ forbidden for zero dipole moment} \tag{15.3-9b}$$

The occurrence of $\Delta v = 0$ allows for the occurrence of rotational spectroscopy, in which the vibrational energy does not change. As will be shown in Exercise 15.8, this case arises from the $\mu(0)$ term in Equation (15.3-8).

The transition dipole moment vanishes for homonuclear diatomics because $\mu(x)$ vanishes for all values of x.

Transitions for which $\Delta v = 0$ give the pure rotational spectrum in the microwave region, which we have already discussed. Transitions for which $\Delta v = \pm 1$ give the vibration-rotation spectrum.

Exercise 15.8   Using the $v = 0$ harmonic oscillator function of Equation (9.6-15), show that a nonzero value of the $\mu(0)$ term on the right-hand side of Equation (15.3-8) leads to a nonzero value of the transition dipole moment for the $v = 0$ to $v = 0$ transition. Use the $v = 0$ and $v = 1$ functions to show that a nonzero value of the second term on the right-hand side of the equation leads to a nonzero value of the transition dipole moment for the $v = 0$ to $v = 1$ transition.

As with the rotational selection rule of Equation (15.3-2b), it is possible to understand qualitatively the selection rule of Equation (15.3-9b) by considering a molecule described by classical mechanics. For such a vibrating molecule to interact with electromagnetic radiation, the molecule must be differently charged at its two ends so that it presents a fluctuating dipole to the radiation as it vibrates.

Figure 15.8 shows the allowed transitions between the ground vibrational state ($v = 0$) and the first excited vibrational state ($v = 1$). Because there are

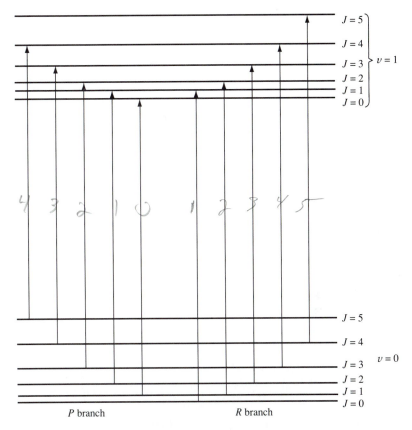

Figure 15.8. Allowed Transitions Leading to a Vibration-Rotation Spectral Band of a Diatomic Substance. In the R branch, the value of J in the upper level is greater than that in the lower level, and in the P branch the opposite is the case. There is no Q branch, which would correspond to the same value of J in the upper and lower energy levels.

different spacings for different values of the rotational quantum number J, a set of spectral lines results, called a **band**. The band corresponding to the allowed transitions shown is called the fundamental band. The spectral lines for which the value of J in the upper state is larger than the value of J in the lower state constitute the **R branch** of the band, and the spectral lines for which the value of J in the upper state is smaller than that in the lower state constitute the **P branch**. If lines occurred for which the values of J in both states were the same, they would constitute the **Q branch**. The rotational selection rules forbid this, and Q branches are not observed in the spectra of diatomic molecules with $^1\Sigma$ electronic states, since the rotational selection rules are well obeyed for these molecules.

Vibrational selection rules are less well obeyed than rotational selection rules, and forbidden vibrational transitions can often be observed, although with lower probabilities than the allowed transitions. A forbidden spectral band with $\Delta v = \pm 2$ is called a **first overtone**, one with $\Delta v = \pm 3$ is called a **second overtone**, etc.

If we neglect the x_e, α, and $\mathscr{D}$ terms in Equation (14.2-40), the reciprocal wavelength of a line of the R branch of the fundamental band is given by

$$\frac{1}{\lambda_R} = \frac{E_{1,J+1} - E_{0,J}}{hc} = \tilde{\nu}_e + 2\tilde{B}_e(J+1) \qquad \textbf{(15.3-10)}$$

where J is the rotational quantum number in the lower ($v = 0$) state. For the absorption spectrum, $\Delta v = 1$ and $\Delta J = 1$ for the R branch.

The R branch consists of a set of roughly equally spaced lines, looking exactly like the rotational spectrum in the microwave region except that they start from the position the Q branch would have (the band center) instead of from zero reciprocal wavelength. The splittings are the same as those of the microwave spectrum.

The reciprocal wavelengths of the P branch are given by

$$\tilde{\nu}_P = \frac{1}{\lambda_P} = \frac{E_{1,J-1} - E_{0,J}}{hc} = \tilde{\nu}_e - 2\tilde{B}_e J \qquad \textbf{(15.3-11)}$$

where J is again the value of J in the lower ($v = 0$) state and must now be at least as large as unity. If the P-branch transitions are observed by absorption spectroscopy, $\Delta J = -1$ while $\Delta v = +1$. The P branch consists of a set of roughly equally spaced lines with the same splitting as the R branch, but the lines are on the other side of the band center. The splitting between the first line of the P branch and the first line of the R branch is equal to twice the splitting between the lines of each branch. Figure 15.9 shows the fundamental band of the HCl molecule. The double lines are due to the two isotopes of chlorine that are present.

In Figure 15.9, it is apparent that the lines are not equally spaced. This is due to the effect of the α term and the $\mathscr{D}$ term in the energy level expression of Equation (14.2-40). In HCl, as in most molecules, the $\mathscr{D}$ term is small, and the effect is due primarily to the α term.

Exercise 15.9   Show that Equations (15.3-10) and (15.3-11) are correct.

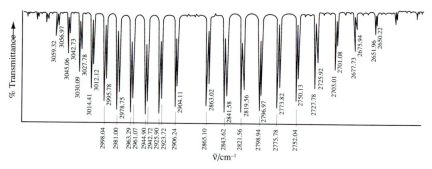

Figure 15.9. The Fundamental Band of the Vibration-Rotation Spectrum of HCl. In this figure we see the R and P branches. The resolution of the spectrum is sufficient to show the lines for $H^{35}Cl$ and $H^{37}Cl$. The ^{35}Cl is the more abundant isotope of chlorine and produces the more intense lines. From N. L. Alpert, W. E. Keiser, and H. A. Szymanski, *IR Theory and Practice of Infrared Spectroscopy*, 2nd ed., Plenum, New York, 1970.

▼   **EXAMPLE 15.4**

a. Using the values of the parameters for CO in Table A16, find the reciprocal wavelength at the band center and the splitting $\Delta\tilde{\nu} = \Delta(1/\lambda)$ for the fundamental band.

b. Find the value of the force constant k for the CO molecule.

c. Assuming that all transition dipole moments are equal and that the temperature is 298 K, find the most intense line in the P branch and in the R branch of the fundamental band.

Solution

a.

$$\tilde{\nu}_{BC} = \frac{1}{\lambda_{BC}} = \tilde{\nu}_e = 2169.81 \text{ cm}^{-1}$$

$$\Delta\tilde{\nu} = \Delta(1/\lambda) = 2\tilde{B}_e = 2(1.93127 \text{ cm}^{-1}) = 3.86254 \text{ cm}^{-1}$$

b. From Equation (14.2-24), and using the value of μ from Example 14.6,

$$\nu = \tilde{\nu}_e c = \frac{1}{2\pi}\sqrt{\frac{k}{\mu}} \quad \text{or} \quad k = 4\pi^2 \tilde{\nu}_e^2 c^2$$

$$k = 4\pi^2(1.1385 \times 10^{-26} \text{ kg})(2169.81 \text{ cm}^{-1})^2(2.9979 \times 10^{10} \text{ cm s}^{-1})^2$$

$$= 1901.8 \text{ kg s}^{-2} = 1901.8 \text{ N m}^{-1} = 1901.8 \text{ J m}^{-2}$$

The force constant of the CO molecule is fairly large because CO has a triple bond, which is quite stiff compared to most single and double bonds.

c. From Example 14.7, the most populated level is that for $J = 7$. This corresponds to the eighth line from the band center in the R branch and the seventh line from the band center in the P branch.

▲

Since the spacing between the vibrational levels is generally quite a bit larger than the spacing between rotational levels, the excited vibrational levels are much less populated than excited rotational levels, as shown in Example 14.8. Transitions from the $v = 1$ and higher levels are generally not seen in absorption spectra near room temperature.

If one or more overtone bands can be observed, it is possible to evaluate the anharmonicity parameter $\tilde{\nu}_e x_e$ from them. The overtone bands consist of a P branch and an R branch, just as does the fundamental. Using the energy level expression of Equation (14.2-40), the reciprocal wavelength of the band center of the fundamental band is

$$\tilde{\nu}_{BC} = \frac{1}{\lambda_{BC}} = \tilde{\nu}_e - \tilde{\nu}_e x_e[(3/2)^2 - (1/2)^2]$$

$$= \tilde{\nu}_e - 2\tilde{\nu}_e x_e \quad \text{(fundamental band)} \qquad (15.3\text{-}12)$$

and the reciprocal wavelength of the band center of the first overtone is

$$\tilde{\nu}_{BC} = \frac{1}{\lambda_{BC}} = 2\tilde{\nu}_e - 6\tilde{\nu}_e x_e \quad \text{(first overtone band)} \qquad (15.3\text{-}13)$$

▼   **EXAMPLE 15.5**

Find the reciprocal wavelength for the band centers of the fundamental and the first overtone of the CO molecule.

Solution

The necessary values are in Table A16. For the fundamental, the reciprocal wavelength of the band center is

$$\tilde{\nu}_{BC} = \frac{1}{\lambda_{BC}} = 2169.8 \text{ cm}^{-1} - 2(13.29 \text{ cm}^{-1}) = 2143.2 \text{ cm}^{-1}$$

For the first overtone,

$$\tilde{\nu}_{BC} = \frac{1}{\lambda_{BC}} = 2(2169.8 \text{ cm}^{-1}) - 6(13.29 \text{ cm}^{-1}) = 4259.9 \text{ cm}^{-1}$$

▲

Exercise 15.10   Find the expression for the reciprocal wavelength of the band center of the $(n + 1)$st harmonic (nth overtone) and find the reciprocal wavelength of the third and fourth overtones for the CO molecule.

If the energy level expression of Equation (14.2-40) is used, the expression for the splitting between the lines of a vibrational band is not the same for the overtones as for the fundamental, as well as depending on the value of J. For example, the reciprocal wavelength of a line in the R branch of the first overtone is

$$\tilde{\nu}_R = \frac{1}{\lambda_R}$$

$$= 2\tilde{\nu}_e - 6\tilde{\nu}_e x_e + 2\tilde{B}_e(J + 1) - 5\tilde{\mathscr{D}}(J + 1)^3 - \tilde{\alpha}(2J^2 + 7J + 5) \quad \textbf{(15.3-14)}$$

Exercise 15.11   **a.** Verify Equation (15.3-14).
b. Obtain the analogue of Equation (15.3-14) for the fundamental band.
c. Find the reciprocal wavelength of each of the first three lines in the R branch of the first overtone of the CO molecule.

15.4   Electronic Spectra of Diatomic Molecules

Transitions of a diatomic molecule from one electronic state to another are complicated by the fact that rotational and vibrational transitions take place simultaneously with the electronic transitions. The selection rules for these transitions are derived much as in the other cases we have considered, by determining which transitions correspond to nonzero transition dipole moments. The following selection rules are obtained:[7]

$$\Delta\Lambda = 0, \pm 1 \quad \textbf{(15.4-1a)}$$

$$\Delta S = 0 \quad \textbf{(15.4-1b)}$$

---

[7] Levine, *op. cit.*, pp. 298ff.

$$\text{Parity of electronic state changes: } (u \to g \text{ or } g \to u) \quad \textbf{(15.4-1c)}$$

$$\Delta J = \pm 1 \quad \textbf{(15.4-1d)}$$

where Λ is the quantum number for the magnitude of the projection of the total electronic orbital angular momentum on the internuclear axis, S is the total electron spin quantum number, and J is the rotational quantum number. There is no restriction on Δv, the change in the vibrational quantum number.

The electronic energy levels are even more widely spaced than the vibrational energy levels, so the electronic spectrum for most diatomic molecules is found in the ultraviolet or visible region.

Exercise 15.12 The lowest-lying excited singlet state of the CO molecule lies 8.0278 eV above the ground state. Find the wavelength of the light absorbed in the transition to this state from the ground state, neglecting changes in rotational and vibrational energy.

Each electronic transition is accompanied by a variety of vibrational transitions and rotational transitions, so it produces a large number of lines, grouped into bands, with one band for each vibrational transition and the lines of each band corresponding to different rotational transitions. Interpretation and measurement of such a band spectrum can yield not only the energy differences between electronic levels but also those between vibrational and rotational levels. This makes it possible to use electronic spectra to determine rotational and vibrational properties of homonuclear diatomic molecules, which have no infrared or microwave spectrum.

In order to understand an electronic spectrum qualitatively, we apply a principle analogous to the Born-Oppenheimer approximation: The **Franck-Condon principle** states that the nuclei do not move appreciably during an electronic transition. Figure 15.10 depicts the situation schematically for a typically diatomic molecule. The two curves are the Born-Oppenheimer energies of two different electronic states. The figure combines three kinds of graphs: The vibrational energy levels are superimposed on the graph in the appropriate positions, and a graph of the square of each vibrational wave function (probability density for internuclear distance) is drawn on the line segment representing its energy level. The vertical scales of these wave function graphs are separate from the energy scale.

In general, excited states have a larger equilibrium internuclear distance and a shallower minimum than ground states.

In the figure, a line segment is drawn to represent a typical electronic transition. The line is vertical, because the Franck-Condon principle requires the internuclear distance after the transition to be equal to that before the transition. The line is drawn from one of the regions of relatively high probability to another region of relatively high probability. Such transitions are the only ones that have appreciable probabilities, since the vibrational wave functions of both states enter in the transition dipole moment integral. One factor in the transition probability is the square of the overlap integral of the two vibrational wave functions. This factor is called the **Franck-Condon factor**.

The vibrational wave functions for the two states must both be significantly different from zero at the same internuclear distances for the Franck-Condon factor to be much different from zero.

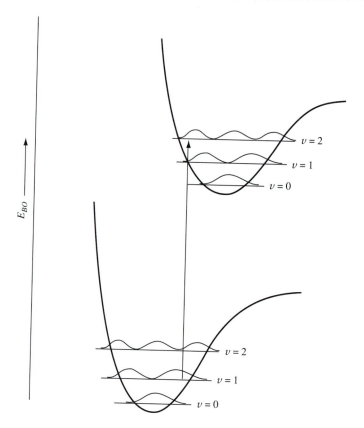

E_{BO}

$v = 2$
$v = 1$
$v = 0$

$v = 2$
$v = 1$
$v = 0$

Figure 15.10. An Electronic Transition According to the Franck-Condon Principle. The vertical lines in this figure illustrate the Franck-Condon principle. The Franck-Condon factor giving the strength of the transition is proportional to the overlap integral for the initial and final vibrational states. In the case of the figure, a transition from the $v = 1$ state of the lower electronic level most likely will lead to the $v = 2$ state of the upper electronic level.

Since the equilibrium internuclear distance of the upper electronic state is somewhat larger than that of the lower state, the wave functions of the lower vibrational states of the upper electronic states are small at the internuclear distance at which the transition takes place, and there is a larger probability that the transition will take place to an excited vibrational state than to the ground vibrational state of the upper electronic state. Because numerous vibrational states can have wave functions with appreciable values at a given internuclear distance, many bands can occur in the electronic system. Figure 15.11 shows the electronic spectrum of nitrogen.

As with atoms, the selection rule forbidding transitions that change the value of S is an important one. However, forbidden transitions between triplet ($S = 1$) states and singlet ($S = 0$) states do occur, but with low probabilities.

If the electronic wave functions are represented by orbital approximations, an electronic transition of a single electron from one orbital to another can be characterized by stating the initial orbital and the final orbital. For example, if the electron raises its energy by going from a pi bonding to a

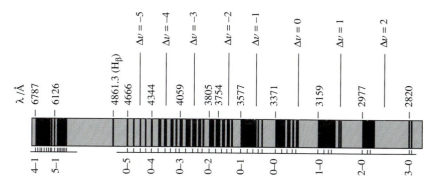

Figure 15.11. The Electronic Spectrum of Diatomic Nitrogen. This spectrum shows a number of bands corresponding to different final vibrational states and sets of lines within each band corresponding to different initial and final rotational states.

pi antibonding orbital, the transition is called a pi to pi-star ($\pi \rightarrow \pi^*$) transition. If the electron goes from a nonbonding orbital to a pi antibonding orbital, the transition is said to be an n to pi-star ($n \rightarrow \pi^*$) transition.

15.5   Spectra of Polyatomic Molecules

The spectra of polyatomic molecules are more complicated than those of atoms or diatomic molecules. However, there are similarities: rotational transitions can occur without vibrational or electronic transitions, vibrational transitions can occur without electronic transitions but are accompanied by rotational transitions, and electronic transitions are accompanied by both vibrational and rotational transitions.

Rotational Spectra of Polyatomic Molecules

The pure rotational transitions of polyatomic molecules produce a microwave spectrum as with diatomic molecules. We have already divided polyatomic molecules into four classes: linear molecules, spherical top molecules, symmetric top molecules, and asymmetric top molecules, We continue to treat a rotating molecule as though it cannot vibrate. Therefore, in all of these classes, a molecule must have a permanent dipole moment in order to produce a microwave spectrum.

Linear polyatomic molecules have rotational wave functions exactly like those of diatomic molecules, so their rotational selection rules and spectra are the same as those of diatomic molecules. Analysis of a spectrum yields the value of the two equal moments of inertia, but not individual bond lengths.

Spherical top molecules are so symmetrical that they cannot have a nonzero dipole moment, which forbids all rotational transitions. There is no microwave spectrum of spherical top molecules. We do not discuss the microwave spectra of symmetric top molecules and asymmetric top molecules, which can be rather complicated.[8] However, a rotational spectrum is always observed, because these molecules must have dipole moments.

Vibrational Spectra of Polyatomic Molecules

As with diatomic molecules, vibrational transitions are accompanied by rotational transitions, giving a band of lines instead of a single vibrational line. The vibrational energy of polyatomic molecules is that of normal modes, each acting more or less like an independent harmonic oscillator. (See Section 14.3.) The selection rules for vibrational transitions are obtained from calculations of the transition dipole moment. If we use approximate vibrational wave functions that are products of harmonic oscillator wave functions, the results are

$$\Delta v_i = 0, \pm 1 \text{ for some one value of } i, \qquad \Delta v = 0 \text{ for } j \neq i \qquad \textbf{(15.5-1a)}$$

$$\text{normal mode } i \text{ must modulate the molecule's dipole moment} \qquad \textbf{(15.5-1b)}$$

The selection rule of Equation (15.5-1b) does not require a molecule to have a nonzero dipole moment in its equilibrium conformation (a permanent dipole). A vibrational normal mode can produce a fluctuating dipole moment that oscillates about zero.

where the indexes i and j identify particular normal modes. The statement that the motion must modulate the dipole moment of the molecule means that the motion, treated classically, must cause the dipole moment to oscillate in value.

According to the selection rule of Equation (15.5-1a), the vibrational quantum number of only one normal mode changes by unity in a vibrational transition. This leads to a fundamental band in the infrared region for each normal mode that modulates the molecule's dipole moment. The case in which all Δv's equal zero gives rise to the rotational spectrum in the microwave region, and this case requires the presence of a permanent dipole moment.

If a polyatomic molecule has a permanent dipole moment, all or most of its normal modes will modulate the dipole and give rise to vibrational bands. For example, in nonlinear triatomic molecules such as H_2O or SO_2, all three of the normal modes shown in Figure 14.7 will give fundamental bands in the infrared spectrum.

Transitions that violate the selection rules occur, as with diatomic molecules. There are overtones like those of diatomic molecules and also **combination bands**, in which two (or more) normal modes change their quantum numbers at once, in violation of Equation (15.5-1a).

▼   **EXAMPLE 15.6**   The infrared spectrum of hydrogen sulfide, H_2S, shows three strong bands at 1290 cm^{-1}, 2610.8 cm^{-1}, and 2684 cm^{-1}. There are weaker bands at 2422 cm^{-1}, 3789 cm^{-1}, and 5154 cm^{-1}. Interpret the spectrum.

---

[8] Davis, *op. cit.*, pp. 322ff.

In Example 15.6 we see a general pattern for triatomic molecules: the bend has the lowest frequency, the symmetric stretch has an intermediate frequency, and the asymmetric stretch has the highest frequency. The weak bands are identified by trial and error, seeing whether their frequencies approximate multiples of a fundamental frequency or are sums of two fundamental frequencies.

Solution

The three strong bands are fundamentals. The lowest frequency, 1290 cm^{-1}, is that of the bend, v_2; the 2610.8 cm^{-1} frequency is that of the symmetric stretch, v_1; and the 2684 cm^{-1} frequency is that of the asymmetric stretch, v_3. The 2422 cm^{-1} frequency is roughly twice that of the bend, so it is the first overtone of the bend. The 3789 cm^{-1} frequency is roughly the sum of 1290 cm^{-1} and 2684 cm^{-1} and is a combination band of the bend and the asymmetric stretch. The 5154 cm^{-1} frequency is roughly twice as large as v_2 and also roughly equal to $v_2 + v_3$. It has been assigned both ways but is said to be the combination band by analogy with the water spectrum.[9]

▲

Even in molecules without a permanent dipole, some of the normal modes can produce a fluctuating dipole that oscillates about zero magnitude and thus produce spectral lines. For example, in CO_2, although each C=O bond is polar, the molecule has no permanent dipole because it is linear and the dipole contributions of the two bonds cancel in the equilibrium position. The normal modes of CO_2 were shown in Figure 14.7. The two bending modes, which have the same frequency, produce temporary dipoles that point perpendicularly to the molecule axis and fluctuate about zero magnitude. These two modes give rise to a single vibrational band called a **perpendicular band**. The asymmetric stretch produces an oscillating dipole parallel to the molecule axis, and the spectral band it produces is called a **parallel band**. The symmetric stretch increases and then decreases both bond dipoles simultaneously, so it does not produce a fluctuating dipole and thus does not give rise to a spectral line. The infrared spectrum of carbon dioxide contains only two fundamental bands, the parallel band at 1340 cm^{-1} and the perpendicular band at 667 cm^{-1}.

For polyatomic molecules the selection rules for rotational transitions that occur during vibrational transitions are not the same as those for pure vibrational transitions, and perpendicular bands exhibit Q branches in addition to P and R branches. The reason is that the two bending modes can produce a motion that is the same as a rotation of a bent molecule, which turns out to permit $\Delta J = 0$ as well as $\Delta J = \pm 1$.[10] Figure 15.12 shows the low-resolution carbon dioxide perpendicular band at 667 cm^{-1}, containing P, Q, and R branches.

To determine whether a given normal mode of a polyatomic molecule will give rise to a vibrational band (be "infrared active"), it is often necessary only to know the pattern of motion of the nuclei in the normal mode and to determine from inspection whether the motion leads to modulation of the dipole moment. In complicated cases, group theory can be used to determine which normal modes will be infrared active and which will not.[11]

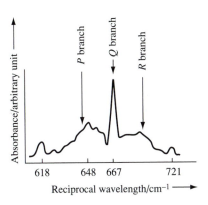

Figure 15.12. The Perpendicular Band of Carbon Dioxide. This spectrum shows a Q branch, which occurs when the normal mode produces a modulation of the dipole moment of the molecule perpendicular to the principal symmetry axis of the molecule. From G. Herzberg, *Molecular Spectra and Molecular Structure,* Vol. II, *Infrared and Raman Spectra of Polyatomic Molecules,* Van Nostrand Reinhold, New York, 1945, p. 273.

[9] G. Herzberg, *Molecular Spectra and Molecular Structure.* Vol. II, *Infrared and Raman Spectra of Polyatomic Molecules,* Van Nostrand Reinhold, New York, 1945, p. 283.

[10] Levine, *op. cit.,* pp. 255ff.

[11] Levine, *op. cit.,* pp. 427ff.

▼   **EXAMPLE 15.7**   The normal modes of cyanogen, C_2N_2, are shown in Example Figure 15.7. (The last two diagrams represent two modes each, of equal frequency.) Tell which are infrared active.

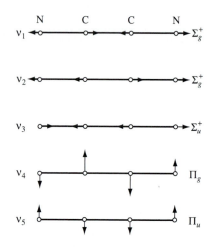

Example Figure 15.7. The Vibrational Normal Modes of Cyanogen, C_2N_2. The arrows show the directions of motion of the nuclei during half of the period of vibration, and the lengths are proportional to the amplitudes of the nuclear oscillations. From G. Herzberg, *Molecular Spectra and Molecular Structure*, Vol. II, *Infrared and Raman Spectra of Polyatomic Molecules*, Van Nostrand Reinhold, New York, 1945, p. 181.

Solution

The molecule is symmetrical and has a zero dipole moment in the equilibrium configuration. Only modes 3 and 5 produce an oscillating dipole moment, since the other modes produce contributions to the dipole moment that cancel.

▲

Exercise 15.13   One of the normal modes of the benzene molecule is the "breathing mode," in which the entire molecule alternately expands and contracts. Will this mode be infrared active?

The selection rules for rotational transitions accompanying vibrational transitions in spherical top, symmetric top, and asymmetric top molecules are slightly different from those for pure rotational transitions. The outcome is that Q branches are sometimes present and P and R branches are always present. Generally, if there is a single symmetry axis, vibrational motions in which the dipole oscillates perpendicular to the axis have a Q branch. For example, in sulfur dioxide the band for the asymmetric stretch has a Q branch.

Electronic Spectra of Polyatomic Molecules

Electronic transitions in polyatomic molecules are similar to those in diatomic molecules. Vibrational and rotational transitions take place along with electronic transitions. The Franck-Condon principle applies just as in diatomic molecules, so the final state is usually an excited vibrational state as well as an excited electronic state. Because any polyatomic molecule has several normal modes, the simultaneous electronic, vibrational, and rotational transitions can give very complicated spectra.

The selection rules for electronic transitions in polyatomic molecules are also more complicated than in diatomic molecules. One rule is the same for all molecules and atoms: the total spin quantum number is the same for the final as for the initial state:

$$\Delta S = 0 \quad \text{(rule for all molecules and atoms)} \qquad \textbf{(15.5-2)}$$

The selection rules for the space part of the electronic wave function can be derived using group theory to investigate the effects of wave function symmetry. We state only the general rule of thumb: the symmetry of the electronic wave function in the final state must be different from the symmetry in the initial state. For example, a transition from a u state to a g state is ordinarily allowed, whereas a transition from a g state to another g state is not allowed.

The molecular electronic selection rules that we have given are not exactly obeyed because they are derived with approximate wave functions. "Forbidden" transitions are often observed. For example, a transition from an excited singlet state to a singlet ground state with emission of a photon is allowed and generally occurs with a mean lifetime in the excited state of a microsecond to a millisecond. A transition from an excited triplet state to a singlet ground state is forbidden and generally occurs with a longer mean lifetime in the excited state, up to about 10 seconds.

If the electronic wave functions of a polyatomic molecule are represented by an orbital approximation, electronic transitions can be classified as in the diatomic case by specifying the initial and final orbitals. In many cases, a whole class of compounds exhibits similar spectral lines. A functional group or other group of atoms in a molecule which exhibits a characteristic absorption is called a **chromophore**. For example, most organic compounds containing a carbonyl group have an absorption near 200 nm due to a pi to pi-star transition and another absorption near 300 nm due to an n to pi-star transition.

15.6   Fluorescence, Phosphorescence, and Photochemistry

In this section we discuss changes of state of molecular systems that involve emission or absorption of photons as well as other kinds of processes, including chemical reactions.

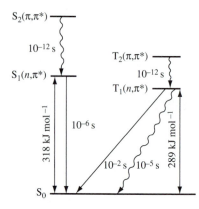

Figure 15.13. Some Energy Levels of the Benzophenone Molecule. The wavy arrows represent radiationless transitions, and the straight arrows represent emissions of photons. The times shown are relaxation times for the transitions. Data from D. L. Pavia, G. M. Lampman, and G. S. Kriz, Jr., *Introduction to Organic Laboratory Techniques*, 2nd ed., Saunders College Publishing, Philadelphia, 1982, p. 364.

Fluorescence and Phosphorescence

Figure 15.13 shows schematically some low-lying levels of the benzophenone molecule due to excitation of electrons in the carbonyl group. An excited level that is reached from the ground level by an n to pi-star transition is labeled (n, π^*) and the excited level reached by a pi to pi-star transition is labeled (π, π^*).

There are two levels labeled (n, π^*) and two levels labeled (π, π^*). These differ in the value of the total spin quantum number S, with singlet levels corresponding to $S = 0$ and triplet levels corresponding to $S = 1$. The ground level is a singlet level, with all electrons paired, and the selection rules allow transitions only to excited singlet levels. Transitions to the two excited singlet levels give rise to two absorptions, one near 330 nm (the n to pi-star) and one near 260 nm (the pi to pi-star). Transitions to the triplet levels from the ground level are forbidden and are not seen.

Instead of giving the term symbol of each level, we follow a common notation and label the singlet levels as S_0, S_1, S_2, etc. in order of increasing energy, and the triplet levels T_1, T_2, etc. in order of increasing energy. The S_0 level is the ground level. We reserve the subscript 0 for the ground level, so there is no T_0 level.

If a molecule absorbs a photon to make a transition to an excited singlet level, there is some probability that it will make the reverse transition and emit a photon of the same wavelength as the photon originally absorbed, but this is not the only thing that can happen. Most likely, the molecule will be in an excited vibrational state after the upward transition, so it can make a transition to a lower-energy vibrational level within the excited electronic level, thus losing some vibrational energy. This energy can be emitted as a photon, which would be in the infrared region. A **radiationless transition** can also occur. The vibrational energy lost by the molecule can be transferred to other vibrational modes in the molecule, or to rotation or translation of the molecule, or to other molecules.

Once the molecule is in a lower vibrational level of the excited electronic level, one possibility is that it can emit a photon and return to the ground electronic level. Such a radiative transition to the ground level from an excited level with the same value of S is called **fluorescence**. Vibrational energy was lost, so the emitted photon will be less energetic than the photon originally absorbed and the emitted light will have a longer wavelength than the absorbed light. Many common objects, including human teeth, certain minerals, and "black-light" posters, can fluoresce, emitting visible light when irradiated with ultraviolet light.

Another possibility is that the molecule will make a radiationless transition to the ground level or to a lower-energy electronic level with the same value of S. Such a radiationless transition is called an **internal conversion**. In our example of a carbonyl compound, an internal conversion could occur from the singlet (π, π^*) level to the singlet (n, π^*) level, followed by fluorescence to the ground level.

Still another possibility is a radiationless transition to an electronic level with a different value of S. For example, to each of the excited singlet levels in Figure 15.13 there corresponds a triplet level with the same electron configuration and a lower energy. A radiationless transition in which the value of S changes is called an **intersystem crossing**.

Do not confuse this use of the word "system" with its common usage to refer to an object under study.

Once a molecule has arrived in a state with a value of S different from that of the ground state, it might make a radiative transition to the ground state, although this is forbidden by the selection rules. This process is called **phosphorescence**. Since phosphorescence is forbidden, it is less probable than fluorescence, and a typical mean time for phosphorescence is from a millisecond to 10 seconds. Phosphorescence can sometimes be observed visually after the radiation being absorbed is turned off. In Figure 15.13 the approximate values of mean transition times are indicated near the arrows.

Photochemistry

In some cases, a molecule in an excited state can undergo a chemical reaction that is inaccessible to a molecule in the ground level. If the excited state was reached directly or indirectly by absorption of radiation, the reaction is a **photochemical reaction**.

Most photochemical reactions are governed by the **Stark-Einstein law of photochemistry**, which states that absorption of one photon can cause reaction of one molecule. This is similar to the statement of the Bohr frequency rule of spectroscopy, that the absorption or emission of a single photon accompanies a transition between atomic or molecular energy levels.

The Stark-Einstein law does not imply that every photon necessarily leads to the reaction of one molecule, since some of the excited molecules might undergo internal conversion, intersystem crossing fluorescence, or phosphorescence leading to unreactive states. Also, a **chain reaction** might occur in which the reaction of one molecule can lead to the reaction of other molecules without absorption of further radiation.

The **quantum yield** of a photochemical reaction, Φ, is defined by

$$\Phi = \frac{\text{total number of molecules reacted}}{\text{number of photons absorbed}} \qquad \textbf{(15.6-1)}$$

In a chain reaction Φ can exceed unity, but in a nonchain reaction unity is its maximum value.

Equation (15.6-1) can be restated in terms of moles of reactant and moles of photons. One mole of photons is an einstein, so

$$\Phi = \frac{\text{amount reacted in moles}}{\text{amount of photons absorbed in einsteins}} \qquad \textbf{(15.6-2)}$$

An example of a photochemical reaction involves benzophenone, whose energy level diagram has been presented in Figure 15.13.[12] Upon irradiation with ultraviolet light in the wavelength range 300 to 350 nm, benzophenone reacts with 2-propanol to form benzpinacol and acetone:

$$\mathrm{Ph_2C{=}O + H(CH_3)_2COH} \xrightarrow{h\nu} \overset{\displaystyle \text{HO} \quad \text{OH}}{\underset{}{\mathrm{Ph_2C{-}CPh_2}}} + \mathrm{(CH_3)_2C{=}O} \qquad \textbf{(15.6-3)}$$

---

[12] D. L. Pavia, G. M. Lampman, and G. S. Kriz, Jr., *Introduction to Organic Laboratory Techniques*, 2nd ed., Saunders College Publishing, Philadelphia, 1982, pp. 362ff.

where Ph stands for the phenyl group. Because 300-nm radiation has photons of insufficient energy to reach the singlet (π, π^*) level and because the radiative transition to a triplet level is forbidden, the first step in the mechanism for this reaction must be absorption of radiation to excite the benzophenone to the singlet (n, π^*) level:

$$Ph_2C{=}O + hv \rightarrow Ph_2\overset{\cdot}{C}{=}\overset{\cdot}{O} \qquad (S_1) \qquad \textbf{(15.6-4)}$$

where the electron in the nonbonding orbital is represented by a dot over the oxygen atom and the electron in the antibonding orbital is represented by a dot over the double bond (which is now a bond with order 3/2).

The next step in the mechanism is an intersystem crossing:

$$Ph_2\overset{\cdot}{C}{=}\overset{\cdot}{O}(S) \xrightarrow{\text{isc}} Ph_2\overset{\cdot}{C}{=}\overset{\cdot}{O} \qquad (T_1) \qquad \textbf{(15.6-5)}$$

This step is followed by abstraction of a hydrogen atom (complete with one electron) from a 2-propanol molecule:

$$Ph_2\overset{\cdot}{C}{=}\overset{\cdot}{O}(T_1) + H{-}\underset{\underset{CH_3}{|}}{\overset{\overset{CH_3}{|}}{C}}{-}OH \rightarrow Ph_2\overset{\cdot}{C}{-}OH + .\underset{\underset{CH_3}{|}}{\overset{\overset{CH_3}{|}}{C}}{-}OH$$

$$\textbf{(15.6-6)}$$

The next step is abstraction of a second hydrogen atom from the 2-propanol molecule, forming another molecule of the radical and an acetone molecule:

$$Ph_2\overset{\cdot}{C}{=}\overset{\cdot}{O}(T_1) + HO{-}\underset{\underset{CH_3}{|}}{\overset{\overset{CH_3}{|}}{C}}. \rightarrow Ph_2\overset{\cdot}{C}{-}OH + O{=}C(CH_3)_2$$

$$\textbf{(15.6-7)}$$

This is followed by combination of two radicals:

$$2Ph_2\overset{\cdot}{C}{-}OH \rightarrow Ph{-}\underset{\underset{Ph}{|}}{\overset{\overset{HO}{|}}{C}}{-}\underset{\underset{Ph}{|}}{\overset{\overset{OH}{|}}{C}}{-}Ph$$

$$\textbf{(15.6-8)}$$

The photochemical reaction can be carried out by use of an ultraviolet lamp, but sunlight contains enough ultraviolet light to produce a significant amount of product in a few days. The reaction will proceed in a borosilicate glass flask.

Exercise 15.14   Borosilicate glass blocks almost all radiation of wavelength less than 300 nm. Calculate the energy per photon and per einstein for radiation of wavelength 300 nm.

If naphthalene is placed in the reaction mixture, no reaction takes place (the reaction is **quenched**). The explanation is that intermolecular energy transfer from an excited benzophenone molecule to a naphthalene molecule returns the benzophenone molecule to its ground level before it can react chemically. Naphthalene has a singlet ground level, a singlet (π, π^*) level at 4.1 eV above the ground level, and a triplet (π, π^*) level at 2.7 eV above the ground level.

A well-obeyed selection rule requires that in an intermolecular energy transfer the sum of the two electron spin quantum numbers remains constant. This means that if the benzophenone molecule makes a transition from a triplet excited level to a singlet ground level, the naphthalene molecule must make a transition from its singlet ground level to a triplet excited level, and if the benzophenone molecule makes a transition from a singlet excited level the naphthalene molecule must make a transition to a singlet excited level.

Because the (π, π^*) singlet excited level of the naphthalene lies higher than the (n, π^*) singlet excited level of benzophenone by 0.8 eV, it cannot be reached by energy transfer from a benzophenone molecule in its (n, π^*) level. However, the (π, π^*) triplet excited level of the naphthalene molecule lies lower than the (n, π^*) triplet level of benzophenone, so there is enough energy to reach this level by energy transfer from a benzophenone molecule. The fact that the naphthalene stops the reaction shows that the triplet (n, π^*) level of benzophenone must be the reactive level.

Photosynthesis is an important set of photochemical reactions in plants that "fix" carbon dioxide and produce carbohydrates and O_2. The cells of photosynthetic plants contain organelles called **chloroplasts**, in which are several chromophores called **chlorophylls**. Chlorophyll molecules absorb blue and red light, giving the chloroplasts their characteristic green color. These chromophores are bound to hydrophobic proteins located in an internal membrane called the thylakoid membrane.

The chlorophylls are structures with four pyrrole rings, quite similar to the heme groups in hemoglobin, myoglobin, and the cytochromes. An important difference is that the chlorophylls contain a magnesium atom instead of the iron atom found in the heme group. It is important to the functioning of the cytochromes that the iron can be oxidized to iron(III) and reduced to iron(II). It is important to the functioning of the chlorophyll that the magnesium occurs in only one oxidation state.

The accepted mechanism of the initial photochemical reaction is as follows:[13] First, the chlorophyll molecule absorbs a photon of red light, making a transition to an excited singlet state from its singlet ground state. Second, the excited chlorophyll molecule, which is a much stronger reducing agent than the ground-state chlorophyll molecule, loses an electron to some other substance involved in the chain of reactions that eventually produce carbohydrates and oxygen. Since the magnesium atom has only a single oxidation state, this electron comes from a delocalized orbital in the ring system, producing a radical cation. Third, some electron donor gives an electron to the chlorophyll. This electron goes into the orbital from which the first electron made its original transition, so the chlorophyll is restored to its ground state.

The whole process depends on the fact that the excited singlet chlorophyll is a much stronger reducing agent than the ground-state chlorophyll, which has a

---

[13] G. Zubay, *Biochemistry*, Addison-Wesley, Reading, MA, 1983, pp. 409ff.

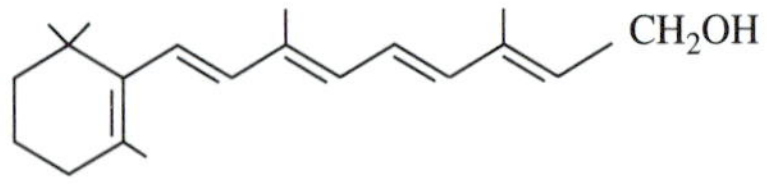

all-*trans*-retinol (vitamin A$_1$)

(a)

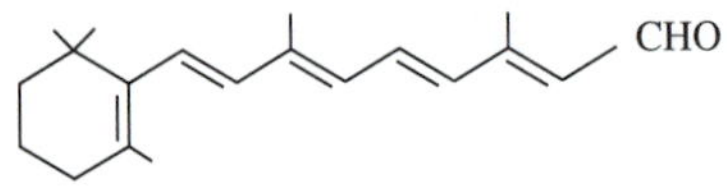

all-*trans*-retinal

(b)

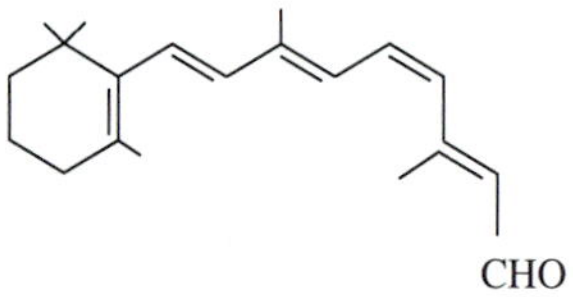

CHO

ll-*cis*-retinal

(c)

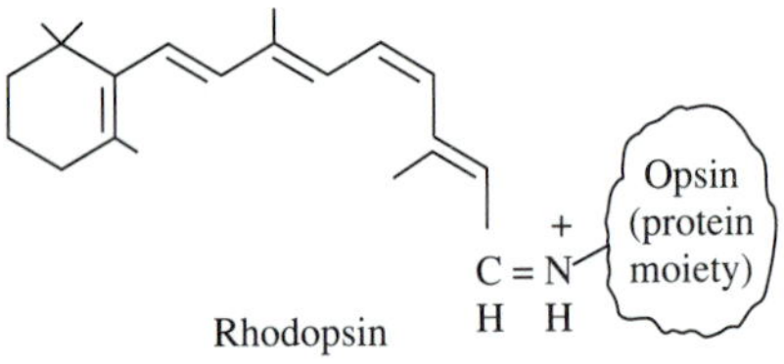

Opsin
(protein
moiety)

Rhodopsin

(d)

Figure 15.14. The Structures of Retinal and Rhodopsin. (a) All-*trans*-retinol (vitamin A). **(b)** All-*trans*-retinal. **(c)** 11-*cis*-retinal. **(d)** Rhodopsin. The absorption of a photon by the 11-*cis*-retinal that is bound to the protein opsin begins the vision process.

half-cell reduction potential under biological conditions of about 0.5 V. Since the excitation energy is about 1.5 eV, the effective half-cell potential of the excited chlorophyll is about -1.0 V. However, reduction of the chlorophyll radical cation is easy because it forms the ground state, not the excited state.

The chlorophyll does not react directly with carbon dioxide. In fact, at least eight photons are required for each oxygen molecule evolved in the photosynthetic process, so the chlorophyll cycles repeatedly from its ground state to its excited state to its oxidized state and back to its ground state in order to produce one oxygen molecule. The later steps in the process occur in the liquid outside the membrane in which the chlorophyll is bound.

Linear polyenes called carotenoids occur in the chloroplasts. In addition to carrying out other functions, they prevent the formation of singlet O_2, which is extremely toxic to the plant cells. The ground level of O_2 is a $^3\Sigma_g^-$ level, and 0.98 eV is required to excite oxygen molecules to a $^1\Delta_g$ excited level. The excited chlorophyll molecule has absorbed 1.5 eV from a photon of red light and can undergo intersystem crossing to a triplet state of slightly lower energy. Because of the selection rule requiring constancy of the sum of the spin quantum numbers, the excited singlet state of chlorophyll cannot excite the oxygen to its singlet state, but the triplet state can do so. The carotenoid molecules have an excited triplet state lower in energy than the excited triplet state of chlorophyll, and they accept the energy from any chlorophyll molecules that have undergone intersystem crossing before significant numbers of singlet oxygen molecules can be produced. The excited triplet states of the carotenoids do not have enough energy to excite the oxygen, preventing formation of singlet oxygen.

Another interesting set of photochemical reactions is involved in vision in vertebrates.[14] The retina of the vertebrate eye has two kinds of light-sensitive cells, called **rods** and **cones**. The rods are primarily responsible for vision in dim light. There are three varieties of cone cells, sensitive to red, green, and blue light respectively, and these require greater illumination to function.

The rod cells contain a protein called **rhodopsin**, which consists of a protein moiety called **opsin** and a polyene called **retinal**. Retinal is related to retinol, which is known as vitamin A and is depicted in Figure 15.14a. Retinal occurs in the eye as the all-*trans* isomer and as the 11-*cis* isomer. The structural formulas of these isomers are shown in Figure 15.14b and 15.14c. The 11-*cis* form attaches to the free $-NH_2$ group of a lysine residue, forming a Schiff base, as shown in Figure 15.14d. The all-*trans* isomer does not bond to the opsin. Each variety of cone cell has one of three proteins that are similar to rhodopsin and function in analogous ways.

Rhodopsin has a broad absorption ranging from 400 nm to 600 nm, with maximum absorption around 500 nm. The corresponding absorption band of 11-*cis*-retinal is centered at 380 nm, in the ultraviolet, so the unbound retinal cannot be responsible for vision in the visible region.

Exercise 15.15

a. Using the structural formulas in Figure 15.14 and the free-electron (particle-in-a-box) model for a conjugated polyene, explain why the absorption maximum of the Schiff base form of rhodopsin absorbs at a longer wavelength than does 11-*cis*-retinal.

---

[14] *ibid.*, pp. 1169ff.

b. Using the free-electron model, calculate the wavelength of maximum absorbance for 11-*cis*-retinal and for rhodopsin, taking an average bond length of 1.39×10^{-10} m and adding one bond length to each end of the conjugated system of bonds.

The accepted mechanism of the photochemical process in rod cells is as follows: First, the rhodopsin absorbs a photon, raising it to an excited state in which a 90-degree rotation has occurred about the double bond between carbons 11 and 12 of the retinal, making the molecule intermediate in shape between the all-*trans* isomer and the 11-*cis* isomer. Second, some of these molecules (about two-thirds) convert into the all-*trans* form, called bathorhodopsin. The retinal is still attached to the opsin, but the protein now undergoes a sequence of transformations producing a sequence of identifiable proteins called lumirhodopsin, metarhodopsin I, and metarhodopsin II. Over a period of several minutes, the metarhodopsin II dissociates into opsin and free all-*trans*-retinal, which can be converted to the 11-*cis* form and attached again to opsin. The time required for this process is related to the time required for the eye to become dark adapted, but the process is much too slow to be involved in the actual process of vision. The process by which a signal is sent into a fiber of the optic nerve is apparently associated with conformational changes in metarhodopsin II.

15.7 Other Types of Spectroscopy

The spectroscopy we have discussed involves emission and/or absorption of ultraviolet, visible, or infrared radiation. There are a number of types of spectroscopy that utilize different processes, and we discuss a few of them in this section.

Raman Spectroscopy

Raman spectroscopy was invented by Chandrasekhara Venkata Raman, 1888–1970, Indian physicist who received the 1930 Nobel Prize in physics for this work.

In Raman spectroscopy, radiation is not absorbed or emitted but is inelastically scattered from the sample substance. Radiation of one wavelength is incident on the sample and radiation of a different wavelength comes out from the sample, ordinarily being observed in a direction at right angles to the incident beam's direction. Figure 15.15 shows schematically a modern Raman spectrometer. Since the scattered beam is much weaker than the incident beam, the use of lasers to provide the incident beam has greatly improved the performance of Raman spectrometers.

In an inelastic scattering process, an amount of energy equal to the difference in the photon energies of the incident and scattered radiation is transferred. This difference must equal the energy difference between two energy levels of the sample molecules. Let the frequency of the incident radiation be denoted by v and the frequency of the scattered radiation by v'. If the radiation loses energy to the molecules,

$$hv - hv' = E_{\text{upper}} - E_{\text{lower}} \tag{15.7-1}$$

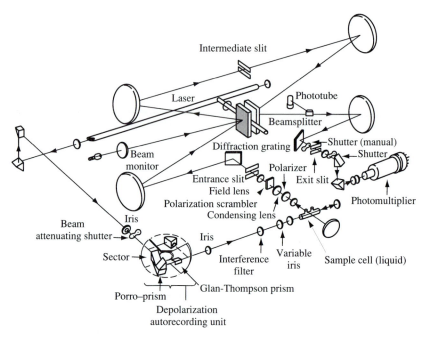

Figure 15.15 Schematic Diagram of a Laser Raman Spectrophotometer. Two distinct beams are shown: the incident beam and the scattered beam at right angles to the incident beam. Courtesy of Jeol, Ltd.

and if the molecules lose energy to the radiation,

$$hv' - hv = E_{\text{upper}} - E_{\text{lower}} \tag{15.7-2}$$

The difference between the scattered and incident frequencies or reciprocal wavelengths is called the **Raman shift**. Spectral lines corresponding to Equation (15.7-1) are called **Stokes lines**, and those corresponding to Equation (15.7-2) are called **anti-Stokes lines**. Incident light of a fixed frequency is used, and the scattered light can be dispersed to determine the wavelengths at which inelastic scattering occurs.

Since Raman spectroscopy uses a different physical process from that of ordinary optical spectroscopy, the selection rules are different, and this makes it possible to observe transitions that are forbidden in emission or absorption spectroscopy. The Raman selection rules for rotational and vibrational transitions are:

Raman Selection Rules

$$\Delta J = 0, \pm 2 \tag{15.7-3a}$$

$$\Delta v = \pm 1 \tag{15.7-3b}$$

The nuclear motion must modulate the polarizability of the molecule

$$\tag{15.7-3c}$$

The **polarizability** is a measure of the tendency of a molecule to acquire an electric dipole in the presence of an electric field. Even if a molecule has no permanent dipole, an electric field will induce a dipole in it that to a good approximation is proportional to the electric field. If the molecule has a permanent dipole moment, the induced dipole and the permanent dipole add as vectors.

For a molecule with the same properties in all directions (an isotropic molecule), the induced moment is proportional to the electric field:

$$\boldsymbol{\mu}_{\text{ind}} = \alpha\mathbf{E} \tag{15.7-4}$$

where α is the **polarizability** and $\mathbf{E}$ is the electric field (a vector quantity). For an anisotropic molecule (with different properties in different directions), the x component of the induced moment is given by

$$\mu_{x,\text{ind}} = \alpha_{xx}E_x + \alpha_{xy}E_y + \alpha_{xz}E_z \tag{15.7-5}$$

with similar equations for the y and z components. The polarizability is now a matrix with nine components (a **tensor**) whose components have two subscripts. Equation (15.7-5) and its analogues become the same as Equation (15.7-4) if

$$\alpha_{xx} = \alpha_{yy} = \alpha_{zz} \tag{15.7-6}$$

and if the other components vanish.

In a molecule, the contributions to the polarizability are due to distortion of the electronic wave function and of the nuclear framework of the molecule. In most molecules, the major contribution is from the electrons.

Modulation of the polarizability by the nuclear motion means that the value of the polarizability oscillates as the motion occurs. In general, the electronic polarizability parallel to a bond is different from the polarizability perpendicular to the bond. Therefore, as a diatomic molecule or linear polyatomic molecule rotates, the components of the polarizability fluctuate periodically. The rotation is **Raman active** (produces a Raman spectrum).

For a nonlinear polyatomic molecule, the polarizabilities of the individual bonds add to make a total polarizability. If the molecule is sufficiently symmetrical, the total polarizability is the same in all directions. A spherical top molecule has no rotational Raman spectrum. Symmetric tops and asymmetric tops have anisotropic polarizabilities and have rotational Raman spectra.

Many vibrational normal modes give rise to Raman spectral lines. The stretching or compression of a bond changes the electronic wave function, changing the polarizability of the electrons in the bond. Diatomic molecules therefore exhibit vibrational Raman spectra.

A polyatomic molecule can have some normal modes in which one bond is stretching at the same time a bond of the same type is shortening. If the changes in polarizabilities of the two bonds cancel, the normal mode is not Raman active. There is a rule of mutual exclusion: *In a molecule with a center of symmetry, a normal mode that is seen in the infrared will not be seen in the Raman, and vice versa.*[15] The normal modes of carbon dioxide are

---

[15] Levine, *op. cit.*, p. 268.

shown in Figure 14.6. The asymmetric stretch, which is seen in the infrared, is not seen in the Raman spectrum, and the symmetric stretch, which is not seen in the infrared, is seen in the Raman. The bending modes, which are seen in the infrared, are not seen in the Raman spectrum, since the bonds do not stretch appreciably as the bond angle bends.

▼   **EXAMPLE 15.8**

Identify which of the normal modes of cyanogen shown in Example Figure 15.7 give rise to Raman lines.

Solution

In modes 1 and 2, the two C—N bonds oscillate in unison, so the polarizability is modulated and the modes are Raman active. In mode 3, the C—C bond does not oscillate and the C—N bonds oscillate out of phase, so this mode is not seen. In mode 4, all bonds stretch in unison, so this mode is seen in the Raman spectrum. In mode 5, the bonds do not change their lengths appreciably, and this mode is not seen in the Raman spectrum. The molecule has a center of symmetry, so the same results could be obtained immediately by using the fact that only modes 3 and 5 are infrared active.

◢

Exercise 15.16

Figure 15.16 shows sketches of some of the vibrational normal modes of ethylene. Tell which are infrared active, and which are Raman active.

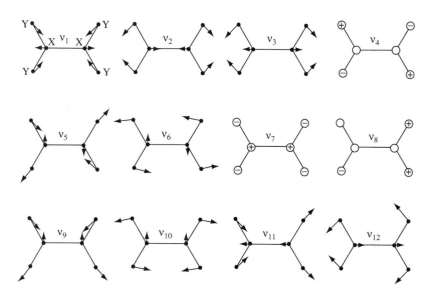

Figure 15.16. The Vibrational Normal Modes of Ethylene, C_2H_4. The arrows show the direction of motion of each nucleus in one half of the period. The length of each arrow is proportional to the amplitude of motion of the nucleus. From G. Herzberg, *Molecular Spectra and Molecular Structure*, Vol. II, *Infrared and Raman Spectra of Polyatomic Molecules*, Van Nostrand Reinhold, New York, 1945, p. 107.

It is possible to deduce the same kinds of structural information from Raman spectra as from infrared and microwave spectra. Using the selection rule for rotation, Equation (15.7-3a), we find for the Raman shift of the Stokes rotational lines of a diatomic or linear polyatomic molecule

$$\tilde{\nu} - \tilde{\nu}' = (E_{J+2} - E_J)/hc = \tilde{B}_e(4J + 6) \tag{15.7-7}$$

where terms in α and $\mathscr{D}$ have been neglected.

EXAMPLE 15.9

Example Figure 15.9 shows the rotational Raman spectrum of $^{12}C^{16}O_2$. From the splitting between the lines, 3.09 cm^{-1}, calculate the equilibrium bond lengths.

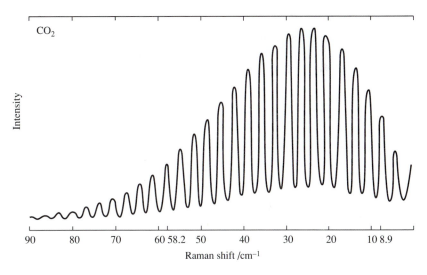

Example Figure 15.9. Rotational Raman Spectrum of Carbon Dioxide. The axis is the difference between the reciprocal wavelength of the incident radiation and that of the scattered radiation. From L. Claron Hoskins, *J. Chem. Educ.* **54**, 642 (1977).

Solution

Carbon dioxide is linear, so Equation (15.7-7) applies, with $\tilde{B}_e$ given by the Equation (14.2-35) divided by c, the speed of light. Since the molecule is symmetrical, with the carbon nucleus at the center of mass, the moment of inertia is

$$I_e = 2m_O r_e^2$$

where r_e is the bond length and m_O is the oxygen nuclear mass, 2.656×10^{-26} kg.
   The CO_2 molecule has symmetry number 2. Since the ^{16}O nuclei have zero spin and the electronic ground state is a sigma state, only even values of J can occur, just as with a homonuclear diatomic molecule with nuclei of zero spin. The first spectral line should occur at $6\tilde{B}_e$, and the splitting between lines should equal $8\tilde{B}_e$. The first full line in the figure is the second line, corresponding to the transition from $J = 2$ to $J = 4$.

Carbon dioxide has no microwave spectrum because it has no permanent dipole moment. The selection rules also rule out a spectrum, because $\Delta J = \pm 1$ for the microwave spectrum and only even values of J are possible for CO_2.

Using the symbol $\Delta \tilde{v}$ for the splitting

$$r_e^2 = \frac{h}{2\pi^2 m_O c \, \Delta \tilde{v}}$$

$$= \frac{6.6261 \times 10^{-34} \text{ J s}}{2\pi^2 (2.656 \times 10^{-26} \text{ kg})(3.09 \text{ cm}^{-1})(2.9979 \times 10^{10} \text{ cm s}^{-1})}$$

$$= 1.364 \times 10^{-20} \text{ m}^2$$

$$r_e = 1.17 \times 10^{-10} \text{ m} = 117 \text{ pm} = 1.17 \text{ Å}$$

This result agrees fairly well with the accepted value, 116.15 pm.

Notice how short the bond length is. It has roughly the same length as the carbon-oxygen triple bond in carbon monoxide. A typical C⚌C double bond has a length near 140 pm = 1.4 Å.

From the vibrational selection rule, $\Delta v = \pm 1$, the Raman shift in a vibrational spectrum will be the same as the frequency of an infrared spectral line if both transitions are allowed.

Exercise 15.17

From data in Table A16 find the Raman shift in reciprocal wavelength for the band center of the Stokes vibrational fundamental band of diatomic oxygen. If the incident light has wavelength 253.7 nm, find the wavelength of this band center.

Photoelectron Spectroscopy

In photoelectron spectroscopy, high-energy electromagnetic radiation is absorbed by the sample substance, causing ejection of an electron:

$$M + \text{photon} \rightarrow M^+ + e^- \qquad \textbf{(15.7-8)}$$

In photoelectron spectroscopy, one can measure the ionization potential of other electrons besides the most easily removed electron.

The calculation of an orbital energy from the photoelectron spectrum depends on the assumption that the other orbitals do not change appreciably when an electron is removed.

where M represents a molecule (or atom) of the sample substance. The kinetic energy of the ejected electron is measured, and the difference in energy between a photon of the incident radiation and the kinetic energy of the electron is taken to be the ionization energy of the particular electron ejected.

If the electronic wave function is represented in the orbital approximation, the wave function of the ion will be like that of the molecule except that one spin orbital will be occupied in the molecule that is not occupied in the ion. The ionization energy will be equal to the orbital energy of this spin orbital, so photoelectron spectroscopy affords a direct means of measuring the orbital energies.[16]

Figure 15.17 shows the photoelectron spectrum of N_2, using 58.4-nm radiation from a helium arc. The kinetic energy of the electrons increases from left to right, so the ionization energy increases from right to left. There are three groups of lines, corresponding to ionization from three different orbitals. The separate lines within each group correspond to different vibra-

---

[16] T. C. Koopmans, *Physica* **1**, 104 (1933).

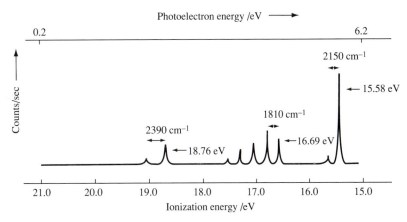

Figure 15.17. Photoelectron Spectrum of Nitrogen. Electrons ejected from three different orbitals are shown. The splittings are for different vibrational states in the product ions. From Ira N. Levine, *Molecular Spectroscopy*, Wiley, New York, 1975, p. 316.

tional states of the ion. Since the sample is at room temperature, only the ground vibrational state of the molecule is significantly occupied.

The ground-level electron configuration of N_2 is

$$(\sigma g 1s)^2 (\sigma^* u 1s)^2 (\sigma g 2s)^2 (\sigma^* u 2s)^2 (\pi u 2p)^4 (\sigma g 2p)^2$$

The rightmost group of lines arises from removal of an electron from the $\sigma g 2p$ bonding orbital, the center group of lines arises from removal of an electron from a $\pi u 2p$ bonding orbital, and the leftmost group of lines arises from removal of an electron from the $\sigma^* u 2s$ antibonding orbital. To determine the orbital energies of other orbitals, shorter-wavelength radiation must be used.

The absorption of a photon with removal of an electron is quite similar to absorption of a photon to make a transition to an excited electronic state, and the Franck-Condon principle applies. The ionization potential determined through photoelectron spectroscopy is referred to as the "vertical" ionization energy, since the ionization is represented by a vertical line in a diagram such as that of Figure 15.10. In the nitrogen spectrum, it appears that ionization to the $v = 1$ vibrational state is the most probable process for the center group of lines, while in the other two groups the transition to the $v = 0$ vibrational state is most probable.

▼   **E X A M P L E   15.10**

The ground-state vibrational frequency for nitrogen is 2359 cm^{-1}. The spacing between the lines in the rightmost group of lines in the photoelectron spectrum corresponds to 2150 cm^{-1}, and the line for the transition from $v = 0$ to $v = 0$ is at 15.58 eV. Find the ionization energy from the minimum of the ground-state potential curve to the minimum in the ion curve.

Solution

The ionization energy from the $v = 0$ vibrational state to the $v = 0$ vibrational state differs from the desired quantity by the difference of the zero-point vibrational energies, given by

$$
\Delta E_{\text{zero point}} = \frac{h\nu_{\text{molecule}}}{2} - \frac{h\nu_{\text{ion}}}{2}
$$

$$
= \frac{hc}{2} (\tilde{\nu}_{\text{molecule}} - \tilde{\nu}_{\text{ion}})
$$

$$
= \frac{(6.6261 \times 10^{-34} \text{ J s})(2.9979 \times 10^{10} \text{ cm s}^{-1})}{2} (209 \text{ cm}^{-1})
$$

$$
= 2.08 \times 10^{-21} \text{ J} = 0.0130 \text{ eV}
$$

$$
\Delta E_{\text{e}-\text{e}} = \Delta E_{0-0} + 0.013 \text{ eV} = 15.58 \text{ eV} + 0.013 \text{ eV} = 15.59 \text{ eV}
$$

In the solution to Example 15.10, notice how small the difference in the zero-point energies is.

▲

Exercise 15.18

Hint: For Exercise 15.18, remember that bonds of higher order are stronger and generally have larger force constants.

Explain why the spacing between the lines in the leftmost group in Figure 15.17 is greater than 2359 cm^{-1}, the vibrational spacing of the ground level, while the spacing between the lines in the other two groups is smaller than 2359 cm^{-1}.

Photoacoustic Spectroscopy

Photoacoustic spectroscopy is a type of absorption spectroscopy in which absorption of energy is detected by the generation of sound waves. A beam of monochromatic radiation is directed on the sample through a "chopper," which is usually a rotating disk with several notches cut in the edge so that the beam is alternately passed and interrupted (chopped).

Photoacoustic spectroscopy was invented by Alexander Graham Bell, 1847–1922, the inventor of the telephone. The method has not been extensively exploited until fairly recently.

Radiation absorbed by the sample heats the surface of the sample and the air next to it during the time that the beam is passed by the chopper. During the time that the beam is interrupted, the sample and the air cool off. The air thus alternately expands and contracts with the frequency of the chopper, producing a sound wave that can be detected by a microphone. If the radiation is not absorbed, no sound wave is generated. The intensity of the sound wave can be measured electronically as the wavelength of the light is varied, giving an absorption spectrum. The frequency of the chopper must be slow enough that the air has time to cool off during the period of beam interruption but fast enough to make a detectable sound wave. A chopper frequency of around 50 hertz is common. The principal advantage of the method is that an opaque sample can be used, such as a strongly colored liquid or a solid.

Circular Dichroism and Optical Rotatory Dispersion

These two types of spectroscopy involve study of optically active substances. An optically active substance is one that rotates the plane of plane-polarized light and is generally one that has molecules without a plane of symmetry.

Plane-polarized light was described in Figure 9.4 as an oscillating electric field remaining in one plane containing the direction of propagation and an oscillating magnetic field in another plane perpendicular to the first, also containing the direction of propagation.

To understand optical activity, one must consider **circularly polarized light**. Consider two plane-polarized rays of equal amplitude and wavelength that are polarized in perpendicular directions and out of phase by a fourth of a wavelength, as in Figure 15.18, which shows only the electric field. The sum of the two electric fields follows a helix, and the light is said to be circularly polarized. If the wave shown in the figure propagates to the right of the figure, a stationary observer facing the source of radiation observes an electric field that rotates clockwise. Such radiation is called **right-polarized** radiation, while radiation that gives a field rotating counterclockwise when looking into the source is called **left-polarized radiation**.

Individual photons correspond to circularly polarized light rather than to plane-polarized light. A photon can have a projection of its spin angular momentum either parallel to its direction of propagation or antiparallel to it, and these two possibilities correspond to the two directions of circular polarization.

Although we depicted a ray of circularly polarized light as being the sum of two plane-polarized rays, it is possible to depict plane-polarized light as the sum of two circularly polarized rays. Figure 15.19a shows the rotation of the electric fields of a right-polarized ray and a left-polarized ray at a fixed location. As the two electric field contributions rotate in opposite directions, their sum remains in a plane if they have the same frequency and the same amplitude. Plane-polarized light should be visualized as made up of equal numbers of left circularly polarized photons and right circularly polarized photons.

As light passes through a transparent or translucent medium, its speed is less than the speed of light in vacuum. This can be explained by absorption

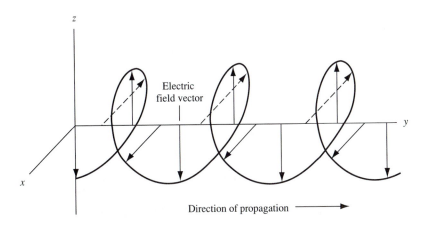

Figure 15.18. Circularly Polarized Light from Plane Polarized Light. The circularly polarized light corresponds to the electric field vector moving in a helical pattern. Plane-polarized light in the y-z and x-z planes can interfere to form circularly polarized light.

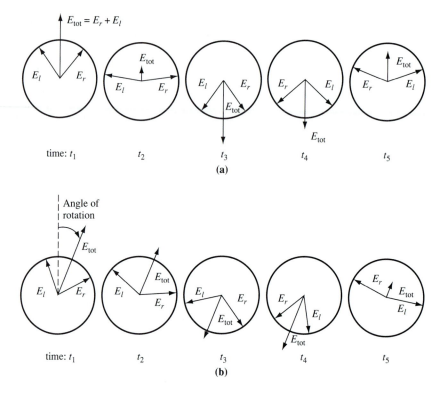

Figure 15.19 Plane Polarized Light (E_{tot}) from Right-Polarized Light (E_r) and Left-Polarized Light (E_l). (a) Plane-polarized light in a vertical plane. This diagram shows how the two vectors from the right- and left-polarized light add to produce a sum vector that oscillates in one plane. **(b)** Plane-polarized light (E_{tot}) from right-polarized light (E_r) and left-polarized light (E_l) that is Delayed Relative to part a. This diagram shows how the two vectors from the right- and left-polarized light add to produce a sum vector that oscillates in one plane. Since one of the beams is delayed, the sum vector oscillates in a different plane from that in part a.

of the light by the molecules or atoms of the medium followed by reradiation in the same direction as before, slowing the propagation of the light. The **refractive index**, n, of a medium is defined as the ratio of the speed of light in a vacuum to the speed of light in the medium:

$$n = \frac{c(\text{vacuum})}{c(\text{medium})} \quad \text{(definition of refractive index)} \quad \textbf{(15.7-9)}$$

The refractive index depends on the wavelength of light as well as on the identity of the medium.

Molecules that have a plane of symmetry interact with photons of both circular polarizations in the same way, since the molecules are identical to their enantiomorphs. However, a molecule without a plane of symmetry appears different to the two kinds of photons, and the speed of light of the two kinds of photons can be different. This phenomenon is called **circular**

birefringence. If it occurs, the rotation of one circularly polarized electric field contribution lags behind the other, and the plane of polarization is rotated, as shown in Figure 15.19b. The angle α through which the plane is rotated is proportional to the length of the sample and to the difference between the refractive indexes of right- and left-polarized light. It is given by

$$\alpha = \frac{\pi(n_R - n_L)L}{\lambda} \tag{15.7-10}$$

where n_R is the refractive index for right-polarized light and n_L is the refractive index for left-polarized light of wavelength λ, and L is the length of the sample.

The rotating power of a substance is commonly expressed as the **specific rotation**, $[\alpha]$, defined by

$$[\alpha] = \frac{\alpha}{\rho L} \quad \text{(definition)} \tag{15.7-11}$$

where α is the angle of rotation, usually measured in degrees, ρ is the density of the substance, and L is the length of the sample. If instead of a pure substance one has a solute in solution, the density is replaced by the concentration of the substance, usually measured in g cm^{-3}. The specific rotation depends on the identity of the substance, the identity of the solvent (if any), the temperature, and the wavelength of light.

The specific rotation may have different signs for two wavelengths between which a spectral line occurs. Specific rotations of many substances have often been tabulated for a single wavelength, usually the yellow sodium "D lines" at 589.0 and 589.6 nm, and the wavelength dependence has been ignored. However, additional information about the stereochemical configuration of molecules can be obtained from the dependence of the specific rotation on wavelength, which is called **optical rotatory dispersion** (ORD).

The absorptivity of an optically active substance can also differ for right-polarized and left-polarized photons. This phenomenon is called **circular dichroism (CD)** and is also studied as a function of wavelength. Until the 1970s, only ultraviolet light and visible light were used for ORD and CD. Since then, however, techniques have been developed for infrared circular dichroism spectroscopy, which is usually called **vibrational circular dichroism (VCD)**. In addition, techniques have been invented for determining the differences in scattering of left- and right-polarized light, and Raman optical activity (ROA) is now being studied.[17]

In addition to the types of spectroscopy we have discussed, many other spectroscopic types and techniques have been developed. Almost any issue of the *Journal of Chemical Physics* contains several articles on spectroscopic techniques that we have not discussed.

---

[17] S. C. Stinson, *Chem. Eng. News* **63**(45), 21 (Nov. 11, 1985).

15.8 • Magnetic Resonance Spectroscopy

When an atom or molecule is in a magnetic field, different spin states of electrons or nuclei in the atom or molecule have different energies. Transitions between the spin states give rise to absorption of radiation at characteristic frequencies depending on the magnetic field and on the properties of the molecule, allowing information about the molecular structure to be deduced.

Magnetic Fields

Note the common usage of the term "magnetic field" for the magnetic induction, not the magnetic field strength.

The strength of a magnetic field is commonly specified by one of two vector quantities, the **magnetic induction B** or the **magnetic field strength H**. These two quantities are proportional to each other. We will generally use the magnetic induction, **B**, and will call it the **magnetic field**.

Ampère's law gives the magnitude of the magnetic field in a vacuum in the vicinity of a long straight wire carrying an electric current I:

$$B = |\mathbf{B}| = \frac{\mu_0 I}{2\pi r} \tag{15.8-1}$$

where r is the perpendicular distance from the wire and μ_0 is the permeability of a vacuum, introduced in Section 9.2:

$$\mu_0 = 4\pi \times 10^{-7} \text{ T m A}^{-1} \quad \text{(exactly, by definition)} \tag{15.8-2}$$

The conventional direction of an electric current is the apparent direction of motion of positive charges. In a current of electrons, the electrons are moving in the opposite direction.

The current is measured in amperes (A), the distance in meters, and the magnetic induction in tesla (T). The direction of the magnetic induction can be obtained from a right-hand rule: If the right thumb points in the direction of the current, then **B** points in the direction of the fingers, tangent to a circle around the wire.

Another unit of magnetic induction, the **gauss**, is defined by

$$1 \times 10^4 \text{ gauss} = 1 \text{ T} \quad \text{(exactly)} \tag{15.8-3}$$

The earth's magnetic field is slightly smaller than 1 gauss in most locations on the earth's surface.

If a particle of charge Q is moving with velocity **v** through a magnetic field **B**, there is a force on the particle

$$\mathbf{F} = Q\mathbf{v} \times \mathbf{B} \tag{15.8-4}$$

where × stands for the vector product of the two vectors.

Magnetic Dipoles

Magnetic dipoles are like bar magnets, with a "north-seeking" pole at one end and a "south-seeking" pole at the other. Just as an electric field applies a torque to an electric dipole, a magnetic field applies a torque to a magnetic dipole. However, an electric dipole can be made up of a positive and a negative charge (two electric monopoles), which can exist separately from each

other. It is not yet certain that magnetic monopoles can exist separately.[18] If they do, they are not commonly observed.

According to classical electromagnetic theory, a magnetic dipole can be produced by an electric current flowing in a closed loop of a conducting material, as shown in Figure 15.20a. The magnetic dipole $\boldsymbol{\mu}$ is a vector whose magnitude is given by the product of the current, I, and the area of the loop, $\mathscr{A}$:

$$\mu = |\boldsymbol{\mu}| = I\mathscr{A} \tag{15.8-5}$$

Do not confuse the symbol μ for a magnetic dipole with the same letter used for the permeability.

The direction of the magnetic dipole vector is as shown in the figure, perpendicular to the plane of the loop. If the fingers of the right hand point in the direction of the current, which is conventionally assigned as the direction of flow of positive charges, the thumb points in the direction of the magnetic dipole.

The potential energy of a magnetic dipole in a magnetic field is given by

$$\boxed{E_{\mathrm{mag}} = -\boldsymbol{\mu} \cdot \mathbf{B} = -|\boldsymbol{\mu}||\mathbf{B}|\cos(\alpha)} \tag{15.8-6}$$

where the dot ($\cdot$) stands for the scalar product of the two vectors and α is the angle between the dipole and the field. The energy is at a minimum if the dipole and the field are parallel and a maximum if the dipole and the field are antiparallel.

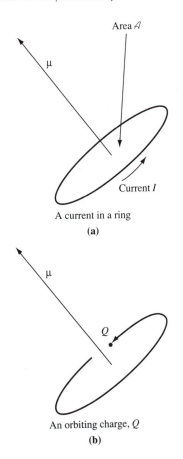

Area $\mathscr{A}$

μ

Current I

A current in a ring

(a)

μ

Q

An orbiting charge, Q

(b)

Figure 15.20. A Magnetic Dipole. (a) From a current. **(b)** From a moving charge. An orbiting charge is equivalent to a current and produces a magnetic dipole just like any other current.

A charged particle orbiting around a fixed point is equivalent to a current, so it produces a magnetic dipole. If a particle of charge Q is moving in a circular orbit, as in Figure 15.20b, the electric current has magnitude equal to the amount of charge passing per unit time, or the particle's charge divided by the time required to make one circuit:

$$I = \frac{Q}{t_{\mathrm{orbit}}} = \frac{Qv}{2\pi r} \tag{15.8-7}$$

where r is the radius of the particle's orbit and v is its speed.

Combining Equations (15.8-5) and (15.8-7) gives

$$\mu = |\boldsymbol{\mu}| = \frac{\pi r^2 Q v}{2\pi r} = \frac{Qvr}{2} \tag{15.8-8}$$

This can be restated in terms of the angular momentum, which for a circular orbit is given by Equation (D-15) of Appendix D as

$$L = |\mathbf{L}| = mvr \tag{15.8-9}$$

so that

$$\mu = |\boldsymbol{\mu}| = \frac{Q}{2m}|\mathbf{L}| \tag{15.8-10}$$

---

[18] J. E. Dodd, *The Ideas of Particle Physics*, Cambridge University Press, 1984, pp. 169ff.

The same relation holds for the vector quantities as for their magnitudes, even for orbits that are not circular (although we do not prove this fact):

$$\boldsymbol{\mu} = \frac{Q}{2m}\, \mathbf{L} \qquad (15.8\text{-}11)$$

Atomic and Molecular Magnetic Dipoles

Equation (15.8-11) is a convenient form to use in discussing quantum mechanical magnetic moments. The operators and eigenvalues for the angular momentum of an orbiting particle are known from Chapters 10 and 11 and allow the operators for the magnetic dipole to be written easily. For an orbiting electron, the charge Q is equal to $-e$, and the operator for the magnetic dipole is

$$\hat{\boldsymbol{\mu}} = -\frac{e}{2m_e}\, \hat{\mathbf{L}} \qquad (15.8\text{-}12)$$

where m_e is the mass of the electron, 9.10939×10^{-31} kg.

Equation (15.8-12) must be modified to hold for spin angular momenta. The operator of the magnetic dipole due to the spin angular momentum of an electron is

$$\hat{\boldsymbol{\mu}} = -g\,\frac{e}{2m_e}\, \hat{\mathbf{S}} \qquad (15.8\text{-}13)$$

In relativistic quantum mechanics, a theory exists that allows the value of this factor to be calculated.

where $\hat{\mathbf{S}}$ is the spin angular momentum operator. The quantity g is a correction factor that accounts for the failure of the electron to obey nonrelativistic mechanics and is known as the **anomalous g factor of the electron**. Its value is $2.0023\ldots$.

The only eigenvalue of the operator $\hat{\mathbf{S}}^2$ is $\hbar^2(1/2)(3/2)$, so the magnitude of the magnetic dipole is

$$|\boldsymbol{\mu}| = g\,\frac{e}{2m_e}\, \hbar[(1/2)(3/2)]^{1/2} = g\beta_e[(1/2)(3/2)]^{1/2} \qquad (15.8\text{-}14)$$

The Bohr magneton is also denoted by the letter μ in some books.

The constant β_e is called the **Bohr magneton**:

$$\beta_e = \frac{e\hbar}{2m_e} = 9.2740 \times 10^{-24}\ \text{J T}^{-1} \qquad (15.8\text{-}15)$$

If a magnetic dipole is placed in a magnetic field, its energy is

$$E_{\text{mag}} = -\boldsymbol{\mu} \cdot \mathbf{B} \qquad (15.8\text{-}16a)$$

For a magnetic field in the direction of the z axis

$$E_{\text{mag}} = -\mu_z B_z \qquad (15.8\text{-}16b)$$

For an electron, $\hat{S}_z$ has two eigenvalues, $m_s\hbar = \pm\hbar/2$, so that

$$E_{\text{mag}} = \pm g\,\frac{e}{2m_e}\,\frac{\hbar}{2}\, B_z = \pm\frac{g\beta_e B_z}{2} \quad \text{(electron)} \qquad (15.8\text{-}17)$$

The difference between the two energies is exploited in **electron spin resonance (ESR) spectroscopy**.

▼ **EXAMPLE 15.11**

For a magnetic field of 0.500 T, find the difference in the energies of the two electron spin states.

Solution

$$\Delta E_{\text{mag}} = g\beta_e B_z = (2.0023)(9.2740 \times 10^{-24} \text{ J T}^{-1})(0.500 \text{ T})$$
$$= 9.28 \times 10^{-24} \text{ J}$$

▲

Exercise 15.19

Find the frequency and wavelength of photons with energy equal to the energy difference in Example 15.11.

Many nuclei have nonzero spin angular momenta and magnetic moments. For a proton, the magnetic dipole operator is

$$\hat{\boldsymbol{\mu}} = g_p \frac{e}{2m_p} \hat{\mathbf{I}} \tag{15.8-18}$$

where $\hat{\mathbf{I}}$ is the spin angular momentum operator of the proton. The proton has the same spin angular momentum properties as the electron, so the only magnitude μ can have is

$$|\boldsymbol{\mu}| = \mu = g_p \frac{e}{2m_p} \hbar[(1/2)(3/2)]^{1/2}$$
$$= g_p \beta_N [(1/2)(3/2)]^{1/2} \tag{15.8-19}$$

The nuclear magneton β_N is also denoted by μ_N in some books.

The constant β_N is analogous to the Bohr magneton and is called the **nuclear magneton**.

In tabulations (such as CODATA 63) the value given for the magnetic moment of the proton is that of the z component, equal to $g_p\beta_N(1/2) = 1.41 \times 10^{-26}$ J T^{-1}.

$$\beta_N = \frac{e\hbar}{2m_p} = 5.050787 \times 10^{-27} \text{ J T}^{-1} \tag{15.8-20}$$

The factor g_p is analogous to the g factor of the electron and is called the **nuclear g factor of the proton**. Its value is 5.58569, so the magnitude of μ is equal to 2.44×10^{-26} J T^{-1}.

Exercise 15.20

Find the difference in the energies of the two spin states of a proton in a magnetic field of 0.500 T. Compare with the result of Example 15.11 for the electron.

The energy required to excite a nucleus to a state other than its ground state is millions of electron volts (or more). Chemists almost never observe such states.

Other nuclei besides the proton have nonzero spin angular momentum. Since we ordinarily encounter nuclei only in their ground states, a given isotope (nuclide) can be taken to have a fixed magnitude of its spin angular momentum:

$$|\mathbf{I}| = \hbar\sqrt{I(I+1)} \tag{15.8-21}$$

where I is a characteristic quantum number for a given nuclide in its ground state. For example, I equals 1 for ^{2}H, 1/2 for ^{13}C, and zero for ^{12}C. Each nuclide also has a characteristic magnetic dipole moment:

$$|\boldsymbol{\mu}| = |g_N|\beta_N \sqrt{I(I+1)} \tag{15.8-22}$$

The nuclear g factor has a different value for each nucleus and must be determined by experiment. The same β_N is used for all nuclei, for convenience. If the mass of the particular nucleus were put into the formula to give a different β_N for each nucleus, it would just change the value of g_N for that nucleus.

where g_N is a characteristic factor for the given nuclide, called the **nuclear g factor of the nuclide**. The nuclear magneton β_N of Equation (15.8-20) containing the mass and charge of the proton is used for all nuclides. The necessary correction for different masses and charges is incorporated into the nuclear g factor g_N. Table A17 lists the nuclear g factors and spin quantum numbers of some common nuclides. Notice the remarkable fact that some nuclides have negative values of the nuclear g factor. In these cases the magnetic dipole has the direction it would be expected to have for a negative particle.

The values I_z can take on are

$$I_z = \hbar M_I \tag{15.8-23}$$

All angular momenta follow the same pattern of magnitudes and projections on one axis. The number of values of M_I is $2I + 1$.

where M_I is a quantum number that ranges in integral steps from I to $-I$. For a proton, M_I can equal $1/2$ or $-1/2$. For a ^{2}H nucleus M_I can equal 1, 0 or -1, etc. The z component of the magnetic dipole can take on values

$$\mu_z = g_N \beta_N M_I \tag{15.8-24}$$

If a nucleus is placed in a magnetic field B_z,

$$E_{\mathrm{mag}} = -g_N \beta_N B_z M_I \tag{15.8-25}$$

For example, a proton could be in either of two energy states.

The electronic, vibrational, and rotational energy levels are determined solely by the nature of the molecules, while the energy levels in Equation (15.8-25) are determined by the externally applied magnetic field as well as by the nature of the molecules.

Electron Spin Resonance Spectroscopy

In electron spin resonance (ESR) spectroscopy, transitions of electrons between the two possible spins are observed in a magnetic field. ESR spectroscopy is also called **electron paramagnetic resonance (EPR)** spectroscopy.

The selection rule for these transitions is

$$\boxed{\Delta m_s = \pm 1} \tag{15.8-26}$$

The "resonance" terminology will seem reasonable if we compare the radiation frequency and the frequency of precession of the angular momentum about its cone of possible directions (see Problem 15.48).

Transitions between the two possible states, $m_s = 1/2$ and $m_s = -1/2$, are allowed. The frequency of radiation absorbed or emitted is

$$v = \frac{E_{\mathrm{mag}}}{h} = \frac{g\beta_e B_z}{h} \tag{15.8-27}$$

Radiation that can be absorbed or emitted is said to be in "resonance" with the electrons.

▼   **EXAMPLE 15.12**   Find the magnetic field necessary to cause ESR absorption or emission of light of wavelength 1.000 cm.

Solution

$$B_z = \frac{hc}{g\beta_e\lambda} = \frac{(6.6261 \times 10^{-34} \text{ J s})(2.9979 \times 10^{10} \text{ cm s}^{-1})}{(2.0023)(9.2740 \times 10^{-24} \text{ J T}^{-1})(1.000 \text{ cm})}$$

$$= 1.070 \text{ T}$$

◢

At first glance, it seems that every substance would absorb radiation at the same frequency if placed in the same magnetic field, since all substances contain electrons. However, most substances do not absorb at all because their electrons all occupy space orbitals in pairs with opposite spins. Such electrons cannot change their spins unless both members of a pair change simultaneously, because of the Pauli exclusion principle. If both change simultaneously, no absorption of energy takes place because one electron gains the same amount of energy that the other electron loses. Only a substance containing unpaired electrons exhibits an electron spin resonance spectrum.

Only substances with unpaired electrons exhibit an ESR spectrum.

It still seems that no useful information about the substance would be obtained except whether it contains unpaired electrons. However, the magnetic field to which an electron is exposed is a vector sum of the externally applied field, $\mathbf{B}_0$, and the contribution from the nuclei in the molecule, $\mathbf{B}_{\text{internal}}$. If the applied field is in the z direction and the molecule has n nuclei, the z component of the field is

$$B_z = B_0 + B_{\text{internal}, z} = B_0 + \sum_{j=1}^{n} a_j M_{Ij} \qquad \textbf{(15.8-28)}$$

where a_j is a **coupling constant** for the jth nucleus and M_{Ij} is the quantum number for the z component of the nuclear spin angular momentum of the jth nucleus. The coupling constants for nuclei in many molecules have values near 1 gauss $(1 \times 10^{-4} \text{ T})$, but they depend on the orbital the electron is in as well as on the identity of the nucleus.

It is found that B_{internal} is a short-range interaction. That is, it has an effect only if the electron comes very close to the nucleus. If an unpaired electron occupies an orbital with a nodal surface at a particular nucleus, its probability of being found at that nucleus is nearly zero and the coupling constant will nearly vanish for that nucleus and that orbital. Since s atomic orbitals are the only atomic orbitals without nodes passing through the nucleus, the coupling constant at a specific nucleus is sometimes said to be a measure of the "s character" of the orbital at that nucleus.

An ESR spectrometer uses microwave radiation, with wavelengths around 1 cm. The microwaves are conducted by waveguides to the sample chamber, which is a cavity with conducting walls in which standing electromagnetic waves can occur. Absorption by the sample ("resonance") is detected by its effect on these standing waves. Since a particular cavity can support standing waves of only a few frequencies, the frequency of the radiation is kept fixed and the applied magnetic field is varied.

Absorption occurs when the magnetic field actually "felt" by an unpaired electron reaches the value B_{res}, the value such that Equation (15.8-27) is satisfied:

$$B_0 + B_{\text{internal}} = B_0 + \sum_{j=1}^{n} a_j M_{Ij} = B_{\text{res}} \qquad (15.8\text{-}29)$$

A sample of a substance with several nuclear dipoles in its molecules can give several spectral lines, or values of B_0 at which resonance occurs, because it has a variety of nuclear spin states. For example, a molecule with an unpaired electron that couples equally with two protons could be in a state with both proton spins up, in either of two states with one proton spin up and one down, or in a state with both proton spins down. All of these states will be represented in the sample, giving a spectrum with three lines in which the middle line is twice as intense as the other two.

In a symmetrical molecule, an unpaired electron often occupies a delocalized orbital that has equal magnitudes at several nuclei of the same element. In this case, the coupling constants are equal and the spectrum has several lines with characteristic patterns.

EXAMPLE 15.13

Assume that the benzene negative ion, $C_6H_6^-$, has its unpaired electron in a delocalized orbital with equal magnitude at each of the protons. Predict the ESR spectrum.

Solution

Since all of the coupling constants are equal, there will be just one line for each value of the sum of M_I's. This sum can equal 3, 2, 1, 0, -1, -2, or -3, so seven lines occur in the spectrum, with a splitting between two adjacent lines equal to the coupling coefficient.

The intensities of the spectral lines will be proportional to the numbers of molecules in each level. For a field of 0.500 T the difference in energy between two nuclear spin states with values of the sum of the M_I's differing by unity is given by Equation (15.8-25):

$$\Delta E_{\text{mag}} = g_N \beta_N B_z = (5.5857)(5.0508 \times 10^{-27}\ \text{J T}^{-1})(0.500\ \text{T})$$
$$= 1.411 \times 10^{-26}\ \text{J}$$

At 300 K the ratio of the populations of two such states is

$$e^{-\Delta E_{\text{mag}}/k_B T} = \exp\left[-\frac{1.411 \times 10^{-26}\ \text{J}}{(1.3807 \times 10^{-23}\ \text{J K}^{-1})(300\ \text{K})}\right]$$
$$= 0.9999966$$

Since the states are nearly equally populated, the population of a level will be nearly proportional to the degeneracy of the level, which is proportional to the number of ways of dividing a set of six spins into a subset of nuclear spins pointing up and a subset of spins pointing down. The number of ways of dividing a set of n objects into a subset of m and a subset of $n - m$ is

$$\text{Number} = \frac{n!}{m!(n-m)!} \qquad (15.8\text{-}30)$$

which is the formula for binomial coefficients. The intensities of the lines are in the ratios 1:6:15:20:15:6:1.

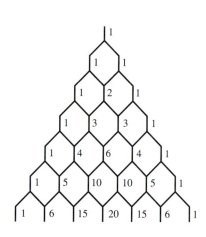

Figure 15.21. Diagram to Find the Splittings in the ESR Spectrum of the Benzene Negative Ion, $C_6H_6^-$. Use of this diagram requires that all of the splitting constants are equal. Otherwise the lines will not coincide, and there will be more than seven lines.

Figure 15.21 shows another way of arriving at the spectrum of Example 15.13. Starting with the spectral line that would occur without any splitting by proton spins, the effect of the first proton is represented by a splitting into two lines, as shown at the top of the diagram. Each of these lines is then split into two lines, representing the effect of the second proton. Since all splitting constants are equal, this gives three lines with relative intensities of 1:2:1, as shown at the second level of the diagram. The third level of the diagram represents the effect of the third proton, etc. The relative intensities at each level are obtained by adding the relative intensities of the two lines at the previous level that combine to produce a given line. The result at each level gives relative intensities proportional to binomial coefficients.

| **Exercise 15.21** | Predict the ESR spectrum of the hydrogen atom. |

Nuclear Magnetic Resonance Spectroscopy

In nuclear magnetic resonance (NMR) spectroscopy, transitions of nuclei from one spin state to another are observed in a magnetic field. A "scanning" NMR instrument is similar to the ESR instruments previously described: A fixed-frequency source of radiation is used, and the magnetic field is varied (scanned). The frequency is smaller than in ESR spectroscopy, and the radio-frequency energy is conducted by coaxial cable to the probe in which a liquid or solid sample is irradiated.

There is only one kind of electron, so there is only one electron g factor and only one Bohr magneton. However, every type of nucleus that has non-zero spin has its characteristic g factor. Because presently available magnets cannot scan over a very large range of magnetic fields without losing the necessary field homogeneity, a single scanning instrument can ordinarily take the spectrum of only one type of nucleus. The most common scanning instruments obtain NMR spectra only of protons.

However, **Fourier transform NMR spectrometers** use pulses of radio-frequency energy that contain many frequencies. Such instruments, which are discussed briefly in the next section of this chapter, can obtain spectra of more than one kind of nucleus and can obtain a spectrum more quickly than can a scanning instrument.

The selection rule for transitions of nuclear spins with absorption or emission of radiation is

$$\Delta M_I = \pm 1 \qquad \textbf{(15.8-31)}$$

so that from Equation (15.8-25),

$$E_{\text{photon}} = h\nu = |g_N|\beta_N B_z \qquad \textbf{(15.8-32)}$$

▼ **EXAMPLE 15.14**

Find the value of the magnetic field necessary for protons to absorb at a frequency of 60.000 megahertz.

Solution

$$B_z = \frac{h\nu}{g_N\beta_N} = \frac{(6.6261 \times 10^{-34} \text{ J s})(60.000 \times 10^6 \text{ s}^{-1})}{(5.5857)(5.050787 \times 10^{-27} \text{ J T}^{-1})} = 1.4092 \text{ T}$$

▲

| **Exercise 15.22** | Find the magnetic field necessary for ^{19}F nuclei to absorb at 60.000 megahertz. |

As in ESR spectroscopy, if absorption is to occur, the magnetic field at the nucleus must equal the value required for the radiation being used. If the externally applied field were the only contribution to this field, every proton would absorb at a fixed value of the field, every ^{13}C nucleus would absorb at another fixed value of the field, etc. No structural information could be obtained except to identify which nuclei with nonzero spins were present. However, the magnetic field at a nucleus has two additional contributions, from the electrons around the nucleus and from other nuclei with magnetic dipoles in the molecule.

The first molecular contribution to the magnetic field at a nucleus comes from the electrons around the nucleus. The externally applied magnetic field, B_0, induces a net magnetic dipole in the electrons of the molecule, antiparallel to the applied field. This phenomenon is called **diamagnetism**, and it causes a substance without unpaired electrons and without nuclear spins to be pushed out of a nonuniform magnetic field. The induced magnetic dipole in the electrons makes a contribution to the magnetic field at a given nucleus that is proportional to the applied field and in the opposite direction. For the jth nucleus,

$$B_{j\,\text{diamagnetic}} = -\sigma_j B_0 \tag{15.8-33}$$

where σ_j is called the **shielding constant** of the jth nucleus. It has a larger value when the probability of finding electrons near the nucleus is larger. Typical values for σ range from 15×10^{-6} to 35×10^{-6} (15 to 35 parts per million).

The magnetic field at the jth nucleus is

$$B_j = (1 - \sigma_j)B_0 \tag{15.8-34}$$

and the applied magnetic field at which absorption by the jth nucleus occurs is

$$B_{0,j} = \frac{h\nu}{g_N \beta_N} + \sigma_j B_{0,j} \tag{15.8-35}$$

An NMR spectrum is observed by scanning B_0 over a range of 10 or 20 parts per million and observing the values of B_0 at which absorption occurs.

It is not possible to use a system of nuclei without electrons as a reference, so it is customary to choose a reference substance to serve as a zero point for our NMR spectra. For proton NMR, the standard reference substance is tetramethyl silane, $Si(CH_3)_4$, which has a single sharp spectral line and a rather large shielding constant, 31 ppm.

The difference between the applied field necessary for the reference substance to absorb and that necessary for a given nucleus to absorb is called the **chemical shift**. One variable used to specify the chemical shift is δ:

$$\delta_j = \frac{B_{0\text{ref}} - B_{0j}}{B_{0\text{ref}}} \times 10^6 \text{ ppm} \tag{15.8-36}$$

Since the first term on the right-hand side of Equation (15.8-35) is the same for both $B_{0\text{ref}}$ and B_{0j},

$$\delta_j = \frac{\sigma_{\text{ref}} B_{0\text{ref}} - \sigma_j B_{0j}}{B_{0\text{ref}}} \times 10^6 \text{ ppm}$$
$$= (\sigma_{\text{ref}} - \sigma_j) \times 10^6 \text{ ppm} \tag{15.8-37}$$

The variable δ ordinarily lies between 0 and 15 ppm for proton NMR. The approximate equality in Equation (15.8-37) holds to four significant digits, because the different values of the applied field differ by only 10 or 20 parts per million.

The values of the shielding constants and of the chemical shifts are related to the structure of the molecule, since a higher electron density around a given nucleus generally corresponds to a larger shielding constant. If a nucleus is close to another nucleus of high electronegativity, it generally has a smaller shielding constant and its peak appears "downfield" at a larger value of δ.

Table A19 provides typical values of δ for different chemical environments for protons. The values will be slightly different in different substances with similar functional groups, but the values in the table are useful as a general guide.

A second variable used to specify the chemical shift is τ:

$$\tau = 10 \text{ ppm} - \delta \tag{15.8-38}$$

Larger values of τ correspond to larger values of the shielding constant and therefore to larger values of the magnetic field.

An NMR spectrum is ordinarily a graph in which a spectral line is represented by a peak and the intensity of the line is proportional to the area under the peak. Since the differences between the energies of nuclear spin states are all much smaller than $k_B T$, the spin states are nearly equally populated.

▼ **EXAMPLE 15.15**

Find the ratio of the populations of the two spin states of protons in a magnetic field of 1.4092 T at 298 K.

Solution

Population ratio $= e^{-\Delta E/k_B T} = e^{-g_p \beta_N B/k_B T}$

$$= \exp\left[\frac{-(5.5857)(5.05079 \times 10^{-27} \text{ J T}^{-1})(1.4092 \text{ T})}{(1.3807 \times 10^{-23} \text{ J K}^{-1})(298 \text{ K})}\right]$$
$$= e^{-9.66 \times 10^{-6}} = 0.99999034$$

▲

Since the different spin states are nearly equally populated, the area under a peak is proportional to the number of nuclei in one molecule producing the peak. This allows us to determine the relative number of nuclei in each type of electronic environment.

▼   **E X A M P L E 15.16**

Example Figure 15.16 shows schematically the low-resolution proton NMR spectrum of 1-propanol. Interpret this spectrum.

Solution

The molecule contains protons in four kinds of electronic environments. The proton on the oxygen is bonded to an electronegative atom and has a smaller shielding constant and a large value of δ. The first peak from the left, with $\delta = 5.8$ ppm and relative area unity, is due to this proton. The protons bonded to the carbinol carbon produce the second peak from the left, with $\delta = 3.6$ ppm and area twice that of the first peak. The two protons on the second carbon are still more distant from the electronegative oxygen atom and produce the third peak from the left, with $\delta = 1.5$ ppm. The protons in the methyl group are most distant from the oxygen and produce the peak with $\delta = 8$ ppm and area three times that of the first peak.

▲

Exercise 15.23

Sketch the low-resolution proton NMR spectrum you would expect from propanal (propionaldehyde).

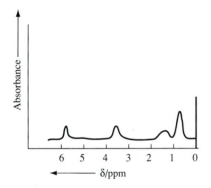

Example Figure 15.16. The Low-Resolution Proton NMR Spectrum of 1-Propanol (simulated). The line near 6 ppm is from the hydroxyl proton, the line near 3.5 ppm from the methylene closest to the hydroxyl group, the line near 1.5 ppm from the other methylene, and the line near 0.8 ppm from the methyl group.

The second molecular contribution to the magnetic field at a given nucleus is due to the presence in the molecule of other nuclei with magnetic dipoles. The interaction between magnetic dipoles is of short range, so the direct interaction between one nuclear dipole and another is usually not important. However, a magnetic dipole in one nucleus induces a magnetic dipole in the electrons of the molecule that move close to both this nucleus and a second nucleus, so the field at the jth nucleus has a **spin-spin coupling** contribution from other nuclei that is transmitted through the electrons:

$$B_{\text{spin-spin},\, j} = \sum_{i=1}^{n} J_{ji} M_{Ii} \qquad (15.8\text{-}39)$$

where the sum includes a term for each other nucleus with nonzero spin. M_{Ii} is the quantum number for the z projection of the spin angular momentum of nucleus i, and the coefficient J_{ij} is called the **spin-spin coupling constant** for nuclei i and j. The spin-spin coupling constant depends on the electronic environments of both nuclei and the distance between them and falls off rapidly for nuclei at larger distances. It is found that nucleus j affects nucleus i in the same way as nucleus i affects nucleus j, so that

$$J_{ij} = J_{ji} \qquad (15.8\text{-}40)$$

The spin-spin coupling constants are sometimes expressed in terms of frequency. The value is the same number of parts per million of the instrument's frequency as the value in Equation (15.8-39) is of the field used in the instrument.

In the case of proton NMR in organic molecules, almost all of the carbons are carbon-12 atoms, which have no magnetic dipole. Almost all of the oxygens are oxygen-16 atoms, which also have no magnetic dipole. If no other atoms with spin are present, only the other protons provide spin-spin coupling for a given proton.

The spin-spin coupling leads to splitting of spectral lines into multiple lines that can be seen in a high-resolution spectrum. For example, in a molecule that contains no magnetic nuclei except for two protons, the field at the first proton is

$$B_1 = B_0(1 - \sigma_1) + J_{12}M_{I2} \qquad \textbf{(15.8-41)}$$

In a sample of many molecules roughly half will have $M_{I2} = 1/2$ and the other half will have $M_{I2} = -1/2$, so there will be two values of B_0 at which some of the molecules have the first proton at the resonant value of the magnetic field. If the chemical shifts of the two protons differ by an amount much larger than J_{12}, both transitions will be observed and two spectral lines will be observed, with a difference in B_0 at resonance equal to J_{12}. This is the **spin-spin splitting** that arises from the spin-spin coupling.

Although there are two lines, each molecule contributes to only one line. Some molecules have the second proton in the spin-up state and some have it in the spin-down state. The second proton will exhibit two lines instead of one in the same way as the first proton. Because of Equation (15.8-40), the splitting will be the same as with the lines of the first proton.

The fact that the spin-spin splitting induced in the absorption of one nucleus by another nucleus is the same as the splitting in the other direction is a considerable help in interpreting NMR spectra.

Here we have coupling without splitting. It is also true that some nuclei generally couple without splitting. For example, ^{35}Cl nuclei and protons couple but do not produce spin-spin splitting.

Another way to state this rule is that two protons will have appreciable spin-spin splitting if the number of bonds from one to the other is no more than three.

If the difference in chemical shift between protons on two different atoms is not large compared with the splitting constant, one does not get a splitting exactly equal to J_{12} and the two lines are not of equal intensity. In the limit that two protons have exactly the same chemical shift, although spin-spin coupling occurs, the transitions are governed by a selection rule that forbids transitions in such a way that spin-spin splitting is not observed.[19]

Spin-spin coupling is a short-range phenomenon. Two protons that are bonded to two atoms that are bonded directly together will ordinarily exhibit spin-spin splittings. Protons in aliphatic compounds that are more distant from each other than this usually do not exhibit significant spin-spin splitting. However, in aromatic compounds spin-spin splitting from meta or para protons is observed because of the delocalized bonding.

If there are n protons on the second carbon atom, there are $n + 1$ possible values for the sum in Equation (15.8-39). There can be n protons with spin up, there can be $n - 1$ with spin up, and so on down to no protons with spin up. This means that the spectral line of protons on the first carbon atom will be split into $n + 1$ lines.

Since the energy differences are much smaller than $k_B T$, the intensities of the lines will be proportional to the degeneracies of the levels. The degeneracy of the level with m protons having spin up out of a set of n protons is the number of ways of choosing a subset of m members out of a set of n members:

$$\text{Degeneracy} = \frac{n!}{m!(n - m)!} \qquad \textbf{(15.8-42)}$$

which is the formula for the binomial coefficients. Two protons on the first atom will thus cause the line of a proton on an adjacent atom to be split into three lines with intensities in the ratio 1:2:1, three protons will produce four lines with intensities in the ratios 1:3:3:1, etc.

---

[19] *ibid.*, pp. 110ff.

If a proton is bonded to an atom that is bonded to two or three adjacent atoms to which protons are bonded, more than one value of the splitting constant can be involved, since the splitting constants depend on the electronic environment of both protons and on the distance between the protons. One way to predict the effect of spin-spin coupling in such a case is to divide the other protons into groups of equal coupling constants. First determine the splitting due to the protons in one group. Then split each resulting line according to the splittings of the protons in the next group, etc.

▼   **EXAMPLE 15.17**

Predict the high-resolution proton NMR spectrum of pure 2-propanol.

Solution

The spectral line of the —OH proton will be split into two lines by the proton on the —CH— carbon. The spectral line of the proton on the —CH— carbon will be split into seven lines by the six protons on the —CH_3— carbons, with relative intensities 1:6:15:20:15:6:1, and each of these lines will be split into two lines by the single —OH proton. Example Figure 15.17 shows the high-resolution spectrum.

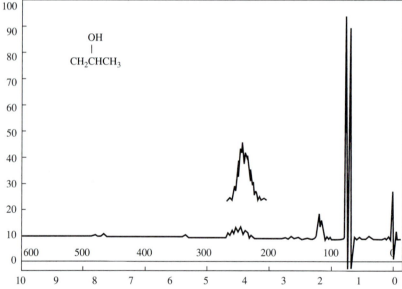

Example Figure 15.17. The High-Resolution Proton NMR Spectrum of 2-Propanol. This spectrum is from a sample dissolved in deuterated chloroform. Because of solvent interaction, the line from the hydroxyl proton occurs near 2 ppm instead of farther downfield. The peak near 4 ppm is from the proton on the carbon attached to the hydroxyl group. It is split into seven lines by the six methyl protons, and each of these lines is split into two lines by the hydroxyl proton. These 14 lines overlap and are not completely resolved. From Charles J. Pouchert and John R. Campbell, *The Aldrich Library of NMR Spectra*, Vol. I, Aldrich Chemical Co., 1974, p. 85.

The shielding term in Equation (15.8-41), which gives rise to the chemical shift, is proportional to the applied field, whereas the spin-spin coupling term is independent of the applied field. Consider two instruments with different frequencies, say a 60-megahertz instrument and a 300-megahertz instrument. In terms of frequency, the spin-spin splitting terms will be the same in both spectra but the chemical shifts will be larger in the higher-frequency instrument, which uses a larger magnetic field. In terms of parts per million of the magnetic field, the chemical shift will be the same and the spin-spin splittings will be smaller in the higher-frequency instrument. Comparing the spectra of the same substance from two such instruments can aid in interpreting a spectrum with a number of overlapping sets of lines.

If a substance is mixed with other substances, they can modify its chemical shifts. Figure 15.22 shows three different high-resolution proton NMR spectra of ethanol. Figure 15.22a shows that of nearly pure ethanol, but apparently with a trace of water present. Figure 15.22b shows the spectrum of carefully purified ethanol, and Figure 15.22c shows the spectrum of a dilute solution of ethanol in deuterated chloroform.

In the first spectrum, the spectral line of the —OH proton is at $\delta = 4.8$ ppm. In the carefully purified ethanol, this peak is in approximately the same place, at $\delta = 5.3$ ppm. In the solution, this spectral line is at $\delta = 2.3$ ppm. Ethanol molecules strongly hydrogen-bond to other ethanol molecules. In the solution in deuterated chloroform, the ethanol molecules are distant from each other and cannot form hydrogen bonds. In the absence of hydrogen bonding, the hydroxyl proton, although adjacent to the electronegative oxygen atom, has a larger shielding constant than the protons on

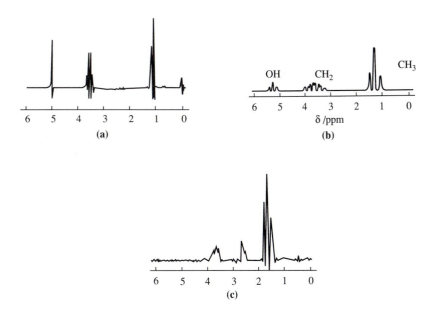

Figure 15.22. The Proton NMR Spectrum of Ethanol. (a) With a trace of water present. Because of exchange of the hydroxyl proton with water protons, the hydroxyl proton shows no spin-spin splitting. From Gilbert W. Castellan, *Physical Chemistry*, 3rd ed., Addison-Wesley, Reading, MA, 1983, p. 606. **(b)** Highly purified ethanol. In the absence of water, the spin-spin splitting of the line from the hydroxyl proton is split into three lines by the methylene protons. From Ira N. Levine, *Molecular Spectroscopy*, Wiley, New York, 1975, p. 353. **(c)** In deuterated chloroform. In this spectrum, the solvent interaction with chloroform moves the hydroxyl line to near 2.4 ppm. The splitting into three lines is not quite resolved. From Charles J. Pouchert and John R. Campbell, *The Aldrich Library of NMR Spectra*, Vol. I, Aldrich Chemical Co., 1974, p. 79.

the carbinol carbon. The O—H bond distance, approximately 96 pm, is so small that the proton is "buried" in the electron density close to the oxygen. When a hydrogen bond forms, the proton is "deshielded," moving to a greater distance from the oxygen nucleus and to a region of lower electron density.

There is another difference between the two spectra. In the spectrum of the carefully purified ethanol, the line of the —OH proton is split into three lines by spin-spin splitting and also splits the —CH$_2$ protons' lines. In the solution spectrum this splitting is also present, although poorly resolved. In the first spectrum, the absence of spin-spin splitting in the —OH proton line is due to exchange of the proton with a small amount of water present in the ethanol. The alcohol molecules, which are slightly acidic, can exchange their —OH protons with protons on water molecules. If the average time for this exchange process is shorter than the period of oscillation of the NMR radiation, the proton NMR spectrum is an average of the spectrum of the proton that is leaving and the proton that is arriving. Since these protons are a random mixture of spin-up and spin-down protons, an average line with no spin-spin splitting is obtained.

15.9   Fourier Transform Spectroscopy

Fourier transform spectroscopy is a technique that is used in both infrared and NMR spectroscopy. A spectrum is obtained without having to disperse radiation and to scan over the different frequencies or magnetic fields. No new physical phenomena are observed, but since it is not necessary to spend time scanning, an instrument can take a spectrum repeatedly in a short time and average the separate spectra, allowing weak spectral lines to be seen and random "noise" to be reduced by the averaging procedure. For example, in ^{13}C NMR, ^{13}C atoms constitute only 1% of the carbon atoms in ordinary carbon and therefore give only a weak spectrum in scanning techniques.

Fourier Series

A **periodic function** is a function that repeats itself at a fixed interval (the period) along the axis of its independent variable. A periodic function of t with a period $2T$ obeys the relation

$$f(t + 2T) = f(t)$$

for any value of t.

A periodic function can be represented by a Fourier series. We have already seen an example of a Fourier sine series in Chapter 9 in connection with a wave in a string. A general Fourier series for a function $I(t)$ of period $2T$ is[20]

$$I(t) = \sum_{n=0}^{\infty} a_n \cos\left(\frac{n\pi t}{T}\right) + \sum_{n=1}^{\infty} b_n \sin\left(\frac{n\pi t}{T}\right) \tag{15.9-1}$$

---

[20] MPC, pp. 123ff.

where the a and b coefficients are called **Fourier coefficients**. The function $I(t)$ is thus represented as a superposition of sine waves and cosine waves, with the strength of each wave being given by the magnitude of its coefficient. The entire set of coefficients contains all of the information contained in the original function, and knowledge of the coefficients allows the function to be reconstructed by summation of the series.

A Fourier series is an example of a linear combination of basis functions as in Equation (10.4-24). Fourier was able to show mathematically that the set of sine and cosine basis functions is complete for any periodic function of period $2T$. That is, if enough terms are taken, an arbitrary function of period $2T$ can be represented to any degree of accuracy. The function does not even have to be continuous.

Exercise 15.24	Using trigonometric identities, show that the series in Equation (15.9-1) is unchanged if t is replaced by $t + 2T$, so the series does represent a periodic function.

Fourier Integrals

If a function is not periodic, it can be considered to be a periodic function in the limit that the period becomes large without limit. In this case, the values of $n\pi t/T$ approach each other more and more closely and the sums are replaced in the limit by integrals, which are called **Fourier transforms** or **Fourier integrals**.

Fourier transform spectrometers operate with a pulse of radiation that contains a large number of frequencies. The intensity of radiation as a function of time can be written as a Fourier transform

$$I(t) = \frac{1}{\sqrt{2\pi}} \int_{-\infty}^{\infty} c(\omega)e^{i\omega t} \, d\omega \qquad \text{(15.9-2)}$$

where $c(\omega)$ is called the Fourier transform of $I(t)$.[21] The variable ω is related to the frequency v of the radiation by

$$\omega = 2\pi v \qquad \text{(15.9-3)}$$

ω is sometimes called the "circular frequency." The Fourier transform $c(\omega)$ contains all of the information contained in $I(t)$, but it gives the intensity as a function of frequency instead of time.

The similarity between Equations (15.9-1) and (15.9-2) is apparent from the identity

$$e^{i\omega t} = \cos(\omega t) + i \sin(\omega t) \qquad \text{(15.9-4)}$$

Substitution of this identity into Equation (15.9-2) gives the Fourier transform in a form more similar to that of Equation (15.9-1).

---

[21] L. Glasser, *J. Chem. Educ.* **64**, A261 (1987) and *J. Chem. Educ.* **64**, A306 (1987).

To determine the function $c(\omega)$, one must "invert" the Fourier transform, which is done by calculating the integral

$$c(\omega) = \frac{1}{\sqrt{2\pi}} \int_{-\infty}^{\infty} I(t)e^{-i\omega t}\, dt \qquad (15.9\text{-}5)$$

Note the similarity between Equations (15.9-2) and (15.9-5). Since they differ only in the sign of the exponent, $I(t)$ is also called the Fourier transform of $c(\omega)$.

EXAMPLE 15.18

Find the Fourier transform of the function

$$f(\omega) = \frac{1}{a^2 + \omega^2}$$

where a is a positive constant.

Solution

$$I(t) = \frac{1}{\sqrt{2\pi}} \int_{-\infty}^{\infty} \frac{e^{i\omega t}}{a^2 + \omega^2}\, d\omega = \frac{1}{\sqrt{2\pi}} \int_{-\infty}^{\infty} \frac{\cos(\omega t) + i\sin(\omega t)}{a^2 + \omega^2}\, d\omega$$

The real part of the integrand is an even function, and the imaginary part is an odd function. The imaginary part will vanish upon integration, and the even part will give twice the value of the integrand from 0 to ∞:

$$I(t) = \frac{2}{\sqrt{2\pi}} \int_{0}^{\infty} \frac{\cos(\omega t)}{a^2 + \omega^2}\, d\omega = \frac{2}{\sqrt{2\pi}} \frac{\pi}{2a} e^{-a|t|} = \sqrt{\frac{\pi}{2}} \frac{1}{a} e^{-a|t|}$$

where we have looked the integral up in a table.[22]

In a Fourier transform infrared spectrometer, the pulse of infrared radiation is passed through an **interferometer**, as depicted in Figure 15.23.[23] The beamsplitter divides the beam into two beams of equal intensity, which are recombined after being reflected by separate mirrors, one of which is movable. As this mirror moves, the detector responds to changes in the intensity as the two beams interfere constructively or destructively. The intensity of the pulse as a function of time (the **interferogram**) is the Fourier transform of the spectrum. As shown in the figure, a single frequency produces a sinusoidal interferogram, and an interferogram corresponding to a number of frequencies encodes the intensities of the various frequencies. A computer that is programmed to carry out a fast Fourier transform algorithm recovers the spectrum from the interferogram.

A spectrum is taken without a sample in position and another is taken with a sample in position, and the difference in the two spectra gives the absorption spectrum of the sample. An interferogram can be taken in less than 1 second using a deuterated triglycine sulfate (DTGS) detector, and as

---

[22] H. B. Dwight, *Tables of Integrals and Other Mathematical Data*, 4th ed., Macmillan, New York, 1961.

[23] W. D. Perkins, *J. Chem. Educ.* **63**, A5 (1986); **64**, A269 (1987); and **64**, A296 (1987).

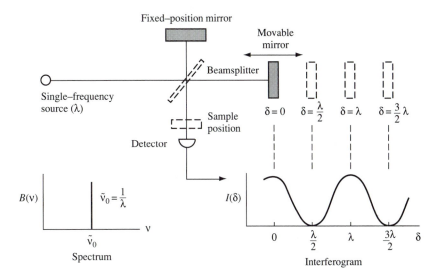

Figure 15.23. The Radiation Path in a Fourier Transform Infrared Spectrophotometer (schematic). If only one wavelength is present, the sinusoidal interferogram at the lower right is observed as the movable mirror in the Michelson-Morley interferometer changes its position. The Fourier transform of this signal is a single sharp line, as shown at the lower left. If many frequencies are present, the Fourier transform will show all frequencies present in the proper intensities. From W. D. Perkins, *J. Chem. Educ.* **63**, A5 (1986).

many as 20 or 30 interferograms can be taken in a second with a detector cooled with mercury–cadmium telluride.

Fourier-transform NMR spectroscopy can be described qualitatively in terms of the semiclassical picture of **Larmor precession**.[24] In a magnetic field, each nuclear spin precesses (moves around the cone of possible directions as shown in Figure 11.14 for an electron). A calculation of this precession frequency is assigned in Problem 15.48. A strong pulse of radiofrequency radiation is delivered to the sample, much like the pulse of infrared radiation used in Fourier transform infrared spectroscopy. The pulse aligns some of the precessing nuclear spins, so there is an oscillating magnetization in the sample. The length of the pulse is carefully controlled so that the magnetization vector becomes perpendicular to the z axis (the direction of the external magnetic field) at the end of the pulse.

After the end of the pulse, the magnetization vector precesses with the Larmor precession of the magnetic moments of the nuclei. This induces an alternating voltage in a coil around the sample. This voltage as a function of time is the signal whose Fourier transform is the NMR spectrum. The detected signal is called the free induction decay spectrum (FIDS).

The interested reader can read more about Fourier transform infrared and NMR spectroscopy in some of the works listed at the end of the chapter, as well as in the references of footnotes 23 and 24.

[24] R. S. Macomber, *J. Chem. Educ.* **62**, 212 (1985).

Summary

Electromagnetic radiation that is absorbed or emitted by atoms or molecules gives information about energy level differences through the Bohr frequency rule

$$E_{\text{photon}} = h\nu = \frac{hc}{\lambda} = E_{\text{upper}} - E_{\text{lower}}$$

However, each pair of energy levels does not necessarily lead to a spectral line for emission or absorption. Selection rules tell whether a transition with emission or absorption of radiation can occur between a given pair of energy levels.

Transitions between rotational states lead to emission or absorption in the microwave region. For diatomic and linear polyatomic molecules with permanent dipole moments, the selection rule is

$$\Delta J = \pm 1$$

which leads to a spectrum of equally spaced lines with a spacing in terms of reciprocal wavelength equal to $2\tilde{B}_e$. Molecules with no permanent dipole moment give no microwave spectrum. The microwave spectra of nonlinear polyatomic molecules are more complicated, but a permanent dipole moment must be present for a microwave spectrum to occur.

Transitions between vibrational levels lead to spectra in the infrared region. For diatomic molecules with permanent dipole moments, the selection rule is

$$\Delta v = 0, \pm 1$$

leading to a "fundamental" band centered at a reciprocal wavelength equal to $\tilde{\nu}_e$. Since the selection rule is only an approximation, "overtone" bands at multiples of this reciprocal wavelength also occur.

The infrared spectra of polyatomic molecules contain one fundamental band for each normal mode whose motion modulates the dipole moment of the molecule. Normal modes that do not modulate the dipole moment of the molecule are not seen in the infrared spectrum. Overtone bands occur as with diatomic molecules, along with combination bands, which are produced when two normal modes make simultaneous transitions.

Atomic and molecular spectra in the visible and ultraviolet regions arise from transitions from one electronic state to another. Vibrational and rotational transitions occur simultaneously with the electronic transitions, producing complicated band spectra. The electronic transitions take place rapidly compared with rotational and vibrational periods and conform to the Franck-Condon principle: the nuclei remain nearly stationary during the transition.

Raman spectroscopy involves inelastic scattering of light instead of absorption or emission. The selection rules for Raman transitions are different from those of absorption and emission spectroscopy, and many transitions that are forbidden in absorption and emission occur in Raman scattering. Raman scattering requires that the motion modulate the polarizability of the molecule. For rotational transitions in diatomic and linear polyatomic molecules

$$\Delta J = 0, \pm 2$$

which leads to a rotational Raman spectrum with lines whose reciprocal wavelengths are equally spaced with a spacing of $4\tilde{B}_e$.

Almost every diatomic molecule has a vibrational Raman spectrum. The vibrational selection rule for diatomic molecules is

$$\Delta v = \pm 1$$

A nonlinear polyatomic molecule exhibits a rotational Raman spectrum only if it has different values of the polarizability in different directions. Most molecules exhibit a rotational Raman spectrum, except for highly symmetric molecules such as spherical tops.

A normal mode must modulate the polarizability to be seen in the vibrational Raman spectrum. The "rule of exclusion" states that in a molecule with a center of symmetry, normal modes not seen in the infrared spectrum will be seen in the Raman spectrum, and those seen in the infrared spectrum will not be seen in the Raman spectrum.

Magnetic resonance spectroscopy is absorption spectroscopy in which the sample substance is placed in a magnetic field, causing spin states that are degenerate in the absence of the field to have different energies. The two principal types of magnetic resonance spectroscopy are electron spin resonance (ESR) and nuclear magnetic resonance (NMR). In ESR spectroscopy, a substance with unpaired electrons absorbs radiation of a fixed frequency when the externally applied magnetic field is such that the magnetic field is

$$B_{\text{res}} = \frac{h\nu}{g\beta_e}$$

where g is known as the g factor of the electron and β_e is called the Bohr magneton. Since the magnetic field "felt" by the electrons includes a term due to nuclear spins in the molecule, structural information can be obtained from the ESR spectrum.

NMR spectroscopy involves transitions of nuclear spins from one state to another in a magnetic field with absorption of radiation. The applied magnetic field must be such that the magnetic field at a given nucleus is

$$B_{\text{res}} = \frac{h\nu}{g_N\beta_N}$$

where g_N is the nuclear g factor for the particular nucleus and β_N is the nuclear magneton (same for every nucleus).

The field at a given nucleus contains two contributions in addition to the externally applied field. One is due to shielding of the electrons around the nucleus and is expressed by the chemical shift. The other is due to the presence of other spins in the molecule and gives rise to spin-spin splitting of spectral lines. Interpretation of the chemical shifts and spin-spin splittings gives structural information.

Additional Reading

R. J. Abraham and P. Loftus, *Proton and Carbon-13 NMR Spectroscopy*, Heyden, London, 1978
This book is an integrated approach to the two principal types of NMR spectroscopy, carbon-13 and proton NMR.

Robert G. Bacher and Samuel Goudsmit, *Atomic Energy States*, McGraw-Hill, New York, 1932
This is a compilation of numerical energy level values.

Gordon M. Barrow, *Introduction to Molecular Spectroscopy*, McGraw-Hill, New York, 1962
A general introductory text at the senior undergraduate/first year graduate level.

Alan Carrington and Andrew D. McLachlan, *Introduction to Magnetic Resonance*, Harper & Row, New York, 1967
A clear and authoritative treatment of both ESR and NMR spectroscopy.

Norman B. Colthup, Lawrence H. Daly, and Stephen W. Wiberley, *Introduction to Infrared and Raman Spectroscopy*, 2nd ed., Academic Press, New York, 1975
This book includes considerable information about instrumentation in addition to spectroscopic theory and is well suited for use by analytical chemists.

Wolfgang Demtroder, *Laser Spectroscopy—Basic Concepts and Instrumentation*, Springer-Verlag, Berlin, 1981
A modern account of various types of spectroscopy using lasers as sources of incident radiation. Fourier transform spectroscopy is included, as is multiphoton spectroscopy.

Russell S. Drago, *Physical Methods in Chemistry*, W. B. Saunders, Philadelphia, 1977
The bulk of this book is devoted to spectroscopy of various sorts, but it also includes chapters on molecular orbital theory, group theory, and X-ray crystallography.

Thomas C. Farrar and Edwin D. Becker, *Pulse and Fourier Transform NMR*, Academic Press, New York, 1971
This brief book on the subject includes information about instrumentation as well as theory.

Gerhard Herzberg, *Molecular Spectra and Molecular Structure*, Vol. I, *Spectra of Diatomic Molecules*, 2nd ed., Van Nostrand, Princeton, NJ, 1950
Vol. II, *Infrared and Raman Spectra of Polyatomic Molecules*, Van Nostrand Reinhold, New York, 1945
Vol. III, *Electronic Spectra and Electronic Structure of Polyatomic Molecules*, Van Nostrand Reinhold, New York, 1966
Vol. IV, *Constants of Diatomic Molecules*, by K. P. Huber and G. Herzberg, Van Nostrand Reinhold, New York, 1979
This set of books is the most authoritative information source for molecular spectroscopy, both in theory and in critically evaluated experimental data on molecular structure.

William G. Laidlaw, *Introduction to Quantum Concepts in Spectroscopy*, McGraw-Hill, New York, 1970

This book at the level of undergraduate physical chemistry interprets spectroscopy in terms of qualitative quantum mechanical concepts.

Ira N. Levine, *Molecular Spectroscopy*, John Wiley & Sons, New York, 1975

A clearly written and authoritative text for a graduate-level course in molecular spectroscopy.

J. A. Pople, W. G. Schneider, and H. J. Bernstein, *High-Resolution Nuclear Magnetic Resonance*, McGraw-Hill, New York, 1959

This book presents an authoritative account of NMR theory as of the date of its publication and is still very useful.

Nicholas J. Turro, *Modern Molecular Photochemistry*, Benjamin/Cummings, Menlo Park, CA, 1978

An authoritative account of photochemistry written from the perspective of physical organic chemistry.

Harvey Elliott White, *Introduction to Atomic Spectra*, McGraw-Hill, New York, 1934

Now primarily useful only as a source of data on atomic energy levels.

See also the references listed at the end of Chapters 9–14.

PROBLEMS

Problems for Section 15.1

*15.25. A solution of phenylalanine in neutral water with concentration of 0.110×10^{-3} mol L^{-1} has an absorbance at 206.0 nm of 1.027 in a 1.000-cm cell. Find the molar absorptivity.

15.26. The absorptivity of hemoglobin at 430 nm is found to be 532 L mol^{-1} cm^{-1}. The molar mass of hemoglobin is 68000 g mol^{-1}.

  a. The concentration of hemoglobin inside red blood cells is approximately 17% by mass. Estimate the absorbance of such a solution in a cell of length 1.000 cm.

  b. Find the concentration in mol L^{-1} and in percent by mass for a solution with an absorbance of 1.00 at 430 nm in a cell of length 1.000 cm.

Problems for Section 15.2

*15.27. A positronium atom consists of a positron, which is an antiparticle with the same mass as an electron and the same charge as a proton, and one electron. Find the wavelengths of the photons emitted in the following "electronic" transitions: (a) $n = 3$ to $n = 2$, (b) $n = 4$ to $n = 2$, (c) $n = 5$ to $n = 2$, and (d) $n = 6$ to $n = 2$. Compare these wavelengths with those of a normal hydrogen atom.

15.28. a. Consider the excited states of the helium atom that arise from the $(1s)(2s)$ and $(1s)(2p)$ configurations. These states were discussed in Chapters 11 and 12. Draw a Grotrian diagram for these states and the ground state of the He atom.

  b. Using energies from Figure 12.3, estimate the wavelengths at which spectral lines would be found from the transitions of part a.

Problems for Section 15.3

15.29. Using an identity from Appendix F, derive the selection rule for a harmonic oscillator, $\Delta v = 0, \pm 1$.

15.30. Find a formula for the rotational frequency (number of revolutions per second) of a rigid diatomic molecule, assuming that classical mechanics holds but that the angular momentum happens to have the magnitude $\hbar\sqrt{J(J+1)}$. Compare this formula with that for the frequency of a photon absorbed when a quantum mechanical molecule makes a transition from J to $J + 1$. Show that the two frequencies are nearly equal for large values of J.

*15.31. Using information on the normal HF molecule from Table A16, predict the reciprocal wavelengths of the absorptions in the microwave spectra of DF, where D is deuterium, ^{2}H. Assuming equal oscillator strengths, find the line of maximum intensity at 298 K.

15.32. Which of the following substances will have a microwave spectrum?

a. CO_2	**d.** $CHCl_3$	**g.** BH_3
b. N_2O	**e.** CH_2Cl_2	**h.** NH_3
c. CCl_4	**f.** CH_3Cl	**i.** C_2H_4

15.33. Using the $\tilde{\nu}_e$ and $\tilde{\nu}_e x_e$ values from Table A16, find the reciprocal wavelength of the band center of the fundamental band, the first overtone band, and the second overtone band of $^1H^{19}F$. Find the reciprocal wavelength of the center of the (high-temperature) band corresponding to the transition from $v = 1$ to $v = 2$.

***15.34.** Using values of parameters in Table A16, find the reciprocal wavelength of the radiation absorbed in the transition from the $v = 0$, $J = 6$ state to the $v = 1$, $J = 7$ state of $^1H^{81}Br$. To which branch does this line belong, and how many lines lie between it and the band center?

Problems for Section 15.5

15.35. The infrared spectrum of HCN shows strong bands at 712.1 cm^{-1} and at 3312.0 cm^{-1}. There is a strong Raman band at 2089.0 cm^{-1}. There are weaker infrared bands at 1412.0 cm^{-1}, 2116.7 cm^{-1}, 2800.3 cm^{-1}, 4004.5 cm^{-1}, 5394 cm^{-1}, and 6521.7 cm^{-1}. Identify these bands as fundamental, overtone, or combination bands and give the shape of the molecule.

15.36. The N_2O molecule has three strong bands in its IR spectrum, at 588.8 cm^{-1}, 1285.0 cm^{-1}, and 2223.5 cm^{-1}. All have been shown to be fundamentals, and the molecule has been shown to be linear. Explain why CO_2, which is also linear, has only two fundamental IR bands while N_2O has three.

***15.37.** Using the frequencies in Problem 15.36, tell where to look for overtone and combination bands in the spectrum of N_2O.

Problems for Section 15.6

15.38. In a photochemical reaction, 0.00100 watt of radiant energy of wavelength 254 nm is incident on the reaction vessel, and 61.2% of this energy is absorbed. If 2.45 mol of the absorbing reactant reacts to form products in 30.0 seconds, find the quantum yield.

***15.39.** Find the frequency and wavelength of light given off by the benzophenone molecule in (a) fluorescence from the S_1 state and (b) phosphorescence from the T_1 state.

15.40. Which of the following substances could quench the photochemical reaction of benzophenone discussed in Section 15.6?

Substance	Energy of First Triplet State
Biphenyl	2.9 eV
Toluene	3.6 eV
Benzene	3.7 eV
9,10-Diphenylanthracene	1.8 eV
trans-1,3-Pentadiene	2.6 eV

Problems for Section 15.7

15.41. The rotational Raman spectrum of H_2 has lines at Raman shifts of 354 cm^{-1}, 586 cm^{-1}, 814 cm^{-1}, and 1034 cm^{-1}. At room temperature, the line of greatest intensity is at 586 cm^{-1}. Explain the relative intensities and calculate the internuclear distance for H_2.

15.42. Which of the following substances will have a rotational Raman spectrum?

   a. CCl_4 **d.** SF_6
   b. BF_3 **e.** PCl_5
   c. CO_2 **f.** NH_3

15.43. Which of the substances in Problem 15.32 will have a rotational Raman spectrum?

Problems for Section 15.8

15.44. The manganese nucleus has a spin quantum number $I = 5/2$. Describe the ESR spectrum of manganese atoms.

15.45. The Hückel molecular orbital method gives the following LCAO-MO for the unpaired electron in the naphthalene negative ion:

$$\psi = 0.42536(-\psi_1 + \psi_4 - \psi_5 + \psi_8)$$
$$+ 0.26286(-\psi_2 + \psi_3 - \psi_6 + \psi_7)$$

where ψ_i is the nonhybridized p orbital on carbon i, numbered as in the diagram:

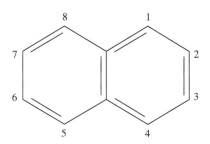

Assume that the coupling constant for each hydrogen is approximately proportional to the square of the coefficient for the carbon on which the hydrogen is bonded, and describe the ESR spectrum of the naphthalene negative ion.

***15.46.** For electrons in a magnetic field such that ESR absorption occurs at a frequency of $9.159 \times 10^9 \text{ s}^{-1}$, calculate the ratio of the populations of the two spin states at 298 K.

15.47. Tell qualitatively what the proton NMR spectrum of carefully purified acetic acid is like at high resolution. Describe the spectrum of acetic acid with a small amount of water present. Describe the spectrum of a dilute anhydrous solution of acetic acid in deuterated chloroform.

15.48. According to classical mechanics, a magnetic dipole μ in a magnetic field **B** precesses about the field direction.

That is, the direction of the dipole vector traces out a cone with the axis of the cone in the direction of the field. This is called the **Larmor precession**. If the magnetic dipole is produced by the orbiting of a particle of charge Q and mass m about a center, its Larmor frequency (number of revolutions about the cone per second) is given by

$$\nu_{Larmor} = \frac{1}{2\pi} \frac{Q}{2m} B$$

The Larmor precession frequency of the magnetic moment of a proton is given by

$$\nu_{Larmor} = \frac{1}{2\pi} g_p \frac{e}{2m_p} B$$

Find the Larmor precession frequency of a proton in a magnetic field of 1.4092 T. Compare it with the frequency of the radiation absorbed by protons in an NMR apparatus at this field. (See Example 15.14 for the radiation frequency.) Comment on why the term "magnetic resonance" is an appropriate name for the interaction of the radiation and the precessing spins.

*15.49. Nuclear magnetic moments are sometimes expressed in terms of the **magnetogyric ratio** γ, defined by

$$|\boldsymbol{\mu}| = \gamma h [I(I+1)]^{1/2}$$

This quantity has a different value for each nuclide. Find the values of the magnetogyric ratio for protons and for ^{13}C nuclei.

Problems for Section 15.9

15.50. a. Find the Fourier transform $c(\omega)$ of the function

$$I(t) = Ae^{-(t-t_0)^2/D}$$

where A, t_0, and D are constants.

b. Sketch graphs of $I(t)$ and $c(\omega)$.

c. Explain in physical terms what $c(\omega)$ represents if $I(t)$ represents the intensity of a pulse of radiation as a function of time. Explain why $c(\omega)$ depends on t_0 as it does, and describe what happens if $t_0 = 0$.

15.51. Find the Fourier transform $I(t)$ of the function $e^{-a|\omega|}$.

15.52. Find the inverse Fourier transform $c(\omega)$ of the function $I(t) = e^{-at^2} \sin(bt)$.

General Problems

*15.53. a. Calculate the reciprocal wavelength of the band center of the fundamental vibrational band of the CO molecule, using the expression for the corrected energy levels in Equation (14.2-40) and information from Table A16. Compare your value with the value from the uncorrected energy level expression.

b. Calculate the reciprocal wavelength of the band center of the first overtone band of the CO molecule, using the

expression for the corrected energy levels and information from Table A16. Compare this reciprocal wavelength with double the reciprocal wavelength from part a. Compare your value with the value from the uncorrected energy level expression.

c. Calculate the reciprocal wavelength of the first line of the microwave spectrum of the CO molecule using the corrected energy level expression. Compare your value with the value from the uncorrected energy level expression.

d. Calculate the reciprocal wavelength of the second line of the microwave spectrum of the CO molecule using the corrected energy level expression. Compare your value with the value from the uncorrected energy level expression.

15.54. a. Write a computer program to calculate the reciprocal wavelength of the spectral lines in the fundamental band of the vibration-rotation spectrum of a diatomic molecule, using the corrected energy level expression of Equation (14.2-40). Use the program to calculate the reciprocal wavelengths of the band center, the first 15 lines in the P branch, and the first 15 lines in the R branch of the HCl fundamental band, using information from Table A16.

b. Use the program to repeat the calculation of part a for the first overtone band of HCl.

c. Modify your program to calculate the reciprocal wavelengths of the lines in the microwave spectrum of a diatomic molecule. Use the program to calculate the reciprocal wavelengths of the first 15 lines of the microwave spectrum of HCl.

15.55. Write a computer program to calculate the relative intensities of the spectral lines in the fundamental band of the vibration-rotation spectrum of a diatomic molecule, assuming that the absorbance is displayed in the spectrum. Set the maximum absorbance of the first line of the P branch equal to 1. Assume the Boltzmann probability distribution and assume that the transition dipole moments for all transitions are equal. Use your program to calculate the relative intensities for the first 15 lines in each branch of the HCl spectrum at 298 K.

15.56. Using information from Table A16, consider the HF molecule.

*a. Find the value of the moment of inertia.

*b. Find the value of the force constant.

c. Draw a simulated microwave spectrum, assuming that the temperature is 298 K and using reciprocal centimeters as the independent variable. Use the uncorrected energy levels. Assume that the transition dipole moments are roughly equal, and show the correct line of maximum intensity.

d. Repeat part c, using the corrected energy levels. If you have done Problems 15.54 and 15.55, use your computer programs to calculate the reciprocal wavelengths and intensities.

e. Draw a simulated infrared spectrum of the fundamental band, assuming that the temperature is 298 K and using reciprocal centimeters as the independent variable. Use the uncorrected energy levels. Assume that the transition dipole moments are roughly equal, and show the line of maximum intensity in each branch.

f. Repeat part e with corrected energy levels.

g. Draw a simulated rotational Raman spectrum of HF, showing the Stokes lines. Assume that the temperature is 298 K, and plot the Raman shift on the horizontal axis, using reciprocal centimeters as the independent variable. Use the uncorrected energy levels.

15.57. Tell how you would distinguish between each pair of substances, using spectroscopic techniques. If possible, specify two different techniques.

a. Dimethyl ether and ethanol

b. 3-Pentanone and 2-pentanone

c. 1-Chloropropane and 2-chloropropane. (The spin-spin coupling between protons and chlorines does not lead to spin-spin splitting.)

d. Methyl acetate and propanoic acid

e. Acetone (propanone) and propionaldehyde (propanal)

f. $^1H^{35}Cl$ and $^2H^{35}Cl$.

g. $HC{\equiv}CH$ and $HC{\equiv}CCl$

*15.58. Label each statement as either true or false. If a statement is true only under special circumstances, label it as false.

a. A forbidden transition cannot occur.

b. A forbidden transition always produces a weaker spectral line than every allowed transition.

c. All forms of spectroscopy require that the radiation is dispersed.

d. All molecules exhibit an absorption spectrum in some part of the electromagnetic spectrum.

e. Any molecule that has a rotational Raman spectrum does not have a microwave absorption spectrum.

f. Any molecule that has a vibrational Raman spectrum does not have an infrared absorption spectrum.

g. Raman spectroscopy requires the use of a laser.

h. All infrared spectroscopy is absorption spectroscopy.

i. Every molecule exhibits an electron spin resonance spectrum.

j. Every molecule exhibits a nuclear magnetic resonance spectrum.

k. A nonpolar molecule cannot have an infrared spectrum.

l. A molecule that is a spherical top has $3n - 6$ vibrations that lead to infrared spectral lines.

m. A molecule that is a spherical top has the same kind of infrared spectrum as a heteronuclear diatomic molecule.

n. Most organic substances are colorless in the visible region and also do not absorb in the ultraviolet region.

o. A linear polyatomic molecule has the same type of infrared spectrum as a diatomic molecule.

p. A linear polyatomic molecule has the same kind of microwave spectrum as a diatomic molecule.

Gas Kinetic Theory: The Molecular Theory of Dilute Gases at Equilibrium

PREVIEW

In the present chapter, we carry out a more careful analysis of the relationship between microscopic states and macroscopic states. This analysis is called kinetic theory, and it is applied to a model system designed to represent a dilute gas.

PRINCIPAL FACTS AND IDEAS

1. The model system that represents a dilute gas is a system of noninteracting molecules.
2. The kinetic theory of gases is the mathematical analysis of the model system and includes averages over microscopic states of the molecules of the system.
3. The probability distribution for molecular velocities is the Maxwell probability distribution:

$$\text{Probability of a state of velocity } v \propto e^{-mv^2/2k_\mathrm{B}T}$$

4. The probability distribution for molecular speeds is obtained from the velocity distribution:

$$\text{Probability of a speed } v \propto v^2 e^{-mv^2/2k_\mathrm{B}T}$$

5. Gas kinetic theory predicts the ideal gas equation of state.
6. Gas kinetic theory predicts the rate of wall collisions and the rate of effusion of a dilute gas.
7. The Boltzmann probability distribution is for a nonuniform gas in which there is a molecular potential energy.
8. The hard-sphere gas is a more realistic representation of a dilute gas

than is a gas of noninteracting molecules, and it allows analysis of molecular collisions.

9. The properties of a liquid can be understood qualitatively in terms of intermolecular forces.

16.1   The Model System for a Dilute Gas

Daniel Bernoulli, 1700–1782, Swiss mathematician best known for Bernoulli's principle, which states that the pressure decreases as the flow velocity of a fluid increases.

James Clerk Maxwell, 1831–1879. The Maxwell of the Maxwell relations of thermodynamics and the Maxwell equations of electrodynamics.

John James Waterston, 1811–1883, who at the time of his work on kinetic theory was teaching in India and was unable to get his work published.

In Chapter 1 we defined the two principal types of states of systems: microscopic or mechanical states and macroscopic or thermodynamic states. We asserted that a macroscopic state corresponds to an average over many microscopic states. One argument was that during the response time of a macroscopic instrument, the system must pass through very many microscopic states. All of these states must correspond to the single macroscopic state that corresponds to the macroscopic value measured by the instrument.

We now study gas kinetic theory, which explicitly averages over molecular states of a system of particles obeying classical mechanics. This theory was originated by Bernoulli, who was the first to test the consequences of assuming that a gas was a mechanical system made up of many tiny moving particles. The theory was brought to an advanced state by Joule and Maxwell. The fundamentals were worked out independently by Waterston, about 15 years prior to the work of Joule and Maxwell.

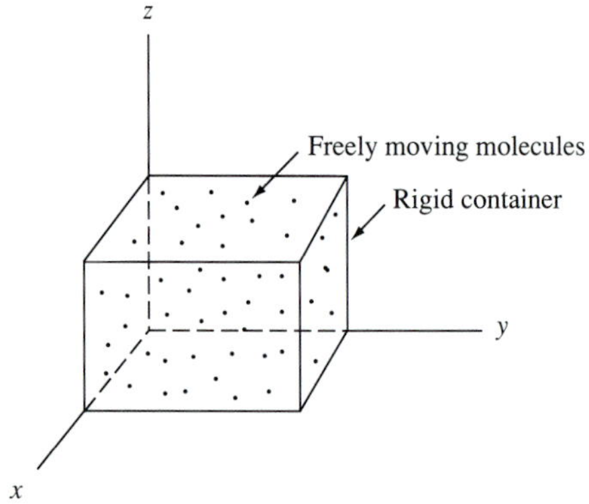

The Model System for a Dilute Gas

A dilute gas is a gas which is at sufficiently low pressure that it nearly obeys the ideal gas equation of state. A model system that represents a dilute gas is depicted in Figure 16.1. It has the properties:

1. It consists of a large number, N, of molecules, moving about in a container. For the present, we assume that all of the molecules have the same mass.
2. The motions of these molecules are governed by classical (Newtonian) mechanics.
3. The molecules are point mass particles that do not exert any forces on each other. A **point mass** is an object of vanishingly small size, so point masses cannot collide with each other.
4. The container confining the gas is a box with smooth hard walls.

Figure 16.1. The Model System to Represent an Ideal Gas. The dots represent point mass molecules moving randomly. Only a few dots are shown, but the actual model system contains many molecules.

The Mechanical States of the Model System

Let us number the particles from 1 to N. The mechanical state (microstate) of the system is specified by separately giving the position and velocity of every particle. The position of particle i is specified by a **position vector** $\mathbf{r}_i$, which has its tail at the origin of our coordinate system and its head at the particle's location. This vector can be represented in terms of its **components** x_i, y_i, and z_i:

$$\mathbf{r}_i = \mathbf{i}x_i + \mathbf{j}y_i + \mathbf{k}z_i \tag{16.1-1}$$

The product of a scalar and a vector is a second vector with the same direction as the first vector if the scalar is positive. If the scalar is negative, the second vector is in the opposite direction.

The unit vector **i** points in the direction of the positive x axis, the unit vector **j** in the direction of the positive y axis, and the unit vector **k** in the direction of the positive z axis. A vector is also denoted by its three components listed in parentheses, as in $\mathbf{r}_i = (x_i, y_i, z_i)$. The components x_i, y_i, and z_i are also the **cartesian coordinates** of the location of the particle. Figure 16.2 shows the vector $\mathbf{r}_i$, the cartesian axes, the unit vectors, and the cartesian components of the vector.

The sum of two vectors can be represented geometrically by moving the second vector so its tail is at the head of the first vector and drawing the sum vector from the tail of the first to the head of the second. Figure 16.3 shows how its three components add to equal the position vector $\mathbf{r}_i$.

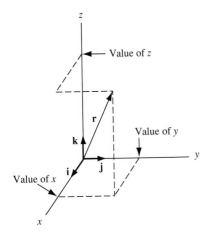

Figure 16.2. A Position Vector r in Three-Dimensional Space. The vector **r** specifies the position of the particle when its tail is at the origin. The unit vectors **i**, **j**, and **k** are also shown.

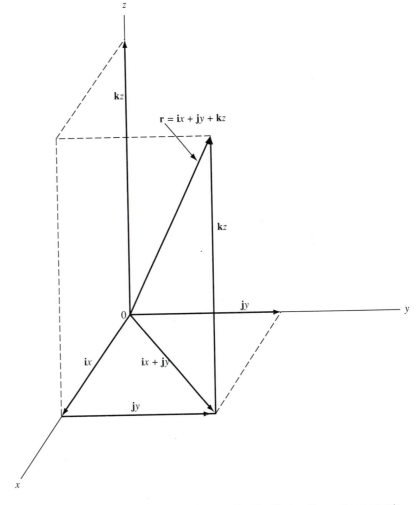

Figure 16.3. The Addition of the Components of a Position Vector. To use the geometric recipe for adding vectors: translate the **j**y vector so that its tail is at the head of the **i**x vector; draw the sum vector **i**x + **j**y from the origin to the head of the **j**y vector. Then translate the **k**z vector so that its tail is at the head of the **i**x + **j**y vector and draw the **r** vector from the origin to the head of the translated **k**z vector.

The velocity of particle i is specified by the **velocity vector**

$$\mathbf{v}_i = \mathbf{i}v_{ix} + \mathbf{j}v_{iy} + \mathbf{k}v_{iz} \qquad \textbf{(16.1-2)}$$

whose components are the rates of change of x_i, y_i, and z_i:

$$v_{ix} = dx_i/dt, \qquad v_{iy} = dy_i/dt, \qquad v_{iz} = dz_i/dt \qquad \textbf{(16.1-3)}$$

The velocity vector can be represented in much the same way as the position vector, with its components represented geometrically as in Figure 16.4. The space of this figure is called **velocity space**. It is not a physical space but is a graph of rates of motion. However, many of the mathematical properties of ordinary space apply to velocity space, including theorems of geometry.

The direction of the velocity vector is the direction of motion of the particle, and its magnitude is the **speed**:

$$v_i = |\mathbf{v}_i| = (v_{ix}^2 + v_{iy}^2 + v_{iz}^2)^{1/2} \qquad \textbf{(16.1-4)}$$

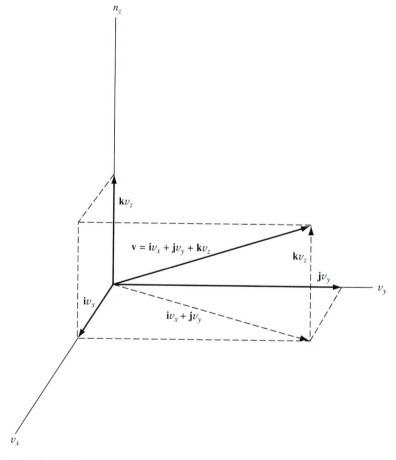

Figure 16.4. A Velocity Vector in Velocity Space. The velocity vector has a magnitude equal to the speed of the particle and a direction which is the instantaneous direction of motion of the particle. The space of this figure is not a physical (coordinate) space, but a graph in which distances on the axes represent velocity components. Cartesian components are shown and can be handled in the same way as components in ordinary space.

We use either of the two notations in Equation (16.1-4) to denote the magnitude of a vector: the boldface letter within vertical bars, and the ordinary letter. The magnitude of a vector is always nonnegative, although its components can be positive, negative, or equal to zero.

Exercise 16.1   **a.** Use the theorem of Pythagoras to verify Equation (16.1-4). (It must be used twice.)
b. Find the speed of a particle with the velocity components:
$$v_x = 400 \text{ m s}^{-1}, \qquad v_y = -600 \text{ m s}^{-1}, \qquad v_z = 750 \text{ m s}^{-1}$$

The instantaneous microstate of the system is specified by giving the values of three coordinates and three velocity components for each of the N particles, a total of $6N$ numbers. Note the vast difference between the amount of information in the specification of the microscopic state and the specification of the macroscopic state of our gas. Specification of the microscopic state requires approximately 3,600,000,000,000,000,000,000,000 values for a system containing one mole. Specification of the equilibrium macroscopic state requires the values of only three variables, such as T, V, and n. This vast difference indicates that many microscopic states must correspond to the same macroscopic state. Gas kinetic theory provides a specific way to average over many microscopic states to represent a single macroscopic state.

There is another significant difference between the microscopic states and the macroscopic states of our system. The macroscopic states we have discussed in Chapters 1 through 8 are equilibrium states, with no tendency to change in time. Our microscopic state is a dynamical state. Even if the system is at macroscopic equilibrium, the molecules are moving and thus constantly changing the microscopic state of the system. The microscopic state is specified in exactly the same way whether the system is at macroscopic equilibrium or not.

The Mechanical Processes of the System—Newton's Laws of Motion

Newton's second law is the most important formula of classical mechanics:

$$\mathbf{F}_i = m_i \mathbf{a}_i = m_i \, d\mathbf{v}_i/dt = m_i \, d\mathbf{r}_i^2/dt^2 \tag{16.1-5}$$

where $\mathbf{F}_i$ is the force on the ith particle, m_i is its mass, and $\mathbf{a}_i$ is its acceleration. Since $\mathbf{r}_i$ and $\mathbf{v}_i$ are vectors, $\mathbf{a}_i$ is also a vector.

Equation (16.1-5) is a vector equation, so it is equivalent to three scalar equations, one of which is

$$F_{ix} = m_i(dv_{ix}/dt) = m_i(dx_i^2/dt^2) \tag{16.1-6}$$

Since the particles in our model system do not interact with each other, the only forces on them are the forces due to interactions at the walls. Away from the walls, the particles move in straight lines at constant speeds.

Mechanical Properties of the System and Mean Values

Just as we recognized state functions of macroscopic states, we recognize state functions of microscopic states. All mechanical properties of the system are functions of the positions and velocities of the particles and therefore are state functions.

The mechanical energy is the sum of the kinetic and the potential energy:

$$E = \mathcal{K} + \mathcal{V} \tag{16.1-7}$$

The kinetic energy of the system, $\mathcal{K}$, is a state function that depends only on the velocities of the particles. It is given by

$$\mathcal{K} = \frac{m}{2}(v_1^2 + v_2^2 + v_3^2 + \cdots + v_N^2) = \frac{m}{2}\sum_{i=1}^{N} v_i^2 \tag{16.1-8}$$

where we have assumed that all the particles of our model system have the same mass.

Forces that are velocity independent can be derived from a potential energy (see Appendix D). The potential energy is a function only of the positions of the particles:

$$\mathcal{V} = \mathcal{V}(\mathbf{r}_1, \mathbf{r}_2, \ldots, \mathbf{r}_N) \tag{16.1-9}$$

Since the molecules of our system have no forces exerted on them except by the walls enclosing the system, their potential energy is constant so long as they remain within these walls, and we can set this potential energy equal to zero inside the container. To represent complete confinement of the particles in the container, we assign an infinite value to the potential energy if any particle is outside the box.

Since the potential energy is zero for possible states of the system, the energy of the system is equal to the sum of the molecular kinetic energies:

$$E = \varepsilon_1 + \varepsilon_2 + \varepsilon_3 + \cdots + \varepsilon_N = \mathcal{K} \tag{16.1-10}$$

where we use the symbol ε_i for the energy of particle i, which is now equal to the kinetic energy of the particle.

The **mean molecular kinetic energy** is given by the sum of the molecular kinetic energies divided by the number of molecules:

$$\langle \varepsilon \rangle = \frac{1}{N}(\varepsilon_1 + \varepsilon_2 + \varepsilon_3 + \cdots + \varepsilon_N) = \frac{\mathcal{K}}{N} = \frac{E}{N} \tag{16.1-11}$$

If p_i is the fraction of the molecules in state j with energy equal to ε_j,

$$p_j = \frac{N_j}{N} \tag{16.1-12}$$

then we can write the mean kinetic energy as

$$\langle \varepsilon \rangle = \sum_j p_j \varepsilon_j \tag{16.1-13}$$

where the sum is over the possible molecular states, not over the molecules.

Since we describe our system with classical mechanics, we must integrate over velocities and coordinates instead of summing over discrete energies.

The sum in Equation (16.1-13) is replaced by an integral like that in Equation (10.4-10), and the probability distribution p_j must be replaced by a probability density.

16.2 The Velocity Probability Distribution

This probability density is analogous to probability densities in Chapter 10 except that it is a probability per unit volume in velocity space instead of ordinary space.

In this section, we obtain the formula that represents the probability distribution for the velocities of molecules in our model system. We begin with a reasonable (but unproved) assumption: *The probability distribution of each velocity component is independent of the other velocity components.* This assumption is sufficient to determine the form of the probability distribution.

Consider the velocity of a representative particle, with components v_x, v_y, and v_z. Let $f(v_x)$ be the probability density for v_x. The probability that v_x lies between v_x' (a particular value of v_x) and $v_x' + dv_x$ is given by

$$\text{Probability} = f(v_x')\, dv_x \qquad (16.2\text{-}1)$$

We must consider the probability density in three dimensions. Let v_x', v_y', and v_z' be particular values of v_x, v_y, and v_z. Let the probability that v_x lies between v_x' and $v_x' + dv_x$, and that v_y is in the range between v_y' and $v_y' + dv_y$, and that v_z lies between v_z' and $v_z' + dv_z$ be given by

$$\text{Probability} = g(v_x', v_y', v_z')\, dv_x\, dv_y\, dv_z = g(\mathbf{v}')\, d^3\mathbf{v} \qquad (16.2\text{-}2)$$

where we abbreviate the product of the three differentials $dv_x\, dv_y\, dv_z$ by the symbol $d^3\mathbf{v}$ and write the dependence on the vector components as dependence on the vector $\mathbf{v}$.

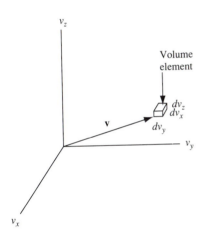

The probability density $g(\mathbf{v})$ is a probability per unit volume in velocity space. That is, the probability for an infinitesimal region in velocity space is equal to the probability density evaluated for the location of the region times the volume of the infinitesimal region, which is $dv_x\, dv_y\, dv_z$. The probability of a finite region is the integral of the probability density over that region. Figure 16.5 shows the velocity space with a volume element, which unfortunately must be drawn with finite size.

Figure 16.5. A Velocity Vector in Velocity Space, with a Cartesian Volume Element. The volume element is a rectangular box that is taken to be infinitesimal, although in the figure it must be drawn with finite size.

If the effects of gravity are negligible, there is no physical difference between any two directions. Therefore, g can depend only on the magnitude of $\mathbf{v}$ (the speed v) and not on its direction. Furthermore, the probability distributions for v_y and v_z must be the same function as for v_x. Finally, we have assumed that the probability densities for v_x, v_y, and v_z are independent of each other. It is a fact of probability theory that the joint probability of three independent events is the product of the probabilities of the three events. Therefore,

Here are some consequences of our assumption that space is isotropic.

$$g(\mathbf{v}) = g(v) = f(v_x)f(v_y)f(v_z) \qquad (16.2\text{-}3)$$

This assumption that space is isotropic and the assumption that the three components are independent are sufficient to produce the Boltzmann probability distribution.

The assumptions made to obtain Equation (16.2-3) are sufficient to determine the mathematical forms of the probability distributions f and g.

We first differentiate g with respect to v_x. By the chain rule, Equation (B-9) of Appendix B,

$$\frac{\partial g}{\partial v_x} = \frac{dg}{dv}\frac{\partial v}{\partial v_x} = \frac{dg}{dv}\frac{\partial}{\partial v_x}[(v_x^2 + v_y^2 + v_z^2)^{1/2}]$$

$$= \frac{dg}{dv}\frac{1}{2}(v_x^2 + v_y^2 + v_z^2)^{-1/2}(2v_x) = \frac{dg}{dv}\frac{v_x}{v} \qquad \textbf{(16.2-4)}$$

We divide this equation by v_x to get

$$\frac{1}{v_x}\frac{\partial g}{\partial v_x} = \frac{1}{v}\frac{dg}{dv} \qquad \textbf{(16.2-5)}$$

Since v_x, v_y, and v_z occur in the expression for the speed v in the same way, the corresponding equation for differentiation by v_y or v_z would be the same except for having v_y or v_z in place of v_x. The right-hand side of each equation would be the same, so that

$$\frac{1}{v_x}\frac{\partial g}{\partial v_x} = \frac{1}{v_y}\frac{\partial g}{\partial v_y} = \frac{1}{v_z}\frac{\partial g}{\partial v_z} = \frac{1}{v}\frac{dg}{dv} \qquad \textbf{(16.2-6)}$$

Since v_y and v_z are treated as constants in the v_x differentiation

$$\frac{\partial g}{\partial v_x} = f(v_y)f(v_z)\frac{df}{dv_x} \qquad \textbf{(16.2-7)}$$

Similar equations for $(\partial g/\partial v_y)$ and $(\partial g/\partial v_z)$ can be written, so Equation (16.2-6) becomes

$$\frac{1}{v_x}f(v_y)f(v_z)\left(\frac{df}{dv_x}\right) = \frac{1}{v_y}f(v_x)f(v_z)\left(\frac{df}{dv_y}\right) = \frac{1}{v_z}f(v_x)f(v_y)\left(\frac{df}{dv_z}\right) \quad \textbf{(16.2-8)}$$

We divide this equation by $f(v_x)f(v_y)f(v_z)$ to obtain

$$\frac{1}{v_x f(v_x)}\left(\frac{df}{dv_x}\right) = \frac{1}{v_y f(v_y)}\left(\frac{df}{dv_y}\right) = \frac{1}{v_z f(v_z)}\left(\frac{df}{dv_z}\right) \qquad \textbf{(16.2-9)}$$

In this equation we have separated the variables v_x, v_y, and v_z, much as we separated independent variables in several Schrödinger equations in Chapters 9 through 14. We have an equation in which each of these variables occurs in one term and only in that term. Since these variables are independent, this can be a correct equation for all values of v_x, v_y, and v_z only if each term of the equation equals a constant. We denote this constant by C. Setting the first term equal to C and multiplying by $v_x f(v_x)$, we obtain

$$\frac{df}{dv_x} = Cv_x f(v_x) \qquad \textbf{(16.2-10)}$$

The method of separation of variables is used in this case to separate an independent variable from a dependent variable. The terms cannot be set equal to constants, but can be integrated to solve the equation.

Equation (16.2-10) is a differential equation that can be solved by separation of variables. The term "separation of variables" has a different meaning here than in the construction of Equation (16.2-9), because now one variable is a dependent variable. We multiply Equation (16.2-10) by dv_x and divide it by $f(v_x)$:

$$\frac{1}{f}\frac{df}{dv_x}dv_x = Cv_x\,dv_x \qquad \textbf{(16.2-11)}$$

We recognize $(df/dv_x)\, dv_x$ as the differential of the dependent variable, df, and write

$$\frac{1}{f}\, df = C v_x\, dv_x \tag{16.2-12}$$

which is an equation with the variables separated. An indefinite integration of both sides of Equation (16.2-12) gives

$$\ln(f) = \frac{C v_x^2}{2} + A \tag{16.2-13}$$

where A is a constant of integration. We take the exponential of each side to obtain the formula for our probability distribution

$$f(v_x) = e^A e^{C v_x^2 / 2} \tag{16.2-14}$$

We will require that f is normalized, which means that the integral of f over all possible values of v_x is equal to unity:

$$\int_{-\infty}^{\infty} f(v_x)\, dv_x = 1 \tag{16.2-15}$$

Since we are using nonrelativistic mechanics, speeds greater than the speed of light are not excluded. The constant C must be negative, since otherwise the integrand in Equation (16.2-15) would grow without bound for large magnitudes of v_x and the integral would diverge. We let $b = -C$, so that b is positive. We now have

$$1 = e^A \int_{-\infty}^{\infty} e^{-b v_x^2 / 2}\, dv_x = e^A \sqrt{\frac{2\pi}{b}} \tag{16.2-16}$$

where we have looked up the definite integral in Appendix C.

Solving Equation (16.2-16) for e^A, we have

$$e^A = \left(\frac{b}{2\pi}\right)^{1/2} \tag{16.2-17}$$

and

$$f(v_x) = \left(\frac{b}{2\pi}\right)^{1/2} e^{-b v_x^2 / 2} \tag{16.2-18}$$

The probability distribution for all three components is

$$g(v) = f(v_x) f(v_y) f(v_z) = \left(\frac{b}{2\pi}\right)^{3/2} e^{-b v_x^2 / 2} e^{-b v_y^2 / 2} e^{-b v_z^2 / 2}$$

$$= \left(\frac{b}{2\pi}\right)^{3/2} e^{-b(v_x^2 + v_y^2 + v_z^2)/2} \tag{16.2-19}$$

The exponent in the probability distribution function is proportional to the kinetic energy of the particle, ε, so we can also write

$$g(v) = \left(\frac{b}{2\pi}\right)^{3/2} e^{-b\varepsilon/m} \tag{16.2-20}$$

The Kinetic Energy of an Ideal Gas

To finish our derivation, we must identify the parameter b. The total energy of a system is the sum of kinetic and potential energy. We have set the potential energy of the system equal to zero, so

The sum of a set of numbers is always equal to the number of members of the set times the mean value of the set.

$$E = \mathcal{K} = N\langle\varepsilon\rangle = Nm\langle v^2\rangle/2 \qquad (16.2\text{-}21)$$

The mean molecular kinetic energy is

$$\langle\varepsilon\rangle = \frac{m}{2}\langle v^2\rangle = \frac{m}{2}\langle v_x^2 + v_y^2 + v_z^2\rangle$$

$$= \frac{m}{2}(\langle v_x^2\rangle + \langle v_y^2\rangle + \langle v_z^2\rangle) \qquad (16.2\text{-}22)$$

Because the x, y, and z velocity component probability distributions are the same function, the three terms in Equation (16.2-22) will be equal to each other after integration, and

$$\langle\varepsilon\rangle = \frac{3m}{2}\langle v_x^2\rangle$$

$$= \left(\frac{3m}{2}\right)\left(\frac{b}{2\pi}\right)^{3/2}\int_{-\infty}^{\infty}\int_{-\infty}^{\infty}\int_{-\infty}^{\infty} v_x^2 e^{-bv_x^2/2}e^{-bv_y^2/2}e^{-bv_z^2/2}\,dv_x\,dv_y\,dv_z$$

$$(16.2\text{-}23)$$

where we have written g in its original factored form. We can factor the multiple integral in Equation (16.2-23), since our limits are constants and the integrand factors:

$$\langle\varepsilon\rangle = \left(\frac{3m}{2}\right)\left(\frac{b}{2\pi}\right)^{3/2}\int_{-\infty}^{\infty} v_x^2 e^{-bv_x^2/2}\,dv_x\int_{-\infty}^{\infty} e^{-bv_y^2/2}\,dv_y\int_{-\infty}^{\infty} e^{-bv_z^2/2}\,dv_z$$

$$(16.2\text{-}24)$$

The integrals over v_y and v_z in Equation (16.2-24) are the same as in Equation (16.2-16), so they produce factors of $(2\pi/b)^{1/2}$. The third integration gives

$$\langle\varepsilon\rangle = \left(\frac{3m}{2}\right)\left(\frac{b}{2\pi}\right)^{1/2}\int_{-\infty}^{\infty} v_x^2 e^{-bv_x^2/2}\,dv_x$$

$$= \left(\frac{3m}{2}\right)\left(\frac{b}{2\pi}\right)^{1/2}(2\pi)^{1/2}b^{-3/2} = \frac{3m}{2b} \qquad (16.2\text{-}25)$$

where we have looked the integral up in Appendix C.

The Maxwell Probability Distribution

In Chapter 2 we asserted as an experimental fact that the thermodynamic energy of a monatomic gas is very nearly given by Equation (2.3-6):

$$U = \frac{3}{2}nRT = \frac{3}{2}Nk_BT \qquad (16.2\text{-}26)$$

where k_B is Boltzmann's constant, introduced in Equation (1.2-10) and equal to R/N_A. We will see later that we could accept the ideal gas equation of state for our system as an experimental fact and then derive Equation (16.2-26).

We assume that the thermodynamic energy and the mechanical energy of the system are equal:

$$U = N\langle\varepsilon\rangle \tag{16.2-27}$$

The mean molecular kinetic energy is now

$$\langle\varepsilon\rangle = \frac{3}{2}k_B T \tag{16.2-28}$$

Comparison of Equations (16.2-25) and (16.2-28) gives

$$b = m/k_B T \tag{16.2-29}$$

Equation (16.2-19) becomes

$$g(v) = \left(\frac{m}{2\pi k_B T}\right)^{3/2} e^{-m(v_x^2 + v_y^2 + v_z^2)/2k_B T} \tag{16.2-30}$$

This function is called the **Maxwell probability distribution** or the **Maxwell-Boltzmann probability distribution**. In terms of the molecular kinetic energy,

$$g(\mathbf{v}) = \left(\frac{m}{2\pi k_B T}\right)^{3/2} e^{-\varepsilon/k_B T} \tag{16.2-31}$$

We can now complete the derivation of the Boltzmann probability distribution of Chapter 1, in which we arbitrarily set the constant c (same as $-b/m$) in Equation (1.7-23) equal to $-1/k_B T$. Equation (16.2-31) is the same as the Boltzmann probability distribution of Equation (1.7-25), in normalized form.

The important qualitative physical facts about the Boltzmann probability distribution were presented in Chapter 1:

1. At a fixed temperature, molecular states of higher energy are less probable than states of lower energy. States with energy much larger than $k_B T$ are quite improbable.
2. A molecular state of high energy will be more probable at a high temperature than at a low temperature. As the temperature approaches infinity, all states approach equal probability.

▼ **EXAMPLE 16.1**

Find the ratio of the probabilities of the following two velocities of neon atoms at 300 K:

First velocity: $v_x = 500$ m s^{-1}, $v_y = -400$ m s^{-1}, $v_z = 250$ m s^{-1}

Second velocity: $v_x = 200$ m s^{-1}, $v_y = 350$ m s^{-1}, $v_z = -275$ m s^{-1}

Solution

Use the average molar mass:

$$m = \frac{0.02018 \text{ kg mol}^{-1}}{6.022 \times 10^{23} \text{ mol}^{-1}} = 3.35 \times 10^{-26} \text{ kg}$$

First velocity:

$$v^2 = (500 \text{ m s}^{-1})^2 + (-400 \text{ m s}^{-1})^2 + (250 \text{ m s}^{-1})^2$$
$$= 472500 \text{ m}^2 \text{ s}^{-2}$$

$$\frac{\varepsilon}{k_\text{B} T} = \frac{mv^2}{2k_\text{B} T} = \frac{(3.35 \times 10^{-26} \text{ kg})(472500 \text{ m}^2 \text{ s}^{-2})}{(2)(1.3807 \times 10^{-23} \text{ J K}^{-1})(300 \text{ K})} = 1.91$$

Second velocity:

$$\frac{\varepsilon}{k_\text{B} T} = 0.963 \quad \text{(calculation similar to first velocity)}$$

$$\text{Probability ratio} = \frac{e^{-1.91}}{e^{-0.963}} = 0.388$$

Exercise 16.2

Find the ratio of the probabilities of the two velocities for argon atoms at 300 K:
First velocity: $v_x = 650 \text{ m s}^{-1}, v_y = 780 \text{ m s}^{-1}, v_z = 990 \text{ m s}^{-1}$
Second velocity: $v_x = 300 \text{ ms}^{-1}, v_y = 290 \text{ m s}^{-1}, v_z = 430 \text{ m s}^{-1}$

We can write the probability density (probability distribution) for any component of the velocity. For example

$$f(v_x) = \left(\frac{m}{2\pi k_\text{B} T}\right)^{1/2} e^{-mv_x^2/2k_\text{B}T} \tag{16.2-32}$$

This probability distribution is a normal (gaussian) probability distribution, as in Equation (10.4-17). It is represented in Figure 16.6 for oxygen molecules at 298 K.

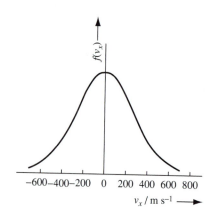

Figure 16.6. The Probability Distribution for a Velocity Component of Oxygen Molecules at 298 K. The most probable value of the velocity component is zero, and most of the molecules have values of the velocity component between -400 m s^{-1} and 400 m s^{-1}.

▼   **EXAMPLE 16.2**

Find the probability that v_x lies between 0 and 500 m s^{-1} for neon atoms at 300 K.

Solution

$$\text{Probability} = \left(\frac{m}{2\pi k_{\mathrm{B}}T}\right)^{1/2} \int_0^{500 \text{ m/s}} e^{-mv_x^2/2k_{\mathrm{B}}T} \, dv_x$$

This integral cannot be worked out in closed form. We change variables to get it into the form of the **error function**, defined in Appendix C by[1]

$$\text{erf}(x) = \frac{2}{\sqrt{\pi}} \int_0^x e^{-t^2} \, dt \tag{16.2-33}$$

In Example 16.2, one must use a table of the error function. However, the integral in the example must be transformed by a change of variables before the desired value can be looked up in a table.

We let

$$t = (mv_x^2/2k_{\mathrm{B}}T)^{1/2}, \qquad dt = (m/2k_{\mathrm{B}}T)^{1/2} \, dv_x$$

At the upper limit, we have

$$t = \sqrt{\frac{(3.35 \times 10^{-26} \text{ kg})(500 \text{ m s}^{-1})^2}{(2)(1.3807 \times 10^{-23} \text{ J K}^{-1})(300 \text{ K})}} = 1.0055$$

so that

$$\text{Probability} = \frac{1}{\sqrt{\pi}} \int_0^{1.0055} e^{-t^2} \, dt = \frac{1}{2} \text{ erf}(1.0055)$$

$$= 0.4225 = 42.25\%$$

where the numerical value was obtained from the table of the error function in Appendix C.

▲

Exercise 16.3

By comparison with Equation (10.4-17), find a formula for the standard deviation σ of the probability distribution of Equation (16.2-32). What fraction of the molecules have x components of the velocity between $-\sigma$ and $+\sigma$? Find the value of the standard deviation for oxgyen molecules at 298 K.

If we are interested in a small finite range Δv_x, we can obtain an approximate value of the probability by replacing the infinitesimal interval dv_x in Equation (16.2-1) by Δv_x.

$$\text{Probability} \approx f(v_x') \, \Delta v_x \tag{16.2-34}$$

where v_x' is a value of v_x within the range Δv_x.

Exercise 16.4

a. Find the probability that v_x for an argon atom in a system at 273.15 K is in the range 650 m/s $< v_x <$ 651 m/s.

b. Find the probability that v_x for an argon atom in a system at 273.15 K is in the range 650 m/s $< v_x <$ 652 m/s.

[1] M. Abramowitz and I. A. Stegun (eds.), *Handbook of Mathematical Functions with Formulas, Graphs, and Mathematical Tables*, U.S. Government Printing Office, Washington, DC, 1964, pp. 297ff.

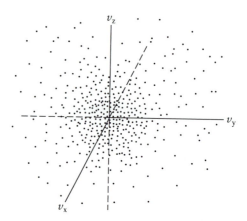

Figure 16.7. Points Representing the Velocity States of Molecules in a System. There is one point for each molecule. Instead of the several hundred points shown here, there might be Avogadro's number of points for an actual system.

Equation (16.2-30) represents the probability density in the three-dimensional velocity space of Figure 16.4. Every point in the velocity space represents a possible velocity of a molecule. We say that there is a one-to-one correspondence between the points of the space and the possible velocity states of a molecule. If we have N molecules, we can represent their velocities by a set of N points, one for each molecule. The density (number of points per unit volume) of this swarm of points at some location, say v, is proportional to $g(v)$, the probability density for the velocity evaluated at that point. Figure 16.7 schematically represents the swarm of points for a system of a few hundred molecules.

16.3   The Distribution of Molecular Speeds

All velocity vectors that have the same length but differ in direction correspond to the same speed, so the probability distribution of speeds is different from the distribution of velocities. We change to a new set of velocity coordinates, in which one of the coordinates is the speed v and the other two coordinates are angles specifying the direction of the velocity vector in velocity space. These coordinates are spherical polar coordinates, analogous to the spherical polar coordinates in coordinate space introduced in Chapter 11.

The volume element is equal to $v^2 \sin(\theta) \, dv \, d\theta \, d\phi$, as shown in Figure 16.8, which also shows the coordinates. We denote the volume element by either of the two symbols

$$d^3\mathbf{v} = v^2 \sin(\theta) \, dv \, d\theta \, d\phi \qquad \textbf{(16.3-1)}$$

Using the symbol $d^3\mathbf{v}$ for both the volume element in cartesian coordinates

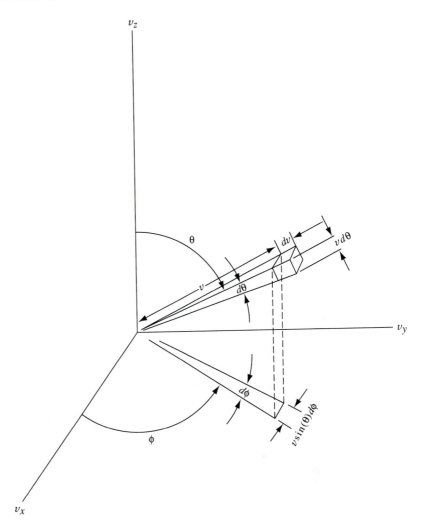

Figure 16.8. Spherical Polar Coordinates in Velocity Space, with a Volume Element. We can choose to specify the location of a point in velocity space and the size of a volume element in velocity space with cartesian coordinates, spherical polar coordinates, or any other coordinate system. The volume element can be considered to be a rectangular box, since it is infinitesimal in size.

and the volume element in spherical polar coordinates enables us to write some equations so that they apply to either set of coordinates.

The probability that the velocity of a randomly chosen molecule lies in the volume element $v^2 \sin(\theta)\, dv\, d\theta\, d\phi$ is

$$\text{Probability} = g(v)\, d^3\mathbf{v} = g(v)v^2 \sin(\theta)\, dv\, d\theta\, d\phi \qquad \textbf{(16.3-2)}$$

To get the probability density for molecular speeds, we integrate the probability over all points in velocity space corresponding to speeds in an infinitesimal range dv. These points make up a spherical shell of thickness dv, as shown in Figure 16.9. We accomplish this integration by integrating Equation (16.3-2) over all values of θ and ϕ for a fixed value of v. Since $g(v)$

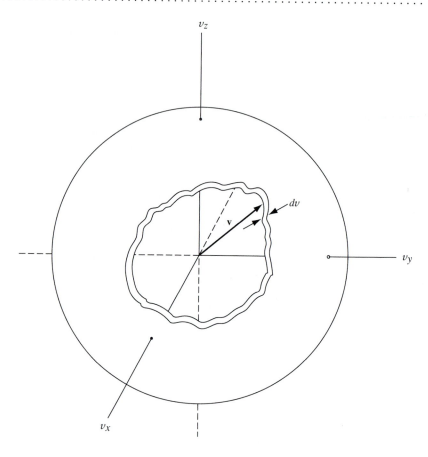

Figure 16.9. A Spherical Shell of Thickness *dv* in Velocity Space. This shell contains all of the points in velocity space that represent velocities with magnitudes (speeds) between *v* and *v* + *dv*.

All the points on a sphere in velocity space correspond to the same speed, since they represent velocities with the same magnitude.

depends only on the speed, it factors out of the integral:

$$\text{Probability} = g(v)v^2\,dv \int_0^\pi \sin(\theta)\,d\theta \int_0^{2\pi} d\phi \qquad \textbf{(16.3-3)}$$

$$= 4\pi v^2 g(v)\,dv \qquad\qquad\qquad \textbf{(16.3-4)}$$

The factor $4\pi v^2$ is the area of one surface of the spherical shell in Figure 16.9, and $4\pi v^2\,dv$ is the volume of the shell. The probability in Equation (16.3-4) is a probability density times the volume of the shell.

Exercise 16.5   Show that the integrals in Equation (16.3-3) lead to the factor 4π in Equation (16.3-4).

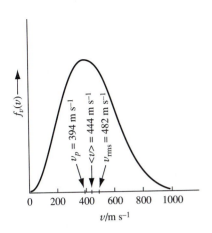

Figure 16.10. The Probability Distribution of Molecular Speeds for Oxygen Molecules at 298 K. The most probable speed, the mean speed, and the root-mean-square speed are labeled on the axis.

Since the probability in Equation (16.3-4) is a probability density times an infinitesimal interval dv, the probability density f_v for molecular speeds is

$$f_v(v) = 4\pi v^2 \left(\frac{m}{2\pi k_B T}\right)^{3/2} e^{-mv^2/2k_B T}$$ **(16.3-5)**

Figure 16.10 shows this probability density for oxygen molecules at 298 K. Compare this with Figure 16.6. The most probable value of any velocity component is zero, whereas the most probable speed is nonzero and the probability of zero speed is zero. This difference is due to the fact that the probability density of a speed is proportional to the area of the spherical shell in velocity space (equal to $4\pi v^2$) times the probability density of the velocities lying in the spherical shell. The area of the spherical shell vanishes for zero speed.

A formula for the **most probable speed**, v_p, is obtained by finding the value of the speed at which the first derivative of the probability density in Equation (16.3-5) vanishes. The result is

$$v_p = \sqrt{\frac{2k_B T}{m}} \quad \text{(most probable speed)}$$ **(16.3-6)**

The derivative also vanishes at $v = 0$ and at $v = \infty$. These points correspond to minimum values of the probability.

Exercise 16.6

a. Verify Equation (16.3-6) by setting the derivative of f_v equal to zero.
b. Find the most probable speed of oxygen molecules at 298 K.
c. Find the most probable speed of helium atoms at 298 K.

The **mean speed** is given by

$$\langle v \rangle = \int_0^\infty v f_v(v) \, dv = 4\pi \left(\frac{m}{2\pi k_B T}\right)^{3/2} \int_0^\infty v^3 e^{-mv^2/2k_B T} \, dv$$

$$\langle v \rangle = \sqrt{\frac{8k_B T}{\pi m}}$$ **(16.3-6a)**

where we have looked up the integral in the table of Appendix C. The mean speed can also be written in terms of the molar mass, M, since k_B is equal to R/N_A:

$$\langle v \rangle = \sqrt{\frac{8RT}{\pi M}}$$ **(16.3-6b)**

▼   **EXAMPLE 16.3**   Find the mean speed of oxygen molecules at 298 K.

Solution

$$\langle v \rangle = \sqrt{\frac{(8)(8.3145 \text{ J K}^{-1} \text{ mol}^{-1})(298 \text{ K})}{(\pi)(0.0320 \text{ kg mol}^{-1})}}$$

$$= 444 \text{ m s}^{-1}$$

▲

The most probable value (the **mode**) and the mean value are two different types of average. A third kind of average is the **root-mean-square** value, which is the square root of the mean-square speed, $\langle v^2 \rangle$:

$$\langle v^2 \rangle = \int_0^\infty v^2 f_v(v) \, dv = 4\pi \left(\frac{m}{2\pi k_B T} \right)^{3/2} \int_0^\infty v^4 e^{-mv^2/2k_B T} \, dv$$

$$= \frac{3k_B T}{m} = \frac{3RT}{M} \tag{16.3-7}$$

where we have looked up the integral in the table of Appendix C. The root-mean-square speed is

$$\boxed{v_{rms} = \langle v^2 \rangle^{1/2} = \sqrt{\frac{3k_B T}{m}} = \sqrt{\frac{3RT}{M}}} \tag{16.3-8}$$

▼   **EXAMPLE 16.4**   Find the root-mean-square speed of oxygen molecules at 298 K.

Solution

$$v_{rms} = \left[\frac{(3)(8.3145 \text{ J K}^{-1} \text{ mol}^{-1})(298 \text{ K})}{0.0320 \text{ kg mol}^{-1}} \right]^{1/2}$$

$$= 482 \text{ m s}^{-1}$$

▲

The most probable speed, the mean speed, and the root-mean-square speed are all shown in Figure 16.10.

Exercise 16.7   Find the most probable speed, the mean speed, and the root-mean-square speed of helium atoms at 298 K.

Exercise 16.8   Show that the two ratios $\langle v \rangle / v_p$ and v_{rms}/v_p have the same two values for all gases at all temperatures, and find those values.

The Pressure of an Ideal Gas

We assume that our model system is confined in a rectangular box whose edges are parallel to our coordinate axes. The walls of the box are assumed to be smooth and impenetrable, so that no particle, however fast it is moving, can escape from the box. Since the walls are smooth, the molecules will collide specularly with them. A **specular collision** has the properties: (1) It is elastic. That is, the kinetic energy of the molecule is the same before and after the collision. (2) No force parallel to the wall is exerted on the particle.

The trajectory of a particle striking the wall at the right end of the box is as shown in Figure 16.11. The x and z components of the velocity do not change, since no force is exerted in these directions. Because the kinetic energy does not change and v_x^2 and v_z^2 do not change, v_y^2 cannot change. The sign of v_y does change. Let $\mathbf{v}(i)$ be the initial velocity and $\mathbf{v}(f)$ be the final velocity:

$$v_x(f) = v_x(i), \qquad v_y(f) = -v_y(i), \qquad v_z(f) = v_z(i) \qquad \textbf{(16.4-1)}$$

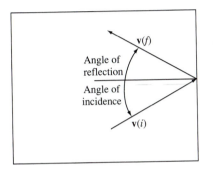

Figure 16.11. The Trajectory of a Particle Colliding Specularly with a Wall. The angle of incidence and the angle of reflection are equal because the wall exerts a force on the molecule only in a direction perpendicular to the wall.

As the particle strikes the wall, the magnitude of the force exerted on the wall rises suddenly and drops just as suddenly. Any measuring instrument used to measure the force on the wall requires a period of time, called the response time, to adjust to a sudden change in the force, The reading of the instrument is a time average over the response time, which we denote by τ. We therefore consider a time average of the force over a period of time τ, instead of the force itself.

The force on the particle is by Newton's third law the negative of the force exerted on the wall. The time average of the y component of the force on the particle over a period of time from 0 to τ is defined by

$$\langle F_y \rangle = \frac{1}{\tau} \int_0^\tau F_y(t)\, dt \quad \text{(definition of time average)} \qquad \textbf{(16.4-2)}$$

We use the $\langle\ \rangle$ symbol to denote either a time average or a mean value computed with a probability distribution. We must determine from the context which type of average is meant in a particular case.

For a particular particle, we choose the starting point of the time average so that the particle will hit the wall during the interval τ. From Newton's second law, Equation (16.1-5), we can replace the force component by the mass times the acceleration component:

$$\langle F_y \rangle = \frac{1}{\tau} \int_0^\tau m \frac{d^2 y}{dt^2}\, dt = \frac{1}{\tau} \int_0^\tau m \frac{dv_y}{dt}\, dt \qquad \textbf{(16.4-3)}$$

We recognize $(dv_y/dt)\, dt$ as the differential dv_y, so that

$$\langle F_y \rangle = \frac{m}{\tau} \int_0^\tau dv_y = \frac{m}{\tau} [v_y(f) - v_y(i)] \qquad \textbf{(16.4-4)}$$

We use Equation (16.4-1) to replace $v_y(f)$ by $-v_y(i)$, so that

$$\langle F_y \rangle = -\frac{2mv_y(i)}{\tau} \tag{16.4-5}$$

By Newton's third law, the force on the wall, $\mathbf{F}_w$, is equal in magnitude to $\mathbf{F}$ and opposite in direction, so that

$$\langle F_{wy} \rangle = -\langle F_y \rangle = 2mv_y/\tau \tag{16.4-6}$$

We use only initial velocities from now on and omit the (i) label on our velocity component.

We now need to sum up the contributions to the force from all of the particles in the system. Consider particles whose velocities lie in the infinitesimal range $dv_x\, dv_y\, dv_z$. The fraction of all particles whose velocities lie in this range is

$$\text{Fraction} = g(\mathbf{v})\, dv_x\, dv_y\, dv_z \tag{16.4-7}$$

All of the particles with velocities in our infinitesimal range are moving nearly in the same direction with nearly the same speed. The particles in this set that will strike an area of size $\mathscr{A}$ in the wall are contained in a prism whose sides are parallel to the direction of motion of the particles, as shown in Figure 16.12.

Here we use the familiar relation distance = rate × time.

The particles that are no farther from the wall in the y direction than a distance equal to τv_y will strike the wall during the time interval τ. The distance in the y direction, not the distance along the surface of the prism, must be used, because we are considering the y component of the velocity, not the speed.

We assume that our system is uniform. The **number density** in any part of the system is therefore equal to N/V, where N is the number of particles and V is the volume of the system. We denote the number density by $\mathscr{N}$.

The volume of the prism is equal to the area of its base times its altitude:

$$V_{\text{prism}} = \mathscr{A} v_y \tau \tag{16.4-8}$$

The number of particles in the prism is

$$\text{Number in prism} = \mathscr{N} \mathscr{A} v_y \tau = \frac{N}{V} \mathscr{A} v_y \tau \tag{16.4-9}$$

so that

$$\begin{array}{l}\text{Number with velocities}\\ \text{in the given range}\\ \text{striking } \mathscr{A} \text{ in time } \tau\end{array} = \mathscr{N} \mathscr{A} v_y \tau g(v)\, dv_x\, dv_y\, dv_z \tag{16.4-10}$$

Let the force on $\mathscr{A}$ be denoted by $\langle F_{\mathscr{A}} \rangle$. Each particle makes a contribution to this force that is given by Equation (16.4-6):

$$\begin{array}{l}\text{Contribution to } \langle F_{\mathscr{A}} \rangle \text{ due}\\ \text{to particles with velocities}\\ \text{in the given range}\end{array} = \left(\frac{2mv_y}{\tau}\right) \mathscr{N} \mathscr{A} v_y \tau g(v)\, dv_x\, dv_y\, dv_z$$

$$= 2m \mathscr{N} \mathscr{A} v_y^2 g(v)\, dv_x\, dv_y\, dv_z$$

$$= 2m \frac{N}{V} \mathscr{A} v_y^2 g(v)\, dv_x\, dv_y\, dv_z \tag{16.4-11}$$

Notice that τ cancels out of this equation.

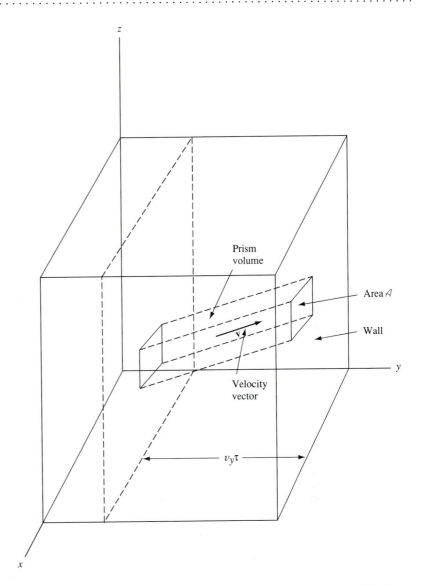

**Figure 16.12. The Prism That Contains Molecules of a Certain Velocity That Will Strike
Area $\mathscr{A}$ in Time τ.** Molecules with velocities not equal to the given velocity lie in
different prisms.

We now add up the contributions for molecules with different velocities
by integrating over the velocity components. Only molecules with positive
values of v_y will strike the wall, and the components parallel to the wall
can be positive or negative. The total time-average force on $\mathscr{A}$ is

$$\langle F_{\mathscr{A}} \rangle = \frac{2mN\mathscr{A}}{V} \int_{-\infty}^{\infty} \int_{0}^{\infty} \int_{-\infty}^{\infty} v_y^2 g(v)\, dv_x\, dv_y\, dv_z \qquad \textbf{(16.4-12)}$$

The integral in this equation is the same as one of the terms in Equation (16.2-22) except that the lower limit of the v_y integration is 0 instead
of $-\infty$. Since the integrand is an even function of v_y, the value of the

integral will be half of that of the integral obtained by changing the lower limit to $-\infty$.

$$\langle F_{\mathscr{A}} \rangle = 2m \frac{N}{V} \mathscr{A} \frac{1}{2} \langle v_y^2 \rangle = 2m \frac{N}{V} \mathscr{A} \frac{1}{6} \langle v^2 \rangle$$

$$= \frac{1}{3} m \frac{N}{V} \mathscr{A} \frac{3k_{\mathrm{B}}T}{m} = \frac{N \mathscr{A} k_{\mathrm{B}} T}{V} = \mathcal{N} \mathscr{A} k_{\mathrm{B}} T \qquad \textbf{(16.4-13)}$$

where we have used Equation (16.3-7) for $\langle v^2 \rangle$ and the fact that $3\langle v_y^2 \rangle = \langle v^2 \rangle$.

The pressure, P, is the force per unit area, so that

$$P = \frac{Nk_{\mathrm{B}}T}{V} = \mathcal{N} k_{\mathrm{B}} T = \frac{nRT}{V} \qquad \textbf{(16.4-14)}$$

where we identify

$$k_{\mathrm{B}} = nR/N = R/N_{\mathrm{A}} \qquad \textbf{(16.4-15)}$$

and where N_{A} is Avogadro's constant. We do not label P as a time average, since it is a time-independent equilibrium quantity.

We have taken a model system whose states and processes are described mechanically on the molecular scale. By averaging over microscopic states, we have found that the macroscopic pressure of the model system is the same as that of an ideal gas. Our analysis depended on our use of Equation (16.2-26), which gave the energy of the gas as proportional to the temperature. If we had wished to do so, we could have assumed that our gas obeys the ideal gas equation of state and then derived Equation (16.2-26).

EXAMPLE 16.5

For an O_2 molecule moving with $v_y = 444$ m s^{-1}, find the impulse (force multiplied by the time over which the force is exerted) on a wall perpendicular to the y axis if the particle collides with the wall.

Solution

$$\langle F_y \rangle \tau = 2mv_y(i) = \frac{(2)(0.0320 \text{ kg mol}^{-1})(444 \text{ m s}^{-1})}{6.0221 \times 10^{23} \text{ mol}^{-1}}$$

$$= 4.72 \times 10^{-23} \text{ kg m s}^{-1} = 4.72 \times 10^{-23} \text{ N s}$$

Exercise 16.9

Find the number of oxygen molecules with $v_y = 444$ m s^{-1} that must strike an area of 1.000 m^2 in 1.000 second for a force of 101325 N to be exerted on the area.

If our system contains a mixture of different gaseous substances and the molecules do not exert forces on each other, the molecules of each substance move as though the other substances were not present. The total pressure is then the sum of the pressures exerted by each set of molecules:

$$P = P_1 + P_2 + P_3 + \cdots + P_c \qquad \textbf{(16.4-16)}$$

where P_1 is the partial pressure of substance 1, or the pressure this substance would exert if it were alone in the container, and similarly for the other substances. Equation (16.4-16) is Dalton's law of partial pressures.

We have found that our model system conforms to the ideal gas equation of state and to Dalton's law of partial pressures. We therefore believe a dilute gas consists of moving molecules that behave in much the same way as the point mass particles of our model system. That is, we believe that the molecules of a dilute gas do not collide with each other frequently enough to make it much different from our model system.

16.5   Wall Collisions and Effusion

Graham's law is named for Thomas Graham, 1805–1869, a British chemist who not only studied diffusion and effusion but also determined the formulas of the various species formed from phosphoric acid.

Effusion is the process by which molecules of a gas pass through a small hole into a vacuum. The hole must be small enough that the gas does not flow through the hole as a fluid, but passes as individual molecules. **Graham's law of effusion** is an empirical law which states that at a given temperature and a given pressure, the rates of effusion of different gases are inversely proportional to the square roots of the densities of the gases and thus to the square roots of the molecular masses.

Refer again to Figure 16.12. Our analysis of effusion will be similar to that of Section 16.4, except that we will now compute the number of molecules striking an area $\mathscr{A}$. If this area is a section of the wall, we obtain the rate of wall collisions. If the area is a hole in the wall, we obtain the rate of effusion.

The number of molecules whose velocities lie in the velocity interval $dv_x\, dv_y\, dv_z$ and which will strike the area $\mathscr{A}$ in time τ is given by Equation (16.4-10):

$$\text{Number striking } \mathscr{A} \text{ in time } \tau = \mathscr{N}\mathscr{A} v_y \tau g(v)\, dv_x\, dv_y\, dv_z \quad \textbf{(16.5-1)}$$

We obtain the total number of molecules striking the area in the time interval τ by integrating over velocities as in Equation (16.4-12):

$$
\begin{aligned}
\begin{matrix}\text{Total number} \\ \text{striking area} \\ \mathscr{A} \text{ in time } \tau\end{matrix}
&= \mathscr{N}\mathscr{A}\tau \int_{-\infty}^{\infty} \int_{0}^{\infty} \int_{-\infty}^{\infty} v_y g(v)\, dv_x\, dv_y\, dv_z \\[2mm]
&= \mathscr{N}\mathscr{A}\left(\frac{m}{2\pi k_\mathrm{B} T}\right)^{3/2} \tau \int_{-\infty}^{\infty} \int_{0}^{\infty} \int_{-\infty}^{\infty} v_y e^{-mv^2/2k_\mathrm{B}T}\, dv_x\, dv_y\, dv_z
\end{aligned}
$$

$$\textbf{(16.5-2)}$$

This equation is factored as was Equation (16.2-24).

$$
\begin{matrix}\text{Total number} \\ \text{striking area} \\ \mathscr{A} \text{ in time } \tau\end{matrix}
= \mathscr{N}\mathscr{A}\left(\frac{m}{2\pi k_\mathrm{B} T}\right)^{1/2} \tau \int_{0}^{\infty} v_y e^{-mv_y^2/2k_\mathrm{B}T}\, dv_y \quad \textbf{(16.5-3)}
$$

The integral in this equation can be performed by the method of substitution. We let $w = v_y^2$, so that the integral becomes

$$
\int_0^\infty v_y e^{-mv_y^2/2k_B T}\, dv_y = \frac{1}{2}\int_0^\infty e^{-mw/2k_B T}\, dw
$$

$$
= \frac{1}{2}\frac{2k_B T}{m}\left(-e^{-w}\right)\Big|_0^\infty
$$

$$
= -\frac{1}{2}\frac{2k_B T}{m}(0 - 1) = \frac{k_B T}{m} \qquad \textbf{(16.5-4)}
$$

Therefore

$$
\begin{array}{l}\text{Total number}\\ \text{striking area}\\ \mathscr{A}\text{ in time }\tau\end{array} = \mathscr{N}\mathscr{A}\tau\left(\frac{k_B T}{2\pi m}\right)^{1/2}
$$

$$
= \frac{1}{4}\mathscr{N}\mathscr{A}\tau\left(\frac{8k_B T}{\pi m}\right)^{1/2} = \frac{1}{4}\mathscr{N}\mathscr{A}\tau\langle v\rangle \qquad \textbf{(16.5-5)}
$$

The number of particles that strike the area $\mathscr{A}$ in time τ is proportional to the area $\mathscr{A}$, proportional to the length of time τ, proportional to the number of particles per unit volume, and proportional to the mean speed of the particles.

The number of molecules striking a unit area per unit time is denoted by v:

$$
\boxed{\; v = \frac{1}{4}\mathscr{N}\langle v\rangle = \frac{\mathscr{N}}{4}\left(\frac{8k_B T}{\pi m}\right)^{1/2} \;} \qquad \textbf{(16.5-6)}
$$

▼ **EXAMPLE 16.6**

Find the number of molecules of air striking a person's eardrum in 1.00 second at 298 K and 1.00 atm. Assume air to be 79 mol % nitrogen and 21 mol % oxygen, and assume the area of the eardrum to equal 0.50 cm^2.

Solution

$$
\langle v(O_2)\rangle = 444 \text{ m s}^{-1} \quad \text{from Example 16.3}
$$

$$
\langle v(N_2)\rangle = 475 \text{ m s}^{-1} \quad \text{by a similar calculation}
$$

$$
\mathscr{N}(N_2) = N(N_2)/V = P(N_2)/k_B T
$$

$$
= \frac{(101325 \text{ N m}^{-2}\text{ atm}^{-1})(0.79 \text{ atm})}{(1.3807 \times 10^{-23} \text{ J K}^{-1})(298 \text{ K})} = 1.9 \times 10^{25} \text{ m}^{-3}
$$

where $P(N_2)$ is the partial pressure of N_2, and $N(N_2)$ is the number of N_2 molecules.

$$
\mathscr{N}(O_2) = N(O_2)/V = 5.2 \times 10^{24} \text{ m}^{-3} \quad \text{by a similar calculation}
$$

$$
v(N_2) = \frac{1}{4}(1.9 \times 10^{25} \text{ m}^{-3})(475 \text{ m s}^{-1}) = 2.3 \times 10^{27} \text{ m}^{-2}\text{ s}^{-1}
$$

$$
v(O_2) = 5.7 \times 10^{26} \text{ m}^{-2}\text{ s}^{-1} \quad \text{by a similar calculation}
$$

$$
\text{Number per second} = [v(N_2) + v(O_2)](5.0 \times 10^{-5} \text{ m}^2) = 1.4 \times 10^{23} \text{ s}^{-1}
$$

Exercise 16.10	**a.** Estimate the number of air molecules striking the palm of your hand in 24 hours. **b.** A certain solid catalyst has a surface area of 55 m^2 per gram of catalyst. A mixture of gases containing 0.5 mol % carbon monoxide passes over the catalyst at 350 K and 1.00 atm. Find the amount of CO in moles striking 1 gram of the catalyst per second.

Equation (16.5-6) also gives the rate of effusion per unit area through a small hole, since nothing in the derivation depended on the fate of the molecules after they struck the area $\mathscr{A}$. The effusion rate predicted by Equation (16.5-6) is inversely proportional to the square root of the mass of the particles, in agreement with Graham's law of effusion. We have now determined that our model system conforms to all of the empirical laws of Section 1.2 as well as to Graham's law of effusion.

16.6   The Probability Distribution of Coordinates

We now consider a system in which the particles have a nonconstant potential energy. We consider only the case that the potential energy of each particle is independent of the positions of the other particles. This case excludes intermolecular forces, which we discuss later, but it includes gravitational potential energy, and it will allow us to derive the barometric pressure equation, already applied in Example 1.7.

Let the potential energy of particle i be denoted by

$$u = u(x_i, y_i, z_i) \qquad \text{(16.6-1)}$$

The total energy of the ith molecule is

$$\varepsilon_i = \frac{m}{2}\left(v_{ix}^2 + v_{iy}^2 + v_{iz}^2\right) + u(x_i, y_i, z_i) \qquad \text{(16.6-2)}$$

The Boltzmann probability distribution applies. From Equations (1.7-25) and (16.2-31), the probability of the state with energy of Equation (16.6-2) is

$$\text{Probability} \propto e^{-m(v_x^2 + v_y^2 + v_z^2)/2k_\text{B}T} e^{-u(x,y,z)/k_\text{B}T} \qquad \text{(16.6-3)}$$

We now normalize this probability distribution in order to write it as an equality instead of a proportionality. Let the probability that the position of a randomly chosen particle lies in the coordinate volume element $dx\,dy\,dz$ and that its velocity lies in the velocity space volume element $dv_x\,dv_y\,dv_z$ be given by a normalized probability density, G:

$$\begin{aligned}\text{Probability} &= G(x', y', z', v_x', v_y', v_z')\,dx\,dy\,dz\,dv_x\,dv_y\,dv_z \\ &= G(\mathbf{r}', \mathbf{v}')\,d^3\mathbf{r}\,d^3\mathbf{v}\end{aligned} \qquad \text{(16.6-4)}$$

where x', y', and z' are coordinates lying in the volume element $dx\,dy\,dz$ and v_x', v_y', and v_z' are velocity coordinates lying in the volume element $dv_x\,dv_y\,dv_z$.

The probability density can be factored into a velocity factor and a co-ordinate factor:

$$G = g(v_x, v_y, v_z)g_c(x, y, z) = g(\mathbf{v})g_c(\mathbf{r}) \tag{16.6-5}$$

where $g(\mathbf{v})$ is the same function as in Equation (16.2-30) and

$$g_c(\mathbf{r}) \propto e^{-u/k_B T} \tag{16.6-6}$$

The probability that a particle is in the coordinate volume element $dx\,dy\,dz$, irrespective of its velocity, is

$$\text{Probability} = \left[\int_{-\infty}^{\infty} \int_{-\infty}^{\infty} \int_{-\infty}^{\infty} g_c(\mathbf{r})g(\mathbf{v})\,dv_x\,dv_y\,dv_z \right] dx\,dy\,dz$$
$$= g_c(r)\,dx\,dy\,dz \tag{16.6-7}$$

The second equation follows from the fact that the velocity probability distribution is separately normalized.

We assume that the system is enclosed in a rectangular box of dimensions a by b by c and write the normalization integral

$$\int_0^c \int_0^b \int_0^a g_c(r)\,dx\,dy\,dx = \int_V g_c(r)\,d^3\mathbf{r} = 1 \tag{16.6-8}$$

where the second version of the integral is an abbreviation for the first version. We define the **configuration integral**, ζ:

$$\zeta = \int_V e^{-u/k_B T}\,d^3\mathbf{r} \tag{16.6-9}$$

The normalized coordinate distribution function is

$$g_c = \frac{1}{\zeta}e^{-u/k_B T} \tag{16.6-10}$$

The barometric pressure for an atmosphere of constant temperature is a special case of this equation:

$$P = P_0 e^{-mgh/k_B T} = P_0 e^{-Mgh/RT} \tag{16.6-11}$$

where h is the altitude, P_0 is the pressure at altitude $h = 0$, g is the acceleration due to gravity, and M is the molar mass of the gas (usually taken as an average value for the atmosphere).

Exercise 16.11   **a.** Show that for a system in a box with $\mathscr{V} = 0$ inside the box, the configuration integral is given by

$$\zeta = V$$

where V is the volume of the box.

b. Find an expression for ζ if $u = mgz$, where z is the vertical coordinate and g is the acceleration due to gravity, 9.80 m s^{-2}. Assume the system to be confined in a box of dimensions a by b by c.

16.7 Intermolecular Forces

In order to discuss a nonideal gas, we must modify our first model system, which consists of point mass particles that exert no forces on each other. Two real molecules attract each other at moderate intermolecular distances and repel each other at smaller intermolecular distances.

The attractive forces are called **London dispersion forces**.[2] They are also called **van der Waals forces**. A classical picture used to visualize the attractive force between neutral molecules is that electron motion produces instantaneous fluctuating dipoles in two molecules, which can become synchronized and attract each other.

Intermolecular repulsions are also visualized in a relatively simple way. One repulsive effect arises from the fact that, according to quantum mechanics, the kinetic energy of a moving object rises when it is confined to a smaller region, as happens to the electrons in two molecules that are pushed close together. Another repulsive effect arises from the fact that the electrons move primarily in the outer regions of a molecule or atom and repel the electrons in another atom or molecule that is close by.

Intermolecular forces can be discussed in the Born-Oppenheimer approximation, presented in Chapter 13. The Born-Oppenheimer energy depends on the internuclear distances but is independent of nuclear velocities. An energy that is independent of velocities is a potential energy. The potential energy function for a pair of argon atoms is shown in Figure 16.13 and is typical of pairs of atoms that do not react chemically. Diatomic and polyatomic molecules exhibit a similar intermolecular potential energy, but with additional dependence on molecular orientations.

A simple model system with intermolecular forces is defined by the following assumptions:

1. The system consists of a number, N, of particles that move according to classical mechanics.
2. The intermolecular forces are independent of the particles' velocities.
3. The force on particle 1 due to particle 2 is unaffected by the position of particle 3, particle 4, etc. and similarly for the other pairs of particles. We say that the forces are **pairwise additive**.
4. The magnitudes of the intermolecular forces depend only on the distances between the particles, not on their orientations.

For pairwise additive forces, the potential energy of the system is a sum of terms, one for each pair of particles:

$$\mathscr{V} = \sum_{i=1}^{N-1} \sum_{j=i+1}^{N} u(r_{ij}) \qquad \textbf{(16.7-1)}$$

Compare the curve in Figure 16.13 with the similar curve in Figures 13.1 and 13.3. The minimum in this curve is at -1.656×10^{-21} J, while that in Figure 13.3 is at -4.5×10^{-18} J. We are dealing now with forces that are much weaker than chemical bonds.

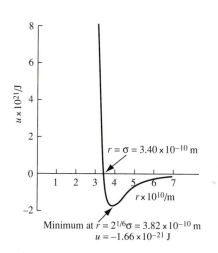

Minimum at $r = 2^{1/6}\sigma = 3.82 \times 10^{-10}$ m
$u = -1.66 \times 10^{-21}$ J

Figure 16.13. The Lennard-Jones Representation of the Intermolecular Potential of a Pair of Argon Atoms. This representation is only qualitatively correct but is widely used as a convenient approximation.

[2] F. London, *Z. Phys. Chem.* **B11**, 222 (1930); *Trans. Faraday Soc.* **33**, 8 (1937).

We used the letter u for a potential energy of one particle in an externally imposed field in Section 16.6 and use it here for a potential energy depending on the distance between a pair of particles. There are, unfortunately, not enough letters in the alphabet to use a different letter for every quantity.

The function u is the **pair potential energy function** of the pair of particles i and j and is a function such as shown in Figure 16.13. The distance between particle i and particle j is denoted by r_{ij}. The limits on the double sum in Equation (16.7-1) are chosen so that the contribution of a single pair of particles is not counted twice and so that $i = j$ does not occur.

Assumptions 3 and 4 limit the systems and states to which we can apply our model. The assumption of pairwise additivity is not very accurate for liquids but is a good approximation for gases.[3] Assumption 4 excludes substances that are not monatomic. However, diatomic substances and polyatomic substances with symmetrical molecules, like methane, are described to a fairly good approximation by this model.

In the gas phase, average nearest-neighbor intermolecular separations are larger than in condensed phases (under ordinary conditions, roughly 10 times as large). The potential energy is nearly constant at these distances, and intermolecular forces are not very important in a gas, except during collisions. A minimum in a potential energy corresponds to a **mechanical equilibrium**, with repulsive forces balancing attractive forces. In a condensed phase (liquid or solid) the average separation of the molecules from their nearest neighbors is approximately equal to the intermolecular distance at the minimum. Since mechanical work must be done either to expand or compress the system, solids and liquids have nearly fixed volumes.

The Lennard-Jones potential is named for J. E. Lennard-Jones, a prominent contemporary British theoretical chemist.

A common approximate representation for the pair potential function is the **Lennard-Jones 6-12 potential**, which was used to draw Figure 16.13:

$$u(r) = 4\varepsilon[(\sigma/r)^{12} - (\sigma/r)^6] \tag{16.7-2}$$

The parameter σ is equal to the intermolecular separation at which the potential energy is zero, and the parameter ε is equal to the depth of the minimum in the curve. The designation 6-12 denotes the choice of the exponents in the formula. Other choices for the exponents are also used, but the Lennard-Jones 6-12 potential is the most common choice and is usually called "the Lennard-Jones potential."

The values of the two parameters σ and ε can be chosen to fit volumetric data, using formulas of statistical mechanics, some of which are discussed in Chapter 20. Table A20 gives some values that were obtained by fitting data on the second virial coefficient, using Equation (20.8-31).[4]

Exercise 16.12

a. Show that for the Lennard-Jones potential,

$$u(\sigma) = 0 \tag{16.7-3}$$

b. Show that the value of r at the minimum in the Lennard-Jones potential is

$$r_{min} = 2^{1/6}\sigma = (1.12246)\sigma \tag{16.7-4}$$

c. Show that

$$u(r_{min}) = -\varepsilon \tag{16.7-5}$$

---

[3] D. R. Williams and L. J. Schaad, *J. Chem. Phys.* **47**, 4916 (1967).
[4] J. O. Hirschfelder, C. F. Curtiss, and R. B. Bird, *Molecular Theory of Gases and Liquids*, Wiley, New York, 1954, pp. 1110ff.

d. The force in the r direction is given by $-du/dr$. Show that the force on one particle due to another particle at distance r is

$$F_r = 4\varepsilon\left(12\frac{\sigma^{12}}{r^{13}} - 6\frac{\sigma^6}{r^7}\right) \tag{16.7-6}$$

e. Show that $F_r = 0$ if $r = 2^{1/6}\sigma$.

A simpler but less realistic representation of the pair potential is the **square well potential**

$$u(r) = \begin{cases} +\infty & \text{if } r < d \\ -\varepsilon & \text{if } d < r < c \\ 0 & \text{if } r > c \end{cases} \tag{16.7-7}$$

This potential function is shown in Figure 16.14. A still simpler representation of the pair potential is the **hard sphere potential**

$$u(r) = \begin{cases} +\infty & \text{if } r < d \\ 0 & \text{if } r > d \end{cases} \tag{16.7-8}$$

The hard sphere representation of a dilute gas is successful primarily because we can choose the effective hard sphere diameter to fit the data and can choose different values at different temperatures if necessary.

which is depicted in Figure 16.15. This is the crudest possible representation of intermolecular potential energy. It completely disregards the attractive forces that are responsible for condensation, so it cannot be used in a realistic description of liquids or solids. However, it is surprisingly successful in the approximate treatment of collisions and transport processes in dilute gases.

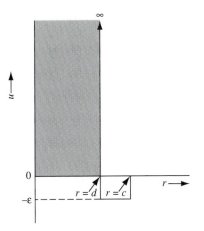

Figure 16.14. The Square Well Representation of the Intermolecular Potential of a Pair of Atoms. This is the crudest representation that contains an attraction.

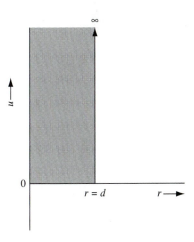

Figure 16.15. The Hard Sphere Representation of the Intermolecular Potential of a Pair of Atoms. This is the crudest of all possible representations of intermolecular potential energy, but it does a surprisingly good qualitative job in describing actual gas behavior.

16.8 The Hard Sphere Gas

The hard sphere gas is the model system with the pair potential function given by Equation (16.7-8). It is like a collection of tiny impenetrable balls. When the appropriate size of the spheres is taken, many of the properties of a real gas can be simulated. Table A21 gives values of the effective hard sphere radius for several gases.

The Equation of State of a Hard Sphere Gas

Consider a one-component gaseous system of N hard sphere molecules of diameter d confined in a volume V. Mechanical states of the system in which the centers of any pair of molecules are closer to each other than a distance d are impossible. We will use this fact to obtain an approximate equation of state for a hard sphere gas.

We first pretend that all of the particles in the system are stationary except particle 1. This moving particle has access to a volume that is smaller than V because of the presence of the other particles. Figure 16.16 shows the volume due to particle 2 into which the center of particle 1 cannot penetrate. This excluded volume is spherical, with radius equal to the sum of the radii of the two particles, or equal to d, the diameter of one particle.

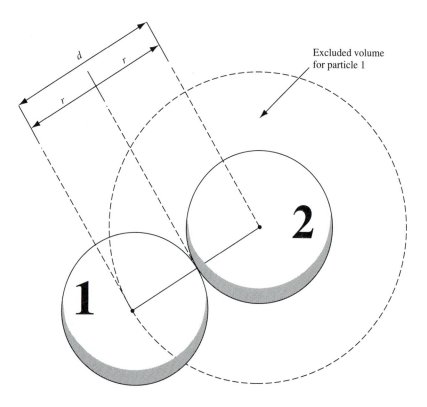

Figure 16.16. The Excluded Volume of a Pair of Hard Spheres. The important fact about this excluded volume is that its radius is the sum of the radii of the two molecules.

The total volume from which particle 1 is excluded is the number of other particles times the volume in Figure 16.16:

$$V_{\text{exc}} = (N - 1)\left(\frac{4}{3}\pi d^3\right) \approx N\frac{4}{3}\pi d^3 \qquad (16.8\text{-}1)$$

where we neglect unity compared with N.

If we now recognize that all of the particles are moving, the excluded volume in Figure 16.16 represents not only the volume from which particle 1 is excluded because of particle 2 but also the volume which particle 2 is excluded because of particle 1. We assign half of this volume to each particle and write for the net excluded volume for a single particle

$$V_{\text{exc}} = \frac{N}{2}\left(\frac{4}{3}\pi d^3\right) = N\frac{2}{3}\pi d^3 \qquad (16.8\text{-}2)$$

We take as our approximate equation of state that of an ideal gas in a container whose volume is equal to the volume in which each particle can move,

$$P(V - V_{\text{exc}}) = Nk_{\text{B}}T = nRT \qquad (16.8\text{-}3)$$

where k_{B} is Boltzmann's constant, n is the amount of the gas measured in moles, and R is the gas constant.

We divide Equation (16.8-3) by n and obtain

$$P(\bar{V} - b) = RT \qquad (16.8\text{-}4)$$

where $\bar{V} = V/n$, the molar volume, and where we define the constant b:

$$b = \frac{N}{n}\frac{2}{3}\pi d^3 = N_{\text{A}}\frac{2}{3}\pi d^3 \qquad (16.8\text{-}5)$$

where N_{A} is Avogadro's constant. Equation (16.8-4) resembles the van der Waals equation of state of Equation (1.3-3) except for the absence of the term containing the parameter a. Since our present system has only repulsive forces, we assert that the parameter b in the van der Waals equation has the same relationship to the effective size of the molecules as in Equation (16.8-5) and represents repulsive forces. The parameter a represents the effect of attractive forces.

If we write Equation (16.8-4) in the form

It has been argued that attractive forces can only slow a particle down just before it strikes a wall, since other molecules are only on the side of the molecule away from the wall as it strikes the wall.

$$P = \frac{RT}{\bar{V} - b} \qquad (16.8\text{-}6)$$

the pressure is greater than that of an ideal gas with the same values of $\bar{V}$ and T. Repulsive forces always make a positive contribution to the pressure. Attractive forces always make a negative contribution to the pressure.

Exercise 16.13   Calculate the radius of an argon atom from the value of the van der Waals parameter b in Table A1. Compare your result with argon's radius in Table A21. Why might the two values differ?

Equation (16.8-4) is only an approximation, because of our assumption that the excluded volumes of the different molecules simply add together. We have ignored, among other things, the effect of finding two particles close enough together that the volumes which they exclude for a third particle partially overlap. The equation of state of a hard sphere fluid has been the subject of considerable research, and far better approximate equations than Equation (16.8-4) have been obtained.[5] One such equation of state is shown in Problem 16.52.

Much of this research has used the technique of **molecular dynamics**, in which solutions to the equations of motion (Newton's second law in appropriate forms) for a system of several hundred particles are numerically simulated by a computer program. Energies, pressures, etc. are then calculated as a function of time from the particles' positions and velocities. The results of these calculations are the closest thing to experimental results for a hard sphere system. The molecular dynamics technique has also been used for other model systems, including those with Lennard-Jones potentials, and has been used to compute nonequilibrium properties.

The hard sphere solid is stabilized by the fact that at high pressure the spheres can lower the volume by packing into a solid lattice.

The results of molecular dynamics calculations indicate that there is no gas-liquid condensation in the hard sphere system, and we would expect no such condensation because there are no attractive intermolecular forces. However, there is considerable evidence from these results that a gas-solid phase transition occurs.[6] This result was originally somewhat surprising.

Molecular Collisions in a Hard Sphere Gas

In the hard sphere model system at fairly low density, a molecule undergoes a number of collisions, moving at a constant velocity between collisions. Study of these collisions will help us to understand transport processes (diffusion, viscous flow, and thermal conduction) and chemical reaction rates in gases.

We now investigate the mean rate of collisions and the mean distance traveled between collisions in a one-component hard sphere gas. For fairly small pressures, the probability that three molecules will collide at the same time is small, so we neglect three-body collisions.

Problem 16.61 involves an estimate of three-body collision rates.

For a first approximate analysis, we assume that only particle 1 is moving, while the others are stationary and distributed throughout the container. The mean number of stationary particles per unit volume is given by

$$\frac{N-1}{V} \approx \frac{N}{V} = \mathscr{N} \tag{16.8-7}$$

where we neglect unity compared with N. The mean number N' of particles in a given portion of the system with volume V' is given by

$$N' = \mathscr{N}V' \tag{16.8-8}$$

---

[5] R. Hoste and W. Van Dael, *J. Chem. Soc. Faraday Trans. 2* **80**, 477 (1984).
[6] H. Reiss and A. D. Hammerich, *J. Phys. Chem.* **90**, 6252 (1986).

As the moving particle travels along, it "sweeps out" a cylindrical volume, as shown in Figure 16.17. The radius of this **collision cylinder** is equal to d, the sum of the radii of two particles, which is called the **collision diameter**. Any molecule whose center is within the cylinder will be struck by the moving particle. The length of the cylinder that contains on the average one stationary particle is equal to the mean distance between collisions, called the **mean free path** and denoted by λ.

The cross-sectional area of the collision cylinder is called the **collision cross section**

$$\text{Collision cross section} = \pi d^2 \tag{16.8-9}$$

The volume containing on the average one stationary particle is equal to

$$\text{Volume containing one particle} = \lambda \pi d^2 \tag{16.8-10}$$

This must equal V/N, the mean volume per molecule:

$$\frac{V}{N} = \frac{1}{\mathcal{N}} = \lambda \pi d^2 \tag{16.8-11}$$

The mean free path is therefore given by

$$\lambda = \frac{1}{\pi d^2 \mathcal{N}} \quad \text{(approximate equation)} \tag{16.8-12}$$

We can now determine the mean rate of collisions for the moving particle in our system. We assume that our moving particle has a speed equal to the

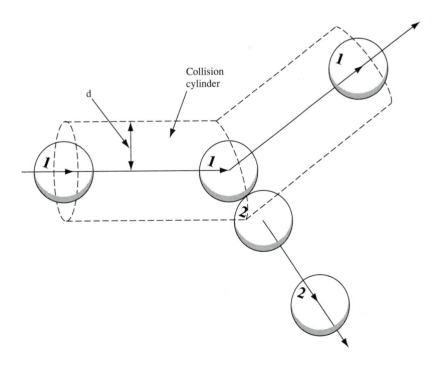

Figure 16.17. A Portion of the Collision Cylinder of Particle 1. Like the excluded volume, the collision cylinder has a radius that is the sum of the radii of the two molecules.

mean molecular speed given by Equation (16.3-6):

$$v_1 = \langle v \rangle = \sqrt{\frac{8k_B T}{\pi m}} \tag{16.8-13}$$

Application of the relationship distance = rate × time.

The mean time between collisions is equal to the mean free path divided by the speed. We denote this time by τ_{coll}.

$$\tau_{\text{coll}} \approx \frac{\lambda}{\langle v \rangle} = \frac{1}{\langle v \rangle \pi d^2 \mathscr{N}}$$

$$= \sqrt{\frac{\pi m}{8k_B T}} \frac{1}{\pi d^2 \mathscr{N}} \quad \text{(approximate equation)} \tag{16.8-14}$$

The rate of collisions (number of collisions per unit time) for our moving particle is denoted by z_1

$$z_1 \approx \frac{1}{\tau_{\text{coll}}} = \pi d^2 \mathscr{N} \sqrt{\frac{8k_B T}{\pi m}} \quad \text{(approximate equation)} \tag{16.8-15}$$

These equations are crude approximations because all of the particles are actually moving. When two particles collide, they might initially be moving toward each other, they might initially be moving roughly at right angles to each other, or they might be moving in the same general direction. We assume that the "average" collision occurs with the particles moving initially at right angles to each other, as in Figure 16.18, and that both particles are moving at the mean speed.

If x is the distance of one particle from the collision site and y is the distance of the other from this site, the separation of the two particles is given by

$$w = (x^2 + y^2)^{1/2} \tag{16.8-16}$$

The **relative speed** is defined as $|dw/dt|$.

If the particles are moving at the same speed, x and y must be equal to each other for the collision to occur. Therefore,

$$w = \sqrt{2}\, x \tag{16.8-17}$$

The relative speed is larger than the speed of each particle by a factor of $\sqrt{2}$:

$$v_{\text{rel}} = \left| \frac{dw}{dt} \right| = \sqrt{2} \left| \frac{dx}{dt} \right| \tag{16.8-18}$$

If both particles are moving at the mean speed, we identify their relative speed as the mean relative speed:

$$\boxed{\langle v_{\text{rel}} \rangle = \sqrt{2} \langle v \rangle = \sqrt{2} \sqrt{\frac{8k_B T}{\pi m}} = \sqrt{\frac{16 k_B T}{\pi m}}} \tag{16.8-19}$$

Our analysis is approximate, but Equation (16.8-19) is the correct formula for the mean relative speed.

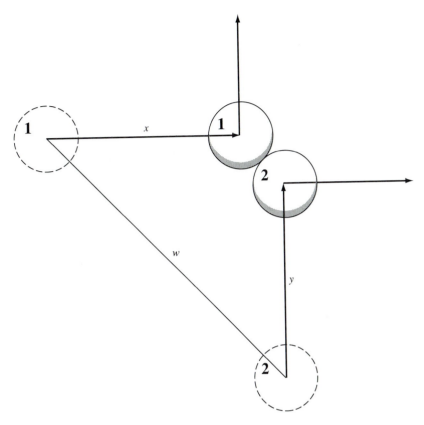

Figure 16.18. An "Average" (Right-Angle) Collision of Two Hard Spheres of the Same Type. Our analysis, based on assuming that this kind of collision is an average that can represent all collisions, leads to the correct formula for the mean relative speed, in spite of its crudity.

If we assume that particle 1 is approaching the other particles at a mean speed of $\sqrt{2}\langle v \rangle$ instead of $\langle v \rangle$, it will meet the other particle in a time that is shorter by a factor of $\sqrt{2}$, so the **mean collision time** becomes

$$\tau_{\text{coll}} = \sqrt{\frac{\pi m}{16 k_{\text{B}} T}} \, \frac{1}{\pi d^2 \mathcal{N}} \qquad \text{(16.8-20)}$$

and the **mean molecular collision rate** becomes

$$z_1 = \frac{1}{\tau_{\text{coll}}} = \pi d^2 \mathcal{N} \sqrt{\frac{16 k_{\text{B}} T}{\pi m}} \qquad \text{(16.8-21)}$$

where we use the subscript 1 to indicate that this is the mean collision rate for substance 1, the only substance present. The mean free path is also shorter by a factor of $2^{1/2}$:

$$\lambda_1 = \frac{1}{\sqrt{2}\,\pi d^2 \mathcal{N}} \qquad\qquad \text{(16.8-22)}$$

EXAMPLE 16.7

For nitrogen gas at 298 K and a molar volume of 24.45 L (approximately at 1 atmosphere pressure):

  a. Find the mean free path.

  b. Find the mean collision time and the mean molecular collision rate.

Solution

$$\mathcal{N} = \frac{N}{V} = \frac{6.022 \times 10^{23}\ \text{mol}^{-1}}{0.02445\ \text{m}^3\ \text{mol}^{-1}} = 2.463 \times 10^{25}\ \text{m}^{-3}$$

$$\langle v \rangle = \left[\frac{8(8.3145\ \text{J K}^{-1}\ \text{mol}^{-1})(298.15\ \text{K})}{\pi(0.0280\ \text{kg mol}^{-1})} \right]^{1/2} = 475\ \text{m s}^{-1}$$

$$\langle v_{\text{rel}} \rangle = \sqrt{2}\,\langle v \rangle = 671\ \text{m s}^{-1}$$

a.

$$\lambda = \frac{1}{\sqrt{2}\,\pi(3.7 \times 10^{-10}\ \text{m}^3)^2(2.463 \times 10^{25}\ \text{m}^{-3})} = 6.7 \times 10^{-8}\ \text{m}$$

b.

$$\tau_{\text{coll}} = \frac{\lambda}{\langle v \rangle} = \frac{6.7 \times 10^{-8}\ \text{m}}{475\ \text{m s}^{-1}} = 1.4 \times 10^{-10}\ \text{s}$$

$$z_1 = \frac{1}{1.4 \times 10^{-10}\ \text{s}} = 7.1 \times 10^9\ \text{s}^{-1}$$

Exercise 16.14

For helium gas at a molar volume of 24.45 L

  a. Find the length of a cube containing on the average one atom.

  b. Find the mean free path.

  c. Why is the mean free path so much larger than the length of the cube of part a?

The total rate of collisions per unit volume is not quite equal to the mean molecular collision rate multiplied by the number of molecules per unit volume, which would count each collision twice. For example, the collision between molecule number 1 and molecule number 37 would be included once for molecule 1 and once for molecule 37. We correct for this double counting by dividing by 2:

$$Z_{11} = \frac{1}{2} z_1 \mathcal{N} = \frac{1}{2} \pi d^2 \langle v \rangle \mathcal{N}^2 = \pi d^2 \sqrt{\frac{4k_\text{B}T}{\pi m}}\,\mathcal{N}^2 \qquad\qquad \text{(16.8-23)}$$

where Z_{11} stands for the **total collision rate per unit volume** of molecules of substance 1 with other molecules of substance 1.

We notice the following important physical facts: (1) The total collision rate per unit volume is proportional to the square of the number density, (2) it is proportional to the collision cross section, and (3) it is proportional to the mean speed and thus to the square root of the temperature. For example, doubling the number density quadruples Z_{11}, while doubling the absolute temperature raises the Z_{11} by the square root of two. These facts will aid in analyzing the rates of chemical reactions in gases.

▼   **EXAMPLE 16.8**

Calculate the total rate of collisions in 1.000 mol of nitrogen gas confined in a volume of 24.45 L at 298 K.

Solution

Using the values from Example 16.7

$$Z_{11} = \frac{1}{2} z_1 \mathcal{N}_1 = \frac{1}{2}(7.1 \times 10^9 \text{ s}^{-1})(2.463 \times 10^{25} \text{ m}^{-3})$$

$$= 8.7 \times 10^{34} \text{ m}^{-3} \text{ s}^{-1}$$

$$\text{Total rate} = (8.7 \times 10^{34} \text{ m}^{-3} \text{ s}^{-1})(0.02445 \text{ m}^3)$$

$$= 2.1 \times 10^{33} \text{ s}^{-1}$$

The total number of collisions per second is about a billion times larger than Avogadro's constant, since each molecule collides about a billion times per second.

▲

Collisions and Free Paths in a Multicomponent Hard Sphere Gas

For two successive collisions of a particular molecule with others of some given substance, the molecule might have collided with other types of molecules between the two collisions of interest if other substances are present. The effect of such collisions will be to put bends in the collision cylinder. As long as there are not too many such bends, the volume of the cylinder will be nearly the same as if it were straight, so the results for a one-component gas can be applied to the collisions between like particles in a multicomponent gas.

The radius of the collision cylinder (collision diameter) for collisions between molecules of substance 1 and substance 2 is equal to the sum of the radii of the molecules, or half the sum of their diameters:

$$d_{12} = \frac{1}{2}(d_1 + d_2) \tag{16.8-24}$$

where d_1 and d_2 are the collision diameters for collisions of like molecules.

Assume that molecule 1 is of substance 1 and is moving at $\langle v_1 \rangle$, the mean speed of molecules of substance 1, and that molecule 2 is of substance 2 and is moving at $\langle v_2 \rangle$, the mean speed of molecules of type 2. Assume again that the average collision takes place at right angles. Figure 16.18 must be modified, as shown in Figure 16.19. For the molecules to collide,

$$x = t_c \langle v_1 \rangle, \qquad y = t_c \langle v_2 \rangle \tag{16.8-25}$$

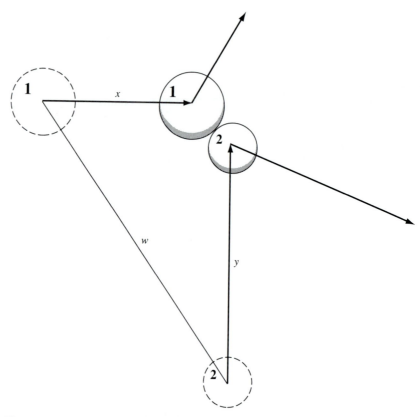

Figure 16.19. An "Average" (Right-Angle) Collision of Two Hard Spheres of Different Types. This is similar to the diagram of Figure 16.18, except that the particles are moving with different speeds and must approach the collision site from different distances.

where t_c is the time yet to elapse before the collision occurs. The molecular separation is given by

$$w = t_c[\langle v_1 \rangle^2 + \langle v_2 \rangle^2]^{1/2} \qquad \textbf{(16.8-26)}$$

and the mean relative speed is given by

$$\langle v_{\mathrm{rel}} \rangle = \langle v_{12} \rangle = \left| \frac{dw}{dt} \right| = \sqrt{\langle v_1 \rangle^2 + \langle v_2 \rangle^2} = \sqrt{\frac{8k_BT}{\pi m_1} + \frac{8k_BT}{\pi m_2}} \quad \textbf{(16.8-27)}$$

$$\boxed{\langle v_{\mathrm{rel}} \rangle = \sqrt{\frac{8k_BT}{\pi \mu_{12}}}} \qquad \textbf{(16.8-28)}$$

The reduced mass is defined in exactly the same way as the reduced mass introduced in Chapter 11. One can also write

$$\frac{1}{\mu_{12}} = \frac{1}{m_1} + \frac{1}{m_2}$$

where m_1 and m_2 are the two molecular masses and μ_{12} is the **reduced mass**:

$$\boxed{\mu_{12} = \frac{m_1 m_2}{m_1 + m_2} \quad \text{(definition)}} \qquad \textbf{(16.8-29)}$$

For a pair of identical particles, μ is equal to $m/2$, so Equation (16.8-28) is valid for that case as well as for two different substances.

▼ **EXAMPLE 16.9** Calculate the mean relative speed of nitrogen and oxygen molecules at 298 K.

Solution

Let nitrogen be substance 1 and oxygen be substance 2

$$\langle v_{12} \rangle = \sqrt{\frac{8k_B T}{\pi \mu_{12}}} = \sqrt{\frac{8RT}{\pi N_A \mu_{12}}} \tag{16.8-30}$$

$$N_A \mu_{12} = \frac{(0.0280 \text{ kg mol}^{-1})(0.0320 \text{ kg mol}^{-1})}{0.0280 \text{ kg mol}^{-1} + 0.0320 \text{ kg mol}^{-1}} = 0.0149 \text{ kg mol}^{-1}$$

$$\langle v_{12} \rangle = \sqrt{\frac{8(8.3145 \text{ J K}^{-1} \text{ mol}^{-1})(298 \text{ K})}{\pi (0.0149 \text{ kg mol}^{-1})}} = 651 \text{ m s}^{-1}$$

▲

When we take account of the fact that the molecules of substance 2 are moving, we obtain the mean free path between collisions of a single particle of substance 1 with particles of substance 2

$$\lambda_{1(2)} = \frac{\langle v_1 \rangle}{\langle v_{12} \rangle} \frac{1}{\pi d_{12}^2 \mathcal{N}_2} = \sqrt{\frac{m_2}{m_1 + m_2}} \frac{1}{\pi d_{12}^2 \mathcal{N}_2} \tag{16.8-31}$$

Collisions with other molecules of substance 1 or of substances other than 2 can intervene between the collisions with molecules of substance 2. The mean free path of a molecule of substance 2 between collisions with molecules of substance 1 is denoted by $\lambda_{2(1)}$ and is obtained by switching the indices 1 and 2 in Equation (16.8-31). It is not necessarily equal to $\lambda_{1(2)}$.

The formula for the mean rate of collisions of one molecule of substance 1 with molecules of substance 2 is analogous to Equation (16.8-21)

$$z_{1(2)} = \frac{1}{\tau_{1(2)}} = \sqrt{\frac{8k_B T}{\pi \mu_{12}}} \pi d_{12}^2 \mathcal{N}_2 \tag{16.8-32}$$

The rate of collisions of a molecule of substance 2 with molecules of substance 1 is obtained by switching the indices 1 and 2 in Equation (16.8-32).

The total rate per unit volume of collisions between molecules of substance 1 and molecules of substance 2 is equal to the collision rate of Equation (16.8-32) times the number density of molecules of substance 1, and vice versa:

$$Z_{12} = z_{1(2)} \mathcal{N}_2 = z_{2(1)} \mathcal{N}_1 = \sqrt{\frac{8k_B T}{\pi \mu_{12}}} \pi d_{12}^2 \mathcal{N}_1 \mathcal{N}_2 \tag{16.8-33}$$

There is no need to divide by 2, as in Equation (16.8-23). The two molecules in a given collision are of different substances, so there is no double counting. An important physical fact is shown in Equation (16.8-33): *the total rate of collisions between molecules of two substances is proportional to the number density of each substance.* This fact, like that of Equation (16.8-23), will aid us in discussing the rates of chemical reactions in gaseous systems.

Exercise 16.15   Assume that 0.800 mol of nitrogen (substance 1) and 0.200 mol of oxygen (substance 2) are contained in 24.45 L at 298 K.
  a. Find $z_{1(2)}$, $z_{2(1)}$, $z_{1(1)}$, and $z_{2(2)}$.
  b. Find Z_{11}, Z_{22}, and Z_{12}.
  c. Find the total number of collisions per second.

16.9   The Molecular Structure of Liquids

The model system of noninteracting particles is usable for a dilute gas because of the large average distances between molecules in a gas. Since liquid molecules are close together, it is more difficult to construct a simple model system for liquids. A great deal of theoretical research has been carried out on the properties of liquids. However, a single comprehensive theory for the liquid state does not exist. The theories that do exist are much more complicated than gas kinetic theory, and we are not prepared to discuss them.

We present a few elementary comments about the equilibrium structure of liquids, based on a simple but important fact: the structure and properties of a liquid are determined primarily by the potential energy of the system, not its kinetic energy. For example, consider liquid argon. The Lennard-Jones representation of the intermolecular potential function for argon was shown in Figure 16.13. If two argon atoms are separated by about 3.8×10^{-10} m (380 pm), they are at a stable equilibrium position, and the potential energy of the pair of atoms must be increased either to separate them or to push them closer together.

Substances such as argon consist of spherical particles. In a solid of spherical atoms, 12 "nearest-neighbor" spheres can surround a given sphere in the solid lattice. Each of these nearest neighbors touches the central sphere and four of the other 11 nearest neighbors. In solid argon, the nearest-neighbor distance is 3.72×10^{-10} m, nearly equal to the equilibrium separation of the Lennard-Jones potential of Figure 16.13. The volume of the solid is nearly fixed because considerable energy is required either to increase or to decrease the nearest-neighbor distance.

In liquid argon, although the density is smaller than in the solid, the nearest neighbors are at very nearly the same distance as in the solid. On the average there are fewer nearest neighbors, because the nearest neighbors are disordered and voids exist between them. In a liquid of spherical molecules, the average number of nearest neighbors is approximately 10 or 11.

Exercise 16.16   Estimate the number of nearest neighbors around an argon atom in the liquid by multiplying 12, the number of nearest neighbors in the solid, by the ratio of the density of the liquid to the density of the solid. The density of solid argon is 1.66 g mL^{-1}, and that of liquid argon is 1.38 g mL^{-1}.

Since molecules in the liquid are surrounded by nearly as many nearest neighbors as are the molecules in a solid, their motions are very different from those of gaseous atoms. The nearest-neighbor molecules form a sort of "cage" in which a given molecule is confined by the repulsive intermolecular forces of the neighbors. Instead of moving considerable distances in nearly straight trajectories between occasional collisions, a molecule of a liquid is almost constantly involved in collisions as it undergoes a kind of zigzag rattling motion.

Intermolecular forces do not depend on velocities, so the velocity distribution of Section 16.2 is valid for a liquid or solid as well as for a gas. The mean speed of molecules of mass m in a liquid at temperature T is given by Equation (16.3-6):

$$\langle v \rangle = \sqrt{\frac{8k_{B}T}{\pi m}} \qquad \textbf{(16.9-1)}$$

Since the molecules in a liquid are much closer together than in a gas and are moving just as rapidly as in a gas at the same temperature, the rate of collisions in a liquid is much greater than in a gas. There is some ambiguity in the definition of a collision between two molecules in a liquid, because the molecules are not exactly like hard spheres and there is no unique instance of contact between them. However, if some definition of a collision is adopted, the rate of collisions between liquid molecules can be calculated theoretically.[7]

▼   **EXAMPLE 16.10**

Estimate the collision frequency of an argon atom in liquid argon at its normal boiling temperature, 87.3 K. Assume that the molecule is moving at the mean speed of gaseous argon molecules at 87.3 K. Find the distance required to travel twice the distance from the minimum in the potential function of Figure 16.13 to a value of r such that the potential energy is equal to the kinetic energy of a particle moving at the mean relative speed.

Solution

$$\langle v \rangle = \sqrt{\frac{8RT}{\pi M}} = \sqrt{\frac{8(8.3145 \text{ J K}^{-1}\text{ mol}^{-1})(87.3 \text{ K})}{\pi(0.039948 \text{ kg mol}^{-1})}} = 215 \text{ m s}^{-1}$$

The kinetic energy corresponding to this speed is 1.53×10^{-21} J. To find the value of r that corresponds to this value of the potential energy function, we set

$$1.53 \times 10^{-21} \text{ J} = 4\varepsilon[(\sigma/r')^{12} - (\sigma/r')^{6}]$$

where r' is the value of r that we seek. This equation must be solved by numerical approximation. With values of ε and σ from Table A20, we find that $r' = 3.3 \times 10^{-10}$ m. The minimum of the potential energy is at 3.8×10^{-10} m. Twice the distance from r' to this value of r is 1.0×10^{-10} m. The time required to traverse this distance at 215 m s^{-1} is equal to 4.7×10^{-13} s, giving a collision rate of 2.1×10^{12} s^{-1}.

▲

The approximate calculation in this example is based on a web of assumptions, but it illustrates the high collision rates in a liquid.

---

[7] P. K. Davis, *J. Chem. Phys.* **57**, 517 (1972).

As seen in Example 16.10, collision rates of a molecule in a typical liquid are roughly a hundred times larger than in a typical gas.

Summary

The macroscopic properties of a model system of noninteracting point mass molecules were deduced by averaging over molecular states. The Maxwell probability distribution of molecular velocities was deduced and used for this averaging process:

$$g(v) = \left(\frac{m}{2k_{\mathrm{B}}T} \right)^{3/2} e^{-mv^2/2k_{\mathrm{B}}T}$$

The probability distribution for molecular speeds is

$$f_v(v) = 4\pi \left(\frac{m}{2k_{\mathrm{B}}T} \right)^{3/2} v^2 e^{-mv^2/2k_{\mathrm{B}}T}$$

Formulas for the mean speed, the most probable speed, and the root-mean-square speed were derived.

It was found that the model system obeyed the ideal gas law and the law of partial pressures. The rate of wall collisions per unit area per unit time was also derived:

$$v = \frac{1}{4} \frac{N}{V} \langle v \rangle = \frac{1}{4} \mathcal{N} \langle v \rangle$$

Beginning with Section 16.7, a model system with interacting molecules was discussed. The hard sphere gas is a special case of this system. An approximate equation of state was derived for this system. Molecular collisions in the hard sphere gas were also discussed. Formulas were obtained for the mean free paths between collisions and for collision rates for both one-component and multicomponent systems. An important result was that the total rate of collisions in a one-component gas is proportional to the square of the number density and to the square root of the temperature. In a multicomponent gas, the rate of collisions between molecules of two different substances was found to be proportional to the number densities of both substances and to the square root of the temperature.

In Section 16.9, we presented a few elementary ideas about the molecular structure of liquids. In a liquid, the shell of nearest neighbors contains voids, so fewer nearest neighbors are present than in the solid. In a typical liquid, a molecule undergoes roughly a hundred times as many collisions per second as does a molecule in a typical gas.

Additional Reading

W. Kauzmann, *Thermal Properties of Matter*, W. A. Benjamin, New York, 1966

This book, written by a prominent theoretical chemist, contains a complete discussion of gas kinetic theory.

E. H. Kennard, *Kinetic Theory of Gases*, McGraw-Hill, New York, 1938
A classic treatment of kinetic theory at the graduate level.

D. Tabor, *Gases, Liquids and Solids*, 2nd ed., Cambridge University Press, Cambridge, 1985
This book, which appears in paperback, is a well-written survey of the theory of all three states of matter at an advanced undergraduate level. It concentrates on simple model systems.

PROBLEMS

Problems for Section 16.1

16.17. A group of people has the following distribution of annual income:

Income/$	Number of People
23000	12
24000	15
25000	18
26000	24
27000	19
28000	20
29000	12
30000	8

a. The **median** of a set of numbers is a kind of average value such that half of the members of the set are greater than or equal to the median, and half are less than or equal to the median. Find the median income of our set of people.

b. Find the mean income.

c. Find the root-mean-square income.

*16.18. Find the mean value and the root-mean-square value of $\sin(x)$ for $0 < x < 2\pi$ radians, assuming a uniform probability distribution.

16.19. Find the mean value and the root-mean-square value of $\sin(x)$ for $-\infty < x < \infty$, assuming the standard normal (gaussian) probability distribution

$$f(x) = \frac{1}{\sqrt{2\pi}} e^{-x^2/2}$$

*16.20. For nitrogen molecules at 300 K, estimate the error in the normalization integral of Equation (16.2-16) that is produced by neglecting relativity. The following asymp-

totic formula[8] gives values of the error function for large arguments:

$$\mathrm{erf}(z) = 1 - \frac{e^{-z^2}}{\sqrt{\pi}\,z}\left[1 + \sum_{m=1}^{\infty} \frac{(1)(3)\cdots(2m-1)}{(2z^2)^m}\right]$$

16.21. a. Use Equation (16.2-34) to estimate the probability that an argon atom in a system at 273.15 K has the x component of its velocity between 0 and 20 m s^{-1}.

b. Find the correct value of the probability and compare it with your result from part a.

Problems for Section 16.2

*16.22. Find the fraction of the molecules in a gas that have

$$v_x^2 > \frac{k_{\mathrm{B}}T}{m}$$

Show that this fraction is the same for all gases at all temperatures.

16.23. The escape velocity is defined as the minimum upward vertical velocity component required to escape the earth's gravity. At the earth's surface, its value is

$$v_{\mathrm{esc}} = 1.12 \times 10^4 \text{ m s}^{-1} = 2.5 \times 10^4 \text{ miles per hour}$$

a. Find the fraction of N_2 molecules at 298 K having an upward vertical velocity component exceeding v_{esc}.

b. Find the fraction of helium atoms at 298 K having an upward vertical velocity component exceeding v_{esc}. The asymptotic formula in Problem 16.20 can be used.

c. Find the fraction of helium atoms at 298 K having a speed exceeding v_{esc}. Note: See the identity in Problem 16.24.

d. Find the temperature at which the mean speed of helium atoms equals the escape velocity.

---

[8] M. Abramowitz and I. A. Stegun, *op. cit.*, p. 298.

Problems for Section 16.3

*16.24. a. Find the fraction of molecules that have speeds greater than $k_B T/m$. Explain the relationship of this fraction to the fraction computed in Problem 16.22. You can use the identity

$$\int_0^x t^2 e^{-at^2}\, dt = \frac{\sqrt{\pi}}{4a^{3/2}}\, \text{erf}(\sqrt{ax}) - \frac{x}{2a} e^{-ax^2}$$

b. Find the fraction of molecules with kinetic energies greater than $k_B T/2$. Explain the relationship of this fraction to that of part a.

16.25. Find the fraction of molecules in a gas that have:

a. Speeds less than the most probable speed

b. Speeds between the most probable speed and the mean speed

c. Speeds between the most probable speed and the root-mean-square speed

d. Speeds greater than the root-mean-square speed
Note: See the identity in Problem 16.24.

Explain why these fractions are independent of the temperature and of the mass of the molecules.

16.26. The speed of sound in air and in other gases is somewhat (about 20%) less than the mean speed of the gas molecules.

a. Explain why you think this is true.

b. How do you think that the speed of sound in air would depend on the temperature?

c. How do you think that the speed of sound in helium gas would compare with its speed in air? Why do you think a person's voice sounds different after a breath of helium?

16.27. Find the fraction of molecules in a gas with speeds greater than 80% of the mean speed.

*16.28. a. The standard deviation of the variable x is defined by Equation (10.4-16).

$$\sigma_x = [\langle x^2 \rangle - \langle x \rangle^2]^{1/2}$$

Find the expression for the standard deviation for the distribution of speeds of gas molecules.

b. Find the value of this standard deviation for oxygen molecules at 298 K.

16.29. If oxygen molecules had mass 32 g instead of 32 amu, what would their mean speed be at 298 K?

16.30. a. Find a general expression for the median speed of molecules in a dilute gas. Hint: See the identity in Problem 16.24.

b. Find the median speed for oxygen molecules at 298 K.

Problems for Section 16.4

16.31. In Section 16.4, it was assumed that the system was contained in a rectangular box. Explain why the pressure of a gas is independent of the shape of the container.

16.32. Show that in a specular collision with a flat wall, the angle of incidence and the angle of reflection are equal. The angle of incidence is the angle between the line perpendicular to the surface and the initial velocity, and the angle of reflection is the angle between the line perpendicular to the surface and the final velocity.

*16.33. Estimate the mass (in pounds and in kilograms) whose gravitational force at the surface of the earth is equal to the force of the atmosphere on your entire body. State any assumptions.

16.34. Derive the ideal gas equation of state without using the velocity distribution function. Hint: Start with Equations (16.2-25) and (16.4-12).

Problems for Section 16.5

16.35. The vapor pressure of solid tungsten (wolfram) at 4763 K is 10.0 torr. A sample of solid at equilibrium with its vapor is maintained at this temperature in a container that has a circular hole of diameter 0.100 mm leading to a vacuum chamber. Find the loss of mass through the hole in 1.00 hour.

*16.36. A spherical vessel of radius 10.0 cm contains argon at 0.980 atm pressure and 298.15 K. Calculate the number of argon atoms striking the container per second.

16.37. In a certain vacuum system, a pressure of 1.0×10^{-10} torr is achieved. Assume that the air inside the vacuum system is 80 mol % N_2 and 20 mol % O_2.

a. Find the number density in molecules per cubic meter for each substance if the temperature is 298 K.

b. Find the number of molecules striking 1.00 square centimeter of wall per second at 298 K.

16.38. A 1.000-liter container full of neon gas and maintained at 298 K is placed in a large vacuum chamber and a circular hole with a diameter of 20 micrometers is punched in the container. If the initial pressure in the container is 1013 Pa, find the time required for half of the neon gas to effuse from the container. Hint: Remember that the pressure will be dropping as the gas escapes.

16.39. Estimate the time required for a bicycle tire to go flat due to a circular hole of 50 micrometers diameter. State your assumptions.

Problems for Section 16.6

16.40. a. Find the probability of drawing the ace of spades from one deck of 52 cards and drawing the eight of diamonds from another deck of 52 cards.

b. Find the probability of drawing the ace of spades and the eight of diamonds (in that order) from a single deck of 52 cards.

c. Find the probability of drawing the ace of spades and the eight of diamonds (in either order) from a single deck of 52 cards.

*16.41. Compute the odds for each of the possible values of the sum of the two numbers showing when two dice are thrown.

16.42. Assume that air is 80% nitrogen and 20% oxygen, by moles, at sea level. Calculate the percentages and the total pressure at an altitude of 20 km, assuming a temperature of $-20°C$ at all altitudes. Calculate the percent error in the total pressure introduced by assuming that air is a single substance with molar mass 0.029 kg mol^{-1}.

16.43. Calculate the difference in the density of air at the top and bottom of a vessel 1.00 m tall at 273.15 K at sea level. State any assumptions.

*16.44. Estimate the difference in barometric pressure between the ground floor of a building and the forty-first floor, assumed to be 400 feet higher. State any assumptions.

16.45. A balloon is filled with helium at sea level and at a temperature of 20°C. The design of the balloon is such that the pressure inside it remains equal to the external atmospheric pressure (the volume can change).

a. If the volume of the balloon at sea level is 1000 m^3 find the mass of helium required to fill the balloon and the mass that can be lifted (including the mass of the balloon).

b. Assuming that the atmosphere has a uniform temperature, find the volume of the balloon at an altitude of 10.0 km and find the mass it can lift, assuming that the same amount of helium is in the balloon as in part a.

Problems for Section 16.7

16.46. Write an expression for the excluded volume of a pair of hard sphere molecules of different sizes.

16.47. Calculate the pressure of carbon dioxide gas at 298.15 K and a molar volume of 24.00 L, assuming (a) the ideal gas equation of state, (b) the van der Waals equation of state, and (c) Equation (16.7-5), taking the same value of the parameter b as for the van der Waals equation of state.

*16.48. For N_2 gas at 298.15 K and 1.000 atm, calculate the ratio of the total volume of the molecules to the volume of the gas, (a) using the hard sphere diameter from Table A20 and (b) using the molecular diameter calculated from the value of the van der Waals parameter b.

16.49. Find the value of the Lennard-Jones representation of the interatomic potential function of argon at interatomic distances equal to each of the effective hard sphere diameters

of argon at different temperatures in Table A20. Explain the temperature dependence of your values.

16.50. Another approximate representation for intermolecular pair potentials is the **exponential-6 potential**

$$u(r) = be^{-ar} - cr^{-6}$$

where a, b, c are parameters to be chosen to fit data for each substance. The function has a nonphysical maximum and approaches negative infinity as r approaches zero. For values of r smaller than the value at the maximum, this expression must be replaced by a different representation. The usual procedure is to define u to be positively infinite in this region. Find the values of a, b, and c in the 6-exponential representation of the interatomic potential of argon such that $c = 4\varepsilon\sigma^6$ and the minimum is at the same value of r as the minimum of the Lennard-Jones representation.

16.51. Calculate the coefficient of the r^{-6} term in the Lennard-Jones potential, $4\varepsilon\sigma^6$, for He, Ne, N_2, O_2, Ar, and CO_2. Make a rough graph of this quantity versus the number of electrons in the atom or molecule. Comment on your result in view of the interpretation that the London attraction is due to synchronized fluctuating dipoles in the electrons of the two attracting atoms or molecules.

Problems for Section 16.8

16.52. A good approximate equation of state for the hard sphere fluid is due to Carnahan and Starling:

$$\frac{PV}{Nk_BT} = \frac{1 + y + y^2 - y^3}{(1 - y)^3}$$

where

$$y = \frac{\pi N d^3}{6V}$$

make a graph of $z = PV/Nk_BT$ as a function of y, and a second graph as a function of $\bar{V} = V/(Nk_B)$, assuming $d = 3.00 \times 10^{-10}$ m.

16.53. Explain why, after the correction for relative motion was made, the formula for the collision time was still

$$\tau_{coll} = \frac{\lambda}{\langle v \rangle} \quad \text{(correct)}$$

instead of

$$\tau_{coll} = \frac{\lambda}{\langle v_{rel} \rangle} \quad \text{(wrong)}$$

*16.54. For a mixture of 2.000 mol of CO and 1.000 mol of O_2 at 292 K and 1.000 atm, calculate:

a. The number of collisions with O_2 molecules suffered by one CO molecule in 1.000 second, taking $d = 2.94 \times 10^{-10}$ m for CO

b. The number of collisions with CO molecules suffered by one O_2 molecule in 1.000 second

c. The time required for one CO molecule to have as many collisions with O_2 molecules as there are O_2 molecules in the system.

16.55. Assume that a certain region of interstellar space contains 1 hydrogen atom (not molecule) per cubic meter at a temperature of 5 K. Estimate the mean time between collisions for a hydrogen atom. You will have to find (or estimate) a value for the atomic diameter.

Problems for Section 16.9

16.56. The density of ice is 0.917 g mL^{-1} at 0°C, and that of liquid water is 1.000 g mL^{-1}. Ice is completely hydrogen-bonded with four nearest neighbors for each water molecule. Estimate the average number of nearest neighbors in liquid water at 0°C. Explain your answer, and explain why liquid water is denser than ice.

*16.57. a. Calculate the density of solid xenon from the Lennard-Jones parameters.

b. The density of liquid xenon is 3.52 g mL^{-1}. Estimate the number of nearest-neighbor atoms in the liquid.

16.58. a. If 1.000 mol of argon atoms is in a perfect crystal lattice such that each atom has 12 nearest neighbors at the interatomic distance equal to the distance at the minimum in the Lennard-Jones potential function, calculate the energy required to turn the crystal into a gas, neglecting all interactions except those of nearest neighbors. Assign half of the interaction energy of a pair to each member of the pair, so that each atom has to break six attractions in the sublimation process. Compare this energy with the actual energy of sub-limitation at 0 K, 8.49 kJ mol^{-1}.

b. Calculate the energy of vaporization of argon, assuming that each argon atom has approximately 10.5 nearest neighbors.

c. Calculate the energy of fusion of argon.

General Problems

16.59. A certain sample of air maintained at 25°C contains dust particles all of which are the same size. The diameter of the dust particles is 2.0 micrometers and their density is 1500 kg m^{-3}.

a. Find the most probable speed, the mean speed, and the root-mean-square speed of the dust particles, treating them as giant molecules.

b. Assuming that the dust particles are described by a Boltzmann distribution, find the ratio of the concentration of dust particles at a height of 1.000 m to the concentration at a height of 0.00 m.

c. Find the rate of collisions of one dust particle with other dust particles if their number density is 1.0×10^9 m^{-3}.

d. Find the total rate of collisions per cubic meter of pairs of dust particles.

e. Find the rate of collisions of one dust particle with nitrogen molecules.

*16.60. Consider a spherical water droplet in a cloud at 25°C. The radius of the droplet is 10.0 micrometers. The equilibrium vapor pressure of water at this temperature is 23.756 torr.

a. Calculate the rate at which water molecules strike the surface of the droplet, assuming that the air is saturated with water vapor (the partial pressure equals the equilibrium vapor pressure).

b. Assume that the air is supersaturated (the water vapor is supercooled) with a water partial pressure of 30.0 torr. Find the rate at which water molecules strike the surface of the droplet.

c. Calculate the rate at which the mass of the droplet in part b is growing. State any assumptions.

16.61. The number of three-body collisions is far smaller than the number of two-body collisions in a dilute gas. Consider three-body collisions in a sample of pure argon at 1.000 bar and 300 K. Assume that a three-body collision occurs when a third body collides with a pair of atoms in the act of colliding,

a. Estimate the number density of colliding pairs by estimating the time during which two colliding atoms are close enough together to be struck by a third body. Take this time as the time for an atom moving at the mean relative speed to travel a distance equal to the collision diameter.

b. Estimate the rate of three-body collisions by estimating the rate of collisions between colliding pairs and third bodies. Take an effective hard sphere diameter of the colliding pair to be twice that of a single atom.

16.62. Assume that a certain sample of polluted air has the following composition by moles: nitrogen, 76.08%; oxygen, 20.41%; water vapor, 2.63%; argon 0.910%; carbon dioxide, 0.0306%; ozone, 0.0004%; carbon monoxide, 0.0005%. The air is maintained at a temperature of 300 K and a pressure of 1.000 bar. Take a value of 4.3×10^{-10} m for the hard-sphere diameter of O_3.

a. Find the number of collisions per second a single ozone molecule undergoes with carbon monoxide molecules.

b. Find the number of ozone–carbon monoxide collisions per cubic meter per second.

c. Find the number of ozone-oxygen collisions per cubic meter per second.

d. Find the number of nitrogen-nitrogen collisions per cubic meter per second.

*16.63. Label each of the following statements as either true or false. If a statement is true only under special circumstances, label it as false.

a. If a given sample of a pure gas is isothermally expanded to twice its original volume, the total rate of collisions in the entire sample drops to one-fourth of its original value.

b. If a given sample of a pure gas is isothermally expanded to twice its original volume, the rate of collisions per unit volume drops to one-fourth of its original value.

c. The mean speed of water molecules at 100°C has the same value in the liquid as in the vapor.

d. The ratio of the most probable speed to the mean speed has the same value for all gases at all temperatures.

e. The ratio of the mean speed to the root-mean-square speed has the same value for all gases at all temperatures.

f. Ordinary gases behave nearly like ideal gases because the molecules are far enough apart on the average that the intermolecular forces are small.

g. In a typical gas under ordinary conditions, the average distance between neighboring molecules is roughly 10 times the distance between neighboring molecules in the liquid.

h. The mean free path in an ordinary gas is roughly equal to the average distance between neighboring molecules.

i. Since the temperature on the Kelvin scale cannot be negative, a state of higher energy cannot have a greater population than a state of lower energy.

j. The mean value of a velocity component is equal to the mean speed of the molecules of a dilute gas.

k. The most probable value of a velocity component is equal to the most probable speed of the molecules of a dilute gas.

l. The mean molecular kinetic energy of a gas is independent of the molecular mass.

17 Transport Processes

PREVIEW

In earlier chapters, we discussed a class of nonequilibrium processes that began and ended with equilibrium states. In this chapter, we discuss more general nonequilibrium processes in a one-phase fluid system in which no chemical reactions can occur. The principal examples of such processes are the transport processes: heat conduction, viscous flow, and diffusion.

PRINCIPAL FACTS AND IDEAS

1. If there is any tendency for a process to occur in it, a system is not at equilibrium.
2. The description of nonequilibrium states of fluid systems requires variables that do not occur in equilibrium thermodynamics: variables to express the rates of processes and variables to specify the extent to which the system deviates from equilibrium.
3. The three principal transport processes are heat conduction, diffusion, and viscous flow.
4. Each transport process is described by an empirical linear law.
5. Theories of transport processes are molecular theories.

OBJECTIVES

After studying this chapter, the student should:

1. understand qualitatively the behavior of a rate of a process governed by a linear law,
2. understand the molecular basis for transport processes in a hard sphere gas,
3. be able to solve a variety of problems related to transport processes in fluids,
4. understand qualitatively the relation of molecular motions in liquids to transport processes in liquids.

17.1 The Macroscopic Description of Nonequilibrium States

We asserted in Chapters 1 and 5 that in a one-phase simple system at equilibrium, the intensive macroscopic state is specified by $s + 1$ variables, where s is the number of independent chemical substances (we use the letter s now rather than c because we will use c for concentrations). These variables could be T, P, and $s - 1$ concentrations or mole fractions.

Processes that take place far from equilibrium, including such things as explosions and turbulent flow, are very difficult to describe mathematically.

Nonequilibrium states are more complicated than equilibrium states, and we will attempt to describe only systems whose states do not deviate too much from equilibrium states. In this chapter, we consider a one-phase simple system containing several substances in which no chemical reactions can occur. The rates of chemical reactions will be treated in Chapters 18 and 19.

The macroscopic variables of a nonequilibrium system can depend on position. However, the definitions of these variables require measurements at equilibrium. In order to define these variables in a nonequilibrium system, we visualize the following process: A small portion of the system (a subsystem) is suddenly removed from the system and allowed to relax adiabatically to equilibrium at fixed volume. Once equilibrium is reached, variables such as the temperature, pressure, density, and concentrations in this subsystem are measured. These measured values are assigned to a point inside the volume originally occupied by the subsystem and to the time at which the subsystem was removed. Each subsystem must contain enough molecules that the macroscopic variables do apply.

This procedure is performed repeatedly at different times and different locations in the system, and interpolation procedures are carried out to obtain smooth functions of position and time to represent the temperature, concentrations, etc.

$$T = T(x, y, z, t) = T(\mathbf{r}, t) \tag{17.1-1a}$$

$$P = P(x, y, z, t) = P(\mathbf{r}, t) \tag{17.1-1b}$$

$$c_i = c_i(x, y, z, t) = c_i(\mathbf{r}, t) \qquad (i = 1, 2, \ldots, s) \tag{17.1-1c}$$

where c_i is the concentration of substance i, measured in moles per unit volume (either L or m^3). We use the symbol $\mathbf{r}$ for the position vector with components x, y, and z.

Since we measure the intensive variables after each subsystem comes to equilibrium, they obey the same relations among themselves as they would in an equilibrium system. Therefore, at any point in a nonflowing system, $s + 1$ intensive variables are independent variables, and all other intensive variables are dependent variables. Specification of the intensive state of a nonflowing system requires specification of $s + 1$ intensive variables at every point of the system. That is, these variables must be expressed as functions of position and time. If the system is flowing, we will assert that, in addition to these variables, the flow velocity must be specified as a function of position and time.

To discuss the state at a point in the system, we do not need to know the state of the entire system. Instead of knowing the function representing a variable at all points of the system, we require only a specification of how rapidly each independent variable changes with position in the vicinity of the point of interest. We will assert as an experimental fact that in the near-equilibrium case, the intensive state at a point is adequately described by $s + 1$ independent intensive variables and the flow velocity plus the gradients of these quantities.

The **gradient** of a scalar function f is defined by Equation (B-27) of Appendix B. It is a vector that points in the direction of the most rapid increase of the function and has a magnitude equal to the derivative with respect to distance in that direction. The **temperature gradient** is denoted by ∇T and is given in cartesian coordinates by

$$\nabla T = \mathbf{i}\,\frac{\partial T}{\partial x} + \mathbf{j}\,\frac{\partial T}{\partial y} + \mathbf{k}\,\frac{\partial T}{\partial z} \qquad (17.1\text{-}2)$$

where $\mathbf{i}$, $\mathbf{j}$, and $\mathbf{k}$ are unit vectors in the directions of the x, y, and z axes, respectively. The concentration gradient of substance number i is

$$\nabla c_i = \mathbf{i}\,\frac{\partial c_i}{\partial x} + \mathbf{j}\,\frac{\partial c_i}{\partial y} + \mathbf{k}\,\frac{\partial c_i}{\partial z} \qquad (17.1\text{-}3)$$

where c_i is the concentration of substance i. Do not confuse the subscript i with the unit vector $\mathbf{i}$. As much as possible, we will discuss systems in which the variables depend on only one coordinate, so that gradients contain only one term.

Exercise 17.1   Assume that the concentration of substance number 2 is represented by the function

$$c_2 = c_2(z, t) = c_0 + a\cos(bz)e^{-t/\tau}$$

where c_0, a, b, and τ are constants. Write the expressions for $(\partial c_2/\partial t)$ and for ∇c_2.

The flow velocity $\mathbf{u}$ can be written as a function of position and time:

$$\mathbf{u} = \mathbf{u}(x, y, z, t) = \mathbf{u}(\mathbf{r}, t)$$
$$= \mathbf{i}u_x(\mathbf{r}, t) + \mathbf{j}u_y(\mathbf{r}, t) + \mathbf{k}u_z(\mathbf{r}, t) \qquad (17.1\text{-}4)$$

Each of the three components of the velocity has a gradient with three components. The gradient of the flow velocity vector, or **rate of shear**, thus has nine components. Such a nine-component quantity is called a **dyadic**, or **cartesian tensor**. Fortunately, we will be able to avoid using more than one component of it at a time. For example, consider

$$(\nabla\mathbf{u})_{zy} = (\partial u_y/\partial z) \qquad (17.1\text{-}5)$$

which is the derivative of the y component of the velocity with respect to z, giving the rate of change of the y component in the z direction. This quantity specifies the rate of shear, or the rate at which one layer of the fluid is sliding (shearing) past an adjacent layer.

Gradients can also be expressed in other coordinate systems. For example, in cylindrical polar coordinates, the three coordinates are z, as in cartesian coordinates, ϕ as in spherical polar coordinates, and ρ, equal to $\sqrt{x^2 + y^2}$. The gradient of a scalar function f is given by

$$\nabla f = \mathbf{e}_\rho\,\frac{\partial f}{\partial \rho} + \mathbf{e}_\phi\,\frac{1}{\rho}\,\frac{\partial f}{\partial \phi} + \mathbf{k}\,\frac{\partial f}{\partial z} \qquad (17.1\text{-}6)$$

where $\mathbf{e}_\rho$ is the unit vector in the ρ direction, $\mathbf{e}_\phi$ is the unit vector in the ϕ direction, and $\mathbf{k}$ is the unit vector in the z direction.

▼ **EXAMPLE 17.1**

The liquid in a pipe with radius R has a velocity that depends on ρ, the distance from the center of the pipe, such that $u_\rho = 0$, $u_\phi = 0$, and

$$u_z(\rho) = A(\rho^2 - R^2)$$

where A is a constant. Find the gradient of this velocity.

Solution

All of the nine components will vanish except for $\partial u_z / \partial \rho$:

$$\nabla f = \mathbf{e}_\rho \frac{\partial u_z}{\partial \rho} = \mathbf{e}_\rho 2A\rho$$

In this case, the gradient of the velocity points at right angles to the velocity. We will see this kind of dependence later in the chapter when we discuss viscous flow.

▲

In the near-equilibrium case that we discuss, all of the gradients of the intensive variables and of the flow velocity are small. We do not attempt to describe turbulent flow and other processes involving states far from equilibrium, in which case the gradients might be large.

17.2 Transport Processes

In the absence of chemical reactions, the principal nonequilibrium processes that can occur in a simple system are heat conduction, diffusion, and viscous flow. These are called **transport processes** because in each process some quantity is transported from one location to another. In heat conduction, energy is transported; in diffusion, molecules are transported; and in viscous flow, momentum is transported. These three processes are important in the chemical process industry, in biological organisms, and in many activities of ordinary life.

Variables to Specify the Rates of Transport Processes

The **heat flux** is a vector $\mathbf{q}$ that has the direction of the flow of heat and magnitude equal to the quantity of heat in joules per square meter per second passing through a plane perpendicular to the direction of heat flow.

The **diffusion flux** of substance i is a vector $\mathbf{J}_i$ that has the direction of the average velocity of the molecules of substance i and a magnitude equal to the net amount of the substance in moles per square meter per second passing through a plane perpendicular to the direction of diffusion. In precise discussions of diffusion, one must specify the velocity of the plane (whether it is stationary in the laboratory or is stationary with respect to the center of mass of a small portion of the fluid in the system, etc.).

Figure 17.1 shows an idealized experimental apparatus for the measurement of viscosity, which is the resistance of a fluid to shearing flow. The fluid is confined between two very large parallel plates. The top plate is dragged along parallel to its surface, in the y direction, and the lower plate is fixed to a stationary object.

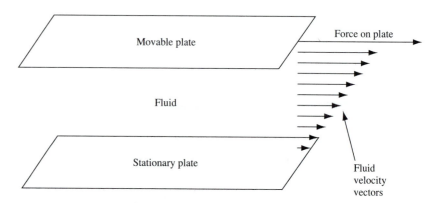

Figure 17.1. An Idealized Viscosity Experiment. The lower plate is stationary, and the upper plate is pulled at a constant speed relative to the lower plate. The fluid between the plates exhibits laminar flow, with a velocity that depends smoothly on the height above the lower plane.

The moving plate drags a layer of fluid along, transferring momentum to it, and this layer of fluid drags another layer along, transferring momentum to it, and so on. The momentum is eventually transferred to the stationary object, whose acceleration is infinitesimal (it must be attached to an object of infinite mass to be stationary). Due to frictional losses, the speed of each layer is a little smaller than the speed of the layer above it, giving a nonzero value to the rate of shear defined in Equation (17.1-5). In Figure 17.1 arrows are used to indicate the magnitude of the flow velocity at different heights. This kind of flow is called **laminar flow** (flow in layers). A flow that is not laminar is called **turbulent**.

Instead of using the momentum flux as a rate variable, it is customary to use the force per unit area that must be exerted on the moving plate by an external agent to keep it at a steady speed. Because Newton's second law relates force to the rate of change of momentum, this variable gives the same information.

The Driving Forces for Transport Processes

Every process needs a variable to express the rate of the process and a variable to express the driving force that produces the process.

To a first approximation each rate variable depends on a single "driving force." The temperature gradient is the driving force for heat conduction, the concentration gradient of substance i is the driving force for diffusion of that substance, and the velocity gradient, or rate of shear, is the driving force for viscous flow. In irreversible thermodynamics, transport processes are treated in a consistent way, using carefully defined driving forces and including the possibility that the driving force for one transport process contributes to the rate of another transport process.[1] An example of such

---

[1] S. deGroot and P. Mazur, *Nonequilibrium Thermodynamics*, North Holland, Amsterdam, 1962.

a "cross-effect" is the production of a diffusion flux by a temperature gradient (thermal diffusion, or the Soret effect). We will not discuss cross-effects.

The important fact is that the rate variables are not separate variables required to specify the state of the nonequilibrium system. They are dependent variables whose values are determined by the state of the system, and the state of the system is specified by $s + 1$ variables plus the flow velocity and the gradients of these quantities.

In some cases, transport processes are studied by measuring the rate at which an isolated system relaxes toward equilibrium. In other cases, a system is maintained in a time-independent nonequilibrium state (a **steady state**). For example, a time-independent temperature gradient might be maintained by keeping one end of a system at one temperature and the other end at another temperature, while heat flows through the system. A steady state is not an equilibrium state, although it is time independent, because an external agent must maintain the steady state.

Linear Phenomenological Laws

Three well-known empirical laws give the dependence of the rates on the driving forces for our three transport processes.

The first is **Fourier's law of heat conduction**:

Fourier's law is named for Jean Baptiste Joseph Fourier, 1768–1830, the famous French mathematician and physicist to whom we referred in Chapters 9 and 15.

$$\mathbf{q} = -\kappa \, \nabla T \quad \text{(Fourier's law)} \qquad (17.2\text{-}1)$$

where κ is called the **thermal conductivity**. For a case in which T varies only in the z direction, this equation is

$$q_z = -\kappa \frac{\partial T}{\partial z} \qquad (17.2\text{-}2)$$

Fourier's law is a **linear phenomenological law**. It is called linear because the rate is directly proportional to, or linearly dependent on, the driving force. The thermal conductivity κ is the proportionality constant and depends on the composition, the temperature, the pressure, and the identities of the substances present, but it does not depend on the gradients of temperature, concentrations, or velocity. The term "phenomenological" means that the law describes the phenomenon. The form of a phenomenological law might be deduced from reasonable assumptions, but within the phenomenological theory there is no theoretical formula for calculating values for coefficients such as the thermal conductivity. Later in the chapter we will discuss a molecular theory for these coefficients in dilute gases.

Although gaseous systems, liquid systems, and solid systems are very different from each other on the molecular level, Fourier's law holds quite accurately for all three. However, in measurements of thermal conductivity in gaseous and liquid systems, one must take extreme pains to eliminate convection contributions (contributions due to thermally driven currents in the fluid). Table A22 gives the values of the thermal conductivity for several pure substances.

▼   **EXAMPLE 17.2**

A cubical cell 0.100 meter on a side is filled with benzene. The top surface is maintained at 25.0°C and the bottom surface is maintained at 15.0°C. Calculate the amount of heat flowing through the benzene per hour after a steady state is achieved, neglecting convection.

Solution

The thermal conductivity of pure benzene at 20°C and 1.00 atm pressure is equal to 0.151 J m^{-1} s^{-1} K^{-1}. Assuming that the temperature depends only on z, the vertical coordinate, the temperature gradient is

$$\frac{dT}{dz} = \frac{\Delta T}{\Delta z} = \frac{10.0 \text{ K}}{0.100 \text{ m}} = 100 \text{ K m}^{-1}$$

The temperature gradient in such a uniform system must be constant because heat would accumulate in some region if it were not, and we would not have a steady state.

We have used an average value for the gradient by replacing dT/dz by the quotient of finite differences. In a steady state, the temperature gradient must be uniform in a system with uniform thermal conductivity and uniform cross-sectional area, and the gradient at any point must equal the average gradient. The heat flux has only a z component

$$q_z = -\kappa(dT/dz) = -(0.151 \text{ J m}^{-1} \text{ s}^{-1} \text{ K}^{-1})(100 \text{ K m}^{-1})$$
$$= -15.1 \text{ J m}^{-2} \text{ s}^{-1}$$

The cross-sectional area is 0.0100 m^2, so the total amount of heat flowing in 1.00 hour is

$$|-(15.1 \text{ J m}^{-2} \text{ s}^{-1})(0.0100 \text{ m}^2)(3600 \text{ s})| = 544 \text{ J}$$

◤

Fick's law is named for Adolf Fick, 1829–1901, German physiologist.

The second linear phenomenological law is **Fick's law of diffusion**

$$\boxed{\mathbf{J}_i = -D_i \nabla c_i \quad \text{(Fick's law)}} \qquad \text{(17.2-3)}$$

If the concentration varies only in the z direction, this equation becomes

$$J_{iz} = -D_i(\partial c_i/\partial z) \qquad \text{(17.2-4)}$$

The quantity D_i is called the **diffusion coefficient** of substance i. It depends on temperature, pressure, composition, and on which other substances are present, but not on the concentration gradient. Fick's law holds quite accurately for most gaseous, liquid and solid systems.

The fact that so many liquid state diffusion coefficients are nearly equal is worth remembering, as is the approximate value of the diffusion coefficients.

Table A23 gives the values of several diffusion coefficients near room temperature. It is remarkable that many liquid substances with "ordinary size" molecules have diffusion coefficients roughly equal to 10^{-9} m^2 s^{-1}. Diffusion coefficients for gases are somewhat larger than this value, and those for solids are much smaller. Diffusion in gases is more rapid because mean free paths in gases are much larger than the distances through which molecules can move quickly in liquids, and diffusion in solids is slow because molecules in a solid are nearly absolutely confined in crystal lattices.

Exercise 17.2	Calculate the value of the steady-state diffusion flux of glucose in water at 25°C if the concentration of glucose at the top of a cell of height 0.100 m is 0.100 mol L^{-1} and at the bottom of the cell is 0.0500 mol L^{-1}.

The version of Fick's law given in Equation (17.2-3) is sometimes called **Fick's first law**. **Fick's second law** is obtained by combining Fick's first law with an equation called the equation of continuity.

We now derive a one-dimensional version of the equation of continuity. Consider a small slab in a fluid system in which properties depend on z but not on x or y. This slab is depicted in Figure 17.2. The area of the large face of the slab is $\mathscr{A}$, and the thickness of the slab is Δz. Assume that a diffusion flux is present in the z direction. The amount of substance i entering the slab per second from below is

$$\text{Influx} = \mathscr{A} J_{iz}(z') \qquad (17.2\text{-}5)$$

where z' is the value of the z coordinate at the bottom of the slab. The amount leaving the slab per second through its top surface is

$$\text{Efflux} = \mathscr{A} J_{iz}(z' + \Delta z) \qquad (17.2\text{-}6)$$

Equation (17.2-7) is a statement of the fact that what goes in minus what comes out is what remains.

so the rate of change of the amount of substance i in the slab is

$$\frac{dn_i}{dt} = \mathscr{A}[J_{iz}(z') - J_{iz}(z' + \Delta z)] \qquad (17.2\text{-}7)$$

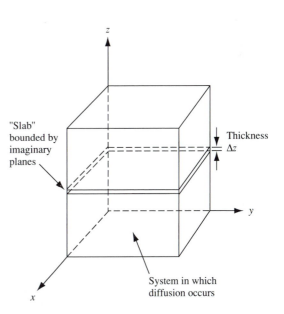

Figure 17.2. A Slab-Shaped Portion of a Fluid System. This figure is used in deriving the one-dimensional version of the equation of continuity by considering the influx and efflux of one component into and out of this slab.

Since the concentration of substance i is the amount of i per unit volume, the rate of change of the concentration is

$$\frac{\partial c_i}{\partial t} = \frac{dn_i/dt}{\mathscr{A}\,\Delta z} = \frac{J_{iz}(z') - J_{iz}(z' + \Delta z)}{\Delta z} \qquad \textbf{(17.2-8)}$$

We write the derivative $(\partial c_i/\partial t)_z$ as a partial derivative because we are taking this derivative at fixed values of the coordinates. Now we take the limit that Δz approaches zero, so that the right-hand-side of Equation (17.2-8) becomes a derivative.

The result is the one-dimensional version of the **equation of continuity** for substance i.

$$\frac{\partial c_i}{\partial t} = -\frac{\partial J_{iz}}{\partial z} \qquad \textbf{(17.2-9)}$$

Its physical content is the conservation of matter, because it is equivalent to saying that the rate of change in the concentration is just the difference between what arrives and what leaves. If a chemical reaction is consuming or producing substance i, this equation must be modified.

The three-dimensional version of the equation of continuity is

$$\frac{\partial c_i}{\partial t} = -\left(\frac{\partial J_{ix}}{\partial x} + \frac{\partial J_{iy}}{\partial y} + \frac{\partial J_{iz}}{\partial z}\right) \qquad \textbf{(17.2-10a)}$$

which is the same as

$$\boxed{\frac{\partial c_i}{\partial t} = -\nabla \cdot \mathbf{J}_i \quad \text{(equation of continuity)}} \qquad \textbf{(17.2-10b)}$$

This physical interpretation of the divergence as a measure of spreading of streamlines explains why the name was chosen and is worth remembering.

where $\nabla \cdot \mathbf{J}_i$ is called the **divergence** of $\mathbf{J}_i$, defined in Equation (B-28) of Appendix B. The divergence is a measure of the rate at which "stream lines" of a vector quantity diverge from each other. If it is positive, the stream lines move away from each other, and in this case the concentration of the substance decreases as one follows the flow.

Exercise 17.3 Derive Equation (17.2-10) by considering a small cube instead of a slab in a fluid system.

We substitute Equation (17.2-4) into Equation (17.2-9) to obtain

$$\frac{\partial c_i}{\partial t} = \frac{\partial}{\partial z}\left(D_i\frac{\partial c_i}{\partial z}\right) \qquad \textbf{(17.2-11)}$$

This is Fick's second law of diffusion for the one-dimensional case. If D_i is independent of composition and is therefore independent of position,

$$\frac{\partial c_i}{\partial t} = D_i\frac{\partial^2 c_i}{\partial z^2} \qquad \textbf{(17.2-12)}$$

The diffusion equation is not valid if D_i is not constant.

This equation is also called the **diffusion equation**. If the concentration depends on all three coordinates,

$$\frac{\partial c_i}{\partial t} = D_i\left(\frac{\partial^2 c_i}{\partial x^2} + \frac{\partial^2 c_i}{\partial y^2} + \frac{\partial^2 c_i}{\partial z^2}\right) = D_i \nabla^2 c_i \qquad \textbf{(17.2-13)}$$

The operator ∇^2 ("del squared") is the same Laplacian operator that occurs in the Hamiltonian operator of quantum mechanics.

Equation (17.2-12) is a partial differential equation that can be solved for some sets of initial conditions. For example, if a solution initially containing substance 2 at concentration c_0 is placed in the bottom half of a cell and pure solvent (substance 1) is carefully layered above it in the top half of the cell, the initial condition is

$$c_2(z, 0) = \begin{cases} c_0 & \text{if } z < 0 \\ 0 & \text{if } z > 0 \end{cases}$$

This initial condition fails to meet our criterion of small gradients at the location $z = 0$, but the state of the system rapidly becomes one in which the gradient of c_2 is fairly small, and Fick's law seems to hold quite accurately for this case.

The solution of Equation (17.2-12) that satisfies this initial condition is

$$c_2(z, t) = (c_0/2)[1 - \text{erf}(z/2\sqrt{D_2 t})] \qquad \textbf{(17.2-14)}$$

where erf( ) is the error function. This solution is shown in Figure 17.3 for several values of t, for a value of D_2 equal to 1.0×10^{-9} m^2 s^{-1}.

Notice the scale on the coordinate axis and the lengths of time of the curves in Figure 17.3. Ordinary liquids require several hours to diffuse a centimeter or so.

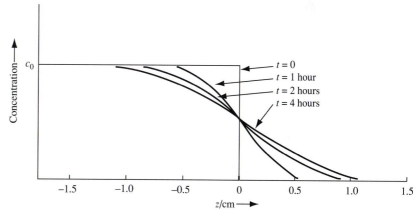

Figure 17.3. Concentration as a Function of Position in a Diffusing System, as in Equation (17.2-14). This figure corresponds to the initial state with a uniform concentration of the diffusion component in the bottom half of the cell and none in the top half. The concentration profile is given for the times indicated, showing that for a typical liquid-state value of the diffusion coefficient, diffusion over a distance of a centimeter requires several hours.

Exercise 17.4 Show by substitution that the function of Equation (17.2-14) satisfies Equation (17.2-12). You will need the identity

$$\frac{d}{dx} \int_0^x f(u)\, du = f(x) \qquad\qquad \textbf{(17.2-15)}$$

Another case that can be analyzed mathematically is obtained when the bottom half of a cell is filled with pure solvent and a very thin layer of solution is layered carefully on it, followed by more pure solvent to fill the cell. Let $z = 0$ at the center of the cell. An idealized representation of this initial condition is

$$c_2(z, 0) = \begin{cases} \infty & \text{if } z = 0 \\ 0 & \text{if } z \neq 0 \end{cases}$$

with the total amount of substance 2 present per unit cell cross-sectional area denoted by n_0. Again, this concentration function fails to meet our criterion of small gradients, as well as being impossible to attain in fact. However, the system will rapidly evolve to a state of small gradients, and Fick's law seems to hold quite well for this case. For a long cell, a solution to Equation (17.2-12) that matches this initial condition is

$$c_2(z, t) = \frac{n_0}{2\sqrt{\pi D_2 t}}\, e^{-z^2/4D_2 t} \qquad\qquad \textbf{(17.2-16)}$$

This function is shown in Figure 17.4 for D_2 equal to 1.0×10^{-9} m² s⁻¹ and for several values of t.

Compare Figure 17.4 with Figure 17.3.

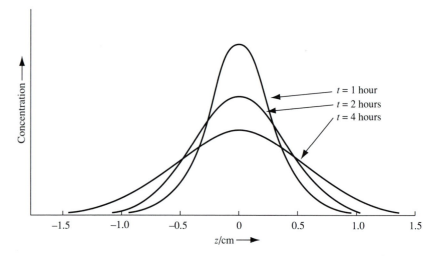

Figure 17.4. Concentration as a Function of Position in a Diffusing System with an Initial Thin Layer of Solute, as in Equation (17.2-16). This figure corresponds to the initial state of a very thin layer of the diffusing substance at $z = 0$.

Exercise 17.5

a. Show by substitution that the function in Equation (17.2-16) satisfies Equation (17.2-12).

b. Show that the same amount of substance 2 is present at a time $t = t_2$ as at time $t = t_1$ by showing that

$$\int_{-\infty}^{\infty} c_2(z, t_2)\, dz = \int_{-\infty}^{\infty} c_2(z, t_1)\, dz = n_0$$

Since the cell is assumed to be long, we use infinite limits for the integrals.

Extending limits of integration to infinity makes little numerical difference if the integrand rapidly approaches zero at the ends of the interval of integration.

Since all of the molecules of substance 2 started out at $z = 0$, we can use Equation (17.2-16) to study their average displacement in the z direction. Consider a little slab of the system lying between z' and $z' + dz$, where z' is some value of z. The fraction of the molecules of substance 2 in the slab at time t is

$$\text{Fraction in slab} = \frac{c_2(z', t)\, dz}{\int_{-\infty}^{\infty} c_2(z, t)\, dz} = \frac{c_2(z', t)\, dz}{n_0} \qquad \textbf{(17.2-17)}$$

where we have used the value of the integral from Exercise 17.5. This fraction is a probability much like those of Chapters 10 and 16. The mean value of the coordinate z at time t is thus, by Equation (10.4-10),

$$\langle z(t) \rangle = \frac{1}{n_0} \int_{-\infty}^{\infty} z c_2(z, t)\, dz = 0 \qquad \textbf{(17.2-18)}$$

The fact that the mean displacement in the z direction is zero means that for every molecule that has moved in the positive z direction, another has moved the same distance in the negative z direction.

The root-mean-square change in the z coordinate is an inherently nonnegative quantity and gives a measure of the magnitude of the distance traveled in the z direction by an average molecule. By analogy with Equation (16.3-7)

The random walk is a model process in which an object repeatedly takes a "step" in a randomly chosen direction. It is discussed in Chapter 22 in connection with polymer conformations.

$$z_{\text{rms}} = \langle z^2 \rangle^{1/2} = \left[\frac{1}{n_0} \int_{-\infty}^{\infty} z^2 c_2(z, t)\, dz \right]^{1/2} = (2 D_2 t)^{1/2} \quad \textbf{(17.2-19)}$$

The root-mean-square displacement is proportional to the square root of the elapsed time and to the square root of the diffusion coefficient. This behavior is typical of a process that can be represented by a "random walk."[2]

Exercise 17.6

Look up the integral and show that Equation (17.2-19) is correct.

[2] L. E. Reichl, *A Modern Course in Statistical Physics*, University of Texas Press, Austin, 1980, pp. 151ff.

▼   **EXAMPLE 17.3**   Find the root-mean-square distance diffused by glucose molecules in 30.0 minutes in water at 25°C.

Solution

The total rms distance traveled in three-dimensional space is

$$r_{rms} = [\langle x^2 \rangle + \langle y^2 \rangle + \langle z^2 \rangle]^{1/2} = [3\langle z^2 \rangle]^{1/2} = \sqrt{3}\, z_{rms}$$

where we use the fact that all three directions are equivalent. From Table A23, the value of D at 25°C is 0.673×10^{-9} m² s⁻¹.

$$r_{rms} = (6D_2 t)^{1/2}$$
$$= [6(0.673 \times 10^{-9} \text{ m}^2 \text{ s}^{-1})(30.0 \text{ min})(60 \text{ s min}^{-1})]^{1/2}$$
$$= 2.70 \times 10^{-3} \text{ m} = 0.270 \text{ cm}$$

◣

Newton's law of viscous flow is named for Sir Isaac Newton, 1642–1727, great British mathematician and physicist, to whom we have repeatedly referred.

P_{zy} is one of nine similar quantities. P_{xx}, P_{yy}, and P_{zz} are pressures (force per unit area) in given directions. P_{yz} and similar quantities are forces per unit area, but they are not pressures because the force is parallel to the area, not perpendicular to it. The nine quantities constitute the "pressure tensor."

Newton's law of viscous flow is the third linear phenomenological law:

$$\boxed{P_{zy} = \eta(\partial u_y/\partial z) \quad \text{(Newton's Law)}} \qquad \textbf{(17.2-20)}$$

where P_{zy} is the force per unit area on the upper plate in Figure 17.1 that is required to maintain a steady speed. The first subscript on P_{zy} indicates that the plane is perpendicular to the z direction, and the second subscript indicates that the force is in the y direction (parallel to the plane). The coefficient η is called the **viscosity coefficient**, or the **viscosity**, and is a constant at constant temperature if Newton's law is obeyed. The derivative $(\partial u_y/\partial z)$ is a component of the velocity gradient, or the rate of shear. Table A24 gives values for viscosity coefficients for a few liquids and gases.

Some liquids, such as blood and polymer solutions, do not obey Newton's law. These fluids are called **non-Newtonian fluids** and are described by a viscosity coefficient that depends on the rate of shear.

▼   **EXAMPLE 17.4**   The viscosity coefficient of water at 20°C equals 0.001002 kg m⁻¹ s⁻¹ (0.001002 Pa s). For an apparatus like that in Figure 17.1, find the force per unit area required to keep the upper plate moving at a speed of 0.250 m s⁻¹ if the tank is 0.500 m deep.

Solution

The component of the velocity gradient has an average value of

$$\frac{\partial u_y}{\partial z} = \frac{0.250 \text{ m s}^{-1}}{0.500 \text{ m}} = 0.500 \text{ s}^{-1}$$

so

$$P_{zy} = (0.001002 \text{ kg m}^{-1} \text{ s}^{-1})(0.500 \text{ s}^{-1})$$
$$= 5.01 \times 10^{-4} \text{ kg m}^{-1} \text{ s}^{-2}$$
$$= 5.01 \times 10^{-4} \text{ N m}^{-2}$$

◣

Exercise 17.7   A certain sleeve bearing in a machine has an area of 26.6 cm². It is lubricated with an oil having a viscosity coefficient of 0.0237 kg m⁻¹ s⁻¹. If the film of oil has a thickness of 0.30 mm and the radius of the bearing is 2.00 cm, find the frictional torque on the shaft if it is turning at 600 revolutions per minute. Assume that the oil does not slip on the metal surfaces.

When laboratory measurements of viscosity coefficients are made, an apparatus like the idealized one of Figure 17.1 is not ordinarily used. One simple apparatus used for liquids consists of a tube of uniform diameter through which the liquid is forced by a pressure difference. For an incompressible liquid, the **volume rate of flow**, dV/dt, is given by **Poiseuille's equation**:

Poiseuille's equation is named for Jean Leonard Marie Poiseuille, 1797–1869, French physician who studied the circulation of blood.

Do not confuse R, the radius of the tube, with the gas constant.

$$\frac{dV}{dt} = \frac{(P_2 - P_1)\pi R^4}{8L\eta} \qquad \text{(17.2-21)}$$

where R is the radius of the tube, L is its length, and $P_2 - P_1$ is the difference in pressure between the ends of the tube.

To derive Poiseuille's equation, we first find how the flow velocity of the fluid in the tube depends on position. Consider a portion of the fluid in the tube that is contained in an imaginary cylinder of radius r that is concentric with the tube walls, as shown in Figure 17.5. When a steady state has been

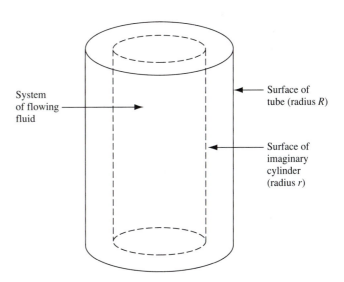

System of flowing fluid

Surface of tube (radius R)

Surface of imaginary cylinder (radius r)

Figure 17.5. The Fluid in a Tube. When fluid undergoes laminar flow in a tube, the velocity at any part of the surface of the imaginary cylinder equals that at any other part of the same cylinder.

reached, the frictional force due to viscosity at the surface of this cylinder exactly balances the hydrostatic force on the liquid in the cylinder. From Newton's law of viscosity

$$\frac{F}{\mathscr{A}} = \eta \frac{du_z}{dr} \tag{17.2-22}$$

where F is the magnitude of the force on the liquid, $\mathscr{A}$ is the area of the cylinder, and du_z/dr is the magnitude of the velocity gradient, which is perpendicular to the axis. We take the axis of the cylinder as the z direction and the perpendicular distance from the axis as the variable r.

The force is equal to the pressure difference times the cross-sectional area of the cylinder, and the area of the cylinder is its circumference times its length:

In this derivation, we assume that the pressure does not depend on r, the distance from the center of the tube.

$$F = (P_2 - P_1)\pi r^2, \qquad \mathscr{A} = 2\pi r L$$

Substitution of this into Equation (17.2-22) gives

$$\frac{(P_2 - P_1)r}{2L} = \eta\left(\frac{du_z}{dr}\right)$$

This is a differential equation that can be solved by separation of variables. We divide both sides of the equation by η and multiply both sides by dr, and the right-hand-side is then equal to the differential du_z. We integrate both sides from $r = R$ to $r = r'$, where r' is some arbitrary value of r:

$$\int_{u_z(R)}^{u_z(r')} du_z = \frac{P_2 - P_1}{2L\eta}\int_{R}^{r'} r\, dr$$

where we have switched sides of the equation. If the liquid adheres to the tube walls, $u_z(R)$ will vanish, and

$$u_z(r') = \frac{P_2 - P_1}{4L\eta}(r'^2 - R^2) \tag{17.2-23}$$

This parabolic dependence of flow velocity on position is shown in Figure 17.6. The length of each arrow is proportional to the flow velocity at its location.

We now find the total rate of flow of the incompressible liquid through the tube. Consider a cylindrical shell of thickness dr and radius r, concentric with the walls of the tube. The volume of the fluid in this shell that flows out of the tube in 1 second is equal to the cross-sectional area of the shell times a length equal to the distance traveled in 1 second:

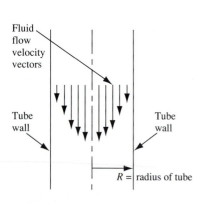

Figure 17.6. The Fluid Velocity in a Tube with Laminar Flow. The fluid velocity is a parabolic function, so the rate of shear is a simple function of position.

$$\text{Contribution of shell to } dV/dt = 2\pi r\, dr\, |u_z(r)|$$

where we put absolute magnitude bars on the velocity u_z since its direction is unimportant. The total volume rate of flow is the integral of the contributions of all such shells. Since the circumference of a shell of radius r is $2\pi r$, the total volume rate of flow is

$$\frac{dV}{dt} = 2\pi \int_0^R r|u_z(r)|\, dr = 2\pi \frac{P_2 - P_1}{4L\eta} \int_0^R (R^2 r - r^3)\, dr$$

$$\boxed{\frac{dV}{dt} = \frac{(P_2 - P_1)\pi R^4}{8L\eta} \quad \text{(Poiseuille's equation)}} \qquad \textbf{(17.2-24)}$$

This is Equation (17.2-21), Poiseuille's equation for an incompressible liquid.

▼ **EXAMPLE 17.5**

Water flows through a tube of length 0.420 m and radius 0.00520 m. If the pressure difference is 0.0500 atmosphere and the temperature is 20°C, find the volume of water that flows in 1.000 hour.

Solution

$$\frac{dV}{dt} = \frac{(0.0500 \text{ atm})(101325 \text{ N m}^{-2} \text{ atm}^{-1})(\pi)(0.00520 \text{ m})^4}{(8)(0.420 \text{ m})(0.001002 \text{ kg m}^{-1}\text{ s}^{-1})}$$

$$= 0.00346 \text{ m}^3 \text{ s}^{-1}$$

$$V = (0.00346 \text{ m}^3 \text{ s}^{-1})(3600 \text{ s}) = 12.4 \text{ m}^3$$

▲

Exercise 17.8

"Blood is thicker than water."

The blood pressure difference across a capillary in a human body is approximately 22 torr. Assume that a human body contains 1×10^{10} capillaries with an average length of 8×10^{-4} m and an average diameter of 7×10^{-6} m. Although blood is a non-Newtonian fluid, assume that a Newtonian viscosity of 0.004 kg m^{-1} s^{-1} can be used. Estimate the volume of blood flowing through the human circulatory system in L min^{-1}, assuming Poiseuille's equation. The actual value is near 5 L min^{-1}. Comment on your result, in view of the fact that the average diameter of red blood cells is near 7×10^{-6} m, the average capillary diameter.

Poiseuille's equation is correct only if the fluid flow is laminar. If the flow is turbulent, the problem is much more complicated, and we will not attempt to discuss it. A dimensionless quantity called the Reynolds number can be used to determine whether flow through a tube is probably laminar. The **Reynolds number** is defined by

$$\mathscr{R} = \frac{R\langle u\rangle \rho}{\eta} \quad \text{(definition of Reynolds number)} \qquad \textbf{(17.2-25)}$$

where R is the radius of the tube, ρ is the density of the fluid, and $\langle u\rangle$ is the mean speed of flow in the tube. It is found experimentally that flow in a tube is nearly always laminar if the Reynolds number is smaller than some value near 2000, no matter what the values of the individual quantities in Equation (17.2-25) are. If the tube is long, smooth, and straight, the flow might be laminar if the Reynolds number is as large as 3000 or 4000, but it is best to assume that the flow is not laminar if $\mathscr{R}$ exceeds 2000.

| **Exercise 17.9** | Estimate the Reynolds number for the flow of blood in a typical human capillary. Is the flow laminar? |

The viscosity can also be measured by using the frictional force that a fluid exerts on a spherical object moving through the fluid. Viscosities can be measured by dropping a spherical ball into the fluid and measuring its rate of descent. For a spherical object moving at a speed v through a fluid with viscosity η, **Stokes' law** is

Stokes' law is named for George Gabriel Stokes, 1819–1903, great Anglo-Irish mathematician and physicist who pioneered the science of hydrodynamics.

$$\mathbf{F}_f = -6\pi\eta r\mathbf{v} \qquad (17.2\text{-}26)$$

where $\mathbf{F}_f$ is the frictional force, r is the radius of the spherical object, and $\mathbf{v}$ is the object's velocity. The negative sign indicates that the friction force is in the opposite direction to the velocity. Stokes' law holds only for velocities small enough that the flow around the sphere is laminar.

▼   **EXAMPLE 17.6**

An iron sphere of density 7.874 g mL^{-1} is falling at a steady speed in glycerol at $20°C$. The density of glycerol is 1.2613 g mL^{-1} and its viscosity is 1.49 Pa s. If the radius of the sphere is 5.00 mm, find the speed.

Solution

The frictional force must be equal in magnitude to the gravitational force, which (corrected for buoyancy) is equal to

$$|\mathbf{F}_g| = \frac{4\pi r^3}{3}(\rho_{Fe} - \rho_{gly})g$$

$$= \frac{4\pi}{3}(5.00 \times 10^{-3}\text{ m})^3(7874\text{ kg m}^{-3} - 1261.3\text{ kg m}^{-3})(9.80\text{ m s}^{-2})$$

$$= 0.0339\text{ kg m s}^{-2} = 0.0339\text{ N}$$

$$v = \frac{|\mathbf{F}_g|}{6\pi\eta r} = \frac{0.0339\text{ kg m s}^{-2}}{6\pi(1.49\text{ kg m}^{-1}\text{ s}^{-1})(5.00 \times 10^{-3}\text{ m})} = 0.241\text{ m s}^{-1}$$

▲

Although Stokes' law was originally observed to hold for macroscopic objects, it is frequently applied as an approximation to molecules moving through a liquid. Since Stokes' law depends on a description of the fluid as a set of flowing layers without recognition of its molecular nature, application of Stokes' law to moving molecules is more meaningful for large molecules such as proteins than for small molecules that are nearly the same size as the solvent molecules. However, even for small molecules we often say that Stokes' law defines an "effective radius."

17.3   Transport Processes in the Hard Sphere Gas

In Chapter 16, we presented a molecular theory for the pressure of a dilute gas. We now discuss an elementary molecular theory for nonequilibrium processes in a hard sphere gas. This theory is based on the simple theory of collisions in a hard sphere gas presented in Chapter 16. We discuss the application of this theory to self-diffusion; but give only the results of its application to heat conduction and viscous flow.

Self-Diffusion

Fick's law of diffusion is given by Equation (17.2-3). It is the objective of our molecular theory to derive this equation for a particular model system, including an expression for the diffusion coefficient.

In order for diffusion to take place, there must be at least two substances present. In **self-diffusion**, the molecules of the two substances have all of their properties in common, but can somehow be distinguished from each other. This situation cannot actually occur, but is approximated in the laboratory by using two substances which differ only by isotopic substitution. The molecular masses will be different, but this difference can sometimes be made small. For example, one might replace in some molecules a carbon-12 atom by a carbon-13 atom and in the other molecules a hydrogen atom by a deuterium atom. Diffusion of two substances that differ only by isotopic substitution is called **tracer diffusion**.

We discuss a gaseous model system with two kinds of hard spherical molecules, both with the same molecular size and molecular mass. The model system is at a uniform temperature and a uniform pressure and is a closed system confined in a rectangular box with four vertical sides. At the start of the experiment, the system is at equilibrium except that the concentrations depend on the z (vertical) coordinate. The sum of the two concentrations is independent of position to avoid a pressure gradient. The process of diffusion will gradually eliminate the concentration gradient as the molecules mix.

This approximate analysis is sometimes presented with a spacing between the planes equal to $2\lambda/3$ or some similar fraction of λ to account for the fact that some molecules approach the center plane in directions other than perpendicular.

Figure 17.7 depicts our model system. In the interior of the system, we have drawn three imaginary horizontal planes. The upper and lower planes are placed at a distance from the center plane equal to λ, the mean free path between two collisions with any kind of particle. These planes specify locations at which we might observe what is going on; there are no physical structures in the system at these locations. We will perform an approximate analysis by counting up the molecules that pass through the center plane, whose value of the z coordinate we call z'.

Since all molecules are of the same size and mass and we are considering all kinds of collisions, the mean free path is given by Equation (16.8-22)

$$\lambda = \frac{1}{\sqrt{2}\,\pi d^2 \mathcal{N}_t} \qquad \textbf{(17.3-1)}$$

where $\mathcal{N}_t$ is the total number density, the sum of the number densities of the two substances:

$$\mathcal{N}_t = \mathcal{N}_1 + \mathcal{N}_2 \qquad \textbf{(17.3-2)}$$

and d is the effective hard sphere diameter of the molecules.

We now assume that all molecules passing upward through the center plane last suffered collisions in the vicinity of the lower plane and were

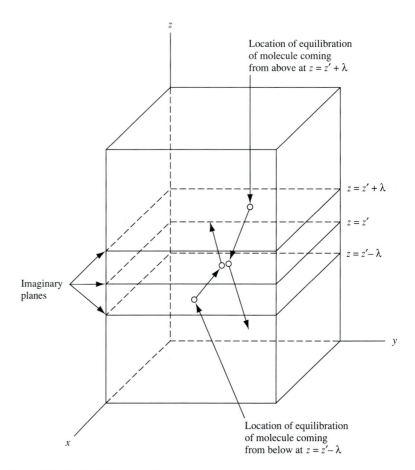

Figure 17.7. The Model System Showing Three Imaginary Planes for Analysis of Self-Diffusion in a Hard Sphere Gas. The spacing between the planes is equal to the mean free path for the gas, and the planes are used in deriving the self-diffusion coefficient for the gas.

equilibrated at that location. Similarly, molecules passing through the center plane from above are assumed to have been equilibrated at the location of the upper plane. This is the crucial assumption of the theory. It is clearly not correct for every molecule, since the z component of every free path is not equal to the mean free path and equilibration might not be complete at each collision. However, it is approximately correct on the average.

Equation (16.5-6) gives the rate of molecules in a gas striking a unit area of a wall or passing through a unit area of an imaginary plane. The number of molecules per unit area per unit time passing upward through the center plane is

The number density must be evaluated at the location at which the molecules had their last collision before approaching the center plane, because this is the location at which we assume they were equilibrated.

$$v_1(\text{up}) = \frac{1}{4}\,\mathcal{N}_1(z' - \lambda)\langle v \rangle \tag{17.3-3}$$

where the argument of $\mathcal{N}_1$ indicates that $\mathcal{N}_1$, which depends on z, is evaluated at $z' - \lambda$. The rate of molecules of substance 1 passing downward through a unit area of the center plane is:

$$v_1(\text{down}) = \frac{1}{4}\,\mathcal{N}_1(z' + \lambda)\langle v \rangle \tag{17.3-4}$$

where $\langle v \rangle$ is the mean speed of the molecules.

The diffusion flux is a vector whose magnitude equals the net amount of the substance passing a unit area of a plane perpendicular to the direction of diffusion per second. Since our system is uniform in the x and y directions, the diffusion flux is in the z direction and is given by

$$J_{1z} = \frac{1}{N_A}\left[v_1(\text{up}) - v_1(\text{down})\right] \tag{17.3-5}$$

where N_A is Avogadro's number and the factor $1/N_A$ is needed to express the flux in moles instead of molecules.

We now substitute our expressions for the upward and downward rates into Equation (17.3-5):

$$J_{1z} = \frac{\langle v \rangle}{4N_A}\left[\mathcal{N}_1(z' - \lambda) - \mathcal{N}_1(z' + \lambda)\right] = \frac{\langle v \rangle}{4}\left[c_1(z' - \lambda) - c_1(z' + \lambda)\right]$$

$$= -\frac{\langle v \rangle}{4}\left[\frac{c_1(z' + \lambda) - c_1(z' - \lambda)}{2\lambda}\right]2\lambda$$

where we have recognized that the concentration $c_1 = \mathcal{N}_1/N_A$ and in the last equation have multiplied and divided by 2λ.

If we were to take the limit of the quantity in square brackets as λ approached zero, it would become equal to the derivative $(\partial c_1/\partial z)$. We cannot take this limit, but if c_1 is nearly a linear function of z over the small region we are considering, the quotient of finite differences is nearly equal to the derivative and we can write, to a good approximation,

$$J_{1z} = -\frac{\langle v \rangle \lambda}{2}\left(\frac{\partial c_1}{\partial z}\right) \tag{17.3-6}$$

For the case of dependence only on the z coordinate, Fick's law is given by Equation (17.2-4):

$$J_{1z} = -D_1(\partial c_1/\partial z) \tag{17.3-7}$$

Comparison of this with Equation (17.3-6) shows that

$$D_1 = \frac{\langle v \rangle \lambda}{2} = \frac{1}{2}\left(\frac{8k_B T}{\pi m}\right)^{1/2}\frac{1}{\sqrt{2}\,\pi d^2\mathcal{N}_t}$$

$$= \frac{1}{\pi d^2\mathcal{N}_t}\left(\frac{k_B T}{\pi m}\right)^{1/2} \quad \text{(appproximate equation)} \tag{17.3-8}$$

where we have used Equation (16.3-6) for the mean speed and Equation (16.8-22) for the mean free path. Since substances 1 and 2 are of the same mass and size, D_1 and D_2 are equal and the subscript is unnecessary.

Our theory was cruder than necessary. A more accurate theory for the hard sphere gas takes into account that the molecules do not all arrive at a given plane from a vertical distance equal to the mean free path. The diffusion coefficient obtained in this treatment has the same dependence on density, temperature, mass, and hard sphere diameter as the expression of Equation (17.3-8) and is only about 18% different from that expression:

$$D = \frac{3\pi}{16} \lambda \langle v \rangle = \frac{3}{8d^2 \mathcal{N}_t} \left(\frac{k_B T}{\pi m} \right)^{1/2} \quad \left(\begin{array}{c} \text{more accurate} \\ \text{equation} \end{array} \right) \qquad (17.3\text{-}9)$$

One should use Equation (17.3-9), not Equation (17.3-8), for any numerical calculations.

Except for the constant, the first version of Equation (17.3-9) is easy to remember because it is so reasonable physically. The farther the molecules travel without collision, the faster they diffuse, and the faster they travel, the faster they diffuse.

EXAMPLE 17.7

Neglecting the mass difference, calculate the diffusion coefficient for the diffusion of isotopically substituted nitrogen molecules in ordinary nitrogen at 298 K and 1.00 atm.

Solution

We use the ideal gas law to calculate the number density:

$$\mathcal{N}_t = \frac{N}{V} = \frac{P}{k_B T} = \frac{101325 \text{ N m}^2}{(1.3807 \times 10^{-23} \text{ J K}^{-1})(298 \text{ K})}$$
$$= 2.46 \times 10^{25} \text{ m}^{-3}$$

From Table A21, $d = 3.7 \times 10^{-10}$ m.

$$D = \frac{3}{8d^2 \mathcal{N}_t} \left(\frac{k_B T}{\pi m} \right)^{1/2} = \frac{3}{8d^2 \mathcal{N}_t} \left(\frac{RT}{\pi M} \right)^{1/2}$$

$$= \frac{3}{8(3.7 \times 10^{-10} \text{ m})^2 (2.46 \times 10^{25} \text{ m}^{-3})} \left[\frac{(8.3145 \text{ J K}^{-1} \text{ mol}^{-1})(298 \text{ K})}{\pi (0.028 \text{ kg mol}^{-1})} \right]^{1/2}$$

$$= 1.9 \times 10^{-5} \text{ m}^2 \text{ s}^{-1}$$

Stop for a moment and consider what we have done. We have taken a mechanical model system and analyzed its behavior. We have compared this behavior to that described by an empirical macroscopic equation, Fick's law of diffusion. We have found that the two give the same behavior and that the molecular equation contains a formula for the diffusion coefficient in terms of molecular properties.

Experimental studies of the dependence of diffusion coefficients on temperature and density agree fairly well with Equation (17.3-9), and reasonable values are obtained when effective hard sphere diameters are calculated from diffusion coefficient data. The calculated hard sphere diameters depend somewhat on temperature. This is explained by the fact that the actual intermolecular repulsive potential is not infinitely steep, as is the hard sphere potential. When two molecules strike together more strongly, as they more often do at higher temperature, the distance of closest approach is smaller and the effective hard sphere diameter is expected to be smaller. Table A25 gives some experimental values for self-diffusion coefficients.

Exercise 17.10	Find the effective hard sphere diameter of argon atoms from each of the self-diffusion coefficient values in Table A25. Comment on your results.

Thermal Conduction

An analysis of heat conduction and viscous flow very similar to that for self-diffusion can be carried out.[3] For heat conduction, instead of counting molecules that pass the central plane, one computes contributions the molecules make to the kinetic energy that is transported (in our hard sphere model, the potential energy vanishes).

The molecular energy is expressed in terms of the heat capacity at constant volume. From Equation (16.2-26), the heat capacity of a monatomic gas is

$$C_V = \left(\frac{\partial U}{\partial T}\right)_{V,n} = \frac{3nR}{2} \quad \text{or} \quad c_V = \frac{3k_B}{2} \tag{17.3-10}$$

where c_V is the heat capacity per molecule. The mean molecular energy is

$$\langle \varepsilon \rangle = c_V T = \frac{3k_B T}{2} \tag{17.3-11}$$

An analysis similar to that leading to Equation (17.3-9) is carried out, with the temperature depending on z and the net transport of $\langle \varepsilon \rangle$ being computed. The result is compared with Fourier's law of thermal conduction, Equation (17.2-1), and an expression for the coefficient of thermal conductivity deduced. For a one-substance hard sphere gas

$$\kappa = \frac{25\pi}{64} c_V \lambda \langle v \rangle \mathcal{N} = \frac{25}{32} \frac{c_V}{d^2} \left(\frac{k_B T}{\pi m}\right)^{1/2} \tag{17.3-12}$$

This equation is at the level of accuracy of Equation (17.3-9).

Exercise 17.11	Carry out the derivation of Fourier's law of thermal conduction and obtain an expression for the thermal conductivity analogous to Equation (17.3-8). A system of one component is used, but assume that the temperature depends on the z coordinate and that the molecules are equilibrated at the temperature of the plane from which they come to approach the central plane. Calculate the percent difference between your result and that of Equation (17.3-12).

For a polyatomic gas with rotational and vibrational energy, the thermal conductivity is more complicated, since only part of the rotational and vibrational energy is transferred in a collision. We do not discuss this case.

---

[3] J. O. Hirschfelder, C. F. Curtiss, and R. B. Bird, *Molecular Theory of Gases and Liquids*, Wiley, New York, 1954, pp. 9ff.

Viscous Flow

In the case of viscous flow in the y direction, with the flow velocity depending on z, the quantity transported is the momentum component mv_y, and it is transported in the z direction. An analysis of the net flow of this quantity in a dilute hard sphere gas gives an expression for the viscosity coefficient:

$$\eta = \frac{5\pi}{32} m\lambda \langle v\rangle \mathcal{N} = \frac{5}{16\sqrt{\pi}} \frac{1}{d^2}\sqrt{mk_{\mathrm{B}}T} \qquad\qquad \textbf{(17.3-13)}$$

where this equation is at the level of accuracy of Equation (17.3-9).

All three transport coefficients are proportional to the mean speed of the molecules, which means that they are proportional to the square root of the temperature. A gas becomes more viscous when the temperature is raised (opposite to the behavior of a liquid). Notice also that the coefficient of viscosity and the coefficient of thermal conductivity are independent of the number density. This interesting behavior was predicted by kinetic theory before it was observed experimentally.

Exercise 17.12   Certain dimensionless ratios of physical constants have been shown to be useful enough that they are called dimensionless groups. Show that for a hard sphere gas each of the following ratios equals a dimensionless constant and find each constant:

$$\frac{\eta}{\mathcal{N}mD}, \qquad \frac{c_V\eta}{\kappa m}, \qquad \frac{Dc_V\mathcal{N}}{\kappa}$$

17.4 The Structure of Liquids and Transport Processes in Liquids

The discussion of this section is designed only to give some qualitative understanding of transport processes in liquids. Keep in mind that the structure and properties of a liquid are determined primarily by potential energy, whereas the properties of gases are determined primarily by kinetic energy.

In Section 16.9, the molecular environment of a typical molecule in a liquid was described as a cage, made up of neighboring molecules, in which the molecule is confined by the repulsive intermolecular forces. This confinement is nearly absolute in a solid, since the nearest neighbors are packed in a rigid crystal lattice.

If a molecule in a liquid were absolutely confined to a cage made up of its nearest neighbors, there could be no diffusion or viscous flow. However, the neighbors are also moving, so there is a chance that after colliding with the neighboring molecules in a given cage a number of times, a molecule in a liquid can move past some of these neighbors into a new cage. Whereas a molecule in a typical gas might undergo a collision every 10^{-10}

See the discussion in Section 16.9.

to 10^{-9} second, a molecule in a typical liquid might undergo a collision with its neighbors every 10^{-12} to 10^{-11} second but then move to a new cage every 10^{-9} to 10^{-8} second.

In some approximate theories of liquid transport, it is found that the motion of a molecule through a fluid is, on the average, impeded by a frictional force representing the retarding effects of the repulsions of the neighboring molecules on the motion from one cage to another. This force turns out to be proportional to the negative of the velocity of the molecule:

$$\mathbf{F}_f = -f\mathbf{v} \tag{17.4-1}$$

where $\mathbf{F}_f$ is the frictional force and f is called the **friction coefficient**. This equation is similar to Stokes' law, Equation (17.2-26). Although Stokes' law was originally derived for a macroscopic sphere moving through a continuous fluid, we can write

$$f = 6\pi\eta r_{\text{eff}} \tag{17.4-2}$$

and use this relation to define r_{eff} as an **effective radius** of the molecule. Reasonable values for effective radii of molecules and hydrated ions are obtained in this way. For macromolecules (molecules of large molecular mass) and colloidal particles (particles roughly 3 to 1000 nm in diameter) that are nearly spherical, Equation (17.2-26) is nearly as accurate as it is for macroscopic spheres.

Around 1905, Einstein devised a theory of **Brownian motion**, the irregular motion of colloidal particles suspended in a liquid. A colloidal particle is a particle that is considerably larger than solvent molecules but small enough to remain suspended without settling out under the force of gravity. Einstein assumed that a colloidal particle is bombarded randomly by the molecules of the solvent and was able to show for a spherical colloidal particle that the mean-square displacement of the particle in the z direction in a time t is given by

Einstein developed this theory, when he was pursuing theoretical physics as a part-time, unpaid pursuit. The number of important theories Einstein developed at this time is remarkable.

$$\langle z^2 \rangle = \frac{k_B T}{3\pi\eta r} t \tag{17.4-3}$$

where k_B is Boltzmann's constant, T is the absolute temperature, r is the radius of the particle, and η is the viscosity of the solvent. Comparison of this equation with Equation (17.2-19) shows that the diffusion coefficient of the colloidal substance is given by

$$\boxed{D_2 = \frac{k_B T}{f} = \frac{k_B T}{6\pi\eta r}} \tag{17.4-4}$$

In 1908, using a dark-field microscope, Perrin repeatedly measured the displacements of colloidal particles and verified Equation (17.4-3) experimentally. Many skeptics considered this the definitive verification of the existence of atoms and molecules, since Einstein's derivation of Equation (17.4-3) depended on the assumption that the colloidal particle was bombarded randomly by solvent molecules. Perrin was able to obtain the value of Boltzmann's constant from Equation (17.4-3) and thus to

Jean Baptiste Perrin, 1870–1942, French physicist.

calculate a value of Avogadro's number, using the known value of the
gas constant.

▼   **EXAMPLE 17.8**

The diffusion coefficient of hemoglobin in water at 20°C is 6.9×10^{-11} m² s⁻¹.
Assuming the molecules to be spherical, calculate their radius. The viscosity coefficient of water at this temperature is 1.002×10^{-3} kg m⁻¹ s⁻¹.

Solution

$$r = \frac{k_B T}{6\pi\eta D} = \frac{(1.3807 \times 10^{-23} \text{ J K}^{-1})(293 \text{ K})}{6\pi(1.002 \times 10^{-3} \text{ kg m}^{-1} \text{ s}^{-1})(6.9 \times 10^{-11} \text{ m}^2 \text{ s}^{-1})}$$

$$= 3.1 \times 10^{-9} \text{ m} = 31 \text{ Å}$$

▲

Exercise 17.13   **a.** Estimate the molar volume of hemoglobin from the molecular size in Example 17.8.

b. Human hemoglobin has a density of 1.335 g mL⁻¹ (a little larger than typical
protein values, which are around 1.25 g mL⁻¹). It has a molar mass of 68000 g
mol⁻¹. Calculate its molar volume and compare with your answer from part a.
State any assumptions.

The Temperature Dependence of Diffusion and Viscosity Coefficients in Liquids

From Equation (17.3-9) we see that the self-diffusion coefficient of a hard
sphere gas is proportional to the square root of the temperature, and from
Equation (17.3-13) we see that the viscosity depends on the temperature
in the same way. This rather weak dependence on temperature is different
from that of liquids, in which coefficients of diffusion and viscosity depend
quite strongly on temperature. Furthermore, liquids are less viscous at
higher temperatures and liquid self-diffusion coefficients increase as the
temperature is raised.

It is found experimentally that diffusion coefficients in liquids are usually
quite well described by the formula

$$D = D_0 e^{-E_{ad}/RT} \tag{17.4-5}$$

where R is the ideal gas constant and T is the absolute temperature. The
quantity D_0 is a parameter that is nearly temperature independent (it is the
limit of D as $1/T \to 0$). E_{ad} is a positive parameter with the dimensions of
energy and is called an activation energy. It is similarly found that liquid
viscosities are quite well described by the formula

$$\eta = \eta_0 e^{E_{a\eta}/RT} \tag{17.4-6}$$

where the symbols have similar meanings as in Equation (17.4-5).

An elementary explanation of Equation (17.4-5) is as follows: For a molecule in a liquid to push past some of its nearest neighbors and move into
the next cage, it must possess a relatively high kinetic energy. If we identify

ε_a as a minimum energy required to break out of a cage (an "activation energy"), then from Equation (16.2-31) we see that the proability that a molecule has a velocity corresponding to this energy is

$$\text{Probability} \propto e^{-\varepsilon_a/k_B T} = e^{-E_a/RT} \qquad (17.4\text{-}7)$$

where k_B is Boltzmann's constant and $E_a = N_A \varepsilon_a$. It is therefore reasonable that a diffusion coefficient in a liquid would obey Equation (17.4-5).

For shearing flow to take place, layers of a liquid must flow past each other. This requires disruption of cages and much the same kind of activation energy as in diffusion. Note that the sign of the exponent in Equation (17.4-6) is opposite from that in Equation (17.4-5). This is because the rate of shear is proportional to the factor in Equation (17.4-7), making the viscosity coefficient inversely proportional to it. It is found that the activation energy for viscosity is roughly equal to the activation energy for self-diffusion in the same liquid, giving further plausibility to this argument.

Although we speak of laminar flow, we think of the molecules as moving individually in this molecular picture of viscous flow.

▼ **EXAMPLE 17.9**

Following are data on the viscosity of carbon tetrachloride:

$T/°C$	0	15	20	30	40	50	60
η/cp	1.329	1.038	0.969	0.843	0.739	0.651	0.585

The viscosity coefficient values are given in centipoise (cp). The poise is an older unit of viscosity, equal to 1 g cm^{-1} s^{-1}; 1 poise = 0.1 kg m^{-1} s^{-1}. Find the values of η_0 and $E_{a\eta}$.

Solution

We linearize Equation (17.4-6) by taking logarithms:

$$\ln\left(\frac{\eta}{\eta_0}\right) = \frac{E_{a\eta}}{RT}$$

A linear least squares fit of $\ln(\eta/1 \text{ cp})$ to $1/T$ gives an intercept of -4.279 and a slope of 1245 K, with a correlation coefficient of 0.9998 (a good fit).

$$E_{a\eta} = (\text{slope}) \times R = (1245 \text{ K})(8.3145 \text{ J K}^{-1} \text{ mol}^{-1})$$
$$= 1.035 \times 10^4 \text{ J mol}^{-1} = 10.4 \text{ kJ mol}^{-1}$$
$$\eta_0 = (1 \text{ cp})e^{-4.279} = 1.39 \times 10^{-2} \text{ cp} = 1.39 \times 10^{-5} \text{ kg m}^{-1} \text{ s}^{-1}$$

This energy of activation is typical of liquids with small, nearly spherical molecules and is a reasonable size for the energy required (per mole) to push through a layer of neighboring molecules. As might be expected, it is somewhat smaller than the energy of vaporization of CCl_4, 33.9 kJ mol^{-1}.

◢

Exercise 17.14

The self-diffusion coefficient of carbon tetrachloride at 25°C is 1.4×10^{-9} m^2 s^{-1}. Estimate the value at 45°C, assuming the same value of the energy of activation as for the viscosity from Example 17.9.

Macromolecules and colloidal particles can sediment though a liquid under the force of gravity or under centrifugal force. Sedimentation experiments provide a second means of determining the size of the particles or molecules through use of Equation (17.4-2). To provide sufficiently rapid sedimentation of protein molecules, one must use an ultracentrifuge, which can turn at speeds of several thousand revolutions per second.

An object of mass m at a distance r from an axis of rotation is maintained in its circular path by a centripetal force that is given by Equation (D-13) of Appendix D:

$$F_c = mr\omega^2 \tag{17.4-8}$$

where ω is the **angular speed**, measured in radians per second. The rate of rotation measured in revolutions per second is equal to ω divided by 2π, since one revolution is equal to 2π radians.

Figure 17.8 shows a schematic diagram of an ultracentrifuge. The rotor chamber must be evacuated to lessen air friction at the rotational speeds that are attained. The sample cell has a transparent top and bottom, so that a beam of light can shine through the cell each time it passes the location of the beam.

When the rotor spins, the macromolecules sediment toward the outside of the rotor if they are denser than the solvent, and the location of those originally closest to the center can be measured by passing a beam of light through the cell and observing the position dependence of the index of refraction, which depends on composition. It is found that after the rotor has been spinning a short time, the sedimentation speed $v_{\text{sed}} = dr/dt$ has attained a steady value. The centripetal force is then equal in magnitude to the frictional force given by Equation (17.4-1).

Since a macromolecule is immersed in a solvent, its centripetal force must be corrected for buoyancy, giving instead of Equation (17.4-8)

$$F_c = (m - m_{\text{solvent}})r\omega^2 \tag{17.4-9}$$

where m_{solvent} is the mass of solvent displaced by one macromolecule. If ρ_1 is the density of the solvent and ρ_2 is the density of the macromolecular substance, we have

$$m_{\text{solvent}} = \left(\frac{\rho_1}{\rho_2}\right)m = \left(\frac{\rho_1}{\rho_2}\right)\frac{M_2}{N_A} \tag{17.4-10}$$

where M_2 is the molar mass of the macromolecular substance and N_A is Avogadro's number. Combination of Equations (17.4-1) and (17.4-9) gives

$$f v_{\text{sed}} = \frac{M_2}{N_A}\left(1 - \frac{\rho_1}{\rho_2}\right)r\omega^2 \tag{17.4-11}$$

The **sedimentation coefficient** S is defined as the ratio of v_{sed} to the centrifugal acceleration, $r\omega^2$:

$$S = \frac{v_{\text{sed}}}{r\omega^2} \quad \text{(definition)} \tag{17.4-12}$$

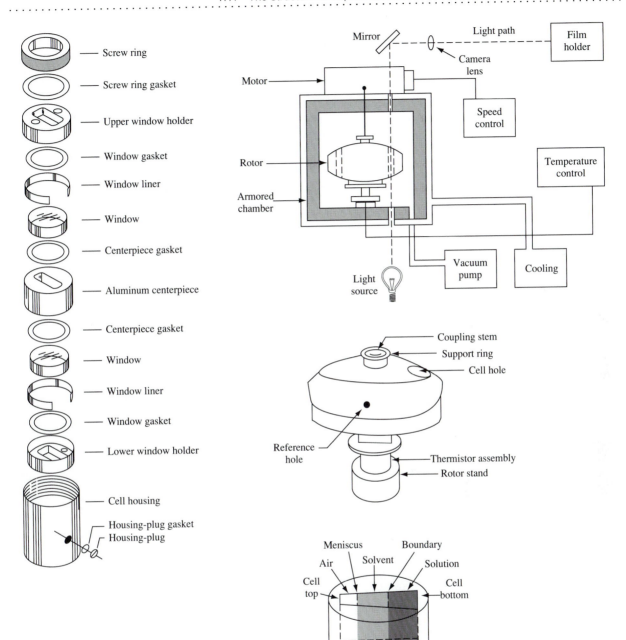

Figure 17.8. An Ultracentrifuge with Its Rotor and Cell. The rotor of an ultracentrifuge spins so rapidly that it is operated in an evacuated chamber to avoid friction with air. Courtesy of Beckman Instruments.

The svedberg unit is named for Theodor (called The, pronounced "Tay") Svedberg, 1884–1971, Swedish biophysical chemist who received the 1926 Nobel Prize for his work on disperse systems (colloids, protein suspensions, etc.).

so that

$$S = \left(\frac{M_2}{N_A}\right)\frac{1 - \rho_1/\rho_2}{f} \tag{17.4-13}$$

The sedimentation coefficient has the units of seconds and is ordinarily approximately equal to 10^{-13} s. The practical unit of sedimentation coefficients is the **svedberg**, defined so that 1 svedberg = 10^{-13} s.

Equation (17.4-4) allows us to express f in terms of D_2, the diffusion coefficient of the macromolecular substance, giving

$$M_2 = \frac{RTS}{D_2(1 - \rho_1/\rho_2)} \tag{17.4-14}$$

Two experimental values are required to solve for the molar mass: the sedimentation coefficient and the diffusion coefficient. It is possible to get both of these values from the same experiment if a concentration profile similar to that of Figure 17.3 can be observed in addition to the sedimentation rate.

where we have replaced $k_B N_A$ by R. This equation has been widely used to obtain molar masses of proteins.

▼ **EXAMPLE 17.10**

The sedimentation coefficient of a sample of human hemoglobin in water is 4.48 svedberg at 20°C, and its density is 1.335 g mL^{-1}. The density of water at this temperature is 0.998 g mL^{-1}. Use the value of the diffusion coefficient from Example 17.8 to determine the molar mass of hemoglobin.

Solution

$$M = \frac{(8.3145 \text{ J K}^{-1} \text{ mol}^{-1})(293.15 \text{ K})(4.48 \times 10^{-13} \text{ s})}{(6.9 \times 10^{-11} \text{ m}^2 \text{ s}^{-1})(1 - 0.998/1.335)}$$
$$= 63 \text{ J m}^{-2} \text{ s}^2 \text{ mol}^{-1} = 63 \text{ kg mol}^{-1}$$

The actual molar mass is near 68 kg mol^{-1}.

◢

17.5 Transport in Electrolyte Solutions

In Sections 16.9 and 17.4, we made some qualitative comments about diffusion and viscous flow in liquids. The motions of molecules in a liquid were crudely described as rattling about in a cage of adjacent molecules with an occasional excursion into a neighboring cage.

In an ionic solution, the motion of an ion is similar to the motion of an uncharged molecule in an ordinary liquid. However, even at concentrations as low as 0.001 mol L^{-1} the electrostatic forces between ions are important, influencing the probability that an ion will move into an adjacent cage and affecting the rate at which an ion will migrate through a solution under the influence of an electric field.

Electric currents consist of migrations of charged particles. Electric currents in conductors and semiconductors are due to the motions of electrons, while electric currents in electrolyte solutions are due to the motions of ions. **Ohm's law** is an empirical law that describes both cases. It states that the current in a conducting system is proportional to the voltage imposed

Ohm's law is named for Georg Simon Ohm, 1787–1854, a German physicist who was a high-school teacher when he discovered Ohm's law.

on the system:

$$I = V/R \qquad (17.5\text{-}1)$$

where V is the voltage, I is the current (equal to the amount of charge passing a given location per second), and R is the **resistance** of the conductor. Ohm's law is obeyed nearly exactly by metallic conductors, to a good approximation by most electrolyte solutions, and less accurately by most semiconductors. Deviations from Ohm's law can be described by using Equation (17.5-1) with a resistance R that is dependent on V.

Ohm found by painstaking experiments with homemade equipment (even including homemade wires) that the resistance of a conductor of uniform cross-sectional area is proportional to its length and inversely proportional to its cross-sectional area. We define the **resistivity** r of a conducting object shaped as in Figure 17.9 by

The units of resistivity are ohm m.

$$r = \frac{R\mathscr{A}}{d} \qquad (17.5\text{-}2)$$

where $\mathscr{A}$ is the cross-sectional area and d is the length of the object. The resistivity is independent of d and $\mathscr{A}$; it depends only on the composition of the object, the temperature, and the pressure (the dependence on the pressure is usually very weak).

The reciprocal of the resistivity is called the **conductance** σ:

The units of conductance are ohm^{-1} m^{-1} = S m^{-1}. The ohm^{-1} has been called the mho, but the SI name for ohm^{-1} is the siemens (S).

$$\sigma = 1/r \qquad (17.5\text{-}3)$$

We define the **current density j** (do not confuse it with the unit vector **j**) as a vector with magnitude equal to the current per unit area and with the

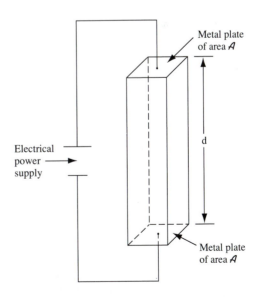

Figure 17.9. An Electrically Conducting System (schematic). A cell of this shape allows us to relate the resistance of the cell to the resistivity and conductivity in a direct fashion, since the resistance is inversely proportional to the area $\mathscr{A}$ and directly proportional to the length d.

The direction of an electric current is by convention the direction of apparent motion of positive charges.

same direction as the current:

$$j = |\mathbf{j}| = \frac{I}{\mathscr{A}} \qquad (17.5\text{-}4)$$

where $\mathscr{A}$ is the cross-sectional area of the conductor. Ohm's law can be written

$$j = \sigma E \qquad (17.5\text{-}5)$$

where E is the magnitude of the electric field, equal to V/d for an object like that of Figure 17.9.

Exercise 17.15   Show that Equation (17.5-5) is correct.

Since the current in an electrolyte solution is due to the motions of ions, we can write the current density as the sum of the cation contribution and the anion contribution. Positive ions no farther from a fixed plane than a distance equal to v_+ times 1 second will pass in 1 second, and similarly for anions:

$$\mathbf{j} = \mathbf{j}_+ + \mathbf{j}_- \qquad (17.5\text{-}6)$$
$$j = |\mathbf{j}| = z_+ N_A e c_+ v_+ + |z_-| N_A e c_- v_- \qquad (17.5\text{-}7)$$

The mean drift velocity is a vector average velocity. At equilibrium, even though the ions are moving about, the mean drift velocity vanishes.

where e is the charge on one proton, 1.6022×10^{-19} C, N_A is Avogadro's number, v_+ is the magnitude of the mean "drift velocity" of cations, and v_- is the magnitude of the mean drift velocity of anions. The quantity z_+ is the valence of the cation (the number of proton charges on one ion) and z_- is the valence of the anion (a negative integer equal in magnitude to the number of electron charges on one ion). The concentrations of cations and anions in mol m^{-3} or mol L^{-1} are denoted by c_+ and c_-, respectively.

If dissociation or ionization is complete and if the solution is made from c moles of electrolyte solute per cubic meter of solution,

$$c_+ = v_+ c \quad \text{and} \quad c_- = v_- c \qquad (17.5\text{-}8)$$

where v_+ is the number of cations per formula unit of the solute and v_- is the number of anions per formula unit. For a single solute that is electrically uncharged, these quantities must obey the electrical neutrality relation

$$v_+ z_+ + v_- z_- = 0 \qquad (17.5\text{-}9)$$

Equation (17.5-7) for j, the magnitude of the current density, can also be written

$$j = F(z_+ c_+ v_+ + |z_-| c_- v_-) \qquad (17.5\text{-}10)$$

where F is Faraday's constant, equal to the charge on 1 mol of protons, 96485 C mol^{-1}.

A direct (nonreversing) electric current can flow only in a closed circuit. If the conductor in Figure 17.9 is an electrolyte solution, the ions that flow cannot pass out of the ends of the container, so the current is carried by electron motion in the rest of the circuit. This means that if a direct current flows, cations will accumulate at one end of the system and anions will accumulate at the other end, unless some chemical reactions consume the ions.

Alternating current is often used to measure the conductivities of electrolyte solutions. In this case the current flows alternately in one direction and then the other. Only small accumulations of ions occur at the ends of the system during one half of the cycle, and these ions move back during the other half of the cycle.

▼ **EXAMPLE 17.11**

For a system such as shown in Figure 17.9, with a solution containing c moles of NaCl per m^3, show that Ohm's law holds if the drift velocities are governed by Equation (17.4-1). Use the relationship between the electric field and the force on a charged particle given in Equations (9.1-1) and (9.1-3). Express the conductance in terms of the friction coefficients f_+ and f_-.

Solution

Using Equations (17.4-1) and (17.5-3)

$$j = FcEe\left(\frac{1}{f_{NA^+}} + \frac{1}{f_{Cl^-}}\right) \tag{17.5-11}$$

Comparison with Equation (17.5-5) shows that

$$\sigma = Fce\left(\frac{1}{f_{NA^+}} + \frac{1}{f_{Cl^-}}\right) \tag{17.5-12}$$

▲

For a general binary electrolyte $M_{v_+}X_{v_-}$, the analogue of Equation (17.5-12) is

$$\sigma = Fe\left(\frac{c_+z_+^2}{f_+} + \frac{c_-z_-^2}{f_-}\right) = Fce\left(\frac{v_+z_+^2}{f_+} + \frac{v_-z_-^2}{f_-}\right) \tag{17.5-13}$$

Exercise 17.16   Show that Equation (17.5-13) is correct.

Since the cations and anions in a given electrolyte do not generally have equal friction coefficients, the two kinds of ions do not necessarily carry the same amount of current. (One substance that comes close is KCl.) The fraction of the current carried by a given type of ion is called its **transference number**, t:

$$t_i = \frac{j_i}{j_{total}} \tag{17.5-14}$$

From Equation (17.5-13), it follows that if only one type of cation and one type of anion are present

$$t_+ = \frac{c_+z_+/f_+}{c_+z_+/f_+ + c_-z_-/f_-} \tag{17.5-15}$$

with a similar equation for t_-.

The mobility is numerically equal to the mean drift velocity if the electric field is 1 volt per meter.

The **mobility** u_i of the ith type of ion is defined by

$$u_i = \frac{|\mathbf{v}_i|}{E} \qquad (17.5\text{-}16)$$

where $\mathbf{v}_i$ is the mean drift velocity of this type of ion. This relation is equivalent to

$$u_i = \frac{|z_i|e}{f_i} \qquad (17.5\text{-}17)$$

where f_i is the friction coefficient for the ith type of ion. Ions with equal valence magnitudes and equal effective radii will have equal mobilities, according to Stokes' law, Equation (17.2-26).

The conductivity can be written in terms of the mobilities:

$$\sigma = F(c_+ z_+ u_+ + c_- |z_-| u_-) \qquad (17.5\text{-}18a)$$
$$\sigma = Fc(v_+ z_+ u_+ + v_- |z_-| u_-) \qquad (17.5\text{-}18b)$$

Exercise 17.17   Write the transference numbers in terms of the ion mobilities.

Ion mobilities are measured by several different techniques. The technique of **electrophoresis** is used in the study of the ion mobilities of protein molecules. In this technique, a solution is placed between electrodes, across which a direct voltage is placed. Most protein molecules contain various weak acidic and basic functional groups, which ionize at different characteristic pH values. The protein molecules have a characteristic average charge that depends on pH. The mobility of the protein thus depends on pH, and electrophoresis experiments at different pH values can be used to separate mixtures of proteins.

It is found that the mobilities and friction coefficients of a given ion depend quite strongly on the concentration of that ion, as well as on the concentrations of other ions. One reason for the strong dependence on concentration is that interionic forces are "long-range" forces. Whereas the forces between uncharged molecules decrease rapidly with distance, electrostatic forces decrease slowly with distance. This means that ions which are at a considerable distance exert significant forces, as we found in our discussion of the Debye-Hückel theory in Chapter 6.

Three important effects contribute to the concentration dependence of the mobilities and friction coefficients. The first is the **electrophoretic effect**, due to the fact that ions of the opposite charge are moving in the opposite direction from a given ion. At nonzero concentration, their attractive forces pull on an ion in the direction of their motion, tending to slow down the given ion.

The second effect is the **relaxation effect**. Every ion repels ions of its own charge and attracts ions of the opposite charge. At nonzero concentration, there is an "ion atmosphere" of excess charge of the opposite sign around every ion. If an ion moves, it is no longer at the center of its ion atmosphere, which must then relax to become centered on the new position of the ion. This effect also slows down the motion of the ion compared with its motion at infinite dilution.

The third effect is that of **solvation**. At high concentrations, ions must compete with each other to attract solvent molecules. Since the effective size of ions can include some of the solvent molecules that are strongly attracted to the ions, any change in the solvation can affect the mobility.

The electrophoretic effect and the relaxation effect vanish in the limit of infinite dilution, and the solvation approaches concentration-independent behavior, so ion mobilities and friction coefficients approach constant values in the limit of infinite dilution. Tabulated values of mobilities are usually given at infinite dilution.

The ions with the largest mobilities in aqueous solutions are the hydrogen ion and the hydroxide ion. The reason for the large mobilities is that hydrogen and hydroxide ions can "exchange" with water molecules. For example, a hydrogen ion can attach itself to a water molecule, making a hydronium ion, H_3O^+, followed by release of a hydrogen ion on the other side of the hydronium ion. This hydrogen ion can attach itself to a second water molecule, after which a different hydrogen ion is released on the other side of the second hydronium ion, etc., providing rapid apparent motion of hydrogen ions. The exchange of hydroxide ions is similar.

Some cations with a small radius, such as lithium, have somewhat lower mobilities and larger effective radii than might be expected. This is attributed to the fact that small cations are more strong hydrated (more strongly bound to water molecules) than larger cations, because water molecules can approach closer to the center of charge of the smaller ion. Table A26 gives values of ion mobilities at infinite dilution in water at 25°C, as well as values for molar conductivities, which we discuss below.

| **Exercise 17.18** | **a.** Calculate the effective radii of the hydrogen and hydroxide ions from the ion mobilities. |

b. Calculate the effective radius of the lithium ion from its ion mobility.

Another quantity that is commonly tabulated is the **molar conductivity**, denoted by Λ, and defined by

$$\Lambda = \sigma/c \qquad \text{(17.5-19a)}$$

where c is the concentration of the electrolyte solute in mol m^{-3} or in mol L^{-1}. From Equations (17.5-13b) and (17.5-18b), we can see that this quantity would be concentration independent if the mobilities were concentration independent:

$$\Lambda = \frac{\sigma}{c} = F(v_+ z_+ u_+ + v_- |z_-| u_-) \qquad \text{(17.5-19b)}$$

At nonzero concentration, it is found that the molar conductivity depends on the square root of the concentration, much like the logarithm of the activity coefficient according to the Debye-Hückel theory:

$$\Lambda = \Lambda_0 - Ac^{1/2} \qquad \text{(17.5-20a)}$$

This equation was discovered empirically by Kohlrausch in 1900. It also represents the results of the Debye-Hückel-Onsager theory, which provides an expression for the parameter A.

In the limit of infinite dilution, Λ approaches a constant limit, just as do the mobility and friction coefficient:

$$\Lambda_0 = \lim_{c \to 0} \Lambda \qquad (17.5\text{-}20b)$$

The value at infinite dilution is called the **limiting molar conductivity**. It is the quantity most often tabulated.

The conductivity contains a term for the cation and a term for the anion, so the molar conductivity can be written as such a sum. For a uni-univalent electrolyte

$$\Lambda = \lambda_+ + \lambda_- \qquad (17.5\text{-}21a)$$
$$\Lambda_0 = (\lambda_+)_0 + (\lambda_-)_0 \qquad (17.5\text{-}21b)$$

where $(\lambda_+)_0$ and $(\lambda_-)_0$ are the limiting molar conductivities of the ions. Equation (17.5-21b) is known as Kohlrausch's law and was discovered empirically around 1875. The limiting molar conductivities for the ions can be tabulated separately, resulting in a shorter table than if each electrolyte were tabulated separately. Table A26 is such a table.

EXAMPLE 17.12

Following are data on the molar conductivity of NaOH as a function of concentration at 25°C:

Concentration/mol L^{-1}	Λ/ohm^{-1} cm^2 mol^{-1}
0.0010	244.5
0.0100	238.0
0.0500	227.6
0.1000	221.2
0.2000	213.0
0.5000	197.6
1.0000	178.8

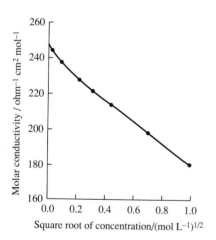

Example Figure 17.12. The Molar Conductivity of Sodium Hydroxide as a Function of the Square Root of the Concentration in mol L^{-1}. This figure shows the extrapolation to zero concentration, which delivers the limiting value of the molar conductivity.

Find the value of Λ_0 for NaOH. Compare it with the value from Table A26.

Solution

As indicated in Equation (17.5-19), a plot of Λ versus $c^{1/2}$ should be nearly linear. Such a plot is shown in Example Figure 17.12. The extrapolation to zero concentration was accomplished by fitting the data to a polynomial of degree 5. The intercept was found to be

$$\Lambda_0 = 247.7 \text{ ohm}^{-1} \text{ cm}^2 \text{ mol}^{-1}$$

From Table A26,

$$\Lambda_0 = \lambda(Na^+) + \lambda(OH^-)$$
$$= 19.8 \times 10^{-3} \text{ ohm}^{-1} \text{ m}^2 \text{ mol}^{-1} + 5.011 \times 10^{-3} \text{ ohm}^{-1} \text{ m}^2 \text{ mol}^{-1}$$
$$= 24.8 \times 10^{-3} \text{ ohm}^{-1} \text{ m}^2 \text{ mol}^{-1} = 248 \text{ ohm}^{-1} \text{ cm}^2 \text{ mol}^{-1}$$

in good agreement with our earlier value.

▲

Because a weak electrolyte is only partially ionized at nonzero concentration, an extrapolation such as that in Example 17.12 is difficult. The percent ionization, as well as the ion mobilities, is changing with concentration. However, one can easily determine the limiting molar conductivity of a weak electrolyte such as acetic acid from the values for the ions.

Exercise 17.19

a. Find the value of the limiting molar conductivity for acetic acid from the values in Table A26.

b. Following are data on the molar conductivity of acetic acid as a function of concentration at 25°C:

Concentration/mol L^{-1}	Λ/ohm^{-1} cm^2 mol^{-1}
0.0001	134.6
0.0010	49.2
0.0100	16.18
0.0500	7.36
0.1000	5.20
0.2000	3.65

This extrapolation is difficult because the degree of ionization of the acid depends on the concentration, and the slope is very steep near zero concentration. It should be possible to divide each value of the molar conductivity by the degree of ionization for each data point to get a more slowly varying quantity.

Attempt an extrapolation as in Example 17.12 to determine the limiting molar conductivity.

Summary

In this chapter, we have discussed nonequilibrium processes that do not involve chemical reactions. We presented the macroscopic description of three transport processes: heat conduction, which is the transport of energy; diffusion, which is the transport of molecules; and viscous flow, which is the transport of momentum. These processes are all described by empirical linear phenomenological laws: heat conduction by Fourier's law, diffusion by Fick's law, and viscous flow by Newton's law of viscous flow.

Transport processes in a hard sphere gas were discussed theoretically. A formula for the self-diffusion coefficient was derived, and similar formulas for thermal conductivities and viscosity coefficients were presented. Each transport coefficient is proportional to the mean free path and to the mean speed and thus is proportional to the square root of the temperature.

A molecule in a liquid was pictured as partially confined in a cage made up of its nearest neighbors. This model and an assumed frictional force were related to diffusion in liquid solutions, viscosity in pure liquids, and sedimentation in solutions of macromolecular substances.

We presented the consequences of assuming that an ion moving through a solution experiences a frictional force proportional to its speed. It was shown that this assumption leads to Ohm's law for an electrolyte solution, with a conductance contribution for each type of ion that is inversely proportional to the friction coefficient.

Additional Reading

P. A. Egelstaff, *An Introduction to the Liquid State*, Academic Press, New York, 1967
This small book is intended to introduce graduate students in physics to the equilibrium and nonequilibrium theory and description of the liquid state.

J. P. Hansen and I. R. MacDonald, *Theory of Simple Liquids*, Academic Press, New York, 1976
This is a theoretically oriented book similar to that of Egelstaff, nearly 10 years more modern.

J. O. Hirschfelder, C. F. Curtiss, and R. B. Bird, *Molecular Theory of Gases and Liquids*, Wiley, New York, 1954
Although it is now about 40 years old, this book is a very useful source of information. It contains a lot of numerical data and carefully presents nearly all of the fluid theory known at the time of its publication. There is a second printing (1965) with corrections and some added material.

E. H. Kennard, *Kinetic Theory of Gases*, McGraw-Hill, New York, 1938
This was the standard reference on gas kinetic theory for many years and is still very useful.

T. M. Reed and K. E. Gubbins, *Applied Statistical Mechanics*, McGraw-Hill, New York, 1973
This book discusses equilibrium and transport processes in fluids for chemical engineers at the graduate level.

R. C. Reid, J. M. Prausnitz, and Thomas K. Sherwood, *The Properties of Gases and Liquids*, 3rd ed., McGraw-Hill, New York, 1977
This book is intended for chemical engineers and contains quite a bit of data, as well as correlations useful in coming up with estimated numerical values when data are not available.

PROBLEMS

Problem for Section 17.1

17.20. Assume that in a two-component solution the temperature is given by

$$T = T_0 + B \cos(a_1 z)e^{-t/b_1}$$

and the concentration of component 2 is given by

$$c_2 = c_0 + C \sin(a_2 z)e^{-t/b_2}$$

where B, C, a_1, a_2, b_1, and b_2 are constants.

a. Write the expressions for the gradients of the temperature and the concentration of component 2.

b. Write the expressions for the time derivatives of the temperature and the concentration of component 2.

Problems for Section 17.2

17.21. a. Show that the concentration expression in Problem 17.20 can satisfy Fick's second law of diffusion, Equation (17.2-12).

b. Find the expression for the constant b_2 in terms of a_2 and D_2, the diffusion coefficient, assuming Fick's law to be valid. What is the physical interpretation of the constant a_2?

c. If $D_2 = 1.00 \times 10^{-9}$ m^2 s^{-1} and $a_2 = 10.0$ m^{-1}, find the value of b_2.

*17.22. At 25.4°C, the diffusion coefficient of methane in glycerol equals 9.5×10^{-10} m^2 s^{-1}. Find the root-mean-square displacement in one direction of a methane molecule in (a) 60 minutes and (b) 120 minutes.

17.23. Estimate the time required for molecules of a neurotransmitter to diffuse across a synapse (the gap between two nerve cells) by calculating the time required for $\langle x^2 \rangle^{1/2}$ to equal 50 nm, a typical synapse spacing, if $D = 5 \times 10^{-10}$ m^2 s^{-1}.

17.24. Find the root-mean-square distance in the z direction traveled by glucose molecules in 1.00 hour in a dilute aqueous solution at 25°C.

*17.25. Liquid water at 20°C is flowing through a tube of radius 4.00 mm with a speed at the center of the tube (at $r = 0$) equal to 5.65 cm s^{-1}.

a. Find the speed at $r = 2.00$ mm.

b. Find $P_2 - P_1$ if the length of the tube is 1.000 m.

c. Find dV/dt, the volume rate of flow through the tube.

d. Find the value of the Reynolds number. Is the flow laminar?

17.26. A glass marble is falling at a steady speed in a swimming pool at 20.0°C. If the density of the marble is 2.2×10^3 kg m^{-3} and the radius of the marble is 0.0065 m, find its speed, assuming laminar flow. If the flow is not laminar, do you think the speed would be greater or less than expected with laminar flow?

17.27. An Ostwald viscometer consists of a capillary tube through which a fixed volume of a liquid is allowed to flow under hydrostatic pressure. If in a certain viscometer the fixed volume of water requires 243.0 s to flow at 20°C, calculate the time required for this volume of sulfuric acid to flow at 20°C. Use the fact that the hydrostatic force in a given viscometer is proportional to the density of the liquid. The density of sulfuric acid is 1.834 g mL^{-1} and that of water is 0.9982 g mL^{-1} at 20°C.

*17.28. In a certain Ostwald viscometer (see previous problem), a fixed volume of water required 162.4 s to flow at 20°C. If the same volume of mercury flows in 18.39 s, find the viscosity of mercury. The density of water is 0.99823 g mL^{-1}, and that of mercury is 13.5462 g mL^{-1}.

17.29. a. A certain garden hose is supposed to be able to deliver 550 gallons of water per hour. If the hose is 50 feet long and has an inside diameter of 5/8 inch, estimate the water pressure necessary to deliver this much water. State any assumptions.

b. Calculate the Reynolds number for the flow in part a, and determine whether the flow is laminar or turbulent.

17.30. a. Using the equation of continuity, show for a diffusing system in a steady state that the concentration gradient must be uniform if the diffusion coefficient is a constant and the cross-sectional area of the system is uniform.

b. Argue that the temperature gradient must be uniform in a system of uniform cross section with a uniform thermal conductivity if a steady state occurs.

Problems for Section 17.3

*17.31. a. Calculate the self-diffusion coefficient of helium gas at STP (Standard Temperature and Pressure).

b. Calculate the self-diffusion coefficient of argon gas at STP.

c. Calculate the self-diffusion coefficient of argon gas at 0.100 atm and 273.15 K. Comment on the comparison with the result of part b.

d. Calculate and compare the rms distances traveled in one direction in 60.0 minutes by He and Ar atoms diffusing at STP in self-diffusion experiments.

17.32. Show that for fixed pressure, the self-diffusion coefficient of a hard sphere gas is proportional to $T^{3/2}$.

17.33. For diffusion in a two-substance hard sphere gas in which the substances have different masses and sizes and neither substance is dilute, one uses the linear phenomeno-logical equation

$$\mathbf{u}_1 - \mathbf{u}_2 = \frac{1}{x_1 x_2} \mathscr{D}_{12} \nabla x_1$$

instead of Fick's law. In this equation, $\mathbf{u}_1$ is the mean drift velocity of molecules of component 1 and $\mathbf{u}_2$ is the same quantity for component 2. The quantities x_1 and x_2 are the two mole fractions and $\mathscr{D}_{12}$ is called the **mutual diffusion coefficient**. For a mixture of hard spheres[4]

$$\mathscr{D}_{12} = \frac{3}{8\pi^{1/2}} \frac{1}{d_{12}^2 \mathscr{N}_t} \left(\frac{k_B T}{2\mu}\right)^{1/2}$$

where μ is the reduced mass, $\mathscr{N}_t$ is the total number density, and d_{12} is the mean collision diameter

$$d_{12} = \frac{d_1 + d_2}{2}$$

Find the mutual diffusion coefficient for helium and argon at 273.15 K and 1.000 atmosphere pressure. Compare with the self-diffusion coefficient of each substance from Problem 17.31.

17.34. At 20.0°C, the viscosity of ammonia gas is equal to 9.82×10^{-6} kg m^{-1} s^{-1}. Find the effective hard sphere diameter of ammonia molecules at this temperature.

*17.35. For a temperature of 298 K and a pressure of 1.00 atm, calculate the viscosities of helium gas and carbon dioxide gas from the hard sphere diameters. Explain why the values compare as they do.

Problems for Section 17.4

17.36. The diffusion coefficient of bovine serum albumin in water at 20.0°C equals 7×10^{-11} m^2 s^{-1}.

a. Assuming the molecule to be spherical, estimate its radius.

b. If the density of the protein molecule equals 1.25 g cm^{-3}, estimate the molecular mass and the molar mass.

c. Estimate the sedimentation coefficient of the protein.

17.37. The diffusion coefficient of horse heart myoglobin in water at 20°C is equal to 1.13×10^{-10} m^2 s^{-1}, and its sedimentation coefficient is equal to 2.04 svedberg. Assume that its density is equal to that of hemoglobin, 1.335 g cm^{-3}, and find its molar mass.

*17.38. Calculate the rms distance diffused in one direction in 30.0 minutes by hemoglobin molecules in water at 20°C. The value of the diffusion coefficient is 6.9×10^{-11} m^2 s^{-1}.

---

[4] *ibid.*, pp. 14, 518.

Problems for Section 17.5

17.39. a. Calculate the force on 1.000 mol of sodium ions (assumed localized in a small volume) at a distance of 1.000 m in a vacuum from 1.000 mol of chloride ions.

b. Find the mass in kilograms and in pounds on which the gravitational force at the earth's surface would equal the force in part a.

17.40. Calculate the transference numbers for H$^+$ and Cl$^-$ ions in a dilute aqueous HCl solution at 25°C.

*17.41. Calculate the effective radii of K$^+$ and Cl$^-$ ions in dilute aqueous solution at 25°C.

17.42. Calculate the transference numbers for K$^+$ and Cl$^-$ in a dilute aqueous KCl solution at 25°C.

17.43. a. Calculate the transference number of each ion in a solution with 0.0010 mol L^{-1} sodium acetate and 0.00050 mol L^{-1} sodium chloride, assuming that infinite-dilution values can be used.

b. Calculate the conductivity of the solution of part a.

c. Find the resistance of a cube-shaped cell with side equal to 1.000 cm, containing the solution of part a and having electrodes on two opposite sides.

17.44. Calculate the transference numbers for each ion in an acetic acid solution, assuming that infinite-dilution values can be used.

General Problems

17.45. The thermal conductivity, viscosity, and self-diffusion coefficient of argon gas are listed for one or more temperatures in Tables A22, A24, and A25.

*a. Calculate the effective hard sphere diameter of argon atoms at 0°C from the viscosity and self-diffusion coefficient values. Compare your two values with each other and with the value in Table A21. Comment on any discrepancies.

*b. Calculate the effective hard sphere diameter of argon atoms at 20°C from the thermal conductivity and viscosity values. Compare your two values with each other and with the value in Table A21. Comment on any discrepancies.

c. Calculate the value of the self-diffusion coefficient of argon at 20°C and 1.00 atm, using Equation (17.3-9) and the effective hard sphere diameter from Table A21.

d. Calculate the self-diffusion coefficient of argon at 1.00 atm and 20°C by interpolation in Table A25. To do the interpolation, divide each value in the table by $T^{3/2}$ and do a linear least-squares fit to the resulting values. Comment on the closeness of your fit. Why is $T^{3/2}$ the correct factor to use?

e. Using your least-squares fit from part d, find the value of the self-diffusion coefficient of argon at 473 K and 1.00 atm.

f. Using your value from part e, calculate the effective hard sphere diameter of argon atoms at 473 K and compare it with the value in Table A21.

17.46. a. Derive an equation for the flow of heat that is analogous to Fick's second law of diffusion, Equation (17.2-12). Assume a constant heat capacity and a constant thermal conductivity.

b. Assume that two pieces of aluminum have been machined so that they fit together perfectly. Assume that one piece is initially at 30°C and the other is initially at 20°C and that they are suddenly placed in contact. Write a formula for the temperature profile as a function of time and of the perpendicular distance from the junction of the pieces.

17.47. Give verbal explanations for each of the following:

a. Each of the formulas for the transport coefficients in a hard sphere gas is proportional to λ and to $\langle v \rangle$.

b. The viscosity of a gas increases with temperature.

c. The diffusion coefficient is inversely proportional to $\mathcal{N}$.

d. The thermal conductivity and viscosity of a hard sphere gas are independent of $\mathcal{N}$.

*17.48. Identify each statement as either true or false. If a statement is true only under certain circumstances, label it as false.

a. An irreversible process always raises the entropy of the universe.

b. An irreversible process always raises the entropy of the system.

c. A temperature gradient can cause a diffusion flow to occur.

d. Ohm's law is exactly obeyed by electrolyte solutions.

e. A smaller ion will always have a smaller effective radius in aqueous solution.

f. If the elapsed time of a diffusion experiment is quadrupled, the root-mean-square distance moved by diffusing molecules is increased by a factor of two.

g. If a mixture of two proteins of equal density is sedimenting in an ultracentrifuge and if one protein has twice the molar mass of the other protein, its sedimentation velocity will be half as large as that of the other protein.

h. The viscosity of a gas increases with increasing temperature, whereas that of a liquid decreases with increasing temperature.

18 The Rates of Chemical Reactions

PREVIEW

This chapter presents the macroscopic description of chemical reaction rates and the techniques of measuring these rates.

PRINCIPAL FACTS AND IDEAS

1. The rate of a chemical reaction is a function of the concentrations of the reactants and products.
2. The rate law of a chemical reaction is the differential equation for the rate of change of the concentration of a reactant or product.
3. Differential rate laws can often be solved to obtain integrated rate laws.
4. Two consecutive reactions constitute a simple "mechanism" for a chemical reaction.
5. Specialized techniques exist for studying fast reactions.

18.1 The Macroscopic Description of Chemically Reacting Systems

In Chapter 17, we described nonequilibrium states of fluid systems in which no chemical reactions could occur. We now discuss fluid systems in which chemical reactions can occur, but which remain uniform during the reaction. Any heat evolved in the reaction must be conducted away from the system as quickly as it is evolved, and the reaction rate must be the same in all parts of the system. These conditions can usually be met by using a sufficiently small system or by stirring the system during the reaction. It is possible to consider reactions in nonuniform systems, such as systems

748

in which the temperature is not uniform.[1] However, we will not discuss such reactions.

Consider a single chemical reaction

$$a\text{A} + b\text{B} \rightarrow d\text{D} + f\text{F} \tag{18.1-1}$$

where the capital letters stand for chemical formulas, and the lowercase letters are stoichiometric coefficients. We specify the rate r by

$$r = -\frac{1}{a}\frac{d[\text{A}]}{dt} = -\frac{1}{b}\frac{d[\text{B}]}{dt} = \frac{1}{d}\frac{d[\text{D}]}{dt} = \frac{1}{f}\frac{d[\text{F}]}{dt} \tag{18.1-2a}$$

where $[\text{A}]$ is the concentration (mol L^{-1} or mol m^{-3}) of substance A, $[\text{B}]$ is the concentration of substance B, etc.

An equation similar to Equation (18.1-2a) can be written for any other reaction, using the appropriate stoichiometric coefficients. In Equation (7.1-1), we wrote a general reaction in the form

$$0 = \sum_{i=1}^{s} v_i \mathscr{F}_i$$

Equation (18.1-2b) gives the same value for the rate for any choice of substance i, because of the definition of the stoichiometric coefficients.

The rate can be written as

$$r = \frac{1}{v_i}\frac{d[\mathscr{F}_i]}{dt} \tag{18.1-2b}$$

This equation is valid in the case that the system is closed and there is only one reaction.

Nonequilibrium states do not possess the same properties as equilibrium states. However, we will assume that the state of the system during a chemical reaction is similar to a metastable state, so that macroscopic variables such as temperature, pressure, and concentrations can be used. We will further assume that the rate of the reaction is a function of the temperature, pressure, and concentrations of the substances in the system.

In most reactions, the rate depends only on the concentrations of the substances occurring in the stoichiometric equation. If a substance that does not occur in the stoichiometric equation affects the rate, it is called **a catalyst** if it speeds up the reaction or an **inhibitor** if it slows it down. We discuss catalysts in the next chapter.

A macroscopic rate is determined by measuring concentrations as a function of time, which makes it a net rate by definition.

The observable rate of a chemical reaction such as that of Equation (18.1-1) is actually a net rate:

$$r = r_{\text{net}} = r_f - r_b \tag{18.1-3}$$

where r_f is the **forward rate**, or the rate at which A and B are being converted to D and F, and r_b is the **reverse rate**, or the rate at which D and F are being converted to A and B.

Chemical reactions generally proceed smoothly toward a macroscopic equilibrium state, in which the forward and reverse rates cancel each other.

---

[1] See, for example, R. G. Mortimer, *J. Phys. Chem.* **67**, 1938 (1963).

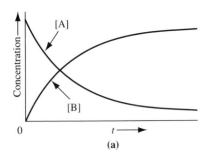

(a)

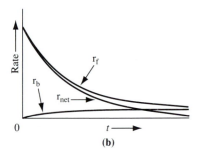

(b)

Figure 18.1. The Approach to Equilibrium of a Hypothetical Reaction. (a) The concentrations of the product B and the reactant A as a function of time. Both of the concentrations approach constant values at large values of the time. **(b)** The forward rate, the reverse rate, and the net rate as a function of time. The net rate approaches zero for large values of the time.

However, some oscillatory reactions do exist[2] and will be discussed briefly in Chapter 21. Figure 18.1 shows schematically how a nonoscillatory chemical reaction, $A \rightarrow B$, approaches equilibrium.

In most reactions, the forward rate depends only on the concentrations of the reactants, so that the forward reaction can be observed in the absence of the reverse reaction by arranging to have small or zero concentrations of the products. In this case, if A and B are reactants

$$r_f = -\frac{1}{a}\frac{d[A]}{dt} = r_f([A], [B]) \quad \text{(no reverse reaction)} \qquad \textbf{(18.1-4)}$$

If the function represented by r_f is known, Equation (18.1-4) is called the **rate law** of the forward reaction.

Similarly, the reverse reaction rate generally depends only on the concentrations of the products. In this case, the reverse reaction can be observed in the absence of the forward reaction by arranging to have small or zero concentrations of the reactants. In this case, if D and F are reactants

$$r_b = \frac{1}{a}\frac{d[A]}{dt} = r_b([D], [F]) \quad \text{(no forward reaction)} \qquad \textbf{(18.1-5)}$$

Equation (18.1-5) is the rate law of the reverse reaction.

There is a large class of chemical reactions in which the reaction rate is proportional to the concentration of each reactant raised to some power. For example, in the absence of a reverse reaction,

$$r_f = -\frac{1}{a}\frac{d[A]}{dt} = k_f[A]^\alpha[B]^\beta \qquad \textbf{(18.1-6)}$$

Equation (18.1-6) is an example of a **rate law with definite orders.** The exponent α is called the **order with respect to substance A** and the exponent β is called the **order with respect to substance B**. These orders are sometimes called **partial orders**. The sum of the orders with respect to the different substances is called the **overall order**. If α and β both equal unity, the reaction is said to be first order with respect to substance A, first order with respect to substance B, and second order overall. Higher orders are similarly assigned. The orders are usually small positive integers, but other cases do occur. Some reactions are not described by Equation (18.1-6). Such reactions are said not to have a definite order.

The proportionality constant k_f in Equation (18.1-6) is independent of the concentrations and is called the **rate constant**. Rate constants depend on temperature and pressure, although the pressure dependence is generally negligible.[3] We will discuss the temperature dependence of reaction rates in Chapter 19 and will discuss only constant-temperature systems in this chapter.

[2] R. J. Field and M. Burger, *Oscillations and Traveling Waves in Chemical Systems*, Wiley, New York, 1985.
[3] R. E. Weston and H. A. Schwarz, *Chemical Kinetics*, Prentice-Hall, Englewood Cliffs, NJ, pp. 181ff.

One of the things one wants to do for a chemical reaction is to determine its rate law. For a reaction with definite orders, this means to determine the orders and the value of the rate constant. This determination is useful for at least two reasons. First, it allows us to predict the rates of the reaction for new values of the concentrations without doing new experiments. Second, the form of the rate law usually provides information about the **mechanism of the reaction**, which is the sequence of molecular steps making up the reaction.

18.2   Forward Reactions with One Reactant

In this section we discuss reactions in which the reverse reaction can be neglected. Inspection of Figure 18.1 shows that when a reaction has nearly come to equilibrium, the reverse rate cannot be neglected. We must avoid applying the results of this section to that case, since the net rate is what can be determined macroscopically. However, many reactions proceed essentially to completion, and in that case we can neglect the reverse reaction for nearly the entire reaction.

The "classical" method for determining the rate law for a reaction is to mix the reactants together and determine the concentration of one of the reactants (or products) as a function of time as the reaction proceeds at constant temperature.

A variety of methods have been used to determine concentrations, including measurement of the following:

1. The absorbance of radiation at some wavelength at which a given product or reactant absorbs
2. The intensity of the emission spectrum of the system at a wavelength at which a given product or reactant emits
3. The volume of a solution required to titrate an aliquot of the system
4. The pressure of the system (for a reaction at constant volume)
5. The volume of the system (for a reaction at constant pressure)
6. The electrical conductance of the system
7. The mass spectrum of the system
8. The ESR or NMR spectrum of the system
9. The dielectric constant or index of refraction of the system

One must convert the concentration information into information about the derivative of the concentration, or one must solve the rate law (rate differential equation) to obtain a formula for the concentration as a function of time. We now proceed to solve the differential rate laws for a number of cases with a single reactant.

First-Order Reactions

Consider a first-order forward reaction:

$$A \rightarrow products \qquad \qquad \textbf{(18.2-1)}$$

The nature of the products is unimportant. The rate law is

$$r = -\frac{d[A]}{dt} = k_f[A] \tag{18.2-2}$$

Once again, we use separation of variables. This time we separate a dependent variable from an independent variable.

Equation (18.2-2) is a differential equation that can be solved by separation of variables. We multiply Equation (18.2-2) by dt and divide by $[A]$:

$$\frac{1}{[A]}\frac{d[A]}{dt}\,dt = \frac{1}{[A]}\,d[A] = -k_f\,dt \tag{18.2-3}$$

We carry out a definite integration from time $t = 0$ to $t = t'$:

$$\ln([A]_{t'}) - \ln([A]_0) = -k_f t' \tag{18.2-4}$$

where the subscript on a concentration indicates the time at which it is measured. Equation (18.2-4) is the same as

$$\boxed{[A]_t = [A]_0 e^{-k_f t} \quad \left(\begin{array}{l}\text{first order}\\\text{no reverse reaction}\end{array}\right)} \tag{18.2-5}$$

The validity of an equation does not depend on the symbols used as long as they are replaced consistently.

where we have written t instead of t'. An indefinite integration can also be carried out, followed by evaluation of the constant of integration.

Exercise 18.1   Carry out an indefinite integration of Equation (18.2-3). Evaluate the constant of integration to obtain Equation (18.2-5).

For ideal gases, the concentration is proportional to the partial pressure.

For a gas-phase reaction, one can use either the concentration or the partial pressure.

▼   **EXAMPLE 18.1**

The gas-phase reaction

$$N_2O_5 \rightarrow 2NO_2 + \frac{1}{2}O_2$$

is found to be first order. The rate constant at 337.6 K is equal to $5.12 \times 10^{-3}\ \text{s}^{-1}$.[4] If the partial pressure of N_2O_5 is 0.500 atm at time $t = 0$, find the partial pressure of N_2O_5 at $t = 60.0$ s, neglecting any reverse reaction and assuming that the system is at constant volume.

Solution

Assuming the gas to be ideal, the partial pressure of a gas is proportional to the concentration. Equation (18.2-5) becomes

$$P_{N_2O_5}(t) = P_{N_2O_5}(0)e^{-k_f t}$$
$$= (0.500\ \text{atm})\exp[-(5.12 \times 10^{-3}\ \text{s}^{-1})(60.0\ \text{s})] = 0.368\ \text{atm}$$

▲

---

[4] H. S. Johnston and Y. Tao, *J. Am. Chem. Soc.* **73**, 2948 (1951).

The **half-life** $t_{1/2}$ of a reaction with a single reactant is defined as the time required for half of the original amount of the reactant to react. If the reverse reaction is negligible for the time interval $0 < t < t_{1/2}$, we can write from Equation (18.2-4)

$$k_f t_{1/2} = -\ln\left(\frac{[A]_{t_{1/2}}}{[A]_0}\right) = -\ln\left(\frac{1}{2}\right)$$

or

$$t_{1/2} = \frac{\ln(2)}{k_f} \quad \left(\begin{array}{l}\text{first order}\\ \text{no reverse reaction}\end{array}\right) \qquad \textbf{(18.2-6)}$$

The **relaxation time** τ is defined as the time for the amount of reactant to drop to a fraction $1/e$ (approximately 0.3679) of its original value. Substitution of this definition into Equation (18.2-4) gives

$$\tau = \frac{1}{k_f} = \frac{t_{1/2}}{\ln(2)} \qquad \textbf{(18.2-7)}$$

The units of a first-order rate constant are s^{-1}, min^{-1}, etc. The units of the rate constant are different for each order.

Exercise 18.2

a. Find the half-life and the relaxation time for the reaction of Example 18.1.
b. Verify Equation (18.2-7).

First-order processes are not confined to chemistry. The decay of radioactive nuclides obeys first-order kinetics, with rate constants that do not appear to depend on temperature or pressure. Half-lives, not rate constants, are tabulated for radioactive nuclides.

Exercise 18.3

The half-life of ^{235}U is equal to 7.1×10^8 years. Find the time required for a sample of ^{235}U to decay to 10.0% of its original amount.

Second-Order Reactions

If the reverse reaction is negligible, the rate law is

$$r = -\frac{d[A]}{dt} = k_f [A]^2 \qquad \textbf{(18.2-8)}$$

With the variables separated,

$$\frac{1}{[A]^2}\frac{d[A]}{dt}\,dt = \frac{d[A]}{[A]^2} = -k_f\,dt \qquad \textbf{(18.2-9)}$$

We carry out a definite integration from time $t = 0$ to time $t = t'$.

$$\int_{[A]_0}^{[A]_{t'}} \frac{d[A]}{[A]^2} = -k_f \int_0^{t'} dt \qquad \textbf{(18.2-10)}$$

*The limits on the two integrals must
correspond to the same state.*
where $[A]_0$ is the concentration of A at time $t = 0$ and $[A]_{t'}$ is the concentration at time $t = t'$. The result of the integration is

$$\frac{1}{[A]_{t'}} - \frac{1}{[A]_0} = k_f t' \quad \left(\begin{array}{c}\text{second order}\\\text{no reverse reaction}\end{array}\right) \qquad \textbf{(18.2-11)}$$

Exercise 18.4   Carry out an indefinite integration and evaluate the constant of integration to obtain Equation (18.2-11) in an alternative way.

▼   **EXAMPLE 18.2**   The gas-phase reaction

$$NO_3 \rightarrow NO_2 + \frac{1}{2}O_2$$

is second order. At a constant temperature of 20°C, it is found that for an NO_3 initial concentration of 0.0500 mol L^{-1} the concentration after 60.0 minutes is 0.0358 mol L^{-1}. Find the forward rate constant, assuming there is no reverse reaction.

Solution

$$k_f = \frac{1}{t}\left(\frac{1}{[A]_t} - \frac{1}{[A]_0}\right)$$

$$= \frac{1}{3600\text{ s}}\left(\frac{1}{0.0358\text{ mol L}^{-1}} - \frac{1}{0.0500\text{ mol L}^{-1}}\right)$$

$$= 2.2 \times 10^{-3}\text{ L mol}^{-1}\text{ s}^{-1}$$

▲

Exercise 18.5   Find the concentration of NO_2 in the experiment of Example 18.2 after a total elapsed time of 120 minutes, assuming there is no reverse reaction.

The half-life of the reaction is defined as before: It is the time for half of the initial amount of reactant to react.

$$\frac{2}{[A]_0} - \frac{1}{[A]_0} = k_f t_{1/2} \qquad \textbf{(18.2-12)}$$

where we have used the definition $[A]_{t_{1/2}} = [A]_0/2$. Equation (18.2-12) leads to

$$t_{1/2} = \frac{1}{k_f[A]_0} \quad \left(\begin{array}{c}\text{second order}\\\text{no reverse reaction}\end{array}\right) \qquad \textbf{(18.2-13)}$$

<table>
<tr><td>

Exercise 18.6

</td><td>

a. Find the half-life of the reaction of Example 18.2 with the given initial concentration.

b. Find the half-life of the reaction of Example 18.2 if the initial concentration is 0.0200 mol L^{-1}.

</td></tr>
</table>

Figure 18.2 shows the concentration of the single reactant for two different hypothetical reactions that happen to have the same half-life for the same initial concentration, but where the reaction of substance B is second order and the reaction of substance A is first order. Note that the concentrations do not differ by very much for $0 < t < t_{1/2}$. If there is considerable experimental error, it might be difficult to tell a first-order reaction from a second-order reaction by inspecting a graph of the concentration versus time if the graph extends only over one half-life.

<table>
<tr><td>

Exercise 18.7

</td><td>

Find expressions for the time required for the concentration of the reactant in each of the reactions of Figure 18.2 to drop to one-sixteenth of its original value, assuming the reverse reaction is absent. Express this time both in terms of $t_{1/2}$ and in terms of the two forward rate constants and the initial concentrations.

</td></tr>
</table>

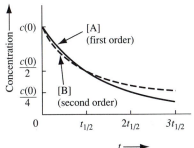

Figure 18.2. Comparison of the Concentrations of the Reactants of a First-Order Reaction and a Second-Order Reaction. This graph shows two hypothetical reactions that have equal initial half-lives for the same initial concentration. During the first half-life, the two curves do not differ very much from each other.

*n*th-Order Reactions

The rate law is

$$r = -\frac{d[A]}{dt} = k_f[A]^n \qquad \text{(18.2-14)}$$

where n is not necessarily an integer but is assumed not to equal unity or zero. The variables can be separated by division by $[A]^n$ and multiplication by dt, giving

$$-\frac{d[A]}{[A]^n} = k_f\,dt \qquad \text{(18.2-15)}$$

We perform a definite integration from $t = 0$ to $t = t'$:

$$\int_{[A]_0}^{[A]_{t'}} \frac{d[A]}{[A]^n} = -k_f \int_0^{t'} dt \qquad \text{(18.2-16)}$$

The result is

$$\frac{1}{n-1}\left(\frac{1}{[A]_{t'}^{n-1}} - \frac{1}{[A]_0^{n-1}}\right) = k_f t' \qquad \text{(18.2-17)}$$

The half-life of an nth-order reaction without a reverse reaction is found by substituting $[A]_{t_{1/2}} = [A]_0/2$ into Equation (18.2-17). The result is

$$t_{1/2} = \frac{2^{n-1} - 1}{(n-1)k_f[A]_0^{n-1}} \qquad \text{(18.2-18)}$$

This formula is not valid if n is equal to unity, in which case Equation (18.2-6) applies.

Exercise 18.8   Verify Equation (18.2-18).

Zero-Order Reactions

In the rare case that a reaction of a single reactant is zero order (independent of the concentration of the reactant), the rate law for the forward rate is

$$r = -\frac{d[A]}{dt} = k_f[A]_0 = k_f \tag{18.2-19}$$

The solution of this equation, assuming no reverse reaction, is

$$[A]_t = \begin{cases} [A]_0 - k_f t & \text{if } 0 < t < [A]_0/k_f \\ 0 & \text{if } [A]_0/k_f < t \end{cases} \tag{18.2-20}$$

where the first line of the solution is obtained from Equation (18.2-19). The second line of the solution is obtained from the physical fact that the reaction stops when the reactant is completely consumed.

Determination of Reaction Order by Comparison with Integrated Rate Laws

One does not ordinarily use a graph of the concentration as a function of time to determine the reaction order, because experimental errors can obscure the differences between orders. Instead, for each order one plots a function of the concentration that will give a linear graph. To test for zero order, one makes a graph of $[A]_t$ as a function of t. For first order, one makes a graph of $\ln([A])$ as a function of t, for second order a graph of $1/[A]$ as a function of t, for third order a graph of $1/2[A]^2$ as a function of t, etc. To test for nonintegral orders, graphs of $1/((n-1)[A]^{n-1})$ for various nonintegral values of n can also be made. The graph that is most nearly linear corresponds to the correct order. Figure 18.3 shows schematic graphs for zero, first, second, and third order.

As an alternative to graphing, one can perform linear least-squares (linear regression) fits to $[A]$ as a function of t, $\ln([A])$ as a function of t, $1/[A]$ as a function of t, etc. The least-squares procedure is generally quicker and more accurate than the graphical procedure if a programmable calculator or computer is used. It is best to use a program that calculates the **correlation coefficient**, which is a measure of the accuracy of the fit.[5]

---

[5] R. G. Mortimer, *Mathematics for Physical Chemistry*, Macmillan, New York, 1981, pp. 295, 381.

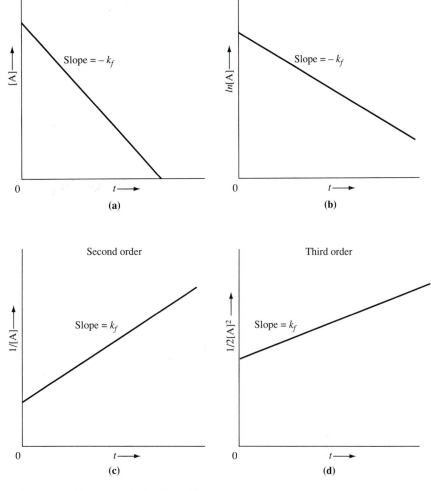

Figure 18.3. Linear Graphs for Zero-, First-, Second-, and Third-Order Reactions.
(a) Zero order. **(b)** First order. **(c)** Second order. **(d)** Third order. In each case, the appropriate function of the concentration is plotted such that the graph is linear.

The Method of Half-Lives

Another way to determine the order of a reaction is to determine how the half-life of the reaction depends on the initial concentration. If the half-life of the reaction is independent of concentration, the reaction is first order. To test for other orders, we take the logarithm of Equation (18.2-18):

$$\ln(t_{1/2}) = \ln\left[\frac{2^{n-1} - 1}{(n-1)k_f}\right] - (n-1)\ln([A]_0) \qquad \textbf{(18.2-21)}$$

To use Equation (18.2-21), one could perform a set of different experiments at the same temperature but with different initial concentrations, determine the half-life for each, and make a plot of $\ln(t_{1/2})$ versus $\ln([A]_0)$. A straight line should result, with slope equal to $-(n-1)$ and intercept equal to the first term on the right-hand side of Equation (18.2-21).

One can also take the data for a single experiment and regard different times during the experiment as "initial" times. One then treats these data in the same way as data from separate experiments. One must be sure that the reverse reaction is negligible for the later half-lives.

▼   **EXAMPLE 18.3**

For the gas-phase reaction at 300°C,

$$C_2F_4 \rightarrow \frac{1}{2}\,\text{cyclo-}C_4F_8$$

assume the following data on the concentration of C_2F_4. Using the half-life method, determine the order of the reaction and the rate constant at this temperature.

Time/min	Concentration/mol L^{-1}
0	0.0500
250	0.0250
750	0.0125
1750	0.00625
3750	0.00312

Solution

Each concentration is half of the previous concentration, so the time interval from a given data point to the next is the half-life for the reaction from the given data point as an "initial state." We have

Half-Life/min	Initial Concentration/mol L^{-1}
250	0.0500
500	0.0250
1000	0.0125
2000	0.00625

so it is apparent that the half-life doubles each time the "initial" concentration is reduced to half its previous value. This behavior indicates a second-order reaction, as confirmed by a linear least-squares fit of the logarithms, which gives order = 2.00 and $k_f = 0.080$ L mol^{-1} min^{-1} = 1.3×10^{-3} L mol^{-1} s^{-1}.

▲

The Method of Initial Rates

In this method, one compares data directly with the differential rate laws instead of the integrated rate laws. The method has two advantages: It is not necessary to integrate the rate law, and there is almost certainly no interference by the reverse reaction.

One monitors the reaction for a short time Δt. Let the change in [A] in time Δt be denoted by

$$\Delta[A] = [A]_{\Delta t} - [A]_0 \qquad (18.2\text{-}22)$$

The time Δt must be short enough that $\Delta[A] \ll [A]$. The initial rate is approximated by a quotient of finite differences:

$$r_{\text{initial}} = -\frac{d[A]}{dt} \approx -\frac{\Delta[A]}{\Delta t} \qquad (18.2\text{-}23)$$

One could alternatively measure the concentration at several times and use a graphical or numerical procedure to obtain a better approximation to the derivative at $t = 0$.

The logarithm of Equation (18.1-6) is

$$\ln(r_{\text{initial}}) = \ln(k_f) + \alpha \ln([A]_0) \qquad (18.2\text{-}24)$$

if there is no second reactant. To determine the order and the rate constant, one carries out several experiments at the same temperature but with different initial concentrations and plots the logarithm of the initial rate as a function of the logarithm of the initial concentration, or carries out a linear least-squares fit. The slope of the line fitting the data points is the order of the reaction, and the intercept is the logarithm of the rate constant.

The method just described has the disadvantage that several experiments must be carried out. A modification to this method would be to determine the value of $\Delta[A]/\Delta t$ and the value of [A] at different times in a single experiment and to use these values in Equation (18.2-24).

▼   **EXAMPLE 18.4**

Assume the following data for the decomposition of ethyl chloride, C_2H_5Cl, at $500°C$. Find the order of the reaction and the rate constant at this temperature.

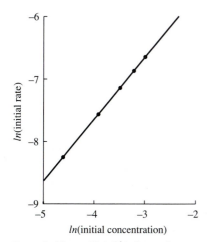

Example Figure 18.4. This figure shows the data points and the least-squares line fitting the natural logarithm of the initial rate to the natural logarithm of the initial concentration.

Initial Concentration/mol L^{-1}	Initial Rate/mol L^{-1} hour^{-1}
0.0500	0.00130
0.0400	0.00104
0.0300	0.00080
0.0200	0.00052
0.0100	0.00026

Solution

A linear least-squares fit of $\ln(r_{\text{initial}})$ against $\ln([C_2H_5Cl]_0)$ gives

$$\alpha = 1.00, \qquad \ln(k_f) = -3.64, \qquad k_f = 0.026 \text{ hour}^{-1}$$

Example Figure 18.4 shows the graph of the data points and the least-squares line. The correlation coefficient squared was equal to 1.000.

▲

18.3   Forward Reactions with More Than One Reactant

Integration of the Rate Law

Consider a reaction that is first order with respect to each of two reactants and second order overall and that has no appreciable reverse reaction:

$$a\text{A} + b\text{A} \rightarrow \text{products} \qquad \text{(18.3-1)}$$

The rate law is

$$r = -\frac{1}{a}\frac{d[\text{A}]}{dt} = k_f[\text{A}][\text{B}] \qquad \text{(18.3-2)}$$

The easiest way to study such a reaction is to carry out experiments in which the reactants are in the stoichiometric ratio:

$$\frac{[\text{A}]_0}{[\text{B}]_0} = \frac{a}{b} \qquad \text{(18.3-3)}$$

The concentrations will remain in this ratio during the reaction.

Exercise 18.9   Show that if Equation (18.3-3) holds,

$$\frac{[\text{A}]_t}{[\text{B}]_t} = \frac{a}{b} \qquad \text{(18.3-4)}$$

for all values of t greater than 0.

Equation (18.3-2) appears to have two dependent variables, $[\text{A}]$ and $[\text{B}]$. However, we can express the equation in terms of a single dependent variable $x(t)$:

$$x(t) = \frac{[\text{A}]_0 - [\text{A}]_t}{a} = \frac{[\text{B}]_0 - [\text{B}]_t}{b} \qquad \text{(18.3-5)}$$

We can now write Equation (18.3-2) in the form

$$\frac{dx}{dt} = k_f([\text{A}]_0 - ax)([\text{B}]_0 - bx) \qquad \text{(18.3-6a)}$$

$$= k_f ab\left(\frac{[\text{A}]_0}{a} - x\right)\left(\frac{[\text{B}]_0}{b} - x\right)$$

$$= k_f ab\left(\frac{[\text{A}]_0}{a} - x\right)^2 \qquad \text{(18.3-6b)}$$

where we have used the fact that $[\text{B}]_0/b = [\text{A}]_0/a$.

The variables can be separated in Equation (18.3-6b) by dividing by $([\text{A}]_0/a - x)^2$ and multiplying by dt:

$$\frac{dx}{\left(\dfrac{[\text{A}]_0}{a} - x\right)^2} = k_f ab\, dt \qquad \text{(18.3-7)}$$

We integrate both sides of Equation (18.3-7) from $t = 0$ to $t = t'$:

$$\frac{1}{\dfrac{[A]_0}{a} - x(t')} - \frac{1}{\dfrac{[A]_0}{a}} = k_f a b t'$$

which, if we replace t' by t, is the same as

$$\frac{1}{[A]_t} - \frac{1}{[A]_0} = k_f b t \qquad (18.3\text{-}8)$$

Equation (18.3-8) is the same as Equation (18.2-11) except for the appearance of the factor b in the right-hand side.

▼   **EXAMPLE 18.5**

One of the reactions implicated in the destruction of the ozone layer of the atmosphere is

$$NO + O_3 \rightarrow NO_2 + O_2$$

The reaction is second order and the rate constant is equal to 1.3×10^6 L mol^{-1} s^{-1} at 298 K. For initial concentrations of NO and O_3 both equal to 1.00×10^{-6} mol L^{-1}, find the concentrations of NO and O_3 at time $t = 2.00$ second.

Solution

$$\frac{1}{[NO]} = \frac{1}{[NO]_0} + k_f t$$

$$= \frac{1}{1.00 \times 10^{-6} \text{ mol L}^{-1}} + (1.3 \times 10^6 \text{ L mol}^{-1} \text{ s}^{-1})(2.00 \text{ s})$$

$$= 3.6 \times 10^6 \text{ L mol}^{-1}$$

$$[NO] = [O_3] = \frac{1}{3.6 \times 10^6 \text{ L mol}^{-1}} = 2.8 \times 10^{-7} \text{ mol L}^{-1}$$

▲

Exercise 18.10

    a. Find the expression for the half-life of the reaction of Equation (18.3-1) for the case of a stoichiometric mixture.

    b. Find the half-life of the reaction in Example 18.5.

In the case that the reactants are not mixed in the stoichiometric ratio, we separate the variables in Equation (18.3-6a):

$$\frac{1}{([A]_0 - ax)([B]_0 - bx)} dx = k_f \, dt \qquad (18.3\text{-}9)$$

This equation can be integrated by the method of partial fractions. We write

$$\frac{1}{([A]_0 - ax)([B]_0 - bx)} = \frac{G}{[A]_0 - ax} + \frac{H}{[B]_0 - bx} \qquad (18.3\text{-}10)$$

where G and H are guaranteed by a theorem of algebra to be constants.

These constants are found to be

$$G = \frac{1}{[B]_0 - b[A]_0/a} \quad \text{and} \quad H = \frac{1}{[A]_0 - a[B]_0/b}$$

Exercise 18.11   Verify the expressions for G and H.

When the expressions for G and H are substituted into Equation (18.3-10) and the resulting expression is substituted into Equation (18.3-9), a definite integration gives

$$\frac{1}{a[B]_0 - b[A]_0} \ln\left(\frac{[B]_t[A]_0}{[A]_t[B]_0}\right) = k_f t \qquad \text{(18.3-11)}$$

Exercise 18.12   Verify Equation (18.3-11).

▼   **EXAMPLE 18.6**   For the reaction of Example 18.5, if the initial concentration of nitric oxide is 1.00×10^{-6} mol L^{-1} and that of ozone is 5.00×10^{-7} mol L^{-1}, find the concentration of ozone after 3.50 seconds.

Solution

Let O_3 be denoted by A and NO be denoted by B. Use of Equation (18.3-11) with $a = b = 1$ gives, with some manipulation:

$$\frac{[B]_0 - x}{[A]_0 - x} = \frac{[B]_0}{[A]_0} \exp\{kt([B]_0 - [A]_0)\}$$

After several steps of algebra, one obtains $x = 4.27 \times 10^{-7}$, so the concentration of O_3 is 0.73×10^{-7} mol L^{-1}.

▲

The Method of Initial Rates for Two or More Reactants

Consider the reaction

$$a\text{A} + b\text{B} + f\text{F} \rightarrow \text{products} \qquad \text{(18.3-12)}$$

with the initial rate

$$\frac{1}{a}\frac{\Delta[A]}{\Delta t} \approx r_{\text{initial}} = k_f[A]_0^\alpha[B]_0^\beta[F]_0^\phi \qquad \text{(18.3-13)}$$

where α, β, and ϕ are the orders with respect to A, B, and F.

Several experiments are carried out at fixed temperature and with fixed values of $[B]_0$ and $[F]_0$, but with different values of $[A]_0$. The initial rate is determined for each experiment. We write

$$\ln(r_{\text{initial}}) = \ln(k_f[B]_0^\beta[F]_0^\phi) + \alpha \ln([A]_0) \qquad \text{(18.3-14)}$$

The first term on the right-hand side of this equation is fixed in these experiments.

A plot of $\ln(r_{initial})$ as a function of $\ln([A]_0)$ should give a straight line with slope equal to α. Additional experiments would then be carried out with the initial concentration of each of the other substances varied in turn. Equations analogous to Equation (18.3-14) allow the order with respect to each substance to be determined. After all of the orders are determined, everything in the right-hand side of Equation (18.3-13) is known but k_f, so k_f can be computed from any one of the experiments.

▼   **EXAMPLE 18.7**

Assume the following data for the gas-phase reaction

$$2NO + Cl_2 \rightarrow 2NOCl$$

at 298 K.

[NO]/mol L^{-1}	[O$_3$]/mol L^{-1}	Initial Rate/mol L^{-1} s^{-1}
0.0200	0.0200	7.1×10^{-5}
0.0400	0.0200	2.8×10^{-4}
0.0200	0.0400	1.4×10^{-4}

Find the order with respect to each reactant and find the rate constant.

Solution

Doubling the initial concentration of NO quadrupled the rate, so the reaction is second order with respect to NO. Doubling the initial concentration of O_3 doubled the rate, so the reaction is first order with respect to O_3. We compute the rate constant from the data of the first experiment. Another choice would be to compute it from all three and to average the three values.

$$k_f = \frac{r_{initial}}{[NO]^2[O_3]} = \frac{(7.1 \times 10^{-5} \text{ mol L}^{-1} \text{ s}^{-1})}{(0.0200 \text{ mol L}^{-1})^2(0.0200 \text{ mol L}^{-1})}$$
$$= 8.9 \text{ L}^2 \text{ mol}^{-2} \text{ s}^{-1}$$

▲

The Method of Isolation

In this method, an experiment is carried out in which the initial concentration of one reactant is made much smaller than the concentrations of the other reactants. During the reaction, the fractional changes in the large concentrations are negligible, and these concentrations are treated as constants. The small concentration is monitored just as in the case of a single reactant. For example, in the reaction of Equation (18.3-12), if [A] is much smaller than [B] and [F], the relative changes in [B] and [F] will be small. We write

$$-\frac{1}{a}\frac{d[A]}{dt} = (k_f[B]^{\beta}[F]^{\phi})[A]^{\alpha} \tag{18.3-15}$$

where the quantity in parentheses is nearly constant. Data from this kind of experiment can be treated like data from reactions with a single reactant. For example, if $\alpha = 2$, Equation (18.2-11) can be transcribed to obtain

$$\frac{1}{[A]_t} = \frac{1}{[A]_0} + (k_f[B]^\beta[F]^\phi)t \qquad \textbf{(18.3-16)}$$

Sets of experiments can also be carried out in which $[B]$ is made much smaller than $[A]$ and $[F]$, and then in which $[F]$ is made much smaller than $[A]$ and $[B]$, in order to determine β, ϕ, and k_f.

If a reaction in a liquid solution includes the solvent as a reactant, the concentration of the solvent is usually much larger than the concentrations of other reactants and is nearly fixed. For example, assume that the solvent S is involved in the reaction:

$$A + S \rightarrow products \qquad \textbf{(18.3-17)}$$

and that the rate law is

$$r = -\frac{d[A]}{dt} = k[S]^\sigma[A]^\alpha = k_{app}[A]^\alpha \qquad \textbf{(18.3-18)}$$

where σ is the order with respect to the solvent. $[S]$ is nearly constant in a fairly dilute solution. The quantity k_{app} is called an **apparent rate constant**. It can be determined by any of the methods we have discussed, as can the order with respect to substance A. However, the actual rate constant k and the order with respect to the solvent cannot be determined unless the concentration of the solvent can be changed and the value of σ found.

If the reaction is first order with respect to substance A and of unknown order with respect to the solvent, the reaction is called a **pseudo first-order reaction**. If the reaction is second order with respect to substance A, it is called **pseudo second-order**, etc.

18.4 Inclusion of a Reverse Reaction; Chemical Equilibrium

Consider the reversible reaction, which comes to equilibrium prior to completion:

$$A \underset{k_b}{\overset{k_f}{\rightleftharpoons}} B \qquad \textbf{(18.4-1)}$$

where the labels indicate that k_f is the forward rate constant and k_b is the backward rate constant. We assume that the reaction is first order in both directions:

Other cases, such as second order in one direction and first order in the other direction, are much more difficult to treat.

The net (observable) rate of the reaction is given by

$$r_{net} = -\frac{d[A]}{dt} = k_f[A] - k_b[B] \qquad \textbf{(18.4-2)}$$

At equilibrium,

$$r_{net(eq)} = 0 = k_f[A]_{eq} - k_b[B]_{eq} \qquad \textbf{(18.4-3)}$$

The equilibrium constant for the reaction of Equation (18.4-1) can be written as in Equation (7.3-3). We assume ideally dilute behavior, and use concentrations instead of activities, so that our equilibrium constants will not necessarily be dimensionless. Equation (18.4-3) is the same as

$$K_{eq} = \frac{[B]_{eq}}{[A]_{eq}} = \frac{k_f}{k_b} \tag{18.4-4}$$

The second equality in Equation (18.4-4) comes from Equation (18.4-3). A large value of the equilibrium constant means that the forward rate constant is large compared with the reverse rate constant, and a small value corresponds to a small forward rate constant. Equation (18.4-4) is an example of a more general relation, as shown in the following exercise:

Exercise 18.13 | Assume that for the reaction

$$aA + bB \overset{k_f}{\underset{k_b}{\rightleftharpoons}} dD + fF \tag{18.4-5}$$

the forward and back rates are given by

$$r = k_f[A]^a[B]^b \quad \text{and} \quad r_b = k_b[D]^d[F]^f \tag{18.4-6}$$

That is, assume that the order with respect to each substance is equal to its stoichiometric coefficient (which might not be the case in a real reaction). Show that

$$K_{eq} = \frac{k_f}{k_b} \tag{18.4-7}$$

Equation (18.4-8) appears to contain two dependent variables. We need to express [B] in terms of [A] before we can solve the equation.

We return to our special case and subtract Equation (18.4-3) from Equation (18.4-2) to obtain

$$-\frac{d[A]}{dt} = k_f([A] - [A]_{eq}) - k_b([B] - [B]_{eq}) \tag{18.4-8}$$

We now assume that $[B]_0 = 0$:

$$[B] = [A]_0 - [A] \quad \text{and} \quad [B]_{eq} = [A]_0 - [A]_{eq}$$

so that

$$[B] - [B]_{eq} = [A]_0 - [A] - ([A]_0 - [A]_{eq}) = -[A] + [A]_{eq} \tag{18.4-9}$$

When Equation (18.4-9) is substituted into Equation (18.4-8), we get

$$-\frac{d[A]}{dt} = (k_f + k_b)([A] - [A]_{eq}) \tag{18.4-10}$$

Since $[A]_{eq}$ is a constant for any particular initial condition, $d[A]/dt = d([A] - [A]_{eq})/dt$. Equation (18.4-10) is exactly like Equation (18.2-2) except for the symbols that occur. The solution is obtained by transcribing Equation (18.2-5):

$$[A]_{t'} - [A]_{eq} = ([A]_0 - [A]_{eq})e^{-(k_f + k_b)t'} \tag{18.4-11}$$

Exercise 18.14   Carry out the separation of variables to obtain Equation (18.4-11).

Figure 18.4 shows the concentration of a hypothetical reactant as a function of time. $[A] - [A]_{eq}$ decays exponentially, as did $[A]$ in the case of Figure 18.2.

We define the half-life of the reversible reaction to be the time required for $[A] - [A]_{eq}$ to drop to half of its initial value. We find that

$$t_{1/2} = \frac{\ln(2)}{k_f + k_b} \qquad (18.4\text{-}12)$$

Exercise 18.15   Verify Equation (18.4-12).

We define the **relaxation time** τ as the time required for $[A] - [A]_{eq}$ to drop to $1/e$ of its original value:

$$\tau = \frac{1}{k_f + k_b} \qquad (18.4\text{-}13)$$

A large value of the backward rate constant is as effective in giving a quick relaxation to equilibrium as is a large value of the forward rate constant, even when starting with no product present.

Exercise 18.16   For a hypothetical isomerization with $k_f = 15.7$ min^{-1} and $k_b = 35.2$ min^{-1}, find the final composition if the initial concentration of A is 0.150 mol L^{-1} and the initial concentration of B is zero. Find the half-life and the relaxation time. Find the composition at time $t = 0.100$ minute.

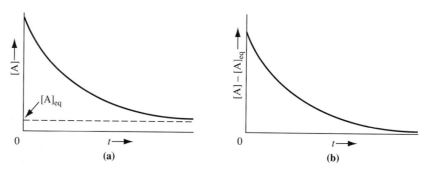

Figure 18.4. Concentration of the Reactant in a Hypothetical Reaction with Forward and Reverse Reactions. (a) $[A]$ as a function of time. **(b)** $[A] - [A]_{eq}$ as a function of time. $[A] - [A]_{eq}$ decays exponentially, like $[A]$ for a reaction without reverse reaction.

18.5 Consecutive Reactions

Consider the case that a single reactant forms a reactive intermediate in a first-order reaction without a reverse reaction, and the reactive intermediate forms a product in another first-order reaction without a reverse reaction:

$$A \overset{k_1}{\to} B \overset{k_2}{\to} F \tag{18.5-1a}$$

where the labels over the arrows indicate that k_1 is the rate constant for the first step and k_2 is the rate constant for the second step. We also use another notation:

From now on, when steps are numbered, the rate constants for the steps are labeled with the number of the step.

$$\begin{align} (1) \quad & A \to B \tag{18.5-1b} \\ (2) \quad & B \to F \tag{18.5-1c} \end{align}$$

It is understood that the rate constant has the same index as the step. If both forward and reverse rate constants occur for step number i, the forward rate constant is called k_i and the reverse rate constant is called k_i'.

Consider the special case that no B or F is present at time $t = 0$. The first step has the same rate law as Equation (18.2-2),

$$\frac{d[A]}{dt} = -k_1 \, dt \tag{18.5-2}$$

The rate law for the second step is

$$\frac{d[B]}{dt} = k_1[A] - k_2[B] \tag{18.5-3}$$

Equations (18.5-2) and (18.5-3) are a set of simultaneous differential equations. We already have the solution to Equation (18.5-2):

$$[A]_t = [A]_0 e^{-k_1 t} \tag{18.5-4}$$

This solution can be substituted into Equation (18.5-3) to obtain a single differential equation for $[B]$:

$$\frac{d[B]}{dt} = k_1[A]_0 e^{-k_1 t} - k_2[B] \tag{18.5-5}$$

We convert this equation into the form

$$dz = M \, d[B] + N \, dt = 0 \tag{18.5-6}$$

where M and N are functions of $[B]$ and t. Equation (18.5-6) is called a **Pfaffian differential equation**. An equation of this type is called an **exact differential equation** if dz is an exact differential.

If dz is exact, M and N must be derivatives of the function z and must obey Equation (B-13) of Appendix B:

$$\frac{\partial M}{\partial t} = \frac{\partial^2 z}{\partial t \, \partial[B]} = \frac{\partial^2 z}{\partial[B] \, \partial t} = \frac{\partial N}{\partial[B]} \tag{18.5-7}$$

We multiply Equation (18.5-5) by dt and recognize that $(d[B]/dt)\, dt = d[B]$.

$$d[B] + (k_2[B] - k_1[A]_0 e^{-k_1 t})\, dt = 0 \qquad \textbf{(18.5-8)}$$

This is not an exact differential equation. However, if it is multiplied by the factor $e^{k_2 t}$, we get the exact differential equation

$$e^{k_2 t}\, d[B] + \{k_2[B]e^{k_2 t} - k_1[A]_0 e^{(k_2 - k_1)t}\}\, dt = 0 \qquad \textbf{(18.5-9)}$$

A factor that converts an inexact Pfaffian differential equation into an exact differential equation is called an **integrating factor**.

Exercise 18.17   Show that Equation (18.5-9) satisfies the condition of Equation (18.5-7) for exactness but Equation (18.5-8) does not.

In this solution, we are using the fact that the line integral of an exact differential is equal to the value of the function at the final point minus the value of the function at the initial point. Since the integral is path independent, we can choose whatever path of integration is convenient.

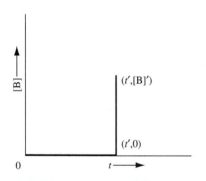

Figure 18.5. The Path of Integration to Solve Equation (18.5-9).

Equations (18.5-8) and (18.5-9) possess the same solution. In order to solve Equation (18.5-9), we perform a line integral on the path shown in Figure 18.5 and set the result equal to zero, since $dz = 0$ satisfies the differential equation.

The $d[B]$ term in Equation (18.5-9) gives no contribution on the first leg of the path. On the second leg, $t = t'$, and we obtain

$$\int_0^{[B]_{t'}} e^{-k_2 t'}\, d[B] = e^{-k_2 t'}[B]_{t'} \qquad \textbf{(18.5-10)}$$

The dt term gives no contribution on the second leg of the path. On the first leg of the path, $[B]$ is equal to zero, so the second term in the integrand vanishes and the result is

$$-\int_0^{t'} k_1[A]_0 e^{(k_2 - k_1)t}\, dt = -\frac{k_1[A]_0}{k_2 - k_1}(e^{(k_2 - k_1)t'} - 1) \qquad \textbf{(18.5-11)}$$

When the contributions of Equations (18.5-10) and (18.5-11) are combined, set equal to zero, and divided by $e^{k_2 t}$, we obtain

$$\boxed{[B]_{t'} = \frac{k_1[A]_0}{k_2 - k_1}(e^{-k_1 t'} - e^{-k_2 t'})} \qquad \textbf{(18.5-12)}$$

Exercise 18.18   Carry out the steps to obtain Equation (18.5-12).

The concentration of the final product F is obtained from

$$[F] = [A]_0 - [A] - [B] \qquad \textbf{(18.5-13)}$$

Figure 18.6a shows the concentrations of all three substances for the case that $k_1 = 0.100\ \text{s}^{-1}$ and $k_2 = 0.500\ \text{s}^{-1}$, and Figure 18.6b shows the concentrations for the case that $k_1 = 0.500\ \text{s}^{-1}$ and $k_2 = 0.100\ \text{s}^{-1}$. Since there is no reverse reaction, the final state is complete conversion to the product

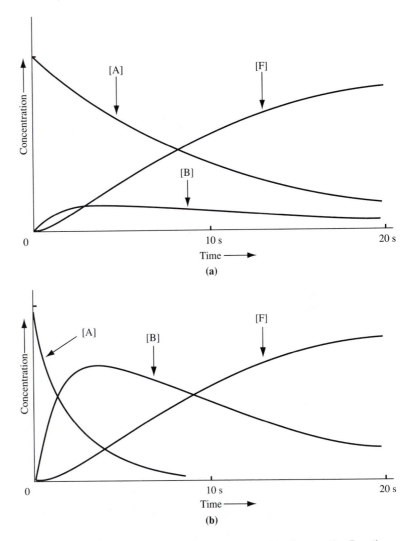

Figure 18.6. The Concentrations of Substances A, B, and F for Consecutive Reactions.
(a) The case that $k_1 = 0.10 \text{ s}^{-1}$ and that $k_2 = 0.50 \text{ s}^{-1}$. The intermediate B is used up
rapidly and [B] remains small. **(b)** The case that $k_1 = 0.50 \text{ s}^{-1}$ and that $k_2 = 0.10 \text{ s}^{-1}$.
[B] rises to a larger value in the intermediate stages of the reaction.

F in both cases. Notice that if $k_1 < k_2$ the amount of B remains relatively
small, but if $k_1 > k_2$ the amount of B becomes fairly large before dropping
eventually to zero.

If reverse reactions are included in both step 1 and step 2 of Equation (18.5-1), we write the reaction equations:

$$(1) \qquad A \rightleftharpoons B \qquad \qquad \textbf{(18.5-14a)}$$

$$(2) \qquad B \rightleftharpoons F \qquad \qquad \textbf{(18.5-14b)}$$

The rate differential equations are

$$\frac{d[A]}{dt} = -k_1[A] + k_1'[B] \tag{18.5-15a}$$

$$\frac{d[B]}{dt} = k_1[A] + k_1'[B] - k_2[B] + k_2'[F] \tag{18.5-15b}$$

$$\frac{d[F]}{dt} = k_2[B] - k_2'[F] \tag{18.5-15c}$$

This set of simultaneous differential equations can be solved,[6] but we do not present the solution.

At equilibrium,

$$\frac{[B]_{eq}}{[A]_{eq}} = \frac{k_1}{k_1'} = K_1 \tag{18.5-16}$$

and

$$\frac{[F]_{eq}}{[B]_{eq}} = \frac{k_2}{k_2'} = K_2 \tag{18.5-17}$$

The equilibrium constant K for the overall reaction is equal to

$$K = \frac{[F]_{eq}}{[A]_{eq}} = \frac{[F]_{eq}}{[B]_{eq}}\frac{[B]_{eq}}{[A]_{eq}} = K_1 K_2 = \frac{k_1}{k_1'}\frac{k_2}{k_2'} \tag{18.5-18}$$

The analogues of Equation (18.5-18) are valid for any stepwise reaction if the orders are equal to the stoichiometric coefficients: *The equilibrium constant is equal to the product of all of the forward rate constants divided by the product of all of the backward rate constants.*

18.6   The Experimental Study of Fast Reactions

The "classical" method of studying reaction rates is to mix the reactants to start the reaction and then determine the concentration of some reactant or product as a function of time.

Flow Techniques

The classical method is clearly inadequate if the reaction time is comparable to or shorter than the time required to mix the reactants. Two common flow methods can be used to speed up the mixing of liquids or gases.

In the **continuous-flow method**, two fluids are forced into a mixing chamber. The newly mixed fluid passes into a transparent tube of uniform diameter. The flow rates into the mixing chamber are kept constant, and the distance along the tube is proportional to the elapsed time since the mixing.

---

[6] T. M. Lowry and W. T. John, *J. Chem. Soc.* **97**, 2634 (1910).

The concentration of a reactant or product is determined spectrophotometrically as a function of position along the tube, using the tube as a spectrophotometer cell. The position dependence of the concentration is translated into time dependence from knowledge of the flow rate.

In the **stopped-flow method**, two fluids are forced into a mixing chamber as in the continuous-flow method. After a steady state is attained, the flow of solutions into the chamber is suddenly stopped, and the concentration of a product or reactant is determined spectrophotometrically as a function of time, using the mixing chamber as a spectrophotometer cell and often using an oscilloscope as a recording device. Figure 18.7 schematically shows a stopped-flow apparatus.

Flow systems have been designed that can mix two liquids in 0.1 ms (0.1 millisecond), so that reactions with half-lives of 1 ms to 1 second can be studied by either of the two flow methods.

Relaxation Techniques

In these techniques, one does not rely on mixing but uses the fact that equilibrium compositions depend on temperature and sometimes on pressure. One begins with a system at equilibrium and suddenly changes the temperature or pressure so that the system is no longer at equilibrium. The relaxation of the system to its new equilibrium state is then monitored.

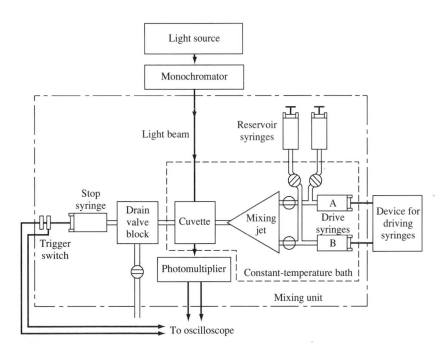

Figure 18.7. A Stopped-Flow Apparatus (schematic). The reaction takes place in the cuvette. A flow of reactants is established and the reaction reaches a steady state. The flow is suddenly stopped, and the reaction relaxes to equilibrium. The concentration of a reactant or a product is monitored spectrophotometrically.

In the **shock-tube method**, a reaction vessel is constructed with two chambers, separated by a diaphragm that can be ruptured suddenly. On one side is a mixture of gaseous reactants and products at equilibrium at a fairly low pressure. On the other side is a "driver" gas at a higher pressure. When the diaphragm is ruptured, the driver gas moves quickly into the low-pressure chamber. Collisions of the driver gas molecules with the other molecules produce a shock wave that propagates through the low-pressure gas, heating it. The reacting system then relaxes to the equilibrium state for the new temperature. The reaction is monitored near the end of the tube until the driver gas reaches the end of the tube. This method is applied to reactions that have half-lives in the range from 1 millisecond to 1 microsecond but is limited to gaseous reactions.

In the **flash photolysis method**[7] a brief (perhaps 1 microsecond) burst of light irradiates the system. The concentration of one of the reactants or products is then measured spectroscopically as a function of time as the system relaxes to its new equilibrium. The emission or absorption is ordinarily measured at right angles to the direction of the photolyzing light. Flash photolysis differs from the shock-wave technique in that the irradiation ordinarily does more than change the temperature of the system. Photochemical processes produce new species, so the system can be far from equilibrium immediately after the irradiation. Figure 18.8 shows schematically an apparatus for flash photolysis.

In the **temperature-jump** ("T-jump") and **pressure-jump** ("P-jump") methods, a gaseous or liquid system is subjected to rapid heating or a rapid change in pressure. A heating pulse can be delivered by a burst of microwave radiation or by passage of a brief pulse of electric current through an electrically conductive system. A rapid change in pressure can be achieved

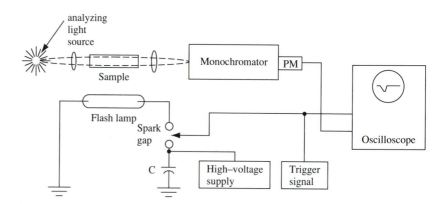

Figure 18.8. A Flash Photolysis Apparatus (schematic). The reaction mixture is subjected to a flash from the flash lamp, and the concentration of a reactant or product (or possibly an intermediate) is monitored spectrophotometrically.

---

[7] See G. Porter, *J. Chem. Soc. Faraday Trans. 2* **82**, 2445 (1986) for a historical survey.

by rupturing a diaphragm. The T-jump technique usually produces a larger effect and is more commonly used than the P-jump technique.

After the temperature or pressure change, the system relaxes to its new equilibrium state. The concentration of a reactant or product is usually monitored spectroscopically, although the reaction of hydrogen ions and hydroxide ions was monitored by measuring the electrical conductivity. Some of the fastest reactions have been studied by the T-jump or P-jump technique. Figure 18.9 shows the range of reaction half-lives for which each of several techniques can be used.

Ordinarily, the system after a T-jump or P-jump method is only slightly different in composition from the equilibrium composition at the new temperature and pressure, because it is usually not possible to change the temperature by more than about 20°C or the pressure by more than a few atmospheres.

Consider a system in which the reaction can occur:

$$(1) \qquad A + B \rightleftharpoons C \tag{18.6-1}$$

Assume that this reaction is second order overall in the forward direction and first order in the reverse direction.

The time to accomplish the T-jump or P-jump should not be greater than 10% of the half-life of the reaction.

We suddenly impose a temperature or pressure jump on the system at time $t = 0$ and allow it to relax to its new equilibrium state under conditions of constant temperature and pressure. Figure 18.10 shows schematically the concentrations of A, B, and C before and after a T-jump.

The initial concentration $[A]_0$ was the equilibrium concentration at the old temperature, but since the equilibrium constant of the reaction depends on temperature, $[A]_0$ is not equal to the new equilibrium concentration, represented by $[A]_{eq}$. The same is true of $[B]$ and $[C]$. We now let

$$\Delta[A] = [A]_t - [A]_{eq} \tag{18.6-2a}$$

$$\Delta[B] = [B]_t - [B]_{eq} \tag{18.6-2b}$$

$$\Delta[C] = [C]_t - [C]_{eq} \tag{18.6-2c}$$

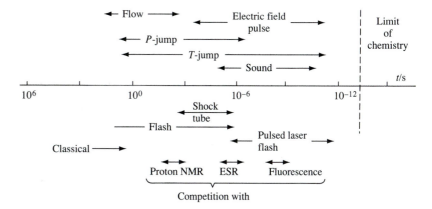

Figure 18.9. Reaction Times for Various Techniques. Classical techniques are those in which the reactants are simply poured together. The fastest technique is the pulsed laser flash technique, followed by the temperature-jump technique.

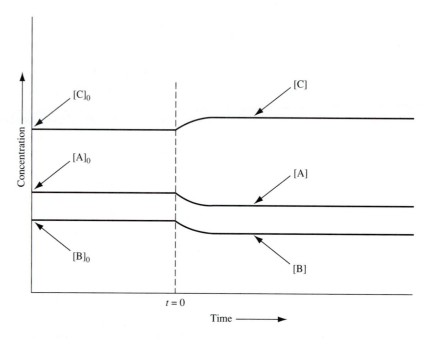

Figure 18.10. The Behavior of a System in a *T*-Jump Experiment. Prior to the temperature jump, the system is at equilibrium at one temperature. After the temperature jump, the system relaxes to its new equilibrium state.

We assume that $|\Delta[A]| \ll [A]$, $|\Delta[B]| \ll [B]$, and $|\Delta[C]| \ll [C]$, since it is not possible to change the equilibrium composition very much with a temperature jump or a pressure jump.

From the stoichiometry of Equation (18.6-1),

$$\Delta[A] = \Delta[B] = -\Delta[C] \qquad (18.6\text{-}3)$$

so that

$$[C] = [C]_{eq} + \Delta[C] \qquad (18.6\text{-}4a)$$
$$[A] = [A]_{eq} - \Delta[C] \qquad (18.6\text{-}4b)$$
$$[B] = [B]_{eq} - \Delta[C] \qquad (18.6\text{-}4c)$$

Equation (18.6-5) appears to have three dependent variables, but Equation (18.6-4abc) can be used to eliminate two of them.

The differential equation for the net rate is

$$\text{Rate} = \frac{d\Delta[C]}{dt} = k_1[A][B] - k_1'[C] \qquad (18.6\text{-}5)$$

Using Equation (18.6-4), we write the rate differential equation in terms of one dependent variable, $\Delta[C]$:

$$\frac{d\Delta[C]}{dt} = k_1([A]_{eq} - \Delta[C])([B]_{eq} - \Delta[C]) - k_1'([C]_{eq} + \Delta[C])$$

$$= k_1[A]_{eq}[B]_{eq} - k_1'[C]_{eq} - k_1([A]_{eq} + [B]_{eq})\,\Delta[C]$$
$$- k_1'\,\Delta[C] + k_1(\Delta[C])^2 \qquad (18.6\text{-}6)$$

The first two terms on the right-hand side of the final version of Equation (18.6-6) cancel, because the first is the forward rate at equilibrium and the second is the reverse rate at equilibrium. The final term on the right-hand side is much smaller than the others, because $\Delta[C]^2 \ll |\Delta[C]|$. We therefore neglect this term, which linearizes the equation:

$$\frac{d\Delta[C]}{dt} = -\left\{ k_1([A]_{eq} + [B]_{eq}) + k_1' \right\} \Delta[C] \qquad \textbf{(18.6-7)}$$

Equation (18.6-7) is exactly like Equation (18.2-2) except for the symbols used, so we can write the solution:

$$\Delta[C] = \Delta[C]_0 e^{-t/\tau} \qquad \textbf{(18.6-8)}$$

where

$$\frac{1}{\tau} = k_1([A]_{eq} + [B]_{eq}) + k_1' \qquad \textbf{(18.6-9)}$$

The relaxation is exponential, in spite of the fact that the reaction is second order in one direction. The quantity τ is the relaxation time for $\Delta[C]$.

Exercise 18.19

a. Verify the steps of algebra leading to Equation (18.6-7).
b. Verify that Equation (18.6-8), with Equation (18.6-9), is a solution to Equation (18.6-7).
c. Write the expressions for $\Delta[A]$ and $\Delta[B]$.

The reaction of hydrogen ions and hydroxide ions in water can be written as

$$H_3O^+ + OH^- \rightleftharpoons 2H_2O \qquad \textbf{(18.6-10)}$$

which is a member of a class of reactions

$$A + B \rightleftharpoons 2C \qquad \textbf{(18.6-11)}$$

Assume that the reaction is second order in both directions. We write

$$[C] = [C]_{eq} + \Delta[C] \qquad \textbf{(18.6-12a)}$$

$$[A] = [A]_{eq} - \frac{1}{2}\Delta[C] \qquad \textbf{(18.6-12b)}$$

$$[B] = [B]_{eq} - \frac{1}{2}\Delta[C] \qquad \textbf{(18.6-12c)}$$

When Equations (18.6-12) are substituted into the rate differential equation and the necessary steps of algebra are carried out with neglect of terms proportional to $(\Delta[C])^2$, we obtain

$$\Delta[C] = \Delta[C]_0 e^{-t/\tau} \qquad \textbf{(18.6-13)}$$

where

$$\frac{1}{\tau} = k_1 \frac{[A]_{eq} + [B]_{eq}}{2} + 2k_1'[C]_{eq} \qquad \textbf{(18.6-14)}$$

▼   **EXAMPLE 18.8**   At 25°C, the forward rate constant of Equation (18.6-10) is equal to 1.4×10^{11} L mol^{-1} s^{-1}. At this temperature, the dissociation equilibrium constant K_w is equal to 1.008×10^{-14} mol^2 L^{-2}.

a. Using Equation (18.5-3), find the value of k_1'.

b. For pure water, find the relaxation time if a T-jump experiment has a final temperature of 25°C.

Solution

a. We convert K_w, for which the H_2O activity is nearly equal to unity, to the equilibrium constant for the reaction of Equation (18.6-10) with all concentrations in terms of molarities:

$$K_1 = \frac{k_1}{k_1'} = \frac{[H_2O]^2}{[H^+][OH^-]} = \frac{(55.35 \text{ mol L}^{-1})^2}{1.008 \times 10^{-14} \text{ mol}^2 \text{ L}^{-2}} = 3.039 \times 10^{17}$$

$$k_1' = \frac{k_1}{K_1} = \frac{1.4 \times 10^{11} \text{ L mol}^{-1} \text{ s}^{-1}}{3.039 \times 10^{17}} = 4.6 \times 10^{-7} \text{ L mol}^{-1} \text{ s}^{-1}$$

From Equation (18.6-14)

$$\frac{1}{\tau} = \frac{(1.4 \times 10^{11} \text{ L mol}^{-1} \text{ s}^{-1})(2 \times 1.004 \times 10^{-7} \text{ mol L}^{-1})}{2}$$

$$+ 2(4.6 \times 10^{-7} \text{ L mol}^{-1} \text{ s}^{-1})(55.35 \text{ mol L}^{-1})$$

$$= 1.4 \times 10^4 \text{ s}^{-1}$$

so that $\tau = 7.1 \times 10^{-5}$ s.

▲

Similar expressions for the relaxation time can be derived for other rate laws.[8]

Exercise 18.20   **a.** Verify Equation (18.6-14).

b. For the reaction equation

$$A + B \rightleftharpoons X + Y$$

assumed second order in both directions, show that the relaxation is exponential with relaxation time given by

$$\frac{1}{\tau} = k_1([A]_{eq} + [B]_{eq}) + k_1'([X]_{eq} + [Y]_{eq}) \qquad (18.6\text{-}15)$$

Summary

A rate law of the form

$$\text{Rate} = k[A]^\alpha[B]^\beta$$

is said to have definite order, and k is called the rate constant. We solved several such differential rate laws to obtain the integrated rate laws.

[8] K. J. Laidler, *Chemical Kinetics*, 3rd ed., Harper & Row, New York, 1987, p. 38.

Some techniques for experimental determination of the rate law involve comparison of the integrated rate equation with concentration data. The method of initial rates allows direct comparison with the experimental data. In the method of isolation, the concentration of one reactant is made much smaller than the concentrations of the other reactants. During the reaction, the fractional changes in the larger concentrations are negligible, and the small concentration behaves like the concentration in a reaction with one reactant.

For a reversible reaction, it was found that the difference between the concentration of the reactant and its equilibrium value relaxes exponentially and that the relaxation time and the half-life are both inversely proportional to the sum of the two rate constants.

The case of two consecutive first-order reactions without reverse reactions was considered. This is the simplest kind of reaction mechanism. It was found that the concentration of the reactive intermediate rises and then falls as the reaction proceeds.

We discussed some techniques for studying fast reactions that cannot be studied by classical experimental techniques. These techniques included continuous-flow and stopped-flow techniques, which are rapid mixing methods, as well as relaxation techniques. The relaxation techniques included shock-tube methods, flash photolysis, and T-jump and P-jump methods. Equations were derived for the relaxation of a reaction after a small perturbation, giving an exponential relaxation for a variety of rate laws.

Additional Reading

J. N. Bradley, *Fast Reactions*, Oxford University Press, Oxford, 1975
This small book concentrates on experimental techniques for studying fast reactions, in both liquid and gas phases.

G. G. Hammes, *Principles of Chemical Kinetics*, Academic Press, New York, 1978
This is a text for a one-semester course at the senior/first-year graduate level.

P. C. Jordan, *Chemical Kinetics and Transport*, Plenum, New York, 1979
This is a clearly written text for a course at the senior/first-year graduate level, including both reaction kinetics and transport processes.

K. J. Laidler, *Chemical Kinetics*, 3rd ed., Harper & Row, New York, 1987
A comprehensive and clearly written book at the first-year graduate level.

K. J. Laidler, *Physical Chemistry with Biological Applications*, Benjamin/Cummings, Menlo Park, CA, 1978
This text for a one-semester course in physical chemistry for biologically oriented students contains a clear discussion of enzyme kinetics.

R. E. Weston and H. A. Schwarz, *Chemical Kinetics*, Prentice-Hall, Englewood Cliffs, NJ, 1972
This is an undergraduate text in chemical kinetics designed to supplement or replace a portion of a comprehensive physical chemistry text.

PROBLEMS

Problems for Section 18.2

18.21. Write the reaction equation of Example 18.1 in the form

$$2N_2O_5 \rightarrow 4NO_2 + O_2$$

with the rate equation

$$-\frac{1}{2}\frac{d[N_2O_5]}{dt} = k''_f[N_2O_5]$$

a. How is k''_f related to the coefficient k_f given in the example?

b. How does the change in the writing of the reaction equation affect the half-life of the reaction? Write the equation relating $t_{1/2}$ to k''_f.

*18.22. The following data were taken for the decomposition of dinitrogen trioxide:

t/s	0	184	526	867	1877
$[N_2O_3]/\text{mol L}^{-1}$	2.33	2.08	1.67	1.36	0.72

Assuming that the reverse reaction is negligible, determine whether the reaction is first, second, or third order and find the value of the rate constant. Proceed by graphing $\ln(c)$, $1/c$, and $1/c^2$ or by making linear least-squares fits to these functions.

18.23. A certain reaction was studied at a fixed temperature. It was found that if the initial concentration of the single reactant was equal to 4.86×10^{-3} mol L^{-1}, the half-life of the reaction was 399 s. When the initial concentration was 2.28×10^{-3} mol L^{-1}, the half-life was 696 s. Find the order of the reaction and the value of the rate constant.

18.24. A certain reaction with a single reactant is found to be fourth order and to have at a certain temperature a half-life of 10.00 s with an initial concentration of 0.100 mol L^{-1}. Find the concentration of the reactant at 20.00 s and 40.00 s for the same initial concentration and temperature.

Problems for Section 18.3

*18.25. The rate constant for the forward reaction

$$H_3O^+ + OH^- \rightarrow 2H_2O$$

is equal to 1.4×10^{11} L mol^{-1} s^{-1}.

a. Assuming instantaneous mixing, find the half-life of the neutralization reaction between a strong acid and a strong base if both the acid and the base have initial concentrations (after mixing) of 0.100 mol L^{-1}.

b. Repeat the calculation for initial concentrations of 1.00×10^{-4} mol L^{-1}.

18.26. The following reaction is first order in each reactant:

$$C_6H_5N(CH_3)_2 + CH_3I \rightarrow C_6H_5N(CH_3)_3^+ + I^-$$

Assume that the reverse reaction is negligible.

a. At 24.8°C with nitrobenzene as the solvent, the rate constant is 8.39×10^{-5} L mol^{-1} s^{-1}. Assuming that both reactants have initial concentrations of 0.100 mol L^{-1}, find the half-life of the reaction at this temperature.

b. Find the time required for 75% of the reactants to react at 24.8°C.

18.27. For the reaction of the previous problem, find the time required at 24.8°C for half of the methyl iodide to react if its initial concentration is 0.100 mol L^{-1} and the initial concentration of the dimethyl phenylamine is 0.0750 mol L^{-1}. Assume that the reverse reaction can be neglected.

18.28. Assume that the reaction

$$A + B \rightarrow \text{products}$$

is second order with respect to A and first order with respect to B. Integrate the rate differential equation

$$-d[A]/dt = k[A]^2[B]$$

to obtain the result

$$kt = \left(\frac{1}{[B]_0 - [A]_0}\right)\left(\frac{1}{[A]} - \frac{1}{[A]_0}\right) + \frac{1}{([B]_0 - [A]_0)^2}\ln\left(\frac{[B]_0[A]}{[A]_0[B]}\right)$$

18.29. For the reaction

$$aA + bB + cC \rightarrow \text{products}$$

assume that the rate law is third order overall:

$$\text{Rate} = -\frac{1}{a}\frac{d[A]}{dt} = k_f[A][B][C]$$

Integrate the rate law for the case of a stoichiometric mixture, such that $[A]:[B]:[C] = a:b:c$.

18.30. For the reaction of the previous problem, integrate the rate law for the case of a nonstoichiometric mixture, using the method of partial fractions.

*18.31. The reaction

$$2H_2O_2 \rightarrow 2H_2O + O_2$$

is catalyzed by iodide ions. Since the iodide ions are not consumed, we can determine the pseudo order with respect to H_2O_2 in the same way as for a reaction with a single reactant. The following data were obtained for the decomposition of hydrogen peroxide in 0.02 mol L^{-1} KI at 25°C:

Time/minutes	Volume of O$_2$ Evolved/mL
0	0
5.00	7.50
10.00	14.00
25.00	28.80
45.00	41.20
65.00	48.30
∞	57.90

Determine the pseudo order with respect to H$_2$O$_2$ and find the value of the apparent rate constant at this temperature. Assume that the reverse reaction can be neglected.

18.32. The decomposition of benzene diazonium chloride in water is given by the reaction equation

$$C_6H_5N_2^+ + 2H_2O \rightarrow C_6H_5OH + N_2 + H_3O^+$$

Since the concentration of water is nearly fixed, we can determine the pseudo order with respect to the diazonium ion as with a reaction having a single reactant.

The reaction was followed at 40°C by monitoring the pressure of the nitrogen evolved by the reaction. The following data were taken, with the pressure in arbitrary units:

Time/s	$P_\infty - P$
0	22.62
60	22.08
120	21.55
240	20.47
360	19.45
480	18.48
600	17.60
900	15.49
1200	13.62
1800	10.54
2400	8.15
3000	6.34
3600	4.88
4800	2.98

Determine the pseudo order with respect to the diazonium ion, and find the value of the apparent rate constant for this temperature.

Problems for Section 18.4

18.33. For the reaction A ⇌ B, which is first order in both directions, the initial concentration of A is 1.000 mol L^{-1} and that of B is 0. Draw a graph of the concentrations of A and B as a function of time for the case that $k_f = 10.0$ s^{-1} and $k_b = 1.00$ s^{-1}.

18.34. Consider the same reaction as in Section 18.4, which is first order in each direction: A ⇌ B. Integrate the rate differential equations for the case that the initial concentrations of both A and B are nonzero.

18.35. Consider the reaction 2A ⇌ B + C. The forward reaction is second order, and the reverse reaction is first order with respect to B and first order with respect to C. Write a computer program using Euler's method to integrate the rate differential equations for the case that the initial concentration of A is nonzero and that of B and C is zero. See problem 21.38 for the method.

Problems for Section 18.5

*18.36. For the consecutive first-order reaction without reverse reactions considered in Section 18.5, assume that $k_1 = 0.01000$ s^{-1} and $k_2 = 0.1000$ s^{-1}. If $[A]_0 = 0.100$ mol L^{-1}, find the value of each concentration at $t = 10.00$ s and at $t = 100.0$ s.

18.37. a. For consecutive first-order reactions without reverse reactions, make a graph of [A], [B], and [F] in the case that $k_1 = 0.0100$ s^{-1} and $k_2 = 0.100$ s^{-1}. Assume that $[A]_0 = 1.00$ mol L^{-1} and that B and F are initially absent.

b. For the values of part a, make a graph of $d[A]/dt$, $d[B]/dt$, and $d[F]/dt$.

18.38. For consecutive first-order reactions without reverse reactions, obtain the expression for [B] in the case that $k_1 = k_2$, assuming that no B or F is present at the beginning of the reaction. Proceed by taking the limit of the expression of Equation (18.5-12) as $k_1 \rightarrow k_2$, using L'Hôpital's rule. Draw a graph showing [A], [B], and [F] as functions of time for the case that $k_1 = k_2 = 0.100$ s^{-1} and $[A]_0 = 1.00$ mol L^{-1}.

Problems for Section 18.6

18.39. a. Find the expression for the relaxation time for the reaction

$$A \rightleftharpoons 2X$$

when the system is subjected to a small perturbation from equilibrium. Assume the forward reaction to be first order and the reverse reaction to be second order.

b. Find the expression for the relaxation time for the reaction

$$2A \rightleftharpoons X$$

when the system is subjected to a small perturbation from equilibrium. Assume the forward reaction to be second order and the reverse reaction to be first order.

c. Find the expression for the relaxation time for the reaction

$$A \rightleftharpoons X$$

when the system is subjected to a small perturbation from equilibrium. Assume both the forward reaction and the reverse reaction to be first order.

*18.40. For the reaction at 298 K,

$$CH_3CO_2^- + H^+ \rightleftharpoons CH_3CO_2H$$

$k_f = 4.5 \times 10^{10}$ L mol^{-1} s^{-1} and $k_b = 8.0 \times 105$ s^{-1}. A solution is made from 0.100 mol of acetic acid and enough water to make 1.000 L. Find the relaxation time if a small perturbation is imposed on the solution such that the final temperature is 298 K.

General Problems

18.41. The reaction

$$NO + H_2 \rightarrow H_2O + \frac{1}{2} N_2$$

has been studied in the gas phase at 826°C by the method of initial rates. The initial rate was monitored by measuring the total pressure of the system, and the data were obtained:

P_{H_2}(initial)/kPa	P_{NO}(initial)/kPa	$-\dfrac{dP}{dt}$ /kPa s^{-1}
53.3	40.0	0.137
53.3	20.3	0.033
38.5	53.3	0.213
19.6	53.3	0.105

a. Find the order with respect to each reactant and find the value of the rate constant. Use the partial pressures in the same way as concentrations are used, since concentrations are proportional to partial pressures in an ideal gas mixture.

b. For P_{H_2}(initial) = 100 kPa and P_{NO}(initial) = 100 kPa find the initial rate.

c. For the initial pressures of part b, find the pressure of each substance 1.00 second after the start of the reaction.

d. For P_{NO}(initial) = 200 kPa and P_{H_2}(initial) = 5.00 kPa, find the time for half of the hydrogen to react.

e. For P_{NO}(initial) = 5.00 kPa and P_{H_2}(initial) = 200 kPa, find the time for half of the nitric oxide to react.

*18.42. Label each of the following statements as either true or false. If a statement is true only under special circumstances, label it false.

a. The rate of a forward reaction can depend on the concentrations of substances other than the reactants in the stoichiometric equation.

b. The order of a reaction with respect to a substance is not necessarily equal to the stoichiometric coefficient of that substance in the reaction equation.

c. The reverse reaction is unimportant during the first half-life of any chemical reaction.

d. The method of initial rates cannot be used in conjunction with the method of isolation.

e. At the equilibrium of a chemical reaction, the rate of the forward reaction and the rate of the reverse reaction cancel.

f. A rate constant is not a true constant, since it depends on temperature and pressure.

g. In the method of initial rates, one must worry about the effect of the reverse reaction.

h. In the case of two consecutive reactions without back reactions, the final state corresponds to zero concentration of the reactant and of the intermediate.

i. In the case of two consecutive reactions without back reactions, the concentration of the intermediate always remains small compared with the initial concentration of the reactant.

j. The linearization of the rate equation that is done in studying the temperature-jump method is a good approximation, because the state immediately after the temperature jump does not deviate very much from the final equilibrium state.

k. First-order processes occur only in chemical processes.

l. All rate laws can be written in a form with definite orders.

19 Mechanisms of Chemical Reactions

OBJECTIVES

After studying this chapter, the student should:

1. understand the relationship between molecularity and order in an elementary reaction,

2. be able to deduce a rate law from a proposed mechanism by either of two approximation schemes,

3. understand the molecular basis of catalysis,

4. be able to solve problems involving the Michaelis-Menten equation for enzyme-catalyzed reactions,

5. be able to solve problems involving the collision theory of gas-phase reactions,

6. understand kinetics experiments that give molecular information.

PREVIEW

In this chapter, we discuss chemical reactions from a molecular point of view.

PRINCIPAL FACTS AND IDEAS

1. An elementary reaction is a chemical process that cannot be broken down into steps. For elementary processes, the overall order equals the molecularity.
2. The sequence of elementary molecular steps that accomplishes a chemical reaction is called the *mechanism* of that reaction.
3. The rate law for a given mechanism can often be deduced approximately from the mechanism.
4. Catalysts provide alternative mechanisms for chemical reactions.
5. The temperature dependence of gas-phase reaction rates can be understood in terms of the collision theory of reactions.
6. In some cases, mechanisms can be verified by detection of reaction intermediates.

19.1 Elementary Processes in Gases

One of the gas-phase reactions that endanger the ozone layer in the earth's upper atmosphere is

$$2NO_2 + O_3 \rightarrow N_2O_5 + O_2 \qquad \text{(19.1-1)}$$

The forward reaction is thought to consist of the following steps:[1]

$$(1) \qquad NO_2 + O_3 \rightarrow NO_3 + O_2 \qquad \text{(19.1-2a)}$$

$$(2) \qquad NO_3 + NO_2 \rightarrow N_2O_5 \qquad \text{(19.1-2b)}$$

---

[1] H. S. Johnston and D. M. Yost, *J. Chem. Phys.* **17**, 386 (1949).

These steps constitute the proposed **mechanism** of the reaction. NO_3 is a **reactive intermediate**. It is produced and then consumed, and it does not occur in the stoichiometric equation.

The steps of Equations (19.1-2a) and (19.1-2b) are said to be **elementary processes**. An elementary process is one that cannot be broken down into smaller steps. All of the reactant molecules in an elementary process must come together at the same time.

We classify an elementary process by its **molecularity**, which is the number of molecules that are involved in it. Both steps in the mechanism of Equation (19.1-2) are **bimolecular**. That is, they involve two particles. **Unimolecular** steps involve a single particle. **Termolecular** processes involve three particles. Termolecular processes are relatively slow because of the smallness of the probability that three molecules will collide or diffuse together at once, so they occur in the mechanisms of fewer reactions than do bimolecular processes. Elementary processes involving four or more particles are not thought to occur in chemical reaction mechanisms.

Bimolecular elementary processes in gases involve the binary collisions we discussed in Chapter 16. According to Equation (16.8-23), the rate of binary collisions of molecules of a single substance in a hard sphere gas is proportional to the square of the number density and thus to the square of the concentration of the substance. From Equation (16.8-33), the rate of binary collisions between unlike molecules is proportional to the product of the concentrations of the two species.

At ordinary pressures the collision rate in a gas is very large, typically several billion collisions per second for a single molecule. If every collision in a reactive mixture led to chemical reaction, gas-phase reactions would be very rapid, coming to completion in a nanosecond. Since gas-phase reactions are almost never this rapid, it is apparent that in general only a small fraction of all collisions lead to chemical reaction.

That the fraction of collisions which lead to reaction depends only on the temperature is the basic assumption in our analysis. We discuss its validity in Section 19.6.

We begin with the assumption that the fraction of binary collisions that lead to chemical reaction in a gas-phase bimolecular elementary process depends only on the temperature. That is, it is constant at a constant temperature. Consider a bimolecular process involving two molecules of substance number 1, whose formula is abbreviated by $\mathscr{F}_1$:

$$\mathscr{F}_1 + \mathscr{F}_1 \rightarrow \text{products} \qquad \textbf{(19.1-3)}$$

since the rate of two-body collisions is proportional to the square of the concentration of the substance. Therefore, the rate of the forward reaction is also proportional to the square of the concentration of the substance:

$$\text{Rate} = -\frac{1}{2}\frac{d[\mathscr{F}_1]}{dt} = \frac{f Z_{11}}{N_A} = f \pi d_1^2 \left(\frac{4 k_B T}{\pi m_1}\right)^{1/2} \mathscr{N}_1^2 \frac{1}{N_A} \qquad \textbf{(19.1-4)}$$

where m_1 is the mass of a molecule of substance 1, d_1 is its effective hard sphere diameter, $\mathscr{N}_1$ is the number density of substance 1, N_A is Avogadro's constant, and f is the fraction of collisions that lead to reaction.

By comparison with Equation (18.1-6) we can write an expression for the bimolecular elementary forward rate constant:

$$ k = \mathscr{f}\pi d_1^2 \left(\frac{4k_{\mathrm{B}}T}{\pi m_1}\right)^{1/2} N_{\mathrm{A}} \quad \text{(one reactant)} \qquad \textbf{(19.1-5)} $$

Avogadro's constant occurs in Equation (19.1-5) in order to express the rate and the concentrations in terms of moles instead of molecules.

The SI units of k in Equation (19.1-5) are $\mathrm{m^3\ mol^{-1}\ s^{-1}}$. If concentrations and rates are to be measured in $\mathrm{mol\ L^{-1}}$, an additional conversion factor is needed, equal to $1000\ \mathrm{L\ m^{-3}}$, to give $\mathrm{L\ mol^{-1}\ s^{-1}}$.

Exercise 19.1 Show that Equation (19.1-5) is correct. Remember that a factor of $1/2$ occurs in the definition of the rate as in Equation (18.1-6), that each reactive collision uses up two molecules of substance 1, and that a factor of $1/2$ was introduced into Equation (16.8-23) to avoid overcounting of collisions.

The quantity πd_1^2 is the cross-sectional area of the collision cylinder depicted in Figure 16.18 and is the area of the "target" with which the center of a molecule can collide. The quantity $\mathscr{f}\pi d_1^2$ is an effective cross-sectional area, or the area of the target that actually leads to reaction. It is called the **reaction cross section**.

In the case of a bimolecular elementary process involving one molecule each of substance 1 and substance 2, the elementary bimolecular process is first order in each substance and second order overall. The rate constant is given by

$$ k = \mathscr{f}\pi d_{12}^2 \left(\frac{8k_{\mathrm{B}}T}{\pi \mu_{12}}\right)^{1/2} N_{\mathrm{A}} \quad \text{(two reactants)} \qquad \textbf{(19.1-6)} $$

where d_{12} is the collision diameter for a collision of a molecule of substance 1 with a molecule of substance 2, as in Equation (16.8-24), and where μ_{12} is the reduced mass of molecules of mass m_1 and m_2, defined in Equation (16.8-29) and in Appendix B:

$$ \mu_{12} = \frac{m_1 m_2}{m_1 + m_2} \qquad \textbf{(19.1-7)} $$

We can now summarize our results for both types of bimolecular processes in gaseous reactions: *A gaseous bimolecular elementary process is second order overall, and for a two-substance reaction it is first order in each substance.*

Exercise 19.2 **a.** Show that Equation (19.1-6) is correct.
b. For a hypothetical bimolecular elementary reaction of a gaseous substance with an effective hard sphere diameter of 3.00×10^{-10} m and a molar mass of $0.060\ \mathrm{kg\ mol^{-1}}$, the rate constant is equal to $6.0 \times 10^{-4}\ \mathrm{m^3\ mol^{-1}\ s^{-1}}$ at 298 K. Find the fraction of collisions that lead to reaction.

A three-body collision can be pictured as a collision of a third particle with a pair of molecules that is undergoing a binary collision. The number of three-body collisions is proportional to the number of pairs of particles that are in the process of colliding with each other and to the number of third particles. The number of colliding pairs is proportional to the square of the number density (or the product of two number densities). If we again assume that the fraction of three-body collisions that leads to reaction is fixed at constant temperature, we obtain a second important result: *Gaseous termolecular elementary processes are third order overall, and the order with respect to any substance is equal to the number of molecules of that substance involved in the three-body collision.*

Unimolecular elementary processes are qualitatively different from bimolecular and termolecular processes, since they involve a single molecule and would at first glance appear to be collision independent. We assert another important result: *Gaseous unimolecular elementary processes are first order.* That is, they have a rate that is proportional to the number of molecules available to react. We will discuss unimolecular processes in Section 19.3 and will find that the above assertion is an oversimplification that applies only at sufficiently high pressure.

In order to summarize our results, we define the molecularity with respect to any substance as the number of molecules of that substance involved in the collision. The overall molecularity is equal to the sum of the molecularities of all substances. Our results are summarized: *For a gaseous elementary process, the order with respect to any substance is equal to the molecularity of that substance.*

This equality of order and molecularity holds only for elementary processes. For example, a first-order reaction could have a multistep mechanism in which there are no unimolecular steps. The order of a process does not imply anything about its molecularity unless it is an elementary process.

19.2 Elementary Reactions in Liquid Solutions

In a liquid, the reacting molecules must diffuse together before reacting chemically and the product molecules diffuse apart after the chemical reaction (the "chemical part" of the process). Such a process in a liquid does not completely satisfy the definition that an elementary process cannot be broken down into steps. However, we will speak of a process in solution that has a one-step chemical part as an elementary process in spite of the presence of the two diffusion processes.

Chemical reactions in solutions can be very rapid. For example, the reaction between hydrogen ions and hydroxide ions in aqueous solution is a second-order process with a rate constant at 25°C equal to 1.4×10^{11} L mol^{-1} s^{-1}. If solutions could be mixed instantaneously to give a solution containing hydrogen ions at 0.10 mol L^{-1} and hydroxide ions at 0.10 mol L^{-1}, this rate constant would correspond to a half-life of 7×10^{-11} s.

| Exercise 19.3 | Verify the half-life value given above. |

The rapidity of some liquid-state reactions surprises some people, since ordinary diffusion processes in liquids take hours or days. The reason for the large difference between the time required for ordinary diffusion processes and for the diffusion processes in liquid-state chemical reactions is the difference between the average distances traveled by the molecules. In an ordinary diffusion process, the root-mean-square distance traveled by molecules might be a few centimeters, whereas the mean distance between reacting molecules in a solution might be a few nanometers.

▼   **EXAMPLE 19.1**

a. Estimate the mean distance between a molecule of substance 2 and the nearest-neighbor molecule of substance 3 in a mixed solution with a concentration of each substance equal to 0.10 mol L^{-1}.

b. For a substance with a diffusion coefficient equal to 1.0×10^{-9} m^2 s^{-1}, find the time for which the root-mean-square distance diffused in one direction is equal to the mean distance in part a.

Solution

a. We estimate the mean distance as the side of a cube in the solution that contains on the average one molecule of each solute.

$$V_{\text{cube}} = \frac{1 \text{ L}}{0.10 \text{ mol}} \frac{1 \text{ mol}}{6.0 \times 10^{23} \text{ mol}^{-1}} \frac{1 \text{ m}^3}{1000 \text{ L}} = 1.7 \times 10^{-26} \text{ m}^3$$

Our solution to Example 19.1 is based on equating the side of a cube to a mean distance. More accurate procedures could have been devised, but we were asked only for an estimate.

Side of cube $= d = (1.7 \times 10^{-26} \text{ m}^3)^{1/3} = 2.6 \times 10^{-9}$ m $= 2.6$ nm

b. From Equation (17.2-19),

$$\langle x^2 \rangle^{1/2} = \sqrt{2Dt}$$

$$t = \frac{d^2}{2D} = \frac{(2.6 \times 10^{-9} \text{ m})^2}{2(1.0 \times 10^{-9} \text{ m}^2 \text{ s}^{-1})} = 3.4 \times 10^{-9} \text{ s} = 3.4 \text{ ns}$$

A solute molecule in a liquid solution can be pictured as being temporarily confined in a cage of other molecules, colliding repeatedly with molecules making up the cage. In Chapter 16, we concluded that collisions of a molecule in a typical liquid are approximately 100 times more frequent than collisions of a molecule in a typical gas, as the molecule rattles around in the cage. Because the molecules making up the cage are also moving, an avenue can occasionally open up for a molecule to move out of a cage into an adjacent cage. A molecule typically undergoes several hundred collisions in one cage before it moves into another cage.

If substance 2 can react with substance 3, a molecule of substance 2 can react only if it is in a cage in which a molecule of substance 3 is one of the cage molecules. Motion of a type 2 molecule into a cage with a type 3 molecule (or vice versa) is called an **encounter**. It is an interesting fact that, over a long time, the average rate of 2-3 collisions in a liquid is approximately the same as the rate of 2-3 collisions in a gas having concentrations equal to those in the liquid. The collision rate when molecules of substance 2 and molecules of substance 3 are in the same cage is much higher, but each molecule of type 2 spends much of its time in cages containing no molecule of type 3, and vice versa.

In a **diffusion-limited** or **diffusion-controlled** bimolecular elementary reaction between substances 2 and 3, the chemical part of the reaction is so rapid that every encounter of a substance 2 molecule with a type 3 molecule leads to reaction. The reaction of hydrogen ions and hydroxide ions in water is an example of a diffusion-limited bimolecular elementary reaction.

A theory for the rate of a bimolecular elementary diffusion-limited process was developed by Smoluchowski.[2] The original version of the theory was based on the assumption that molecules of type 2 are diffusing toward fixed molecules of type 3. On the average, the motion of the substance 2 molecules toward the fixed molecules of type 3 constitutes a diffusion flux that obeys Fick's law of diffusion. It is assumed that the first collision of any encounter will produce reaction, so the reaction occurs as soon as the center of the type 2 molecule reaches a critical distance d_{23} from the center of the type 3 molecule. The concentration of type 2 molecules closer to the center of the substance 3 molecule than this distance vanishes. The distance d_{23} is called the **reaction diameter**. It is not necessarily equal to the hard sphere collision diameter of the reacting molecules.

A solution to Fick's law of diffusion is sought such that the concentration is time independent and vanishes at distances less than or equal to d_{23} from the fixed type 3 molecule. The diffusion flux gives the rate of reaction, since all molecules that diffuse up to the sphere of radius d_{23} are assumed to react.

The diffusion flux is proportional to the bulk concentration of substance 2 (the concentration at distances far from any type 3 molecule), and the rate of reaction is also proportional to the concentration of (fixed) type 3 molecules, so the reaction is second order overall: first order with respect to 2 and first order with respect to 3. We do not present the mathematical analysis. The result is that the rate constant is given by

$$k = 4\pi N_A D_2 d_{23} \qquad \textbf{(19.2-1)}$$

where D_2 is the diffusion coefficient of substance 2 and N_A is Avogadro's number.

When the fact that the type 3 molecules are also moving is included, the result is[3]

$$\boxed{k = 4\pi N_A(D_2 + D_3)d_{23} \qquad \begin{array}{l}\text{diffusion-limited} \\ \text{reaction with} \\ \text{two reactants}\end{array}} \qquad \textbf{(19.2-2)}$$

where D_3 is the diffusion coefficient of substance 3.

---

[2] M. V. Smoluchowski, *Z. Phys. Chem.* **92**, 129 (1927). See K. J. Laidler, *Chemical Kinetics*, Harper & Row, New York, 1987, pp. 212ff.
[3] K. J. Laidler, *op. cit.*, pp. 212ff.

If two molecules of substance 2 react in a diffusion-controlled reaction, the reaction is second order in that substance. By an analysis that is analogous to that leading to Equation (19.2-2),

$$k = \frac{1}{2} 4\pi N_A (2D_2) d_{22} = 4\pi N_A D_2 d_{22} \quad \begin{array}{l} \text{diffusion-limited} \\ \text{reaction with} \\ \text{one reactant} \end{array} \qquad \textbf{(19.2-3)}$$

where d_{22} is the reaction diameter for two type 2 molecules. The factor $1/2$ in Equation (19.2-3) is included because of the factor of $1/2$ in the definition of the rate in Equation (18.1-6) for a substance with stoichiometric coefficient equal to 2.

▼ **EXAMPLE 19.2**

Calculate the rate constant for the diffusion-limited reaction

$$2I \rightarrow I_2$$

in carbon tetrachloride solution at 298 K. Take the reaction diameter equal to 4.0×10^{-10} m and the diffusion coefficient of iodine atoms in CCl_4 as 4.2×10^{-9} m^2 s^{-1}.

Solution

$$k = 4\pi(6.02 \times 10^{23} \text{ mol}^{-1})(4.2 \times 10^{-9} \text{ m}^2 \text{ s}^{-1})(4.0 \times 10^{-10} \text{ m})$$
$$= 1.3 \times 10^7 \text{ m}^3 \text{ mol}^{-1} \text{ s}^{-1} = 1.3 \times 10^{10} \text{ L mol}^{-1} \text{ s}^{-1}$$

▲

Equations (19.2-2) and (19.2-3) are not valid for ions. According to a theory of Debye,[4] the right-hand side of Equation (19.2-2) must be modified by multiplication by the **electrostatic factor**, f:

$$f = \frac{y}{e^y - 1} \qquad \textbf{(19.2-4)}$$

where

$$y = \frac{z_2 z_3 e^2}{4\pi\varepsilon d_{23} k_B T} \qquad \textbf{(19.2-5)}$$

The respective valences of the two ions are represented by z_2 and z_3. The symbol e stands for the charge on a proton and ε stands for the permittivity of the solvent (equal to the permittivity of a vacuum multiplied by the dielectric constant of the solvent).

The electrostatic factor f is greater than unity if z_2 and z_3 have opposite signs, corresponding to enhancement of the rate due to electrostatic attraction of the ion pair. If z_2 and z_3 have the same sign, the electrostatic factor is smaller than unity, corresponding to a decrease in the rate due to electrostatic repulsion.

---

[4] P. Debye, *Trans. Electrochem. Soc.* **82**, 265 (1942).

▼   **EXAMPLE 19.3**

The diffusion coefficient of H^+ in water at 25°C is 9.31×10^{-9} m^2 s^{-1}, and that of OH^- is 5.26×10^{-9} m^2 s^{-1}. Estimate the reaction diameter d. The value of the dielectric constant for water is 78.4 at 25°C. Use the value of k given above.

Solution

From Equation (19.2-5), we first obtain the relation between y and d:

$$y = \frac{(-1)(1.6022 \times 10^{-19} \text{ C})^2}{4\pi(78.4)(8.85 \times 10^{-12} \text{ C}^2 \text{ N}^{-1} \text{ m}^{-2})(1.38 \times 10^{-23} \text{ J K}^{-1})(298 \text{ K})d}$$

$$= -\frac{7.16 \times 10^{-10} \text{ m}}{d}$$

Example 19.3 is solved by successive approximations, a method that we have used previously.

Since the reaction diameter d occurs in both the expression for k in Equation (19.2-2) and the expression for f in Equation (19.2-4), we solve by successive approximation. We guess a value for d and calculate the corresponding value of k, then guess another value of d and calculate k again, etc., until an adequate approximation for k is obtained. By this procedure, the correct value of k is obtained with a value of d equal to 8.64×10^{-10} m.

▲

The reaction diameter for the reaction between hydrogen ions and hydroxide ions is somewhat larger than the sum of the radii of these ions. The explanation is that water is the solvent for the reaction, and water ionizes to form hydrogen and hydroxide ions. Water molecules can exchange hydrogen and hydroxide ions. When a hydrogen ion approaches within 8.5×10^{-10} m of a hydroxide ion, it can begin a chain reaction in which a water molecule combines with a hydrogen ion to form a hydronium ion, H_3O^+, which can then lose a different hydrogen ion to a second water molecule, which then passes a still different hydrogen ion to a third water molecule, etc. This exchange process also explains the large magnitudes of the diffusion coefficients of hydrogen and hydroxide ions in water and the large mobilities of these ions in water.

Exercise 19.4

Calculate the values of the electrostatic factor f for $z_2 z_3$ equal to 2, 1, 0, -1, and -2 at 298.15 K, assuming a dielectric constant of 78.4 and a reaction diameter of 0.50 nm. Comment on your results.

If the diffusion coefficients in Equation (19.2-3) are related to the molecular radii and the viscosity of the solvent through Equation (17.4-4), we obtain

$$k = \frac{4N_A k_B T}{6\eta}\left(\frac{1}{r_2} + \frac{1}{r_3}\right)(r_2 + r_3) = \frac{2RT}{3\eta}\left(2 + \frac{r_2}{r_3} + \frac{r_3}{r_2}\right) \quad \textbf{(19.2-6)}$$

If r_2 and r_3 are nearly equal, Equation (19.2-6) becomes

$$k = \frac{8RT}{3\eta} \quad \text{(two different reacting substances)} \quad \textbf{(19.2-7a)}$$

so that k is independent of the identity of the reactants. If the two reacting molecules are of the same substance, we must divide by 2 in order to avoid counting the same encounter twice:

$$k = \frac{4RT}{3\eta} \quad \text{(single reacting substance)} \qquad \textbf{(19.2-7b)}$$

▼ **EXAMPLE 19.4**

Use the value of the viscosity of water at 25°C to estimate the value of the rate constant for any bimolecular diffusion-controlled reaction of uncharged molecules in water at that temperature.

Solution

Use Equation (19.2-7) as an approximation and assume that the reactants are uncharged:

$$k = \frac{8(8.3145 \text{ J K}^{-1} \text{ mol}^{-1})(298 \text{ K})}{3(8.904 \times 10^{-4} \text{ kg m}^{-1} \text{ s}^{-1})}$$

$$= 7.4 \times 10^6 \text{ m}^3 \text{ mol}^{-1} \text{ s}^{-1} = 7.4 \times 10^9 \text{ L mol}^{-1} \text{ s}^{-1}$$

▲

Example 19.4 illustrates the remarkable fact that, to this approximation, the rate constant of the reaction is independent of the identity of the reactants.

There are solution reactions in which the observed rate is smaller than the rate of diffusion of the reactant molecules toward each other. Reactions that are only slightly slower than the rate of diffusion are known as **partially diffusion-limited reactions**. Solution reactions that are much slower than diffusion-limited reactions are called **activation-limited reactions**.

In an activation-limited process, the reaction will not occur until the motions of the molecules produce a high-energy collision. The molecules might have to undergo many encounters before a collision occurs with relative kinetic energy larger than the critical value. Since the diffusion processes for a diffusion-limited reaction and an activation-limited reaction have roughly equal rates, a diffusion-limited reaction is in general faster than an activation-limited reaction.

We will assume, as with gaseous reactions, that in the case of reactions that are not diffusion-limited, only a fraction of collisions lead to reaction. We assume that this fraction depends only on the temperature. For an activation-limited reaction between type 2 molecules and type 3 molecules, the molecules must undergo a 2-3 encounter to begin a set of collisions, and the rate must therefore be proportional to the number of encounters if a fixed fraction of collisions leads to reaction. Since the number of 2-3 encounters is proportional to the number of type 2 molecules and the number of type 3 molecules, such an activation-limited bimolecular elementary reaction is first order with respect to 2, first order with respect to 3, and second order overall, just as with a diffusion-limited reaction.

The rates of unimolecular elementary processes are proportional to the number of molecules that are involved, so the reactions are first order, just as in the gas phase. The rates of diffusion-limited termolecular elementary processes in liquid phases are proportional to the number of encounter pairs (pairs of molecules in the midst of an encounter) and also to the number of "third" molecules present to diffuse into the same cage as the encounter pair. Therefore, diffusion-limited termolecular elementary processes are third order, just as in the gas phase. Activation-limited termolecular elementary reactions are also third order if the fraction of collisions that lead to reaction is independent of the concentration.

We can now summarize the facts for elementary processes in both liquids and gases: *The molecularity of a substance in an elementary process is equal to its order.*

19.3   Reaction Mechanisms and Rate Laws

Unfortunately, there is no way to take an experimentally determined rate law for a given reaction and deduce the correct mechanism of the reaction from it. For example, the reaction

$$H_2 + I_2 \rightarrow 2HI$$

is second order overall. This rate law could indicate that the reaction is a bimolecular elementary reaction, and such was once thought to be the case. However, other mechanisms also conform to the same rate law, and the reaction is now thought to proceed by several competing mechanisms, including the elementary mechanism.[5]

Although we cannot directly deduce a mechanism from a rate law, we can often deduce the rate law that corresponds to a given assumed mechanism and can compare it with the experimental rate law. If the two do not match, the mechanism must be incorrect. If they do match, the mechanism *might* be correct, but we cannot say for sure that it is. Fortunately, other types of experiments can sometimes be done, including direct detection of reaction intermediates, molecular beam experiments, and radioactive tracer experiments. The results can sometimes verify a possible mechanism. However, we must regard even a well-accepted mechanism as tentative.

A mechanism is a theory, and even a well-accepted mechanism might be wrong.

Since a proposed mechanism consists of elementary steps, we can deduce a rate differential equation for each step from the fact that for an elementary process the order equals the molecularity. Every multistep mechanism leads to a set of simultaneous differential equations analogous to those for the consecutive reactions of Equation (18.5-1). For example, consider the reaction of Equation (19.1-1). If the reaction were elementary, the forward termolecular reaction would be third order overall, with the rate law:

$$\text{Rate} = d[N_2O_5]/dt = k[NO_2]^2[O_3] \quad \left(\begin{matrix} \text{rate law} \\ \text{for one-step} \\ \text{mechanism} \end{matrix} \right) \quad \textbf{(19.3-1)}$$

On the other hand, for the mechanism of Equation (19.1-2) we write the set of simultaneous differential equations

$$d[O_3]/dt = -k_1[NO_2][O_3] \qquad \textbf{(19.3-2a)}$$

$$d[N_2O_5]/dt = k_2[NO_3][NO_2] \qquad \textbf{(19.3-2b)}$$

---

[5] K. J. Laidler, *op. cit.*, pp. 297ff.

It is always true that there are as many independent differential rate equations as there are steps in the mechanism.

There are as many independent differential rate equations as there are steps in the mechanism. However, we have some choice as to which concentration time derivatives are used.

Exercise 19.5   Write the differential equation for $d[NO_3]/dt$ (it will have two terms on the right-hand side). Show that the right-hand side of this equation is equal to a linear combination (weighted sum or difference) of the right-hand sides of Equations (19.3-2a) and (19.3-2b) and that the equation is therefore not independent of the other two equations.

We do not attempt a solution of the set of Equation (19.3-2), but apply one of two common approximation schemes that reduce the set of differential equations to a single rate law.

The Rate-Limiting Step Approximation

Let us assume that the elementary reaction of Equation (19.3-2b) is much more rapid than that of Equation (19.3-2a). We mean by this assumption that $k_2 \gg k_1$, not that the actual rate of the second step is greater during the reaction (which cannot be, since one of its reactants is furnished by the first step). The inherently rapid second step uses up a molecule of NO_3 very quickly after it is produced by the slow first step.

We call the slow first step the **rate-limiting** or **rate-determining** step. Since the products of the first step are immediately used up, the rate of the reaction is controlled by the first step, and the rate law of the forward reaction is Equation (19.3-2a), the rate differential equation of the slow first step. The rapid second step plays no role in determining the rate law. The reaction is predicted to be first order with respect to nitrogen dioxide, first order with respect to ozone, and second order overall, which agrees with the experimental rate law. The mechanism of Equation (19.1-2) is possibly correct, although other mechanisms can be found that predict the same rate law.

If a step other than the first step is much slower than all other steps, the slow step will still be the rate-limiting step. In this case, the steps prior to the rate-determining step will play a role in determining the rate law. Any steps after the rate-limiting step will play no role in determining the rate law.

Consider the gaseous reaction

$$2O_3 \rightarrow 3O_2 \tag{19.3-3}$$

This reaction is thought to proceed by the mechanism

$$(1) \quad O_3 + M \rightleftharpoons O_2 + O + M \quad \text{(fast)} \tag{19.3-4a}$$

$$(2) \quad O + O_3 \rightarrow 2O_2 \qquad\qquad \text{(slow)} \tag{19.3-4b}$$

where M stands for any molecule, such as an O_2 molecule or a molecule of another substance (if other substances are present). An inelastic collision with the molecule M is needed to provide the energy to break the bond in the O_3 molecule.

The second step is assumed to be the rate-determining step. Because of the slowness of the second step, a reverse reaction for the first step is included, to allow for the possibility that this reverse reaction is faster than the slow second step. We do not include the reverse reaction for the second step, which is negligible until the reaction nearly reaches equilibrium. We obtain only the forward rate law. Only if reverse steps are included for all steps can we obtain a rate law that includes the reverse reaction.

The rate differential equation for the second step is

$$\text{Rate} = \frac{1}{2}\frac{d[O_2]}{dt} = k_2[O_3][O] \tag{19.3-5}$$

This equation cannot be solved by itself because of the presence of $[O]$, a concentration that is not known.

We invoke the **equilibrium approximation**, which is the assumption that all steps prior to the rate-limiting step are at equilibrium. These equilibria are sometimes called **pre-equilibria**. This assumption is not strictly correct while the reaction is proceeding, but if the rate-limiting step really is much slower than all previous steps, it gives nearly correct relations among the concentrations.

The equilibrium expression for the first step is

$$K_1 = \frac{k_1}{k_1'} = \frac{[O_2][O][M]}{[O_3][M]} \tag{19.3-6}$$

where we use the relation between the equilibrium constant K_1 and the rate constants, Equation (18.4-4) or (18.4-7). Equation (19.3-6) is solved for $[O]$ (canceling $[M]$ top and bottom), and the result is substituted into Equation (19.3-5) to obtain the final rate law:

$$\text{Rate} = k_2 K_1 \frac{[O_3]^2}{[O_2]} \tag{19.3-7}$$

The order with respect to O_2, a product, is -1. Equation (19.3-7) holds for the forward rate only if some O_2 is present.

Exercise 19.6 Find the rate law for the forward rate of the gaseous reaction

$$2NO + 2H_2 \rightarrow N_2 + 2H_2O$$

assuming the mechanism

(1) $2NO + H_2 \rightarrow N_2 + H_2O_2$ (slow)

(2) $H_2O_2 + H_2 \rightarrow 2H_2O$ (fast)

The Steady-State Approximation

This is the second approximation scheme that is used to produce a single rate law from a set of differential equations. It consists of the assumption that the rate of change of the concentration of one or more reactive intermediates is negligibly small. It is used when no step is sufficiently slow compared with the others that the rate-limiting step approximation can be applied accurately.

The steady-state approximation is generally a good approximation only if the concentration of the intermediate is small.

The nature of the steady-state approximation can be seen by considering the successive reactions of Section 18.5, which constitute a simple mechanism if the two successive reactions are elementary. If the first step is fast compared with the second step, as in Figure 18.6b, the concentration of the intermediate B becomes large for intermediate values of the time. Since the concentration is large, its time derivative is fairly large for most values of the time.

If the second step is fast compared with the first step, the concentration of the intermediate B remains small, as in Figure 18.6a. Unless it oscillates, a small quantity likely has a small time derivative. The time derivative of the concentration of B will therefore have a small magnitude compared with the time derivatives of reactants' and products' concentrations. Numerical solutions to sets of simultaneous rate differential equations have shown that the approximation is often quite nearly correct.[6]

The Lindemann Mechanism

There is a class of thermal decomposition reactions in the gas phase that can be represented by

$$A \rightarrow B + C \qquad \text{(19.3-8)}$$

For example, A could stand for cyclopentene, C_5H_8, in which case B stands for cyclopentadiene, C_5H_6, and C stands for H_2. This example reaction occurs spontaneously when cyclopentene is heated to about 500°C. We use this reaction both as an example of the application of the steady-state approximation and as a discussion of unimolecular processes.

The forward reaction for this class of reactions is found experimentally to be described by the rate law:

$$\text{Rate} = \frac{d[B]}{dt} = \frac{k[A]^2}{k' + k''[A]} \qquad \text{(19.3-9)}$$

where k, k', and k'' are temperature-dependent coefficients. This rate law does not correspond to a definite order, since it is not of the form of Equation (18.1-6).

Lindemann[7] proposed the following mechanism

$$(1) \qquad A + A \rightleftharpoons A^* + A \qquad \text{(19.3-10a)}$$

$$(2) \qquad A^* \rightarrow B + C \qquad \text{(19.3-10b)}$$

Step 1 is not a chemical reaction in the usual sense, since no new substance is created. The symbol A^* stands for a molecule of A that is in an excited state due to energy gained through the inelastic collision of step 1 and is able to decompose according to step 2. If another substance M is present, the second molecule of A in step 1 could be replaced by an M molecule.

---

[6] L. A. Farrow and D. Edelson, *Int. J. Chem. Kinet.* **6**, 787 (1974).
[7] F. A. Lindemann, *Trans. Faraday Soc.* **17**, 598 (1922).

When you write a differential rate equation for each step, write all but one of the equations as the time derivative of a reactive intermediate. These time derivatives will be set approximately equal to zero, giving an algebraic equation. Be sure not to invoke the steady-state approximation for a reactant or a product.

We write two differential equations,

$$\frac{d[A^*]}{dt} = k_1[A]^2 - k_1'[A][A^*] - k_2[A^*] \approx 0 \qquad \textbf{(19.3-11)}$$

$$\text{Rate} = d[B]/dt = k_2[A^*] \qquad \textbf{(19.3-12)}$$

It is important to choose the concentration of the reactive intermediate A* as one of the concentrations whose time derivatives are written. Imposing the steady-state approximation and setting this time derivative equal to zero will give an algebraic equation, the second equality of Equation (19.3-11). This equation is solved for $[A^*]$ to obtain

$$[A^*] = \frac{k_1[A]^2}{k_2 + k_1'[A]} \qquad \textbf{(19.3-13)}$$

We substitute this into Equation (19.3-12):

$$\text{Rate} = \frac{d[B]}{dt} = \frac{k_1 k_2[A]^2}{k_2 + k_1'[A]} \qquad \textbf{(19.3-14)}$$

This equation agrees with Equation (19.3-9) if $k = k_2 k_1$, $k' = k_2$, and $k'' = k_1'$. As with the equilibrium approximation, we have reduced a set of differential equations to a single rate law by replacing a differential equation by an algebraic equation.

For some reactions, it is possible to make the pressure or concentration of substance A small enough that $k_2 \gg k_1'[A]$. If we neglect the term $k_1'[A]$, the rate law is second order. It is generally possible to make the pressure of substance A large enough that $k_2 \ll k_1'[A]$, and in this case the rate law is first order.

▼ **EXAMPLE 19.5**

a. Find the rate law predicted by the mechanism of Equation (19.3-10) if the second step is rate determining. What condition turns the expression of Equation (19.3-14) into this result?

b. Find the rate law predicted by the mechanism of Equation (19.3-10) if the forward reaction of the first step is rate determining.

Solution

a. The equilibrium approximation gives

$$K_1 = \frac{k_1}{k_1'} = \frac{[A][A^*]}{[A]^2} = \frac{[A^*]}{[A]} \qquad \textbf{(19.3-15)}$$

When Equation (19.3-15) is substituted into Equation (19.3-12), we obtain the first-order equation

$$\text{Rate} = \frac{d[B]}{dt} = k_2 K_1[A] = \frac{k_2 k_1}{k_1'}[A] \qquad \textbf{(19.3-16)}$$

which is the same as Equation (19.3-14) if the k_2 term in the denominator of Equation (19.3-14) is deleted. This deletion corresponds to the case in which the forward rate of the second step is much smaller than the reverse rate of the first step (the second step is rate limiting).

b. The rate law is second order in the case that $k_2 \gg k_1'[A]$:

$$\text{Rate} = -\frac{d[A]}{dt} = k_1[A]^2$$

▲

For the mechanism of Equation (19.3-10), the steady-state approximation is more general than the rate-limiting step approximation, since the result of the latter is a special case of the steady-state result.

As noted above, the first step of the mechanism of Equation (19.3-10) is not a chemical reaction but a transfer of energy from translational kinetic energy to kinetic energy of intramolecular motions. In fact, it appears that gaseous unimolecular processes generally proceed by mechanisms like that of Equation (19.3-10) and are therefore not truly elementary.[8]

Prior to 1921, it was thought that the energy to produce gaseous unimolecular processes in polyatomic molecules came from the surroundings by radiation. This "radiation theory" of chemical reactions was abandoned upon the acceptance of the mechanism of Lindemann, which was also advanced by Christiansen.[9]

It is now accepted that the energy to provide excited molecules for unimolecular processes generally comes from collisions and that, if the pressure is sufficiently low (the "fall-off region"), such processes become second order instead of first order. However, most unimolecular processes are observed in the first-order (high-pressure) region, and we will continue to assume first-order kinetics for unimolecular steps in multistep mechanisms.

Not very many reactions are known to be unimolecular. The first one discovered was the isomerization of cyclopropane to propene. Others are the dissociation of molecular bromine and the decomposition of sulfuryl chloride.[10]

Mechanisms with More Than Two Steps

If a proposed mechanism consists of three steps, three independent simultaneous differential equations can be written. As in the case of two steps, we can apply either approximation scheme to obtain a single rate law. If the third step is rate limiting, algebraic equations are written for the equilibria of the first two steps and are used to eliminate the concentrations of reactive intermediates from the differential equation for the rate-limiting step. If the rate-limiting step approximation cannot be made, the steady-state approximation is applied to the concentrations of two reactive intermediates in order to replace two differential equations by algebraic equations.

▼   **EXAMPLE 19.6**   Find the rate law for the liquid-state mechanism[11]

$$(1) \quad H^+ + HNO_2 + NO_3^- \rightleftharpoons N_2O_4 + H_2O \qquad \text{(fast)}$$

$$(2) \quad N_2O_4 \rightleftharpoons 2NO_2 \qquad \text{(fast)}$$

$$(3) \quad NO_2 + Fe(CN)_6^{3-} \rightarrow \text{further intermediates} \quad \text{(slow)}$$

using the rate-limiting step approximation.

[8] K. J. Laidler, *op. cit.*, pp. 150ff.

[9] J. S. Christiansen, Ph.D. thesis, University of Copenhagen, 1921.

[10] K. J. Laidler, *ibid.*

[11] M. V. Twigg, *Mechanisms of Inorganic and Organometallic Reactions*, Plenum, New York, 1983, p. 39.

Since the third step is rate limiting, we must write two chemical equilibrium expressions. These equations must be solved simultaneously to obtain expressions for concentrations that can be substituted into the rate differential equation for step 3.

Solution

We apply the equilibrium approximation to steps 1 and 2:

$$K_1 = \frac{k_1}{k_1'} = \frac{[N_2O_4][H_2O]}{[H^+][HNO_2][NO_3^-]} \tag{19.3-17}$$

$$K_2 = \frac{k_2}{k_2'} = \frac{[NO_2]^2}{[N_2O_4]} \tag{19.3-18}$$

The rate differential equation for the third step is

$$\text{Rate} = k_3[NO_2][\text{fer}] \tag{19.3-19}$$

where [fer] is the concentration of the ferricyanide ion, $Fe(CN)_6^{3-}$. Equation (19.3-18) is solved to give the concentration of NO_2,

$$[NO_2] = (K_2[N_2O_4])^{1/2} \tag{19.3-20}$$

However, this equation still contains the concentration of N_2O_4, which is also a reactive intermediate. We solve Equation (19.3-17) to obtain

$$[N_2O_4] = \frac{K_1[H^+][HNO_2][NO_3^-]}{[H_2O]} \tag{19.3-21}$$

Equation (19.3-21) is substituted into Equation (19.3-20) and the resulting equation is substituted into Equation (19.3-19). The result is

$$\text{Rate} = \left(\frac{k_3 K_2 K_1[H^+][HNO_2][NO_3^-]}{[H_2O]}\right)^{1/2}[\text{fer}]$$

$$= k_{app}([H^+][HNO_2][NO_3^-])^{1/2}[\text{fer}] \tag{19.3-22}$$

We have incorporated the concentration of H_2O into k_{app}, the apparent rate constant, since the concentration of water is nearly constant in aqueous solution.

▲

Exercise 19.7   Apply the steady-state approximation to the mechanism of Example 19.6. Omit the reverse reaction in step 2.

▼

E X A M P L E   19.7   The following is a famous gas-phase reaction, with a mechanism proposed by Ogg.[12] Stoichiometry:

$$2N_2O_5 \rightarrow 4NO_2 + O_2$$

Mechanism:

(1) $N_2O_5 \rightleftharpoons NO_2 + NO_3$

(2) $NO_2 + NO_3 \rightarrow NO + O_2 + NO_2$

(3) $NO + NO_3 \rightarrow 2NO_2$

Find the rate law using the steady-state approximation.

---

[12] R. A. Ogg, Jr., *J. Chem. Phys.* **15**, 337, 613 (1947).

Solution

We write the three differential equations:

$$\text{Rate} = -\frac{d[N_2O_5]}{dt} = k_1[N_2O_5] - k_1'[NO_2][NO_3] \quad \textbf{(19.3-23)}$$

$$\frac{d[NO_3]}{dt} = k_1[N_2O_5] - k_1'[NO_2][NO_3]$$

$$\qquad\qquad - k_2[NO_2][NO_3] - k_3[NO][NO_3]$$

$$= 0 \quad \textbf{(19.3-24)}$$

$$\frac{d[NO]}{dt} = k_2[NO_2][NO_3] - k_3[NO][NO_3] = 0 \quad \textbf{(19.3-25)}$$

The second equalities in Equations (19.3-24) and (19.3-25) give two simultaneous algebraic equations, which can be solved for $[NO_3]$ and $[NO]$. The results are

$$[NO] = \frac{k_2[NO_2]}{k_3}$$

and

$$[NO_3] = \frac{k_1[N_2O_5]}{(k_1' + 2k_2)[NO_2]}$$

When these two equations are substituted into Equation (19.3-23), we obtain

$$\text{Rate} = \frac{2k_1k_2}{k_1' + 2k_2}[N_2O_5] = k_{app}[N_2O_5] \quad \textbf{(19.3-26)}$$

so the reaction is first order.

▲

Exercise 19.8   **a.** Verify Equation (19.3-26). Assume that the second step of the mechanism of Example 19.7 is rate limiting. Find the rate law for the reaction. What assumptions will cause the steady-state result to become the same as this result?

In the examples of this section, we have seen that it is a fairly routine matter to deduce a rate law from a proposed mechanism if either the rate-limiting step approximation or the steady-state approximation can be used. It is not always a routine matter to decide what mechanism to propose. However, inspection of the experimental rate law can provide some guidance.[13]

If in the experimental rate law there is a denominator with two or more terms, one should attempt to find a mechanism to which the steady-state approximation applies. If the rate law shows definite orders, one should first attempt to find a mechanism to which the rate-limiting step approximation applies.

---

[13] J. O. Edwards, E. F. Greene, and J. Ross, *J. Chem. Educ.* **45**, 381 (1968).

In order to propose a reasonable mechanism to which the rate-limiting step approximation can be applied, one can use some of the following rules:

1. If some reactants do not appear in the rate law for the forward reaction, these substances occur only in steps after the rate-limiting step.
2. If no products appear in the rate law for the forward reaction, a possible mechanism is that the first step is rate limiting.
3. If negative orders or fractional orders occur in the rate law, the rate-limiting step cannot be the first step.
4. Substances with positive orders have a larger sum of stoichiometric coefficients on the left-hand sides of step equations up to and including the rate-limiting step, and substances with negative orders have a larger sum of stoichiometric coefficients on the right-hand sides of step equations prior to the rate-limiting step.

Let us examine a few hypothetical cases. If the rate-limiting step is step number i,

$$(i) \qquad a\text{A} + b\text{B} \rightarrow \text{further intermediates or products} \qquad \textbf{(19.3-27)}$$

the rate is given by

$$\text{Rate} = k_i[\text{A}]^a[\text{B}]^b \qquad \textbf{(19.3-28)}$$

where a and b are the molecularities of the substances *in the rate-limiting step*. If A and B are both reactants, then the first step can be the rate-limiting step and Equation (19.3-28) can be the rate law.

If the rate-limiting step is not the first step and either A or B is a reactive intermediate, then the concentration of the reactive intermediate is deduced from equilibrium expressions for previous steps. For example, say that the mechanism is

$$(1) \qquad \text{B} + \text{F} \rightleftharpoons \text{A} + \text{D} \qquad (\text{fast})$$
$$(2) \qquad a\text{A} + b\text{B} \rightarrow \text{products} \qquad (\text{slow})$$

In this event,

$$[\text{A}] = K_1 \frac{[\text{B}][\text{F}]}{[\text{D}]} \qquad \textbf{(19.3-29)}$$

so the rate expression is

$$\text{Rate} = k_2 \left(K_1 \frac{[\text{B}][\text{F}]}{[\text{D}]} \right)^a [\text{B}]^b = k_{\text{app}}[\text{B}]^{a+b}[\text{F}]^a[\text{D}]^{-a} \qquad \textbf{(19.3-30)}$$

If the rate-limiting step is the third step, two equilibrium expressions can be used to replace concentrations of reactive intermediates in Equation (19.3-28). For example, if the mechanism is

$$(1) \qquad \text{B} + \text{F} \rightleftharpoons \text{A} + \text{D} \qquad (\text{fast})$$
$$(2) \qquad \text{F} + \text{H} \rightleftharpoons \text{B} \qquad (\text{fast})$$
$$(3) \qquad a\text{A} + b\text{B} \rightarrow \text{products} \qquad (\text{slow})$$

then the overall rate law is

$$\text{Rate} = k_3(K_1[\text{B}][\text{F}][\text{D}]^{-1})^a(K_2[\text{F}][\text{H}])^b$$
$$= k_{\text{app}}[\text{B}]^a[\text{F}]^{a+b}[\text{D}]^{-a}[\text{H}]^b \qquad \textbf{(19.3-31)}$$

Exercise 19.9   Verify Equation (19.3-31).

▼ **EXAMPLE 19.8**   For the gaseous reaction

$$4\text{HNO}_3 \rightarrow 4\text{NO}_2 + 2\text{H}_2\text{O} + \text{O}_2$$

the rate law is found to be

$$\text{Rate} = k_{\text{app}}[\text{HNO}_3]^2[\text{NO}_2]^{-1}$$

Propose a possible mechanism.

Solution

We must have NO_2 on the right-hand side of some step prior to a rate-limiting step. One possibility is

    (1) $\text{HNO}_3 \rightleftharpoons \text{HO} + \text{NO}_2$ (fast)

    (2) $\text{HO} + \text{HNO}_3 \rightarrow$ further intermediates/products   (slow)

with step 2 assumed to be rate limiting. Other steps following step 2 do not affect the rate law.

                                                                                    ▲

Exercise 19.10   **a.** Show that the proposed mechanism in Example 19.8 leads to the correct rate law.
b. If H_2O and NO_3 are the products of step 2 and no further HNO_3 enters in later steps, propose steps 3 and 4 to complete the mechanism and give the correct stoichiometry.

Fractional orders can occur if one of the substances in the rate-limiting step is produced in a previous step with a stoichiometric coefficient greater than unity, as in Example 19.6.

19.4   Some Additional Mechanisms, Including Chain and Photochemical Mechanisms; Competing Mechanisms

Chain reactions are characterized by a mechanism in which one or more steps produce reactive intermediates (**chain carriers**) in addition to products. The chain carriers react further, producing more products and still more chain carriers, which react further, and so forth. A chain mechanism often consists of four steps: an **initiation step**, in which chain carriers are formed;

two **chain propagation steps**, in which products are formed and chain carriers are produced as well as being used up; and a **chain termination step**, in which chain carriers are consumed.

A thoroughly studied chain reaction is the gas-phase reaction

$$H_2 + Br_2 \rightarrow 2HBr \tag{19.4-1}$$

The empirical rate law for the forward reaction in the presence of some HBr is

$$-\frac{d[H_2]}{dt} = \frac{k_a[H_2][Br_2]^{1/2}}{1 + k_b[HBr]/[Br_2]} \tag{19.4-2}$$

where k_a and k_b are temperature-dependent parameters.

The accepted mechanism for the forward reaction is[14]

$$(1) \qquad Br_2 \rightleftharpoons 2Br \tag{19.4-3a}$$
$$(2) \qquad Br + H_2 \rightleftharpoons HBr + H \tag{19.4-3b}$$
$$(3) \qquad H + Br_2 \rightarrow HBr + Br \tag{19.4-3c}$$

The reverse reaction of step 3 is omitted because its rate is small.

The forward reaction of step 1 is the initiation step, which produces Br, one of the two chain carriers. The forward reactions of steps 2 and 3 are propagation steps, producing the two chain carriers, Br and H. The reverse reaction of step 1 is the termination step. The reverse reactions of steps 2 and 3 regenerate chain carriers, so they are not termination steps. They do consume the product and are called **inhibition processes**.

Once Br atoms are formed in the initiation step, the reaction can proceed almost indefinitely without further initiation. The **chain length** γ is defined as the average number of times the cycle of the two propagation steps is repeated for each initiation step. It is possible to have a chain length as large as 10^6. In this reaction, the initiation step gives two Br atoms and each of these gives two molecules of HBr per cycle, so the average number of molecules of product for each initiation step is four times the chain length.

The rate law is obtained by use of the steady-state approximation. We write differential equations for the time derivatives of the concentrations of H_2, H and Br. We choose H_2 over Br_2 to specify the rate of the overall reaction because H_2 occurs in only one elementary step and will give a simpler differential equation. We obtain the simultaneous equations:

We have a choice of the concentration whose time derivative expresses the rate. When faced with such a choice, we always try to see which alternative will make things easier.

$$-\frac{d[H_2]}{dt} = k_2[Br][H_2] - k_2'[HBr][H] \tag{19.4-4a}$$

$$\frac{d[Br]}{dt} = 2k_1[Br_2] - k_2[Br][H_2] + k_3[H][Br_2]$$
$$+ k_2'[HBr][H] - 2k_1'[Br]^2$$
$$= 0 \tag{19.4-4b}$$

$$\frac{d[H]}{dt} = k_2[Br][H_2] - k_2'[HBr][H] - k_3[H][Br_2] = 0 \tag{19.4-4c}$$

---

[14] K. J. Laidler, *op. cit.*, pp. 291ff.

where we have applied the steady-state approximation and set the time derivatives of the concentration of the chain carriers equal to zero. To solve the algebraic versions of Equations (19.4-4b) and (19.4-4c), we add these equations to give

$$k_1[Br_2] - k_1'[Br]^2 = 0 \tag{19.4-5a}$$

which is the same as

$$[Br] = \left(\frac{k_1}{k_1'}\right)^{1/2}[Br_2]^{1/2} \tag{19.4-5b}$$

Equation (19.4-5b) is the same as the equation that would result from assuming that step 1 is at equilibrium.

Equation (19.4-5b) is substituted into Equation (19.4-4b) or Equation (19.4-4c) to obtain (after several steps of algebra)

$$[H] = \frac{k_2(k_1/k_1')^{1/2}[H_2][Br_2]^{1/2}}{k_3[Br_2] + k_2'[HBr]} \tag{19.4-6}$$

We now simplify Equation (19.4-4a) by noticing that the first two terms in Equation (19.4-4c) are the same as the two terms on the right-hand side of Equation (19.4-4a), so

$$-d[H_2]/dt = k_3[H][HBr_2]$$

When Equation (19.4-6) is substituted into this equation, we have

$$-\frac{d[H_2]}{dt} = \frac{k_2(k_1/k_1')^{1/2}[H_2][Br_2]^{1/2}}{1 + k_2'[HBr]/k_3[Br_2]} \tag{19.4-7}$$

which reproduces the empirical rate law with the following expressions for the empirical parameters:

$$k_a = k_2\left(\frac{k_1}{k_1'}\right)^{1/2} \tag{19.4-8}$$

$$k_b = \frac{k_2'}{k_3} \tag{19.4-9}$$

Exercise 19.11   Verify Equations (19.4-6) and (19.4-7).

The Photochemical Reaction

The initiation step of the chain mechanism of Equation (19.4-3) requires 190 kJ mol^{-1} to break the Br—Br bond. This energy can be supplied by ultraviolet light instead of inelastic molecular collisions.

Exercise 19.12   Calculate the minimum frequency and maximum wavelength of light with sufficient energy per photon to break a Br—Br bond.

The accepted mechanism for the photochemically initiated reaction is

(1a)	$Br_2 + h\nu \rightarrow 2Br$	**(19.4-10a)**
(1')	$2Br \rightarrow Br_2$	**(19.4-10b)**
(2)	$Br + H_2 \rightleftharpoons HBr + H$	**(19.4-10c)**
(3)	$H + Br_2 \rightarrow HBr + Br$	**(19.4-10d)**

where we use the expression for the energy of a photon, $h\nu$, as a symbol for the photon itself. The mechanism is just as in Equation (19.4-3) except for replacing process 1 by the photochemical process 1a. The termination reaction is the same as before and is labeled 1'.

Photochemical reactions were discussed in Chapter 15. They are described by two empirical laws. The **Grotthuss-Draper law** states that only the absorbed radiation is effective in producing a photochemical change. That is, a large intensity of incident light will not produce a photochemical effect if none of it is absorbed. The **Stark-Einstein law of photochemical equivalence** states that for each photon absorbed, one molecule undergoes the photochemical process. With high-intensity laser light, a molecule can absorb several photons in a single photochemical process, providing an exception to the Stark-Einstein law.[15]

The hydrogen-bromine reaction conforms to the laws of photochemistry, so the rate of the initiation step of Equation (19.4-10a) is proportional to the rate at which photons are absorbed. By measuring incident and transmitted intensities of light, one can measure the amount of radiation absorbed in the range of wavelengths that can produce Br atoms. We let J be the average rate of absorption of light, measured in einsteins per unit volume per second. The rate of photochemical production of Br atoms is equal to $2J$, so that the steady-state equation for Br atoms is now

An einstein is a mole of photons.

$$\frac{d[Br]}{dt} = 2J - k_2[Br][H_2] + k_3[H][Br_2] + k_2'[HBr][H] - 2k_1'[Br]^2$$

$$= 0 \qquad \textbf{(19.4-11)}$$

We neglect the collisional (thermal) production of Br atoms, which is much slower than the photochemical production. The steady-state equation for H atoms is still Equation (19.4-4c): When this equation is added to Equation (19.4-11), we obtain

$$J - k_1'[Br]^2 = 0 \qquad \textbf{(19.4-12)}$$

Equation (19.4-12) is combined with Equation (19.4-5) and substituted into Equation (19.4-7) to obtain

$$-\frac{d[H_2]}{dt} = \frac{k_2(2/k_1')^{1/2}[H_2]J^{1/2}}{1 + k_2'[HBr]/k_3[Br_2]} \qquad \textbf{(19.4-13)}$$

which agrees with experiment.

---

[15] See for example L. Li, M. Wu, and P. M. Johnson, *J. Chem. Phys.* **86**, 1131 (1987).

The **quantum yield Φ** of a photochemical reaction is defined as the number of molecules of product produced per photon absorbed. It is also equal to the number of moles of product per einstein of photons absorbed. The quantum yields of photochemical reactions range from nearly zero to about 10^6. Quantum yields greater than unity ordinarily indicate a chain reaction. The quantum yield of the hydrogen-bromine reaction is equal to four times the chain length and therefore can greatly exceed unity.

Hydrogen reacts in the gas phase with chlorine or with iodine much as with bromine, but there are differences between the three reactions. The study of the hydrogen-chlorine reaction is complicated by the fact that it is inhibited by the presence of oxygen, even in very small concentrations. An approximate empirical rate law for the photochemical reaction is[16]

$$\frac{d[HCl]}{dt} = \frac{k_a J[H_2][Cl_2]}{k_b[Cl_2] + [O_2]([H_2] + k_c[Cl_2])} \tag{19.4-14}$$

where k_a, k_b, and k_c are temperature-dependent parameters. In the complete absence of oxygen, the reaction becomes first order in hydrogen and zero order in chlorine (except for the dependence of J on the concentration of chlorine). However, k_b is sufficiently small that partial pressures of oxygen down to a few hundredths of a torr are effective in inhibiting the reaction.

The principal difference between the hydrogen-chlorine reaction and the hydrogen-bromine reaction is that the recombination of chlorine atoms, analogous to process 1′ in Equation (19.4-10), is unimportant, but termination of chains by combination of chlorine atoms with the surface of the reaction vessel and with other molecules (such as oxygen) is important. Another difference is that the hydrogen-chlorine reaction gives off enough heat to the reaction mixture to speed up the reaction and cause an explosion.

The hydrogen-iodine reaction was mentioned at the beginning of the previous section:

$$H_2 + I_2 \rightarrow 2HI \tag{19.4-15}$$

It is different from both of the other reactions, since it is not primarily a chain reaction, except at high temperatures. It obeys second-order kinetics and was thought for a long time to be a bimolecular elementary reaction. However, it is now thought that several mechanisms compete, including the elementary mechanism, and that under different conditions of temperature and pressure different mechanisms dominate. The chain mechanism analogous to Equation (19.4-10) is dominant above 750 K but is unimportant below 600 K. The following nonchain mechanism appears to be dominant below 600 K:[17]

$$(1) \quad I_2 \rightleftharpoons 2I \qquad \text{(fast)} \tag{19.4-16a}$$

$$(2) \quad 2I + H_2 \rightarrow 2HI \quad \text{(slow)} \tag{19.4-16b}$$

[16] K. J. Laidler, *op. cit.*, pp. 295ff.

[17] *ibid.* See also J. H. Sullivan, *J. Chem. Phys.* **46**, 73 (1967).

Exercise 19.13. **a.** Find the rate law for the mechanism of Equation (19.4-16) using the rate-limiting step approximation.

b. Find the rate law for the mechanism of Equation (19.4-16) using the steady-state approximation.

The Principle of Detailed Balance

For any reaction that proceeds by two competing mechanisms, a fundamental physical principle governs the rate constants for the two mechanisms. This is called the **principle of detailed balance**: *All mechanisms for the same reaction must give the same value of the equilibrium constant at the same temperature and pressure.* Another statement is: *At equilibrium, each mechanism must separately be at equilibrium, with canceling forward and reverse rates.*

Consider the elementary mechanism

$$(1a) \qquad H_2 + I_2 \rightleftharpoons 2HI \qquad (19.4\text{-}17)$$

and the mechanism of Equation (19.4-16) with inclusion of a back reaction in step 2. Figure 19.1 shows the two pathways. The principle of detailed balance states that at equilibrium it is not possible for the rate of the forward reaction of one mechanism to be canceled by the reverse reaction of the other mechanism.

If it were possible at equilibrium for the forward rate of one mechanism to cancel the reverse rate of a second mechanism, it would be possible to violate the laws of thermodynamics. For example, let us assume that equilibrium of the reaction of Figure 19.1 corresponds to a large forward rate of the two-step mechanism and a large reverse rate of the one-step mechanism, with smaller rates for the other two processes.

If a solid substance can be found which absorbs iodine atoms, insertion of a sample of this substance into the reaction vessel would slow down the forward rate of the two-step reaction but would do nothing to the rate of the one-step mechanism. The system would no longer be at equilibrium, and it would have to change its composition to reestablish equilibrium. Removal of the solid material would cause the system to return to its original equilibrium state, so that one could at will change the equilibrium composition back and forth. It might be possible to harness some of the Gibbs energy change of the forward reaction, and one would have a perpetual motion machine of the second kind, in violation of the second law of thermodynamics.

From Equation (18.4-4), the equilibrium constant for reaction 1a is given by

$$K_{1a} = \frac{k_{1a}}{k'_{1a}} \qquad (19.4\text{-}18)$$

By analogy with Equation (18.5-18), the equilibrium constant for the mechanism of Equation (19.4-16) is

$$K = \frac{k_1 k_2}{k'_1 k'_2} \qquad (19.4\text{-}19)$$

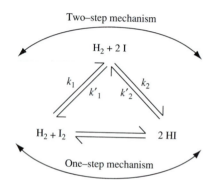

Two–step mechanism

H₂ + 2 I

k_1 k'_1 k'_2 k_2

H₂ + I₂ ⇌ 2 HI

One–step mechanism

Figure 19.1. Two Mechanisms for the H₂ + I₂ Reaction. Since both mechanisms lead from the same reactants to the same products, the equilibrium constants for the two mechanisms must be equal.

Since both equilibrium constants refer to the same reaction, they must be equal by the principle of detailed balance:

$$\frac{k_1 k_2}{k_1' k_2'} = \frac{k_{1a}}{k_{1a}'} \qquad \text{(19.4-20)}$$

The principle of detailed balance follows from a more fundamental principle, the **principle of microscopic reversibility**, which states that all mechanical processes are time reversible. That is, the equations governing these processes must be unchanged if the value of the time is replaced by its negative, so that time appears to run backward. (The preceding statement is correct for classical mechanics. In quantum mechanics, other transformations must be carried out in addition to time reversal, but the overall situation is much the same.)

If it were possible to take a moving picture of the motions of molecules, the mechanical laws of motion would still seem to apply if the movie were run backward. Macroscopic (thermodynamic) processes do not obey this principle. A movie of a diffusion process or a chemical reaction run backward would appear to violate the second law of thermodynamics, since the entropy of the universe would seem to decrease. One of the most interesting tasks of science is to answer the question: How can irreversible macroscopic processes result from time-reversible molecular processes? Although much progress has been made in understanding how to average over microstates to represent a macrostate, and it is apparent that the answer to the question lies in this averaging, the question is not yet completely answered.

Branching-Chain Reactions

The combustion of hydrogen with oxygen is also a chain reaction, but it appears to proceed by a branching-chain mechanism, which means that some propagation steps produce more chain carriers than they consume, accelerating the reaction and sometimes producing an explosion. A simplified version of the accepted mechanism of the hydrogen-oxygen reaction is[18]

(1)	$H_2 + \text{wall} \rightarrow H(\text{adsorbed}) + H$	(initiation)
(2)	$H + O_2 \rightarrow OH + O$	(branching)
(3)	$O + H_2 \rightarrow OH + H$	(branching)
(4)	$OH + H_2 \rightarrow H_2O + H$	(propagation)
(5)	$H + O_2 + M \rightarrow HO_2 + M$	(termination)
(6)	$H + \text{wall} \rightarrow \text{stable species}$	(termination)
(7)	$HO + \text{wall} \rightarrow \text{stable species}$	(termination)
(8)	$HO_2 + \text{wall} \rightarrow \text{stable species}$	(termination)

In step 5, M represents any molecule that can collide with the H and O_2. This could be an H_2 or an O_2 molecule, but it could be an impurity molecule

---

[18] S. W. Benson, *The Foundations of Chemical Kinetics*, McGraw-Hill, New York, 1960, pp. 454ff.

if one is present. Steps 2 and 3 use up one chain carrier but provide two chain carriers, so as these steps occur the reaction rate is accelerated and an explosion can occur.

Because several processes at the walls of the vessel are included in the mechanism, the ratio of the surface area to the volume is important, as are the temperature and pressure, in determining whether an explosion will occur. It is possible to predict from the values of pressure, temperature, and surface-to-volume ratio whether the system will explode or will react smoothly.

19.5 Catalysis

Jons Jakob Berzelius, 1779–1848, great Swedish chemist who dominated the field of chemistry for several decades and who invented the present system of chemical symbols for the elements and compounds.

A substance that affects the rate of a chemical reaction but does not appear in the chemical equation for the reaction is either a **catalyst** or an **inhibitor**. A substance that increases the rate of the reaction is called a catalyst, and a substance that decreases the rate of the reaction is called an inhibitor. The term "catalyst" was coined in 1836 by Berzelius from the Greek words "kata" (wholly) and "lyein" (to loosen).[19]

Catalysis is divided into two principal classes: In **homogeneous catalysis**, all substances involved in the reaction, including the catalyst, occur in the same phase. In **heterogeneous catalysis**, the catalyzed reaction occurs at the boundary between two phases. Most heterogeneous catalysts are solids, such as those in the catalytic converters in automobile exhaust systems.

In general, a catalyst provides an alternative mechanism that competes with the uncatalyzed mechanism. If the catalyzed mechanism is faster than the uncatalyzed mechanism, the observed rate of the reaction is due mostly to the catalyzed mechanism, although the reaction is also still proceeding by the uncatalyzed mechanism.

Heterogeneous Catalysis

Absorption involves a substance being taken into the bulk of a phase, while adsorption involves a substance being taken onto a surface.

We consider mechanisms involving molecules adsorbed from a gas or liquid phase onto the surface of a solid catalyst. For example, we could have an uncatalyzed unimolecular process:

$$A(gas) \rightarrow \text{products or intermediates} \qquad \textbf{(19.5-1a)}$$

and a competing catalyzed process:

$$A(gas) + \text{surface} \rightarrow A(adsorbed)$$

$$A(adsorbed) \rightarrow \text{same products or intermediates} \qquad \textbf{(19.5-1b)}$$

The rate of the first process depends on the concentration of substance A in the gas phase, while the rate of the second process depends on the amount of substance A adsorbed on the solid surface.

---

[19] K. J. Laidler, *op. cit.*, p. 229.

The Theory of Adsorption

*Irving Langmuir, 1881–1957, American
industrial chemist who won the 1932
Nobel Prize in chemistry for his work
on surface chemistry.*

The study of adsorption was pioneered by Langmuir, who derived an expression for the equilibrium fraction of a solid surface covered by an adsorbed substance as a function of the concentration of the substance in the gas or liquid phase.

The adsorption process is represented by

$$(1) \qquad A + \text{surface site} \rightleftharpoons A(\text{adsorbed})$$

It is assumed that the surface contains a set of sites at each of which a molecule of A can be adsorbed, and that only a single layer of molecules of A (a **monolayer**) can be adsorbed on the surface. The sites might include all of the atoms of the solid surface, or might be special locations such as a "step" between two layers of atoms, as schematically depicted in Figure 19.2. It is assumed that the total number of surface sites is fixed for a fixed amount of catalyst. The fraction of the surface sites occupied by adsorbed A molecules is denoted by θ.

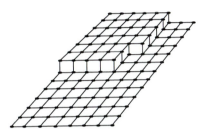

**Figure 19.2. Idealized Sites on a Solid
Surface.** A real crystal surface is not a
perfect plane of molecules, but contains
imperfections such as the step imperfec-
tion shown in this figure. The adsorption
probably takes place primarily at the
imperfections.

The adsorption process is assumed to be an elementary process, so the rate of adsorption is first order in the concentration of A in the fluid phase and is also proportional to $1 - \theta$, the fraction of surface sites available for adsorption:

$$\text{Rate of adsorption} = k_1[A](1 - \theta) \qquad \textbf{(19.5-2)}$$

The desorption is also assumed to be an elementary process, so

$$\text{Rate of desorption} = k_1'\theta \qquad \textbf{(19.5-3)}$$

At equilibrium, the rate of desorption equals the rate of adsorption, and we can write

$$k_1'\theta = k_1[A](1 - \theta) \qquad \textbf{(19.5-4)}$$

which can be solved for θ to give

$$\theta = \frac{k_1[A]}{k_1' + k_1[A]} = \frac{K[A]}{1 + K[A]} \qquad \textbf{(19.5-5)}$$

where K is a type of equilibrium constant and has the units of reciprocal concentration (L mol^{-1} or m^3 mol^{-1}).

$$K = \frac{k_1}{k_1'} \qquad \textbf{(19.5-6)}$$

The function of Equation (19.5-5) is known as the **Langmuir isotherm**. The name "isotherm" is used because the formula gives the fraction of the surface covered as a function of the concentration of A at a fixed temperature. Figure 19.3 schematically depicts the Langmuir isotherm for a hypothetical system.

The value of K can be determined from a graph of the Langmuir isotherm by determining the value of the concentration of A corresponding to $\theta = 1/2$.

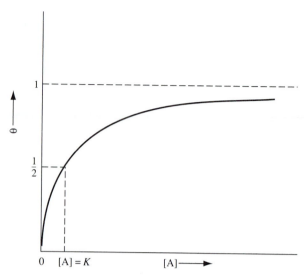

Figure 19.3. The Langmuir Isotherm. This curve has the property that when $[A] = K$, the surface adsorption sites are half occupied. The asymptote corresponds to full occupation.

Exercise 19.14   Show that $1/K$ equals the value of $[A]$ corresponding to $\theta = 1/2$.

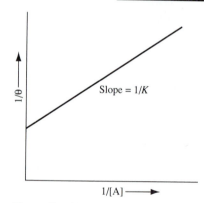

Figure 19.4. Linear Plot of the Langmuir Isotherm. The plot of $1/\theta$ against $1/[A]$ is linear according to the Langmuir isotherm.

It is easier to make a linear plot or a linear least-squares fit than a plot of or a fit to any other curve. If such a transformation can be done, it is worth the trouble required to transform an equation into a linear form.

Since the total area of an adsorbing surface and the area occupied by an adsorbed molecule will probably not be known, the value of θ is generally not directly measurable. However, the mass adsorbed is proportional to θ, and a graph of the mass adsorbed will have the same shape as the graph of Figure 19.3; the location of the asymptote corresponds to $\theta = 1$, allowing one to determine where $\theta = 1/2$ is located on the graph.

Locating an asymptote on a graph is difficult if the data suffer from experimental errors, so it is desirable to make a graph that gives a linear plot, as shown in Figure 19.4. In this graph, $1/\theta$ is plotted as a function of $1/[A]$, corresponding to the version of Equation (19.5-5):

$$\frac{1}{\theta} = \frac{1 + K[A]}{K[A]} = \frac{1}{K[A]} + 1 \qquad \textbf{(19.5-7)}$$

A linear least-squares procedure can be used as an alternative to graphing.

A plot of the reciprocal of the mass adsorbed is proportional to a plot of $1/\theta$. Since a plot of $1/\theta$ as a function of $1/[A]$ has an intercept equal to unity, it is possible to determine the proportionality constant and thus calculate the slope of the $1/\theta$ plot, from which we can obtain the value of K.

▼   **EXAMPLE 19.9**   Chloroethane from the gas phase is adsorbed on a sample of charcoal at 273.15 K. The mass adsorbed for each concentration in the gas phase is:

$[C_2H_5Cl]$/mol L^{-1}	0.00117	0.00294	0.00587	0.0117	0.0176
mass/g	3.0	3.8	4.3	4.7	4.8

a. Find the value of θ for each concentration and the value of K.

b. If each chloroethane molecule occupies an area of 2.60×10^{-19} m^2, find the area of the sample of charcoal.

Solution

a. Since m, the mass absorbed, is proportional to θ if the area is fixed, we write from Equation (19.5-7)

$$\frac{1}{m} = \frac{B}{\theta} = \frac{B}{K[A]} + B \qquad (19.5\text{-}8)$$

where B is a proportionality constant which we evaluate from the data. For each data point, we calculate $1/m$ and $1/[A]$, using grams as the unit of mass and mol L^{-1} (M) as the unit of concentration:

In the solution of Example 19.9, the mass adsorbed is given instead of the value of θ. The proportionality constant B has to be introduced. However, K and B can both be found from the data.

$\dfrac{1}{m/\mathrm{g}}$	0.333	0.263	0.233	0.213	0.208
$\dfrac{1}{[A]/\mathrm{M}}$	855	340	170	85.5	56.8

Example Figure 19.9 shows a graph of $1/m$ as a function of $1/[A]$, with the linear least-squares line drawn in. The slope of this line is equal to 1.55×10^{-4} M g^{-1} and its intercept is equal to 0.203 g^{-1}. The parameter B is therefore equal to 0.203 g^{-1}, and the value of K is equal to 0.203 g$^{-1}/1.55 \times 10^{-4}$ M g^{-1} = 1310 M^{-1}.

The value of θ for each concentration is given by $\theta = Bm$, giving the set of values:

0.609	0.771	0.873	0.954	0.974

The asymptotic value of m is equal to $1/B$, or 4.93 g.

b. The effective area of the sample is obtained from the assumption that the asymptotic amount adsorbed corresponds to a full monolayer. Since the molar mass of chloroethane is 64.515 g mol^{-1}, the asymptotic amount adsorbed is 0.0764 mol. The area is

$$\mathscr{A} = (0.260 \text{ nm}^2 \text{ molecule}^{-1})(6.022 \times 10^{23} \text{ molecule mol}^{-1})(0.0764 \text{ mol})$$
$$= 1.20 \times 10^{22} \text{ nm}^2 = 1.20 \times 10^4 \text{ m}^2$$

Although this corresponds to the area of a macroscopic square 110 m on a side, charcoal can be so finely divided that this sample of charcoal might have a mass of only 1 gram.

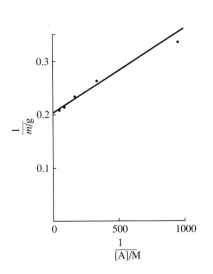

Example Figure 19.9. Plot of 1/m vs. 1/[A]. The value of the Langmuir constant K and the total surface area are deduced from fitting these data to a straight line.

Adsorption processes are divided into two classes: **physical adsorption** and **chemical adsorption (chemisorption)**. In physical adsorption, the binding forces are London dispersion forces, dipole-dipole attractions, etc. In chemisorption, covalent chemical bonds are formed between the atoms or molecules of the surface and the atoms or molecules of the adsorbed substance. The Langmuir isotherm applies to both classes, so long as only a monolayer of atoms or molecules can be adsorbed on the surface and the adsorbed molecules do not dissociate. Other isotherms apply to the case of multiple layers.[20]

In some cases of chemisorption, the molecules of the adsorbed substance dissociate and are bonded to the surface as atoms or free radicals. For example, when hydrogen is adsorbed on platinum, it dissociates into individual hydrogen atoms. Platinum is an effective catalyst for hydrogenation reactions for this reason.

For this case, we obtain a different isotherm from that of Equation (19.5-5). If a substance A_2 dissociates to form two A atoms, it occupies two sites on the surface, so we write for the adsorption process

$$A_2(g) + 2 \text{ surface sites} \rightleftharpoons 2A(\text{adsorbed})$$

If the adsorption is elementary, its rate is

$$\text{Rate of adsorption} = k_1[A_2](1 - \theta)^2 \qquad \textbf{(19.5-9)}$$

The rate of desorption is proportional to the square of the number of adsorbed atoms per unit area and thus to θ^2:

$$\text{Rate of desorption} = k_1'\theta^2 \qquad \textbf{(19.5-10)}$$

This dependence of the rate of desorption follows from the fact that the substance desorbs as diatomic molecules, so the rate of desorption is second order in adsorbed atoms.

When the rates of adsorption and desorption are equated, we obtain the equilibrium isotherm:

$$\theta = \frac{K^{1/2}[A_2]^{1/2}}{1 + K^{1/2}[A_2]^{1/2}} \qquad \textbf{(19.5-11)}$$

The Rate of a Heterogeneously Catalyzed Reaction

Consider the mechanism in which the reactant does not dissociate upon adsorption:

(1) $A + \text{surface site} \rightleftharpoons A(\text{adsorbed})$ (fast)

(2) $A(\text{adsorbed}) \rightarrow \text{further intermediates or products}$ (slow)

We assume that the second step is rate limiting.

Since θ is proportional to the amount of adsorbed substance A and we can assume the first step to be at equilibrium, the rate is given by

To obtain Equation (19.5-12), we have used the Langmuir isotherm expression.

$$\text{Rate} = k_2\theta = \frac{k_2K[A]}{1 + K[A]} \qquad \textbf{(19.5-12)}$$

For sufficiently small values of $[A]$, this equation becomes first order in A, but for large enough values of $[A]$ it is zero order in A. This limit corresponds to the fully covered catalytic surface.

The catalyzed reaction of a substance that dissociates upon adsorption can also be studied:

---

[20] K. J. Laidler, *op. cit.*, p. 234.

Exercise 19.15 Find the rate law for the mechanism

(1) A_2 + 2 surface sites $\rightleftharpoons$ 2A(adsorbed) (fast)

(2) A(adsorbed) $\rightarrow$ further intermediates or products (slow)

where the second step is rate limiting.

For the case of two different substances reacting at a solid surface, we consider two possible mechanisms. If only one of the reactants is adsorbed, the mechanism is called the **Langmuir-Rideal mechanism.** For example, we might have

(1) A + surface site $\rightleftharpoons$ A(adsorbed)

(2) A(adsorbed) + B $\rightarrow$ further intermediates or products

Since molecules must collide to react, this mechanism means that the B molecules from the fluid phase must strike the adsorbed A molecules. If the second step is rate limiting, the rate law is

$$\text{Rate} = \frac{k_2 K_1 [B][A]}{1 + K[A]} \qquad (19.5\text{-}13)$$

If both of the reacting molecules are adsorbed and if at least one of them can move around on the surface, the reaction between two adsorbed molecules can occur. This mechanism is called the **Langmuir-Hinshelwood mechanism.** It apparently occurs more commonly than the Langmuir-Rideal mechanism. It can be represented by

(1) A + surface site $\rightleftharpoons$ A(adsorbed)

(2) B + surface site $\rightleftharpoons$ B(adsorbed)

(3) A(adsorbed) + B(adsorbed) $\rightarrow$ further intermediates or products

We assume both substances adsorb on the same set of sites, so that the fraction of free sites is equal to $1 - \theta_A - \theta_B$, where θ_A is the fraction of sites with adsorbed A molecules and θ_B is the fraction of sites with adsorbed B molecules:

$$\text{Rate of adsorption of A} = k_1[A](1 - \theta_A - \theta_B)$$
$$\text{Rate of adsorption of B} = k_2[B](1 - \theta_A - \theta_B)$$

The rates of desorption are

$$\text{Rate of desorption of A} = k_1' \theta_A$$
$$\text{Rate of desorption of B} = k_2' \theta_B$$

When the rate of adsorption is equated to the rate of desorption for each substance and the resulting equations are solved simultaneously for θ_A and θ_B, we get the equilibrium relations

$$\theta_A = \frac{K_1[A]}{1 + K_1[A] + K_2[B]} \qquad (19.5\text{-}14)$$

$$\theta_B = \frac{K_2[B]}{1 + K_1[A] + K_2[B]} \qquad (19.5\text{-}15)$$

| Exercise 19.16 | Verify Equation (19.5-15). |

If step 3 in the Langmuir-Hinshelwood mechanism is rate-limiting, the rate law is

$$\text{Rate} = \frac{k_3 K_1 K_2 [A][B]}{(1 + K_1[A] + K_2[B])^2} \tag{19.5-16}$$

Figure 19.5 shows a schematic plot of the rate as a function of [A] for a fixed value of [B]. For small values of [A], the rate is roughly proportional to [A], but as [A] is increased, the rate passes through a maximum and then drops, becoming proportional to 1/[A] for large values of [A]. This decline in the rate corresponds to a value of $K_1[A]$ that is larger than the other two terms in the denominator, so the denominator becomes proportional to $[A]^2$.

The physical reason for the decline is that as the A molecules compete more and more successfully for the surface sites, there are fewer B molecules adsorbed on the surface. The reaction then slows down because of the scarcity of adsorbed B molecules. In the reaction of CO with O_2 on platinum (one of the reactions carried out in an automobile's catalytic converter), the CO is bonded much more strongly on the catalyst surface than is the O_2, and the rate is inversely proportional to [CO] for nearly all cases,[21] corresponding to the case that $K_1[CO]$ is much larger than the other terms.

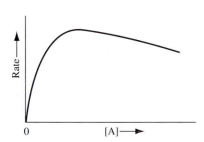

Figure 19.5. Schematic Plot of the Rate of a Catalyzed Reaction A + B → Products as a Function of [A] with Fixed [B]. The decrease in the rate comes from the fact that most of the adsorption sites are occupied by A molecules, and there are too few B molecules adsorbed to sustain the maximum rate of the reaction.

Homogeneous Catalysis

This class of catalysis occurs in a gas or a liquid phase. One subclass of homogeneous catalysis in aqueous solutions is acid-base catalysis. One type of acid catalysis is called **general acid catalysis**, defined as catalysis depending on the concentration of undissociated weak acid. If the catalysis depends on the concentration of hydrogen ions, irrespective of the strong or weak acid from which the hydrogen ions come, it is called **specific hydrogen ion catalysis**.

We illustrate acid and base catalysis by an example,[22] the isomerization of α-D-glucose to β-D-glucose (or vice versa). This is a famous reaction and was one of the first reactions shown to exhibit generalized acid catalysis. It is sometimes called the "mutarotation" of glucose because of the change in optical rotation of the solution as the reaction proceeds. The reaction is

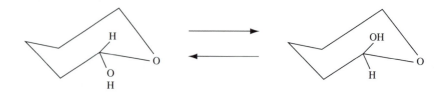

---

[21] K. J. Laidler, *op. cit.*, p. 249.
[22] S. W. Benson, *op. cit.*, pp. 558ff.

where the structural formulas are abbreviated by omission of some H's and OH's.

When the reaction is carried out in pure water, both the forward and back reactions are found to be first order. The rate law for the forward reaction of the alpha isomer is

$$\text{Forward rate} = k_0[\alpha] \tag{19.5-17}$$

where $[\alpha]$ stands for the concentration of the alpha isomer. The rate constant k_0 has the value 0.0054 min^{-1} at 18°C.

In the presence of a strong acid, the rate law is

$$\text{Forward rate} = k_0[\alpha] + k_{H^+}[H^+][\alpha] \tag{19.5-18}$$

where k_0 has the same value as before and k_{H^+} is equal to 0.0040 L mol^{-1} min^{-1} at 18°C. The uncatalyzed mechanism gives rise to the first term and the catalyzed mechanism gives rise to the second term, showing the competition between the two mechanisms. The second term corresponds to specific hydrogen ion catalysis, since hydrogen ions from any strong acid give the same contribution to the rate.

In basic solution, the same reaction exhibits **base catalysis**, with the rate law

$$\text{Forward rate} = k_0[\alpha] + k_{H^+}[H^+][\alpha] + k_{OH^-}[OH^-][\alpha] \tag{19.5-19}$$

where $k_{OH^-} = 3800$ L mol^{-1} min^{-1} at 18°C. In basic solution the concentration of hydrogen ions will be small, and in acidic solution the concentration of hydroxide ions will be small, so that only one of the last two terms will make an important contribution in a given case. However, both terms are still present, because introducing another mechanism does not shut down an existing mechanism. In fact, if weak acids and bases are also present, each acid and base contributes an additional term to the rate law, corresponding to general acid catalysis.

The mechanism proposed to explain acid catalysis is that either the alpha or beta pyranose ring isomer of glucose is converted to the open-chain form, which can then close the ring to form either ring isomer. The mechanism for forming the open-chain form is thought to be the following for general acid catalysis:[23]

(1)

(2)

---

[23] *ibid.*

(3)

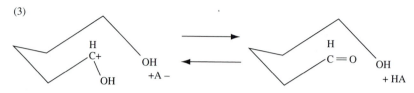

The alpha isomer is shown, but the mechanism for forming the open-chain structure from the beta isomer is analogous, and the mechanism for forming either pyranose ring isomer from the open-chain structure is the reverse of this mechanism.

Step 1 is thought to be rate limiting, so the forward reaction is predicted to be first order in α-glucose and first order in the acid HA:

$$\text{Rate} = k_1[\alpha][\text{HA}] \qquad (19.5\text{-}20)$$

Other examples of homogeneous catalysis correspond to similar mechanisms. For example, the gas-phase decomposition of ozone

$$2O_3 \rightarrow 3O_2$$

is catalyzed by N_2O_5. The proposed mechanism is[24]

$$
\begin{array}{ll}
(1) & N_2O_5 \rightarrow NO_3 + NO_2 \\
(2) & NO_2 + O_3 \rightarrow NO_3 + O_2 \\
(3) & 2NO_3 \rightarrow 2NO_2 + O_2 \\
(4) & NO_3 + NO_2 \rightarrow N_2O_5
\end{array}
$$

Step 1 is the same as the first step of the mechanism of Example 19.7. Step 4 is the reverse of step 1. We have written it separately to emphasize that the N_2O_5 is regenerated.

▼   **EXAMPLE 19.10**

Find the rate law for the forward reaction of the O_3 decomposition according to the above mechanism.

Solution

We write differential equations for the rate of change of $[O_3]$, which gives the rate, and for the rates of change of $[NO_3]$ and $[NO_2]$, which are the reactive intermediates whose concentrations can be considered to be constant in the steady-state approximation:

$$\text{Rate} = -\frac{1}{2}\frac{d[O_3]}{dt} = k_2[NO_2][O_3] \qquad (19.5\text{-}21a)$$

$$\frac{d[NO_2]}{dt} = k_1[N_2O_5] - k_1'[NO_3][NO_2] + 2k_3[NO_3]^2 - k_2[NO_2][O_3] = 0$$

$$(19.5\text{-}21b)$$

$$\frac{d[NO_3]}{dt} = k_1[N_2O_5] - k_1'[NO_3][NO_2] - 2k_3[NO_3]^2 + k_2[NO_2][O_3] = 0$$

$$(19.5\text{-}21c)$$

[24] H. S. Johnston, *Gas Phase Reaction Rate Theory*, Ronald Press, New York, 1966.

Subtraction of Equation (19.5-21c) from Equation (19.5-21b) gives an equation that is solved to obtain

$$[NO_2] = \frac{2k_3[NO_3]^2}{k_2[O_3]} \qquad (19.5\text{-}22)$$

Adding Equations (19.5-21b) and (19.5-21c) gives an equation that is combined with Equation (19.5-22) and solved to give

$$[NO_3] = \left(\frac{k_1}{2k'_1 k_3}[N_2O_5][O_3]\right)^{1/3} \qquad (19.5\text{-}23)$$

Equation (19.5-23) is substituted into Equation (19.5-22), which is substituted into Equation (19.5-20a) to give our solution:

$$\text{Rate} = -\frac{1}{2}\frac{d[O_3]}{dt} = k_2 k_3 \left(\frac{k_1}{2k'_1 k_3}[N_2O_5][O_3]\right)^{2/3}$$
$$= k_{app}[N_2O_5]^{2/3}[O_3]^{2/3} \qquad (19.5\text{-}24)$$

It is unusual to find a reaction that is four-thirds order. However, the rate depends on both the concentration of the reactant and that of the catalyst, as expected.

▲

Enzyme Catalysis

In cellular biological organisms nearly all reactions are catalyzed by **enzymes**. Most enzymes are proteins, although some ribonucleic acids have been found to exhibit catalytic activity.[25] Enzymes are usually named by adding the ending -ase to the name of the reactant or the reaction that is catalyzed. For example, the enzyme urease catalyzes the hydrolysis of urea.

Enzymes generally exhibit **specificity**. That is, each enzyme catalyzes only a single reaction or a set of similar reactions. Three kinds of specificity are recognized. **Absolute specificity** means that the enzyme catalyzes the reaction of only one substance. Urease exhibits this kind of specificity, since it does not catalyze the reaction of anything other than urea. **Group specificity** means that the enzyme catalyzes any of a group of reactions. For example, protease catalyzes the hydrolysis of various kinds of proteins, although it does not catalyze the hydrolysis of fats or carbohydrates. **Stereochemical specificity** means that the enzyme catalyzes the reaction of one optical isomer but not its enantiomorph. This case occurs with protease, which catalyzes the hydrolysis of polypeptides made of L-amino acids, but not polypeptides made of D-amino acids.

A typical enzyme molecule has an **active site** at which a reactant molecule can attach itself. The active site is often like a socket into which the reactant molecule fits, like a key in a lock, as shown schematically in Figure 19.6.

Once situated in the active site, the reactant molecule is rendered more reactive, like an adsorbed molecule on a heterogeneous catalyst. A molecule in an active site does not ordinarily dissociate into reactive fragments, but through conformational changes or polarizations produced by interaction with the enzyme it is put into a state of greater reactivity.

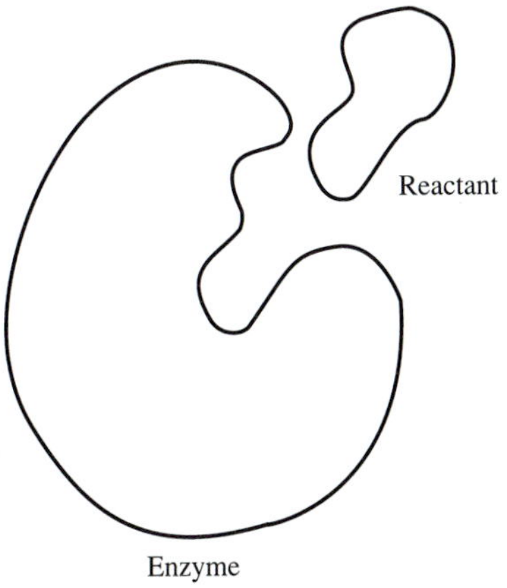

Figure 19.6. The Active Site on a Hypothetical Enzyme. The reactant molecule fits into the active site in such a way that a bond is stretched or compressed, or the molecule is made more reactive when it is combined with the active site.

Enzyme

Reactant

[25] T. R. Cech, *Science* **236**, 1532 (1987).

Many biochemistry textbooks denote the reactant by S (for substrate) and its concentration by [S] instead of [R].

The first accepted mechanism for enzyme catalysis was proposed by Michaelis and Menten.[26] For the case of a single reactant R and a single product P, this mechanism is

(1) $E + R \rightleftharpoons ER$ **(19.5-25a)**

(2) $ER \rightarrow E + P$ **(19.5-25b)**

where E stands for the enzyme, R stands for the reactant, ER stands for the enzyme-reactant complex, and P stands for the product. In addition to this mechanism, there is presumably an uncatalyzed mechanism proceeding at the same time but with a smaller rate, so that the catalyzed mechanism dominates.

The application of the steady-state approximation to obtain the rate law was first carried out by Briggs and Haldane.[27] The two differential rate equations are

$$\frac{d[ER]}{dt} = k_1[E][R] - k_1'[ER] - k_2[ER] \qquad \textbf{(19.5-26a)}$$

$$\frac{d[P]}{dt} = k_2[ER] \qquad \textbf{(19.5-26b)}$$

The steady-state approximation is invoked by setting Equation (19.5-26a) equal to zero. This equation can be solved for [ER] and the expression for [ER] can be substituted into Equation (19.5-26b). However, in a typical case [E], the concentration of uncombined enzyme, is not known. An unknown fraction of the enzyme is in the combined form ER, and [E] will differ significantly from the total concentration of enzyme, given by $[E]_{total} = [E] + [ER]$. However, since the concentration of reactant is much larger than the enzyme concentration and is thus much larger than [ER], [R] can be considered approximately equal to both the total concentration of reactant and the concentration of uncombined reactant.

When we substitute $[E]_{total} - [ER]$ into the right-hand side of Equation (19.5-26a) in place of [E], set the result equal to zero, and solve for [ER], we obtain

$$[ER] = \frac{k_1[E]_{total}[R]}{k_1' + k_2 + k_1[R]} \qquad \textbf{(19.5-27)}$$

The final form of our equation is determined by the fact that we must express the rate in terms of the total enzyme concentration, not just the concentration of uncombined enzyme.

The rate is obtained from Equation (19.5-26b):

$$\text{Rate} = \frac{d[P]}{dt} = \frac{k_2[E]_{total}[R]}{K_m + [R]} \quad \left(\begin{array}{l} \text{Michaelis-Menten} \\ \text{equation} \end{array} \right) \qquad \textbf{(19.5-28)}$$

where

$$K_m = \frac{k_1' + k_2}{k_1} \qquad \textbf{(19.5-29)}$$

---

[26] L. Michaelis and M. L. Menten, *Biochem. Z.* **49**, 333 (1913).
[27] G. E. Briggs and J. B. S. Haldane, *Biochem. J.* **19**, 338 (1925).

The coefficient K_m is called the **Michaelis-Menten constant** or the **Michaelis constant**. Like other rate constants, it is a constant only at constant temperature. Since the Michaelis-Menten mechanism of Equation (19.5-26) does not include a back reaction for step 2, it does not apply near equilibrium. (See Problem 19.44.)

The method of initial rates is commonly used to apply Equation (19.5-28). A number of experiments with the same concentration of enzyme, but with different concentrations of reactant, are carried out, and the initial rate in each case is determined. Figure 19.7 shows the initial rate given by Equation (19.5-28) as a function of reactant concentration. Note the resemblance of Equation (19.5-28) to Equation (19.5-5) and to Equation (19.5-11) and the resemblance of Figure 19.7 to Figure 19.3. The initial rate increases monotonically as the reactant concentration is increased, approaching the value $k_2[E]_{total}$ asymptotically. The value of K_m can be determined by locating the asymptote and equating K_m to the value of $[R]$ at which the initial rate is equal to one-half of the asymptotic value, as indicated in the figure.

The resemblance of the Langmuir isotherm to the Michaelis-Menten equation is striking.

The number of reactant molecules that react per enzyme molecule per second is called the **turnover number**. Its maximum value is equal to k_2. It can be as large as 10^6 s^{-1}.

Exercise 19.17

a. Show that the initial rate approaches the value $k_2[E]_{total}$ for large values of $[R]$.

b. Show that K_m is equal to the value of $[R]$ at which the initial rate is equal to half of its asymptotic value.

c. Show that the maximum value of the turnover number is equal to k_2.

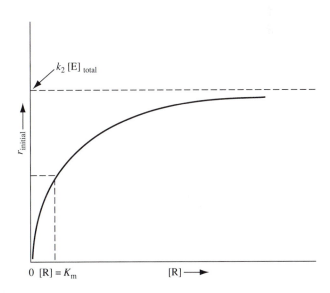

Figure 19.7. The Initial Rate as a Function of Reactant Concentration for the Michaelis-Menten Mechanism. Compare this graph with the graph of the Langmuir isotherm in Figure 19.3.

Just as with Equation (19.5-5), accurate location of the asymptote in Figure 19.7 from experimental data on initial rates is difficult if there is considerable experimental error. To avoid this problem, plots of initial rate data analogous to that of Figure 19.4 can be made. If Equation (19.5-29) is solved for the reciprocal of the initial rate, we obtain

$$\frac{1}{r_i} = \frac{K_m}{k_2[E]_{total}[R]} + \frac{1}{k_2[E]_{total}} \qquad \text{(19.5-30)}$$

where $r_i = (d[P]/dt)_{initial}$ is the initial rate. Data on initial rates should give a straight line when $1/r_i$ is plotted as a function of $1/[R]$, as in Example Figure 19.11. This type of plot is called a **Lineweaver-Burk plot**.[28] The slope of the line is equal to $K_m/k_2[E]_{total}$, the intercept on the vertical axis is equal to $1/k_2[E]_{total}$, and the intercept on the horizontal axis is equal to $-1/K_m$.

▼   **EXAMPLE 19.11**

The following data were gathered for the myosin-catalyzed hydrolysis of ATP at 25°C and pH 7.0:

[ATP]/μmol L^{-1}	Initial Rate/μmol L^{-1} s^{-1}
7.5	0.067
12.5	0.095
20.0	0.119
43.5	0.149
62.5	0.185
155.0	0.191
320.0	0.195

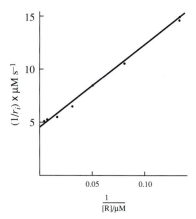

Example Figure 19.11. The Lineweaver-Burk Plot. Compare this plot with the linear plot in Figure 19.4.

In the solution to Example 19.11, an unweighted least-squares fit was used. If the experimental errors in the initial rates are all nearly equal, the experimental errors in the reciprocals are not equal and a weighted least-squares procedure would be preferable.

Determine the value of the Michaelis-Menten constant.

Solution

The Lineweaver-Burk plot of the data is shown in Example Figure 19.11. The line drawn in the figure is determined by an unweighted linear least-squares procedure. The slope of this line is equal to 76.58 s, and the intercept on the vertical axis is equal to 4.547 μmol L^{-1} s^{-1} (μM s^{-1}). The correlation coefficient for the least-squares fit is equal to 0.9975. The intercept on the horizontal axis is equal to -0.0595 μM^{-1}, so the Michaelis-Menten constant is equal to 16.8 μM.

▲

---

[28] H. Lineweaver and D. Burk, *J. Am. Chem. Soc.* **56**, 658 (1934).

| Exercise 19.18 |

An alternative linear plot is the **Eadie plot**,[29] for which Equation (19.5-28) is put into the form

$$\frac{r_i}{[R]} = -\frac{r_i}{K_m} + \frac{k_2 [E]_{total}}{K_m} \qquad (19.5\text{-}31)$$

Use the data of Example 19.11 to make a plot or a linear least-squares fit of $r_i/[R]$ as a function of r_i. Find the value of the Michaelis-Menten constant for the reaction of Example 19.11.

Inhibition

Many enzyme-catalyzed reactions are subject to **inhibition**. That is, the rate of the process is decreased by the presence of some substance. The **degree of inhibition** is defined as

$$i = 1 - \frac{r}{r_0} \qquad (19.5\text{-}32)$$

where r is the rate of the catalyzed reaction in the presence of the inhibitor and r_0 is the rate in the absence of the inhibitor.

A **competitive inhibitor** is defined as one for which the degree of inhibition decreases if the reactant concentration is increased with constant concentration of the inhibitor. A **noncompetitive inhibitor** is one for which the degree of inhibition is independent of the reactant concentration, and an **anticompetitive inhibitor** is one for which the degree of inhibition increases if the concentration of the reactant is increased.

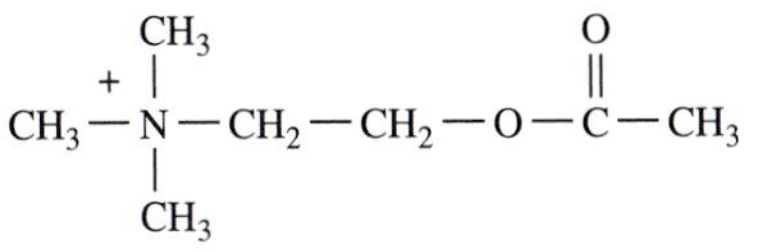

Acetylcholine is a **neurotransmitter**, a substance that diffuses across the synapse between two nerve cells and causes a signal to pass along the second nerve cell. This substance is hydrolyzed by the enzyme cholinesterase. This hydrolysis causes the signal to be interrupted by lowering the concentration of acetylcholine.

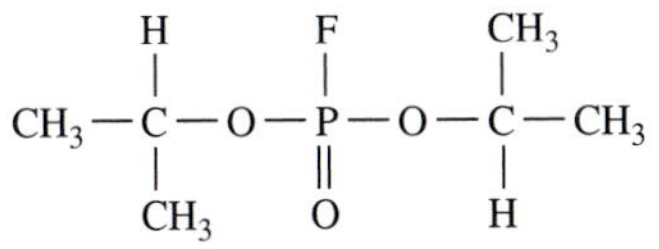

Diisopropyl fluorophosphate is a competitive inhibitor for this hydrolysis. Figure 19.8 shows the structural formulas of acetylcholine and diisopropyl fluorophosphate. Note the similarity in molecular size and shape. The diisopropyl fluorophosphate molecules compete with the acetylcholine molecules for the active sites. They inhibit the catalyzed reaction, because enzyme molecules with diisopropyl fluorophosphate molecules in their active sites are not available for acetylcholine hydrolysis. The diisopropyl fluorophosphate molecules are strongly attracted to the active sites, and a fairly small dose can nearly completely shut down the hydrolysis. In this case, the neurotransmitter remains in the synapse and the nerve cell transmits a signal repeatedly. If the nerve cell repeatedly stimulates a muscle to contract, the muscle soon succumbs to fatigue. Various substances similar to diisopropyl fluorophosphate have been prepared as insecticides and as chemical warfare agents ("nerve gases").

Acetylcholine

Diisopropyl fluorophosphate

Figure 19.8. The Structural Formulas of Acetylcholine and Diisopropyl Fluorophosphate. Diisopropyl fluorophosphate is a competitive inhibitor of cholinesterase, the enzyme that catalyzes the reaction of acetylcholine. It is apparent that both molecules might fit into the same enzyme active site.

---

[29] G. S. Eadie, *J. Biol. Chem.* **146**, 85 (1942).

The simplest Michaelis-Menten type of mechanism for competitive inhibition includes a process to form the enzyme-inhibitor complex EI:

$$(1) \qquad E + R \rightleftharpoons ER \qquad\qquad \textbf{(19.5-33a)}$$

$$(3) \qquad E + I \rightleftharpoons EI \qquad\qquad \textbf{(19.5-33b)}$$

$$(2) \qquad ER \rightarrow E + P \qquad\qquad \textbf{(19.5-33c)}$$

The rate law for this mechanism is obtained by application of the steady-state approximation for both the enzyme-reactant complex ER and the enzyme-inhibitor complex EI.[30]

$$\frac{d[ER]}{dt} = k_1[E][R] - (k_1' + k_2)[ER] = 0 \qquad \textbf{(19.5-34a)}$$

$$\frac{d[EI]}{dt} = k_3[E][I] - k_3'[EI] = 0 \qquad \textbf{(19.5-34b)}$$

$$\text{Rate} = \frac{d[P]}{dt} = k_2[ER] \qquad \textbf{(19.5-34c)}$$

Equation (19.5-34b) is the same as

$$\frac{[E][I]}{[EI]} = \frac{k_3'}{k_3} = K_I \qquad\qquad \textbf{(19.5-35)}$$

K_I is not the equilibrium constant for the formation of the complex but is the constant for its dissociation.

where K_I is the equilibrium constant for the *dissociation* of the enzyme-inhibitor complex.

We write

$$[E]_{\text{total}} = [E] + [ER] + [EI] \qquad \textbf{(19.5-36)}$$

We replace [EI] by its expression from Equation (19.5-35), and replace [E] by its expression from Equation (19.5-34a), obtaining

$$[E]_{\text{total}} = [ER]\left[1 + \frac{k_1' + k_2}{k_1[R]}\left(1 + \frac{[I]}{K_I} \right) \right] \qquad \textbf{(19.5-37)}$$

When Equation (19.5-37) is solved for [ER] and the result substituted into Equation (19.5-34c), the result is

$$\frac{d[P]}{dt} = \frac{k_2[E]_{\text{total}}[R]}{[R] + K_m(1 + [I]/K_I)} \qquad \textbf{(19.5-38)}$$

where K_m is still defined as in Equation (19.5-29).

Exercise 19.19

a. Verify Equation (19.5-38).

b. Derive the formula for the degree of inhibition for Equation (19.5-38) and show that it decreases if the reactant concentration is increased.

A proposed mechanism for a noncompetitive inhibitor is that an enzyme has a second site, other than the catalytic active site, to which the inhibitor can bind. The inhibited enzyme molecule is assumed unable to catalyze the reaction. The

---

[30] K. J. Laidler, *Physical Chemistry with Biological Applications*, Benjamin/Cummings, Menlo Park, CA, 1978, pp. 436ff.

only difference between this mechanism and that of Equation (19.5-33) is that the inhibitor can bind to the enzyme-reactant complex as well as to the uncombined enzyme. The mechanism is

$$\text{(1)} \quad E + R \rightleftharpoons ER \qquad\qquad \textbf{(19.5-39a)}$$

$$\text{(3)} \quad E + I \rightleftharpoons EI \qquad\qquad \textbf{(19.5-39b)}$$

$$\text{(4)} \quad ER + I \rightleftharpoons ERI \qquad\qquad \textbf{(19.5-39c)}$$

$$\text{(2)} \quad ER \rightarrow E + P \qquad\qquad \textbf{(19.5-39d)}$$

When the steady-state approximation is applied to EI, ER, and ERI, the rate law is obtained:[31]

$$\frac{d[P]}{dt} = \frac{k_2[E]_{\text{total}}[R]}{K_m(1 + [I]/K_I) + [R](1 + [I]/K_I')} \qquad \textbf{(19.5-40)}$$

where K_I is defined as in Equation (19.5-35) for the enzyme-inhibitor complex without the reactant and K_I' is the dissociation constant for the enzyme-inhibitor complex which also has a molecule of the reactant bound to it:

$$K_I' = \frac{[ER][I]}{[ERI]} \qquad\qquad \textbf{(19.5-41)}$$

Exercise 19.20   Find the expression for the degree of inhibition for the mechanism of Equation (19.5-39), and show that it is independent of reactant concentration in the case that $K_I = K_I'$.

19.6

The Temperature Dependence of Rate Constants; The Collision Theory of Gaseous Reactions

Reaction rates depend strongly on temperature, nearly always increasing when the temperature is raised. A rule of thumb is that the rate of a reaction doubles if the temperature is raised by 10°C.

The first quantitative generalizations about the temperature dependence of rate constants were published in the last half of the nineteenth century, and various empirical formulas were proposed.[32] The most widely used empirical relation is that of Arrhenius, which was proposed in 1889. This formula has gained wide acceptance because it is based on a physical picture of elementary processes.

Svante Arrhenius, 1859–1927, Swedish chemist who won the 1905 Nobel Prize in chemistry for his theory of dissociation and ionization of substances in solution.

Arrhenius pointed out that typical rate constants for gaseous reactions are much smaller than they would be if every collision led to reaction and that the typical temperature dependence of reaction rate constants is much too strong to be explained by the temperature dependence of collision rates. He postulated that "activated" molecules (with high energy) must exist in

---

[31] Laidler, *op. cit.*, p. 440.
[32] K. J. Laidler, *Chemical Kinetics*, Harper & Row, New York, 1987, pp. 40ff.

order to react and that the numbers of such activated molecules would be governed by the Boltzmann probability distribution of Equation (1.7-25) or Equation (16.2-31). This assumption leads to the **Arrhenius relation**:

$$k = Ae^{-\varepsilon_a/k_B T} \qquad\qquad (19.6\text{-}1)$$

The quantity ε_a is the energy relative to the ground-state energy which the molecules must have in order to react and is called the **activation energy**. The temperature-independent factor A is called the **pre-exponential factor**.

It is common to express Equation (19.6-1) in terms of a **molar activation energy** E_a which we usually call the activation energy:

Although Arrhenius arrived at his formula by some analysis, we regard it as an empirical relation.

$$k = Ae^{-E_a/RT} \qquad\qquad (19.6\text{-}2)$$

where $E_a = N_A \varepsilon_a$ is the molar activation energy. Experimental molar activation energy values are usually in the range 50 to 200 kJ mol^{-1}, somewhat smaller than energies required to break chemical bonds.

▼   **EXAMPLE 19.12**

For the reaction

$$H_2 + I_2 \rightarrow 2HI$$

at 373.15 K, the rate constant is 8.74×10^{-15} L mol^{-1} s^{-1}. At 473.15 K it is 9.53×10^{-10} L mol^{-1} s^{-1}. Find the value of the activation energy and of the pre-exponential factor.

Solution

For any two temperatures T_1 and T_2, Equation (19.6-1) gives

$$E_a = \frac{R \ln[k(T_2)/k(T_1)]}{1/T_1 - 1/T_2} \qquad\qquad (19.6\text{-}3)$$

Notice that $\ln(k)$ is a linear function of $1/T$, so if more than two data points were given a plot of $\ln(k)$ versus $1/T$ could be used. Substitution of the values gives

$$E_a = \frac{(8.3145 \text{ J K}^{-1} \text{ mol}^{-1}) \ln(9.53 \times 10^{-10}/8.74 \times 10^{-15})}{1/373.15 \text{ K} - 1/473.15 \text{ K}}$$

$$= 1.70 \times 10^5 \text{ J mol}^{-1} = 170 \text{ kJ mol}^{-1}$$

$$A = ke^{E_a/RT}$$

$$= (8.74 \times 10^{-15} \text{ L mol}^{-1} \text{ s}^{-1}) \exp\left(\frac{1.70 \times 10^5 \text{ J mol}^{-1}}{(8.3145 \text{ J K}^{-1} \text{ mol}^{-1})(373.15 \text{ K})}\right)$$

$$= 5.47 \times 10^9 \text{ L mol}^{-1} \text{ s}^{-1}$$

▲

Exercise 19.21   **a.** Find the value of the activation energy if a rate constant doubles in value between 20°C and 30°C.

b. Find the value of the activation energy if a rate constant doubles in value between 50°C and 60°C.

c. A common definition of the activation energy of a reaction is

$$E_a = RT^2\left(\frac{d\ln(k)}{dT}\right) \quad \text{(definition of } E_a) \tag{19.6-3a}$$

Show that if k is given by Equation (19.6-2), Equation (19.6-3a) gives the same E_a as in Equation (19.6-2) if A is temperature-independent.

The Collision Theory of Bimolecular Elementary Gaseous Reactions

This theory will justify our assumption that a fixed fraction of collisions lead to reaction at fixed temperature, and it will not only explain the temperature dependence of elementary bimolecular rate constants but also predict their values.

The basic assumption of the collision theory of bimolecular elementary reactions is that the initiation of the reaction involves an inelastic collision in which energy is transferred from translational kinetic energy to energy of internal motions. The energy that can be transferred to internal motion is not the total kinetic energy of the two particles. For example, if two rapidly moving molecules happen to be moving in nearly the same direction with nearly the same speed, they can have rather large kinetic energies but not transfer much of their kinetic energies if they collide. It is the energy of motion of one particle relative to the other, determined by the relative speed, that determines the amount of energy transferred.

We first assume that the probability of reaction equals zero if the relative speed is smaller than a certain critical value (different for every reaction) and that it equals unity for relative speeds larger than this value, as shown in Figure 19.9a. This corresponds to a reaction cross section equal to zero

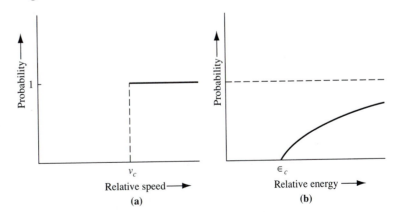

Figure 19.9. Probability of Reaction as a Function of Relative Speed. (a) First assumption. The probability of reaction is assumed to rise suddenly from zero to unity at a critical relative speed. **(b)** Assumed probability of reaction as a function of relative kinetic energy according to Equation (19.6-16). This curve gives a more reasonable assumed behavior than does the curve of part (a). The reaction probability is assumed to rise gradually from zero above a critical relative energy and to approach unity asymptotically.

The assumption of Figure 19.9a will be improved on later.

for relative speeds smaller than the critical value and equal to the collision cross section for speeds larger than the critical value.

Consider a bimolecular elementary reaction between molecules of substance 1 and substance 2. From the discussion of Chapter 2, we can write the expression for the total rate of collisions per unit volume that involve molecules of type 1 whose velocities lie in the infinitesimal range $d^3\mathbf{v}_1$ and molecules of type 2 whose velocities lie in the infinitesimal range $d^3\mathbf{v}_2$. We write this collision rate as dZ_{12}:

$$dZ_{12} = \pi d_{12}^2 \mathscr{N}_1 \mathscr{N}_2 |\mathbf{v}| f_1(\mathbf{v}_1)\, d^3\mathbf{v}_1\, f_1(\mathbf{v}_2)\, d^3\mathbf{v}_2 \qquad \textbf{(19.6-4)}$$

where $\mathscr{N}_1$ is the number density of molecules of type 1, $\mathscr{N}_2$ is the number density of molecules of type 2, and d_{12} is the collision diameter of the molecule pair. The relative velocity $\mathbf{v}$ is given by

$$\mathbf{v} = \mathbf{v}_2 - \mathbf{v}_1 \qquad \textbf{(19.6-5)}$$

The magnitude of the relative velocity is the relative speed.

In order to obtain the total rate of collisions that lead to reaction, we integrate over all velocities that satisfy the condition

$$v = |\mathbf{v}_2 - \mathbf{v}_1| > v_c \qquad \textbf{(19.6-6)}$$

where v_c is the minimum relative speed that can lead to reaction.

In order to carry out this integration, we change variables, expressing the kinetic energy of the pair of particles in terms of the relative velocity of Equation (19.6-5) and the velocity of the center of mass of the two molecules, which is given in Equation (D-26) of Appendix D:

We have used this expression for the kinetic energy in the solution of the Schrödinger equation for the central-force problem in Chapter 11 and in the analysis of the relative velocity in Chapter 16.

$$\mathscr{K} = \frac{1}{2} M V^2 + \frac{1}{2} \mu v^2 \qquad \textbf{(19.6-7)}$$

where $M = m_1 + m_2$ and where the reduced mass of the particles $\mu = m_1 m_2 / M$. The first term on the right-hand side of Equation (19.6-7) is the kinetic energy of the center of mass and the second term is the relative kinetic energy.

Using Equation (16.2-30) for the probability distributions, we can write

$$dZ_{12} = \pi d_{12}^2 \mathscr{N}_1 \mathscr{N}_2 \left(\frac{m_1}{2\pi k_B T} \right)^{3/2} \left(\frac{m_2}{2\pi k_B T} \right)^{3/2}$$
$$\times\, v e^{-MV^2/2k_B T} e^{-\mu v^2/2k_B T}\, d^3\mathbf{V}\, d^3\mathbf{v} \qquad \textbf{(19.6-8)}$$

Exercise 19.22   Verify Equation (19.6-8).

We integrate Equation (19.6-8) over all values of $\mathbf{V}$ and over the values of $\mathbf{v}$ that satisfy Equation (19.6-6). Integration over $\mathbf{V}$ is just like the integration in Equation (16.2-16) and gives a factor of $(2\pi k_B T/M)^{3/2}$. We now have the expression for the total number of reactive collisions per unit volume per unit time:

$$
Z_{12}(\text{reactive}) = \pi d_{12}^2 \mathcal{N}_1 \mathcal{N}_2 \left(\frac{2\pi k_B T}{M} \right)^{3/2} \left(\frac{m_1}{2\pi k_B T} \right)^{3/2} \left(\frac{m_2}{2\pi k_B T} \right)^{3/2}
$$

$$
\times \int v e^{-\mu v^2 / 2k_B T} \, d^3\mathbf{v} \tag{19.6-9}
$$

This integration is carried out in spherical polar coordinates. Integration over the angles θ and ϕ gives a factor of 4π. The integration must include only values of v satisfying Equation (19.6-6). We use a tabulated indefinite integral to obtain

$$
\int_{v_c}^{\infty} e^{-\mu v^2 / 2k_B T} v^3 \, dv = \frac{1}{2} e^{-\mu v_c^2 / 2k_B T} \left(\frac{2k_B T}{\mu} \right) \left(v_c^2 + \frac{2k_B T}{\mu} \right)
$$

The final result is

$$
Z_{12}(\text{reactive}) = \pi d_{12}^2 \mathcal{N}_1 \mathcal{N}_2 \left(\frac{8k_B T}{\pi \mu} \right)^{1/2} \left(1 + \frac{\mu v_c^2}{2k_B T} \right) e^{-\mu v_c^2 / 2k_B T} \tag{19.6-10}
$$

Comparison of this equation with Equation (19.1-4) shows that we have an expression for the fraction of collisions that lead to reaction in a bimolecular elementary process:

$$
f = \left(1 + \frac{\mu v_c^2}{2k_B T} \right) e^{-\mu v_c^2 / 2k_B T} \tag{19.6-11}
$$

Exercise 19.23   Verify Equation (19.6-10).

The critical value of the relative kinetic energy is given by

$$
\varepsilon_c = \frac{1}{2} \mu v_c^2 \tag{19.6-12}
$$

so that Equation (19.6-10) can be written

$$
Z_{12}(\text{reactive}) = \pi d_{12}^2 \mathcal{N}_1 \mathcal{N}_2 \left(\frac{8k_B T}{\pi \mu} \right)^{1/2} \left(1 + \frac{\varepsilon_c}{k_B T} \right) e^{-\varepsilon_c / k_B T} \tag{19.6-13}
$$

which corresponds to

$$
k = N_A \pi d_{12}^2 \left(\frac{8k_B T}{\pi \mu} \right)^{1/2} \left(1 + \frac{E_c}{RT} \right) e^{-E_c / RT} \tag{19.6-14}
$$

where N_A is Avogadro's number, where we let

$$
E_c = N_A \varepsilon_c \tag{19.6-15}
$$

and use the fact that $R = N_A k_B$.

Equation (19.6-14) is not quite the same as Equation (19.6-1), since the pre-exponential factor in Equation (19.6-14) depends on T. However, the exponential factor depends so much more strongly on temperature than does the pre-exponential factor in Equation (19.6-14) that the difference between the two equations is numerically small over a limited range of temperature, and E_c can be approximately identified with E_a, the Arrhenius activation energy.

▼ **EXAMPLE 19.13**

For the reaction of Example 19.12, calculate the fractional change in the exponential factor and in the pre-exponential factor in Equation (19.6-14) if T is changed from 20°C to 30°C.

Solution

The value of E_c/R is 2.045×10^4 K. The ratio of the exponential factors is

$$\frac{\exp(-2.045 \times 10^4 \text{ K}/303.15 \text{ K})}{\exp(-2.045 \times 10^4 \text{ K}/293.15 \text{ K})} = 9.99$$

The ratio of the pre-exponential factors is

$$\frac{(303.15 \text{ K})^{1/2}(1 + 2.045 \times 10^4 \text{ K}/303.15 \text{ K})}{(293.15 \text{ K})^{1/2}(1 + 2.045 \times 10^4 \text{ K}/293.15 \text{ K})} = 0.984$$

The change in the pre-exponential factor is negligible.

◢

The condition of Equation (19.6-6) corresponds to the probability of reaction shown in Figure 19.9a. In a more sophisticated version of the collision theory, it is assumed that the probability of reaction is given by

$$\text{Probability} = \begin{cases} 0 & \text{if } E_r < E_c \\ 1 - E_c/E_r & \text{if } E_r > E_c \end{cases} \qquad \textbf{(19.6-16)}$$

which is shown in Figure 19.9b. This assumption can be defended as follows: A collision with a relative kinetic energy barely great enough to initiate a reaction should have a lower probability of producing a reaction than a collision with a higher relative kinetic energy, since some of the translational energy could be transferred into "inactive" internal motions of the molecules that do not lead to reaction. Collisions with a large relative kinetic energy should provide plenty of energy even if some is lost in inactive internal motions.

When the probability of Equation (19.6-16) is introduced into the integration of Equation (19.6-9), the result is[33]

$$k = N_A \pi d_{12}^2 \left(\frac{8k_B T}{\pi \mu}\right)^{1/2} e^{-E_c/RT} \qquad \textbf{(19.6-17)}$$

▼ **EXAMPLE 19.14**

Assuming the activation energy and the value of the rate constant at 373 K from Example 19.12, find the effective collision diameter of a hydrogen molecule and an iodine molecule.

[33] K. J. Laidler, *op. cit.*, pp. 85ff.

Solution

Assuming that E_c and E_a can be identified with each other and using Equation (19.6-17),

$$\pi d^2 = \left(\frac{\pi\mu}{8k_{\mathrm{B}}T}\right)^{1/2}\frac{k}{N_{\mathrm{A}}}e^{E_a/RT} = \frac{k}{\langle v_{\mathrm{rel}}\rangle N_{\mathrm{A}}}e^{E_a/RT}$$

The value of $\langle v_{\mathrm{rel}}\rangle$ is 1988 m s^{-1}, so that

$$d^2 = \frac{8.74 \times 10^{-18}\text{ m}^3\text{ mol}^{-1}\text{ s}^{-1}}{\pi(1988\text{ m s}^{-1})(6.022 \times 10^{23}\text{ mol}^{-1})}e^{54.794} = 1.46 \times 10^{-21}\text{ m}^2$$

$$d = 3.8 \times 10^{-11}\text{ m}$$

The value of the collision diameter in Example 19.14 is too small by a factor of about 10, which is typical of the collision theory. There is a simple explanation for the smallness of the collision diameter: not only do the molecules have to collide with at least a minimum relative energy, but in many reactions they must also be oriented properly with respect to each other in order to react. For example, an organic molecule with a functional group would be much more likely to react if struck on the functional group than if struck on the hydrocarbon portion of the molecule.

To account for the orientation dependence, an additional factor, called the **steric factor**, was introduced into the collision theory. The steric factor is defined as the fraction of collisions in which the orientation of the molecules is appropriate for reaction. If this factor is denoted by φ, Equation (19.6-17) becomes

$$k = N_{\mathrm{A}}\varphi\pi d_{12}^2\left(\frac{8k_{\mathrm{B}}T}{\pi\mu}\right)^{1/2}e^{-E_c/RT} \qquad \text{(19.6-18)}$$

Exercise 19.24   Find the value of the steric factor for the reaction of Example 19.14 that will give a value for the collision diameter equal to the mean of the values for H_2 and I_2 in Table A21.

Liquid-State Reactions

The temperature dependence of rate constants for both gaseous and liquid-state reactions is reasonably well described by the Arrhenius formula, Equation (19.6-2). For activation-limited reactions in liquid phases, the activation energies are roughly equal to those for gas-phase reactions. This is as expected, since the collisional activation is very similar to that of gaseous reactions. For diffusion-limited reactions, the activation energies are somewhat smaller, often near the energies of activation for diffusion processes, which again is what we would expect.

| Exercise 19.25 | For the reaction |

$$2I \rightarrow I_2$$

in carbon tetrachloride, the rate constant at 23°C is 7.0×10^{-6} m^3 mol^{-1} s^{-1}. At 30°C, the value is 7.7×10^{-6} m^3 mol^{-1} s^{-1}. Find the activation energy and compare it with the activation energy for the viscosity of carbon tetrachloride in Example 17.9.

We will continue our discussion of the theories of chemical reaction rates in Chapter 21.

19.7   Experimental Molecular Study of Chemical Reactions

The "classical" study of chemical reaction rates involves determination of concentrations of reactants or products and delivers information only about net rates. Such information cannot lead directly to knowledge of a reaction mechanism. However, there exist techniques that deliver molecular information about a reaction mechanism.

Observation of Reaction Intermediates

If a reactive intermediate included in a proposed mechanism can be detected in the experimental system, that mechanism becomes more plausible, and if the intermediate's concentration can be determined as a function of time, individual rate constants for elementary steps can sometimes be evaluated.

The most direct technique for detecting reactive intermediates is spectroscopy. An early example of spectroscopic detection of a reactive intermediate was a study of the decomposition of N_2O_5.[34] According to the mechanism of Example 19.7, the first step is the formation of NO_2 and NO_3 from N_2O_5. Schott and Davidson carried out shock tube studies, using the reaction tube as a spectrophotometer cell. They monitored the absorption of light at 546 nm and 652 nm, at which wavelengths NO_3 absorbs much more strongly than NO_2; at 366 nm, at which NO_2 absorbs more strongly than NO_3; and at 436 nm, at which NO_2 and NO_3 absorb nearly equally.

Schott and Davidson were able to determine the concentration of NO_3 as a function of time and to calculate values for the elementary rate constants in the mechanism of Example 19.7, finding for example that the pre-exponential factor in k_2 is equal to 1.66×10^8 L mol^{-1} s^{-1} and that the activation energy is approximately equal to 16 kJ mol^{-1} over the temperature range 300 K to 820 K.

---

[34] G. Schott and N. Davidson, *J. Am. Chem. Soc.* **80**, 1841 (1958).

Another technique for detecting reactive intermediates is mass spectrometry. In a mass spectrometer, molecules are converted into positive ions, often undergoing fragmentation in the process. The resulting ions are accelerated by an electrical field, attaining a speed depending on their charge/ mass ratio. They are then passed through electrical and magnetic fields (or other analyzing devices) and separated, so that the number of ions with each charge/mass ratio can be determined. Very often, the identity of the original substance can be deduced not only from its molecular mass but also from its fragmentation pattern.

For mass spectrometric detection of reactive intermediates, one carries out a gas-phase reaction in a vessel that adjoins the ionization chamber of a mass spectrometer. The total pressure in the ionization chamber must be small, around 10^{-4} torr. A small aperture allows the reacting gases to pass into the ionization chamber of the mass spectrometer. The mass spectrum of a reactive intermediate can sometimes be found in the mass spectrum of the reacting mixture of reactants, intermediates, and products.

An advantage of the mass spectrometric method of detecting reactive intermediates is that the pressure in the mass spectrometer is very small, both the analyzing chamber and in the ionizing chamber. The low pressure in both chambers lowers the collision rate of reactive intermediates, prolonging their lifetimes and making it easier to detect short-lived species.

In addition to direct detection techniques, it is possible to infer the presence of certain kinds of reactive intermediates, especially free radicals, from their chemical effects. In the **mirror technique**, a reacting gas is passed through a tube with a metallic mirror deposited on its inner surface. If free radicals are present, they can combine with the metal to form volatile products that can be trapped at low temperature and analyzed. For example, a lead mirror will combine with methyl radicals to form tetramethyl lead, $Pb(CH_3)_4$, a stable substance that can be condensed in a trap.[35]

Another technique that can be used to detect the presence of free radical intermediates is based on the fact that almost any free radical catalyzes the conversion of ortho-H_2 to para-H_2 and vice versa. At room temperature, equilibrium H_2 consists of 75% ortho-H_2 and 25% para-H_2, whereas at low temperatures the equilibrium mixture is nearly 100% para-H_2. (See Section 14.2 for more information about ortho- and para-H_2.) If para-H_2 prepared at low temperature is brought in contact with free radicals, the rate of conversion to the equilibrium mixture is a measure of the amount of free radicals present.

Still another way to detect free radical intermediates is to investigate the effect of adding a substance known to combine with free radicals. Molecular oxygen is such a substance, and if addition of oxygen to a reacting system inhibits the rate of reaction, it is likely that some kind of free radical intermediate is present in the reaction mechanism. However, these three techniques for detecting free radical intermediates do not distinguish one free radical intermediate from another.

---

[35] S. W. Benson, *op. cit.*, p. 101, gives a table of free radicals and metals that had been studied as of the 1950s.

A final technique is the use of isotopic substitution. For example, the decomposition of acetaldehyde

$$CH_3CHO \rightarrow CH_4 + CO$$

was thought to proceed by the mechanism

(1) $CH_3CHO \rightarrow CH_3 + CHO$
(2) $CH_3 + CH_3CHO \rightarrow CH_4 + CH_3CO$
(3) $CH_3CO \rightarrow CH_3 + CO$

A mixture of CH_3CHO and CD_3CDO was reacted, where D stands for deuterium, ^2H.[36] The product mixture contained the statistically expected mixture of randomly isotopically substituted methanes, which would not have been the case with a mechanism other than a free radical mechanism.

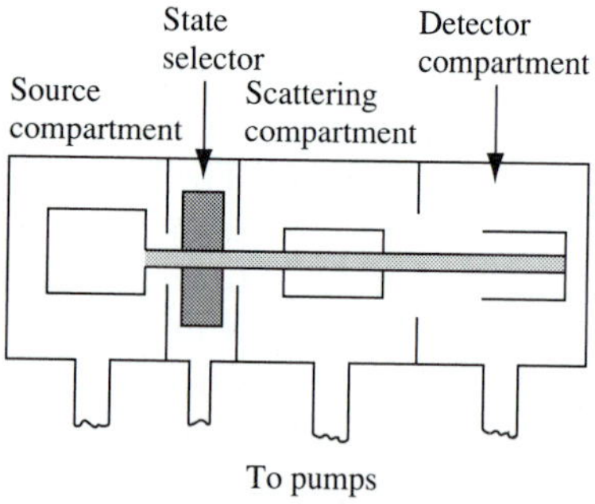

Molecular Beam Reactions

In this technique, the reaction is carried out by forming beams of reactants in an otherwise evacuated chamber, instead of by mixing the reactants. Figure 19.10 shows schematically an apparatus for generating a molecular beam from a solid or liquid meterial. The material is vaporized in an oven, and the molecules exit from a small aperture into an evacuated chamber. The diagram shows a second chamber through which the molecules pass, which is separately evacuated in order to lower the pressure even further. The molecules then pass into a third chamber, in which the beam can be observed or reacted with a second beam. Only molecules moving in nearly the same direction can pass through both barriers, producing a nearly unidirectional (collimated) beam.

If it is desirable to select only molecules in a narrow range of speeds, a velocity selector can be used. Figure 19.11 shows schematically one type of velocity selector. The rotating disks have slots through which molecules can pass. When the set of disks is rotating, it allows molecules to pass only if their speed is such that they reach the second disk when its slot is in the beam position and reach the third disk when its slot is in the beam position, etc. Varying the speed of rotation of the disks allows different speeds to be selected.

Chemical reactions can be carried out with two crossed molecular beams, as schematically depicted in Figure 19.12. The beams are generally brought together at right angles, and the product molecules that are scattered away from the collision region are detected by a movable detector, allowing the angular distribution of products to be determined. Since the product molecules are scattered away from the reaction region into a region where further collisions are unlikely, reactive intermediates can be detected and identified. The detector is usually a mass spectrometer, but molecules containing alkali metal atoms can be detected by a surface ionization detector.[37]

Figure 19.10. An Apparatus for Generating a Molecular Beam (schematic). The molecular beam is generated thermally in one compartment and passed through a velocity selector in the second compartment before impinging on another substance in a third compartment.

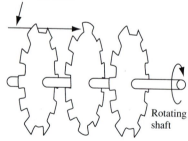

Slotted disks to pass molecules within a certain range of speeds (depending on rotational speed)

Figure 19.11. A Velocity Selecting Apparatus (schematic). Such a rotating apparatus will allow molecules to pass only if they have a speed in a fairly narrow range. The speed of the molecules depends on the rate of rotation.

[36] S. W. Benson, *op. cit.*, p. 108.
[37] G. G. Hammes, *Principles of Chemical Kinetics*, Academic Press, New York, 1978, pp. 113ff.

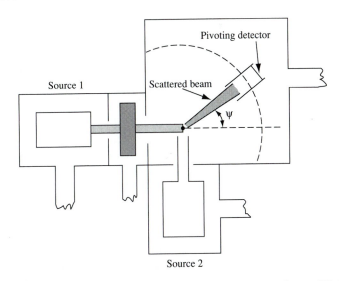

Figure 19.12. An Apparatus for Carrying out a Chemical Reaction in Crossed Molecular Beams. In this type of apparatus, two molecular beams impinge on each other in the reaction chamber. A movable detector allows the products to be determined as a function of scattering angle.

It is also possible to carry out a reaction by bringing a beam of molecules into a stationary gaseous sample. If every collision leads to reaction, the reaction cross section is equal to the collision cross section, but if some collisions do not lead to reaction, the reaction cross section is equal to the fraction of collisions that are reactive times the collision cross section. The reaction cross section generally depends on the states of the reactants and products and on the collision energy.

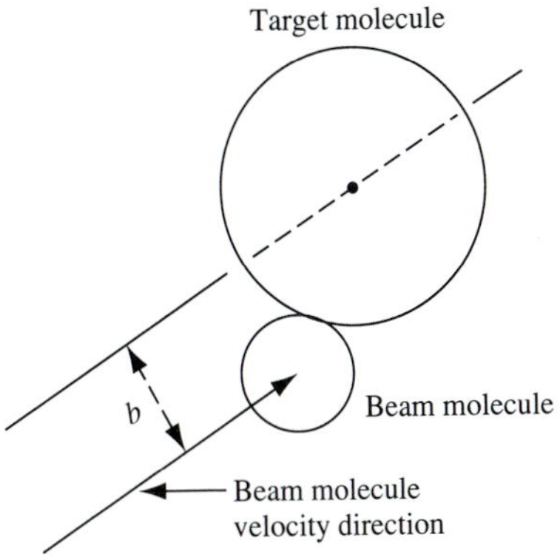

The probability that a collision will lead to reaction depends also on the **impact parameter**, b, which is defined as the perpendicular distance between the extrapolated path of the beam molecule and the center of the target molecules, as depicted in Figure 19.13. The **opacity function**, $P(b)$, is defined as the fraction of collisions with impact parameter b that lead to collision. The reaction cross section is given by

$$\sigma_R = 2\pi \int_0^\infty bP(b)\,db \tag{19.7-1}$$

Note that σ_R and $P(b)$ can depend on the relative energy of the collision.

In order to test theoretical calculations, it is useful to have information on a "state-to-state" reaction in which the reactant and product molecules are in known rotational, vibrational, and electronic states. For example, consider the gas-phase reaction[38]

$$H(^2S) + ICl(v', J') \to HI(v'', J'') + Cl(^2P_{1/2}) \tag{19.7-2}$$

Figure 19.13. The Impact Parameter. This diagram illustrates the definition of the impact parameter b.

Target molecule

Beam molecule

Beam molecule velocity direction

b

---

[38] R. D. Levine and R. B. Bernstein, *Molecular Reaction Dynamics and Chemical Reactivity*, Oxford University Press, New York, 1987, p. 208.

where we have specified the electronic states of the atoms and the rotational and vibrational states of the molecules, which are assumed to be in their electronic ground states. An ordinary chemical reaction is a sum of such reactions, since various states of the reactant and product molecules are represented in a system of many molecules.

The information on the state-to-state reaction probability is given in terms of the reaction cross section for the reaction with reactants and products in definite states. The total reaction cross section is a sum of the cross sections for all of the different possible states of the reactants and products, weighted with the probabilities of the states.

The "ideal" molecular beam kinetics experiment would give the reaction cross section for different values of the collision energy and for different states of the reactants, the angular distribution of products (the angle relative to the original molecular beam at which the product molecules leave the collision region), the velocity distribution of the product molecules, and the distribution of electronic, vibrational, and rotational states of the products. No single experiment has given all of these pieces of information, but each of them has been obtained in at least one kind of experiment.

Some types of molecular beam experiments that are used to obtain molecular kinetic information are:[39]

1. Chemiluminescence. In this method, radiation emitted by excited products is spectroscopically analyzed as it is emitted. The intensities of radiation due to various transitions and the Franck-Condon factors for these transitions can be used to determine the population distribution for product states. Modern techniques also allow time-resolved spectra to be observed (intensity as a function of time as well as of wavelength), and measurements in the picosecond region are becoming common.

2. Chemical lasers. Some reactions produce product molecules with an inverted population distribution. That is, the population of some state of higher energy is larger than that of some state of lower energy in the products. In this case, a **chemical laser** is possible, in which incident radiation produces stimulated emission. For example, the flash photolysis of trifluoroiodomethane in the presence of hydrogen and a buffer gas can produce excited HF molecules with a population inversion:

   (1) $CF_3I \xrightarrow{\text{uv flash}} F + CF_2I$

   (2) $F + H_2 \rightarrow H + HF^*$

   (3) $HF^* \xrightarrow{\text{stimulated emission}} HF + h\nu$

   Figure 19.14 shows the laser emission as a function of time for a number of transitions in this system.

3. Laser pump and laser probe. In this technique, one laser is trained on a beam of reactant molecules, essentially using photons as one of the

---

[39] *ibid.*, p. 210ff.

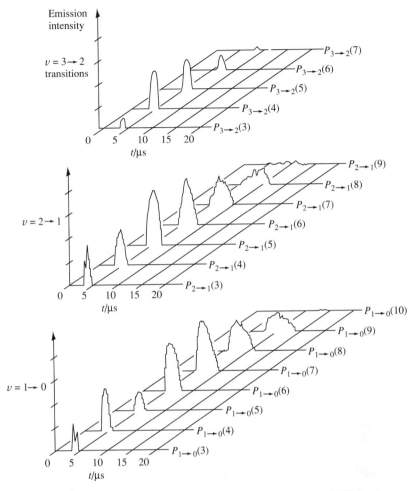

Figure 19.14. Intensity of Laser Pulses as a Function of Time for Excited HF Molecules.
In each plot, the vertical axis represents the intensity of a laser pulse corresponding to
emission from HF formed in a flash-initiated reaction. The horizontal axis represents time
measured in microseconds. Each curve is labeled with both the initial and final values
of the vibrational quantum number v. The number in parentheses is the final value of J,
the rotational quantum number (the initial value of J is smaller than this value by unity).
From R. D. Levine and R. B. Bernstein, *Molecular Reaction Dynamics and Chemical
Reactivity*, Oxford University Press, New York, 1987, p. 213.

reagents. A second laser is trained on the beam in the product region,
raising product molecules to excited states, from which they fluoresce.
Spectroscopic analysis of the fluorescent radiation gives information
about the distribution of products and their states.

4. Crossed molecular beams. This method is illustrated in Figure 19.15,
which shows an apparatus for the reaction of chlorine atoms with
molecules of the other halogens. The detector is a mass spectrometer
that can be moved to different angles, so that the angular distribution
of the products can be studied. The TOF (time-of-flight) chopper allows determination of the velocity distribution of the products, giving

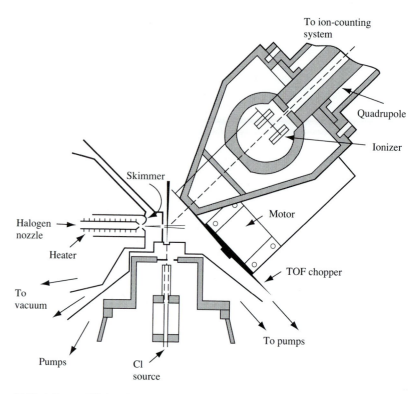

Figure 19.15. A Crossed Molecular Beam Apparatus (schematic). From R. D. Levine and R. B. Bernstein, *Molecular Reaction Dynamics and Chemical Reactivity*, Oxford University Press, New York, 1987, p. 234.

information on the distribution of energy between translational and internal degrees of freedom. By use of chemiluminescence or laser excitation fluorescence, the vibrational and rotational state distribution can be determined, giving the distribution of energy among the translational, rotational, and vibrational degrees of freedom.

Summary

An elementary process is one that cannot be divided into steps. A liquid-state "elementary" process is actually preceded and followed by diffusion processes. If the diffusion of the reactant molecules is a slow process compared to the chemical part, the reaction is called a diffusion-limited or diffusion-controlled reaction. In either a gaseous or liquid-state elementary process, the molecularity of a substance is equal to its order. The mechanism of a reaction is the sequence of elementary steps making up the reaction.

Two approximation schemes, the steady-state approximation and the rate-limiting step approximation, were introduced to deduce the rate law that corresponds to a given mechanism. Example mechanisms were studied, including chain reactions, in which propagation steps are included in the mechanism.

Catalysis involves an alternative reaction mechanism in which the catalyst is involved but not consumed. Catalysis is divided into two classes, heterogeneous catalysis and homogeneous catalysis. Heterogeneous catalysis at the surfaces of solids was discussed by using expressions for adsorption.

The accepted theory of biological catalysis is that an enzyme has an active site into which the reactant molecule fits in such a way that it is more reactive in the active site than out of it. A rate law for the simplest mechanism, due Michaelis and Menten, was obtained. Inhibition of enzyme-catalyzed reactions was discussed.

The empirical Arrhenius formula for the temperature dependence of rate constants was presented. This empirical formula was based on an idea that "activated" molecules with high energy are necessary for the reaction to occur and that the population of molecules with a characteristic activation energy is given by the Boltzmann probability distribution. The collision theory of bimolecular reaction rates was presented, using first the assumption that all collisions with a relative kinetic energy greater than a critical value would lead to reaction.

Various techniques for direct detection of reaction intermediates were discussed. These techniques included direct observation of the reaction intermediates and study of their effects.

Additional Reading

R. B. Bernstein, *Chemical Dynamics via Molecular Beam and Laser Techniques*, Oxford University Press, New York, 1982
This book is based on a set of lectures given at Oxford University in 1980 and was designed to bring advanced undergraduate students up to date with research in this area.

A. A. Frost and R. G. Pearson, *Kinetics and Mechanism*, 2nd ed., Wiley, New York, 1961
A useful book containing a number of reaction mechanisms.

H. S. Johnston, *Gas Phase Reaction Rate Theory*, Ronald Press, New York, 1966
This book was designed for a one-semester course at the senior/first-year graduate level and is still very useful.

K. J. Laidler, *Physical Chemistry with Biological Applications*, Benjamin/Cummings, Menlo Park, CA, 1978
A text for a one-semester course in physical chemistry for biologically oriented students.

See also the books listed at the end of Chapter 18.

PROBLEMS

Problems for Section 19.2

*19.26. The reaction

$$2I \rightarrow I_2$$

is diffusion controlled in carbon tetrachloride and also in water. Estimate the rate constant of the reaction in water at 20°C from data in Exercise 19.25 and viscosities in Table A24.

19.27. a. The reaction

$$2CH_3 \rightarrow C_2H_6$$

in toluene is diffusion controlled. The viscosity of toluene at 30°C is 5.236×10^{-4} kg m^{-1} s^{-1}. Estimate the rate constant of the reaction. State any assumptions.

b. The viscosity of toluene at 20°C is 5.9×10^{-4} kg m^{-1} s^{-1}. Estimate the activation energy of the reaction of part a and the value of the rate constant at 40°C. State any assumptions.

19.28. Compute the reaction diameter d_{12} for the reaction at 298.15 K

$$CH_3CO_2^- + H^+ \rightarrow CH_3CO_2H$$

for which $k = 4.5 \times 10^{10}$ L mol^{-1} s^{-1}. Use the values of the ion mobilities from Table A26.

Problems for Section 19.3

19.29. The formation of phosgene

$$CO + Cl_2 \rightleftharpoons COCl_2$$

is thought to proceed by the mechanism[40]

(1) $Cl_2 \rightleftharpoons 2Cl$
(2) $Cl + CO \rightleftharpoons COCl$
(3) $COCl + Cl_2 \rightleftharpoons COCl_2 + Cl$

where step 3 is rate-limiting.

a. Find the forward rate law.

b. Find the reverse rate law.

c. Show that these expressions are consistent with the equilibrium constant expression.

19.30. An alternative mechanism for the reaction of the previous problem is

(1) $Cl_2 \rightleftharpoons 2Cl$
(2) $Cl + Cl_2 \rightleftharpoons Cl_3$
(3) $Cl_3 + CO \rightleftharpoons COCl_2 + Cl$

where step 3 is assumed to be rate limiting. Repeat parts a, b, and c of the previous problem for this mechanism, and compare with the results of the previous problem.

---

[40] K. J. Laidler, *op. cit.*, pp. 301ff.

*19.31. The reaction equation for the nitration of an aromatic compound in a strong acid (usually sulfuric acid) can be written

$$ArH + HNO_3 \rightarrow ArNO_2 + H_2O$$

where Ar stands for an aromatic group such as the phenyl group, C_6H_5. A proposed mechanism is[41]

(1) $HNO_3 + HA \rightleftharpoons H_2NO_3^+ + A^-$
(2) $H_2NO_3^+ \rightleftharpoons H_2O + NO_2^+$
(3) $NO_2^+ + ArH \rightarrow ArNO_2H^+$
(4) $ArNO_2H^+ + A^- \rightarrow ArNO_2 + HA$

a. Find the rate law assuming that step 2 is rate limiting.

b. Find the rate law assuming that step 3 is rate limiting.

c. Find the rate law assuming the steady-state approximation.

19.32. For the reaction

$$2ICl + H_2 \rightarrow I_2 + 2HCl$$

the forward rate law is found to be

$$Rate = \frac{d[I_2]}{dt} = k[ICl][H_2]$$

Propose a mechanism that predicts this rate law.

19.33. The reaction

$$H_2O_2 + 3I^- + 2H^+ \rightarrow 2H_2O + I_3^-$$

obeys the rate law[42]

$$-\frac{d[H_2O_2]}{dt} = k_1[H_2O][I^-] + k_2[H_2O][I^-][H^+]$$

The two terms presumably correspond to two competing mechanisms. Propose a reasonable mechanism for the first term and two different reasonable mechanisms for the second term.

Problems for Section 19.4

*19.34. The chain mechanism for the $H_2 + I_2$ reaction is

(1) $I_2 \rightleftharpoons 2I$
(2) $I + H_2 \rightarrow HI + H$
(3) $H + I_2 \rightarrow HI + I$

Note that no reverse reaction is included in step 3.

---

[41] R. J. Gillespie, E. D. Hughes, C. K. Ingold, and R. I. Reed, *Nature* **163**, 599 (1949).

[42] D. Benson, *Mechanisms of Inorganic Reactions in Solution*, McGraw-Hill, New York, 1968, pp. 6ff.

a. Find the rate law using the steady-state approximation.

b. Find the rate law using a hybrid of the rate-limiting step approximation and the steady-state approximation: Assume that step 1 is at equilibrium, and assume a steady state for the concentration of H.

19.35. Insert a reverse reaction in step 3 of the previous problem. Write the expression for the equilibrium constant in terms of the rate constants. Write the relation between the rate constants for the chain mechanism and the other mechanisms in Section 19.4, using the principle of detailed balance.

19.36. The thermal decomposition of acetaldehyde follows the reaction equation

$$CH_3CHO \rightarrow CH_4 + CO$$

The following mechanism is proposed:[43]

(1) $CH_3CHO \rightarrow CH_3 + CHO$
(2) $CH_3CHO + CH_3 \rightarrow CH_4 + CO + CH_3$

with the following termination steps, which make only traces of C_2H_6 and H_2 and are not included in the stoichiometric equation:

(3) $2CH_3 \rightarrow C_2H_6$
(4) $2CHO \rightarrow 2CO + H_2$

a. Find the rate law assuming that the concentration of CH_3 is steady. Ignore the termination steps.

b. Add a reverse reaction to step 1 and find the rate law in the steady-state approximation.

c. Johnston[44] gives a mechanism in which step 2 is replaced by the steps

(2a) $CH_3 + CH_3CHO \rightarrow CH_3CO + CH_4$
(2b) $CH_3CO \rightarrow CH_3 + CO$

Repeat the steady-state solution using this mechanism. What must be assumed to bring the two results into agreement?

Problems for Section 19.5

19.37. a. The hydrogenation of ethylene on copper appears to follow the Langmuir-Hinshelwood mechanism, with the rate law

$$Rate = \frac{k_a[H_2][C_2H_4]}{(1 + k_b[C_2H_4])^2}$$

What conclusion can you draw from this rate law?

*b. The same reaction on nickel appears to follow the Langmuir-Rideal mechanism. Write the rate law you would expect to find for this case.

19.38. Apply the steady-state approximation to obtain a rate law for the unimolecular reaction on a catalytic surface:

(1) $A + surface \rightleftharpoons A(adsorbed)$
(2) $A(adsorbed) \rightarrow product$

19.39. a. Derive the rate law for the reaction of a substance A that undergoes a unimolecular reaction on a catalytic surface in the case that a nonreacting substance C also adsorbs on the surface. Assume that each substance obeys the Langmuir isotherm.

b. In the case that C absorbs more strongly than A, the catalyst can be "poisoned" by C. Poisoning can occur if lead is adsorbed on the surface of the catalyst in an automobile's catalytic converter. Find the rate law if C is much more strongly adsorbed than A.

19.40. The decomposition of nitrous oxide, N_2O, on a platinum surface is apparently inhibited by the adsorption of the O_2 produced. For the case of zero initial partial pressure of O_2, the observed rate law is

$$Rate = \frac{kP(N_2O)}{1 + b[P_0(N_2O) - P(N_2O)]}$$
$$= \frac{kP(N_2O)}{1 + bP(O_2)}$$

where k and b are temperature-dependent parameters and $P_0(N_2O)$ is the initial pressure of N_2O.

a. Assuming that both substances obey the Langmuir isotherm, derive this rate law, stating any necessary assumptions. Identify the parameters k and b.

b. Integrate the rate law to obtain

$$\frac{1 + ba}{t} \ln\left(\frac{a}{a - x}\right) = k + \frac{bx}{t}$$

where $a = \frac{1}{2}P_0(N_2O)$ and $x = P(O_2)$.

c. Following are data of Hinshelwood and Pritchard for this reaction at 741°C and for $a = 95$ torr.

t/s	315	750	1400	2250	3450	5150
x/torr	10	20	30	40	50	60

Using a graphical or a least-squares method, fit these data to the integrated rate law and determine the values of k and b. How well does the formula fit the data points?

[43] J. L. Latham, *Elementary Reaction Kinetics*, 2nd ed., Butterworths, London, 1969, p. 128.
[44] H. S. Johnston, *op. cit.*, p. 36.

19.41. The platinum-catalyzed oxidation of methanol apparently proceeds through formation of formaldehyde and then carbon monoxide.[45] The following mechanism is proposed for the oxidation of the carbon monoxide:

(1) $CO(g)$ + surface site $\rightleftharpoons$ CO(adsorbed)

(2) O_2 + surface site $\rightleftharpoons$ O_2(adsorbed)

(3) O_2(adsorbed) + surface site $\rightarrow$ 2O(adsorbed)

(4)   CO(adsorbed) + O(adsorbed) $\rightarrow$ $CO_2(g)$

Use the steady-state approximation, assuming that the rates of change of θ_{CO}, θ_{CO_2}, and θ_O are negligible. Find the rate law for the oxidation of CO. Assume that

$$1 - \theta_{CO} - \theta_{O_2} - \theta_O \approx 1 - \theta_{CO}$$

and neglect k'_2 compared with $k_3(1 - \theta_{CO})$.

19.42. The gas-phase thermal decomposition of acetaldehyde is catalyzed by iodine. The proposed mechanism is[46]

(1) $CH_3CHO + I_2 \rightleftharpoons CH_3I + HI + CO$

(2) $CH_3I + HI \rightarrow CH_4 + I_2$

a. Find the rate law, assuming the steady-state approximation.

b. Find the rate law, assuming that the back reaction in step 1 is negligible.

c. Tell how you would decide which assumption is preferable.

19.43. Obtain the rate law for the simple Michaelis-Menten mechanism of Equation (19.5-25), assuming the second step to rate limiting instead of assuming the steady-state approximation.

19.44. a. Add the reverse reaction to the second step of the Michaelis-Menten mechanism of Equation (19.5-25) and obtain the rate law.

b. Take the limit of the rate expression as the concentration of product approaches zero.

c. Take the limit of the rate expression as the concentration of reactant approaches zero.

d. Write an expression for the equilibrium constant of the reaction.

19.45. Derive the rate law for the forward rate of the enzyme-catalyzed reaction with the mechanism

(1) $E + R \rightleftharpoons ER$

(2) $ER \rightarrow ER' + P_1$

(3) $ER' \rightarrow E + P_2$

where P_1 and P_2 are two different products. Comment on the relationship of your answer to the ordinary Michaelis-Menten formula.

Problems for Section 19.6

*19.46. The first-order rate constant for the thermal decomposition of 3-methylcyclobutanone has the values:[47]

T/K	$k/10^{-4}$ s^{-1}	T/K	$k/10^{-4}$ s^{-1}
552.24	0.4259	589.05	6.459
561.81	0.8936	596.96	11.201
570.41	1.707	606.14	20.83
579.35	3.207		

Find the value of the activation energy and the value of the pre-exponential factor.

19.47. The gas-phase decomposition of acetaldehyde, CH_3CHO, obeys second-order kinetics. Some values of the rate constant are:

T/K	k/L mol^{-1} s^{-1}	T/K	k/L mol^{-1} s^{-1}
703	0.0110	811	0.79
733	0.0352	836	2.14
759	0.105	865	4.95
791	0.343		

a. Find the activation energy and the pre-exponential factor.

b. Find the value of the rate constant at 500°C.

c. If the initial pressure of pure acetaldehyde is equal to 0.500 atm at 500°C, find the time for 50.0% of the acetaldehyde to react.

*19.48. a. The rate constant for the bimolecular elementary gaseous reaction

$$CO + O_2 \rightarrow CO_2 + O$$

is equal to 1.22×10^5 L mol^{-1} s^{-1} at 2500 K and 3.66×10^5 L mol^{-1} s^{-1} at 2800 K. Find the activation energy and the pre-exponential factor.

[45] R. W. McCabe and D. F. McCready, *J. Phys. Chem.* **90**, 1428 (1986).

[46] J. L. Latham, *op. cit.*, p. 108.

[47] H. M. Frey, H. P. Watts, and I. D. R. Stevens, *J. Chem. Soc. Faraday Trans. 2* **83**, 601 (1987).

b. Assuming a hard sphere diameter of 350 pm for O_2 and of 360 pm for CO, calculate the value of the steric factor in the collision theory.

General Problems

19.49. For the reaction

$$H_2 + I_2 \rightarrow 2HI$$

the value of ΔU is -8.2 kJ mol^{-1}. Find the value of the activation energy for the reverse reaction, using information in the chapter.

19.50. The decomposition of ethyl bromide, C_2H_5Br, in the gas phase is observed to be a first-order reaction.

a. Write the steps of the mechanism, assuming the Lindemann mechanism. The products are ethene, C_2H_4, and hydrogen bromide, HBr. What must be the case for the first-order rate law to be observed?

b. The rate constant at 527°C is equal to 0.0361 s^{-1}. Find the half-life of the reaction at this temperature. Neglect any reverse reaction.

c. If the original pressure of pure ethyl bromide is equal to 1.00 atm, find the partial pressure of each substance after an elapsed time of 60 seconds. Neglect any reverse reaction.

d. The rate constant is equal to 1.410 s^{-1} at 627°C. Find the value of the Arrhenius activation energy and of the pre-exponential factor for the reaction.

19.51. The gas-phase recombination of iodine atoms proceeds in the presence of a second substance, M:

$$I + I + M \rightarrow I_2 + M$$

Two mechanisms are proposed.[48] The first is

(1) $I + I \rightleftharpoons I_2^*$
(2) $I_2^* + M \rightarrow I_2 + M$

where I_2^* represents a high-energy molecule. The second mechanism is

(1) $I + M + M \rightleftharpoons IM + M$
(2) $IM + I \rightarrow I_2 + M$

where IM represents a loosely bound molecule that is not necessarily capable of permanent existence.

a. Find the rate differential for each mechanism, using the rate-limiting step approximation and assuming that the second step in each mechanism is rate limiting.

*b. It is found that the activation energy for the overall reaction is negative. If M is argon, $k = 8.3 \times 10^{-33}$ cm^6 s^{-1} at 300 K and $k = 1.3 \times 10^{33}$ cm^6 s^{-1} at 1300 K. Note that the concentrations are expressed in atoms cm^{-3} for these values of the rate constants. Find the value of the activation energy.

c. Explain the fact that the activation energy is negative.

*19.52. Label each statement as either true or false. If a statement is true only under special circumstances, label it as false.

a. A useful rule of thumb is that the rate of a chemical reaction doubles for each increase in temperature of 10°C.

b. Every chemical reaction proceeds by a multistep mechanism.

c. Every reaction mechanism is a sequence of steps that follow each other sequentially.

d. In a typical gas-phase chemical reaction, only a small fraction of molecular collisions lead to reaction.

e. Termolecular steps are relatively rare in chemical reaction mechanisms.

f. One way to verify a reaction mechanism is to observe the presence of reactive intermediates.

g. Diffusion-limited reactions are rapid because diffusion is an inherently rapid process.

h. Only mechanisms with branching chains can lead to explosions.

i. An inhibitor for an enzyme-catalyzed reaction must compete with a reactant for an active site.

j. A reaction can be catalyzed heterogeneously if only one of the reactants is adsorbed.

---

[48] H. S. Johnston, *op. cit.*, pp. 253ff.

20 Equilibrium Statistical Mechanics

PREVIEW

In Chapter 1 of this book, we stated that a macroscopic state of a system of many molecules corresponds to many microscopic states and that a single macroscopic state can be represented as an average over many microscopic states. In this chapter, we study the mathematical procedures by which this averaging is accomplished.

PRINCIPAL FACTS AND IDEAS

1. Statistical mechanics is the connection between the microscopic states and macroscopic states of a system of many molecules.
2. Statistical mechanics identifies macroscopic states with averages over microscopic states.
3. For a system representing a dilute gas, we can average over molecular states instead of averaging over system microstates.
4. The average over molecular states is done with a probability distribution.
5. Equilibrium thermodynamic variables can be expressed in terms of a sum over molecular states called the molecular partition function.
6. Explicit formulas can be derived for molecular partition functions.
7. Equilibrium constants for chemical reactions in dilute gases can be calculated theoretically from molecular partition functions.
8. Statistical mechanics can be applied to systems other than dilute gases by using ensemble theory.
9. Statistical mechanics can be based on either classical mechanics or quantum mechanics.

The Quantum Statistical Mechanics of a Sample System of Four Molecules

In this chapter, we will analyze the interrelationships among three different kinds of states: (1) the mechanical states of single molecules or atoms, (2) the mechanical states (microstates) of systems of many molecules, and (3) the thermodynamic states (macrostates) of systems of many molecules.

Gas kinetic theory, which we discussed in Chapter 16, connects micro-states and macrostates for dilute gaseous systems whose molecules are assumed to obey classical mechanics. Statistical mechanics is much like gas kinetic theory, in that it uses averages over microstates to represent mac-rostates. However, statistical mechanics can treat any kind of a system, whether it is solid, liquid, or gas, and can provide representations of all thermodynamic functions. Both classical and quantum versions of statistical mechanics exist, as well as equilibrium and nonequilibrium versions. In this chapter, we will discuss only equilibrium statistical mechanics, which is sometimes called statistical thermodynamics, and will focus on quantum statistical mechanics.

Discussion of a Simple Example System

To introduce the principles of statistical mechanics, we examine a simple example system: a dilute gas containing four hydrogen molecules. This system is small enough that its microstates can be enumerated.

We assume that the molecules are sufficiently far apart on the average that the intermolecular potential energy can be ignored. In this case the Hamiltonian operator of the system is

$$\hat{H}_s(1, 2, 3, 4) = \hat{H}(1) + \hat{H}(2) + \hat{H}(3) + \hat{H}(4) \tag{20.1-1}$$

where $\hat{H}(1)$ is the hydrogen molecule Hamiltonian operator of molecule number 1, etc. The indexes in the parentheses stand for all of the coordinates of both nuclei and both electrons of the specified hydrogen molecules. The molecules can be far apart from each other if the gas is dilute, so that the neglect of intermolecular interactions can be a good approximation.

This system Hamiltonian operator gives a time-independent Schrödinger equation in which the variables can be separated by use of the trial solution

$$\Psi(1, 2, 3, 4) = \psi_1(1)\psi_2(2)\psi_3(3)\psi_4(4) \tag{20.1-2}$$

Bosons are defined in Section 11.3. They require symmetric wave functions instead of the antisymmetric wave functions that electrons and other fermions require. A symmetrized wave function can be written as a permanent, which is like a determinant except that all positive signs are written when it is expanded.

where ψ_1, ψ_2, etc. are individual hydrogen molecule energy eigenfunctions and Ψ is a system wave function. The subscripts on the ψ factors are abbreviations for all of the quantum numbers of a hydrogen atom. For now, we ignore the fact that the hydrogen molecules are bosons and proceed as though they were distinguishable, with a system wave function that is neither antisymmetrized nor symmetrized.

| Exercise 20.1 | **a.** Since a hydrogen molecule consists of four fermions, show that it is a boson. **b.** Symmetrize the wave function of Equation (20.1-2) by writing 24 terms. All terms must have the same sign to produce a symmetric wave function. |

The energy eigenvalues corresponding to Equation (20.1-2) are given by

$$E = \varepsilon_1 + \varepsilon_2 + \varepsilon_3 + \varepsilon_4 \tag{20.1-3}$$

where ε_1, ε_2, etc. are hydrogen molecule energy eigenvalues and E is a system energy eigenvalue.

| Exercise 20.2 | Using separation of variables, show that Equation (20.1-3) follows from Equation (20.1-1). |

To the approximation of Chapter 14, each molecule wave function is a product of factors:

$$\psi_i = \psi_{i,\,\mathrm{tr}}\psi_{i,\,\mathrm{rot}}\psi_{i,\,\mathrm{vib}}\psi_{i,\,\mathrm{elec}} \tag{20.1-4}$$

and each molecule energy eigenvalue is a sum of terms

$$\varepsilon_i = \varepsilon_{i,\,\mathrm{tr}} + \varepsilon_{i,\,\mathrm{rot}} + \varepsilon_{i,\,\mathrm{vib}} + \varepsilon_{i,\,\mathrm{elec}} \tag{20.1-5}$$

where the subscripts stand for the quantum numbers that enumerate the energy eigenfunctions and eigenvalues for one molecule. For example, the subscript i, elec stands for all electronic quantum numbers of a hydrogen molecule.

We focus our attention on the vibrational states of the molecules, assuming the harmonic oscillator approximation. The vibrational energy of the system is the sum of the vibrational energies of the molecules:

$$E_{\mathrm{vib}} = \varepsilon_{v_1} + \varepsilon_{v_2} + \varepsilon_{v_3} + \varepsilon_{v_4} \tag{20.1-6a}$$
$$= h\nu(v_1 + v_2 + v_3 + v_4) \tag{20.1-6b}$$

where ν is the classical vibration frequency, given by Equation (9.1-10), v_1 is the vibrational quantum number of molecule number 1, v_2 is the vibrational quantum number of molecule number 2, etc. We include the vibrational zero-point energies in the electronic energies, so that the ground-state vibrational energy is equal to zero.

Specification of the equilibrium macrostate of the system requires three variables, such as the energy of the system, the volume of the system, and the number of molecules in the system. Since we are considering only the vibrational energy, which is independent of the volume, the macrostate is specified by the number of particles in the system and value of the vibrational energy of the system.

Consider the case that the vibrational energy of the system equals $4h\nu$. All of the possible system vibrational microstates are listed in Table 20.1. There are 35 possible vibrational microstates for our system of four hydrogen molecules, comprising a system energy level with a degeneracy equal to 35.

Table 20.1. Vibrational States for a System of Four Diatomic Molecules

v_1	v_2	v_3	v_4	v_1	v_2	v_3	v_4
1	1	1	1	2	2	0	0
0	1	1	2	0	0	1	3
0	1	2	1	0	0	3	1
0	2	1	1	0	1	0	3
1	0	1	2	0	3	0	1
1	0	2	1	1	0	0	3
2	0	1	1	3	0	0	1
2	1	0	1	0	1	3	0
1	2	0	1	0	3	1	0
1	1	0	2	1	0	3	0
2	1	1	0	3	0	1	0
1	2	1	0	1	3	0	0
1	1	2	0	3	1	0	0
0	0	2	2	0	0	0	4
0	2	0	2	0	0	4	0
2	0	0	2	0	4	0	0
0	2	2	0	4	0	0	0
2	0	2	0				

If we possess only macroscopic information (the value of the total vibrational energy), we have no information about which of the 35 microstates the system occupies at a particular instant. We make an assumption, which is the first postulate of statistical mechanics:

Postulate I: *A thermodynamic property of the system can be equated to the average of the corresponding property over the system microstates that might be occupied.*

To define the type of average to be taken, we make another fundamental assumption, the second postulate of statistical mechanics:

Postulate II: *All system microstates of equal energy are equally probable.*

These postulates are an assumption with a minimal amount of arbitrariness. Any assumption of unequal probabilities would be more arbitrary than this assumption, corresponding to a greater amount of assumed information about the microstate of the system.

Since the system can make transitions from one microstate to another, in some period of time the system might pass through all of the 35 microstates. Therefore, a measurement of a macroscopic property that takes a long time to perform should give the same value as would some kind of average over the microstates, making Postulate I plausible. Postulate II is equivalent to assuming that over some long period of time the system would spend equal amounts of time in each of the 35 microstates. For a system of many particles, there are very many possible microstates and there is some doubt that all of them would be occupied in any period of time, no matter how long, so our postulates must be regarded as hypotheses that have no *a priori* justification.

We now determine the consequences of our postulates for the four-molecule system. We seek first the probability of finding a particular molecular vibrational state if a molecule is randomly selected from the system. We look at the state of molecule number 1. All of the molecules' states are alike, so it doesn't matter which one we examine. Of the 35 microstates, there are 15 states with $v_1 = 0$. The probability that a randomly selected molecule will have its vibrational quantum number equal to zero is therefore

$$p_0 = \frac{15}{35} = 0.42857 \qquad \textbf{(20.1-7a)}$$

There are 10 states for which $v_1 = 1$, so

$$p_1 = \frac{10}{35} = 0.28571 \qquad \textbf{(20.1-7b)}$$

The other probabilities are

$$p_2 = \frac{6}{35} = 0.17143 \qquad \textbf{(20.1-7c)}$$

$$p_3 = \frac{3}{35} = 0.08571 \qquad \textbf{(20.1-7d)}$$

$$p_4 = \frac{1}{35} = 0.02857 \qquad \textbf{(20.1-7e)}$$

In statistical mechanics, an average nearly always denotes a mean.

Since these probabilities are obtained by averaging over system microstates, we call them an **average probability distribution**. These probabilities will be used in calculating average molecular properties, but since they were obtained by averaging over microstates of the whole system, every molecular average quantity will be a double average: an average over molecular states using a probability distribution that was itself obtained by averaging over system microstates.

The average vibrational energy per molecule is given by the general formula for a mean value, Equation (16.1-13):

$$\langle \varepsilon_{\text{vib}} \rangle = \sum_{v=0}^{4} p_v \varepsilon_v$$
$$= p_0 0 + p_1 hv + p_2 2hv + p_3 3hv + p_4 4hv \qquad \textbf{(20.1-8)}$$

Exercise 20.3 Calculate the value of the mean molecular vibrational energy in Equation (20.1-8) and show that to five significant digits it equals hv (as it must, since four molecules share $4hv$ of energy).

The probabilities of Equation (20.1-7) decrease smoothly with increasing energy, just as did the Boltzmann probability distribution of Equations (16.2-31) and (1.7-25).

▼   **EXAMPLE 20.1**   Compare the probability distribution of Equation (20.1-7) with the Boltzmann probability distribution for equally spaced energy levels

$$\varepsilon = vh\nu \qquad (v = 0, 1, 2, \dots)$$

and for the mean energy $\langle \varepsilon \rangle = h\nu$.

Solution

The probability distribution is that used by Planck, Equation (9.3-7), and the corresponding formula for the mean energy is given in Equation (9.3-8). To have

$$\langle \varepsilon \rangle = h\nu$$

we must have a value of the temperature given by

The first part of the solution of Example 20.1 is to determine the temperature that corresponds to our average energy.

$$e^{h\nu/k_B T} - 1 = 1 \quad \text{or} \quad \frac{h\nu}{k_B T} = \ln(2)$$

which gives, from Equation (9.3-7) and the result of Problem 9.46,

$$p_{v(\text{Boltzmann})} = \frac{1}{2} e^{-v \ln(2)}$$

This formula gives the values shown in Table 20.2. The Boltzmann distribution does not stop at $v = 4$, because it applies to a large system, not a system of four molecules.

Table 20.2. Average and Boltzmann Probability Distributions for the Vibrational States of Diatomic Molecules in a System of Four Diatomic Molecules

Value of v	p_v	$p_v(\text{Boltzmann})$
0	0.42857	0.50000
1	0.28571	0.25000
2	0.17143	0.12500
3	0.08571	0.06250
4	0.02857	0.03125
5	0	0.01562
6	0	0.00781
...		

▲

If the number of molecules in a system is increased, the average probability distribution becomes more and more similar to the Boltzmann distribution. In the limit that the number of molecules in the system becomes infinite, the average distribution becomes the same as the Boltzmann distribution, although we do not prove this fact.

We will not treat the translational, rotational, and electronic states of our four-particle system, since these states do not lend themselves to a simple treatment. However, the same principles would apply. Averaging over all system microstates of the same energy with equal weight leads to a molecular probability distribution in which the probability of a molecular state depends on the energy in a characteristic way, with molecular states of lower energy being more probable.

The Probability Distribution for a Dilute Gas

Statistical mechanics is not restricted to dilute gases. However, the most productive elementary applications of statistical mechanics are to dilute gases, and we focus our discussion on dilute gases. Consider a closed dilute gaseous system in which all molecules are identical. Let the thermodynamic energy U have a known value, the number of particles N have a known value, and the system be confined in a rectangular box of known volume V. The equilibrium macrostate of the system is thus known.

We assume that our molecules are on the average far enough apart that the intermolecular potential energy is small enough to be neglected. The system Hamiltonian operator can be written as in Equation (20.1-1):

$$\hat{H}_{sys} = \sum_{i=1}^{N} \hat{H}(i) \tag{20.2-1}$$

where $\hat{H}(i)$ is the molecular Hamiltonian operator for molecule number i. Solution of the Schrödinger equation with this Hamiltonian operator by separation of variables leads to energy eigenvalues

$$E_n = \sum_{i=1}^{N} \varepsilon_{ni} \tag{20.2-2}$$

and to energy eigenfunctions

$$\Psi_n = \prod_{i=1}^{N} \psi_{ni}(i) \tag{20.2-3}$$

where (i) stands for the coordinates of all of the particles in molecule i and the subscript ni stands for the values of the quantum numbers needed to specify the molecular state for molecule i that corresponds to system microstate number n.

Since our molecules are identical, they are indistinguishable from each other. If the molecules are fermions (like electrons), the system wave function must be antisymmetrized, which leads to an exclusion principle for the fermion molecules. If the particles are bosons, the wave function must be symmetrized. A symmetrized product wave function does not vanish if two factors are identical, and there is no exclusion principle for bosons.

Let us consider molecular energy levels instead of molecular states. If we knew the system microstate, we would know all of the factors in Equation (20.2-3) and would know the number of molecules occupying states in molecular level j. We denote this number by $N_j(k)$ for system microstate k. If the molecules are fermions, $N_j(k)$ can be no larger than g_j, the degeneracy of level j. If the molecules are bosons, $N_j(k)$ can be any nonnegative integer that is no larger than N, the number of molecules in the system.

The symbol $\{N\}$ stands for an entire set of numbers.

The set of numbers $N_1(k)$, $N_2(k)$, $N_3(k)$, etc. is a **distribution**. We will denote the distribution for system microstate k by the single symbol $\{N(k)\}$ and a general distribution by the symbol $\{N\}$. We are not particularly interested in the distribution for a particular system microstate. What we want is a distribution that applies to a given system macrostate, which must correspond to some kind of average over many system microstates.

We specify the macrostate of the system by the values of N, V, and U. We must consider all microstates of the system that correspond to a given value of the energy E equal to U, number of molecules equal to a given value N, and system volume equal to a given value of V, because the system might be in any of these microstates, so far as we know. Let the number of system microstates corresponding to these values of E, N, and V be denoted by Ω. This quantity is the degeneracy of a system energy level. For example, in the four-particle system, if there were only vibrational energy, Ω would equal 35 for a system vibrational energy $E = 4h\nu$. For a many-particle system, Ω is a very large number.

The Average Distribution

We invoke the same two postulates as with the four-molecule system. That is, we equate observed values of system variables with an average over all of the system microstates that correspond to the system macrostate, and we give all these system microstates equal weight. If we average the numbers $N_j(1)$, $N_j(2)$, $N_j(3),\ldots, N_j(k),\ldots$, over all of the system microstates, we will have an average number of molecules in molecular level j:

$$\bar{N}_j = \frac{1}{\Omega} \sum_{k=1}^{\Omega} N_j(k) \tag{20.2-4}$$

where the sum includes all system microstates with the correct values of N, V, and E, which we number from 1 to Ω. The set of average numbers for all molecular levels is the average distribution, denoted by $\{\bar{N}\}$. Unfortunately, the sum in Equation (20.2-4) cannot actually be carried out, because there are too many system microstates. There are techniques for approximating the sum, but we do not discuss them.

The Most Probable Distribution

There can be a number of system microstates that correspond to the same distribution. Let $W(\{N\})$ be the number of system microstates that correspond to a particular distribution $\{N\}$. The total number of system microstates with the correct values of E, V, and N is the sum of the W's for all distributions that correspond to the correct values of E, V, and N:

$$\sum_{\{N\}} W(\{N\}) = \Omega \tag{20.2-5}$$

where the sum includes one term for each distinct distribution that corresponds to the correct values of E, V, and N.

The distribution with the largest value of W is the **most probable distribution**, since all system microstates considered are equally probable. Instead of seeking to determine what the average distribution is like, we seek the most probable distribution, assuming that these two distributions will resemble each other.[1] The average distribution and the most probable distribution are identical if the system has a very large number of molecules, but we omit the proof of this fact.

We now obtain an expression for W for a given distribution. Consider a single molecular energy level, number j. We consider the special case that for nearly every level

$$N_j \ll g_j \tag{20.2-6}$$

even though we will later assume that N_j is a rather large number. This case is called the case of **dilute occupation**. It applies for a sufficiently dilute gas, since for a system in a very large volume, the translational energy levels are very close together and their degeneracies are very large. In the limit of an infinitely large system, which is called the **thermodynamic limit**, the degeneracies become infinite and the spacings between levels approach zero.

We need an expression for the number of distinct ways to choose molecular states from the g_j states in level j to be occupied by N_j molecules. If the molecules are fermions, no more than one molecule can occupy each state. Therefore, one needs to count up the ways to divide the g_j states of the level into N_j occupied levels and $g_j - N_j$ unoccupied levels. This number is

$$t_j = \frac{g_j!}{(g_j - N_j)! N_j!} \tag{20.2-7}$$

where $N_j!$ stands for N_j factorial:

$$N_j! = N_j(N_j - 1)(N_j - 2) \cdots (3)(2)(1) \tag{20.2-8}$$

We consider the case that $g_j \gg N_j$, so that the quotient $g_j!/(g_j - N_j)!$ consists of N_j factors that are all nearly equal to g_j:

$$t_j \approx \frac{g_j^{N_j}}{N_j!} \tag{20.2-9}$$

Equation (20.2-9) can also be derived by noting that there are nearly g_j choices for each occupied state and that there are $N_j!$ different orders to pick the occupied states.

Exercise 20.4

Find the percent error in approximating Equation (20.2-7) by Equation (20.2-9) if (a) $g_j = 100$ and $N_j = 3$ and (b) $g_j = 1000000$ and $N_j = 3$.

In the approximation that $N_j \ll g_j$, Equation (20.2-9) also applies to a system of bosons. If our gas is sufficiently dilute, the distributions for both fermions and bosons are the same.

---

[1] Our discussion follows that of N. Davidson, *Statistical Mechanics*, McGraw-Hill, New York, 1962, Chapter 6.

For two levels, the total number of ways to choose states is the product of the number of ways of choosing states for each level, since each state in the first level can be combined with every state in the second level. For the entire distribution (many levels) the number of system microstates is given by a product containing one factor as in Equation (20.2-9) for each molecular energy level:

$$W(\{N\}) = \prod_j t_j = \prod_j \frac{g_j^{N_j}}{N_j!} \qquad \text{(20.2-10)}$$

If any level is unoccupied, it contributes a factor of unity, since 0! is defined to equal unity. We do not indicate the upper limit for the product, but there are infinitely many molecular energy levels.

If a distribution did not satisfy Equations (20.2-11) and (20.2-12), it would not correspond to the correct macrostate.

We consider only those distributions which conform to the given macrostate of the system, a state of fixed E, V, and N. The distribution must obey the conditions

$$\sum_j N_j = N \qquad \text{(20.2-11)}$$

$$\sum_j N_j \varepsilon_j = E \qquad \text{(20.2-12)}$$

The problem of finding the correct distribution is now the problem of finding the distribution $\{N\}$ that maximizes the value of $W(\{N\})$ subject to the constraints of Equations (20.2-11) and (20.2-12). In order to use the methods of calculus, we assume that the N_j's are able to take on any real value, instead of just integral values. A value of N_j corresponding to the maximum can be rounded to the closest integer to give the desired value.

The maximum value subject to constraints is found by Lagrange's method of undetermined multipliers.[2] For example, to find a maximum or minimum of a function $f = f(x, y, z)$ subject to two constraints, the constraints are first expressed by equations of the form

$$g(x, y, z) = 0 \quad \text{and} \quad h(x, y, z) = 0$$

One then proceeds by finding a point at which

$$\frac{\partial}{\partial x} (f + \alpha g + \beta h) = 0$$

$$\frac{\partial}{\partial y} (f + \alpha g + \beta h) = 0$$

$$\frac{\partial}{\partial z} (f + \alpha g + \beta h) = 0$$

The quantities α and β are constants called **undetermined multipliers**. The values of these constants must be found from the equations for the constraints. If there are more variables and more constraints, there is an undetermined multiplier for each constraint and there is an equation for each independent variable.

---

[2] H. Anton, *Calculus*, 3rd ed., Wiley, New York, 1988, pp. 1032ff.

It is easier to find a maximum in the logarithm of W than in W itself. The logarithm is a monotonic function of its argument, so the largest value of $\ln(W)$ occurs with the largest value of W. At the constrained maximum there is a set of simultaneous equations, one for each variable. We write the equation for N_i, the number of molecules in level number i:

$$\frac{\partial}{\partial N_i}\left[\ln(W) + \alpha\left(\sum_j N_j - N\right) - \beta\left(\sum_j N_j\varepsilon_j - E\right)\right] = 0 \quad \textbf{(20.2-13)}$$

We use α and $-\beta$ as symbols for the undetermined multipliers. The method is not changed by using $-\beta$ instead of β for one multiplier.

From Equation (20.2-10),

$$\ln[W(\{N\})] = \sum_j [N_j \ln(g_j) - \ln(N_j!)] \quad \textbf{(20.2-14)}$$

We use **Stirling's approximation** for the factorial:

$$N! \approx (2\pi N)^{1/2} N^N e^N \quad \textbf{(20.2-15)}$$

For $\ln(N_j!)$ we use a simpler version of the approximation, which is valid for large values of N_j:

$$\ln(N_j!) \approx N_j \ln(N_j) - N_j \quad \textbf{(20.2-16)}$$

Exercise 20.5	Using a table of factorials or a calculator that computes factorials, find the percent error in Equation (20.2-16) for $N_j = 10$ and $N_j = 60$. Take the logarithm of Equation (20.2-15) and find the percent error for the same values of N_j. Find the difference between the results using Equation (20.2-16) and the logarithm of Equation (20.2-15) for $N_j = 1 \times 10^9$.

Use of Equations (20.2-14) and (20.2-16) in Equation (20.2-13) gives

$$\frac{\partial}{\partial N_i} \sum_j [N_j \ln(g_j) - N_j \ln(N_j) + N_j + \alpha N_j - \beta N_j\varepsilon_j] = 0$$

where we have omitted the terms N and E because these quantities are constants and have zero derivatives.

Since N_1, N_2, N_3, etc. are all independent variables, only the term with $j = i$ has a nonzero derivative, and we have at the constrained maximum in $\ln(W)$

$$\ln(g_i) - \ln(N_i) - 1 + 1 + \alpha - \beta\varepsilon_i = 0 \quad \textbf{(20.2-17)}$$

for each value of i. The set of simultaneous equations is a set of separate equations with one variable in each equation. Equation (20.2-17) is the same as

$$N_j = g_j e^\alpha e^{-\beta\varepsilon_j} \quad \textbf{(20.2-18)}$$

Switching the index in the equation from i to j makes no difference so long as all occurrences of the index are changed.

where we return to the subscript j to designate the energy level. Equation (20.2-18) is one equation for each value of j and represents the most probable distribution.

The probability that a randomly selected molecule is found in energy level j is

$$p_j = \frac{N_j}{N} = \frac{1}{N} g_j e^{\alpha} e^{-\beta \varepsilon_j} \quad \text{(probability of level } j) \qquad \textbf{(20.2-19)}$$

This probability is independent of N and proportional to the degeneracy of the level, as expected. We assume that all of the states of a level are equally probable, so the probability of a molecular state is

$$p_i = \frac{N_i}{N} = \frac{1}{N} e^{\alpha} e^{-\beta \varepsilon_i} \quad \text{(probability of state } i) \qquad \textbf{(20.2-20)}$$

Note the resemblance of our probability distribution to the Boltzmann probability distribution of kinetic theory, Equation (16.2-31). Once we show that the parameter β is equal to $1/(k_B T)$, our probability distribution will be the same as the Boltzmann distribution.

Equation (20.2-20) is for a gas with dilute occupation, where the distinction between fermions and bosons is unimportant. The expression for t_j for fermions was stated in Equation (20.2-7). The corresponding expression for bosons is

$$t_j = \frac{(g_j + N_j - 1)!}{(g_j - 1)! N_j!} \quad \text{(bosons)} \qquad \textbf{(20.2-21)}$$

If Equation (20.2-7) or (20.2-21) is used in the expression for W instead of the expression in Equation (20.2-10), the results are

$$N_j = \frac{g_j e^{\alpha} e^{-\beta \varepsilon_j}}{1 + e^{\alpha} e^{-\beta \varepsilon_j}} \quad \text{(fermions)} \qquad \textbf{(20.2-22)}$$

$$N_j = \frac{g_j e^{\alpha} e^{-\beta \varepsilon_j}}{1 - e^{\alpha} e^{-\beta \varepsilon_j}} \quad \text{(bosons)} \qquad \textbf{(20.2-23)}$$

Consider now the behavior of both of these distributions for large values of the molecular energy. For sufficiently large energies, the second terms in the denominators of both the expressions in Equations (20.2-22) and (20.2-23) become negligible, and both expressions approach

$$N_j = g_j e^{\alpha} e^{-\beta \varepsilon_j} \qquad \textbf{(20.2-24)}$$

which is the same as Equation (20.2-20). Therefore, the parameters α and β must have the same meaning for the fermion and boson distributions as for the dilute occupation distribution.

20.3 The Probability Distribution and the Molecular Partition Function

To make our probability distribution of Equation (20.2-18) usable, we need to evaluate the parameters α and β. In this section, we will carry out this evaluation and will introduce the molecular partition function, a function that can be used to calculate thermodynamic quantities without an explicit averaging process.

The Molecular Partition Function and the Parameter α

The parameter α is evaluated by substituting the distribution of Equation (20.2-18) into Equation (20.2-11):

$$\sum_j g_j e^{\alpha} e^{-\beta \varepsilon_j} = N \tag{20.3-1}$$

Equation (20.3-1) is the same as

$$e^{\alpha} = N \left(\sum_j g_j e^{-\beta \varepsilon_j} \right)^{-1} = \frac{N}{z} \tag{20.3-2}$$

where we define the **molecular partition function** z:

$$z = \sum_j g_j e^{-\beta \varepsilon_j} \quad \text{(sum over molecular levels)} \tag{20.3-3}$$

The name "partition function" was chosen because z is related to the way in which the molecules are partitioned among the possible molecular states. The German name for a partition function is **Zustandsumme**, translated as **sum over states** or **state sum**. This name is more descriptive than "partition function" and is occasionally used in the English language, with or without translation. The German name is the reason for using the letter z to denote the partition function, but the letter q is also used.

The partition function plays a central role in equilibrium statistical mechanics. We will show that the values of all thermodynamic variables of any system can be obtained from the partition function. There are also several kinds of system partition functions, in which the sum is over microstates of a macroscopic system instead of over molecular states. We will discuss one such system partition function in Section 20.7.

The probability of finding a randomly selected molecule in molecule energy level j is equal to the fraction of all molecules of the system that are in this level:

$$p_j = \frac{N_j}{N} = \frac{1}{z} g_j e^{-\beta \varepsilon_j} \quad \text{(probability of level } j) \tag{20.3-4}$$

In the partition function there is one term for each molecular energy level. Each term is proportional to the number of states comprising that level, so we can write the molecular partition function as a sum over states instead of a sum over levels:

$$z = \sum_j e^{-\beta \varepsilon_j} \quad \text{(sum over molecular states)} \tag{20.3-5}$$

Each term in this sum is proportional to the probability of finding a randomly selected molecule in the state corresponding to that term, with the term for a state of zero energy equal to unity. The molecular partition function is therefore a sum of probabilities of states relative to a state of zero energy, and its value can be thought of as an effective total number of states accessible to a molecule under the given conditions.

The Parameter β

In order to determine what the parameter β is, we substitute the expression for N_j into the constraint of Equation (20.2-12):

$$\frac{1}{z}\sum_j \varepsilon_j g_j e^{-\beta \varepsilon_j} = \frac{E}{N} = \frac{U}{N} \tag{20.3-6}$$

where we have again equated the mechanical energy of the system, E, to the thermodynamic energy, U.

Equation (20.3-6) can be written in another way, using a mathematical trick:

$$\frac{U}{N} = \frac{1}{z}\sum_j \varepsilon_j g_j e^{-\beta \varepsilon_j} = -\frac{1}{z}\sum_j \frac{\partial}{\partial \beta}\left(g_j e^{-\beta \varepsilon_j}\right)$$

$$= -\frac{1}{z}\frac{\partial}{\partial \beta}\left(\sum_j g_j e^{-\beta \varepsilon_j}\right) = -\frac{1}{z}\frac{\partial z}{\partial \beta} = -\frac{\partial \ln(z)}{\partial \beta} \tag{20.3-7}$$

where all variables on which the energy eigenvalues depend are held fixed in the differentiation.

In order to use Equation (20.3-7) to identify the parameter β, we must know how the thermodynamic energy U depends on temperature and any other variables for at least one specific system. Once we identify β for this system, we will assert that the same relation for β exists for any system of independent molecules, since nothing we have done so far depends on the nature of the system. There are other schemes for identifying β, and we present one of them in Section 20.5.

The Molecular Partition Function for a Dilute Monatomic Gas

For our specific system we choose a dilute monatomic gas whose atoms remain in their ground electronic state, with the system confined in a rectangular box of dimensions a by b by c. If we choose the potential energy to have value zero in the box, the energy eigenvalues are given by Equation (9.5-33), which is the same as Equation (14.1-5):

$$\varepsilon_{tr} = \varepsilon_{n_x n_y n_z} = \frac{h^2}{8m}\left(\frac{n_x^2}{a^2} + \frac{n_y^2}{b^2} + \frac{n_z^2}{c^2}\right) \tag{20.3-8}$$

where m is the molecular mass. These energy eigenvalues depend on a, b, and c. However, in a fluid system without surface tension, the thermodynamic energy depends only on the volume of the container, and not on its shape. We therefore assume that the box is cubical ($a = b = c$) so that the translational energy eigenvalues depend only on the volume of the box, and we specify that the partial derivative in Equation (20.3-7) is taken at constant volume.

We assume that the energy of our monatomic gas depends in the same way on temperature as that of the classical gas of point mass particles

The gas kinetic theory result is based on classical mechanics, and our present discussion is based on quantum mechanics. It is a fact, called the **correspondence principle**, *that quantum results become the same as classical results in the limit that energies and/or particle masses are large enough.*

treated in gas kinetic theory, which also had only translational energy:

$$U_{tr} = \frac{3}{2} N k_B T \qquad (20.3\text{-}9)$$

where k_B is Boltzmann's constant and T is the temperature.

It is easier in this case to sum over states than over levels, so we sum over all values of n_x, n_y, and n_z to obtain the partition function:

$$z_{tr} = \sum_{n_x=1}^{\infty} \sum_{n_y=1}^{\infty} \sum_{n_z=1}^{\infty} \exp\left[\frac{-\beta h^2(n_x^2/a^2 + n_y^2/b^2 + n_z^2/c^2)}{8m} \right] \qquad (20.3\text{-}10)$$

This partition function is labeled z_{tr} and is called the **translational partition function**.

A multiple sum can always be factored if the limits are constants and if the summand can be factored.

Each value of n_x can occur with every value of n_y, etc., so that the sum factors:

$$z_{tr} = z_x z_y z_z \qquad (20.3\text{-}11)$$

where

$$z_x = \sum_{n_x=1}^{\infty} \exp\left(\frac{-\beta h^2 n_x^2}{8ma^2} \right) \qquad (20.3\text{-}12)$$

and where z_y and z_z are given by analogous formulas.

Exercise 20.6

By writing all the terms on both sides of the equation, show that

$$\sum_{i=1}^{3} \sum_{j=1}^{3} a_i b_j = \left(\sum_{i=1}^{3} a_i \right) \left(\sum_{j=1}^{3} b_j \right)$$

where a_i and b_j represent arbitrary quantities.

The sum can also be depicted as the sum of the heights of the rectangles, but this is the same as the area under the rectangles because each rectangle has unit width.

Figure 20.1 shows a graphical representation of the sum of Equation (20.3-12), in which each term is represented by the area of a rectangle with height equal to the value of the term and width equal to unity. The rectangle for a given value of n_x is drawn to the left of that value on the horizontal axis, so the area begins at $n_x = 0$ while the sum begins at $n_x = 1$.

Also shown in the figure is a curve representing the function

$$f(n_x) = \exp\left(\frac{-\beta h^2 n_x^2}{8ma^2} \right) \qquad (20.3\text{-}13)$$

where n_x now takes on any real value, not just integral values. This function is equal to the terms in the sum of Equation (20.3-12) for integral values of n_x and interpolates between these values for nonintegral values of n_x. The area under the curve is approximately equal to the combined areas of the rectangles, so when we look up the integral,

$$z_x \approx \int_0^{\infty} \exp\left(\frac{-\beta h^2 n_x^2}{8ma^2} \right) dn_x = \left(\frac{2\pi m}{h^2 \beta} \right)^{1/2} a \qquad (20.3\text{-}14)$$

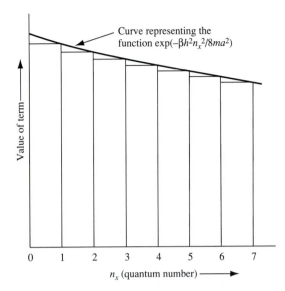

Figure 20.1. A Graphical Representation of the Translational Partition Function (schematic). The translational partition function is equal to the area under the bar graph, and the integral approximation to it is equal to the area under the curve. The error due to the small roughly triangular areas is negligible for ordinary temperatures.

with analogous results for z_y and z_z. The approximation of Equation (20.3-14) is accurate if the areas between the rectangles and the curve in Figure 20.1 are small.

The translational partition function is now

$$z_{tr} = \left(\frac{2\pi m}{h^2 \beta}\right)^{3/2} abc = \left(\frac{2\pi m}{h^2 \beta}\right)^{3/2} V \tag{20.3-15}$$

where $V = abc$, the volume of the box containing the gas. The translation of any molecule in a box is essentially the same as that of an atom, as discussed in Chapter 14, so that Equation (20.3-15) is valid for the translation of any gaseous substance.

We substitute Equation (20.3-15) into Equation (20.3-7):

$$\frac{U}{N} = -\frac{3}{2}\frac{d\ln(1/\beta)}{d\beta} = \frac{3}{2\beta} \tag{20.3-16}$$

If Equation (20.3-16) is to agree with Equation (20.3-9), we must let

$$\boxed{\beta = \frac{1}{k_B T}} \tag{20.3-17}$$

The translational partition function is now

$$\boxed{z_{tr} = \left(\frac{2\pi m k_B T}{h^2}\right)^{3/2} V \quad \text{(any gaseous substance)}} \tag{20.3-18}$$

▼ **EXAMPLE 20.2**

The solution to Example 20.2 is instructive in showing us the large number of translational states that are effectively accessible to a molecule in a dilute gas.

Find the value of z_{tr} for argon atoms confined in a box of volume 25 L at 298.15 K.

Solution

$$z_{tr} = \left(\frac{(2\pi)(6.6 \times 10^{-26} \text{ kg})(1.38 \times 10^{-23} \text{ J K}^{-1})(298.15 \text{ K})}{(6.6 \times 10^{-34} \text{ J s})^2} \right)^{3/2} (0.025 \text{ m}^3)$$

$$= 6.1 \times 10^{30}$$

▲

The large magnitude of the translational partition function in Example 20.2 is typical. Very many translational states are effectively accessible to an atom or molecule at room temperature.

Exercise 20.7

Find the value of z_{tr} for helium atoms confined in a box of volume 25 L at 298.15 K. Compare this value with that of Example 20.2 and comment on the difference.

We can now judge the accuracy of replacing the sum of Equation (20.3-13) by the integral of Equation (20.3-14).

▼ **EXAMPLE 20.3**

a. Assuming that n_x, n_y, and n_z are equal to each other and that our gas is confined in a cubical box 1.00 m on a side, find the value of n_x that will make the energy of a neon atom equal to $3k_BT/2$ at 300 K.

b. Find the change in energy if n_x is increased by unity from the value in part a, and show that this change is small compared with $3k_BT/2$ at 300 K.

Solution

a.

$$\frac{3}{2} k_B T = \frac{h^2}{8ma^2} (3n_x^2)$$

$$n_x^2 = \frac{4ma^2 k_B T}{h^2}$$

$$= \frac{4(0.02018 \text{ kg mol}^{-1})(1.00 \text{ m})^2(1.3807 \times 10^{-23} \text{ J K}^{-1})(300 \text{ K})}{(6.022 \times 10^{23} \text{ mol}^{-1})(6.6261 \times 10^{-34} \text{ J s})^2}$$

$$= 1.2 \times 10^{21}$$

$$n_x = 3.5 \times 10^{10}$$

We expect that the integral approximation to the translational partition function is accurate to about 9 significant digits.

b.

$$\frac{\partial \varepsilon}{\partial n_x} \Delta n_x = \frac{h^2}{8ma^2} 6n_x \Delta n_x = \frac{2\varepsilon}{n_x} \Delta n_x = \frac{2\varepsilon}{n_x}$$

The difference in energy is roughly equal to $\varepsilon/10^{10}$.

▲

Exercise 20.8

For $T = 300$ K, evaluate the term in Equation (20.3-12) corresponding to the value of n_x in Example 20.3. Show that the difference between this term and the term with n_x increased by unity is small compared with the term, and the approximation of Equation (20.3-14) is nearly exact.

In our original derivation of the probability distribution, there was no restriction to a particular kind of dilute gaseous system. Therefore, since $\beta = 1/(k_B T)$ for a monatomic gas with no electronic energy, we assert that $\beta = 1/(k_B T)$ for all dilute gases. With this identification, the probability distribution for molecular energy levels of any substance in the dilute gas phase is

$$p_j = f\left(\frac{N_j}{N}\right) = \frac{1}{z}\, g_j e^{-\varepsilon_j/k_B T} \quad \text{(level } j\text{)} \tag{20.3-19a}$$

and the probability distribution for molecular states is

$$p_i = \frac{N_i}{N} = \frac{1}{z}\, e^{-\varepsilon_i/k_B T} \quad \text{(state } i\text{)} \tag{20.3-19b}$$

The **molecular partition function** is given by either of the equations:

$$z = \sum_j g_j e^{-\varepsilon_j/k_B T} \quad \text{(sum over levels)} \tag{20.3-20a}$$

$$z = \sum_i e^{-\varepsilon_i/k_B T} \quad \text{(sum over states)} \tag{20.3-20b}$$

The probability distribution of Equations (20.3-19a) and (20.3-19b) is the probability distribution that corresponds to a given system macrostate. Although we specified the system macrostate by giving the values of U, V, and N, the value of the temperature is determined macroscopically if the macrostate is given for a particular system. The probability distribution could be restated in terms of U.

Exercise 20.9   Using the relationship of Equation (20.3-9) for a monatomic gas, write the expression for p_i and for z in terms of U instead of T.

Our probability distribution for quantum molecular states is identical to the classical Boltzmann probability distribution of Equations (16.2-31) and (1.7-25). It has two important qualitative properties. First, at a finite temperature the equilibrium probability decreases exponentially as higher values of the energy are considered. Only states with energy no larger than $2k_B T$ or $3k_B T$ will be significantly populated. In the limit of zero temperature, only the lowest-energy level is occupied. Second, if the temperature is increased, the probabilities of states of high energy increase. In the limit of infinite temperature, all states are equally probable.

Exercise 20.10   **a.** Calculate the ratio of the population of the $v = 1$ vibrational state of HF to the $v = 0$ vibrational state at (a) 298.15 K, (b) 500 K, (c) 1000 K, and (d) 5000 K. What is the limit as $T \to \infty$? Data are found in Table A16.

b. Calculate the ratio of the population of the $J = 1$ rotational level of HF to the $J = 0$ rotational level at (a) 298.15 K and (b) 1000 K. What is the limit as $T \to \infty$? Don't forget the degeneracies of the rotational levels.

c. Find the J value that has the largest population for HF at 298.15 K.

Nonequilibrium population distributions can occur in which a state of higher energy has a higher population than a state of lower energy. This situation occurs in lasers. Such a distribution is said to correspond to a "negative temperature." If so defined, negative temperatures are "hotter" than positive temperatures. However, temperature is an equilibrium concept and must be nonnegative. Negative temperatures are parameters used to describe nonequilibrium distributions and are not thermodynamic temperatures.

Exercise 20.11 In a certain laser, the population of a state of energy 2.08 eV is 50% larger than the population of a state of energy zero. Find the negative temperature corresponding to this situation.

We can rewrite Equation (20.3-7) in terms of T:

$$U = -N\left(\frac{\partial \ln(z)}{\partial T}\right)_V \left(\frac{dT}{d\beta}\right) = Nk_B T^2 \left(\frac{\partial \ln(z)}{\partial T}\right)_V \quad \begin{pmatrix} \text{any} \\ \text{dilute} \\ \text{gas} \end{pmatrix} \qquad \textbf{(20.3-21)}$$

Exercise 20.12 Verify Equation (20.3-21).

Let us investigate the effect of changing the zero of energy, which means adding a constant to the value of the potential energy. Let ε_j'' represent the new value of the energy eigenvalue of level number j:

$$\varepsilon_j'' = \varepsilon_j + \mathcal{V}'' \qquad \textbf{(20.3-22)}$$

where $\mathcal{V}''$ is a constant.

Denote the new value of the partition function by z'':

$$z'' = \sum_j g_j e^{-\varepsilon_j''/k_B T} = \sum_j g_j e^{-(\varepsilon_j + \mathcal{V}'')/k_B T}$$
$$= e^{-\mathcal{V}''/k_B T} \sum_j g_j e^{-\varepsilon_j''/k_B T} = z e^{-\mathcal{V}''/k_B T} \qquad \textbf{(20.3-23)}$$

The partition function depends on the choice of the zero of energy because it is a sum of Boltzmann factors relative to zero energy. Its interpretation as the effective number of available molecular states is relative to a Boltzmann factor of unity for a state of zero energy.

The probability distribution is

$$p'' = \frac{1}{z''} g_j e^{-(\varepsilon_j + \mathcal{V}'')/k_B T} = \frac{1}{z e^{-\mathcal{V}''/k_B T}} g_j e^{-\varepsilon_j/k_B T} e^{-\mathcal{V}''/k_B T}$$
$$= \frac{1}{z} g_j e^{-\varepsilon_j/k_B T} \qquad \textbf{(20.3-24)}$$

The probability distribution is independent of the choice of the zero of energy, as we would expect, but the value of the partition function changes by the factor $e^{-\mathscr{V}''/k_BT}$.

Exercise 20.13

Using Equation (20.3-21), show that the thermodynamic energy using the new zero of energy is given by

$$U'' = U + N\mathscr{V}'' \tag{20.3-25}$$

In later sections, we will see that a change in the zero of energy has the same effect on all other energy quantities (the enthalpy, Gibbs energy, and Helmholtz energy) as it does on the thermodynamic energy, while the entropy and the heat capacity are independent of the choice of the zero of energy.

20.4 The Calculation of Molecular Partition Functions

In this section, we obtain formulas for the molecular partition function for all types of dilute gases and carry out example calculations of partition functions.

Monatomic Gases

To an excellent approximation, the translational energy eigenvalues of any atom are the same as those of point mass molecules, and the electronic and translational energy are separated, as in Equation (14.1-2):

$$\varepsilon_j = \varepsilon_{jtr} + \varepsilon_{jelec} \tag{20.4-1}$$

where ε_{jtr} is the same as in Equation (20.3-8) and where we include any constant potential energy in the electronic energy.

Since any electronic state can occur with any translational state, the degeneracy of energy level number j is

$$g_j = g_{jtr}g_{jelec} \tag{20.4-2}$$

The molecular partition function is a double sum:

$$z = \sum_{jtr} \sum_{jelec} g_{jtr}g_{jelec}e^{-(\varepsilon_{jtr} + \varepsilon_{jelec})/k_BT} \tag{20.4-3}$$

We can factor this double sum just as we factored the triple sum in Equation (20.3-10):

$$z = \sum_{jtr} g_{jtr}e^{-\varepsilon_{jtr}/k_BT} \sum_{jelec} g_{jelec}e^{-\varepsilon_{jelec}/k_BT} = z_{tr}z_{elec} \tag{20.4-4}$$

The factor z_{tr} is called the **translational factor** of the partition function, or the **translational partition function**. Since the translational energy levels are the same as in Equation (20.3-8), the translational partition function is the same as that of Equation (20.3-18). The factor z_{elec} is called the **electronic factor of the partition function**, or the **electronic partition function**:

$$z_{\text{elec}} = g_0 e^{-\varepsilon_0/k_B T} + g_1 e^{-\varepsilon_1/k_B T} + \cdots \qquad \textbf{(20.4-5)}$$

Except for hydrogen-like atoms, there is no simple formula for atomic electronic energy eigenvalues. We cannot obtain a general formula for the electronic partition function.

For a typical monatomic gas such as helium or argon, the ground electronic energy level is a nondegenerate ^1S state, and the first excited state has an energy at least 1 eV higher than that of the ground level. Most molecular substances also have first excited states at least 1 eV higher than the ground state. In this case, all terms past the first term are negligible at ordinary temperatures:

$$\boxed{z_{\text{elec}} \approx g_0 e^{-\varepsilon_0/k_B T} \quad \text{(most substances)}} \qquad \textbf{(20.4-6)}$$

If we choose the zero of energy at the ground-state energy and if the ground level is nondegenerate,

$$z_{\text{elec}} \approx 1 \qquad (\varepsilon_0 = 0 \text{ chosen}, g_0 = 1) \qquad \textbf{(20.4-7)}$$

If Equation (20.4-7) can be used, the partition function is equal to the translational factor, and we have the case of Section 20.3.

▼   **EXAMPLE 20.4**   For the case that $g_0 = 1$ and $\varepsilon_1 = 1$ eV, show that $z_{\text{elec}} \approx 1$ is an adequate approximation at $T = 300$ K.

Solution

$$\frac{\varepsilon_1}{k_B T} = \frac{(1.00 \text{ eV})(1.602 \times 10^{-19} \text{ J eV}^{-1})}{(1.3807 \times 10^{-23} \text{ J K}^{-1})(300 \text{ K})} = 38.7$$

$$e^{-38.7} = 1.6 \times 10^{-17}$$

which is negligible compared with unity.

Diatomic Gases

In addition to translational and electronic motion, a diatomic gas has rotational and vibrational motion. To the approximation of Equation (20.1-5),

$$\varepsilon_j = \varepsilon_{j\text{tr}} + \varepsilon_{j\text{rot}} + \varepsilon_{j\text{vib}} + \varepsilon_{j\text{elec}} \qquad \textbf{(20.4-8)}$$

Equation (20.4-8) is a fairly good approximation, but such things as the α cross-term in Equation (14.2-37) have been ignored.

where $j\text{tr}$ is an abbreviation for the translational quantum numbers, $j\text{rot}$ is an abbreviation for the rotational quantum numbers, etc.

Equation (20.4-8) requires some comment. The electronic and vibrational energies are separated through use of the Born-Oppenheimer approximation. The Born-Oppenheimer energy depends on relative nuclear positions and acts as a potential energy for the vibration of the molecule, as depicted schematically in Figure 20.2, which is equivalent to one of the curves in Figure 14.1. Figure 20.2 also shows several of the vibrational energy levels.

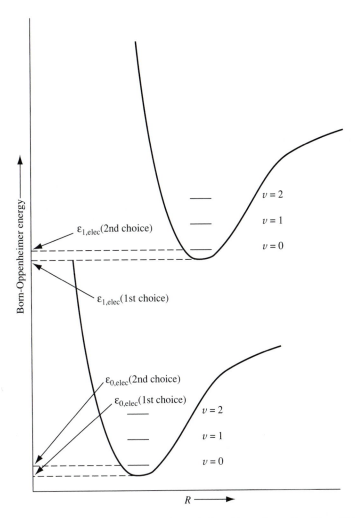

Figure 20.2. The Electronic Energy (Born-Oppenheimer Energy) as a Function of Internuclear Distance for Two Electronic States. The vibrational energy levels are superimposed for both states. The two possible choices for the zero of electronic energy are shown. The second choice, with the electronic energy set equal to the constant energy of the $v = 0$ vibrational state, is more convenient.

There are two common choices for dividing the Born-Oppenheimer energy into electronic and vibrational energies. The first is to assign the minimum value of the Born-Oppenheimer energy as the electronic energy and assign the remainder to be the vibrational potential energy, as was done in Chapter 14. The second choice is to assign the zero-point vibrational energy as part of the electronic energy, making the vibrational ground-state energy equal to zero. Both these choices are depicted in Figure 20.2. With either choice, the electronic energy is taken as a constant.

With the second choice, the energy is, in the harmonic oscillator–rigid rotor approximation,

$$\varepsilon_j = \frac{h^2}{8ma^2}\left(n_x^2 + n_y^2 + n_z^2\right) + hB_e J(J + 1) + h\nu_e v + \varepsilon_{j\text{elec}} \tag{20.4-9}$$

where m is the molecular mass and v_e is the vibrational frequency, given by Equation (14.2-24)

$$v_e = \frac{1}{2\pi}\sqrt{\frac{k}{\mu}}$$

where k is the force constant and μ is the reduced mass of the molecule. The rotational constant, B_e, is given by Equation (14.2-35)

$$B_e = \frac{h}{8\pi^2\mu R_e^2} = \frac{h}{8\pi^2 I_e}$$

where I_e is the equilibrium moment of inertia and R_e is the equilibrium internuclear distance (the distance at which no force is exerted in the R direction).

The subscript j in Equation (20.4-9) is an abbreviation for the entire set of quantum numbers needed to specify the energy level. The translational, rotational, and vibrational quantum numbers are denoted by their usual symbols, and the subscript $j$elec stands for the electronic quantum numbers.

If the first excited electronic level is high enough in energy that only a single electronic state needs to be included, z_{elec} is given by Equation (20.4-6), and Equation (20.4-9) leads to a factoring of the molecular partition function similar to that of Equation (20.4-4):

$$z = z_{\text{tr}} z_{\text{rot}} z_{\text{vib}} z_{\text{elec}} \quad \left(\begin{array}{l}\text{diatomic or polyatomic}\\ \text{substances}\end{array}\right) \qquad \textbf{(20.4-10)}$$

Exercise 20.14   Verify the factoring in Equation (20.4-10).

One can imagine that a rotating diatomic molecule will bounce crookedly off a wall. However, when one averages over a lot of molecules, the effect of the rotation should be in various directions and should average to zero.

The rotational and vibrational motions can affect the details of individual collisions of diatomic molecules with the walls of the container, as discussed in Chapter 14. However, due to the small size of the molecules compared with size of the box containing the gas, use of the translational energy eigenvalues of point mass molecules is an excellent approximation, and Equation (20.3-18) gives an essentially exact result for the translational partition function.

The rotational partition function is given by

$$z_{\text{rot}} = \sum_{J=0}^{\infty} (2J+1)e^{-hB_e J(J+1)/k_B T} \qquad \textbf{(20.4-11)}$$

since the rotational degeneracy is $2J+1$ (see Section 14.2).

Figure 20.3 shows the sum of Equation (20.4-11) represented by the combined areas of a set of rectangles, as was done with the translational partition function in Figure 20.1. Since $J = 0$ is the lower limit of the sum, we draw the rectangle for a given value of J to the right of that value of J. Figure 20.3 also shows a curve representing the function that is obtained by allowing J to take on all real values. The area under the curve is an inte-

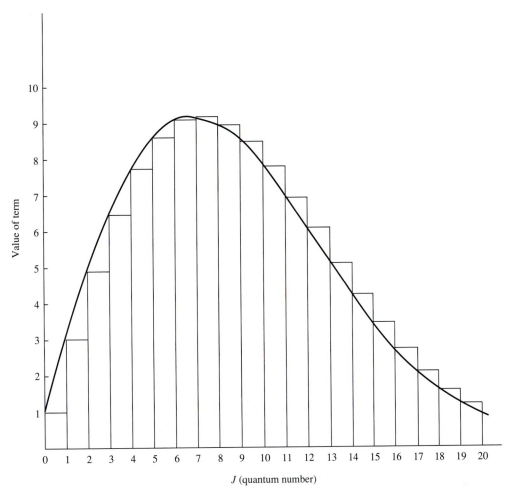

Figure 20.3. A Graphical Representation of the Rotational Partition Function (Drawn for CO at 298.15 K). This figure is analogous to Figure 20.1 for the translational partition function. The area under the bar graph is equal to the rotational partition function for a diatomic molecule, and the area under the curve is equal to the integral approximation to the rotational partition function.

gral that is approximately equal to the sum:

$$z_{\text{rot}} \approx \int_0^\infty (2J + 1)e^{-hB_eJ(J+1)/k_{\text{B}}T} \, dJ$$

To work out the integral, let $u = J(J + 1)$, so that $du = (2J + 1) \, dJ$:

$$z_{\text{rot}} \approx \int_0^\infty e^{-hB_e u/k_{\text{B}}T} \, du = \frac{k_{\text{B}}T}{hB_e} = \frac{8\pi^2 I_e k_{\text{B}}T}{h^2} = \frac{8\pi^2 \mu R_e^2 k_{\text{B}}T}{h^2} \quad \textbf{(20.4-12)}$$

The formula of Equation (20.4-12) is acceptable for heteronuclear diatomic molecules but must be modified for homonuclear diatomic molecules, for which either odd values of J or even values of J are not allowed, as discussed in Section 14.3. This restriction corresponds to omitting half of the rectangles in Figure 20.3, approximately cutting the value of the sum in half.

The symmetry number σ is defined as in Section 14.3. For diatomic molecules,

$$\sigma = \begin{cases} 2 & \text{homonuclear diatomic molecules} \\ 1 & \text{heteronuclear diatomic molecules} \end{cases} \qquad \textbf{(20.4-13)}$$

so that for any diatomic molecule, to some approximation,

$$z_{\text{rot}} = \frac{1}{\sigma}\frac{8\pi^2 I_e k_B T}{h^2} = \frac{1}{\sigma}\frac{k_B T}{h B_e} \quad \text{(diatomic substances)} \qquad \textbf{(20.4-14)}$$

In the translational partition function, the replacement of a sum by an integral was essentially exact. In the rotational partition function it is a good approximation for molecules containing heavy atoms, a fair approximation at ordinary temperatures for molecules containing atoms of moderate mass, and a fairly poor approximation for molecules containing light atoms such as hydrogen. Corrections can be made.[3]

▼ **EXAMPLE 20.5**

Calculate the rotational partition function for $^{35}Cl_2$ at 298.15 K.

Solution

$$\mu = \frac{0.03497 \text{ kg mol}^{-1}}{2(6.0221 \times 10^{23} \text{ mol}^{-1})} = 2.903 \times 10^{-26} \text{ kg}$$

From Table A16, $R_e = 1.988 \times 10^{-10}$ m.

$$I_e = (2.903 \times 10^{-26} \text{ kg})(1.988 \times 10^{-10} \text{ m})^2 = 1.147 \times 10^{-45} \text{ kg m}^2$$

$$z_{\text{rot}} = \frac{1}{2}\frac{8\pi^2(1.147 \times 10^{-45} \text{ kg m}^2)(1.3807 \times 10^{-23} \text{ J K}^{-1})(298.15 \text{ K})}{(6.6261 \times 10^{-34} \text{ J s})^2}$$

$$= 424.6$$

▲

Larger values of z_{rot} indicate that the replacement of the sum by an integral in Equation (20.4-12) is a better approximation. The value in Example 20.5 is typical for molecules of moderate size near room temperature and is large enough to be a fairly good approximation.

Molecules of small mass (especially hydrogen and molecules with one hydrogen atom) have rotational partition functions that are smaller than that of Example 20.5 and are less well approximated by Equation (20.4-14). In either event, far fewer rotational states than translational states are accessible near room temperature.

Exercise 20.15

a. Calculate the rotational partition function for H_2 at 298.15 K, using Equation (20.4-14).

b. Obtain an adequate approximation for the rotational partition function of H_2 by adding up terms explicitly, assuming that the hydrogen is all para hydrogen (only even values of J occur). Use a programmable calculator or computer if possible.

---

[3] N. Davidson, *op cit.*, pp. 111ff; R. S. McDowell, *J. Chem. Phys.* **88**, 356 (1988). See also Exercise 20.12.

Continue the sum until one additional term in the sum over levels does not change the first four digits of the sum. Calculate the percent error in the result of Example 20.5 and the difference of its natural logarithm from that of the correct value.

  c. Repeat part b for $^{35}Cl_2$.

The vibrational partition function of a diatomic molecule is

$$z_{vib} = \sum_{v=0}^{\infty} e^{-hvv/k_B T} = \sum_{v=0}^{\infty} a^v \qquad \text{(20.4-15)}$$

This sum is a **geometric progression** with infinitely many terms, given by a well-known formula

$$\sum_{v=0}^{\infty} a^v = \frac{1}{1-a} \qquad \text{(20.4-16)}$$

where we must require that $|a| < 1$, which is satisfied by $a = \exp(-hv/k_B T)$. Therefore,

$$z_{vib} = \frac{1}{1 - e^{-hv/k_B T}} \quad \text{(diatomic substances)} \qquad \text{(20.4-17)}$$

EXAMPLE 20.6

Calculate the vibrational partition function of $^{35}Cl_2$ at 298.15 K. The vibrational frequency is 1.6780×10^{13} s^{-1}.

Solution

Let

$$x = \frac{hv}{k_B T} = \frac{(6.6261 \times 10^{-34} \text{ J s})(1.6780 \times 10^{13} \text{ s}^{-1})}{(1.3807 \times 10^{-23} \text{ J K}^{-1})(298.15 \text{ K})} = 2.701$$

$$z_{vib} = \frac{1}{1 - e^{-x}} = \frac{1}{1 - e^{-2.701}} = 1.0720$$

Only the ground vibrational state is significantly accessible at room temperature, as shown by the closeness of the value of z_{vib} to unity. For a typical diatomic molecule at room temperature, very many translational states are effectively accessible, about a hundred rotational states are effectively accessible, but only one vibrational state and one electronic state are accessible.

Exercise 20.16

Calculate the vibrational partition functions for H_2 and for I_2 at 298.15 K. Explain your results.

Exercise 20.17

Show that if the vibrational energy is written as

$$\varepsilon_{vib}'' = hv\left(v + \frac{1}{2}\right)$$

the vibrational partition function is

$$q''_{vib} = \frac{e^{-h\nu/2k_BT}}{1 - e^{-h\nu/k_BT}} \qquad \text{(20.4-18)}$$

Calculate the value of this vibrational partition function for $^{35}Cl_2$, using information in Example 20.6.

The factoring of Equation (20.4-10) depends on the assumption that only the ground electronic level is needed in the electronic partition function. If additional electronic terms must be included, the same rotational and vibrational partition functions do not apply to different electronic states, because not all states have the same force constant and internuclear distance.

If different values of B_e and ν_e must be used for different electronic states, we must compute a different rotational and vibrational partition function for each electronic state. The partition function becomes

$$z = z_{tr}(g_0 e^{-\varepsilon_0/k_BT} z_{rot,\,0} z_{vib,\,0} + g_1 e^{-\varepsilon_1/k_BT} z_{rot,\,1} z_{vib,\,1} + \cdots) \quad \text{(20.4-19)}$$

where $z_{rot,\,0}$ and $z_{vib,\,0}$ are the rotational and vibrational partition functions for electronic state 0, $z_{rot,\,1}$ and $z_{vib,\,1}$ are the corresponding quantities for electronic state 1, etc.

If as an approximation we can assume that all electronic states have the same values of B_e and ν_e, we can write the partition function in factored form as in Equation (20.4-10) even if more than one term is included in z_{elec}. The values of B_e and ν_e of the ground electronic level are used, since this term makes the largest contribution.

Exercise 20.18

The ground electronic level of NO is a $^2\Pi_{1/2}$ term. The first excited level is a $^2\Pi_{3/2}$ term with an energy 2.380×10^{-21} J above the ground level. All other electronic states are more than 7×10^{-19} J above the ground level. The value of ν_e, the vibrational frequency, is 5.7086×10^{13} s^{-1}, and the value of the rotational constant, B_e, is 5.0123×10^{10} s^{-1} for the ground level.

a. Find the molecular partition function of NO at 298.15 K and at a volume equal to the molar volume at 101325 Pa. Assume that the ground-level values of B_e and ν_e apply to both electronic states.

b. Find the molecular partition function for NO at the conditions of part a, using the actual values for the first excited state, $\nu_e = 5.7081 \times 10^{13}$ s^{-1}, $B_e = 5.1569 \times 10^{10}$ s^{-1}.

Polyatomic Gases

Polyatomic molecules obey Equation (20.4-8) to roughly the same approximation as do diatomic molecules. Therefore, the molecular partition function of a polyatomic substance factors into the same four factors as in Equation (20.4-10).

As with diatomic molecules, we assume that the translational energy levels of a polyatomic gas are like those of point mass molecules. Therefore, the translational partition function is given by Equation (20.3-18). The electronic partition function is given by the same formula as in Equation (20.4-5), and the approximation of Equation (20.4-7) is often valid if the molecule does not have unpaired electrons.

The Rotational Partition Function

For a linear polyatomic molecule like acetylene, the rotational energy levels are given by the same formula as those of a diatomic molecule in terms of the moment of inertia, and Equation (20.4-14) can be used for the rotational partition function, with the appropriate definition of the symmetry number and the moment of inertia.

As mentioned in Section 14.3, the rotational energy levels of nonlinear polyatomic molecules are more complicated than those of diatomic molecules. The derivation of the rotational partition function for nonlinear molecules is complicated, and we merely cite the result, which is valid for all types of nonlinear polyatomic molecules:

$$
z_{rot} = \frac{1}{\sigma} \sqrt{\pi} \left(\frac{8\pi^2 k_B T}{h^2} \right)^{3/2} (I_A I_B I_C)^{1/2} \quad \begin{pmatrix} \text{nonlinear} \\ \text{polyatomic} \\ \text{substances} \end{pmatrix} \qquad \textbf{(20.4-20)}
$$

where I_a, I_b, and I_c are the principal moments of inertia, defined in Section 14.4, and σ is the symmetry number, also defined in Section 14.4. The symmetry number is equal to the number of ways a model of the molecule can be oriented with each nuclear location occupied by the same chemical species as in the other orientations.

Equation (20.4-20) is accurate to the same degree of approximation as Equation (20.4-14), equivalent to having replaced a sum by an integral. However, z_{rot} for polyatomic molecules is generally large enough that Equation (20.4-20) is a fairly good approximation. An accurate expression for spherical top molecules has been derived.[4]

▼ **EXAMPLE 20.7**

Calculate the rotational partition function of CCl_4 at 298.15 K, assuming all chlorine atoms to be ^{35}Cl. The molecule is a spherical top, with bond lengths equal to 1.766×10^{-10} m and with all moments of inertia equal to 4.829×10^{-45} kg m^2.

Solution

$$
z_{rot} = \frac{\sqrt{\pi}}{12} \left(\frac{8\pi^2 (1.3807 \times 10^{-23} \text{ J K}^{-1})(298.15 \text{ K})}{(6.6261 \times 10^{-34} \text{ J s})^2} \right)^{3/2}
$$
$$
\times (4.829 \times 10^{-45} \text{ kg m}^2)^{3/2}
$$
$$
= 3.157 \times 10^4
$$

▲

The value of the rotational partition function for a typical polyatomic molecule is considerably larger than that for a diatomic molecule, since the

4 R. S. McDowell, *J. Quant. Spectrosc. Radiat. Transfer.* **38**, 337 (1987).

additional degree of freedom (additional quantum number or additional classical rotational coordinate) increases the number of accessible states. Moments of inertia are also generally larger.

Exercise 20.19

Hint: All three of the principal moments of inertia are equal. It is necessary to calculate only one of them.

The C—H bond distances in methane are equal to 1.091×10^{-10} m. Calculate the rotational partition function for methane at 298.15 K. Comment on the comparison between your value and that for carbon tetrachloride in Example 20.7.

The Vibrational Partition Function

In the harmonic oscillator approximation, the vibrational energy of a polyatomic molecule is the sum of the energies of several normal modes, each of which is like a harmonic oscillator, as in Equation (14.3-19). A nonlinear polyatomic molecule has $3n - 6$ such normal modes, and a linear polyatomic molecule has $3n - 5$ such normal modes, where n is the number of nuclei in the molecule. Using the choice that the energy of the ground vibrational state is equal to zero,

$$\varepsilon_{\text{vib}} = \sum_{i=1}^{3n-5(6)} h\nu_i v_i \tag{20.4-21}$$

where ν_i represents the classical vibration frequency of normal mode number i, and v_i is the quantum number for this normal mode.

The vibrational partition function factors, with one factor for each normal mode:

We have skipped several steps in obtaining Equation (20.4-22). The summand factors because it is an exponential of a sum of terms. The multiple sum factors because the limits of the sums are independent of each other.

$$z_{\text{vib}} = \prod_{i=1}^{3n-5(6)} \frac{1}{1 - e^{-h\nu_i/k_B T}} \quad \text{(polyatomic substances)} \tag{20.4-22}$$

▼ **EXAMPLE 20.8**

The vibrational frequencies of carbon dioxide are 4.162×10^{13} s^{-1}, 2.000×10^{13} s^{-1} (two bends at this frequency), and 7.043×10^{13} s^{-1}. Find the vibrational partition function at 298.15 K.

Solution

Let

$$x_1 = \frac{h\nu_1}{k_B T} = \frac{(6.6261 \times 10^{-34} \text{ J s})(4.162 \times 10^{13} \text{ s}^{-1})}{(1.3807 \times 10^{-23} \text{ J K}^{-1})(298.15 \text{ K})}$$

$$= 6.699$$

Similarly, $x_2 = 3.219$ and $x_3 = 11.337$. The vibrational partition function is

$$z_{\text{vib}} = \frac{1}{1 - e^{-6.699}} \left(\frac{1}{1 - e^{-3.219}} \right)^2 \frac{1}{1 - e^{-11.34}}$$

$$= 1.086$$

The lowest-frequency vibrations make the largest contributions to z_{vib}.

◣

Exercise 20.20	Calculate the vibrational partition function of sulfur dioxide at 298.15 K. The vibrational frequencies are 1.556×10^{13} s^{-1}, 3.451×10^{13} s^{-1}, and 4.080×10^{13} s^{-1}.

The four factors of the molecular partition function that we have discussed are not the complete partition function for molecules of a dilute gas. First, internal nuclear energy has been neglected. However, excitation of a typical nucleus to its first excited state requires millions or even billions of electron volts. Excited nuclear states are unpopulated at room temperature, and the factor for the nuclear excitation is equal to the degeneracy of the nuclear ground state.

In addition, there are sometimes degenerate nuclear spin states. For example, if a molecule of ortho hydrogen (see Section 14.2) is not in an external magnetic field, the three values of the projection of the total nuclear angular momentum have the same energy. There is only one value of the spin projection for para hydrogen. Ortho hydrogen should have a factor of 3 included for its nuclear spin degeneracy, while para hydrogen (which has the same ground-state energy) requires a factor equal to unity. In the presence of a catalyst that can dissociate the hydrogen molecules, ortho and para hydrogen can interconvert rapidly, and a factor of 4 should be included for the nuclear spin degeneracy.

Furthermore, most elements occur in the earth's crust as isotopic mixtures. There is a degeneracy due to the near equality of the energy of molecules with different isotopes present, corresponding to an entropy of isotopic mixing.

However, all of these factors cancel between products and reactants in an ordinary chemical process, as well as in the process of heating a sample of material from 0 K to an ordinary temperature. It is customary to omit them and to calculate partition functions and thermodynamic variables just as though these factors did not exist.

For a better approximation than the integral approximation with a symmetry number, a direct summation can be carried out, with the ortho (odd J) terms multiplied by a factor of 3.

This policy of ignoring nuclear degeneracy is just an arbitrary convention and should be regarded as such.

20.5	## Calculations of Thermodynamic Functions of Dilute Gases

In Section 20.4, we obtained formulas for the calculation of molecular partition functions for any dilute gas, and also a formula for the thermodynamic energy of a dilute gas in terms of the molecular partition function. In this section, we obtain formulas for the other thermodynamic variables in terms of the molecular partition function. We also obtain statistical mechanical interpretations of heat and work and present an alternative derivation for the relationship $\beta = 1/(k_B T)$.

The Entropy

The entropy is not a mechanical quantity, like the energy or the pressure, so we cannot calculate an average value using a probability distribution.

To proceed, we adopt the definition of the statistical entropy in Equation (3.4-3):

$$S_{st} = k_B \ln(\Omega) + \text{constant} \quad \text{(definition)}$$

(20.5-1)

where Ω is the number of system microstates in which the system might be found, so far as we know. It is the same as the quantity in Equation (20.2-5) if the macrostate of the system is specified through known values of U, V, and N. Since all of the system microstates of equal energy are assumed equally probable, Ω is sometimes called the **thermodynamic probability**. We will set the additive constant in Equation (20.5-1) equal to zero, which corresponds to setting the statistical entropy equal to zero for a system known to be in a single microstate.

▼   **EXAMPLE 20.9**

Estimate the value of Ω for 1.000 mol of helium at 298.15 K and 1.000 atm, using the third-law entropy, 126 J K^{-1}.

Solution

$$\Omega = e^{S/k_B} = \exp\left(\frac{126 \text{ J K}^{-1}}{1.38 \times 10^{-23} \text{ J K}^{-1}}\right)$$
$$= e^{9.13 \times 10^{24}} = 10^{3.96 \times 10^{24}}$$

Ten raised to an exponent roughly as large as Avogadro's number is a *very* large number.

▲

The total number of possible system microstates is equal to the sum of the number of microstates corresponding to each possible distribution:

$$\Omega = \sum_{\{N\}} W(\{N\})$$

(20.5-2)

where the summation index $\{N\}$ means that there is one term in the sum for each possible distribution.

It is really surprising to most of us that such an approximation can be used. It works because the logarithm of a very large number is a very slowly varying function of its argument.

We now make what seems like a drastic approximation: we replace the entire sum in Equation (20.5-2) by its largest term (the term corresponding to the most probable distribution). This replacement is a very bad approximation for the sum, but a very good approximation for the logarithm of the sum.

We illustrate this strange situation as follows: Assume that we want to have 12 significant digits correct in the logarithm of Ω and that Ω has roughly the value in Example 20.9. This means that we can have an error in $\ln(\Omega)$ that is no larger in magnitude than 10^{13}. Call the magnitude of this error $\ln(x)$, so that

$$\ln(x) \leq 10^{13} \quad \text{or} \quad x \leq e^{10^{13}}$$

This equation is the same as

$$\Omega(\text{exact}) = x\Omega(\text{approximate})$$

(20.5-3)

An error in Ω that makes it incorrect by a factor of $e^{10^{13}}$ is permissible. The exponent in this expression is larger than the federal debt of the United States in dollars, and e raised to that power is beyond imagination. However, this tremendous error in Ω still produces a value of $\ln(\Omega)$ that is correct to 12 significant digits. Such is the strange behavior of very large numbers.

Although we do not prove this fact, the largest term in the sum of Equation (20.5-2) is smaller than the entire sum by roughly the same factor as that in Equation (20.5-3), giving about 12 significant digits in approximating $\ln(\Omega)$ by $\ln(W_{max})$.

We have already studied the probability distribution corresponding to W_{max}. From Equations (20.5-1) and (20.2-10)

$$S = k_B \ln(W_{max}) = k_B \ln\left(\prod_j \frac{g_j^{N_j}}{N_j!}\right)$$

$$= k_B \sum_j \left[N_j \ln\left(\frac{g_j}{N_j}\right) + N_j \right] \qquad \textbf{(20.5-4)}$$

where we have used the fact that the logarithm of a product is the sum of the logarithms of the factors and have used Stirling's approximation, Equation (20.2-16).

We now replace N_j within the parentheses in Equation (20.5-4) by its expression from Equation (20.3-19a):

$$S = k_B \sum_j \left[N_j \ln\left(\frac{z}{N} e^{\varepsilon_j/k_B T}\right) + N_j \right]$$

$$= k_B \sum_j \left(N_j \ln\left(\frac{z}{N}\right) + \frac{N_j \varepsilon_j}{k_B T} + N_j \right) = k_B \left(N \ln\left(\frac{z}{N}\right) + \frac{E}{k_B T} + N \right)$$

$$\boxed{S = \frac{U}{T} + Nk_B \ln\left(\frac{z}{N}\right) + Nk_B} \qquad \textbf{(20.5-5a)}$$

where we have identified the mechanical energy E with the thermodynamic energy U.

Equation (20.5-5a) is a working equation that gives the entropy of a dilute gas in terms of the molecular partition function. As will be the case in all formulas for thermodynamic variables, it is the logarithm of the partition function that occurs in the formula.

▼ **EXAMPLE 20.10**

Calculate the molar entropy of helium gas at 1.000 bar pressure and 298.15 K.

Solution

We assume that there is no significant electronic excitation, so that $z = z_{tr}$. Using the fact that the molar energy $\bar{U} = 3RT/2$,

$$\bar{S} = \frac{5R}{2} + R \ln\left(\frac{z}{N_A}\right)$$

Using Equation (20.3-18) and the ideal gas equation of state,

$$\frac{z}{N_A} = \left(\frac{2\pi m k_B T}{h^3}\right)^{3/2} \frac{V}{N_A} = \left(\frac{2\pi m k_B T}{h^3}\right)^{3/2} \frac{k_B T}{P} = \frac{(2\pi m)^{3/2}(k_B T)^{5/2}}{h^3 P}$$

$$= \frac{(2\pi/N_A)^{3/2} k_B^{5/2} M^{3/2} T^{5/2}}{h^3 P^\circ}\left(\frac{P^\circ}{P}\right)$$

where M is the molar mass of the gas. This equation is valid for any monatomic gas without electronic excitation. Substituting in the values of the constants and using $P^\circ = 1.000$ bar, we obtain

$$\frac{z}{N_A} = (820.61 \ \text{kg}^{-3/2} \ \text{mol}^{3/2} \ \text{K}^{-5/2}) M^{3/2} T^{5/2}\left(\frac{P^\circ}{P}\right)$$

and

$$\boxed{\bar{S} = R\left[9.2100 + \frac{3}{2}\ln\left(\frac{M}{1 \ \text{kg mol}^{-1}}\right) + \frac{5}{2}\ln\left(\frac{T}{1 \ \text{K}}\right) - \ln\left(\frac{P}{P^\circ}\right)\right]} \qquad \textbf{(20.5-5b)}$$

where the divisors $1 \ \text{kg mol}^{-1}$ and $1 \ \text{K}$ are inserted to make the arguments of the logarithms dimensionless. Equation (20.5-5b) is known as the **Sackur-Tetrode equation**.

The molar mass of helium is $M = 0.0040026 \ \text{kg mol}^{-1}$. Substitution of this value, $T = 298.15$ K, and $P = 1.000$ bar gives $\bar{S}^\circ = 126.15 \ \text{J K}^{-1} \ \text{mol}^{-1}$ for helium at 298.15 K. This value of the entropy corresponds to the choice that the energy of the electronic ground state of the atoms is set equal to zero and that the arbitrary constant in Equation (20.5-1) is equal to zero. This value agrees well with the third-law entropy value in Table A9.

▲

Exercise 20.21   Using Equation (20.3-23), show that the value of the entropy of a dilute gas is unchanged if the zero of energy is changed.

The Pressure

Since the molecules of a dilute gas act independently, the pressure is equal to the sum of the pressures exerted by the individual molecules. If the system is in microstate number i, its pressure is

$$P_i = P_{i,1} + P_{i,2} + P_{i,3} + \cdots + P_{i,N} \qquad \textbf{(20.5-6)}$$

where $P_{i,1}$ is the pressure exerted by the first molecule when the system is in microstate i, etc.

The sum in Equation (20.5-6) can be written in another way by summing over molecular states instead of over molecules. We number the molecular states $1, 2, 3, \ldots$:

$$P = N_1 P_1 + N_2 P_2 + N_3 P_3 \cdots \qquad \textbf{(20.5-7)}$$

where N_1 is the number of molecules in molecular state 1, N_2 is the number of molecules in state 2, etc. Equation (20.5-7) also applies when the system microstate is not known, if the distribution is known.

To find P_1, consider a system consisting of a single molecule in the box that contains our entire system. From Equation (4.2-6),

$$P = -\left(\frac{\partial U}{\partial V}\right)_{S,n} \qquad \text{(20.5-8)}$$

If the single molecule is known to be in molecular state 1, its statistical entropy is fixed, since $\Omega = 1$. The amount of substance is fixed at one molecule, so that

$$P_1 = -\frac{d\varepsilon_1}{dV}$$

where ε_1 is the energy eigenvalue of molecular state 1, which we have already assumed to depend only on V. A similar formula applies for each other molecular state.

Using the probability distribution of Equation (20.3-20), we obtain

$$P = \frac{N}{z}\sum_i P_i e^{-\varepsilon_i/k_B T} = -\frac{N}{z}\sum_i \left(\frac{d\varepsilon_i}{dV}\right)e^{-\varepsilon_i/k_B T} \qquad \text{(20.5-9)}$$

The sum in the final version of Equation (20.5-9) can be written in the form

$$\boxed{P = Nk_B T\left(\frac{\partial \ln(z)}{\partial V}\right)_T \quad \left(\begin{array}{l}\text{any dilute}\\\text{gaseous substance}\end{array}\right)} \qquad \text{(20.5-10)}$$

Exercise 20.22   Verify Equation (20.5-10).

Using the factored version of the molecular partition function,

$$P = Nk_B T\left(\frac{\partial \ln(z_{\mathrm{tr}})}{\partial V}\right)_T + Nk_B T\left(\frac{\partial \ln(z_{\mathrm{rot}})}{\partial V}\right)_T$$
$$+ Nk_B T\left(\frac{\partial \ln(z_{\mathrm{vib}})}{\partial V}\right)_T + Nk_B T\left(\frac{\partial \ln(z_{\mathrm{elec}})}{\partial V}\right)_T$$

The translational partition function is the only factor that depends on V, so that

$$P = Nk_B T\left(\frac{\partial \ln[(2\pi m k_B T/h^2)^{3/2}V]}{\partial V}\right)_T$$

$$\boxed{P = Nk_B T\left(\frac{d \ln(V)}{dV}\right) = \frac{Nk_B T}{V} \quad \text{(any dilute gas)}} \qquad \text{(20.5-11)}$$

Other Thermodynamic Functions

A formula for the Helmholtz energy can be obtained from its definition:

$$A = U - TS = U - T[U/T + Nk_B + Nk_B \ln(z/N)]$$

$$\boxed{A = -Nk_B T \ln(z/N) - Nk_B T} \qquad \text{(20.5-12)}$$

This change in the definition of the chemical potential is not necessary for our discussion. If we did not make the change, a factor of Avogadro's number would occur in the formulas for μ.

In statistical mechanics, it is customary to change the definition of the chemical potential from a derivative with respect to the number of moles to a derivative with respect to the number of molecules. For example,

$$\mu = \left(\frac{\partial A}{\partial N}\right)_{T,V} \qquad \text{(20.5-13)}$$

The new chemical potential is just the old chemical potential divided by Avogadro's number. Of course, the number of molecules is an integer, and the derivative cannot actually be taken. Therefore, we use a quotient of finite differences to approximate the derivative

$$\mu = \frac{A_N - A_{N-1}}{1} = A_N - A_{N-1} \qquad \text{(20.5-14)}$$

where the subscripts indicate the number of molecules in the system. Since z is independent of the number of molecules in the system,

$$\mu = -Nk_B T \ln(z/N) - Nk_B T - [-(N-1)k_B T \ln(z/N-1) - (N-1)k_B T]$$
$$= -Nk_B T \ln(z) + Nk_B T \ln(N) - Nk_B T + (N-1)k_B T \ln(z)$$
$$\quad - (N-1)k_B T \ln(N-1) + Nk_B T - k_B T$$
$$= -k_B T \ln(z) + Nk_B T \ln(N/N-1) - k_B T \ln(N-1) - k_B T$$

Since N is a large number (near 10^{24} in most systems), $\ln(N-1)$ can be replaced by $\ln(N)$, and

$$-\ln(N/N-1) = \ln[(N-1)/N] = \ln(1 - 1/N) \approx 1/N$$

so that

$$\boxed{\mu = -k_B T \ln(z/N)} \qquad \text{(20.5-15)}$$

Exercise 20.23   Equation (20.5-15) can also be obtained by pretending that N can take on any real value and performing the differentiation in Equation (20.5-13). Carry out the differentiation to obtain this equation.

The Gibbs energy of a dilute gas is

$$G = A + PV = -Nk_B T \ln(z/N) - Nk_B T + Nk_B T$$

$$\boxed{G = -Nk_B T \ln(z/N)} \qquad \text{(20.5-16)}$$

which is equal to $N\mu$, as required by Euler's theorem.

The enthalpy of a dilute gas is

$$H = U + PV = Nk_{\text{B}}T^2\left(\frac{\partial \ln(z)}{\partial T}\right)_V + Nk_{\text{B}}T \qquad \text{(20.5-17)}$$

The heat capacity at constant volume is given by

$$C_V = \left(\frac{\partial U}{\partial T}\right)_{N,V} = \left(\frac{\partial}{\partial T}\left[Nk_{\text{B}}T^2\left(\frac{\partial \ln(z)}{\partial T}\right)_V\right]\right)_{N,V}$$

$$\boxed{C_V = 2Nk_{\text{B}}T\left(\frac{\partial \ln(z)}{\partial T}\right)_V + Nk_{\text{B}}T^2\left(\frac{\partial^2 \ln(z)}{\partial T^2}\right)_V} \qquad \text{(20.5-18)}$$

The heat capacity at constant pressure can be obtained, using a thermodynamic identity, by adding Nk_{B} to the expression of Equation (20.5-18).

All of the equations that relate thermodynamic quantities to the molecular partition function contain the logarithm of the partition function. If the partition function is factored as in Equation (20.4-10), the logarithm is a sum of terms:

$$\boxed{\ln(z) = \ln(z_{\text{tr}}) + \ln(z_{\text{rot}}) + \ln(z_{\text{vib}}) + \ln(z_{\text{elec}})} \qquad \text{(20.5-19)}$$

There are thus four additive contributions to any thermodynamic function for diatomic and polyatomic substances and two contributions for a monatomic substance, for which z_{rot} and z_{vib} do not occur. For example, the thermodynamic energy is given by Equation (20.3-21):

$$\begin{aligned}
U &= Nk_{\text{B}}T^2\left(\frac{\partial \ln(z)}{\partial T}\right)_V \\
&= Nk_{\text{B}}T^2\left(\frac{\partial \ln(z_{\text{tr}})}{\partial T}\right)_V + Nk_{\text{B}}T^2\left(\frac{\partial \ln(z_{\text{rot}})}{\partial T}\right)_V \\
&\quad + Nk_{\text{B}}T^2\left(\frac{\partial \ln(z_{\text{vib}})}{\partial T}\right)_V + Nk_{\text{B}}T^2\left(\frac{\partial \ln(z_{\text{elec}})}{\partial T}\right)_V \\
&= U_{\text{tr}} + U_{\text{rot}} + U_{\text{vib}} + U_{\text{elec}}
\end{aligned} \qquad \text{(20.5-20)}$$

The translational energy is the same for all dilute gases:

$$\boxed{U_{\text{tr}} = \frac{3}{2}Nk_{\text{B}}T = \frac{3}{2}nRT} \qquad \text{(20.5-21)}$$

The rotational energy is, for a diatomic substance or a polyatomic substance with linear molecules,

$$U_{\text{rot}} = Nk_{\text{B}}T^2\left(\frac{d \ln(8\pi^2 I k_{\text{B}}T/\sigma h^2)}{dT}\right) = Nk_{\text{B}}T^2\frac{1}{T}$$

$$\boxed{U_{\text{rot}} = Nk_{\text{B}}T = nRT \quad \begin{pmatrix}\text{diatomic or}\\\text{linear molecules}\end{pmatrix}} \qquad \text{(20.5-22)}$$

and for a polyatomic substance with nonlinear molecules,

$$U_{rot} = \frac{3}{2} Nk_BT = \frac{3}{2} nRT \quad \text{(nonlinear molecules)} \qquad \textbf{(20.5-23)}$$

Exercise 20.24　Show by differentiation that Equation (20.5-23) is correct.

The vibrational energy of a diatomic substance is, with the choice of the zero of energy used in Equation (20.5-3),

$$U_{vib} = Nk_BT^2 \frac{d}{dT}\left[\ln\left(\frac{1}{1 - e^{-h\nu/k_BT}}\right)\right]$$

$$U_{vib} = \frac{Nh\nu}{e^{h\nu/k_BT} - 1} \quad \text{(diatomic substance)} \qquad \textbf{(20.5-24)}$$

Exercise 20.25　**a.** Carry out the steps to obtain Equation (20.5-24).

b. Find the expression for the vibrational energy if the zero-point energy is included in the vibrational energy, giving the vibrational partition function of Equation (20.4-18).

For a polyatomic substance, the vibrational energy is a sum of contributions, one for each normal mode:

$$U_{vib} = \sum_{i=1}^{3n-5(6)} Nk_BT^2 \frac{d}{dT}\left[\ln\left(\frac{1}{1 - e^{-h\nu_i/k_BT}}\right)\right]$$

$$U_{vib} = \sum_{i=1}^{3n-5(6)} \frac{Nh\nu_i}{e^{h\nu_i/k_BT} - 1} \quad \left(\begin{array}{c}\text{polyatomic}\\\text{substances}\end{array}\right) \qquad \textbf{(20.5-25)}$$

▼ **EXAMPLE 20.11**　The vibrational frequency of CO is 6.5048×10^{13} cm^{-1}.

a. Find the vibrational partition function at 298.15 K.

b. Find the vibrational contribution to the molar energy at 298.15 K.

Solution

We use the zero of vibrational energy at the lowest vibrational state, as in Equation (20.4-15). We let

$$x = \frac{h\nu}{k_BT} = \frac{(6.6261 \times 10^{-34} \text{ J s})(6.5048 \times 10^{13} \text{ s}^{-1})}{(1.3807 \times 10^{-23} \text{ J K}^{-1})(298.15 \text{ K})} = 10.47$$

a.

$$z_{\text{vib}} = \frac{1}{1 - e^{-x}} = 1.0000284$$

b.

$$U_{\text{vib}} = \frac{Nh\nu}{e^x - 1}$$

$$= \frac{(6.022 \times 10^{23} \text{ mol}^{-1})(6.6261 \times 10^{-34} \text{ J s})(6.5048 \times 10^{13} \text{ s}^{-1})}{e^{10.47} - 1}$$

$$= 0.737 \text{ J mol}^{-1}$$

The closeness of the partition function to unity and the small value of the vibrational energy show that the ground vibrational state is virtually the only occupied vibrational state.

▲

Exercise 20.26

a. Repeat the calculation of Example 20.11 with the zero-point energy included in the vibrational energy, so that the vibrational partition function is given by Equation (20.4-18).

b. Find the value of the vibrational partition function and the vibrational energy for 1.000 mol of gaseous iodine at 500 K, using both choices of the zero of energy.

The electronic contribution to the energy is

$$U_{\text{elec}} = Nk_BT^2 \left(\frac{d \ln(z_{\text{elec}})}{dT} \right)$$

$$U_{\text{elec}} = Nk_BT^2 \frac{d}{dT} \left[\ln(g_0 e^{-\varepsilon_0/k_BT} + g_1 e^{-\varepsilon_1/k_BT} + \cdots) \right]$$

$$= Nk_BT \frac{1}{z_{\text{elec}}} \left[g_0 \frac{\varepsilon_0}{k_BT} e^{-\varepsilon_0/k_BT} + g_1 \frac{\varepsilon_1}{k_BT} e^{-\varepsilon_1/k_BT} + \cdots \right] \quad \textbf{(20.5-26)}$$

In most atoms and molecules, the ground electronic level is nondegenerate and the first excited level is sufficiently high in energy that at room temperature the electronic partition function is very nearly equal to unity, and

$$U_{\text{elec}} = N\varepsilon_0 \quad \left(\begin{array}{l} \text{molecules with high} \\ \text{first excited level} \end{array} \right) \quad \textbf{(20.5-27)}$$

However, some substances, such as NO, have an excited level that is near the ground level, so that Equation (20.5-27) does not apply.

Exercise 20.27

Find the electronic contribution to the thermodynamic energy of 1.000 mol of gaseous NO at 1.000 atm and 298.15 K. See Exercise 20.18 for the value of ε_1.

The entropy is a sum of contributions, like the energy. From Equation (20.4-10)

$$
\begin{aligned}
S = \frac{U_{tr}}{T} + Nk_B \ln\!\left(\frac{z_{tr}}{N}\right) + Nk_B + \frac{U_{rot}}{T} + Nk_B \ln(z_{rot}) \\
+ \frac{U_{vib}}{T} + Nk_B \ln(z_{vib}) + \frac{U_{elec}}{T} + Nk_B \ln(z_{elec}) \\
= S_{tr} + S_{rot} + S_{vib} + S_{elec}
\end{aligned}
\tag{20.5-28}
$$

where

$$
S_{tr} = \frac{U_{tr}}{T} + Nk_B \ln\!\left(\frac{z_{tr}}{N}\right) + Nk_B
\tag{20.5-29}
$$

and where the other terms are apparent by inspection.

There is only one additive term, Nk_B, and there is only one divisor, N, in the first version of Equation (20.5-27), and these are arbitrarily placed with the translational entropy. The same policy is followed with the formulas for the other thermodynamic functions. This is the only policy possible with atomic substances and is continued for the other substances.

The contributions to the Helmholtz energy are

$$
A_{tr} = -Nk_B T \ln\!\left(\frac{z_{tr}}{N}\right) - Nk_B T
\tag{20.5-30a}
$$

$$
A_{rot} = -Nk_B T \ln(z_{rot})
\tag{20.5-30b}
$$

$$
A_{vib} = -Nk_B T \ln(z_{vib})
\tag{20.5-30c}
$$

$$
A_{elec} = -Nk_B T \ln(z_{elec})
\tag{20.5-30d}
$$

It is important that the single Nk_B term and the single divisor N be included only once. If you mistakenly put them in the other terms as well as in the translational term, you will get the wrong value for the thermodynamic function. However, if you put them in any other term and not in the translational term, you would get the correct total value for the thermodynamic function.

▼   **EXAMPLE 20.12**

Usually, when you look up vibrational frequencies, you will find values of frequencies divided by the speed of light, usually expressed in cm^{-1} and denoted by $\tilde{\nu}$. Multiply these values by the speed of light to get the frequencies.

Find the vibrational contributions to the thermodynamic energy, Helmholtz energy, and entropy of 1.000 mole of water vapor at 1.00 atm and 100°C. There are three vibrational normal modes, with frequencies $1.0947 \times 10^{14}\ s^{-1}$, $4.7817 \times 10^{13}\ s^{-1}$, and $1.1260 \times 10^{14}\ s^{-1}$.

Solution

Let $x_i = h\nu_i/k_B T$, so that at 373.15 K, $x_1 = 14.079$, $x_2 = 6.1498$, and $x_3 = 14.482$. The vibrational energy is

$$
\begin{aligned}
U_{vib} &= \sum_{i=1}^{3} \frac{Nh\nu_i}{e^{h\nu_i/k_B T} - 1} = Nh \sum_{i=1}^{3} \frac{\nu_i}{e^{x_i} - 1} \\
&= (1.000\ \text{mol})(6.022 \times 10^{23}\ \text{mol}^{-1})(6.6261 \times 10^{-34}\ \text{J s}) \\
&\quad \times \left(\frac{1.0947 \times 10^{14}\ s^{-1}}{e^{14.079} - 1} + \frac{4.7817 \times 10^{13}\ s^{-1}}{e^{6.1498} - 1} + \frac{1.1260 \times 10^{14}\ s^{-1}}{e^{14.482} - 1} \right) \\
&= 40.86\ \text{J}
\end{aligned}
$$

The vibrational contribution to the Helmholtz energy is

$$
\begin{aligned}
A_{vib} &= Nk_B T \ln(z_{vib}) \\
&= nRT \ln\!\left[\left(\frac{1}{1 - e^{-14.079}} \right)\!\left(\frac{1}{1 - e^{-6.1498}} \right)\!\left(\frac{1}{1 - e^{-14.482}} \right) \right] \\
&= (1.000\ \text{mol})(8.3145\ \text{J K}^{-1}\ \text{mol}^{-1})(373.15\ \text{K})(2.137 \times 10^{-3}) \\
&= 6.632\ \text{J}
\end{aligned}
$$

$$S_{\text{vib}} = \frac{U - A}{T} = \frac{40.86 \text{ J} - 6.63 \text{ J}}{373.15 \text{ K}} = 0.0917 \text{ J K}^{-1}$$

▲

Equations can be written for the other thermodynamic functions. For example, the vibrational contribution to the heat capacity of a dilute diatomic gas is

$$C_{V\text{vib}} = Nk_{\text{B}} \left(\frac{hv}{k_{\text{B}}T}\right)^2 \frac{e^{hv/k_{\text{B}}T}}{(e^{hv/k_{\text{B}}T} - 1)^2} \qquad \textbf{(20.5-31)}$$

Exercise 20.28
a. Verify Equation (20.5-31) by differentiating Equation (20.5-24).
b. Calculate C_V for 1.000 mol of $^{35}\text{Cl}_2$ at 298.15 K. See Table A16 for needed information.

The chemical potential is given by

$$\mu_{\text{tr}} = -k_{\text{B}}T \ln(z_{\text{tr}}/N) \qquad \textbf{(20.5-32a)}$$
$$\mu_{\text{rot}} = -k_{\text{B}}T \ln(z_{\text{rot}}) \qquad \textbf{(20.5-32b)}$$
$$\mu_{\text{vib}} = -k_{\text{B}}T \ln(z_{\text{vib}}) \qquad \textbf{(20.5-32c)}$$
$$\mu_{\text{elec}} = -k_{\text{B}}T \ln(z_{\text{elec}}) \qquad \textbf{(20.5-32d)}$$

The divisor N is placed with the translational contribution, as usual.

Exercise 20.29
Calculate the four contributions to the chemical potential of 1.000 mol of $^{35}\text{Cl}_2$ at 298.15 K and 1.000 atm. See Table A16 for needed information.

Now that we have the expression for the chemical potential in terms of the molecular partition function, we can write the molecular probability distribution for a dilute gas in a slightly different way. We write Equation (20.5-15) in the form

$$\frac{N}{z} = e^{\mu/k_{\text{B}}T} \qquad \textbf{(20.5-33)}$$

The equation for the molecular probability distribution for a dilute gas can now be written from Equation (20.3-19)

$$N_j = g_j e^{\mu/k_{\text{B}}T} e^{-\varepsilon_j/k_{\text{B}}T} \qquad \textbf{(20.5-34)}$$

This equation gives the explicit expression for the Lagrange multiplier, α, of Equation (20.2-13):

$$\alpha = \mu/k_{\text{B}}T \qquad \textbf{(20.5-35)}$$

Heat and Work in Statistical Mechanics

From thermodynamics, we have the relation for a closed simple system

$$dU = T\,dS - P\,dV \tag{20.5-36}$$

Since we identify the mechanical energy of the system, E, with its thermodynamic energy, U, dE is identified with dU.

$$dE = d\left(\sum_i N_i\varepsilon_i\right) = \sum_i \varepsilon_i\,dN_i + \sum_i N_i\,d\varepsilon_i \tag{20.5-37}$$

Since we have assumed that the molecule energy eigenvalues depend only on the volume of the system, we write

$$dE = \sum_i \varepsilon_i\,dN_i + \sum_i N_i\frac{d\varepsilon_i}{dV}\,dV \tag{20.5-38}$$

From the expression for the pressure in Equation (20.5-9) we can write

$$dE = \sum_i \varepsilon_i\,dN_i - P\,dV = \sum_i \varepsilon_i\,dN_i + dw_{\text{rev}} \tag{20.5-39}$$

where we have used the fact that $dw_{\text{rev}} = -P\,dV$ in a simple system such as a dilute gas.

From the first law of thermodynamics, we now have

$$dq_{\text{rev}} = \sum_i \varepsilon_i\,dN_i \tag{20.5-40}$$

This identification of work is reasonable because the energy levels depend on the volume and, for a simple system such as a dilute gas, work involves changes in the volume of the system. The other term is identified as heat because it is what is left over after the work is identified.

For a dilute gas, we now have statistical mechanical interpretations of work and heat for reversible processes. Work is a change in energy that corresponds to shifts of the molecular energy levels, and heat is a change in energy that corresponds to changes in the numbers of molecules occupying the energy levels.

More on the Entropy

Using the definition of the statistical entropy in Equation (20.5-1) and replacing $\ln(\Omega)$ by its largest term, we can write

$$dS_{\text{st}} = k_{\text{B}}\,d\ln(W_{\text{max}}) = k_{\text{B}}\sum_i \left(\frac{\partial\ln(W_{\text{max}})}{\partial N_i}\right)dN_i \tag{20.5-41}$$

From Equation (20.2-13),

$$\frac{\partial\ln(W_{\text{max}})}{\partial N_i} = -\alpha + \beta\varepsilon_i \tag{20.5-42}$$

so that

$$dS_{\text{st}} = k_{\text{B}}\sum_i (-\alpha + \beta\varepsilon_i)\,dN_i = k_{\text{B}}\beta\sum_i \varepsilon_i\,dN_i = k_{\text{B}}\beta\,dq_{\text{rev}} \tag{20.5-43}$$

where we have used the fact that since the number of particles is fixed, the dN's must sum to zero, and have used Equation (20.5-40).

We could do two things with Equation (20.5-43). If we assume that the statistical entropy is the same as the thermodynamic entropy, we can rec-

ognize that $\beta = 1/(k_B T)$. If we take $\beta = 1/(k_B T)$ as given, we have a demonstration that the statistical entropy is the same as the thermodynamic entropy except for an additive constant.

20.6 Chemical Equilibrium in Dilute Gases

In this section, we obtain the relationship between the equilibrium constant for a chemical reaction in a dilute gas mixture and the partition functions of the reactants and products. There are two ways to derive this relationship. In the first procedure, one begins with the thermodynamic relation of Equation (7.1-10). In the other procedure, one starts from scratch without assuming anything from thermodynamics.[5] We use the first procedure.

We write a chemical reaction equation as in Equation (7.1-1)

$$0 = \sum_{j=1}^{c} v_j \mathscr{F}_j \tag{20.6-1}$$

where the v's are stoichiometric coefficients (not vibrational frequencies) and the $\mathscr{F}$'s stand for chemical formulas. The number of chemical substances involved is denoted by c. The thermodynamic criterion for chemical equilibrium is given by Equation (7.1-10):

$$0 = \sum_{j=1}^{c} v_j \mu_j \quad \text{(at equilibrium)} \tag{20.6-2}$$

We will use the statistical mechanical expression for the chemical potential in terms of the molecular partition function. The molecular partition function is given by

$$z = \sum_i e^{-\varepsilon_i / k_B T} \tag{20.6-3}$$

where ε_i denotes the molecular energy eigenvalue corresponding to molecular state i.

Let us rewrite Equation (20.6-3)

$$z = \sum_i e^{-(\varepsilon_i - \varepsilon_0 + \varepsilon_0)k_B T} = e^{-\varepsilon_0 / k_B T} \sum_i e^{-(\varepsilon_i - \varepsilon_0)k_B T} \tag{20.6-4}$$

where we have factored a common factor out of every term in the sum. The sum in Equation (20.6-4) is denoted by z':

$$z = e^{-\varepsilon_0 / k_B T} z' \tag{20.6-5}$$

We must use a consistent zero of energy for all substances. Equation (20.6-5) makes it easy to do so. We simply measure ε_0 relative to the same zero for all substances.

The sum z' is independent of the choice of the zero of energy, since it contains only energy differences. It is equal to the molecular partition function if the choice is made that the energy of the ground state, ε_0, is equal to zero. We can make this choice for a single substance if we wish to do so, but cannot do so simultaneously for two or more substances. For several substances, the factor in front of the sum in Equation (20.6-4) must be included for each, with a consistent zero of energy chosen for all substances.

---

[5] N. Davidson, *op. cit.*, pp. 97ff.

The formula for the chemical potential of a substance in a system of N particles is Equation (20.5-15). We use the chemical potential per molecule. Equation (20.6-2) is still valid. From Equation (20.5-15)

$$\mu = -k_B T \ln(z/N) = -k_B T[-\varepsilon_0/k_B T + \ln(z')]$$
$$= \varepsilon_0 - k_B T \ln(z'/N) \qquad (20.6\text{-}6)$$

We substitute Equation (20.6-6) into Equation (20.6-2), labeling the chemical species by the subscript j:

$$0 = \sum_{j=1}^{c} v_j[\varepsilon_{0j} - k_B T \ln(z'_j/N_j)] \qquad (20.6\text{-}7)$$

The sum of the terms $v_j \varepsilon_{j0}$ is a difference because the v's are negative for reactants. One molecule of reaction corresponds to a number of molecules of product j equal to v_j being produced and a number of molecules of reactant j equal to $|v_j|$ being consumed.

The sum of the terms $v_j \varepsilon_{0j}$, which we denote by $\Delta\varepsilon_0$, is the difference in the energies of the ground states of the product and reactant molecules.

We divide Equation (20.6-7) by $k_B T$. We use the fact that the sum of logarithms is the logarithm of a product and use the identity

$$a \ln(x) = \ln(x^a) \qquad (20.6\text{-}8)$$

The result is

$$0 = \frac{\Delta\varepsilon_0}{k_B T} - \ln\left[\prod_{i=1}^{c} \left(\frac{z'_j}{N_j}\right)^{v_j}\right] \qquad (20.6\text{-}9)$$

The molecular partition functions are of the form of Equation (20.4-4) or Equation (20.4-10):

$$z = z_{tr} z_{elec} \qquad \text{(atoms)} \qquad (20.6\text{-}10a)$$
$$z = z_{tr} z_{rot} z_{vib} z_{elec} \quad \text{(molecules)} \qquad (20.6\text{-}10b)$$

In either event, z_{tr} is given by Equation (20.3-18):

$$z_{tr} = \left(\frac{2\pi m k_B T}{h^2}\right)^{3/2} V \qquad (20.6\text{-}11)$$

where V is the volume of the system. We use the abbreviation

$$z''_j = \frac{z'_j}{V} \qquad (20.6\text{-}12)$$

Since our dilute gases obey the ideal gas law,

$$\frac{V}{N_j} = \frac{k_B T}{P_j} = \frac{k_B T}{P^\circ} \frac{P^\circ}{P_j} = \frac{\bar{V}^\circ}{N_A} \frac{P^\circ}{P_j} \qquad (20.6\text{-}13)$$

where P_j is the partial pressure of substance j, N_A is Avogadro's number, and $\bar{V}^\circ$ is the volume occupied by 1.000 mol of ideal gas at pressure P° and temperature T. Equation (20.6-9) can now be written

$$0 = \frac{\Delta\varepsilon_0}{k_B T} - \ln\left[\prod_{j=1}^{c} \left(\frac{z''_j \bar{V}^\circ}{N_A}\right)^{v_j}\right] + \ln\left[\prod_{j=1}^{c} \left(\frac{P_j}{P^\circ}\right)^{v_j}\right] \qquad (20.6\text{-}14)$$

Since the partial pressures in the last term are the equilibrium partial pressures, this term is the logarithm of the equilibrium constant K_P, and we can write

$$\ln(K_P) = -\frac{\Delta\varepsilon_0}{k_B T} + \ln\left[\prod_{j=1}^{c}\left(\frac{z_j''\bar{V}°}{N_A}\right)^{v_j}\right] \qquad \textbf{(20.6-15a)}$$

$$K_P = e^{-\Delta\varepsilon_0/k_B T}\prod_{j=1}^{c}\left(\frac{z_j'°}{N_A}\right)^{v_j} \quad \left(\begin{array}{c}\text{reaction}\\ \text{in dilute gas}\end{array}\right) \qquad \textbf{(20.6-15b)}$$

where $z_j'°$ is the full molecular partition function of substance number j, given that the volume in which the molecules are found is equal to the molar volume at pressure $P°$ and that the energy of the ground state of the molecules is set equal to zero.

EXAMPLE 20.13   Calculate the equilibrium constant for the reaction

$$H_2 \rightleftharpoons 2H$$

at 500 K. Assume that the vibrational partition function of the molecule and all electronic partition functions can be approximated by unity.

Solution

From Equation (20.4-14),

$$z_{\text{rot, H}_2} = \frac{1}{2}\frac{8\pi^2 I_e k_B T}{h^2} = \frac{1}{2}\frac{8\pi^2\mu r_e^2 k_B T}{h^2}$$

$$= \frac{1}{2}\frac{8\pi^2(1.674\times10^{-27}\text{ kg})(0.741\times10^{-10}\text{ m})^2(1.3807\times10^{-23}\text{ J K}^{-1})(500\text{ K})}{(6.6261\times10^{-34}\text{ J s})^2}$$

$$= 5.7$$

For atomic hydrogen, from Equation (20.3-18),

$$\frac{z'°}{N_A} = \frac{z_{\text{tr}}'°}{N_A} = \left(\frac{2\pi m k_B T}{h^2}\right)^{3/2}\frac{\bar{V}°}{N_A}$$

$$= \left[\frac{2\pi(1.674\times10^{-27}\text{ kg})(1.3807\times10^{-23}\text{ J K}^{-1})(500\text{ K})}{(6.6261\times10^{-34}\text{ J s})^2}\right]^{3/2}$$

$$\times\frac{0.04103\text{ m}^3\text{ mol}^{-1}}{6.022\times10^{23}\text{ mol}^{-1}}$$

$$= 1.449\times10^5$$

The translational partition function of the molecule is given by the same expression except for the mass, and is equal to

$$\frac{z_{\text{tr}}'°(H_2)}{N_A} = 2.049\times10^5$$

From Chapter 13,

$$\Delta\varepsilon_0 = 4.75\text{ eV} = 7.61\times10^{-19}\text{ J}$$

$$e^{-\Delta\varepsilon_0/k_B T} = \exp\left[\frac{-7.61\times10^{-19}\text{ J}}{(1.3807\times10^{-23}\text{ J K}^{-1})(500\text{ K})}\right] = 1.3\times10^{-48}$$

$$K_P = e^{-\Delta\varepsilon_0/k_B T}(z_{\text{trH}}'°)^2(z_{\text{trH}_2}'°)^{-1}(z_{\text{rotH}_2}'°)^{-1}$$

$$= \frac{(1.3\times10^{-48})(1.449\times10^5)^2}{(2.049\times10^5)(5.7)} = 2.3\times10^{-44}$$

Exercise 20.30 Without making detailed calculations, estimate the equilibrium constant for the reaction at 298.15 K:

$$H_2 + D_2 \rightleftharpoons 2HD$$

20.7 The Canonical Ensemble

Thus far, our discussion of statistical mechanics has been based on the probability distribution for molecular states in a single macroscopic system. This approach can be used only for a dilute gas. If intermolecular interactions must be included, the microstates of the system cannot be represented by the individual states of the molecules.

An alternative approach to statistical mechanics is more general and can in principle be applied to any kind of system. This approach is called the ensemble approach. An **ensemble** is an imaginary collection of many replicas of the actual system. In a statistical mechanical ensemble, each replica is in the same macrostate as the real system, but different members of the ensemble occupy different system microstates. For example, the system we discussed in the first part of this chapter had fixed values of U, V, and N. The ensemble constructed to represent such a system is called the **microcanonical ensemble**. It consists of very many replicas of this system, all with the same value of U, V, and N, but occupying different microstates compatible with this macroscopic state.

The reason for defining an ensemble into existence is to visualize the averaging processes of statistical mechanics. The first postulate of statistical mechanics now becomes:

Postulate I. *Ensemble average values and observed values of thermodynamic functions can be equated.*

The second postulate of statistical mechanics can be restated:

Postulate II. *In the microcanonical ensemble, each system microstate corresponding to the given values of E, N and V is occupied by an equal number of systems, and no systems in the ensemble occupy other system microstates.*

The principal application of the microcanonical ensemble is to systems of noninteracting particles (dilute gases). We have already carried out this analysis, but without using the ensemble terminology. Use of the ensemble terminology would not change the results of Sections 1 through 6 of this chapter.

The **canonical ensemble** represents a closed system with given values of T, V, and N. The **grand canonical ensemble** represents an open system with given values of T, V, and μ (the chemical potential). Discussions of the grand canonical ensemble can be found in the statistical mechanics textbooks listed at the end of the chapter, but we do not discuss it. There are also additional ensembles that are less commonly used.

It is possible to use an alternative approach to the statistical mechanics of dilute gases, based on the canonical ensemble.

A canonical ensemble can be used to represent any kind of system. If the canonical ensemble is applied to dilute gases, the working equations for the calculation of thermodynamic quantities are identical to those of previous sections.

A system to be represented by a canonical ensemble is a closed system of constant volume, held at a fixed temperature. Instead of placing each system of the ensemble in an individual constant-temperature bath, the systems of the ensemble are placed in thermal contact with each other, so that a given system of the ensemble is in a constant-temperature bath consisting of the other systems of the ensemble. Since the nature of the constant-temperature bath is unimportant so long as it remains at constant temperature, this replacement is acceptable. Figure 20.4 schematically depicts a canonical ensemble.

The Schrödinger equation for a system of many particles cannot be solved. However, the solutions exist even if we cannot find them, and we can write equations containing them so long as we do not have to have explicit expressions.

Imagine that the Schrödinger equation for the entire system has been solved and that we have a list of the very many energy eigenstates of the real physical system corresponding to the correct volume and correct number of molecules. The energy eigenfunctions are numbered Ψ_1, Ψ_2, Ψ_3, etc., with energy eigenvalues E_1, E_2, E_3, etc. All of these very many microstates occur in the ensemble. We let p_i be the fraction of the systems of the ensemble that are found in system microstate i.

States of equal energy can differ from each other in the values of some other properties. If all states of equal energy are equally probable, the probability cannot depend on these other properties.

We apply a version of Postulate II to the ensemble and assert that in our canonical ensemble all system microstates of the same energy are equally probable, with equal values of p_i. The probabilities of system microstates of different energies are not assumed to be equal. If all system microstates of equal energy are equally probable, the probability depends only on the system energy:

$$p_i = p(E_i) \tag{20.7-1}$$

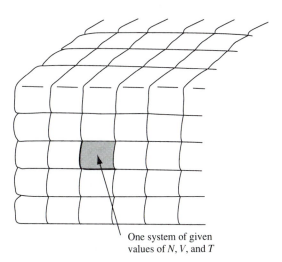

One system of given
values of N, V, and T

Figure 20.4. A Canonical Ensemble (schematic). Although there is only one actual system, the ensemble consists of very many imaginary replicas of this system. We imagine that every possible state of the system is occupied by a number of systems of the ensemble.

where p without a subscript represents a mathematical function, which we now seek to determine. We make the further assumption that p is the same function of its argument for all different kinds of systems.

Consider a system that is made up of two parts, subsystem I and subsystem II. A canonical ensemble that represents the system is made up of many replicas of the system, each with two parts. Each part is closed and at constant volume, but the two parts are in thermal contact with each other and maintained at a fixed temperature. We assume that the two subsystems occupy mechanical states independently of each other and the system energy is a sum of the subsystem energies:

$$E_i = E_{I,i} + E_{II,i} \tag{20.7-2}$$

Consider a particular microstate of subsystem I, denoted as state number I,i. The fraction of the replicas of subsystem I in this microstate is

$$p_{I,i} = p(E_{I,i}) \tag{20.7-3a}$$

and the fraction of the replicas of subsystem II in microstate II, i is

$$p_{II,i} = p(E_{II,i}) \tag{20.7-3b}$$

If subsystem I is in microstate I,i and subsystem II is in microstate II,i, we denote the microstate of the entire system by i. It is a fact of probability theory that the joint probability of two independent occurrences is equal to the product of the probabilities of the two occurrences. Therefore

$$p(E_i) = p(E_{I,i} + E_{II,i}) = p(E_{I,i})p(E_{II,i}) \tag{20.7-4}$$

where the three probabilities denoted by p are the same function of their arguments.

Only one function has this property. This is the exponential function, as shown in Section 1.7:

$$p(E) = Ae^{-\beta E} \tag{20.7-5}$$

where A and β do not depend on E. If β is positive, the negative sign must be chosen to prevent the probability from increasing indefinitely for large energy. By our assumptions, Equation (20.7-5) represents the probability distribution for states of any kind of system represented by a canonical ensemble. It is not restricted to dilute gases or any other particular kind of system, nor is it restricted to a system made up of two separate subsystems.

Exercise 20.31   **a.** Choose several different functions of E and show that they do not satisfy Equation (20.7-4), and show that the function of Equation (20.7-5) does satisfy this relation.

b. One can carry out an analysis similar to that of Section 1.7 to show that Equation (20.7-5) follows from Equation (20.7-4). Let x and y represent the two subsystem energies, so that

$$p(x + y) = p(x)p(y)$$

Differentiate both sides of this equation once with respect to x and once with respect to y to obtain two equations. Relate these equations, separate the variables, and complete the analysis.

We evaluate the quantity A in the same way that we evaluated the parameter α in Equation (20.2-20). We require the sum of the probabilities of all possible states to equal unity:

$$\sum_i p_i = A \sum_i e^{-\beta E_i} = 1 \tag{20.7-6}$$

which is equivalent to

$$\frac{1}{A} = \sum_i e^{-\beta E_i} = Z \tag{20.7-7}$$

Equation (20.7-7) defines Z, the **canonical partition function** or **canonical Zustandsumme**. It is a sum over all possible system states of the system under study.

Each term in the canonical partition function is proportional to the probability of finding the state corresponding to that term, with the term for a state of zero energy set equal to unity. The partition function is therefore a measure of the number of system microstates that are effectively accessible to the system. The canonical partition function at nonzero temperature is always a very large number.

The canonical probability distribution can now be written

$$p_i = \frac{1}{Z} e^{-\beta E_i} \tag{20.7-8}$$

The ensemble average energy can be written

$$\langle E \rangle = U = \sum_i E_i p_i = \frac{1}{Z} \sum_i E_i e^{-\beta E_i} \tag{20.7-9}$$

where we use the symbol $\langle \ \rangle$ for an ensemble average value. This equation can be rewritten in exactly the same way as was Equation (20.3-7):

$$\langle E \rangle = U = -\frac{1}{Z} \left(\frac{\partial Z}{\partial \beta} \right)_{V,N} = -\left(\frac{\partial \ln(Z)}{\partial \beta} \right)_{V,N} \tag{20.7-10}$$

We can also write an equation for the pressure that is analogous to Equation (20.5-10). Let P_i be the pressure of the system, given that it is in system microstate i. If the microstate is known to be state number i, the statistical entropy is constant and we can write

$$P_i = \left(\frac{\partial U}{\partial V} \right)_{S,N} = \left(\frac{\partial E_i}{\partial V} \right)_N \tag{20.7-11}$$

where we assume that the energy eigenvalue E_i depends only on V and N. The ensemble average pressure is

$$\langle P \rangle = \sum_i P_i p_i = \frac{1}{Z} \sum_i \left(\frac{\partial E_i}{\partial V} \right)_N e^{-\beta E_i} = \frac{1}{\beta} \left(\frac{\partial \ln(Z)}{\partial V} \right)_{\beta,N} \tag{20.7-12}$$

Exercise 20.32   Carry out the steps to obtain Equation (20.7-10) and (20.7-12).

To identify the parameter β, we use the properties of a particular system, a dilute gas of noninteracting particles. In a dilute gas, we can neglect the intermolecular potential energy and assume that the particles are independent. The system energy is a sum of the energies of the molecules:

$$E_i = \varepsilon_{1i} + \varepsilon_{2i} + \varepsilon_{3i} + \varepsilon_{4i}\cdots \tag{20.7-13}$$

where the double subscript $1i$ stands for whatever values of the molecular quantum numbers for molecule 1 correspond to system state i, etc.

The canonical partition function for our dilute gaseous system is given by

$$Z = \sum_{1i} \sum_{2i} \sum_{3i} \cdots \sum_{Ni} \exp[-\beta(\varepsilon_{1i} + \varepsilon_{2i} + \varepsilon_{3i} + \cdots)] \tag{20.7-14}$$

If each of the sums were independent of the others, so that every set of values of the quantum numbers represented by $1i$ could combine with every set of values of $2i$, etc., then the multiple sum could be factored. Each sum would give a factor of z, leading to $Z = z^N$, where z is the molecular partition defined in Equation (20.3-3).

However, the molecules of our gas are indistinguishable, and two sets of quantum numbers that differ only by having the values of the quantum numbers for one particle exchanged for those of another particle cannot represent different system states. For example, if molecule 4 is in molecular state 478 and molecule 17 is in molecular state 12, the system microstate is not different from the system state with particle 4 in molecular state 12 and molecule 17 in molecular state 478, if all other molecules are in the same states. That is, since the system wave function must be antisymmetrized for a system of fermions or symmetrized for a system of bosons, both states will be identical after symmetrization or antisymmetrization. Furthermore, for a system of fermions, we must delete any set of quantum numbers in which two or more molecules have the same values for all of their quantum numbers, since the wave function for such a state will vanish upon antisymmetrization.

We assume the case of dilute occupation, in which the total number of molecular states available to be occupied is much larger than the number of molecules in the system. In this case, the forbidden sets of quantum numbers for a system of fermions occur in only a small fraction of the terms of the sum, and it is not a serious error to leave these sets in the sum.

We must still take account of the indistinguishability of the particles. If we fail to do this and sum independently over all values of $1i$, $2i$, etc., we will include many terms that would be identical after symmetrization or antisymmetrization. The number of ways of arranging N objects in N boxes, one to a box, is called the number of permutations of N objects and is equal to $N!$. For any given set of quantum numbers in which every molecule has a distinct set of molecule quantum numbers, this is the number of ways of assigning the molecules to the sets of quantum numbers. If we sum independently over $1i$, $2i$, etc., we have therefore included $N!$ terms where there should be only one term. We can make correction for this by writing

$$Z = \frac{z^N}{N!} \quad \text{(gas of independent molecules)} \tag{20.7-15}$$

If the molecules were distinguishable, instead of Equation (20.7-15) we would have $Z = z^N$. In Chapter 22, we will discuss the Einstein crystal model, in which atoms in a crystal are treated as distinguishable and $Z = z^N$.

For a monatomic gas without electronic excitation, Equation (20.3-15) is

$$z = z_{tr} = \left(\frac{2\pi m}{h^2 \beta}\right)^{3/2} V \qquad \text{(20.7-16)}$$

From Equation (20.7-12), the pressure of such a gas is

$$\langle P \rangle = \frac{1}{\beta}\left(\frac{\partial \ln(Z)}{\partial V}\right)_{\beta, N} = \frac{N}{\beta}\left(\frac{\partial \ln(z)}{\partial V}\right)_{\beta, N}$$
$$= \frac{N}{\beta}\frac{d \ln(V)}{dV} = \frac{N}{\beta V} \qquad \text{(20.7-17)}$$

In order to make our dilute gas obey the ideal gas equation of state, we must let $\beta = 1/(k_B T)$. The canonical probability distribution, Equation (20.7-8), becomes

$$\boxed{p_i = \frac{1}{Z}e^{-E_i/k_B T}} \qquad \text{(20.7-18)}$$

Although we have established that $\beta = 1/(k_B T)$ only for a dilute gas, we assert that β cannot have a different interpretation for different systems and that Equation (20.7-18) is valid for any kind of system.

We can now write Equation (20.7-10) in terms of the temperature:

$$\langle E \rangle = U = k_B T^2 \left(\frac{\partial \ln(Z)}{\partial T}\right)_{V,N} \qquad \text{(general system)} \qquad \text{(20.7-19)}$$

$$\langle E \rangle = U = N k_B T^2 \left(\frac{\partial \ln(z)}{\partial T}\right)_V \qquad \text{(dilute gas)} \qquad \text{(20.7-20)}$$

Equation (20.7-20) is identical with Equation (20.3-21).

We can also write Equations (20.7-12) and (20.7-17) for the pressure in terms of the temperature:

$$\langle P \rangle = k_B T \left(\frac{\partial \ln(Z)}{\partial V}\right)_{T,N} \qquad \text{(general system)} \qquad \text{(20.7-21)}$$

$$\langle P \rangle = N k_B T \left(\frac{\partial \ln(z)}{\partial V}\right)_T \qquad \text{(dilute gas)} \qquad \text{(20.7-22)}$$

In order to obtain formulas for other thermodynamic variables, we must have a formula for the entropy. One procedure is to begin with a formula from thermodynamics, Equation (4.2-3). For a closed system,

$$dU = T\,dS - P\,dV \qquad \text{(20.7-23)}$$

We divide this equation by T:

$$dS = \frac{1}{T}dU + \frac{P}{T}dV = d\left(\frac{U}{T}\right) + \frac{U}{T^2}dT + \frac{P}{T}dV \qquad \text{(20.7-24)}$$

Using Equations (20.7-19) and (20.7-21) for U and P,

$$dS = d\left(\frac{U}{T}\right) + k_B\left(\frac{\partial \ln(Z)}{\partial T}\right)_V dT + k_B\left(\frac{\partial \ln(Z)}{\partial V}\right)_T dV$$
$$= d\left(\frac{U}{T}\right) + k_B d \ln(Z) \tag{20.7-25}$$

since Z is a function only of T and V if N is fixed.

We now perform an indefinite integration:

$$S = \frac{U}{T} + k_B \ln(Z) + C \tag{20.7-26}$$

where C is a constant of integration. We choose to let $C = 0$. The entropies thus calculated agree with third-law entropies and correspond to zero statistical entropy if the system is known to be in a specific microstate.

There is an alternative definition of the statistical entropy, in addition to the definition of Equation (3.4-3):

$$\boxed{S_{st} = -k_B \sum_i p_i \ln(p_i) \quad \text{(alternative definition)}} \tag{20.7-27}$$

where the sum is over all microstates of the system.

If we apply Equation (20.7-27) to the microcanonical ensemble (the ensemble representing the averaging process used in Sections 20.2–20.6), all of the p_i values are equal to $1/\Omega$, where Ω is the number of possible microstates:

$$S_{st} = -k_B \sum_{i=1}^{\Omega} \frac{1}{\Omega} \ln\left(\frac{1}{\Omega}\right) = k_B \ln(\Omega) \tag{20.7-28}$$

which is the same as Equation (3.4-3) and Equation (20.5-1).

To apply Equation (20.7-27) to the canonical ensemble, we substitute Equation (20.7-18) into the argument of the logarithm:

$$S_{st} = -k_B \sum_i p_i \left[\frac{-E_i}{k_B T} - \ln(Z)\right]$$
$$= \frac{\langle E \rangle}{T} + k_B \ln(Z) = \frac{U}{T} + k_B \ln(Z) \tag{20.7-29}$$

Equation (20.7-29) is the same as Equation (20.7-26) with the choice $C = 0$.

We can now write formulas for the other thermodynamic functions in terms of the canonical partition function. The results are:

$$\boxed{U = k_B T^2 \left(\frac{\partial \ln(Z)}{\partial T}\right)_{V,N}} \tag{20.7-30a}$$

$$\boxed{P = k_B T \left(\frac{\partial \ln(Z)}{\partial V}\right)_{T,N}} \tag{20.7-30b}$$

$$S = \frac{U}{T} + k_{\text{B}} \ln(Z)$$

(20.7-30c)

$$A = U - TS = -k_{\text{B}} T \ln(Z)$$

(20.7-30d)

$$C_V = k_{\text{B}} T^2 \left(\frac{\partial^2 \ln(Z)}{\partial T^2} \right)_{V,N} + 2k_{\text{B}} T \left(\frac{\partial \ln(Z)}{\partial T} \right)_{V,N}$$

(20.7-30e)

$$G = A + PV = k_{\text{B}} T \ln(Z) + V k_{\text{B}} T \left(\frac{\partial \ln(Z)}{\partial V} \right)_{T,N}$$

(20.7-30f)

$$\mu = \left(\frac{\partial A}{\partial N} \right)_{T,V} = -k_{\text{B}} T \ln\left(\frac{Z_N}{Z_{N-1}} \right)$$

(20.7-30g)

$$H = U + PV = k_{\text{B}} T^2 \left(\frac{\partial \ln(Z)}{\partial T} \right)_{V,N} + V k_{\text{B}} T \left(\frac{\partial \ln(Z)}{\partial V} \right)_{T,N}$$

(20.7-30h)

where we have repeated the equations for U, P, and S for completeness. In Equation (20.7-30g), we have replaced a derivative with respect to the number of particles by a finite difference, as in Equation (20.5-14), and have attached a subscript to the canonical partition function to indicate the number of particles in the system. The formulas of Equation (20.7-30) are valid for any kind of system, whether solid, liquid, or gas.

Exercise 20.33   Verify the formulas in Equation (20.7-30).

The formulas of Equation (20.7-30) can be applied to a dilute gaseous system by using Equation (20.7-15) to express the canonical partition function in terms of the molecular partition function. The formulas of Section 20.5 are recovered.

Exercise 20.34   Show that Equation (20.7-30) leads to the same formulas for the thermodynamic functions of a dilute gas in terms of the molecular partition function as in Section 20.5.

The thermodynamic functions for a dilute gaseous system are given by the same formulas for a canonical ensemble and a microcanonical ensemble. In the "thermodynamic limit" of a very large system, the thermodynamic variables of any kind of system have the same values in both ensembles. We do not prove this assertion, but it is a remarkable fact, since the canonical ensemble includes system states with all possible values of the energy, while the microcanonical ensemble includes only states of a given fixed energy. Following are some comments on this fact:

Consider Equation (20.7-18) for the canonical distribution function. At ordinary temperatures, $k_B T$ is a very small quantity of energy compared with the energy of an entire macroscopic system, being roughly equal to the energy of one molecule. Therefore, except for the very lowest-energy states, $E/k_B T$ is a very large quantity and $e^{-E/k_B T}$ is a small and rapidly decreasing function of E. Therefore, all but the lowest-energy states of a macroscopic system are nearly unpopulated.

The probability of a system energy level, made up of all of the states of a given energy E, is given by

$$p_E \propto \Omega(E)e^{-E/k_B T} \qquad\qquad \textbf{(20.7-31)}$$

where $\Omega(E)$ is the number of system microstates of energy E (the degeneracy of the system energy level).

In Section 20.5 we estimated a value of Ω and found it to be a very large number. (See Example 20.9.) Furthermore, Ω increases rapidly as a function of E. The situation is depicted schematically in Figure 20.5. Figure 20.5a shows the probability distribution of states, $e^{-E/k_B T}$, and Figure 20.5b shows $\Omega(E)$. Unfortunately, the extremely small size of $e^{-E/k_B T}$ and the extremely large size of $\Omega(E)$ cannot be shown accurately. The large factor is rising so steeply and the small factor falling so rapidly that there is a very small range of values of E where the product of the two factors is much larger than for other values of E, as shown in Figure 20.5c. In fact (although we do not prove this) for a system of N molecules, the width of the energy probability distribution is proportional to $\sqrt{N}$, while the energy itself is proportional to N. For large values of N, the width of the distribution in E becomes very small compared to the value of E. This is why the results of the canonical and microcanonical ensembles are so similar. The energy probability distribution of the microcanonical ensemble, which is nonzero for only a single energy, as shown schematically in Figure 20.5d, is virtually indistinguishable from that of the canonical ensemble.

The small width of the canonical energy probability distribution is related to a very interesting question: why is one macroscopic variable, say U, given as a mathematical function of other thermodynamic variables such as T, V, and N? The energy of almost every system in the ensemble is nearly equal to the ensemble average energy, which we equate with the thermodynamic energy. Average quantities in very large populations behave regularly and predictably, even if the properties of each member of the population do not. This regular behavior leads to the fact that the ensemble average energy is a single-valued mathematical function of T, V, and N. We do not discuss the details of this assertion.

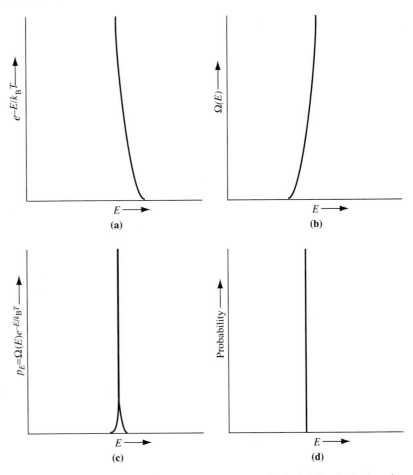

Figure 20.5. The Probability of System States and Levels. (a) Probability distribution of system states in the canonical ensemble as a function of energy (schematic). The probability of system states in the canonical ensemble is a very rapidly decreasing function of the energy. **(b)** The degeneracy of system energy levels as a function of energy (schematic). This degeneracy is a very large and very rapidly increasing function of the energy. **(c)** The canonical probability of system energy levels as a function of energy (schematic). The product of a rapidly decreasing function of the energy and a rapidly increasing function of the energy turns out to be a very sharply peaked function of the energy. **(d)** The microcanonical probability of system energy levels as a function of energy (schematic). This probability distribution is nonzero for only a single value of the energy. However, the canonical distribution is so sharply peaked that the small difference between the micro- canonical and canonical distributions does not produce any significant differences in thermodynamics functions for the two ensembles.

20.8   Classical Statistical Mechanics

Classical statistical mechanics averages over classically described system micro- states to represent system macrostates. Classical mechanics is an approximation to the actual behavior of objects that applies when particle masses and ener- gies are sufficiently large. It is an adequate approximation for the transla- tional motion of heavy atoms and molecules near room temperature, and it is a fairly good approximation for the rotational motion of heavy molecules near room temperature.

Consider first the classical states of a system consisting of a single particle confined in a one-dimensional box of length a. We construct a two-dimensional state space with the coordinate plotted on one axis and the momentum on the other axis.

The point representing the instantaneous state of the system moves along a trajectory in this space, which is called a **phase space**. If undisturbed, the particle moves back and forth in the box with constant speed. The trajectory corresponding to this motion is depicted in Figure 20.6. The trajectory jumps suddenly from a positive value of the momentum to a negative value and vice versa as the particle hits the walls of the box.

For a system of one point mass particle moving in three dimensions, the phase space has six dimensions, three for coordinates and three for momentum components. The phase space for a system of N atoms without electronic excitation has $6N$ dimensions.

We represent our system by a canonical ensemble of replicas of the system, just as in quantum statistical mechanics. We plot all of the phase points for the many systems of the ensemble in a single phase space, so there is a swarm of very many phase points in this phase space, one for each system in the ensemble. This swarm of points moves about in phase space in a way that is analogous to fluid flow in three dimensions.[6]

The probability distribution (probability density) f for the ensemble is defined such that $f(\mathbf{r}_1, \mathbf{p}_1, \mathbf{r}_2, \mathbf{p}_2, \ldots, \mathbf{r}_N, \mathbf{p}_N) \, d^3\mathbf{r}_1 \, d^3\mathbf{p}_1 \, d^3\mathbf{r}_2 \, d^3\mathbf{p}_2 \cdots d^3\mathbf{r}_N \, d^3\mathbf{p}_N$ is the probability that the phase point of a randomly selected system of the ensemble will lie in the $6N$-dimensional volume element $d^3\mathbf{r}_1 \, d^3\mathbf{p}_1 \, d^3\mathbf{r}_2 \, d^3\mathbf{p}_2 \cdots d^3\mathbf{r}_N \, d^3\mathbf{p}_N$. This probability is equal to the fraction of the ensemble's phase points that lie in the volume element.

Classical mechanical formulas must be equal to those obtained by taking a limit of quantum mechanical formulas as masses and energies become large. The equilibrium canonical probability density must therefore be the same function of the energy as that of quantum statistical mechanics, given by Equation (20.7-18). For a one-component system of atoms without electronic excitation, but with intermolecular forces, the classical energy (classical Hamiltonian function) is

$$\mathscr{H} = \mathscr{K} + \mathscr{V} = \frac{1}{2m} \sum_{i=1}^{N} (p_{xi}^2 + p_{yi}^2 + p_{zi}^2) + \mathscr{V}(q) \qquad \textbf{(20.8-1)}$$

where $\mathscr{K}$ is the kinetic energy and $\mathscr{V}$ is the potential energy. The canonical probability density is

$$f = f(p, q) = f[\mathscr{H}(p, q)] = \frac{1}{Z_{\mathrm{cl}}} e^{-\mathscr{H}(p,q)/k_{\mathrm{B}}T} \qquad \textbf{(20.8-2)}$$

where p and q stand for all of the momenta and coordinates.

The probability density is normalized so that its integral over all of phase space equals unity. The **classical partition function** Z_{cl}, which is also called the

Other mathematical spaces are also called phase spaces, but all of them have time-dependent quantities plotted on all axes.

The limit in which quantum mechanics must produce classical results is called the correspondence limit.

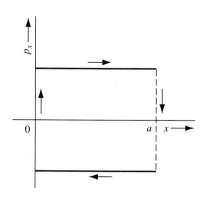

Figure 20.6. A Trajectory in the Two-Dimensional Phase Space for a Particle in a Hard One-Dimensional Box. This figure conveys the same information as does Figure 9.20: the particle moves back and forth at constant speed, changing its direction suddenly at the ends of the box.

[6] D. A. McQuarrie, *Statistical Mechanics*, Harper & Row, New York, 1976, pp. 119ff.

phase integral, accomplishes this normalization:

$$Z_{cl} = \int \int \cdots \int e^{-\mathscr{H}(p,q)/k_B T}\, d^3\mathbf{r}_1\, d^3\mathbf{p}_1\, d^3\mathbf{r}_2\, d^3\mathbf{p}_2 \cdots d^3\mathbf{r}_N\, d^3\mathbf{p}_N$$

$$= \int e^{-\mathscr{H}(p,q)/k_B T}\, dq\, dp \tag{20.8-3}$$

where dp stands for all of the momentum differentials and dq stands for all of the coordinate differentials.

The integrations range over all values of the coordinates inside the box containing the system, and from negative infinity to positive infinity for each momentum component. The integral over the momentum components factors into a product of integrals, one for each momentum component, giving the result

$$Z_{cl} = (2\pi m k_B T)^{3N/2} \int e^{-\mathscr{V}(q)/k_B T}\, dq = (2\pi m k_B T)^{3N/2} \zeta \tag{20.8-4}$$

where ζ is called the **configuration integral**, an integral over the $3N$ coordinates:

$$\zeta = \int e^{-\mathscr{V}(q)/k_B T}\, dq \tag{20.8-5}$$

Exercise 20.35   Carry out the integrations to obtain Equation (20.8-4).

Consider the case that the system is a dilute gas, so that intermolecular forces can be neglected. Let the system be confined in a box of volume V, inside which the potential energy is a constant, taken to equal zero. In this case, the configuration integral is

$$\zeta = \int e^0\, d^3\mathbf{r}_1\, d^3\mathbf{r}_2 \cdots d^3\mathbf{r}_N = V^N \tag{20.8-6}$$

The classical partition function is

$$Z_{cl} = (2\pi m k_B T)^{3N/2} V^N \tag{20.8-7}$$

Let $f_1(\mathbf{p}_1, \mathbf{r}_1)\, d^3\mathbf{p}_1\, d^3\mathbf{r}_1$ be the probability that the phase point of molecule number 1 will be found in the volume element $d^3\mathbf{p}_1\, d^3\mathbf{r}_1$. The probability density f_1 is equal to the full probability density in phase space integrated over all possible states of the $N - 1$ other particles:

$$f_1(\mathbf{p}_1, \mathbf{r}_1) = \int \cdots \int f(q, p)\, d^3\mathbf{r}_2\, d^3\mathbf{p}_2\, d^3\mathbf{r}_3\, d^3\mathbf{p}_3 \cdots d^3\mathbf{r}_N\, d^3\mathbf{p}_N \tag{20.8-8a}$$

If there are no intermolecular forces, each of the integrations is the same as the integrations in Z_{cl}, so that

$$f_1(\mathbf{p}_1, \mathbf{r}_1) = \frac{1}{Z_{cl}}(2\pi m k T)^{3(N-1)/2} V^{N-1} e^{-p_1^2/2mkT}$$

$$= (2\pi m k T)^{-3/2} V^{-1} e^{-p_1^2/2mkT} \quad \text{(dilute gas)} \tag{20.8-8b}$$

where

$$p_1^2 = p_{x1}^2 + p_{y1}^2 + p_{z1}^2 \tag{20.8-9}$$

The function in Equation (20.8-8) can be written:

$$f_1(\mathbf{p}_1, \mathbf{r}_1) = \frac{1}{z_{\mathrm{cl}}} e^{-p_1^2/2mk_\mathrm{B}T} = \frac{1}{z_{\mathrm{cl}}} e^{-\varepsilon/k_\mathrm{B}T} \qquad (20.8\text{-}10)$$

where the energy of the molecule (which is all kinetic energy) is denoted by ε, and we define the **classical molecular partition function**, or **classical one-particle phase integral**:

$$z_{\mathrm{cl}} = \int e^{-\varepsilon/k_\mathrm{B}T} \, d^3\mathbf{p}_1 \, d^3\mathbf{r}_1 = \int e^{-p_1^2/2mk_\mathrm{B}T} \, d^3\mathbf{p}_1 \, d^3\mathbf{r}_1 \qquad (20.8\text{-}11)$$

$$= (2\pi m k_\mathrm{B}T)^{3/2} V \qquad (20.8\text{-}12)$$

Compare the classical molecular partition function and canonical partition function of the monatomic dilute gas with their quantum mechanical counterparts:

$$z_{\mathrm{qm}} = \left(\frac{2\pi m k_\mathrm{B}T}{h^2}\right)^{3/2} V^3 = \frac{z_{\mathrm{cl}}}{h^3} \qquad (20.8\text{-}13)$$

$$Z_{\mathrm{qm}} = \frac{z_{\mathrm{qm}}^N}{N!} = \frac{(2\pi m k_\mathrm{B}T/h^2)^{3N/2} V^N}{N!} = \frac{Z_{\mathrm{cl}}}{h^{3N}N!} = \frac{1}{N!}\left(\frac{z_{\mathrm{cl}}}{h^3}\right)^N \qquad (20.8\text{-}14)$$

The divisor $N!$ that occurs in the quantum partition function but not in the classical partition function arises from the indistinguishability of the particles, which is not recognized in classical mechanics. The factors of Planck's constant indicate a relationship between a volume in phase space and a quantum state. Consider, for example, the two-dimensional phase space of Figure 20.6. Figure 20.7 shows the same diagram, with several trajectories indicated. The first trajectory is for the motion of a classical particle whose energy happens to be ε_1, the $n = 1$ energy eigenvalue of a quantum mechanical particle of the same mass. The second trajectory is for the motion of a classical particle with energy equal to ε_2, etc.

Let us calculate the area in the figure that lies between any two adjacent trajectories. For n', a given value of the quantum number, the energy is

$$\varepsilon_{\mathrm{cl}} = \frac{p^2}{2m} = \varepsilon_{\mathrm{qm}} = \varepsilon_n' = \frac{h^2 n'^2}{8ma^2} \qquad (20.8\text{-}15)$$

When Equation (20.8-15) is solved for the momentum, we obtain

$$|p| = \frac{hn'}{2a} \qquad (20.8\text{-}16)$$

The area between the trajectory for n' and that for $n' + 1$ is

$$\text{Area} = \left[\frac{h(n'+1)}{2a} - \frac{hn'}{2a}\right](2a) = h \qquad (20.8\text{-}17)$$

This equation is an example of a general relationship. For each two dimensions in phase space (one coordinate and one momentum component) an area equal to h corresponds to one quantum state. For a six-dimensional phase space, a volume equal to h^3 corresponds to one quantum state, and in a $6N$-dimensional phase space, a volume equal to h^{3N} corresponds to one quantum state.

The molecular phase integral is an integral of the probability of states relative to a state of zero energy, taken over the entire phase space, and it is a measure

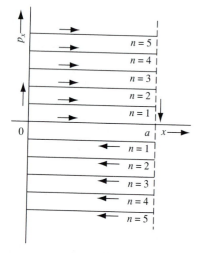

Figure 20.7. Several Trajectories in a Two-Dimensional Phase Space for a Particle in a Hard One-Dimensional Box. This figure shows the classical trajectories in phase space for several energies, each one of which is equal to one of the quantum-mechanical quantized energies. The area between each pair of successive trajectories is equal to Planck's constant.

of the volume in phase space that is effectively accessible to a molecule, just as the quantum molecular partition function is a measure of the number of states effectively accessible to a molecule. Dividing the molecular phase integral by h^3 converts the phase integral into an approximation for the quantum molecular partition function.

The canonical partition function is similar. Division by a factor of h^{3N} is the same as dividing by the volume per quantum state. This division and a division by $N!$ to account for indistinguishability convert a canonical phase integral into an approximation for a quantum statistical canonical partition function.

For diatomic and polyatomic molecules, the molecular phase integral is a product of translational, rotational, and vibrational factors, and factors of h have to be included for all factors. The rotational factor in the classical molecular partition function of a substance with diatomic molecules is given by

$$z_{\text{cl, rot}} = 8\pi^2 I_e k_B T \tag{20.8-18}$$

Exercise 20.36

In Exercise 20.36, the integrations over the angles range from 0 to π for θ and 0 to 2π for ϕ, while the integrations over the conjugate momenta range from $-\infty$ to ∞.

Derive Equation (20.8-18). You must express the rotational energy in terms of the conjugate momenta for the angles θ and ϕ, as given in Appendix D. The integration is taken over both the angles and the conjugate momenta.

The formula in Equation (20.8-18) is turned into that for the quantum mechanical rotational partition function by division by h^2, since there are two rotational coordinates (two angles are required to specify the orientation of a diatomic molecule), and by division by the symmetry number, which is needed to account for the indistinguishability of the nuclei in a homonuclear molecule.

A classical partition function that has been corrected for the quantum effects as above described is called a **semiclassical partition function**. The semiclassical canonical partition function of a dilute monatomic gas without electronic excitation is

$$Z_s = \frac{Z_{\text{cl}}}{h^{3N}N!} = \frac{1}{N!}\left(\frac{2\pi^2 m k_B T}{h^2}\right)^{3N/2} V^N \tag{20.8-19}$$

which is the same as the quantum canonical partition function of Equations (20.7-15) and (20.7-16).

For a diatomic molecule, the vibrational factor in the semiclassical partition function is

$$z_{s,\text{vib}} = \frac{1}{h}\int \exp\left[-\frac{p^2/2\mu + kx^2/2}{k_B T}\right] dp\, dx$$

$$= \frac{1}{h}(2\pi\mu k_B T)^{1/2}(2\pi k_B T/k)^{1/2} = \frac{1}{h}(2\pi k_B T)\sqrt{\frac{\mu}{k}} = \frac{k_B T}{h\nu} \tag{20.8-20}$$

where μ is the reduced mass, k is the force constant, and ν is the frequency of the oscillator, from Equations (9.1-10) and (14.2-24).

The semiclassical vibrational partition function of Equation (20.8-20) approaches the quantum harmonic oscillator vibrational partition function only for large values of T or small values of ν. However, inspection of Figure 14.4 shows that for large values of the vibrational quantum number (which are occupied at high temperature), the semiclassical harmonic approximation to the potential energy function is poor, and the harmonic oscillator partition function will be a poor approximation to the actual partition function.

Exercise 20.37	Show that for large values of T, the vibrational partition function of Equation (20.4-17) approaches the semiclassical vibrational partition function of Equation (20.8-20).

The semiclassical electronic partition function is not used, because the mass of the electron is so small that classical mechanics cannot describe the motions of electrons in molecules to any usable approximation.

The Energy in Classical Statistical Mechanics

The ensemble average energy is given for a general system by

$$U = \langle E \rangle = \frac{1}{Z_{cl}} \int \mathscr{H}(p, g) e^{-\mathscr{H}(p,q)/k_B T} \, dp \, dq$$

$$= \frac{1}{Z_{cl}} k_B T^2 \int \left(\frac{\partial}{\partial T} \left[e^{-\mathscr{H}(p,q)/k_B T} \right] \right) dp \, dq$$

$$= k_B T^2 \left(\frac{\partial \ln(Z_{cl})}{\partial T} \right)_V \qquad \text{(20.8-21)}$$

where we have indicated that V is held constant in the differentiation, since the potential energy can depend on V.

If the system is a dilute gas,

$$U = \langle E \rangle = N \langle \varepsilon \rangle = N k_B T^2 \left(\frac{\partial \ln(z_{cl})}{\partial T} \right)_V \qquad \text{(20.8-22)}$$

Exercise 20.38	**a.** Verify Equation (20.8-22). **b.** Using Equation (20.8-22), show that the energy of a classical dilute monatomic gas without electronic excitation is $$U = \frac{3N k_B T}{2} \qquad \text{(20.8-23)}$$

Equipartition of Energy

The classical **theorem of equipartition of energy** states that if a molecular variable occurs in the classical energy in a quadratic form, the contribution to the ensemble average system energy corresponding to that variable is equal to $N k_B T/2$. Equation (20.8-23) is an example of this theorem, since each momentum component occurs in a quadratic form in the energy expression of Equation (20.8-1). The equipartition of energy was used by Rayleigh and Jeans in their analysis of blackbody radiation, discussed in Chapter 9.

A proof of the theorem follows: Let x be some variable, such as a coordinate or a momentum component. The contribution of this variable to the energy of one molecule is assumed to be

$$\varepsilon = ax^2 \qquad \text{(20.8-24)}$$

where a is a constant. The average energy contribution is

$$\langle E \rangle = N \langle \varepsilon \rangle = N \frac{\int ax^2 e^{-ax^2/k_B T}\, dx}{\int e^{-ax^2/k_B T}\, dx} \qquad \textbf{(20.8-25a)}$$

where the limits on the integrals are assumed to be from negative infinity to positive infinity. When the integrals are looked up in a table such as that of Appendix C, we obtain

$$\langle E \rangle = \frac{N a \pi^{1/2} (k_B T/a)^{3/2}/2}{(\pi k_B T/a)^{1/2}} = \frac{N k_B T}{2} \qquad \textbf{(20.8-25b)}$$

The result is independent of a, so a translational energy contribution is independent of the mass of the molecules and a rotational energy contribution is independent of the moment of inertia.

In classical statistical mechanics, the vibration of a diatomic molecule makes a contribution of $N k_B T$ to the energy, since there is one coordinate and one momentum component, both of which enter quadratically in the energy expression.

Exercise 20.39   Show that the vibrational energy given by Equation (20.5-24) approaches the equipartition value of $N k_B T$ for large values of T.

Other Thermodynamic Functions in Classical Statistical Mechanics

The formulas for thermodynamic functions in terms of the classical canonical partition function are the same as those in terms of the quantum canonical partition function. The classical version of the definition of the statistical entropy given in Equation (20.7-27) is

$$S_{\text{st}} = -k_B \int f(p, q) \ln[f(p, q)]\, dp\, dq \qquad \textbf{(20.8-26)}$$

$$= -k_B \int f(p, q) \left[-\frac{E}{k_B T} - \ln(Z_{\text{cl}}) \right] dp\, dq = \frac{U}{T} + k_B \ln(Z_{\text{cl}}) \qquad \textbf{(20.8-27)}$$

Equation (20.8-27) is analogous to Equation (20.7-29). If the classical partition function is used without conversion into a semiclassical partition function, values are obtained for the entropy that differ by a constant from the semiclassical values.

Exercise 20.40   **a.** Evaluate the standard-state entropy of 1.000 mol of argon at 298.15 K using Equation (20.8-27) and the classical canonical partition function of Equation (20.8-7).
b. Evaluate the standard-state entropy of 1.000 mol of argon at 298.15 K using Equation (20.8-27) and the semiclassical partition function of Equation (20.8-19).

Now that we have the same formulas for the energy and the entropy as in the quantum canonical ensemble, all of the formulas for the other thermodynamic functions must be the same, and all of the formulas of Equation (20.7-30) hold for the classical canonical ensemble.

a. Show that all of the formulas in Equation (20.7-30) are valid for the classical canonical ensemble.

b. Obtain a formula for the entropy of a dilute diatomic gas in the semiclassical approximation. Include translation and rotation, but no vibrational or electronic energy.

Classical Statistical Mechanics of Nonideal Gases and Condensed Phases

Gas nonideality and nearly all of the important properties of liquids and solids are due to intermolecular forces. The application of classical statistical mechanics to systems of particles with intermolecular forces is complicated, although it is more tractable than the quantum statistical mechanics of the same systems. We provide only a few comments.

The simplest intermolecular potential function is a sum of terms that depend only on the distances between pairs of molecules, introduced in Chapter 16:

$$\mathscr{V} = \sum_{i=1}^{N-1} \sum_{j=i+1}^{N} u(r_{ij}) \tag{20.8-28}$$

where r_{ij} is the distance between the centers of molecules i and j.

The momentum integrations of the phase integral in Equation (20.8-4) still apply, but the configuration integral becomes

$$\zeta = \int \exp\left[\frac{-\sum_{i=1}^{N-1}\sum_{j=i+1}^{N} u(r_{ij})}{k_B T} \right] dq \tag{20.8-29}$$

Mayer[7] showed that the configuration integral can be expressed as a sum of integrals called **cluster integrals** and also that the virial coefficients can be expressed as sums of cluster integrals of another type. The second virial coefficient of a nonideal gas is given by a single integral involving the coordinates of only two particles:

$$B_2 = -\frac{N_A}{2V} \int [e^{-u(r_{12})/k_B T} - 1]\, d^3\mathbf{r}_1\, d^3\mathbf{r}_2 \tag{20.8-30}$$

where both integrations are over all positions in the box containing the gas.

The variables in this integration can be changed from $\mathbf{r}_1$ and $\mathbf{r}_2$ to $\mathbf{r}_1$ and $\mathbf{r}_{12} = \mathbf{r}_2 - \mathbf{r}_1$:

$$B_2 = -\frac{N_A}{2V} \int d^3\mathbf{r}_1 \int [e^{-u(r_{12})/k_B T} - 1]\, d^3\mathbf{r}_{12}$$

Integration over $\mathbf{r}_{12}$ gives a result that is independent of $\mathbf{r}_1$ unless particle 1 is very near the walls of the box, so that the integration over $\mathbf{r}_1$ will then just give a factor of V, the box volume, giving

$$B_2 = -\frac{N_A}{2} \int_0^\infty [e^{-u(r)/k_B T} - 1] 4\pi r^2\, dr \tag{20.8-31}$$

[7] J. E. Mayer and M. G. Mayer, *Statistical Mechanics*, Wiley, New York, 1940. See T. Hill, *Statistical Thermodynamics*, Addison-Wesley, Reading, MA, 1960, pp. 261ff, for a readable discussion.

where the integral was transformed to spherical polar coordinates and the integrations over θ and ϕ have been carried out. The subscript on r_{12} has been omitted, and the limit of the r_{12} integration has been extended to infinity. This change in limit produces negligible error, since the integrand function rapidly approaches zero for large values of r_{12}.

▼ **EXAMPLE 20.14**

Calculate the second virial coefficient for the hard sphere gas.

Solution

For the hard sphere gas with molecular diameter a,

$$u(r) = \begin{cases} \infty & \text{if } r < a \\ 0 & \text{if } r > a \end{cases}$$

so that

$$e^{-u(r)/k_{\mathrm{B}}T} - 1 = \begin{cases} -1 & \text{if } r < a \\ 0 & \text{if } r > a \end{cases}$$

The second virial coefficient of the hard sphere gas is

$$B_2 = -\frac{N_A}{2} \int_0^a (-1) 4\pi r^2 \, dr = \frac{2\pi N_A a^3}{3} \tag{20.8-32}$$

Unlike the virial coefficients of real gases, this virial coefficient is independent of temperature.

▲

Exercise 20.42

Obtain a formula for the second virial coefficient of a "square well" gas, for which

$$u(r) = \begin{cases} \infty & \text{if } r < a \\ -u_0 & \text{if } a < r < b \\ 0 & \text{if } r > b \end{cases}$$

where u_0 is a positive constant. Explain why this virial coefficient depends on temperature while that of a hard sphere gas does not.

Summary

The properties of macroscopic systems are determined by the behavior of molecules making up the system. Through statistical mechanics we can in principle calculate these properties from molecular properties.

The two principal postulates of statistical mechanics are (I) observed macroscopic properties of a system held at a fixed energy can be equated to an average over system mechanical states corresponding to that energy and (II) all system states of the same energy are given equal weight in this average.

For a dilute gaseous system, we can average over molecular states with a molecular probability distribution. The most probable distribution was found as an approximation to the average distribution. It is the Boltzmann distribution

$$p_j = \frac{1}{z} g_j e^{-\varepsilon_j/k_B T}$$

where p_j is the probability that a randomly selected molecule would be found in molecule energy level j of degeneracy g_j and energy ε_j.

The quantity z is the molecular partition function. It can be written as a sum over levels,

$$z = \sum_j g_j e^{-\varepsilon_j/k_B T}$$

and can also be written as a sum over states.

For atomic gases,

$$z = z_{tr} z_{elec}$$

and for diatomic or polyatomic gases

$$z = z_{tr} z_{rot} z_{vib} z_{elec}$$

All factors except for the electronic partition function can be expressed with general formulas. The electronic partition function must be summed up term by term but can often be approximated by a single term.

The thermodynamic functions of a dilute gas can be calculated from the molecular partition function of the gas. The necessary formulas are based on the postulates of statistical mechanics and on the definition of the statistical entropy

$$S_{st} = k_B \ln(\Omega)$$

where Ω is the total number of system mechanical states that might be occupied, given the information we possess about the state of the system.

A theoretical expression for the equilibrium constant for a chemical reaction in a dilute gas mixture was obtained:

$$K_P = e^{-\Delta\varepsilon_0/k_B T} \prod_{j=1}^{c} \left(\frac{z_j'^{\circ}}{N_A}\right)^{v_j}$$

where z_j' is the molecular partition function of substance number j, given that the volume of the system is equal to the molar volume of a dilute gas at pressure P° and at the temperature of the system.

Statistical mechanics can also be studied through an ensemble, an imaginary collection of many replicas of the physical system. All of the systems in the ensemble are in the same macroscopic state as the physical system, but members of the ensemble occupy all possible microscopic states compatible with the macroscopic state. The probability distribution for the canonical ensemble is

$$p_i = \frac{1}{Z} e^{-E_i/k_B T}$$

where Z is the canonical partition function, which is a sum over system states. In the case that the system is a dilute gas,

$$Z = \frac{z^N}{N!}$$

where z is the molecular partition function.

In classical statistical mechanics, the partition functions are integrals over states, called phase integrals, instead of sums over states. The formulas for the thermodynamic functions in terms of the canonical partition function are the same as those for the quantum canonical ensemble.

Additional Reading

D. Chandler, *Introduction to Modern Statistical Mechanics*, Oxford University Press, New York, 1987
This small book is a text for a one-quarter or one-semester course in statistical mechanics for chemists at the senior/first-year graduate level. It is written by one of the prominent practitioners of statistical mechanics and emphasizes modern applications.

N. Davidson, *Statistical Mechanics*, McGraw-Hill, New York, 1962
This book presents a complete and clear exposition of the principles of statistical mechanics for chemists at the first-year graduate level.

E. A. Desloge, *Statistical Physics*, Holt, Rinehart, & Winston, New York, 1966
This is a textbook for a senior/first-year graduate course in statistical mechanics and kinetic theory for physics students. It is clearly written and useful.

H. L. Friedman, *A Course in Statistical Mechanics*, Prentice-Hall, Englewood Cliffs, NJ, 1985
This is a text for a one-semester course in statistical mechanics for chemists at the graduate level. It is mathematically oriented and emphasizes the liquid state.

T. Hill, *Statistical Thermodynamics*, Addison-Wesley, Reading, MA, 1960
A clear and useful introduction to equilibrium statistical mechanics for chemistry students at the first-year graduate level.

C. Kittel and H. Kroemer, *Thermal Physics*, 2nd ed., W. H. Freeman, San Francisco, 1980
This text for an undergraduate physics course in thermodynamics and statistical mechanics is clear and useful.

D. A. McQuarrie, *Statistical Mechanics*, Harper & Row, New York, 1976
A complete and carefully written account of statistical mechanics for first-year graduate students in physical chemistry. It is probably the leading modern textbook in this area.

PROBLEMS

Problems for Section 20.1

*20.43. a. Find the total number of system vibrational states for a system of three hydrogen molecules with a total energy of $3hv$.

b. Find the molecular probability distribution for the system of part a.

20.44. Consider a system of four distinguishable rigid rotating diatomic molecules with a total energy of $20hB$, where B is the rotational constant.

a. Make a list of the possible system states analogous to that of Table 20.1. Don't forget the degeneracies.

b. Calculate the average distribution of molecular levels, analogous to that of Equation (20.1-7).

c. Calculate the Boltzmann distribution for the same total energy.

Problems for Section 20.2

*20.45. It would be impossible to write down the antisymmetrized wave function of a system containing more than a few fermions. Estimate the number of terms in an antisymmetrized wave function for a system containing a number of fermions equal to Avogadro's number, using Stirling's approximation.

20.46. The number of distinct ways to choose states for N_j bosons from a level with degeneracy g_j is given by Equation (20.2-21). Find the percent differences between the result of this formula, the result of Equation (20.2-7), and the result of Equation (20.2-9) for $g_j = 1000$ and $N_j = 5$. Repeat the calculation for $g_j = 1000000$ and $N_j = 5$.

Problems for Section 20.3

20.47. Calculate the ratio of the population of the $v = 1$ vibrational state of CO to the $v = 0$ vibrational state at (a) 298.15 K, (b) 500 K, (c) 1000 K, and (d) 5000 K. (e) What is the limit as $T \to \infty$?

*20.48. Calculate the ratio of the population of the $J = 1$ rotational level of CO to the $J = 0$ rotational level at (a) 298.15 K and (b) 1000 K. (c) What is the limit as $T \to \infty$?

20.49. Find the J value with the largest population for CO at 500 K. The equilibrium internuclear distance is equal to 1.13×10^{-10} m.

Problems for Section 20.4

*20.50. Calculate the value of the molecular partition function of argon at 298.15 K for a volume of 0.00100 m^3 and also for a volume of 1.00 m^3. Explain in words what the difference between the two values means.

20.51. Calculate the molecular partition functions of neon and krypton at 298.15 K, assuming each gas is confined in a volume of 24.45 L. Explain the difference in the two results in terms of effective number of accessible states.

20.52. Calculate the molecular partition function of argon at 300. K and at 500. K, assuming the gas is confined in a volume of 100. L. Explain the difference in the two results in terms of the effective number of accessible states.

*20.53. a. Calculate the four factors in the partition function of N_2 at 298.15 K if 1.000 mol is confined at 1.000 atm.

b. Calculate the value of z/N for the system of part a. Calculate the value of z/N for 2.000 mol of N_2 at 298.15 K and 1.000 atm. Explain why z/N does not depend on the size of the container so long as the pressure is unchanged.

20.54. A formula for the rotational partition function of a diatomic substance that gives corrections to the formula of Equation (20.4-14) is[8]

$$z_{rot} = \frac{T}{\sigma \Theta_{rot}} \left[1 + \frac{1}{3} \left(\frac{\Theta_{rot}}{T} \right) + \frac{1}{15} \left(\frac{\Theta_{rot}}{T} \right)^2 + \frac{4}{315} \left(\frac{\Theta_{rot}}{T} \right)^3 + \cdots \right]$$

where Θ_{rot} is sometimes called the "rotational temperature":

$$\Theta_{rot} = \frac{h^2}{8\pi^2 I_e k_B} = \frac{hB_e}{k_B}$$

a. Calculate the rotational partition function of H_2 at 298.15 K and compare your answer with that of Exercise 20.15.

b. Calculate the rotational partition function of Cl_2 at 298.15 K and compare your result with the value obtained using Equation (20.4-14).

*20.55. The bond distances in H_2O are equal to 95.8 pm, and the bond angle is equal to 104.45°. Find the location of the center of mass and the principal moments of inertia. Calculate the rotational partition function at 298.15 K. Don't forget the symmetry number.

20.56. The bond distances in NH_3 are equal to 101.4 pm, and the bond angles are equal to 107.3°. Find the rotational partition function of NH_3 at 500 K.

*20.57. Represent the electron in a hydrogen atom as an electron in a cubic box 1.00×10^{-10} m on a side.

a. Find the electronic partition function at 298.15 K, using Equation (20.3-18).

b. Find the electronic partition function at 298.15 K by summing the terms explicitly. Use a programmable calculator or a computer if possible. Continue the sum until one

---

[8] N. Davidson, *op. cit.*, p. 118.

added term does not change the first four digits of the sum. Explain any difference between the results of parts a and b.

20.58. In calculating rotational and vibrational partition functions, the upper limit of the sum is taken as infinity. However, states with very high values of the quantum numbers do not occur because the molecule would dissociate before reaching such a high energy. Explain why this fact does not produce a serious error in calculating the partition function.

Problems for Section 20.5

20.59. a. Using a thermodynamic calculation, find the standard-state molar entropy of helium at 323.15 K. State any assumptions and approximations.

b. Using the value from part a, estimate the value of Ω for 1.000 mol of helium at 323.15 K. Find the ratio of this value to the value of Example 20.9 and comment on this ratio.

*20.60. Calculate the standard-state molar entropy of argon at 298.15 K. Compare your result with the value in Table A9. Compare your result with the value for helium at the same temperature and comment on the difference.

20.61. Calculate the standard-state energy, entropy, and Gibbs energy of 1.000 mol of xenon at 298.15 K.

20.62. Calculate the translational, rotational, and vibrational contributions to the constant-volume heat capacity of SO_2 gas at 298.15 K and 1.000 bar. The vibrational frequencies are in Figure 14.6.

*20.63. Calculate the standard-state molar entropy of CO at 298.15 K. Compare your result with the value in Table A9.

20.64. Using data in Exercise 20.18, make a graph of the electronic contribution to the molar energy of NO and a graph of the electronic contribution to the molar heat capacity of NO between 0 K and 500 K. The behavior of this electronic heat capacity as a function of temperature is known as the Schottky effect.

20.65. Calculate the molar entropy of water vapor at 100°C and 1.000 bar. The bond distances and bond angle are in Problem 20.55. The vibrational frequencies are 4.7817×10^{13} s^{-1}, 1.0947×10^{14} s^{-1}, and 1.1260×10^{14} s^{-1}.

20.66. Calculate the value of $\mu - \varepsilon_0$ for water vapor at 100°C and 1.000 bar, where ε_0 is the ground-state energy of the molecule.

*20.67. Calculate the molar heat capacity at constant pressure for carbon dioxide at 298.15 K. Compare your result with the value in Table A9.

Problems for Section 20.6

20.68. Calculate the equilibrium constant for the dissociation of hydrogen at 5000 K.

20.69. Calculate the value of the equilibrium constant at 900 K for the reaction

$$CO_2 + H_2 \rightleftharpoons CO + H_2O$$

Carbon dioxide is linear, with bond lengths of 1.161×10^{-10} m. The vibrational frequencies are in Example 20.8. Information about the H_2O molecule is in Problems 20.55 and 20.65. The value of ΔU_0 for the reaction is equal to 40.33 kJ mol^{-1}.[9] The experimental value of the equilibrium constant is 0.46. Calculate your per cent error.

*20.70. Calculate the value of the equilibrium constant at 750 K for the reaction

$$H_2 + I_2 \rightleftharpoons 2HI$$

The value of ΔU_0° is equal to -8.461 kJ mol^{-1}.[10]

20.71. Without doing any detailed calculations, estimate the equilibrium constant for each of the following gas-phase reactions. State any assumptions or approximations.

a. $NH_3 + D_2O \rightleftharpoons NH_2D + HDO$

b. $^{35}Cl_2 + {}^{37}Cl \rightleftharpoons {}^{35}Cl^{37}Cl + {}^{35}Cl$

20.72. Using statistical mechanics, show that the equilibrium constant for a racemization reaction equals unity.

20.73. The ionization energy of the hydrogen atom is equal to 13.60 eV. Find the equilibrium constant for the dissociation of a hydrogen atom into a proton and an electron at 10000 K.

Problems for Section 20.7

20.74. Derive a formula for the change in the canonical partition function when the zero of energy is changed.

*20.75. From the result of Example 20.2, calculate the value of the canonical partition function for 1.000 mol of argon confined in a volume of 25 L at 298.15 K.

20.76. a. Using the value of third-law entropy from Table A9 and the statistical mechanical value for the energy, calculate the experimental value of the Helmholtz energy of 1.000 mol of oxygen gas at 298.15 K and 1.000 atm. Take the energy of supercooled oxygen gas at 0 K to equal zero.

b. Find the value of the canonical partition function for this system from the result of part a.

c. Find the value of the molecular partition function from the result of part b.

[9] G. Herzberg, *Molecular Spectra and Molecular Structure. II. Infrared and Raman Spectra of Polyatomic Molecules*, Van Nostrand Reinhold, New York, 1945, p. 429.

[10] K. P. Huber and G. Herzberg, *Molecular Spectra and Molecular Structure. IV. Constants of Diatomic Molecules*, Van Nostrand Reinhold, New York, 1979, pp. 240, 330, 324.

20.77. Calculate the value of the molecular partition function of oxygen gas at 298.15 K if 1.000 mol of the gas is confined at a pressure of 1.000 atm. Compare your result with the result of the previous problem.

Problems for Section 20.8

20.78. a. Sketch a diagram in a two-dimensional phase space for a harmonic oscillator that is analogous to Figure 20.6 for a particle in a hard box. Show the trajectories for energies equal to the $v = 0$, $v = 1$, and $v = 2$ quantum states.

b. Calculate the area between any two such trajectories.

*20.79. Calculate the vibrational contribution to the molar heat capacity of CO at 500 K, using the semiclassical approximation for the vibration. Compare your result with that obtained by using Equation (20.5-31). Find the minimum temperature at which the two results agree within 5% and the minimum temperature at which they agree within 1%.

20.80. The intermolecular potential of argon can be represented as a square well with $a = 3.162 \times 10^{-10}$ m, $b = 5.850 \times 10^{-10}$ m, and $u_0 = 9.58 \times 10^{-22}$ J. Evaluate the second virial coefficient of argon at 173 K, 273 K, and 373 K. Compare your values with those in Table A2. Find the temperature at which the second virial coefficient is equal to zero.

General Problems

20.81. a. Choose a diatomic gas and compute its translational, rotational, vibrational, and electronic partition functions at 298.15 K and 1.000 bar, looking up parameters as needed in Table A16 or in some more complete table.[11] Unless you choose NO or a similar molecule with an unpaired electron, assume that only the ground electronic state needs to be included.

------

[11] K. P. Huber and G. Herzberg, *op. cit.*

b. Compute the standard-state energy, entropy, Helmholtz energy, Gibbs energy, and heat capacity of your gas at 298.15 K. It might be interesting for members of a class to choose different substances and to make a table of their collective results.

*20.82. Identify the following statements as either true or false. If a statement is true only under special circumstances, label it as false.

a. Dilute gases are the only systems that can be treated with statistical mechanics.

b. Dilute gases are the only systems for which molecular partition functions apply.

c. It is impossible for a state of higher energy to have a larger population than a state of lower energy.

d. The statistical mechanical approach used in the first seven sections of this chapter is equivalent to using the microcanonical ensemble.

e. An isomerization reaction must have an equilibrium constant equal to unity.

f. The value of the molecular partition function can be interpreted as the effective number of states available to the molecule at the temperature and pressure of the system.

g. Sums over states and sums over levels can be used interchangeably.

h. For a system of macroscopic size, the average distribution of molecular states and the most probable distribution are essentially the same.

i. Fermion and boson probability distributions become more and more like the Boltzmann distribution as the occupation becomes more dilute.

j. Fermion and boson probability distributions become more and more like the Boltzmann distribution as the energy increases.

k. Values of standard-state thermodynamic functions of gases calculated with statistical mechanics are always less reliable than experimental values.

l. Molecular partition functions are exactly factored into translational, rotational, vibrational, and electronic factors.

21 Theories of Nonequilibrium Processes

PREVIEW

This chapter is devoted to the theoretical discussion of nonequilibrium processes. It treats a collection of loosely related topics, but all of the discussions involve macroscopic consequences of molecular processes.

PRINCIPAL FACTS AND IDEAS

1. The activated complex theory provides a means of calculating rate constants of chemical reactions.
2. Very few reactions have been thoroughly studied on the molecular level.
3. Transport processes in liquids can be discussed in terms of the activated complex theory.
4. Nonequilibrium electrochemistry can be studied with the same theories as other chemical reactions.
5. Electrical conductivity in solids is described by the Drude theory.
6. Oscillatory chemical reactions can provide an example of deterministic chaotic behavior.

OBJECTIVES

After studying this chapter, the student should:

1. understand the assumptions underlying the activated complex theory of chemical reactions,
2. be able to solve a variety of problems involving the activated complex theory,
3. understand the collisional activation of unimolecular reactions,
4. understand the molecular basis of transport processes in liquids,
5. understand the principles of nonequilibrium electrode processes,
6. understand the assumptions behind the Drude theory of electrical conduction and be able to solve simple problems related to it,
7. understand how deterministic equations of evolution can produce seemingly chaotic behavior.

21.1 Theories of Chemical Reactions

In the collision theory of bimolecular gaseous reactions, we treated molecules as though they were hard spheres without attractive intermolecular forces. In this section, we discuss more realistic rate theories that include the effect of intermolecular potential energy. For example, consider a hypothetical bimolecular elementary gas phase reaction

$$CD + F \rightarrow C + DF \qquad (21.1\text{-}1)$$

Potential Energy Surfaces

In the Born-Oppenheimer approximation, the electronic Schrödinger equation for a molecule is solved with the nuclei assumed to be stationary. The study of electronic motion in our reacting system of three nuclei and several electrons is not fundamentally different from solving the Schrödinger equation for a stable triatomic molecule. The only difference is that we now must consider nuclear positions in which one nucleus is a large distance from the other two, as well as nuclear positions that are close to each other.

The solution of the Schrödinger equation in the Born-Oppenheimer approximation is a set of electronic wave functions and a set of electronic energy eigenvalues. Both the wave functions and the energy eigenvalues depend on the positions of the three nuclei. The position-dependent electronic energy, when added to the electrostatic repulsion of the nuclei, is the Born-Oppenheimer energy. This energy acts as a potential energy for nuclear motion.

If the three nuclei are constrained to be collinear (on the same line), with zero "bonding angle," the nuclear potential energy is a function only of two internuclear distances:

The collinear case is assumed in order to simplify the potential energy function. The calculation of the potential energy surface in the general case has been carried out in some cases.

$$\mathscr{V} = \mathscr{V}(r_{CD}, r_{DF}) \tag{21.1-2}$$

where r_{CD} is the distance from nucleus C to nucleus D, and r_{DF} is the distance from nucleus D to nucleus F.

The earliest calculation of a potential energy function for a three-atom reacting system was a semiempirical calculation for the reaction of a hydrogen atom with a hydrogen molecule at zero bonding angle.[1] This system has been repeatedly studied ever since.[2] Figure 21.1a shows a schematic graph of the potential energy function for this reaction for zero bonding angle. This graph is called a **potential energy surface**. Figure 21.1b shows the same information in a different way, by giving contours that represent positions of equal potential energy.

A trough runs from the point labeled "a" at the upper left of the diagram, representing the reactants, to the point labeled "c" at the lower right, representing the products. In a region near the lower left of the diagram, the potential energy surface is shaped like a saddle or a mountain pass. A curve is drawn along the bottom of the trough and over the saddle at point "b" in Figure 21.1b, and the energy as a function of position along this curve is shown in Figure 21.1c. Progress along the curve in Figure 21.1b (along the axis in Figure 21.1c) represents the accomplishment of the reaction, and we call the distance along this curve the **reaction coordinate** ζ.

The **activation energy** is the energy that must be added to the reactants to bring them to the state of maximum potential energy. The height of the maximum in the curve of Figure 21.1c above the energy of the reactants is

[1] H. Eyring and M. Polanyi, *Z. Phys. Chem.* **B12**, 279 (1931).
[2] T. J. Park and J. C. Light, *J. Chem. Phys.* **91**, 974 (1989); see D. G. Truhlar and C. J. Horowitz, *J. Chem. Phys.* **68**, 2466 (1978) for an accurate potential energy function.

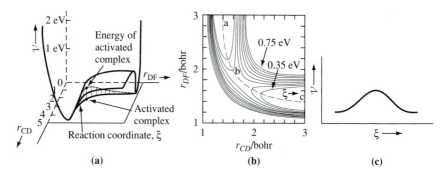

Figure 21.1. Potential Energy Surface of H_3 for Linear Geometry. (a) Perspective view of the surface (schematic). For the linear geometry, the potential energy depends on only two coordinates, r_{CD} and r_{DF}. The potential energy surface in this figure represents this function. Note the two valleys separated by a saddle, similar to a mountain pass. **(b)** Contours of constant potential energy for H_3. This figure presents the same information as in (a) in quantitative form. From R. D. Levine and R. B. Bernstein, *Molecular Reaction Dynamics and Chemical Reactivity*, Oxford University Press, New York, 1987, p. 124. **(c)** Potential energy along the reaction coordinate. The reaction coordinate measures the distance along the first valley, over the saddle surface, and into the second valley.

approximately equal to the activation energy, since except for zero-point energy it is the minimum excitation energy required for the reactants to pass over the maximum to form products. Calculations show that the activation energy for the hydrogen reaction is lower for zero bonding angle than for any nonzero bonding angle, and many calculations of three-atom reactions include only the linear geometry. A more complete discussion would take an average over the possible values of the bonding angle.

Trajectory Calculations

Once the potential energy surface is obtained, the direct way to study the reaction is to solve the Schrödinger equation or the classical equations for nuclear motion along the reaction coordinate. Most nuclei are sufficiently massive that classical mechanics is thought to be an adequate approximation for some purposes, so the classical equations of motion are sometimes used. However, quantum mechanical calculations are required for reactions involving hydrogen atoms and are often carried out for other reactions. Neither the classical nor quantum equations can be solved exactly, and the motions are numerically simulated using computer programs.

For the classical equations, it is not generally an adequate approximation to solve only for motion along the single curve in Figure 21.1b. The equations of motion are solved for a number of initial conditions in the vicinity of the point labeled "a" in Figure 21.1b, and the results are averaged. The fraction of the trajectories that pass over the saddle and the time required for the passage are used to calculate a rate constant.[3]

---

[3] R. N. Porter, *Ann. Rev. Phys. Chem.* **25**, 317 (1974).

The Activated Complex Theory

Henry Eyring, 1900–1981, American physical chemist who made contributions in almost every area of theoretical physical chemistry; Michael Polanyi, 1891–1976, Hungarian-born chemist who was originally trained as a physician and who later became a professor of social sciences.

Rather than calculating trajectories and averaging the results, one can use statistical mechanics to perform a theoretical average. The oldest such approximate theory was pioneered in the 1930s by Eyring and Polanyi and is called the **activated complex theory** or the **transition state theory**.[4]

For a gaseous bimolecular reaction, the theory begins with the assumptions: (1) When the nuclei are in the state of maximum potential energy along the reaction coordinate (point "b"), they are in a **transition state**, or form an **activated complex**, which has some of the properties of a triatomic molecule and can be identified as a distinct chemical species. (2) The concentration of activated complexes can be obtained by assuming that they are in chemical equilibrium with the reactants. (3) The rate of the chemical reaction is equal to the concentration of activated complexes times the frequency of passage of an activated complex over the maximum:

$$\text{Rate} = \nu_{\text{passage}}[\text{CDF}^\ddagger] = \nu_{\text{passage}} K_c[\text{CD}][\text{F}] \qquad \textbf{(21.1-3)}$$

where ν_{passage} is the frequency of passage over the maximum, $[\text{CDF}^\ddagger]$ is the concentration of the activated complex, and K_c is the equilibrium constant for the reaction of the reactants to form the activated complex:

$$K_c = \frac{[\text{CDF}^\ddagger]}{[\text{CD}][\text{F}]} \qquad \textbf{(21.1-4)}$$

The activated complex is not a stable molecule, so it is customary to give it a special symbol, $\text{CDF}^\ddagger$.

In view of the apparent lack of justification of the assumption of equilibrium between the activated complex and the reactants, the theory is more successful than one might expect.

The assumptions of the activated complex theory are not exactly correct. In most reactions, there is too little time for the activated complex to equilibrate with the reactants. However, the theory gives useful results, and it is commonly used to provide qualitative interpretations of experimental results.

We consider only gaseous reactions and assume the gases to be ideal. We first write an equation for K_c in terms of the molecular partition functions. From Equation (20.6-9),

$$0 = -\frac{\Delta\varepsilon_0}{k_B T} + \ln\left[\prod_{j=1}^{c}\left(\frac{z_j'}{N_i}\right)^{\nu_j}\right]$$

$$= -\frac{\Delta\varepsilon_0}{k_B T} + \ln\left[\prod_{j=1}^{c}\left(\frac{z_j'}{N_A V}\right)^{\nu_j}\right] - \ln\left[\prod_{j=1}^{c}[\mathscr{F}_j]^{\nu_j}\right] \qquad \textbf{(21.1-5)}$$

where z' denotes a partition function with the ground state of the substance taken as zero, as in Section 20.6. We denote the concentration of substance number j by $[\mathscr{F}_j]$:

$$[\mathscr{F}_j] = \frac{N_j}{N_A V} \qquad \textbf{(21.1-6)}$$

[4] H. Eyring, *J. Chem. Phys.* **3**, 107 (1935); M. Evans and M. Polanyi, *Trans. Faraday Soc.* **312**, 875 (1935). See K. J. Laidler and M. C. King, *J. Phys. Chem.* **87**, 2657 (1983) for a history of the theory.

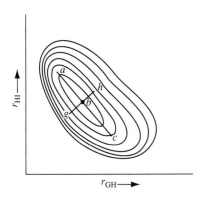

Figure 21.2. A Potential Energy Surface for a Stable Diatomic Molecule (schematic). A stable triatomic molecule differs from the activated complex in that the point "b" is located at a relative minimum rather than at a saddle point.

The shape of the potential energy surface in the vicinity of the saddle point shows that the activated complex is not a stable triatomic molecule.

Equation (21.1-5) leads to

$$K_c = \prod_{j=1}^{c} [\mathscr{F}_j]^{\nu_j} = e^{-\Delta\varepsilon_0/k_B T} \prod_{j=1}^{c} \left(\frac{z'_j}{N_A V} \right)^{\nu_j} \qquad \textbf{(21.1-7)}$$

We omit the factors c° that would make the equilibrium constant dimensionless. For the formation of the activated complex

$$K_c = e^{-\Delta\varepsilon_0^\ddagger/k_B T} \frac{(z'_{CDF}/N_A V)}{(z'_{CD}/N_A V)(z'_F/N_A V)} \qquad \textbf{(21.1-8)}$$

where $\Delta\varepsilon_0^\ddagger$ is the change in energy to form one activated complex in its ground state from one CD molecule and one F atom in their ground states. This is the activation energy for the reaction at 0 K and, except for zero-point energy, is equal to the height of the maximum in Figure 21.1c above the energy of the reactants.

The activated complex resembles an ordinary triatomic molecule, except that a stable linear triatomic molecule (call it GHI) would have a potential energy surface as shown in Figure 21.2, with a relative minimum at point "b" instead of a saddle point. Motion along path abc corresponds to the asymmetric stretch, and motion along path gbh corresponds to the symmetric stretch (see Section 14.3).

The activated complex has an ordinary symmetric stretch, because the saddle in the potential energy corresponds to a minimum in the gbh direction. We also assume that the two bending modes of the activated complex are ordinary oscillations. However, the asymmetric stretch of the activated complex corresponds to motion along the reaction coordinate and leads to dissociation, either to form products or to form reactants, since the potential energy is at a maximum instead of a minimum along the reaction coordinate.

We write the partition function of the activated complex in the form

$$z'_{CDF} = z^\ddagger_{CDF} z_{rc} \qquad \textbf{(21.1-9)}$$

where z_{rc} is the factor in the partition function for motion on the reaction coordinate and $z^\ddagger_{CDF}$ represents all the other factors in the partition function, including a translational factor, a rotational factor, an electronic factor, and three vibrational factors (for the symmetric stretch and the two bends).

We now adopt a nonrigorous approach to obtain a formula for z_{rc}. We first pretend that we can change the potential energy surface so that the activated complex CDF is like the triatomic molecule GHI. The partition function for the reaction coordinate is now that for an asymmetric stretch:

$$z_{rc} = \frac{1}{1 - e^{-h\nu_{as}/k_B T}} \qquad \textbf{(21.1-10)}$$

where ν_{as} is the frequency of oscillation. We now let the potential energy surface become flatter in the abc direction. As its curvature in the direction of the reaction coordinate becomes small, ν becomes small, and we can use the approximation $e^{-x} = 1 - x$, which gives

$$z_{rc} = \frac{k_B T}{h\nu_{as}} \qquad \textbf{(21.1-11)}$$

The frequency of the asymmetric stretch, ν_{as}, gives the number of oscillations per second in the direction of the reaction coordinate. We assume that we can identify ν_{as} with $\nu_{passage}$, and write

$$\nu_{passage} z_{rc} = k_B T/h \qquad \textbf{(21.1-12)}$$

This process of distorting the potential energy surface is not very satisfactory, because in order to recover the actual surface, the process must be continued until the curvature is not just small but becomes negative. There are other ways to derive Equation (21.1-12).

The final result for the rate is

$$\text{Rate} = \frac{k_B T}{h} e^{-\Delta \varepsilon_0^{\ddagger}/k_B T} \frac{(z_{CDF}^{\ddagger}/N_A V)}{(z'_{CD}/N_A V)(z'_F/N_A V)} [CD][F] \qquad \textbf{(21.1-13)}$$

The rate constant is

$$k = \frac{k_B T}{h} e^{-\Delta \varepsilon_0^{\ddagger}/k_B T} \frac{(z_{CDF}^{\ddagger}/N_A V)}{(z'_{CD}/N_A V)(z'_F/N_A V)} \qquad \textbf{(21.1-14)}$$

The transmission coefficient is commonly used as a correction factor to make the theory agree with experiment. It is probably better just to admit that the theory is approximate and cannot be expected to give exact agreement with experiment.

Sometimes the expression of Equation (21.1-14) is multiplied by a **transmission coefficient** κ, which represents the fraction of activated complexes that actually react. The transmission coefficient cannot be evaluated theoretically, and in view of the approximate nature of the theory, we omit it.

Equation (21.1-14) can also be written in a "thermodynamic" version:

$$k = \frac{k_B T}{h} \frac{1}{c^{\circ}} K_c^{\ddagger} = \frac{k_B T}{h} \frac{1}{c^{\circ}} e^{-\Delta G^{\ddagger\circ}/RT} \qquad \textbf{(21.1-15)}$$

where the equilibrium constant $K_c^{\ddagger}$ and the standard-state Gibbs energy change $\Delta G^{\ddagger\circ}$ are for the formation of the activated complex without the motion along the reaction coordinate. The factor $1/c^{\circ}$, where $c^{\circ} = 1$ mol L^{-1}, takes care of the fact that the equilibrium constant in Equation (21.1-15) is dimensionless.

The Relation of the Activated Complex Theory to the Arrhenius Formula and the Collision Theory

To compare Equation (21.1-14) with the Arrhenius equation, Equation (19.6-2), and with the collision theory equation, Equation (19.6-18), we write

$$k = \frac{k_B T}{h} \frac{1}{c^{\circ}} e^{\Delta S^{\ddagger\circ}/R} e^{-\Delta H^{\ddagger\circ}/RT}$$

$$= \frac{k_B T}{h} \frac{1}{c^{\circ}} e^1 e^{\Delta S^{\ddagger\circ}/R} e^{-\Delta U^{\ddagger\circ}/RT} \qquad \textbf{(21.1-16)}$$

where we have used Equations (2.8-1)–(2.8-5) with $\Delta \nu_g = -1$.

The standard-state entropy and enthalpy changes of activation also do not include motion on the reaction coordinate.

If we identify $\Delta U^{\ddagger\circ}$ with the Arrhenius activation energy, then the Arrhenius pre-exponential factor is

$$A = \frac{k_B T}{h} \frac{1}{c^{\circ}} e e^{\Delta S^{\ddagger\circ}/R} \qquad \textbf{(21.1-17)}$$

▼   **EXAMPLE 21.1**   From Example 19.12 the value of the pre-exponential factor for the reaction

$$H_2 + I_2 \rightarrow 2HI$$

is equal to 1.65×10^7 L mol^{-1} s^{-1}. Find the value of $\Delta S^{\ddagger \circ}$ at 298.15 K.

Solution

From Equation (21.1-17)

$$\Delta S^{\ddagger \circ} = R \ln\left(\frac{hAc^{\circ}}{ek_B T}\right)$$

$$\frac{hAc^{\circ}}{ek_B T} = \frac{(6.62 \times 10^{-34} \text{ J s})(1.65 \times 10^7 \text{ L mol}^{-1} \text{ s}^{-1})(1 \text{ mol L}^{-1})}{e(1.38 \times 10^{-23} \text{ J K}^{-1})(298 \text{ K})}$$

$$= 9.77 \times 10^{-7}$$

$$\Delta S^{\ddagger \circ} = (8.3145 \text{ J K}^{-1} \text{ mol}^{-1}) \ln(9.77 \times 10^{-7}) = -115 \text{ J K}^{-1} \text{ mol}^{-1}$$

▲

Exercise 21.1   **a.** Explain why the entropy change of activation for a bimolecular gaseous reaction is generally negative.

b. Estimate the pre-exponential factor for a typical bimolecular elementary reaction in which two atoms form a diatomic activated complex. Assume typical values of masses and internuclear distance.

We now show that the activated complex theory and the collision theory of Section 19.6 give the same result for a bimolecular reaction of hard spherical atoms. For the hypothetical reaction

$$C + D \rightarrow CD^{\ddagger} \rightarrow \text{products} \qquad \text{(21.1-18)}$$

we assume an activated complex consisting of the two spheres in contact, with internuclear distance equal to the sum of the two hard sphere radii.

Since the activated complex is a diatomic species, the single vibration is replaced by motion along the reaction coordinate. The partition functions are, assuming that there is no electronic excitation,

$$\frac{z_{CD}^{\ddagger}}{N_A V} = \frac{z_{trCD} z_{rotCD}}{N_A V}$$

$$= \frac{[2\pi(m_C + m_D)k_B T]^{3/2}\left[8\pi^2 \dfrac{m_C m_D}{m_C + m_D}(r_C + r_D)^2 k_B T\right]h^{-5}}{N_A}$$

$$\qquad \text{(21.1-19a)}$$

$$\frac{z'_C}{N_A V} = \frac{z_{trC}}{N_A V} = \left(\frac{2\pi m_C k_B T}{h^2}\right)^{3/2} \qquad \text{(21.1-19b)}$$

$$\frac{z'_D}{N_A V} = \frac{z_{trD}}{N_A V} = \left(\frac{2\pi m_D k_B T}{h^2}\right)^{3/2} \qquad \text{(21.1-19c)}$$

When these formulas are substituted into Equation (21.1-14) for the rate constant, we obtain

$$k = N_A \left(\frac{8k_B T}{\pi \mu} \right)^{1/2} \pi (r_C + r_D)^2 e^{-\Delta \varepsilon_0^{\ddagger}/k_B T} \qquad (21.1-20)$$

which is identical to Equation (19.6-18) if $\Delta \varepsilon_0^{\ddagger}/k_B$ is identified with E_c/R, and if the steric factor in Equation (19.6-18) is omitted.

The assumptions of the collision theory and the activated complex theory seem quite different. The fact that two quite different theories give identical results for a fairly ordinary case is interesting, and shows that there might be some common ground between the theories.

Activated Complex Theory of Termolecular Reactions

The same kind of analysis can be carried out for a termolecular reaction as for a bimolecular reaction and can be applied to explicit calculations if the potential energy function can be approximated in some way. For the reaction

$$C + D + G \rightarrow CDG^{\ddagger} \rightarrow products$$

the activated complex result is similar to Equation (21.1-14):

$$k = \frac{k_B T}{h} e^{-\Delta \varepsilon_0^{\ddagger}/k_B T} \frac{(z_{CDG}^{\ddagger}/N_A V)}{(z_C'/N_A V)(z_D/NV)(z_G'/N_A V)} \qquad (21.1-21)$$

Exercise 21.2   Write the termolecular analogues of Equations (21.1-16) and (21.1-17).

Theories of Unimolecular Reactions

Unimolecular reactions are fairly rare, and many reactions once thought to be unimolecular actually occur by multistep mechanisms that produce first-order kinetics. Three reactions accepted as unimolecular are[5]

$$Br_2 \rightarrow 2Br \qquad (21.1-22a)$$

$$SO_2Cl_2 \rightarrow SO_2 + Cl_2 \qquad (21.1-22b)$$

$$cyclo\text{-}C_3H_6 \rightarrow CH_3CH{=}CH_2 \qquad (21.1-22c)$$

The Lindemann-Christiansen mechanism discussed in Section 19.3 apparently applies to most unimolecular reactions. For the unimolecular reaction of a substance C, this mechanism is given in Equation (19.3-10):

$$(1) \qquad C + M \rightleftharpoons C^* + M \qquad (21.1-23a)$$

$$(2) \qquad C^* \rightarrow products \qquad (21.1-23b)$$

---

[5] K. J. Laidler, *Chemical Kinetics*, 3rd ed., Harper & Row, New York, 1987, p. 151.

where C* stands for an excited or activated molecule of C and we now assume that there is a second substance, M, present with which molecules of C collide.

From Equation (19.3-14),

$$\frac{d[C]}{dt} = -k_{uni}[C] \qquad (21.1\text{-}24)$$

where

$$k_{uni} = \frac{k_1 k_2 [M]}{k_2 + k_1'[M]} = \frac{k_2 k_1}{(k_2/[M]) + k_1'} \qquad (21.1\text{-}25)$$

In the limit of large concentration of M (high pressure), k_{uni} becomes nearly constant and the reaction is first order.

Once the C molecule becomes activated, even though it has enough energy to dissociate or to isomerize, this process does not occur immediately. If the dissociation or isomerization occurred immediately upon collision, the process would be an elementary bimolecular process and would be second order. The principal place for "storage" of the energy of excitation is in the vibrational degrees of freedom. One picture of what happens is that a "strong" collision excites all or most of the vibrational normal modes of the reactant molecule. This excited vibration goes on for a time, during which the normal modes exchange energy or change their relative phases until the vibrational amplitude of a specific bond becomes so large that the molecule reacts.

The Activated Complex Theory of Unimolecular Reactions

This theory can be applied to some unimolecular reactions in a fairly direct way. For example, in a *cis-trans* isomerization such as occurs in 11-*cis*-retinal (see Section 15.6), the reaction is a 180° internal rotation of the atoms at one end of the double bond relative to those at the other end.

In the original planar conformation, the overlap between the unhybridized *p* orbitals from which we construct the pi bonding orbital is at a maximum. Upon a rotation of 90°, the positive overlap cancels the negative overlap, the carbon-carbon bond is approximately a single bond instead of a double bond, and the molecule is in a transition state of high electronic energy. Further rotation to an angle of 180° restores the overlap and produces the other stable isomer. The activation energy for this process roughly equals the difference between the C—C single bond energy and the C=C double bond energy.

Application of the activated complex theory to the unimolecular process gives the expression analogous to Equation (21.1-14) for the rate constant:

$$k = \frac{k_B T}{h} e^{-\Delta\varepsilon_0^{\ddagger}/k_B T} \frac{z^{\ddagger}}{z_r} \qquad (21.1\text{-}26)$$

where z_r is the partition function of the reactant and $z^{\ddagger}$ is the partition function of the transition state without the factor for the reaction coordinate.

Since the activated complex and the reactant molecule have the same atoms and nearly the same chemical bonding, all factors in the partition function might nearly cancel except for the factor of the "missing" vibrational normal mode. Since vibrational partition functions near room temperature are generally nearly equal to unity, the ratio $z^{\ddagger}/z_r$ is nearly unity in cases in which the activated complex resembles the reactant molecule. In this case, the pre-exponential factor in the rate constant is roughly $k_B T/h$, which at room temperature is approximately 6×10^{12} s^{-1}.

In other cases, the activated complex is "looser" than the reactant molecule, having lower vibrational frequencies and larger moments of inertia, and in this case the pre-exponential factor is larger than 6×10^{12} s^{-1}, since $z^{\ddagger}$ is then larger than z_r.

Exercise 21.3	Estimate the rate constant and the half-life of a *cis-trans* isomerization at (a) 300 K and (b) 310 K. State any assumptions.

Other Theories of Unimolecular Reactions

Modern theories of unimolecular reactions are generally based on a theory due to Rice, Ramsperger, and Kassel[6] that predated the activated complex theory. It is based on the Lindemann-Christiansen mechanism, and includes a mathematical formulation of the "storage" of energy in the vibrational degrees of freedom.

We assume that the energy of a particular vibrational normal mode must have at least a threshold energy ε_0^* in order to react. That is, if a molecule undergoes a strong collision such that its vibrational energy after the collision is equal to ε^*, then there is a chance for reaction to occur if ε^* is at least as large as ε_0^* and if that energy can become concentrated in the reactive normal mode before deactivation occurs by another collision.

The expression for the rate constant in the Lindemann-Christiansen mechanism is modified by assuming a different rate constant for each value of ε^* and writing the total rate constant as the integral over all values of ε^* that are at least as large as ε_0^*:

$$dk_{uni} = \frac{k_2(\varepsilon^*)f(\varepsilon^*)}{1 + k_2(\varepsilon^*)/k_1'[M]} \, d\varepsilon^* \qquad \text{(21.1-27a)}$$

$$k_{uni} = \int_{\varepsilon_0^*}^{\infty} dk_{uni} = \int_{\varepsilon_0^*}^{\infty} \frac{k_2(\varepsilon^*)f(\varepsilon^*)}{1 + k_2(\varepsilon^*)/k_1'[M]} \, d\varepsilon^* \qquad \text{(21.1-27b)}$$

where we have replaced the quotient k_1/k_1' by $f(\varepsilon^*)$. This quantity is the equilibrium constant for step 1, and it is taken as equal to the Boltzmann probability for energy ε^*. When this is expressed as a function of ε^*, the result for s vibrational normal modes is[7]

$$f(\varepsilon^*) \, d\varepsilon^* = \frac{1}{(s-1)!} \left(\frac{\varepsilon^*}{k_B T}\right)^{s-1} \frac{1}{k_B T} e^{\varepsilon^*/k_B T} \, d\varepsilon^* \qquad \text{(21.1-28)}$$

---

[6] O. K. Rice and H. C. Ramsperger, *J. Am. Chem. Soc.* **49**, 1616 (1927); **50**, 617 (1928); L. S. Kassel, *J. Phys. Chem.* **32**, 225 (1928). See K. J. Laidler, *Chemical Kinetics*, 2nd ed., Harper & Row, New York, 1987, pp. 150ff, for a clear discussion.

[7] K. J. Laidler, *op. cit.*, pp. 77, 159.

The k_2 is the rate constant for the vibrational energy of excitation to accumulate in the reactive degree of freedom and for the reaction to occur. Rice, Ramsperger, and Kassel made the approximation that all of the vibrational degrees of freedom have the same frequency, so their quanta of energy are of the same size. Consider a molecule with s vibrational normal modes. The number of ways to distribute j quanta of energy among s oscillators, with no restriction on the number of quanta in each oscillator, is

$$w = \frac{(j + s - 1)!}{j!(s - 1)!}$$ (21.1-29a)

and the number of ways to have m quanta in one oscillator if there are $j - m$ quanta in the other oscillators is

$$w' = \frac{(j - m + s - 1)!}{(j - m)!(s - 1)!}$$ (21.1-29b)

The probability of having m quanta in one oscillator if there are j quanta in all s oscillators is the ratio of these quantities:

$$r = \frac{w'}{w} = \frac{(j - m + s - 1)! \, j!}{(j - m)!(j + s - 1)!}$$ (21.1-30)

where we denote the ratio by the letter r.

If we apply Stirling's approximation in the form

Stirling's approximation is not very good for the value of n that we have.

$$n! \approx \left(\frac{n}{e}\right)^n$$

the factors of e cancel. If we then make the assumption that $j - m \gg s - 1$, we get the result

$$r = \left(\frac{j - m}{j}\right)^{s-1} = \left(\frac{\varepsilon^* - \varepsilon_0^*}{\varepsilon^*}\right)^{s-1}$$ (21.1-31)

where we have used the fact that the number of quanta is proportional to the vibrational energy. The rate constant k_2 is now written in the form

$$k_2 = k^{\ddagger}\left(1 - \frac{\varepsilon_0^*}{\varepsilon^*}\right)^{s-1}$$ (21.1-32)

where $k^{\ddagger}$ is the rate constant for "free passage" over the "dividing surface" that marks the locations on the potential energy surface beyond which reaction is certain. Notice the resemblance of Equation (21.1-32) to Equation (19.6-16).

When Equation (21.1-32) is substituted into Equation (21.1-28), the result is

$$k_{\text{uni}} = \frac{k^{\ddagger} e^{-\varepsilon_0^*/k_B T}}{(s - 1)!} \int_0^{\infty} \frac{x^{s-1} e^{-x}}{1 + \dfrac{k^{\ddagger}}{k_1'[\text{M}]}\left(\dfrac{x}{x + b}\right)^{s-1}} \, dx$$ (21.1-33)

where the substitution has been made

$$x = \frac{\varepsilon^* - \varepsilon_0^*}{k_B T}$$ (21.1-34)

In the limit of high pressure (large values of [M]), the rate constant approaches the limit:

$$\lim_{c \to \infty} k_{\text{uni}} = k^{\ddagger} e^{-\varepsilon_0^*/k_B T}$$ (21.1-35)

Exercise 21.4	Verify Equation (21.1-35).

In the original RRK theory, $k^{\ddagger}$ was considered to be a constant. However, the results of numerical calculations with Equation (21.1-33) did not always agree well with experiment, often giving a value for the pre-exponential factor that was too small. However, if the high-pressure limit is made to agree with the activated complex result, Equation (21.1-27), then

$$k^{\ddagger} = \frac{k_{\mathrm{B}} T}{h} \frac{z^{\ddagger}}{z_r} \tag{21.1-36}$$

This allows large pre-exponential factors in the case that the activated complex is "looser" than the reactant.

The RRK theory has been modernized by Marcus.[8] The resulting theory is called the RRKM theory and is the principal modern theory of unimolecular reactions.

21.2 The Molecular Case History of a Chemical Reaction

A reaction that has been thoroughly studied is[9]

$$\mathrm{F} + \mathrm{H}_2 \rightarrow \mathrm{H} + \mathrm{HF} \tag{21.2-1}$$

Figure 21.3a shows the potential energy along the reaction coordinate, based on a semiempirical London-Eyring-Polanyi-Sato (LEPS) calculation. As can be seen, $\Delta E_0 = -29.2$ kcal mol^{-1} (-122 kJ mol^{-1}). The height of the energy barrier is small, 0.9 kcal mol^{-1} (4 kJ mol^{-1}), while the zero-point energies of the H_2 and F_2 are fairly large, respectively equal to 6.2 kcal mol^{-1} and 5.85 kcal mol^{-1} (26 kJ mol^{-1} and 24.5 kJ mol^{-1}). The zero-point energy of the activated complex is 6.9 kcal mol^{-1} (29 kJ mol^{-1}), so that the activation energy $\Delta E_0^{\ddagger}$ is equal to 1.6 kcal mol^{-1} (6.7 kJ mol^{-1}). The smallness of the activation energy correlates with an activated complex that is similar in structure to the reactants, in agreement with **Hammond's postulate**, which states that "the more exoergic the reaction, the more the transition state will resemble the reagents." The reaction is sufficiently exoergic that the first four vibrational states of the product HF can be populated, as shown in Figure 21.3b.

Levine and Bernstein review the experimental study of this reaction up to the time of publication of their book. The first stage was a chemical laser study, which verified that there is a population inversion in the products, with the $v = 2$ vibrational state more highly populated than the other states, so that laser action can occur to the $v = 1$ state and the $v = 0$ state.

---

[8] R. Marcus, *J. Chem. Phys.* **20**, 359 (1952).

[9] R. D. Levine and R. B. Bernstein, *Molecular Reaction Dynamics and Chemical Reactivity*, Oxford University Press, New York, 1987, pp. 306ff, 396ff.

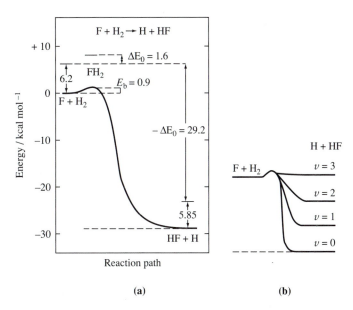

Figure 21.3. The Potential Energy on the Reaction Coordinate for Colinear FH$_2$. This is analogous to Figure 21.1c. The activation energy for the forward reaction is rather small. Part (b) shows the surface for various final vibrational states. From R. D. Levine and R. B. Bernstein, *Molecular Dynamics and Chemical Reactivity*, Oxford University Press, New York, 1987, p. 127.

The next stage was a crossed molecular beam study, in which the angular distribution of products was determined and it was found that most of the products were "backscattered" (scattered generally in the direction from which the reactants had come).

Next came a classical trajectory calculation, in which the large backscattering was also found to occur. After this, an ab initio calculation of the potential energy surface was carried out, giving a surface in fairly good agreement with the LEPS surface from which Figure 21.3 was drawn. Next came a quantum mechanical calculation showing that "resonances" or "quasi-bound states" occurred at the transition state, which means that the activated complex existed longer than expected, as it would if there were a "basin" in the potential energy surface at the transition state. This phenomenon was explained by the fact that the potential energy surface has a saddle that is fairly broad in the direction of the symmetric stretch, lowering the energy of the activated complex and causing its relative persistence.

Finally, a detailed chemiluminescence study gave the distribution of product states, showing the most backscattering in the lower four vibrational states, but significant forward scattering in the $v = 4$ vibrational state, and determining that on the average 66% of the energy of reaction goes into vibrational energy, 8% into rotational energy, and the remainder into translational energy. The calculations did not agree well with the new experimental results.

Since the publication of the book by Levine and Bernstein, detailed quantum mechanical calculations of state-to-state reaction probabilities have been carried out.[10] Several different potential energy surfaces were used in these calculations, but only states of zero angular momentum were included. Better agreement with experiment was attained. Further research is focused on finding better potential energy surfaces. A recent calculation gave a barrier height of 0.089 eV.[11] It still cannot be said that this reaction is completely understood, although it involves only small molecules in the gas phase and is among the most intensively studied reactions.

<table>
<tr><td>21.3</td></tr>
</table>

Theories of Transport Processes in Fluid Systems

In Chapter 16, we presented an elementary kinetic theory of transport processes in gases, based on formulas for collision rates and mean free paths. As in most areas of theoretical science, there are several layers of theories for transport processes in gases and in liquids, each layer more complicated than the previous one but capable of giving better agreement with experiment.

Here we see an example of the fact that chemical and physical processes are really not completely distinct from each other.

The first and simplest theory for transport processes in liquids is based on the activated complex theory of Eyring and Polanyi. It assumes a quasi-lattice model system that resembles a disordered crystal, much as in the cage model described in Chapters 16, 17, and 19.

We assume that vacancies occur in the liquid and that these vacancies occasionally become as large as a molecule through random molecular motions. For a molecule to move into a vacancy, it must push some neighboring molecules aside, thereby moving through a state of high potential energy, as depicted schematically in Figure 21.4. This potential energy maximum is analogous to the maximum along the reaction coordinate for a chemical reaction. We therefore treat the coordinate representing motion into the vacancy as a reaction coordinate and treat the transition state of high potential energy as an activated complex.

In the thermodynamic formulation of the activated complex theory, the rate constant is given by

$$k = \frac{k_B T}{h} e^{-\Delta G^{\ddagger\circ}/RT} \tag{21.3-1}$$

where $\Delta G^{\ddagger\circ}$ is the Gibbs energy change per mole to form the activated complex in its standard state, not including motion along the reaction coordinate.

Assume a two-component liquid system in which the concentrations depend on the vertical coordinate z, with a larger concentration of com-

---

[10] C. Yu, Y. Sun, D. J. Kouri, P. Halvick, D. G. Truhlar, and D. W. Schwenke, *J. Chem. Phys.* **90**, 7608 (1989); C. Yu, D. J. Kouri, M. Zhaoo, D. G. Truhlar, and D. W. Schwenke, *Chem. Phys. Lett.* **157**, 491 (1989).

[11] G. E. Scuseria, *J. Chem. Phys.* **95**, 7426 (1991).

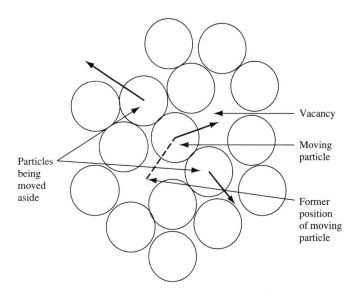

Figure 21.4. Motion of a Molecule into a Vacancy in a Liquid. This figure attempts to illustrate the position of maximum potential energy of a molecule moving from one shell of neighbors to another.

ponent 2 at the bottom of the system. Consider a vacancy at height $z = z'$. On the average, there will be slightly smaller concentration of component 2 above the vacancy than below. The rate at which molecules move into the vacancy from below is equal to

$$\text{Rate}_b = kc_2(z' - a) \qquad (21.3\text{-}2)$$

where a is an average z component of the displacement into the vacancy and $c_2(z' - a)$ is the concentration of component 2 evaluated at $z' - a$. The distance a will presumably be somewhat smaller than the nearest-neighbor distance, since some of the displacements are not parallel to the z axis.

The rate at which molecules move into the vacancy from above is equal to

$$\text{Rate}_a = kc_2(z' + a) \qquad (21.3\text{-}3)$$

since the molecules coming from above should on the average have the same z component and the same rate constant as those coming from below.

Since the average z component of the displacement of the particles is equal to a, the net contribution to the diffusion flux is

$$J_{2z} = k[c_2(z' - a) - c_2(z' + a)]a \qquad (21.3\text{-}4)$$

Let us write $c_2(z' + a)$ as a Taylor series around the point $z' - a$:

$$c_2(z' + a) = c_2(z' - a) + \frac{\partial c_2}{\partial z}(2a) + \cdots \qquad (21.3\text{-}5)$$

where the derivative is evaluated at $z = z' - a$. If we neglect the terms not shown in Equation (21.3-5), Equation (21.3-4) becomes

$$J = -\frac{k_B T}{h}(2a^2)e^{-\Delta G^{\ddagger\circ}/RT}\left(\frac{\partial c_2}{\partial z}\right) \qquad \text{(21.3-6)}$$

Comparison with Fick's law, Equation (17.2-4), gives an expression for the diffusion coefficient,

$$D = 2\frac{k_B T}{h}a^2 e^{-\Delta G^{\ddagger\circ}/RT} \qquad \text{(21.3-7)}$$

EXAMPLE 21.2

One should always check the reasonableness of the result of any calculation. An activation Gibbs energy change equal to a fraction of the energy of vaporization is reasonable, since vaporization corresponds to breaking the attraction with all of the neighbors of a given molecule, while moving into a vacancy involves interactions with part of the neighbors.

Many liquids with molecules of ordinary size have diffusion coefficients approximately equal to 10^{-9} m^2 s^{-1}. Assume that a is equal to 1×10^{-10} m and estimate the value of $\Delta G^{\ddagger\circ}$ for $T = 300$ K.

Solution

$$\Delta G^{\ddagger\circ} = -RT \ln\left(\frac{Dh}{2a^2 k_B T}\right)$$

$$= -(8.3 \text{ J K}^{-1} \text{ mol}^{-1})(300 \text{ K}) \ln\left[\frac{(10^{-9} \text{ m}^2 \text{ s}^{-1})(6.6 \times 10^{-34} \text{ J s})}{2(10^{-10} \text{ m})^2(1.4 \times 10^{-23} \text{ J K}^{-1})(300 \text{ K})}\right]$$

$$= 1 \times 10^4 \text{ J mol}^{-1} = 10 \text{ kJ mol}^{-1}$$

This value is reasonable, being about one-fourth of a typical energy change of vaporization.

The diffusion coefficient is sometimes written in the Arrhenius form, analogous to Equation (19.6-2):

$$D = A_d e^{-\Delta E_{ad}/RT} \qquad \text{(21.3-8)}$$

where E_{ad} is the activation energy and A_d is the pre-exponential factor for the diffusion process.

To compare Equation (21.3-7) with this equation, we recognize that the enthalpy change of activation and the energy change of activation are nearly equal, since there is little change in PV in a liquid-state process. Therefore,

$$D = 2\frac{k_B T}{h}a^2 e^{-\Delta S^{\ddagger\circ}/R}e^{-\Delta U^{\ddagger\circ}/RT} \qquad \text{(21.3-9)}$$

Exercise 21.5

The diffusion coefficient for 1,1,1-trichloroethane in a mixed solvent of 2,2-dichloropropane and carbon tetrachloride was measured to be 1.41×10^{-9} m^2 s^{-1} at 25°C and 2.02×10^{-9} m^2 s^{-1} at 45°C.
  a. Find the value of the apparent energy of activation.
  b. Estimate the value of $\Delta S^{\ddagger}$, assuming that $a = 4 \times 10^{-10}$ m. Comment on the magnitude and sign of your answer.

The motion of a single molecule into a molecule-size hole is not the only molecular diffusion process that can be treated in a simple model theory. Various processes, including the exchange in position of two molecules of different species, have been considered.[12] This exchange would require a very high activation energy unless there is a vacancy adjacent to the pair of molecules, in which case the activation energy could be small enough that this process might make a large contribution compared with the one-body motion into a hole.

In a three-component system, the exchange in position of two molecules of different solute species means that for a molecule of one species to move in one direction, a molecule of another species must move in the opposite direction. This process can lead to "cross-effects," so that Fick's law must be written in an extended form in which the concentration gradient of one species makes a contribution to the diffusion flux of another species:

$$\mathbf{J}_i = - \sum_{j=1}^{c} D_{ij} \nabla c_j \qquad \textbf{(21.3-10)}$$

In the theory of nonequilibrium thermodynamics, such cross-effects are systematically studied and thermodynamic theorems relating the cross-coefficients are proved.[13]

Viscosity

In viscous (shearing) flow in a liquid, one layer of molecules flows at a greater speed than an adjacent layer. If we use a moving coordinate system, one layer of molecules can be considered stationary, with the adjacent layer moving relative to it. In the simple model theory of transport processes, this motion is accomplished by the motion of individual molecules into holes in the liquid, rather than by concerted motion of a whole sheet of molecules at once. Therefore, the rate of shear is proportional to the rate constant in Equation (21.3-1). By Equation (17.2-20), the viscosity is inversely proportional to the rate of shear, so a formula of the Arrhenius type is

$$\eta = A_\eta e^{+E_{a\eta}/RT} \qquad \textbf{(21.3-11)}$$

where $E_{a\eta}$ is an activation energy and A_η is a pre-exponential factor for viscous flow. According to the simple theory, this activation energy should nearly equal that for diffusion in the same liquid.

| Exercise 21.6 | The viscosity of carbon tetrachloride at 20°C is 9.69×10^{-4} kg m^{-1} s^{-1} and at 50°C is 6.51×10^{-4} kg m^{-1} s^{-1}. Calculate the Arrhenius activation energy for viscosity, and compare it with the value for the activation energy for diffusion in Exercise 21.5a. Try to explain any difference in the activation energies. |

---

[12] R. G. Mortimer and N. H. Clark, *Ind. Eng. Chem. Fundam.* **10**, 604 (1971).
[13] S. R. DeGroot and P. Mazur, *Nonequilibrium Thermodynamics*, North Holland, Amsterdam, 1962.

More Advanced Theories of Transport in Gases

For a nonequilibrium monatomic gas, the equilibrium classical one-body probability density of Equation (20.8-8) must be replaced by a nonequilibrium probability density that depends on time as well as on coordinates and velocity. The **Boltzmann equation** is used to solve approximately for this time-dependent probability density.

In a gas without collisions, the probability density changes by "streaming." That is, the points representing molecule states move around in the six-dimensional phase space, as described by the equations of motion of the individual molecules. This motion in six dimensions is similar to that of a fluid flowing in three dimensions. If there are no collisions, the flow is like an incompressible (although nonuniform) fluid, so that the one-particle probability density is a constant if we move along with a point representing the state of any given molecule:

$$\frac{df}{dt} = 0 \quad \text{(along a stream line, no collisions)} \qquad \textbf{(21.3-12)}$$

Since f is a function of coordinates, velocity, and time, we can use the chain rule to write

$$\frac{df}{dt} = \frac{\partial f}{\partial t} + \frac{\partial f}{\partial x}\frac{dx}{dt} + \frac{\partial f}{\partial y}\frac{dy}{dt} + \frac{\partial f}{\partial z}\frac{dz}{dt} + \frac{\partial f}{\partial v_x}\frac{dv_x}{dt} + \frac{\partial f}{\partial v_y}\frac{dv_y}{dt} + \frac{\partial f}{\partial v_z}\frac{dv_z}{dt}$$

$$= \frac{\partial f}{\partial t} + \nabla_r f \cdot \frac{d\mathbf{r}}{dt} + \nabla_v f \cdot \frac{d\mathbf{v}}{dt} = 0 \quad \begin{pmatrix}\text{along a stream line} \\ \text{with no collisions}\end{pmatrix} \qquad \textbf{(21.3-13)}$$

where we use the symbol $\mathbf{r}$ for the position vector (x, y, z) and the symbol $\mathbf{v}$ for the velocity vector (v_x, v_y, v_z), and where ∇_r means the gradient with respect to the coordinates and ∇_v means the gradient with respect to the velocity. The symbol $\cdot$ stands for the scalar product of two vectors. The partial derivative $(\partial f/\partial t)$ is the time derivative that would be observed by an observer who is stationary in the six-dimensional phase space.

Equation (21.3-13) can be rewritten using the fact that $\mathbf{v} = d\mathbf{r}/dt$ and using Newton's second law:

$$\mathbf{F} = m\frac{d\mathbf{v}}{dt} \qquad \textbf{(21.3-14)}$$

Since we have not yet included intermolecular forces, the only force on the molecules is an external force, such as gravity, or in the case of charged particles, an external electric field. We denote an external force by $\mathbf{X}$.

$$\frac{df}{dt} = \frac{\partial f}{\partial t} + \mathbf{v} \cdot \nabla_r f + \frac{\mathbf{X}}{m} \cdot \nabla_v f = 0 \quad \begin{pmatrix}\text{along a stream line} \\ \text{with no collisions}\end{pmatrix} \quad \textbf{(21.3-15)}$$

The **Boltzmann equation** is obtained by adding two terms to the right-hand side of Equation (21.3-15). One represents the rate at which collisions bring molecules into the molecular state being considered (specified by the values of $\mathbf{v}$ and $\mathbf{r}$), and the other represents the rate at which collisions remove molecules from this state. Assuming now that we have a system containing several components, we write Equation (21.3-15) for component number j and add the two collision terms:

$$\frac{df_j}{dt} = \frac{\partial f_j}{\partial t} + \mathbf{v}_j \cdot \nabla_{rj} f + \frac{\mathbf{X}_j}{m_j} \cdot \nabla_{vj} f = \left(\frac{\partial f}{\partial t}\right)_{\text{coll}} = \sum_{i=1}^{c} [\Gamma_{ji}^{(+)} - \Gamma_{ji}^{(-)}] \quad \textbf{(21.3-16)}$$

where $\Gamma_{ji}^{(+)}$ is the rate of increase of f_j due to collisions with molecules of component i such that molecules of component j are brought into the state being considered, and $\Gamma_{ji}^{(-)}$ is the rate of decrease of f_j due to collisions with molecules of component i that remove molecules of component j from the state being considered. Equation (21.3-16) is the Boltzmann equation.

At equilibrium, the two collision terms cancel, and the Boltzmann equation is satisfied by the equilibrium probability density.

| **Exercise 21.7** | For a uniform gravitational field in the z direction, |

$$X_{jz} = -m_j g$$

where g is the acceleration due to gravity. Show that the equilibrium probability density of Equation (16.6-3) satisfies Equation (21.3-16).

In order to obtain a usable equation from Equation (21.3-16), we must have explicit expressions for the collision terms. Any collision between a molecule in the given state and any other molecule almost certainly removes the first molecule from the given state. In addition, the time-reversed version of any collision will bring a molecule into the given state.

The assumption is made that the rate of collisions can be computed as though the probabilities of the positions and velocities of the two colliding molecules are completely independent of each other, or are **uncorrelated**. The nonequilibrium analogue of the radial distribution function is equal to unity for all intermolecular separations. This assumption was called the **Stosszahlansatz** by Boltzmann and in English is usually called the assumption of **molecular chaos**. It is clearly not exactly correct, since the intermolecular forces will cause the molecules to affect each other's motions.

With these assumptions, the collision terms become[14]

$$\left(\frac{\partial f_j}{\partial t}\right)_{\text{coll}} = 2\pi \sum_{i=1}^{c} \int [f'_i f'_j - f_i f_j]|\mathbf{v}_i - \mathbf{v}_j| b\, db\, d^3\mathbf{v} \qquad \textbf{(21.3-17)}$$

where b is the impact parameter defined in Figure 19.13. We do not derive this equation. The two probability densities labeled f' are the probability densities for the velocities after the collision and give the rate at which particles are scattered into the state of interest. The Boltzmann equation cannot be solved exactly, but there is a procedure for obtaining approximate solutions, called the Chapman-Enskog method.[15]

Formal Theories of Transport Processes in Fluids

Further classical nonequilibrium theories for both gases and liquids consist of various approximation schemes for studying the evolution of the system probability distribution. Most of these theories begin as formal theories, which means that the equations derived are valid but cannot be used directly to make calculations without introducing approximations or assumptions.

---

[14] D. A. McQuarrie, *Statistical Mechanics*, Harper & Row, New York, 1976, pp. 409ff.
[15] S. Chapman and T. G. Cowling, *The Mathematical Theory of Non-Uniform Gases*, Cambridge University Press, Cambridge, 1952.

There is also a quantum mechanical version of the Liouville equation.

The equation that governs the time dependence of a system probability distribution is the **Liouville equation**. For a monatomic gas of N particles, this equation is, in cartesian coordinates,

$$\frac{\partial f}{\partial t} + \sum_{i=1}^{N}\left(\frac{p_i}{m_i}\cdot\nabla_{ri}f + \mathbf{F}_i\cdot\nabla_{pi}f\right) = 0 \tag{21.3-18}$$

where ∇_{pi} and ∇_{ri} are the gradients with respect to the momentum and coordinates of particle i and where $\mathbf{F}_i$ is the total force on particle i. The force $\mathbf{F}_i$ is the total force on particle i, including all intermolecular forces.

The Liouville equation is similar to the collisionless version of the Boltzmann equation, except that it corresponds to streaming of the probability density in the $6N$-dimensional phase space instead of a six-dimensional phase space. The Liouville equation cannot be solved exactly, but there are several approximate approaches to it. One approach is to define reduced distribution functions, which are a set of one-body, two-body, ..., n-body functions obtained by integrating the N-body distribution function f over the coordinates and momenta of all particles but one, or all particles but two, etc.

The Liouville equation is converted into a set of equations for the reduced distribution functions, called the **Bogoliubov-Born-Green-Yvon-Kirkwood (BBGYK) hierarchy** of equations. This set of equations is obtained by integrating over the coordinates and momenta of all of the particles except one to obtain one equation, integrating over the coordinates and momenta of all of the particles except two to obtain another equation, etc. The first one or two of the equations are separated from the others by approximations and solved in various approximate ways.

Another approach is to derive formal expressions for transport coefficients and then seek approximation schemes to evaluate these expressions. This approach was pioneered by Kubo and by Green, and the formulas for transport coefficients are sometimes called **Kubo formulas**. For example, the coefficient of self-diffusion is given by[16]

$$D = \frac{1}{3}\int_{0}^{\infty}\langle\mathbf{v}(0)\cdot\mathbf{v}(t)\rangle\,dt \tag{21.3-19}$$

where $\mathbf{v}$ is the velocity of a molecule and where $\langle\ \rangle$ denotes an ensemble average. This kind of ensemble average of a product of two quantities evaluated at different times is called a **time-correlation function**. At $t = 0$, the time-correlation function in Equation (21.3-19) is equal to the average of the square of the velocity, which in a classical system is $3k_BT/m$. As time passes, the time correlation function approaches zero, representing the decay of the correlation of the velocity of a molecule with its initial velocity. This behavior allows the integral in Equation (21.3-19) to converge. Frequency-dependent transport coefficients are given as Fourier transforms of the time-correlation functions instead of time integrals of the time-correlation functions, which are equivalent to zero-frequency Fourier transforms.

▼ **EXAMPLE 21.3**

Assume that the velocity time correlation function of a molecule in a liquid is equal to

$$\langle\mathbf{v}(0)\cdot\mathbf{v}(t)\rangle = \frac{3k_BT}{m}e^{-t/\tau} \tag{21.3-20}$$

---

[16] D. A. McQuarrie, *op. cit.*, pp. 467ff.

where τ is a **correlation time**, or a time for correlation of a particle's velocity with its initial velocity to be lost. The self-diffusion coefficient of CCl_4 is equal to $1.30 \times 10^{-9} \text{ m}^2 \text{ s}^{-1}$ at 298.15 K. Find the value of τ.

Solution

$$D = \frac{k_B T}{m} \int_0^\infty e^{-t/\tau} dt = \frac{k_B T \tau}{m}$$

$$\tau = \frac{mD}{k_B T} = \frac{(2.56 \times 10^{-25} \text{ kg})(1.3 \times 10^{-9} \text{ m}^2 \text{ s}^{-1})}{(1.38 \times 10^{-23} \text{ J K}^{-1})(298 \text{ K})} = 8.2 \times 10^{-14} \text{ s}$$

▲

Exercise 21.8

If a carbon tetrachloride molecule is moving with a speed equal to the rms speed of carbon tetrachloride molecules at 298 K, find the distance it travels in a time equal to the correlation time from Example 21.3. Comment on the size of your answer.

Other transport coefficients are expressed in terms of different time-correlation functions. The exact calculation of a time-correlation function cannot be carried out, since this would involve solution of the equations of motion of the entire system, but various techniques have been developed to obtain approximate results.[17]

21.4   Nonequilibrium Electrochemistry

In Chapter 8, we discussed the cell voltage of electrochemical cells at equilibrium. If a nonzero current is being drawn from a galvanic cell or if a nonzero current is being passed through an electrolytic cell, the cell voltage can differ from the equilibrium value. This deviation is called the **polarization** of the cell. The extent of polarization for a given current is determined not only by the nature of the reactions but also by the concentrations of the electrolyte solutes, the surface area of the electrodes, stirring in the solution, etc.

An electrode that passes no current, even for a large potential, is called an **ideal polarized electrode**. An electrode that does not change its potential no matter what current is passing is called a **nonpolarizable electrode** or an **ideal depolarized electrode**. All real electrodes are intermediate between these two nonexistent extremes.

Electrolytic Cells near Equilibrium

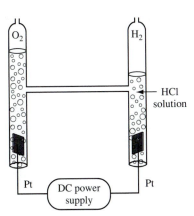

Figure 21.5. An Electrolytic Cell. This is a cell in which water can be electrolyzed by passage of a direct electric current. The volume of H_2 formed is twice that of the O_2 formed.

Figure 21.5 schematically shows a particular electrolytic cell. The two electrodes are made of platinum, and the electrolyte solution is an aqueous HCl solution with unit activity. The external power supply drives electrons through the external circuit from left to right. In the solution, positive ions move from left to right, and negative ions move in the opposite direction.

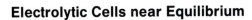

[17] *ibid.*

▼   **EXAMPLE 21.4**

Find the voltage required to produce an infinitesimal electrolysis current at 298.15 K in the cell of Figure 21.5. Determine what products will be produced. Assume that any gaseous products are maintained at pressure $P°$ and that the hydrogen and chloride ions are at unit activity.

Solution

The only possible cathode half-reaction is

$$2H^+ + 2e^- \rightarrow H_2(g) \qquad E° = 0.0000 \text{ V}$$

If the hydrogen gas is maintained at the electrode at pressure $P°$, the cathode is a standard hydrogen electrode.

There are two conceivable anode half-reactions:

$$2H_2O \rightarrow O_2(g) + 4H^+ + 4e^- \qquad E° = 1.229 \text{ V}$$
$$2Cl^- \rightarrow Cl_2(g) + 2e^- \qquad E° = -1.3583 \text{ V}$$

To produce oxygen at pressure $P°$, the reversible cell voltage would be -1.229 V. To produce chlorine at pressure $P°$, the reversible cell voltage would have to be -1.3583 V. Therefore, water, not chloride ions, will be oxidized at the anode at an infinitesimal rate if a voltage slightly larger than 1.229 V is applied to the cell.

▲

In the solution of Example 21.4, we determine what half-reaction will occur at infinitesimal current. We will later see that at finite current the situation is more complicated.

If the products and reactants in Example 21.4 were not at unit activity, the Nernst equation would be used to find the voltage required to produce each possible product.

Exercise 21.9

a. Determine what products will be formed if an aqueous solution containing chloride ions with activity on the molality scale equal to 5.000 is electrolyzed with an infinitesimal current, and find the voltage required. Assume that any gaseous products are maintained at pressure $P°$. Write the half-reaction equations and the cell reaction equation.

b. Pure sodium chloride can be electrolyzed above its melting temperature of 801°C. Write the half-reaction equations and the cell reaction equation.

The Electrical Double Layer

In order to discuss electrochemical cells with finite currents, we must investigate the molecular processes at the electrodes. If a metallic phase such as an electrode is negatively charged, the extra electrons will repel each other and move to the surface of the phase, where they are as far from each other as possible. Similarly, the excess positive charges will be found at the surface of a positively charged conductor.

Ions and neutral molecules can also be adsorbed on an electrode surface. Adsorbed ions are divided into two classes. If the ion is adsorbed directly on the surface, it is said to be **specifically adsorbed**. Specifically adsorbed ions are generally adsorbed without a complete "solvation sphere" of attached solvent molecules. Every electrolyte solution contains at least one type of cation and one type of anion and the cations are generally not adsorbed to the same extent as the anions, so adsorbed ions contribute to the charge at the surface.

In addition to specifically adsorbed ions, an excess of oppositely charged ions will be attracted to the vicinity of a charged electrode surface. However, these ions will be fully solvated, and a solvated ion cannot approach the surface as closely as specifically adsorbed ions. Such ions are not held tightly to the surface and are distributed in a diffuse layer. Such ions are sometimes said to be **nonspecifically adsorbed**.

The region near an electrode surface thus contains two layers of charge, the electrode surface (including specifically adsorbed ions) with one sign of charge and a more diffuse layer made up of nonspecifically adsorbed ions of the opposite charge, as shown schematically in Figure 21.6a. This structure is called the **electrical double layer**, or sometimes the **diffuse double**

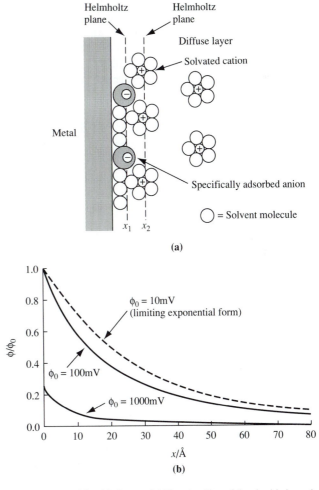

(a)

(b)

Figure 21.6. The Electrical Double Layer. (a) The structure of the double layer (schematic). The double layer consists of a layer of ions of one sign specifically adsorbed on the electrode surface, and a diffuse layer of hydrated ions of the opposite charge further from the electrode surface. **(b)** The electric potential in the diffuse double layer. (From A. J. Bard and L. R. Faulkner, *Electrochemical Methods—Fundamentals and Applications*, Wiley, New York, 1980, p. 505.)

layer. The layer of specifically adsorbed ions is called the **compact layer**, the **Helmholtz layer**, or the **Stern layer**. The plane of closest approach of nonspecifically adsorbed ions is sometimes called the **outer Helmholtz plane**, or the **Guoy plane**.

Guoy and Chapman developed a theory of the charge distribution in the diffuse double layer about 10 years before Debye and Hückel developed their theory of ionic solutions, which is quite similar to it.[18] If one neglects nonelectrostatic contributions to the potential energy of an ion of valence z, the concentration of ions in a region of electric potential φ is given by the Boltzmann probability formula, Equation (1.7-25):

$$c = c_0 e^{-ze\varphi/k_B T} = c_0 e^{-zF\varphi/RT} \qquad \textbf{(21.4-1)}$$

where c is the concentration of the particular type of ion and c_0 is the concentration at a location where $\varphi = 0$. In the second version of the equation, F is Faraday's constant, $F = N_A e = 96485$ C mol^{-1}.

Guoy and Chapman combined Equation (21.4-1) with the Poisson equation of electrostatics and found (by an analysis similar to that of Debye and Hückel) that if the potential is taken equal to zero at large distances, the electric potential in the diffuse double layer is given by

$$\varphi = \varphi_0 e^{-\kappa x} \qquad \textbf{(21.4-2)}$$

where φ_0 is the value of the potential at the surface of the electrode. The parameter k is the same parameter as in the Debye-Hückel theory, given by Equation (6.4-3):

$$\kappa = e\left(\frac{2N_A \rho_1 I}{\varepsilon k_B T}\right)^{1/2} \qquad \textbf{(21.4-3)}$$

where the symbols are defined in Section 6.4. Figure 21.6b shows this potential as a function of distance from the electrode surface for an ionic strength of 0.010 mol kg^{-1} and a temperature of 298.15 K.

The **Debye length** is defined to equal $1/\kappa$. It is the distance from the surface of the electrode at which the potential is equal to $1/e$ (about 0.37) of its value at the surface and is a measure of the effective thickness of the diffuse double layer.

The electrical double layer resembles a capacitor, which is a pair of parallel conducting plates separated by a dielectric medium or a vacuum and which can carry charges of opposite sign on the two plates. Typical values of the capacitance of the electrical double layer range from 10 to 40 microfarads per square centimeter.[19]

Remember that the farad is a very large unit of capacitance. Typical capacitors found in electrical circuits prior to the invention of integrated circuits have capacitances of small fractions of a microfarad (usually measured in nanofarads or picofarads).

Exercise 21.10   Find the Debye length for a solution of HCl with a molality of 0.0200 mol kg^{-1}.

---

[18] G. Guoy, *J. Phys. (Paris)* **9**, 457 (1910); D. L. Chapman, *Philos. Mag.* **25**, 475 (1913).
[19] A. J. Bard and L. R. Faulkner, *Electrochemical Methods—Fundamentals and Applications*, Wiley, New York, 1980, p. 8.

The **charge density** (charge per unit volume) ρ_c is given by

$$\rho_c = F(z_+ c_+ + z_- c_-) \tag{21.4-4}$$

where z_+ and z_- are the valences of the cation and anion, respectively, and c_+ and c_- are the concentrations of the cation and anion, respectively. The charge density can be obtained by combining Equations (21.4-1), (21.4-2) and (21.4-4).

Exercise 21.11   Write the equation for the charge density in the Guoy-Chapman theory.

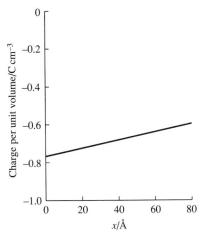

Figure 21.7. The Distribution of Charge in the Diffuse Double Layer. This distribution was calculated with the Guoy-Chapman theory.

Figure 21.7 shows the charge density in an aqueous 1-1 electrolyte solution of molality 0.010 mol kg^{-1} at 298.15 K in the vicinity of a positive electrode with $\varphi_0 = 10$ mV, according to the Guoy-Chapman theory. Compare this figure with Figure 6.11 for the charge density near a negative ion in an electrolyte solution.

Rates of Electrode Processes

When any chemical reaction comes to equilibrium, the forward and reverse processes cancel. At an electrode, these processes are an oxidation (anodic) half-reaction and a reduction (cathodic) half-reaction. Either of these half-reactions if occurring alone would produce a current, and the half-reactions can be thought of as producing anodic and cathodic currents that cancel at equilibrium. The **exchange current** is defined as the equilibrium magnitude of each of these currents. The exchange current per unit area is denoted by i_0 and is commonly expressed in A cm^{-2} (amperes per square centimeter). In SI units it is expressed in A m^{-2}.

If the exchange current is large, a relatively small change in either the anodic or cathodic current can produce a sizable net current, so that the electrode can approximate a nonpolarizable electrode. However, if the exchange current is small, only a small net current is likely to occur, and the electrode can approximate an ideal polarizable electrode.

The magnitude of the exchange current depends on the temperature and is different for different electrode materials and different solution compositions. Typical values range from 5.4×10^{-3} A cm^{-2} for 0.02 mol L^{-1} Zn^{2+} against a Zn amalgam with mole fraction 0.010 down to 10^{-10} A cm^{-2} for the oxygen electrode with a platinum surface in acid solution.[20] The zinc electrode is said to be a reversible electrode because it is possible in practice to reverse the direction of the net current at this electrode with a small change in potential. The oxygen electrode is called an **irreversible electrode**, since a small change in potential produces a negligible change in net current because of the small size of the exchange current.

---

[20] H. A. Laitinen and W. E. Harris, *Chemical Analysis*, 2nd ed., McGraw-Hill, New York, 1975, p. 233.

The Overpotential

There are three sources of **back e.m.f.** which oppose the passage of an electrolytic current through a cell. The first is the **reversible back e.m.f.** due to the cell reaction. For example, in a Daniell cell with unit activities the reversible back e.m.f. is the equilibrium cell potential of 1.100 V. For activities other than unit activities, the reversible back e.m.f. can be calculated from the Nernst equation. The second source of back e.m.f. is the "IR drop" in the electrolyte, the voltage difference across the electrolyte solution due to its electrical resistance. In many cases, we will be able to neglect this contribution. The other source of back e.m.f. is the **overpotential**, the change in potential beyond the reversible cell potential that is needed to produce a given net current in an electrochemical cell (neglecting the IR drop). The overpotential is a measure of the polarization of the electrode.

There are two principal contributions to the overpotential.[21] The first contribution is the **concentration overpotential**, which is due to changes in concentration near the surface of the electrodes. The second contribution is the **activation overpotential**, related to the activation energy of the chemical reaction at the electrode. It exists in at least some systems because the reactive species have to pass through a transition state of high potential energy in order to react at an electrode.

The Concentration Overpotential

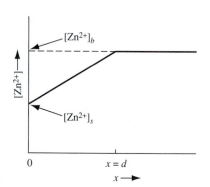

Figure 21.8. The Assumed Concentration Profile in a Boundary Layer Near an Electrode. This concentration profile is assumed in order to obtain a relatively simple theory of the rates of electrode processes.

If electrolysis (charging) is occurring in a Daniell cell at a nonzero rate, zinc ions are being reduced at the surface of the zinc electrode. If there is no stirring, the zinc ions are replaced from the bulk solution only by diffusion. As an oversimplification, let us assume that the concentration of zinc ions near the zinc electrode is as represented in Figure 21.8, where d is the effective thickness of a boundary layer. The value of d depends on the shape of the electrodes, the concentration of the solutes, the value of the diffusion coefficient, etc., so we will not attempt to evaluate it directly but will express it in terms of a limiting current.

For a planar electrode, Fick's law of diffusion, Equation (17.2-4), gives the diffusion flux of zinc ions as

$$J_{Zn^{2+}} = D \frac{[Zn^{2+}]_b - [Zn^{2+}]_s}{d} \tag{21.4-5}$$

where D is the diffusion coefficient of Zn^{2+} ions, $[Zn^{2+}]_b$ is the concentration in the bulk of the solution, and $[Zn^{2+}]_s$ is the concentration at the surface. The **current density** j (current per unit area of electrode surface) is

$$j = nFD \frac{[Zn^{2+}]_b - [Zn^{2+}]_s}{d} \tag{21.4-6}$$

where n is the number of electrons reacting per ion (2 in this case).

---

[21] *ibid.*, pp. 258ff.

If an electrolytic current flows, the concentration of zinc ions at the surface becomes smaller. The current density approaches a limit for large counter e.m.f., when the surface concentration approaches zero:

$$j_{lim} = \frac{nFD[Zn^{2+}]_b}{d} \tag{21.4-7}$$

Of course, if the potential is increased enough, water can be reduced to form hydrogen gas, so the limiting value in Equation (21.4-7) must be estimated from potentials that are not sufficient to reduce water or any reducible species other than zinc ions.

If there is stirring, the zinc ions are brought to the electrode by convection as well as diffusion. If convection predominates, the quotient D/d in Equation (21.4-7) is replaced by the rate of convection (volume of solution brought to unit area of electrode per second). With a combination of convection and diffusion D/d is replaced by a weighted sum of D/d and the rate of convection, called m, the **mass transport coefficient**.

If we regard the boundary layer as a concentration cell, the analogue of Equation (8.3-7) gives the electric potential difference across the boundary layer. This potential difference is the **concentration overpotential**, denoted by η_{conc}:

$$\eta_{conc} = \frac{RT}{nF} \ln\left(\frac{[Zn^{2+}]_b}{[Zn^{2+}]_s}\right) = \frac{RT}{nF} \ln\left(1 - \frac{j}{j_{lim}}\right) \tag{21.4-8}$$

where we have used Equations (21.4-6) and (21.4-7). We assume activity coefficients equal to unity and use the concentration description instead of the molality description.

Exercise 21.12

a. Show that Equation (21.4-8) follows from Equations (21.4-6) and (21.4-7).

b. Evaluate the concentration overpotential at a cadmium amalgam electrode, the diffusion flux of Cd^{2+} ions, and the current per square centimeter for the assumption that $d = 200\ \mu m$, that the concentration at the electrode is $0.0100\ mol\ L^{-1}$, and that the concentration in the bulk is $0.0200\ mol\ L^{-1}$. The diffusion coefficient for Cd^{2+} ions equals $8.7 \times 10^{-10}\ m^2\ s^{-1}$.[22]

Equation (21.4-8) gives the concentration overpotential for one electrode. The concentration overpotential of the Daniell cell also contains a contribution from the other electrode, from which newly formed copper ions diffuse into the bulk of the solution.

It is sometimes possible to study the behavior of a single electrode by the use of a third electrode, called a **reference electrode**. A common choice for a reference electrode is a silver–silver chloride electrode in a saturated potassium chloride solution. The electrode is built with a protruding tube containing a porous plug at the end, forming a liquid junction. The liquid

---

[22] A. J. Bard and L. R. Faulkner, *op. cit.*, p. 154.

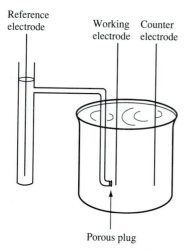

Reference electrode   Working electrode   Counter electrode

Porous plug

Figure 21.9. An Electrochemical Cell with a Third Electrode. The function of the third electrode (the counter electrode) is to provide a means of passing a current through the cell without passing a current between the reference electrode and the working electrode.

The single reaction coordinate represents a complicated process: the ion moves toward the electrode, the electrons detach from the ion, and the electrons move into the electrode.

junction potential is presumably fairly small and should be very nearly constant if the KCl solution is much more concentrated than the cell solution. The liquid junction is placed close to the surface of the electrode to be studied (the working electrode), as in Figure 21.9. The potential difference between the reference electrode and the working electrode is measured without allowing a current to pass between the reference electrode and the working electrode, but allowing a controlled current to pass between the working electrode and the other electrode, called the **counter electrode**.

The Activation Overpotential

This part of the overpotential is due to chemical processes at the electrode. For example, consider a cation of valence z that can be oxidized by losing n electrons at an inert electrode such as a platinum electrode. An example of such a half-reaction is

$$Fe^{2+} \rightleftharpoons Fe^{3+} + e^-$$

We write the general version of such a half-reaction equation:

$$R^{z+} \rightleftharpoons O^{(z+n)+} + ne^-$$

where R^{z+} stands for the reduced species and $O^{(z+n)+}$ stands for the oxidized species. In the Fe^{2+}/Fe^{3+} half-reaction, $z = 2$ and $n = 1$.

We assume that the progress of the reaction can be represented by a potential energy depending on a reaction coordinate as shown in Figure 21.10a. The reaction coordinate changes from left to right in this figure as the reactant ion approaches the electrode and n electrons are pulled from the ion by the electric field near the electrode. The region near an electrode surface can be a region of very high electric fields; a potential difference of 1 volt across an interface region with a thickness of 10 nm corresponds to a field of 1×10^8 volts m^{-1}.

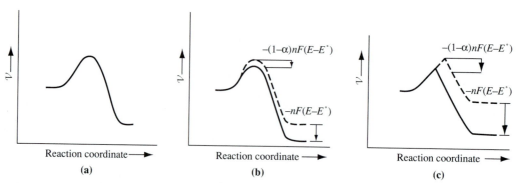

(a)   (b)   (c)

Figure 21.10. The Potential Energy as a Function of a Reaction Coordinate. (a) Without applied potential. Here, the potential energy is that for the oxidation of a cation by the loss of n electrons. The ion must approach the electrode, and the electrons must be removed. **(b)** With a negative applied potential. This is for the same process, but with a negative applied potential at the electrode. The applied potential changes the potential energy of the negative electrons in their final state (in the electrode) and changes the height of the maximum in the curve by a smaller amount. **(c)** The effect of the symmetry factor. This shows an oversimplified analysis to determine the shift in the maximum of the potential energy curve. After A. J. Bard and L. R. Faulkner, *Electrochemical Methods—Fundamentals and Applications*, Wiley, New York, 1980, p. 94.

The maximum in the potential energy is partly due to the increase of the potential energy of the positive ion as it approaches the positive electrode and partly due to the energy required to remove n electrons from the ion. This increase in potential energy is related to the ionization potential of the ion, but differs in that the electron is not removed into a vacuum but into a region between the ion and the electrode that contains solvent molecules and other ions.

The decrease of potential energy as the reaction proceeds past the maximum in Figure 21.10a is due to the binding of the electrons into the electrode. This amount of energy is related to the work function of the electrode material, which is the minimum energy required to remove an electron from the electrode material into a vacuum, but differs from it because the electrons move into the electrode from the solution near the electrode.

The reverse (reduction) half-reaction mentioned here is not the same as the reduction half-reaction occurring at the other electrode. It is the reverse of the oxidation reaction we are discussing.

We regard the state of high potential energy as a transition state and interpret the potential energy to reach the maximum from the left as the activation energy for the forward (oxidation) half-reaction. The potential energy required to reach the maximum from the right is the activation energy for the reverse (reduction) half-reaction. We write the forward and reverse rate constants for our unimolecular reaction in the "thermodynamic form" analogous to Equation (21.1-15)

$$k = \frac{k_B T}{h} e^{\Delta S^{\ddagger \circ}/R} e^{-\Delta H^{\ddagger \circ}/RT} = \frac{k_B T}{h} e^{\Delta S^{\ddagger \circ}/R} e^{-\Delta U^{\ddagger \circ}/RT} \qquad \textbf{(21.4-9)}$$

where we neglect the difference between $\Delta H^{\ddagger}$ and $\Delta U^{\ddagger}$. As in Equation (21.1-17), we identify $\Delta U^{\ddagger}$ with the Arrhenius activation energy, E_a, and identify the pre-exponential factor

$$A = \frac{k_B T}{h} e^{\Delta S^{\ddagger \circ}/R} \qquad \textbf{(21.4-10)}$$

The rate constants of the forward and reverse half-reactions can be written in the Arrhenius form:

$$k_{ox} = A_{ox} e^{-E_{a(ox)}/RT} \qquad \textbf{(21.4-11a)}$$

$$k_{red} = A_{red} e^{-E_{a(red)}/RT} \qquad \textbf{(21.4-11b)}$$

Although the activation energy E_a is a total energy, including kinetic energy as well as potential energy, we make the assumption that any variation in E_a is due to variation in the potential energy of the ion at the maximum in Figure 21.10a. Let us assume that the potential energy of the reacting species is as represented in Figure 21.10a when the cell is at equilibrium in its standard state. For example, if the counter electrode is the standard hydrogen electrode and our electrode half-reaction is the Fe^{2+}/Fe^{3+} half-reaction, the electrode will be 0.771 V more positive than the standard hydrogen electrode, and the counter e.m.f. is equal in magnitude to the cell voltage, which is equal to the equilibrium standard-state voltage E°. If activity coefficients can be assumed to equal unity, the concentrations of the oxidized and reduced species will be equal to each other in this state.

If the system is at equilibrium and each half-reaction is first order in its reactant, we can write for unit area of the electrode

Oxidation rate per unit area $= k_{ox}[R]_{eq}$

$$= \text{reduction rate per unit area} = k_{red}[O]_{eq}$$

where $[O]$ is the concentration of the oxidized species (Fe^{3+} in our example) and $[R]$ is the concentration of the reduced species at the surface of the electrode (we omit the subscript s).

At equilibrium at the standard-state voltage the concentrations are equal and $k_{red} = k_{ox}$. We denote this value of the rate constants by k°:

$$k^{\circ} = A_{ox} \exp(-E^{\circ}_{aox}/RT) = A_{red} \exp(-E^{\circ}_{ared}/RT) \qquad \textbf{(21.4-12)}$$

where the superscript $^{\circ}$ means that the values apply in the standard state at equilibrium.

Let us now change the counter e.m.f. in the external circuit, thus changing the electric potential in the electrode from φ° (the value of φ when $E = E^{\circ}$) to a new value φ. The potential energy of n moles of electrons in the electrode is changed by an amount equal to $-nF(\varphi - \varphi^{\circ})$. If the maximum in the potential energy curve remains at the same position, the activation energy for the reduction increases by $nF(\varphi - \varphi^{\circ})$ while the activation energy of the oxidation remains unchanged. However, the ordinary case is that, as shown schematically in Figure 21.10b, the position of the maximum is lowered but by a magnitude smaller than $|nF(\varphi - \varphi^{\circ})|$.

We define a parameter α such that the position of the minimum is lowered by $(1 - \alpha)nF(\varphi - \varphi^{\circ})$, lowering the activation energy of the oxidation by the same amount. The activation energy of the reduction is increased by the amount $\alpha nF(\varphi - \varphi^{\circ})$. The parameter α is called the **transfer coefficient** or the **symmetry factor**. The name "symmetry factor" is used because its value, which generally lies between 0 and 1, is related to the shape (symmetry) of the curve in Figure 21.10a.

Figure 21.10c shows in an oversimplified way how the shape of the curve affects the value of α. For this illustration we assume that the position of the right side of the curve is determined solely by the electric potential at the electrode, while the left side of the curve is determined solely by chemical factors that are unaffected by the potential of the electrode, and assume also that the two sides of the curve meet at a cusp. When the electric potential is increased by an amount $\varphi - \varphi^{\circ}$, the entire right side is lowered by an amount $nF(\varphi - \varphi^{\circ})$. The peak drops by an amount $nF(\varphi - \varphi^{\circ})/2$ (corresponding to $\alpha = 1/2$) if the slopes of the two sides are equal in magnitude and by a different amount if the slopes have different magnitudes. The parameter α is thus a measure of the asymmetry of the curve in the diagram.

| **Exercise 21.13** | Show by sketching graphs similar to Figure 21.10c that $\alpha < 1/2$ if the left side of the curve is steeper than the right and that $\alpha > 1/2$ if the right side is steeper. |

Assuming that the pre-exponential factors in Equation (21.4-11) do not depend on the electric potential in the electrode, we write

$$k_{\text{ox}} = A_{\text{ox}} \exp\left[-\frac{E_{a\text{ox}} - (1-\alpha)nF(\varphi - \varphi^\circ)}{RT} \right]$$
$$= k^\circ \exp\left[\frac{(1-\alpha)nF(\varphi - \varphi^\circ)}{RT} \right] \qquad \textbf{(21.4-13a)}$$

Similarly,

$$k_{\text{red}} = k^\circ \exp[-\alpha nF(\varphi - \varphi^\circ)/RT] \qquad \textbf{(21.4-13b)}$$

Let us now investigate the case of equilibrium at a potential not necessarily equal to φ°. The magnitudes of the anodic and cathodic currents per unit area of electrode are given by

$$|j_a| = nFk_{\text{ox}}[\text{R}] \qquad \textbf{(21.4-14a)}$$
$$|j_c| = nFk_{\text{red}}[\text{O}] \qquad \textbf{(21.4-14b)}$$

At equilibrium, the net current is zero and the surface concentrations are equal to the bulk concentrations, so that

$$nFk^\circ \exp\left[\frac{(1-\alpha)nF(\varphi_{\text{eq}} - \varphi^\circ)}{RT} \right][\text{R}]_b$$
$$= nFk^\circ \exp\left[-\frac{\alpha nF(\varphi_{\text{eq}} - \varphi^\circ)}{RT} \right][\text{O}]_b \qquad \textbf{(21.4-15)}$$

Each side of Equation (21.4-15) is equal to the **exchange current density**, the exchange current per unit area. When equal factors on the two sides of this equation are canceled, the symmetry factor disappears from the equation:

$$\exp\left[\frac{nF(\varphi_{\text{eq}} - \varphi^\circ)}{RT} \right] = \frac{[\text{O}]_b}{[\text{R}]_b} \qquad \textbf{(21.4-16)}$$

It is always important to ensure that any result has the correct behavior in limiting cases. In this treatment, we see that as we approach equilibrium the equilibrium Nernst equation is recovered.

This equation is equivalent to the Nernst equation if the reactants and products at the counter electrode are at unit activities, which means that our treatment is consistent with equilibrium electrochemical theory.

Exercise 21.14 Carry out the algebraic steps to put Equation (21.4-16) into the standard form of the Nernst equation.

If both sides of Equation (21.4-16) are raised to the $-\alpha$ power, we obtain

$$\exp\left[\frac{-\alpha nF(\varphi_{\text{eq}} - \varphi^\circ)}{RT} \right] = \left(\frac{[\text{O}]_b}{[\text{R}]_b} \right)^{-\alpha} \qquad \textbf{(21.4-17)}$$

When Equation (21.4-17) is substituted into Equation (21.4-15), we obtain an expression for the exchange current density, j_0:

$$j_0 = nFk^\circ[\text{R}]_b^{1-\alpha}[\text{O}]_b^{\alpha} \qquad \textbf{(21.4-18)}$$

If the voltage is changed from the equilibrium value, the resulting current density is

$$
j = j_a - |j_c| = nFk^\circ \exp\left[\frac{(1 - \alpha)nF(\varphi - \varphi^\circ)}{RT}\right][R]
$$
$$
- nFk^\circ \exp\left[-\frac{\alpha nF(\varphi - \varphi^\circ)}{RT}\right][O] \tag{21.4-19}
$$

where we count the current as positive if the electrode half-reaction proceeds in the oxidation direction.

Equation (21.4-19) gives the dependence of the current on the potential for any value of the potential. We now express this dependence in terms of the **overpotential**, η, defined by

$$
\eta = \varphi - \varphi_{eq} \tag{21.4-20}
$$

We divide the first term on the right-hand side of Equation (21.4-19) by the left-hand side of Equation (21.4-15) and divide the second term by the right-hand side of Equation (21.4-15) and obtain

$$
\frac{j}{j_0} = \frac{[R]}{[R]_b} \exp\left[\frac{(1 - \alpha)nF\eta}{RT}\right] - \frac{[O]}{[O]_b} \exp\left(\frac{-\alpha nF\eta}{RT}\right) \tag{21.4-21}
$$

Exercise 21.15   Carry out the steps of algebra to obtain Equation (21.4-21).

Equation (21.4-21) is a fundamental equation of electrode kinetics. It indicates that the current is proportional to the exchange current but that it has a fairly complicated dependence on the overpotential. Figure 21.11 shows the anodic current, the cathodic current, and the net current for a hypothetical electrode reaction. Ohm's law is not obeyed except for small values of the overpotential.

If stirring is so efficient that the bulk and surface concentrations are equal, the concentrations cancel out of Equation (21.4-21), which is then known as the **Butler-Volmer equation**:

$$
\frac{j}{j_0} = \exp\left[\frac{(1 - \alpha)nF\eta}{RT}\right] - \exp\left(\frac{-\alpha nF\eta}{RT}\right) \tag{21.4-22}
$$

There are two important limiting cases. One is the case of small overpotential, in which case the exponents in Equation (21.4-21) or Equation (21.4-22) are small, and the approximation

$$
e^x \approx 1 + x
$$

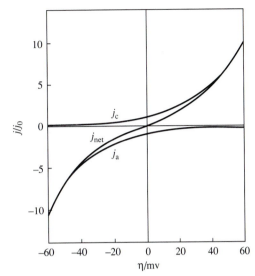

Figure 21.11. The Anodic Current, the Cathodic Current, and the Net Current at an Electrode as a Function of the Overpotential (schematic). This figure shows how the net current is composed of the two contributions, which cancel at zero overpotential.

can be used. Application of this approximation to the Butler-Volmer equation gives

$$j \approx j_0 \left[1 + \frac{(1 - \alpha)nF\eta}{RT} - 1 + \frac{\alpha nF\eta}{RT} \right] \approx \frac{j_0 nF\eta}{RT} \qquad \textbf{(21.4-23)}$$

so that Ohm's law is obeyed. The factor $RT/j_0 nF$ has the dimensions of resistance times area. This factor divided by the area is sometimes called the **charge transfer resistance**.

The second limiting case is that of large overpotential. Consider the case that the overpotential is large and negative. The second term on the right-hand side of the Butler-Volmer equation, Equation (21.4-22), is much larger than the first term, and

$$|j| \approx j_0 e^{-\alpha nF\eta/RT} \qquad \textbf{(21.4-24)}$$

Solving this equation for η gives

$$\eta = \frac{RT}{\alpha nF} \ln(j_0) - \frac{RT}{\alpha nF} \ln(|j|) \qquad \textbf{(21.4-25)}$$

Equation (21.4-25) is of the form of the **Tafel equation**, an empirical equation:

$$|\eta| = a + b \log_{10}(|j|) \qquad \textbf{(21.4-26)}$$

Exercise 21.16   Derive the version of Equation (21.4-22) that applies for large positive values of the overpotential.

One of the electrode reactions that has been extensively studied is the reduction of hydrogen ions at various metal electrode surfaces. The mechanism of the electrode reaction can depend on the material of the electrode. One possible mechanism is[23]

$$(1)\qquad H^+ + e^- + M \rightleftharpoons M\text{—}H \qquad\qquad \textbf{(21.4-27a)}$$

$$(2)\qquad 2M\text{—}H \rightarrow 2M + H_2 \qquad\qquad \textbf{(21.4-27b)}$$

where M denotes the metal of the electrode, and M—H denotes a chemisorbed hydrogen atom. Another possible mechanism is[24]

$$(1)\qquad H^+ + e^- + M \rightleftharpoons M\text{—}H \qquad\qquad \textbf{(21.4-28a)}$$

$$(2)\qquad M\text{—}H + H^+ + e^- \rightarrow M + H_2 \qquad\qquad \textbf{(21.4-28b)}$$

In either of these mechanisms, it is possible for different metals that either the first step or the second step is rate limiting. If the first step of the mechanism of Equation (21.4-27) is rate limiting in both directions,

$$\text{Forward rate} = k_1[H^+]_G(1 - \theta)$$

where $[H^+]_G$ denotes the concentration of hydrogen ions at the Guoy plane (outer Helmholtz plane) and θ denotes the fraction of surface sites occupied by hydrogen atoms. If the electric potential in the bulk of the solution is called zero, then by the Boltzmann probability

$$[H^+]_G = [H^+]_b e^{-F\varphi_1/RT} \qquad\qquad \textbf{(21.4-29)}$$

where $[H^+]_b$ denotes the concentration of H^+ in the bulk of the solution and φ_1 is the electric potential at the Guoy plane.

The rate at equilibrium gives the exchange current. Using Equation (21.4-13a) for the rate constant,

$$j_0 = F(1 - \alpha)[H^+]_b \exp\left(\frac{-F\varphi_1}{RT}\right) k^\circ \exp\left[\frac{(1 - \alpha)nF(\varphi - \varphi^\circ)}{RT}\right] \qquad \textbf{(21.4-30)}$$

With the expression in Equation (21.4-30) for the exchange current, the Butler-Volmer equation, Equation (21.4-22), and Equations (21.4-23) through (21.4-25) can be used, since it is of the same form for a reduction as for an oxidation half-reaction.

For a current density of 0.01 ampere per square centimeter and a hydrogen ion concentration of 1.0 mol L^{-1}, the overpotentials for several electrodes are[25]

0.035 V for platinized platinum

0.56 V for iron

0.76 V for silver

1.10 V for mercury

---

[23] K. J. Laidler, *J. Chem. Educ.* **47**, 600 (1970).

[24] *ibid.*

[25] W. C. Gardiner, Jr., *Rates and Mechanisms of Chemical Reactions*, W. A. Benjamin, New York, 1969, p. 197.

Overpotentials for different reactions at the same electrode will also differ from each other, and it is even possible to react one substance at a nonzero current when another substance in the solution would preferentially react at an infinitesimal current. An example that is exploited in polarography is the reduction of a fairly active metal onto a mercury electrode. The overpotential for production of hydrogen from water on a mercury electrode is so large that almost no hydrogen is evolved at a potential difference sufficient to plate out the metal, even if hydrogen would preferentially be evolved at an infinitesimal current.

21.5 Electrical Conductivity in Solids

The Drude model system[26] consists of conduction electrons and positively charged "cores," which are the ions produced if the conduction electrons are removed from the atoms of the crystal. The electrons are assumed to obey classical mechanics, which is obviously a bad approximation for the motion of one electron but might be a usable approximation for the average motion of many electrons.

Since electrons are fermions, it is necessary to assume dilute occupation, so that most of the available electron states are vacant. Because of this assumption, the Drude model cannot be applied at low temperatures, when fermions occupy states of lowest possible energy, as assumed in the Aufbau principle.

As the electrons move about in the crystal, they interact with the nuclei. We assume that this interaction can be described adequately as a sequence of collisions with the nuclei, and define

$$\text{Collision probability per unit time} = \frac{1}{\tau} \qquad \text{(21.5-1)}$$

The quantity $1/\tau$ is like a first-order rate constant, and the rate of collisions per unit volume is the number of electrons per unit volume times $1/\tau$, so by analogy with the first-order chemical kinetics result

$$n(t) = n(0)e^{-t/\tau} \qquad \text{(21.5-2)}$$

where $n(t)$ is the number of electrons per unit volume that have not yet collided with a core at time t.

Now let us impose an external electric field $\mathbf{E}$ on the crystal. The field will accelerate the conduction electrons. Since the force on an electron is $-e\mathbf{E}$, (constant), the change in velocity of an electron in time t is, from Newton's second law,

$$\Delta \mathbf{v}(t) = -\frac{e\mathbf{E}t}{m} \qquad \text{(21.5-3)}$$

where e is the magnitude of the charge on the electron and m is its mass.

This change in velocity is imposed on whatever velocity the electron originally had. Before the field was imposed no current was flowing, so the original velocities

---

[26] See J. S. Blakemore, *Solid State Physics*, 2nd ed., W. B. Saunders, Philadelphia, 1974, p. 158ff, and D. Tabor, *Gases, Liquids and Solids*, 2nd ed., Cambridge University Press, Cambridge, England, 1979, pp. 188ff.

The velocity in Equation (21.5-3) is a contribution to the velocity in addition to a part that averages to zero. It is similar to the drift velocity in our discussions of diffusion and ionic conductivity.

of all conduction electrons vanish on the average. We treat the initial velocity as zero and take the right-hand side of Equation (21.5-3) as the final velocity. The distance traveled (in excess of the distance due to the original velocity) from time 0 to time t' is obtained by integrating Equation (21.5-3) from time 0 to time t':

$$\mathbf{r}(t) = -\frac{e\mathbf{E}t'^2}{2m} \tag{21.5-4}$$

We assume that with each collision the electron velocity is again randomized, so that on the average the velocity returns to zero value after the collision.

The rate at which electrons are undergoing collisions at time t is

$$\frac{dn}{dt} = -\frac{n(t)}{\tau} = \frac{n(0)}{\tau} e^{-t/\tau} \tag{21.5-5}$$

The contribution of all electrons to the "electron transport" (equivalent to the electron flux times the time of transport) is

$$\int_0^\infty \mathbf{r}(t) \frac{dn}{dt} \, dt = \frac{-e\mathbf{E}n(0)}{2m\tau} \int_0^\infty t^2 e^{-t/\tau} \, dt = -\frac{e\mathbf{E}n(0)\tau^2}{m} \tag{21.5-6}$$

This formula for the electron transport is the same as if $n(0)$ electrons all had the "drift velocity" given by Equation (21.5-3) for an acceleration time equal to τ,

$$\mathbf{v}_{\text{drift}} = -\frac{e\mathbf{E}\tau}{m} \tag{21.5-7}$$

and traveled for a time equal to τ. The electric current per unit area, $\mathbf{j}$, is equal to the charge on one electron, $-e$, times the electron flux

$$\mathbf{j} = (-e)\left(\frac{-e\mathbf{E}n(0)\tau}{m}\right) = \frac{ne^2\tau}{m}\mathbf{E} \tag{21.5-8}$$

This equation is the same as Ohm's law, Equation (17.5-5), if the conductivity is given by

$$\boxed{\sigma = \frac{ne^2\tau}{m}} \tag{21.5-9}$$

EXAMPLE 21.5

From the density of gold and the assumption that one conduction electron comes from each atom, the density of conduction electrons in gold is equal to 5.90×10^{28} m^{-3}. The resistivity (reciprocal of the conductivity) is equal to 2.24 microohm cm at 20°C. Find the value of τ that corresponds to these values.

Solution

$$\sigma = \frac{1}{2.24 \text{ microohm cm}} \left(\frac{10^6 \text{ microohm}}{1 \text{ ohm}}\right)\left(\frac{100 \text{ cm}}{1 \text{ m}}\right)$$

$$= 4.46 \times 10^7 \text{ ohm}^{-1} \text{ m}^{-1}$$

$$\tau = \frac{m\sigma}{ne^2} = \frac{(9.1 \times 10^{-31} \text{ kg})(4.46 \times 10^7 \text{ ohm}^{-1} \text{ m}^{-1})}{(5.90 \times 10^{28} \text{ m}^{-3})(1.60 \times 10^{-19} \text{ C})^2} = 2.7 \times 10^{-14} \text{ s}$$

Exercise 21.17

a. Verify the units of the answer of Example 21.5, obtaining the SI units of ohms from Ohm's law.

b. If a current per unit area of 1.00×10^6 A m^{-2} is flowing in a sample of gold at 20°C, find the mean drift speed. Find the ratio of the mean drift speed to the root-mean-square speed of electrons (using the classical formula) at this temperature.

The conductivity can also be expressed in terms of the mean free path between collisions. The mean free path is approximately given by

$$\lambda = \langle v \rangle \tau \tag{21.5-10}$$

where $\langle v \rangle$ is an average speed of the electrons. This is not the average drift speed, which is a small speed superimposed on a large speed. It is an average of the actual speeds, which is nearly the same as the equilibrium value, due to the smallness of the drift velocity. If we use the root-mean-square speed of Equation (16.3-8), the conductivity is

$$\sigma = \frac{ne^2 \lambda}{(3mk_B T)^{1/2}} \tag{21.5-11}$$

EXAMPLE 21.6

Find the mean free path for electrons in gold at 20°C.

Solution

$$\lambda = \frac{(3mk_B T)^{1/2} \sigma}{ne^2}$$

$$= \frac{[3(9.11 \times 10^{-31} \text{ kg})(1.38 \times 10^{-23} \text{ J K}^{-1})(293 \text{ K})]^{1/2}(4.46 \times 10^7 \text{ ohm}^{-1} \text{ m}^{-1})}{(5.90 \times 10^{28} \text{ m}^{-3})(1.60 \times 10^{-19} \text{ C})^2}$$

$$= 3.1 \times 10^{-9} \text{ m}$$

This value seems reasonable, as it is roughly 10 lattice distances, and an electron would be expected to pass a number of cores before it collides with one.

The Drude model is crude, but it contains the accepted mechanism for electrical resistance in solids, which is the effect of collisions with the cores of the crystal. One problem is that the conductivities of most common metals are found experimentally to be nearly inversely proportional to the temperature, instead of being inversely proportional to the square root of the temperature as in Equation (21.5-11). One can rationalize this by arguing that the mean free path should decrease as the temperature rises, due to the increased vibrational amplitude of the cores, making them larger effective targets.

There are a number of more sophisticated theories than the Drude theory. However, the results of these theories are similar in general form to Equation (21.5-9). The major differences are in the interpretation of the quantities n, τ, and m.[27]

---

[27] J. S. Blakemore, *op. cit.*, pp. 162ff.

Superconductivity

This phenomenon was discovered in 1911 by Onnes,[28] who found that the conductivity of mercury suddenly rose to a value at least as large as 10^{15} ohm^{-1} m^{-1} (and possibly infinite) when the mercury was cooled below a **transition temperature**, T_c, of 4.2 K. Onnes coined the name "superconductivity" for the phenomenon. Since that time, numerous substances have been found to exhibit superconductivity, generally with a transition temperature below 23 K.

The transition to the superconducting state is found to be a second-order phase transition (see Section 5.4). It is also found that a superconducting material rejects a magnetic field. That is, within the surface of the sample of material, a compensating magnetic field is generated that exactly cancels the magnetic field within the material except for a surface layer of thickness 10 to 100 nm. This effect is called the **Meissner effect** and can cause the levitation of a magnet above a superconductor. However, if the magnetic field is increased above a certain critical value, which depends on temperature and on the substance, the superconductivity disappears.

Beginning in 1986, a number of ceramic compounds were discovered that exhibit superconductivity with transition temperatures as high as 125 K, and the search is on for a superconducting compound with a transition temperature at or above room temperature. The commercial possibilities of such a compound have not been lost on hopeful potential inventors and the communications media.[29]

The "high-temperature" semiconductors are oxides, containing copper along with two or three other metals, such as barium and yttrium, or thallium, barium, and calcium. The first compound to exhibit a transition above that of boiling liquid nitrogen was $YBa_2Cu_3O_x$, where x appears to range in value from about 6.5 to 7.2. If x were equal to 9, the substance could have the crystal structure of a perovskite, with six oxygens surrounding each copper and the copper atoms in layers. The crystal structure of the superconducting material was determined by X-ray diffraction and corresponds to four oxygen atoms around each copper atom, all in the plane of the copper atoms.[30]

In 1988, $Tl_2Ca_2Ba_2Cu_3O_{10+y}$, where y is smaller than unity, was found to have a transition temperature of 125 K. The crystal structure of this compound was determined and found to have planes containing copper and oxygen atoms, as well as planes containing thallium and oxygen atoms.[31] The relationship of this structure to the superconductivity of the compounds is apparently not yet clear.

There is a generally accepted theory of superconductivity in metals,[32] which however appears to be inadequate for the ceramic materials. This theory is based on the notion that at sufficiently low temperature some electrons can "condense" into pairs. Under certain conditions the electrons interact with the lattice of the solid and modify the lattice vibrations in such a way that two electrons form a pair with opposite spins having a lower energy than two single uncorrelated electrons. The pair of electrons is called a **Cooper pair**.

---

[28] H. Kamerlingh Onnes, *Akad. van Wetenschappen (Amsterdam)* **14**, 113 (1911).

[29] A. M. Thayer, *Chem. Eng. News* **67**(48), 9 (Nov. 27, 1989).

[30] R. Dagani, *Chem. Eng. News* **65**(19), 7 (May 11, 1987).

[31] R. Dagani, *Chem. Eng. News* **66**(20), 24 (May 16, 1988); S. S. P. Parkin, *Phys. Rev. Lett.* **61**, 750 (1988).

[32] J. Bardeen, L. N. Cooper, and J. R. Schrieffer, *Phys. Rev.* **108**, 1175 (1957). This theory is known as the BCS theory.

> At sufficiently low temperature, the electrons occupy the lowest-energy states compatible with the Pauli exclusion principle and a few of the highest-energy electrons form Cooper pairs. Unless there is an input of energy to break up the pair, it is not possible for one of the electrons to be scattered by a nucleus unless the other electron is also scattered or the pair is broken up. At sufficiently low temperature, there is not enough energy to break up the pair, so scattering does not occur. The electrons remain as a pair, even though they do not necessarily remain close to each other. A small voltage can now induce a current of Cooper pairs, conducting electrical current without observable resistance.

21.6   Oscillatory Chemical Reactions and Chemical Chaos

Until recently, it had been thought that the concentrations of substances involved in chemical reactions evolved in a relatively simple way. Concentrations of reactants were supposed to decay smoothly, concentrations of products were supposed to rise smoothly, and concentrations of reactive intermediates were supposed to rise and fall no more than once. However, we now recognize more complicated behavior, including oscillations and apparently chaotic behavior.

Oscillatory Chemical Reactions

The first known oscillatory chemical reaction was the iodate-catalyzed decomposition of hydrogen peroxide, discovered in 1920. In this reaction, the color of the solution (due to iodine) and the evolution of oxygen can vary in a nearly periodic way. For many years, most chemists believed that the oscillations were due to some factor other than ordinary chemical reactions and thought that no purely chemical reaction could oscillate.

The most famous oscillatory reaction is the Belousov-Zhabotinskii (BZ) reaction, which is the reaction of citric acid, bromate ion, and ceric ion in acidic solution. This reaction produces not only oscillations in time but also chemical waves, consisting of roughly concentric or spiral rings of different colors moving outward from various centers.[33] The first article describing this reaction was rejected for publication because of the common belief that chemical oscillations could not occur. However, since the 1960s chemical oscillations have been widely studied and have even become common lecture demonstrations. Most of the articles published on oscillatory reactions have concerned the BZ reaction, although numerous other oscillatory reactions have been discovered.

A mechanism proposed for the BZ reaction has 18 steps and involves 21 different chemical species.[34] A computer simulation of the 18 simultaneous rate differential equations for the mechanism has been carried out and does produce oscillatory behavior.

---

[33] See I. R. Epstein, *Chem. Eng. News* **65**(13), 24 (March 30, 1987) for pictures and a history.

[34] R. J. Field, E. Körös, and R. M. Noyes, *J. Am. Chem. Soc.* **94**, 8649 (1972). This mechanism is called the FKN mechanism. See also the previous footnote.

If A is continually replenished, the following simple mechanism can show oscillatory behavior:

$$
\begin{array}{lll}
(1) & A + X \rightarrow 2X & \textbf{(21.6-1a)} \\
(2) & X + Y \rightarrow 2Y & \textbf{(21.6-1b)} \\
(3) & Y \rightarrow P & \textbf{(21.6-1c)}
\end{array}
$$

This mechanism corresponds to the stoichiometric equation

$$ A \rightarrow P $$

Exercise 21.18   Write the three rate differential equations for the mechanism of Equation (21.6-1).

The mechanism of Equation (21.6-1) has been used in ecology as a simple predator-prey model, in which A represents the food supply (grass) for prey animals (hares), represented by X. Predators (wolves) are represented by Y, and dead wolves are represented by P. The consumption of grass (A) by the hares (X) allows them to reproduce as in step 1, and the consumption of hares by wolves (Y) allows them to reproduce as in step 2. However, the consumption of the hares and the death of the wolves mean that the overall reaction corresponds to the conversion of grass into dead wolves.

For the case that [A] is held fixed, a solution to the set of differential equations is

$$ k_2([X] + [Y]) - k_3 \ln([X]) - k_1[A] \ln([Y]) = C \qquad \textbf{(21.6-2)} $$

where C is a constant.[35]

Exercise 21.19   Show that the function of Equation (21.6-2) satisfies the differential equations obtained in Exercise 21.18. Hint: First differentiate Equation (21.6-2) with respect to time. Then substitute the differential equations of Exercise 21.18 into this equation.

The form of the solution in Equation (21.6-2), with the time not explicitly appearing, suggests that the results can be displayed by a plot of [Y] as a function of [X]. This plot shows that the solution is periodic, with a closed curve that is retraced over and over again as time passes. The two-dimensional space spanned by [X] and [Y] is a **phase space**. Two time-dependent quantities are plotted on the axes, and a curve in the space is traced out as time passes. The oscillations predicted by the mechanism resemble the actual oscillations in predator and prey populations in ecosystems.

---

[35] I. R. Epstein, *op. cit.*

Exercise 21.20 Plot the curve in phase space representing Equation (21.6-2) for $C = 1.00 \text{ s}^{-1}$, $[A] = 1.000 \text{ mol L}^{-1}$, $k_1 = 1.000 \text{ L mol}^{-1} \text{ s}^{-1}$, $k_2 = 0.500 \text{ L mol}^{-1} \text{ s}^{-1}$, and $k_3 = 0.1 \text{ s}^{-1}$. One way to proceed is to write a computer program to solve for $[Y]$ for each value of $[X]$, using Newton's method or the method of bisection.

The known mechanisms that produce oscillatory behavior have two characteristics in common. The first is **autocatalysis**: the product of a step must catalyze that step, as in Equations (21.6-1a) and (21.6-1b). Second, nonlinear differential equations occur. That is, second- or third-order steps must occur.

Deterministic Chaos

The theories of the time-dependent behavior of chemical systems contain various differential equations, such as the time-dependent Schrödinger equation, Newton's laws, Liouville's equation and its quantum analogue, and the sets of differential equations that are derived from assumed mechanisms in reaction kinetics. All of these equations are **deterministic**. That is, if an initial condition is given, the state of the system evolves in a unique way that is determined by the equation and the initial condition. It would seem that there is no room for chaos, if what we mean by chaos is random, unpredictable behavior. However, it has been found both by experiment and by theory that a kind of nonperiodic deterministic behavior can occur that is not random, although under casual inspection it may appear to be random.[36] This kind of behavior is called **deterministic chaos.**

Chaos in Chemically Reacting Systems

Deterministic chaos in chemically reacting systems can occur in oscillatory reactions, which under some circumstances go into nonperiodic oscillations.[37] Figure 21.12 shows experimental results on the electrode potential for a bromide ion-specific electrode[38] in a continuously stirred tank reactor in which the BZ reaction was carried out. Figure 21.12a shows periodic behavior obtained with one flow rate of the reactants, and Figure 21.12c shows another periodic behavior with a different flow rate. Figure 21.12b shows nonperiodic (chaotic) oscillatory behavior for a flow rate intermediate between the other two.

One way in which chaos appears is **period doubling**, which is illustrated in Figure 21.13. In this behavior, as the flow rate of reactants is changed the period doubles, as in going from Figure 21.13a to Figure 21.13b. There are still oscillations that are as rapid as before, but twice as many oscillations must occur before the behavior repeats itself exactly. A further change in

---

[36] R. M. May, *Nature* 261 (1976).
[37] F. Arnoul, A. Arneodo, P. Richett, J. C. Rous, and H. L. Swinney, *Accounts Chem. Res.* **20**, 436 (1987).
[38] J. L. Hudson, M. Hart, and D. Marinko, *J. Chem. Phys.* **71**, 1601 (1979).

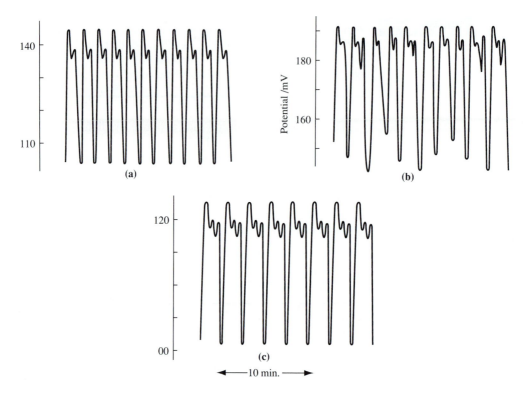

Figure 21.12. The Electrode Potential of a Bromide Ion-Specific Electrode in a BZ Reaction in a Continuously Stirred Tank Reactor. The residence time (reactor volume divided by total reactant flow rate) was **(a)** 0.104 hour, **(b)** 0.098 hour, and **(c)** 0.097 hour. Periodic behavior is apparent. From J. L. Hudson, M. Hart, and D. Marinko, *J. Chem. Phys.* **71**, 1601 (1979).

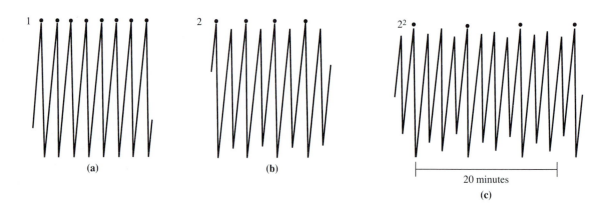

Figure 21.13. An Example of Period Doubling (schematic). Although the oscillations remain of the same frequency, successive oscillations are not identical. First, two oscillations are necessary to make a period, then four oscillations are required, etc. From F. Argoul, A. Arneodo, P. Richett, J. C. Rous, and H. L. Swinney, *Accounts Chem. Res.* 20, 436 (1987).

the flow rate doubles the period again, as in going to Figure 21.13c. It is found that the period doubles repeatedly with an ever-decreasing change in the flow rate, so that at an **accumulation point** the period becomes infinite. The behavior becomes nonperiodic, or chaotic.[39]

For a reaction like the BZ reaction, with several concentrations, a multi-dimensional phase space can be constructed with one axis for each concentration. Periodic behavior is described by a closed curve called a **limit cycle**. Chaotic behavior is described by an irregular trajectory, called a **strange attractor**, which can lie in a kind of torus (doughnut-shaped region) but does not retrace itself exactly in a finite length of time.

A general characteristic of a strange attractor is "exponential separation of nearby points on the attractor."[40] This separation means that if one chooses two nearby points in the portion of phase space occupied by the attractor and follows the natural trajectories of the two points, the distance between them grows larger in an exponential fashion as time passes. In practice, since an initial state can be determined only within a certain experimental error, the long-term behavior of a chaotic system cannot be predicted with certainty. Small errors in specification of an initial state could lead to very large errors in predicting a state at a much later time. If chaotic behavior exists, the fact that our equations of motion are mathematically deterministic does not permit practical prediction of later states.

Chaos in an Artificial System

Mathematicians have constructed various models in order to study chaotic behavior in simple cases. An interesting example of chaotic behavior arises in the iteration of the equation

$$x_{n+1} = ax_n(1 - x_n) \qquad \text{(21.6-3)}$$

The behavior of the logistic difference equation is chosen to illustrate the origin of deterministic chaos in a simple way. It has no direct connection with any chemical phenomena.

which is sometimes called the "logistic difference equation." Iteration of the equation means that the function $ax(1 - x)$ is evaluated for one value of x, and that value is assigned to be the next value of x, which is substituted into the function to obtain a third value of x, etc.

If $a < 4$, a graph of the function is as shown in Figure 21.14a. The maximum in the parabola is lower than unity, and iteration of the equation beginning with any value of x in the closed interval [0, 1] gives a value of x in that interval no matter how many times the equation is iterated.

However, if $a > 4$, a graph of the function appears as in Figure 21.14b, and there are values of x in the interval [0, 1] for which iteration leads to a value of x greater than 1 or less than 0, whereupon further iteration leads to values of x with larger and larger magnitudes. Chaos arises from the fact that there are values of x in the interval [0, 1] that will eventually lead out of the interval upon sufficiently many iterations and are arbitrarily close to values of x that will not lead out of the interval, no matter how many iterations are done.[41]

---

[39] Arnoul *et al., op. cit.*

[40] *ibid.*

[41] R. L. Devaney, *An Introduction to Chaotic Dynamical Systems*, Addison-Wesley, Reading MA, 1987, pp. 31ff. See also footnote 36.

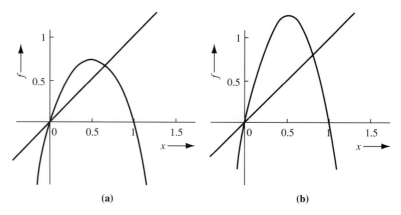

Figure 21.14. The Function $f = ax(1 - x)$. (a) The function $f = 3x(1 - x)$. This version of the function, with $a = 3$, does not lead to deterministic chaos. **(b)** The function $f = 5x(1 - x)$. Iteration of this version of the function, for which $a = 5$, illustrates a kind of deterministic chaos, as do other versions of the function with $a > 4$.

For example, consider $a = 5$, which is the case of Figure 21.14b. The curve representing the function $f = ax(1 - x)$ and the line representing the function $f = x$ cross at $x = 4/5$. This point is called an **attractor**, since if $x = 4/5$, further iterations of the equation continue to give this value. However, a value of x nearly but not exactly equal to $4/5$ leads upon enough iterations outside the interval $[0, 1]$, no matter how close in value to $4/5$ it is. Two points arbitrarily close to each other thus lead to different behaviors: one stays in the interval for arbitrarily many iterations, and the other leads out of the interval. A similar behavior occurs for any point that after a certain number of iterations leads to $x = 4/5$. It will lead only to points in the interval but will also have points arbitrarily close to it that lead out of the interval.

Exercise 21.21

a. Write a computer program to iterate the equation of Equation (21.6-3). Put a branch in the program so that you can iterate arbitrarily many times, but have the program display the new value of x for each iteration.

b. Run the program for $a = 1$, various initial values of x, and various number of iterations.

c. Run the program for $a = 3$ and various initial values of x. What kind of point is $x = 2/3$? What kind of point is $x = 1$? What kind of point is $x = 0$?

d. Run the program for $a = 5$ and various initial values of x. See if you can find some points that lead to values in the interval $[0, 1]$ no matter how many iterations you do. Try iterating $x = 4/5$ and see what round-off error in your computer does. If possible, run the program a second time in double precision.

e. For $a = 5$, find a value of x other than $4/5$ that leads to $x = 4/5$ after two iterations.

f. What is the difference between $x = 2/3$ for $a = 3$ and $x = 4/5$ for $a = 5$?

Chaos in Other Systems

The chemical chaos described above in the case of oscillating chemical reactions is a chaos in a macroscopic description, in which concentrations and times are the variables used. Another example of chaos in a macroscopic description is turbulence in hydrodynamic (fluid) flow.[42] Fluid flow is governed by the Navier-Stokes equation, which is a differential equation for the velocity of a fluid. For an incompressible fluid, the Navier-Stokes equation is[43]

$$\frac{d\mathbf{u}}{dt} = -\nabla P + \rho \mathbf{X} + \eta \, \nabla^2 \mathbf{u} \qquad \text{(21.6-4)}$$

where $\mathbf{X}$ is the external force (gravity, etc.), P is the pressure, η is the viscosity, ρ is the density, and $\mathbf{u}$ is the fluid velocity, which depends on position and time.

When a fluid is flowing with laminar flow (low Reynolds number), the solutions of the Navier-Stokes equation correspond to streaming along smooth curves. However, when the Reynolds number increases beyond a certain value, turbulence can suddenly set in and the fluid streams along chaotic curves.

In addition to chaos in macroscopic descriptions, chaos on the molecular level is now being studied extensively. Two current areas of interest are the study of trajectories on potential energy surfaces, which we mentioned in Section 21.1,[44] and the study of intramolecular energy flow.[45]

The study of trajectories in the phase space of classical statistical mechanics, described in Section 20.8, also has some relation to chaos. The Liouville equation, Equation (21.3-18), gives the rate of change of the classical probability density for an ensemble. The equation is equivalent to saying that the probability density is fixed if one travels along with the phase point specifying the state of one of the systems of the ensemble. This motion is somewhat similar to that of an incompressible (but nonuniform) fluid in hydrodynamics, as given by the Navier-Stokes equation, Equation (21.6-4).

The question arises whether the motion in phase space is chaotic, analogous to turbulent flow in a fluid. This question is related to the fundamental postulate of statistical mechanics, that an ensemble average can be equated with a measured value. In classical statistical mechanics, this postulate is known as the **ergodic hypothesis**.

A measured value of a physical quantity is a time average over the response time of the measuring instrument. During this time, the phase point representing the state of the system moves about in the $6N$-dimensional phase space of the system, sampling some part of the phase space. On the other hand, the ensemble average samples all parts of the $6N$-dimensional space. During a short response time, the phase space trajectory of the physical system cannot pass through all parts of phase space, but if we measure only equilibrium variables, we can think of using instruments with infinitely slow response times without changing the measured values. The question is then whether the phase space trajectory passes through all parts of phase space in an infinite time.

If the motion has a finite period with a closed trajectory, there must be parts of phase space that are not visited by the trajectory, since the trajectory retraces the closed curve over and over again. However, if the motion is nonperiodic (chaotic), it might be that the trajectory will sooner or later pass arbitrarily close to any point in the phase space. Unfortunately, it is apparently not known

---

[42] L. P. Kadanoff, *Phys. Today*, Dec. 1983, p. 46.

[43] D. A. McQuarrie, *Statistical Mechanics*, Harper & Row, New York, 1976, p. 391ff.

[44] See for example N. DeLeon and B. J. Berne, *J. Chem. Phys.* **75**, 3495 (1981), and B. J. Berne, N. DeLeon, and R. O. Rosenberg, *J. Phys. Chem.* **86**, 2166 (1982).

[45] See for example W. P. Reinhardt, *J. Phys. Chem.* **86**, 2158 (1982). The issue in which this article appears contains several articles on related topics.

whether the motion is chaotic or not, and the ergodic hypothesis remains an unproved postulate. A number of theorems are related to the ergodic hypothesis, and the interested reader is referred to advanced works on statistical mechanics.[46]

Summary

In this chapter, we have examined several theories of nonequilibrium processes at the molecular level. The principal theory discussed for chemical reactions is the activated complex theory, in which it is assumed that the reactants in an elementary process form an activated complex, which can be assumed to be at chemical equilibrium with the reactants.

A case history of the gas-phase chemical reaction was presented:

$$F + H_2 \rightarrow H + HF$$

Although this reaction has been extensively studied on the molecular level, both experimentally and theoretically, it is still the subject of ongoing research and cannot be said to be completely understood.

The activated complex theory was also applied to diffusion in liquids, in which the motion of a molecule from one "cage" into an adjacent cage is treated similarly to motion along a reaction coordinate of a chemical reaction, and during which an activated complex is assumed to form. More advanced theories of transport in liquids and gases were also discussed.

Nonequilibrium electrochemistry was discussed. The important physical fact is that with a finite current flowing, the cell voltage can be different from its equilibrium value. The difference is partly due to the overpotential, which was discussed.

Electrical conduction in metals was briefly discussed in terms of the Drude model, in which electrical resistance is assumed to be due to collisions of conducting electrons with atomic cores (ions produced when conduction electrons are removed from atoms). Superconductivity was discussed briefly.

A brief description of deterministic chaos was presented. This phenomenon corresponds to irregular but theoretically predictable behavior. However, if initial conditions cannot be specified to arbitrary precision, theoretically predictable behavior becomes chaotic in practice.

Additional Reading

R. B. Bernstein, *Chemical Dynamics via Molecular Beam and Laser Techniques*, Oxford University Press, New York, 1982
This book comes from the Hinshelwood Lectures, given at Oxford University in 1980.

---

[46] See for example, A. I. Khinchin, *Mathematical Foundations of Statistical Mechanics*, Dover, New York, 1949, Chapter III.

P. R. Brooks and E. F. Hayes (eds.), *State-to-State Chemistry*, ACS Symposium Series, No. 56. American Chemical Society, Washington, DC, 1977
A report on the state of the art in state-to-state chemistry as of 1977.

R. L. Devaney, *An Introduction to Chaotic Dynamical Systems*, Addison-Wesley, Reading MA, 1987
A treatment of the mathematics of models that can produce chaotic behavior. The models are discussed with mathematical precision but are generally far removed from physical reality.

A. I. Khinchin, *Mathematical Foundations of Statistical Mechanics*, Dover Publications, New York, 1949
A highly mathematical treatment of some theoretical aspects of classical statistical mechanics. Chapter 3 treats the ergodic hypothesis.

K. J. Laidler, *Chemical Kinetics*, 3rd ed., Harper & Row, New York, 1987
This is the latest edition of a standard textbook in reaction kinetics at the senior/first-year graduate level. It is authoritative and clearly written.

R. D. Levine and R. B. Bernstein, *Molecular Dynamics and Chemical Reactivity*, Oxford University Press, New York, 1987
An excellent source of information about the molecular study of reaction kinetics. It is clearly written by authors who are experts in the field.

PROBLEMS

Problems for Section 21.1

21.22 Moss and Coady[47] give information about approximate potential energy surfaces for several reactions, calculated with the London-Eyring-Polanyi-Sato (LEPS) method. For the reaction

$$Br + H_2 \rightarrow HBr + H$$

their value of $\Delta \varepsilon_0^{\ddagger}$ is 9.0×10^{-21} J, from a barrier height of 9.3×10^{-21} J. They give the following values for the activated complex at 298 K: $z'^{\ddagger}_{rot} = 118.25$, $z'^{\ddagger}_{vib} = (1.00)(1.199)^2$ (for the symmetric stretch and the two degenerate bends). Find the value of the rate constant at 298 K. Compare your value with that from the experimental Arrhenius activation energy, 9 kJ mol^{-1}, and the Arrhenius pre-exponential factor, 6.3×10^7 mol^{-1} s^{-1}. What is an explanation for the difference between the theoretical and experimental values?

*21.23. For the gas-phase reaction

$$CH_3 + CH_4 \rightarrow CH_4 + CH_3$$

the Arrhenius pre-exponential factor equals 5×10^4 m^3 mol^{-1} s^{-1}. Estimate $\Delta S^{\ddagger \circ}$ at 298 K. Explain your value of $\Delta S^{\ddagger \circ}$.

21.24. Calculate the forward rate constant at 300 K and at 1000 K for the reaction

$$H + H_2 \rightarrow H_2 + H$$

using the activated complex theory. Assume a linear activated complex with all vibrational partition functions equal to unity. The intermolecular distance of hydrogen is 0.741×10^{-10} m. For the activated complex take each bond as 0.942×10^{-10} m. The value of $\Delta \varepsilon_0^{\ddagger}$ is 6.8×10^{-20} J.

Problem for Section 21.2

*21.25. a. Using the C—C and C=C bond energies, estimate the activation energy for a *cis-trans* isomerization around a C=C bond.

b. Estimate the ratio of the rate of a *cis-trans* isomerization at 310 K to the rate at 300 K. State any assumptions.

Problems for Section 21.3

21.26. The self-diffusion coefficient of liquid CCl$_4$ is given by Rathbun and Babb:[48]

$t/°C$	25	40	50	60
$D/10^{-9}$ m^2 s^{-1}	1.30	1.78	2.00	2.44

Find the value of A_d and the value of E_{ad} in Equation (21.3-8). Assuming that $a = 4 \times 10^{-10}$ m, find $\Delta S^{\ddagger \circ}$ and $\Delta G^{\ddagger \circ}$ at 40°C.

[47] S. J. Moss and C. J. Coady, *J. Chem. Educ.* **60**, 445 (1983).

[48] R. E. Rathbun and A. L. Babb, *J. Phys. Chem.* **65**, 1072 (1961).

*21.27. The viscosity of water at 20°C is 0.001005 kg m^{-1} s^{-1}, and at 50°C it is 0.0005494 kg m^{-1} s^{-1}. Calculate the Arrhenius activation energy for viscosity in water.

21.28. Derive an integrodifferential equation for $f^{(2)}$, the two-body reduced distribution function, by integrating the Liouville equation, Equation (21.3-18), over the coordinates of all particles except two. Assume that the potential energy is pairwise additive, as in Equation (16.7-1). There will be an integral remaining in the equation which contains $f^{(3)}$, the three-body reduced distribution function.

*21.29. Using the data in Problem 21.26, calculate the value of the correlation time τ in Equation (21.3-20) for carbon tetrachloride at 50°C. Explain the temperature dependence of this parameter.

21.30. The representation of Example 21.3 for a velocity time correlation function is not the only crude approximation that can be used.

a. Find the value of the parameter a in the representation

$$\langle \mathbf{v}(t) \cdot \mathbf{v}(0) \rangle = \frac{3 k_B T}{m} e^{-t^2/2a^2}$$

for carbon tetrachloride at 25°C.

b. Draw rough graphs of the two representations of $\langle \mathbf{v}(t) \cdot \mathbf{v}(0) \rangle$ on the same sheet of graph paper.

c. How do you think the parameter a will depend on temperature?

Problems for Section 21.4

21.31. In the Hall-Heroult process for the production of aluminum, aluminum oxide is dissolved in molten cryolite, Na_3ClF_6, which melts at 1000°C, and electrolyzed. The cathode is molten aluminum, and the anode is made of graphite. The anode product is carbon dioxide. Write the half-reaction equations and the cell reaction equation.

*21.32. Find the activity of chloride ion necessary so that chlorine instead of oxygen will be evolved when an infinitesimal current flows at 298.15 K in the electrolysis cell of Example 21.4.

21.33. Find the value of the charge transfer resistance at 298.15 K for the reduction of Zn^{2+} ions from a solution of 0.0200 mol L^{-1} on a zinc amalgam electrode with Zn mole fraction of 0.0100 and an area of 1.00 cm^2.

21.34. For the formation of hydrogen gas from a dilute sulfuric acid solution on a mercury electrode at 25°C, the current density was 4.8×10^{-6} A cm^{-2} at an overpotential of 0.60 V and was 3.7×10^{-7} A cm^{-2} at an overpotential of 0.50 V. Find the values of the parameters in the Tafel equation, Equation (21.4-26).

Problems for Section 21.5

*21.35. The resistivity of silver at 20°C is 1.59 microohm cm and its density is 10.5 g cm^{-3}. Assume that there is one conduction electron per atom.

a. Find the value of τ.

b. Find the mean free path for electrons in silver.

c. Give a simple explanation for the comparison of the mean free path of electrons in silver and gold (see Examples 21.3 and 21.4).

d. Without looking up the value, estimate the resistivity of copper. Look up the correct value and compare your estimate to it.

21.36. The resistivity of solid mercury at -39.2°C is 25.5 microohm cm, and that of liquid mercury at -36.1°C is 80.6 microohm cm. Give a qualitative explanation for this behavior.

*21.37. a. Assume that an electric field can be instantaneously turned off. Estimate the length of time for a current to drop to $1/e$ of its initial value when an electric field is instantaneously turned off in gold at 20°C.

b. A current has been flowing in a superconducting ring since about 1940, without an applied electric field. Estimate a minimum value of the conductivity. State any assumptions.

Problems for Section 21.6

21.38. Equation (21.6-3) resembles the differential equation

$$dx/dt = ax(1 - x)$$

which is used to describe population growth and some other phenomena. When x is small compared with unity, x grows nearly exponentially, but when x is larger the rate decreases and x approaches an asymptotic value for large values of t.

a. At what value of x is the rate dx/dt at a maximum?

b. What is the asymptotic value of x?

c. Sketch a rough graph of $x(t)$ for $a = 1$.

d. Write a computer program to integrate the equation using Euler's method, in which x as a function of t is approximated by iterating

$$x(t + \Delta t) = x(t) + (dx/dt) \, \Delta t$$
$$= x(t) + ax(t)[1 - x(t)] \, \Delta t$$

where dx/dt is evaluated at time t. This method delivers a usable approximation if Δt is chosen to be small enough. Run the program for several values of a, making sure your value of Δt is small enough for adequate accuracy. Note that you cannot start with $x = 0$ at time zero. Make an accurate graph of x as a function of t for $a = 1$ and compare it with your graph of part b. Use computer graphics if possible to make the graph.

e. Look up the Runge-Kutta method[49] and write a computer program to integrate the equation. Run the program for several values of *a*.

21.39. a. Write a computer program to integrate the three simultaneous differential equations for the mechanism of Equation (21.6-1) using Euler's method (see Problem 21.38). Assume that the concentration of A is kept fixed. Make your value of Δt small enough for adequate accuracy. Run the program for several different sets of values of the initial concentrations and the rate constants. Try to determine how the behavior depends on these values.

b. Determine the period of the oscillation for one or more of your sets of initial conditions.

c. Using the Runge-Kutta method (see the footnote to Problem 21.38) write a computer program to integrate the differential equations of part a. Run the program for some of the same sets of values as used in part a to test the accuracy of the Euler method.

General Problem

*21.40. Label each of the following statements as either true or false. If a statement is true only under special circumstances, label it as false.

a. The activated complex generally has sufficient time to equilibrate with the reactants of a chemical reaction.

b. The activated complex is a type of reaction intermediate and can sometimes be isolated from the reaction mixture.

---

[49] See for example R. L. Burden, J. D. Faires, and A. C. Reynolds, *Numerical Analysis*, 2nd ed., Prindle, Weber and Schmidt, Boston, 1981, pp. 200ff.

c. The collision theory of chemical reactions is the best relatively simple way to study unimolecular reactions.

d. The usual source of energy to populate the transition state in a unimolecular reaction is an inelastic collision.

e. The RRKM theory of unimolecular reactions is outmoded and no longer useful.

f. The mechanisms of most chemical reactions are not well characterized and understood.

g. Nonequilibrium electrochemical reactions can be understood with the same theories as other chemical reactions in the liquid phase.

h. By changing the amount of current flowing in an electrolytic cell, it is possible to change the products of the reaction.

i. The electrical conductivity of a typical metal increases with increasing temperature.

j. The Boltzmann equation is an exact equation.

k. The Liouville equation is an exact equation, if classical mechanics is assumed to be valid.

l. A deterministic equation of motion can lead to apparently chaotic results.

m. Zinc should be a better electrical conductor than copper because it has two $4s$ electrons while copper has only one.

n. It is reasonable that the activated complex should resemble the reactants more than the products if the reaction is highly exoergic.

o. It is reasonable that the activated complex should resemble the reactants more than the products if the activation energy of a given reaction is small.

22 The Structure of Condensed Phases

PREVIEW

The properties of gases are primarily determined by the kinetic energy of the molecules. In condensed phases (liquids and solids), the molecules are close to their nearest neighbors, and the properties of condensed phases are primarily determined by the potential energy. In this chapter we discuss some aspects of the structure of solids and liquids.

PRINCIPAL FACTS AND IDEAS

1. The properties of liquids and solids are primarily determined by intermolecular forces.
2. The study of crystalline solids is based on their regular lattice structure.
3. The Einstein and Debye theories describe vibrations in crystals.
4. The band theory explains the electronic structure of solids.
5. Liquids can be studied as very dense gases or as disordered solids.
6. Polymers are made up of long chain-like molecules.

22.1 General Features of Solids and Liquids

Liquids and solids have several things in common. The most important is that in both phases, the molecules (or other formula units) are close together. Because of this proximity, the intermolecular forces are strong. In fact, the potential energy is more important than the kinetic energy in determining their properties, whereas in gases the kinetic energy is more important. However, the most important distinction between solids and liquids is that solids generally possess a crystal lattice structure, while liquids possess only vestiges of such a structure at short range and have no long-range order.

Solids

We group solids into four classes, according to the principal cohesive (attractive) force that holds the solid together. The first class is **molecular solids**, which consist of molecules that are attracted to each other by forces other than chemical bonds. Two examples of molecular solids are solid water (ice) and solid carbon dioxide (dry ice). Atomic solids such as solid argon and solid krypton are also included in this class.

The principal cohesive (attractive) force in molecular solids is the London dispersion force. However, some molecular solids also exhibit other attractive forces. In solid water, the principal intermolecular attraction is hydrogen bonding, which holds the solid in a structure with a tetrahedral arrangement of four nearest-neighbor water molecules around each water molecule.

A rough measure of the strength of the cohesive forces in a solid is its melting temperature. The stronger the attractions, the higher the melting temperature. Because of the relative weakness of their intermolecular attractions, molecular solids generally melt at fairly low temperatures.

The second class is **ionic solids**. For example, sodium chloride consists of sodium ions and chloride ions. Each ion is surrounded by six ions of the opposite charge, which are closer to the particular ion than are the closest ions of like charge. The attractive electrostatic forces thus exceed the repulsive electrostatic forces in magnitude and provide the cohesive force that holds the solid together. However, the net cohesive force is a sum of attractive and repulsive electrostatic forces, and there is considerable variation in melting temperatures among ionic solids, although melting temperatures are generally higher than those of molecular solids. Sodium chloride melts at 801°C, and ammonium nitrate melts at 170°C.

The third class is **covalent solids**, in which the principal cohesive forces are provided by covalent bonds. A sample of a covalent solid can be considered to be one giant molecule. An example is solid carbon, which in both of its naturally occurring allotropic forms, diamond and graphite, is a network of atoms bonded together by covalent bonds. In diamond, each carbon atom is bonded to four other carbon atoms in a tetrahedral geometry. In graphite, the atoms are arranged in sheets in which each carbon is bonded to three other carbons at bond angles of 120°, forming a network of hexagons. There are weaker London forces between the sheets, which can fairly easily be cleaved from each other.

Due to the strength of the covalent bonds, covalent solids generally melt at very high temperatures. For example, graphite sublimes at 1 atm at 3367°C (its triple point is at a higher pressure than 1 atm, like that of carbon dioxide). Quartz (one of the naturally occurring allotropes of silicon dioxide) melts at 1610°C.

The fourth class consists of **metallic solids**. The bonding in metallic solids is primarily delocalized covalent bonding. This bonding can be represented by molecular orbitals that are linear combinations of atomic orbitals from many atoms. The electrons occupying these orbitals move over large regions of the solid and can easily conduct electrical currents.

Some solids are **crystalline**. A crystalline solid consists of a regular geometric array of repeating identical molecular units. Most samples of crystalline substance are **polycrystalline**, which means that they are made up of many pieces of crystal lattice stuck together in various orientations. For example, if you break a piece of cast iron, you can usually see grains, which might be single crystals. If a single crystal can be grown, it usually has greater mechanical strength than a polycrystalline sample of the same material.

Solids that are not crystalline are **amorphous**. They do not have a regular array of many repeating identical units, but have a more disordered structure. **Glasses** are amorphous materials that soften gradually as they are heated, becoming liquid without a definite melting temperature. They are often considered to be supercooled liquids, although they can be very rigid. Silica (silicon dioxide) readily forms a glass called "fused silica" or "fused quartz." Silica also has several crystalline forms, including quartz and cristobalite, in which each silicon atom is covalently bonded to four tetrahedrally arranged oxygen atoms. Each oxygen atom is also bonded to another silicon atom, which is bonded to four oxygen atoms, and so forth to form the crystal.

| Exercise 22.1 | **a.** Without trying to represent the correct three-dimensional structure, sketch a section of a network of Si and O atoms in which every silicon atom is bonded to four oxygen atoms and every O atom is bonded to two Si atoms. |

a. Without trying to represent the correct three-dimensional structure, sketch a section of a network of Si and O atoms in which every silicon atom is bonded to four oxygen atoms and every O atom is bonded to two Si atoms.

b. What kind of hybrid orbitals on the silicon atoms and on the oxygen atoms would be used to construct localized molecular orbitals to represent the bonding in crystalline silicon dioxide?

In fused silica, the silicon tetrahedra are disordered. Some of the silicon atoms may be bonded to fewer than four oxygen atoms, and there may exist sizable voids in the structure into which atoms or molecules of other substances can be placed. There are vestiges of a crystal lattice at short range, but the geometric regularity is not complete and does not persist over large distances.

Polymers have large molecules made up of smaller units, called **monomers**, covalently bonded together. A commonly encountered polymer is polyethylene, which is made up of linear molecules constructed by polymerization of ethene (ethylene), $CH_2=CH_2$. A portion of the structural formula for polyethylene is

$$\cdots-CH_2-CH_2-CH_2-CH_2-CH_2-CH_2-CH_2-CH_2-\cdots$$

Polymer molecules with thousands or tens of thousands of monomer units are common. Solid polymers are nearly always amorphous, although some polymers with linear molecules, like polyethylene, can be partially crystalline, having sets of parallel molecules.

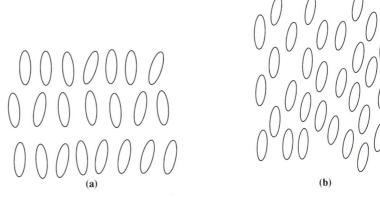

Figure 22.1. Liquid Crystals. (a) A smectic phase. This kind of phase is a type of liquid crystal in which layers of molecules occur. **(b)** A nematic phase. This kind of phase is a type of liquid crystal in which the molecules are aligned, but not in layers.

Liquids

The same intermolecular or chemical forces that hold a solid together act in the liquid of the same substance, and at short range most liquids exhibit partial retention of their crystal structures. Substances with large intermolecular attractions generally have high boiling temperatures. For example, liquid water has considerable hydrogen bonding and exhibits vestiges of the tetrahedral structure of the solid. Water has the unusual property that the solid is less dense than the liquid, due to partial collapse of the crystal structure upon melting.

Liquid crystals are liquids that sometimes form from substances with rodlike molecules. In liquid crystals, the molecules cluster in nearly parallel sets in the liquid phase, attracted to each other by London dispersion forces or other intermolecular forces. In a **smectic phase**, the molecules are aligned in layers, with the long axes of the molecules perpendicular to the layer. In a **nematic phase**, the molecules are still parallel, but are not arranged in layers. In a **cholesteric phase**, the molecules are arranged in layers, with the molecules in a given layer having roughly the same angle with the boundary of the layer but with molecules in different layers having different angles, so that a generally helical arrangement of the molecules over several layers is found. Smectic and nematic liquid crystal phases are depicted schematically in Figure 22.1.

22.2   Crystals

Crystals are solid phases consisting of repetitive geometric arrays of identical molecular units. The **basis** of a crystal is the smallest set of atoms, ions, or molecules with fixed bond distances and angles and with identical orientation and molecular environment that repeats again and again to make up the crystal. For example, the basis of the sodium chloride crystal consists of one sodium ion and one chloride ion. The crystal could be reproduced by stacking replicas of the basis, all with the same orientation.

The basis is not necessarily the same as the formula unit of the substance. For example, the basis of a carbon dioxide crystal contains four molecules. Even though the molecules all have the same bond distances and angles, the four molecules have different molecular orientations and environments.

Crystal Lattices

The **crystal lattice** is generated by placing a **lattice point** at the same location in each basis. Usually a point at the center of an atom or ion is chosen, but any point will do. The crystal lattice is the set of these points and is not the same as the crystal, which is a real piece of matter.

The crystal lattice can be divided into identical **unit cells**. The unit cell is the smallest region in space that can reproduce the lattice by translations in three directions. That is, the lattice is reproduced by stacking replicas of the unit cell in straight rows, files, and columns, with no spaces between them. The contents of the unit cell must have the same stoichiometry as the whole crystal and must have the same symmetry properties as the entire lattice.

Figure 22.2 shows the unit cell of the sodium chloride lattice. This unit cell is a cube with side equal to 5.64×10^{-10} m. It contains four sodium ions: a one-eighth share of each of eight sodium ions at the corners of the unit cell and a one-half share of each of six sodium ions in the faces. It also contains four chloride ions: a one-fourth share of the 12 ions at the centers of the edges, plus the chloride ion at the center of the cell. It contains four basis units.

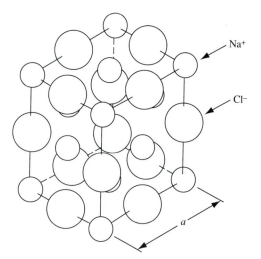

Figure 22.2. The Sodium Chloride Unit Cell. The sodium chloride lattice consists of a face-centered cubic lattice of sodium ions (the small spheres in the figure) and another face-centered cubic lattice of chloride ions a position shifted by a distance a/2 relative to the sodium ion lattice.

Table 22.1. Crystal Systems and Lattices

System	Defining Conditions	Bravais Lattices
Cubic	$a = b = c$ $90° = \alpha = \beta = \gamma$	P, I, F
Tetragonal	$a = b \neq c$ $90° = \alpha = \beta = \gamma$	P, I
Orthorhombic	$a \neq b \neq c$ $90° = \alpha = \beta = \gamma$	P, C, I, F
Monoclinic	$a \neq b \neq c$ $90° = \alpha = \gamma \neq \beta$	P, I
Triclinic	$a \neq b \neq c$ $\alpha \neq \beta = \gamma$	P
Trigonal (rhombohedral)	$a = b = c$ $90° \neq \alpha = \beta = \gamma < 120°$	P
Hexagonal	$a = b \neq c$ $90° = \alpha = \beta, \gamma = 120°$	P

The edges of a unit cell and their lengths are denoted by the letters a, b, and c. The angle between a and b is called γ, the angle between a and c is called β, and the angle between b and c is called α. The directed line segments a, b, and c define the axes along which the unit cell is translated repeatedly to reproduce the lattice. In some lattices, these axes are not perpendicular to each other.

There are seven different **crystal systems**, or unit cell shapes, which are listed in Table 22.1. The unit cells are depicted in Figure 22.3 with the lattice points indicated. If one is willing to allow rotation of replicas of the unit cell while stacking them, the hexagonal unit cell can be cut into three unit cells, one of which is drawn darker than the others in Figure 22.3.

Some of the crystal systems correspond to more than one kind of lattice. A **primitive** lattice (denoted by P) is one in which lattice points occur only at the corners of the unit cell. A unit cell of a primitive lattice contains one basis unit: one-eighth of the basis unit at each corner. A **body-centered** lattice (denoted by I, for German *innenzentriert*) is one in which there is a lattice point at the center of the unit cell as well as at the corners. A **face-centered** lattice (denoted by F) is one in which there is a lattice point at the center of each face of the unit cell as well as at the corners. A **base-centered** or **end-centered** lattice (denoted by C) is one in which there is a lattice point at the center of only one pair of opposite faces as well as at the corners. The table and figure also show the 14 possible lattices, which are called the **Bravais lattices**.

Exercise 22.2 For each of the 14 Bravais lattices, list the number of basis units in the unit cell.

The sodium chloride lattice is a face-centered cubic lattice. In Figure 22.2, the lattice points have been chosen at the centers of the sodium ions. If the

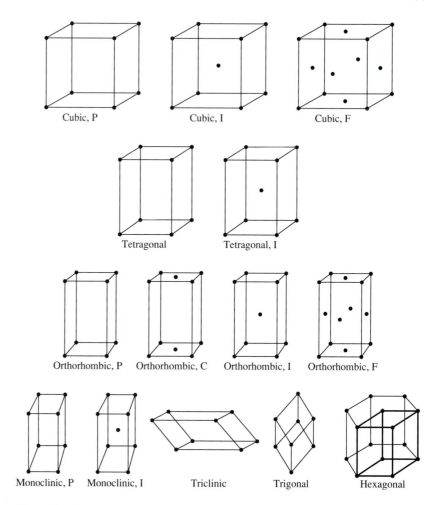

Figure 22.3. The Unit Cells of the Fourteen Bravais Lattices. These are all of the space-filling unit cells. The designations are as follows: P means primitive (no lattice points except at the corners of the unit cell); I (for innenzentriert) means body centered; F means face centered; and C means end centered.

centers of the chloride ions had been chosen, there would be another lattice of exactly the same properties, with the corner of the unit cell for one lattice at the center of the unit cell for the other lattice.

The location of a point within a unit cell is specified by three fractions, which indicate the distances from the origin in the directions of the three axes. For example, if the first fraction is $\frac{1}{2}$, it means a distance of $a/2$ along the a axis; if the second fraction is $\frac{1}{4}$, it means a distance of $b/4$ along the b axis, etc. The center of a unit cell is denoted by $\frac{1}{2}\frac{1}{2}\frac{1}{2}$.

X-Ray Diffraction

One of the most powerful tools for the study of crystal lattices is X-ray diffraction. Since X-rays are electromagnetic radiation with wavelengths of the same general size as crystal lattice spacings, crystals can act as diffraction gratings for X-rays. Study of the angles and intensities of diffracted X-ray

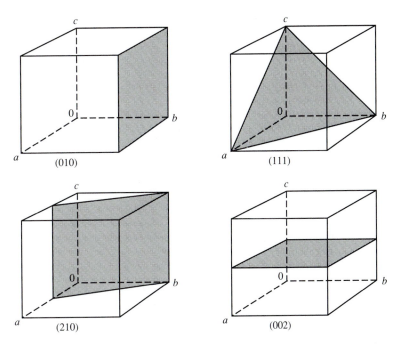

Figure 22.4. Some Planes and Their Miller Indices. All of these planes intersect the edges of the unit cell at points that are distances a/h, b/k, and c/ℓ from the origin where h, k, and ℓ are integers (the Miller indices).

beams can reveal unit cell dimensions and positions of atoms within the unit cells.

In order to describe the diffraction of X-rays, it is necessary to specify the directions of planes that fit into a crystal lattice in such a way that they contain a repeating regular pattern of lattice points. The planes in which we are interested intersect with one or more of the axes within a unit cell. We consider any plane that intersects with the a axis at a distance a/h from the origin, with the b axis at a distance b/k from the origin, and with the c axis at a distance c/ℓ from the origin, where h, k, and ℓ are integers. Many planes intersect the axes in other ways, but they are not included in our discussion. The planes we consider all contain repeating patterns of lattice points.

The three integers h, k, and ℓ are called **Miller indices** and are denoted by their values inside parentheses, as in $(hk\ell)$. The Miller indices are not required to be positive, and a zero value for an index means that the plane is parallel to that axis. Negative values are often denoted by putting the negative sign above the digit, as in $(11\bar{1})$. Figure 22.4 shows a unit cell with several planes and their Miller indices. Some planes are parallel to other planes. For example, the $(00\bar{1})$ plane is parallel to the (001) plane, and the (222) plane is parallel to the (111) plane. However, two parallel planes do not necessarily contain the same pattern of lattice points.

Figure 22.5 shows schematically how a crystal acts like a diffraction grating. Two planes of lattice points are shown, in which lattice points of the first plane are perpendicularly opposed to lattice points of the second plane. We assume that there is an atom to act as a scattering center at each

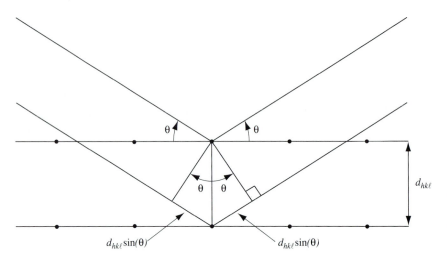

Figure 22.5. Diagram Showing the Bragg Condition for X-Ray Diffraction by a Crystal Lattice. The path lengths of the two rays must differ by an integral number of wavelengths.

lattice point. Each atom diffracts incident electromagnetic radiation, sending a spherical electromagnetic wave out in all directions. The condition for constructive interference in a particular direction is that two waves from adjacent diffraction centers have crests and troughs at the same locations. If the distance between the planes is equal to d and the wavelength of the radiation is λ, the condition is that the extra distance traveled by the wave diffracted from the second layer is an integral number of wavelengths. Trigonometry gives the condition

$$n\lambda = 2d \sin(\theta) \qquad \textbf{(22.2-1)}$$

where θ is the angle between the plane and the direction of the radiation. Equation (22.2-1) is called the **Bragg equation**.

After Sir William Henry Bragg and his son William Lawrence Bragg, who received the 1915 Nobel Prize in physics for their studies in X-ray diffraction.

At first glance, it would seem that the two angles labeled θ in the diagram would not have to be equal. However, one cannot consider just two atoms as scattering centers. For the scattering from other pairs of atoms in the same two planes to produce constructive interference, the two angles must be equal, so the diffraction condition is similar to a reflection from the planes of atoms. Diffracted X-ray beams are therefore sometimes called "reflections."

Exercise 22.3	By drawing a replica of Figure 22.5 and drawing incident and diffracted rays from other pairs of atoms, show that if the two angles labeled θ in Figure 22.5 are equal, all of the diffracted beams interfere constructively if the Bragg condition is satisfied.

It is not necessary that a plane have lattice points directly across from corresponding lattice points in a parallel plane to diffract X-rays. Any one of the sets of planes specified by Miller indices can diffract X-rays, so we specify the Miller indices of the plane in Equation (22.2-1)

$$n\lambda = 2d_{hk\ell} \sin(\theta) \qquad \textbf{(22.2-2)}$$

This equation is the same as

$$\lambda = 2d_{nh, nk, n\ell} \sin(\theta) \tag{22.2-3}$$

For example, the distance between the (200) planes is half as great as the distance between the (100) planes, so the second-order ($n = 2$) diffraction from the (100) planes is at the same wavelength as the first-order ($n = 1$) diffraction from the (200) planes. For a cubic lattice the distance between any two parallel planes is given by

$$d_{hk\ell} = \frac{a}{(h^2 + k^2 + \ell^2)^{1/2}} \tag{22.2-4}$$

where a is the unit cell dimension. We omit the derivation of this equation.

The diffraction of X-rays by a crystal lattice is more complicated than we have indicated. In some cases, the diffracted beams interfere destructively and are not seen. This destructive interference is called **extinction**. Some extinction rules are:[1]

For a primitive lattice: No extinctions

For a face-centered lattice: All three Miller indices must be even or all three must be odd to avoid extinction.

For a body-centered lattice: The sum of the three Miller indices must be an even integer to avoid extinction.

The simplest kind of X-ray diffraction experiment is carried out with a finely powdered sample of the crystalline material placed in a small container and placed in an X-ray beam. Since there are very many small crystals, an impinging collimated X-ray beam strikes some crystals at any given angle, and a number of diffracted beams come from the sample in cones of directions concentric with the incident beam. A photographic plate or film is placed to intercept these beams and record their positions, allowing one to calculate the diffraction angles.

Analysis of the pattern of the diffraction angles allows one to determine from the extinction conditions whether one has a primitive lattice, a face-centered lattice, or a body-centered lattice. From the wavelength of the radiation and the diffraction angles, one can determine the unit cell dimensions.

Other techniques of X-ray diffraction are more complicated. They usually require a small single crystal with minimal imperfections. Most of them involve recording the intensities of the reflections with radiation detectors instead of film, and many of them involve synchronized angular motions of the crystal and the detector. We do not discuss these techniques.

We have discussed the diffraction of X-rays up to now as though each atom in the crystal lattice were a point from which the X-rays are scattered. In fact, the atoms and molecules consist of nuclei and electrons, and the scattering takes place over the entire unit cell. Electrons scatter X-rays far more strongly than do nuclei. The scattering from the different parts of the

---

[1] M. J. Buerger, *Contemporary Crystallography*, McGraw-Hill, New York, 1970, chapter 5.

unit cell interferes constructively and destructively in ways that are determined by the electron density in the unit cell.

In some cases, analysis of the relative intensities of the different diffracted beams allows reconstruction of the electron density. The electron density is a periodic function of position, since it repeats in each unit cell. It therefore can be represented as a Fourier series, and it has been shown that the coefficients in the Fourier series are related to the intensities of the diffracted beams. Reconstructing the electron density is a very complicated process, which we cannot describe in detail. The first such structure determinations were done before the advent of programmable computers, with many hours of hand calculation. Present-day calculations are done almost automatically by sophisticated computer programs, using intensity data taken with automated computer-driven diffractometers.

Crystal Habits

When a crystal grows, it adds more and more basis units in each of three directions. The crystal will have the same shape as a unit cell if unit cells are added at the same rate in each direction. However, crystals can exhibit various **habits**, which are shapes corresponding to different growth rates in the three lattice axis directions. For example, a cubic lattice could give rise to a crystal that is rectangular but not cubic. Furthermore, the faces of the crystal do not have to be the (001), (010), and (100) planes corresponding to the faces of the unit cell. Other planes, such as the (111) plane, can form boundaries of the crystal.

Modern studies in surface catalysis often use single crystals with an exposed face whose Miller indices are known,[2] and it is sometimes found that different planes have different catalytic activities. A real crystal can also have a variety of defects, some of which are schematically depicted in Figure 22.6 and in Figure 19.2. The presence of defects often increases the catalytic activity.

A number of monatomic substances, including the inert gases and most metals, crystallize in **close-packed lattices.** If a collection of spheres of equal size is packed as closely as possible, each sphere is in contact with 12 other spheres, which surround it symmetrically. We say that the **coordination number** equals 12. There are two ways to accomplish this closest packing, as depicted in Figure 22.7. If one layer of spheres is laid down in a plane, another layer can be placed on it, as shown in Figure 22.7a (the lower layer is drawn in broken curves). There are now two choices for laying down a third layer: either in the locations marked h (directly over the spheres of the first layer) or in the locations marked c. If the third layer is laid down in the h locations, a lattice with a hexagonal unit cell results, as shown in Figure 22.7b. If the third layer is laid down in the c locations, a lattice with a face-centered cubic (fcc) unit cell results, as shown in Figure 22.7c. The cubic unit cell is tilted with respect to the original layers, which are parallel to the (111) plane.

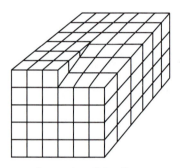

A screw-type dislocation

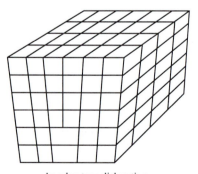

An edge-type dislocation

Figure 22.6. Some Crystal Defects (schematic). In addition to the step defect pictured in Figure 19.2, these are the most common simple types of crystal defects.

---

[2] D. W. Goodman, *Ann. Rev. Phys. Chem.* **37**, 425 (1986).

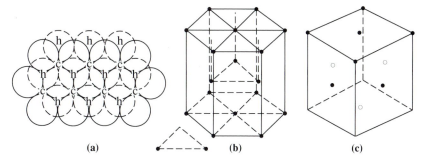

Figure 22.7. Closest Packing of Spheres. (a) Layers of spheres. Two layers of spheres are shown, with the second layer resting in hollows of the first layer. The third layer can be placed in either of two positions. The positions labeled h give the hexagonal close-packed lattice, and the positions labeled c give the cubic close-packed lattice. **(b)** The hexagonal close-packed unit cell. Here the third layer has each sphere directly over a sphere of the first layer. **(c)** The face-centered cubic close-packed unit cell. This unit cell is oriented so that the layers of atoms in (a) are parallel to a plane that passes through the lattice point at the top rear of the cell, the lower left of the cell, and the lower right of the cell.

Both the hexagonal and the fcc close-packed lattices give the same **packing fraction**, which is the fraction of the unit cell volume occupied by the spheres. (It is also called the "filling fraction" or the "packing efficiency.") The nearest-neighbor distance and the number of nearest neighbors are also identical. However, the neighbors beyond the shell of nearest neighbors are not at the same distances in both lattices, and a given substance at equilibrium will crystallize in only one of these lattices. Nickel crystallizes in the face-centered cubic lattice, whereas cobalt crystallizes in the hexagonal lattice. All of the inert gases crystallize in the face-centered cubic lattice except for helium, which crystallizes in the hexagonal lattice.

Exercise 22.4

a. From the formula for the volume of a sphere and the fact that the diagonal of the face of the unit cell is equal to four times the sphere radius, show that the packing fraction for the fcc close-packed lattice is equal to 0.74.

b. Iron crystallizes in the body-centered cubic lattice. Give the coordination number for this lattice and calculate the packing fraction for this lattice, assuming spheres that touch.

22.3 Crystal Vibrations

In Section 22.2 we discussed crystal lattices as though the atoms of the crystal were permanently fixed at certain locations in the unit cell. These positions are actually equilibrium positions about which the atoms vibrate. Sound waves in a crystal correspond to organized collective vibrations, but random vibrations occur at any nonzero temperature. At high temperatures, the atoms can even diffuse at measurable rates through the crystal. The motions of individual atoms are difficult to study, since the potential energy of a crystal is a complicated function of all of the atoms' coordinates and the motion of one atom affects the motions of the other atoms. However, there are simple model systems that represent the dynamics of a crystal in an approximate way.

The Einstein Crystal Model

This model system represents an atomic crystal, such as a solidified inert gas, a metal, or a covalent crystal like diamond. It is assumed that the potential energy of each atom depends only on the displacement of that atom from its equilibrium position. This assumption is equivalent to pretending that all of the other atoms in the crystal are fixed at their equilibrium positions while we study the motion of one atom.

The Einstein crystal model is a model system in which one pretends that each atom moves independently of the other atoms. Therefore, we can apply the statistical mechanical treatment of independent particles to it.

For an Einstein crystal of N atoms of a single element, the Hamiltonian operator is

$$\hat{H}_{\text{sys}} = \sum_{i=1}^{N} \hat{H}_1(i) \tag{22.3-1}$$

where $\hat{H}_1(i)$ is a one-atom Hamiltonian operator for atom number i

$$\hat{H}_1(i) = -\frac{\hbar^2}{2m}\nabla_i^2 + \mathscr{V}(x_i, y_i, z_i) \tag{22.3-2}$$

and where x_i, y_i, and z_i are the cartesian components of the displacement of atom i from its equilibrium position. That is, x_i is the distance in the x direction from the atom's equilibrium position to its present position, etc.

The Hamiltonian operator of Equation (22.3-1) is a sum of one-particle operators. It gives a product wave function and a system energy eigenvalue that is a sum with one term for each atom:

$$E_j = \sum_{i=1}^{N} \varepsilon_{ij} \tag{22.3-3}$$

where the double subscript ij stands for the values of the quantum numbers of atom i if the whole crystal is in state j.

In the dilute gas, we must take account of the indistinguishability of the molecules and symmetrize or antisymmetrize the wave function. Two product wave functions that differ only in having the factors for two atoms exchanged cannot represent different microstates. In the crystal, although the atoms are indistinguishable, the lattice sites at which the atoms are confined are in principle distinguishable from each other. We regard the atoms as fixed at particular sites and treat them as distinguishable. The system wave function does not have to be symmetrized or antisymmetrized, and two product wave functions that differ only in having the factors for two atoms exchanged do represent different microstates of the crystal.

Calculation of the Partition Function for the Einstein Crystal Model

We represent the crystal by a canonical ensemble. The canonical partition function is

$$Z = \sum_{1j}\sum_{2j}\cdots\sum_{Nj} e^{-(\varepsilon_{1j}+\varepsilon_{2j}+\varepsilon_{3j}\cdots+\varepsilon_{Nj})/k_{\text{B}}T} \tag{22.3-4}$$

In this multiple sum, each individual sum is performed independently, since each state of one atom can occur with any state of every other atom. The canonical partition function factors:

$$Z = z^N \tag{22.3-5}$$

where z is the partition function of one atom:

$$z = \sum_j e^{-\varepsilon_j / k_B T} \tag{22.3-6}$$

The index j denotes a state of one atom. Equation (22.3-5) is different from Equation (20.7-15). The $N!$ divisor in Equation (20.7-15) comes from the indistinguishability of the molecules of the dilute gas, which does not apply here.

We can now write formulas for the thermodynamic functions of the Einstein crystal model, using Equation (20.7-30):

Compare these equations with those for a dilute gas in Section 20.6. The differences are due to the distinguishability of the atoms in the Einstein model.

$$U = k_B T^2 \left[\frac{\partial \ln(Z)}{\partial T} \right]_V = N k_B T^2 \left[\frac{\partial \ln(z)}{\partial T} \right]_V \tag{22.3-7a}$$

$$S = \frac{U}{T} + k_B \ln(Z) = \frac{U}{T} + N k_B \ln(z) \tag{22.3-7b}$$

$$A = -k_B T \ln(Z) = -N k_B T \ln(z) \tag{22.3-7c}$$

$$\mu = A_N - A_{N-1} = -k_B T \ln(z) \tag{22.3-7d}$$

$$P = k_B T \left[\frac{\partial \ln(Z)}{\partial V} \right]_T = N k_B T \left[\frac{\ln(z)}{\partial V} \right]_T \tag{22.3-7e}$$

As usual, we have replaced a derivative by a finite difference in the equation for the chemical potential, which is chosen to be the chemical potential per molecule.

There is a difficulty with the pressure of the Einstein crystal model. For a one-component system, G is given by Euler's theorem as

$$G = N\mu = -N K_B T \ln(z) \tag{22.3-8}$$

so that $G = A$, and PV is predicted to vanish, since G equals $A + PV$. This result is a shortcoming of a crude model, but for a crystal the value of PV is small compared with G and A and the numerical effect of this shortcoming is not very important.

This potential energy function is the simplest reasonable assumption. However, the real function must depend in a complicated way on the positions of the surrounding particles, and this feature is completely omitted in our function.

As a first approximation, we assume that the potential energy is an isotropic quadratic function of the displacement from the equilibrium position of the central atom:

$$\mathscr{V}(x_i, y_i, z_i) = \mathscr{V}_e + \frac{k}{2}(x_i^2 + y_i^2 + z_i^2) \tag{22.3-9}$$

where k is a force constant and $\mathscr{V}_e$ is the potential energy at the equilibrium position.

This function is the potential energy function of a three-dimensional harmonic oscillator, similar to three harmonic oscillators of the same frequency. The energy eigenfunctions are products of three factors, and the energy eigenvalues are

Once again, we have recognized a problem that has already been solved, so we can use the solution from that problem.

$$\varepsilon = h\nu(v_x + v_y + v_z) + \mathscr{V}_0 \tag{22.3-10}$$

where v is the harmonic oscillator frequency, from Chapter 9,

$$v = \frac{1}{2\pi}\sqrt{\frac{k}{m}} \tag{22.3-11}$$

and the three quantum numbers v_x, v_y, and v_z can take on nonnegative integral values. We let $\mathcal{V}_0$ include the zero-point energy, so that $\mathcal{V}_0 = \mathcal{V}_e + 3hv/2$.

The molecular partition function is now

$$z = e^{-\mathcal{V}_0/k_\mathrm{B}T}\sum_{v_x=0}^{\infty}\sum_{v_y=0}^{\infty}\sum_{v_z=0}^{\infty}e^{-hv(v_x+v_y+v_z)/k_\mathrm{B}T} \tag{22.3-12}$$

This sum can be factored like that of Equation (20.3-10). Each factor is just like the vibrational partition function in Equation (20.4-17), so that

$$z = e^{-\mathcal{V}_0/k_\mathrm{B}T}\left(\frac{1}{1-e^{-hv/k_\mathrm{B}T}}\right)^3 \tag{22.3-13}$$

We can now write formulas for the thermodynamic functions of the Einstein crystal model. For example,

$$U = N\mathcal{V}_0 + \frac{3Nhv}{e^{hv/k_\mathrm{B}T}-1} \tag{22.3-14a}$$

$$C_V = 3Nk_\mathrm{B}\left(\frac{hv}{k_\mathrm{B}T}\right)^2\frac{e^{hv/k_\mathrm{B}T}}{(e^{hv/k_\mathrm{B}T}-1)^2} \tag{22.3-14b}$$

$$\mu = \mathcal{V}_0 + 3k_\mathrm{B}T\ln(1-e^{-hv/k_\mathrm{B}T}) \tag{22.3-14c}$$

where we use the chemical potential per molecule, as in Chapter 20. Since there is no simple way to determine the value of the force constant, it is customary to choose that value of v which gives the best fit to heat capacity data.

The formulas for the thermodynamic functions can be restated in terms of the parameter

$$\Theta_\mathrm{E} = \frac{hv}{k_\mathrm{B}} \quad \text{(definition)} \tag{22.3-15}$$

which has the dimensions of temperature, and which is called the **Einstein temperature** or **characteristic temperature**.

Exercise 22.5   Derive the formula for the entropy of an Einstein crystal.

Figure 22.8 shows the heat capacity of diamond as a function of temperature, as well as the heat capacity of the Einstein crystal model with an Einstein temperature of 1320 K, which gives the best fit to these experimental data.

In the limit of high temperature

$$\lim_{T\to\infty}C_V = 3Nk_\mathrm{B} \tag{22.3-16}$$

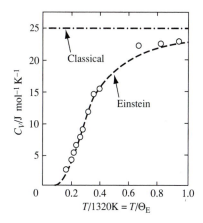

Figure 22.8. The Heat Capacity of Diamond Fit to the Einstein Crystal Model Result. The horizontal line corresponds to the classical result. From J. S. Blakemore, *Solid State Physics*, W. B. Saunders, Philadelphia, 1974, p. 121.

This formula agrees with the empirical **law of Dulong and Petit**, which states that the molar heat capacity of atomic crystals is approximately equal to $3R$.

Exercise 22.6   Derive Equation (22.3-16).

Most atomic solids have a sufficiently small Einstein temperature that the law of Dulong and Petit applies quite well near room temperature.

The Debye Crystal Model

This model is a physically motivated improvement over the Einstein crystal model. The Einstein crystal model has a single vibrational frequency, which is far from the actual situation, as seen by the fact that crystals can conduct sound waves of almost any frequency. Debye treated the crystal like a single giant molecule, so that the vibrational motions of the atoms in the crystal correspond to normal modes, with numerous atoms moving together in any given mode. Debye assumed that these normal modes could be approximately represented by the vibrations of a continuous solid (a model solid system with no molecular structure). These vibrations are sound waves, and their quanta of energy are called **phonons**.

Consider a cubical sample of the crystal with side L, and assume that the surfaces cannot vibrate. A standing wave must have zero amplitude at the surfaces, so that the amplitude of the wave is the same function of position as the wave function of a particle in a three-dimensional box:

The assertion that the standing waves must vanish at the surface of the crystal is an approximation. Clearly, there are no atoms to vibrate beyond the surface, but the atoms on the surface can vibrate.

$$\text{Amplitude} = \sin(s_x \pi x/L)\,\sin(s_y \pi y/L)\,\sin(s_z \pi z/L)$$

where s_x, s_y, and s_z are positive integers. We let

$$s^2 = s_x^2 + s_y^2 + s_z^2 \tag{22.3-17}$$

The wavelength is given by

$$\lambda = 2L/s \tag{22.3-18}$$

and the frequency of the wave is given by

$$v = \frac{c}{\lambda} = \frac{cs}{2L} \tag{22.3-19}$$

where c is the speed of propagation of the waves in the solid (the speed of sound in the solid, which is generally much larger than the speed of sound in a gas).

The number of sets of integers such that s lies between s and $s + ds$ is approximately equal to

$$\text{Number of waves in } (s, s + ds) = \frac{\pi s^2}{2} ds \tag{22.3-20}$$

which is the same relation as in Equation (9.3-2). The result in Equation (22.3-20) must be multiplied by a factor of 3, since in a given direction in a continuous solid there can be two transverse waves (polarized at right angles to each other) and one longitudinal wave. In a given solid, the speed of sound can depend on the frequency but is found to approach a constant value for small frequency. Furthermore, the longitudinal and transverse waves might not move at the same speed, in which case we regard the constant value of c as some kind of average speed, which is different for each solid substance.

Exercise 22.7   Show that Equation (22.3-20) is correct by constructing a space in which s_x, s_y, and s_z are plotted on three cartesian axes. The number of points inside a given region of this space corresponding to sets of integral values is nearly equal to the volume of that region, since there is one such point per unit volume. Consider a spherical shell of thickness ds. Only one octant of the coordinate system is included, corresponding to the requirement that all of the integers are positive.

Equations (22.3-19) and (22.3-20) can be combined:

$$
\begin{aligned}
\text{Number of waves in } ds &= \frac{3\pi}{2}\left(\frac{2Lv}{c}\right)^2 ds = g(v)\, dv \\
&= \frac{3\pi}{2}\left(\frac{2Lv}{c}\right)^2 \frac{2L}{c}\, dv = 12\pi \frac{V}{c^3} v^2\, dv
\end{aligned}
\tag{22.3-21}
$$

The function $g(v)$ is called the **frequency distribution**.

In crystal of N atoms, the total number of vibrational modes is $3N - 6$, which can be approximated by $3N$. All possible frequencies cannot be included, since this would give infinitely many normal modes. Debye chose a maximum frequency v_D such that he had the correct number of modes:

$$3N = \int_0^{v_D} g(v)\, dv = \frac{12\pi V}{c^3} \int_0^{v_D} v^2\, dv = \frac{4\pi V v^3}{c^3} \tag{22.3-22}$$

$$v_D^3 = \frac{3Nc^3}{4\pi V} \tag{22.3-23}$$

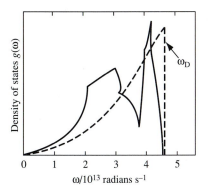

Figure 22.9. The Debye Distribution of Frequencies, with the Experimental Distribution of Frequencies for Copper. The experimental distribution is obviously complicated enough that a theory to reproduce such a distribution would likely be difficult to produce. From J. S. Blakemore, *Solid State Physics*, W. B. Saunders, Philadelphia, 1974, p. 126.

Debye's frequency distribution is shown in Figure 22.9, along with the experimental distribution of frequencies for copper. It is given in terms of the "circular frequency" ω, equal to $2\pi v$.

The **Debye temperature**, Θ_D, is a parameter with the dimensions of temperature defined by

$$\Theta_D = \frac{h v_D}{k_B} \quad \text{(definition)} \qquad \textbf{(22.3-24)}$$

The appropriate value of Θ_D for a given crystal is usually chosen by fitting heat capacity data to the Debye formula in Equation (22.3-26), rather than by using a measured value for the speed of sound in the crystal.

The vibrational energy of a normal mode of frequency v is given by the analogue of Equation (20.5-24), so that

$$\begin{aligned} U &= \int_0^{v_D} \frac{h v}{e^{h v/k_B T} - 1} g(v)\, dv \\ &= \frac{9N}{v_D^3} \int_0^{v_D} \frac{h v^3}{e^{h v/k_B T} - 1}\, dv \\ &= \frac{9N k_B T}{u_D^3} \int_0^{u_D} \frac{u^3}{e^u - 1}\, du \end{aligned} \qquad \textbf{(22.3-25)}$$

where $u = h v/k_B T$ and $u_D = h v_D/k_B T$. The integral in Equation (22.3-25) cannot be evaluated in closed form, but can be evaluated numerically to any desired accuracy.

The heat capacity is given by

$$C_V = \frac{9N k_B}{v_D^3} \int_0^{v_D} \left(\frac{h v}{k_B T}\right)^2 \frac{e^{h v/k_B T} v^2}{e^{h v/k_B T} - 1}\, dv \equiv 3N k_B D\!\left(\frac{\Theta_D}{T}\right) \quad \textbf{(22.3-26)}$$

which defines the function D, called the **Debye function**. It depends only on the ratio Θ_D/T. Tables of values of this function are available,[3] and the function can easily be evaluated with a computer program to any desired accuracy.

The Debye function generally fits experimental data on the heat capacity of atomic solids quite well if an optimum choice of Θ_D is made. Figure 22.10 shows the heat capacities of several elements, along with curves representing the Debye function for the Debye temperatures given. At high temperatures, the Debye expression conforms to the law of Dulong and Petit.

Exercise 22.8 Show that the energy expression in Equation (22.3-25) reduces to $U = 3N k_B T$ for high temperatures, so that $C_V = 3N k_B$. Hint: for high temperatures, e^u can be approximated by $1 + u$.

The Debye temperature of diamond is equal to 1890 K. Compare this with the Einstein temperature of diamond, equal to 1320 K. Since these

---

[3] N. Davidson, *Statistical Mechanics*, McGraw-Hill, New York, 1962, p. 359.

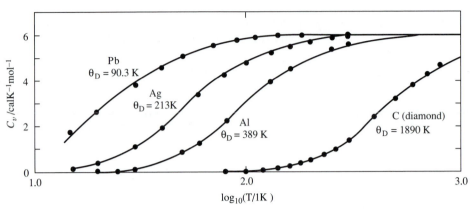

Figure 22.10. The Heat Capacity of Several Elements as a Function of Temperature, with Debye Curves. The value of the Debye temperature is chosen to give the best fit for each substance.

From G. N. Lewis and M. Randall, *Chemical Thermodynamics,* 2nd ed., rev. by K. S. Pitzer and L. Brewer, McGraw-Hill, New York, 1961, p. 56.

If the Einstein frequency is a kind of average frequency, it should lie between zero and the Debye frequency.

temperatures are proportional to the frequency and the Einstein temperature should be equal to some kind of average frequency in the Debye model, these values seem reasonable.

Exercise 22.9   Obtain a formula for an average Debye frequency

$$v_{av} = \frac{\int_0^{v_D} v g(v) \, dv}{\int_0^{v_D} g(v) \, dv}$$

Compare the average Debye frequency for diamond with the Einstein frequency for diamond.

For small temperatures (T smaller than $\Theta_D/10$), u_D is large and the upper limit of the integral can be extended to infinity without serious error. The integral is then equal to $\pi^4/15$, and

$$U = \frac{3\pi^4 N k_B T^4}{5 \Theta_D^3} \tag{22.3-27}$$

For small values of the temperature, the heat capacity is

$$C_V = \frac{12\pi^4}{5} \frac{N k_B T^3}{\Theta_D} \tag{22.3-28}$$

Equation (22.3-28) is a useful result. The approximations of the Debye theory become more nearly exact at low temperature. Heat capacities are hard to measure at low temperatures, and data below 15 K are quite rare. Equation (22.3-28) is commonly used as a substitute for data between 0 K and 15 K in calculating third-law entropies. (See Section 3.4.) In metals, there is also a significant contribution to the heat capacity from the electronic motion, which we discuss in the next section.

Formulas for the other thermodynamic functions can be obtained as integrals over the frequencies, which are analogous to sums over the normal modes of a polyatomic molecule. For example,

$$A = -Nk_B T \int_0^{v_D} \ln\left(\frac{1}{1 - e^{-hv/k_B T}}\right) g(v) \, dv \qquad \textbf{(22.3-29)}$$

Exercise 22.10　Write a formula for the vibrational entropy of a Debye crystal.

Modifications to the Debye theory have been devised that use a temperature-dependent Debye temperature and give improved agreement with experiment.[4]

22.4 The Electronic Structure of Solids

In any type of solid, the cohesive forces that hold the solid together are due to the Coulomb attractions between electrons and nuclei. These forces manifest themselves in different types of solids as chemical bonds, London dispersion forces, dipole-dipole interactions, hydrogen bonds, etc.

Electronic Structure of Molecular Crystals

Molecular crystals are held together by intermolecular forces, such as London dispersion forces, dipole-dipole forces, and hydrogen bonds. The electronic wave function of the crystal can be approximated as a product of wave functions of individual molecules, which are similar to those of the same molecules in the gas phase.

Electronic Structure of Ionic Crystals

As with molecular crystals, the electronic wave function of an ionic crystal can be approximated as a product of the wave functions of the ions, which are much the same as those of the separated ions.

Electronic Structure of Covalent and Metallic Crystals

Covalently bonded crystals, including metallic crystals, can be thought of as giant molecules. There can be localized and delocalized covalent bonding, which is similar to the molecular bonding discussed in Chapter 13.

---

[4] See J. S. Blakemore, *Solid State Physics*, 2nd ed., W. B. Saunders, Philadelphia, 1974, pp. 128ff.

▼ **EXAMPLE 22.1**

Describe the bonding in a gold crystal of N atoms.

Solution

A gold atom in its ground state has one electron in the $6s$ space orbital and 78 electrons in filled subshells. For a first approximation to the wave function of the crystal, we construct a wave function that is a product of orbitals. We use the atomic orbitals of the filled subshells as nonbonding orbitals. The $6s$ space orbitals from the N atoms can be made into N different LCAO molecular orbitals, which will be delocalized and extend over the entire crystal. These LCAO-MOs will have slightly different energies.

Each of these N space orbitals can combine with either the α spin function or the β spin function, giving $2N$ spin orbitals. At low temperature, the N spin orbitals of lowest energy will be occupied. That is, the $N/2$ space orbitals will be occupied by pairs of electrons with opposite spins. The occupied orbitals will have fewer nodes than the unoccupied orbitals and will have more bonding than antibonding character. The crystal is therefore held together very strongly, so that gold melts at 1063°C.

▲

Exercise 22.11

Explain in simple terms why mercury, which has one more electron per atom than gold, melts at a low temperature (-38.9°C).

The crude description of the bonding in crystalline gold of Example 22.1 is somewhat analogous to a description of the bonding in C_2, in which the electron configuration can be taken as

$$(1sA)^2(1sB)^2(2sA)^2(2sB)^2(\pi u2px)^2(\pi u2py)^2$$

where one nucleus is denoted as A and the other as B. Another description uses the electron configuration

$$(\sigma g1s)^2(\sigma *u1s)^2(\sigma g2s)^2(\sigma *u2s)^2(\pi u2px)^2(\pi u2py)^2$$

In homonuclear diatomic molecules such as C_2, two atomic orbitals of equal orbital energies from the two atoms produce two linear combinations: a bonding orbital and an antibonding orbital. Similarly, in the delocalized pi bonding of benzene, six pi LCAO-MOs are formed from six $2p$ atomic orbitals on the six carbons. The energies of these orbitals span a fairly narrow range, as shown in Figure 13.21. In the ground state of the molecule, the six pi electrons occupy the three lowest-energy delocalized pi space orbitals, leaving the other orbitals vacant. The occupied molecular orbitals have more bonding character than antibonding character, contributing to the stability of the molecule.

As in the second configuration of C_2, we can use delocalized molecular orbitals for all the electrons of a gold crystal instead of using nonbonding orbitals. We can make one set of linear combinations from the $1s$ orbitals from the N atoms of the crystal, another set of linear combinations from the $2s$ orbitals, a third set of linear combinations from the $2p$ orbitals, etc.

Here we are comparing a multielectron system with two systems having a few electrons.

The Band Theory of Solid Electronic Structure

In a crystal of N atoms, N atomic orbitals of equal energy form N LCAO delocalized orbitals of different orbital energies. These energies are confined to a fairly narrow range, as were the pi orbital energies of benzene. This range is called a **band**. One band comes from the $1s$ orbitals, another from the $2s$ orbitals, a third from the $2p$ orbitals, etc. In many cases, these bands do not overlap, leaving a **band gap** between two adjacent bands. Figure 22.11 shows the X-ray photoelectron spectrum of a gold foil, in which several of the bands can be seen, as well as the $1s$ band from a carbon impurity. The subscript on each band label is the value of j, the quantum number for the total (orbital plus spin) angular momentum for the electrons.

If a band is created from orbitals that are filled in the ground state of the separated atoms, there are as many electrons to fill the band as there are spin orbitals in the band, so the band is completely filled in the ground state. If a band is created from orbitals that are only partly filled in the separated atoms, as in the $6s$ band of gold in Example 22.1, the resulting band is only partly filled.

If the highest occupied band in a crystal is completely filled, an electron can move from one orbital to another in the band only if another electron vacates the second orbital. If there is no way for the orbital to be vacated, the crystal is an insulator, unable to conduct electricity.

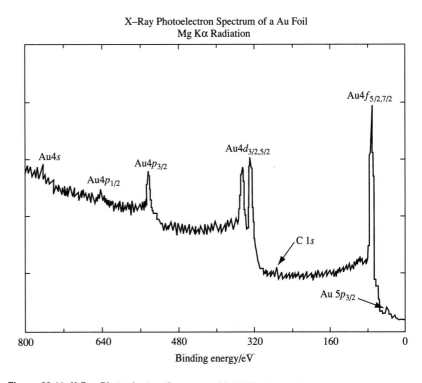

Figure 22.11. X-Ray Photoelectron Spectrum of Gold. Each peak is labeled with the subshell and the value of j, the quantum number for the total angular momentum. Courtesy of Dr. Kevin Ogle.

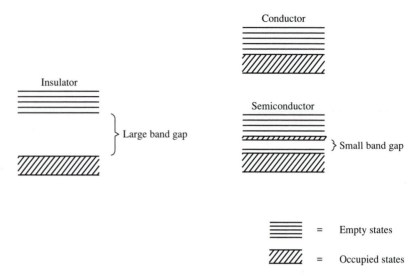

Figure 22.12. Bands of Orbital Energies in a Hypothetical Insulator, Conductor, and Semiconductor. The top figure, for an insulator, shows the band gap, which is large compared with $k_B T$. The middle figure, for a conductor, shows a partly filled band, so that there is no band gap. The bottom figure, for a semiconductor, shows a band gap that is not large compared with $k_B T$.

If the highest occupied band in a crystal is roughly half filled, there are plenty of unoccupied orbitals for electrons to move into and plenty of movable electrons in the occupied orbitals. The crystal is an electrical conductor.

If the highest occupied band has only a few electrons, there are plenty of unoccupied orbitals but only a few movable electrons, and the crystal is a semiconductor, with a fairly large resistance. If the highest occupied band is nearly filled, there are plenty of movable electrons but only a few unoccupied orbitals to move into, and this crystal is also a semiconductor. Semiconductors with only a few electrons in the highest band are called n-type semiconductors. Those with only a few vacant orbitals (referred to as "holes") are called p-type semiconductors. The designation p stands for positive, since the positive holes are thought of as moving, and the designation n stands for negative, since electrons are thought of as moving. Figure 22.12 schematically depicts the band occupations in insulators, conductors, and semiconductors.

The behavior of semiconductors is strongly temperature dependent. From Equation (20.2-22), the probability of finding a fermion in state i with energy ε_i is

$$\frac{N_i}{g_i} = f(\varepsilon_i) = \frac{1}{e^{-\alpha + \beta \varepsilon_i} + 1} \tag{22.4-1}$$

where g_i is the degeneracy of the level.

Since we established that the parameters α and β are the same for fermions as for the dilute occupation case, we can substitute $1/k_B T$ for β. For the dilute occupation case, Equation (20.3-2) gives

$$\alpha = \ln(N/z) \tag{22.4-2}$$

where z is the molecular partition function. Equation (20.5-15) gives

$$\mu = -k_{\mathrm{B}}T \ln(z/N) \qquad \textbf{(22.4-3)}$$

so that

$$\alpha = \mu/k_{\mathrm{B}}T \qquad \textbf{(22.4-4)}$$

where μ is the chemical potential. It is not permissible to use the molecular partition function for fermions or bosons if the dilute occupation case cannot be used. However, α must be the same parameter for all cases, so Equation (22.4-4) can be used for noninteracting fermions or bosons even if dilute occupation does not apply. The fermion distribution now becomes

$$f(\varepsilon_i) = \frac{1}{e^{(\varepsilon_i - \mu)/k_{\mathrm{B}}T} + 1} \qquad \textbf{(22.4-5)}$$

Here we argue that the parameter α can be identified with $\mu/k_{\mathrm{B}}T$ even if the dilute occupation case does not apply.

Electrons do interact strongly with each other in a solid, but in order to understand the states of the system qualitatively, we pretend that they are noninteracting fermions and obey Equation (22.4-5).

The value of the chemical potential of the electron is called the **Fermi level** and is also denoted by ε_{F}. Figure 22.13 shows the fermion probability distribution for 0 K and for two nonzero temperatures. At 0 K, all of the states with energies up to the Fermi level are fully occupied and those above the Fermi level are vacant. As the temperature is increased, more and more of the states with energies just below the Fermi level become unpopulated and more and more of the states just above the Fermi level become populated. The range of energy over which the probability distribution changes from approximately unity to nearly zero is approximately equal to $k_{\mathrm{B}}T$.

If the Fermi level lies within a band, the crystal will be a conductor. If the Fermi level lies at the top of a band or between two bands, the crystal will be either an insulator or a semiconductor. If the band gap to the next higher band is small, then at ordinary temperature some of the highest-energy states in the filled band will be vacant and some of the low-lying states in the first vacant band will be occupied, and the crystal will be a semiconductor. If the band gap is large, there will be little chance that electrons can move to the vacant band, the highest occupied band will be completely filled, and the crystal will be an insulator.

Silicon is the most widely used semiconductor. The silicon crystal is similar to the diamond crystal, with each silicon atom tetrahedrally bonded to four other silicon atoms. To a first approximation, the bonding orbitals in both crystals can be approximated as localized bonding molecular orbitals made from two sp^3 hybrid orbitals on adjacent silicons. The antibonding molecular orbitals lie somewhat higher in energy and are vacant.

In silicon, there is a band from the $3d$ orbitals. The band gap between the occupied $3sp^3$ and the $3d$ band is small enough that silicon is a semiconductor at room temperature. Diamond is an insulator because there is no $2d$ subshell. The lowest vacant band of diamond is the $3s$ band, which has a large band gap from the occupied $2sp^3$ band. Figure 22.14 shows a rough energy level diagram. At 0 K, no electrons in the silicon crystal could occupy orbitals in the $3d$ band and silicon would be an insulator like diamond.

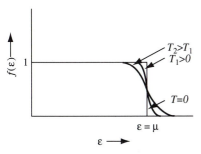

Figure 22.13. The Fermion Probability Distribution. At $T = 0$, the electron states are completely filled up to the Fermi level and completely empty above the Fermi level. At higher temperatures, some of the states below the Fermi level are empty and some of them above the Fermi level are occupied.

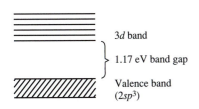

Figure 22.14. The Electron Bands of Silicon (schematic). The band gap, 1.17 eV, is roughly 43 times as large as kT at room temperature. Although silicon is a semiconductor, it must be "doped" by addition of impurities, making either holes or extra electrons, to conduct much electricity.

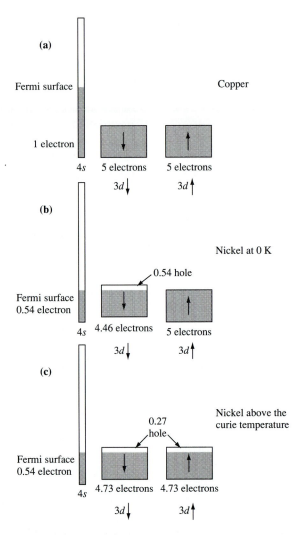

Figure 22.15. The 4s and 3d Bands of Nickel and Copper. The vertical axis in each diagram represents the electronic energy. The fact that the 3d spin-up electrons in nickel have a slightly lower energy than the 3d spin-down electrons gives an excess of spin-up electrons, which can align in domains to make nickel ferromagnetic. From N. B. Hannay, *Solid-State Chemistry*, Prentice-Hall, Englewood Cliffs, NJ, 1967, p. 38.

In some applications, silicon is "doped" with small amounts of other substances. For example, if a few aluminum atoms are introduced into the lattice, there are "holes" in the available orbitals, since aluminum has 13 electrons and silicon has 14. This makes the doped silicon into a *p*-type semiconductor, which would conduct electricity even at 0 K.

Nickel is ferromagnetic (can be permanently magnetized).[5] Figure 22.15 shows the 4s and 3d bands for both nickel and copper. In both elements, the two bands overlap in energy, since the 3d and 4s orbitals in the isolated

---

[5] N. B. Hannay, *Solid-State Chemistry*, Prentice-Hall, Englewood Cliffs, NJ, 1967, p. 38.

atoms are nearly at the same energy. However, the 4s band covers a wider range of energy. In copper, which has one 4s electron and ten 3d electrons in the isolated atom, the Fermi level is halfway up the 4s band. The 4s band is half full and the 3d band is full.

In nickel, which has two 4s electrons and eight 3d electrons, the Fermi level is lower and lies below the top of the 3d band. The spin-up states of the 3d band have slightly lower energy than the spin-down states, due to "exchange interaction," and at 0 K there is an average of 0.54 hole per atom in the spin-down states of the 3d band and an average of 0.54 electron per atom in the 4s band. The spins of the excess spin-up electrons interact strongly with each other and tend to form domains in the crystal in which all of the excess spins are aligned parallel to each other. These domains can be aligned to produce a permanent macroscopic magnetic moment, characteristic of ferromagnetism. Above 633 K, the **Curie temperature** for nickel, thermal energy overrides the exchange interaction and the ferromagnetism disappears, as shown in the last part of the figure.

The Curie temperature of a substance is the temperature above which ferromagnetism does not occur for that substance.

The Free-Electron Theory

This simple theory provides an approximate description of the electron states within the occupied bands. The electrons are represented as a gas of noninteracting fermions. This approximation is a three-dimensional analogue of the free-electron approximation of Equation (13.5-18), in which the delocalized pi orbitals for a conjugated chain of carbon atoms were represented by one-dimensional particle-in-a-box wave functions.

We represent the orbitals for the mobile electrons in our crystal by free-particle wave functions, representing traveling waves, as in Section 9.5:

$$\psi = e^{i\mathbf{k}\cdot\mathbf{r}} = e^{i(k_x x + k_y y + k_z z)} \tag{22.4-6}$$

where the vector $\mathbf{k}$ is called the **wave vector**.

We treat a very large crystal and consider a portion of the crystal that is a cube with dimension L. We impose "periodic boundary conditions:"

$$\psi(x + L, y, z) = \psi(x, y, z) \tag{22.4-7}$$

with similar equations for y and z. To satisfy this condition,

$$k_x = 2\pi n_x/L, \qquad k_y = 2\pi n_y/L, \qquad k_z = 2\pi n_z/L \tag{22.4-8}$$

where n_x, n_y, and n_z are integers. The situation is similar to that in the Debye crystal model, except the integers are now not required to be positive, since we are dealing with free particles rather than particles in a box, and the particles can move in any direction. The energy of a wave with wave vector $\mathbf{k}$ is

$$\varepsilon_k = \frac{\hbar^2 k^2}{2m} = \frac{\hbar^2}{2m}(k_x^2 + k_y^2 + k_z^2) \tag{22.4-9}$$

The number of sets of integers in the range dn is given by Equation (22.3-20):

$$\text{Number of sets in } (n, n + dn) = 4\pi n^2\, dn \tag{22.4-10}$$

This number is 8 times as large as that in Equation (22.3-20) to account for the possibility that negative integers can occur. There are two possible spin states for each electron, so we double this expression to get the number of states. Using Equation (22.4-8), we obtain for the number of states in the range dk

$$\text{Number of states in } dk = g(k)\,dk = \frac{L^3 k^2}{\pi^2}\,dk$$

Using Equation (22.4-9),

$$\text{Number of states in } d\varepsilon = \frac{L^3 \sqrt{2}\,m^{3/2}}{\pi^2 \hbar^3}\,\varepsilon^{1/2}\,d\varepsilon \qquad \textbf{(22.4-11a)}$$

$$\begin{array}{l}\text{Number of states} \\ \text{in } d\varepsilon \text{ per unit volume}\end{array} = g(\varepsilon)\,d\varepsilon = \frac{\sqrt{2}\,m^{3/2}}{\pi^2 \hbar^3}\,\varepsilon^{1/2}\,d\varepsilon \qquad \textbf{(22.4-11b)}$$

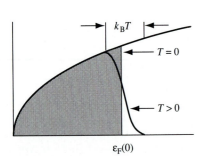

This degeneracy is depicted in Figure 22.16. At 0 K, electrons occupy the available states from the state of lowest energy up to the Fermi level. This occupation is shown by the shaded area in Figure 22.16. At nonzero temperatures, the occupation of states is given by Equation (22.4-6), and the occupation of the levels is given by the curve in the figure.

The number of electrons per unit volume is

$$n = \int_0^\infty g(\varepsilon) f(\varepsilon)\,d\varepsilon \qquad \textbf{(22.4-12)}$$

Figure 22.16. The Degeneracy of Energy Levels in the Free-Electron Theory. The monotonically rising curve is the degeneracy. The vertical cutoff gives the degeneracy times the fermion state distribution at 0 K. That is, it gives the distribution in terms of levels. The curve labeled $T > 0$ K is the product of the degeneracy and the fermion state distribution for a nonzero temperature. From N. B. Hannay, *Solid-State Chemistry*, Prentice-Hall, Englewood Cliffs, NJ, 1967, p. 26.

where $f(\varepsilon)$ is the fermion probability distribution of Equation (22.4-5). At 0 K, each of the states with energy less than the Fermi level is occupied by one electron and all of the states above the Fermi level are vacant. The upper limit of the integral can be moved from infinity to the Fermi level without error and $f(\varepsilon)$ can be replaced by unity without error:

$$n = \frac{\sqrt{2}\,m^{3/2}}{\pi^2 \hbar^3} \int_0^{\mu_0} \varepsilon^{1/2}\,d\varepsilon = \frac{1}{3\pi^2} \left(\frac{2m\mu_0}{\hbar^2}\right)^{3/2} \qquad \textbf{(22.4-13)}$$

The Fermi level at 0 K is

$$\mu_0 = \varepsilon_{\text{F0}} = (3\pi^2 n)^{2/3}\,\frac{\hbar^2}{2m} \qquad \textbf{(22.4-14)}$$

A typical metal has a density of mobile electrons approximately equal to 10^{28} m^{-3}, corresponding to a value of the Fermi level equal to several electron volts.

Exercise 22.12   **a.** Find the value of the constant A when Equation (22.4-14) is written in the form $\varepsilon_{\text{F0}} = A n^{2/3}$.
b. The density of copper is 8960 kg m^{-3}. Find the density of mobile electrons, assuming one electron from each atom, and find the zero-temperature value of the Fermi level in joules and in electron volts.

The Fermi level for nonzero temperature is approximately[6]

$$\mu = \varepsilon_F = \varepsilon_{F0}\left[1 - \frac{(\pi k_B T)^2}{12\varepsilon_{F0}^2}\right] \tag{22.4-15}$$

For fairly low temperatures, this formula gives nearly the same value for the Fermi level as Equation (22.4-14).

The energy per unit volume of the free-electron gas at 0 K is

$$U_0 = \int_0^{\varepsilon_{F0}} \varepsilon g(\varepsilon)\, d\varepsilon = \frac{1}{5\pi^2}\left(\frac{2m}{h^2}\right)\varepsilon_{F0}^{5/2} = \frac{3n\varepsilon_{F0}}{5} \tag{22.4-16}$$

The energy at a nonzero temperature is approximately[7]

$$U = U_0 + \frac{n\pi^2 k_B^2 T^2}{4\varepsilon_F} \tag{22.4-17}$$

if $T \ll \varepsilon_F/k_B$. Equation (22.4-17) leads to a formula for the electronic contribution to the heat capacity per unit volume

$$C_{\text{elec}} = \frac{\pi^2 n k_B^2 T}{2\varepsilon_F} \tag{22.4-18}$$

We obtain a formula for the molar heat capacity by replacing n by the number of mobile electrons per mole.

If the electron gas obeyed classical mechanics, the heat capacity would be $3k_B/2$ per electron, so that

$$C_{\text{elec}} = \frac{\pi^2 k_B T}{3\varepsilon_F} C_{\text{class}} \tag{22.4-19}$$

This meaning of the word "degenerate" is different from the usage in previous chapters, where it applied to an energy level consisting of more than one state.

Although the classical heat capacity is independent of T, the quantum mechanical heat capacity is proportional to T. The quantum mechanical electron gas is sometimes called the **degenerate electron gas** because its heat capacity is "degenerated" by the factor given in Equation (22.4-19).

Exercise 22.13   **a.** Find the electronic contribution to the heat capacity of copper at 15 K, using the same assumptions as in Exercise 22.12.

b. Find the ratio of the electronic contribution to the classical prediction of the electronic contribution.

c. Find the ratio of the electronic contribution to the lattice vibration contribution for copper at 15 K, using the Debye theory result with the Debye temperature 315 K.

---

[6] J. S. Blakemore, *op. cit.*, p. 176.
[7] *ibid.*

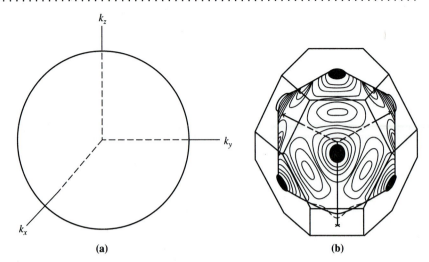

Figure 22.17. The Fermi Surface. (a) The Fermi surface in the free-electron theory. This surface is a sphere in the space of the wave vectors denoted by **k**. **(b)** The Fermi surface of copper. This surface is a distorted sphere. It is shown imbedded in a polyhedron whose surfaces coincide with the dark areas of the Fermi surface. From G. E. Smith, in *The Fermi Surface*, edited by W. A. Harrison and M. B. Webb, Wiley, New York, 1960.

The simple free-electron theory has been modified by Bloch.[8] The uniform potential energy function of the free-electron theory is replaced by a potential function that is periodic, with the same period as the crystal lattice. The wave function is like that of Equation (22.4-6) except for the presence of another factor u, called the Bloch function:

$$\psi = e^{i\mathbf{k} \cdot \mathbf{x}} u(\mathbf{x}) \tag{22.4-20}$$

The energy levels are found to lie in bands, much like the bands of the LCAO molecular orbitals discussed above. Each band is said to correspond to a **Brillouin zone**.

The Fermi level is sometimes depicted as a surface in a space in which k_x, k_y, and k_z are plotted on the axes. For the free-electron theory, all states below the Fermi surface are contained within a sphere, as schematically depicted in Figure 22.17a. For the Bloch theory, the Fermi surface is not spherical. Figure 22.17b shows the Fermi surface of copper.

22.5   The Structure of Liquids

Like solids, liquids can be divided into classes according to their attractive forces. The first class consists of molecular liquids, in which the molecules of the liquid are similar to isolated (gaseous) molecules. The cohesive forces holding the liquids together are intermolecular forces, including London forces, hydrogen bonds, etc. The second class consists of liquid

[8] F. Bloch, *Z. Phys.* **52**, 555 (1928). See Hannay, *op. cit.*, pp. 27ff, or Blakemore, *op. cit.*, pp. 204ff.

metals, in which delocalized electrons move about in the liquid much as in solid metals. The third class consists of fused salts, which are somewhat like the ionic crystals which melt to produce them, and are held together by electrostatic forces.

When a solid melts to form a liquid, a considerable change in properties takes place with a small change in temperature. The rigidity of the solid suddenly disappears. The molar volume increases slightly (typically by 5% to 15%, although it decreases by about 8% in the case of water). The isothermal compressibility increases, by a few percent in some cases and by a factor of two to three in others. The thermal conductivity drops, but diffusion coefficients increase, often by a factor of a thousand.

The change in properties upon melting is due to a significant change in molecular structure. The lattice structure of the solid, which extends over long distances compared with molecular dimensions, suddenly collapses. Vestiges of this structure exist in the liquid, but only at short distances. For example, in most atomic solids, each atom has 12 nearest neighbors. When the solid melts, the average nearest-neighbor distance changes only slightly, but the average number of nearest neighbors decreases. In liquid argon at the melting temperature, the average nearest-neighbor distance is only 1% larger than in the solid at the same temperature, but the average number of nearest neighbors lies between 10 and 11. The lower density of the liquid is due to the presence of voids around molecules, and "hole theories" of liquids have been devised to explain the properties of liquids on this basis.[9]

The void in the shell of nearest neighbor molecules does not necessarily represent one missing molecule with all others in the same locations as in the solid. Instead, there are generally numerous small vacant spaces in the shell, which move around and change their sizes as the nearest neighbor molecules move. When this disorder is passed on to more distant shells, the long-range order of the solid is absent.

There are three principal approaches to the theoretical study of liquids. The first approach is by fundamental statistical mechanics. When this approach is used, the liquid is usually treated as though it were a very dense nonideal gas. The second approach is by the use of approximate model systems. When this approach is used, the liquid is usually treated as though it were a somewhat disordered solid. The third approach is through numerical simulation, in which computer programs calculate liquid properties from the motions and positions of a collection of molecules.

The Fundamental Approach to Liquid Structure

It is generally assumed that the motions of the molecules of most liquids except for liquid hydrogen and helium can be described to an adequate approximation by classical mechanics. Let us assume that we have an atomic liquid of N atoms that do not exhibit electronic excitation. The equilibrium probability density of

---

[9] H. Eyring and M. S. Jhon, *Significant Liquid Structures*, John Wiley & Sons, Inc., New York, 1969

classical statistical mechanics is given by Equation (20.8-2):

$$f(p, q) = \frac{1}{Z_{cl}} e^{-[\mathscr{K}(p) + \mathscr{V}(q)]/k_B T} \tag{22.5-1}$$

where $\mathscr{K}$ is the kinetic energy, $\mathscr{V}$ is the potential energy, p stands for the momenta, and q stands for the coordinates of all of the molecules.

As in Section 16.7, we assume that the potential energy is a sum of two-body terms (is **pairwise additive**) and depends only on the distances between pairs of particles:

$$\mathscr{V} = \sum_{i=1}^{N-1} \sum_{j=1+1}^{N} u(r_{ij}) \tag{22.5-2}$$

where r_{ij} is the distance between atom number i and atom number j. This assumption means, for example, that the contribution of a pair of molecules to the potential energy is unchanged if a third molecule is moved close to them. It is apparently a fairly good approximation for simple liquids.

Equation (20.8-3) gives the classical canonical partition function for our system:

$$Z_{cl} = (2\pi m k_B T)^{3N/2} \int e^{-\mathscr{V}(q)/k_B T} \, dq \tag{22.5-3}$$

where the integral is called the configuration integral. From Equation (20.8-21), the ensemble average energy of the liquid is

$$U = \langle E \rangle = k_B T^2 \left[\frac{\partial \ln(Z_{cl})}{\partial T} \right]_V = \frac{3N k_B T^2}{2} \frac{d \ln(2\pi m k_B T)}{dT} + \frac{1}{\zeta} \int \mathscr{V} e^{-\mathscr{V}/k_B T} \, dq$$

$$= \frac{3}{2} N k_B T + \langle \mathscr{V} \rangle \tag{22.5-4}$$

Exercise 22.14   Verify Equation (22.5-4).

The kinetic energy of the liquid is the same as for an ideal gas, but the potential energy cannot be computed exactly. We can write

$$\langle \mathscr{V} \rangle = \frac{1}{\zeta} \sum_{i=1}^{N-1} \sum_{j=i+1}^{N} \int u(r_{ij}) e^{-\mathscr{V}/k_B T} \, dq \tag{22.5-5}$$

where we have exchanged the order of summation and integration.

Every pair of particles in our system is just like every other pair, so that all of the $N(N-1)/2$ terms in the sum of Equation (22.5-2) will be equal after integration. Therefore,

$$\langle \mathscr{V} \rangle = \frac{N(N-1)}{2} \frac{1}{\zeta} \int u(r_{12}) e^{-\mathscr{V}/k_B T} \, dq \tag{22.5-6}$$

Reduced Distribution Functions

The integration in Equation (22.5-6) is over all of the coordinates of the N particles. We can divide up the integrations

$$\int u(r_{12}) e^{-\mathscr{V}/k_B T} \, dq = \int d^3 \mathbf{r}_1 \, d^3 \mathbf{r}_2 \, u(r_{12}) \int e^{-\mathscr{V}/k_B T} \, dq_{N-2} \tag{22.5-7}$$

This distribution function is for coordinates only. It is a probability density in the six-dimensional coordinate space of the two particles. A two-body reduced distribution function for both momenta and coordinates can also be defined as in Equation (20.8-8)

Our two-body reduced distribution function is only for atoms. For molecules, the two-body reduced distribution function depends on molecular orientations and vibrational positions.

The radial distribution function has the same name as the probability density for finding an electron at a specified distance from the nucleus in an atom, but is a different quantity.

where $\int \cdots dq_{N-2}$ stands for integrating over the coordinates of all particles except numbers 1 and 2. This much of the integration would be the same if we were averaging any quantity that depends only on the coordinates of two particles at a time, so it is useful to define the **two-body reduced distribution function**.

$$n^{(2)}(\mathbf{r}_1, \mathbf{r}_2) = n^{(2)}(1, 2) = \frac{1}{\zeta} \int e^{-\mathscr{V}/k_B T} \, dq_{N-2} \qquad \textbf{(22.5-8)}$$

The two-body reduced distribution function $n^{(2)}(1, 2)$ is the probability density for finding particles 1 and 2: The probability that particle 1 is in a volume element $d^3\mathbf{r}_1$ and that particle 2 is simultaneously in a volume element $d^3\mathbf{r}_2$, irrespective of where the other $N - 2$ particles are, is

$$\text{Probability} = n^{(2)}(1, 2) \, d^3\mathbf{r}_1 \, d^3\mathbf{r}_2$$

In an equilibrium gas without intermolecular forces, the two-body distribution function is

$$n^{(2)} = \frac{1}{V^2} \quad \text{(ideal gas)} \qquad \textbf{(22.5-9)}$$

where V is the volume of the system. In any equilibrium system, $n^{(2)}$ depends only on r_{12}, the distance between particles 1 and 2.

Reduced distribution functions for one particle, three particles, etc., are also defined. The one-particle reduced distribution function is independent of position in any equilibrium fluid system if the effects of gravity can be ignored.

$$n^{(1)}(1) = \frac{1}{V} \qquad \textbf{(22.5-10)}$$

The **radial distribution function**, or **pair correlation function**, $g(r)$, is defined by

$$g(r_{12}) = \frac{n^{(2)}(1, 2)}{n^{(1)}(1)n^{(1)}(2)} = \frac{n^{(2)}(1, 2)}{V^2} \qquad \textbf{(22.5-11)}$$

The radial distribution, $g(r)$, is the probability of finding a second molecule at distance r from a given molecule, divided by the probability of finding a molecule in the bulk phase (far from the given molecule). In an ideal gas, $g(r)$ is equal to unity for all distances. In other words, the positions of the particles are uncorrelated. In an ordinary liquid, $g(r)$ vanishes at $r = 0$ due to the short-range repulsions between molecules. It approaches unity at large values of r. There are several maxima representing shells of neighboring molecules.

If one averages over different directions, the radial distribution function can also be defined for a solid. In a solid, the radial distribution function has large "blips" at distances equal to distances between lattice points and vanishes for other distances. Figure 22.18 shows the radial distribution function for liquid and solid mercury. The vestiges of the solid structure can be seen in the layers of neighbors at very nearly the distances between some pairs of lattice points.

The pressure can be expressed in terms of the radial distribution function[10]

$$P = \frac{Nk_B T}{V} \left[1 - \frac{N}{Vk_B T} \int r \left(\frac{du}{dr} \right) g(r) \, d^3\mathbf{r} \right] \qquad \textbf{(22.5-12)}$$

---

[10] P. A. Egelstaff, *An Introduction to the Liquid State*, Academic Press, New York, 1967, p. 20.

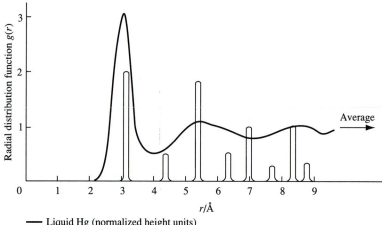

Figure 22.18. The Radial Distribution Function of Solid and Liquid Mercury. Since the solid has a lattice structure, the positions of neighboring atoms give narrow "blips" in the radial distribution function. In the liquid, the disorder that is present makes the function into a smooth curve, which shows vestiges of the crystal lattice. From D. Tabor, *Gases, Liquids, and Solids*, 2nd ed., Cambridge University Press, Cambridge, 1979, p. 197.

Exercise 22.15   Show that Equation (22.5-12) gives the correct pressure of an ideal gas.

We will not discuss them, but there are a number of theories used to calculate approximate radial distribution functions for liquids.[11] Some of the theories involve integral equations satisfied by approximate radial distribution functions. Others are "perturbation" theories somewhat like quantum mechanical perturbation theory (see Section 12.2). These theories take a hard sphere fluid or other fluid with purely repulsive forces as a zero-order system and consider the attractive part of the forces to be a perturbation. This procedure was originally suggested by the fact that the repulsive forces are the principal determining factor for the short-range order in a liquid.

Any perturbation theory divides the energy into two parts: a zero-order part and a perturbation, which is assumed small compared to the zero-order part.

The radial distribution function can be determined by neutron diffraction or X-ray diffraction. Neutrons are scattered in much the same way as X-rays, since they exhibit de Broglie wavelengths roughly equal to intermolecular spacings in liquids when moving at speeds near their thermal average speeds.

Exercise 22.16   **a.** Find the speed of a neutron such that its de Broglie wavelength is equal to 2.00×10^{-10} m.

b. Find the temperature such that the root-mean-square speed of thermally equilibrated neutrons is equal to the speed of part a, using Equation (16.3-8). Neutrons that have been equilibrated near room temperature are called **thermal neutrons**.

---

[11] H. L. Friedman, *A Course in Statistical Mechanics*, Prentice-Hall, Englewood Cliffs, NJ, 1985, Ch. 7, 8, and 9, and D. A. McQuarrie, *Statistical Mechanics*, Harper & Row, New York, 1976, Ch. 13 and 14.

Neutron diffraction is more commonly used to determine radial distribution functions in liquids than is X-ray diffraction, because X-rays are scattered primarily by electrons and neutrons are scattered primarily by nuclei. This makes the neutron diffraction simpler to interpret, since the nuclei are localized while the electrons are "smeared out" over the entire molecule.

Solid-Like Model Approaches to Liquid Structure

Several model theories treat a liquid like a disordered solid. These model theories are sometimes called "quasi-lattice" theories of liquids.

The cell model is similar to the Einstein crystal model, in that the motion of a given atom or molecule is assumed to be independent of the motions of the surrounding atoms or molecules. In the cell theory, the potential energy is taken to be uniform in the cell instead of being a harmonic oscillator potential.

In the **cell model**[12] each atom is assumed to be confined in a cell whose walls are made up of its nearest neighbors. In the simple cell model, this cell is approximated as a spherical cavity inside which the potential energy of the moving atom is constant.

Since each atom moves independently, the classical canonical partition function is like that of Equation (20.8-14) except that each atom moves in a small cell instead of the entire volume of the system:

$$Z_{cl} = (2\pi m k_B T)^{3N/2} (V_f e^{-u_0/k_B T})^N \tag{22.5-13}$$

where V_f is the "free volume" in which the center of the atom can move and u_0 is the constant potential energy of an atom in a cell. We consider the atoms in the solid at 0 K to be hard spheres in contact with the centers of nearest neighbors at a distance d. At nonzero temperature, the thermal expansion of the lattice moves the nearest neighbors to an average distance b that is larger than d. To a reasonable approximation, the center of the atom can now move about in a small sphere at the center of the cell that has radius equal to $b - d$.

The free volume is

$$V_f = \frac{4}{3}\pi(b - d)^3 \tag{22.5-14}$$

so that the partition function and thermodynamic variables can be found from Equation (22.5-13). If we assume a close-packed lattice, the distance b can be expressed in terms of the molar volume.

Exercise 22.17   In a face-centered cubic close-packed lattice, the unit cell dimension is $\sqrt{2}\,b$ and there are four atoms in a unit cell. Show that

$$b^3 = \frac{\sqrt{2}\,\bar{V}}{N_A} \tag{22.5-15}$$

where $\bar{V}$ is the molar volume and N_A is Avogadro's number.

---

[12] T. L. Hill, *Statistical Thermodynamics*, Addison-Wesley, Reading, MA, 1960, Ch. 16, and H. Eyring and M. S. Jhon, *op. cit.*, Ch. 2.

Using Equation (22.5-15), the classical canonical partition function of an atomic liquid of N atoms is

$$Z_{cl} = (2\pi m k_B T)^{3N/2} \left\{ \frac{4}{3}\pi \left[\left(\frac{\sqrt{2}\,\bar{V}}{N_A} \right)^{1/3} - \left(\frac{\sqrt{2}\,\bar{V}_s}{N_A} \right)^{1/3} \right]^3 e^{-u_0/k_B T} \right\}^N \quad (22.5\text{-}16)$$

where $\bar{V}_s$ is the molar volume of the solid at 0 K.

The pressure can be calculated from Equation (20.7-12)

$$P = k_B T \left[\frac{\partial \ln(Z)}{\partial V} \right]_T = \frac{N k_B T}{V} \frac{1}{1 - (\bar{V}_s/\bar{V})^{1/3}} \quad (22.5\text{-}17)$$

Exercise 22.18   Carry out the differentiation to obtain Equation (22.5-17).

The simple cell theory result for the pressure given in Equation (22.5-17) approaches the ideal gas value for large molar volume and diverges as the molar volume approaches that of the solid at 0 K. This behavior is qualitatively correct, but the cell model does not predict accurate values of the pressure. Lennard-Jones and Devonshire[13] developed an improved version in which they explicitly summed up the potential energy contributions for the nearest neighbors, obtaining better results.

There is a problem with all quasi-lattice theories of the liquid state. In calculating the entropy and any entropy-related thermodynamic variables, such as the Gibbs energy, we must make the semiclassical correction to the partition function of Equation (22.5-13). If we really were discussing a crystal, we would divide by h^{3N}, where h is Planck's constant. There would be no correction for the indistinguishability of the particles, since the lattice sites are distinguishable, as in the discussion of the Einstein crystal model in Section 22.3.

If we were discussing a gas, as in Section 14.8, we would divide by $N!$ as well as by h^{3N}, as in Equation (20.8-14). Our present model is an intermediate case. There is insufficient order to claim that the quasi-lattice sites are distinguishable, but our model is still not as completely disordered as a gas. The simplest procedure is to treat the system like a gas and divide by $N!$. Omission of this division makes a contribution to the entropy that is called the **communal entropy**. There has been some discussion of how much of the communal entropy should be included, but in view of the crudity of the cell model and the fact that nearly all current research concentrates on the fundamental statistical mechanical approach rather than quasi-lattice approaches, this discussion is not very important.

Exercise 22.19   Write expressions for the energy and entropy of the simple cell model.

---

[13] J. E. Lennard-Jones and A. F. Devonshire, *Proc. R. Soc. London, Ser. A* **163**, 53 (1937) and A165, 1 (1938).

The Significant Structures Theory

The model system defined in the significant structures theory is not realistic, since a single large vacancy is less likely than several small vacancies in the nearest-neighbor shell. However, the numerical effect of this difference might be fairly small.

This is another quasi-lattice liquid theory.[14] The model system for this theory is assumed to have atom-size vacancies in the shell of nearest neighbors. For each vacancy, one atom can move back and forth in a cavity roughly the size of two atoms. The atom is considered to have a gas-like motion in this cavity, and the canonical partition function of the system is written as a product of gas-like factors and solid-like factors. The fraction of atoms that can have gas-like motion is equal to the fraction of vacancies, which is assumed to be

$$\text{Fraction of vacancies} = \frac{\bar{V} - \bar{V}_s}{\bar{V}} \qquad \textbf{(22.5-18)}$$

where $\bar{V}$ is the molar volume of the liquid under the conditions of interest and $\bar{V}_s$ is the molar volume of the solid at 0 K.

The partition function for a system of N molecules is therefore

$$Z_N = z_g^{N_g} z_s^{N_s} \qquad \textbf{(22.5-19)}$$

where z_s is the molecular partition function for solid-like motion, obtained by use of the Einstein crystal model, and z_g is the molecular partition function for gas-like motion. N_g is the number of atoms with gas-like motion and N_s is the number of atoms with solid-like motion:

$$N_G = N\,\frac{\bar{V} - \bar{V}_s}{\bar{V}} \qquad \textbf{(22.5-20a)}$$

$$N_s = N\,\frac{\bar{V}_s}{\bar{V}} \qquad \textbf{(22.5-20b)}$$

Eyring and co-workers have obtained excellent agreement with experiment in calculations of various properties of liquids using the significant structures theory,[15] but the theory is not widely used because it is based on assumptions and approximations introduced at the beginning of the theory and cannot easily be related to more fundamental theories.

Numerical Simulations of Liquid Structure

With the advent of fast programmable computers, numerical simulations of liquid structure have become practicable. The two principal types of simulation are the Monte Carlo method and the molecular dynamics method.

The **Monte Carlo** method is so named because it uses a random number generator, reminiscent of the six-sided random number generators (dice) used in gambling casinos. This method was pioneered by Metropolis.[16] It is a modification of the original Monte Carlo method, which is used to evaluate integrals by randomly choosing points within the interval of integration. The integrand is evaluated at these points, and summing these values with equal weight gives an approximation to the integral.

---

[14] Eyring and Jhon, *op. cit.*, Ch. 3.
[15] *ibid.*
[16] N. Metropolis, A. W. Rosenbluth, M. N. Rosenbluth, A. H. Teller, and E. Teller, *J. Chem. Phys.* **21**, 1087 (1953).

In the Monte Carlo method of Metropolis and co-workers, a system of several hundred or a few thousand molecules is considered. Such a system is, of course, much smaller than a macroscopic liquid sample, but the error so introduced can be estimated. A sample set of coordinate states for all the molecules is generated, and the average of mechanical quantities (energy, pressure, etc.) is taken over all of these states. These states are generated as follows: an initial coordinate state of the system is assigned in some way, and a random number generator is used to choose a possible displacement of one particle. That is, a random number generator is used to pick a number, b, between -1 and 1. The particle is moved a distance $x = ab$, where a is a predetermined maximum displacement. The change in potential energy of the system, $\Delta \mathscr{V}$, is calculated. If $\Delta \mathscr{V} < 0$, the particle is left at the new location. If $\Delta \mathscr{V} > 0$, the particle is given a probability of staying at the new location which is equal to $\exp(-\Delta \mathscr{V}/k_B T)$. This is done by choosing a new random number, c, between 0 and 1. If $c > \exp(-\Delta \mathscr{V}/k_B T)$, the particle is left at the new location. Otherwise, it is returned to its old location. Similar displacements are taken in the y and z directions for the first particle and then in all three directions for all other particles.

Each time a new set of locations is obtained (including a set obtained by returning a particle to its old position), the value of the quantity to be averaged is calculated and added to the sum that is producing the average value. It was shown by Metropolis et al. that if sufficient terms are taken, this procedure produces averages that are correctly weighted by the canonical probability distribution, proportional to $\exp(-\mathscr{V}/k_B T)$.

To make the system act somewhat like a piece of a larger system, **periodic boundary conditions** are applied: If a move carries a particle out of the system, which is usually contained in a cube, the particle is reintroduced into the system through the same location on the opposite wall.

Early applications of the Monte Carlo method were limited to systems of hard spheres because of the limited computer power available. Systems of square well particles and Lennard-Jones particles have now been studied in detail. Present-day computers have advanced to the point that useful calculations can even be done on personal computers.

The second simulation technique is **molecular dynamics**. In this technique, which was pioneered by Alder,[17] the classical equations of motion of all the particles of a system of several hundred particles are numerically integrated by a computer program. The method is somewhat similar to the Monte Carlo method, except that the displacements of the particles are determined by approximately solving the classical equations of motion. The positions and velocities are given increments corresponding to a time step of about 1 fs (1 femtosecond). Periodic boundary conditions are applied, as in the Monte Carlo method.

The equations of motion are followed for a period of time that is short on a macroscopic time scale, a small fraction of a second. However, this

---

[17] B. J. Alder and T. E. Wainwright, *J. Chem. Phys.* **31**, 459 (1959).

much calculation requires a large amount of computer time. The first molecular dynamics calculations were done on systems of hard spheres, but the method has been applied to systems of square well particles and Lennard-Jones particles.

Although the Monte Carlo method is limited to the calculation of equilibrium properties, both equilibrium and nonequilibrium information can be obtained by the molecular dynamics technique. After a sufficient time, the molecules settle into motions that simulate motion of molecules in equilibrium liquids, and properties of the equilibrium state can then be studied. However, the principal virtue of the method is that the relaxation of the system toward equilibrium can be studied, giving direct information on transport properties.[18]

22.6 Polymer Formation and Conformation

Polymer molecules are often called **macromolecules**. The word polymer comes from **poly**, meaning many, and **meros**, meaning parts. Polymer molecules are formed from smaller molecules (monomers), which react together to form covalently bonded chains or networks. Examples of synthetic polymers are found in almost all manufactured products, and polymer chemistry is probably the most important area of industrial chemistry. There are also many naturally occurring polymers, including natural rubber, proteins, starches, celluloses, and nucleic acids. As with all other substances, the properties of polymers are determined by their molecular structures, and the properties of synthetic polymers can often be tailored to specific applications.

The simplest polymers have linear chain-like molecules. These materials are usually **thermoplastic** substances. That is, the material gradually softens as the temperature is raised. Other polymers have networks instead of chains. Some of these are made up of long chains with short chains (cross-links) fastening two or more chains together, and others are three-dimensional networks that are bonded in much the same way in all three dimensions. Network polymers are sometimes called **thermosetting** substances because they are commonly cross-linked at elevated temperatures.

Synthetic polymers are also classified by the type of reaction that forms them. The two major classes are **condensation polymers** and **addition polymers**. When a monomer unit is added to a condensation polymer chain, there is another product (often water) besides the lengthened chain. In an addition polymer, there is no other product, so the polymer has the same empirical formula as the monomer. Two common examples of condensation polymers are nylon and polyester, and two common examples of addition polymers are polyethylene and polystyrene.

---

[18] See W. G. Hoover, *Ann. Rev. Phys. Chem.* **34**, 103 (1983) for a review of work in this field.

Polymerization Kinetics

We discuss only the reaction kinetics of condensation polymerization, but addition polymerization can be discussed in much the same way.[19] Consider the formation of a polyester from a diacid, HOOC—X—COOH, and a dialcohol, HO—Y—OH, where X and Y represent two hydrocarbon chains. The first step in the polymerization is

$$HOOC—X—COOH + HO—Y—OH$$
$$\rightarrow HOOC—X—COO—Y—OH + H_2O$$

The resulting dimer has one carboxyl group and one hydroxyl group, so it is available for reaction with further monomer units. The next step can be the reaction with a diacid or a dialcohol. Two chains of any length can also bond together to form a longer chain. In any event, a long chain with the repeating unit —OOC—X—COO—Y— is eventually formed, with a carboxyl group at one end and a hydroxyl group at the other.

Solving the rate differential equations for the polymerization reactions is difficult, because we have a large number of different kinds of molecules and different rate coefficients. To simplify the problem, we assume that the rate coefficients for all condensation reactions have the same value. This approximation is commonly justified by the assertion that the reaction is not diffusion limited and that the behavior of a functional group in a "cage" of neighboring molecules or groups is nearly independent of the length of the chain to which the functional group is attached.

The variable c decreases smoothly as time passes.

Assume that we begin with a stoichiometric mixture, having equal concentrations of diacid and dialcohol. We simplify the problem further by using a single dependent variable, c, the sum of the concentrations of all types of molecules other than catalyst or solvent molecules. At $t = 0$ when the polymerization reaction begins, the diacid concentration is equal to $c_0/2$. There are two carboxyl groups on each diacid molecule, so the initial concentration of carboxyl groups is c_0. We assume that the water is removed as it is formed, so that each time a condensation reaction occurs, one free carboxyl group disappears and the number of molecules decreases by unity; thus c remains equal to the concentration of free (unesterified) carboxyl groups and also to the concentration of free hydroxyl groups.

These orders are taken as experimental fact. Since both the concentration of carboxyl groups and the concentration of alcohol groups equal c, the reaction rate is proportional to c².

If the polyesterification reaction is carried out with an acid catalyst, the reaction is first order in the diacid (or other molecule with a carboxyl group), first order in the dialcohol (or other molecule with a hydroxyl group), and first order in the catalyst. We assume that there is no back reaction and that the uncatalyzed mechanism can be neglected. The forward rate differential equation is

$$\frac{dc}{dt} = -k_f[\mathrm{H^+}]c^2 = -k'c^2 \qquad \textbf{(22.6-1)}$$

---

[19] See H. R. Allcock and F. W. Lampe, *Contemporary Polymer Chemistry*, Prentice-Hall, Englewood Cliffs, NJ, 1981, pp. 245ff, or C. Tanford, *Physical Chemistry of Macromolecules*, Wiley, New York, 1961, pp. 588ff for the standard treatments of both types of polymerization.

where

$$k' = k_f[H^+] \tag{22.6-2}$$

Since a catalyst is not consumed, $[H^+]$ is a constant.

Equation (22.6-1) is the same as Equation (18.2-8). Equation (18.2-11) gives its solution:

$$\frac{1}{c} = \frac{1}{c_0} + k't \tag{22.6-3}$$

We define the variable p, the fraction of carboxyl groups that have reacted:

$$p = 1 - \frac{c}{c_0} \tag{22.6-4}$$

Equation (22.6-3) can be written

$$p = \frac{k'c_0 t}{1 + k'c_0 t} \tag{22.6-5}$$

which is equivalent to

$$\frac{1}{1-p} = 1 + k'c_0 t \tag{22.6-6}$$

Exercise 22.20   Verify Equations (22.6-5) and (22.6-6).

The **degree of polymerization**, x, is defined to equal the number of monomer units in a molecule. In the present example, if x is even, there are $x/2$ diacid molecules and $x/2$ dialcohol molecules combined in the polymer molecule. If x is odd, there is an extra diacid or an extra dialcohol. The **number-average degree of polymerization** is called $\bar{x}_n$ and is given by

$$\bar{x}_n = \frac{c_0}{c} = \frac{1}{1-p} \tag{22.6-7}$$

This is a mean value with each molecule given equal weight. The result of Equation (22.6-3) can be expressed in terms of $\bar{x}_n$:

$$\bar{x}_n = 1 + k'c_0 t \tag{22.6-8}$$

Equation (22.6-8) agrees with experiment fairly well for the later stages of polymerization, although not so well for the early stages. Figure 22.19 shows data for two polyesterification reactions, beginning with 82% of the carboxyl groups esterified.

Since there is a distribution of degrees of polymerization, there is a distribution of molecular masses. If x is even, the molecular mass is

$$M_x = \frac{x}{2} M_r + 18 \text{ amu} = xM_0 + 18 \text{ amu} \approx xM_0 \tag{22.6-9}$$

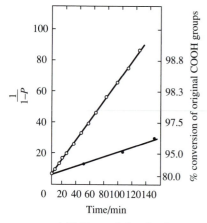

• 1,10–Decamethylen glycol
  at 161° C [S.D. Hamann, D.H.
  Solomon, and J.D.Swift, J.
  Macromol Sci. Chem. **A2**,
  153 (1968)]
• Diethylene glycol at 109°C
  [P.J. Flory, *J. Am. Chem.
  Soc.* **61**, 3334 (1939)]

Figure 22.19. Polyesterification of Adipic Acid with Two Polyalcohols, Catalyzed with *p*-Toluene Sulfonic Acid. The left axis shows the value of $1/(1 - p)$ where p is the fraction of carboxyl groups that have reacted. The right axis is labeled with the percent of original carboxyl groups that have reacted. From H. R. Allcock and F. W. Lampe, *Contemporary Polymer Chemistry*, Prentice-Hall, Englewood Cliffs, NJ, 1981, p. 254.

We frequently replace an upper limit of an integral or a sum by infinity when this approximation produces negligible error.

where M_r is the mass of the repeating unit of the polymer and M_0 is equal to $M_r/2$. The 18 amu is added because each molecule has an OH group at one end and an H atom at the other end that are not part of the repeating unit. Equation (22.6-9) can be modified for odd values of x. For either odd or even x, the final approximate equality is valid for large values of x, for which 18 amu can be neglected.

Let X_x be the **number fraction** of molecules with degree of polymerization equal to x:

$$X_x = \frac{N_x}{N} \tag{22.6-10}$$

where N_x is the number of molecules with degree of polymerization equal to x and N is the total number of molecules.

The **number-average molecular mass** is the mean molecular mass with each molecule given equal importance in the averaging process:

$$\bar{M}_n = \frac{1}{N} \sum_{x=1}^{\infty} N_x M_x = \sum_{x=1}^{\infty} X_x M_x \tag{22.6-11}$$

The number average molecular mass is equal to the total mass divided by the total number of molecules.

A formula for X_x can be derived, either from statistical reasoning[20] or from reaction kinetics.[21] For the assumption that all rate coefficients are equal, the formula is obtained by the following simple statistical argument, for an even value of x: The polymer molecule consists of $x - 1$ units with esterified carboxyl groups and one unit with a free carboxyl group. The fraction of all carboxyl groups that are free is $1 - p$, and the fraction that are esterified is equal to p. The probability of having a given value of x is

$$\text{Probability of } x = p^{x-1}(1 - p) \tag{22.6-12}$$

The number-average molecular mass can now be written

$$\bar{M}_n = \sum_{x=1}^{\infty} X_x x M_0 = M_0(1 - p) \sum_{x=1}^{\infty} x p^{x-1} \tag{22.6-13}$$

where a term equal to 18 amu has been omitted.

The sum in Equation (22.6-13) should have a finite upper limit because the system is of finite size. However, since $p < 1$, the terms become smaller and smaller for large values of x and we approximate the sum by extending the limit to infinity. This sum can be found in tables,[22] giving

$$\bar{M}_n = \frac{M_0}{1 - p} \tag{22.6-14}$$

From Equations (22.6-6) and (22.6-14), we can write an equation for the time dependence of $\bar{M}_n$:

$$\bar{M}_n = M_0(1 + c_0 k' t) \tag{22.6-15}$$

[20] Allcock and Lampe, *op. cit.*
[21] Tanford, *op. cit.*
[22] See, for example, H. B. Dwight, *Tables of Integrals and Other Mathematical Data*, 4th ed., Macmillan, New York, 1961, p. 8.

Equation (22.6-16) expresses a simple relationship: the number-average molecular mass increases linearly with time.

where we continue to neglect 18 amu compared with $\bar{M}_n$. For fairly large values of the time, the unity inside the parentheses can also be neglected, giving

$$\bar{M}_n = M_0 c_0 k' t \qquad (22.6\text{-}16)$$

The time dependence of the number fraction of any degree of polymerization can be obtained by substituting Equation (22.6-5) into Equation (22.6-12).

Exercise 22.21 Obtain an expression for the time dependence of X_x.

The **mass fraction** of molecules with degree of polymerization x is

$$W_x = \frac{\text{mass of molecules of degree of polymerization } x}{\text{mass of all molecules}}$$

$$= \frac{N_x M_x}{\sum_{x=1}^{\infty} N_x M_x} \qquad (22.6\text{-}17)$$

The **mass-average molecular mass** (usually called the "weight-average molecular weight") is defined by

$$\bar{M}_w = \sum_{x=1}^{\infty} W_x M_x = \frac{\sum_{x=1}^{\infty} N_x M_x^2}{\sum_{x=1}^{\infty} N_x M_x} \qquad (22.6\text{-}18)$$

The required sum can be found in tables,[23] giving

$$\bar{M}_w = (1-p)^2 M_0 \sum_{x=1}^{\infty} x^2 p^{x-1} = M_0 \frac{1+p}{1-p} \qquad (22.6\text{-}19)$$

The mass-average molecular mass is always equal to or larger than the number-average molecular mass, since the heavier molecules are given larger weight in the average. In our case, if p is nearly equal to unity, the mass-average molecular mass is nearly twice as large as the number-average molecular mass.

The evolution in time of the mass-average molecular mass can be expressed as a function of time, similar to Equation (22.6-15):

Exercise 22.22 Show that

$$\bar{M}_w = M_0(1 + 2k'c_0 t) \qquad (22.6\text{-}20)$$

It is also possible to derive an expression for the time dependence of the mass fraction of any degree of polymerization. Figure 22.20a shows the distribution of mass in a polyester, according to Equation (22.6-12). Figure 22.20b shows the evolution in time of several mass fractions during a condensation polymerization.

---

[23] *ibid.*

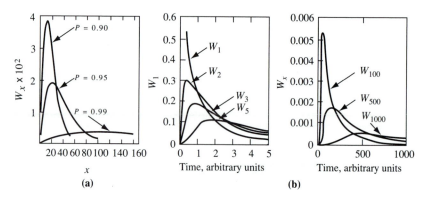

Figure 22.20. A Condensation Polymerization. (a) Distribution by mass in a polyester for conversions of 90, 95, and 99%. The horizontal axis is labeled with x, the degree of polymerization, or the number of monomer units in the polymer molecule. The vertical axis is labeled with the mass fraction of molecules with x monomer units. From H. R. Allcock and F. W. Lampe, *Contemporary Polymer Chemistry*, Prentice-Hall, Englewood Cliffs, NJ, 1981, p. 261. **(b)** The evolution of mass fractions. Note that all of the mass fractions shown except for W_1 rise and then decline as molecules of larger size are formed. From C. Tanford, *Physical Chemistry of Macromolecules*, Wiley, New York, 1961, p. 595.

Polymer Conformation

A typical polymer molecule is relatively flexible, since most bonds are able to rotate even if the bond angles are fixed. The principal elementary piece of information about the conformation of a polymer molecule is the end-to-end distance. Even if we had a monodisperse sample of a polymer (one in which all molecules had the same molecular mass), there would be a distribution of end-to-end distances, since each molecule would coil up differently from the others.

Let us approximate a polymer molecule by a **freely jointed chain**, which is a set of links of fixed length a fastened together end-to-end in a chain such that each joint can rotate into any position, even folding back at a 180° angle. Of course, a real polymer molecule is not this flexible.

To simplify the problem further, let us suppose that each link of the chain can be directed in one of only six directions, parallel to the x, y, and z axes of a cartesian coordinate system. The more general problem of arbitrary orientations can also be solved, and the solution is similar to the one we will obtain. We place the beginning of the chain at the origin, so that the ends of the links can fall only on the lattice points of a simple cubic lattice with lattice spacing a, which is very much like a crystal lattice.

We are constructing a model system with a web of assumptions that are fairly reasonable but obviously not exactly like the real system.

Let us now write an equation for $p(n + 1, x, y, z)$, the probability that the end of link number $n + 1$ is at a particular lattice point, denoted by the cartesian coordinates (x, y, z). This probability will give us our distribution of end-to-end distances for a chain of $n + 1$ links, since the distance from the origin to (x, y, z) can easily be calculated. If the end of link $n + 1$ is at (x, y, z), then the end of link n can be at one of only six possible locations: $(x + a, y, z), (x - a, y, z), (x, y + a, z), (x, y - a, z), (x, y, z + a)$, or $(x, y, z - a)$. We assume that the probabilities of the six possible directions of a link are equal, so that

$$p(n + 1, x, y, z) = \frac{1}{6} \left[p(n, x + a, y, z) + p(n, x - a, y, z) \right.$$

$$+ p(n, x, y + a, z) + p(n, x, y - a, z)$$

$$\left. + p(n, x, y, z + a) + p(n, x, y, z - a) \right] \quad \textbf{(22.6-21)}$$

Equation (22.6-21) is a difference equation that can be solved.[24] We apply the condition that the beginning of the chain is known to be at the origin:

$$p(0, x, y, z) = \begin{cases} 1 & \text{if } x = 0, y = 0, \text{ and } z = 0 \\ 0 & \text{otherwise} \end{cases} \quad \textbf{(22.6-22)}$$

The difference equation will maintain the normalization

$$\sum_x \sum_y \sum_z p(n, x, y, z) = 1 \quad \textbf{(22.6-23)}$$

where the summations are over all values of x, y, and z corresponding to lattice points.

We define a one-dimensional probability by summing $p(n + 1, x, y, z)$ over all values of y and z:

$$p(n + 1, x) = \sum_y \sum_z p(n + 1, x, y, z) \quad \textbf{(22.6-24)}$$

which is the probability that the end of link $n + 1$ is at x, irrespective of the y and z values. Equation (22.6-21) is now summed over all values of y and z:

In Equation (22.6-25) we have used the fact that it makes no difference whether y or y + a enters in the argument of a p term as long as all values of y are summed over, and similarly with z.

$$p(n + 1, x) = \frac{1}{6} \left[p(n, x + a) + 4p(n, x) + p(n, x - a) \right] \quad \textbf{(22.6-25)}$$

where we have recognized that the y and z directions are equivalent, so that four terms are equal to each other after the summation.

The principal variable that characterizes the distribution of end-to-end distances is the mean of the square of the distance, which is called the **second moment** of the distribution or the square of the standard deviation of the distribution. The x component of this quantity for link $n + 1$ is given by Equation (22.6-25)

The second moment is the same as the variance.

$$\langle x^2 \rangle_{n+1} = \sum_x x^2 p(n + 1, x)$$

$$= \frac{1}{6} \sum_x x^2 p(n, x + a) + \frac{2}{3} \sum_x x^2 p(n, x)$$

$$+ \frac{1}{6} \sum_x x^2 p(n, x - a) \quad \textbf{(22.6-26)}$$

---

[24] F. T. Wall, *Chemical Thermodynamics*, 2nd ed., W. H. Freeman, San Francisco, 1974, pp. 341ff.

In the first term let $x + a = x'$, and in the third sum let $x - a = x''$. The second sum is equal to $\langle x^2 \rangle_n$, so that

$$\langle x^2 \rangle_{n+1} = \frac{1}{6} \sum_{x'} (x'^2 - 2ax' + a^2) p(n, x') + \frac{2}{3} \langle x^2 \rangle_n$$

$$+ \frac{1}{6} \sum_{x''} (x''^2 + 2ax'' + a^2) p(n, x'') \qquad \textbf{(22.6-27)}$$

The sums over x' and x'' range over all values from $-\infty$ to $+\infty$, so there is no distinction between a sum over x, over x', or over x'' after the summation is done. Therefore, we can replace x' or x'' by x in the sums. The two sums containing $2ax$ cancel. The two sums containing x^2 give $\langle x^2 \rangle_n$, and the two sums containing a^2 can be combined:

$$\langle x^2 \rangle_{n+1} = \left(\frac{2}{3} + \frac{2}{6} \right) \langle x^2 \rangle_n + \frac{2}{3} a^2 \sum_x p(n, x)$$

$$= \langle x^2 \rangle_n + \frac{a^2}{3} \qquad \textbf{(22.6-28)}$$

where we have used the fact that the distribution is normalized as in Equation (22.6-23). Equation (22.6-28) is a recursion relation, somewhat similar to one encountered in the solution of the Schrödinger equation for the harmonic oscillator. If the value for $n = 0$ is known, the value for $n = 1$ can be calculated, and from this the value for $n = 2$ can be calculated, etc.

A recursion relation relates one member of a sequence to a later member. A recursion relation was used in solving the Schrödinger equation for the harmonic oscillator.

From the initial condition in Equation (22.6-22)

$$\langle x^2 \rangle_0 = 0 \qquad \textbf{(22.6-29)}$$

so that

$$\langle x^2 \rangle_1 = \frac{a^2}{3} \qquad \textbf{(22.6-30)}$$

Each iteration of Equation (22.6-28) adds a term $a^2/3$, so that

$$\langle x^2 \rangle_n = \frac{na^2}{3} \qquad \textbf{(22.6-31)}$$

The three directions are all equivalent, so that $\langle x^2 \rangle_n = \langle y^2 \rangle_n = \langle z^2 \rangle_n$. By Pythagoras' theorem, the mean square end-to-end distance in three dimensions is

$$\langle r^2 \rangle_n = \langle x^2 \rangle_n + \langle y^2 \rangle_n + \langle z^2 \rangle_n = na^2 \qquad \textbf{(22.6-32)}$$

and the root-mean-square distance is

$$r_{\text{rms}} = \langle r^2 \rangle^{1/2} = n^{1/2} a \qquad \textbf{(22.6-33)}$$

As expected, this distance is proportional to the length of a link. However, it is proportional to the square root of the number of links in the polymer chain, not to the number of links. This behavior arises physically from the fact that a longer chain has more ways to coil up than a short chain, so adding more links increases the root-mean-square distance less rapidly than in proportion to the number of links.

Our discussion of the end-to-end distance in a freely jointed chain is analogous to a **random walk** or **random flight** problem, in which the time t plays the same role as n, the number of links. It is applied to various model systems, including a model for diffusion in a liquid, in which a diffusing molecule is assumed to jump a distance equal to a at regular intervals in time. The analysis of this model leads to Equation (17.2-16).

In order to solve for the full distribution, we approximate Equation (22.6-21) by a differential equation.[25] We expand the function p in four different Taylor series in n, x, y, and z, treating n as though it could take on nonintegral values:

$$p(n + 1, x, y, z) = p(n, x, y, z) + \frac{\partial p}{\partial n} + \cdots \qquad \text{(22.6-34a)}$$

$$p(n, x \pm a, y, z) = p(n, x, y, z) \pm \frac{\partial p}{\partial x} a + \frac{1}{2} \frac{\partial^2 p}{\partial x^2} a^2 \pm \cdots \qquad \text{(22.6-34b)}$$

with equations like Equation (22.6-34b) for y and z. These series are substituted into Equation (22.6-21). The lowest-order terms that do not cancel are kept, and the higher-order terms are discarded:

$$\frac{\partial p}{\partial n} = \frac{a^2}{6} \left(\frac{\partial^2 p}{\partial x^2} + \frac{\partial^2 p}{\partial y^2} + \frac{\partial^2 p}{\partial z^2} \right) \qquad \text{(22.6-35)}$$

Equation (22.6-35) is valid in the case that n is large compared with unity and a is small compared with the values of x, y, and z that are important. Compare this equation with Fick's second law of diffusion, Equation (17.2-13).

We transform to spherical polar coordinates:

$$\frac{\partial p}{\partial n} = \frac{a^2}{6r^2} \frac{\partial}{\partial r} \left(r^2 \frac{\partial p}{\partial r} \right) \qquad \text{(22.6-36)}$$

When we transform to spherical polar coordinates, p becomes a function of r, θ, and φ. However, we apply a physical argument to assert that all directions in space are equivalent, so that p cannot depend on direction, and we write p = p(n, r).

The derivatives with respect to the angles θ and ϕ do not occur because p depends only on the end-to-end distance and not on direction. The solution to Equation (22.6-36) is[26]

$$p(n, r) = \left(\frac{3}{2\pi na^2} \right)^{3/2} e^{-3r^2/2na^2} \qquad \text{(22.6-37)}$$

where the constant provides for normalization:

$$\int_0^\infty p(n, r) 4\pi r^2 \, dr = 1 \qquad \text{(22.6-38)}$$

Compare Equation (22.6-37) to Equation (17.2-16), which can be derived from a one-dimensional random walk.

---

[25] *ibid.*
[26] *ibid.*

Exercise 22.23

a. Carry out the substitution of the Taylor series into Equation (22.6-21) to obtain Equation (22.6-35).
b. Verify that the function of Equation (22.6-37) satisfies Equation (22.6-36).
c. Verify that the function of Equation (22.6-37) is normalized.

The freely jointed chain, which we have discussed, is only a crude first approximation for real polymers. Every real polymer has some rigidity built into its bonds, so the chain is not freely jointed. Furthermore, we have ignored the problem of **excluded volume**, which means that two parts of a polymer chain or of two different polymer chains cannot occupy the same location at the same time. In addition, the effect of intermolecular attractions on the conformation can be considerable. Any of the books on polymer chemistry contain discussions of more elaborate theories than the simple freely jointed chain theory.

Most biological macromolecules have far more regular molecular conformations than synthetic polymers. Proteins, which are condensation polymers of amino acids, can form intramolecular hydrogen bonds that generally hold the chains in a helical conformation or a pleated sheet conformation. The proper conformation is essential to the biological function of the molecule, and if the molecule is disrupted into a more random conformation, it is said to be **denatured** and loses its biological function.

Nucleic acids are polymers of five-carbon sugars (either ribose or deoxyribose), phosphoric acid residues, and certain ring-containing molecules called bases. Deoxyribonucleic acid (DNA) is held in a double helix of two chains by hydrogen bonds between certain pairs of bases: cytosine (C) hydrogen-bonds to guanine (G) and adenine (A) hydrogen-bonds to thymine (T), so in an intact DNA molecule a C must be opposite every G on the other chain and a T must be opposite every A on the other chain.

22.7   Rubber Elasticity

Rubber is a naturally occurring polymer that was first used as pencil erasers (this use is the origin of the name). Rubber was shown by Staudinger to be macromolecular in nature.

Rubber is an addition polymer of isoprene (2-methyl-1,3-butadiene). Since each molecule of the monomer has two double bonds, there remains one double bond per monomer. Natural rubber has the *cis* configuration at all of these bonds. A portion of the structural formula for a rubber molecule is (hydrogens have been omitted):

Hermann Staudinger, 1881–1965, German chemist who received the 1953 Nobel Prize in chemistry for his pioneering work in the chemistry of macromolecules.

Prior to 1955, various kinds of synthetic rubber were invented, but they were found to be inferior to natural rubber for making automobile tires. It was possible to make polyisoprene, but the double bonds were in a random mixture of the *cis* and *trans* configurations. In 1955, catalysts were developed that can produce a product identical to natural rubber.

In 1839, after 10 years of trial-and-error experimentation in his kitchen, Goodyear invented the **vulcanization** process, in which sulfur is reacted with natural rubber. From natural rubber, which is a soft, sticky, and semifluid thermoplastic, a more nearly solid and elastic product is obtained. Vulcanized rubber can be made in varying degrees of hardness, from flexible rubber like that in inner tubes, through that used in tires, to the hard rubber used in combs.

Charles Goodyear, 1800–1860, American inventor whose patent was widely infringed upon and who died in debt.

In the vulcanization process, the sulfur reacts with double bonds in two chains, forming short chains of sulfur atoms between the polymer chains. The extent of this cross-linking and the presence of various additives such as "carbon black" (finely powdered carbon) determine the physical properties of the rubber.

A piece of vulcanized rubber is not a simple system in the thermodynamic sense, since work other than compression work ($P\,dV$ work) can be done on it. When it is stretched reversibly, work is done on it:

$$dw = -P\,dV + f\,dL \qquad \textbf{(22.7-1)}$$

where L is the length of the piece of rubber and f is called the **tension force**.

It is found experimentally that the stretching of rubber approximately obeys three properties: (1) the volume remains constant, (2) the tension force is proportional to the absolute temperature, and (3) the energy is independent of the length at constant temperature. An **ideal rubber** is defined as one that exactly obeys these three properties.

Since the volume is constant, the first law of thermodynamics for reversible processes in a closed system made of ideal rubber is

$$dU = T\,dS + f\,dL \qquad \textbf{(22.7-2)}$$

This equation is completely analogous to Equation (4.2-3). We define an enthalpy-like variable, K:

$$K = U - fL \quad \text{(definition)} \qquad \textbf{(22.7-3)}$$

The Helmholtz energy is defined in the standard way:

$$A = U - TS \quad \text{(definition)} \qquad \textbf{(22.7-4)}$$

We denote the analogue to the Gibbs energy by J:

$$J = K - TS = U - fL - TS \quad \text{(definition)} \qquad \textbf{(22.7-5)}$$

Just as in Chapter 4, we can write differential expressions:

$$dK = T\,dS - L\,df \qquad \textbf{(22.7-6a)}$$
$$dA = -S\,dT + f\,dL \qquad \textbf{(22.7-6b)}$$
$$dJ = -S\,dT - L\,df \qquad \textbf{(22.7-6c)}$$

Compare the Maxwell relation in Equation (22.7-7) with the Maxwell relations in Section 4.2.

and can write Maxwell relations from these. For example, we will need the Maxwell relation that can be obtained from Equation (22.7-6b):

$$\left(\frac{\partial f}{\partial T}\right)_L = -\left(\frac{\partial S}{\partial L}\right)_T \tag{22.7-7}$$

Exercise 22.24 Write the other three Maxwell relations from Equations (22.7-2), (22.7-6a), and (22.7-6c).

Using Equations (22.7-2) and (22.7-7), we can derive a **thermodynamic equation of state** analogous to Equation (4.5-1):

$$\left(\frac{\partial U}{\partial L}\right)_T = T\left(\frac{\partial S}{\partial L}\right)_T + f$$

$$= f - T\left(\frac{\partial f}{\partial T}\right)_L \tag{22.7-8}$$

We can now show that property 3 of an ideal rubber follows from property 2. If f is proportional to T, it can be written

$$f = T\varphi(L) \tag{22.7-9}$$

where φ is some function of L and possibly of V, which we cannot determine from thermodynamics. We now have

$$T\left(\frac{\partial f}{\partial T}\right)_L = T\varphi = f \tag{22.7-10}$$

so that

$$\left(\frac{\partial U}{\partial L}\right)_T = f - f = 0 \tag{22.7-11}$$

Equation (22.7-11) shows the difference between a rubber band and a spring. When a spring is stretched at constant temperature, the energy increases as work is done on the spring. When a rubber band is stretched at constant temperature, doing work on the system, heat must flow out if the energy remains constant. (This behavior is easily observed by stretching a rubber band in contact with the upper lip, which is well supplied with temperature sensors.) Stretching the rubber band at constant temperature must therefore decrease its entropy. This fact seems reasonable from a molecular point of view, since the polymer molecules are more nearly parallel and more nearly ordered in the stretched state than in the relaxed state. From Equations (22.7-8) and (22.7-11) we can derive a relation for this decrease in entropy:

$$f = -T\left(\frac{\partial S}{\partial L}\right)_T \tag{22.7-12}$$

A Molecular Theory of Rubber Elasticity

We can derive an equation of state for an ideal rubber that gives the tension force as a function of T and L. We now define a model system[27] which has the following properties: (1) The equilibrium system is a rectangular piece of rubber, made up of a set of polymer molecules with an equilibrium distribution of end-to-end lengths given by the freely jointed chain formula of Equation (22.6-37). (2) A certain number, N, of randomly selected polymer chains are cross-linked. For simplicity, we assume that they are cross-linked only at their ends and that all of the cross-linked molecules have the same number of links, n. (3) When the rubber is stretched in the x direction, the y and z dimensions change so that the volume remains constant and the x, y, and z components of all end-to-end vectors change in the same ratio as the x, y, and z dimensions of the rubber.

At equilibrium, there will be N_i molecules with an end-to-end vector $\mathbf{r}_i = (x_i, y_i, z_i)$. After the deformation, the end-to-end vector of these molecules will be (x_i', y_i', z_i'):

$$x_i' = \alpha x_i, \qquad y_i' = \frac{y_i}{\alpha^{1/2}}, \qquad z_i' = \frac{z_i}{\alpha^{1/2}} \qquad \textbf{(22.7-13)}$$

where α is the degree of elongation, equal to L/L_0, the elongated x dimension divided by the original x dimension. The number of molecules with this new end-to-end vector is still equal to N_i. The changes in the y and z directions preserve the original volume.

To calculate the entropy change on elongation, we use the definition of the statistical entropy, Equation (3.4-3):

$$S_{\text{st}} = k_{\text{B}} \ln(\Omega) + \text{const} \qquad \textbf{(22.7-14)}$$

where Ω is the thermodynamic probability, equal to the number of system mechanical states that the system might occupy, so far as we know. To use this formula, we calculate the probability, P, that the elongation would occur spontaneously, equal to the probability that N_1 chains will have end-to-end vector $\mathbf{r}_1'$, N_2 chains will have end-to-end vector $\mathbf{r}_2'$, etc. This probability is proportional to Ω for this state and is equal to

In Equation (22.7-15) we assume that the chains act independently and use the fact that the joint probability of several independent events is the product of the probabilities of the events. This probability is then multiplied by the number of ways to divide the set of polymer molecules into the specified subsets.

$$P' = N! \prod_i \frac{1}{N_i!} \, p(n, x_i', y_i', z_i')^{N_i} \qquad \textbf{(22.7-15)}$$

where the factor $N!/\prod_i N_i!$ is the number of ways to divide the N chains into the required subsets. With Stirling's approximation,

$$\ln(P') = N \ln(N) - N + \sum_i [N_i \ln(p_i') - N_i \ln(N_i) + N_i]$$

$$= \sum_i N_i \ln\left(p_i' \frac{N}{N_i} \right) \qquad \textbf{(22.7-16)}$$

where we have abbreviated $p(n, x_i', y_i', z_i')$ by p_i'. We now write the same equation for the equilibrium distribution:

$$\ln(P) = \sum_i N_i \ln\left(p_i \frac{N}{N_i} \right) \qquad \textbf{(22.7-17)}$$

---

[27] Wall, *op. cit.*, Ch. 16; Flory, *op. cit.*, pp. 464ff.

Since P is proportional to Ω,

$$\Delta S = S(\text{stretched}) - S(\text{equilibrium}) = k[\ln(P') - \ln(P)]$$

$$= k_B \sum_i N_i \ln\left(\frac{p_i'}{p_i}\right) = Nk_B \sum_i p_i \ln\left(\frac{p_i'}{p_i}\right) \qquad \text{(22.7-18)}$$

where we have used the fact that $p_i = N_i/N$, since the randomly selected cross-linked chains obey the equilibrium distribution.

If we can approximately consider x, y, and z to range continuously, we can replace the sum by an integral:

$$\Delta S = Nk_B \int p \ln\left(\frac{p'}{p}\right) dx \, dy \, dz \qquad \text{(22.7-19)}$$

where the integral is over all values of x, y, and z.

From Equation (22.6-37),

$$\ln\left(\frac{p'}{p}\right) = \frac{3}{2na^2}\left[-x^2(\alpha^2 - 1) - (y^2 + z^2)\left(\frac{1}{\alpha} - 1\right)\right] \qquad \text{(22.7-20)}$$

When Equation (22.7-20) is substituted into Equation (22.7-19),

$$\Delta S = \frac{3Nk_B}{2na^2}\left[-\langle x^2\rangle_n(\alpha^2 - 1) - (\langle y^2\rangle_n + \langle z^2\rangle_n)\left(\frac{1}{\alpha} - 1\right)\right]$$

$$= -\frac{Nk_B}{2}\left(\alpha^2 + \frac{2}{\alpha} - 3\right) \qquad \text{(22.7-21)}$$

where we have used Equation (22.6-31) for the equilibrium value of $\langle x^2\rangle_n$, which is also equal to $\langle y^2\rangle_n$ and $\langle z^2\rangle_n$.

Using Equations (22.7-21) and (22.7-12), we can write an equation of state for ideal rubber:

$$f = -T\left(\frac{\partial S}{\partial L}\right)_T = -\frac{T}{L_0}\left(\frac{\partial S}{\partial \alpha}\right)_T = \frac{Nk_B T}{L_0}\left(\alpha - \frac{1}{\alpha^2}\right) \qquad \text{(22.7-22)}$$

This equation of state agrees fairly well with experiment for values of α up to 3 or 4.[28]

EXAMPLE 22.2   Derive an expression for the reversible work done in stretching a piece of ideal rubber at constant temperature.

Solution

Let α' be the final value of the extent of elongation.

$$dw = f \, dL = \frac{Nk_B T}{L}\left(\alpha - \frac{1}{\alpha^2}\right) dL = Nk_B T\left(\alpha - \frac{1}{\alpha^2}\right) d\alpha$$

$$w = Nk_B T \int_1^{\alpha'}\left(\alpha - \frac{1}{\alpha^2}\right) d\alpha = \frac{Nk_B T}{2}\left(\alpha'^2 + \frac{2}{\alpha'} - 3\right)$$

---

[28] *ibid.*

Exercise 22.25	**a.** Write the formula for q for the elongation of a piece of ideal rubber.

b. For a piece of ideal rubber large enough that 1.00×10^{-6} mole of cross-linked polymer chains occur in it, find the value of q and w for stretching it to 3.00 times its original length at 298.15 K.

22.8 Polymers in Solution

Most polymers are made from relatively nonpolar monomer molecules and are nearly completely insoluble in water. However, polymers such as polyvinyl alcohol, made from polar monomers, can be somewhat soluble in water, and other polymers can dissolve in nonpolar solvents.

The conformation of polymer molecules in solution is similar to that of pure polymers, except that there is an additional effect due to intermolecular forces between polymer molecules and solvent molecules and between solvent molecules. A nonpolar polymer molecule attracts polar solvent molecules less strongly than the polar solvent molecules attract each other and tends to form a tight ball in a polar solvent. Similarly, polar polymer molecules tend to form tight balls in nonpolar solvents. However, molecules of a polar polymer such as polyvinyl alcohol attract water molecules and can swell in water. Similarly, nonpolar polymers can swell in nonpolar solvents (try placing a rubber object in benzene).

Just as in a pure polymer, we use the mean-square end-to-end distance as a measure of the conformation of the polymer molecules. The **expansion coefficient** α is defined such that

The same symbol, α is unfortunately used for the coefficient of thermal expansion.

$$\langle r^2 \rangle = \alpha^2 \langle r^2 \rangle_0 \tag{22.8-1}$$

where $\langle r^2 \rangle_0$ is the mean-square end-to-end distance in the pure polymer and $\langle r^2 \rangle$ is the mean-square end-to-end distance in the solution.

A solvent in which polymer molecules adopt on the average the same conformation as in the pure polymer is called a **theta solvent**. In a theta solvent, $\alpha = 1$. In this kind of solvent, the polymer molecules should approximately conform to the freely jointed chain model of Section 22.6. In a poor solvent for the particular polymer, α is smaller than unity, and in a good solvent, which causes the polymer to swell, α exceeds unity.

In a typical polymer solution, a polymer molecule and its associated solvent molecules occupy a roughly spherical region in space, with a diameter approximately equal to the molecule's end-to-end distance, and move through a solution in much the same way as a rigid object of that size. The presence of the polymer molecules determines the viscosity of the solution, so the viscosity can be used as an indirect means of determining the molecular mass of the polymer.

Einstein solved the hydrodynamic equations for flow around a hard sphere in a viscous fluid and found that for a dilute suspension of spheres, the viscosity is given by[29]

$$\eta = \eta_0 \left(1 + \frac{5}{2} \phi \right) \tag{22.8-2}$$

where η_0 is the viscosity of the pure solvent and ϕ is the volume fraction of the spheres in the suspension:

$$\phi = v/V \tag{22.8-3}$$

The total volume of the spheres is denoted by v and the volume of the entire suspension is denoted by V.

Equation (22.8-2) can be written in the form

$$\eta_{sp} = \frac{\eta}{\eta_0} - 1 = \eta_r - 1 = \frac{5}{2} \phi \tag{22.8-4}$$

where η_{sp} is called the **specific viscosity** and η_r is called the **relative viscosity**.

The theory of Einstein is valid only for a dilute suspension, since it requires that the flow around one sphere is not perturbed by the flow around other spheres. In the limit of small concentration, a polymer solution should behave approximately like such a suspension of spheres.

Since the volume of a set of spherical particles is proportional to the number of the spheres, the specific viscosity of a dilute suspension of spheres is proportional to the concentration of the spheres. We define the **intrinsic viscosity**, $[\eta]$, also called the **limiting viscosity number**,

$$[\eta] = \lim_{c \to 0} \left(\frac{1}{c} \eta_{sp} \right) = \lim_{c \to 0} \left[\frac{1}{c} \left(\frac{\eta}{\eta_0} - 1 \right) \right] \tag{22.8-5}$$

where c is the concentration of the polymer, expressed as mass per unit volume (often grams per deciliter). The reason for using mass per unit volume is that one cannot specify the concentration in moles per unit volume until the molar mass is known.

For a dilute suspension of hard spheres of the same size, the specific viscosity divided by c is independent of the concentration. Therefore,

$$[\eta] = \frac{1}{c} \eta_{sp} = \frac{5}{2} \frac{1}{c} \frac{v}{V} = \frac{5}{2} \frac{1}{M} \frac{4}{3} \pi r^3 \tag{22.8-6}$$

where M and r are the mass and radius of one of the spheres.

Exercise 22.26   Verify Equation (22.8-6).

We assume that the radius of the sphere occupied by a polymer molecule and associated solvent molecules is proportional to the root-mean-square end-to-end distance, which itself is proportional to the square root

---

[29] A. Einstein, *Ann. Phys.* **19**, 289 (1906).

of the molecular mass. Therefore, the volume is proportional to $M^{3/2}$ and the intrinsic viscosity is proportional to $M^{1/2}$.

$$[\eta] = K'M^{1/2} \tag{22.8-7}$$

where K' depends on temperature and on the identities of the solvent and polymer, but not on M.

Equation (22.8-7) is known as the **Mark-Houwink equation** and is considered to be valid for a polymer in a theta solvent. For other solvents, we write a modified version

$$[\eta] = K''\alpha^3 M^{1/2} \tag{22.8-8}$$

Since K' and K'' are parameters that will be experimentally evaluated, Equation (22.8-8) is no different from Equation (22.8-7) if the expansion coefficient is a constant for a given polymer and a given solvent.

It is found experimentally that the expansion coefficient increases when M increases, so that Equations (22.8-7) and (22.8-8) are replaced by the relation

$$[\eta] = KM^a \tag{22.8-9}$$

where the exponent a is determined by experiment. Flory and Leutner prepared nearly monodisperse samples of polyvinyl alcohol and found that for aqueous solutions at 25°C,[30]

$$[\eta] = (2.0 \times 10^{-4} \text{ dL g}^{-1})(M/1 \text{ amu})^{0.76} \tag{22.8-10}$$

For a typical polydisperse sample of a single polymer, one can apply Equation (22.8-9) to each molecular mass that is present, multiply each equation by W_i, the mass fraction for molecular mass M_i, and sum the equations over all possible molecular masses. Recognizing that the intrinsic viscosity is a sum of all contributions (since it applies to infinite dilution) and that K has the same value for all molecules of a given polymer, we write

$$[\eta] = K \sum_i W_i M_i^a \tag{22.8-11}$$

We define the **viscosity-average molecular mass**:

$$\bar{M}_v = \left(\sum_i W_i M_i^a \right)^{1/a} \tag{22.8-12}$$

and Equation (22.8-11) becomes

$$[\eta] = K\bar{M}_v^a \tag{22.8-13}$$

The viscosity-average molecular mass is not defined in the same way as the number-average and mass-average molecular masses, being more like a geometric mean than an arithmetic mean. However, it can be shown that if $a = 1$ the viscosity-average molecular mass is the same as the mass-average molecular mass. It can also be shown that the viscosity-average molecular

---

[30] P. J. Flory and F. S. Leutner, *J. Polym. Sci.* **5**, 267 (1950).

mass is the same as the number-average molecular mass if $a = -1$. However, $a = 1$ is conceivable, but $a = -1$ is not. The viscosity-average molecular mass is more nearly equal to the mass-average value than to the number-average value.

| Exercise 22.27 | Using the fact that the mass fractions sum to unity, show that the viscosity-average molecular mass is the same as the mass-average molecular mass in the case that $a = 1$. |

Summary

In this chapter, we have studied the molecular structure of solids and liquids. Many solids are crystalline, with molecular units arranged in a regular three-dimensional lattice. We have discussed the principal types of solid lattices.

There are two principal theories for the vibrations of lattices of atoms, the Einstein and Debye theories. In the Einstein model, each atom is assumed to vibrate independently of the other atoms, with all atoms vibrating with the same frequency. In the Debye model, the lattice is assumed to vibrate with the same distribution of frequencies as would a structureless solid. In each theory, the principal result is a formula for the heat capacity of the solid lattice. In each case, the high-temperature limit of the formula conforms to the law of Dulong and Petit.

The electronic structure of solids was discussed through the band theory, in which the electrons are assumed to occupy delocalized orbitals that comprise bands of energy levels. The differences between conductors, semiconductors, and insulators were discussed in terms of the band theory.

The structure of liquids is more difficult to discuss than the structure of solids because liquids are more disordered than solids but not completely disordered as are gases. Some elementary comments on the structure of liquids were presented, including the definition of the radial distribution function, which gives the probability of finding a second molecule at a given distance from another molecule.

The kinetics of the formation of addition polymers was discussed. An approximation solution of the rate law was presented, and distributions of molecular mass were discussed. The conformation of a simple polymer model, a freely jointed chain, was also presented. Rubber elasticity was discussed at the same level of approximation.

Additional Reading

A. W. Adamson, *Physical Chemistry of Surfaces*, 2nd ed., Interscience Publishers, New York, 1967
This is a textbook for a graduate-level course in surface chemistry.

H. R. Allcock and F. W. Lampe, *Contemporary Polymer Chemistry*, Prentice-Hall, Englewood Cliffs, NJ, 1981
A complete, clearly written, and useful textbook for a course in polymer chemistry.

J. S. Blakemore, *Solid-State Physics*, 2nd ed., W. B. Saunders, Philadelphia, 1974
This is one of the standard textbooks for an undergraduate course in solid-state physics and is quite clear and readable.

C. A. Croxton, *Liquid State Physics—A Statistical Mechanical Introduction*, Cambridge University Press, 1974
A reference book in the statistical mechanical theory of liquids at an advanced graduate level.

P. A. Egelstaff, *An Introduction to the Liquid State*, Academic Press, New York, 1967
An introduction to the theory of the liquid state at the level of beginning graduate students in physics. Although it grew out of a course, it is more of a reference book than a textbook.

P. J. Flory, *Principles of Polymer Chemistry*, Cornell University Press, Ithaca, NY, 1953
This classic textbook in polymer chemistry is still very useful.

N. B. Hannay, *Solid-State Chemistry*, Prentice-Hall, Englewood Cliffs, NJ, 1967
This is a paperback book intended for advanced undergraduates. It includes chemical reactions as well as structure and is nicely written and very informative.

J. P. Hansen and I. R. McDonald, *Theory of Simple Liquids*, Academic Press, New York, 1976
This is a reference book, not a textbook. It is at a rather advanced level and is quite complete for its time.

C. Kittel, *Introduction to Solid State Physics*, 6th ed., Wiley, New York, 1986
One of the best-known textbooks of solid state physics at the undergraduate level.

W. Moore, *Seven Solid States*, W. A. Benjamin, New York, 1967
This is a small paperback book, written at a fairly elementary undergraduate level, in which the principles of solid state chemistry and physics are illustrated by study of seven example solids.

R. B. Seymour and C. E. Carraher, Jr., *Polymer Chemistry—An Introduction*, Marcel Dekker, New York, 1981
A nonmathematical survey of polymer chemistry.

D. Tabor, *Gases, Liquids and Solids*, 2nd ed., Cambridge University Press, Cambridge, 1979
This is a fairly small paperback book. It is a clear and valuable reference and derives approximate formulas for nearly all properties of gases, liquids, and solids from simple model systems.

C. Tanford, *Physical Chemistry of Macromolecules*, Wiley, New York, 1961
This book is written by a prominent biophysical chemist but gives good coverage to synthetic polymers as well as to biological macromolecules.

See also the books listed at the end of Chapter 20, most of which contain applications of statistical mechanics to solids and liquids.

PROBLEMS

Problems for Section 22.1

22.28. Without looking up any properties, classify the following solids into the four classes of Section 22.1. Rearrange the list in order of increasing melting temperature without looking up any data.

Diamond	Cesium nitrate
Neon	Sodium
Copper	Krypton

22.29. Explain why a single crystal has greater mechanical strength than a polycrystalline sample of the same material.

Problems for Section 22.2

*22.30. What is the basis for a crystal of argon (fcc cubic lattice)? What is the number of atoms per unit cell? What is the number of bases per unit cell?

22.31. The CsCl crystal has a cubic unit cell with a cesium ion at each corner and a chloride ion at the center. To which Bravais lattice does it belong? How many bases are there in a unit cell? The density of CsCl is 3988 kg m^{-3}. Find the unit cell dimension.

22.32. Barium crystallizes in the body-centered cubic lattice.

a. Find the number of atoms per unit cell.

b. The density of barium is 3.5 g cm^{-3}. Find the unit cell dimension and the radius of a barium atom.

c. Calculate the molar volume of solid barium and the volume of empty space in 1.00 mol of barium.

*22.33. a. Find the perpendicular distance between 111 planes in the molybdenum carbide lattice, which is face-centered cubic with $a = 4.28 \times 10^{-10}$ m.

b. If the $n = 1$ reflection from the 111 plane gives $\theta = 36.5°$, find the wavelength of the X-rays.

22.34. a. The stretching of a uniform bar due to a tensile force is described by **Young's modulus**, E, defined by

$$E = \frac{\text{stress}}{\text{strain}} = \frac{F/A}{\Delta L/L}$$

where F is the magnitude of the tensile force, A is the cross-sectional area, L is the length, and ΔL is the change in the length due to the force F. Derive an approximate expression for Young's modulus for a perfect crystalline substance with

a simple cubic lattice and an intermolecular potential energy given by

$$u(r) = u(a) + \frac{k}{2}(r - a)^2$$

where k is a constant, r is the lattice spacing, and a is its equilibrium value. Include only nearest-neighbor interactions and assume that each unit cell stretches in the same ratio as the entire bar.

b. The value of E for fused quartz is 7.17×10^{10} N m^{-2}. Estimate the force constant k for the Si—O bond, assuming (contrary to fact) that quartz has a simple cubic lattice. Take $a = 1.5 \times 10^{-10}$ m, the approximate Si—O bond distance. Comment on your value in view of the fact that force constants for most single bonds in molecules are roughly equal to 500 N m^{-1}. Try to explain any discrepancy.

Problems for Section 22.3

*22.35. The value of Θ_E that fits the Einstein crystal model heat capacity formula to aluminum data is 240 K.

a. What is the vibrational frequency corresponding to this value of the parameter?

b. Draw a graph of the molar heat capacity of aluminum from 0 K to 300 K, according to the Einstein model.

22.36. Derive a formula for the vapor pressure of an Einstein crystal. Take the potential energy of the gas equal to zero, and call the energy of an atom of the crystal in its ground vibrational state $-u_0$ (a constant).

22.37. Write a computer program to evaluate the Debye function, using Simpson's rule.[31]

22.38. Consider a modified Einstein crystal model with two frequencies. There are N atoms in the crystal. One-third of them oscillate in three dimensions with frequency v, and two-thirds of them oscillate in three dimensions with frequency $2v$.

a. Write a formula for C_V.

---

[31] S. I. Grossman, *Calculus*, 3rd ed., Academic Press, Orlando, FL, 1984, pp. 518ff, or any standard calculus text.

b. Draw a graph of C_V versus T for $\nu = 3.94 \times 10^{12}$ s^{-1} (one-half of the Debye frequency for germanium).

c. If a table of the Debye function is available,[32] find the ratio of your result to the Debye result for several values of T.

Problems for Section 22.4

22.39. Make an accurate graph of the fermion distribution, Equation (22.4-5), using ε/k_BT as the independent variable and assuming (a) that $T = \mu/10k_B$, (b) that $T = \mu/k_B$, and (c) that $T = 10\mu/k_B$.

*22.40. Evaluate the Fermi level for copper at 298 K using Equation (22.4-15). Find the percent error of using Equation (22.4-14).

22.41. The Fermi level for sodium is equal to 3.1 eV.

a. Find the number of mobile electrons per cubic meter and per mole. The density of sodium is 0.971 g mL^{-1}.

b. Find the speed of electrons that have kinetic energy equal to the Fermi level for sodium.

22.42. In silicon, the band gap between the highest filled band (the valence band) and the lowest vacant band (the conduction band) is equal to 1.17 eV.

Assuming the Boltzmann distribution, find the ratio of the population of the lowest conduction band states and the highest valence band states at 300 K.

*22.43. Calculate the electronic contribution to the heat capacity of copper at 298.15 K. Find the percent contribution to the total heat capacity, using the law of Dulong and Petit for the vibrational contribution.

Problems for Section 22.5

22.44. The density of ice at 0°C is 0.917 g ml^{-1} and that of liquid water is 1.000 g ml^{-1}. The water molecules in the ice crystal have four nearest neighbors. Estimate the number of nearest neighbors in liquid water at 0°C.

22.45. Show that the pressure of a system of hard spheres can be obtained from the value of the radial distribution at only one value of r. What value is it?

*22.46. At 1.00 atm and 84 K, the normal melting temperature, the density of solid argon is 1.82 g ml^{-1} and that of liquid argon is 1.40 g ml^{-1}. Find the pressure according to the cell model result, Equation (22.5-17), and compare it with the correct value, 1.00 atm.

Problems for Section 22.6

22.47. The sum in Equation (22.6-13) is similar to the geometric progression of Equation (20.4-15). Show that the result used in Equation (22.6-14) can be obtained by differentiating the sum

$$S = \sum_{n=0}^{\infty} p^x$$

with respect to p.

*22.48. Find the number-average and mass-average molecular masses for a sample of a polyvinyl alcohol that has equal numbers of molecules with molecular masses 5000 amu, 10000 amu, 15000 amu, 20000 amu, 25000 amu, 30000 amu, and 35000 amu.

22.49. The hydroxy acid HO—(—CH$_2$—)$_5$—CO$_2$H forms a polyester by condensation polymerization.

a. Show that the equations used in Section 22.6 apply for the condensation polymerization of a diacid and a dialcohol.

b. If the mass-average molecular mass is reported to 20000 amu, find the fraction of carboxyl groups that have been esterified.

c. Find the number-average molecular mass for part b.

d. Find the average degree of polymerization.

22.50. The amino acid 7-aminoheptanoic acid forms a polyamide (nylon) in m-cresol solution, with the second-order rate coefficients:[33]

temperature/°C	150	187
k/kg mol^{-1} min^{-1}	1.0×10^{-3}	2.74×10^{-2}

a. Find the times required at 150°C to attain a value of p equal to 0.950 and a value of p equal to 0.990 for an initial molality of 0.50 mol kg^{-1}.

b. Find the number-average and mass-average molecular masses at $p = 0.99$.

c. Find the activation energy of the reaction and find the time required to reach a value of p equal to 0.95 at 175°C.

22.51. Carry out the solution of the kinetic equation for an uncatalyzed polyesterification which is assumed to be second order in carboxyl groups and first order in hydroxyl groups:

$$\frac{dc}{dt} = -k_f c^3$$

---

[32] N. Davidson, *Statistical Mechanics*, McGraw-Hill, New York, 1962, p. 359.

[33] Zhubanov et al., *Izv. Akad. Nauk. SSR. Ser. Khim.* **17**, 69 (1967), cited in Allcock and Lampe, *op. cit.*, p. 268.

22.52. Show that Equation (22.6-30) is compatible with the fact that a "chain" of one link is rigid.

*22.53. For a freely jointed chain with links 3×10^{-10} m long, find the root-mean-square end-to-end distance and the ratio of this length to the sum of the link lengths for (a) 100 links and (b) 10000 links.

Problems for Section 22.7

22.54. a. Write a general expression for $C_f - C_L$, the difference between the heat capacity at constant f and that at constant L, for rubber.

b. Write the expression for part a for ideal rubber.

22.55. a. Write equations for q and w for each step of a Carnot cycle using a piece of ideal rubber instead of a working fluid.

b. Write an expression for the efficiency of a Carnot engine using a piece of ideal rubber instead of a working fluid.

Problems for Section 22.8

*22.56. Assume that the diameter of a molecule of polyvinyl alcohol is equal to the root-mean-square end-to-end distance. Assume that (since six carbon atoms can make a ring) each set of three carbons constitutes a link in a freely jointed chain, of length 2.57×10^{-10} m. (Each link constitutes 1.5 monomer unit.)

a. Find the volume of a sphere containing one molecule of polyvinyl alcohol of mass 60000 amu, assuming the diameter of the sphere to equal the root-mean-square end-to-end distance.

b. Find the viscosity at 20.0°C of a solution of such molecules (all assumed identical) with 1.00 g L^{-1}, using the Einstein equation, Equation (22.8-2).

c. Find the viscosity at 20.0°C of the solution of part a using Equation (22.8-10) and assuming that the limiting value of the specific viscosity can be used at this concentration.

22.57. Find the viscosity-average molecular mass of the polyvinyl alcohol sample of Problem 22.48.

General Problems

22.58. Write a computer program to evaluate the configuration integral for a system of eight hard spheres of radius 1.00×10^{-10} m in a box 5.00×10^{-10} m on a side, using the Monte Carlo method. Start with the particles at the centers of the octants of the box, and take 1.00×10^{-10} m as your maximum displacement.

*22.59. Label each of the following statements as either true or false. If a statement is true only under special circumstances, label it as false.

a. All solids are crystalline.

b. The melting temperature is an indicator of the strength of the forces holding a solid together.

c. Liquid crystals are intermediate in their properties between liquids and solids.

d. The properties of gases arise primarily from the molecular kinetic energy, while the properties of liquids and solids arise primarily from the molecular potential energy.

e. The stretching of a rubber band can be treated thermodynamically just like the stretching of a coil spring.

f. The Einstein temperature of a crystal is generally smaller than the Debye temperature of the same crystal.

g. The effective volume of a polymer molecule in a theta solvent is approximately proportional to $N^{3/2}$, where N is the number of monomer units.

h. All solids obey the law of Dulong and Petit at sufficiently high temperature.

i. The molar mass of a polymer determined by a viscosity experiment will always be different from the molar mass determined by an osmotic pressure measurement.

j. The heat capacity of a nonconductor is approximately proportional to T^3 at sufficiently low temperature.

Tables of Numerical Data

Table A1. Parameters for Some Equations of State
(Chapter 1, page 16)

Parameters for the van der Waals Equation of State

$$\left(P + \frac{a}{\bar{V}^2}\right)(\bar{V} - b) = RT$$

Substance	$a/\text{Pa m}^6 \text{ mol}^{-2}$	$b \times 10^5/\text{m}^3 \text{ mol}^{-1}$
Ammonia	0.4225	3.707
Argon	0.1363	3.219
Carbon dioxide	0.3640	4.267
Helium	0.003457	2.370
Hydrogen	0.2476	2.661
Methane	0.2283	4.278
Neon	0.02135	1.709
Nitrogen	0.1408	3.913
Oxygen	0.1378	3.183
Xenon	0.4250	5.105
Water	0.5536	3.049

Parameters for the Berthelot Equation of State

$$\left(P + \frac{a}{T\bar{V}^2}\right)(\bar{V} - b) = RT$$

Substance	$a/\text{Pa m}^6 \text{ mol}^{-2} \text{ K}$	$b \times 10^5/\text{m}^3 \text{ mol}^{-1}$
Ammonia	171	3.70
Argon	20.5	3.20
Carbon dioxide	111	4.28
Helium	0.019	2.41
Methane	43.6	4.27
Neon	0.98	1.77
Nitrogen	17.3	3.87
Oxygen	21.3	3.18
Xenon	121	5.13
Water	357	3.04

(*continues*)

Table A1. (*Continued*)
Parameters for the Dieterici Equation of State

$$Pe^{a/\bar{V}RT}(\bar{V} - b) = RT$$

Substance	a/Pa m^6 mol^{-2}	$b \times 10^5$/m^3 mol^{-1}
Ammonia	0.540	4.00
Argon	0.174	3.47
Carbon dioxide	0.468	4.63
Helium	0.0046	2.60
Hydrogen	0.031	2.83
Methane	0.293	4.62
Neon	0.028	1.91
Nitrogen	0.176	4.19
Oxygen	0.177	3.45
Xenon	0.536	5.56
Water	0.709	3.29

Parameters for the Redlich-Kwong Equation of State

$$P = RT/(\bar{V} - b) - a/[T^{1/2}\bar{V}(\bar{V} + b)]$$

Substance	a/Pa m^6 mol^{-2} K$^{1/2}$	$b \times 10^5$/m^3 mol^{-1}
Ammonia	8.59	2.56
Argon	1.69	2.22
Carbon dioxide	6.44	2.96
Helium	0.00835	1.67
Hydrogen	0.14195	1.813
Methane	3.20	2.96
Neon	0.149	1.22
Nitrogen	1.56	2.68
Oxygen	1.74	2.21
Xenon	7.20	3.56
Water	14.24	2.11

From R. C. Weast (ed.), *Handbook of Chemistry and Physics*, 69th ed., CRC Press, Boca Raton, FL, 1988–1989, p. D188.

Table A2. Second Virial Coefficients
(Chapter 1, page 15)

	$B_2 \times 10^5$/m^3 mol^{-1}					
	Temperature/°C					
Substance	−100	−50	0	50	100	150
Argon	−6.43	−3.74	−2.15	−1.12	−0.42	0.11
Carbon dioxide			−15.4	−10.3	−7.3	−5.1
Helium	1.17	1.19	1.18	1.16	1.14	1.10
Nitrogen	−5.19	−2.64	−1.04	−0.04	0.63	1.19
Water			−45.0	−28.4		
Xenon			−8.12			

Data from D. P. Shoemaker, C. W. Garland, and J. I. Steinfeld, *Experiments in Physical Chemistry*, 4th ed., McGraw-Hill, New York, 1981, p. 64, and J. O. Hirschfelder, C. F. Curtiss, and R. B. Bird, *Molecular Theory of Gases and Liquids*, Wiley, New York, 1954, pp. 167, 227.

Table A3. Isothermal Compressibilities of Liquids
(Chapter 1, page 19)

Substance	Pressure/atm	$\kappa_T \times 10^{10}/Pa^{-1}$ Temperature/°C			
		25	**45**	**65**	**85**
Aniline	1	4.67	5.22	5.84	6.56
	1000	3.23	3.48	3.76	4.04
Benzene	1	9.67	11.32	13.39	
	1000		5.07	5.50	5.98
Bromobenzene	1	6.68	7.52	8.50	9.65
	1000	4.09	4.39	4.72	5.06
Carbon tetrachloride	1	10.67	12.54	14.87	
	1000	5.30	5.75	6.22	
Chlorobenzene	1	7.51	8.55	9.76	11.23
	1000	4.39	4.73	5.10	5.49
Nitrobenzene	1	5.03	5.59	6.24	6.99
	1000	3.39	3.64	3.91	4.20
Water	1	4.57	4.41	4.48	4.65
	1000	3.48	3.40	3.42	3.53

From R. C. Weast (ed.), *Handbook of Chemistry and Physics*, 69th ed., CRC Press, Boca Raton, FL, 1988–1989, pp. F12–15.

Table A4. Coefficients of Thermal Expansion at 20°C
(Chapter 1, page 19)

Substance	$\alpha \times 10^3/K^{-1}$
Benzene	1.237
Carbon disulfide	1.218
Carbon tetrachloride	1.236
Chloroform	1.273
Phenol	1.090
Sulfuric acid	0.558
Water	0.207

From C. D. Hodgman (ed.), *Handbook of Chemistry and Physics*, 33rd ed., Chemical Rubber Publishing Co., Cleveland, 1951, p. 1855.

Table A5. Values of the Constant C and the Function $L(T)$ for the Tait Equation
(Chapter 1, page 20)

| | | $L(T)/\text{kbar}$ | | | |
| | | Temperature/°C | | | |
Substance	C	25	45	65	85
Aniline	0.2159	2.007	1.798	1.606	1.429
Benzene	0.2159	0.970	0.829	0.701	
Bromobenzene	0.2159	1.404	1.247	1.103	0.972
Carbon tetrachloride	0.2129	0.867	0.737	0.622	
Chlorobenzene	0.2159	1.249	1.098	0.961	0.835
Ethylene glycol	0.2176	2.544	2.363	2.186	2.011
Nitrobenzene	0.2159	1.865	1.679	1.504	1.342
Water	0.3150	2.996	3.081	3.052	2.939

From R. E. Gibson and O. H. Loeffler, *J. Am. Chem. Soc.* **61**, 2575 (1939); 63, 898 (1941).

Table A6. Critical Constants
(Chapter 1, page 23)

Substance	T_c/K	P_c/bar	$\bar{V}_c \times 10^6/\text{m}^3\ \text{mol}^{-1}$	$P_c\bar{V}_c/RT$
Ammonia	405.6	114.0	70.4	0.238
Argon	151	49	75.2	0.291
Carbon dioxide	304	73.9	98.2	0.287
Helium	5.3	2.29	57.8	0.300
Methane	190.7	46.4	99.0	0.290
Neon	44.5	26.2	41.7	0.296
Nitrogen	126.1	33.9	90.1	0.292
Oxygen	154.4	50.4	74.4	0.292
Xenon	289.81	58.66	120.2	0.293
Water	647.2	221.2	54.5	0.224

From J. O. Hirschfelder, C. F. Curtiss, and R. B. Bird, *Molecular Theory of Gases and Liquids*, Wiley, New York, 1954, p. 245, and R. C. Weast (ed.), *Handbook of Chemistry and Physics*, 69th ed., CRC Press, Boca Raton, FL, 1988–1989, pp. F66–F67.

Table A7. Molar Heat Capacities
(Chapter 2, page 70)

Gases

$$\bar{C}_P = a + bT + cT^{-2}$$

(applicable between 298 K and 2000 K)

Substance	a/J K^{-1} mol^{-1}	$b \times 10^3$/J K^{-2} mol^{-1}	$c \times 10^{-5}$/J K^{-1} mol^{-1}
H_2	27.3	3.3	0.50
O_2	30.0	4.18	-1.67
N_2	28.6	3.8	-0.50
CO	28.4	4.1	-0.46
CO_2	44.2	8.79	-8.62
H_2O	30.5	10.3	0
CH_4	23.6	47.86	-1.8

Liquids

Substance	T/K	$\bar{C}_P$/J K^{-1} mol^{-1}
H_2O	273–373	75.48
CO	80	60.7
C_2H_6	100	68.6
Hg	234	28.4

Solids

Substance	T/K	$\bar{C}_P$/J K^{-1} mol^{-1}
Hg	234	28.4
Cu	1357	31.0
Zn	693	29.7
Sn	505	30.5

From K. S. Pitzer and L. Brewer, *Thermodynamics*, 2nd. ed., McGraw-Hill, New York, 1961, pp. 63, 66.

Table A8. Specific Enthalpy Changes of Fusion and Vaporization
(Chapter 2, page 53)

Substance	M.P./°C	ΔH_{fus}/J g^{-1}	B.P./°C	ΔH_{vap}/J g^{-1}
Ammonia	66.7		-33.35	1372
Benzene	5.5	127.	80.1	549.24
Carbon dioxide	-56.6^a	180.7	-78.5^b	526.6
Carbon monoxide	-199	29.8	-191.5	240.98
Ethane	-183.3	95.10	-88.6	520.32
Ethanol	-117.3	109.0	78.5	878.58
Methane	-182	58.41	-164	555.19
Water	0.0	333.5	100.0	2257.5

From R. C. Weast, *Handbook of Chemistry and Physics*, 69th ed., CRC Press, Boca Raton, FL, 1988–89, pp. B224ff, C666ff, C672ff.

[a] At 5.2 atm pressure.
[b] Sublimation temperature at 1.000 atm pressure.

Table A9. Values of Thermodynamic Functions
(Chapter 2, page 70, 79)

T	$\bar{C}_P^\circ$	$\bar{S}^\circ$	Quantities $-\dfrac{\bar{G}^\circ - \bar{H}_{298}^\circ}{T}$	$\bar{H}^\circ - \bar{H}_{298}^\circ$	$\Delta\bar{H}_f^\circ$	$\Delta\bar{G}_f^\circ$
K	$J\ K^{-1}\ mol^{-1}$	$J\ K^{-1}\ mol^{-1}$	$J\ K^{-1}\ mol^{-1}$	$kJ\ mol^{-1}$	$kJ\ mol^{-1}$	$kJ\ mol^{-1}$
			Ag(*cr*)			
0				−5.745		
298	25.351	42.55		0	0	0
			Ag$^+$(*ao*)			
298	21.8	72.78			105.579	77.107
			AgCl(*cr*)			
298	50.79	96.2			−127.068	−109.78
			Ar(*g*)			
0	0	0		−6.197	0	0
298	20.786	154.845	154.845	0	0	0
500	20.786	165.591	157.200	4.196	0	0
1000	20.786	179.999	165.410	14.589	0	0
2000	20.786	194.407	176.719	35.375		
			Br$^-$(*ao*)			
298	−141.8	82.4			−121.55	−103.96
			Br$_2$(*g*)			
0	0	0		−9.722	45.697	45.697
298	36.048	245.394	245.394	0	30.910	3.126
500	37.077	264.329	249.526	7.402	0	0
1000	37.787	290.293	264.143	26.150	0	0
2000	38.945	316.785	283.569	64.432	0	0
			Br$_2$(*l*)			
298	75.674	152.206	152.206	0	0	0
			C(*cr*, graphite)			
0	0	0		−1.051	0	0
298	8.517	5.740	5.740	0	0	0
500	14.623	11.662	6.932	2.365	0	0
1000	21.610	24.457	12.662	11.795	0	0
2000	25.094	40.771	23.008	35.525	0	0
			C(*cr*, diamond)			
0				−0.523	2.423	
298	6.113	2.377		0	1.895	2.900

Table A9. (*Continued*)

T	$\bar{C}_P^\circ$	$\bar{S}^\circ$	Quantities $-\dfrac{\bar{G}^\circ - \bar{H}_{298}^\circ}{T}$	$\bar{H}^\circ - \bar{H}_{298}^\circ$	$\Delta \bar{H}_f^\circ$	$\Delta \bar{G}_f^\circ$
K	J K^{-1} mol^{-1}	J K^{-1} mol^{-1}	J K^{-1} mol^{-1}	kJ mol^{-1}	kJ mol^{-1}	kJ mol^{-1}
			C(g)			
0	0	0		−6.536	711.185	711.185
298	20.838	158.100	158.100	0	716.170	671.244
500	20.804	168.863	160.459	4.202	718.507	639.906
1000	20.791	183.278	168.678	14.600	719.475	560.654
2000	20.952	197.713	179.996	35.433	716.577	402.694
			CH$_4$(g)			
0	0	0		−10.024	−66.911	−66.911
298	35.639	186.251	186.251	0	−74.873	−50.768
500	46.342	207.014	190.614	8.200	−80.802	−32.741
1000	71.795	247.549	209.370	38.179	−89.849	19.492
2000	94.399	305.853	244.057	123.592	−92.709	130.802
			CO(g)			
0	0	0		−8.671	−113.805	−113.805
298	29.142	197.653	0	0	−110.527	−137.163
500	39.794	212.831	200.968	5.931	−110.003	−155.414
1000	33.183	234.538	212.848	21.690	−111.983	−200.275
2000	36.250	258.714	230.342	56.744	−118.896	−286.034
			CO$_2$(g)			
0	0	0		−9.364	−393.151	−393.151
298	37.129	213.795	213.795	0	−393.522	−394.389
500	44.627	234.901	281.290	8.054	−393.676	−394.939
1000	54.308	269.299	235.901	33.397	−394.623	−395.886
2000	60.350	309.293	263.574	91.439	−396.784	−396.333
			CO$_2$(ao)			
298		117.6			−413.80	−385.98
			CH$_3$OH(g)			
298	43.89	289.81		0	−201.17	−162.46
			C$_2$H$_2$(g)			
0	0	0		−10.012	235.755	235.755
298	44.095	200.958	200.958	0	226.731	248.163
500	54.869	226.610	206.393	10.108	220.345	264.439
1000	68.275	269.192	227.984	41.208	202.989	315.144
2000	81.605	321.335	262.733	117.203	166.980	441.068

(*continues*)

Table A9. (*Continued*)

T	$\bar{C}_P^\circ$	$\bar{S}^\circ$	Quantities $-\dfrac{\bar{G}^\circ - \bar{H}_{298}^\circ}{T}$	$\bar{H}^\circ - \bar{H}_{298}^\circ$	$\Delta\bar{H}_f^\circ$	$\Delta\bar{G}_f^\circ$
K	J K^{-1} mol^{-1}	J K^{-1} mol^{-1}	J K^{-1} mol^{-1}	kJ mol^{-1}	kJ mol^{-1}	kJ mol^{-1}
			$C_2H_4(g)$			
0	0	0		-10.518	60.986	60.986
298	42.886	219.330	219.330	0	52.467	68.421
500	62.477	246.215	224.879	10.668	46.641	80.933
1000	93.899	300.408	249.742	50.665	38.183	119.122
2000	118.386	374.791	295.101	159.381	34.894	202.070
			$C_2H_6(g)$			
0				-11.950	-69.132	
298	52.63	229.60		0	-84.68	-32.82
			Ethanol, $C_2H_5OH(l)$			
298	111.46	160.7			-277.69	-174.78
			Ethanol, $C_2H_5OH(ao)$			
298		148.5			-288.3	-181.64
			Acetic acid, $CH_3CO_2H(l)$			
298	124.3	282.5			-484.5	-389.9
			Acetic acid, $CH_3CO_2H(ai)$			
298	-6.3	86.6			-486.01	-369.31
			Acetic acid, $CH_3CO_2H(ao)$			
298		178.7			-485.76	-396.46
			Acetate ion, $CH_3CO_2^-(ao)$			
298	-6.3	86.6			-486.01	-369.31
			$C_3H_8(g)$			
298	73.51	270.02			-103.85	-23.27
			n-Butane, $C_4H_{10}(g)$			
298	97.45	310.23		0	-201.17	-162.46
			Benzene, $C_6H_6(g)$			
298	81.67	269.31		0	82.93	129.73
			Benzoic acid, $C_6H_5COOH(cr)$			
298	146.8	167.6		0	-385.1	-245.3
			Octane, $C_8H_{18}(l)$			
298		361.1			-249.9	6.4
			Isooctane, $C_8H_{18}(l)$			
298					-255.1	

Table A9. (*Continued*)

T	$\bar{C}_P^\circ$	$\bar{S}^\circ$	Quantities $-\dfrac{\bar{G}^\circ - \bar{H}_{298}^\circ}{T}$	$\bar{H}^\circ - \bar{H}_{298}^\circ$	$\Delta \bar{H}_f^\circ$	$\Delta \bar{G}_f^\circ$
K	**J K^{-1} mol^{-1}**	**J K^{-1} mol^{-1}**	**J K^{-1} mol^{-1}**	**kJ mol^{-1}**	**kJ mol^{-1}**	**kJ mol^{-1}**
			Ca(*cr*)			
298	25.951	43.070	43.070	0	1.056	0.61
			Ca^{2+}(*ao*)			
298		−53.1			−542.83	−553.58
			CaCl$_2$(*cr*)			
298	72.59	104.6			−795.8	−748.1
			Calcite, CaCO$_3$(*cr*)			
298	81.88	92.9			−1206.92	−1128.79
			Aragonite, CaCO$_3$(*cr*)			
298	81.25	88.7			−1207.13	−1127.75
			CaO(*cr*)			
298	42.120	38.212	38.212	0	−635.089	−603.501
			Cl(*g*)			
0	0	0		−6.272	119.621	119.621
298	21.838	165.189	165.189	0	121.302	105.306
500	22.744	176.752	167.708	4.522	122.272	94.203
1000	22.233	192.430	176.615	15.815	124.334	65.288
2000	21.341	207.505	188.749	37.512	127.058	5.081
			Cl$^-$(*g*)			
298	22.958	167.5567	167.556	0	1378.801	1355.845
500	23.706	179.700	170.205	4.748	1384.192	1338.
			Cl$^-$(*ao*)			
298	−136.4	56.5			−167.159	−131.228
			Cl$_2$(*g*)			
0	0	0		−9.181	0	0
298	33.949	223.079	223.079	0	0	0
500	36.064	241.228	227.020	7.104	0	0
1000	37.438	266.767	241.203	25.565	0	0
2000	38.428	293.033	261.277	63.512	0	0
			CuO(*cr*)			
298	42.246	42.594	42.594	0	−156.063	−128.292

(*continues*)

Table A9. (Continued)

T	$\bar{C}_P^\circ$	$\bar{S}^\circ$	$-\dfrac{\bar{G}^\circ - \bar{H}_{298}^\circ}{T}$	$\bar{H}^\circ - \bar{H}_{298}^\circ$	$\Delta \bar{H}_f^\circ$	$\Delta \bar{G}_f^\circ$
K	J K^{-1} mol^{-1}	J K^{-1} mol^{-1}	J K^{-1} mol^{-1}	kJ mol^{-1}	kJ mol^{-1}	kJ mol^{-1}
			$Cu_2O(cr)$			
298	35.693	234.617	234.617	0	306.269	276.788
			$F(g)$			
0	0	0		−6.518	77.284	77.284
298	22.746	158.750	158.750	0	79.390	62.289
500	22.100	170.363	161.307	4.528	80.597	50.350
1000	21.266	185.362	170.038	15.324	82.403	19.317
2000	20.925	199.963	181.778	36.369	84.387	−44.635
			$F_2(g)$			
0	0	0		−8.825	0	0
298	31.302	202.789	202.789	0	0	0
500	34.255	219.738	206.452	6.743	0	0
1000	37.057	244.552	219.930	24.622	0	0
2000	38.846	270.904	239.531	62.745	0	0
			$Fe(cr)$			
0				−4.489		
298	25.10	27.28		0	0	0
			$Fe^{2+}(ao)$			
298		−137.7			−89.1	−78.90
			$Fe^{3+}(ao)$			
298		−315.9			−48.5	−4.7
			$H_2(g)$			
0	0	0		−8.467	0	0
298	28.836	130.680	130.680	0	0	0
500	29.260	145.737	133.973	5.882	0	0
1000	30.205	166.216	145.536	20.680	0	0
2000	34.280	188.418	161.943	52.951	0	0
			$H^+(ao)$			
298	0	0			0	0
			$H_2CO_3(ao)$			
298		187.4			−699.65	−623.08
			$HF(g)$			
298	29.138	173.780	173.780	0	−272.546	−274.646

Table A9. (*Continued*)

T	$\bar{C}_P^\circ$	$\bar{S}^\circ$	Quantities $-\dfrac{\bar{G}^\circ - \bar{H}_{298}^\circ}{T}$	$\bar{H}^\circ - \bar{H}_{298}^\circ$	$\Delta \bar{H}_f^\circ$	$\Delta \bar{G}_f^\circ$
K	$\mathrm{J\ K^{-1}\ mol^{-1}}$	$\mathrm{J\ K^{-1}\ mol^{-1}}$	$\mathrm{J\ K^{-1}\ mol^{-1}}$	$\mathrm{kJ\ mol^{-1}}$	$\mathrm{kJ\ mol^{-1}}$	$\mathrm{kJ\ mol^{-1}}$
			$\mathrm{HCl}(g)$			
0	0	0		-8.640	-92.127	-92.127
298	29.136	186.901	186.901	0	-92.312	-95.300
500	29.304	201.898	190.205	5.892	-92.912	-97.166
1000	31.628	222.903	201.857	21.046	-94.388	-100.799
2000	35.600	246.246	218.769	54.953	-95.590	-106.631
			$\mathrm{HI}(g)$			
0	0	0		-8.656	28.535	28.535
298	29.156	206.589	206.589	0	26.359	1.560
500	29.736	221.760	209.905	5.928	-5.622	-10.088
1000	33.135	243.404	221.763	21.641	-6.754	-14.006
2000	36.623	267.680	239.248	56.863	-7.589	-21.009
			$\mathrm{HNO_3}(g)$			
0				-11.778	-125.27	
298	53.35	266.38		0	-135.06	-74.72
			$\mathrm{HNO_3}(ai)$			
298	-86.6	146.4		0	-207.36	-111.25
			$\mathrm{H_2O}(g)$			
0	0	0		-9.904	-238.921	-238.921
298	33.590	188.834	188.834	0	-241.826	-228.582
500	35.226	206.534	192.685	6.925	-243.826	-219.051
1000	41.268	232.738	206.738	26.000	-247.857	-192.590
2000	51.180	264.769	228.374	72.790	-251.575	-135.528
			$\mathrm{H_2O}(l)$			
298	75.351	69.950	69.950	0	-285.830	-237.141
500	83.694	109.898	78.579	15.659	-279.095	-206.002
			$\mathrm{H_2S}(g)$			
298	34.192	205.757	205.757	0	-20.502	-33.329
500	37.192	224.102	209.726	7.188	-27.762	-40.179
1000	45.786	252.579	224.599	27.980	-90.024	-20.984
			$\mathrm{H_2SO_4}(l)$			
298	83.761	298.796	298.796	0	-735.129	-653.366
			$\mathrm{H_2SO_4}(ai)$			
298	-293	20.1		0	-909.27	-744.53

(*continues*)

Table A9. (*Continued*)

T	$\bar{C}_P^\circ$	$\bar{S}^\circ$	Quantities $-\dfrac{\bar{G}^\circ - \bar{H}_{298}^\circ}{T}$	$\bar{H}^\circ - \bar{H}_{298}^\circ$	$\Delta \bar{H}_f^\circ$	$\Delta \bar{G}_f^\circ$
K	$J\,K^{-1}\,mol^{-1}$	$J\,K^{-1}\,mol^{-1}$	$J\,K^{-1}\,mol^{-1}$	$kJ\,mol^{-1}$	$kJ\,mol^{-1}$	$kJ\,mol^{-1}$
			He(g)			
0	0	0		−6.197	0	0
298	20.786	126.152	126.152	0	0	0
500	20.786	136.899	128.507	4.196	0	0
1000	20.786	151.306	136.718	14.589	0	0
2000	20.786	165.714	148.027	35.375	0	0
			Hg(l)			
298	27.978	76.028	0	0	0	0
			Hg(g)			
298	20.786	174.970	174.970	0	61.380	31.880
			HgO(cr, red orthorhombic)			
298	44.06	70.29			−90.83	−58.539
			I$_2$(cr)			
298	54.436	116.142	116.142	0	0	0
			I$_2$(g)			
298	36.887	260.685	260.685	0	62.421	19.325
			I(g)			
0				−6.197	107.240	
298	20.786	180.791		0	106.838	70.250
			I$^-$(ao)			
298	−142.3	111.3			−55.19	−51.57
			K$^+$(ao)			
298	21.8	102.5			−252.38	−283.27
			KOH(cr)			
298	64.9	78.9		12.150	−424.764	−379.08
			Li$^+$(ao)			
298	68.6	13.4			−278.49	−293.31
			Mg(cr)			
0				−5.000		
298	24.89	32.68		0	0	0
			MgO(cr, macrocrystal)			
0				−5.167	−597.530	
298	37.15	26.74		0	−601.70	−569.43

Table A9. (*Continued*)

T	$\bar{C}_P^\circ$	$\bar{S}^\circ$	Quantities $-\dfrac{\bar{G}^\circ - \bar{H}_{298}^\circ}{T}$	$\bar{H}^\circ - \bar{H}_{298}^\circ$	$\Delta \bar{H}_f^\circ$	$\Delta \bar{G}_f^\circ$
K	J K^{-1} mol^{-1}	J K^{-1} mol^{-1}	J K^{-1} mol^{-1}	kJ mol^{-1}	kJ mol^{-1}	kJ mol^{-1}
			$N_2(g)$			
0	0	0		−8.670	0	0
298	29.124	191.609	191.609	0	0	0
500	29.580	206.739	194.917	5.911	0	0
1000	32.697	228.170	206.708	21.463	0	0
2000	35.971	252.074	224.006	56.137	0	0
			$NH_3(g)$			
0	0	0		−10.045	−38.907	−38.907
298	35.652	192.774	192.774	0	−45.898	−16.367
500	42.048	212.659	197.021	7.819	−49.857	4.800
1000	56.491	246.486	213.849	32.637	−55.013	61.910
2000	72.833	291.525	242.244	98.561	−54.833	179.447
			$NH_3(ao)$			
298		111.3			−80.29	−26.50
			$NH_4^+(ao)$			
298	79.9	113.4			−132.51	−79.31
			$NO(g)$			
298	29.845	210.758	210.758	0	90.291	86.600
			$NO_2(g)$			
298	36.874	240.034	240.034	0	33.095	51.258
			$N_2O_4(g)$			
298	77.256	304.376	304.376	0	9.179	97.787
			$N_2O(g)$			
298	38.617	219.957	219.957	0	82.048	104.179
			$NO_3^-(ao)$			
298	−86.6	146.4			−205.0	−108.74
			$N_2O_5(g)$			
298	96.303	346.548	346.548	0	11.297	118.013
			$NOCl(g)$			
0				−11.364	53.60	
298	44.69	261.69		0	51.71	66.08
			$Na^+(ao)$			
298	46.4	59.0			−240.12	−261.905

(continues)

Table A9. (*Continued*)

T	$\bar{C}_P^\circ$	$\bar{S}^\circ$	Quantities $-\dfrac{\bar{G}^\circ - \bar{H}_{298}^\circ}{T}$	$\bar{H}^\circ - \bar{H}_{298}^\circ$	$\Delta \bar{H}_f^\circ$	$\Delta \bar{G}_f^\circ$
K	$\text{J K}^{-1}\,\text{mol}^{-1}$	$\text{J K}^{-1}\,\text{mol}^{-1}$	$\text{J K}^{-1}\,\text{mol}^{-1}$	kJ mol^{-1}	kJ mol^{-1}	kJ mol^{-1}
			NaOH(cr)			
0	0	0		−10.487	−421.396	−421.396
298	59.530	64.445	64.445	0	−425.931	−379.741
500	75.157	98.172	71.595	13.288	−427.401	−347.767
			Ne(g)			
0	0	0		−6.197	0	0
298	20.786	146.327	146.327	0	0	0
500	20.786	157.074	148.683	4.196	0	0
1000	20.786	171.482	156.893	14.589	0	0
2000	20.786	185.889	168.202	35.375	0	0
			O(g)			
0	0	0		−6.725	246.790	246.790
298	21.911	161.058	161.058	0	249.173	231.736
500	21.257	172.197	163.511	4.343	250.474	219.549
1000	20.915	186.790	171.930	14.860	252.682	187.681
2000	20.826	201.247	183.391	35.713	255.299	121.552
			O$_2$(g)			
0	0	0		−8.683	0	0
298	29.376	205.147	205.147	0	0	0
500	31.091	220.693	208.524	6.084	0	0
1000	34.870	243.578	220.875	22.703	0	0
2000	37.741	268.748	239.160	59.175	0	0
			O$_3$(g)			
0	0	0		−10.351	145.348	145.348
298	39.238	238.932	238.932	0	142.674	163.184
500	47.262	261.272	243.688	8.792	142.340	177.224
1000	55.024	297.048	262.228	34.819	143.439	211.759
2000	58.250	336.469	290.533	91.873	145.784	279.089
			OH$^-$(ao)			
298	−148.5	−10.75			−229.994	−157.244
			PCl$_3$(g)			
0				−15.932		
298	71.581	311.682	311.682	0	−288.696	−269.610
			PCl$_5$(g)			
0				−22.852		
298	111.890	364.288	364.288	0	−360.184	−290.271

Table A9. (*Continued*)

T	$\bar{C}_P^\circ$	$\bar{S}^\circ$	Quantities $-\dfrac{\bar{G}^\circ - \bar{H}_{298}^\circ}{T}$	$\bar{H}^\circ - \bar{H}_{298}^\circ$	$\Delta \bar{H}_f^\circ$	$\Delta \bar{G}_f^\circ$
K	J K^{-1} mol^{-1}	J K^{-1} mol^{-1}	J K^{-1} mol^{-1}	kJ mol^{-1}	kJ mol^{-1}	kJ mol^{-1}
			Pb(*cr*)			
0				−6.878		
298	26.44	64.81		0	0	0
			PbO$_2$(*cr*)			
298	64.64	68.6			−277.4	−217.33
			PbSO$_4$(*cr*)			
0				−20.062		
298	103.207	148.57		0	−919.94	(−813.14)
			Pt(*cr*)			
0				−5.740		
298	25.86	41.63		0	0	0
			S(*cr*, rhombic)			
0				−4.410		
298	22.64	31.80		0	0	0
			S(*cr*, monoclinic)			
298					0.33	
			SO$_2$(*g*)			
0	0	0		−10.552	−294.299	−294.299
298	39.878	248.212	248.212	0	−296.842	−300.125
500	46.576	270.495	252.979	8.758	−302.736	−300.871
1000	54.484	305.767	271.339	34.428	−361.940	−288.725
2000	58.229	345.007	299.383	91.250	−360.981	−215.929
			SO$_3$(*g*)			
0	0	0	−11.697	−390.025	−390.025	
298	50.661	256.769	256.769	0	−395.765	−371.016
500	63.100	286.152	262.992	11.580	−401.878	−352.668
1000	75.968	334.828	287.768	47.060	−459.581	−293.639
2000	81.140	389.616	326.421	126.390	−454.351	−129.768
			SO$_4^{2-}$(*ao*)			
298	−293.		20.1		−909.27	−744.53
			Sn^{2+}(*ao*) (μ(NaClO$_4$) = 3.0)			
298		−17			−8.8	−27.2

(*continues*)

Table A9. (*Continued*)

T	$\bar{C}_P^\circ$	$\bar{S}^\circ$	Quantities $-\dfrac{\bar{G}^\circ - \bar{H}_{298}^\circ}{T}$	$\bar{H}^\circ - \bar{H}_{298}^\circ$	$\Delta\bar{H}_f^\circ$	$\Delta\bar{G}_f^\circ$
K	$J\ K^{-1}\ mol^{-1}$	$J\ K^{-1}\ mol^{-1}$	$J\ K^{-1}\ mol^{-1}$	$kJ\ mol^{-1}$	$kJ\ mol^{-1}$	$kJ\ mol^{-1}$
			$Sn^{4+}(ao)$ (in $HCl + \infty\ H_2O$)			
298		−117			30.5	2.5
			$Zn(cr)$			
0				−5.669		
298	25.287	41.717		0	0	0
			$Zn^{2+}(ao)$			
298	46.	−112.1			−153.89	−147.06
			$ZnO(cr)$			
298	40.25	43.64			−348.28	−318.30
			Sphalerite, $ZnS(cr)$			
298	46.0	57.7			−205.98	−201.29
			Wurtzite, $ZnS(cr)$			
298					−192.63	

From M. W. Chase, Jr., C. A. Davies, J. R. Downey, Jr., D. J. Frurip, R. A. McDonald, and A. N. Syverud, *JANAF Thermochemical Tables*, 3rd ed., *J. Phys. Chem. Ref. Data*, Vol. 14, 1985, Supplement No. 1, published by the American Chemical Society and the American Institute of Physics for the National Bureau of Standards; D. D. Wagman, W. H. Evans, V. B. Parker, R. H. Schymm, I. Halow, S. M. Bailer, K. L. Churney, and R. L. Nuttall, *The NBS Tables of Chemical Thermodynamic Properties—Selected Values for Inorganic and C_1 and C_2 Organic Substances in SI Units, J. Phys. Chem. Ref. Data.*, Vol. 11, 1982, Supplement No. 2, published by the American Chemical Society and the American Institute of Physics for the National Bureau of Standards.

    In the case of small discrepancies between Chase et al. and Wagman et al., the values from Chase et al. have been taken. Values for organic substances with more than two carbons are from D. R. Stull, E. F. Westrum, and G. C. Sinke, *The Chemical Thermodynamics of Organic Compounds*, Wiley, New York, 1969.

    Abbreviations: *cr* = crystal; *l* = liquid; *g* = gas; *ao* = aqueous, molality standard state without (further) ionization; *ai* = aqueous, molality standard state with ionization.

Table A10. Average Bond Energies
(Chapter 2, page 89)

Bond	Bond Energy/$kJ\ mol^{-1}$
Br—Br	193
C—C	343
C=C	615
C≡C	812
C—Cl	326
C—F	490
C—H	416
C—N	290
C≡N	891
C—O	351
C=O	724

(continues)

Table A10. (*Continued*)

Bond	Bond Energy/kJ mol^{-1}
C$=$O (in CO$_2$)	799
Cl—Cl	326
F—F	158
H—Br	366
H—Cl	432
H—F	568
H—H	436
H—I	298
H—N	391
H—S	367
N—N	160
N$\equiv$N	946
N—O	176
O—H	464
O—O	144
O$=$O (in O$_2$)	498

From F. T. Wall, *Chemical Thermodynamics*, 3rd ed., W. H. Freeman, San Francisco, 1974, p. 63.

Table A11. Surface Tension Values
(Chapter 5, page 191)

Substance	In Contact With	Temperature/°C	Value/N m^{-1}
Acetone	air	0	0.02621
	air	20	0.02370
	air	40	0.02116
Argon	vapor	−188	0.0132
Benzene	air	20	0.02885
	vapor	20	0.02889
Carbon dioxide	vapor	−25	0.00913
	vapor	20	0.00116
Carbon tetrachloride	vapor	20	0.02695
Ethanol	vapor	20	0.02275
Glycerol	air	20	0.0634
Gold	H$_2$, Ar	1120	1.128
Mercury	air	20	0.4355
Water	air	20	0.07275
	air	25	0.07197
	air	30	0.07118

From R. C. Weast, *Handbook of Chemistry and Physics*, 69th ed., CRC Press, Boca Raton, FL, 1988–89, pp. F34ff.

Table A12. Mean Ionic Activity Coefficients of Aqueous Electrolytes at 25°C (Chapter 6, page 237)

Molality	0.1	0.2	0.5	1.0	1.6	2.5
			1–1 Electrolytes			
Debye-Hückel	0.754	0.696	0.615	0.556	0.519	0.487
Davies	0.781	0.747	0.733	0.791	0.912	1.175
HCl	0.796	0.767	0.757	0.809	0.916	1.147
NaOH	0.764	0.725	0.688	0.677	0.690	0.741
NaCl	0.778	0.735	0.681	0.657	0.655	0.688
KCl	0.770	0.718	0.649	0.604	0.580	0.569
Sodium acetate	0.791	0.757	0.735	0.757	0.809	0.914
NH_4NO_3	0.740	0.677	0.582	0.504	0.447	0.391
			1–2 and 2–1 Electrolytes			
Debye-Hückel	0.436	0.359	0.275	0.226	0.199	0.179
Davies	0.538	0.547	0.790	1.868	5.865	35.27
$MgCl_2$	0.528	0.488	0.480	0.569	0.802	1.538
Na_2SO_4	0.452	0.371	0.270	0.204	0.168	0.144
K_2CrO_4	0.466	0.356	0.298	0.240	0.212	0.194
			2–2 Electrolytes			
Debye-Hückel	0.162	0.109	0.064	0.044	0.035	0.026
Davies	0.284	0.336	1.069	12.24	284.5	—
$CuSO_4$	0.150	0.104	0.0620	0.0432	—	—
$ZnSO_4$	0.150	0.104	0.0630	0.0435	0.0363	0.0367
$CdSO_4$	0.150	0.103	0.0615	0.0415	0.0338	0.0317

From R. A. Robinson and R. H. Stokes, *Electrolyte Solutions*, 2nd ed., Butterworths, London, 1968.

Table A13. Partial Vapor Pressure of Hydrogen Halides in Aqueous Solution at 298.15 K (Chapter 7, page 273)

	$P/$torr		
$m/$mol kg^{-1}	HCl	HBr	HI
4.0	0.0182		
5.0	0.0530		
6.0	0.140	0.00151	0.00057
7.0	0.348	0.00370	0.00182
8.0	0.844	0.0089	0.0065
9.0	1.93	0.0226	0.0295
10.0	4.20	0.059	0.132
11.0		0.151	

From S. J. Bates and H. D. Kirschman, *J. Am. Chem. Soc.* **41**, 1991 (1919).

Table A14. Some Standard Half-Cell Potentials in Aqueous Solution at 25°C and $P° = 1$ bar
(Chapter 8, page 308)

Half-Cell Symbol	Half-Reaction	$E°/V$
$F^-|F_2|Pt$	$F_2(g) + 2e^- \rightarrow 2F^-$	$+2.87$
$H_2SO_4|PbSO_4|PbO_2$	$PbO_2 + SO_4^{2-} + 4H^+ + 2e^- \rightarrow PbSO_4 + 2H_2O$	$+1.685$
$Mn^{2+}, MnO_4^-|Pt$	$MnO_4^- + 8H^+ + 5e^- \rightarrow Mn^{2+} + 4H_2O$	$+1.491$
$Ce^{3+}, Ce^{4+}|Pt$ (in 1 M HNO_3)	$Ce^{4+} + e^- \rightarrow Ce^{3+}$	$+1.61$
$Au^{3+}|Au$	$Au^{3+} + 3e^- \rightarrow Au$	$+1.42$
$Cl^-|Cl_2|Pt$	$Cl_2 + 2e^- \rightarrow 2Cl^-$	$+1.3583$
$Hg_2^{2+}|Hg(l)|Pt$	$Hg_2^{2+} + 2e^- \rightarrow 2Hg(l)$	$+0.7961$
$Br^-|Br_2|Pt$	$Br_2 + 2e^- \rightarrow 2Br^-$	$+1.065$
$Ag|Ag^+$	$Ag^+ + e^- \rightarrow Ag$	$+0.7986$
$H^+|O_2(g)|Pt$	$O_2(g) + 4H^+ + 4e^- \rightarrow 2H_2O$	$+1.229$
$Fe^{2+}|Fe^{3+}|Pt$	$Fe^{3+} + e^- \rightarrow Fe^{2+}$	$+0.770$
$MnO_2|MnO_4^-|Pt$	$MnO_4^- + 2H_2O + 3e^- \rightarrow MnO_2 + 4OH^-$	$+0.588$
$I^-|I_2(s)|Pt$	$I_2(s) + 2e^- \rightarrow 2I^-$	$+0.535$
$I^-|I_2(aq)|Pt$	$I_2(aq) + 2e^- \rightarrow 2I^-$	$+0.6197$
$I^-, I_3^-|Pt$	$I_3^- + 2e^- \rightarrow 3I^-$	$+0.5338$
$OH^-|O_2(g)|Pt$	$O_2 + 2H_2O + 4e^- \rightarrow 4OH^-$	$+0.401$
$Cu(s)|Cu^{2+}$	$Cu^{2+} + 2e^- \rightarrow Cu(s)$	$+0.3402$
$Cl^-|Hg_2Cl_2(s)|Hg(l)$	$Hg_2Cl_2(s) + 2e^- \rightarrow 2Hg(l) + 2Cl^-$	$+0.268$
$Cl^-|AgCl(s)|Ag(s)$	$AgCl(s) + e^- \rightarrow Ag(s) + Cl^-$	$+0.2223$
$Sn^{4+}, Sn^{2+}|Pt$	$Sn^{4+} + 2e^- \rightarrow Sn^{2+}$	$+0.15$
$Br^-|AgBr(s)|Ag(s)$	$AgBr(s) + e^- \rightarrow Ag(s) + Br^-$	$+0.0713$
$H^+|H_2|Pt$	$2H^+ + 2e^- \rightarrow H_2$	0.000
$Fe^{3+}|Fe(s)$	$Fe^{3+} + 3e^- \rightarrow Fe(s)$	-0.036
$Pb^{2+}|Pb(s)$	$Pb^{2+} + 2e^- \rightarrow Pb(s)$	-0.1263
$I^-|AgI(s)|Ag(s)$	$AgI(s) + e^- \rightarrow Ag(s) + I^-$	-0.1519
$Cd^{2+}|Cd(Hg)$	$Cd^{2+} + 2e^- \rightarrow Cd(Hg)$	-0.3521
$SO_4^{2-}|PbSO_4(s)|Pb(s)$	$PbSO_4^{2-} + 2e^- \rightarrow Pb(s) + SO_4^{2-}$	-0.356
$Cd^{2+}|Cd(s)$	$Cd^{2+} + 2e^- \rightarrow Cd(s)$	-0.4026
$Fe^{2+}|Fe(s)$	$Fe^{2+} + 2e^- \rightarrow Fe(s)$	-0.44
$S^{2-}|Ag_2S(s)|Ag(s)$	$Ag_2S(s) + 2e^- \rightarrow 2Ag(s) + S^{2-}$	-0.7051
$Zn^{2+}|Zn(s)$	$Zn^{2+} + 2e^- \rightarrow Zn(s)$	-0.7628
$OH^-|H_2(g)|Pt$	$2H_2O + 2e^- \rightarrow H_2 + 2OH^-$	-0.8277
$Mg^{2+}|Mg(s)$	$Mg^{2+} + 2e^- \rightarrow Mg(s)$	-2.375
$Na^+|Na(s)$	$Na + e^- \rightarrow Na(s)$	-2.7109
$Ca^{2+}|Ca(s)$	$Ca^{2+} + 2e^- \rightarrow Ca(s)$	-2.868
$K^+|K(s)$	$K^+ + e^- \rightarrow K(s)$	-2.925
$Li^+|Li(s)$	$Li^+ + e^- \rightarrow Li(s)$	-3.045

Table A15. Electronegativities on the Pauling Scale
(Chapter 13, page 536)

Element	Value	Element	Value
H	2.1	K	0.8
He	—	Ca	1.0
Li	1.0	Sc	1.3
Be	1.5	Ti	1.5
B	2.0	V	1.6
C	2.5	Cr	1.6
N	3.0	Mn	1.5
O	3.5	Fe	1.8
F	4.0	Co	1.8
Ne	—	Ni	1.8
Na	0.9	Cu	1.9
Mg	1.2	Zn	1.6
Al	1.5	Ga	1.6
Si	1.8	Ge	1.8
P	2.1	As	2.0
S	2.5	Se	2.4
Cl	3.0	Br	2.8
Ar	—	Kr	—

Table A16. Properties of Diatomic Molecules
(Chapter 14, page 578)

Molecule/Term	D_e/hc cm^{-1}	$\tilde{\nu}_e$ cm^{-1}	$\tilde{\nu}_e x_e$ cm^{-1}	$\tilde{B}_e$ cm^{-1}	R_e Å	$\tilde{\alpha}_e$ cm^{-1}	$\mathscr{D}_e$ 10^{-4} cm^{-1}
$^{79}Br_2/^1\Sigma_g^+$		325.32	1.077	0.0821	2.281	0.0003187	0.00021
$^{12}C^{16}O/^1\Sigma^+$	90544	2169.8	13.29	1.931	1.128	0.0175	0.0612
$^{35}Cl_2/^1\Sigma_g^+$		559.7	2.68	0.2440	1.988	0.0015	0.000186
$^1H_2/^1\Sigma_g^+$	38297	4401.2	121.3	60.85	0.741	3.06	0.00471
$^1H^{81}Br/^1\Sigma^+$		2648.98	45.22	8.4649	1.414	0.233	3.46
$^1H^{35}Cl/^1\Sigma^+$	37240	2990.9	52.8	10.593	1.275	0.307	5.32
$^1H^{19}F/^1\Sigma^+$		4138.3	89.9	20.956	0.9168	0.798	21.51
$^1H^{127}I/^1\Sigma^+$		2309.01	39.64	6.4264	1.609	0.1689	2.069
$^{127}I_2/^1\Sigma_g^+$	12550	214.5	0.615	0.03737	2.666	0.000114	0.000042
$^{14}N_2/^1\Sigma_g^+$	79890	2358.6	14.3	1.998	1.098	14.3	0.0576
$^{14}N^{16}O/^2\pi_r$		1904.20	14.075	1.67195	1.15077	0.0171	0.54
$^{16}O_2/^3\Sigma_g^-$		1580.2	11.98	1.4456	1.208	0.0159	0.0048
$^{19}F_2/^1\Sigma_g^+$		916.64	11.236	0.89019	1.41193	0.013847	0.033

Data primarily from K. P. Huber and G. Herzber, *Molecular Spectra and Molecular Structure*, Vol. IV, *Constants of Diatomic Molecules*, Van Nostrand Reinhold, New York, 1979.

Table A17. Properties of Some Nuclei in Their Ground States
(Chapter 14, page 579, 640)

Nuclide	I	g_N	Abundance/% (Earth's Crust)	Atomic Mass/amu
^1H	1/2	5.5856	99.985	1.007825
^2H	1	0.8574	0.015	2.014102
^7Li	3/2	2.171	92.44	7.01600
^{12}C	0	—	98.89	12.00000
^{13}C	1/2	1.4048	1.11	13.00335
^{14}N	1	0.403	99.63	14.003074
^{15}N	1/2	−0.567	0.37	15.00011
^{16}O	0	—	99.759	15.994915
^{17}O	5/2	−0.757	0.037	16.99913
^{19}F	1/2	5.257	100	18.99840
^{23}Na	3/2	1.478	100	22.98977
^{29}Si	1/2	−1.111	4.70	28.97650
^{31}P	1/2	2.263	100	30.97376
^{33}S	3/2	0.429	0.76	32.97146
^{35}Cl	3/2	0.548	75.77	34.96885
^{37}Cl	3/2	0.456	24.23	36.96590
^{63}Cu	3/2	1.484	69.17	62.92959
^{65}Cu	3/2	1.590	30.83	64.92779

Data on abundances and masses from Norman E. Holden and F. William Walker, *Chart of the Nuclides,* General Electric Co., Schenectady, NY, 1968.

Table A18. Characteristic Vibrational Frequencies
(Chapter 14, page 588)

Group	Frequency/cm^{-1}
C—H (stretch)	2850–3000
C—H (bend)	1350–1450
O—H (stretch)	3600–3700
C=C	1600–1680
C≡C	2200–2260
C=O	1660–1870
C—Cl	600–800
C≡N	ca. 2250

Table A19. Characteristic Proton Chemical Shifts Relative to Tetramethyl Silane (TMS)
(Chapter 15, page 645)

Group	δ/ppm
$-CO_2H$	11–13
$-CHO$	9–10
$Ar-H$	7–8
$-C\equiv CH$	5–7
ROH	3–5
$-CO_2CH_3$	4
$RCOCH_3$	3–4
$Ar-CH_3$	2–3
$-CH_2-$	1–5
$R-CH_3$	1–2
$Si(CH_3)_4$	0.00

R stands for an aliphatic radical, Ar for an aromatic radical.

Table A20. Parameters for the Lennard-Jones Potential
(Chapter 16, page 688)

Substance	$\sigma \times 10^{10}$/m	(ε/k_B)/K
Argon	3.40	120
Carbon dioxide	4.5	199
Helium	2.56	10.22
Krypton	3.60	190
Neon	2.75	33
Nitrogen	3.7	91
Oxygen	3.5	109
Xenon	4.1	222

From J. O. Hirschfelder, C. F. Curtiss, and R. Byron Bird, *Molecular Theory of Gases and Liquids*, Wiley, New York, 1954, pp. 1110ff.

Table A21. Effective Hard-Sphere Diameters
(Chapter 16, page 690)

Substance	Temperature/K	$d \times 10^{10}$/m
Ammonia	293	4.38
Argon	273	3.65
	293	3.61
	373	3.48
	473	3.37
Carbon monoxide	295	3.72
Carbon dioxide	293	4.53
Chlorine	293	5.39
Ethane	293	5.26
Ethylene	293	4.90
Helium	293	2.17
Hydrogen	294	2.72
Iodine	397	6.35
Methane	293	4.10
Nitrogen	301	3.71
Oxygen	292	3.58
Water	373	4.18
Xenon	293	4.81

Calculated from viscosity values in R. C. Weast (ed.), *Handbook of Chemistry and Physics*, 69th ed., CRC Press, Boca Raton, FL, 1988–1989, pp. F41ff.

Table A22. Thermal Conductivities
(Chapter 17, page 714)

Substance	Temperature/°C	κ/J s^{-1} m^{-1} K^{-1}
Gases		
Air, 1 atm	20	0.02353
Argon, 1 atm	20	0.01625
Carbon dioxide, 1 atm	20	0.01409
Helium, 1 atm	20	0.13954
Liquids		
Benzene	22.5	0.1582
Carbon tetrachloride	20	0.1033
Octane	30	0.1451
Water	27	0.6092
Solids		
Aluminum	25	237
Iron	25	80.4

From R. C. Weast (ed.), *Handbook of Chemistry and Physics*, 69th ed., CRC Press, Boca Raton, FL, 1988–1989, pp. E3ff.

Table A23. Diffusion Coefficients for Dilute Solutes in Liquid Solutions
(Chapter 17, page 714)

Solute	Solvent	Temperature/°C	$D \times 10^9/m^2\ s^{-1}$
Acetylene	Water	0	1.10
α-Alanine	Water	25	0.91
Carbon dioxide	n-Pentanol	25	1.91
Ethane	n-Heptane	30	5.60
Glucose	Water	25	0.673
Glycine	Water	25	1.064
Potassium chloride	Water	25	1.917
Sodium chloride	Water	25	1.545

From R. C. Weast (ed.), *Handbook of Chemistry and Physics*, 69th ed., CRC Press, Boca Raton, FL, 1988–1989, pp. F47–F48.

Table A24. Coefficients of Viscosity
(Chapter 17, page 720)

Substance	State	Temperature/°C	$\eta \times 10^3/kg\ m^{-1}\ s^{-1}$
Benzene	Liquid	20	0.652
Carbon tetrachloride	Liquid	20	0.969
Ethanol	Liquid	20	1.200
		40	0.834
Glycerol	Liquid	0	12110
		20	1490
Hexane	Liquid	0	0.401
		20	0.326
Mercury	Liquid	0	1.685
		20	1.554
Sulfuric acid	Liquid	20	25.4
Water	Liquid	0	1.787
		20	1.002
		100	0.2818
Argon	Gas	0	0.02096
		20	0.02217
Helium	Gas	20	0.01941
		100	0.02281
Nitrogen	Gas	27.4	0.01781
		127.2	0.02191
Oxygen	Gas	0	0.0189
		19.1	0.2018
Water	Gas	100	0.01255
		200	0.01635

From R. C. Weast (ed.), *Handbook of Chemistry and Physics*, 69th ed., CRC Press, Boca Raton, FL, 1988–1989, pp. F41ff.

Table A25. Self-Diffusion Coefficients at 1.000 atm
(Chapter 17, page 728)

Substance	Temperature/K	$D \times 10^4/\text{m}^2 \text{ s}^{-1}$
Argon	77.7	0.0134 ± 0.0002
	273.2	0.157 ± 0.0003
	353.2	0.249 ± 0.0003
Carbon dioxide	273.2	0.0970
Hydrogen	273	1.285 ± 0.002
Methane	273.2	0.206 ± 0.006
Nitrogen	353.2	0.287 ± 0.009
Oxygen	353.2	0.301 ± 0.004
Xenon	300.5	0.0576 ± 0.0009

Data from J. O. Hirschfelder, C. F. Curtiss, and R. Byron Bird, *Molecular Theory of Gases and Liquids*, Wiley, New York, 1954, p. 581.

Table A26. Molar Conductivities and Ion Mobilities at Infinite Dilution in Water at 298.15 K
(Chapter 17, page 741)

Ion	$\lambda_0 \times 10^3/\text{m}^2 \text{ ohm}^{-1} \text{ mol}^{-1}$	$u \times 10^8/\text{m}^2 \text{ s}^{-1} \text{ V}^{-1}$
H^+	34.982	36.25
K^+	7.352	7.619
Na^+	5.011	5.193
Li^+	3.869	4.010
NH_4^+	7.34	7.61
Ca^{2+}	11.90	6.166
OH^-	19.8	20.5
Cl^-	7.634	7.912
Br^-	7.84	8.13
I^-	7.685	7.96
NO_3^+	7.144	7.404
Acetate, $C_2H_3O_2^-$	4.09	4.24
ClO_4^-	6.80	7.05
SO_4^{2-}	15.94	8.27

From A. J. Bard and L. R. Faulkner, *Electrochemical Methods*, Wiley, New York, 1980, p. 67.

B Calculus with Several Variables

Differential Calculus   The **fundamental equation of differential calculus** for a function f with independent variables x, y, and z is

$$df = (\partial f/\partial x)_{y,z}\, dx + (\partial f/\partial y)_{x,z}\, dy + (\partial f/\partial z)_{x,y}\, dz \qquad \text{(B-1)}$$

where $(\partial f/\partial x)_{y,z}$, $(\partial f/\partial y)_{x,z}$, and $(\partial f/\partial z)_{x,y}$ are **partial derivatives**. A partial derivative with respect to one independent variable is obtained by the ordinary procedures of differentiation, treating all other independent variables as though they were constants.

An example of Equation (B-1) is

$$dP = (\partial P/\partial T)_{V,n}\, dT + (\partial P/\partial V)_{T,n}\, dV + (\partial P/\partial n)_{T,V}\, dn \qquad \text{(B-2)}$$

This equation represents the value of an infinitesimal change in pressure that is produced when we impose arbitrary infinitesimal changes dT, dV, and dn on the system.

An approximate version of Equation (B-2) can be written for finite increments in P, T, V, and n, equal respectively to ΔP, ΔT, ΔV, and Δn:

$$\Delta P \approx (\partial P/\partial T)_{V,n}\, \Delta T + (\partial P/\partial V)_{T,n}\, \Delta V + (\partial P/\partial n)_{T,V}\, \Delta n \qquad \text{(B-3)}$$

where $\approx$ means "is approximately equal to." Equation (B-3) will usually be more nearly correct if the finite increments ΔT, ΔV, and Δn are small and less nearly correct if the increments are large.

An Identity for a Change of Variables   The expression for dU is

$$dU = (\partial U/\partial T)_{V,n}\, dT + (\partial U/\partial V)_{T,n}\, dV + (\partial U/\partial n)_{T,V}\, dn \qquad \text{(B-4)}$$

if T, V, and n are used as the independent variables. If T, P, and n are used as the independent variables, then dU is given by

$$dU = (\partial U/\partial T)_{P,n}\, dT + (\partial U/\partial P)_{T,n}\, dP + (\partial U/\partial n)_{T,P}\, dn \qquad \text{(B-5)}$$

In a nonrigorous fashion, we "divide" Equation (B-4) by dT and specify that P and n are fixed. Of course, you cannot correctly do this, since dT is infinitesimal, but it gives the correct relationship between the derivatives. Each "quotient" such as dU/dT is interpreted as a partial derivative with the same variables fixed in each "quotient." The result is, holding T and n fixed:

$$
\begin{aligned}
(\partial U/\partial T)_{P,n} = {} & (\partial U/\partial T)_{V,n}(\partial T/\partial T)_{P,n} + (\partial U/\partial V)_{T,n}(\partial V/\partial T)_{P,n} \\
& + (\partial U/\partial n)_{T,V}(\partial n/\partial T)_{P,n}
\end{aligned}
\qquad \text{(B-6)}
$$

The derivative of T with respect to T is equal to unity, and the derivative of n with respect to anything is equal to zero if n is fixed, so that:

$$(\partial U/\partial T)_{P,n} = (\partial U/\partial T)_{V,n} + (\partial U/\partial V)_{T,n}(\partial V/\partial T)_{P,n} \qquad \textbf{(B-7)}$$

Equation (B-7) is an example of the **variable-change identity**. The version for any particular case can be obtained by systematically replacing each letter by the letter for any desired variable.

The Reciprocal Identity   An example of this identity is

$$\left(\frac{\partial V}{\partial P}\right)_{T,n} = \frac{1}{(\partial P/\partial V)_{T,n}} \qquad \textbf{(B-8)}$$

The same variables must be held fixed in both derivatives in the reciprocal identity. This identity has the same form as though the derivatives were simple quotients, instead of limits of quotients.

The Chain Rule   An example is

$$(\partial U/\partial T)_{V,n} = (\partial U/\partial P)_{V,n}(\partial P/\partial T)_{V,n} \qquad \textbf{(B-9)}$$

The same quantities must be held fixed in all of the derivatives in the identity.

We can obtain a derivative of a quantity that is expressed as a function of one variable that is in turn given as a function of other variables. For example, if $f = f(u)$ and $u = u(x, y, z)$:

$$\left(\frac{\partial f}{\partial x}\right)_{y,z} = \left(\frac{df}{du}\right)\left(\frac{\partial u}{\partial x}\right)_{y,z} \qquad \textbf{(B-10)}$$

The differential of f can be written

$$df = \frac{df}{dx}\left[\left(\frac{\partial u}{\partial x}\right)_{y,z} dx + \left(\frac{\partial u}{\partial y}\right)_{x,z} dy + \left(\frac{\partial u}{\partial z}\right)_{x,y} dz\right] \qquad \textbf{(B-11)}$$

Euler's Reciprocity Relation or the Second-Derivative Identity   A second derivative is the derivative of a first derivative. If f is a differentiable function of two independent variables, x and y, there are four second derivatives:

$$\frac{\partial^2 f}{\partial y\,\partial x} = \left(\frac{\partial}{\partial y}\left(\frac{\partial f}{\partial x}\right)_y\right)_x \qquad \textbf{(B-12a)}$$

$$\frac{\partial^2 f}{\partial x\,\partial y} = \left(\frac{\partial}{\partial x}\left(\frac{\partial f}{\partial y}\right)_x\right)_y \qquad \textbf{(B-12b)}$$

$$\left(\frac{\partial^2 f}{\partial x^2}\right)_y = \left(\frac{\partial}{\partial x}\left(\frac{\partial f}{\partial x}\right)_y\right)_y \qquad \textbf{(B-12c)}$$

$$\left(\frac{\partial^2 f}{\partial y^2}\right)_x = \left(\frac{\partial}{\partial y}\left(\frac{\partial f}{\partial y}\right)_x\right)_x \qquad \textbf{(B-12d)}$$

We refer to the second partial derivatives in Equations (B-12a) and (B-12b) as **mixed second partial derivatives**.

The **Euler reciprocity relation** is a theorem of mathematics. We sometimes call it the **second-derivative identity**: If f is differentiable, then the two mixed second partial derivatives in Equations (B-12a) and (B-12b) are the same function:

$$\frac{\partial^2 f}{\partial y\,\partial x} = \frac{\partial^2 f}{\partial x\,\partial y} \tag{B-13}$$

For a function of three variables, there are nine second partial derivatives, six of which are mixed derivatives. The mixed second partial derivatives obey relations exactly analogous to Equation (B-13). For example, one case of the second-derivative identity is

$$\left(\frac{\partial^2 V}{\partial T\,\partial P}\right)_n = \left(\frac{\partial^2 V}{\partial P\,\partial T}\right)_n \tag{B-14}$$

In these derivatives, the same third independent variable is held fixed in both derivatives, as shown by the subscript.

The Cycle Rule   If x, y, and z are related so that any two of them can be considered as independent variables and the third is then a dependent variable,

$$\left(\frac{\partial z}{\partial x}\right)_y\left(\frac{\partial x}{\partial y}\right)_z\left(\frac{\partial y}{\partial z}\right)_x = -1 \tag{B-15}$$

We can obtain this identity in a nonrigorous way. The differential dz can be written

$$dz = \left(\frac{\partial z}{\partial x}\right)_y dx + \left(\frac{\partial z}{\partial y}\right)_x dy \tag{B-16}$$

We consider the special case in which z is held fixed so that $dz = 0$, and "divide" Equation (B-16) nonrigorously by dy. The "quotient" dx/dy at constant z is interpreted as a partial derivative at constant z, and the "quotient" dy/dy equals unity. We obtain

$$0 = \left(\frac{\partial z}{\partial x}\right)_y\left(\frac{\partial x}{\partial y}\right)_z + \left(\frac{\partial z}{\partial y}\right)_x \tag{B-17}$$

We multiply by $(\partial y/\partial z)_x$ and apply the reciprocal identity to obtain

$$\left(\frac{\partial z}{\partial x}\right)_y\left(\frac{\partial x}{\partial y}\right)_z\left(\frac{\partial y}{\partial z}\right)_x = -1 \tag{B-18}$$

which is equivalent to Equation (B-15).

Exact and Inexact Differentials   Equation (B-1) gives the differential of a function, which is called an **exact differential**. We can also write a general differential in terms of dx, dy, and dz:

$$du = L(x, y, z)\,dx + M(x, y, z)\,dy + N(x, y, z)\,dz \tag{B-19}$$

where L, M, and N are some functions of x, y, and z. If these functions are not the appropriate partial derivatives of the same function, then the differential du is an **inexact differential** and has some different properties from an exact differential. A general differential form like that of Equation (B-19) is sometimes called a **Pfaffian form**.

To test the differential du for exactness, we can see if the appropriate derivatives of L, M, and N are mixed second derivatives of the same function and obey the Euler reciprocity relation:

$$\left(\frac{\partial L}{\partial y}\right)_{x,z} = \left(\frac{\partial M}{\partial x}\right)_{y,z} \qquad \textbf{(B-20a)}$$

$$\left(\frac{\partial L}{\partial z}\right)_{x,y} = \left(\frac{\partial N}{\partial x}\right)_{y,z} \qquad \textbf{(B-20b)}$$

$$\left(\frac{\partial M}{\partial z}\right)_{x,y} = \left(\frac{\partial N}{\partial y}\right)_{x,z} \qquad \textbf{(B-20c)}$$

If the conditions of Equation (B-20) are not obeyed, du is an inexact differential, and if they are obeyed, du is an exact differential.

Integral Calculus   There are two principal types of integrals of functions of several variables: the line integral and the multiple integral.

Line Integrals   For a differential with two independent variables,

$$du = M(x, y)\,dx + N(x, y)\,dy$$

the line integral is denoted by

$$\int_c du = \int_c [M(x, y)\,dx + N(x, y)\,dy] \qquad \textbf{(B-21)}$$

where the letter c denotes a curve in the x-y plane. This curve gives y as a function of x and x as a function of y, as in Figure B-1. We say that the integral is carried out along this curve (or path). To carry out the integral, we substitute the expression for y as a function of x in M and substitute the expression for x as a function of y in N, using the functions specified by the curve. We represent these functions by $y(x)$ and $x(y)$:

$$\int_c du = \int_{x_1}^{x_2} M(x, y(x))\,dx + \int_{y_1}^{y_2} N(x(y), y)\,dy \qquad \textbf{(B-22)}$$

where x_1 and y_1 are the coordinates of the point on the curve designated as the initial point of the line integral and x_2 and y_2 are the coordinates of the final point. Each integral is now an ordinary integral and can be carried out by the methods of the calculus of functions of one variable.

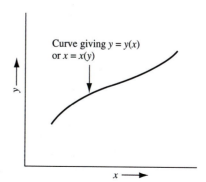

Figure B-1 A curve giving y as a function of x or x as a function of y.

If the differential form has three or more independent variables, the procedure is analogous. The curve must be a curve in a space of all independent variables, giving each of the independent variables except one as a function of that one variable.

There is an important theorem of mathematics concerning the line integral of an exact differential: If dz is an exact differential, say the differential of $z = z(T, V, n)$,

$$\int_c dz = \int_c \left[\left(\frac{\partial z}{\partial T} \right)_{V,n} dT + \left(\frac{\partial z}{\partial V} \right)_{T,n} dV + \left(\frac{\partial z}{\partial n} \right)_{T,V} dn \right]$$
$$= z(T_2, V_2, n_2) - z(T_1, V_1, n_1) \tag{B-23}$$

where T_2, V_2, and n_2 are the values of the independent variables at the final point of the curve and T_1, V_1, and n_1 are the values at the initial point of the curve. Since many different curves can have the same initial and final points, Equation (B-23) means that the line integral depends only on the initial point and the final point and is independent of the curve between these points. It is said to be **path independent**.

The line integral of an inexact differential is path dependent. That is, one can always find two or more paths between a given initial point and a given final point for which the line integrals are not equal.

Multiple Integrals   If $f = f(x, y, z)$ is an integrand function, a multiple integral with constant limits is denoted by

$$I(a_1, a_2, b_1, b_2, c_1, c_2) = \int_{a_1}^{a_2} \int_{b_1}^{b_2} \int_{c_1}^{c_2} f(x, y, z)\, dz\, dy\, dx \tag{B-24}$$

This integral is carried out as follows: z is first integrated from c_1 to c_2 in the ordinary manner, treating x and y as constants during the integration. The result is a function of x and y, which is the integrand for the other two integrations. Then y is integrated from b_1 to b_2 in the ordinary manner, treating x as a constant. The result is a function of x, which is the integrand for the final integration. Then x is integrated from a_1 to a_2. The leftmost differential and the rightmost integral sign belong together, and this integration is done first, and so on.

In the integral of Equation (B-24), the limits of the z integration can be replaced by functions of x and y, and the limits of the y integration can be replaced by functions of x. The limit functions are substituted into the indefinite integral in exactly the same way as are constants when the indefinite integral is evaluated at the limits.

If the variables are cartesian coordinates and the limits are constants, the region of integration is inside a rectangular parallelepiped (box) as shown in Figure B-2. If the limits for the first two integrations are not constants, the region of integration can have a more complicated shape.

We can visualize the integration process as in Figure B-3. The product $dx\, dy\, dz$ is considered to be a "volume element," which is depicted in the figure as a little box. If (x, y, z) represents a point in the element, then the contribution of the element of volume to the integral is equal to the value of the function at (x, y, z) times the volume of the volume element:

$$\text{Contribution} = f(x, y, z)\, dx\, dy\, dz$$

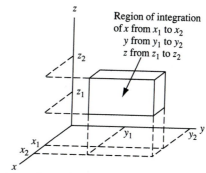

Figure B-2 An integration region in cartesian coordinates with constant limits.

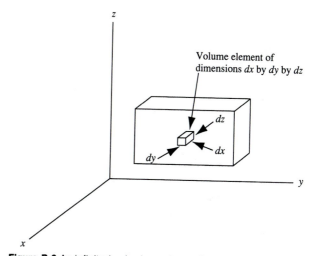

Figure B-3 An infinitesimal volume element in cartesian coordinates.

The integral is the sum of the contributions of all elements of volume in the region of integration.

If an integral over a volume in three-dimensional space is needed and spherical polar coordinates are used, the volume element is as depicted in Figure B-4. The length of the volume element in the r direction is equal to dr. The length of the box in the θ direction (the direction in which an infinitesimal change in θ carries a point in space) is equal to $r\,d\theta$ if θ is measured in radians. This fact comes from the definition of the measure of an angle in radians, which is the ratio of the arc length to the radius. The length of the volume element in the ϕ direction is $r\sin(\theta)\,d\phi$, which comes from the fact that the projection of r in the x-y plane has length $r\sin(\theta)$, as shown in the figure. The volume of the element of volume is thus

$$d^3\mathbf{r} = r^2\sin(\theta)\,d\phi\,d\theta\,dr \qquad\qquad \textbf{(B-25)}$$

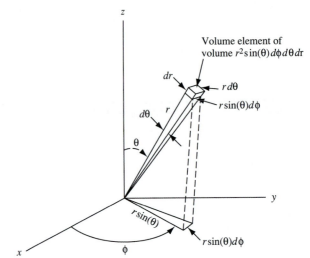

Figure B-4 An infinitesimal volume element in spherical polar coordinates.

where $d^3\mathbf{r}$ is a general abbreviation for the volume element. An integral over all of space using spherical polar coordinates is

$$I = \int_0^\infty \int_0^\pi \int_0^{2\pi} f(r, \theta, \phi)r^2 \sin(\theta)\, d\phi\, d\theta\, dr \qquad \textbf{(B-26)}$$

Since the limits are constants, this integral is carried out in the same way as that of Equation (B-24).

For other coordinate systems, a factor analogous to the factor $r^2 \sin(\theta)$ must be used. This factor is called a **Jacobian**. For example, for cylindrical polar coordinates, where the coordinates are z, ϕ (the same angle as in spherical polar coordinates, and ρ (the projection of r into the x-y plane), the Jacobian is the factor ρ, so the element of volume is $\rho\, d\rho\, dz\, d\phi$.

Vectors   A vector is a quantity with both magnitude and direction. The vector $\mathbf{A}$ can be represented by its cartesian components, A_x, A_y, and A_z:

$$\mathbf{A} = \mathbf{i}A_x + \mathbf{j}A_y + \mathbf{k}A_z \qquad \textbf{(B-27)}$$

where $\mathbf{i}$, $\mathbf{j}$, and $\mathbf{k}$ are unit vectors in the x, y, and z directions, respectively.

The **dot product**, or **scalar product**, of two vectors is a scalar quantity equal to the product of the magnitudes of the two vectors times the cosine of the angle between them:

$$\mathbf{A} \cdot \mathbf{B} = |\mathbf{A}||\mathbf{B}| \cos(\alpha) = AB \cos(\alpha) \qquad \textbf{(B-28)}$$

where α is the angle between the vectors. The scalar product is commutative:

$$\mathbf{A} \cdot \mathbf{B} = \mathbf{B} \cdot \mathbf{A} \qquad \textbf{(B-29)}$$

The scalar product of a vector with itself is the square of the magnitude of the vector:

$$\mathbf{A} \cdot \mathbf{A} = |\mathbf{A}|^2 = A^2 \qquad \textbf{(B-30)}$$

The dot products of the unit vectors are

$$\mathbf{i} \cdot \mathbf{j} = \mathbf{i} \cdot \mathbf{k} = \mathbf{j} \cdot \mathbf{k} = 0 \qquad \text{(B-31)}$$
$$\mathbf{i} \cdot \mathbf{i} = \mathbf{j} \cdot \mathbf{j} = \mathbf{k} \cdot \mathbf{k} = 1$$

In terms of cartesian components, it follows from Equation (B-31)

$$\mathbf{A} \cdot \mathbf{B} = A_x B_x + A_y B_y + A_z B_z \qquad \text{(B-32)}$$

The **cross product**, or **vector product**, of two vectors is a vector quantity that is perpendicular to the plane containing the two vectors with magnitude equal to the product of the magnitudes of the two vectors times the sine of the angle between them.

$$|\mathbf{A} \times \mathbf{B}| = |\mathbf{A}||\mathbf{B}| \sin(\alpha) \qquad \text{(B-33)}$$

The product vector is perpendicular to the plane containing **A** and **B** and in the direction in which an ordinary (right-handed) screw moves if it is rotated in the direction in which the vector on the left must be rotated to coincide with the vector on the right, through an angle less than or equal to 180°.

The cross product is not commutative:

$$\mathbf{A} \times \mathbf{B} = -\mathbf{B} \times \mathbf{A} \qquad \text{(B-34)}$$

The cross products of the unit vectors are

$$\mathbf{i} \times \mathbf{i} = \mathbf{0}, \qquad \mathbf{j} \times \mathbf{j} = \mathbf{0}, \qquad \mathbf{k} \times \mathbf{k} = \mathbf{0} \qquad \text{(B-35)}$$
$$\mathbf{i} \times \mathbf{j} = \mathbf{k}, \qquad \mathbf{i} \times \mathbf{k} = -\mathbf{j}, \qquad \mathbf{j} \times \mathbf{k} = \mathbf{i}$$

In terms of cartesian components, we can deduce from Equation (B-35) that

$$\mathbf{A} \times \mathbf{B} = \mathbf{i}[A_y B_z - A_z B_y] + \mathbf{j}[A_z B_x - A_x B_z] + \mathbf{k}[A_x B_y - A_y B_x] \qquad \text{(B-36)}$$

The product of a vector and a scalar is a vector whose magnitude is equal to the magnitude of the vector times the magnitude of the scalar. Its direction is the same as the direction of the first vector if the scalar is positive and is the opposite of the direction of the first vector if the scalar is negative.

Vector Derivatives The **gradient** is a vector derivative of a scalar function. If f is a function of x, y, and z, its gradient is given by

$$\nabla f = \mathbf{i}(\partial f/\partial x) + \mathbf{j}(\partial f/\partial y) + \mathbf{k}(\partial f/\partial z) \qquad \text{(B-37)}$$

The symbol for the gradient operator, ∇, is called "del." At a given point in space, the gradient points in the direction of most rapid increase of the function, and its magnitude is equal to the rate of change of the function in that direction.

The gradient of a vector function is also defined, and the gradient of each component is as defined in Equation (B-37). There are nine components of this quantity, which is called a **dyadic** or a **cartesian tensor**.

The **divergence** is a three-dimensional derivative of a vector function. The divergence of a vector function **F** is defined by

$$\nabla \cdot \mathbf{F} = (\partial F_x/\partial x) + (\partial F_y/\partial y) + (\partial F_z/\partial z) \qquad \text{(B-38)}$$

The divergence is a scalar quantity. If the vector function represents the flow velocity of a fluid, the divergence is a measure of the spreading out of the streaming curves along which small elements of the fluid flow. A positive value of the divergence corresponds to a decrease in density along a curve following the flow.

The **curl** is a vector derivative of a vector function. The curl of **F** is analogous to the vector product (cross product) of two vectors and is given by

$$\mathbf{V} \times \mathbf{F} = \mathbf{i}[(\partial F_z/\partial y) - (\partial F_y/\partial z)] + \mathbf{j}[(\partial F_x/\partial z) - (\partial F_z/\partial x)]$$
$$+ \mathbf{k}[(\partial F_y/\partial x) - (\partial F_x/\partial y)] \tag{B-39}$$

The curl of a vector function is a kind of measure of the rotation, or turning of the vector as a function of position, and has also been called the "rotation."

The divergence of the gradient is also called the **Laplacian**. The Laplacian of a scalar function f is given by

$$\nabla^2 f = \frac{\partial^2 f}{\partial x^2} + \frac{\partial^2 f}{\partial y^2} + \frac{\partial^2 f}{\partial z^2} \tag{B-40}$$

The vector derivative operators can be expressed in other coordinate systems. In spherical polar coordinates, the gradient is

$$\nabla f = \mathbf{e}_r \frac{\partial f}{\partial r} + \mathbf{e}_\theta \frac{1}{r} \frac{\partial f}{\partial \theta} + \mathbf{e}_\phi \frac{1}{r \sin(\theta)} \frac{\partial f}{\partial \phi} \tag{B-41}$$

where $\mathbf{e}_r$ is the unit vector in the r direction (the direction of motion if $\mathbf{r}$ is increased keeping θ and ϕ fixed), $\mathbf{e}_\theta$ is the unit vector in the θ direction, and $\mathbf{e}_\phi$ is the unit vector in the ϕ direction. In spherical polar coordinates, the Laplacian is

$$\nabla^2 f = \frac{1}{r^2} \left[\frac{d}{dr} \left(r^2 \frac{df}{dr} \right) + \frac{1}{\sin(\theta)} \frac{\partial}{\partial \theta} \left(\sin(\theta) \frac{\partial f}{\partial \theta} \right) + \frac{1}{\sin^2(\theta)} \frac{\partial^2 f}{\partial \phi^2} \right] \tag{B-42}$$

C A Short Table of Integrals

Indefinite integrals

$$\int \sin^2(ax)\, dx = \frac{x}{2} - \frac{\sin(2x)}{4}$$

$$\int \cos^2(ax)\, dx = \frac{x}{2} + \frac{\sin(2x)}{4}$$

$$\int x \sin^2(x)\, dx = \frac{x^2}{4} - \frac{x \sin(2x)}{4} - \frac{\cos(2x)}{8}$$

$$\int x \cos^2(x)\, dx = \frac{x^2}{4} + \frac{x \sin(2x)}{4} + \frac{\cos(2x)}{8}$$

$$\int e^{ax}\, dx = \frac{1}{a} e^{ax}$$

Definite integrals

$$\int_0^{\pi} \sin(mx) \sin(nx)\, dx = \begin{cases} \dfrac{\pi}{2} & \text{if } m = n \\ 0 & \text{if } m \neq n \end{cases} \quad (m \text{ and } n \text{ integers})$$

$$\int_0^{\pi} \cos(mx) \cos(nx)\, dx = \begin{cases} \dfrac{\pi}{2} & \text{if } m = n \\ 0 & \text{if } m \neq n \end{cases} \quad (m \text{ and } n \text{ integers})$$

$$\int_0^{\pi} \sin(mx) \cos(nx)\, dx = \begin{cases} 0 & \text{if } m = n \\ 0 & \text{if } m \neq n \quad (m + n \text{ odd}) \\ \dfrac{2m}{m^2 - n^2} & \text{if } m \neq n \quad (m + n \text{ even}) \end{cases}$$

$$\int_0^{\infty} e^{-ax}\, dx = \frac{1}{a}, \quad a > 0$$

$$\int_0^{\infty} x^n e^{-ax}\, dx = \frac{1}{a^{n+1}} n!, \quad a > 0, n = 0, 1, 2, \ldots$$

$$\int_0^{\infty} e^{-ax^2}\, dx = \frac{1}{2} \sqrt{\frac{\pi}{a}}, \quad a > 0$$

$$\int_0^{\infty} x e^{-ax^2}\, dx = \frac{1}{2a}, \quad a > 0$$

$$\int_0^\infty x^2 e^{-ax^2}\, dx = \frac{1}{4}\sqrt{\frac{\pi}{a^3}}, \qquad a > 0$$

$$\int_0^\infty x^3 e^{-ax^2}\, dx = \frac{3}{a^2}, \qquad a > 0$$

$$\int_0^\infty x^{2n} e^{-ax^2}\, dx = \frac{(1)(3)(5)\cdots(2n-1)}{2^{n+1}a^{n+1/2}}\sqrt{\pi}, \qquad a > 0$$

$$\int_0^\infty x^{2n+1} e^{-ax^2}\, dx = \frac{n!}{2a^{n+1}}, \qquad a > 0$$

The error function

The error function is defined by

$$\text{erf}(z) = \frac{2}{\sqrt{\pi}}\int_0^z e^{-t^2}\, dt$$

So that for $z = 0$ and $z = \infty$,

$$\text{erf}(0) = 0$$
$$\text{erf}(\infty) = 1$$

The following identity is sometimes useful:

$$\int_0^x t^2 e^{-at^2}\, dt = \frac{\sqrt{\pi}}{4a^{3/2}}\text{erf}(\sqrt{ax}) - \frac{x}{2a}e^{-ax^2}$$

The error function cannot be expressed in closed form for values of z other than 0 or ∞. The following is a short table of values:

x		0	1	2	3	4	5	6	7	8	9
0.0	0.0	000	113	226	338	451	564	676	789	901	*013
1	0.1	125	236	348	459	569	680	790	900	*009	*118
2	0.2	227	335	443	550	657	763	869	974	*079	*183
3	0.3	286	389	491	593	694	794	893	992	*090	*187
4	0.4	284	380	475	569	662	755	847	937	*027	*117
5	0.5	205	292	379	465	549	633	716	798	879	959
6	0.6	039	117	194	270	346	420	494	566	638	708
7		778	847	914	981	*047	*112	*175	*238	*300	*361
8	0.7	421	480	538	595	651	707	761	814	867	918
9		969	*019	*068	*116	*163	*209	*254	*299	*342	*385
1.0	0.8	427	468	508	548	586	624	661	698	733	768
1		802	835	868	900	931	961	991	*020	*048	*076
2	0.9	103	130	155	181	205	229	252	275	297	319
3		340	361	381	400	419	438	456	473	490	507
4	0.95	23	39	54	69	83	97	*11	*24	*37	*49
5	0.96	61	73	84	95	*06	*16	*26	*36	*45	*55
6	0.97	63	72	80	88	96	*04	*11	*18	*25	*32
7	0.98	38	44	50	56	61	67	72	77	82	86
8		91	95	99	*03	*07	*11	*15	*18	*22	*25

x		0	1	2	3	4	5	6	7	8	9
9	0.99	28	31	34	37	39	42	44	47	49	51
2.0	0.995	32	52	72	91	*09	*26	*42	*58	*73	*88
1	0.997	02	15	28	41	53	64	75	85	95	*05
2	0.998	14	22	31	39	46	54	61	67	74	80
3		86	91	97	*02	*06	*11	*15	*20	*24	*28
4	0.999	31	35	38	41	44	47	50	52	55	57
5		59	61	63	65	67	69	71	72	74	75
6		76	78	79	80	81	82	83	84	85	86
7		87	87	88	89	89	90	91	91	92	92
8	0.9999	25	29	33	37	41	44	48	51	54	56
9		59	61	64	66	68	70	72	73	75	77

D Classical Mechanics

Newton's Laws The first law is the **law of inertia**: *If not acted upon by a force, a stationary object remains stationary, and a moving object continues to move in a straight line at a constant speed.*

Newton's second law, the most important of the three laws, is the **law of acceleration**:

$$\mathbf{F} = m\mathbf{a} = m\frac{d\mathbf{v}}{dt} = \frac{d^2\mathbf{r}}{dt^2}$$

or

$$\mathbf{i}F_x + \mathbf{j}F_y + \mathbf{k}F_z = m\left(\mathbf{i}\frac{d^2x}{dt^2} + \mathbf{j}\frac{d^2y}{dt^2} + \mathbf{k}\frac{d^2z}{dt^2}\right) \tag{D-1}$$

where m is the mass of the particle and $\mathbf{i}$, $\mathbf{j}$, and $\mathbf{k}$ are unit vectors in the direction of the x, y, and z coordinate axes, respectively. The vector $\mathbf{a}$, the second time derivative of the position vector, is called the **acceleration**.

Newton's third law is the **law of action and reaction**: *If one object exerts a force on a second object, the second object exerts a force on the first object which is equal in magnitude to the first force and opposite in direction.*

If the force is a known function of position, Equation (D-1) is an **equation of motion**, which determines the particle's position and velocity for all values of the time if the position and velocity are known for a single time. Classical mechanics is thus said to be **deterministic**.

All mechanical quantities have values that are determined by the values of coordinates and velocities. The **kinetic energy** of a point mass particle is

$$\mathscr{K} = \frac{1}{2}mv^2 = \frac{1}{2}m(v_x^2 + v_y^2 + v_z^2) \tag{D-2}$$

If the forces on the particles of a system depend only on the particles' positions, these forces can be derived from the **potential energy**. Consider motion in the z direction. If z' and z'' are two values of z, the difference in the potential energy $\mathscr{V}$ between these two locations is defined to equal the reversible work done on the system by an external agent to move the particle from z' to z''.

$$\mathscr{V}(z'') - \mathscr{V}(z') = \int_{z'}^{z''} F_{\text{ext(rev)}}(z)\, dz \tag{D-3}$$

The external force $F_{\text{ext(rev)}}$ must exactly balance the force due to the other particles in order for the process to be reversible:

$$F_{\text{ext(rev)}}(z) = -F_z \tag{D-4}$$

By the principles of calculus, the integrand in Equation (D-3) is equal to the derivative of the function $\mathscr{V}$, so that

$$F_z = -d\mathscr{V}/dz \tag{D-5}$$

In the case of motion in three dimensions, analogous equations for the x and y components can be written, and the vector force is given by

$$\begin{aligned}\mathbf{F} = \mathbf{i}F_x + \mathbf{j}F_y + \mathbf{k}F_z &= \mathbf{i}(\partial\mathscr{V}/\partial x) + \mathbf{j}(\partial\mathscr{V}/\partial y) + \mathbf{k}(\partial\mathscr{V}/\partial z)\\ &= -\nabla\mathscr{V}\end{aligned} \tag{D-6}$$

where the symbol ∇ stands for the three-term **gradient operator** in the right-hand side of the first line of Equation (D-6).

Since only the difference in potential energy is defined in Equation (D-3), we have the option of deciding at what location we want to have the potential energy equal zero. This is always the case in nonrelativistic quantum or classical mechanics.

A system in which no forces occur except those derivable from a potential energy is called a **conservative system**. It is a theorem of mechanics that the energy of such a system is constant, or **conserved**.

Lagrangian and Hamiltonian Mechanics The methods of Lagrange and the method of Hamilton are designed for problems in which cartesian coordinates cannot conveniently be used.

We denote the positions of the particles in a system by the coordinates $q_1, q_2, q_3, \ldots, q_n$, where n is the number of coordinates, equal to three times the number of particles if they are point mass particles that move in three dimensions. These coordinates can be of any kind—spherical polar coordinates, cylindrical polar coordinates, etc. To specify the state of the system, some measures of the particles' velocities are needed in addition to coordinates. The Lagrangian method uses the time derivatives of the coordinates:

$$\dot{q}_i = dq_i/dt \qquad (i = 1, 2, \ldots, n) \tag{D-7}$$

We use a symbol with a dot over it to represent a time derivative.

The Lagrangian function is defined by

$$\boxed{\mathscr{L} = \mathscr{K} - \mathscr{V} \quad \text{(definition of the Lagrangian)}} \tag{D-8}$$

where $\mathscr{K}$ and $\mathscr{V}$ are expressed in terms of the $\dot{q}$'s and the q's. The **Lagrangian equations of motion** are

$$\boxed{\frac{d}{dt}\frac{\partial\mathscr{L}}{\partial\dot{q}_i} - \frac{\partial\mathscr{L}}{\partial q_i} = 0 \qquad (i = 1, 2, \ldots, n)} \tag{D-9}$$

These equations are equivalent to Newton's second law. The convenient thing about them is that they have exactly the same form for all coordinate systems, so the work of transforming Newton's second law into a particular coordinate system can be avoided.[1]

One application of Lagrange's equations of motion is to a particle orbiting about a fixed point in a plane. Using plane polar coordinates ρ and ϕ (measured in radians), with the origin at the fixed point, the component of the velocity parallel to the position vector is $\dot{\rho}$ and the component perpendicular to this direction is $\rho\dot{\phi}$, so the Lagrangian is

$$\mathscr{L} = \frac{1}{2}m\dot{\rho}^2 + \frac{1}{2}m\rho^2\dot{\phi}^2 - \mathscr{V}(\rho, \phi) \tag{D-10}$$

and the Lagrangian equations of motion are

$$\frac{d(2m\dot{\rho})}{dt} - m\rho\dot{\phi}^2 + \frac{\partial \mathscr{V}}{\partial \rho} = 0 \tag{D-11}$$

$$\frac{d(m\rho^2\dot{\phi})}{dt} + \frac{\partial \mathscr{V}}{\partial \phi} = 0 \tag{D-12}$$

The second term in Equation (D-11) produces a rate of change in $\dot{\rho}$ if $\mathscr{V}$ does not depend on ρ. This is the **centrifugal force**, which is not a force but an expression of the natural tendency of an orbiting particle to move off in a straight line. To maintain a circular orbit about the origin of coordinates, the second term must be canceled by a **centripetal force**:

$$F(\text{centripetal}) = -\frac{\partial \mathscr{V}}{\partial \rho} = -m\rho\dot{\phi}^2 = -\frac{mv^2}{\rho} \tag{D-13}$$

where the speed v in a circular orbit equals $\rho\dot{\phi}$.

An important quantity for an orbiting object is the **angular momentum** around a fixed center. This is the vector

$$\mathbf{L} = m\mathbf{r} \times \mathbf{v} \tag{D-14}$$

where $\mathbf{r}$ is the position vector from the fixed center to the particle, $\mathbf{v}$ is the velocity vector, and $\times$ stands for the cross product of two vectors. For a circular orbit, the angular momentum vector has the magnitude

$$L = m\rho v = m\rho^2\dot{\phi} \tag{D-15}$$

The method of Hamilton is similar to that of Lagrange in that it provides equations of motion that have the same form in any coordinate system. However, it uses variables called **conjugate momenta** instead of coordinate time derivatives as state variables. The **momentum conjugate to the ith coordinate** is defined by

$$p_i = \frac{\partial \mathscr{L}}{\partial \dot{q}_i} \qquad (i = 1, 2, \ldots, n) \quad (\text{definition}) \tag{D-16}$$

---

[1] This fact is proved in J. C. Slater and N. Frank, *Mechanics*, McGraw-Hill, New York, 1947, pp. 69ff, or in any other book on the same subject.

The momentum conjugate to a cartesian coordinate is a component of the ordinary (linear) momentum:

$$p_x = mv_x, \qquad p_y = mv_y, \qquad p_z = mv_z \qquad \textbf{(D-17)}$$

As with the angular momentum, the vector sum of the momenta of a set of interacting particles is conserved if no external forces act on the particles. That is, the total vector momentum is conserved.

The **Hamiltonian function**, also called **Hamilton's principal function**, is defined by

$$\mathscr{H} = \sum_{i=1}^{n} p_i \dot{q}_i - \mathscr{L} \qquad \textbf{(D-18)}$$

The Hamiltonian must be expressed as a function of coordinates and conjugate momenta. It is equal to the total energy of the system (kinetic plus potential).[2]

$$\mathscr{H} = \mathscr{K} + \mathscr{V} \qquad \textbf{(D-19)}$$

The Hamiltonian equations of motion are

$$\dot{q}_i = \frac{\partial \mathscr{H}}{\partial p_i}, \qquad \dot{p}_i = -\frac{\partial \mathscr{H}}{\partial q_i} \qquad (i = 1, 2, \ldots, n) \qquad \textbf{(D-20)}$$

There is one pair of equations for each value of i, as indicated.

The Two-Body Problem   Consider a two-particle system with a potential energy that depends only on the distance between the particles. This case applies, for example, to the hydrogen atom and to the nuclei of a rotating diatomic molecule.

We first examine the case in which there is motion only in the x direction. The Lagrangian of the system is

$$\mathscr{L} = \mathscr{K} - \mathscr{V}$$
$$= \frac{1}{2} m_1 \dot{x}_1^2 + \frac{1}{2} m_2 \dot{x}_2^2 - \mathscr{V}(x_2 - x_1) \qquad \textbf{(D-21)}$$

where x_1 and m_1 are the coordinate and the mass of the first particle and x_2 and m_2 are the coordinate and the mass of the second particle.

We now change to a different set of coordinates:

$$X = \frac{m_1 x_1 + m_2 x_2}{m_1 + m_2} \qquad \textbf{(D-22)}$$

$$x = x_2 - x_1 \qquad \textbf{(D-23)}$$

The coordinate X is the location of the **center of mass** of the system, and the coordinate x is called the **relative coordinate**.

---

[2] Slater and Frank, *op. cit.*, pp. 74ff.

These two coordinates have separate equations of motion. We solve Equations (D-22) and (D-23) for x_1 and x_2:

$$x_1 = X - \frac{m_2 x}{M} \tag{D-24}$$

$$x_2 = X + \frac{m_1 x}{M} \tag{D-25}$$

where $M = m_1 + m_2$.

Taking the time derivatives of Equations (D-24) and (D-25), we can write the Lagrangian as

$$
\begin{aligned}
\mathscr{L} &= \frac{1}{2} m_1 \left(\dot{X} - \frac{m_2 \dot{x}}{M} \right)^2 + \frac{1}{2} m_2 \left(\dot{X} + \frac{m_1 \dot{x}}{M} \right)^2 - \mathscr{V}(x) \\
&= \frac{1}{2} m_1 \left[\dot{X}^2 - 2m_2 \frac{\dot{X}\dot{x}}{M} + \left(\frac{m_2 \dot{x}}{M} \right)^2 \right] \\
&\quad + \frac{1}{2} m_2 \left[\dot{X}^2 + 2m_1 \frac{\dot{X}\dot{x}}{M} + \left(\frac{m_1 \dot{x}}{M} \right)^2 \right] - \mathscr{V}(x) \\
&= \frac{1}{2} M\dot{X}^2 + \frac{1}{2} \mu \dot{x}^2 - \mathscr{V}(x) \tag{D-26}
\end{aligned}
$$

where we have introduced the **reduced mass** of the pair of particles, defined by

$$\mu = \frac{m_1 m_2}{m_1 + m_2} = \frac{m_1 m_2}{M} \tag{D-27}$$

Since the variables are separated, we obtain separate equations of motion for X and x:

$$M \frac{d\dot{X}}{dt} = \frac{\partial \mathscr{L}}{\partial X} = 0 \tag{D-28}$$

$$\mu \frac{d\dot{x}}{dt} = \frac{\partial \mathscr{L}}{\partial x} = -\frac{d\mathscr{V}}{dx} \tag{D-29}$$

These equations imply that the center of mass of the two particles moves like a particle of mass M that has no forces acting on it, while the relative coordinate changes like the motion of a particle of mass μ moving at a distance x from a fixed origin and subject to the potential energy $\mathscr{V}(x)$. The motion of the two-particle system has been separated into two one-body problems.

For motion in three dimensions, the separation is completely analogous. *The fictitious particle of mass μ moves around the origin of its coordinate in the same way that particle 2 moves relative to particle 1, while the center of mass moves like a free particle of mass M.*

E Euler's Theorem

Statement: A function f that depends on $n_1, n_2, n_3, \ldots, n_c$ is said to be **homogeneous of degree m** in the n's if

$$f(\lambda n_1, \lambda n_2, \ldots, \lambda n_c) = \lambda^m f(n_1, n_2, \ldots, n_c) \qquad \textbf{(E-1)}$$

for every positive real value of λ.

Euler's theorem states that for such a function

$$mf = \sum_{i=1}^{c} n_i \left(\frac{\partial f}{\partial n_i} \right)_{n'} \qquad \textbf{(E-2)}$$

where the subscript n' means that all of the n's are held fixed except for n_i.

Proof: Differentiate Equation (E-1) with respect to λ:

$$\sum_{i=1}^{c} \left(\frac{\partial f}{\partial(\lambda n_i)} \right)_{n'} \left(\frac{\partial(\lambda n_i)}{\partial(\lambda)} \right)_{n'} = m\lambda^{m-1} f(n_1, n_2, \ldots, n_c) \qquad \textbf{(E-3)}$$

We use the fact that

$$\left(\frac{\partial(\lambda n_i)}{\partial(\lambda)} \right)_{n'} = n_i \qquad \textbf{(E-4)}$$

and set λ equal to unity to obtain Equation (E-2), and the theorem is proved.

Thermodynamic Applications of Euler's Theorem If T and P are held fixed, then U, H, A, G, V, etc. are homogeneous of degree 1 in the n's. That is, they are extensive quantities. If T and P are held fixed, then the molar quantities $\bar{U}, \bar{H}, \bar{A}, \bar{G}, \bar{V}$, etc. are homogeneous of degree 0. That is, they are intensive quantities. For example,

$$G = \sum_{i=1}^{c} n_i \left(\frac{\partial G}{\partial n_i} \right)_{T,P,n'} \qquad \textbf{(E-5)}$$

E. A. Desloge, *Statistical Physics*, Holt Rinehart & Winston, New York, 1966, Appendix 10.

Special Mathematical Functions Occurring in Energy Eigenfunctions

Hermite Polynomials   The energy eigenfunctions of a harmonic oscillator are written in Chapter 9 as

$$\psi(z) = e^{-az^2/2}S(z) \tag{F-1}$$

where $S(z)$ is a power series

$$S(z) = c_0 + c_1 z + c_2 z^2 + c_3 z^3 + \cdots = \sum_{n=1}^{\infty} c_n z^n \tag{F-2}$$

with constant coefficients $c_1, c_2, c_3, \ldots$. These coefficients are governed by the recursion relation given in Chapter 9:

$$c_{n+2} = \frac{2an + a - b}{(n+2)(n+1)} c_n \qquad (n = 0, 1, 2, \ldots) \tag{F-3}$$

where a and b are parameters defined in the chapter. The series is found to terminate, making it a polynomial, containing one term, two terms, or three terms, etc. This termination puts a condition on the parameter b, determining the energy eigenvalues. The polynomials are called Hermite polynomials, and each contains either odd powers or even powers of z, but not both. The first three normalized energy eigenfunctions are

$$\psi_0 = \left(\frac{a}{\pi}\right)^{1/4} e^{-az^2/2} \tag{F-4a}$$

$$\psi_1 = \left(\frac{4a^3}{\pi}\right)^{1/4} ze^{-az^2/2} \tag{F-4b}$$

$$\psi_2 = \left(\frac{a}{4\pi}\right)^{1/4} (2az^2 - 1)e^{-az^2/2} \tag{F-4c}$$

corresponding to the first three Hermite polynomials

$$H_0(x) = 1 \tag{F-5a}$$
$$H_1(x) = 2x \tag{F-5b}$$
$$H_2(x) = 4x^2 - 2 \tag{F-5c}$$

where in order to correspond to our wave functions, $x = \sqrt{a}\,z$. The normalization must be accomplished by an additional factor in the formula for the wave function.

All of the Hermite polynomials can be generated by the formula

$$H_n(x) = (-1)^n e^{-x^2} \frac{d^n}{dx^n}(e^{x^2}) \qquad \text{(F-6)}$$

or the formula[3]

$$H_n(x) = n! \sum_{m=0}^{[n/2]} \frac{(-1)^m (2x)^{n-2m}}{m!(n-2m)!} \qquad \text{(F-7)}$$

where $[n/2]$ stands for $n/2$ if n is even and for $(n-1)/2$ if n is odd.

A number of identities are obeyed by Hermite polynomials, and some of these can be found in reference 3. One useful identity is

$$x H_n(x) = n H_{n-1}(x) + \frac{1}{2} H_{n+1}(x) \qquad \text{(F-8)}$$

An important fact is that if n is even, $H_n(x)$ is an even function of x, and if n is odd, $H_n(x)$ is an odd function of x.

$$H_n(-x) = H_n(x) \qquad (n \text{ even}) \qquad \text{(F-9)}$$
$$H_n(-x) = -H_n(x) \qquad (n \text{ odd}) \qquad \text{(F-10)}$$

The Hydrogen-like Energy Eigenfunctions   The eigenfunctions are written as

$$\psi_{n\ell m} = R_{n\ell}(r) Y_{\ell m}(\theta, \phi) = R_{n\ell}(r) \Theta_{\ell m}(\theta) \Phi_m(\phi) \qquad \text{(F-11)}$$

where the $Y_{\ell m}$ functions are called **spherical harmonic functions**. The Y functions are the same for any central-force system, including the rigid rotor and the rotating diatomic molecule in the Born-Oppenheimer approximation.

The Φ functions are

$$\Phi_m = \frac{1}{\sqrt{2\pi}} e^{im\phi} \qquad \text{(F-12)}$$

where m is an integer.

The $\Theta_{\ell m}$ functions obey the equation

$$\sin(\theta) \frac{d}{d\theta} \sin(\theta) \frac{d\Theta}{d\theta} - m^2 \Theta + K \sin^2(\theta) \Theta = 0 \qquad \text{(F-13)}$$

where K is a constant. With a change of variables, $y = \cos(\theta)$, $P(y) = \Theta(\theta)$, the equation becomes, after quite a bit of manipulation,

$$(1 - y^2) \frac{d^2 P}{dy^2} - 2y \frac{dP}{dy} - \frac{m^2}{1 - y^2} P + KP = 0 \qquad \text{(F-14)}$$

---

[3] Erdelyi *et al., Higher Transcendental Functions*, Vol. II, McGraw-Hill, New York, 1953, pp. 192ff.

Equation (16-4) is the same as the **associated Legendre equation** if $K = \ell(\ell + 1)$, where ℓ is an integer that must be at least as large as $|m|$. The set of solutions is known as the **associated Legendre functions**, given for non-negative values of m by[4]

$$P_\ell^m(y) = (1 - y^2)^{m/2} \frac{d^m P_\ell(y)}{dy^m} \tag{F-15}$$

where $P_\ell(y)$ is the Legendre polynomial

$$P_\ell(y) = \frac{1}{2^\ell \ell!} \frac{d^\ell}{dy^\ell} (y^2 - 1)^\ell \tag{F-16}$$

With suitable normalization,

$$\Theta = \Theta_{\ell m} = \left(\frac{(2\ell + 1)(\ell - m)!}{2(\ell + m)!} \right)^{1/2} P_\ell^m(\cos(\theta)) \tag{F-17}$$

This equation is valid only for nonnegative values of the integer m, but Equation (F-13) contains only the square of m, so the solution of the Schrödinger equation is the same for any integral value of m and for the negative of that value. Therefore

$$\Theta_{\ell m} = \Theta_{\ell, -m} \tag{F-18}$$

We insert a comma to avoid confusing two subscripts having values ℓ and $-m$ with a single subscript having a value equal to $\ell - m$. Equations (F-15)–(F-17) can be used to construct all of the Θ functions in normalized form.

The R functions obey the equation

$$r^2 \frac{d^2 R}{dr^2} + 2r \frac{dR}{dr} - \frac{2\mu r^2}{\hbar^2} \left(E + \frac{Ze^2}{4\pi\varepsilon_0 r} \right) R + \ell(\ell + 1)R = 0 \tag{F-19}$$

We make the following substitutions:

$$\alpha^2 = -\frac{2\mu E}{\hbar^2}, \qquad \beta = \frac{Z^2 e}{4\pi\varepsilon_0 \alpha\hbar^2}, \qquad \rho = 2\alpha r \tag{F-20}$$

The resulting equation is divided by ρ^2, giving the **associated Laguerre equation**:

$$\frac{d^2 R}{d\rho^2} + \frac{2}{\rho} \frac{dR}{d\rho} - \frac{R}{4} + \frac{\beta R}{\rho} + \frac{\ell(\ell + 1)R}{\rho^2} = 0 \tag{F-21}$$

This has the solution

$$R(\rho) = G(\rho)e^{-\rho/2} \tag{F-22}$$

where $G(\rho)$ is a power series

$$G(\rho) = \sum_{j=1}^{\infty} a_j \rho^j \tag{F-23}$$

---

[4] J. C. Davis, Jr., *Advanced Physical Chemistry*, Ronald Press, New York, 1965, pp. 596ff.

with constant coefficients $a_1, a_2, a_3, \ldots$. This series must terminate after a finite number of terms in order to keep the wave function from becoming infinite for large values of ρ.[5] In order for the series to terminate, making it into a polynomial, the parameter β must equal an integer, n. The variable ρ is given by

$$\rho = 2\alpha r = \frac{2Zr}{na} \tag{F-24}$$

where a is the Bohr radius, given by

$$a = \frac{\hbar^2 4\pi\varepsilon_0}{\mu_e^2} \tag{F-25}$$

The polynomial G is given in terms of associated Laguerre functions:

$$G(\rho) = N_{n\ell}\rho^\ell L_{n+1}^{2\ell+1}(\rho) \tag{F-26}$$

where $N_{n\ell}$ is a normalizing factor,

$$N_{n\ell} = \left[\left(\frac{2Z}{na}\right)^3 \frac{(n-\ell-1)!}{2n[(n+\ell)!]^3}\right]^{1/2} \tag{F-27}$$

The associated Laguerre functions are

$$L_u^s(\rho) = \frac{d^s}{d\rho^s} L_u(\rho) \tag{F-28}$$

where L_u is the Laguerre polynomial

$$L_u(\rho) = e^\rho \frac{d^u}{d\rho^u} (\rho^u e^{-\rho}) \tag{F-29}$$

---

[5] *ibid.*

G Symbols Used in This Book

Efforts have been made to avoid using the same letter for two or more different variables in the same section, but use of the same letter for several quantities in different sections is unavoidable. In some cases, the section in which the quantity first appears is listed with the symbol. In addition to the quantities in this list, various letters are used for constants and variables that appear in only one section.

Symbol	Quantity
a	parameter in the van der Waals equation of state, as well as in other equations of state (Section 1.4)
a	distance of closest ionic approach in Debye-Hückel theory (Section 6.4)
a	Bohr radius (Sections 9.3 and 11.2)
a_0	Bohr radius with nucleus assumed infinitely massive (Section 11.2)
a_i	activity of substance i (Section 6.3)
a	absorptivity (Section 15.1)
a_j	coupling constant (Section 15.8)
$\mathbf{a}$	acceleration
$\mathscr{A}$	area
A	Helmholtz energy
A_n	nth pressure virial coefficient (Section 1.4)
$A(\lambda)$	absorbance of an absorbing solution at wavelength λ (Section 15.1)
A	pre-exponential factor in Arrhenius expression (Section 19.6)
$\hat{A}$	general symbol for an operator (Section 10.2)
$[\hat{A}, \hat{B}]$	commutator of two operators (Section 10.2)
b	parameter in the van der Waals equation of state, as well as in other equations of state
b	parameter in extended Debye-Hückel theory (Section 6.4)
b	impact parameter for a molecular collision (Section 19.7)
B_n	nth virial coefficient (Section 1.4)
$\mathbf{B}$	magnetic induction (also called magnetic field) (Sections 9.2 and 15.8)
B	rotational constant (Section 14.2)
$\tilde{B}$	rotational constant $= B/c$ (Section 14.2)
c	speed of light

c	speed of propagation of any wave
c	number of components (Section 5.2)
c_i	concentration of substance i, also denoted by $[\mathscr{F}_i]$ (Section 6.2)
c°	1 mol L^{-1} (exactly) (Section 6.3)
C_V	heat capacity at constant volume
C_P	heat capacity at constant pressure
$\hat{C}_n$	a rotation operator (Section 13.1)
d	collision diameter (Section 16.8)
d	reaction diameter (Section 19.2)
d	distance between crystal planes (Section 22.2)
$\mathbf{D}$	electric displacement vector (Section 9.2)
D_e	dissociation energy (Section 14.2)
$\tilde{\mathscr{D}}$	parameter in molecular energy level (Section 14.2)
D_i	diffusion coefficient of substance i (Section 17.2)
D	Debye function (Section 22.3)
e	base of natural logarithms, 2.71828182846...
e	charge on a proton
E	general symbol for a mechanical energy
E_c	center-of-mass energy (Section 11.1)
E_r	relative energy (Section 11.1)
$E^{(n)}$	perturbation correction to an energy (Section 12.2)
E	cell voltage (Section 8.2)
E°	cell voltage in the standard state (Section 8.2)
$\mathbf{E}$	electric field (Section 8.1)
E_{AB}	average energy of a bond between elements A and B
E_a	activation energy in Arrhenius expression (Section 19.6)
F	Faraday constant (Section 8.1)
f	number of independent intensive variables, or variance, given by the Gibbs phase rule (Section 5.2)
f	usual symbol for a probability density, especially for molecular velocity components (Section 16.2)
$f_v(v)$	probability density for molecular speeds (Section 16.3)
f	friction coefficient (Section 17.4)
f	electrostatic factor for diffusion-controlled reaction (Section 19.2)
f'	fraction of collisions leading to reaction (Section 19.1)
f	tension force of a sample of rubber (Section 22.7)
$\mathbf{F}$	force
$\mathscr{F}_i$	abbreviation for the chemical formula of substance i
g	probability density for molecular velocity in three dimensions (Section 16.2)
g	acceleration due to gravity
g_i	degeneracy of level i
$g(\lambda)\,d\lambda$	number of possible standing waves in wavelength range $d\lambda$ (Section 9.3)
g	electron g-factor (Section 15.8)
g_N	nuclear g-factor (Section 15.8)

g_c	coordinate probability distribution (Section 16.6)
$g(r)$	radial distribution function (Section 22.5)
G	probability density for coordinates and velocities (Section 16.6)
G	Newtonian constant of gravitation
G	Gibbs energy
$G^{(\sigma)}$	surface Gibbs energy (Section 5.5)
ΔG_{mix}	Gibbs energy change of mixing (Section 6.3)
G^E	excess Gibbs energy (Section 6.3)
h	Planck's constant
$\hbar = h/2\pi$	Planck's constant divided by 2π
$h, k,$ and ℓ	Miller indices (Section 22.2)
H	enthalpy
$\mathbf{H}$	magnetic field strength (Section 9.2)
$\Delta H_{int, i}$	integral heat of solution of substance i (Section 6.3)
$\Delta H_{diff, i}$	differential heat of solution of substance i (Section 6.3)
$\mathcal{H}(p, q)$	classical Hamiltonian, equal to the energy E, expressed in terms of coordinates and conjugate momenta
$\hat{H}$	Hamiltonian operator
$\hat{i}$	inversion operator (Section 13.1)
i	degree of inhibition (Section 19.6)
$\mathbf{i}$	unit vector in x direction
I	ionic strength (Section 6.4)
I	moment of inertia (Sections 14.2 and 14.3)
$\mathbf{I}$	nuclear spin angular momentum (Section 15.8)
I	quantum number for a nuclear spin angular momentum
I	electric current
$\mathbf{j}$	unit vector in y direction
$\mathbf{j}$	electric current density (Section 17.5)
$\mathbf{J}$	total angular momentum (Section 11.2)
J	quantum number for a total angular momentum (Section 11.3)
J	quantum number for the rotation of a molecule (Section 14.2)
J_{ij}	spin-spin coupling constant for nuclei i and j (Section 15.8)
J	rate of absorption of photons (Section 19.4)
$\mathbf{J}_i$	diffusion flux of substance i (Section 17.2)
J	analogue to Gibbs energy (Section 22.7)
k	compression (Section 1.5)
k_i	Henry's law constant (Section 6.2)
k	force constant (Section 9.1)
k	usual symbol for a rate constant (Section 18.1)
$\mathbf{k}$	unit vector in z direction
k_B	Boltzmann's constant
$k_{\pm}^{(m)}$	proportionality constant for the vapor pressure of a volatile electrolyte (Section 7.4)
$\mathcal{K}$	kinetic energy
K	parameter in the Tait equation (Section 1.5)
K_d	Nernst Distribution law constant (Section 6.2)

K_b	boiling point elevation constant (Section 6.6)
K_f	freezing point depression constant (Section 6.6)
K	equilibrium constant (Section 7.1)
K	analogue to enthalpy (Section 22.7)
K	proportionality constant in relation between intrinsic viscosity and average molecular mass (Section 22.8)
K_a	acid ionization constant (Section 7.5)
K_m	Michaelis-Menten constant (Section 19.5)
ℓ	quantum number for the magnitude of the orbital angular momentum in a hydrogen-like atom
L	usual symbol for a length
$L(T)$	function in the Tait equation (Section 1.5)
$\mathbf{L}$	angular momentum, usually the orbital angular momentum (Section 11.1)
L	quantum number for the magnitude of an orbital angular momentum
m	general symbol for a mass
m_e	electron rest mass
m_i	molality of substance i (Section 6.2)
$m_\pm$	mean ionic molality (Section 7.4)
m°	1 mol kg^{-1} (exactly) (Section 6.2)
m_N	neutron rest mass
m	quantum number for the z projection of the orbital angular momentum in a hydrogen-like atom
m_s	quantum number for the z projection of the spin angular momentum of a single particle
M	molecular mass of a polymer (Section 22.6)
$\bar{M}_n$	number-average molecular mass of a polymer (Section 22.6)
$\bar{M}_w$	mass-average molecular mass of a polymer (Section 22.6)
$\bar{M}_v$	viscosity-average molecular mass of a polymer (Section 22.6)
M_i	molar mass of substance i
M_i	effective mass of normal mode i (Section 14.3)
M	symbol sometimes used for a sum of masses
M_I	quantum number for the s component of a nuclear spin angular momentum (Section 15.8)
M_J	quantum number for the z component of a total angular momentum of an atom or molecule (Section 11.3)
M_L	quantum number for the z component of an orbital angular momentum of an atom or molecule (Section 11.3)
M_S	quantum number for the z component of a spin angular momentum of an atom or molecule (Section 11.3)
n_i	the amount of substance i (measured in moles)
$n_j(r)$	charge density around a given ion (Section 6.4)
$n_i^{(\sigma)}$	surface amount of substance i (Section 5.5)
n	principal quantum number of a hydrogen-like atom
n	refractive index
$n^{(2)}$	two-body reduced distribution function (Section 22.5)
N_A	Avogadro's constant

N_i	number of molecules of substance i
$\{N\}$	notation for a distribution $N_1, N_2, N_3, \ldots$
$\mathcal{N}_i$	number density of substance i
$\mathbf{p}$	momentum vector
p	a general abbreviation for conjugate momenta
p	a general symbol for a probability
p_i	probability of state i (Section 1.7)
p	number of phases (Section 5.2)
p	fraction of functional groups reacted (Section 22.8)
pH	$-\log_{10}[a(\mathrm{H}^+)]$
pK_a	$-\log_{10}(K_a)$
P	pressure
P_i	partial pressure of substance i (Section 6.1)
P_c	critical pressure (Section 1.6)
P_r	reduced pressure (Section 1.6)
P_{ext}	externally imposed pressure (Section 2.1)
P_i^*	vapor pressure of pure substance i (Section 6.3)
P°	standard pressure, equal to 1 bar
q	an amount of heat
q	a general abbreviation for coordinates
$\mathbf{q}$	heat flux (Section 17.2)
Q	electric charge (Section 6.4)
Q	activity quotient of a chemical reaction (Section 7.1)
r	usual symbol for a radius or a scalar distance
$\mathbf{r}$	position vector
r	rate of a chemical reaction (Section 18.1)
r	resistivity (Section 17.5)
R	gas constant
$\mathscr{R}$	Rydberg constant
R_{nl}	radial factor in the wave function of a hydrogen-like atom
R	electrical resistance (Section 17.5)
R	internuclear distance (also denoted by r)
$\mathscr{R}$	Reynolds number (Section 17.2)
S	entropy
$\mathbf{S}$	Poynting's vector (Section 9.2)
$\mathbf{S}$	spin angular momentum vector (Section 11.2)
S	quantum number for the magnitude of a spin angular momentum (Section 11.2)
S	overlap integral (Section 13.2)
S	sedimentation coefficient (Section 17.4)
t	time
$t_{1/2}$	half-life of a chemical reaction (Section 18.2)
t_C	Celsius temperature
T	absolute temperature
T_c	critical temperature (Section 1.6)
T	tension force on a vibrating string (Section 9.2)

$T(\lambda)$	transmittance of an absorbing solution at wavelength λ (Section 15.1)
T_r	reduced temperature (Section 1.6)
$u(x, y, z)$	potential energy of one molecule (Section 16.6)
$u(r)$	potential energy function for a pair of molecules (Section 16.7)
$\mathbf{u}$	flow velocity (Section 17.1)
$u(r)$	pair potential energy (Section 16.7)
$u(x)$	Bloch function (Section 22.4)
U	thermodynamic energy (also called internal energy)
U	energy density of an electromagnetic field (Section 9.3)
$\mathbf{v}$	velocity vector
v	speed (magnitude of a velocity)
$\langle v \rangle$	mean speed (Section 16.3)
v_p	most probable speed (Section 16.3)
v_{rms}	root-mean-square speed (Section 16.3)
v	vibrational quantum number
V	volume
$\bar{V}$	molar volume
$\bar{V}_c$	critical molar volume (Section 1.6)
V_r	reduced volume (Section 1.6)
v_{rel}	relative speed (Section 16.8)
V	voltage (Section 17.4)
$\mathscr{V}$	potential energy
w	an amount of work
w_{net}	work other than compression work (Section 4.1)
w_{rev}	reversible work (Section 2.1)
w_i	mass of substance i (Section 6.2)
$w(\lambda)\,d\lambda$	energy per unit volume in wavelength range $d\lambda$ (Section 9.3)
W	variational energy
$W(\{N\})$	number of system states corresponding to the distribution $\{N\}$ (Section 20.2)
W_x	mass fraction of polymer molecules with degree of polymerization x (Section 22.6)
x_i	mole fraction of substance i
X_i	electronegativity of substance i (Section 13.4)
x_e	parameter in molecule energy level (Section 14.2)
x	degree of polymerization (Section 22.6)
$\mathbf{X}$	external force (Section 21.3)
X_x	number fraction of polymer molecules with degree of polymerization x (Section 22.6)
Y	letter used to stand for a general extensive thermodynamic variable such as G, H, U, or V
$\bar{Y}$	symbol used to stand for a general molar or mean molar quantity, such as $\bar{G}$ or $\bar{V}$ (Section 4.4)
$\Delta\bar{Y}_f(i)$	symbol used to stand for a general molar quantity of formation, such as $\Delta\bar{G}_f(i)$ or $\Delta\bar{H}_f(i)$ (Section 2.7 and 7.1)
$\bar{Y}_i$	symbol used to stand for a general partial molar quantity, such as $\bar{G}_i$ or $\bar{V}_i$ (Section 4.4)

$\bar{Y}_i^\circ$	symbol used to stand for a general partial molar quantity, such as $\bar{G}_i$ or $\bar{V}_i$ in the standard state (Section 4.4)
$\bar{Y}_i^*$	symbol used to stand for a general partial molar quantity, such as $\bar{G}_i$ or $\bar{V}_i$ in the pure state (Section 6.1)
ΔY_{mix}	symbol used to stand for a general mixing quantity such as the change in Gibbs energy on mixing, ΔG_{mix} (Section 6.3)
$Y(\theta, \phi)$	spherical harmonic function (Section 11.1)
$z_{i(j)}$	rate of collisions of one molecule of substance i with molecules of substance j (Section 16.8)
z_i	valence of ion i (Section 8.1)
z^*	complex conjugate of z (Section 10.2)
z	molecular partition function (Section 20.3)
Z	compression factor of a gas (Section 1.4)
Z_{ij}	total rate of collisions per unit volume of molecules of substance i with molecules of substance j (Section 16.6)
Z	atomic number; number of protons in a nucleus
Z	canonical partition function (Section 20.7)
α	coefficient of thermal expansion (Section 1.5)
α_L	coefficient of linear thermal expansion (Section 1.5)
α, α'	critical exponent for the heat capacity (Section 5.4)
α	parameter in Debye-Hückel theory (Section 6.4)
α	parameter in hydrogen atom wave function (Section 11.2)
α	spin function (spin up) (Section 11.2)
$\tilde{\alpha}$	parameter in molecule energy level (Section 14.2)
α	polarizability (Section 15.7)
α	angle of rotation of polarized light (Section 15.7)
$[\alpha]$	specific rotation (Section 15.7)
α	Lagrange multiplier (Section 20.2)
α	transfer coefficient or symmetry factor (Section 21.4)
α	degree of elongation of a sample of rubber (Section 22.8)
β	critical exponent for the density (Section 5.4)
β	parameter in Debye-Hückel theory (Section 6.4)
β	spin function (spin down) (Section 11.2)
β_e	Bohr magneton (Section 15.8)
β_N	nuclear magneton (Section 15.8)
β	Lagrange multiplier (Section 20.2)
γ, γ'	critical exponent for the compressibility (Section 5.4)
γ	surface or interfacial tension (Section 5.5)
γ_i	activity coefficient of substance i (Section 6.3)
$\gamma_\pm$	mean ionic activity coefficient (Section 6.4)
δ	critical exponent for the pressure (Section 5.4)
δ	chemical shift in NMR spectroscopy (Section 15.8)
Δ	symbol for a difference or an increment
ε	permittivity of a medium
ε_0	permittivity of the vacuum (Section 8.1)
ε_i	energy eigenvalue of molecule state i (Section 20.2)

ε_F	Fermi level (Section 22.4)
$\zeta(t)$	time factor in a wave function (Section 9.4)
ζ	configuration integral (Section 20.8)
ζ	reaction coordinate (Section 21.1)
η	efficiency or coefficient of performance (Section 3.1)
η	spectral radiant emittance (Section 9.3)
η	viscosity coefficient (Section 17.2)
η_{sp}	specific viscosity (Section 22.8)
$[\eta]$	intrinsic viscosity (Section 22.8)
η	overpotential (Section 21.4)
θ	thermodynamic temperature (Section 3.1)
θ	angle coordinate in spherical polar coordinates. Also often used for an arbitrary angle
θ	contact angle (Section 5.5)
θ	fraction of a surface covered by adsorbed molecules (Section 19.5)
$\Theta(\theta)$	angular factor in an atomic wave function (Section 11.1)
Θ_D	Debye temperature (Section 22.3)
Θ_E	Einstein temperature (Section 22.3)
κ_S	adiabatic compressibility (Section 4.5)
κ_T	isothermal compressibility (Section 1.5)
κ	parameter in Debye-Hückel theory, reciprocal of Debye length (Section 6.4)
κ	thermal conductivity (Section 17.2)
λ	mean free path (Section 16.8)
$\lambda_{i(j)}$	mean free path between collisions of one molecule of substance i with molecules of substance j (Section 16.8)
λ	wavelength
λ	perturbation parameter (Section 12.2)
Λ_i	molar conductivity of substance i (Section 17.5)
Λ	quantum number for the magnitude of the z component of the orbital angular momentum (Section 13.3)
μ	reduced mass (Section 11.1 and Appendix D)
μ	usual symbol for the mean value of a distribution (Section 16.1)
μ_i	chemical potential of component i
μ_i°	chemical potential of component i in the standard state
μ_i^*	chemical potential of component i in the pure liquid or solid state
$\mu_{i(\text{chem})}$	chemical part of the chemical potential (Section 8.1)
μ	permeability of a medium
μ_0	permeability of the vacuum
$\boldsymbol{\mu}$	electric dipole vector (Section 13.4)
μ_B	Bohr magneton (Section 15.8)
μ_N	nuclear magneton (Section 15.8)
ν_i	stoichiometric coefficient of substance i (Section 2.7)
$\nu_\pm$	mean ionic stoichiometric coefficient (Section 6.4)
ν	frequency of oscillation of a photon or an oscillating object
$\tilde{\nu}$	reciprocal wavelength; photon frequency divided by the speed of light

$\tilde{\nu}_e$	vibrational parameter in molecule energy level (Section 14.3)
ν	rate of wall collisions (Section 16.5)
ν_D	Debye frequency (Section 22.3)
ξ	reaction progress variable (Section 7.1)
π	3.14159265359...
Π	symbol for a product
Π	osmotic pressure (Section 6.6)
ρ	density
ρ	mass per unit length of a vibrating string (Section 9.2)
ρ	radial variable in hydrogen atom wave function (Section 11.2)
ρ_c	charge density (Section 21.4)
σ	Stefan-Boltzmann constant (Section 9.3)
σ	standard deviation (Section 10.4)
σ	electrical conductivity (Section 17.5)
$\hat{\sigma}$	a reflection operator (Section 13.1)
σ	shielding constant (Section 15.8)
σ	symmetry number (Section 14.3)
Σ	symbol for a sum
τ	a period of time, especially a relaxation time or a period of an oscillation
$\tau_{i(j)}$	mean time between collisions of one molecule of substance i with molecules of substance j (Section 16.8)
τ	tension force (Section 22.7)
φ	electrical potential (Section 8.1)
φ	steric factor (Section 19.6)
ϕ	angular coordinate in spherical polar and cylindrical polar coordinates
ϕ	osmotic coefficient (Section 8.4)
$\varphi(L)$	function of L in theory of rubber elasticity (Section 22.7)
ϕ	volume fraction of a suspension of spheres or polymer molecules (Section 22.8)
Φ	quantum yield of a photochemical reaction (Section 15.6)
$\Phi(\phi)$	an angular factor in a central-force wave function (Section 11.1)
χ	center of mass wave function (Section 11.2)
ψ	usual symbol for a coordinate wave function; usual symbol for an orbital
Ψ	usual symbol for a time-dependent wave function or for a coordinate wave function of a multiparticle system
ω	angular speed (Section 17.4)
Ω	number of possibly occupied mechanical states (thermodynamic probability) (Sections 3.4 and 20.5)
∇	gradient operator (Section 17.1)
$\nabla \cdot$	divergence operator (Section 17.1)
∇^2	Laplacian operator (Section 9.2)

Answers

CHAPTER 1

Answers to Numerical Exercises

Exercise 1.1

$R = 0.082058$ L atm K^{-1} $mol^{-1} = 1.9872$ cal K^{-1} mol^{-1}
$ = 82.058$ cm^3 atm K^{-1} mol^{-1}
$ = 83145$ cm bar K^{-1} mol^{-1}

Exercise 1.2b

$P = 4.8157 \times 10^6$ Pa $= 48.1547$ bar $= 47.527$ atm
Ideal value: 4.9579×10^6 Pa $= 48.931$ atm, for a difference of 2.87%

Exercise 1.3

$$V = n\bar{V}_0(1 - \kappa'(P - P_0) + \alpha'(T - T_0))$$

where κ' and α' are constants.

Exercise 1.4

a. $\dfrac{\Delta V}{V} = 2.9 \times 10^{-4}$

b. $\kappa_T = \dfrac{1}{P}$, $\alpha = \dfrac{1}{T}$

c. $\kappa_T = 9.869 \times 10^{-6}$ $Pa^{-1} = 0.9869$ $bar^{-1} = 1.000$ atm^{-1}

$\dfrac{\kappa_T(\text{ideal gas})}{\kappa_T(\text{water})} = 21595$

d. $\alpha = 3.354 \times 10^{-3}$ K^{-1}

$\dfrac{\alpha(\text{ideal gas})}{\alpha(\text{water})} = 16.2$

Exercise 1.6

a. $T_{\text{Boyle}} = \dfrac{a}{bR}$

b. $T_{\text{Boyle}} = 505$ K

d. $\bar{V} = 1.06 \times 10^{-4}$ m^3 $mol^{-1} = 0.106$ L mol^{-1}
$ P = 211$ atm

Exercise 1.7

a. $\dfrac{\Delta V}{V} = 0.080 = 8.0\%$

b. $\kappa_T = -0.0138 = -1.38\%$

Exercise 1.9

$\kappa_T = 9.75 \times 10^{-5}$ $bar^{-1} = 9.75 \times 10^{-10}$ Pa^{-1}

Exercise 1.12

c. $a = 4bRT_c = 0.176$ J m^3 mol^{-2}

$b = \dfrac{RT_c}{e^2 P_c} = 4.19 \times 10^{-5}$ m^3 mol^{-1}

Exercise 1.13

$a = 1.56$ Pa $K^{1/2}$ m^6 mol^{-2}
$b = 2.68 \times 10^{-5}$ J mol^{-1} $Pa^{-1} = 2.68 \times 10^{-5}$ m^3 mol^{-1}

Exercise 1.16 $\langle w \rangle = 78.8$

Exercise 1.17

$$\langle w^2 \rangle = 6348$$
$$w_{\text{rms}} = 79.7$$

Exercise 1.18 $P = 0.55$ atm

Answers to Selected Problems

Problem 1.20 $V = 0.02514$ $m^3 = 25.14$ L

Problem 1.25

a. $\dfrac{\Delta V}{V} = -1.081 \times 10^{-4} = -0.01081\%$

b. $\dfrac{\Delta V}{V} = 1.2 \times 10^{-2} = 1.2\%$

Problem 1.31b $\kappa_T = 9.88 \times 10^{-6}$ Pa^{-1}

Problem 1.37
Van der Waals: $\quad Z = 1.1434 \quad$ (0.86% error)
Dieterici: $\quad\quad\quad Z = 1.1255 \quad$ (0.71% error)
Redlich-Kwong: $\quad Z = 1.1153 \quad$ (1.6% error)

Problem 1.40 You could save about 0.6%.

Problem 1.43 $\bar{V} = 39.02 \times 10^{-6}$ m^3 mol^{-1}

Problem 1.48

a. Probability $= \dfrac{1}{52} \dfrac{1}{52} = 0.0003698$

b. Probability $= \dfrac{1}{52}\dfrac{1}{51} = 0.0003771$

c. Probability $= 2\dfrac{1}{52}\dfrac{1}{51} = 0.0007541$

Problem 1.52
Assume that P_0, the pressure at the ground floor, equals
1.000 atm. Assume that the temperature is 298 K. Assume
that air is a single gas with $M = 0.029$ kg mol^{-1}.
$$\Delta P = -0.014 \text{ atm}$$

CHAPTER 2

Answers to Numerical Exercises

Exercise 2.1 $w = 912$ J

Exercise 2.2

b. $\displaystyle\int dP = -3.10 \times 10^5$ Pa $= -3.06$ atm

Exercise 2.3
a. $w_{\text{surr}} = 507$ N m $= 507$ J
b. $\Delta P = -3.10 \times 10^5$ Pa $= -3.06$ atm

Exercise 2.4
b. $w_{\text{rev}} = -2145$ J   **c.** $w = -506.6$ J

Exercise 2.5
a. $q = 13567$ J $= 1.357$ kJ   **b.** $T = 350.6$ K, $t_C = 77.5°$C

Exercise 2.6 $m = 674$ g

Exercise 2.7 $\Delta T = 0.0351$ K

Exercise 2.8
a. $E_{\text{rest-mass}} = 3.5903 \times 10^{15}$ J
b. $U = 3718$ J
ratio $= 1.036 \times 10^{-12}$
$\Delta m = 4.14 \times 10^{-14}$ kg

Exercise 2.9 $q = 5066$ J

Exercise 2.12
a. $\Delta U = 83$ J, $w = -1824$ J, $q = 1907$ J
For an ideal gas, $\Delta U = 0$, $w = -1824$ J, $q = 1824$ J
b. $\Delta T = -6.66$ K

Exercise 2.13
a. $w = 5690$ J, $q = 6708$ J
b. $\Delta U = 78$ J, $q = 8151$ J, $w = -7138$ J
c. $\Delta U = 1013$ J

Exercise 2.14
a. $T_2 = 179.4$ K   **b.** $V = 21.00$ L

Exercise 2.18 $T = (1.062)(298 \text{ K}) = 316$ K or $t_C = 33°$C

Exercise 2.19
a.
At 298.15 K, $\bar{C}_P = 29.37$ J K^{-1} mol^{-1} (0.93% difference
from $7R/2$)
At 500 K, $\bar{C}_P = 31.422$ J K^{-1} mol^{-1} (7.98% difference
from $7R/2$)

b. For iron: $\bar{C}_P = 25.12$ J K^{-1} mol^{-1}
For copper: $\bar{C}_P = 24.56$ J K^{-1} mol^{-1}
(0.7% difference for iron, -1.5% difference for copper.)

Exercise 2.22
a. $\Delta U = 935$ J   **b.** $\Delta H = 1559$ J, $q = 3086$ J, $w = -2151$ J

Exercise 2.23
Assume that the $CaCO_3(s)$ is calcite, not aragonite.
$$\Delta H° = 178.31 \text{ kJ mol}^{-1}$$

Exercise 2.24 $\Delta H°_{473\,\text{K}} = 177.85$ kJ mol^{-1}

Exercise 2.25 $T_f = 5229$ K

Exercise 2.26 $T_f = 7263$ K

Exercise 2.27 $\Delta(PV) = -7437$ J mol^{-1}

Exercise 2.28 $\Delta U° = 175.83$ kJ mol^{-1}

Exercise 2.29 $\Delta T = 4.360$ K

Exercise 2.30
$$\Delta H \approx -794 \text{ kJ mol}^{-1}$$
Compare with correct value of -802.301 kJ mol^{-1}.
The first answer is incorrect by only 8 kJ mol^{-1}.

Answers to Selected Problems

Problem 2.33

a. $w = -RT \ln\left(\dfrac{\bar{V}_2 - b}{\bar{V}_1 - b}\right) + \dfrac{a}{\bar{V}_2} - \dfrac{a}{\bar{V}_1}$

b. For CO_2, $a = 0.3640$ Pa m^6 mol^{-2},
$b = 4.267 \times 10^{-5}$ m^3 mol^{-1}, $w = -3507$ J mol^{-1}
c. $w_{\text{surr}} = 1520$ J

Problem 2.37
$$q = 203300 \text{ J}$$
$$w = -15.51 \text{ kJ}$$
$$\Delta U = 187.8 \text{ kJ}$$

Problem 2.40
a. $T_2 = 171.2$ K   **b.** $T_2 = 170.8$ K

Problem 2.46
a. $t_{\text{final}} = -0.88°$C

b. Assume that $\bar{C}_P = \dfrac{7}{2}R$: $\left(\dfrac{\partial \bar{H}}{\partial P}\right)_T = -1.28$ J mol^{-1} atm^{-1}

Problem 2.49
b. $\Delta H° = -197.846$ kJ mol^{-1}, $\Delta U° = -195.367$
d. $\Delta H° = 1236.79$ kJ mol^{-1}, $\Delta U° = 1234.31$ kJ mol^{-1}

Problem 2.52
a. $\Delta \bar{H}°_{f\,373\,\text{K}}(CH_4) = -76.56$ kJ mol^{-1}
$\Delta \bar{H}°_{f\,373\,\text{K}}(CO_2) = -392.98$ kJ mol^{-1}
$\Delta \bar{H}°_{f\,373\,\text{K}}(H_2O) = -283.443$ kJ mol^{-1}
b. $\Delta H° = -883.306$ kJ mol^{-1}
c. $\Delta H° = -883.282$ kJ mol^{-1}

Problem 2.55
a. $\Delta H^\circ \approx -2796$ kJ mol^{-1}
b. $\Delta H^\circ_{298\,K} = -2855.68$ kJ mol^{-1}
$\Delta \bar{H}^\circ_{f\,373\,K} = -2852.74$ kJ mol^{-1}

Problem 2.58
a. Assume constant volume. $\Delta H = 180$ J, $\Delta U = 0$, $q = 0$,
$w = 0$
b. Assume constant volume. $\Delta H = 180$ J, $\Delta U = 0$, $q = 0$,
$w = 0$

CHAPTER 3

Answers to Numerical Exercises

Exercise 3.2 $\eta_c = 0.254$

Exercise 3.3
a. $\eta_{hp} = 17.6$ b. $\eta_r = 15.4$

Exercise 3.6
a. $\Delta S = 40.14$ J K^{-1} b. $\Delta S_{surr} = -\Delta S = -40.14$ J K^{-1}

Exercise 3.7
$$\Delta H_{vap} = 87.86 \text{ kJ}$$
$$q = 87.86 \text{ kJ}$$
$$\Delta S = 249.8 \text{ J K}^{-1}$$

Exercise 3.8
$$\Delta S = na \ln\left(\frac{T_2}{T_1}\right) + nb(T_2 - T_1) + \frac{nc}{2}\left(\frac{1}{T_2^2} - \frac{1}{T_1^2}\right)$$

Exercise 3.9
Assume that argon is an ideal gas, and assume that $\bar{C}_V$ is
constant and equal to $3R/2$.
$$\Delta S = 24.31 \text{ J K}^{-1}$$

Exercise 3.10
$$\text{entropy production} = \frac{dS_{univ}}{dt} = 0.00293 \text{ J K}^{-1} \text{ s}^{-1}$$

Exercise 3.11
$$\Delta S = 23.05 \text{ J K}^{-1}$$
$$\Delta S_{surr} = 0$$
$$\Delta S_{univ} = 23.05 \text{ J K}^{-1}$$

Exercise 3.12
a. $\Delta S_{mix} = 4.64$ J K^{-1} b. $\Delta S_{mix} = 3.57$ J K^{-1}

Exercise 3.13
b. $N = 1296$ c. $N = 21$

Exercise 3.14 $\Omega = 10^{2.5 \times 10^{25}}$

Exercise 3.18
$$\Delta S_{surr} = 1898.3 \text{ J K}^{-1}$$
$$\Delta S_{univ} = 1725.5 \text{ J K}^{-1}$$

Exercise 3.19 $S_{st} = 14.90$ J K^{-1} mol^{-1}

Answers to Selected Problems

Problem 3.20
$$P_4 = 9.61 \text{ atm}$$
$$P_3 = 2.40 \text{ atm}$$
$$V_1 = 0.205 \text{ L}$$
$$V_2 = 0.821 \text{ L}$$
$$V_3 = 1.28 \text{ L}$$
$$V_4 = 0.318 \text{ L}$$

Problem 3.24
$\Delta S_1 = 1.15$ J K^{-1}, $\Delta S_2 = 0$, $\Delta S_3 = -1.15$ J K^{-1}, $\Delta S_4 = 0$
$\Delta S_{cycle} = 0$

Problem 3.28
a. $\Delta S = 9.370$ J K^{-1}
b. $\Delta S = 9.370$ J K^{-1}
c. $\Delta S_{surr} = -9.370$ J K^{-1}
d. $\Delta S = 9.370$ J K^{-1}
$\Delta S_{surr} = -7.275$ J K^{-1}

Problem 3.32
a. $\Delta \bar{S}^\circ = -326.607$ J K^{-1} mol^{-1}
c. Assume that the $CaCO_3$ is calcite, not aragonite.
$\Delta \bar{S}^\circ = 159.1$ J K^{-1} mol^{-1}

Problem 3.35 $\Delta S^\circ_{473\,K} = -97.8$ J K^{-1} mol^{-1}

Problem 3.37
a. $\bar{S}^\circ(\text{solid}, 231.49 \text{ K}) = 111.04$ J K^{-1} mol^{-1}
b. $\bar{S}^\circ(\text{liquid}, 231.49 \text{ K}) = 146.80$ J K^{-1} mol^{-1}
$\bar{S}^\circ(\text{liquid}, 298.15 \text{ K}) = 178.75$ J K^{-1} mol^{-1}

Problem 3.41
a. F b. F c. T d. F e. F f. T g. T h. F

CHAPTER 4

Answers to Numerical Exercises

Exercise 4.2 $\left(\frac{\partial P}{\partial S}\right)_{V,n} = 8124$ K m^{-3}

Exercise 4.3 $\left(\frac{\partial V}{\partial S}\right)_{P,n} = \frac{nR}{C_V + nR}\frac{T}{P} = 0.00118$ K Pa^{-1}

Exercise 4.4
a. $\left(\frac{\partial S}{\partial P}\right)_{T,n} = -\frac{nR}{P}$
b. $\left(\frac{\partial S}{\partial P}\right)_{T,n} = 0.0001641$ J K^{-1} Pa^{-1}
c. $\Delta S = 23.05$ J K^{-1}

Exercise 4.5 $\bar{G}^{\circ(atm)} - \bar{G}^\circ = 32.62$ J mol^{-1}

Exercise 4.6 $f = 4.976$ atm $= 504200$ Pa $= 504.2$ kPa

Exercise 4.7 $\bar{G} - \bar{G}^\circ = 7.092 \times 10^{-2}$ J mol^{-1}

Exercise 4.8 $\Delta G_{383.15\,K} = -1090$ J mol^{-1}

The negative sign means that the process is spontaneous at 1.000 atm and 383.15 K.

Exercise 4.9 $\left(\dfrac{\partial G}{\partial n_i}\right)_{T,V,n'} = \mu_i + V\left(\dfrac{\partial P}{\partial n_i}\right)_{T,V,n'}$

Exercise 4.11 $\bar{A} = \mu^\circ + RT \ln\left(\dfrac{P}{P^\circ}\right) - RT$

Exercise 4.12
$$\mu_i - \mu_i^\circ = -11550 \text{ J mol}^{-1} = -11.55 \text{ kJ mol}^{-1}$$

Exercise 4.13

b. $\left(\dfrac{\partial U}{\partial V}\right)_{T,n} = 309$ J m^{-3}

Exercise 4.14

b. $\left(\dfrac{\partial H}{\partial P}\right)_{T,n} = -9.0 \times 10^{-5}$ m$^3 = -9.0$ J Pa^{-1}

Exercise 4.15

a. $\bar{C}_V = 74.509$ J K^{-1} mol^{-1} b. $\bar{C}_P - \bar{C}_V = 0$

Exercise 4.16 $C_V = 0.4418$ J K^{-1} g^{-1}

Answers to Selected Problems

Problem 4.25
a. $\Delta G = 1718$ J   b. $\Delta G = 1005$ J

Problem 4.28 $\bar{G}(T, P') - \bar{G}^\circ(T) = 31.03$ J mol^{-1}

Problem 4.31 $\bar{V} = 56.72 \pm 0.11$ cm^3 mol^{-1}

Problem 4.35
b. $v_s = 346$ m s^{-1}
c. $v_s = 1016$ m s^{-1}
d. For air, the ratio $= 0.741$. For helium, the ratio $= 0.809$.

Problem 4.39
a. Homogeneous, degree 2
b. Homogeneous, degree 0
e. Homogeneous, degree 4

Problem 4.41
a. F   b. F   c. F   d. F   e. F

CHAPTER 5

Answers to Numerical Exercises

Exercise 5.2
a. $c = 3$ b. $c = 3$ c. $c = 1$
d. $c = 1$ e. $c = 2$ f. $c = 1$

Exercise 5.4 $P_2 - P_1 = 1.3496 \times 10^6$ Pa $= 13.32$ atm

Exercise 5.6 $P = 0.339$ atm $= 258$ torr $= 34.4$ kPa

Exercise 5.7 $P = 0.678$ atm $= 515$ torr $= 68.7$ kPa

Exercise 5.8 $P = 1367197$ Pa $= 13.49$ atm $= 10252$ torr

Exercise 5.11
a. We assume that the three nearest neighbors in the surface are in a triangular arrangement, at the corners of hexagons that completely cover the surface. Each side of the hexagons is equal to 3.0×10^{-10} m. $\gamma = 0.142$ J m^{-2}
b. Ratio $= 2.4 \times 10^{-8}$

Exercise 5.12
a. $h = 4.2$ cm   b. $h = 0.21$ mm

Exercise 5.13 $h = -6.55$ mm

Answers to Selected Problems

Problem 5.16
$$\bar{G}^{(l)} - \bar{G}^{(s)} = 0$$
$$\bar{A}^{(l)} - \bar{A}^{(s)} = -0.167 \text{ J mol}^{-1}$$
$$\bar{H}^{(l)} - \bar{H}^{(s)} = 6006 \text{ J mol}^{-1}$$
$$\bar{U}^{(l)} - \bar{U}^{(s)} = 6006 \text{ J mol}^{-1}$$
$$\bar{S}^{(l)} - \bar{S}^{(s)} = 21.99 \text{ J K}^{-1} \text{ mol}^{-1}$$

Problem 5.19
a. $f = 3$ b. $f = 2$ c. $f = 1$ d. $f = 2$ e. $f = 0$

Problem 5.23
a. $\Delta\bar{H}_{vap} = 5.10 \times 10^4$ J mol^{-1}
b. $\Delta\bar{H}_{vap} = 4.46 \times 10^4$ J mol^{-1}
c. $\Delta\bar{H}_{fus} = 6.4 \times 10^3$ J mol^{-1}

Problem 5.27 $\Delta H_{vap} = 259400$ J mol^{-1}

Problem 5.32 $\gamma = 1.05$ J m^{-2}

Problem 5.36 $P = 44.4$ torr

Problem 5.38
a. 842 torr, which is 11% higher than the correct value.
b. 749 torr, which is much closer to the correct value than the value of part a.
c. $\Delta H = 40.88$ kJ mol^{-1}, which is 0.5% higher than the correct value.
d. $P = 451$ atm. This value is too high because ΔH is not constant.
e. $P = 138$ atm. This value is too low because ΔC_p is not constant.
f. $\Delta H = 29.4$ kJ mol^{-1}, assuming that ΔC_p is constant.
$T = 1352$ K, which is about 108% higher than the actual critical temperature.

CHAPTER 6

Answers to Numerical Exercises

Exercise 6.1
First term $= -8.323$ J mol^{-1}
Second term $= -7937$ J mol^{-1}
The first term is 0.1% of the second term.

Exercise 6.5

Any two of o-xylene, m-xylene, p-xylene
ethyl benzene and toluene
1-propanol and 2-propanol
Any two of naphthalene, anthracene, and phenanthrene
2-methylpentane and 3-methylpentane

Exercise 6.7 $x_{benzene} \approx 0.86$, boiling temperature ≈ 357 K.

Exercise 6.9 $k_1 = 985$ torr

Exercise 6.10

b. $x_{i(H_2O)}^{(eq)} = 0.000022$, $x_{i(CCl_4)}^{(eq)} = 0.00998 \approx 0.010$

Exercise 6.12 $c_{suc} = 0.0999$ mol L^{-1}

Exercise 6.14

$$(P - P^\circ) = 1.37 \times 10^6 \text{ Pa} = 13.7 \text{ bar} = 13.5 \text{ atm}$$
$$P = 14.5 \text{ atm}$$

Exercise 6.15

$$a_i^{(l)} = 0.9513$$
$$\gamma_i^{(l)} = 1.312$$

Exercise 6.16

$$a_i^{(l)} = 0.9232$$
$$\gamma_i^{(l)} = 1.026$$

Exercise 6.19 $I = 0.9$ mol kg^{-1}

Exercise 6.24

$$\gamma_\pm = 0.902$$
$$\% \text{ difference} = 0.32\%$$

Exercise 6.29   5.36%, 0.50%, 0.05%, 0.005%

Exercise 6.31

a. $K_{b1} = 0.513$ K kg mol^{-1} **b.** $t_b = 100.015°C$

Exercise 6.32

a. $P_1 = 0.99947$ atm $= 101272$ Pa   **b.** $\Delta P_{vap} = P_1^* M_1 x_2$

Exercise 6.33

b. $\Pi = 7.24 \times 10^4$ Pa $= 0.715$ atm

Answers to Selected Problems

Problem 6.36 $\gamma_2^{(m)} = 0.9275$, $\gamma_1^{(c)} = 1.036$

Problem 6.40

a. $K_d^{(c)} = 0.012$

b. $V_{thio} = \dfrac{0.000090 \text{ mol}}{0.1000 \text{ mol L}^{-1}} = 0.0090 \text{ L} = 9.0 \text{ mL}$

Problem 6.45

$P - P^\circ = 1.50 \times 10^9$ Pa $= 1.50 \times 10^4$ bar $= 1.48 \times 10^4$ atm
$a(\text{diamond}) = 7.98$, $a(\text{graphite}) = 25.9$

Problem 6.49

a. $\gamma_2^{(m)} = 1.056$ **b.** $\gamma_1^{(l)} = \dfrac{a_1^{(l)}}{x_1} = \dfrac{0.669}{0.7090} = 0.944$

Problem 6.53

a. For KCl, $\gamma_\pm = 0.926$. For FeSO$_4$, $\gamma_\pm = 0.926$

b. For KCl, $\gamma_\pm = 0.927$, % difference $= 0.1\%$
For FeSO$_4$, $\gamma_\pm = 0.871$, % difference $= 0.7\%$

Problem 6.65

a. $\Delta T_f = 0.0788$ K

b. $\Delta T = 0.0786$ K

c. $\Delta T = 0.0789$ K

Problem 6.68

a. $\Pi = 3.63 \times 10^5$ Pa, $h = 36.3$ m

b. $\Pi = 3.75 \times 10^5$ Pa, $h = 37.5$ m

c. $\Pi = 2.71 \times 10^6$ Pa, $h = 271$ m

Problem 6.70

a. F **b.** F **c.** F **d.** T **e.** T
f. T **g.** T **h.** F **i.** T **j.** T

CHAPTER 7

Answers to Numerical Exercises

Exercise 7.1

a. $\Delta G^\circ = 20.661$ kJ mol^{-1}

b. $\Delta G^\circ = 20.665$ kJ mol^{-1}; the small difference is due to small inaccuracies in the data.

Exercise 7.2

a. $\Delta H^\circ = 71.488$ kJ mol^{-1}, $\Delta S^\circ = 170.473$ J K^{-1} mol^{-1}

b. $\Delta G^\circ = 20.661$ kJ mol^{-1}

Exercise 7.4 $\left(\dfrac{\partial G}{\partial \xi}\right)_{T,P} = 24.390$ kJ mol^{-1}

Exercise 7.5

a. $P(C_3H_8) = 7.4 \times 10^{-58}$ Pa, $P(O_2) = 3.7 \times 10^{-57}$ Pa

b. $V = 5.6 \times 10^{36}$ m^3. This is a very large volume. The volume of a sphere with radius equal to the radius of the earth's orbit is 1.36×10^{34} m^3.

Exercise 7.6

$$\Delta G^{\circ(II)} = -558519 \text{ J mol}^{-1} = -558.519 \text{ kJ mol}^{-1}$$

Exercise 7.10

a. $\Delta G^\circ = 79885$ kJ mol^{-1}, $K = 1.01 \times 10^{-14}$

b. $m(H^+)/m^\circ = m(OH^-)/m^\circ = 1.005 \times 10^{-7}$

Exercise 7.11   pH $= 1.63$

Exercise 7.13

a. $\dfrac{a(H^+)a(HCO_3^-)}{a(H_2CO_3) + a(CO_2^{(aq)})} = 4.37 \times 10^{-7}$

b. pH $= 3.91$

c. pH $= 3.91$

Exercise 7.17

b. $K = 14.8$; difference $= 1.3\%$.
$\Delta G^\circ = 8.37$ kJ mol^{-1}; difference $= 0.2\%$.

Exercise 7.18 $\alpha = 0.126$

Exercise 7.20 $\left(\dfrac{\partial G}{\partial \xi}\right)_{T,P} = -16.03 \text{ kJ mol}^{-1}$

Exercise 7.21
a. $K = 1.81 \times 10^4$
b. $c(\text{ADP})/c^\circ = c(\text{PEP})/c^\circ = 7.41 \times 10^{-3}$

Exercise 7.22
b. $\Delta G^{\circ\prime} = 19.4 \text{ kJ mol}^{-1}$, $K = 48$

Answers to Selected Problems

Problem 7.24 $K = 11.25$

Problem 7.26
a.
$$\Delta H^\circ = 181.66 \text{ kJ mol}^{-1}$$
$$\Delta G^\circ = 117.078 \text{ kJ mol}^{-1}$$
$$\Delta S^\circ = 216.607 \text{ J K}^{-1}\text{ mol}^{-1}$$
$$\Delta H^\circ - T\,\Delta S^\circ = 117.08 \text{ kJ mol}^{-1}$$

Problem 7.29
a. $\Delta G^\circ = -32.734 \text{ kJ mol}^{-1}$, $K = 5.43 \times 10^5$
b. $P(\text{N}_2) = 0.01555 \text{ atm}$, $P(\text{H}_2) = 0.04665 \text{ atm}$, $P(\text{NH}_3) = 0.9378 \text{ atm}$

Problem 7.34a
$$\Delta H^\circ = -223.94 \text{ kJ mol}^{-1}$$
$$\Delta S^\circ = -216.537 \text{ J K}^{-1}\text{ mol}^{-1}$$
$$\Delta G^\circ = -159.316 \text{ kJ mol}^{-1}$$
$$\Delta H^\circ - T\,\Delta S^\circ = -159.378 \text{ kJ mol}^{-1}$$

Problem 7.37 $K_{\text{sp}} = 1.19 \times 10^{-5}$

Problem 7.42
a. 0.0863 mol of NaOH.   b. 0.0890 mole of NaOH

Problem 7.46
a. $K_{400\,\text{K}} = 43.6$
b. $x_{\text{N}_2} = 0.121$, $x_{\text{H}_2} = 0.363$, $x_{\text{NH}_3} = 0.516$

Problem 7.52 $K = 262$

Problem 7.56
d. $\Delta G^\circ = -5.248 \text{ kJ mol}^{-1}$; $K_P = 1.65$ at 1259 K
e. $\Delta G^\circ_{1259\,\text{K}} = -5988 \text{ J mol}^{-1}$; $K_P = 1.77$

CHAPTER 8

Answers to Numerical Exercises

Exercise 8.4b $E = 1.091 \text{ V}$

Exercise 8.8 $E = 1.1131 \text{ V}$

Exercise 8.9 $E = -0.0464 \text{ V}$

Exercise 8.11   pH $= 1.100$

Exercise 8.13 $E^{(\text{II})} - E^{(\text{I})} = 0.385 \text{ V}$

Exercise 8.14 $K = 7.6 \times 10^{36}$

Answers to Selected Problems

Problem 8.15 $Vc = 4 \times 10^7 \text{ mol L}^{-1}\text{ m}^{-1}$

Problem 8.18
a. Pt | Pb | Pb(NO$_3$)$_2$(aq) ‖ HCl(aq) | AgCl(s) | Ag(s) | Pt
c. $E = 0.2777 \text{ V}$

Problem 8.23
$$E = 1.0424 \text{ V}$$
$$\text{pH} = -\log_{10}[a(\text{H}^+)] = 12.88$$

Problem 8.25
a. $K = 1.42 \times 10^{-18}$ c. $K = 1.72 \times 10^{-8}$

Problem 8.29
a. $w_{\text{surr, max}} = 474.282 \text{ kJ}$
b. $w = 342 \text{ kJ}$
c. Wasted heat $= 142 \text{ kJ}$

CHAPTER 9

Answers to Numerical Exercises

Exercise 9.3
$$\sin(1.000^\circ) = \sin(0.017453) = 0.017452$$
$$\tan(1.000^\circ) = \tan(0.017453) = 0.017455$$

Exercise 9.7
Smaller by a factor of 1/16. Increased by a factor of 4.

Exercise 9.8 $\lambda = 0.768 \text{ m}$

Exercise 9.13 $T = 4.14 \times 10^3 \text{ K}$

Exercise 9.15 $\sigma = 5.671 \times 10^{-8} \text{ J m}^{-2}\text{ s}^{-1}\text{ K}^{-4}$

Exercise 9.16
a. $\lambda_{\text{threshold}} = 2.5 \times 10^{-7} \text{ m} = 250 \text{ nm}$
b. $v_{\text{max}} = 6.5 \times 10^5 \text{ m s}^{-1}$

Exercise 9.17 $|F| = 29.6 \text{ N}$

Exercise 9.20
$$\lambda = 656.467 \text{ nm}$$
$$v = 4.5667 \times 10^{14} \text{ s}^{-1} \text{ (red light)}$$

Exercise 9.21 $v = 3.38 \times 10^6 \text{ m s}^{-1}$

Exercise 9.23
If the length of the box is doubled, the energy decreases by a factor of 1/4. If the mass of the particle is doubled, the energy decreases by a factor of 1/2.

Exercise 9.26

$v = 2.728 \times 10^{14} \ s^{-1}$

$\lambda = 1.099 \times 10^{-6} \ m = 1099 \ nm$

Exercise 9.27

For the $n = 1$ state, $v = 9.09 \times 10^{13} \ s^{-1}$

For the $n = 2$ state, $v = 3.64 \times 10^{14} \ s^{-1}$

The first frequency is one-third as large as the frequency of the photon in Exercise 9.24, and the second frequency is four-thirds as large.

Exercise 9.29

State	E divided by $\dfrac{h^2}{8ma^2}$	Degeneracy
1, 1, 1	3	1
1, 1, 2; 1, 2, 1; 1, 1, 2	6	3
1, 2, 2; 2, 1, 2; 2, 2, 1	9	3
1, 1, 3; 1, 3, 1; 3, 1, 1	11	3
2, 2, 2	12	1

Exercise 9.30 $B = D + F, C = i(D - F)$

Exercise 9.37

$v = \dfrac{1}{2\pi} \sqrt{\dfrac{k}{m}}$. The frequency v is identical to the classical frequency of the oscillator.

Exercise 9.38 $v_{\text{de Broglie}} = \dfrac{3}{2} v$

Exercise 9.39

$z_t = 1.23 \times 10^{-11} \ m$, 16.6% of the bond length.

Answers to Selected Problems

Problem 9.40

$v_a = 440 \ Hz, v_b = 880 \ Hz, v_c = 1320 \ Hz, v_d = 1760 \ Hz$

Problem 9.43

$c(\text{in water}) = 2.25 \times 10^8 \ m \ s^{-1}$,

$\varepsilon = 1.57 \times 10^{-11} \ N^{-1} \ C^2 \ m^{-2}$

Problem 9.45 $\lambda = 0.00106 \ m = 1.06 \ mm$

Problem 9.48

For $n_2 = 3, \lambda = 6.5646 \times 10^{-7} \ m = 656.46 \ nm$ (red)

For $n_2 = 4, \lambda = 4.8627 \times 10^{-7} \ m = 486.27 \ nm$ (green-blue)

For $n_2 = 5, \lambda = 4.34317 \times 10^{-7} \ m$

$= 434.17 \ nm$ (blue-violet)

For $n_2 = 6, \lambda = 4.1029 \times 10^{-7} \ m = 410.29 \ nm$ (violet)

Problem 9.51 $\lambda = 2.307 \times 10^{-11} \ m$

Problem 9.55

a. $E_1 = 1.247 \times 10^{-19} \ J, E_2 = 4.989 \times 10^{-19} \ J$,

$E_3 = 1.1226 \times 10^{-18} \ J$

b. $v = 9.41 \times 10^{14} \ s^{-1}, \lambda = 3.19 \times 10^{-7} \ m = 319 \ nm$

Problem 9.58

a. $z_t = \sqrt{\dfrac{3}{a}}$, where $a = \dfrac{\sqrt{km}}{\hbar}$

b. Probability $= 0.1116 = 11.16\%$

Problem 9.61

a. $\mu = 0.99544 \ amu = 1.653 \times 10^{-27} \ kg$

b. $k = 411.8 \ kg \ s^{-2} = 411.8 \ N \ m^{-1}$

Problem 9.64 $\lambda = 1.33 \times 10^{-9} \ m$

Problem 9.66

a. F   b. F   c. T   d. T   e. F

f. T   g. F   h. T   i. T   j. F

CHAPTER 10

Answers to Numerical Exercises

Exercise 10.3 $\left[x^2, \dfrac{d}{dx} \right] = -2x$

Exercise 10.5   The commutator is $\left[x, i\dfrac{d}{dx} \right] = -i$

Exercise 10.6b

Eigenfunction $= Ae^{bx}$ where A and b are constants.

Eigenvalue $= b^4$.

Exercise 10.14 $\langle x \rangle = \dfrac{a}{2}$

Exercise 10.15   Probability $= 8.000 \times 10^{-6}$

Exercise 10.16   Probability $= 0.1955$

Exercise 10.17 $\sigma_x = 0.26583a$, probability $= 0.5225$

Exercise 10.18   Ratio $= 1/e = 0.36788$

Exercise 10.19 $\Delta x \, \Delta p_x \geq \dfrac{\hbar}{2} = \dfrac{h}{4\pi}$

Exercise 10.22 $\Delta t \geq 5 \times 10^{-14} \ s$

Exercise 10.23

b. $\langle E \rangle = \dfrac{3}{2} hv, \sigma_E^2 = 0$

Exercise 10.24 $\langle E \rangle = 3E_1, \sigma_E^2 = 2E_1^2$

Answers to Selected Problems

Problem 10.36

a.

$$\text{For } n = 1, \langle p_x^2 \rangle = \frac{h^2}{4a^2}; \text{ for } n = 2, \langle p_x^2 \rangle = \frac{h^2}{a^2};$$

$$\text{for } n = 3, \langle p_x^2 \rangle = \frac{9h^2}{4a^2}$$

b. $\langle p_x^2 \rangle = \dfrac{h^2 n^2}{4a^2}$

Problem 10.39

a. $\langle E \rangle = \dfrac{3}{2} h\nu$ **b.** $\sqrt{\dfrac{2}{3}} h\nu$

c. In a set of many measurements of E, the only values that occur are E_0, E_1, and E_2. Each of these values will occur one-third of the time.

Problem 10.42

a. For general n: $\sigma_x \sigma_{p_x} = \dfrac{h}{2} \left(\dfrac{n^2}{12} - \dfrac{1}{2\pi^2} \right)^{1/2}$

For $n = 1$: $\sigma_x \sigma_{p_x} = \left(\dfrac{1}{12} - \dfrac{1}{2\pi^2} \right)^{1/2} h = 0.09038h$

For $n = 2$: $\sigma_x \sigma_{p_x} = \left(\dfrac{4}{12} - \dfrac{1}{2\pi^2} \right)^{1/2} h = 0.26538h$

For $n = 3$: $\sigma_x \sigma_{p_x} = \left(\dfrac{9}{12} - \dfrac{1}{2\pi^2} \right)^{1/2} h = 0.41813h$

b. Same as part a, since the uncertainty product does not depend on the length of the box.

Problem 10.44

a. F   **b.** F   **c.** F   **d.** T   **e.** T
f. F   **g.** T   **h.** T   **i.** T   **j.** T

CHAPTER 11

Answers to Numerical Exercises

Exercise 11.3b
Distance to nucleus $= 5.44 \times 10^{-14}$ m
Distance to electron $= 9.995 \times 10^{-11}$ m

Exercise 11.8
a. For $J = 0$, $E_J = 0$; for $J = 1$, $E_J = 7.677 \times 10^{-23}$ J
b. $\nu = 1.159 \times 10^{11}$ s^{-1}; $\lambda = 2.588 \times 10^{-3}$ m

Exercise 11.11   0.054% error in both

Exercise 11.12
In n, l, m order: 4, 0, 0; 4, 1, 1; 4, 1, 0; 4, 1, -1; 4, 2, 2;
4, 2, 1; 4, 2, 0; 4, 2, -1; 4, 2, -2; 4, 3, 3; 4, 3, 2; 4, 3, 1;
4, 3, 0; 4, 3, -1; 4, 3, -2; 4, 3, -3. For each of these sets,
$m_s = \pm \frac{1}{2}$

Exercise 11.14
a. $\langle r \rangle = 3a/(2Z)$

b. $\langle r^2 \rangle = 3a^2/Z^2$
c. $r(\text{most probable}) = a/Z$

Exercise 11.24
b. $^4S, ^2S, ^2S$ **c.** $E^{(0)} = -166.6$ eV

Answers to Selected Problems

Problem 11.29
The energy of a deuterium atom is lower (larger in magnitude) by 0.027%

Problem 11.31 $35.264°, 65.905°, 90°, 114.095°, 144.736°$

Problem 11.33
For the 2s state, $\langle r \rangle = 6a/Z$
For the 2p states, $\langle r \rangle = 5a/Z$

Problem 11.36 $\langle z \rangle = 0$; $\sigma_z = a = 0.529 \times 10^{-10}$ m

Problem 11.39
For the 1s state, $r(\text{most probable}) = a$
For the 2s state, $r(\text{most probable}) = 5.23606a$
For the 2p state, $r(\text{most probable}) = 4a$

Problem 11.42
The vectors can either cancel or give a vector parallel to the y axis, with length $\hbar$. This corresponds to $M_S = 0$, not for the correct magnitude of the vector for $S = 1$. The situation is physically impossible.

Problem 11.45
a. $(1s)^2(2s)^2(2p)^2$; $^1D, ^3P, ^1S$
b. $(1s)^2(2s)^2(2p)^6(3s)^2(3p)^6(4s)^2(3d)^{10}(4p)^4$; $^1D, ^3P, ^1S$
c. $(1s)^2(2s)^2(2p)^6(3s)^2(3p)^6$; ^1S
d. $(1s)^2(2s)^2(2p)^6(3s)^2$; ^1S

Problem 11.49
a. T   **b.** F   **c.** F   **d.** T   **e.** T   **f.** F   **g.** F   **h.** F

CHAPTER 12

Answers to Numerical Exercises

Exercise 12.7
a. Se: $(1s)^2(2s)^2(2p)^6(3s)^2(3p)^6(4s)^2(3d)^{10}(4p)^4$
b. Nb: $(1s)^2(2s)^2(2p)^6(3s)^2(3p)^6(4s)^2(3d)^{10}(4p)^6(5s)^2(4d)^3$
c. Pb: $(1s)^2(2s)^2(2p)^6(3s)^2(3p)^6(4s)^2(3d)^{10}(4p)^6(5s)^2(4d)^{10}$
$(5p)^6(6s)^2(4f)^{14}(5d)^{10}(6p)^2$

Exercise 12.8
a. ^2P **b.** ^3P **c.** ^3P

Answers to Selected Problems

Problem 12.11
a. $W = \dfrac{5h^2}{4\pi^2 ma^2} = 1.0132 E_1$, or 1.32% high.

b. $W = \dfrac{6h^2}{4\pi^2 m a^2} = 1.2158 E_1$, or 21.58% high.

c. This function is unusable. It is not differentiable at $x = a/2$.

Problem 12.14 $E \approx E^{(0)} + E^{(1)} = \dfrac{\hbar}{2}\sqrt{\dfrac{k}{m}} + \dfrac{3c\hbar^2}{4km}$

Problem 12.17
a. $(1s)^2(2s)^2(2p)^6(3s)^2(3p)^6(4s)^2(3d)^6$
b. $(1s)^2(2s)^2(2p)^6(3s)^2(3p)^6(4s)^2(3d)^{10}(4p)^6(5s)^2(4d)^{10}$
   $(5p)^6(6s)^2(4f)^{14}(5d)^{10}(6p)^6$
d. $(1s)^2(2s)^2(2p)^6(3s)^2(3p)^6(4s)^2(3d)^{10}(4p)^6(5s)^1$

Problem 12.21 $Z = 1.837$

Problem 12.23
H: 1, He: 0, Li: 1, Be: 0, B: 0, C: 2, N: 3, O: 2, F: 1, Ne: 0,
Na: 1, Mg: 0, Al: 1, Si: 2, P: 3, S: 2, Cl: 1, Ar: 0

Problem 12.28
e. According to part a, $W = 5.225 \times 10^{-20}$ J
   According to part c, $W = 7.546 \times 10^{-20}$ J
f. According to part d, $W = 5.304 \times 10^{-20}$ J for $n = 1$.

Problem 12.29
a. F **b.** F **c.** F **d.** T **e.** F **f.** T
g. F **h.** T **i.** F **j.** F **k.** T

CHAPTER 13

Answers to Numerical Exercises

Exercise 13.1
a. $(-1, -2, -3)$ **b.** $(4, -2, 2)$ **c.** $(-7, -6, 3)$

Exercise 13.2
a. $(1, -2, -3)$ **b.** $(0.37, 1, 1.37)$

Exercise 13.3
a. $(r, 180° - \theta, 180° + \phi) = (r, \pi - \theta, \pi + \phi)$;
$(r, 180° - \theta, \phi) = (r, \pi - \theta, \phi)$

Exercise 13.14   Bond order $= 1$

Exercise 13.16
a. $\psi_{3\sigma} \approx -0.99\psi_{2sp2Li} + 0.124\psi_{1sH}$ (slightly bonding)
b. $\psi_{4\sigma} \approx -1.34\psi_{2sp1Li} + 1.24\psi_{1sH}$ (antibonding)

Exercise 13.20 $c_C \approx 0.575, c_O \approx 0.425$

Exercise 13.22
$C_{VB} \approx 0.41$; $C_I \approx 0.77$. Another term $\psi_{2spLi}(3)\psi_{2spLi}(4)$ is needed with coefficient 0.22 to make the two wave functions equal.

Exercise 13.23
a. 1.7; compare with 1.9 from the table.
b. 1.1; compare with 1.0 from the table.
c. 0.6; compare with 0.5 from the table.

Exercise 13.24
a. Polar covalent **b.** Polar covalent **c.** Polar covalent
d. Primarily ionic **e.** Primarily ionic **f.** Purely covalent

Exercise 13.36 $\beta \approx -5.5 \times 10^{-19}$ J

Exercise 13.38
$\dfrac{1}{\lambda} = 22300$ cm^{-1}, 7% lower than the experimental value.

Answers to Selected Problems

Problem 13.39 $\hat{C}_{2z}$ (180° rotation about z axis). Yes.

Problem 13.46
a. $2px$ **b.** $2py$ **c.** $2pz$

Problem 13.49
$(\sigma g1s)^2(\sigma^*u1s)^2(\sigma g2s)^2(\sigma^*u2s)^2(\pi u2p)^4(\sigma^*g2p)^2(\pi^*g2p)^4$
$(\sigma^*u2p)(\sigma g3s)$
Term symbol $^3\Sigma_u^+$. Bond order $= 1$. Other configurations are possible.

Problem 13.55 $c_I \approx 0.425, c_{VB} \approx 0.91$ (unnormalized);
18.1% ionic.

Problem 13.59   Nonzero dipole moments: (c), (f)

Problem 13.64
In CH_2, sp hybrids are used in the MOs. In CCl_2, sp^2 hybrids are used. CH_2 has two unpaired electrons in the unhybridized 2p carbon orbitals.

Problem 13.68
$E_{photon} = 2.21 \times 10^{-18}$ J;
$\lambda = 89.8$ nm, 50% lower than 180 nm.

Problem 13.73
Assume C uncharged, assume tetrahedral bond angles.
$Q_{cl} = -2.9 \times 10^{-20}$ C $= -0.18e$;
$Q_H = 9.7 \times 10^{-21}$ C $= 0.060e$

Problem 13.77
a. F **b.** F **c.** T **d.** T

CHAPTER 14

Solutions to Numerical Exercises

Exercise 14.1
$\Delta E = 7.55 \times 10^{-41}$ J $= 4.71 \times 10^{-22}$ eV; smaller than 8.315 eV by a factor of 5.67×10^{-23}.

Exercise 14.2
a. 9.07×10^9 **b.** 4.57×10^{-31} J

Exercise 14.4
For translation, $E_{211} - E_{111} = 8.86 \times 10^{-22}$ eV
For rotation, $E_1 - E_0 = 4.84 \times 10^{-24}$ J $= 3.02 \times 10^{-5}$ eV
For vibration, $E_1 - E_0 = 1.12 \times 10^{-20}$ J $= 0.070$ eV

Exercise 14.6
a. $\lambda = 4657.6$ nm, $v = 6.4365 \times 10^{13}$ s^{-1}
b. $\lambda = 4657.6$ nm
c. $\lambda = 4600.5$ nm

Exercise 14.8 J is even

Exercise 14.10
a. 3 **b.** 2 **c.** 12 **d.** 2

Exercise 14.11 $J_{mp} = 2$

Exercise 14.12 0.208

Exercise 14.14 4.6×10^{-38}

Answers to Selected Problems

Problem 14.15 $\Delta E = 6.14 \times 10^{-22}$ eV

Problem 14.17 273.90 cm^{-1}

Problem 14.20
a. Rotational: 48.6% decrease; vibrational: 28.3% decrease
b. Rotational: 0.2% decrease; vibrational: 0.1% decrease

Problem 14.22
a. 966 N m^{-1} **b.** 575 N m^{-1}

Problem 14.25
a. 3 **b.** 4 **c.** 9 **d.** 9 **e.** 9 **f.** 9 **g.** 30 **h.** 18 **i.** 18
j. 12 **k.** 7 **l.** 12

Problem 14.27
$$I_A = 1.94 \times 10^{-47} \text{ kg m}^2$$
$$I_B = 2.96 \times 10^{-47} \text{ kg m}^2$$
$$I_C = 1.02 \times 10^{-46} \text{ kg m}^2$$

Problem 14.32
a. 5.9×10^{-10} **b.** 0.355

Problem 14.34
a. $E_1 = 2.2 \times 10^{-23}$ J, $g_1 = 1$; $E_2 = 4.4 \times 10^{-23}$ J, $g_2 = 3$;
$E_3 = 6.6 \times 10^{-23}$ J, $g_3 = 3$
b. $\lambda = 9.0 \times 10^{-3}$ m, $v = 3.3 \times 10^{10}$ s^{-1}
c. 2.94

Problem 14.36
a. T **b.** T **c.** F **d.** F **e.** T
f. F **g.** T **h.** F **i.** T **j.** T

CHAPTER 15

Answers to Numerical Exercises

Exercise 15.1
a. 2.0×10^{-23} J, 12 J mol^{-1}
b. 2.0×10^{-15} J, 1.2×10^9 J mol^{-1}

Exercise 15.2 3.2×10^{-173}

Exercise 15.3 1.23×10^4 L mol^{-1} cm^{-1}

Exercise 15.6 H^{35}Cl: 21.17 cm^{-1}, H^{37}Cl: 21.13 cm^{-1}

Exercise 15.7 $J_{mp} = 3$

Exercise 15.11
c. 4263.6 cm^{-1}, 4267.3 cm^{-1}, 4271.0 cm^{-1}

Exercise 15.12 154.4 nm

Exercise 15.14 6.62×10^{-19} J, 399 kJ mol^{-1}

Exercise 15.15b 831 nm

Exercise 15.17 1556.2 cm^{-1}; 264.1 nm

Exercise 15.19 1.40×10^{10} s^{-1}; 0.0214 m

Exercise 15.20
1.41×10^{-26} J, smaller than electron energy difference by a factor of 1.52×10^{-3}.

Exercise 15.22 1.497 T

Answers to Selected Problems

Problem 15.25 9.34×10^3 L mol^{-1} cm^{-1}

Problem 15.27
For $n_2 = 3$, $\lambda = 1312.2$ nm, larger than H value of 656.47 nm by a factor of 1.999.
For $n_2 = 4$, $\lambda = 972.01$ nm, larger than H value of 486.27 nm by a factor of 1.999.
For $n_2 = 5$, $\lambda = 867.87$ nm, larger than H value of 434.17 nm by a factor of 1.999.
For $n_2 = 6$, $\lambda = 820.14$ nm, larger than H value of 410.29 nm by a factor of 1.999.

Problem 15.31
The spectrum has equally spaced lines with a spacing of 22.03 cm^{-1}. The third line (for J going from 2 to 3) has the greatest intensity.

Problem 15.34 2661.89 cm^{-1}, R branch, 6 lines

Problem 15.37
Overtones: near 1178 cm^{-1}, 2570 cm^{-1}, and 4447 cm^{-1}
Combination bands: near 1874 cm^{-1}, 2228 cm^{-1}, and 3508 cm^{-1}

Problem 15.39
a. $v = 7.97 \times 10^{14}$ s^{-1}, $\lambda = 376$ nm
b. $v = 7.24 \times 10^{14}$ s^{-1}, $\lambda = 414$ nm

Problem 15.46 0.99853

Problem 15.49
Proton: 2.67513×10^8 s^{-1} T^{-1}; ^{13}C: 6.7282×10^7 s^{-1} T^{-1}

Problem 15.53
a. 2143.2 cm^{-1}, 1.23% smaller than uncorrected value, 2169.8 cm^{-1}.

b. 4259.9 cm^{-1}, 1.84% smaller than uncorrected value, 4339.6 cm^{-1}.

c. 3.844 cm^{-1}, 0.45% smaller than uncorrected value, 3.862 cm^{-1}.

d. 7.688 cm^{-1}, 0.46% smaller than uncorrected value, 7.724 cm^{-1}.

Problem 15.56
a. 1.3358×10^{-47} kg m^2 **b.** 965.7 N m^{-1}

Problem 15.58
a. F **b.** F **c.** F **d.** F **e.** F **f.** F **g.** F **h.** F
i. F **j.** F **k.** F **l.** F **m.** F **n.** F **o.** F **p.** T

CHAPTER 16

Answers to Numerical Exercises

Exercise 16.1b $v = 1040$ m s^{-1}

Exercise 16.2 $\dfrac{p_1}{p_2} = 1.80 \times 10^{-6}$

Exercise 16.3 Fraction = 0.683; $\sigma = 278$ m s^{-1}

Exercise 16.4
a. Probability $\approx 4 \times 10^{-5}$ **b.** Probability $\approx 8 \times 10^{-5}$

Exercise 16.6
b. $v_p = 394$ m s^{-1} **c.** $v_p = 1112$ m s^{-1}

Exercise 16.7
$v_p = 1112$ m s^{-1}, $\langle v \rangle = 1256$ m s^{-1}, $v_{rms} = 1363$ m s^{-1}

Exercise 16.8 $\dfrac{\langle v \rangle}{v_p} = 1.1284$, $\dfrac{v_{rms}}{v_p} = 1.2247$

Exercise 16.9 2.15×10^{27}

Exercise 16.10
a. Assume that air is one substance with $M = 0.029$ kg mol^{-1}, that $\mathscr{A} \approx 80$ cm^2, and that $T = 298$ K. $N = 1.6 \times 10^{30}$
b. $= 1.2 \times 10^3$ mol s^{-1}

Exercise 16.11
b. $\zeta = \dfrac{abk_B T}{mg}(1 - e^{-mgc/k_B T})$

Exercise 16.13
Diameter = 2.94×10^{-10} m, radius = 1.47×10^{-10} m.

Exercise 16.14
a. $L = 3.44 \times 10^{-9}$ m
b. $\lambda = 1.94 \times 10^{-7}$ m

c. $\dfrac{\lambda}{L} = 56.5$

Exercise 16.15
a. $z_{1(2)} = 1.34 \times 10^9$ s^{-1}, $z_{2(1)} = 5.34 \times 10^9$ s^{-1}
$z_{1(1)} = 5.72 \times 10^9$ s^{-1}, $z_{2(2)} = 1.25 \times 10^9$ s^{-1}

b. $Z_{12} = 2.63 \times 10^{34}$ s^{-1} m^{-3}, $Z_{11} = 5.63 \times 10^{34}$ s^{-1} m^{-3}
$Z_{22} = 3.07 \times 10^{33}$ s^{-1} m^{-3}
c. Total number of collisions = 2.10×10^{33} s^{-1}

Exercise 16.16 Number ≈ 9.2

Answers to Selected Problems

Problem 16.18
Mean value = 0, rms value = $\dfrac{1}{\sqrt{2}} = 0.70711$

Problem 16.20 Fractional error $\approx 10^{-2.11 \times 10^{11}}$

Problem 16.22 Fraction = 0.210

Problem 16.24
a. Fraction = 0.318 **b.** Fraction = 0.318

Problem 16.28
a. $\sigma_v = \left[\left(3 - \dfrac{8}{\pi}\right)\dfrac{RT}{M}\right]^{1/2}$ **b.** $\sigma_v = 187.4$ m s^{-1}

Problem 16.33
Assume that the body area is 2.0 m^2 and that the atmospheric pressure is 1.0 bar.
$m = 2.0 \times 10^4$ kg $\approx 4.4 \times 10^4$ pounds

Problem 16.36 3.01×10^{26} s^{-1}

Problem 16.41
$2: \dfrac{1}{36}$, $3: \dfrac{2}{36}$, $4: \dfrac{3}{36}$, $5: \dfrac{4}{36}$, $6: \dfrac{5}{36}$, $7: \dfrac{6}{36}$, $8: \dfrac{5}{36}$, $9: \dfrac{4}{36}$,

$10: \dfrac{3}{36}$, $11: \dfrac{2}{36}$, $12: \dfrac{1}{36}$

Problem 16.44 $P - P_0 = -0.014$ atm

Problem 16.48
a. ratio = 6.5×10^{-4}
b. ratio = 4.0×10^{-4}

Problem 16.54
a. $z_{CO(O_2)} = 1.80 \times 10^9$ s^{-1}
b. $z_{O_2(CO)} = 3.60 \times 10^9$ s^{-1}
c. time = 1.06×10^7 years

Problem 16.57
a. $\rho = 3100$ kg m^{-3}
b. No. of nearest neighbors ≈ 10.6

Problem 16.60
a. rate = 1.43×10^{17} s^{-1}
b. rate = 1.81×10^{17} s^{-1}
c. rate of mass growth = 1.12×10^{-9} kg s^{-1}

Problem 16.63
a. F **b.** T **c.** T **d.** T **e.** T **f.** T **g.** T **h.** F
i. F **j.** F **k.** F **l.** T

CHAPTER 17

Answers to Numerical Exercises

Exercise 17.2 3.36×10^{-11} mol m^{-2} s^{-1}

Exercise 17.7   Torque = 0.00528 N m

Exercise 17.8
32 L min^{-1}, larger than the correct value by a factor of about 6. Blood cannot flow as a structureless fluid through capillaries that are no larger than the blood cells.

Exercise 17.9
$\mathcal{R} \approx 0.001$. The flow would be laminar, except for the presence of blood cells.

Exercise 17.10
77.7 K: 4.61×10^{-10} m; 273.2 K: 3.46×10^{-10} m; 353.2 K: 3.33×10^{-10} m

Exercise 17.12
$$\frac{\eta}{mD\mathcal{N}} = \frac{5}{6}, \frac{c_V\eta}{\kappa m} = \frac{2}{5}, \frac{Dc_V\mathcal{N}}{\kappa} = \frac{12}{25}$$

Exercise 17.13
a. 7.5×10^{-2} m^3 mol^{-1} = 75 L mol^{-1}
b. 5.1×10^{-2} m^3 mol^{-1} = 51 L mol^{-1}

Exercise 17.14 $D = 1.8 \times 10^{-9}$ m^2 s^{-1} at 45°C

Exercise 17.18
a. $r_{\text{eff}}(\text{H}^+) = 0.263 \times 10^{-10}$ m,
    $r_{\text{eff}}(\text{OH}^-) = 0.466 \times 10^{-10}$ m
b. $r_{\text{eff}}(\text{Li}^+) = 2.38 \times 10^{-10}$ m

Answers to Selected Problems

Problem 17.22
a. 0.0026 m   b. 0.0037 m

Problem 17.25
a. $u = 0.0424$ m s^{-1} = 4.24 cm s^{-1} at r = 2.00 mm.
b. $P_2 - P_1$ = 14.15 Pa
c. $dV/dt = 1.42 \times 10^{-6}$ m^3 s^{-1}
d. $\mathcal{R}$ = 113. The flow is laminar.

Problem 17.28 $\eta = 1.54 \times 10^{-3}$ kg m^{-1} s^{-1}

Problem 17.31
a. $D = 1.26 \times 10^{-4}$ m^2 s^{-1}
b. $D = 1.41 \times 10^{-5}$ m^2 s^{-1}
c. $D = 1.41 \times 10^{-4}$ m^2 s^{-1}
d. He: $z_{\text{rms}} = 0.952$ m for t = 1.000 hour
    Ar: $z_{\text{rms}} = 0.319$ m for t = 1.000 hour

Problem 17.35
    He: $\eta = 1.96 \times 10^{-5}$ kg m^{-1} s^{-1};
    CO$_2$: $\eta = 1.49 \times 10^{-5}$ kg m^{-1} s^{-1}

Problem 17.38 5.0×10^{-4} m

Problem 17.41   K$^+$: 1.25×10^{-10} m; Cl$^-$: 1.21×10^{-10} m

Problem 17.45
a. At 0°C: d = 346 pm from diffusion coefficient, d = 365 pm from viscosity coefficient, d = 365 pm from the table.
b. At 20°C: d = 373 pm from viscosity coefficient, d = 373 pm from thermal conductivity, d = 361 pm from the table.

Problem 17.48
a. T   b. F   c. T   d. F   e. F   f. T   g. F   h. T

CHAPTER 18

Answers to Numerical Exercises

Exercise 18.2
a. $t_{1/2}$ = 135 s, τ = 195 s

Exercise 18.3 2.4×10^9 years

Exercise 18.5   0.028 mol L^{-1}

Exercise 18.6
a. 150 minutes   b. 380 minutes

Exercise 18.7
a. $t = 4t_{1/2}$ b. $t = 15t_{1/2}$

Exercise 18.10
b. $t_{1/2}$ = 0.77 s

Exercise 18.16
$[B]_{\text{eq}} = 0.104$ mol L^{-1}, $[A]_{\text{eq}} = 0.046$ mol L^{-1}
τ = 0.0196 minute, $t_{1/2}$ = 0.0136 minute
At t = 0.100 minute, $[B]$ = 0.103 mol L^{-1},
$[A]$ = 0.047 mol L^{-1}

Answers to Selected Problems

Problem 18.22   First order, $k = 6.25 \times 10^{-4}$ s^{-1}

Problem 18.25
a. 7.1×10^{-11} s   b. 7.1×10^{-8} s

Problem 18.31   pseudo first order, k_{app} = 0.0276 min^{-1}

Problem 18.36
At t = 10.0 s, $[A]$ = 0.0905 mol L^{-1}, $[B]$ = 0.0060 mol L^{-1}, $[F]$ = 0.0035 mol L^{-1}.
At t = 100.0 s, $[A]$ = 0.0370 mol L^{-1}, $[B]$ = 0.0041 mol L^{-1}, $[F]$ = 0.0589 mol L^{-1}.

Problem 18.40

$\tau = 8.3 \times 10^{-9}$ s, assuming activity coefficients equal to unity

Problem 18.42
a. T   b. T   c. F   d. F   e. T   f. T
g. F   h. T   i. F   j. T   k. F   l. F

CHAPTER 19

Answers to Numerical Exercises

Exercise 19.4   0.17, 0.45, 1, 1.88, 3.03

Exercise 19.6   Rate $= \dfrac{d[N_2]}{dt} = k_1[NO_2]^2[H_2]$

Exercise 19.12 $v = 4.86 \times 10^{14}$ s^{-1}, $\lambda = 617$ nm

Exercise 19.18 $K_m = 16.3$ μmol L^{-1}

Exercise 19.21
a. 51 kJ mol^{-1} b. 62 kJ mol^{-1}

Exercise 19.24 $\varphi = 7.0 \times 10^{-3}$

Exercise 19.25 $E_a = 10.2$ kJ mol^{-1}

Answers to Selected Problems

Problem 19.26

$k = 6.8 \times 10^6$ m^3 mol^{-1} s^{-1} $= 6.8 \times 10^9$ L mol^{-1} s^{-1}

Problem 19.31

a. Rate $= \dfrac{k_2 K_1[HNO_3][H^+]}{K_a}$, where $K_a = \dfrac{[H^+][A^-]}{[HA]}$

b. Rate $= \dfrac{k_3 K_1 K_2[HNO_3][H^+][ArH]}{K_a[H_2O]}$

Problem 19.34

a. Rate $= \dfrac{-d[H_2]}{dt} = k_2\left(\dfrac{k_1}{k_1'}\right)^{1/2}[I_2]^{1/2}[H_2]$

b. Same as part (a)

Problem 19.37

b. Rate $= \dfrac{k_2 K_1[H_2][C_2H_4]}{1 + K_1[C_2H_4]}$

Problem 19.46 $E_a = 201$ kJ mol^{-1}, $A = 4.01 \times 10^{14}$ s^{-1}

Problem 19.48
a. 213 kJ mol^{-1}, $A = 3.5 \times 10^9$ L mol^{-1} s^{-1}
b. $\varphi = 0.0077$

Problem 19.51
b. $E_a = -6.0$ kJ mol^{-1} (note negative value)

Problem 19.52
a. T   b. F   c. F   d. F   e. T
f. T   g. F   h. F   i. F   j. T

CHAPTER 20

Answers to Numerical Exercises

Exercise 20.4
a. 3.09% error   b. 0.00030% error

Exercise 20.5
For Equation (20.2-16), there is 13.76% error for $N = 10$.
For Equation (20.2-16), there is 1.573% error for $N = 60$.
For the logarithm of Equation (20.2-15), there is 0.0552% error for $N = 10$.
For the logarithm of Equation (20.2-15), there is 0.00074% error for $N = 60$.
For $N = 1 \times 10^9$, there is no difference to nine digits.

Exercise 20.7 $z_{tr} = 1.937 \times 10^{29}$

Exercise 20.8
Term $= 0.61609$, difference $= 1.705 \times 10^{-11}$.

Exercise 20.10
a. At 298.15 K, population ratio $= 2.12 \times 10^{-9}$. At 500 K, population ratio $= 6.74 \times 10^{-6}$. At 1000 K, population ratio $= 2.60 \times 10^{-3}$. At 5000 K, population ratio $= 0.304$. In the limit as $T \to \infty$, population ratio $\to 1$.
b. At 298.15 K, population ratio $= 2.451$. At 1000 K, population ratio $= 2.824$. In the limit as $T \to \infty$, population ratio $\to 3$.
c. $J_{mp} = 3$

Exercise 20.11 $T = -5.95 \times 10^4$ K

Exercise 20.15
a. $z_{rot} = 1.703$
b. $z_{rot} = 1.8840$
c. $z_{rot} = 424.8$, larger than the value of the example by 0.1

Exercise 20.16   H$_2$: $z_{vib} = 1.000000001$; I$_2$: $z_{vib} = 1.5508$

Exercise 20.17 $z_{vib} = 0.2778$

Exercise 20.18
a. $z = 1.505 \times 10^{33}$ b. $z = 1.490 \times 10^{33}$

Exercise 20.19   36.42

Exercise 20.20 $z_{vib} = 1.095$

Exercise 20.26
a. $z_{vib} = 0.005327$, $\bar{U} = 12797$ J mol^{-1}
b. With zero-point energy excluded, as in Example 20.11, $z_{vib} = 2.17125$, $\bar{U} = 3005.42$ J mol^{-1}. With zero-point energy included, as in Example 20.11, $z_{vib} = 1.5947$, $\bar{U} = 4288.42$ J mol^{-1}.

Exercise 20.27 $\bar{U} = 515.1$ J mol^{-1}

Exercise 20.28
b. $\bar{C}_V = 25.466$ J K^{-1} mol^{-1}

Exercise 20.29

$$\mu_{tr} = -6.698 \times 10^{-20} \text{ J}$$
$$\mu_{rot} = -2.491 \times 10^{-20} \text{ J}$$
$$\mu_{vib} = -2.862 \times 10^{-22} \text{ J}$$
$$\mu_{elec} \approx 0$$

Exercise 20.30 $K_P = 4.24$

Exercise 20.40
a. -1304 J K^{-1} **b.** 146.53 J K^{-1}

Exercise 20.42 $B_2 = \left[\dfrac{2\pi b^3}{3} - \dfrac{2\pi}{3}(b^3 - a^3)e^{u_0/k_B T} \right] N_A$

Answers to Selected Problems

Problem 20.43
a. 10 states
b and **c.** The Boltzmann distribution is exactly as in Table 20.2.

Value of v	p_v	p_v(Boltzmann)
0	0.4	0.50000
1	0.3	0.25000
2	0.2	0.12500
3	0.1	0.06250
4	0	0.03125
5	0	0.01562
6	0	0.00781

Problem 20.45 $10^{1.406 \times 10^{25}}$

Problem 20.48
a. 2.6389   **b.** 2.8874   **c.** 3

Problem 20.50
For $V = 0.00100$ m^3, $z = 2.44 \times 10^{29}$
For $V = 1.00$ m^3, $z = 2.44 \times 10^{32}$

Problem 20.53
a. $z_{tr} = 3.509 \times 10^{30}$, $z_{rot} = 51.86$, $z_{vib} = 1.00001$,
   $z_{elec} = 1$
b. $z/N = 3.022 \times 10^8$ for both cases.

Problem 20.55
The center of mass is 0.06567×10^{-10} m from the O atom on the line bisecting the bond angle. This line is the B axis; the axis perpendicular to the molecular plane is the C axis. $I_A = 1.0234 \times 10^{-47}$ kg m^2, $I_B = 1.919 \times 10^{-47}$ kg m^2, and $I_C = 2.9427 \times 10^{-47}$ kg m^2.

$$z_{rot} = 42.91$$

Problem 20.57
a. $z_{elec} = 1.24 \times 10^{-5}$
b. $z_{elec} = 1$ (to roughly 3800 significant digits)

Problem 20.60
$\bar{S} = 154.847$ J K^{-1} mol^{-1}, in good agreement with the table's value of 154.845 J K^{-1} mol^{-1} and 22.7% higher than helium's value of 126.155 J K^{-1} mol^{-1}.

Problem 20.63
$\bar{S} = 197.60$ J K^{-1} mol^{-1}, 0.025% lower than the value in the table.

Problem 20.67
$\bar{C}_P = 37.0508$ J K^{-1} mol^{-1}, 0.21% lower than the value in the table, 37.129 J K^{-1} mol^{-1}

Problem 20.70 $K_P = 53.5$

Problem 20.75 $Z = 10^{4.48 \times 10^{24}}$

Problem 20.79
$\bar{C}_{vib} = 0.6321$ J K^{-1} mol^{-1} at 500 K. $\bar{C}_{vib} = 0.95R$ at 3960 K, $\bar{C}_{vib} = 0.99R$ at 8985 K.

Problem 20.82

a. F   **b.** F   **c.** F   **d.** T   **e.** F   **f.** T
g. T   **h.** T   **i.** T   **j.** T   **k.** F   **l.** F

CHAPTER 21

Answers to Numerical Exercises

Exercise 21.1
b. Assume reaction diameter $= 5 \times 10^{-10}$ m,
$\mu = 4 \times 10^{-26}$ kg.
$A = 2.4 \times 10^{11}$ L mol^{-1} s^{-1}

Exercise 21.3
Assume $z^{\ddagger}/z_{reactant} \approx 1$, $E_a \approx$ difference between C=C and C—C bond energies. $k = 2.7 \times 10^{-35}$ s^{-1}, $t_{1/2} = 2.6 \times 10^{34}$ s.

Exercise 21.5
a. $E_a = 14.2$ kJ mol^{-1} **b.** $\Delta S^{\ddagger\circ} = -13$ J K^{-1} mol^{-1}

Exercise 21.6 $E_{a\eta} = 11.9$ kJ mol^{-1}

Exercise 21.8   Distance $= 1.8 \times 10^{-11}$ m

Exercise 21.10

Debye length $= \dfrac{1}{\kappa} = 2.15 \times 19^{-9}$ m $= 21.5$ Å

Exercise 21.12b
$\eta_{conc} = -3.6 \times 10^{-6}$ V; $J = 4.4 \times 10^{-5}$ mol m^{-2} s^{-1};
$j = 8.4 \times 10^{-4}$ A cm^{-2}

Exercise 21.17b

$v_{drift} = 1.06 \times 10^{-4}$ m s^{-1}; $\dfrac{v_{drift}}{v_{rms}} = 9.2 \times 10^{-10}$

Answers to Selected Problems

Problem 21.23 $\Delta S^{\pm\circ} = -160 \text{ J K}^{-1} \text{ mol}^{-1}$

Problem 21.25
b. $k(310 \text{ K})/k(300 \text{ K}) = 34$

Problem 21.27 $E_{a\eta} = 15.8 \text{ kJ mol}^{-1}$

Problem 21.29 $\tau = 1.15 \times 10^{-13} \text{ s}$

Problem 21.32 $a_{\text{Cl}^-} = 153$

Problem 21.35
a. $\tau = 3.81 \times 10^{-14} \text{ s}$
b. $\lambda = 4.40 \times 10^{-9} \text{ m}$
d. $r \approx 8 \times 10^{-9}$ ohm m, 50% lower than the correct value, 1.72×10^{-8} ohm m

Problem 21.37
a. Relaxation time = $\tau = 2.7 \times 10^{-14}$ s, from Example 21.5.
b. $\sigma_{\text{min}} \approx 4 \times 10^{30}$ ohm^{-1} m^{-1}

Problem 21.40
a. F **b.** F **c.** F **d.** T **e.** F **f.** T **g.** T **h.** T
i. T **j.** F **k.** T **l.** T **m.** F **n.** T **o.** T

CHAPTER 22

Answers to Numerical Exercises

Exercise 22.2
Cubic, P: 1; cubic, I: 2; cubic, F: 4; tetragonal, P: 1; tetragonal, I: 2; orthorhombic, P: 1; orthorhombic, C: 2; orthorhombic, I: 2; orthorhombic, F: 4; monoclinic, P: 1; monoclinic, I: 2; triclinic: 1, hexagonal (small unit cell): 1

Exercise 22.4
b. Coordination number = 8; packing fraction = 0.68

Exercise 22.12
a. $A = 2.7236 \times 10^{-38}$ J m^{-2}
b. $n = 8.491 \times 10^{28}$ m^{-3}; $\varepsilon_{F0} = 5.26 \times 10^{-19}$ J

Exercise 22.13
a. $C_{\text{elec}} = 2.3 \times 10^3$ J K^{-1} m^{-3};
$\bar{C}_{\text{elec}} = 0.0165$ J K^{-1} mol^{-1}

b. Ratio = 0.0013
c. Ratio = 0.077

Exercise 22.16
a. $v = 1.98 \times 10^3$ m s^{-1} **b.** $T = 158$ K

Exercise 22.25
b. $q = -0.00083$ J; $w = 0.00083$ J

Answers to Selected Problems

Problem 22.30
Basis = one argon atom
4 atoms per unit cell
4 bases per unit cell

Problem 22.33
a. $d_{111} = 2.47 \times 10^{-10}$ m **b.** $\lambda = 2.94 \times 10^{-10}$ m

Problem 22.35
a. $v = 5.00 \times 10^{12}$ s^{-1}

Problem 22.40
$\varepsilon_F = 5.26 \times 10^{-19}$ J; no error to 3 significant digits

Problem 22.43 $\bar{C}_{\text{elec}} = 0.321$ J K^{-1} mol^{-1}; 1.27%

Problem 22.46
$P = 2888$ atm, in very poor agreement with 1.00 atm

Problem 22.48
$$\bar{M}_n = 20000 \text{ amu}$$
$$\bar{M}_m = 25000 \text{ amu}$$

Problem 22.53
a. Ratio = 0.1   **b.** Ratio = 0.01

Problem 22.56
a. $V = 5.4 \times 10^{-25}$ m^3
b. $\eta = 1.015 \times 10^{-3}$ kg m^{-1} s^{-1}
c. $\eta = 1.011 \times 10^{-3}$ kg m^{-1} s^{-1}

Problem 22.59
a. F **b.** T **c.** T **d.** T **e.** F
f. T **g.** T **h.** F **i.** F **j.** T

Index

Note: In the case of several page numbers for one entry, a bold-face page number indicates the primary reference to that entry. Appendix pages are labeled with the letter A.